SIPRI Yearbook 2002
Armaments, Disarmament and International Security

sipri
Stockholm International Peace Research Institute

SIPRI is an independent international institute for research into problems of peace and conflict, especially those of arms control and disarmament. It was established in 1966 to commemorate Sweden's 150 years of unbroken peace.

The Institute is financed mainly by the Swedish Parliament. The staff and the Governing Board are international. The Institute also has an Advisory Committee as an international consultative body.

The Governing Board is not responsible for the views expressed in the publications of the Institute.

Governing Board
Ambassador Rolf Ekéus, Chairman (Sweden)
Dr Alexei G. Arbatov (Russia)
Dr Willem F. van Eekelen (Netherlands)
Dr Nabil Elaraby (Egypt)
Sir Marrack Goulding (United Kingdom)
Professor Helga Haftendorn (Germany)
Dr Catherine M. Kelleher (United States)
Professor Ronald G. Sutherland (Canada)
The Director

Director
Professor Adam Daniel Rotfeld (Poland)

Adam Daniel Rotfeld, Director, *Yearbook Editor and Publisher*
Ian Anthony, *Executive Editor*
Connie Wall, *Managing Editor*
Coordinators
Taylor B. Seybolt, Elisabeth Sköns, Shannon N. Kile
Editors
Pamela Boston, Jetta Gilligan Borg, Andy Mash

sipri
Stockholm International Peace Research Institute
Signalistgatan 9, SE-169 70 Solna, Sweden
Cable: SIPRI
Telephone: 46 8/655 97 00
Telefax: 46 8/655 97 33
Email: sipri@sipri.org
Internet URL: http://www.sipri.org

SIPRI Yearbook 2002

Armaments, Disarmament and International Security

Stockholm International Peace Research Institute

OXFORD UNIVERSITY PRESS
2002

OXFORD
UNIVERSITY PRESS

Great Clarendon Street, Oxford OX2 6DP
Oxford University Press is a department of the University of Oxford.
It furthers the University's objective of excellence in research, scholarship,
and education by publishing worldwide in

Oxford New York
Athens Auckland Bangkok Bogotá Buenos Aires
Cape Town Chennai Dar es Salaam Delhi Florence Hong Kong Istanbul
Kolikata Karachi Kuala Lumpur Madrid Melbourne Mexico City Mumbai
Nairobi Paris São Paulo Singapore Taipei Tokyo Toronto Warsaw
and associated companies in Berlin Ibadan

Oxford is a registered trade mark of Oxford University Press
in the UK and certain other countries

Published in the United States
by Oxford University Press Inc., New York

© SIPRI 2002

Yearbooks before 1987 published under title
'World Armaments and Disarmament:
SIPRI Yearbook [year of publication]'

All rights reserved. No part of this publication may be reproduced,
stored in a retrieval system, or transmitted, in any form or by any means,
without the prior permission in writing of SIPRI or as expressly permitted by law,
or under terms agreed with the appropriate reprographics rights organizations.
Enquiries concerning reproduction outside the scope of the above should be sent to
SIPRI, Signalistgatan 9, SE-169 70 Solna, Sweden

You must not circulate this book in any other binding or cover
and you must impose the same condition on any acquirer

British Library Cataloguing in Publication Data

Data available
ISSN 0953–0282
ISBN 0-19-925176-2

Library of Congress Cataloging in Publication Data

Data available
ISSN 0953–0282
ISBN 0-19-925176-2

Typeset and originated by Stockholm International Peace Research Institute
Printed and bound in Great Britain by
Biddles Ltd., Guildford and King's Lynn

Contents

Dedication	xv
Preface	xvii
Glossary	xix
Christer Berggren and Connie Wall	

Introduction: Global security after 11 September 2001		1
Adam Daniel Rotfeld		
	I. Introduction	1
	II. Global terrorism and global responsibility	1
	III. The change of perception	4
	IV. The new nature of the threat	5
	V. New security building blocks	6
	A new agenda for NATO—The NATO–Russia relationship—A new agenda for the European Union—The West and the Islamic world	
	VI. SIPRI findings	11
	VII. Concluding remarks	16

Part I. Security and conflicts, 2001

1. Major armed conflicts		21
Taylor B. Seybolt		
	I. Introduction	21
	II. Conflicts in Africa	24
	Algeria—Angola—Burundi—The Democratic Republic of the Congo—Sierra Leone—Somalia—Sudan	
	III. Conflicts in Asia	39
	Afghanistan—India: Kashmir—Indonesia—The Philippines—Sri Lanka	
	IV. Conflicts in Europe	52
	Russia: Chechnya	
	V. Conflicts in the Middle East	55
	Israel–Palestinians	
	VI. Conflicts in South America	58
	Colombia	
	VII. Conclusions	61
Appendix 1A. Patterns of major armed conflicts, 1990–2001		63
Mikael Eriksson, Margareta Sollenberg and Peter Wallensteen		
	I. Global patterns	63
	II. Regional patterns	65
	III. Changes in the table of conflicts for 2001	67
	The new conflict in 2001—Conflicts recorded in 2000 that were not recorded for 2001—Changes in intensity of conflict	
Table 1A.1.	Regional distribution, number and types of major armed conflicts, 1990–2001	64
Table 1A.2.	Regional distribution of locations with at least one major armed conflict, 1990–2001	64
Table 1A.3.	Table of conflict locations with at least one major armed conflict in 2001	69
Figure 1A.	Regional distribution and total number of major armed conflicts, 1990–2001	66

vi SIPRI YEARBOOK 2002

Appendix 1B. Definitions, sources and methods for the conflict data   77
    I. Definitions   77
    II. Sources   78
    III. Methods   79

Appendix 1C. Measuring violence: an introduction to conflict data sets   81
Taylor B. Seybolt
    I. Introduction   81
    II. Variations in purpose, definitions and methodology   82
        Purposes—Definitions—Methods and coding rules
    III. The leading data sets   88
        Patterns of conflict occurrence—Causes and processes of conflict—
        Costs of conflict and conflict early warning
    IV. Conclusions   96

2. Conflict prevention   97
Renata Dwan
    I. Introduction   97
    II. The rise of conflict prevention   97
    III. The UN and EU reports on conflict prevention in 2001   100
        The UN Secretary-General's Report—The EU documents
    IV. Prevention in practice: West Africa and Zimbabwe   107
        Regional conflict in West Africa—Civil conflict in Zimbabwe
    V. Prevention after 11 September 2001   121
  Figure 2.1. Map of West Africa: the 15 member states of ECOWAS   110

Appendix 2A. Multilateral peace missions, 2001   124
Renata Dwan, Thomas Papworth and Sharon Wiharta
    I. Peace missions in 2001   124
        The International Security Assistance Force in Afghanistan—Peace
        operations in the Balkans—Peace operations in Africa
    II. Table of multilateral peace missions   129
Table 2A.   Multilateral peace missions   130

3. The military dimension of the European Union   151
Zdzislaw Lachowski
    I. Introduction   151
    II. From 1998 to 2000   152
        A credible security actor
    III. Towards operational capabilities   155
        The ESDP structures and procedures—Military capabilities—EU–NATO
        cooperation—Cooperation with non-members—Cooperation with
        international organizations
    IV. Challenges to politico-military integration   166
        The EU–US relationship—EU access to NATO capabilities: Turkish
        opposition—The financial hurdle
    V. Conclusions   172

4. The challenges of security sector reform   175
Dylan Hendrickson and Andrzej Karkoszka
    I. Introduction   175
    II. The policy agenda   177
        Defining the security sector—Defining security sector reform—
        Approaches to security sector reform

III.	Relevance to international security Instruments of security sector reform—The implications of 11 September 2001	183
IV.	Security sector reform in a regional context Africa—Asia—Eastern Europe—Latin America	188
V.	Challenges to security sector reform Operationalizing concepts—Unfavourable environments—Conflicting objectives—Integrated programmes	194
VI.	Conclusions	200

5. Sanctions applied by the European Union and the United Nations 203
Ian Anthony

I.	Introduction	203
II.	The United Nations sanctions experience The UN Working Group on the General Issues on Sanctions	204
III.	The EU sanctions experience Sanctions as an instrument of EU policy: the targeted sanctions against Yugoslavia after 1998	210
IV.	Sanctions against Iraq The debate in the Security Council in 2001—Revising the sanctions regime	215
V.	Sanctions against Afghanistan UN sanctions against Afghanistan—European Union sanctions against Afghanistan	221
VI.	Sanctions against terrorism EU sanctions related to terrorism	225
VII.	Conclusions	227

Part II. Military spending and armaments, 2001

6. Military expenditure 231
Elisabeth Sköns, Evamaria Loose-Weintraub, Wuyi Omitoogun and Petter Stålenheim

I.	Introduction	231
II.	World military expenditure: trends and pattern The global pattern—Economic burden	232
III.	The war on terrorism The costs of the war in Afghanistan—The impact on future budgets	239
IV.	Regional trends in military expenditure Africa—The Middle East—Central Asia—South Asia—East Asia—South America—Western Europe—Russia	245
V.	Conclusions	264
Table 6.1.	Changes in world and regional military expenditure estimates, 1992–2001	234
Table 6.2.	The 15 major spenders, military expenditure in 2001	235
Table 6.3.	Global pattern of military expenditure, population and gross national income, 2000	236
Table 6.4.	Countries with the highest defence burden in 1999: social and military expenditure as a share of gross domestic product, 1995–2000	238
Table 6.5.	Western European and NATO military expenditure, 1992–2001	254
Table 6.6.	Burden sharing within NATO	256
Table 6.7.	The Russian Federation: military expenditure, 1992–2002	260
Table 6.8.	The Russian defence budget for 2002, as adopted on 30 December 2001	261

Appendix 6A. Tables of military expenditure		266
Elisabeth Sköns, Evamaria Loose-Weintraub, Wuyi Omitoogun and Petter Stålenheim		
Table 6A.1.	Military expenditure by region, in constant US dollars, 1992–2001	266
Table 6A.2.	Military expenditure by region and country, in local currency, 1992–2001	270
Table 6A.3.	Military expenditure by region and country, in constant US dollars, 1992–2001	276
Table 6A.4.	Military expenditure by region and country, as a percentage of gross domestic product, 1992–2000	282
Appendix 6B. Tables of NATO military expenditure by category		292
Table 6B.1.	NATO military expenditure on personnel and equipment, 1992–2001	292
Table 6B.2.	Military equipment expenditure of France, 1992–2001	295
Appendix 6C. Sources and methods for military expenditure data		297
	I. Purpose of the data	297
	II. Coverage of the data	297
	Definition of military expenditure	
	III. Methods	298
	Estimation—Calculations	
	IV. Limitations of data	300
	V. Sources	301
Appendix 6D. Official data on military expenditure		302
Evamaria Loose-Weintraub		
	I. Introduction	302
	II. International reporting of military expenditure	302
	III. Reporting of military expenditure data to SIPRI and the UN	306
	IV. Conclusions	307
Table 6D.	SIPRI and UN requests for military expenditure data, 2001	305
Appendix 6E. US military expenditure and the 2001 Quadrennial Defense Review		309
David Gold		
	I. Introduction	309
	II. The US defence budget in the long-term perspective	311
	The decline in the defence burden—Pressures to increase the defence budget	
	III. The Quadrennial Defense Review and US military strategy	314
	IV. US military spending in FY 2002 and beyond	316
	The investment accounts—Personnel—Operations and maintenance—Homeland security—Transformation	
	V. Conclusions	321
Table 6E.	US military expenditure, FYs 1955–2003	310
7. Arms production		323
Elisabeth Sköns and Reinhilde Weidacher		
	I. Introduction	323
	II. Concentration, internationalization and privatization	325
	Concentration—Internationalization—Privatization	
	III. Russia	346
	Quantitative trends—Privatization of the Russian arms industry—Industrial structure	
	IV. Conclusions	352
Table 7.1.	National arms sales and arms exports, Western Europe, 1990–2000	324
Table 7.2.	Change in concentration ratios, SIPRI Top 100 companies, 1990–1998	327

Table 7.3.	Major acquisitions of arms-producing companies in North America and Western Europe, 2001	329
Table 7.4.	International, West European and transatlantic joint ventures and mergers among arms producing companies, established in 2000–2001	334
Table 7.5.	Company acquisitions linked to arms deals with minor producer countries, 2000–2001	338
Table 7.6.	Major cases of company privatization, 1990–2001	342
Table 7.7.	Selected companies providing services to the military, 2000	345
Table 7.8.	Ownership in the Russian arms industry, as of 31 December 2000	349
Table 7.9.	Major arms-producing companies in Russia, 2000	350

Appendix 7A. Arms industry data — 354
Reinhilde Weidacher and the SIPRI Arms Industry Network

	I. Selection criteria and sources of data	354
	II. Definitions	354
	III. Coverage	355
	IV. The 100 largest arms-producing companies in the OECD and developing countries, 2000	356
Table 7A.1.	Regional/national shares of arms sales for the top 100 arms-producing companies in the OECD and developing countries in 2000	356
Table 7A.2.	The 100 largest arms-producing companies in the OECD and developing countries, 2000	357
Table 7A.3.	A tentative list of the 20 largest arms-producing companies in Russia, 2000	364

Appendix 7B. Available government and industry data on the arms industry — 366
Reinhilde Weidacher

	I. Introduction	366
	II. Reporting practices	367
	Industry associations—Government organization	
	III. Conclusions	372
Table 7B.	Open and regular sources for national data on the arms industry in the 20 largest arms-producing countries	369

8. International arms transfers — 373
Björn Hagelin, Pieter D. Wezeman, Siemon T. Wezeman and Nicholas Chipperfield

	I. Introduction	373
	II. The suppliers and recipients	374
	The major suppliers—The major recipients	
	III. Arms transfers to areas of conflict	380
	Arms imports by India and Pakistan—The significance of small transfers of arms—International arms embargoes	
	IV. Arms trade and competition	390
	The value of the arms trade—Competition	
	V. New weapons and transatlantic cooperation: the JSF	395
	The arguments for the JSF—The JSF as a model for transatlantic cooperation—Long-term considerations	
	VI. Arms transfer reporting and transparency	400
	International transparency—National transparency	
	VII. Conclusions	402
Table 8.1.	Transfers of major conventional weapons from the 10 leading suppliers to the 38 leading recipients, 1997–2001	376
Table 8.2.	International arms embargoes, 1997–2001	388
Table 8.3.	Arms deliveries according to national reporting, 1996–2000	392

Table 8.4	Examples of reportedly offered or demanded offsets	394
Table 8.5.	International participation in the Joint Strike Fighter competitive development phase, 1995–2001	398
Figure 8.1.	The trend in international transfers of major conventional weapons, 1987–2001	374

Appendix 8A. The volume of transfers of major conventional weapons: by recipients and suppliers, 1997–2001 — 403
Björn Hagelin, Pieter D. Wezeman, Siemon T. Wezeman and Nicholas Chipperfield

Table 8A.1.	The recipients of major conventional weapons	403
Table 8A.2.	The suppliers of major conventional weapons	407

Appendix 8B. The volume of transfers of major conventional weapons: by regions and other groups of recipients and suppliers, 1992–2001 — 409
Björn Hagelin, Pieter D. Wezeman, Siemon T. Wezeman and Nicholas Chipperfield

Table 8B.1.	Volume of imports of major conventional weapons	409
Table 8B.2.	Volume of exports of major conventional weapons	410

Appendix 8C. Register of the transfers and licensed production of major conventional weapons, 2001 — 413
Björn Hagelin, Pieter D. Wezeman, Siemon T. Wezeman and Nicholas Chipperfield

Table 8C.1.	Register of transfers and licensed production of major conventional weapons, 2001, by recipients	413
Table 8C.2.	Index of suppliers of major conventional weapons and their recipients and licensees, 2001	454

Appendix 8D. Sources and methods for arms transfers data — 456
    I. Selection criteria and coverage — 456
        Selection criteria—Major conventional weapons: the coverage
    II. The SIPRI trend indicator — 457
    III. Sources — 459

Appendix 8E. Government and industry data on national arms exports — 460
Pieter D. Wezeman

Table 8E.	Government and industry data on national arms exports, 1996–2000	461

Part III. Non-proliferation, arms control and disarmament, 2001

9. Arms control after the attacks of 11 September 2001 — 469
Ian Anthony
    I. Introduction — 469
    II. Salient characteristics of arms control — 470
        Compliance and effectiveness in arms control
    III. The Bush Administration and arms control — 476
        Arms control and missile defence
    IV. International effects of the Bush Administration approach — 482
    V. The impact of the attacks of 11 September — 485
    VI. Conclusions — 487

10. Ballistic missile defence and nuclear arms control — 489
Shannon N. Kile
    I. Introduction — 489
    II. Ballistic missile defence and the future of the ABM Treaty — 490
        The US missile defence debate—CBO cost estimates—Missile defence funding and programme changes—US–Russian discussions on the future of the ABM Treaty—The US decision to withdraw from the ABM Treaty

III.	US–Russian strategic nuclear arms control	511
	Implementation of the START I Treaty—Towards deeper reductions in strategic nuclear arms	
IV.	Cooperative nuclear security initiatives	518
V.	The Comprehensive Nuclear Test-Ban Treaty	522
VI.	Conclusions	523
Table 10.1.	Funding of US ballistic missile defence programmes, FY 2002	496
Table 10.2.	Soviet/Russian and US strategic offensive nuclear forces, by delivery vehicles and START-accountable warheads, September 1990 and December 2001	514
Table 10.3.	Summary of funding for principal DOD and DOE non-proliferation programmes in the former Soviet Union, February 2002	520

Appendix 10A. World nuclear forces 525
Hans M. Kristensen and Joshua Handler

I.	Introduction	525
II.	US nuclear forces	526
	The 'enduring nuclear stockpile'— 'Phantom arsenals'—Strategic bombers—Intercontinental ballistic missiles—Ballistic missile submarines—Non-strategic nuclear weapons—Nuclear command and control	
III.	Russian nuclear forces	539
	Strategic aviation—Intercontinental ballistic missiles—Strategic submarines and SLBMs—The Russian nuclear stockpile and non-strategic nuclear weapons	
IV.	British nuclear forces	545
	Strategic submarines—Nuclear warhead maintenance	
V.	French nuclear forces	548
	Nuclear strike aircraft—Ballistic missile submarines—Naval nuclear aviation	
VI.	Chinese nuclear forces	552
	Nuclear aviation—Land-based ballistic missiles—Ballistic missile submarines—Non-strategic weapons	
VII.	Indian nuclear forces	557
	Strike aircraft—Land-based missiles—Naval weapons	
VIII.	Pakistani nuclear forces	562
	Strike aircraft—Ballistic missiles	
IX.	Israeli nuclear forces	565
	Strike aircraft—Ballistic missiles—Other weapon systems	
Table 10A.1.	World nuclear forces, January 2002	526
Table 10A.2.	US nuclear warhead status and modifications, January 2002	527
Table 10A.3.	US nuclear forces, January 2002	536
Table 10A.4.	Russian SSBN and SSN/SSGN patrols per year, 1991–2001	542
Table 10A.5.	Russian nuclear forces, January 2002	544
Table 10A.6.	British nuclear forces, January 2002	548
Table 10A.7.	French nuclear forces, January 2002	550
Table 10A.8.	Chinese nuclear forces, January 2002	556
Table 10A.9.	Indian nuclear forces, January 2002	560
Table 10A.10.	Pakistani nuclear forces, January 2002	564
Table 10A.11.	Israeli nuclear forces, January 2002	566

Appendix 10B. Tactical nuclear weapons 568
Nicholas Zarimpas

    I. Introduction 568
    II. Definitions, history and current status 568
        Definitions—A brief history—The current status
    III. Risks and challenges 575
        Recent developments concerning tactical nuclear weapons—Arms control developments—The Kaliningrad controversy
    IV. Prospects and means for further reductions 580
        Nuclear warhead transparency and control
    V. Conclusions 584

Appendix 10C. The changing Russian and US nuclear warhead production complexes 585
Oleg Bukharin

    I. Introduction 585
    II. The Russian complex 587
        Nuclear weapon complex reductions: the early years—The 1998 programme and its implementation—Beyond planned reductions
    III. The US complex 593
    IV. Conclusions 596

Table 10C.1. The Russian Minatom nuclear warhead production complex, 2001 588
Table 10C.2. The US DOE nuclear warhead production complex, 2001 594

Appendix 10D. Efforts to improve nuclear material and facility security 598
George Bunn and Lyudmila Zaitseva

    I. Introduction 598
    II. The attacks of 11 September and threats to nuclear facilities 598
    III. Illicit trafficking in nuclear and other radioactive material 601
    IV. International efforts after 11 September to improve security against terrorists 606
        Nuclear security in the largest nuclear weapon states—International efforts to strengthen physical protection
    V. Conclusions 612

Figure 10D.1. Incidents of illicit trafficking in nuclear and other radioactive material, 1991–2001 602
Figure 10D.2. Incidents of illicit trafficking involving nuclear material, other radioactive material and both, 1991–2001 603

11. The military uses of outer space 613
John Pike

    I. Introduction 613
    II. The United States 614
        Communications satellites—Navigation satellites—Weather satellites—Early-warning satellites—Ocean-surveillance satellites—Signals intelligence satellites—Imagery intelligence satellites—Anti-satellite systems—Technology development
    III. Russia 627
        Navigation satellites—Communications satellites—Weather satellites—Early-warning satellites—Electronic intelligence satellites—Ocean-surveillance satellites—Imagery intelligence satellites—Anti-satellite systems

	IV. Other countries	635
	Australia—China—Europe—India—Israel—Japan—South Korea— Taiwan—Turkey	
	V. Commercial operators	642
	Space Imaging—DigitalGlobe—Orbimage—ImageSat International	
	VI. Prevention of an arms race in outer space	645
	VII. Ballistic missile defence	647
	The Terminal Defense Segment—The Midcourse Defense Segment— The Boost Defense Segment	
	VIII. Conclusions	654
	IX. Tables of operational military satellites	655
Table 11.1.	US operational military satellites, as of 31 December 2001	656
Table 11.2.	Russian operational military satellites, as of 31 December 2001	660
Table 11.3.	Rest of the world, operational military satellites, as of 31 December 2001	664

12. Chemical and biological weapon developments and arms control 665
Jean Pascal Zanders, John Hart and Frida Kuhlau

	I. Introduction	665
	II. Biological weapon disarmament	666
	Developments in the Ad Hoc Group—The suspended protocol—US objections to the projected protocol—The Fifth Review Conference of the Parties to the BTWC	
	III. Biotechnology, biological defence research and the BTWC	677
	The US BW defence research programme—The BTWC and BW defence	
	IV. Chemical weapon disarmament	683
	Implementing the CWC—Destruction of chemical weapons and related facilities—Old and abandoned chemical weapons	
	V. Terrorism with mail-delivered anthrax spores	696
	Anthrax bacteria as a biological warfare agent—Letters as a means of delivering anthrax bacteria—Proliferation implications of the anthrax attacks	
	VI. CBW proliferation	704
	US proliferation allegations—Iraq's CBW programmes and their elimination—The past South African CBW programme	
	VII. Conclusions	707
Table 12.1.	Type, location and amount of Russian CW destruction assistance	692

13. Conventional arms control 709
Zdzislaw Lachowski

	I. Introduction	709
	II. European arms control	711
	The Treaty on Conventional Armed Forces in Europe—Regional arms control in Europe	
	III. European CSBMs	719
	The fourth Vienna military doctrine seminar—The CSBM Agreement in Bosnia and Herzogovina—Negotiations under Article V of the Agreement on Regional Stabiliztion—New bilateral CSBM accords— The Treaty on Open Skies	
	IV. Non-European CSBM arrangements	727
	The ASEAN Regional Forum—Arms control in Central Asia—The OSCE–Korea CSBM seminar	
	V. Landmines and certain conventional weapons	730
	Landmines—The CCW Convention	

xiv SIPRI YEARBOOK 2002

	VI. Conclusions	734
Table 13.1.	CFE and CFE-1A ceilings and holdings in the ATTU zone, as of 1 January 2002	710
Table 13.2.	Russian entitlements and holdings in the flank zone under the 1999 Agreement on Adaptation, 1999–2002	716
Table 13.3.	The status of the APM Convention, as of 1 January 2002	732

Appendix 13A. The UN conference on the illicit trade in small arms and light weapons 736
Pieter D. Wezeman

    I. Introduction 736
    II. The UN conference 736
    III. Conclusions 739

Appendix 13B. Documents on conventional arms control 740
Concluding Document of the Negotiations under Article V of Annex 1B of the General Framework Agreement For Peace in Bosnia and Herzegovina

14. Multilateral export controls 743
Ian Anthony

    I. Introduction 743
    II. The Missile Technology Control Regime 745
        MTCR compliance issues—The International Code of Conduct and efforts to control ballistic missile proliferation
    III. The Nuclear Suppliers Group 751
    IV. The Wassenaar Arrangement 755
    V. The impact of the 11 September terrorist attacks on multilateral export control 756
    VI. Conclusions 758

Table 14.1. Membership of multilateral weapon and technology export control regimes, as of 1 January 2002   744

Annexes

Annex A. Arms control and disarmament agreements 761
Christer Berggren

Annex B. Chronology 2001 785
Christer Berggren

About the contributors 799

Abstracts 805

Errata 811

Index 812

To
Billie Bielckus
1946–2001
SIPRI Editor 1978–2001,
colleague and friend

Preface

The Stockholm International Peace Research Institute presents in this volume the 33rd edition of the SIPRI Yearbook. It contains the results of SIPRI's traditional research as well as new studies, reflecting the dramatic changes in the world security environment over the past 12 months.

The terrorist attacks on the United States on 11 September 2001 sent shock waves throughout the world and influenced the way in which many processes of international life are perceived, particularly those of security and arms control. The authors of this volume present comprehensive data sets and analyses, all based on open sources, which illuminate the new—or newly prominent—risks and threats to international security. Their findings provide a powerful resource for politicians, diplomats, analysts, journalists and the academic community.

SIPRI's research is organized around three main areas, reflected in the titles of the three parts of this volume: security and conflicts; military spending and armaments; and non-proliferation, arms control and disarmament. Three SIPRI research project leaders coordinated the three parts of this Yearbook—Taylor B. Seybolt, Elisabeth Sköns and Shannon N. Kile.

Among the special studies in this Yearbook, SIPRI's own research staff have contributed chapters on the qualitative change in arms control and on the sanctions applied by the European Union and the United States. Another chapter describes the new military dimension of the European Union, and an appendix provides a thorough examination of the existing conflict databases.

Several special studies were written for SIPRI by external experts. Dylan Hendrickson, King's College, London, and Andrzej Karkoszka, of the Geneva Centre for the Democratic Control of Armed Forces (DCAF), contributed a chapter on security sector reform. David Gold, of Rutgers University, analysed the trends in US military expenditure and the 2001 Quadrennial Defense Review. The chapter on ballistic missile defence and nuclear arms control is supplemented by appendices on world nuclear forces (contributed by Hans M. Kristensen, Nautilus Institute, and Joshua Handler, Princeton University); tactical nuclear weapons; the changing Russian and US nuclear warhead production complexes (by Oleg Bukharin, Princeton University); and efforts to improve the security of nuclear material and facilities in the light of the attacks of 11 September (George Bunn and Lyudmila Zaitseva, Stanford University). The chapter by John Pike (globalsecurity.org) provides a comprehensive overview of world spacecraft and the efforts to control the military uses of outer space. Finally, the Conflict Data Project of Uppsala University, under the direction of Peter Wallensteen, provided the annual data on major armed conflicts.

All the chapters and appendices provide sober, well-documented assessments of world security affairs. It is this research tradition that has built up SIPRI's high reputation over the past 37 years. The SIPRI Yearbook is also published annually in Russian, Ukrainian and other languages, several supported by generous contributions from DCAF. By making SIPRI's research results accessible to new professional and general readers in these countries, the Institute is offering an important contribution to openness, transparency, and democratic control over the military and the security sector.

A volume of this size could not have been produced in a year of such immense change—both in world affairs and in SIPRI—without the diligent work and support of the entire SIPRI staff, to whom we are deeply grateful. Particular appreciation is due to the editors with long professional experience of Yearbook production, Jetta Gilligan Borg and Connie Wall, and those who recently joined the staff, Pamela Boston and Andy Mash. Eve Johansson kindly spent time at the Institute checking the texts and Anna Lundeborg provided technical assistance at the final stage of production. Our appreciation also extends to the library staff, under the direction of Christine-Charlotte Bodell. Ian Anthony provided helpful guidance during the period of transition between directors. Connie Wall, Managing Editor, provided invaluable assistance to the authors.

Finally, we gratefully acknowledge the support of Carol Barta, Secretary for the Director's Office; the SIPRI IT Department, under the direction of Gerd Hagmeyer-Gaverus; Cynthia Loo, project assistant for the military expenditure, arms production and arms transfers research projects; and Peter Rea, indexer.

Adam Daniel Rotfeld
Director of SIPRI 1991–2002

Alyson J. K. Bailes
Director of SIPRI from July 2002

May 2002

Glossary

CHRISTER BERGGREN and CONNIE WALL

Acronyms

ABACC	Brazilian–Argentine Agency for Accounting and Control of Nuclear Materials	CBM	Confidence-building measure
ABL	Airborne laser	CBO	Congressional Budget Office
ABM	Anti-ballistic missile	CBW	Chemical and biological weapon/warfare
ACM	Advanced Cruise Missile		
ACP	African, Caribbean and Pacific (countries)	CD	Conference on Disarmament
		CEE	Central and Eastern Europe
ACV	Armoured combat vehicle	CEI	Central European Initiative
ACW	Abandoned chemical weapon	CFE	Conventional Armed Forces in Europe (Treaty)
AEMI	Annual Exchange of Military Information	CFSP	Common Foreign and Security Policy
AG	Australia Group	CICA	Conference on Interaction and Confidence-Building Measures in Asia
AIAM	Annual Implementation Assessment Meeting		
ALA	Agreement-limited armaments	CIS	Commonwealth of Independent States
ALCM	Air-launched cruise missile	CPC	Conflict Prevention Centre
APM	Anti-personnel mine	CSBM	Confidence- and security-building measure
ARF	ASEAN Regional Forum		
ASEAN	Association of South-East Asian Nations	CSCAP	Council for Security Cooperation in the Asia Pacific
ATTU	Atlantic-to-the Urals (zone)	CTBT	Comprehensive Nuclear Test-Ban Treaty
AU	African Union		
BMDO	Ballistic Missile Defence Organization	CTR	Co-operative Threat Reduction
BMDS	Ballistic Missile Defence System	CW	Chemical weapon/warfare
		CWC	Chemical Weapons Convention
BSEC	Organization of Black Sea Economic Cooperation	DRC	Democratic Republic of the Congo
BTWC	Biological and Toxin Weapons Convention	DTRA	Defense Threat Reduction Agency
BW	Biological weapon/warfare		
C^3I	Command, control, communications and intelligence	EAEC	European Atomic Energy Community (Euratom)
		EAPC	Euro-Atlantic Partnership Council
CALCM	Conventional ALCM		

ECOMOG	ECOWAS Monitoring Group	ICRC	International Committee of the Red Cross
ECOWAS	Economic Community of West African States	IGAD	Intergovernmental Authority on Development
EDIG	European Defence Industries Group	IMF	International Monetary Fund
		INF	Intermediate-range Nuclear Forces (Treaty)
EDP	Engineering, manufacturing and development	IRBM	Intermediate-range ballistic missile
EPC	European Political Cooperation	ISTC	International Science and Technology Centre
ERRF	European rapid reaction force	JCG	Joint Consultative Group
ESDP	European Security and Defence Policy	JSF	Joint Strike Fighter
		KFOR	Kosovo Force
ETAP	European Technology Acquisition Plan	MPC&A	Material Protection, Control and Accounting
EU	European Union	MSF	Multinational Stabilization Force
FRY	Federal Republic of Yugoslavia	MTCR	Missile Technology Control Regime
FSC	Forum for Security Co-operation	NAM	Non-Aligned Movement
FY	Fiscal year	NATO	North Atlantic Treaty Organization
FYROM	Former Yugoslav Republic of Macedonia	NBC	Nuclear, biological and chemical (weapons)
G7	Group of Seven		
G8	Group of Eight	NGO	Non-governmental organization
GBI	Ground-based interceptor	NMD	National missile defence
GCS	Global Control System	NNSA	National Nuclear Security Agency
GDP	Gross domestic product		
GEMI	Global Exchange of Military Information	NPT	Non-Proliferation Treaty
GLCM	Ground-launched cruise missile	NSG	Nuclear Suppliers Group
		OAS	Organization of American States
GNI	Gross national income		
GNP	Gross national product	OAU	Organization of African Unity
HEU	Highly enriched uranium		
IAEA	International Atomic Energy Agency	OCHA	Office for the Coordination of Humanitarian Affairs
IANSA	International Action Network on Small Arms	OCW	Old chemical weapons
		OECD	Organisation for Economic Co-operation and Development
ICBM	Intercontinental ballistic missile		
ICOC	International Code of Conduct	OEF	Operation Enduring Freedom

OIC	Organization of the Islamic Conference	SLCM	Sea-launched cruise missile
OIP	Office of the Iraq Programme	SNDV	Strategic nuclear delivery vehicle
OMB	Office of Management and Budget	SRBM	Short-range ballistic missile
		SSM	Surface-to-surface missile
OPCW	Organisation for the Prohibition of Chemical Weapons	SSP	Stockpile Stewardship Program
		START	Strategic Arms Reduction Treaty
OSCC	Open Skies Consultative Commission	THAAD	Theater High-Altitude Area Defense
OSCE	Organization for Security and Co-operation in Europe	TLE	Treaty-limited equipment
		TMD	Theatre missile defence
PC	Permanent Council	TNF	Theatre nuclear forces
PCASED	Programme for Coordination and Assistance for Security and Development	UN	United Nations
		UNDP	UN Development Programme
PFP	Partnership for Peace		
PNI	Presidential Nuclear Initiative	UNHCR	UN High Commissioner for Refugees
PPP	Purchasing power parity	UNMOVIC	UN Monitoring, Verification and Inspection Commission
R&D	Research and development		
RACVIAC	Regional Arms Control Verification and Implementation Assistance Centre	UNSCOM	UN Special Commission on Iraq
		UTLE	Unaccounted-for and uncontrolled treaty-limited equipment
RD&E	Research, development and engineering		
		WA	Wassenaar Arrangement
RDT&E	Research, development, testing and evaluation	WEU	Western European Union
		WFP	World Food Programme
SADC	Southern African Development Community	WMD	Weapon of mass destruction
		WTO	Warsaw Treaty Organization (Warsaw Pact)
SALW	Small arms and light weapons		
		WTO	World Trade Organization
SAM	Surface-to-air missile		
SANDF	South African National Defence Forces		
SBIRS	Satellite-based infrared system		
SCC	Standing Consultative Commission		
SCO	Shanghai Cooperation Organization		
SLBM	Submarine-launched ballistic missile		

Intergovernmental bodies and international organizations

The main organizations discussed in this Yearbook are described in the glossary. Members or participants of the organizations are listed on pages xxix–xxxv. For acronyms that appear here, see pages xix–xxi; for the arms control and disarmament agreements mentioned in the glossary, see annex A.

African Union (AU)	The Constitutive Act of the African Union entered into force on 26 May 2001, formally establishing the African Union, with headquarters in Addis Ababa, Ethiopia, and open to membership of all African states. Its objective will be to coordinate and harmonize the promotion of enhanced unity, security, democracy, human rights, economic conditions and sustainable development in Africa. The AU will replace the OAU after a transitional period of one year or more, and all the OAU member states are expected to be members of the AU when it becomes operational. *See* Organization of Africa Unity.
Agency for the Prohibition of Nuclear Weapons in Latin America and the Caribbean (OPANAL)	Established by the 1967 Treaty of Tlatelolco to resolve, together with the IAEA, questions of compliance with the treaty.
Arab League	The League of Arab States, established in 1945, with Permanent Headquarters in Cairo. Its principal objective is to form closer union among Arab states and foster political and economic cooperation. An agreement for collective defence and economic cooperation among the members was signed in 1950. *See* the list of members.
Association of South-East Asian Nations (ASEAN)	Established in 1967 to promote economic, social and cultural development as well as regional peace and security in South-East Asia. The seat of the Secretariat is in Jakarta, Indonesia. The ASEAN Regional Forum (ARF) was established in 1993 to address security issues. The ASEAN Post Ministerial Conference (ASEAN–PMC) was established in 1979 as a forum for discussions of political and security issues with Dialogue Partners. *See* the lists of the members of ASEAN, ARF and ASEAN–PMC.
Australia Group (AG)	Group of states, formed in 1985, which meets informally each year to monitor the proliferation of chemical and biological products and to discuss chemical and biological weapon-related items which should be subject to national regulatory measures. *See* the list of members.
Black Sea Economic Cooperation (BSEC)	*See* Organization of Black Sea Economic Cooperation.
Central European Initiative (CEI)	Established in 1989 to promote cooperation among members in the political and economic spheres. It provides support to its non-EU members in their process of accession to the EU. The seat of the Executive Secretariat is in Trieste, Italy. *See* the list of members.

GLOSSARY xxiii

Commonwealth of Independent States (CIS)	Established in 1991 as a framework for multilateral cooperation among former Soviet republics. *See* the list of members.
Commonwealth of Nations	An organization established as the Imperial Conference which adopted the present name in 1971, with its Secretariat in London. It is an organization of 54 developed and developing states whose aim is to advance democracy, human rights, and sustainable economic and social development within its member countries and beyond.
Comprehensive Nuclear Test-Ban Treaty Organization (CTBTO)	Established by the 1996 CTBT to resolve questions of compliance with the treaty and as a forum for consultation and cooperation among the states parties. Its seat is in Vienna.
Conference on Disarmament (CD)	A multilateral arms control negotiating body, set up in 1961 as the Eighteen-Nation Committee on Disarmament; it has been enlarged and renamed several times and has been called the Conference on Disarmament since 1984. The CD is based in Geneva, Switzerland, and is today composed of states representing all the regions of the world, including the permanent members of the UN Security Council. It reports to the UN General Assembly. *See* the list of members under United Nations.
Conference on Interaction and Confidence-Building Measures in Asia (CICA)	Initiated in 1992, and established by the 1999 Declaration on the Principles Guiding Relations among the CICA Member States, as a forum to enhance security cooperation and confidence-building measures among the member states. It also promotes economic, social and cultural cooperation. *See* the list of members.
Council for Security Cooperation in the Asia Pacific (CSCAP)	Established in 1993 as an informal, non-governmental process for regional confidence building and security cooperation through dialogue and consultation in Asia–Pacific security matters. *See* the list of members.
Council of Europe	Established in 1949, with its seat in Strasbourg, France, the Council is open to membership of all the European states that accept the principle of the rule of law and guarantee their citizens human rights and fundamental freedoms. Among its organs is the European Court of Human Rights and the Council of Europe Development Bank (CEB). *See* the list of members.
Economic Community of West African States (ECOWAS)	A regional organization established in 1975, with its Executive Secretariat in Lagos, Nigeria, to promote trade and cooperation and contribute to development in West Africa. In 1981 it adopted the Protocol on Mutual Assistance in Defence Matters. The ECOWAS Monitoring Group (ECOMOG) was established in 1990. *See* the list of members.
Euro-Atlantic Partnership Council (EAPC)	Established in 1997, the EAPC provides the overarching framework for cooperation between NATO and its PFP partners. *See* the list of members under North Atlantic Treaty Organization.

European Atomic Energy Community (Euratom or EAEC)
: Created by the 1957 Treaty Establishing the European Atomic Energy Community (Euratom Treaty) to promote the development of nuclear energy for peaceful purposes and to administer the multinational regional safeguards system covering the EU member states. Euratom is located in Brussels. The members of Euratom are the EU member states

European Union (EU)
: Organization of European states, with its headquarters in Brussels. The 1992 Treaty on European Union (Maastricht Treaty), which created the EU, entered into force in 1993. The 1997 Treaty of Amsterdam Amending the Treaty on European Union (Amsterdam Treaty), which entered into force in 1999, strengthens the political dimension of the EU. The 2000 Treaty of Nice settles institutional issues not dealt with in the Treaty of Amsterdam and prepares the EU for further enlargement. The three EU pillars are: cooperation in economic and monetary affairs and Euratom; the Common Foreign and Security Policy (CFSP); and cooperation in justice and home affairs. *See also* European Atomic Energy Community, and *see* the list of members.

Group of Seven/Eight (G7/G8)
: Group of the seven leading industrialized nations which have met informally, at the level of heads of state or government, since the 1970s; from 1997 Russia has participated with the G7 in meetings of the G8. It holds annual meetings and has agreed practical initiatives on conflict prevention and has been involved in various peace settlements. *See* the list of members.

Group of 21 (G-21)
: Originally 21, now over 30, non-aligned CD member states which act together on proposals of common interest. It is also known as the Group of Non-Aligned Countries.

Intercontinental ballistic missile (ICBM)
: Ground-launched ballistic missile with a range longer than 5500 km.

Intergovernmental Authority on Development (IGAD)
: Established in 1996 to promote peace and stability in the Horn of Africa and to create mechanisms for conflict prevention, management and resolution. Its Secretariat is in Djibouti. *See* the list of members.

International Atomic Energy Agency (IAEA)
: An intergovernmental organization within the UN system, with headquarters in Vienna. The IAEA is endowed by its Statute, which entered into force in 1957, to promote the peaceful uses of atomic energy and ensure that nuclear activities are not used to further any military purpose. It has cooperated with UNSCOM (and is requested in UN Security Council Resolution 1284 (1999) to assist UNMOVIC) in carrying out the removal of nuclear weapon-usable material from Iraq. Under the NPT and the nuclear weapon-free zone treaties, non-nuclear weapon states must accept IAEA nuclear safeguards to demonstrate the fulfilment of their obligation not to manufacture nuclear weapons. *See* the list of IAEA members under United Nations.

GLOSSARY xxv

Joint Consultative Group (JCG)	Established by the 1990 CFE Treaty to promote the objectives and implementation of the treaty by reconciling ambiguities of interpretation and implementation. Under the 1999 Agreement on Adaptation of the CFE Treaty (not in force) the JCG will also address issues arising from the intentions of states to revise their TLE ceilings, consider cooperative measures to enhance the verification regime, consider requests to accede to the treaty and conduct any further negotiations.
Joint Compliance and Inspection Commission (JCIC)	The forum established by the 1991 START I Treaty to resolve questions of compliance, clarify ambiguities and discuss ways to improve implementation of the START treaties. It convenes at the request of at least one of the parties.
Minsk Group	Group of states created in 1992 which act together in the OSCE for political settlement of the conflict in the Armenian enclave of Nagorno-Karabakh in Azerbaijan. *See* the list of members under Organization for Security and Co-operation in Europe.
Missile Technology Control Regime (MTCR)	An informal military-related export control regime, established in 1987, which in 1987 produced the Guidelines for Sensitive Missile-Relevant Transfers (revised in 1992). Its goal is to limit the spread of weapons of mass destruction by controlling ballistic missile delivery systems. *See* the list of members.
NATO–Russia Permanent Joint Council (PJC)	Established by the 1997 NATO–Russia Founding Act on Mutual Relations, Cooperation and Security for regular exchanges of information and consultation. In 2001 the NATO and Russian foreign ministers decided to form a new cooperation mechanism within a council of 20 (instead of 19 plus Russia), to elaborate and implement joint decisions and actions and replace the PJC. The procedures of the new council will be worked out during the spring of 2002.
NATO–Ukraine Commission	Established by the 1997 NATO–Ukraine Charter on a Distinctive Partnership, the commission meets for consultations on political and security issues, conflict prevention and resolution, non-proliferation, arms exports and technology transfers, and other subjects of common concern.
Non-Aligned Movement (NAM)	Established in 1961 as a forum for consultations and coordination of positions in the United Nations on political, economic and arms control issues among non-aligned states. *See* the list of members.
North Atlantic Treaty Organization (NATO)	Established in 1949 by the North Atlantic Treaty (Washington Treaty) as a Western defence alliance. Article 5 of the treaty defines the members' commitment to respond to an armed attack on any party to the treaty. The 1999 NATO Strategic Concept states that the alliance will seek to prevent conflict or, should a crisis arise, contribute to its effective management, consistent with international law, including through the possibility of conducting non-Article 5 crisis response operations. Its headquarters are in Brussels. *See* the list of members.

Nuclear Suppliers Group (NSG)	Also known as the London Club and established in 1975, the NSG coordinates multilateral export controls on nuclear materials. In 1977 it agreed the Guidelines for Nuclear Transfers (London Guidelines, revised in 2000), which contain a 'trigger list' of materials that should trigger IAEA safeguards when exported for peaceful purposes to any non-nuclear weapon state. In 1992 the NSG agreed the Guidelines for Transfers of Nuclear-Related Dual-Use Equipment, Material, Software and Related Technology (Warsaw Guidelines, revised in 2000). *See* the list of members.
Open Skies Consultative Commission (OSCC)	Established by the 1992 Open Skies Treaty to resolve questions of compliance with the treaty.
Organisation for Economic Co-operation and Development (OECD)	Established in 1961, its objectives are to promote economic and social welfare by coordinating policies among the member states. Its headquarters are in Paris. *See* the list of members.
Organisation for the Prohibition of Chemical Weapons (OPCW)	Established by the 1993 Chemical Weapons Convention as a body for the parties to oversee implementation of the convention and resolve questions of compliance. Its seat is in The Hague, the Netherlands.
Organisme Conjoint de Coopération en Matière d'Armement (OCCAR)	Established in 1996, with headquarters in Bonn, Germany, as a management structure for international cooperative armaments programmes between France, Germany, Italy and the UK. It is also known as the Joint Armaments Cooperation Organization (JACO).
Organization for Security and Co-operation in Europe (OSCE)	Initiated in 1973 as the Conference on Security and Co-operation in Europe (CSCE), which adopted the Helsinki Final Act in 1975. The 1990 Charter of Paris for a New Europe set up several standing institutions and regular summit meetings. The new mandate included the implementation of human rights, pluralistic democracy (election monitoring), and economic and environmental security. In 1995 it was renamed the OSCE and transformed into an organization, with headquarters in Vienna, as a primary instrument for early warning, conflict prevention and crisis management. Its Forum for Security Co-operation (FSC), located in Vienna, Austria, deals with arms control and CSBMs. The OSCE comprises several institutions, all located in Europe. *See* the list of members.
Organization of African Unity (OAU)	A union of African states established in 1963 to promote African international cooperation and harmonization of *inter alia* defence policies. It has mediated in conflicts and prepared peace and arms control agreements. The seat of the Secretary-General is in Addis Ababa, Ethiopia. The African Union will replace the OAU after a transitional period. *See* African Union, and *see* the list of members.
Organization of American States (OAS)	Group of states in the Americas which adopted a charter in 1948, with the objective of strengthening peace and security in the western hemisphere. The General Secretariat is in Washington, DC. *See* the list of members.

GLOSSARY xxvii

Organization of Black Sea Economic Cooperation (BSEC)	Established in 1992 as Black Sea Economic Cooperation and renamed the Organization of Black Sea Economic Cooperation in 1999, with its Permanent Secretariat in Istanbul, Turkey. Its aims are to ensure peace, stability and prosperity in the Black Sea region and to promote and develop economic cooperation and progress.
Organization of the Islamic Conference (OIC)	Established in 1971 by Islamic states to promote cooperation among the members and to support peace, security and the struggle of the people of Palestine and all Muslim people. Its Secretariat is in Jedda, Saudi Arabia. *See* the list of members.
Partnership for Peace (PFP)	Launched in 1994, the PFP is the programme for political and military cooperation between NATO and its partner states within the framework of the EAPC. It is open to all OSCE states able to contribute to the programme. The Enhanced PFP programme, adopted in 1997, is intended to strengthen political consultation, develop a more operational role, and provide for greater involvement of partners in PFP decision making and planning. *See* the list of members under North Atlantic Treaty Organization.
Shanghai Cooperation Organization (SCO)	Established in 1996 as the Shanghai Five and later called the Shanghai Forum. In 2001 it was renamed the Shanghai Cooperation Organization and opened to membership of all states that support its aims. The member states cooperate on confidence-building measures and regional security and in the economic sphere. *See* the list of members.
South Pacific Forum	A group of South Pacific states which hold high-level meetings. It proposed the South Pacific Nuclear Free Zone, embodied in the 1985 Treaty of Rarotonga. The Secretariat is in Suva, Fiji. *See* the list of members.
Southern African Development Community (SADC)	Established in 1992 to promote regional economic development and fundamental principles of sovereignty, peace and security, human rights and democracy. The Secretariat is in Gaborone, Botswana. *See* the list of members.
Stability Pact for South Eastern Europe	Initiated by the EU at the 1999 Conference on South Eastern Europe, convened in Cologne, Germany, and placed under OSCE auspices. The facilitating states, organizations and institutions endorsed the Stability Pact through the 1999 Sarajevo Summit Declaration. The Pact is to promote political and economic reforms, development and enhanced security, and facilitate the integration of south-east European countries into the Euro-Atlantic structures. Its activities are coordinated by the South Eastern Europe Regional Table, chaired by the Special Co-ordinator of the Stability Pact, appointed by the EU after consultations with the OSCE Chairman-in-Office. The seat of the Special Co-ordinator is in Brussels. *See* the list of facilitating states, organizations and institutions.
Standing Consultative Commission (SCC)	Established by the 1972 ABM Treaty as the body to which parties may refer issues regarding implementation of the treaty.

Sub-Regional Consultative Commission (SRCC)	Established by the 1996 Agreement on Sub-Regional Arms Control concerning Yugoslavia (Florence Agreement) as the forum for the parties to resolve questions of compliance with the agreement.
United Nations (UN)	The world intergovernmental organization, open to membership of all states, with headquarters in New York, founded in 1945 through the adoption of its Charter at San Francisco, California. Its six principal organs are the General Assembly, the Security Council, the Economic and Social Council (ECOSOC), the Trusteeship Council, the International Court of Justice (ICJ) and the Secretariat. It also has a large number of specialized agencies and other autonomous bodies. *See* the list of members.
Wassenaar Arrangement (WA)	The Wassenaar Arrangement on Export Controls for Conventional Arms and Dual-Use Goods and Technologies was formally established in 1996. It aims to prevent the acquisition of armaments and sensitive dual-use goods and technologies for military uses by states whose behaviour is cause for concern to the member states. *See* the list of members.
Western European Union (WEU)	Established by the 1954 Modified Brussels Treaty. The seat of the WEU is in Brussels. After the transfer of the WEU's operational activities (the 'Petersberg tasks') to the EU in 2000, the residual organization is essentially intended to ensure the respect of obligations stemming from Article V (mutual assistance in case of aggression) of the Modified Brussels Treaty and the continuation of the activities of the Western European Armaments Group (WEAG). The Parliamentary Assembly, seated in Paris, also continues to function.
Zangger Committee	Established in 1971, the Nuclear Exporters Committee, called the Zangger Committee after its first chairman, is a group of nuclear supplier countries that meets informally twice a year to coordinate export controls on nuclear materials. *See* the list of members.

Membership of intergovernmental bodies and international organizations as of 1 January 2002

The UN member states and organizations within the UN system are listed first, followed by all other organizations in alphabetical order. Note that not all members or participants of the organizations are UN member states. Information is as of 1 January 2002 unless otherwise indicated in a note.

United Nations members (189) and year of membership

Afghanistan, 1946
Albania, 1955
Algeria, 1962
Andorra, 1993
Angola, 1976
Antigua and Barbuda, 1981
Argentina, 1945
Armenia, 1992
Australia, 1945
Austria, 1955
Azerbaijan, 1992
Bahamas, 1973
Bahrain, 1971
Bangladesh, 1974
Barbados, 1966
Belarus, 1945
Belgium, 1945
Belize, 1981
Benin, 1960
Bhutan, 1971
Bolivia, 1945
Bosnia and Herzegovina, 1992
Botswana, 1966
Brazil, 1945
Brunei Darussalam, 1984
Bulgaria, 1955
Burkina Faso, 1960
Burundi, 1962
Cambodia, 1955
Cameroon, 1960
Canada, 1945
Cape Verde, 1975
Central African Republic, 1960
Chad, 1960
Chile, 1945
China, 1945
Colombia, 1945
Comoros, 1975
Congo, Democratic Republic of the (DRC), 1960
Congo, Republic of, 1960
Costa Rica, 1945
Côte d'Ivoire, 1960
Croatia, 1992
Cuba, 1945
Cyprus, 1960

Czech Republic, 1993
Denmark, 1945
Djibouti, 1977
Dominica, 1978
Dominican Republic, 1945
Ecuador, 1945
Egypt, 1945
El Salvador, 1945
Equatorial Guinea, 1968
Eritrea, 1993
Estonia, 1991
Ethiopia, 1945
Fiji, 1970
Finland, 1955
France, 1945
Gabon, 1960
Gambia, 1965
Georgia, 1992
Germany, 1973
Ghana, 1957
Greece, 1945
Grenada, 1974
Guatemala, 1945
Guinea, 1958
Guinea-Bissau, 1974
Guyana, 1966
Haiti, 1945
Honduras, 1945
Hungary, 1955
Iceland, 1946
India, 1945
Indonesia, 1950
Iran, 1945
Iraq, 1945
Ireland, 1955
Israel, 1949
Italy, 1955
Jamaica, 1962
Japan, 1956
Jordan, 1955
Kazakhstan, 1992
Kenya, 1963
Kiribati, 1999
Korea, Democratic People's Republic of (North Korea), 1991

Korea, Republic of (South Korea), 1991
Kuwait, 1963
Kyrgyzstan, 1992
Lao People's Democratic Republic (Laos), 1955
Latvia, 1991
Lebanon, 1945
Lesotho, 1966
Liberia, 1945
Libya, 1955
Liechtenstein, 1990
Lithuania, 1991
Luxembourg, 1945
Macedonia, Former Yugoslav Republic of (FYROM), 1993
Madagascar, 1960
Malawi, 1964
Malaysia, 1957
Maldives, 1965
Mali, 1960
Malta, 1964
Marshall Islands, 1991
Mauritania, 1961
Mauritius, 1968
Mexico, 1945
Micronesia, 1991
Moldova, 1992
Monaco, 1993
Mongolia, 1961
Morocco, 1956
Mozambique, 1975
Myanmar (Burma), 1948
Namibia, 1990
Nauru, 1999
Nepal, 1955
Netherlands, 1945
New Zealand, 1945
Nicaragua, 1945
Niger, 1960
Nigeria, 1960
Norway, 1945
Oman, 1971
Pakistan, 1947
Palau, 1994
Panama, 1945

Papua New Guinea, 1975
Paraguay, 1945
Peru, 1945
Philippines, 1945
Poland, 1945
Portugal, 1955
Qatar, 1971
Romania, 1955
Russia, 1945
Rwanda, 1962
Saint Kitts and Nevis, 1983
Saint Lucia, 1979
Saint Vincent and the
  Grenadines, 1980
Samoa, Western, 1976
San Marino, 1992
Sao Tome and Principe, 1975
Saudi Arabia, 1945
Senegal, 1960
Seychelles, 1976
Sierra Leone, 1961
Singapore, 1965
Slovakia, 1993
Slovenia, 1992
Solomon Islands, 1978
Somalia, 1960
South Africa, 1945
Spain, 1955
Sri Lanka, 1955
Sudan, 1956
Suriname, 1975
Swaziland, 1968
Sweden, 1946
Syria, 1945
Tajikistan, 1992
Tanzania, 1961
Thailand, 1946
Togo, 1960
Tonga, 1999
Trinidad and Tobago, 1962
Tunisia, 1956
Turkey, 1945
Turkmenistan, 1992
Tuvalu, 2000
Uganda, 1962
UK, 1945
Ukraine, 1945
United Arab Emirates, 1971
Uruguay, 1945
USA, 1945
Uzbekistan, 1992
Vanuatu, 1981
Venezuela, 1945
Viet Nam, 1977
Yemen, 1947
Yugoslavia, 2000
Zambia, 1964
Zimbabwe, 1980

Notes: In a referendum of 3 Mar. 2001, Switzerland voted in favour of joining the UN; it is expected to submit a formal application for membership during the UN General Asssembly session in Sep. 2002. On 14 Mar. 2002 it was proposed that the name of Yugoslavia be changed to Serbia and Montenegro, subject to the approval of the Serbian, Montenegran and Yugoslavian parliaments.

UN Security Council

Permanent members (the P5): China, France, Russia, UK, USA

Non-permanent members in 2001 (elected by the UN General Assembly for two-year terms; the year in brackets is the year at the end of which the term expires): Bangladesh (2001), Colombia (2002), Ireland (2002), Jamaica (2001), Mali (2001), Mauritius (2002), Norway (2002), Singapore (2002), Tunisia (2001), Ukraine (2001)

Note: Bulgaria, Cameroon, Guinea, Mexico and Syria were elected non-permanent members for 2002–2003.

Conference on Disarmament (CD)

Algeria, Argentina, Australia, Austria, Bangladesh, Belarus, Belgium, Brazil, Bulgaria, Cameroon, Canada, Chile, China, Colombia, Congo (Democratic Republic of), Cuba, Ecuador, Egypt, Ethiopia, Finland, France, Germany, Hungary, India, Indonesia, Iran, Iraq, Ireland, Israel, Italy, Japan, Kazakhstan, Kenya, Korea (North), Korea (South), Malaysia, Mexico, Mongolia, Morocco, Myanmar (Burma), Netherlands, New Zealand, Nigeria, Norway, Pakistan, Peru, Poland, Romania, Russia, Senegal, Slovakia, South Africa, Spain, Sri Lanka, Sweden, Switzerland, Syria, Tunisia, Turkey, UK, Ukraine, USA, Venezuela, Viet Nam, Yugoslavia*, Zimbabwe

* The Federal Republic of Yugoslavia does not participate in the work of the CD.

International Atomic Energy Agency (IAEA)

Afghanistan, Albania, Algeria, Angola, Argentina, Armenia, Australia, Austria, Azerbaijan, Bangladesh, Belarus, Belgium, Bolivia, Bosnia and Herzegovina, Brazil, Bulgaria, Burkina Faso, Cambodia, Cameroon, Canada, Central African Republic, Chile, China, Colombia, Congo (Democratic Republic of), Costa Rica, Côte d'Ivoire, Croatia, Cuba, Cyprus, Czech Republic, Denmark, Dominican Republic, Ecuador, Egypt, El Salvador, Estonia, Ethiopia, Finland, France, Gabon, Georgia, Germany, Ghana, Greece, Guatemala, Haiti, Holy See, Honduras*, Hungary, Iceland, India, Indonesia, Iran, Iraq, Ireland, Israel, Italy, Jamaica, Japan, Jordan, Kazakhstan, Kenya, Korea (South), Kuwait, Latvia, Lebanon, Liberia, Libya, Liechtenstein, Lithuania, Luxembourg, Macedonia (Former Yugoslav Republic of), Madagascar, Malaysia,

Mali, Malta, Marshall Islands, Mauritius, Mexico, Moldova, Monaco, Mongolia, Morocco, Myanmar (Burma), Namibia, Netherlands, New Zealand, Nicaragua, Niger, Nigeria, Norway, Pakistan, Panama, Paraguay, Peru, Philippines, Poland, Portugal, Qatar, Romania, Russia, Saudi Arabia, Senegal, Sierra Leone, Singapore, Slovakia, Slovenia, South Africa, Spain, Sri Lanka, Sudan, Sweden, Switzerland, Syria, Tajikistan, Tanzania, Thailand, Tunisia, Turkey, Uganda, UK, Ukraine, United Arab Emirates, Uruguay, USA, Uzbekistan, Venezuela, Viet Nam, Yemen, Yugoslavia, Zambia, Zimbabwe

* Honduras has been approved as a member by the General Conference but has not yet deposited its instrument of acceptance of the IAEA Statute.

Note: North Korea was a member of the IAEA until Sep. 1994.

Arab League

Algeria, Bahrain, Comoros, Djibouti, Egypt, Iraq, Jordan, Kuwait, Lebanon, Libya, Mauritania, Morocco, Oman, Palestine, Qatar, Saudi Arabia, Somalia, Sudan, Syria, Tunisia, United Arab Emirates, Yemen

Arab Maghreb Union (AMU)

Algeria, Libya, Mauritania, Morocco, Tunisia

Association of South-East Asian Nations (ASEAN)

Brunei Darussalam, Cambodia, Indonesia, Laos, Malaysia, Myanmar (Burma), Philippines, Singapore, Thailand, Viet Nam

ASEAN Regional Forum (ARF)

The 10 ASEAN states plus Australia, Canada, China, European Union, India, Japan, Korea (North), Korea (South), Mongolia, New Zealand, Papua New Guinea, Russia, USA

ASEAN Post Ministerial Conference (ASEAN–PMC)

The 10 ASEAN states plus Australia, Canada, China, European Union, India, Japan, Korea (South), New Zealand, Russia, USA

Australia Group (AG)

Argentina, Australia, Austria, Belgium, Bulgaria, Canada, Cyprus, Czech Republic, Denmark, Finland, France, Germany, Greece, Hungary, Iceland, Ireland, Italy, Japan, Korea (South), Luxembourg, Netherlands, New Zealand, Norway, Poland, Portugal, Romania, Slovakia, Spain, Sweden, Switzerland, Turkey, UK, USA

Observer: European Commission

Black Sea Economic Co-operation (BSEC)

Albania, Armenia, Azerbaijan, Bulgaria, Georgia, Greece, Moldova, Romania, Russia, Turkey, Ukraine

Central European Initiative (CEI)

Albania, Austria, Belarus, Bosnia and Herzegovina, Bulgaria, Croatia, Czech Republic, Hungary, Italy, Macedonia (Former Yugoslav Republic of), Moldova, Poland, Romania, Slovakia, Slovenia, Ukraine, Yugoslavia

Commonwealth of Nations

Antigua and Barbuda, Australia, Bahamas, Bangladesh, Barbados, Belize, Botswana, Brunei Darussalam Darussalam, Cameroon, Canada, Cyprus, Dominica, Fiji, Gambia, Ghana, Grenada, Guyana, India, Jamaica, Kenya, Kiribati, Lesotho, Malawi, Malaysia, Maldives, Malta, Mauritius, Mozambique, Namibia, Nauru, New Zealand, Nigeria, Pakistan, Papua New Guinea, Saint Kitts and Nevis, Saint Lucia, Saint Vincent and the Grenadines, Samoa, Seychelles, Sierra Leone, Singapore, Solomon Islands,

South Africa, Sri Lanka, Swaziland, Tanzania, Tonga, Trinidad and Tobago, Tuvalu, Uganda, UK, Vanuatu, Zambia, Zimbabwe*

* On 19 Mar. 2002 Zimbabwe was suspended from membership of the Commonwealth for 12 months.

Commonwealth of Independent States (CIS)

Armenia, Azerbaijan, Belarus, Georgia, Kazakhstan, Kyrgyzstan, Moldova, Russia, Tajikistan, Turkmenistan, Ukraine, Uzbekistan

Conference on Interaction and Confidence-Building Measures in Asia (CICA)

Afghanistan, Azerbaijan, China, Egypt, India, Iran, Israel, Kazakhstan, Kyrgyzstan, Pakistan, Palestinian Authority, Russia, Tajikistan, Turkey, Uzbekistan

Council for Security Cooperation in the Asia Pacific (CSCAP)

Member committees: Australia, Cambodia, Canada, China, Europe, India, Indonesia, Japan, Korea (North), Korea (South), Malaysia, Mongolia, New Zealand, Papua New Guinea, Philippines, Russia, Singapore, Thailand, USA, Viet Nam

Associate member committee: Institute for Defence Studies and Analyses (IDSA), India

Council of Europe

Albania, Andorra, Armenia, Austria, Azerbaijan, Belgium, Bulgaria, Croatia, Cyprus, Czech Republic, Denmark, Estonia, Finland, France, Germany, Georgia, Greece, Hungary, Iceland, Ireland, Italy, Latvia, Liechtenstein, Lithuania, Luxembourg, Macedonia (Former Yugoslav Republic of), Malta, Moldova, Netherlands, Norway, Poland, Portugal, Romania, Russia, San Marino, Slovakia, Slovenia, Spain, Sweden, Switzerland, Turkey, UK, Ukraine

Observers to the Committee of Ministers: Canada, Holy See, Japan, Mexico, USA

Observers to the Parliamentary Assembly: Canada, Israel, Mexico

Special guests to the Parliamentary Assembly: Bosnia and Herzegovina, Yugoslavia

Economic Community of West African States (ECOWAS)

Benin, Burkina Faso, Cape Verde, Côte d'Ivoire, Gambia, Ghana, Guinea, Guinea-Bissau, Liberia, Mali, Niger, Nigeria, Senegal, Sierra Leone, Togo

European Union (EU)

Austria, Belgium, Denmark, Finland, France, Germany, Greece, Ireland, Italy, Luxembourg, Netherlands, Portugal, Spain, Sweden, UK

Group of Seven/Eight (G7/G8)

Canada, France, Germany, Italy, Japan, UK, USA. As the G8, the members of the G7 plus Russia.

Intergovernmental Authority on Development (IGAD)

Djibouti, Eritrea, Ethiopia, Kenya, Somalia, Sudan, Uganda

Missile Technology Control Regime (MTCR)

Argentina, Australia, Austria, Belgium, Brazil, Canada, Czech Republic, Denmark, Finland, France, Germany, Greece, Hungary, Iceland, Ireland, Italy, Japan, Korea (South), Luxembourg, Netherlands, New Zealand, Norway, Poland, Portugal, Russia, South Africa, Spain, Sweden, Switzerland, Turkey, UK, Ukraine, USA

Non-Aligned Movement (NAM)

Afghanistan, Algeria, Angola, Bahamas, Bahrain, Bangladesh, Barbados, Belarus, Belize, Benin, Bhutan, Bolivia, Botswana, Brunei Darussalam, Burkina Faso, Burundi, Cambodia, Cameroon, Cape Verde, Central African Republic, Chad, Chile, Colombia, Comoros, Congo (Democratic Republic of), Congo (Republic of), Côte d'Ivoire, Cuba, Cyprus, Djibouti, Ecuador, Egypt, Equatorial Guinea, Eritrea, Ethiopia, Gabon, Gambia, Ghana, Grenada, Guatemala, Guinea, Guinea-Bissau, Guyana, Honduras, India, Indonesia, Iran, Iraq, Jamaica, Jordan, Kenya, Korea (North), Kuwait, Laos, Lebanon, Lesotho, Liberia, Libya, Madagascar, Malawi, Malaysia, Maldives, Mali, Malta, Mauritania, Mauritius, Mongolia, Morocco, Mozambique, Myanmar (Burma), Namibia, Nepal, Nicaragua, Niger, Nigeria, Oman, Pakistan, Palestine, Panama, Papua New Guinea, Peru, Philippines, Qatar, Rwanda, Saint Lucia, Sao Tome and Principe, Saudi Arabia, Senegal, Seychelles, Sierra Leone, Singapore, Somalia, South Africa, Sri Lanka, Sudan, Suriname, Swaziland, Syria, Tanzania, Thailand, Togo, Trinidad and Tobago, Tunisia, Turkmenistan, Uganda, United Arab Emirates, Uzbekistan, Vanuatu, Venezuela, Viet Nam, Yemen, Yugoslavia*, Zambia, Zimbabwe

* The Federal Republic of Yugoslavia has not been permitted to participate in NAM activities since 1992.

North Atlantic Treaty Organization (NATO)

Belgium, Canada, Czech Republic, Denmark, France*, Germany, Greece, Hungary, Iceland, Italy, Luxembourg, Netherlands, Norway, Poland, Portugal, Spain, Turkey, UK, USA

* France is not in the integrated military structures of NATO.

Euro-Atlantic Partnership Council (EAPC)

The 19 NATO states plus Albania, Armenia, Austria, Azerbaijan, Belarus, Bulgaria, Croatia, Estonia, Finland, Georgia, Ireland, Kazakhstan, Kyrgyzstan, Latvia, Lithuania, Macedonia (Former Yugoslav Republic of), Moldova, Romania, Russia, Slovakia, Slovenia, Sweden, Switzerland, Tajikistan, Turkmenistan, Ukraine, Uzbekistan

Partnership for Peace (PFP)

Albania, Armenia, Austria, Azerbaijan, Belarus, Bulgaria, Croatia, Czech Republic, Estonia, Finland, Georgia, Hungary, Ireland, Kazakhstan, Kyrgyzstan, Latvia, Lithuania, Macedonia (Former Yugoslav Republic of), Moldova, Poland, Romania, Russia, Slovakia, Slovenia, Sweden, Switzerland, Turkmenistan, Ukraine, Uzbekistan

Note: Tajikistan joined the PFP on 20 Feb. 2002.

Nuclear Suppliers Group (NSG)

Argentina, Australia, Austria, Belarus, Belgium, Brazil, Bulgaria, Canada, Cyprus, Czech Republic, Denmark, Finland, France, Germany, Greece, Hungary, Ireland, Italy, Japan, Korea (South), Latvia, Luxembourg, Netherlands, New Zealand, Norway, Poland, Portugal, Romania, Russia, Slovakia, Slovenia, South Africa, Spain, Sweden, Switzerland, Turkey, UK, Ukraine, USA

Organisation for Economic Co-operation and Development (OECD)

Australia, Austria, Belgium, Canada, Czech Republic, Denmark, Finland, France, Germany, Greece, Hungary, Iceland, Ireland, Italy, Japan, Korea (South), Luxembourg, Mexico, Netherlands, New Zealand, Norway, Poland, Portugal, Slovakia, Spain, Sweden, Switzerland, Turkey, UK, USA

The European Commission participates in the work of the OECD.

Organization for Security and Co-operation in Europe (OSCE)

Albania, Andorra, Armenia, Austria, Azerbaijan, Belarus, Belgium, Bosnia and Herzegovina, Bulgaria, Canada, Croatia, Cyprus, Czech Republic, Denmark, Estonia, Finland, France, Georgia, Germany, Greece, Holy See, Hungary, Iceland, Ireland, Italy, Kazakhstan, Kyrgyzstan, Latvia, Liechtenstein, Lithuania, Luxembourg, Macedonia (Former Yugoslav Republic of), Malta, Moldova, Monaco, Netherlands, Norway, Poland, Portugal, Romania, Russia, San Marino, Slovakia, Slovenia, Spain, Sweden, Switzerland, Tajikistan, Turkey, Turkmenistan, UK, Ukraine, USA, Uzbekistan, Yugoslavia

Members of the Minsk Group in 2001: Armenia, Austria, Azerbaijan, Belarus, France, Germany, Italy, Portugal, Romania, Russia, Sweden, Turkey, USA

Organization of African Unity (OAU)

Algeria, Angola, Benin, Botswana, Burkina Faso, Burundi, Cameroon, Cape Verde, Central African Republic, Chad, Comoros, Congo (Democratic Republic of), Congo (Republic of), Côte d'Ivoire, Djibouti, Egypt, Equatorial Guinea, Eritrea, Ethiopia, Gabon, Gambia, Ghana, Guinea, Guinea-Bissau, Kenya, Lesotho, Liberia, Libya, Madagascar, Malawi, Mali, Mauritania, Mauritius, Mozambique, Namibia, Niger, Nigeria, Rwanda, Western Sahara (Saharawi Arab Democratic Republic, SADR*), Sao Tome and Principe, Senegal, Seychelles, Sierra Leone, Somalia, South Africa, Sudan, Swaziland, Tanzania, Togo, Tunisia, Uganda, Zambia, Zimbabwe

* Western Sahara was admitted in 1982, but its membership was disputed by Morocco and other states. Morocco withdrew from the OAU in 1985.

Organization of American States (OAS)

Antigua and Barbuda, Argentina, Bahamas, Barbados, Belize, Bolivia, Brazil, Canada, Chile, Colombia, Costa Rica, Cuba*, Dominica, Dominican Republic, Ecuador, El Salvador, Grenada, Guatemala, Guyana, Haiti, Honduras, Jamaica, Mexico, Nicaragua, Panama, Paraguay, Peru, Saint Kitts and Nevis, Saint Lucia, Saint Vincent and the Grenadines, Suriname, Trinidad and Tobago, Uruguay, USA, Venezuela

* Cuba has been excluded from participation since 1962.

Permanent observers: Algeria, Angola, Armenia, Austria, Azerbaijan, Belgium, Bosnia and Herzegovina, Bulgaria, Croatia, Cyprus, Czech Republic, Denmark, Egypt, Equatorial Guinea, Estonia, European Union, Finland, France, Germany, Ghana, Greece, Holy See, Hungary, India, Ireland, Israel, Italy, Japan, Kazakhstan, Korea (South), Latvia, Lebanon, Morocco, Netherlands, Norway, Pakistan, Philippines, Poland, Portugal, Romania, Russia, Saudi Arabia, Spain, Sri Lanka, Sweden, Switzerland, Thailand, Tunisia, Turkey, UK, Ukraine, Yemen

Organization of the Islamic Conference (OIC)

Afghanistan, Albania, Algeria, Azerbaijan, Bahrain, Bangladesh, Benin, Brunei Darussalam, Burkina Faso, Cameroon, Chad, Comoros, Côte d'Ivoire, Djibouti, Egypt, Gabon, Gambia, Guinea, Guinea-Bissau, Indonesia, Iran, Iraq, Jordan, Kazakhstan, Kuwait, Kyrgyzstan, Lebanon, Libya, Malaysia, Maldives, Mali, Mauritania, Morocco, Mozambique, Niger, Nigeria, Oman, Pakistan, Palestine, Qatar, Saudi Arabia, Senegal, Sierra Leone, Somalia, Sudan, Suriname, Syria, Tajikistan, Togo, Tunisia, Turkey, Turkmenistan, Uganda, United Arab Emirates, Uzbekistan, Yemen

Observers: Bosnia and Herzegovina, Central African Republic, Thailand

Shanghai Cooperation Organization (SCO)

China, Kazakhstan, Kyrgyzstan, Russia, Tajikistan, Uzbekistan

South Pacific Forum

Australia, Cook Islands, Fiji, Kiribati, Marshall Islands, Micronesia, Nauru, New Zealand, Niue, Palau, Papua New Guinea, Samoa (Western), Solomon Islands, Tonga, Tuvalu, Vanuatu

Southern African Development Community (SADC)

Angola, Botswana, Congo (Democratic Republic of), Lesotho, Malawi, Mauritius, Mozambique, Namibia, Seychelles, South Africa, Swaziland, Tanzania, Zambia, Zimbabwe

Stability Pact for South Eastern Europe

The European Union member states and the European Commission

The countries of the region and their neighbours: Albania, Bosnia and Herzegovina, Bulgaria, Croatia, Czech Republic, Hungary, Macedonia (Former Yugoslav Republic of), Moldova, Poland, Romania, Slovakia, Slovenia, Yugoslavia and Turkey

Non-EU members of the G8: USA, Canada, Japan and Russia

Other countries: Norway and Switzerland

International organizations: United Nations (UN), Organization for Security and Co-operation in Europe (OSCE), Council of Europe, UN High Commissioner for Refugees (UNHCR), North Atlantic Treaty Organization (NATO), Organisation for Economic Co-operation and Development (OECD)

International financial institutions: World Bank, International Monetary Fund (IMF), European Bank for Reconstruction and Development (EBRD) and European Investment Bank (EIB), Council of Europe Development Bank (CEB)

Regional initiatives: Black Sea Economic Co-operation (BSEC), Central European Initiative (CEI), South East European Co-operative Initiative (SECI), South East Europe Co-operation Process (SEECP)

Wassenaar Arrangement (WA)

Argentina, Australia, Austria, Belgium, Bulgaria, Canada, Czech Republic, Denmark, Finland, France, Germany, Greece, Hungary, Ireland, Italy, Japan, Korea (South), Luxembourg, Netherlands, New Zealand, Norway, Poland, Portugal, Romania, Russia, Slovakia, Spain, Sweden, Switzerland, Turkey, UK, Ukraine, USA

Western European Union (WEU)

Belgium, France, Germany, Greece, Italy, Luxembourg, Netherlands, Portugal, Spain, UK

Associate Members: Czech Republic, Hungary, Iceland, Norway, Poland, Turkey

Associate Partners: Bulgaria, Estonia, Latvia, Lithuania, Romania, Slovakia, Slovenia

Observers: Austria, Denmark, Finland, Ireland, Sweden

Members of WEAG and WEAO: Austria, Belgium, Czech Republic, Denmark, Finland, France, Germany, Greece, Hungary, Italy, Luxembourg, Netherlands, Norway, Poland, Portugal, Spain, Sweden, Turkey, UK

Zangger Committee

Argentina, Australia, Austria, Belgium, Bulgaria, Canada, China, Czech Republic, Denmark, Finland, France, Germany, Greece, Hungary, Ireland, Italy, Japan, Korea (South), Luxembourg, Netherlands, Norway, Poland, Portugal, Romania, Russia, Slovakia, Slovenia, South Africa, Spain, Sweden, Switzerland, Turkey, UK, Ukraine, USA

Conventions

. .	Data not available or not applicable
–	Nil or a negligible figure
( )	Uncertain data
b.	Billion (thousand million)
km	Kilometre (1000 metres)
kt	Kiloton (1000 tonnes)
m.	Million
Mt	Megaton (1 million tonnes)
th.	Thousand
tr.	Trillion (million million)
$	US dollars, unless otherwise indicated
€	Euros

Introduction
Global security after 11 September 2001

ADAM DANIEL ROTFELD

I. Introduction

The 11 September 2001 terrorist attacks on the United States marked a watershed in the international security process. From the ruins of the World Trade Center in New York and the destroyed Pentagon building grew the catalyst for the shaping of a new security system.[1] Unresolved issues, clashes of interests and conflicts began to be analysed from a different perspective, with new issues and priorities. The term 'post-cold war period' is no longer adequate.[2]

Although the unprecedented terrorist acts of 2001 were directed against the United States, they were the impulse for a reassessment and redefinition of the security policies by practically all states and major international security institutions. The attacks helped precipitate the shaping of a new global security system.

II. Global terrorism and global responsibility

In many parts of the world, the relationship between globalization and massscale terrorism is still ignored or underestimated, and globalization phenomena are seen there as 'Americanization'.[3] Two major forces are driving the evolution of the world today—modernization and globalization accompanied by fragmentation. The terrorist attacks of 11 September are perceived as the start of the first major war of the age of globalization.[4] However, although globalization has changed the way in which war is waged, the attacks were not a war on globalization. They were not a war of the poor against the rich, although the blatant disproportion in development and income between a handful of high-income countries of the Organisation for Economic Co-operation and

---

[1] On 11 Sep. 2001 a group of al-Qaeda terrorists hijacked 4 US aircraft on domestic flights and deliberately crashed 2 of them into the World Trade Center in New York and 1 into the Pentagon in Arlington, Virginia, across the Potomac River from Washington, DC. The fourth aircraft, heading for Washington, crashed in Pennsylvania. About 3000 people died in the attacks.

[2] During the cold war, the term 'bipolar' security system denoted the division of the world into 2 antagonistic poles or blocs. The term 'post-cold war era', as put by the US Under Secretary of State for Political Affairs, described our environment not for what was, but for what it was not. Grossman, M., 'Global trends for the coming decade and the formulation of US foreign policy', Speech delivered to the National Newspaper Association, 21 Mar. 2002, *Washington File*, 21 Mar. 2002, URL <http://usinfo.state.gov/topical/pol/nato/02032705.htm>.

[3] Bauman, Z., 'Imiona cierpienia, imiona wstydu' [Names of anguish, names of shame], *Tygodnik Powszechny*, no. 38 (23 Sep. 2001), p. 9.

[4] Campbell, K. M., 'Globalization's first war?', *Washington Quarterly*, vol. 25, no. 1 (winter 2002), pp. 7–14.

Development (OECD) and the rest of the world is of great importance for understanding the essence of conflicts.[5] The terrorist attacks were not a war of civilizations, although they were inspired, masterminded and executed under the banner of Islam. In this regard, the question arises how the experience of 11 September and the war on global terrorism[6] will affect the shaping of a new global security system.

In the security field, the policies and mutual relations of the United States, Russia and many other states have changed. It is widely acknowledged that combating terrorism has become a matter of the highest priority.[7] However, the transatlantic community is confronted with a disagreement on what should be the main aim: whether to focus on disrupting and defeating the al-Qaeda network (as urged by the USA) or eliminating the roots of terrorism with a broader range of policies, also stressing non-military measures (as preferred by the European states). As a result of the intervention against the al-Qaeda network and its allies in Afghanistan initiated later in 2001, the processes of the internal transformation and enlargement of the North Atlantic Treaty Organization (NATO) and the European Union (EU) have accelerated. The states of Central Asia have gained in significance, and the policies of, for example, Kazakhstan, Kyrgyzstan, Pakistan, Tajikistan, Turkey and Uzbekistan are now more salient than those of many European states.

Meanwhile, the new US defence policy document reflects a change in the philosophy of US defence planning—from the 'threat-based' pattern that has dominated thinking in the past to a 'capabilities-based' model for the future.[8] Indeed, the premises of defence planning today need to be rapidly and flexibly adapted to the new, volatile circumstances—'adapting to surprise', in the

---

[5] Illustrative here are the data presented in the *Human Development Report 2001*: 'The richest 1% of the world's people received as much income as the poorest 57%. The richest 10% of the US population (around 25 million people) has a combined income greater than that of the poorest 43% of the world's people (around 2 billion people). Around 25% of the world population received 75% of the world's income (in PPP US dollars)'. UN Development Programme (UNDP), *Human Development Report 2001: Making New Technologies Work for Human Development* (Oxford University Press: New York and Oxford, 2001), p. 19. See also chapter 6 in this volume.

[6] Although there is no agreed definition of terrorism, the UN Secretary-General has identified the common denominator as 'the calculated use of deadly violence against civilians for political purposes'. UN document SG/SM/8021, 12 Nov. 2001, URL <http://www.un.org/News/Press/docs/2001/sgsm8021.doc.htm>.

[7] The UN Security Council, the General Assembly and Secretary-General Kofi Annan have condemned terrorism in all its forms in many statements and documents. United Nations Security Council Resolution 1368, 12 Sep. 2001; UN document SG/SM/7948, 11 Sep. 2001; and UN document SG/SM/7951, 12 Sep. 2001. In an address to the UN General Assembly, Annan expressed his support for a global coalition against terrorism: 'This Organization is the natural forum in which to build such a universal coalition. It alone can give global legitimacy to the long-term struggle against terrorism'. UN document SG/SM/7965, 24 Sep. 2001. Decisions were aimed chiefly at developing a long-term strategy to ensure global legitimacy for combating terrorism. Annan urged the UN member states to ratify the 12 conventions and protocols on international terrorism adopted under UN auspices. UN document SG/SM/7944, 1 Oct. 2001.

[8] 'We can be clear about trends, but uncertain about events.' US Department of Defense (DOD), *Quadrennial Defence Review Report* (DOD: Washington, DC, 30 Sep. 2001), p. III, URL <http://www.defenselink.mil/pubs/qdr2002.pdf>.

words of US Defense Secretary Donald Rumsfeld.⁹ The fundamental change of circumstances is seen by many as an enormous opportunity: 'we have to build a new trans-Atlantic relationship that can be a central pillar in the war on terrorism and the constructive prospects for peace, which will follow. Unfortunately, this is an opportunity that, thus far, neither side of the Atlantic has enthusiastically welcomed'.¹⁰ The causes of this lost opportunity are manifold.

As a rule, the deliberations on globalization have an economic and financial bias and often lose sight of the fact that globalization processes embrace all aspects of life—political, cultural and civilizational.

The attacks on the main global power produced a powerful response of global solidarity. However, although almost all states of the world defined a common problem and signed up for the anti-terrorist campaign, the world is still in the initial stage of building a global security system. The strong sense of common responsibility and interdependence has not been translated into new norms, principles, procedures, forms and operating mechanisms.

The issue of the strengthening of global responsibility for the prevention of terrorism became topical after 11 September.¹¹ However, in spite of the many declarations and UN Security Council resolutions, expectations fell flat—both globally, in the United Nations, and regionally, in NATO. There is a lack of internationally recognized legal instruments for effectively tackling situations in which states have traditionally exercised their discretionary power and/or justified their actions as self-defence. Although the issue concerns both domestic and external security, the need for common responses in the security field has not been accepted globally. The interventions in Kosovo (Federal Republic of Yugoslavia) and Afghanistan reflect the new aspiration to establish international rules for protecting and defending respect for the basic principles and norms of international order.

The methods for the prevention of international terrorism are not adequate to the scale of the threat. Rather than bringing an end to unilateralism, the reaction of the United States to this reality has been to increase its unilateral actions.¹²

The USA has a position that is unprecedented in history in terms of its military, economic and technological capabilities, including military superiority in

---

⁹ '[A]dapting to surprise—adapting quickly and decisively—must be a condition of 21st century military planning.' Rumsfeld, D. H., 'Beyond this war on terrorism', 2 Nov. 2001 (published in *The Washington Post* on 1 Nov. 2001), URL <http://usa.or.th/news/press/2001/nrot117.htm>.

¹⁰ Lugar, R. G. (Sen., Indiana), 'NATO after 9/11: crisis or opportunity?', Speech to the Council on Foreign Relations, Washington, DC, 4 Mar. 2002, URL <http://www.cfr.org/public/resource.cgi?pub!4379>.

¹¹ This issue was raised by the UN Secretary-General on 12 Nov. 2001: 'In addition to measures taken by individual Member States, we must now strengthen the global norms against the use of and proliferation of weapons of mass destruction. We must also strengthen controls over other types of weapons that pose grave dangers through terrorist use. This means doing more to ensure a ban on the sale of small arms to non-State groups; making progress in eliminating landmines; improving the physical protection of sensitive industrial facilities, including nuclear and chemical plants; and increased vigilance against cyber-terrorist threats'. UN document SG/SM/8021 (note 6).

¹² Miller, S. E., 'The end of unilateralism or unilateralism redux?', *Washington Quarterly*, vol. 25, no. 1 (winter 2002), pp. 15–29. Two examples illustrate the US policy of unilateralism: its opposition to the 1997 Kyoto Protocol to the UN Convention on Climate Change, aimed at preventing global warming; and its refusal to ratify the Statute of the International Criminal Court.

conventional arms, nuclear weapons and missiles.[13] In the economic sphere, the US national product accounts for 31 per cent of world product—equal to the next four countries combined (Japan, Germany, the United Kingdom and France). It also leads in science and technology. It is this preponderance that both tempts and permits the USA to act unilaterally. However, in the age of new information technologies, security is based on interdependence rather than independence or preponderance. While this understanding is reflected in official US statements, in practice the US tendency towards unilateralism in decision making prevails.

What are the reasons for this course of events? Why has the USA not moved from unilateralism to multilateralism after 11 September?

III. The change of perception

In his speech to the Munich Conference on Security Policy, US Deputy Secretary of Defense Paul Wolfowitz explained the nature of the anti-terrorist campaign. First, 'the mission must determine the coalition, the coalition must not determine the mission'. Second, 'there will not be a single coalition, but rather different coalitions for different missions, "flexible" coalitions'.[14] Third, the USA will take the actions it deems appropriate, and states that wish to cooperate may join such actions.

There is a dichotomy in US policy regarding multilateral action. From a US perspective, those who do not join the struggle against terrorism are seen as anti-American. In its war on terrorism the USA received the political support it needed from its allies and most other states, but it neither expected or requested military support. Only a few states, principally the United Kingdom, played a significant role in the Enduring Freedom military campaign in Afghanistan.[15] The aims of the multinational coalition were to root out the terrorist al-Qaeda network from Afghanistan, overthrow the Taliban regime which hosted it and deter other states from supporting or sheltering terrorist organizations. While the Taliban regime was overthrown sooner than

---

[13] The USA leads both in military spending and in the Revolution in Military Affairs. Nye, J. S. Jr, 'The new Rome meets the new barbarians', *The Economist*, 23–29 Mar. 2002, pp. 23–35. See also Nye, J. S. Jr, *The Paradox of American Power: Why the World's Only Superpower Can't Go It Alone* (Oxford University Press: Oxford, 2002).

[14] 'This means that the coalition will not "unravel" if some country stops doing something or fails to join in some missions . . . Some will join us publicly; others will choose quiet and discrete forms of cooperation.' Wolfowitz, P., Speech at the Munich Conference on Security Policy, 2 Feb. 2002, URL <http://www.defenselink.mil/speeches/2002/s20020202-depsecdef1.html>.

[15] The International Security Assistance Force Operation (ISAF) was launched on a British initiative on the basis of the mandate of UN Security Council Resolution 1386, 20 Dec. 2001, URL <http://www.un.org/Docs/scres/2001/res1386e.pdf>. The operation was envisaged to last 6 months (until 13 July 2002). Some 4700 military from 17 European countries (more than 90% from NATO members) and New Zealand are taking part in the ISAF. The operation is under the command of the UK, which is to hand it over to Turkey in Apr. 2002. At various stages of the operation, more than 30 states have made military contributions. The following countries have been engaged in the military activities: 15 NATO member states (only 3 allied states did not participate—Iceland, Luxembourg and Hungary), 5 Partnership for Peace states (Finland, Russia, Romania, Sweden and Uzbekistan) and 12 non-Euro-Atlantic states (Australia, Bahrain, Cambodia, Egypt, Japan, Jordan, South Korea, New Zealand, Pakistan, the Philippines, Qatar and the United Arab Emirates).

expected, a much more difficult task facing the international community and calling for a qualitatively new approach is the eradication of terrorism globally.

It is therefore important to address a few basic questions. Did the terrorist attacks on the USA indeed provide a new catalyst in the process of international security and, if so, why? What are the short- and long-term implications of the tragic attacks? Can new instruments meet the new threats and risks?

IV. The new nature of the threat

Four features of the attacks of 11 September 2001 are qualitatively new.

The scale of the attacks. More people perished in the attacks of 2001 than in the 1941 Japanese attack on Pearl Harbor or in all the other terrorist acts that have been directed at the United States.

The nature of the attacks. The attacks were sudden, unexpected and not predicted. Although planned and directed from abroad, they were launched from US territory. No weapons or complex technologies were used (acting as missiles, passenger planes with innocent civilians were targeted at buildings with thousands of innocent people), and no state claimed responsibility for the acts.

The new type of adversary. The attacks were carried out by a non-state criminal network. The USA was the victim of aggression, but the attacks were outside any of the definitions of aggression in international law.

The target. The target of the attacks was the mightiest world power, which had considered its territory the safest, even a kind of sanctuary in security terms.

From the strategic perspective, the attacks can be seen as a part of a *sui generis* civil war in the Islamic world. The terrorists sought to compel the USA to pull back from the Muslim world by inciting a neo-isolationist mood and thus attempting to get the American people to put pressure on its government. Were this to have happened, it would have opened the way to the fall of existing regimes in the Islamic countries and the seizing of power by aggressive, extreme fundamentalist groups in most of the states where Islam is the ruling religion, including Pakistan. It would have affected the entire Arab world, whose autocratic regimes do not have popular support and have not come to power through free elections.

Islamic fundamentalists also attempted to diminish the impact of modernization by posing a serious threat to the very foundation of open, democratic society. Faced with the choice between security and democratic freedoms, societies, as a rule, tend to opt for the former. In this regard, the attacks have already adversely affected such values as privacy, freedom of movement and other civil liberties. However, the main aim of the terrorists was not so much to undermine the democratic system in the Western countries as to prevent the spread of democracy to the Islamic countries and to create the conditions for seizing power there.

V. New security building blocks

The attacks on the USA have shown that the norms, procedures, mechanisms and institutions of the current security system are not sufficient for dealing with this kind of threat effectively. The international community—states, the United Nations and regional structures, including NATO—was caught off guard, in spite of the fact that there had been some warnings of an impending attack.

In seeking the correct responses to this new challenge, it is important to avoid solutions based on a belief that the sources of threats stem from specific arrangements within states (e.g., lax airline safety) or fundamental problems such as low living standards and inequitable distribution of wealth between the rich North and the poor South. The logical conclusion of the former analysis is that the attacks could have been prevented by enhancing controls at airports. The latter approach accepts the argument that a gang of desperate idealists sacrificed their lives and those of several thousand victims in order to defend the rights of the poor and underprivileged. It does not explain why the great mass of those suffering in the world neither become nor support terrorists.

Four premises are of key importance in shaping a new global security system. The first is that the development and spread of the technologies of 'the network age', particularly information technology, are a part of the process of globalization. The second is that a growing number of states are too weak to control developments on their territory; consequently, they have become a base and an asylum for international crime and terrorist networks. The phenomenon of failed states is due to various factors: the emergence of new states after the collapse of multinational federations, such as the Soviet Union and Yugoslavia; the exposure of poor states to the globalization and modernization processes; and the higher standards of governance called for by the international community. Paradoxically, at the beginning of the 21st century, threats to international security are generated less by strong and rich powers than by poor and powerless states and non-state actors. The third is the blurring of the distinction between domestic and external security. The fourth is the growing importance of non-military aspects of state security. The stability and efficacy of the state and respect for the norms and rules of law are more important for the maintenance of international order than a state's military potential.[16]

---

[16] This is illustrated by a statement by President George W. Bush regarding the second phase of the war on terrorism: 'America encourages and expects governments everywhere to help remove the terrorist parasites that threaten their own countries and peace of the world'. 'No neutrality, warns Bush', 11 Mar. 2002, URL <http://news.bbc.co.uk/hi/english/world/americas/newsid_1867000/1867101.stm>. With this objective, the USA has initiated programmes of assistance for the training of anti-terrorist units (in Georgia, the Philippines and Yemen) and monitoring of capital transfers.

These premises offer a starting point for governments' operational decisions. The consequences of globalization in decision-making processes have to be taken into account in practically all the spheres of activity of states, both in their external relations and in ensuring internal stability.[17] The international system can only function if it is organized around the principles of democratization, the rule of law, and respect for human rights and the rights of minorities.[18] The emerging new system is predicated in equal measure on the harmonization of interests, common values, and the norms and tenets of law.

A new agenda for NATO

The close interrelationship between security and freedom is the basis of the North Atlantic Alliance. President George W. Bush referred to this in an address on the further enlargement of the Alliance.[19] In 2001 four issues were in focus: NATO defence capabilities; transatlantic relations; NATO enlargement eastward; and new principles of cooperation with Russia within the '20' format.[20] In 2002 decisions will be taken with regard to NATO's new capabilities, new members and new relationships. It will be a period of external enlargement and internal transformation.

Underlying NATO's transformation agenda is a search for answers to the question 'what is the essence of collective defence?'. Some see it as the need for NATO to develop its capacities for counter-terrorism. They also call for a reform of the Alliance command structures to make it 'leaner, more streamlined and more cost efficient, and, above all, more flexible'.[21] Others see the problems in a broader perspective, not confined to military aspects. They demand 'an unrestricted, comprehensive and transnational strategy that focuses attention beyond the immediate and towards the future new horizon'.[22]

In 2001 a formal gap in the structure of transatlantic cooperation was filled with an unprecedented NATO–EU agreement on cooperation in the field of

---

[17] Marc Grossman has rightly observed: 'Globalization's challenges extend beyond international terrorism. The current failure of the Argentine economy, the self-enforced isolation of some countries such as North Korea and the deepening digital divide all tell us we have to do a better job integrating the world into a contemporary economic and political system'. Grossman (note 2). See also the speech by NATO Secretary General Lord Robertson, 'NATO and the future of global security', delivered at the inauguration of the Potsdam Center for Transatlantic Security and Military Affairs, Potsdam, 4 Mar. 2002.

[18] See also Rotfeld, A. D., 'The organizing principles of global security', *SIPRI Yearbook 2001: Armaments, Disarmament and International Security* (Oxford University Press: Oxford, 2001), pp. 1–12.

[19] 'All of Europe's democracies, from the Baltic to the Black Sea and all that lie between, should have the same chance for security and freedom—and the same chance to join the institutions of Europe—as Europe's old democracies have.' President George W. Bush, Remarks by the president in address to faculty and students of Warsaw, Warsaw University, 15 June 2001, URL <http://www.whitehouse.gov/news/releases/2001/06/20010615-1.html>.

[20] Following the decisions adopted at the meeting of NATO foreign ministers in Brussels on 6–8 Dec. 2001, in the course of talks conducted with Russia in the spring of 2002 a document on a qualitatively new format for Russia–NATO cooperation 'at 20' was elaborated. Matser, W., 'Towards a new strategic partnership', *NATO Review*, Dec. 2001–Jan./Feb. 2002 (winter 2001/2002), p. 21. It will replace the existing '19 + 1' forum for cooperation (the NATO–Russia Permanent Joint Council), established by the 1997 NATO–Russia Founding Act.

[21] Wolfowitz (note 14).

[22] Hall, R. and Fox, C., 'Rethinking security', *NATO Review*, Dec. 2001–Jan./Feb. 2002, p. 11.

crisis management and European defence. In transatlantic relations, collaboration between the USA and its European allies should continue to pursue the following goals: (*a*) ensuring the close interrelationship between US and European security; (*b*) maintaining US leadership in the Alliance; and (*c*) taking into account security interests and guaranteeing the sovereign equality of all NATO members. Considering the new situation and new needs for the future of the Alliance, the build-up of mobile rapid reaction forces, adjustment of the European states' military capabilities to new tasks, collaboration in identifying new threats through the cooperation of the intelligence branches and preventing new non-military threats are all of key importance. The emphasis is not on increasing military spending, but on restructuring the existing capabilities in Europe. The capabilities of the new NATO allies are contingent on the rational use of existing resources to enable the effective fulfilment of defence-related and security tasks.

NATO's new political philosophy incorporates the concept of inclusive security that was adopted in the late 1990s and is addressed to those states that are not members and do not intend to join the alliance in the coming years. This political concept was reinforced in the face of a universal terrorist threat and helped align NATO both with Russia and other post-Soviet states as well as the European non-NATO countries.

The NATO–Russia relationship

An essential change has taken place in NATO and US relations with Russia. Although Russia was considered a source of potential tension and rivalry even after the end of the cold war, since 11 September 2001 there has been a profound transformation of their relations. Confrontation and distrust are being replaced by a spirit of collaboration and the gradual engagement of Russia in the Western security structures and arrangements. The new common priorities are cooperation in combating terrorism, crisis management, arms control and confidence-building measures, non-proliferation of weapons of mass destruction, theatre missile defence, search and rescue at sea, military-to-military cooperation and civil emergencies.[23]

In 2001 Russia's incipient new policy was visible at the US–Russian summit meeting in Ljubljana, Slovenia, on 16 June and at the meeting of the Group of Eight (G8) in Genoa, Italy, on 22 July. The opinions and recommendations offered by prominent Russian analysts and researchers were also indicative.[24] While the changes in Russian policy were prompted by the 11 September events, they had much deeper motives and there are strong indications that the reorientation of policy is of a strategic and durable character. As the USA and other Western states do not harbour any ill intentions, and in particular have

---

[23] See note 20.
[24] See the presentation by Alexei G. Arbatov in Rotfeld, A. D. (ed.), SIPRI, *The New Security Dimensions. Europe After the NATO and EU Enlargements* (SIPRI: Stockholm, June 2001), pp. 40–41; and Trenin, D., *The End of Eurasia: Russia on the Border Between Geopolitics and Globalization* (Carnegie Moscow Center: Washington, DC, 2001), p. 319.

no aggressive schemes vis-à-vis Russia, there is no reason to retain the political instruments that determined the cold war confrontation. However, many Russian diplomatic and military officials, including some of President Vladimir Putin's close advisers, neither share the new philosophy nor support the new premises for Russian–Western relations which it requires. The Russian president will have to overcome the conservative stereotypes prevailing within the Russian establishment.[25]

The change in US–Russian relations enabled Russia to stop seeing NATO enlargement as an insurmountable obstacle to closer cooperation with the Alliance. Lord Robertson quoted President Putin: 'If NATO and Russia can develop a true partnership, our past differences over NATO enlargement *will cease to be a relevant issue*'. The NATO Secretary General pointed out the permanent qualitative change in the relationship between NATO and Russia.[26]

The implementation of an 'open door' political philosophy will promote the evolution and transformation of the Alliance. The membership of Bulgaria Estonia, Latvia, Lithuania, Romania, Slovakia and Slovenia would expand the zone of political stability in Europe. Such a decision would better serve the goals of political than purely military security. Although none of the newcomers would bring militarily significant assets, the entry of some candidates (such as Bulgaria and Romania) would carry NATO infrastructure and access closer to several areas of current concern. The decision about the scale and time frame of further enlargement will be taken at the Prague NATO summit meeting in November 2002. A significant extension of NATO should be seen as an important element in the process of changing and reforming the Alliance.[27]

Other new elements in the building of a global cooperative security system include the closure of Russian bases in Cuba and Viet Nam and Russian consent to the presence of the USA and other Western states in Central Asia—where time-limited US engagement is regarded by Russia as in its security interests.

A new agenda for the European Union

Ten days after the terrorist attacks on the USA, the European Council and the High Representative for the Common Foreign and Security Policy (CFSP),

[25] Unable to hamper the process of improvement of NATO–Russia relations, opponents seek to slow it down by suggesting new bases on which it might rest. Illustrative in this context is the position of Dmitri Rogozin, Chairman of the State Duma Foreign Affairs Committee: 'We should not allow any rush towards accommodating the talks on creating new cooperation mechanisms in order to meet the artificially set political terms'. Rogozin, D., 'Rossiya i NATO: Sleduyet-li speshit s "dvadtsatkoi"'? [Russia and NATO: should we rush for the '20'?], *Nezavisimoye Voyennoye Obozreniye*, 12 Apr. 2001. At the Dec. 2001 North Atlantic Council meeting in Brussels, with the participation of Russian Foreign Minister Igor Ivanov, it was agreed that the document on the new format of cooperation with Russia should be prepared before the Reykjavik NATO ministerial meeting in May 2002.

[26] Lord Robertson, 'NATO and Russia: a special relationship', Speech by NATO Secretary General Lord Robertson at the Volgograd Technical University, 22 Nov. 2001, available at URL <http://www.nato.int/docu/speech/2001/s011121a.htm>.

[27] Hopkinson, W., *Enlargement: A New NATO*, Chaillot Papers, no. 49 (Institute for Security Studies, WEU: Paris, 2001).

Javier Solana, elaborated a programme and agreed on specific measures of action.[28] It included combating terrorism within the European Union and cooperation with the USA in its anti-terrorism programme. Judicial cooperation, cooperation between police and intelligence services, preventing the financing of terrorism and enhanced border controls were among the specific measures. The decisions became an important constituent part of the Union's common security policy and accelerated the development of international legal instruments. The European Union supported the Indian proposal for framing within the United Nations a general convention against international terrorism. The convention should enhance the impact of the measures taken over the past 25 years under UN auspices. The new, decisive aspects of EU activities are focused on: (*a*) stopping the funding of terrorism, including money laundering, and freezing assets; (*b*) strengthening air security, including quality control of security measures applied by member states, classification of weapons, technical training for crews; checking and monitoring of hold luggage; and protection of cockpit access.[29]

In the Union's efforts to combat terrorism, two elements deserve attention. First, the implementation of the task has accelerated the development of the CFSP and facilitates making the European Security and Defence Policy more operational. Second, it brought home to the EU states that the fight against the scourge of terrorism will be the more effective if it is based on an in-depth political dialogue with those countries and regions of the world in which terrorism comes into being.

The West and the Islamic world

The overthrow of the Taliban regime in Afghanistan and the response of states in Central Asia call for a redefinition of the policy of the West as a whole vis-à-vis the Islamic world.[30] A new strategy should also address the significance of Asian states, particularly Pakistan, India and China.

The states of Central Asia joined the global anti-terrorist coalition and helped produce an agreed position within the Euro-Atlantic Partnership Council (EAPC) by declaring their unconditional condemnation of the attacks and by pledging to make all efforts to combat terrorism. Their political declarations were supported by concrete commitments concerning military cooperation, including allowing the stationing on their territory of troops of the coalition of states participating in the armed intervention in Afghanistan. More

---

[28] Conclusions adopted by the Council (Justice and Home Affairs), Brussels, 20 Sep. 2001, SN 3926/6/01 Rev. 6, 20 Sep. 2001.

[29] Conclusions and Plan of Action of the Extraordinary European Council Meeting, SN 140/01, 21 Sep. 2001.

[30] A Turkish security analyst has recalled that, on 8 Sep. 2000, a year before the 11 Sep. attacks, Uzbek President Islam Karimov warned the UN General Assembly that 'Afghanistan has turned into a training ground and a hotbed of international terrorism' and is posing 'a threat to the security of not only the states of the Central Asian region, but to the whole world'. Yavuzalp, O., 'On the front line', *NATO Review*, vol. 49 (Dec. 2001/Jan./Feb. 2002), p. 24.

important than military contributions is their understanding of the essence of the threat and non-military ways of preventing terrorism within states.[31]

The threat has been described as a conflict of values within a civilization, not between civilizations.[32] It stems not from the incompatibility of values between religions in general, and Islam and the West in particular, but from the modern challenges to traditional autocratic structures in many Muslim countries, especially in the Arab world. There are no evil religions, but in many conflicts great religions have been misused for evil purposes.[33] Indeed, the clash that is taking place has little to do with religion *per se*. It is about the conflict between values such as individualism and collectivism, conservatism and liberalism, modernism and traditionalism.

It is important to distinguish between Islam as a religion and the attempts to forge it into an instrument for aggressive and extremist groups to seize power in Islamic countries. In the Middle East conflict, it is not ideology or religion that is at stake but the existential interests and the right of Israelis to live within secure boundaries and of Palestinians to obtain statehood.

VI. SIPRI findings

The main findings of the studies presented in this volume, based on original data, facts and analyses of the developments in 2001, particularly against the background of the tragic events of 11 September and the ongoing change in the international security process, are the following.

Armed conflicts.[34] In 2001 there were 24 major armed conflicts in 22 locations throughout the world. The only interstate conflict that was active in 2001 occurred between India and Pakistan. All of the 15 most deadly conflicts in 2001—those that caused 100 or more deaths during the year—were intra-state conflicts. Most commonly, they threatened to destabilize neighbouring states through the burden of refugees, cross-border movement of rebels (and occasionally national military forces), and the undermining of legitimate economic and political structures through the illicit trade in resources and arms. Eleven

---

[31] At the Munich Conference on Security Policy, the Deputy Prime Minister and Minister for Finance of Singapore made an interesting comment: 'Terrorism, especially by extremists invoking the name of Islam, or what Francis Fukuyama called "Islamo-Fascism", is a long-term problem. . . . In countries with large Muslim populations, there are small minorities at the fringes who are inclined toward extremist views and terrorist methods. Globalisation and easy foreign travel enabled them to come into contact with the extremist teachings of militants abroad, and to become part of an international terrorist network. . . . Terrorism, therefore, is not indigenous to South East Asia. It is a problem imported from abroad, but the danger is that it may become endemic to the region'. Lee Hsien Loong, 'Global war on terrorism', 2 Feb. 2002, URL <http://www.securityconference.de/konferenzen/default.asp?jahr=2002&topic=reden>. Similar views were offered by other Asian representatives at the conference. The US representative noted that 'to win the war against terrorism we have to reach out to the hundreds of millions of moderate and tolerant people in the Muslim world, including the Arab world'. Wolfowitz (note 14).

[32] Heller, M. A., 'The clash *within* civilizations', *Strategic Assessment* (Jaffee Center for Strategic Studies), vol. 4, no. 4 (Feb. 2002), pp. 3–6.

[33] See the views expressed by Bronislaw Geremek, Tadeusz Mazowiecki and Adam Michnik in a discussion sponsored by *Gazeta Wyborcza* (Warsaw): 'Czy Bog uprawia polityke?' [Does God engage in politics?], 27–28 Apr. 2002, p. 21.

[34] See chapter 1 and appendix 1A in this volume.

of the 15 conflicts have lasted for eight or more years, leading to extensive destruction of economic and social infrastructure. One of the reasons for their endurance is the inability of either side to prevail by force. In the vast majority of the conflicts, rebels used a guerrilla military strategy but failed to win wide popular support. Governments were unable to use their full strength against small and mobile opponents.

Conflict prevention, management and resolution.[35] The war against terrorism has led to the forging of new relationships between states that were formerly at odds with each other. The global effort against international terrorism marks the appearance of a new paradigm in international politics. It is important that it does not undermine the norms that have so recently been established. There were no new United Nations peace missions in 2001, for the first time since 1996. However, five new multilateral missions were initiated—one of them, the International Security Assistance Force in Afghanistan, on the authority of the UN Security Council. There were 51 multilateral peace missions under way in 2001. The shock generated by the attacks and the responses that have followed them carry serious repercussions for the international adoption and practice of conflict prevention. Prevention of terrorism, as currently practised, consists of measures taken to stop international terrorism, cut off the financial, political and military sources of terrorist support and, where possible, apprehend terrorists before they commit acts of terror. The attacks on Afghanistan have inevitably focused attention on the military elements and given vent to the idea of a global military engagement against terrorism.

The military dimension of the European Union.[36] Almost a decade passed after the end of the cold war before the EU developed the concept of a 'second pillar' European Security and Defence Policy (ESDP). Efforts have been made to meet security risks and challenges in Europe and elsewhere by carrying out the full range of military crisis management tasks. Events in 2001 served as a mid-course test for the success of these efforts.

Democratic control of the security sector.[37] Although security sector reform has become established as part of the international security agenda, there is still not a shared understanding at the international level of what this term means. The response of states to the terrorist attacks on the USA of 11 September 2001 may slow down the development of a security sector reform agenda—that is, demobilization and demilitarization, spending cuts and the promotion of transparency. Increased importance is being placed on developing cooperation with the armed forces, intelligence services and law enforcement services of other states to identify and eliminate groups and individuals engaged in terrorist acts.

Sanctions applied by the European Union and the United Nations.[38] Sanctions, recognized as a necessary and important instrument for conflict resolution, have not been effective in bringing about a change in the behaviour of

---

[35] See chapter 2 and appendix 2A in this volume.
[36] See chapter 3 in this volume.
[37] See chapter 4 in this volume.
[38] See chapter 5 in this volume.

their targets when applied as a 'stand-alone' measure. Efforts have been directed to improving sanctions design and implementation as well as considering how they might be used as part of a wider strategy. In the light of experience the United Nations has modified its approach away from the use of comprehensive sanctions towards the use of measures targeted only at those held responsible for wrong or illegal behaviour. The European Union has also elaborated its approach to the use of sanctions as part of its common foreign and security policy. In addition to giving effective expression to decisions of the United nations, the EU has also developed a distinctive approach to the use of sanctions in foreign policy areas where the UN has not provided direction, notably to support the parts of the CFSP aimed at improving human rights.

Military expenditure.[39] World military expenditure in 2001 is estimated at $839 billion in current prices, an increase of 2 per cent in real terms over 2000. This corresponds to an average of 2.6 per cent of world gross domestic product (GDP) and $137 per capita. The post-cold war decline in military expenditure lasted 11 years (1988–98). Since 1998 world military expenditure has been increasing again. The combined increase in world military expenditure since 1998 is 7 per cent in real terms. More than half of the world total is spent by five high-income countries in the West: in 2001 the USA accounted for 36 per cent of the world total, followed by Japan with 6 per cent and France, Russia and the UK, with 4 per cent each. Most of the countries with the heaviest economic burden in terms of military expenditure as a share of GDP are located in the Middle East, where official military expenditure accounts for more than 5 per cent in several countries. Countries involved in armed conflict in Africa also have a high defence burden, even according to their official figures, although these are likely to understate actual expenditure by a significant amount.

Arms production.[40] The restructuring of arms production that has taken place during the post cold war period has resulted in an increased level of concentration, in particular among the major arms-producing companies in the USA. In Europe, efforts continue to achieve increased concentration into larger companies on the European level. During 2001 several new European joint ventures were created for this purpose. While internationalization at the transatlantic level has been subject to even more political challenges than within Europe, the participation of the UK and British companies in the US-led JSF programme marked an important step in transatlantic military industrial relations. The approval in June 2001 by the French and US governments of a major joint venture may also serve as a test case for future transatlantic military industrial links. The Russian arms industry is still characterized by an extreme degree of over capacity and dependence on arms exports.

Arms transfers.[41] In 2001 the volume of international transfers of major conventional weapons and military technology for foreign licensed production of such weapons was $16.2 billion, in constant 1990 prices, slightly higher than

[39] See chapter 6 and appendices 6A–6E in this volume.
[40] See chapter 7 and appendices 7A and 7B in this volume.
[41] See chapter 8 and appendices 8A–8E in this volume.

in 2000. Calculated as a five-year moving average, however, the level continued to decline in 2001 (the total volume in 1997–2001 was $100.7 billion). The USA, accounting for 44.5 per cent of global arms transfers, remained the largest supplier in the period 1997–2001 despite a 65 per cent reduction in its arms deliveries since 1998. Russia was the second largest supplier in the period 1997–2001, accounting for 17 per cent of total arms transfers. As the result of a 24 per cent increase in arms transfers between 2000 and 2001 it became the largest supplier in 2001, with 30.7 per cent of the total. The five major recipients in the period 1997–2001 were China, India, Saudi Arabia, Taiwan and Turkey. China was by far the largest recipient in 2001, with 19 per cent of the total, after an increase of 44 per cent from 2000. Imports by India increased by 50 per cent, making it the third largest recipient in 2001 after the UK.

Arms control after 11 September.[42] While restoring the traditional linkage between arms control and military security, the Bush Administration underlined that agreements and arrangements need to be adapted to the contemporary strategic environment. This is not an abandonment of arms control. However, at present arms control can be seen as primarily a framework in which structured dialogue can be organized around armaments policy.

Nuclear arms control and missile defence.[43] On 13 December President Bush gave formal notice that the USA would withdraw from the ABM Treaty. There was disagreement between the USA and Russia over the form and substance of the reductions in their countries' strategic nuclear force levels. Specifically, there was a dispute over whether they would be made within the framework of a 'traditional' arms control treaty or as parallel, non-legally binding instruments. There was also a disagreement over whether the nuclear warheads scheduled to be removed from delivery vehicles should be dismantled, as insisted upon by Russia, or could be placed in storage, as advocated by the USA.

The military uses of outer space.[44] At the end of 2001 the USA, as the only superpower, had nearly 110 operational military spacecraft—well over two-thirds of all the military spacecraft orbiting the earth. Russia was a distant second, with about 40, and the rest of the world had only about 20 satellites in orbit. The issue of the military uses of space has resurfaced on the arms control agenda. China, France and Russia, supported by Canada, Sri Lanka and other states, have called for the negotiation of a new multilateral treaty to prohibit the deployment of weapons in space.

Chemical and biological weapon developments and arms control.[45] The attacks of 11 September increased the sense of vulnerability to indiscriminate mass-casualty terrorism. This sense of vulnerability was further augmented by a series of letters containing very high-grade anthrax spores sent to representatives of the US media and politicians. Although the mailed anthrax spores

---

[42] See chapter 9 in this volume.
[43] See chapter 10 in this volume.
[44] See chapter 11 in this volume.
[45] See chapter 12 in this volume.

did not result in the mass casualties predicted by some analysts, the letter attacks nevertheless demonstrated the potential for widespread social and economic disruption. The major events in biological weapon arms control were the US rejection of a draft protocol to strengthen the 1972 Biological and Toxin Weapons Convention (BTWC) and to reject the negotiating mandate of the body that had drafted the protocol. The main US objections were that the protocol would be an ineffective verification tool, that it would allow 'proliferators' political cover by allowing them to claim to be in compliance with the protocol and that confidential business information belonging to biotechnology firms and information relating to national biodefence facilities could be unnecessarily compromised.

Conventional arms control.[46] In 2001 the international community focused its attention on regional and domestic sources of conflict and relevant arms control measures, particularly those of an operational character. In Europe the focus was on the implementation of agreed measures and the search for new approaches to the politico-military dialogue. The 1999 Agreement on Adaptation of the 1990 Treaty on Conventional Armed Forces in Europe (Agreement on Adaptation of the CFE Treaty) is being implemented, but Russia's non-compliance has hindered its entry into force. A second review conference was held in 2001. Russia has made insufficient progress towards complying with its obligations with regard to agreed flank levels, although it has met its commitments regarding troop withdrawals from Moldova. In Georgia the future of one Russian military base and the continued presence of Russian forces remain to be resolved. The Balkan arms control regimes worked well, and the agreement on regional stabilization 'in and around Yugoslavia' was successfully concluded. Regional and bilateral confidence- and security-building measures (CSBMs) continued to work smoothly, and new bilateral CSBMs were introduced in Europe. The Organization for Security and Co-operation in Europe (OSCE) military doctrine seminar evaluated new threats and challenges and identified possible additional directions for the work of the OSCE in the zone of application. After years of deadlock the 1992 Treaty on Open Skies entered into force on 1 January 2002, after Russia and Belarus ratified it in 2001.

Multilateral export control.[47] After the terrorist attacks on the USA, certain decisions that were difficult to take in the framework of the export control regimes may have become possible because of a heightened awareness of the need to reduce the risk that entities planning terrorist acts will acquire non-conventional weapons. Particular attention is being paid to the following questions: the development of procedures for sharing information related to licensing and enforcement; the development of a more harmonized approach to risk assessment and the identification of programmes of concern; the development of common approaches to end-user controls in countries where pro-

---

[46] See chapter 13 in this volume.
[47] See chapter 14 in this volume.

grammes of concern are located; and to the question of how to apply controls to new types of commercial practices in a changing market.

VII. Concluding remarks

In 2001—in the age of globalization—terrorism was transformed from a local into a global problem and the entire structure of the international security system was called into question.

One paradox of the globalization era is that, although the threat of external aggression is non-existent in many parts of the world, national security cannot be taken for granted in any state, including the United States. National security is dependent in greater measure on the developments within states and an effective international security system. Because of its position in the world, the USA plays the most crucial role in shaping this new system. In the search for a new 'grand strategy' for the USA, various concepts have emerged after 11 September 2001. Arguments based on a balance of power are anachronistic and are not reflected in the recent decisions of the US Government. A new and serious threat to international security is the revival of nationalism, quite often in extreme forms, including the risk of possible re-nationalization of security policies. The process of modernization is often perceived as Westernization or Americanization—an attempt to impose the values of the Western world, in particular the United States, on other regions, cultures and civilizations. This misunderstanding has to do with the fact that values such as democracy, accountability, human rights, the rule of law, and social systems based on tolerance of diversity, individual freedoms and gender equality are identified with Western liberal, secular ideas. These are then pitted against traditionalist cultures that put the laws of closed communities above the rights of individuals and the tenets and norms of religious beliefs as well as customs and traditions above the law.

The qualitatively novel phenomena and changes in the world call for a new, unconventional approach. Since the risks are global, the responses should be global as well. This, in turn, requires a system that fosters and generates cooperation rather than balancing influence and threats resulting from competition and rivalry among powers and other actors. The world is interdependent. Positive and negative processes and phenomena are of a global character.[48] The greatest challenge of the contemporary world is not so much the rivalry over power or territorial expansion—motives that dominated in the colonial era—as it is dealing with the new threats of global terrorism, proliferation of weapons of mass destruction and organized crime, on the one hand, and local and regional conflicts, on the other. The inconsistent reactions to the 11 September events which will affect future arms control are reflected both in

[48] NATO Secretary General Lord Robertson stated: 'This is a networked world. . . . The dark side to globalization is that security threats, too, are networking. They are going global. . . . And, as a result, they pose a qualitatively new menace to the international community'. 'A new security network for the 21st century', Speech by NATO Secretary General, Lord Robertson, at the Economist Conference, Athens, 17 Apr. 2002, URL <http://www.nato.int/docu/speech/2002/s020417a.htm>.

the re-prioritization of arms export controls and in the *sui generis* 'de-prioritization' of nuclear concerns among the states combating terrorism (e.g., with respect to India and Pakistan).

There is a close relationship between global and local or regional threats. The main regions of conflict—Africa, Asia, the Balkans and the Middle East—have become a fertile ground for organized crime, terrorists and arms smugglers. The task ahead for the international community is to dismantle these networks. 'To accomplish this, we need an international security network.'[49] This is one of the new global tasks which face the North Atlantic community.

The current situation can be characterized not only by the diversity of threats, but also by the diversity of opportunities. A new NATO must play a central role in addressing the central security challenge. A US–European strategic partnership is essential for winning a global war on terrorism.[50]

The very notion of *cooperative* or *inclusive security* is more about political philosophy than a concrete programme of action. Depending on specific needs, cooperative security has different contents and includes different forms of operational collaboration. It is reflected in the transatlantic debate between the United States, which is inclined towards a 'pro-active' approach based on a 'single truth', and the European states, which emphasize the existential aspects of security—mutual restraint and respect for universal regulation.

There is nothing deterministic about the shaping of a new security system. Nothing is prejudged; institutions and procedures should be transformed. Nothing is shaped forever, and even joint decisions are often interpreted differently.[51]

In the search for a new security system, two common institutional norms of behaviour are dominant in the present circumstances: (*a*) democratic institutions, human rights and the protection of minorities; and (*b*) the bringing to justice under international criminal law of political leaders who have committed massive crimes against humanity. These two elements constitute a new, inalienable part of the evolving international security system. A security

---

[49] Robertson (note 48).

[50] The same idea was put forward by US Secretary of State Colin Powell, who observed that the 'process of participating in this great global campaign against terrorism may well open the door for us to strengthen or reshape international relationships and expand or establish areas of cooperation'. Powell, C. L., 'Seizing the moment', *US Foreign Policy Agenda* (US Department of State), Nov. 2001.

[51] In Aug. 1993, in Warsaw, Russian President Boris Yeltsin agreed with Polish President Lech Walesa that each country should determine for itself and choose the optimal road towards ensuring its own security. This was interpreted to mean that Russia would not oppose the enlargement of NATO eastward, but it was the wrong interpretation. After his return to Moscow, Yeltsin wrote to US President Clinton, and 3 other Western leaders, to try to persuade him to abandon NATO enlargement. Yeltsin's main motive was the domestic situation in Russia. 'We do not see NATO as a bloc opposing us. But it is important to take into account how our public opinion may react to such a step. Not only the opposition, but the moderates, too, would no doubt see this as a sort of neo-isolation of the country as opposed to its natural introduction into the Euro-Atlantic space.' 'And generally', he went on, 'we favor a situation where the relations between our country and NATO would be by several degrees warmer than those between the Alliance and Eastern Europe'. Russian President Boris Yeltin's letter to US President Bill Clinton, 15 Sep. 1993, reproduced in *SIPRI Yearbook 1994* (Oxford University Press: Oxford, 1994), pp. 249–50.

regime based on shared values should in equal measure take into account the specific features of states and regions and the needs of the global community.

The adaptation of the cooperative security system to new tasks calls for the elaboration of new principles and norms adequate to the requirements of the contemporary world. International structures, institutions and organizations are also being reassessed, since they have so far not been able to address effectively the needs and challenges of global processes and the accelerated modernization in the world today.

The USA can play or is playing a decisive role. Enlightened and engaged US leadership is crucial. This leadership should be recognized by the coalition of democratic states within a common framework, not by imposition but by their willing acceptance based on moral consensus and common values.[52]

It is less important to debate the merits of unilateralism against those of multilateralism in US policy than to provide the means to overcome the 'democratic deficit' in international relations. Joseph Nye has suggested that governments

can do several things to respond to the concerns about this phenomenon. They can try to design international institutions that preserve as much space as possible for domestic political processes to operate. There is no single and simple answer to the question of how to reconcile the necessary global institutions with democratic accountability. There is a need to think more about the norms and procedures for the governance of globalization.[53]

The fact that in the era of globalization and information technology the USA cannot act independently in pursuing its international security goals meets with a growing understanding among various groups in the US establishment. As Nye rightly noted, 'the paradox of American power in the 21st century is that the largest power since Rome cannot achieve its objectives unilaterally in a global information age'.[54] The world needs the United States as never before, but the United States needs the rest of the world, too. Neither domination and hegemony nor neo-isolationism offer an adequate response to the new challenges.

---

[52] Kissinger, H., *Does America Need a Foreign Policy? Toward a Diplomacy for the 21st Century* (Simon & Schuster: New York, 2001), p. 288.

[53] Nye, J. S. Jr, 'Globalization's democratic deficit: how to make international institutions more accountable'. *Foreign Affairs*, July/Aug. 2001.

[54] Joseph Nye follows Henry Kissinger in his argument that the test of history for this generation of American leaders will be whether they can turn the current predominant power into an international consensus and widely-accepted norms. 'And that cannot be done unilaterally.' Nye (note 13), p. 25.

Part I. Security and conflicts, 2001

Chapter 1. Major armed conflicts

Chapter 2. Conflict prevention

Chapter 3. The military dimension of the European Union

Chapter 4. The challenges of security sector reform

Chapter 5. Sanctions applied by the European Union and the United Nations

1. Major armed conflicts

TAYLOR B. SEYBOLT*

I. Introduction

In 2001 the security challenges faced by most countries of the world were similar to those that have prevailed since the end of the cold war. All of the 15 major armed conflicts reviewed in this chapter were intra-state conflicts. Although they developed according to their own dynamics, they were often influenced by outside events and actors. Most of the conflicts also had an impact on neighbouring states, as certain effects were felt across international borders in what is commonly called 'spillover'. The potential for spillover from intra-state conflict to lead to interstate conflict was highlighted by the increased tensions between India and Pakistan caused by the violence emanating from the disputed territory of Kashmir.

The conflicts in 2001 pitted rebels using guerrilla tactics against regimes using repressive counter-insurgency strategies, in some cases combined with conventional military tactics. It was rarely the case that either side displayed much interest in winning the support of the population. Civilians were regularly the victims of violence perpetrated by both sides.

Most conflicts were sustained by revenue from the sale of natural resources, on the one hand, and the purchase of arms and ammunition, on the other hand.[1] In general, the trade in resources and arms by governments was considered to be legitimate, while the same trade by rebel groups was considered to be illegal.[2] Regardless of the legal status, the flow of money and arms enabled both sides to continue to pursue their objectives on the battlefield. While governments usually had considerably more wealth and military power than the rebel groups they opposed, the nature of guerrilla warfare meant that they could not take advantage of the disparity to end the conflicts through military means.

The conflict in Sierra Leone was the only one that appeared to end in 2001, mainly because the rebels lost the support of neighbouring Liberia. As a consequence, they abandoned their control of diamond-mining areas in accordance with a peace plan implemented by the United Nations. In the long-running conflict in Afghanistan, the overwhelming military power of the United States led to the rapid defeat of the Taliban regime. However, occa-

---

[1] In some cases, such as Sri Lanka, an important source of revenue for rebels was remittances from supporters living abroad.
[2] Some countries were the subject of economic sanctions or arms embargoes or both. In those cases, some forms of trade by the government were also considered to be illegal.

* Hannes Baumann assisted in the research for several sections of this chapter.

SIPRI Yearbook 2002: Armaments, Disarmament and International Security

sional intense battles continued, traditional rivalries re-emerged, and the ultimate outcome of the conflict has yet to be determined.

These two examples demonstrate the importance of external influence on internal conflicts. In most cases, the supply of military *matériel* by state and sub-state actors and overt military intervention by states served to prolong and intensify the conflicts. Other states and intergovernmental organizations attempted to counteract this type of external influence through mediation and promotion of the peaceful settlement of disputes.

Negotiations in six of the conflicts reviewed in this chapter held out the possibility of peace. New agreements were signed or existing agreements were implemented in Afghanistan, Burundi, the Democratic Republic of the Congo (DRC), the Philippines and Sierra Leone. In Sri Lanka, the two sides engaged in informal negotiations at the end of the year. Three of the conflicts deteriorated considerably in 2001. In Colombia, Indonesia (Aceh) and the conflict between Israel and the Palestinians, fighting intensified and peace talks collapsed or were non-existent. In the remaining six conflicts there was no marked increase in violence or in the likelihood of negotiated settlements.[3]

The intra-state conflicts not only were influenced by external actors but also affected their external environments. Of the 15 most deadly conflicts in 2001, 11 spilled over international borders.[4] Most commonly, they threatened to destabilize neighbouring states through the burden of refugees, cross-border movement of rebels (and occasionally national military forces), and the undermining of legitimate economic and political structures through the illicit trade in resources and arms. The regional dimension of conflict was most clearly visible in the African Great Lakes region, where the conflicts in Angola, Burundi and the DRC are interlinked and have a destabilizing effect on Rwanda, Tanzania and Uganda. Intra-state conflicts may also have a direct impact beyond their immediate region, as demonstrated by the 11 September attacks in the United States by the Afghanistan-based al-Qaeda network.

Although the general pattern of conflict worldwide in 2001 was consistent with that of previous years, the priorities and perceptions of many states changed as a result of the terrorist attacks in the United States. The campaign against terrorism by the United States and its allies in the latter part of the year directly influenced a small number of conflicts, as noted throughout the chapter in the appropriate places. Beyond its immediate ramifications, the campaign against terrorism has brought to the fore a number of conflict-related issues, the full scope of which remains to be seen. These include the militarization of responses to terrorism, the global role of violent sub-state actors, and the connection between intra- and interstate conflict.

Before 2001, terrorism was perceived largely as a type of violent political action waged on a limited scale in the location of a dispute. It was usually addressed using the tools of criminal investigation, policing and the criminal

---

[3] These 6 conflicts are Algeria, Angola, Kashmir, Russia (Chechnya), Somalia and Sudan.
[4] There was minimal spillover of the conflicts in Algeria, Indonesia, the Philippines and Sri Lanka. The last 3 are island states, which makes spillover to other countries less likely because of the natural barrier.

justice system. In 2001 terrorism came to be regarded as politically or religiously motivated violence that could be perpetrated with few limitations and that required the waging of a military campaign against terrorist groups throughout the world. The militarization of efforts to control terrorism holds the potential to intensify ongoing conflicts as governments use the rhetoric of counter-terrorism to overcome the international diplomatic constraints on the use of force. It may also increase the number of major armed conflicts, as seen in the apparent US interest in attacking Iraq in retaliation for its alleged sponsorship of terrorism.

The global role of sub-state actors was previously recognized in the context of their efforts to mitigate conflict. For example, Médecins sans Frontières, the International Campaign to Ban Landmines and the Pugwash Conferences on Science and World Affairs have all received the Nobel Peace Prize in recent years. The use of commercial aircraft by the al-Qaeda network to obliterate a global financial centre brought into shocking relief the destructive potential of sub-state groups. It also highlighted the effectiveness of the classic guerrilla tactic of surprise strikes against unprotected targets. Although 11 September was not the first time in which guerrilla tactics had been used on an international scale, the high profile of the attacks has caused some states to re-examine their national security needs and has drawn attention to sub-state groups as potential threats to international peace and stability.

The preoccupation with terrorism and the predominance of intra-state conflicts diverted attention from the danger of potential interstate conflict in 2001. The rapidity with which an interstate conflict can arise was demonstrated in October, when the USA attacked Afghanistan in retaliation for the September terrorist attacks. A reminder of the potential for catastrophic interstate conflict came in December, when India and Pakistan came close to a significant military escalation of their dispute over the territory of Kashmir in the wake of terrorist attacks in India by extremists based in Kashmir and Pakistan. Both countries possess nuclear weapons and neither has ruled out their use in the event of a war.

Sections II–VI of this chapter review major armed conflicts in the regions of Africa, Asia, Europe, the Middle East and South America. For each conflict, information is provided on the parties, their objectives, the major events in 2001, the human costs and the regional impact.[5] Section VII highlights the major findings of this review.

For the purposes of this chapter, a 'major armed conflict' is defined as the use of armed force between two or more organized armed groups, resulting in the battle-related deaths of at least 1000 people in any single calendar year and in which the incompatibility concerns control of government, territory or

---

[5] Additional information on these conflicts can be found in previous editions of the Yearbook. See, e.g., Seybolt, T. B., 'Major armed conflicts', *SIPRI Yearbook 2001: Armaments, Disarmament and International Security* (Oxford University Press: Oxford, 2001), pp. 15–51; and Seybolt, T. B. and Uppsala Conflict Data Project, 'Major armed conflicts', *SIPRI Yearbook 2000: Armaments, Disarmament and International Security* (Oxford University Press: Oxford, 2000), pp. 15–49.

communal identity.[6] The conflicts reviewed below conform to this definition and, in addition, caused over 100 deaths in 2001. Major armed conflicts that were not intense enough to cause 100 deaths are not described in this chapter but do appear in table 1A.3 in appendix 1A.

II. Conflicts in Africa

The biggest changes in the conflicts in Africa in 2001 took place in Burundi, the DRC and Sierra Leone. A peace accord was signed by most parties to the conflict in Burundi; a new leader in the DRC helped to revive the stalled peace accord signed in 2000; and the Sierra Leonean Government and rebel groups implemented the peace accord signed in 2000. Two regions demonstrated the ways in which a conflict in one country can increase violence and sustain conflicts in neighbouring countries. In the Great Lakes region, progress towards peace in the DRC threatened to escalate the level of violence in Burundi as rebel forces moved across borders. The DRC conflict also served as a location for the Rwandan Government to fight its opponents, many of whom fled Rwanda for the DRC in 1994. In West Africa, the presence of Sierra Leonean rebels along the borders of Guinea and Liberia threatened to ignite a region-wide conflict, as Liberia supported Sierra Leonean and Guinean rebels and Guinea supported Liberian rebels. In spite of these developments, there were no new major armed conflicts in Africa in 2001.

Algeria

The conflict between the government and Islamic rebels that began in 1992 continued in 2001. An estimated 100 000–150 000 people, most of whom were civilians,[7] have been killed in fighting between the Algerian Government and two rebel groups—the Front Islamique du Salut (FIS, Islamic Salvation Front), the Groupe Islamique Armé (GIA, Armed Islamic Group) and the Jamiyy'a Islamiyya Da'wa wal Jihad (variously translated as the Islamic Group for Mission and Holy War, the Islamic Group for Call and Combat, and the Salafist Group for Preaching and Salvation, usually referred to by the French acronym GSPC).[8]

Throughout 2001, attacks on civilian and military targets consisted of ambushes along mountain roads and hit-and-run operations in towns. The incidents occurred in all of the heavily populated northern regions.[9] Most of

---

[6] This definition is based on the one developed by the Uppsala Conflict Data Project but differs from it in that it does not require a government to be one of the parties to the conflict and in that it takes into account conflicts that are motivated by communal identity and not clearly about government or territory. See appendix 1B for information on the definitions, sources and methods used by the Uppsala Conflict Data Project to generate table 1A, appendix 1A.

[7] Khalaf, R., 'Algeria reaps rewards of anti-terrorist war', *Financial Times*, 15 Jan. 2002, p. 4, and Blanche, E., 'Algeria's civil war flares up again', *Jane's Intelligence Review*, Dec. 2001, pp. 6–7.

[8] In French, the group is known as the Groupe Salafiste pour la Prédication et le Combat (GSPC) or Group Salafiyyste de Da'wa et Jihad (GSDJ).

[9] Agence France-Presse (Domestic Service), 10 Mar. 2001, in 'Over 60 killed in latest Islamist-linked violence in Algeria', Foreign Broadcast Information Service, *Daily Report–West Europe* (*FBIS-WEU*),

the killing took place in the outskirts of major towns, allowing guerrillas an easy retreat to the mountains.[10] While it is likely that the rebels are responsible for many of the attacks, a group of dissident soldiers and some activists in France have accused the Algerian military of massacring civilians for political reasons and then blaming the rebels.[11]

Neither the GIA nor the GSPC is strong enough to hold territory. Their weakness was revealed during the summer, when they did not capitalize on the political upheaval created by violent protests among the Berber community, who make up about one-third of Algeria's population of 30 million.[12] Nevertheless, the rebels show no sign of stopping their violent activities and the government does not appear to be able to bring an end to the conflict through either political or military means.[13] The government estimated that a total of 700–800 rebels were still active in 2001, a low figure compared to the 6000 who have surrendered since 1999. The number of killings has consequently dropped, but there were nonetheless about 2300 deaths in 2001.[14]

Angola

Since 1975 the Movimento Popular de Libertação de Angola (MPLA, Popular Movement for the Liberation of Angola) and the União Nacional Para a Independência Total de Angola (UNITA, National Union for the Total Independence of Angola) have fought for control of the country, with the MPLA holding governmental power since 1975.[15] Throughout 2001, the conflict remained deadlocked. Although the government military is about 90 000 strong and has ample oil revenue to pay for weapons and operations, most observers consider it unlikely that the government can achieve a military victory over the approximately 8000 guerrilla fighters of UNITA.[16]

Despite the international sanctions on 'conflict diamonds' coming from unlicensed sources in Angola, a thriving illicit market persists. The UN Moni-

---

FBIS-WEU-2001-0310, 13 Mar. 2001; Agence France-Presse (Domestic Service), 8 Apr. 2001, in 'Algeria: one officer killed, 27 soldiers injured, 33 Islamists reported killed', FBIS-WEU-2001-0408, 9 Apr. 2001; '17 Algeria farmers killed in attack', *International Herald Tribune*, 14 Aug. 2001, p. 5; and Agence France-Presse (Domestic Service), 4 Nov. 2001, in 'Algeria: two policemen, seven armed Islamists die in incidents', FBIS-WEU-2001-1104, 5 Nov. 2001.

[10] Richburg, K., 'The Algeria challenge: growing social upheaval', *International Herald Tribune*, 16–17 June 2001, p. 7.

[11] 'Algerian dissidents claim massacre', BBC Online Network, 27 Feb. 2001, URL <http://news.bbc.co.uk/hi/english/world/>; and Salah, H., 'France moves to boost Algeria links', BBC Online Network, 13 Feb. 2001, URL <http://news.bbc.co.uk/hi/english/world/>.

[12] 'Algeria: the Berbers rise', *The Economist*, 5 May 2001, p. 39; 'Berbers lead huge rally for democracy in Algiers', *International Herald Tribune*, 15 June 2001, p. 4; Blanche (note 7); and Agence France-Presse (Domestic Service), in 'Algeria: riots spread to more towns; Islamist group urges Kabyles to join it', 17 June 2001, in FBIS-WEU-2001-0617, 19 June 2001.

[13] There were rumours of secret talks between the government and the GSPC in Aug. but there has been no evident movement towards negotiations since then. Agence France-Presse (Domestic Service), 18 Aug. 2001, in 'Algerian rebel group said to hold secret talks with government', FBIS-WEU-2001-0818, 20 Aug. 2001.

[14] Blanche (note 7).

[15] A third group, the Frente Nacional de Libertação de Angola (NFLA, National Front for the Liberation of Angola), also fought for power and then became a legal opposition party.

[16] 'Angola: who blinks first?', *Africa Confidential*, 1 June 2001, pp. 1–3.

toring Mechanism on Sanctions against UNITA reported that embargoed diamonds valued at about $350–420 million left Angola during the year. UNITA probably sold 25–30 per cent of the recently mined diamonds leaving the country, in addition to diamonds in its stockpile.[17] The income of $100–150 million enabled the rebels to maintain their supplies of small arms, communications equipment and other *matériel* needed to avoid military defeat.

The fighting in 2001 consisted of guerrilla attacks by UNITA and counter-insurgency efforts by the government, the main consequence of which was to add to the already high number of refugees and internally displaced persons (IDPs).[18] In May, hundreds of UNITA troops attacked strategic towns, including one only 60 km from the capital Luanda.[19] At the end of the year the government made a strong effort to rout UNITA in Moxico province, where it was believed that UNITA leader Jonas Savimbi was operating.[20] Moxico, located on the border with Zambia, allows the rebels access to supply and escape routes. There was also frequent fighting in the northern province of Uige and the central province of Bie. Uige allows UNITA access to supply and escape routes through the DRC.

The diamond and arms trades are two of the transmission belts by which the Angolan conflict spills over into neighbouring countries. The main reason why the Angolan military has been engaged in the war in the DRC on the side of the DRC Government is to try to shut down supply routes that UNITA has maintained there for decades. Relations between Angola and Zambia, to the east, were strained in 2001 by several incursions by Angolan troops into Zambian territory in pursuit of rebels. For years the Zambian Government has hosted a large population of Angolan refugees, some of whom trade diamonds, arms and fuel with UNITA and whose movement across the border the Zambian Government is unable to control.[21] Zambia does not allow Angola to attack UNITA members on Zambian territory. The Zambian military killed Angolan army troops along the border soon after an alleged incursion by Angola that killed several Zambian civilians.[22] In contrast, Namibia, Angola's close ally to the south, allows cross-border attacks by the Forças Armadas de

[17] 'Supplementary report of the Monitoring Mechanism on Sanctions against UNITA', contained in United Nations, Letter dated 12 October 2001 from the Chairman of the Security Council Committee established pursuant to Resolution 864 (1993) concerning the situation in Angola addressed to the President of the Security Council, UN document S/2001/996, 12 Oct. 2001.

[18] United Nations Office for the Coordination of Humanitarian Affairs (OCHA), Integrated Regional Information Network for Southern Africa (IRIN-SA), 'Angola: Savimbi wanted—dead or alive', IRIN-SA weekly round-up 5, 3–9 Feb. 2001; and IRIN-CEA, 'Angola: thousands of Angolan refugees arriving', IRIN-CEA weekly round-up 63, 3–9 Mar. 2001. News items from all the IRIN offices—Central Asia, IRIN-CA (Islamabad, Pakistan), Central and Eastern Africa, IRIN-CEA (Nairobi, Kenya), Horn of Africa, IRIN-HOA (Nairobi, Kenya), Western Africa, IRIN-WA (Abidjan, Côte d'Ivoire) and Southern Africa, IRIN-SA (Johannesburg, South Africa)—are archived on ReliefWeb, URL <http://www.irinnews.org/>.

[19] Brittain, V., 'Unita attacks block progress on Angola peace talks', *Guardian Weekly*, 17–23 May 2001, p. 4.

[20] IRIN-SA, 'Angola: humanitarian impact of government offensive', 12 Dec. 2001, URL <http://www.irinnews.org>.

[21] IRIN-SA, 'Zambia–Angola: IRIN focus on border trade', IRIN-SA weekly round-up 31, 4–10 Aug. 2001; and 'Bitter borders', *Africa Confidential*, 7 Dec. 2001, p. 4.

[22] IRIN-SA, 'Zambia admits killing 10 Angolan soldiers', IRIN-SA weekly round-up 46, 17–23 Nov. 2001.

Angola (FAA, Armed Forces of Angola) and occasionally sends its own troops into Angola in pursuit of UNITA fighters.[23]

The continued violence and lawlessness throughout much of Angola prolonged the suffering of civilians. The UN Office for the Coordination of Humanitarian Affairs (OCHA) reported in July that malnutrition and mortality rates in Bie province exceeded emergency levels.[24] The situation was not unique. A lack of housing, food and medical assistance plagued the majority of the population of 12 million. The most common estimate is that over 500 000 people have been killed in the 26-year war and 3.6–3.8 million people have been displaced, including tens of thousands who were forced to flee in 2001.[25] Revenue from rich diamond and oil deposits supports the war effort rather than basic social services, which are in a state of neglect.[26]

Burundi

The main development in Burundi in 2001 was the establishment on 1 November of a transitional government in accordance with the Arusha Peace and Reconciliation Agreement, signed in Arusha, Tanzania, on 28 August 2000. The achievement of former South African President Nelson Mandela, with strong support from the UN Security Council, the European Union and the Regional Peace Initiative,[27] in bringing 19 parties together in a compromise agreement was welcomed as a significant step towards ending the civil war that began in 1993. To date, the conflict has killed over 200 000 people, about half of them during the first year. In a population of 6.7 million, it has caused the prolonged displacement of nearly 900 000 people, about one-third of whom are refugees, and the destruction of the social, economic and physical infrastructure. Thousands of people were killed and tens of thousands more were displaced by the fighting in 2001.[28]

In early February heavy clashes occurred in the south between the army and rebel groups.[29] At the end of the month, the Parti pour la Libération du Peuple Hutu–Forces Nationales de Libération (PALIPEHUTU–FNL, Party for the Liberation of the Hutu People–National Liberation Forces, known as the FNL) launched one of the biggest offensives of the war on the northern outskirts of

---

[23] 'Bitter borders' (note 21); and IRIN-SA, 'Namibia–Angola: Namibian forces cross into Angola in hot pursuit of UNITA', IRIN-SA weekly round-up 9, 3–9 Mar. 2001.
[24] IRIN-SA, 'Angola: mortality, malnutrition rates in Bie still high', IRIN-SA weekly round-up 26, 7–13 July 2001.
[25] Save the Children Fund, 'Angola emergency update Nov. 2001', 5 Nov. 2001; and IRIN-SA, 'IDPs increase, UNITA commander is killed', IRIN-SA weekly round-up 43, 27 Oct.–2 Nov. 2001.
[26] Save the Children Fund (note 25); and 'Oil, diamonds and danger in Angola', *The Economist*, 13 Jan. 2001, p. 38.
[27] The members of the Regional Peace Initiative are Burundi, the Democratic Republic of the Congo, Ethiopia, Gabon, Kenya, Malawi, Rwanda, Tanzania, Uganda and Zambia. The UN, the Organization of African Unity and the Implementation Monitoring Committee are represented in the initiative.
[28] United Nations, Interim report of the Secretary-General to the Security Council on the situation in Burundi, UN document S/2001/1076, 14 Nov. 2001; Strindberg, A., 'Burundi heats up as Congo cools down', *Jane's Intelligence Review*, July 2001, pp. 40–41; and Amnesty International, *Burundi: Between Hope and Fear*, document AFR 16/007/2001, 22 Mar. 2001.
[29] IRIN-CEA, 'Burundi: heavy fighting in southern provinces', IRIN-CEA weekly round-up 59, 3–9 Feb. 2001.

the capital Bujumbura. The battle lasted for nearly two weeks, killed hundreds and displaced as many as 50 000 civilians before the rebels retreated to the surrounding hills.[30] There was also an upsurge in fighting in the central and southern provinces that caused massive destruction and displacement.[31] The rebel groups involved were the FNL and the Conseil National pour la Défense de la Démocratie–Forces pour la Défense de la Démocratie (CNDD–FDD, National Council for the Defence of Democracy–Forces for Defence of Democracy, known as the FDD).

The increase in violence in the first half of the year appeared to be related to early efforts to implement the 1999 Lusaka Ceasefire Accord in the DRC (see below). Many of the rebels engaged in the attacks had recently entered Burundi from the DRC.[32] For several years the DRC Government allowed the FDD and the FNL to operate in the DRC in an attempt to disrupt the Congolese rebels who control the part of the DRC that borders Burundi.[33] As the situation in the DRC became more stable and the possibility increased of disarming or expelling 'negative forces' in compliance with the Lusaka Accord, the Burundian rebels returned to Burundi by the thousands.[34] Other rebels who left the DRC went to Tanzania, where the Tanzanian Government has either no means or no desire to prevent the recruitment and training of rebels or cross-border attacks into Burundi.[35]

In this volatile environment, Peace Agreement Facilitator Mandela introduced the power-sharing proposal that would ultimately serve as the basis for an agreement, although most parties to the talks rejected the proposal at the time.[36] An agreement on a transitional arrangement was reached among 19 political parties on 23 July 2001.

On 1 November a new, three-year government was put in place. Its primary task is to end the civil war. For the first 18 months, President Pierre Buyoya, a Tutsi, will serve as president. His vice-president is Domitien Ndayizeye, the Hutu head of the main opposition party, the Front pour la Démocratie au Burundi (Frodebu, Democratic Front of Burundi). During the second 18-month period, Ndayizeye will serve as president and the Tutsi parties will designate a vice-president. An Implementation Monitoring Commission was established under UN leadership to oversee the transition. A special protection unit was agreed by the parties and endorsed by the UN Security Council to

---

[30] IRIN-CEA, 'Burundi: heavy fighting in Bujumbura', IRIN weekly round-up 63, 3–9 Mar. 2001; 'Burundi fighting leaves 200 dead', BBC Online Network, 12 Mar. 2001, URL <http://news.bbc.co.uk/hi/english/world/>; and United Nations, Spokesman for the Secretary-General, Secretary-General dismayed at outbreak of renewed fighting in Burundi, UN document SG/SM/7731 AFR/305, 1 Mar. 2001.

[31] IRIN-CEA, 'Burundi: thousands displaced by fighting in the southeast', IRIN weekly round-up 66, 24–30 Mar. 2001; IRIN-CEA, 'Burundi: thousands temporarily displaced by fighting', IRIN weekly round-up 70, 21–27 Apr. 2001; and IRIN-CEA, 'Burundi: security "restored" in Kayanza province', IRIN weekly round-up 73, 12–18 May 2001.

[32] IRIN-CEA (note 29).

[33] 'Central Africa: new year hopes', *Africa Confidential*, 11 Jan. 2002, p. 6.

[34] Turner, M., 'Fears grow over the build up of rebel forces in Burundi', *Financial Times*, 26 June 2001, p. 6; and 'Rwanda/Burundi: negating the negatives', *Africa Confidential*, 28 Sep. 2001, p. 8.

[35] Strindberg (note 28); and 'Burundi–Tanzania security talks', BBC Online Network, 17 June 2001, URL <http://news.bbc.co.uk/hi/english/world/>.

[36] 'Burundi talks fail amid new fighting', *International Herald Tribune*, 27 Feb. 2001, p. 8.

protect returning exiled politicians and possibly assist the reform of the military.[37] South Africa was the main contributor of troops, with 1500 soldiers in Burundi by the end of the year.[38]

In response to the transitional agreement, the FNL and the FDD declared that they would continue fighting.[39] Battles with government forces persisted for the rest of the year, particularly in the north-east, where rebel soldiers had arrived from the DRC.[40] The intensification of fighting made it impossible to carry out the provisions for the transitional phase, such as the return of displaced persons and reform of the security institutions.[41]

The focus at the end of the year was to broker ceasefire agreements between the transitional government and the two rebel groups. The FNL and the FDD both gave conditional approval for negotiations with the government. Normally rivals, they announced a joint negotiating position for future talks, to be chaired by Gabon.[42] The prospects for successful talks seemed poor. The FDD was estimated to have about 10 000 troops and the FLN about 5000 arrayed against the government's 40 000 soldiers.[43] The rebels also received support from the DRC, Tanzania and Zimbabwe.[44] However, although they have the ability to be a potent fighting force, the rebels are politically fractured.[45] This makes a consistent rebel negotiating position difficult, undermines the transitional government's trust in the rebel negotiators and raises the possibility that some rebel factions will continue to fight even if others decide to make peace.

The Democratic Republic of the Congo

The assassination of President Laurent Kabila on 16 January 2001 and the succession of his son, Joseph Kabila, shifted the political ground in the DRC and gave new life to the effort to end the war that began in 1998.[46] The elder Kabila had refused to cooperate with the United Nations or with the facilitator of the Inter-Congolese Dialogue—the primary diplomatic mechanism for implementing the Lusaka Peace Accord—former Botswanan President

---

[37] United Nations, Interim report of the Secretary-General to the Security Council on the situation in Burundi, UN document S/2001/1076, 14 Nov. 2001.

[38] Cauvin, H. E., 'Burundi rivals take office in bid to end war', *International Herald Tribune*, 3–4 Nov. 2001, p. 9.

[39] IRIN-CEA, 'Burundi: mixed reactions to Buyoya nomination, rebels vow to fight on', IRIN weekly round-up 81, 7–13 July 2001.

[40] IRIN-CEA, 'Burundi: army claims capture of rebel base, killing hundreds', IRIN weekly round-up 104, 22–28 Dec. 2001.

[41] Interim report of the Secretary-General (note 37).

[42] IRIN-CEA, 'Burundi: army launches push against rebels', IRIN weekly round-up 101, 24–30 Nov. 2001.

[43] Strindberg (note 28); and Römer-Heitman, H., 'South Africa may deploy troops to Burundi', *Jane's Defence Weekly*, 22 Aug. 2001, p. 16.

[44] 'Burundi/Congo-Kinshasa: piecemeal', *Africa Confidential*, 26 Oct. 2001, pp. 6–7.

[45] 'Burundi/Congo-Kinshasa: piecemeal' (note 44).

[46] There is no agreement about who is responsible for killing the president. The DRC Government blames Rwanda, Uganda and their rebel allies, but there are also rumours of a plot by the DRC military, Angola or Zimbabwe. 'Congo leader shot by guard: Belgium says Kabila died in an attempted coup', *International Herald Tribune*, 17 Jan. 2001, pp. 1, 8; 'Congolese enemies accused in slaying', *International Herald Tribune*, 24 May 2001, p. 4; and 'Kabila is dead, long live Kabila', *The Economist*, 27 Jan. 2001, p. 41.

Ketumile Masire, who was appointed by the Organization for African Unity (OAU).[47] The new president won the approval of foreign governments by reshuffling his cabinet and military leadership, recognizing the legitimate role of Masire and taking steps to reform government economic practices.[48]

For the first time since the 1999 Lusaka Peace Accord was signed, there was substantial progress towards its implementation.[49] The accord set out four steps: a ceasefire; disarmament, demobilization and reintegration (DDR); a dialogue to set up new political arrangements; and the withdrawal of foreign troops. The multitude of armed groups and national militaries involved in the conflict makes each one of the Lusaka Accord steps difficult to achieve.[50]

The two sides in the civil war consist of the DRC Government and its allies Angola, Namibia and Zimbabwe versus a divided set of rebel groups and their foreign allies. The Mouvement de Libération Congolais (MLC, Congolese Liberation Movement), the Rassemblement Congolais pour la Démocratie–Mouvement de Libération (RCD-ML, Congolese Rally for Democracy–Liberation Movement) and the Rassemblement Congolais pour la Démocratie–Goma (RCD-Goma, Congolese Rally for Democracy–Goma) each control a portion of the eastern half of the country. The MLC and the RCD-ML are in the north-east and are backed by Uganda. The RCD-Goma is in the central, eastern and south-eastern parts of the country and is backed by Rwanda.

The ceasefire between the government and the rebels was largely respected in 2001. There was no substantial fighting despite the distrust between them and the slow deployment of UN ceasefire monitors, the Mission d'Observation des Nations Unies au Congo (MONUC, UN Observation Mission in the Democratic Republic of the Congo).[51] UN Secretary-General Kofi Annan blamed the countries of the region for the slow deployment of the mission, claiming that they had not demonstrated an interest in ending the war.[52] The UN contingent had fewer than 200 personnel at the beginning of 2001 but was at almost full strength by the end of the year.[53]

Almost all the parties implemented a disengagement and redeployment plan that was agreed in February 2001. It stipulated that all the signatories of the

[47] The OAU member states adopted the Constitutive Act of the African Union on 12 June 2000; it entered into force on 26 May 2001, formally establishing the African Union (AU), with headquarters in Addis Ababa. The AU will replace the OAU after a transitional period.

[48] Shaxson, N., 'Kabila fosters diplomatic drive to end Congo war', *Financial Times*, 6 Apr. 2001, p. 9.

[49] The Lusaka Ceasefire Agreement is contained in United Nations, Letter dated 23 July 1999 from the permanent representative of Zambia to the United Nations addressed to the President of the Security Council, UN document S/1999/815, 23 July 1999, available at URL <http://www.un.org/search/>.

[50] For a detailed account of the war and the peace process in 1998–99 see Seybolt, T. B., 'The war in the Democratic Republic of Congo', *SIPRI Yearbook 2000* (note 5), pp. 59–75.

[51] IRIN-CEA, 'DRC: "Ceasefire and disengagement continuing to hold"—UN', IRIN weekly round-up 84, 28 July–3 Aug. 2001.

[52] Crossette, B., 'Annan scolds Congo parties', *International Herald Tribune*, 22 Feb. 2001, p. 2.

[53] United Nations, Monthly summary of military and CivPol personnel deployed in current United Nations operations as of 31 Dec. 2000, United Nations Information Centre (UNIC), Copenhagen, 16 Jan. 2001; and United Nations, Monthly summary of military and CivPol personnel deployed in current United Nations operations as of 31 Dec. 2001, United Nations Information Centre (UNIC), Copenhagen, available at URL <http://www.un.org/Depts/dpko/dplo/contributors/31-13-01.pdf>. See also appendix 2A in this volume.

Lusaka Accord withdraw 15 km from about 100 key positions along the 2400-km front line to create a 30-km buffer zone.⁵⁴ The MLC complied in July.⁵⁵ The RCD-Goma did not leave the city of Kisangani, strategically located on the Congo river, despite repeated demands from the UN.⁵⁶ The disengagement provided the countries with troops in the DRC with an opportunity to repatriate their soldiers. By September, Namibia had completely withdrawn its troops, which had numbered about 2000 at their peak strength.⁵⁷ Angola was expected to withdraw many of its estimated 2000–2500 troops by the end of the year.⁵⁸ Zimbabwe made the withdrawal of most of its estimated 11 000 troops conditional on the outcome of the Inter-Congolese Dialogue.⁵⁹ On the opposing side, Uganda withdrew an estimated 7000 troops from most of the DRC but maintained a presence in the region bordering on Uganda.⁶⁰ Rwanda initially withdrew a substantial number of troops but maintained a presence in the DRC of possibly tens of thousands.⁶¹

The parties to the Lusaka Accord are not the only armed groups in the DRC. Several tribal and rebel groups referred to in the accord as 'negative forces' are also active. Of these groups, only the Mayi-Mayi are Congolese. They are fighting primarily to expel Tutsis and their Rwandan backers. As noted above, the Burundian FDD and FNL rebels use the DRC as a rear base. They cooperate with the Armée pour la liberation du Rwanda (ALiR, Army for the Liberation of Rwanda),⁶² which wants to overthrow the Rwandan Government.

In 2001 fighting occurred sporadically between the RCD-Goma rebels and 'negative forces'.⁶³ Most of the fighting appeared to be part of an effort by the RCD-Goma and the Rwandan Army to repress the Mayi-Mayi and the ALiR.⁶⁴ The most sustained and deadly engagement occurred in September around the town of Fizi, on the eastern border of the DRC. Up to 4000 Mayi-Mayi and

⁵⁴ McGreal, C. and Brittain, V., 'Troops pull back in Congo', *Guardian Weekly*, 5–11 Apr. 2001, p. 4; and IRIN-CEA, 'DRC: military disengagement begins', IRIN weekly round-up 64, 10–16 Mar. 2001.

⁵⁵ IRIN-CEA, 'DRC: MLC withdrawal confirmed by UN military observers', IRIN weekly round-up 70, 2–8 June 2001; and United nations, Press statement by Security Council President on Democratic Republic of Congo, UN document SC/7097, 6 July 2001.

⁵⁶ IRIN-CEA, 'DRC: Annan visits rebel-held Kisangani', IRIN weekly round-up 89, 1–7 Sep. 2001.

⁵⁷ IRIN-SA, 'Namibia: troops pull-out applauded', IRIN-SA weekly round-up 35, 1–7 Sep. 2001.

⁵⁸ 'Ugandan troops leaving DR Congo', BBC Online Network, 8 May, URL <http://news.bbc.co.uk/hi/english/world/>; and IRIN-CEA, 'Angola–DRC: Angola announces "substantial" troops withdrawal', IRIN weekly round-up 99, 10–16 Nov. 2001.

⁵⁹ IRIN-CEA, 'DRC: Zimbabwe troops begin leaving', IRIN weekly round-up 67, 31 Mar.–6 Apr. 2001; and IRIN-CEA, 'DRC: Zimbabwe's withdrawal depends on outcome of dialogue', IRIN weekly round-up 85, 4–10 Aug. 2001.

⁶⁰ IRIN-CEA, 'DRC: Uganda to withdraw an additional 7,000 soldiers', IRIN weekly round-up 77, 9–15 June 2001.

⁶¹ Turner, M., 'Upbeat mood over Congo peace prospects', *Financial Times*, 7 Mar. 2001, p. 10; and IRIN-CEA, 'Rwanda: troops arrive back from DRC', IRIN weekly round-up 65, 17–23 Mar. 2001.

⁶² The ALiR is composed of the Rwandan Interahamwe militia and former Rwandan Army Forces (ex-FAR, Forces Armées Rwandaises), who have been in the DRC since they were driven out of Rwanda after they committed genocide in 1994.

⁶³ IRIN-CEA, 'DRC: heavy fighting in Shabunda', IRIN weekly round-up 60, 10–16 Feb. 2001. Because the 'negative forces' are not parties to the Lusaka Accord, battles with them were not considered to be violations of the agreement.

⁶⁴ IRIN-CEA, 'DRC: RCD increases efforts to quash Mayi-Mayi', IRIN weekly round-up 76, 2–8 June 2001; and IRIN-CEA, 'DRC: rebels claim to seize "strategic" town', IRIN weekly round-up 84, 28 July–3 Aug. 2001.

Burundian and Rwandan rebels captured the town, only to be expelled again by the RCD-Goma a month later.[65] The interlocking of the conflicts in the African Great Lakes region was revealed by the timing of the initial offensive by the militia, which occurred after Burundian government forces withdrew from the DRC in response to increased rebel activity in Burundi (see above).[66]

A critical stumbling block came with the second step of the Lusaka Accord: the disarmament, demobilization and reintegration of the armed rebel groups. Successful DDR is necessary to convince the Rwandan Government that it is no longer threatened by the Hutu ALiR. If it is not convinced, it will not withdraw from the DRC. If the DRC Government does not believe that all Burundian, Rwandan and Ugandan troops will leave its territory, then it will not consider sharing power with the anti-government rebels. A power-sharing arrangement is the ultimate objective of the Inter-Congolese Dialogue.[67]

Following the first successful preparatory meeting for the Inter-Congolese Dialogue, held in August, the government of Joseph Kabila demonstrated its willingness to remove Rwandan ALiR fighters when it handed over 1600 of them to MONUC in September. It also pledged to disarm and hand over all of the about 6000 ALiR fighters in its area of control. Rwanda was not satisfied, claiming that there were 20 000–40 000 ALiR fighters in the DRC.[68] However, as the government pointed out, the 'negative forces' had regrouped during the year along the borders of Burundi and Rwanda to avoid being disarmed. Those areas are under the control of the RCD-Goma and Rwanda, not the DRC Government. By mid-summer Rwandan troops had captured about 2000 ALiR fighters.[69] The approach to the Mayi-Mayi was far more conciliatory. After the August preparatory meeting, the DRC Government and the rebel RCD-ML both supported the inclusion of the Mayi-Mayi in the Inter-Congolese Dialogue, on the grounds that they are Congolese rather than foreign. The Rwandan Government appeared to be sympathetic to this position.[70]

On 15 October 2001, two years after the Lusaka Accord was signed, facilitator Masire convened in Addis Ababa, Ethiopia, the first meeting of the Inter-Congolese Dialogue in an effort to fulfil the third step of the Lusaka Accord. The talks broke down immediately, however, and were postponed until an

[65] Vesperini, H., 'DR Congo ceasefire under threat', BBC Online Network, 30 Sep. 2001, URL <http://news.bbc.co.uk/hi/english/world/>; and IRIN-CEA, 'DRC: RCD-Goma claims recapture of Fizi in east', IRIN weekly round-up 94, 8–12 Oct. 2001.

[66] IRIN-CEA, 'DRC: rebel coalition captures strategic town', IRIN weekly round-up 92, 22–28 Sep. 2001.

[67] United Nations, President of Rwanda outlines conditions for withdrawal of forces from Democratic Republic of the Congo, UN document SC/7008, 7 Feb. 2001; Smith, S., 'There can't be dialogue in the presence of invaders', *Guardian Weekly*, 26 Apr.–2 May 2001, p. 29; and International Crisis Group (ICG), *Disarmament in the Congo: Jump-starting DDRRR to Prevent Further War*, ICG Africa Report no. 38 (ICG: Nairobi/Brussels, 14 Dec. 2001).

[68] 'Congo-Kinshasa: dialogue in Addis', *Africa Confidential*, 28 Sep. 2001, pp. 5–6; IRIN-CEA, 'DRC: 6,000 Interahamwe to be handed over', IRIN weekly round-up 91, 17–21 Sep. 2001; and 'Congo-Kinshasa: friends abroad, foes at home', *Africa Confidential*, 13 July 2001, pp. 4–5.

[69] ICG (note 67); IRIN-CEA, 'Great Lakes: Hutu rebels reportedly regrouping', IRIN weekly round-up 78, 16–22 June 2001; and IRIN-CEA (note 68).

[70] IRIN-CEA, 'DRC: government supports Mayi-Mayi participation in talks', IRIN weekly round-up 91, 17–21 Sep. 2001; and IRIN-CEA, 'DRC: RCD-ML rebels seek to incorporate Mayi-Mayi in dialogue', 11 Sep. 2001.

unspecified date because of the Kabila Government's objections that the delegations were too small, that the Mayi-Mayi were not included and that funding for the planned 45-day meeting was uncertain.[71]

Although all the parties appear to be tired of fighting, each has something to lose if the peace process succeeds. The government will have to share power with the rebels in an as yet unspecified way. There is a chance that new political arrangements will recognize the de facto partition of the country and give the rebels federated control of half the territory. Under a different possible scenario, the rebel groups might have to give up control of the east in exchange for their participation in the national government. Angola, Rwanda and Uganda will lose access to the rear operating areas of their own rebels, thus possibly jeopardizing their own security.[72]

All of the countries with troops in the DRC will also lose control over the DRC's lucrative natural resources. Military and political leaders from these countries have grown rich through the extraction of and trade in minerals, and the trade in timber and coffee. The government's adversaries flatly deny plundering the DRC.[73] A United Nations Panel of Experts submitted two reports in 2001 on the extraction of wealth from the DRC. The panel confirmed that continued resource exploitation provided all the parties with a disincentive to settle the DRC conflict. It also found that contests for control of wealth constituted a major source of the violence among all the armed groups and factions. Although security concerns initially drove foreign countries to become involved in the DRC conflict, the primary motive for remaining involved is the material benefits, according to the panel.[74] In short, resolution of the political problems will have to address the economic interests of the parties.

Foreign plunder of wealth is not the only cost of conflict to the DRC. A report on the human cost of the conflict in the eastern part of the country estimated that from August 1998 to the end of March 2001 the war had caused 2.5 million deaths in excess of the number of people who would have died in peacetime. About 350 000 of those deaths were due to violence and the rest to disease and malnutrition.[75] A separate investigation of the access to health services supported the assertion that the conflict has completely disrupted the minimal medical infrastructure that once existed and left many people with no

---

[71] 'Walk out', *Africa Confidential*, 26 Oct. 2001, p. 7; and ICG, *The Inter-Congolese Dialogue: Political Negotiation or Game of Bluff?*, ICG Africa Report no. 37 (ICG: Brussels/Nairobi/Kinshasa, 16 Nov. 2001).

[72] Burundi would lose the same advantage, but it is not part of the Lusaka process.

[73] 'The military–financial complex', *Africa Confidential*, 15 June 2001; and IRIN-CEA, 'DRC: Burundi, Rwanda, Uganda deny plundering resources', IRIN weekly round-up 69, 14–20 Apr. 2001.

[74] 'Report of the Panel of Experts on the illegal exploitation of natural resources and other forms of wealth of the Democratic Republic of the Congo', contained in United Nations, Letter dated 12 April 2001 from the Secretary-General to the President of the Security Council, UN document S/2001/357, 12 Apr. 2001; and 'Addendum to the report of the Panel of Experts on the illegal exploitation of natural resources and other forms of wealth of the Democratic Republic of the Congo', contained in United Nations, Letter dated 13 November 2001 from the Secretary-General to the President of the Security Council, UN document S/2001/1072, 13 Nov. 2001.

[75] Roberts, L. et al., *Mortality in Eastern Democratic Republic of Congo: Results from Eleven Mortality Surveys* (International Rescue Committee: New York, 2001), available at URL <http://www.theIRC.org/docs/mortality_2001/mortII_report.pdf>.

medical care at all.[76] In September 2001, an estimated 2.1 million people remained displaced, about 350 000 of whom had fled across an international border and gained the status of refugees.[77] Looting and human rights abuses by all the armed elements made it impossible for most people to return home and caused several hundred thousand more to flee.[78]

Sierra Leone

At the beginning of 2001 there was grave pessimism concerning the prospect of advancing the 1999 Lomé Peace Agreement, which was intended to end the civil war.[79] The rebel Revolutionary United Front (RUF) and the government-allied Civil Defence Forces (CDF) showed little sign of adhering to the DDR programme, which is an essential part of the peace process. The UN Assistance Mission to Sierra Leone (UNAMSIL) lost much of its military strength in January, when the Indian and Jordanian contingents withdrew. UNAMSIL forces were only positioned around the capital Freetown. The Government of Ahmed Tejan Kabbah and the Sierra Leone Army (SLA) controlled only the territory around the capital and needed the support of 600 British troops to do even that. Although a ceasefire signed in November 2000 was generally respected, there was fighting along the borders with Guinea and Liberia and clashes between the RUF and the CDF in northern districts.[80]

By the end of the year, the picture was entirely different. A joint committee on disarmament had overseen the successful implementation of the programme in 10 of the 12 districts of the country, including diamond-producing districts and former RUF strongholds. The RUF had declared its intention to participate in the political process. UNAMSIL personnel, numbering 17 500, extended their presence throughout the country and the government was in the process of extending its authority. In January 2002 the disarmament programme was completed and a formal end to the war, which began in 1991, was declared. Elections were scheduled for May 2002.[81]

The turning point came in May 2001 when the RUF, increasingly constrained and under pressure, began to abide by the terms of the Lomé Agree-

---

[76] Médecins Sans Frontières, *Violence and Access to Health in Congo (DRC): Results of Five Epidemiological Surveys* (Médecins Sans Frontières: Brussels, Dec. 2001), available at URL <http://www.msf.org/source/countries/africa/drc/2001/healthreport/report.doc>.

[77] UN OCHA, 'Flash OCHA RDC situation humanitaire au 30 Septembre 2001', 30 Sep. 2001.

[78] US Committee for Refugees, 'Current country update: Congo Kinshasa', 2 Oct. 2001, available at URL <http://www.refugees.org/world/countryrpt/africa/Mid_countryrpt01/congokinshasa.htm>; and Oxfam UK, Save the Children UK and Christian Aid, *No End in Sight: The Human Tragedy of the Conflict in the Democratic Republic of Congo* (Oxfam UK: London, Aug. 2001).

[79] United Nations, Peace Agreement between the Government of Sierra Leone and the Revolutionary United Front of Sierra Leone, UN document S/1999/777, 7 July 1999, annex. For a detailed account of the war and the troubled peace process in Sierra Leone from 1991 to 2000 see Reno, W., 'War and the failure of peacekeeping in Sierra Leone', *SIPRI Yearbook 2001* (note 5), pp. 149–62.

[80] United Nations, Ninth report of the Secretary-General on the United Nations Mission in Sierra Leone, UN document S/2001/228, 14 Mar. 2001.

[81] United Nations, Twelfth report of the Secretary-General on the United Nations Mission in Sierra Leone, UN document S/2001/1195, 13 Dec. 2001; IRIN-WA 'Sierra Leone: disarmament officially ends', IRIN-WA weekly round-up 105, 5–11 Jan. 2002; and 'The president's speech', BBC Online Network, 18 Jan. 2002, URL <http://news.bbc.co.uk/hi/english/world/>.

ment. The RUF lacked direction since its founder, Foday Sankoh, remained imprisoned after being captured in 2000. Nor did the rebels have the support of a political constituency. Since 2000, the RUF's only political demands were that the SLA disarm and that all British troops leave the country.[82] In meetings in early May, they did not even make those demands.[83] More importantly, the RUF lost the support of Liberian President Charles Taylor, who had provided financial and military assistance since 1991.[84] Taylor faced international condemnation and increasingly tight economic sanctions for his role in sustaining the Sierra Leone conflict by trading 'conflict diamonds' that originated in RUF-controlled parts of Sierra Leone.[85] In March 2001 the UN Security Council called on Liberia to end all support for the RUF and imposed an arms embargo on Liberia.[86] Taylor also faced an insurgency in the north of Liberia that put demands on the meagre resources his government controlled and reportedly obliged him to ask RUF fighters for support.[87]

The final element in the RUF's demise as a military force was its disastrous incursion into Guinea, at Taylor's request, in support of Guinean rebels. It is unclear why the RUF acceded to the request. The Parrots Beak region, where Guinea, Liberia and Sierra Leone meet, had suffered increasingly violent battles for months in an escalation of the long-standing animosity between the leaders of Guinea and Liberia. Taylor supported the Guinean rebels, with the help of the RUF, and Guinean President Lansana Conté supported the Liberian rebels. The RUF lost hundreds of soldiers and suffered heavy attacks by the Guinean military on its bases in eastern Sierra Leone, which the Sierra Leone Government tacitly welcomed.[88]

On 15 May the pro-government CDF and the rebel RUF signed an additional ceasefire agreement that provided for simultaneous disarmament in accordance with the DDR programme. The implementation was to be monitored by a joint ad hoc committee consisting of government, UNAMSIL and RUF observers.[89] The agreement was a breakthrough as thousands of fighters turned in their weapons and registered for reintegration programmes according to a plan that progressed methodically from district to district. The RUF even

---

[82] 'Sierra Leone: precarious calm', *Africa Confidential*, 29 June 2001, pp. 1–4; and ICG, *Sierra Leone: Time for a New Military and Political Strategy*, ICG Africa Report no. 28 (ICG: Freetown/London/Brussels, 11 Apr. 2001).

[83] 'Sierra Leone: precarious calm' (note 82).

[84] On the role of Liberia and Charles Taylor in supplying arms to the rebels see chapter 8 in this volume.

[85] 'Report of the Panel of Experts appointed pursuant to Security Council Resolution 1306 (2000), paragraph 19, in relation to Sierra Leone', contained in United Nations, Note by the President of the Security Council, UN document S/2000/1195, 20 Dec. 2000; 'Liberia moves to avert sanctions', BBC Online Network, 22 Jan. 2001, URL <http://news.bbc.co.uk/hi/english/world/>; United Nations, Security Council demands that Liberia immediately cease support for Sierra Leone's RUF and other armed rebel groups, UN document SC/7023, 7 Mar. 2001; and UN Security Council Resolution 1343 (7 Mar. 2001).

[86] United Nations Security Council Resolution 1343, 7 Mar. 2001.

[87] Hirsch, J. L., 'War in Sierra Leone', *Survival*, vol. 43, no. 3 (autumn 2001), pp. 145–162; and Astill, J., 'A promise of peace in Sierra Leone', *Guardian Weekly*, 31 May–6 June 2001, p. 3.

[88] IRIN-WA, 'Sierra Leone: government won't condemn Guinean attacks', IRIN-WA weekly round-up 65, 24–30 Mar. 2001; and 'Sierra Leone: precarious calm' (note 82).

[89] IRIN-WA, 'Sierra Leone: rivals agree to stop fighting', IRIN-WA weekly round-up 72, 12–18 May 2001; and 'Sierra Leone: precarious calm' (note 82).

demobilized in the diamond districts, where access to the gems had provided an incentive and the means to continue the war in years past. By December, over 36 700 combatants had turned in their arms to UNAMSIL, about one-third of them members of the RUF and two-thirds members of the CDF. The number far exceeded the initial expectation of 28 000.[90]

The success of the voluntary disarmament programme surprised many observers, who recalled the history of the RUF making agreements only to break them. Four factors seem to have contributed to the success: the rebel group was leaderless after Sankoh was jailed; its patron was no longer able to provide material support; it had suffered military losses at the hands of Guinea; and there was a peace agreement in place that offered an alternative to continued fighting.

During 11 years of war, approximately 43 000 people were killed and about 2 million displaced, out of a population of some 4.5 million.[91] The RUF was infamous for using child soldiers and for mutilating people. The hope for successful elections in 2002 and a lasting peace was tempered by a desperate need for international funding to support the training and reintegration of demobilized fighters, many of whom only knew how to make a living through violence and plunder. There was also grave concern that the peace process in Sierra Leone was too compartmentalized and did not take into account the overlapping conflicts in the West African region.[92] Weak governments, the prevalence of rebel movements and the propensity of rival governments to undermine each other, combined with the abundance of small arms and easily mined gems, remained a recipe for continued insecurity in the region.[93]

Somalia

In 2001 the Transitional National Government (TNG) of Somalia was able to exert control over only part of the capital Mogadishu and a strip of territory along the coast. The northern regions of Somaliland and Puntland maintained their self-declared independent status, although no other state has recognized them.[94] In the central and southern regions of Somalia, there were occasional violent clashes between clan-based armed groups that sought to maintain dominance over local areas, including the capital. Some groups supported the TNG, but most did not.[95] The fighting caused thousands of refugees to flee to

---

[90] United Nations (note 81).

[91] 'Sierra Leone: the spreading battleground', *The Economist*, 7 Apr. 2001, p. 46.

[92] See chapter 2 in this volume.

[93] 'Toward a comprehensive approach to durable and sustainable solutions to priority needs and challenges in West Africa: report of the Inter-Agency Mission to West Africa', contained in United Nations, Letter dated 30 April 2001 from the Secretary-General addressed to the President of the Security Council, UN document S/2001/434, 2 May 2001; and IRIN-WA, 'West Africa: some progress, but region remains volatile, UN says', IRIN-WA weekly round-up 103, 15–21 Dec. 2001.

[94] 'Somaliland votes on independence', BBC News Online, 31 May 2001, URL <http://news.bbc.co.uk/hi/english/world/>.

[95] United Nations, Report of the Secretary-General on the situation in Somalia, UN document S/2001/963, 11 Oct. 2001; IRIN-HOA, 'Somalia: 19 killed in fighting in the southwest', IRIN-HOA weekly round-up 39, 26 May–1 June 2001; and IRIN-HOA, 'Somalia: Mogadishu fighting escalates', 16 July 2001.

Sudan

The war in Sudan escalated in 2001 and the prospects for peace appeared to be more remote than ever. Since 1983, the National Islamic Front government in Khartoum has fought several groups in the south of the country, the largest of which is the Sudanese People's Democratic Front (SPDF). The complex of objectives that drive the war include disputes over religion, governance, autonomy and resources. The government is opposed to giving African animist and Christian groups autonomy, either as a separate state or within a federated union.[98] It is estimated that over 2 million people have died during the war as a result of violence, famine and disease. Approximately 4 million people have been displaced within Sudan and an additional 420 000 are refugees. Over 150 000 people were displaced in 2001.[99]

The Inter-Governmental Authority on Development (IGAD) sub-regional organization has tried since 1994 to initiate peace talks, but without success. The IGAD initiative accepts the concept of self-determination for the south, which the government flatly rejects.[100] The Sudanese Government favours an alternative peace proposal made in 1999, known as the Libyan–Egyptian Initiative. The initiative reportedly calls for a transitional government of all the political parties, revision of the constitution, general elections, recognition of diversity, guarantee of basic rights, a cessation of violence and a single unified Sudanese state.[101] A tentative third peace initiative began in November 2001 when the US Special Envoy to Sudan, John Danforth, held talks with the government, the National Democratic Alliance (NDA) and the Sudan People's Liberation Army (SPLA).[102] All three efforts failed to seriously address the

---

[96] IRIN-HOA, 'Kenya–Somalia: More than 10,000 refugees in Mandera', IRIN-HOA weekly round up 33, 14–20 Apr. 2001; and IRIN News, 'Somalia: Ethiopia says it has no reason to deploy troops', IRIN News, 10 Jan. 2002.

[97] IRIN-HOA, 'Somalia: Nairobi peace talks to go ahead', IRIN-HOA weekly round-up 67, 8–14 Dec. 2001.

[98] ICG, *God, Oil and Country: Changing the Logic of War in Sudan* (Brussels: ICG, 2002).

[99] US Committee for Refugees, 'Current country update: Sudan', 2 Oct. 2001.

[100] 'Sudan: delusions of peace', *Africa Confidential*, 10 Aug. 2001, pp. 1–4; and Kenya Broadcasting Corporation, 4 Oct. 2001, in 'Sudan to give regional body "another chance" to resolve conflict' Foreign Broadcast Information Service, *Daily Report–Sub-Saharan Africa (FBIS-AFR)*, FBIS-AFR-2001-1004, 5 Oct. 2001.

[101] IRIN-HOA, 'Sudan: Khartoum accepts Libyan–Egyptian peace initiative', IRIN-HOA weekly round-up 44, 30 June–4 July 2001.

[102] al-Hasan Ahmad, M., 'Ignoring the initiatives and giving the principal role in peace to the clergy', *Al-Sharq al-Awast*, 20 Nov. 2001, in 'Writer assesses US envoy Danforth's mediation in Sudanese crisis', FBIS-AFR-2001-1120, 27 Nov. 2001; IRIN-HOA, 'Sudan: US hopes to be "a catalyst for peace"', IRIN-HOA weekly round-up 64, 17–23 Nov. 2001; and Allen, M. and Mufson, S., 'US presses Sudan to end persecution', *International Herald Tribune*, 5–6 May 2001, p. 4.

two central issues of the relationship between the north and the south and the relationship between religion and the state.[103]

The extraction of oil begun by several international corporations in 1999 caused a tremendous increase in violence in the south-central provinces. In 2001 the government stepped up its scorched-earth campaign in an effort to clear all residents from the areas of the oilfields. The result was hundreds of deaths and tens of thousands of displaced people, adding to the population of IDPs that was already the largest in the world.[104] Airstrips and roads built by the oil companies gave the military better access to remote areas to continue its campaign. In addition, in recent years the Sudanese Government has increased its military spending. These factors transformed the military's strategy from holding garrison towns and launching dry-season attacks to systematically taking territory and destroying everything in its path.[105] The government and oil companies deny that there is any increase in deaths or human rights abuses, but numerous humanitarian aid and human rights organizations claim that the devastation is overwhelming.[106]

There was occasional heavy fighting in the Nuba mountains of central Sudan as government and SPLA forces battled for control of the region that lies just north of the oilfields. The rebels hope to be able to attack the oil companies' assets in an attempt to shut them down.[107] In May, the government launched its largest offensive since 1992, committing some 7500 troops. The SPLA repulsed them and despite continued fighting the situation had not significantly changed by the end of the year.[108] The humanitarian crisis in the area worsened and famine loomed on the horizon.[109]

The SPLA concentrated its war effort to the west of the oilfields, in Bar el-Ghazal province, where it attacked garrison towns held by the military and pro-government militia. In February the SPLA began an attack that paved the way for a major push in May.[110] By early June the rebels claimed to control all of Bar al-Ghazal for the first time. Independent sources reported that the fighting was not heavy, as local forces that were supposed to defend the government's position switched sides.[111] Nevertheless, the fighting caused tens of

[103] 'Sudan: delusions of peace' (note 100).
[104] US Committee for Refugees (note 99); and IRIN-HOA, 'Sudan: largest internally displaced population in the world', IRIN-HOA weekly round-up 42, 16–22 June 2001.
[105] Christian Aid, *The Scorched Earth: Oil and War in Sudan* (Christian Aid: London, Mar. 2001); and Janney, H., 'Oil reserves transform', *Jane's Intelligence Review*, June 2001, pp. 12–15.
[106] 'War, famine and oil in Sudan', *The Economist*, 14 Apr. 2001, p. 41; and IRIN-HOA, 'Sudan: government denies oil "atrocities"', IRIN-HOA weekly round-up 29, 17–23 Mar. 2001.
[107] Agence France-Presse (World Service), 22 Aug. 2001, in 'Sudan: rebels claim capture of two garrisons in Nuba Mountains', FBIS-AFR-2001-0822, 23 Aug. 2001.
[108] Flint, J., 'Jihad aims at starving out Nuba rebels', *The Guardian*, 4 June 2001, in *Horn of African Bulletin*, vol. 11, no. 3 (May/June 2001), p. 25; IRIN-HOA, 'Sudan: government offensive in Nuba Mountains offensive', IRIN-HOA weekly round-up 40, 4–8 June 2001; and IRIN-HOA, 'Sudan: rebels tell of "fierce fighting" in Nuba Mountains', IRIN-HOA weekly round-up 67, 8–14 Dec. 2001.
[109] 'War, famine and oil in Sudan' (note 106); and Flint (note 107).
[110] IRIN-HOA, 'Sudan: displaced need urgent assistance in Bhar al-Ghazal', IRIN-HOA weekly round-up, 27, 5–9 Mar. 2001.
[111] Reuters, 'SPLA capture two more garrisons', 30 May 2001, in *Horn of Africa Bulletin*, vol. 13, no. 3 (May/June 2001), p. 27; and 'Sudan rebels take government region', BBC Online Network, 4 June 2001, URL <http://news.bbc.co.uk/hi/english/world/>.

thousands of people to flee at a time when they would normally have been planting seeds in a region that has usually supplied food to much of the country.[112] In October, the government regained control of the provincial capital Raga. An intensified bombing and ground offensive in October and November triggered calls of alarm from humanitarian organizations.[113]

III. Conflicts in Asia

The most dramatic event in Asia in 2001 was the entry of the USA into the conflict in Afghanistan after the terrorist attacks in the USA. The US action changed the conflict from a deadlock that favoured the Taliban regime to a dynamic war in which the Taliban were defeated. It also shifted political relationships throughout Central and South-East Asia. The conflicts in India (Kashmir) and Indonesia remained intractable, while conflicts in the Philippines and Sri Lanka showed signs of moving towards negotiated settlements.

Afghanistan

The reaction of the United States to the terror attacks of 11 September completely transformed the conflict in Afghanistan between the Taliban and the United Islamic Front for the Salvation of Afghanistan (UIFSA), also called the Northern Alliance. The Taliban, under Supreme Leader Mullah Muhammad Omar, and the Northern Alliance, led by military chief Ahmad Shah Massoud and political head Burhanuddin Rabbani, had fought over control of the state since 1994. That conflict was a continuation of the war that began in 1978, when the Soviet Union invaded Afghanistan in an attempt to ensure a pro-Soviet government. For most of 2001 the Taliban controlled the capital Kabul and 90–95 per cent of the country. Although the Northern Alliance only held territory in the north-east, it was recognized by the United Nations as the legitimate government. By the end of the year the Taliban had been militarily and politically defeated by the USA and the Northern Alliance. A new, temporary government was set up in Kabul, headed by Hamid Karzai and composed of a spectrum of political leaders but dominated by members of the Alliance.

In January 2001 UN diplomatic and economic sanctions and an arms embargo on the Taliban (but not the UIFSA) came into effect because of their support for and training of international terrorists and their role in drug trafficking.[114] The impact of the economic sanctions was minimal since 80 per cent of the Afghan economy was dependent on the production of and trade in

---

[112] United Nations, Secretary-General says recent offensives in Sudan cause 'massive disruption', UN document SG/SM/7881, 6 July 2001; IRIN-HOA, 'Sudan: insecurity brings IDP concentration in Awoda', IRIN-HOA weekly round-up 46, 14–20 July 2001; and IRIN-HOA, 'Sudan: Raga depopulated after fighting, bombings', IRIN-HOA weekly round-up 45, 7–13 Jul. 2001.
[113] IRIN-HOA, 'Sudan: army recaptures Raga', IRIN-HOA weekly round-up 59, 15–19 Oct. 2001; and IRIN-HOA, 'Sudan: NGO reports civilians suffering in Aweil offensive', IRIN-HOA weekly round-up 61, 27 Oct.–2 Nov. 2001.
[114] UN Security Council Resolution 1333, 19 Dec. 2000.

illegal drugs, according to the UN.[115] The arms embargo helped the Northern Alliance on the battlefield, since the Taliban had trouble resupplying their troops and the Northern Alliance was able to buy arms and helicopters from India, Iran and Russia.[116] The Northern Alliance was thought to pay for the aid by selling gems, as well as profiting from the production of opium.[117]

The Northern Alliance's new *matériel* probably helped it avoid further losses to the Taliban, but it did not significantly change the balance between the sides.[118] The Northern Alliance was estimated to have 12 000–15 000 fighters under the command of Massoud.[119] Additional anti-Taliban groups which were not part of the Northern Alliance numbered in the thousands and operated in pockets of northern, central and western Afghanistan.[120] These fighters were under the command of local leaders who have traditionally fought each other for influence. The Taliban were thought to have a fighting strength of 40 000–45 000, of whom an estimated 4000–12 000 came from Chechnya, Pakistan, Uzbekistan and various Arab countries.[121]

There was considerable fighting during the first eight months of 2001, but the conflict remained in the stagnant state that had characterized it for several years.[122] In particular there was frequent fighting in the middle of the country and in the north-eastern province of Takhar, near the border with Tajikistan, where the Taliban were trying to crush the Northern Alliance's stronghold.[123]

On 9 September UIFSA military leader Massoud was mortally wounded by suicide bombers posing as journalists.[124] The loss of their strongest general might have turned the tide against the Northern Alliance, had it not been for the terrorist attacks on the World Trade Center and the Pentagon in the United

[115] Interfax, 11 July 2001, in 'Afghanistan still a major narcotics threat according to CIS border chiefs', Foreign Broadcast Information Service, *Daily Report–Central Eurasia (FBIS-SOV)*, FBIS-SOV-201-0711, 12 July 2001. The 3 countries that had diplomatic relations with the Taliban prior to Sep. 2001 were Pakistan, Saudi Arabia and the United Arab Emirates.

[116] Davis, A., 'Pakistan in quandary over new sanctions against the Taliban', *Jane's Intelligence Review*, Feb. 2001, pp. 34–37; and Davis, A., 'India increasing aid to Afghan opposition', *Jane's Defence Weekly*, 28 Mar. 2001, p. 14.

[117] Davis, A., 'Interview: Ahmad Shah Massoud, commander of Afghanistan's United Front', *Jane's Defence Weekly*, 4 July 2001, p. 32.

[118] Davis, A., 'Afghanistan reorganises to counter Taliban push', *Jane's Defence Weekly*, 4 July 2001, p. 14.

[119] Davis (note 117).

[120] Davis (note 117); and Wesolowsky, T., 'Afghanistan talk of strikes throws lifeline to opposition', Radio Free Europe/Radio Liberty, available on Eurasianet at URL <http://www.eurasianet.org/departments/insight/articles/pp091901.shtml>.

[121] Davis, A., 'The Taliban tinderbox', *Jane's Defence Weekly*, 18 July 2001, pp. 18–19; and RIA, 24 July 2001, in 'Russia: Uzbeks help Taleban fighting in North Afghanistan', FBIS-SOV-2001-0724, 25 July 2001.

[122] Interfax, 12 May 2001, in 'Afghanistan: Northern Alliance sources say offensive against Taleban continues', FBIS-SOV-2001-0512, 14 May 2001.

[123] Clark, K., 'Conflicting claims in Afghan fighting', BBC Online Network, 22 Jan. 2001, URL <http://news.bbc.co.uk/hi/english/world/>; Agence France-Presse (Hong Kong), 13 Jan. 2001, in 'Opposition says "heavy" fighting between rival Afghan factions erupted in Takhar', Foreign Broadcast Information Service, *Daily Report–NearEast/South Asia (FBIS-NES)*, FBIS-NES-2001-0113, 16 Jan. 2001; and ITAR-TASS, 10 Mar. 2001, in 'Russian border guards report intensive fighting in Afghanistan', FBIS-SOV-2001-0310, 13 Mar. 2001.

[124] 'Blast injures Afghan opposition chief', BBC Online Network, 9 Sep. 2001, URL <http://news.bbc.co.uk/hi/english/world/>; and 'Taliban air offensive reported near Kabul', *International Herald Tribune*, 12 Sep. 2001, p. 6.

States two days later. The USA immediately blamed the al-Qaeda network and its leader Osama bin Laden, who was known to live in Afghanistan under the protection of the Taliban. The Taliban leadership refused to hand over bin Laden in the face of US threats to use force, just as it had refused to comply with UN Security Council resolutions demanding that it do so. In response, the USA began building up its naval forces in the Indian Ocean and negotiating limited access rights to airbases in Pakistan, Kazakhstan, Kyrgyzstan, Tajikistan and Uzbekistan.[125] Emboldened by the prospect of US military strikes on the Taliban, the Northern Alliance intensified its fighting and made small advances.[126]

The UN Security Council condemned the terrorist attacks, interpreted them as a threat to international peace and security, and interpreted member states' right of self-defence to include retaliatory action against the perpetrators.[127] This opened the legal avenue for the US war in Afghanistan.

The Pakistani Government, under President General Pervez Musharraf, made a decision to end its support for the Taliban and cooperate with the US effort to topple the regime of its former ally. The decision eliminated any hope the Taliban had of receiving outside assistance in their impending war against the USA, as Saudi Arabia and the United Arab Emirates cut off their diplomatic relations with the Taliban.[128] Musharraf's decision also enabled US air operations by allowing flights over Pakistani territory. Without that permission, US aircraft would not have been able to reach Afghanistan from the Indian Ocean, since Iran would not allow overflights. Aircraft based on aircraft carriers, in Kuwait and on the Indian Ocean island of Diego Garcia provided most of the firepower used by the USA.

The UK and the USA began military operations against the Taliban and the al-Qaeda on 7 October, when they struck targets in Afghanistan with submarine- and air-launched cruise missiles and other munitions.[129] For about the first two weeks the attacks focused on gaining uncontested control of the air, destroying terrorist training camps, and destroying the Taliban's command, control and communications capacity.[130] By mid-October US special forces were in Afghanistan. They coordinated Northern Alliance forces and gave US aircraft more accurate target information so that they could strike troop formations rather than just fixed targets.[131] A number of other countries

---

[125] Norton-Taylor, R., 'Former Soviet states offer bases for US strikes,' *The Guardian* (Internet edn), 26 Sep. 2001, URL <http://www.guardian.co.uk/guardian/>.

[126] Agence France-Presse (Hong Kong), 24 Sep. 2001, in 'Afghanistan: Taliban confirms opposition take over of strategic town', FBIS-NES-2001-0924, 26 Sep. 2001.

[127] UN Security Council Resolution 1368, 12 Sep. 2001.

[128] 'UAE cuts ties with Taleban', BBC News Online, 22 Sep. 2001, URL <http://news.bbe.co.uk/hi/english/world/>.

[129] 'Bush: "battle joined", US and UK bomb targets in Afghanistan', *International Herald Tribune*, 8 Oct. 2001, pp. 1, 4.

[130] Gordon, M. R., 'US military campaign in Afghanistan outpaces political plan', *International Herald Tribune*, 16 Oct. 2001, p. 4.

[131] Mufson, S. and Ricks, T. E., 'Bolstering rebel offensive, special forces open new western front', *International Herald Tribune*, 12 Nov. 2001, p. 4; and Fitchett, J., 'Gunship use signals start of attacks on troops', *International Herald Tribune*, 17 Oct. 2001, pp. 1, 4.

eventually contributed ground, air and naval forces, most of them in support rather than combat roles.[132]

Two changes at the end of October opened the way for a sudden breakthrough on the ground. First, the number of daily air attacks increased considerably, many of which focused on the Taliban front lines.[133] Second, Russia provided the Northern Alliance with tanks, armoured vehicles, artillery, ammunition and other equipment.[134] The Taliban collapsed. On 9 November Northern Alliance forces took control of the northern city of Mazar-i-Sharif.[135] On 12 November the Northern Alliance captured the northern city of Taloqan in Takhar province, where there had been so much indeterminate fighting earlier in the year.[136] They also captured the crossroads city of Herat in the west near the border with Iran.

On 13 November the Taliban fled Kabul. Despite requests by many countries that the Northern Alliance not enter the city and re-establish themselves as the government, about 2000 military and police from the Northern Alliance took control.[137] International misgivings were overcome by the establishment of calm and order in an environment that had threatened to become anarchic. The next day the Taliban surrendered Jalalabad, east of Kabul, near the border with Pakistan.[138] Two Taliban strongholds remained. Many non-Afghan fighters held out in the northern city of Kunduz but surrendered under intense military pressure on 26 November.[139] The seat of Taliban leader Mullah Omar was in the southern city of Kandahar, in a part of the country dominated by Pashtun tribes, from which the Taliban drew most of their support. The focus of the air war shifted south, US Marines established a base at an airfield south of Kandahar on 26 November, and the Northern Alliance captured the city on 6 December.[140]

Although the fight against the Taliban was successful, another US objective was not achieved. Neither Mullah Omar nor Osama bin Laden were captured and it did not appear that either had been killed. In December the USA focused its operations on a network of caves in the region of Tora Bora on the Pakistani border, where it believed bin Laden was hiding. However, a military

---

[132] Australia and the UK supplied combat forces. Ground and air support forces were provided or promised by Canada, France, Germany, Italy, Japan, Jordan, the Netherlands, Poland, Russia and Turkey. 'Operation Enduring Freedom: a day-by-day account of the war in Afghanistan', *Air Forces Monthly*, Nov. 2001, pp. 35–50; and Willis, D., 'Afghanistan: the second month', *Air Forces Monthly*, Dec. 2001, pp. 74–82.

[133] Kock, A., Sirak, M. and Burger K., 'Pressure pushes US bomb raids into overdrive', *Jane's Defence Weekly*, 7 Nov. 2001, p. 2.

[134] 'War and politics in Afghanistan: now for an equally hard part', *The Economist*, 17 Nov. 2001, pp. 15–17; and Willis (note 132).

[135] Rishburg, K. and Branigan, W., 'Afghan alliance claims capture of strategic city', *International Herald Tribune*, 10–11 Nov. 2001, pp. 1, 4.

[136] Knowlton, B., 'Anti-Taliban forces continue their march', *International Herald Tribune*, 12 Nov. 2001, pp. 1, 4.

[137] Rohde, D., 'Taliban flee as Northern Alliance takes Kabul', *International Herald Tribune*, 14 Nov. 2001, pp. 1, 4.

[138] 'Northern Alliance consolidates power; Jalalabad captured', *International Herald Tribune*, 15 Nov. 2001, p. 1.

[139] Willis (note 132).

[140] Mallet, V. and Wolffe, R., 'Taliban to give up Kandahar', *Financial Times*, 7 Dec. 2001, p. 1.

assault on the cave complex did not result in the capture or death of al-Qaeda leaders, and US officials admitted that they had lost all trace of bin Laden.[141]

When the US attack on Afghanistan began in October, there was a widespread expectation that the fight would be long and difficult. Analogies were drawn with the experiences of the UK in the 19th century and the Soviet Union in the late 1970s and early 1980s, in which both countries expended great effort but were unable to dominate the Afghan people. The swiftness of the US-led military campaign can be attributed to several factors.[142] First, the US forces utterly dominated the Taliban and al-Qaeda fighters with overwhelming firepower and a near monopoly on battlefield information. Second, the Northern Alliance proved to be willing and able allies who pressed the ground campaign forward as the USA operated from the sky. Third, the USA used special operations forces to great advantage by attaching them to Northern Alliance units so that they could coordinate ground and air attacks, pinpoint targets for aircraft, and coordinate logistics and supply help for the Afghan forces. Fourth, the Taliban were weak. Their lack of popularity and a tradition of switching allegiance in Afghanistan meant that many local commanders joined the Northern Alliance or fled without fighting. In addition, the number of Taliban troops was small compared to the size of the territory they protected, so when a defensive line gave way the advancing Alliance could rapidly capture territory before meeting further resistance. Finally, many commentators forgot that, while Soviet troops also captured Kabul relatively quickly, it was when they tried to stay and impose a government that they encountered difficult problems.

On 27 November–5 December 2001, the Special Representative of the Secretary-General for Afghanistan, Lakhdar Brahimi, brought together delegates from four Afghan factions in Bonn, Germany, to discuss arrangements for the replacement of the Taliban regime. The delegates agreed to establish an Interim Authority—a power-sharing council to govern for six months, starting from 22 December, under the leadership of Pashtun tribal leader Hamid Karzai.[143] Some Northern Alliance leaders threatened not to recognize the interim government, but in the end they agreed to respect the decision.[144] The interim government is intended to give way to a two-year transitional government established by a traditional council of elders at the end of the six-month transitional period, in mid-June 2002.

---

[141] 'Hunt for Omar stepped up', BBC Online Network, 18 Dec. 2001, URL <http://news.bbc.co.uk/hi/english/world/>.

[142] Bener, B., Burger, K. and Koch, A., 'Afghanistan: first lessons', *Jane's Defence Weekly*, 19 Dec. 2001, pp. 18–21; Frantz, D., 'Exiled Afghan leaders see hope for defections', *International Herald Tribune*, 19 Oct. 2001, p. 2; and 'War and politics in Afghanistan' (note 134).

[143] The Agreement on Provisional Arrangements in Afghanistan pending the Re-establishment of Permanent Government Institutions (the Bonn Agreement), contained in United Nations, Letter dated 5 December 2001 from the Secretary-General addressed to the President of the Security Council, UN document S/2001/1154, 5 Dec. 2001, available at URL <http://www.uno.de/frieden/afghanistan/talks/agreement.htm>.

[144] Hoyos, C. and Fidler, S., 'Afghans sign accord to form interim government', *Financial Times*, 6 Dec. 2001, p. 1.

As requested in the Bonn Agreement, the UN Security Council approved the International Security Assistance Force in Afghanistan (ISAF) on 20 December.[145] A force of up to 3000 troops was authorized for six months to operate in and around Kabul. It was to provide stability in the area, assist the Afghan Interim Authority in developing future security structures and help administer reconstruction assistance.[146]

Reconstruction was desperately needed after 23 years of war. There was an even more urgent need for emergency humanitarian assistance. Before the escalation of violence in October, war and a three-year drought had driven 2.2 million refugees into Pakistan and 1.5 million into Iran.[147] About 1 million were internally displaced at the end of September.[148] The UN aid agencies estimated in September that 5 million people needed humanitarian assistance to survive.[149] The increase in violence exacerbated the situation. During the US attacks, large proportions of the population of cities and towns fled.[150]

The number of civilians killed by US actions was difficult to determine. The Taliban alleged that the casualty rates were high, but few people regarded this claim as credible. The US Government acknowledged that the operations had killed civilians but insisted that the number of casualties was low and offered no total figure. Independent estimates of the number of Afghan civilians killed by early December range from a conservative 1000–1300[151] to 3767–5000.[152]

India: Kashmir

India and Pakistan fought wars over the territory of Kashmir in 1965 and 1971, as well as clashing in the Kargil region in 1999.[153] An intra-state conflict arose in 1989, when several groups in the Indian province of Jammu and Kashmir began to use violence in their effort to gain independence for the Muslim-dominated province from Hindu-dominated India. Some groups wanted the province to become part of Pakistan, while others wanted to establish an independent state. Since then, most of the militant groups, composed of indigenous Kashmiris who fought for locally defined objectives, have turned away from violence. They have been replaced by militant groups composed largely of Afghans, Pakistanis and other foreigners motivated by a desire to

---

[145] United Nations Security Council Resolution 1386, 20 Dec. 2001.
[146] Military Technical Agreement between the International Security Assistance Force (ISAF) and the Interim Administration of Afghanistan, 30 Dec. 2001, p. 3. See appendix 2A in this volume.
[147] IRIN-Asia, 'Afghanistan: fact-sheet on displaced and refugees', 18 May 2001.
[148] US Committee for Refugees, 'Pakistan: Afghan refugees shunned and scorned', 28 Sep. 2001.
[149] United Nations, In Afghanistan, a population in crisis, UN document AFG/145, ORG/1336, 24 Sep. 2001.
[150] IDP Project, 'Country profile: Afghanistan', Oct. 2001.
[151] Conetta, C., *Operation Enduring Freedom: Why a Higher Rate of Civilian Bombing Casualties*, Briefing Report no. 11 (Project on Defense Alternatives: Cambridge, Mass., 24 Jan. 2002), available at URL <http://www.comw.org/pda/0201oef.html>.
[152] IRIN News, 'Afghanistan: mounting concern over civilian casualties', 7 Jan. 2002, URL <http://www.irinnews.org/>.
[153] See Widmalm, S., 'The Kashmir conflict', *SIPRI Yearbook 1999: Armaments, Disarmament and International Security* (Oxford University Press: Oxford, 2000), pp. 34–46.

establish Islamic rule across the entire region.[154] The Indian Government has long insisted that the status of Kashmir is an internal issue and has refused to seek a negotiated solution with Pakistan.

Pakistan is widely believed to support the rebel groups, although it denies doing so. The Inter-Services Intelligence (ISI) Directorate, which had ties to the Taliban in Afghanistan, provides assistance that includes training and logistical, financial and doctrinal support. In addition, fighters attend religious schools in Pakistan that preach a violent interpretation of jihad.[155] The five main groups in Kashmir that benefit from Pakistani support are Hizbul Mujahideen, Lashkar-e-Toyeba, Harkat-ul-Mujahideen, al Babr and Jaish-e-Mohammed. Only Hizbul Mujahideen is composed mainly of Kashmiris.[156] In mid-2001 there were an estimated 3500–4000 rebel fighters and 350 000–400 000 Indian military and paramilitary in Kashmir.[157] The number of Islamic fighters in Kashmir probably increased at the end of the year, as they fled from Afghanistan across Pakistan to Kashmir.

In the first half of 2001 there was movement towards the establishment of peace talks, but it led nowhere. An Indian unilateral ceasefire declared in November 2000 was extended twice by Prime Minister Atal Bihari Vajpayee, ending in May 2001.[158] The militant groups claimed that the ceasefire was a ploy to attract international support and did not reciprocate. The All Parties Hurriyat (Freedom) Conference—a collection of 20 parties, most of which disavow violence—expressed its interest in talks but required consultation with Pakistan first. The Indian Government refused to issue passports to several people in the Hurriyat's delegation, so the negotiations never opened.[159] An expression of interest by the Democratic Freedom Party in early April also came to nothing.[160]

Surprisingly, when the ceasefire expired in May, Vajpayee invited Pakistani President Musharraf to India for talks in July about several aspects of their countries' relations, including Kashmir.[161] The talks, held in Agra, India, were the first between the countries since they fought at Kargil in 1999.[162] The summit meeting ended without a joint statement because the two sides could not agree on how to refer to the sensitive issue of Kashmir.[163]

---

[154] Chalk, P., 'Pakistan's role in the Kashmir insurgency', *Jane's Intelligence Review*, Sep. 2001, pp. 26–27.

[155] Chalk (note 154).

[156] Chalk (note 154).

[157] Gardner, D., 'A search for peace in a divided valley', *Financial Times*, 14–15 July 2001, p. 7.

[158] Dugger, C. W., 'India vows 3rd month for Kashmir cease-fire', *International Herald Tribune*, 24 Jan. 2001, p. 4; and Donald, A., 'New Delhi to extend ceasefire in Kashmir', *Financial Times*, 22 Feb. 2001, p. 7.

[159] 'Kashmir: muted applause', *The Economist*, 24 Feb. 2001, p. 66.

[160] 'Kashmir separatist chief wants talks', BBC Online Network, 30 Apr. 2001, URL <http://news.bbc.co.uk/hi/english/world/>; and 'Slow progress in Kashmir talks', BBC Online Network, 28 May 2001, URL <http://news.bbc.co.uk/hi/english/world/>.

[161] Gardner (note 157).

[162] Bedi, R., 'War of words over Kashmir', *Jane's Defence Weekly*, 25 July 2001, p. 23.

[163] Constable, P., 'India–Pakistan summit collapses over Kashmir', *International Herald Tribune*, 17 July 2001, p. 1.

Throughout 2001, small attacks and suicide bombings by militant groups occurred weekly.[164] The Indian security forces responded aggressively to the rebel actions, particularly after May. The violence caused more than 2000 deaths in 2001.[165] About 35 000 people have been killed since 1989.[166] The largest attack since 1999 took place in October 2001 when militants drove a car bomb into the legislature building in Jammu and Kashmir's summer capital Srinagar, killing 38 people.[167] The attack generated pressure by Indian hardliners to strike at Pakistan, a country they believe harbours terrorists. Despite a visit by US Secretary of State Colin Powell to the two countries in October, they exchanged artillery fire across the Line of Control in October and November, as they had done earlier in the year.[168] In November, India accused Pakistan of provocative troop movements, to which Pakistan replied that India had been massing troops near the Line of Control.[169]

The political stakes escalated dramatically in December, when suicide bombers and gunmen attacked the Indian Parliament building in New Delhi. No group claimed responsibility but Indian authorities were certain that one of the Kashmiri groups was to blame, most likely Lashkar-e-Toyeba or Jaish-e-Mohammed.[170] India demanded that Pakistan disband the militant groups. Under intense pressure, President Musharraf took steps to control the militants, whom he called 'freedom fighters', including the arrest of many of their leaders. Not satisfied with Musharraf's actions, India built up its military forces along the Line of Control in a game of brinkmanship designed to induce Pakistan to fully disband the militant groups. The year ended with Pakistan matching Indian moves in a tit-for-tat military escalation. Both leaders said that they were prepared to go to war.[171]

Indonesia

In a country consisting of 17 000 islands and over 210 million people, the Indonesian Government faces separatist movements in Aceh and Irian Jaya and violent communal conflicts in West Kalimantan, the Moluccas and Central Sulawesi. Aceh was the only location that experienced sustained, politically motivated violence in 2001. Gerakan Aceh Merdeka (GAM, Free Aceh Movement) has sought an independent state on the northern tip of Sumatra since 1976. Violence between the GAM and government security forces has

[164] Bearak, B., '11 die in shoot-out at Kashmir airport', *International Herald Tribune*, 17 Jan. 2001, p. 4; 'Kashmir attack kills five', BBC Online Network, 25 May 2001, URL <http://news.bbc.co.uk/hi/english/world/>; and 'Militants killed in Kashmir', *Jane's Defence Weekly*, 21 Nov. 2001, p. 14.

[165] Bedi, R., 'Renewed fighting in Kashmir after talks fail', *Jane's Intelligence Review*, Sep. 2001, p. 6.

[166] Gardner (note 157).

[167] 'Rebels attack Kashmir assembly', *International Herald Tribune*, 2 Oct. 2001, p. 4.

[168] Dugger, C. W., 'India rejects Pakistan offer to hold talks', *International Herald Tribune*, 24 Oct. 2001, p. 5; and 'Kashmir clashes leave 35 dead', *International Herald Tribune*, 5 Nov. 2001, p.1.

[169] 'Pakistan move angers India', *International Herald Tribune*, 2 Nov. 2001, p. 7.

[170] Luce, E., 'Terrorists attack Indian Parliament', *Financial Times*, 14 Dec. 2001, p. 4.

[171] Chandrasekaran, R., 'India arrays missiles along border', *International Herald Tribune*, 27 Dec. 2001, p. 1; and Chandrasekaran, R., 'Pakistan and India spar with sanctions', *International Herald Tribune*, 28 Dec. 2001, p. 1.

caused 5000–6000 deaths, most of them civilians, in 25 years of conflict.[172] In 2001 the situation deteriorated considerably, with a complete breakdown in talks between the government and the rebels. The Indonesian Red Cross reported at least 1500 deaths in 2001.[173]

The year began with a one-month extension of the May 2000 ceasefire. The stated intention was to create conditions conducive to discussion. However, as in 2000, both sides tried to strengthen their positions, resulting in an increase in violence rather than a decrease.[174] In February the two sides met in Switzerland for peace talks mediated by the Henry Dunant Centre for Humanitarian Dialogue. On the day the talks opened the Indonesian Government announced that it was sending an additional 3000 troops to the province, bringing the total number of military, paramilitary and police forces to over 30 000.[175] The talks were fruitless and broke down completely in July when the government arrested members of the GAM negotiating team.[176]

Small-scale attacks by the GAM and counter-insurgency operations by government forces were nearly daily events in 2001. With an estimated strength of about 3000 fighters and even fewer military-type weapons, the GAM planted bombs and launched surprise attacks on police, military and industrial targets, and civilians of Javanese background.[177] (The central government's practice of settling people from the island of Java on other, less densely populated, islands is a source of tension throughout Indonesia.) With popular support apparently growing in the countryside, the separatists managed to prevent the government from exercising effective administrative control of about 80 per cent of the province.[178] Government forces, oblivious to the concept of winning the support of the population, frequently burned entire villages to the ground and killed suspected rebels and their sympathizers.[179]

One of the GAM's grievances is that Jakarta siphons off Aceh's resources and returns little benefit to the province. In March GAM attacks on natural gas-producing facilities run by ExxonMobil caused the company to shut down production until it resumed on a limited basis in July. The government, which exports an estimated $1.3 billion worth of oil and gas from Aceh each year,

---

[172] Olson, E., 'Indonesia and Aceh rebels agree to extend cease-fire', *International Herald Tribune*, 11 Jan. 2001, p. 5; and 'Aceh violence brings troops out', BBC Online Network, 17 Aug. 2001, URL <http://news.bbc.co.uk/hi/english/world/>.

[173] Soekanto, S. W. E., 'Autonomy policy no solution to Aceh conflict', *Jakarta Post*, 27 Dec. 2001, URL <http://globalarchive.ft.com/globalarchive/aritcle.html?id=011227001091&query=autonomy+policy+no+solution+to+aceh+conflict>.

[174] 'Indonesia: even nastier', *The Economist*, 20 Jan. 2001, p. 55.

[175] 'New round of Aceh peace talks', BBC Online Network, 15 Feb. 2001, URL <http://news.bbc.co.uk/hi/english/world/>; and Harris, P., 'Indonesia prepares for assault on GAM', *Jane's Intelligence Review*, Apr. 2001, p. 4.

[176] Human Rights Watch (HRW), *Indonesia: the War in Aceh*, HRW Report, vol. 13, no. 4(c), (Aug. 2001), p. 2.

[177] Gunaratna, R., 'The structure and nature of GAM', *Jane's Intelligence Review*, Apr. 2001, pp. 33–35.

[178] ICG, *Aceh: Why Military Force Won't Bring Lasting Peace*, ICG Asia Report no. 17 (ICG: Jakarta/Brussels, 12 June 2001), p. ii; and Gunaratna (note 177).

[179] Soekanto (note 173); 'Troops kill 7 in Aceh, rebels say', *International Herald Tribune*, 26 Mar. 2001, p. 5; and McBeth, J., 'Hard hearts, bitter minds', *Far East Economic Review*, 30 Aug. 2001, pp. 26–27.

secured the installation with thousands of troops.[180] In response to the continued violence, closure of the gas fields and failed talks, President Abdurrahman Wahid issued Presidential Instruction (Inpres) no. 4 in April. The decree authorized increased military and police operations in Aceh and allowed military reinforcements to be sent to the province.[181] The largest clashes of the year occurred in June, when a GAM attack on Javanese settlers brought a strong counter-attack by the military and local militia, causing hundreds of deaths, hundreds of houses to be burned and thousands of people to flee.[182]

The parliament elected Megawati Sukarnoputri as president on 23 July after it impeached former President Abdurrahman Wahid for corruption.[183] In her first state of the nation address, President Sukarnoputri apologized for past abuses by security forces in Aceh and Irian Jaya. However, she is a strong nationalist who is opposed to Wahid's moves towards decentralization and has flatly stated her opposition to independence for restive provinces.[184] In August she sent additional national police trained in counter-insurgency techniques.[185] She also signed a law giving Aceh a measure of autonomy. The action, which she made the centrepiece of her Aceh policy, did not seem to impress the people of Aceh, who were not consulted about it, and did not meet the GAM's demand for complete independence.[186] The year ended with continued and frequent clashes.[187]

The Philippines

In January 2001 Gloria Macapagal Arroyo replaced Joseph Estrada as president of the Philippines. She turned away from the bellicose stance adopted by the government in 2000 towards the Moro Islamic Liberation Front (MILF) and the New People's Army (NPA). Her strategy was to engage these two groups in dialogue, maintain a peaceful relationship with the Moro National Liberation Front (MNLF) and try to eradicate the smallest and most violent group, the Abu Sayyaf.

The NPA has fought for a Marxist government since 1968. In 2001 it launched occasional assaults on government forces and installations.[188] The

[180] Haseman, J., 'Indonesian troops poised to extend operations', *Jane's Defence Weekly*, 28 Mar. 2001, p. 13; Harris (note 175); and McBeth, J., 'Indonesia: pumping from day to day', *Far Eastern Economic Review*, 16 Aug. 2001, pp. 24–25.
[181] 'Wahid orders a crackdown', *International Herald Tribune*, 13 Apr. 2001, p. 5; and Human Rights Watch, *World Report 2002: Events of 2001* (Human Rights Watch: New York, 2002), p. 232.
[182] Human Rights Watch (note 181).
[183] Chandrasekaran, R., 'President's ouster grips Indonesia', *International Herald Tribune*, 24 July 2001, p. 1, 4.
[184] Haseman, J. and Karniol, R., 'Test of endurance: country briefing—Indonesia', *Jane's Defence Weekly*, 29 Aug. 2001, pp. 20–25; and 'Megawati to visit Aceh on Saturday', BBC Online Network, 3 Sep. 2001, URL <http://news.bbc.co.uk/hi/english/world/>.
[185] Haseman, J., 'More personnel deploy to Aceh', *Jane's Defence Weekly*, 1 Aug. 2001, p. 14.
[186] HRW (note 181), p. 233; and Schultz, K., 'Indonesia strives to restore order in Aceh', *Jane's Intelligence Review*, Sep. 2001, pp. 28–30.
[187] Soekanto (note 173).
[188] 'Rebels kill Philippine soldiers', *International Herald Tribune*, 13 Feb. 2001, p. 5; 'Philippine president seeks peace talks', *Jane's Defence Weekly*, 21 Feb. 2001, p. 6; and Williamson, H., 'Manila set to resume talks with rebels', *Financial Times*, 7 Mar. 2001, p. 4.

government has estimated that the NPA has about 6000 fighters, operating mostly in the northern regions although they are present throughout the country. Previous talks between the NPA and the government had broken down in 1999. In February President Arroyo appointed a panel to reopen talks with the rebels and their political arm, the National Democratic Front (NDF).[189] In March she ordered a ceasefire that was reciprocated by the NPA.[190] Peace talks between the government and the NDF began in Norway on 27 April.[191] After minor progress, the government suspended the talks in June when the NPA killed a member of parliament.[192] An offensive by the military in November resulted in a month of clashes that were the worst battles between the government and the NPA in a decade.[193] The government and rebels called reciprocal truces in December.[194]

The MILF has fought for an independent Muslim state on the southern island of Mindanao since 1984, when it broke off from the MNLF. It maintains a military presence and has popular support in a sizeable portion of the island. On 20 February 2001 President Arroyo ordered a suspension of the military operations against the MILF that had begun the previous year,[195] allowing about two-thirds of the displaced population to return home. With the mediation of Malaysian officials, the government and rebels reached an agreement in March on plans for a joint ceasefire and arrangements to hold peace talks.[196] The two sides made progress during talks held in Libya in June and signed a joint ceasefire agreement on 7 August in Malaysia.[197] The agreement outlined plans for the deployment of monitoring teams involving observers from the government, the MILF, Indonesia, Libya and Malaysia. A follow-up agreement was signed in October. The question at the end of the year was whether the peace talks will be accompanied by economic development in one of the poorest parts of the country. Economic development is one of the MILF's demands.[198]

---

[189] 'Philippine president seeks peace talks' (note 188).
[190] 'Philippine rebels ready to set truce', *International Herald Tribune*, 13 Mar. 2001, p. 8.
[191] Burgonio, T. J., 'Oslo tlaks start today', *Philippine Daily Inquirer* (online edn), 27 Apr. 2001, URL <http://www.inq7.net/nat/2001/apr/21/nat_8-1.htm>.
[192] AXF (AP) Asia, 'Philippines to resume talks with communist rebels in August', 17 June 2001, URL <http://globalarchive.ft.com>.
[193] 'Guide to Philippines conflict', BBC News Online, 6 Dec. 2001, URL <http://news.bbc.co.uk/hi/english/world/>.
[194] 'Philippines truce with communists', BBC News Online, 6 Dec. 2001, URL <http://news.bbc.co.uk/hi/english/world/>.
[195] 'Arroyo orders ceasefire with rebels', BBC Online Network, 20 Feb. 2001, URL <http://news.bbc.co.uk/hi/english/world/>.
[196] 'Philippines rebels agree truce', BBC Online Network, 27 Mar. 2001, URL <http://news.bbc.co.uk/hi/english/world/>; and Davis, A., 'Arroyo seeks to "heal and build"', *Jane's Intelligence Review*, Apr. 2001, p. 36.
[197] 'Philippine rebels sign ceasefire', BBC Online Network, 7 Aug. 2001, URL <http://news.bbc.co.uk/hi/english/world/>; and Karniol, R., 'Philippines government agrees ceasefire with MILF', *Jane's Defence Weekly*, 15 Aug. 2001, p. 12.
[198] Davis, A., 'Prospects for peace emerge in southern Philippines', *Jane's Intelligence Review*, Sep. 2001, p. 7; Sheehan, D., 'A price for peace', *Far East Economic Review*, 30 Aug. 2001, p. 17; and '"Big step" for Philippines peace talks', BBC Online Network, 18 Oct. 2001, URL <http://news.bbc.co.uk/hi/english/world/>.

In November a small element of the MNLF took up arms again, under the direction of former MNLF leader Nur Misuari. About 200 MNLF members attacked army positions on the southern island of Jolo. More than a week of fighting on Jolo and Mindanao left about 200 people dead, most of them rebels.[199] The attack was a surprise because the MNLF had agreed to stop fighting the government in 1996. It is thought that Misuari, who had recently been ousted as the MNLF leader, turned to violence in an attempt to stop the November 2001 gubernatorial election. The political settlement between the government and the main body of the MNLF remained in place. The election proceeded but with very low participation. Misuari was arrested in Malaysia and extradited to the Philippines.[200]

The government intensified its effort to eradicate the Abu Sayyaf, which it sees as a criminal organization rather than a political group.[201] The Abu Sayyaf claims to be fighting for an independent Muslim state, but its actions reveal a far greater interest in capturing hostages for ransom. It appears to number in the hundreds and have no ability to capture and hold territory.[202] Nevertheless, the Abu Sayyaf attracted the attention of the United States for its alleged connections to the al-Qaeda terrorist network. Before the 11 September terrorist attacks, the USA had suggested the holding of joint military exercises with Philippine forces, and at the end of the year the Philippine Government accepted the proposal. The USA sent several hundred military advisers to the Philippines and promised additional military and economic assistance.[203]

Sri Lanka

The level of violence in Sri Lanka was considerably lower in 2001 than in 2000. The Liberation Tigers of Tamil Eelam (LTTE) have waged a struggle for a separate Tamil state in the northern part of Sri Lanka since 1983, in a conflict that has claimed over 62 000 lives.[204] LTTE leader Vellupillai Prabhakaran declared a one-month unilateral ceasefire from 24 December 2000 and then extended it several times until 24 April 2001. He indicated his readiness to engage in peace talks on the conditions that the government reciprocate the ceasefire, lift the ban on the LTTE as a legal political organization, and lift the economic embargo on the Jaffna peninsula, held by the rebels.[205] The

[199] Kirk, D., 'Philippine rebels to free captives', *International Herald Tribune*, 28 Nov. 2001, p. 5.
[200] 'Voters shun Philippines poll', BBC Online Network, 26 Nov. 2001, URL <http://news.bbc.co.uk/hi/english/world/>; and 'Philippine leader arrested', BBC Online Network, 24 Nov. 2001, URL <http://news.bbc.co.uk/hi/english/world/>.
[201] 'Arroyo declares war on Abu Sayyaf', BBC Online Network, 2 Apr. 2001, URL <http://news.bbc.co.uk/hi/english/world/>.
[202] 'Philippine troops battle kidnappers', BBC Online Network, 1 June 2001, URL <http://news.bbc.co.uk/hi/english/world/>; and 'Manila says rebels are trapped', *International Herald Tribune*, 10 Oct. 2001, p. 10.
[203] 'Terrorism in the Philippines: the Jolo conundrum', *The Economist*, 24 Nov. 2001, p. 56.
[204] Dugger, C. W., 'Sri Lankan President turns up the pressure', *International Herald Tribune*, 6 Dec. 2001, p. 2.
[205] 'Sri Lanka rebels extend ceasefire', *International Herald Tribune*, 24 Jan. 2001, p. 4; 'Norwegian peace envoy in Sri Lanka', BBC News Online, 4 Apr. 2001, URL <http://news.bbc.co.uk/hi/english/world/>; and 'Talks with Tigers "cordial"', BBC News Online, 9 Apr. 2001, URL <http://news.bbc.co.uk/hi/english/world/>.

rebels seemed to be trying to take advantage of the territorial gains they had made in fierce military engagements during the past two years as well as the good offices of Norwegian envoy Erik Solheim, who began talks with both sides in 1999.

President Chandrika Bandaranaike Kumaratunga expressed scepticism that the LTTE were interested in peace and found their conditions unacceptable. Despite pressure from international economic donors, in January the government launched a military offensive to recapture important transport links on the Jaffna peninsula. The government forces, using tanks, artillery and aircraft, gained control of the road connecting the two main towns on the peninsula, but they were not able to advance towards the Elephant Pass, which connects the peninsula with the rest of the country. The battle ended after several days of strong resistance from the LTTE.[206]

After January, fighting remained at a low level for the duration of the LTTE ceasefire, with only occasional government air attacks and clashes at sea.[207] On 25 April government forces tried to capture the town of Pallai as a prelude to advancing on the Elephant Pass. Several days of fighting left hundreds of soldiers dead and did not change the front-line positions. The offensive was launched by the government's best troops with new equipment, and its failure raised concerns in the government about its ability to maintain a presence on the peninsula.[208]

Solheim's efforts to bring the two sides together foundered in May on the rebels' conditions and the government's insistence on unconditional talks. Beyond the procedural problems, President Kumaratunga has argued in favour of a new constitution that would give the Tamils autonomy, but the LTTE insists on independence.

The main action shifted from the northern peninsula to the capital Colombo in the summer. In July LTTE suicide commandos attacked the country's only international airport and an adjacent military airbase outside Colombo. The attack was significant because of the highly visible target and because it was the first attack by the LTTE on a military target so far away from the Jaffna peninsula. The LTTE destroyed three passenger aircraft, amounting to half the national airline fleet, as well as six fighter planes and two helicopters. In addition to its direct military and economic costs, the raid undermined Sri Lanka's tourism industry. The military immediately resumed its bombing of LTTE positions.[209] For the remainder of the year, fighting consisted of sporadic raids

---

[206] 'Poker game', *The Economist*, 13 Jan. 2001, p. 54; 'Sri Lanka Army in major push', BBC News Online, 16 Jan. 2001, URL <http://news.bbc.co.uk/hi/english/world/>; and 'Jaffna battle rages on', BBC News Online, 17 Jan. 2001, URL <http://news.bbc.co.uk/hi/english/world/>.

[207] AFP (Hong Kong), 22 Mar. 2001, in 'Sri Lankan jets bomb LTTE targets as 20 killed in 21 March sea battle', FBIS-NES-2001-0322, 23 Mar. 2001; and 'Upsurge of fighting in Sri Lanka', BBC News Online, 28 Mar. 2001, URL <http://news.bbc.co.uk/hi/english/world/>.

[208] 'Tiger teeth', *The Economist*, 5 May 2001, p. 5.

[209] Dugger, C. W., 'Tamil suicide bombers strike Colombo airport', *International Herald Tribune*, 25 July 2001, pp. 1, 5; and 'The Tigers pounce', *The Economist*, 28 July 2001, p. 54.

by the LTTE on army bases, police stations and ships in the waters around Jaffna and similarly sporadic government responses.[210]

President Kumaratunga's October call for parliamentary elections to be held in December ignited a divisive political campaign. The president led the People's Alliance Party campaign and was stridently intransigent towards the LTTE.[211] While she accused the main opposition party of secretly planning to grant the LTTE control over the north and east of the country, her opponent, Ranil Wickremesinghe of the United National Party (UNP), advocated a return to the peace talks under Norwegian auspices.[212] The campaign was marred by numerous election-related acts of violence and murder, but it appeared as though the perpetrators were supporters of various political parties rather than the LTTE.[213] The UNP won the election and Wickremesinghe became the new prime minister.[214] Kumaratunga remained the president.

By the end of the year the peace process had slowly resumed. In November, LTTE leader Prabhakaran said for the first time that the group would consider settling for something less than total independence. In December the new prime minister agreed to a truce and lifted an economic embargo on rebel-held areas, thus meeting two of the LTTE's three conditions for talks. The legal ban on the LTTE remained in place and the government would only communicate through the Norwegian intermediary. After elections in Norway brought a new government into office, Deputy Foreign Minister Vidor Helgeson was appointed as the new mediator.

The new anti-terrorism climate abroad led Australia, Canada and the UK to declare the LTTE a terrorist group, as India, Sri Lanka and the USA had done in previous years. Since many Tamils who support the LTTE live in those countries, the designation was bound to put financial and political pressure on the LTTE to resume peace negotiations.[215]

IV. Conflicts in Europe

The only active major armed conflict in Europe in 2001 was in the Russian republic of Chechnya. The conflict had transnational aspects in that non-Chechen Islamists fought in Chechnya, Chechens fought in Afghanistan, and the presence of Chechen refugees and rebels in the neighbouring state of Georgia threatened to exacerbate separatist conflicts there. Russia has consistently referred to the Chechen rebels as terrorists. After September, the interest of other states in stopping terrorism turned their criticism of Russian human

[210] Athas, I., 'Sri Lankan Navy in battle with Tigers', *Jane's Defence Weekly*, 26 Sep. 2001, p. 10; and Harrison, F., 'Tigers attack army camps', BBC News Online, 23 Aug. 2001, URL <http://news.bbc.co.uk/hi/english/world/>.
[211] 'Start again', *The Economist*, 20 Oct. 2001, p. 65; 'Sri Lanka sets elections as coalition fails', *International Herald Tribune*, 11 Oct. 2001, p. 9; and Dugger (note 204).
[212] Jayasinghe, A., 'Sri Lankan opposition claims election victory', *Financial Times*, 7 Dec. 2001, p. 3; and Dugger (note 204).
[213] 'Vote ends with killings and a curfew', *International Herald Tribune*, 6 Dec. 2001, p. 2.
[214] Jayasinghe (note 212).
[215] Dugger, C. W., 'Truce and new leader raise hope for Sri Lanka peace', *International Herald Tribune*, 26 Dec. 2001, p. 4.

rights violations in Chechnya into support for the struggle against extremists. Other militarized disputes in the Caucasus remained at a low level of violence. The Former Yugoslav Republic of Macedonia faced armed attack by ethnic Albanian rebel forces, but the fighting did not reach the level of a major armed conflict.

Russia: Chechnya

The conflict between the Russian Government and separatist rebels in the republic of Chechnya that began in 1999 continued to grind on with little change over the year.[216] Although the government claimed to control the entire country, it controlled only the major population centres.[217] The security situation was tenuous, even in the Chechen capital Grozny, as rebels infiltrated and attacked military posts at night and Russian forces occasionally tried to completely seal off the city.[218] According to official figures, there were 80 000 troops from the Russian defence and interior ministries in Chechnya at the beginning of the year.[219]

A plan to transform the Russian operation from a military to a police operation led to the transfer of responsibility from the army to the Federal Security Service (Federal'nya Sluzhba Bezopasnosti, FSB).[220] Reflecting its confidence that the rebels were almost defeated, the General Staff announced that it would reduce its forces in Chechnya to 50 000.[221] Although the withdrawal of troops started in March, it was halted in May and an independent estimate of the total number of Russian forces in Chechnya in September was 90 000.[222]

The rebels were strongest in the southern mountains, especially in the Argun and Vedeno gorges, where there were sporadic clashes with federal forces.[223] In the rest of the country the rebels employed guerrilla tactics, such as ambushing Russian troops, detonating bombs and assassinating military leaders and Chechens who worked for the Russian district administration.[224] The

---

[216] Chufrin, G., 'Russia: separatism and conflicts in the North Caucasus', *SIPRI Yearbook 2000* (note 5), pp. 157–80.
[217] Scott, R., 'Chechnya strains Russian resolve', *Jane's Intelligence Review*, Sep. 2001, p. 21.
[218] 'Fearing raids, Russians seal off Chechen capital', *International Herald Tribune*, 6 Aug. 2001, p. 7; and ITAR-TASS (Moscow), 4 Sep. 2001, in 'Russian troops kill 30 rebels in Chechnya over past two days', FBIS-SOV-2001-0904, 7 Sep. 2001.
[219] Interfax (Moscow), 22 Jan. 2001, in 'Total Russian strength in Chechnya 80,000', FBIS-SOV-2001-0122, 23 Jan. 2001.
[220] Beal, C., 'Putin appoints FSB to counter Chechen rebels', *Jane's Defence Weekly*, 31 Jan. 2001, p. 2.
[221] 'Russia to reduce forces in Chechnya', *Jane's Defence Weekly*, 14 Feb. 2001, p. 5.
[222] 'Russia begins Chechnya pullout', BBC News Online, 13 Mar. 2001, URL <http://news.bbc.co.uk/hi/english/world/>; Galeotti, M., 'FSB also fails in Chechnya', *Jane's Intelligence Review*, July 2001, p. 8; and Scott (note 217).
[223] Scott (note 217); ITAR-TASS (Moscow), 27 Nov. 2001, in 'Russians claim to have killed over 50 Chechens in a day', FBIS-SOV-2001-1127, 28 Nov. 2001; and ITAR-TASS (Moscow), 27 Dec. 2001, in 'Russians carry out 20 special operations in Chechnya in December', FBIS-SOV-2001-1227, 28 Dec. 2001.
[224] 'Chechen rebels attack convoy', *International Herald Tribune*, 7-8 Apr. 2001, p. 2; AFP (Paris), 18 Aug. 2001, in 'Chechen rebels say "dozens" of Russian troops killed in attack near Vedeno', FBIS-SOV-2001-0818, 20 Aug. 2001; 'Mass arrests after Chechnya attacks', BBC News Online, 18 Sep. 2001, URL <http://news.bbc.co.uk/hi/english/world/>; Tyler, P. E., 'Rebels kill a Chechen on Russia's side',

only figures for the number of rebels come from the Russian Government, which estimated in March that there were 3000–5000 rebels.[225] By November its estimate had declined to 1500–2000 rebels.[226]

The reduction in the government's estimated number of rebels is probably due to battle casualties. Russia claimed to kill 80–90 rebels each month, although there was no independent verification.[227] Casualties on the Russian side were a politically sensitive issue. There were claims that 40–45 soldiers had been killed each month in 2001.[228] In December the interior ministry said that 800 of its troops had died since the start of the conflict, including 168 in 2001.[229] The defence ministry said that 2355 of its soldiers had been killed and 6000 wounded.[230] In September 2001 the Association of Soldiers' Mothers protest group claimed that the total number of government forces killed was 10 500.[231]

The conflict is not only costly for the fighters; it is also devastating for civilians. Grozny has been destroyed, as has the economy of the republic and much of its infrastructure. Chechens in the Russian-sanctioned administration complain that Moscow has done little to rebuild the republic.[232] The number of civilians who have been killed is not known, but according to Russian authorities there were 225 000 IDPs in Chechnya and 180 000 in Ingushetia.[233] Human rights organizations accused the Russian troops and Chechen rebels of torture, summary executions, 'disappearances' and other abuses, but laid most blame on the Russian side.[234] The head of the district administration in Chechnya also criticized federal troops for abusing civilians, and a Russian general admitted that his forces had tortured and abused civilians on at least one occasion.[235] Moscow appointed a special presidential human rights commissioner, but officials from the Organization for Security and Co-operation in Europe were sceptical about whether the appointment had led to any change.[236]

International Herald Tribune, 21 Aug. 2001, p. 5; and 'Missile kills 10 of Russian General Staff', *International Herald Tribune*, 18 Sep. 2001, p. 9.

[225] 'More of the same', *The Economist*, 31 Mar. 2001, p. 32.

[226] Interfax (Moscow), 1 Nov. 2001, in 'Russian authorities say up to 2,000 rebels active in Chechnya', FBIS-SOV-2001-1101, 2 Nov. 2001.

[227] Jack, A., 'Ruins of a republic still at war', *Financial Times*, 8 Dec. 2001, p. 4.

[228] Jack (note 227).

[229] Interfax (Moscow), 18 Dec. 2001, in 'Russian interior troops losses in Chechnya disclosed', FBIS-SOV-2001-1218, 19 Dec. 2001.

[230] ITAR-TASS (Moscow), 25 Dec. 2001, in 'Russian minister issues federal casualty figures in Chechnya', FBIS-SOV-2001-1225, 26 Dec. 2001.

[231] Interfax (Moscow), 'Russian committee claims "twice as heavy" losses in Chechnya', 27 Sep. 2001, in 'Russia: Group claims losses in Chechnya "twice as heavy" as officially reported', FBIS-SOV-2001-0927, 29 Sep. 2001.

[232] Tyler, P., 'Grozny could be the most dangerous capital on earth', *International Herald Tribune*, 27 Apr. 2001, p. 4.

[233] IDP Project, 'Country profile: Russian Federation', Oct. 2001, URL <http://www.db.idpproject.org/Sites/idpSurvey/.nsf/wCountires/Russian+Federation>.

[234] Human Rights Watch (note 181), pp. 340–45.

[235] 'Chechen governor criticizes military', *International Herald Tribune*, 31 Aug. 2001, p. 6; and Tyler, P., 'Russia admits troops' crimes in Chechnya', *International Herald Tribune*, 12 July 2001, pp. 1, 5.

[236] Interfax, 12 Sep. 2001, in 'OSCE official concerned over lack of control of troops in Chechnya', FBIS-SOV-2001-0912, 13 Sep. 2001.

The conflict spilled over into neighbouring Georgia in the form of refugees, 7000–8000 of whom remained there at the end of 2001. Moscow frequently accused Georgia of harbouring Chechen fighters, a charge the Georgian Government has denied.[237] In 2001 Georgian officials accused Russia of launching air strikes into Georgia in October and November but Russia denied the charge.[238] At the end of the year Chechens in the breakaway Georgian region of Abkhazia were reported to be assisting Georgian irregulars against the Abkhaz Government.[239] In another example of external influence on an intrastate conflict, a Taliban official said that the Taliban and other Muslim governments had supplied the Chechen rebels with money and arms.[240] There were also reports of non-Chechens fighting for the rebels.[241] The presence of Chechen fighters in Afghanistan was confirmed by Northern Alliance officials.[242] These reports corroborated long-standing Russian claims that Chechen rebels had ties with radical Islamists outside Chechnya.

V. Conflicts in the Middle East

The worsening conflict between the Israeli Government and various Palestinian groups dominated Middle Eastern politics in 2001. Al-Qaeda leader Osama bin Laden declared that the September terrorist attacks in the USA were meant *inter alia* to induce the US Government to withdraw its military presence in the Middle East and end its support for Israel.[243] There was no indication of a US intention to withdraw from the Middle East.

Israel–Palestinians

The Intifada conflict between Israel and Palestinians that began in September 2000 became more violent and intractable during 2001. On 6 February Ariel Sharon of the right-wing Likud Party defeated incumbent Ehud Barak of the moderate Labour Party to become prime minister of Israel. Barak's defeat was widely interpreted as a rejection of his attempts to negotiate a peace agreement with Yasser Arafat, Chairman of the Palestinian Authority (PA).[244] Sharon came into office having denounced as no longer relevant the 1995 Israeli–

---

[237] Billingsley, D., 'Caucasus wars collide in the Kodori corridor', *Jane's Intelligence Review*, Dec. 2001, pp. 8–9.
[238] 'Georgian officials condemn bombing on Kodori', Radio Free Europe/Radio Liberty Newsline, 10 Oct. 2001, URL <http://rferl.org/newsline/2001/10/101001.asp>; and 'Georgia protests Russian air raid', Radio Free Europe/Radio Liberty Newsline, 29 Nov. 2001, URL <http://rferl.org/newsline/2001/11/291101.asp>.
[239] 'Abkhazia opens attack on rebels', *International Herald Tribune*, 18 Dec. 2001, p. 7.
[240] RIA (Moscow), 17 Jan. 2001, in 'Russia: Afghan minister vows to continue assisting Chechen separatists', FBIS-SOV-2001-0117, 19 Jan. 2001.
[241] Galeotti, M., 'Al-Qaeda and the Chechens', *Jane's Intelligence Review*, Dec. 2001, p. 48.
[242] Interfax, 3 July 2001, in 'Afghan official confirms presence of Chechen rebels among Taleban fighters', FBIS-SOV-2001-0703, 6 July 2001.
[243] 'Bin Laden video condemned', BBC Online Network, 27 Dec. 2001, URL <http://news.bbc.co.uk/hi/english/world/>.
[244] Hockstader, L., 'Sharon routs Barak in Israeli election', *International Herald Tribune*, 7 Feb. 2001, pp. 1, 6.

Palestinian Interim Agreement on the West Bank and the Gaza Strip, which ended the first Intifada (1987–93) and led to the establishment of PA control over the Gaza Strip and parts of the West Bank.[245] The Palestinians responded to the election result with intensified attacks, to which the Israeli military responded even more strongly.[246]

Although the level of violence increased, the pattern of attacks and counter-attacks remained unchanged throughout the year. Palestinians detonated car bombs and launched suicide attacks in Israeli towns using bombs strapped to their bodies.[247] They also attacked Israeli troops and settlers living in the West Bank. These attacks escalated in 2001 from primarily stone throwing to the use of assault rifles, mortars and grenade launchers, which they received in several small shipments during the year.[248] Israeli troops retaliated, armed with assault rifles, armoured personnel carriers, tanks and helicopter gunships. Many observers, including UN Secretary-General Kofi Annan, characterized the Israeli responses as disproportionate and called for restraint, but to no avail.[249]

There were nearly constant attempts to broker peace talks by diplomats from Egypt, Jordan, the UK and the USA with the support of the United Nations and the European Union. Although some of the efforts led to ceasefire agreements, none resulted in stopping the violence for more than a day and none led to substantive talks between the two sides.[250]

In May 2001 the Sharm el-Sheikh Fact-Finding Committee released its report, known as the Mitchell Report, on the outbreak of violence between Israelis and Palestinians.[251] The report called for an immediate and unconditional cessation of violence, recommended a number of steps towards the rebuilding of confidence between the two sides, and called for a resumption of negotiations to solve the underlying causes of the conflict. Among the confidence-building steps were a full effort by the PA to stop terrorist acts and punish the perpetrators and an Israeli freeze on the building of Jewish settlements in the West Bank. Although the report was accepted in principle by the PA and the Israeli Government, it did not lead to any pragmatic measures.[252]

---

[245] Sontag, D., 'Oslo peace agreement repudiated by Sharon', *International Herald Tribune*, 11 Jan. 2001, pp. 1, 5.
[246] Hockstader, L., 'Palestinian Authority described as "terrorist"', *International Herald Tribune*, 1 Mar. 2001, p. 4.
[247] Hockstader, L., 'Suicide bomb kills 15 and wounds 100 in Jerusalem', *International Herald Tribune*, 10 Aug. 2001, pp. 1, 10.
[248] 'Palestinians again shell Israel from Gaza Strip', *International Herald Tribune*, 19 Apr. 2001, pp. 1, 8; and 'Israel sounds the alarm', BBC Online Network, 8 May 2001, URL <http://news.bbc.co.uk/hi/english/world/>
[249] United Nations, Question of Palestine: the situation in the Middle East, report of the Secretary-General, UN documents A/56/642 and S/2001/110, 23 Nov. 2001.
[250] Richburg, K., 'Israel giving short shrift to peace proposal', *International Herald Tribune*, 28–29 Apr. 2001, p. 4; Hoge, W., 'Arafat joins Blair's call for "reinvigorated" talks', *International Herald Tribune*, 16 Oct. 2001, p. 1; and Erlanger, S., 'Arafat and Peres meet as EU pushes for peace', *International Herald Tribune*, 7 Nov. 2001, p. 8.
[251] *Report of the Sharm el-Sheikh Fact-Finding Committee*, 30 Apr. 2001, available at the Meridian International Centre Internet site at URL <http://www.meridian.org/pdfs/report.pdf>.
[252] 'Israel and the Palestinians: yes to a ceasefire, no to a halt on settlements', *The Economist*, 26 May 2001, pp. 47–48; and 'Israeli-Palestinian fighting: prospects for war', *The Economist*, 21 July 2001, pp. 33–34.

By July an upsurge in violence that coincided with a diplomatic impasse led to a tit-for-tat cycle of violence that continued for the rest of the year. The Israeli military used ground and air assaults to attack the PA and to destroy homes and farmland.[253] Israel also assassinated individuals in a practice that began in November 2000 and led to about 60 murders by the end of 2001.[254] After one such attack in October, the Popular Front for the Liberation of Palestine (PFLP) assassinated the Israeli tourism minister, who was an extreme nationalist. The action marked an escalation of the conflict as it was the first time a cabinet minister had been murdered in this conflict. In response Prime Minister Sharon pledged 'an all-out war' on terrorists and their collaborators.[255] Israeli forces occupied six towns in the West Bank for a month.[256]

The current Intifada is more violent than the first one and became more violent as it entered its second year. During the first 12 months of the 1987–93 Intifada, over 300 Palestinians and 11 Israelis were killed. Between September 2000 and September 2001, 560 Palestinians and 177 Israelis were killed, according to the Palestinian Human Rights Monitoring Group and the Israeli Government, respectively.[257] By the end of November 2001, the violence had killed 725 Palestinians and 192 Israelis and had devastated the economy in the Palestinian territories of the Gaza Strip and the West Bank.[258]

Initially, the new US Administration of President George W. Bush put pressure on the Israeli Government to take a less militaristic approach, calling Israeli actions provocative and agreeing on the desirability of international observers in the region, which the PA had frequently requested and Israel had opposed.[259] For several months after the September terrorist attacks in the USA, the US Government put pressure on the Israeli Government to agree to a ceasefire, as part of its strategy to convince Arab governments to join its global effort against terrorists. The Bush Administration also endorsed the idea of a Palestinian state and began to refer to 'Palestine', marking the first time a US administration had used the term the Palestinians use to designate their homeland.[260] Israel strongly resisted US pressure and by the end of the year the

---

[253] Sontag, D., 'Israelis quickly end a push into Gaza', *International Herald Tribune*, 18 Apr. 2001, pp. 1, 4; 'Israel strikes back after bombing', BBC Online Network, 9 Aug. 2001, URL <http://news.bbc.co.uk/hi/english/world/>; and Hockstader, L. and Williams, D., 'Israel raids town, killing 5, despite US plea for calm', *International Herald Tribune*, 25 Oct. 2001, p. 1.

[254] Hockstader, L., 'A top Palestinian is killed', *International Herald Tribune*, 28 Aug. 2001, p. 1; and Blanche, E., 'Israel uses Intifada informers to abet assassination campaign', *Jane's Intelligence Review*, Dec. 2001, pp. 22–25.

[255] Hockstader, L., 'Israeli rightist slain in revenge attack', *International Herald Tribune*, 18 Oct. 2001, p. 1.

[256] 'Israel pulls back in "goodwill gesture"', BBC Online Network, 27 Nov. 2001, URL <http://news.bbc.co.uk/hi/english/world/>.

[257] Bennet, J., 'After a year of Intifada, hearts have hardened', *International Herald Tribune*, 29–30 Sep. 2001, p. 4.

[258] 'The Palestinians and Hamas: Hamas has the people's heart', *The Economist*, 1 Dec. 2001, p. 41.

[259] 'US condemns "provocative" Israel', BBC Online Network, 10 July 2001, URL <http://news.bbc.co.uk/hi/english/world/>; and 'US backs Mideast observers', BBC Online Network, 19 July 2001, URL <http://news.bbc.co.uk/hi/english/world/>.

[260] Hardy, R., 'Bush "endorses" Palestinian state', BBC Online Network, 2 Oct. 2001, URL <http://news.bbc.co.uk/hi/english/world/>.

USA once again put full responsibility on Arafat for taking the first steps towards a ceasefire.²⁶¹

It was impossible for Arafat to completely end the violence, even if he had wanted to.²⁶² He was prevented from leaving the town of Ramallah by Israeli tanks, the PA's infrastructure had been destroyed and the PA had effectively lost control of the Palestinian side. The far more radical Hamas and Islamic Jihad groups were ascendant.²⁶³ Prime Minister Sharon's cabinet formally pronounced Chairman Arafat as 'irrelevant', although Foreign Minister Shimon Peres believed that he was still a possible negotiating partner.²⁶⁴ The year ended with Israelis and Palestinians radicalized by fear and desperation.

VI. Conflicts in South America

Colombia was the location of the only major armed conflict in South America in 2001. The conflict took on a regional aspect as neighbouring countries became concerned that the violence would spill over their borders in the wake of a large US infusion of military funding and equipment to the government. The peace process was internationalized with the appointment of a UN envoy.

Colombia

The Colombian Government faces two leftist rebel groups and a right-wing paramilitary group. The Fuerzas Armadas Revolucionarias de Colombia (FARC, Revolutionary Armed Forces of Colombia), motivated by Marxist ideology, has been fighting since the late 1960s in an effort to overthrow the Colombian Government. Composed of small farmers and day labourers, FARC has about 18 000 fighters, of whom 6000 are lightly armed urban militia. It operates throughout the country, but its main strength is in the south. It claims to have a broad base of public support.²⁶⁵

The Ejército de Liberación Nacional (ELN, National Liberation Army) is smaller, with about 3500 fighters and about the same number of urban militia. Its main strength is in the central region of the country, but it has been seriously weakened by paramilitary attacks on its civilian base.²⁶⁶ It is motivated by a Marxist-inspired concept of governance and economic reform. The ELN and the FARC see each other as rivals, although pressure by the paramilitary and government forces has led them to launch occasional joint operations.

---

²⁶¹ Plett, B., 'Don't sacrifice Israel, warns Sharon', BBC Online Network, 4 Oct. 2001, URL <http://news.bbc.co.uk/hi/english/world/>.

²⁶² 'The Palestinians and Hamas' (note 258); and 'Profile: PFLP', BBC Online Network, 17 Oct. 2001, URL <http://news.bbc.co.uk/hi/english/world/>.

²⁶³ Williams, D., 'Arafat calls for end to all attacks', *International Herald Tribune*, 17 Dec. 2001, pp. 1, 5; and 'The Palestinians and Hamas' (note 258).

²⁶⁴ Bennet (note 257); Bennet, J., 'The Israeli left is splintered as Mideast violence grows', *International Herald Tribune*, 12 Nov. 2001, p. 5; and Hockstader, L., 'Israelis attack Arafat's authority by tank and jet', *International Herald Tribune*, 14 Dec. 2001, p. 1.

²⁶⁵ Reid, M., 'Talk of peace, acts of war', in 'A survey of Colombia', *The Economist*, 21 Apr. 2001, p. 10.

²⁶⁶ Reid, M., 'The curse of the vigilantes', *The Economist* (note 265), p. 13.

The fastest growing armed group is the right-wing paramilitary umbrella organization Autodefensas Unidas de Colombia (AUC, United Self-Defence Forces of Colombia), led by Carlos Castaño. The AUC began as self-defence forces for large landowners and continues to state that its purpose is to fight the guerrillas. However, the paramilitaries are increasingly motivated by the desire for profit from the drug trade, extortion and kidnappings. They use terror to win territory from the guerrillas and are responsible for the worst atrocities against civilians committed in the country. The government says that it is taking action against the paramilitaries, but human rights groups have accused the Colombian military of complicity in the AUC atrocities. They have also criticized the US Government for ignoring or playing down evidence of the ties between the military and paramilitaries.[267] Castaño appears to be interested in political influence and has built the AUC into a national organization. In 2001 the AUC's fighting strength stood at over 8000, up from about 1200 in 1993 and 4500 in 1998.[268] In 2001 the US State Department classified the AUC as a terrorist organization.[269] It classified FARC and the ELN as terrorist organizations in 1997.

The Colombian military benefited tremendously in 2000 and 2001 from becoming the third largest recipient of US military aid, behind Israel and Egypt, as part of Plan Colombia. Out of a total US aid package of $1.3 billion, the Colombian military and police were to receive $952.3 million.[270] Most of the money went to the military to pay for three mobile attack battalions, armed with combat helicopters. By May 2001 all three battalions had been formed.[271] At the end of 2001 the Bush Administration contemplated providing the means to train a fourth mobile battalion.[272]

The Plan Colombia aid is supposed to be spent on the 'war on drugs' by contributing to the eradication of coca cultivation.[273] In practice, much of the coca is planted in regions that are controlled by FARC or the AUC, who earn hundreds of millions of dollars per year from the drug trade. Therefore, sending military units into coca-growing areas to aerial-spray pesticides and to seize and control territory makes the anti-drug programme hard to distinguish from a counter-insurgency strategy.[274] Despite the infusion of military aid, the government is not strong enough to overcome the rebels or the paramilitary by force. The army, air force and navy together number about 140 000 and the police about 100 000.[275] Conscript soldiers who have received a secondary

---

[267] Human Rights Watch, *The 'Sixth Division': Military–Paramilitary Ties and US Policy in Colombia* (HRW: New York, 2001), p. 89.
[268] Reid (note 266).
[269] 'AUC is formally a "terrorist" organization', *Latin American Weekly Report*, 18 Sep. 2001, p. 435.
[270] Centre for International Policy, 'US aid to Colombia', URL <http://www.ciponline.organization/colombia/aid/aid0001.htm>.
[271] Human Rights Watch (note 267).
[272] Sipress, A., 'US contemplates anti-drug options as Powell heads to Colombia', *International Herald Tribune*, 11 Sep. 2001, p. 3.
[273] Reid, M., 'Spraying misery', *The Economist* (note 265), p. 7; and Reid, M., 'Colombia: drugs, war and democracy', *The Economist* (note 265), p. 4.
[274] LeoGrande, W. M. and Sharpe, K. E., 'Two wars or one? drugs, guerrillas and Colombia's new Violencia', *World Policy Journal*, fall 2000, pp. 1–11; and Reid (note 266).
[275] Reid (note 265).

education are prohibited by law from being sent into battle, which keeps the country's elite somewhat insulated from the conflict.[276]

Fighting took place throughout the country in 2001 but was most frequent in the north and south. The armed groups and government forces rarely engaged each other directly, preferring to try to weaken each other's support base by attacking civilians thought to be sympathetic to one group or another.[277] However, there were direct clashes at times, the most intense of which took place in March and April in Antioquia province in the north. Government, AUC and FARC forces were all involved in ambushes, attacks on villages and counter-attacks.[278] In a departure from the past, fighting between the AUC and rebels also took place in cities on several occasions.[279] The frequent battles and massacres made 2001 one of the deadliest years of the conflict. Over the past decade, the conflict is estimated to have caused over 35 000 deaths.[280]

Violence also began to spread into Ecuador. Ecuadorian forces discovered coca plantations and cocaine-processing laboratories controlled by the AUC. In February AUC members fired on Ecuadorian villagers who allegedly gave Ecuadorian troops information that led to the destruction of a cocaine plant and the death of Colombians.[281] The Ecuadorian military has also discovered camps belonging to FARC in an area close to the Colombian province of Putumayo, where FARC and AUC forces are fighting each other.[282]

Colombian President Andrés Pastrana was elected in 1998 on a platform centred around peace talks with the rebels. In a highly controversial move, he removed all government forces from a 42 000-square kilometre zone in southern Colombia and turned it over to FARC for a limited period of time. The rebels had demanded such a safe zone as a precondition for their participation in negotiations. President Pastrana has repeatedly extended the time period for the FARC zone, even though the rebels have not always been willing to negotiate. Talks were formally restarted in February 2001, having been suspended by FARC in November 2000. President Pastrana and FARC leader Manuel Marulanda Vélez agreed on a 13-point plan for further talks, including the involvement of foreign third parties.[283] After little progress, the talks almost

[276] Reid (note 265).
[277] Wilson, J., 'Colombia violence soars as politicians falter', *Financial Times*, 21–22 Apr. 2001, p. 3; Caracol Colombia Radio (Bogota), 1 June 2001, in 'Toll of FARC Action in Cordoba reportedly increases to 80 peasants killed', Foreign Broadcast Information Service (FBIS), *Daily Report–The Americas*, FBIS-LAT-2001-0601, 4 June 2001; and 'Paramilitary massacres across the land', *Latin American Weekly Report*, 16 Oct. 2001, p. 486.
[278] Wilson (note 277); Caracol Colombia Radio (Bogota), 27 Mar. 2001, in 'Colombian Army launches offensive to recover La Llorona Canyon', FBIS-LAT-2001-0327, 28 Mar. 2001; and Caracol Colombia Radio (Bogota), 16 Apr. 2001, in 'Colombia: violent Holy Week fighting reportedly leaves more than 80 dead', FBIS-LAT-2001-0416, 17 Apr. 2001.
[279] Forero, J., 'Colombian paramilitaries adjust attack strategies', *New York Times* on the Web, 22 Jan. 2001, URL <http://www.nytimes.com>.
[280] Wilson (note 277); and McDermott, J., 'Fresh massacre in Colombia', BBC Online Network, 16 Apr. 2001, URL <http://news.bbc.co.uk/hi/english/world/>.
[281] NOTIMEX (Mexico City), 8 Feb. 2001, in 'Colombian paramilitary groups reportedly attack Ecuadorian indigenous town', FBIS-LAT-2001-0208, 12 Feb. 2001.
[282] 'Discovery of a major FARC base in Ecuador sets the alarm bells ringing', *Latin American Weekly Report*, 4 Sep. 2001, p. 409.
[283] 'Pastrana & Tirofijo pull Colombia back from brink and salvage talks—for now', *Latin American Weekly Report*, 13 Feb. 2001, p. 73.

collapsed in October 2001, when FARC killed a former government minister and the government stepped up military patrols around the zone.[284] Although Pastrana stated that FARC were terrorists and drug traffickers, he again extended the period for the FARC zone, until January 2002, and the talks carried on under growing pressure from conservatives to end them.[285]

The ELN has sought a similar zone free of government control as a place to hold its own negotiations with the Pastrana Administration. A zone was agreed in 2000, but the AUC has blocked its establishment because it would be located in a region where the AUC controls the towns and coca industry.[286]

VII. Conclusions

The central contention in all of the conflicts described above is control over either government or territory. However, the diverse state and non-state actors reveal multiple and overlapping objectives related to political power, economic gain and ideological belief. In 2001, economic incentives and ideological belief sustained many of the conflicts to the point that it was difficult in some cases to distinguish where politics left off and other considerations became primary. The role of mineral wealth in West Africa and the African Great Lakes region provided clear examples of the importance of economic gain in sustaining conflicts once they have started.

The spillover of intra-state conflicts into neighbouring states was common, but its impact was varied. In some cases, the cross-border movement of rebels and arms caused conflicts in neighbouring states to intensify, as in Burundi when rebels entered from the DRC. In other cases, conflicts were not significantly affected by spillover from neighbouring states. For example, the Ugandan Government's fight against several rebel groups was not noticeably affected by changes in the conflicts in Sudan and the DRC. Spillover threatened to turn the minor conflicts in Guinea and Liberia into a larger, regional conflict. Fortunately, the region has not yet suffered the fate of the Great Lakes region in the 1990s. Some countries at peace were threatened by spillover, for instance, when rebels and right-wing paramilitaries from Colombia entered Ecuador. Other countries did not appear to be destabilized, for instance, when refugees and armed elements crossed into Kenya from Somalia. Research into the reasons for the wide variation in the impact of conflict spillover could have significant implications for policy.

Eleven of the 15 conflicts reviewed have lasted for at least eight years. One of the reasons for their endurance is the inability of either side to prevail by force. In the vast majority of the conflicts reviewed in this chapter, rebels used a guerrilla military strategy. They supported their military effort through the

[284] 'Colombia peace talks rescued', BBC News Online, 6 Oct. 2001, URL <http://news.bbc.co.uk/hi/english/world/>; and 'Fresh talks with guerrillas on the cards', *Latin American Andean Group Report*, RA-01-10, 11 Dec. 2001, p. 2.

[285] 'Fresh talks with the guerrillas on the cards' (note 284); and Agence France-Presse, 'Chronology of the strained Colombia–FARC peace process', 10 Jan. 2002, available from ReliefWeb at URL <http://wwww.reliefweb.int/>.

[286] Reid (note 265).

sale of minerals, timber and narcotics and through remittances from supporters abroad. However, very few groups tried to win the loyalty of the population through political, economic or social programmes. Historically, such programmes have been important elements of successful insurgencies. From the perspective of governments, it is very difficult to win a guerrilla war militarily. It is difficult to use the military's full strength against small and mobile opponents, and even a military victory does not solve the problem that led to the insurgency. The longer a conflict lasts, the more it creates the economic and social conditions in which rebel organizations can recruit disaffected people into their forces. This combination of factors means that both sides can fight for a long time without one defeating the other or debilitating it through attrition. Long conflicts, where weak antagonists often attack even weaker targets, cause a large number of civilian casualties and destroy economic and social infrastructure.

Events in several countries in 2001 demonstrate that the duration of a conflict alone is not an indication of whether it is 'ripe for resolution' or ready for escalation. Long-running conflicts in the Philippines and Sri Lanka showed signs of resolution. In contrast, the protracted conflicts in Colombia and between Israel and the Palestinians intensified during the year. Analysts and policy makers often conceive of conflicts progressing through a series of phases, from a low level of violence to more intense violence and back to a low level before they are resolved. The concept of conflict phases is a useful one, but it is important to bear in mind that violent conflicts rarely progress in a linear fashion. More often they are cyclical, passing several times through phases of more and less violence.

The full impact of the United States' response to the terrorist attacks of 11 September 2001 will take several years to develop. It is likely that changes in political alliances and flows of military aid brought about by the new US security agenda will have an effect on many conflicts in the future, particularly in those states where the government successfully makes the case that its opponents are terrorists. In addition, the USA has indicated its increased willingness to engage in direct military confrontation now that it perceives its national security to be in danger. Its ability to dominate battlefields will have an immediate and dramatic effect on any conflict in which it becomes involved. However, the extent of future US military engagement throughout the world and whether a military campaign can lead to victory in the 'war on terrorism' are open questions.

Appendix 1A. Patterns of major armed conflicts, 1990–2001

MIKAEL ERIKSSON, MARGARETA SOLLENBERG AND
PETER WALLENSTEEN*

I. Global patterns

In 2001 there were 24 major armed conflicts in 22 locations throughout the world. The number of major armed conflicts and the number of conflict locations in 2001 are slightly lower than in 2000, when there were 25 major armed conflicts in 23 locations.[1] The conflicts and locations for 2001 are presented in table 1A.3. For the definition of a major armed conflict, see appendix 1B.

The only interstate conflict that was active in 2001 occurred between India and Pakistan. Other states contributed regular troops to internal conflicts in Angola, where Namibia contributed troops to the Angolan Government; in the Democratic Republic of the Congo (DRC), where Angola, Namibia and Zimbabwe contributed troops to the DRC Government, and Rwanda and Uganda were involved on the side of the opposition; in Afghanistan, where a multinational coalition contributed troops to the opposition forces; and in the USA, where the multinational coalition supported the government.[2]

In the 12-year post-cold war period 1990–2001, there were 57 different major armed conflicts in 45 different locations. The number of conflicts in 2001 was lower than in the period 1990–95, when the yearly number of major armed conflicts ranged between 28 and 33. The number of locations in 2001 was lower than in the period 1990–95: there were conflicts in 26 locations in 1990–93, 25 in 1994 and 23 in 1995. The lowest number of conflicts for the period 1990–2001 was recorded for 1997, when there were 19.

All but three of the major armed conflicts registered for 1990–2001 were internal; that is, the issue concerned control over the government or territory of one state. The three interstate conflicts in this period were Iraq versus Kuwait, India versus Pakistan and Ethiopia versus Eritrea. Other states contributed regular troops to one side or the other in 15 of the internal conflicts.

---

[1] A location may have 1 or more conflicts over territory; it may also have a conflict over territory and a conflict over government. There can be only 1 conflict over government in each location because, by definition, there can be only 1 government in each location.

[2] See section III of this appendix for an elaboration of the preliminary assessment of this case, particularly the discussion of the incompatibility and the ways in which it differs from other incompatibilities recorded by the Uppsala Conflict Data Project.

* Uppsala Conflict Data Project, Department of Peace and Conflict Research, Uppsala University. For table 1 A.3, Ylva Blondel was responsible for the conflict location Algeria; Erika Forsberg for the Philippines and Turkey; Helena Grusell for Colombia; Kristine Höglund for Burundi; Desiree Nilsson for Sudan; Isak Svensson for Sri Lanka; and Mimmi Söderberg for the Democratic Republic of the Congo. Mikael Eriksson and Margareta Sollenberg were responsible for the remaining conflict locations.

SIPRI Yearbook 2002: Armaments, Disarmament and International Security

Table 1A.1. Regional distribution, number and types of major armed conflicts, 1990–2001

Region	1990 G	1990 T	1991 G	1991 T	1992 G	1992 T	1993 G	1993 T	1994 G	1994 T	1995 G	1995 T	1996 G	1996 T	1997 G	1997 T	1998 G	1998 T	1999 G	1999 T	2000 G	2000 T	2001 G	2001 T
Africa	8	3	8	3	6	1	6	1	5	1	4	1	2	1	4	–	10	1	10	1	8	1	7	–
America	4	–	4	–	3	–	3	–	3	–	3	–	3	–	2	–	2	–	2	–	2	–	3[a]	–
Asia	4	9	3	8	4	8	4	6	4	6	4	7	4	6	3	6	3	6	2	7	2	7	2	7
Europe	–	–	–	1	–	3	–	5	–	4	–	3	–	1	–	–	–	1	–	2	–	1	–	1
Middle East	1	3	2	4	2	3	2	4	2	4	2	4	2	4	2	2	2	2	1	2	2	2	2	2
Total	17	15	17	16	15	15	15	16	14	15	13	15	11	12	11	8	17	10	15	12	14	11	14[a]	10
Total	32		33		30		31		29		28		23		19		27		27		25		24[a]	

G = government and T = territory, the two types of incompatibility.

Table 1A.2. Regional distribution of locations with at least one major armed conflict, 1990–2001

Region	1990	1991	1992	1993	1994	1995	1996	1997	1998	1999	2000	2001
Africa	10	10	7	7	6	5	3	4	11	11	9	7
America	4	4	3	3	3	3	3	2	2	2	2	3[a]
Asia	8	7	9	8	8	8	9	8	8	7	7	7
Europe	–	1	3	4	3	3	1	–	1	2	1	1
Middle East	4	4	4	4	5	4	4	4	4	3	4	4
Total	26	26	26	26	25	23	20	18	26	25	23	22[a]

[a] Note that this number includes the conflict between the USA and al-Qaeda. See section III for an elaboration of this case and its ambiguities.

Source: The Uppsala Conflict Data Project.

The proportion of major armed conflicts waged over territory as compared to those concerning government was roughly the same for the period 1990–2001. Conflicts concerning government were slightly more numerous than those concerning territory in 1990–91 and clearly more numerous in 1997–2001. In 1992 there was the same number of conflicts over government and territory. Conflicts concerning territory were marginally more numerous than those over government in 1993–96.

II. Regional patterns

In 2001 there were seven conflicts in Africa and nine conflicts in Asia. As during most of the 12-year period, the vast majority of the conflicts in 2001 occurred in these regions. There were three major armed conflicts in America,[3] one in Europe and four in the Middle East. The regional distribution of major armed conflicts and locations over the period 1990–2001 is shown in tables 1A.1 and 1A.2. Figure 1A shows the regional distribution and total number of conflicts for each year in this period.

For *Africa,* 19 major armed conflicts were registered for the period 1990–2001.[4] There was a marked decline in the number of conflicts in Africa from 1991 to 1996. The events in the Great Lakes Region of Central Africa led to a sharp increase from three conflicts in 1996 to 11 in 1998 and 1999, the same number registered for the beginning of the period (1990 and 1991). The number of conflicts decreased by two in 2000 and 2001, although the figure still remained higher than in the period 1994–97. Eritrea and Ethiopia fought the only interstate conflicts that took place in Africa during the period 1990–2001. In at least seven of the conflicts that were active in 1990–2001 there was military involvement by other states in intra-state conflicts: Angola, the Republic of Congo, the DRC, Guinea-Bissau, Liberia, Sierra Leone and Somalia. The vast majority of the conflicts in Africa concerned government in each year of the period.

For *America,* a total of five major armed conflicts were registered for 1990–2001.[5] Of the four major armed conflicts in America recorded for 1990, two remained, and one new major armed conflict was registered for 2001. There have been no interstate major armed conflicts in America in the period. In the intra-state conflict in the USA, other states contributed regular troops. All the conflicts in America concerned government.

For *Asia,* a total of 16 major armed conflicts were registered for 1990–2001.[6] Asia had the highest number of major armed conflicts for most years of the period, including 2001. In 1991 and 1998–2000, Africa had as many conflicts as or more than Asia. There has been an overall reduction in armed conflicts in Asia since 1990. As in the Middle East, most of the conflicts in Asia have been active since well before the 1990s. There was one interstate conflict in Asia during the period, between India and

[3] In previous years, 'Central and South America' was treated as a separate region; for 2001, the region 'America' includes North, Central and South America and the Caribbean states.
[4] The 19 conflicts in Africa are Algeria, Angola, Burundi, Chad, the Republic of Congo, the Democratic Republic of the Congo (formerly Zaire) (2 conflicts), Ethiopia, Ethiopia (Eritrea), Eritrea–Ethiopia, Guinea-Bissau, Liberia, Morocco, Mozambique, Rwanda, Somalia, South Africa, Sudan and Uganda.
[5] The 5 conflicts in America are Colombia, El Salvador, Guatemala, Peru and the USA (the conflict between the US Government and al-Qaeda).
[6] The 16 conflicts in Asia are Afghanistan, Cambodia, India (Kashmir), India (Punjab), India (Assam), India–Pakistan, Indonesia (East Timor), Indonesia (Aceh), Myanmar (Kachin), Myanmar (Karen), Myanmar (Shan), the Philippines, the Philippines (Mindanao), Sri Lanka, Sri Lanka (Tamil Eelam) and Tajikistan.

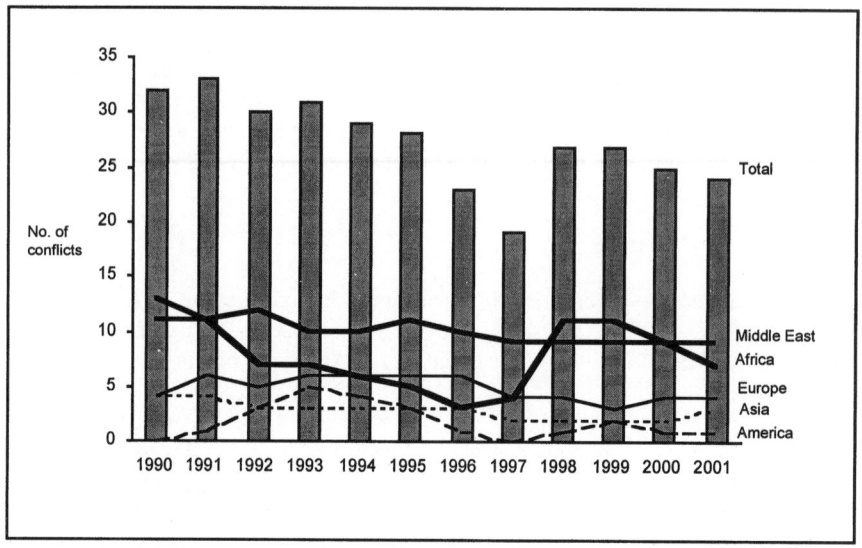

Figure 1A. Regional distribution and total number of major armed conflicts, 1990–2001

Pakistan. It was active in 1990, 1992 and 1996–2001. In the intra-state conflicts in Tajikistan, active in 1992–96, and Afghanistan in 2001, other states contributed regular troops. The vast majority of the conflicts in Asia have concerned territory.

For *Europe*, a total of eight major armed conflicts were registered for the period 1990–2001.[7] The yearly number of conflicts has declined since the peak year 1993, when there were five major armed conflicts. The only active major armed conflict in Europe in 2001 was the Chechnya conflict in Russia. All the new conflicts in Europe during the period emanated from the breakup of the Soviet Union and Yugoslavia. There were no interstate major armed conflicts. In five of the eight conflicts in Europe in 1990–2001, other states contributed regular troops: Azerbaijan, the two conflicts in Bosnia and Herzegovina, Croatia and Yugoslavia (Kosovo). All the major armed conflicts in Europe have concerned territory.

For *the Middle East*, nine major armed conflicts were registered for 1990–2001.[8] The number of conflicts increased from 1990 to 1991, remained fairly constant until the drop in 1997, and remained fairly constant again for the rest of the period. The four major armed conflicts in the Middle East in 2001 were active during almost the entire period. There was one interstate conflict, between Iraq and Kuwait, in 1991. In the intra-state conflict in Lebanon, which was active in 1990, other states contributed troops. Most conflicts in the Middle East have concerned territory, but in 1997–98 and 2001 the number of incompatibilities concerning government and territory was the same.

---

[7] The 8 conflicts in Europe are Azerbaijan, Bosnia and Herzegovina (Republika Srpska), Bosnia and Herzegovina (Herceg-Bosna), Croatia, Georgia, Russia, the Socialist Federal Republic of Yugoslavia (Croatia), and the Federal Republic of Yugoslavia (Kosovo).

[8] These 9 conflicts are Iran, Iran (Kurdistan), Iraq, Iraq (Kurdistan), Iraq–Kuwait, Israel, Lebanon, Turkey and Yemen.

III. Changes in the table of conflicts for 2001

The new conflict in 2001

The only new conflict registered for 2001 was that between the United States and the al-Qaeda network. It was a complex, unusual conflict, warranting an elaboration of the factors considered when classifying and including it in the conflict data for 2001.

The 11 September 2001 attacks on the mainland United States were the most dramatic conflict developments of 2001. They caused the deaths of over 3000 people in a matter of hours and changed the foreign policy agenda of the USA, the world's leading actor. From an analytical point of view, the attacks and the subsequent military campaign against the al-Qaeda network in Afghanistan pose challenges to established definitions of armed conflict because of the transnational nature of the situation.

In the preliminary assessment made in this appendix, the case is registered as a separate conflict over government involving the United States and a multinational coalition[9] versus the al-Qaeda network. Although closely intertwined with the conflict in Afghanistan in 2001, this conflict has a history of its own, involving attacks by al-Qaeda on US military and diplomatic posts in Saudi Arabia (1995 and 1996), Kenya (1998), Tanzania (1998) and Yemen (2000). Geographically, the military action after 11 September took place in Afghanistan since the al-Qaeda network was based there. As the USA began strikes on al-Qaeda bases in Afghanistan, it also joined the United Islamic Front for the Salvation of Afghanistan (UIFSA, also known as the Northern Alliance) in its attempt to overthrow the Taliban Government of Afghanistan. Thus, two wars were fought simultaneously: the UIFSA, the USA and the multinational coalition, on the one side, and the Taliban and al-Qaeda, on the other side.

Although the conflict between al-Qaeda and the USA was militarily inseparable from the conflict in Afghanistan in 2001, it was not fought exclusively for political power in Afghanistan nor was it a traditional interstate conflict between the United States and Afghanistan. The conflict between al-Qaeda and the USA is therefore treated as a separate, intra-state conflict because significant aspects of the conflict differ from the intra-state conflict in Afghanistan.

The incompatibility in the conflict between al-Qaeda and the USA is the global influence of the USA and its political system; that is, it was fought over government. The statements that have been made by al-Qaeda leaders are general, but they mention the objective of the destruction of the United States, in particular its military and economic system.[10] Other issues concern the US influence in the Middle East and the existence of Israel. Thus, the incompatibility is somewhat difficult to identify in the terms of the standard definition used for the present data set.[11] It is unique in that al-Qaeda does not strive for a specific alternative to the status quo, which is normally

---

[9] For the states which contributed troops and other military assets to the coalition, see table 1A.3.

[10] See, e.g., Associated Press, 'Text of Osama bin Laden's statement, broadcast after US–British strikes', 7 Oct. 2001, available at the *Boston Herald* Internet site, URL <http://www.bostonherald.com/attack/world_reaction/ausbintext10072001.htm> (originally aired by Al-Jazeera on 7 Oct. 2001); 'Transcript of Bin Laden's October interview', CNN.com, 5 Feb. 2002, URL <http://www.cnn.com/2002/WORLD/asiapcf/south/02/05/binladen.transcript/> (interview with bin Laden originally made by Al-Jazeera in Oct. 2001); and 'Osama claims he has nukes: if US uses N-arms it will get same response', *Dawn* (Internet edn), 10 Nov. 2001, URL <http://www.dawn.com/2001/11/10/top1.htm> (interview with bin Laden by Hamid Mir, on 7 Nov. for *Dawn* and *Ausaf*).

[11] See appendix 1B for the definition of an incompatibility for the present data set.

the case, but for the destruction of a political system. Nevertheless, the case is listed as a separate conflict because it meets all the other criteria of a major armed conflict. In tables 1A.1–3 this conflict is clearly marked, drawing attention to the fact that researchers' own judgement may be exercised in the compilation of their data sets.

Conflicts recorded in 2000 that were not recorded for 2001

Two major armed conflicts registered in 2000 do not appear in the table for 2001.

In the interstate war between Eritrea and Ethiopia, a ceasefire agreement that ended the violence was signed in June 2001. The war was settled by the Comprehensive Peace Agreement, signed in Algiers, Algeria, on 12 December 2000; it stipulated that the contested border should be demarcated by an independent commission, leaving it basically where it was before the war began.

In Sierra Leone, the 11 November 2000 ceasefire agreement, signed in Abuja, Nigeria, between the Government of Sierra Leone and the Revolutionary United Front (RUF), seems largely to have been implemented in 2001. A major step for the peace process was taken when UN troops were deployed in rebel-held areas. During the year there were reports of clashes between former RUF fighters and pro-government militias. However, these fighters could not be definitely linked to the main RUF organization. During the year the RUF was transformed into a political party, the Revolutionary United Front Party.[12]

Changes in intensity of conflict

Four of the 24 major armed conflicts in 2001 showed a higher intensity compared to 2000. In one of these four conflicts, the battle-related deaths increased by more than 50 per cent.[13] Six conflicts showed reduced intensity in 2001 compared to 2000. In three of these cases intensity levels declined by more than 50 per cent.[14] Six conflicts showed no change in intensity from 2000.[15] Of the remaining eight conflicts, one was new in 2001 and the intensity of the others was difficult to compare with 2000 because of the lack of reliable information on battle-related deaths.

Eleven of the major armed conflicts in 2001 caused at least 1000 deaths during the year: Afghanistan, Algeria, Angola, Burundi, Colombia, India (Kashmir), Russia (Chechnya), Rwanda, Sri Lanka, Sudan and the USA. For nine of them, over 1000 battle-related deaths were also recorded for 2000: Afghanistan, Algeria, Angola, Burundi, Colombia, India (Kashmir), Russia, Sri Lanka and Sudan.

---

[12] Chapter 1 reviews the fighting that took place in Sierra Leone and the steps towards implementation of the disarmament, demobilization and reintegration plan that was part of the ceasefire agreement.

[13] These 4 conflicts are Burundi, India (Kashmir), Israel–Palestinians (increased by > 50%) and Sudan.

[14] These 6 conflicts are India–Pakistan, the Philippines (NPA), the Philippines (MILF), Russia (Chechnya), Sri Lanka and Turkey, of which the Philippines (MILF), Russia (Chechnya) and Sri Lanka declined by > 50%.

[15] These 6 conflicts are Algeria, Colombia, India (Assam), Indonesia (Aceh), Peru and Myanmar.

Table 1A.3. Table of conflict locations with at least one major armed conflict in 2001

Location	Incompat-ibility[a]	Year formed/ year joined[b]	Warring parties[c]	No. of troops in 2001[d]	Total deaths[e] (incl. 2000)	Deaths in 2001	Change from 2000[f]
Africa							
Algeria	Govt	1993/1993	Govt of Algeria vs. GIA	300 000* ..	40 000– 100 000**	> 1 000	0

GIA: Groupe Islamique Armé (Armed Islamic Group)
* Including the Gendarmerie, the National Security Forces and Legitimate Defence Groups (local militias).
** Note that these figures include deaths in the fighting since 1992 in which parties other than those listed above participated, notably the Front Islamique du Salut (FIS), or Islamic Salvation Front.

| Angola | Govt | 1975/1998 | Govt of Angola, Namibia vs. UNITA | 130 000 .. 10 000–30 000 | .. | > 1 000* | ..* |

UNITA: União Nacional para a Independência Total de Angola (National Union for the Total Independence of Angola)
* This death figure serves only as an indication of the absolute minimum number of battle-related deaths; the actual figure may be much higher. In this case, the uncertainty also means that comparison with the figure for deaths in 2000 is not meaningful.

| Burundi | Govt | 1998/1998 ../.. | Govt of Burundi vs. CNDD–FDD vs. Palipehutu | 40 000 10 000–16 000 2 000–3 000 | > 5 000* | > 1 000 | + |

CNDD: Conseil national pour la défense de la démocratie–Forces pour la défense de la démocratie (National Council for the Defence of Democracy–Forces for the Defence of Democracy)
Palipehutu: Parti pour la libération du peuple Hutu (Party for the Liberation of the Hutu People)
* Political violence in Burundi since 1993 is reported to have claimed a total of at least 200 000 lives. This figure includes deaths incurred by groups other than those listed above that are no longer active, deaths in intra-group fighting, as well as deaths that have not been classified as battle-related.

Location	Incompat-ibility[a]	Year formed/ year joined[b]	Warring parties[c]	No. of troops in 2001[d]	Total deaths[e] (incl. 2000)	Deaths in 2001	Change from 2000[f]
Congo, Democratic Republic of the			Govt of Dem. Rep. of Congo, Angola, Namibia, Zimbabwe	45 000–55 000 .. 1 500–2 000** 10 000–12 000	..	..	..
	Govt	1998 /1998	vs. RCD, Rwanda	15 000–20 000 10 000–20 000			
			vs. RCD–ML,* MLC,* Uganda	2 500 10 000–15 000 8 000–9 000			

RCD: Rassemblement Congolais pour la démocratie (Congolese Rally for Democracy)
RCD–ML: Rassemblement Congolais pour la démocratie–Mouvement de libération (Congolese Rally for Democracy–Liberation Movement)
MLC: Mouvement de libération Congolais (Congolese Liberation Movement)

* A merger of the RCD-ML and the MLC into the FLC (Front de Libération Congolais) was reported in Jan. 2001. However, there were indications that the new organization dissolved into its constituent organizations during the year.
** Namibian troops were withdrawn during 2001; by October no troops remained in the DRC.

| Rwanda | Govt | 1994/1994 | Govt of Rwanda vs. Opposition alliance* | 40 000–60 000 30 000–50 000 | .. | > 1000 | .. |

* Consisting of former government troops of the Forces Armées Rwandaises (the former Rwandan Armed Forces, ex-FAR) and the Interahamwe militias. There are contradictory reports on whether the alliance is identical to the Peuples en armes pour la libération du Rwanda (People in Arms for the Liberation of Rwanda).

| Somalia | Govt | 1991/1991 | Govt of Somalia vs. SRRC | .. .. | .. | .. | .. |

SRRC: Somali Reconciliation and Restoration Council.

MAJOR ARMED CONFLICTS   71

| Sudan | Govt | 1980/1983 | Govt of Sudan vs. NDA** | 110 000–120 000*
30 000–50 000 | ..*** | > 2 000 | + |

NDA: National Democratic Alliance
* Including police forces.
** The June 1995 Asmara Declaration forms the basis for the political and military activities of the NDA. The NDA is an alliance of several southern and northern opposition organizations, of which the SPLM (Sudan People's Liberation Movement) is the largest. SPLM leader John Garang is also the commander of the Unified Military Command for the NDA.
*** Total military deaths until 1991 are estimated at 37 000–40 000.

America

| Colombia | Govt | 1949/1978
1965/1978 | Govt of Colombia
vs. FARC
vs. ELN | 280 000*
15 000–20 000
3 000–5 000 | > 40 000** | > 1000*** | 0*** |

FARC: Fuerzas Armadas Revolucionarias Colombianas (Revolutionary Armed Forces of Colombia)
ELN: Ejército de Liberación Nacional (National Liberation Army)
* Including police forces.
** This figure includes deaths in the fighting since 1964 in which parties other than those listed above also participated.
*** The total number of deaths from political violence in 2001, also involving right-wing paramilitary groups, is at least 2000–3000.

| Peru | Govt | 1980/1981 | Govt of Peru vs. Sendero Luminoso | 190 000*
200–600 | > 28 000 | < 25 | 0 |

Sendero Luminoso: Shining Path
* Including paramilitary forces.

Location	Incompat- ibility[a]	Year formed/ year joined[b]	Warring parties[c]	No. of troops in 2001[d]	Total deaths[e] (incl. 2000)	Deaths in 2001	Change from 2000[f]
USA*	Govt*	2001/2001***	Govt of USA, Multinational coalition** vs. al-Qaeda	.. .. ..	> 3 000	> 3 000	n.a.

* See section III for an elaboration of the preliminary assessment of this case, particularly the discussion of the incompatibility and the ways in which it differs from other incompatibilities recorded by the Uppsala Conflict Data Project.
** Including troops from Australia, the UK and the USA. Military contributions were also made by Canada, France, Germany, Italy, Japan, Jordan, the Netherlands, Poland, Russia and Turkey.
*** Before 2001 the incompatibility stated by al-Qaeda concerned US policy on the Middle East. In 2001 statements indicated that the incompatibility had developed into the larger issue of government, i.e., the destruction of the US political system.

Asia

Location	Incompat- ibility[a]	Year formed/ year joined[b]	Warring parties[c]	No. of troops in 2001[d]	Total deaths[e] (incl. 2000)	Deaths in 2001	Change from 2000[f]
Afghanistan	Govt	1992/1992	Govt of Afghanistan vs. UIFSA,* Multinational coalition**	20 000–40 000 10 000–20 000	..	> 1 000***	..***

UIFSA: United Islamic Front for the Salvation of Afghanistan

* A military alliance, the SCDA (Supreme Council for the Defence of Afghanistan), was formed in Oct. 1996 by the Jamiat-i-Islami, Hezb-i-Wahdat and Jumbish-i Milli-ye Islami. The SCDA changed its name to the UIFSA in June 1997.
** Including troops from Australia, the UK and the USA. Military contributions were also made by Canada, France, Germany, Italy, Japan, Jordan, the Netherlands, Poland, Russia and Turkey.
*** This death figure serves only as an indication of the absolute minimum number of battle-related deaths; the actual figure may be much higher. In this case, the uncertainty also means that comparison with the figure for deaths in 2000 is not meaningful.

MAJOR ARMED CONFLICTS 73

India	Territory (Kashmir)	../1989	Govt of India vs. Kashmir insurgents*	1 300 000 5 000	> 23 000	> 3 000	+
	Territory (Assam)	1982/1988 ../1986	vs. ULFA vs. NDFB	.. ..		> 200	0

ULFA: United Liberation Front of Assam
NDFB: National Democratic Front of Bodoland
* Several groups are active, some of the most important being the Hizb-ul-Mujahideen, the Harkat-ul-Mujahideen, the Lashkar-e-Toiba and the Jesh-e-Mohammadi.

India–Pakistan	Territory	1947/1996	Govt of India vs. Govt of Pakistan	1 300 000 600 000	..	> 100	–
Indonesia	Territory	1976/1989	Govt of Indonesia vs. GAM	500 000* 2000–5000	> 2 000	100–200	0

GAM: Gerakan Aceh Merdeka (Free Aceh Movement)
* Including paramilitary forces. Some 30 000 troops were used in Aceh.

Myanmar	Territory	1948/1948	Govt of Myanmar vs. KNU	400 000* 2 000–4 000	1948–50: 8 000 1981–88: 5 000–8 000	50–200	0

KNU: Karen National Union
* Including paramilitary forces.

Philippines	Govt	1968/1968	Govt of the Philippines vs. NPA	100 000 11 000–13 000	21 000– 25 000	> 100	–
	Territory	1984/1987	vs. MILF	10 000–15 000	> 2 000	>25	– –

NPA: New People's Army
MILF: Moro Islamic Liberation Front

Location	Incompatibility[a]	Year formed/ year joined[b]	Warring parties[c]	No. of troops in 2001[d]	Total deaths[e] (incl. 2000)	Deaths in 2001	Change from 2000[f]
Sri Lanka	Territory	1976/1983	Govt of Sri Lanka vs. LTTE	100 000–110 000 6 000–7 000	> 60 000	> 1 000	− −

LTTE: Liberation Tigers of Tamil Eelam

Europe

Russia	Territory	1991/1999	Govt of Russia vs. Republic of Chechnya	1 000 000* 8 000–25 000	40 000– 70 000	> 1 000	− −

* Some 70 000–80 000 troops, including paramilitary forces, were used in Chechnya.

Middle East

Iran	Govt	1970/1991	Govt of Iran vs. Mujahideen e-Khalq	600 000* . .	. .	. .	.

* Including the Revolutionary Guard.

Iraq	Govt	1980/1991	Govt of Iraq vs. SAIRI	430 000 . .	. .	. .	.

SAIRI: Supreme Assembly for the Islamic Revolution in Iraq

Israel	Territory	1964/1964	Govt of Israel vs. Palestinian organizations**	160 000–170 000 . .	1948–: > 13 000*	150 (mil.) 350–400 (civ.)	+ +

* Note that this figure also covers the period 1948–63, in which parties other than those listed above participated.
** Mainly Fatah/Tanzim, but also Hamas and Islamic Jihad.

| Turkey | Territory | 1974/1984 | Govt of Turkey vs. PKK | 800 000* 3 000–5 000 | >30 000 | 100–200 |

PKK: Partiya Karkeren Kurdistan, Kurdish Worker's Party, or Apocus
* Including the Gendarmerie/National Guard.

The following notes apply to table 1A.3. Note that, although some countries are also the location of minor armed conflicts, the table lists only the major armed conflicts in those countries. For the definitions, methods and sources used, see appendix 1B.

The conflicts listed in table 1A.3 are listed by location, in alphabetical order, within 5 geographical regions: *Africa*—excluding Egypt; *America*—including North, Central and South America and the states in the Caribbean; *Asia*—including Oceania, Australia and New Zealand; *Europe*—including the states in the Caucasus; and *Middle East*—Egypt, Iran, Iraq, Israel, Jordan, Kuwait, Lebanon, Syria, Turkey and the states of the Arabian peninsula.

[a] The stated general incompatible positions. 'Govt' and 'Territory' refer to contested incompatibilities concerning government (type of political system or a change of central government or its composition) and territory (control of territory [interstate conflict], secession or autonomy), respectively. Each location may have 1 or more incompatibilities over territory if the disputed territories are different entities. There can be only 1 incompatibility over government in each location as, by definition, there can be only 1 government in each location. For each incompatibility there may be more than 2 parties.

[b] 'Year formed' is the year in which the incompatibility was stated. 'Year joined' is the year in which use of armed force began or recommenced.

[c] The non-governmental warring parties are listed by the name of the parties using armed force. Only those parties and alliances which were active during 2001 are listed in this column. Alliances are indicated by a comma between the names of the warring parties.

[d] The figures for 'No. of troops in 2001' are for total armed forces (rather than for army forces, as in the *SIPRI Yearbooks 1988–1990*) of the government warring party (i.e., the government of the conflict location) and for forces of non-governmental parties from the conflict location. Non-government parties supporting a government with troops are not included as part of the government forces unless specifically noted. For government and non-governmental parties from outside the location, the figures in this column are for total armed forces within the country that is the location of the armed conflict. Deviations from this method are indicated by a note (*) and explained.

[e] The figures for deaths refer to total battle-related deaths, that is, those deaths that were caused by the warring parties and which can be directly connected to the incompatibility, during the conflict. 'Mil.' and 'civ.' refer, where figures are available, to *military* and *civilian* deaths, respectively; where there is no such indication, the figure refers to total military and civilian battle-related deaths in the period or year given. Information which covers a calendar year is necessarily more tentative for the last months of the year. Experience has also shown that the reliability of figures improves over time; they are therefore revised each year.

f The 'change from 2000' is measured as the increase or decrease in the number of battle-related deaths in 2001 compared with the number of battle-related deaths in 2000. Although based on data that cannot be considered totally reliable, the symbols represent the following changes:

++   increase in battle deaths of > 50%
+   increase in battle deaths of > 10 to 50%
0   stable rate of battle deaths ($\pm$10%)
−   decrease in battle deaths of > 10 to 50%
− −   decrease in battle deaths of > 50%
n.a.   not applicable, since the major armed conflict was not recorded for 2000.

Note: In the last 3 columns ('Total deaths', 'Deaths in 2001 and 'Change from 2000'), '..' indicates that no reliable figures, or no reliable disaggregated figures, were given in the sources consulted.

Appendix 1B. Definitions, sources and methods for the conflict data

This appendix explains the definitions and methods and describes the sources used for the data on major armed conflicts compiled by the Uppsala Conflict Data Project of the Department of Peace and Conflict Research, Uppsala University. These data are presented in appendix 1A.

I. Definitions

The Uppsala Conflict Data Project defines a major armed conflict as a contested incompatibility that concerns government and/or territory over which the use of armed force between the military forces of two parties, of which at least one is the government of a state, has resulted in at least 1000 battle-related deaths in any single year.[1]

The separate elements of this definition are defined as follows.[2]

1. *Incompatibility that concerns government and/or territory*. The incompatibility must concern government and/or territory, and it refers to the stated generally incompatible positions. An *incompatibility that concerns government* refers to the type of political system, the replacement of the central government or change of the composition of the current government. An *incompatibility that concerns territory* refers to the status of a territory, for example, a change of the state in control of a certain territory (interstate conflict), secession or autonomy (intra-state conflict).

2. *Use of armed force*. This refers to the use of arms by the military forces of the parties in order to promote the parties' general position in the conflict, resulting in deaths. Arms are defined as any material means of combat, for example, manufactured weapons as well as sticks, stones, fire, water, and so on.

3. *Party*. This refers to the government of a state or an opposition organization or alliance of opposition organizations. The *government of a state* is that party which is generally regarded as being in central control, even by those organizations seeking to take over power. If this criterion is not applicable, the government is the party controlling the capital of the state. In most cases where there is a government, the two criteria coincide. An *opposition organization* is any non-governmental group which has announced a name for the group and its political goals and has used armed force to achieve them.

4. *State*. A state is an internationally recognized sovereign government controlling a specified territory or an internationally non-recognized government controlling a

---

[1] This definition of major armed conflict differs slightly from the definition applied to the data of the Uppsala Conflict Data Project published in *SIPRI Yearbooks 1988–1999* (Oxford University Press: Oxford, 1988–99). The requirement that a conflict must cause 1000 or more battle-related deaths in a single year, rather than over the entire course of the conflict, ensures that only conflicts that reach a high level of intensity, as measured by deaths, are included. The tables and figure in appendix 1A have been retroactively adjusted to reflect this new definition.

[2] Sollenberg, M. (ed.), *States in Armed Conflict 2000*, Report no. 60 (Department of Peace and Conflict Research, Uppsala University: Uppsala, 2000), appendix 2; and Heldt, B. (ed.), *States in Armed Conflict 1990–91*, 2nd edn, Report no. 35 (Department of Peace and Conflict Research, Uppsala University: Uppsala, 1992), pp. 31–34, available at URL <http://www.peace.uu.se>.

specified territory whose sovereignty is not disputed by an internationally recognized sovereign government which previously controlled the same territory.

5. *Battle-related deaths.* This refers to those deaths that are caused by the warring parties and which can be directly related to combat over the contested incompatibility.

Once a conflict has reached the threshold of 1000 battle-related deaths, it continues to appear in the annual tables of conflicts until the contested incompatibility has been resolved and/or until there is no recorded use of armed force, resulting in deaths, between the parties and concerning the same incompatibility during the year. The same conflict may reappear in subsequent years if there is renewed use of armed force between the same parties, resulting in deaths and concerning the same incompatibility.

There is frequently international involvement of various types in intra-state conflicts. Only one type of international involvement is included in appendix 1A: another state or multinational coalition is considered as a party to a conflict if, and only if, it contributes regular troops to one of the warring parties and shares the goals of that party. A traditional peacekeeping operation is not considered to be a party to the conflict but rather an impartial part of a consensual peace process. It should also be noted that rebel groups operating from a base in a neighbouring state are listed as parties to the conflict in the location where the government is challenged, regardless of their nationality or where they are based.

The object of study is not political violence per se but incompatibilities that are contested by the use of armed force. Thus, the project registers one major class of political violence, battle-related deaths, which serves as a measure of the magnitude of the conflict. Other types of political violence are excluded. Examples of such other types of violence are: unilateral use of armed force, for example, government repression, massacres, ethnic cleansing and genocide; unorganized or spontaneous violence, for example, violent demonstrations and communal violence; and violence which is not directed at the state, for example, non-governmental organizations fighting each other. It is argued that these categories of political violence are expressions of phenomena other than armed conflict as defined here—that is, reciprocal, organized, political and deliberate in nature. For example, reciprocal violence is different from unilateral violence, that is, war is different from genocide. This is not to say that such other types of violence or violent conflict are not as important, but the distinction between them is important. There are other projects that collect data on these other types of violence.[3]

II. Sources

The data presented in appendix 1A are based on information taken from a selection of publicly available sources, printed as well as electronic. The sources include news agencies, newspapers, periodical journals, research reports, and documents of international, multinational and non-governmental organizations. The latter include docu-

---

[3] See, e.g., the work of Ted Robert Gurr *et al.*, Center for International Development and Conflict Management, University of Maryland, for data on ethno-political rebellion and communal violence. Several data sets on various types of violence and conflict are available on the Internet site of the Inter-university Consortium for Political and Social Research (ICPSR) of the Institute for Social Research, University of Michigan, at URL <http://www.icpsr.umich.edu>.

ments of the warring parties (governments and opposition organizations) when such sources are available, since they serve as a crucial complement when identifying statements about the parties' incompatible positions. Global, regional and country-specific sources are used.

Independent news sources that have been selected over several years form the basis of the source collection. Two major general news sources consulted for the data collection are Reuters News Service and the BBC World Service. The project also uses region- and country-specific sources extensively. However, they are not comparable between regions. This means that for some countries several sources are consulted, whereas for other countries and regions only a few high-quality region- or country-specific sources are used. Regional sources include, for example in the case of Africa, *Africa Confidential*, *Africa Research Bulletin* and the Integrated Information Regional Network (IRIN). For Asia, more country-specific news sources are consulted since few reliable regional sources are publicly available. For example, for the countries in South Asia articles from a large number of national and local newspapers are used.

The project consistently scrutinizes and revises the selection and combination of sources in order to maintain a high level of reliability and comparable coverage of all regions and states. One of the priorities is to arrive at a balanced combination of sources of different origins in order to avoid a bias.

The reliability of the sources is judged by using the expertise within the project together with advice from a global network of experts. Of highest priority is the general reputation and expertise of the sources as judged by regional experts. If possible, the members of the project discuss the sources with academics in the respective countries in conflict. The independence of the sources is crucial as well as the transparency of the origins of the sources. Each source is judged according to the context in which it is published, that is, according to the potential interests of the source in misrepresenting political or violent events. In the case of biased sources which are used to identify statements by the parties, they must be official sources issued by the parties. Since most sources are secondary sources, the project attempts to trace reports back to the primary source in order to decide whether they are reliable. In addition to deciding the level of reliability of available sources, the project strives to identify the existence of censorship. Thus, other sources than regular news sources must be used to establish what is occurring in a country. Documents and reports issued by international, multinational and non-governmental organizations are consulted for this purpose.

III. Methods

The data on major armed conflicts are compiled by calendar year. They include data on conflict location, type of incompatibility, year the incompatibility was formed, year the warring party began its use of armed force, warring parties, number of troops, total battle-related deaths, battle-related deaths during the year, and the change in battle-related deaths compared to the previous year.[4]

The data on battle-related deaths constitute the largest part of the data collection. Figures for battle-related deaths are produced through a comprehensive review of reports on individual violent incidents in each conflict which are then aggregated.

---

[4] See also the notes to table 1A.3 in appendix 1A.

Ideally, these individual figures are corroborated by two or more independent sources. The aggregated figures are also compared to total figures that appear in official documents, special reports and the news media. Regional experts—for example, researchers, diplomats and journalists—are often consulted during the process of the data collection. Their contribution is mainly clarification of the contexts in which events occur, thus facilitating proper interpretation of the reporting in published sources.

Little information on the exact number of deaths in armed conflicts is publicly available. The project therefore in many instances presents these figures as ranges or approximations, and they are best estimates. The numbers of battle-related deaths are based on conservative estimates. Experience shows that, as more information on an armed conflict becomes available, the conservative estimates based on information about each individual event are more often correct than the less conservative, higher estimates. If no figures are available or if published figures are too contradictory to establish even a minimum reliable figure, no figure is given. Figures are revised retroactively each year as new information becomes available.

Appendix 1C. Measuring violence: an introduction to conflict data sets

TAYLOR B. SEYBOLT

I. Introduction

The systematic study of violent conflict was pioneered by Quincy Wright in *A Study of War*, first published in 1942.[1] The book brought quantitative analysis of conflicts to the attention of a wide audience for the first time by drawing on numerical and statistical research, in addition to historical and legal material.[2] Following Wright, the cornerstone of modern, systematic, quantitative studies of conflict was set in place in the mid-1960s by J. David Singer and Melvin Small with their Correlates of War project.[3] Since the 1980s, with the advent of the widespread use of computers, a multitude of conflict data-collection projects have emerged.[4] As the number of systematic data collections has increased, the field has become increasingly diverse and complex.[5] As a result, there is disagreement on some of the most basic questions. Is the world more or less violent today than in the past? Are wars more or less destructive than they used to be? Are modern violent conflicts different from earlier ones? What are the causes of conflict initiation, continuation and termination?

Most of the differences in opinion about these and other questions can be traced to differences in the collection and use of data. Variations in purpose, definitions and coding rules can lead to significant divergence on basic parameters, such as the number, frequency, duration and dispersal of armed conflicts in the world.[6] The purpose of this appendix is to introduce the users of conflict data to important methodological, theoretical and policy-related questions that face researchers in the systematic study of conflict. The world's primary English-language data-collection projects are presented, as are reasons for the differences between them.[7]

---

[1] Wright, Q., *A Study of War*, 2nd edn (University of Chicago Press: Chicago, Ill., 1965).
[2] Richardson, L. F., 'Generalized foreign politics', *British Journal of Psychology: Monograph Supplements*, vol. 23 (1934); and Sorokin, P., *Social and Cultural Dynamics*, vol. 3, *Fluctuation of Social Relationships, War and Revolution* (American Book Company: New York, 1937).
[3] Singer, J. D. (ed.), *The Correlates of War I: Research Origins and Rationale* (Free Press: New York, 1979). SIPRI published a list of 11 studies of international wars and armed conflicts in its first Yearbook. The list gave the definition of armed conflict used and publication information for each study. The studies were used to generate a table of conflicts, noting their type, size, parties, and beginning and end dates for the period from 1820 to 1968. Naidu, H., 'Conflicts', *SIPRI Yearbook of World Armaments and Disarmament 1968/69* (Almqvist & Wiksell: Stockholm, 1969), pp. 359–73.
[4] Two excellent central sources of databases on many aspects of social, economic and political life are the Inter-University Consortium for Social and Political Research, available at URL <http://www.icpsr.umich.edu/>, and the Harvard–MIT Data Center, available at URL <http://data.fas.harvard.edu/>. Access to a wide range of data collections on several aspects of international affairs is provided by Paul Hensel at URL <http://garnet.acns.fsu.edu/~phensel/intldata.html> and Richard Tucker at URL <http://www.vanderbilt.edu/~rtucker/methods/data/>.
[5] Identifying Wars: Systematic Conflict Research and its Utility in Conflict Resolution and Prevention, Conference held at Uppsala University, Uppsala, Sweden, 8–9 June 2001. The conference agenda, list of participants and papers are available at URL <http://www.pcr.uu.se/>.
[6] Gleditsch, N. P. *et al.*, 'Armed conflict 1946–99: a new dataset', Paper presented at the Identifying Wars Conference (note 5).
[7] Restricting this survey to English-language projects may under-represent the variety of work being done, but it captures almost all of the work in the sub-field. The vast majority of conflict data-collection

Section II reviews variations in purpose, definitions, methods and coding rules as sources of difference between data sets. Section III summarizes the primary data-gathering projects whose data are publicly available and provides contact information for them. Internet access to all the data projects reviewed here is available from the SIPRI Internet site at URL <http://projects.sipri.se/conflictstudy/index.html>. The last section summarizes the conclusions of this examination of conflict data.

II. Variations in purpose, definitions and methodology

Purposes

One of the primary explanations for the differences between data sets is that they are intended to shed light on different aspects of conflict. Data-collection projects are categorized in section III according to whether they are primarily concerned with the global pattern of conflict, processes of conflict initiation and termination, or conflict prediction and cost.[8]

The type of data-collection project that attempts to facilitate the understanding of the patterns of conflict occurrence, rather than the processes of conflict development, is the most well established. These projects measure the frequency, location and severity of conflicts. Most of them determine when a conflict should be included in the data set by the estimated number of people killed.[9] A few projects in this category use qualitative, rather than quantitative, determinants of what to consider as a conflict.[10] Data projects in this category cover both interstate and intra-state conflicts and distinguish between them. Data sets that focus on counting incidents are useful for investigating the prevalence and locations of various types of conflict. Some projects include data on the characteristics of the antagonists in violent conflicts, such as their geographical location, relative capabilities and ethnic identity.[11] This information allows for testing hypotheses on the causes of conflict. Data sets of this type also estimate the costs of conflicts in terms of the number of deaths. They can potentially assist cost-of-conflict studies by providing data on population displacement, economic performance, military expenditure and other factors, if they are combined with non-conflict data sets.[12] The geographical data offered are usually too general to allow

projects are conducted in the United States, probably because of the behavioural approach to studying social phenomena that is prominent in the USA but not elsewhere. In addition, some projects that reside in other countries publish some or all of their material in English. The only notable data project that provides data sets in another language is Arbeitsgemeinschaft Kriegsursachenforschung (AKUF). The AKUF Internet site provides data sets in German; some English-language papers are available at URL <http://www.sozialwiss.uni-hamburg.de/Ipw/Akuf/home.html>.

[8] These categories overlap considerably in practice. The projects are not monolithic and can be used to study several different aspects of conflict.

[9] These are the Correlates of War project, the Conflict Data Project and the Major Episodes of Political Violence project. Full information is provided in section III for all the data-collection projects named in the footnotes, unless otherwise noted.

[10] The Conflict Simulation Model (Konflikt-Simulations-Modell, KOSIMO) and AKUF projects.

[11] E.g., the Correlates of War project.

[12] Information on population displacement is available from the UN High Commissioner for Refugees at URL <http://www.unhcr.ch>, the US Committee for Refugees at URL <http://www.refugees.org/> and the Norwegian Refugee Council at URL <http://www.nrc.no/engindex.htm>. Economic data are available from *inter alia* the World Bank at URL <http://www.worldbank.org/> and the UN Development Programme (UNDP) at URL <http://www.undp.org/>. Military expenditure data are available from the SIPRI Military Expenditure Project, URL <http://projects.sipri.se/milex/mex_data_index.html>. Data on the dollar value of transfers of major conventional weapons are available from the SIPRI Arms Transfers Project at URL <http://projects.sipri.se/armstrade/atfproj/html>. Data on legal transfers

greater specificity than an entire country, which leads to a distorted picture of zones of conflict and zones of peace.

A second type of data-collection effort is concerned with studying conflict dynamics by examining the factors that are thought to have an impact on the outbreak, continuation and termination of conflicts. Data projects in this category tend to include either intra-state or interstate conflicts, but not both. Projects that are primarily concerned with the causes of conflict look at a range of factors, from historically common catalysts, to characteristics of groups vulnerable to crisis behaviour.[13] Some data sets that are concerned with the dynamics of ongoing conflicts focus on a particular kind of event, such as intervention by outside states.[14] Other projects attempt to capture the complexity of societies in conflict.[15] A few data projects are primarily concerned with when and how conflicts end.[16] Some data-collection projects that do not examine conflict are frequently used by researchers in combination with conflict data sets to test, for example, hypotheses on the relationship between the type of regime and violent conflict.[17] The diverse data sets that illuminate possible causes of conflict initiation, continuation and termination are useful for addressing questions related to conflict management and resolution. The current weakness of these data sets is that they are only just beginning to provide detailed information on the particular actions and motives of the antagonists.[18]

A third type of effort, to collect data known as 'events data', gathers very detailed information on the impact of violence and factors thought to influence the outbreak of violence. Event data projects attempt to record all interactions chronologically, without grouping them into distinct cases of conflict. When these data are combined with the knowledge of area specialists, the output can provide warning of potential outbreaks of violence.[19] The greatest shortcoming of events data projects is that they rely on a very limited number of media sources, so their content is skewed by the editorial decisions of media outlets, which are beyond their control. Also, owing to the quantity of material handled, coding is done by computers and must be checked by the user.[20]

---

of small arms are available from the Norwegian Initiative on Small Arms Transfers (NISAT) at URL <http://www.nisat.org/>. The Facts on International Relations and Security Trends (FIRST) project at SIPRI, available at URL <http://first.sipri.org/>, combines information from a number of databases to provide security-related country profiles. The Armed Conflict and Intervention Project (see section III) is currently working on the provision in a single location of information on conflicts, displaced populations, trade flows, political interactions, membership of international organizations and military interventions.

[13] These 3 topics are treated by the Issue Correlates of War project, the Minorities at Risk Project, and the International Crisis Behavior Project and Behavioral Correlates of War projects, respectively.

[14] The Third Party Interventions in Intrastate Conflicts project.

[15] The Armed Conflict and Intervention Project and the Minorities at Risk project.

[16] The Violent Intrastate Nationalist Conflicts project.

[17] Political Regime Characteristics and Transitions, 1800–1999 (Polity IV), available at URL <http://www.bsos.umd.edu/cidcm/polity/>.

[18] Collier, P. and Hoeffler, A., 'Data issues in the study of conflict', Paper presented at the Identifying Wars Conference (note 5).

[19] The Swiss Peace Foundation sponsors Früh-Analyse von Spannungen und Tatsachenermittlung (FAST). Its objective is early recognition of impending or potential crisis situations for the purpose of early action and conflict prevention. The project combines multiple methods, including events data analysis, field investigation and consultation of experts. URL <http://www.swisspeace.ch/html/navigation/fr_program_fast.html>.

[20] There are also systematic, computerized, text-oriented projects intended to assist conflict prevention and management. They are distinct from data-collection projects in that they are organized around case studies. The University of Southern California provides access to 3 such projects (SHERFACS, a database compiled by Frank Sherman; Computer-Aided System for Analyzing Conflict (CASCON); and

Definitions

Even data sets of the same type can present different views of conflict. For example, among the projects that study the global pattern of conflict, a comparison of the Correlates of War, Conflict Data Project, Conflict Simulation Model (Konflikt-Simulations-Modell, KOSIMO) and Arbeitsgemeinschaft Kriegsursachenforschung (AKUF) data sets demonstrates a low correlation between them on the basic matter of the number of interstate and intra-state wars between 1950 and 1999.[21] However, the pattern of rise and decline in the number of conflicts over time is similar in each of the four projects.[22] Disagreements among data sets of the same type are usually the result of different definitions of variables and coding rules.[23] As the definitions of the dependent variables for the 16 data projects presented in section III indicate, there are considerable differences between them. Depending on one's perspective, variation among definitions is either a significant problem or an opportunity.

Lack of consistency is a problem if the reader wants to consult a single authoritative source: there is none. More importantly, some analysts are concerned that the variation indicates basic theoretical differences. For example, does a state have to be one of the antagonists for a violent event to have political significance? These differences are often not made explicit, so data projects cannot be compared on the basis of full information. This undermines the possibility of the progressive accumulation of knowledge, which is the essential purpose of systematic studies of human behaviour.[24]

Variation among data sets of the same type is an asset if one wants to test the robustness of a hypothesis by subjecting it to two different sets of data. If a hypothesis yields the same answer with two different data sets, it can be considered stronger than if it is tested against only one. If the two tests yield different answers, then both the hypothesis and the data sets need to be examined more closely.[25] This process can lead to stronger hypotheses as well as to more refined and accurate data-collection projects.

It is instructive to look at fatality thresholds in this regard.[26] Several projects use a fatality count to determine the degree of violence.[27] Once a specified fatality threshold is crossed, the event is counted in the data set. This method of operationalizing the definition of violent conflict raises several questions. How many people must be killed before an event is considered significant enough to be included? Should the count be on an annual or overall basis? The Correlates of War project requires

HAAS, a database compiled by Ernst Haas) through its Prototype Action Recommenders' Information Support (PARISinLA) project, available at URL <http://www.usc.edu/dept/ancntr/Paris-in-LA/>.

[21] Eberwein, W.-D. and Chojnacki, S., 'Scientific necessity and political utility: a comparison of data on violent conflicts', Discussion paper P01-304, International Politics Working Group, Wissenschaftszentrum-Berlin (Sep. 2001), available at URL <http://www.wz-berlin.de/ip/pub.en.htm>.

[22] Gleditsch et al. (note 6).

[23] The use of different sources is another likely reason for disagreements. The limited amount of information about the sources used by projects is discussed in the next section.

[24] Eberwein and Chojnacki (note 21).

[25] Collier and Hoeffler (note 18).

[26] Sambanis, N., 'A note on the death threshold in coding civil war events', Conflict Processes Newsletter online (June 2001), available at URL <http://mailer.fsu.edu/~whmoore/cps/newsletter/june2001/Sambanis.html>.

[27] The Conflict Data Project, whose data have been published in the SIPRI Yearbook since 1987, uses a threshold measure of fatalities. Among the data projects summarized in section III, fatality thresholds are used by 3 out of 5 projects in the 'patterns of conflict occurrence' category and by 2 out of 8 projects in the 'causes and processes of conflict' category.

1000 battle-related deaths over the course of the conflict for interstate wars and 1000 battle-related deaths in a single year for intra-state and extra-systemic (imperial and anti-colonial) wars. The Major Episodes of Political Violence data set uses a fatality measure of 500 'directly related' fatalities over the course of the conflict, together with substantial destruction of infrastructure and population displacement. The Conflict Data Project distinguishes three levels of violence that combine annual and total battle-related deaths. Minor conflicts cause at least 25 battle-related deaths in a year, but fewer than 1000 overall; intermediate conflicts cause more than 1000 battle-related deaths overall, but fewer than 1000 in any single year; and wars cause at least 1000 battle-related deaths in a single year. The lack of direct comparability between the three projects is obvious.

The seemingly simple question of what to count as a conflict fatality is also open to interpretation. These three projects count only 'battle-related' deaths, which excludes civilian massacres as well as any other deaths that were not directly caused by battle, even if they were brought on by military actions.[28] Other projects use a broader definition of deaths but sometimes do not clearly specify what they include.[29] Differences over what to count as a conflict death have a direct bearing on whether and when a fatality threshold is reached. Then there is the question of whether the threshold should be defined in absolute or relative (for example, per capita) terms. Violence that killed hundreds of people in the early 1980s shook the foundations of the state in Surinam (population $c.$ 500 000). The violent death of hundreds of people in India (population $c.$ 1008 million), as has occurred many times, is a tragedy but not a political crisis. The point is that absolute measures fail to capture the cost differences for small and large countries.[30]

All of this demonstrates that data sets are open to the challenge that the number of deaths may not be the best way to determine the severity of a conflict.[31] The context of the violence may be critically important for determining its severity. For example, the approximately 3000 people killed by the terrorist attack on the World Trade Center in New York City on 11 September 2001 had a far larger impact on world politics than the 60 000 killed in Sri Lanka's civil war over the past 18 years. An inability to incorporate the contextual dimension of behaviour is a fundamental problem for all quantitative research, and conflict data projects can do little to resolve it. Researchers who use data sets would be well advised to keep in mind the inherent limitations of such projects. As an illustration, nearly all data projects, even those that use several measures of severity, treat war as a distinct phenomenon recognizable by the absolute number of people killed. Yet this begs the question of whether war really is a phenomenon distinct from other types of political violence. Wars usually begin as lesser conflicts that escalate. In what ways are the dynamics of war and war termination different from the dynamics and termination of lower levels of violence? Does treating war as distinct reduce our ability to understand it? The projects summarized in the section below on the causes and processes of conflict attempt to address these questions.

---

[28] In an extreme example of the distortion that a definition can impose, the Conflict Data Project did not register the deaths from the genocide in Rwanda, even though it registered the existence of a major armed conflict in that country in 1994. The genocide deaths were not considered to be battle-related.
[29] The KOSIMO project.
[30] Collier and Hoeffler (note 18).
[31] The data projects located at the Center for International Development and Conflict Management (CIDCM), University of Maryland, are particularly innovative in using other measures of severity in addition to or in place of fatalities.

The definitions a data project uses, like the project's purpose, can determine the value of any particular data set for answering a research question. For reasons of consistency over time, data-collection projects must adhere to a predetermined set of variables and definitions for those variables. As a consequence, they reveal a consistent view (or frame) of conflict, with variation occurring within the frame.[32] They do not reveal changes outside the frame. One example is that most projects use definitions of conflict that require a state to be one of the antagonists.[33] Violent conflicts in which a state is not an antagonist, such as inter-communal violence, are excluded. If one wants to investigate whether the role of the state in violent conflict is in decline, it would be useful to know the relative frequencies of conflicts involving states and not involving states. Whether or not a data set can provide the information depends on the definition of conflict used.

Methods and coding rules

In broad terms, all data producers follow the same method. They gather and examine copious amounts of information on actual or potential conflicts and conflict actors, pick out material related to their study variables and code the information according to a set of rules. Exactly where project researchers get their information is somewhat of a mystery, since virtually none of them offers a specific listing of sources used to create its database.[34] Given the limited resources available to them and the global scope of all the projects listed here, it is safe to assume that project personnel do not have a network of colleagues in the field who feed them first-hand information. It is likely that they rely primarily on news media reports and secondary sources.[35] In some cases this type of source is supplemented by consulting regional experts and using field reports produced by non-governmental and international organizations. This situation raises a potential problem when combined with the fact that the vast majority of data projects are located in the United States, and apparently all the projects are Western. In all likelihood, different projects use largely the same set of sources, such as major US and European newspapers and reports from respected organizations. Any bias in news reporting will influence the information in every data project, making the entire enterprise a less reliable reflection of reality. This situation is in the process of changing with the advent of the Internet. Lexis-Nexis Academic Universe is a widely used Internet-based service that provides subscribers with access to several hundred news sources from around the world.[36]

---

[32] In practice, data projects occasionally adjust their definitions and refine their categories. Sarkees, M. R., 'The Correlates of War data on war: an update to 1997', *Conflict Management and Peace Science*, vol. 18, no. 1 (fall 2000), pp. 123–44. SIPRI adjusted the way it uses data collected by the Conflict Data Project, starting in 2000 (in this volume, see appendices 1A and 1B). A major armed conflict is now defined as one that caused at least 1000 deaths in a year rather than over the entire course of the conflict.

[33] Exceptions are the Major Episodes of Political Violence project, the Minorities at Risk project and the Violent Intrastate Nationalist Conflicts project. The Correlates of War project is currently expanding its typology to include conflicts with no state party. Sarkees, M. R. and Singer, J. D., 'Armed conflict past and future: a master typology?', Paper presented at the Identifying Wars Conference (note 5).

[34] Events data projects explicitly rely on news media and do reveal their original sources.

[35] Some of the projects reviewed here combine data from other projects. The Major Episodes of Political Violence project is an example, as is the Conflict Data Project's extension of its data set from 1989 back to 1946. In cases such as these, the projects cite the data collections they use, but the sources used for the original data sets are still not known.

[36] Hensel, P., Private communication with the author, 6 Mar. 2002.

Discussion of the exact methods used by each project and of the differences between them is beyond the scope of this survey. Instead, the remainder of this section makes three brief points, all of which carry the message that quantitative analytical understanding of conflict is highly subject to coding rules.

First, the duration of conflicts is a topic of interest to many researchers and policy makers. Are the durations of conflicts longer or shorter than in the past? Why do some conflicts last longer than others? Are the factors that explain the outbreak of a conflict the same as those that explain its duration? If a conflict appears to subside and then flares up again, is this a recurrence of the same conflict or a new conflict? Analyses of duration and related questions require that the start and end of conflicts be dated.[37] Some projects mark the beginning of a conflict from the initiation of sustained fighting that leads to the fatality threshold. Projects that do not use thresholds count the start of a conflict from the point of observable incompatibility. Still others wait until a conflict has crossed a threshold and then backdate it to the point of stated incompatibility.[38] In similar fashion, various coding rules are used to determine the end of a conflict and whether additional fighting is a new conflict or the continuation of an old one.[39]

Second, increased attention has been paid to conflict management and resolution during the past decade. A common analytical approach divides conflicts into phases of escalation and de-escalation. What factors lead from one phase to the next? Are some types of conflict more likely to escalate than others? Which actions by outside actors lead to de-escalation (or escalation) of the violence and under what circumstances? Analysis of conflict dynamics requires the demarcation of phases.[40] Some projects do not distinguish levels of intensity at all, so a conflict is either 'on' or 'off'.[41] Projects that do measure changes in intensity use additive coding rules so that, once a conflict has accumulated enough fatalities to cross into the next category of violence, it cannot be re-coded in a lower category if it diminishes in intensity.[42] Some projects use an ordinal scale to incorporate several measures to indicate the impact of a conflict on society.[43] These have great potential for the study of conflict management, but they are often used to mark the highest level of violence reached.[44]

Third, a fundamental principle of systematic, or scientific, research projects is that they are objective, transparent and consistent. These standards are not always maintained by conflict data-collection projects since they work to maintain the balance between validity (relevance) and reliability. The definitions of some variables require personal interpretation by the coder. For example, when determining the number of battle-related deaths, projects that count civilians must identify the line between civilians killed by cross-fire (battle-related) and massacres (not battle-related).[45] Some projects attempt to avoid definitional restrictions by setting

---

[37] Collier and Hoeffler (note 18).
[38] An example of each approach is the Correlates of War project, the KOSIMO project and the Conflict Data Project, respectively.
[39] The Violent Intrastate Nationalist Conflicts project pays special attention to conflict endings.
[40] A project that explicitly addresses conflict escalation in terms of phases is SHERFACS (note 20).
[41] The Correlates of War project.
[42] The Conflict Data Project allows movement from 'war' down to 'intermediate' but not to 'minor' conflict.
[43] The Minorities at Risk and the Major Episodes of Political Violence projects.
[44] The Violent Intrastate Nationalist Conflicts project.
[45] The various Correlates of War projects do not count civilian deaths, even if they occur in a battle.

qualitative rather than quantitative parameters for variables.[46] However, variables that are not strictly defined require personal interpretation and may result in the establishment of categories that overlap or are not mutually exclusive. In another example, grievances are often viewed as a cause of conflict, but few objective measures of grievance are available.[47] Even when there is a precise definition for a variable, it may be difficult in practice to gather reliable information. The number of people killed in a conflict is a central piece of information for many projects, but it is notoriously hard to determine with certainty.[48] This is not to say that systematic data collections are inherently unreliable, only that they are not foolproof. Most researchers are aware of these problems, and the projects listed below appear to do an admirable job of collecting, coding and verifying the information they use.

III. The leading data sets

The 16 data sets described below are grouped according to whether they focus primarily on the patterns of conflict occurrence, the causes and processes of conflict, or conflict early warning. The categories are not exclusive and are intended only as a rough guide. This list is not comprehensive, but an attempt has been made to make it complete within certain parameters. Every data set is directly concerned with conflict, provides worldwide coverage, is publicly available in English and is widely judged to be reputable. A large number of data sets created for specific publications and made available by authors are not included here because they constitute data use rather than data collection.[49]

Patterns of conflict occurrence[50]

Correlates of War (COW and COW²)

Location: University of Michigan, Ann Arbor, Michigan, USA, and Pennsylvania State University, State College, Pennsylvania, USA.

Principal investigators: J. David Singer, University of Michigan; and Stuart Bremer, Pennsylvania State University.

Purpose: To promote and support the scientific study of the causes of war and the conditions of peace by collecting and processing large quantities of historical information in an attempt to identify and explain empirical regularities that lead to war.

---

[46] The KOSIMO project, e.g., defines a conflict as consisting of 'some duration' and 'magnitude'.

[47] Collier and Hoeffler (note 18).

[48] The 1999 NATO bombing in the Federal Republic of Yugoslavia is a case in point. It was a short-duration event in a small territory with a large number of independent observers. Nevertheless, a careful study could only estimate the number of people killed by all sides during the period as in the range of 7449–13 627 (with 95% confidence). Projects that use a single number are likely to adopt the study's best estimate of 10 500, but there is a small chance that it could be incorrect. American Bar Association and American Academy for the Advancement of Science, *Political Killings in Kosova/Kosovo, March–June 1999* (ABA Central and Eastern European Law Initiative: Washington, DC, 2000), pp. 7–8. The mortality in most conflicts has not been so exhaustively analysed.

[49] A notable source for research on conflicts that uses data extensively but is not a data-collection project is the World Bank project The Economics of Civil Wars, Crime and Violence, available at URL <http://www.worldbank.org/research/conflict/index.htm>.

[50] In the interest of accuracy, the wording of the descriptions of the purpose and dependent variables of each project is as close as possible to that of the project's own presentation. The information in this section was gathered from project responses to a questionnaire, supplemented with information from project Internet sites.

Interstate conflict is the special focus of the project, with emphasis on those conflicts that involve the threat, use or display of force. Intra-state and extra-systemic conflicts are also studied.

Current coverage: 1816–1997.

Dependent variables: (*a*) Interstate war is sustained combat between the regular military forces of two or more state members of the international system in which there is a total of at least 1000 battle-related fatalities. (*b*) Intra-state war is sustained armed combat between two armed forces within the boundaries of a state, in which there are at least 1000 battle-related fatalities per year. (*c*) Extra-systemic war is sustained armed combat between a state member of the international system and a non-system-member political entity outside its territorial boundaries, in which there are at least 1000 battle-related fatalities per year.

Availability: URL <http://cow2.la.psu.edu> provides access to the most recent data sets, code books, history and contact information. URL <http://www.umich.edu/~cowproj/> provides earlier data sets, code books, publications and contact information.

Conflict Data Project

Location: Department of Peace and Conflict Research, Uppsala University, Uppsala, Sweden.

Principal investigators: Peter Wallensteen and Margareta Sollenberg.

Purpose: To collect information on selected variables relating to armed conflict, primarily to be used in research on various aspects of the origins, dynamics and resolution of conflict. Data have been collected on a global and yearly basis since 1989.

Current coverage: 1989–2001. The project has recently collaborated with others to extend the coverage from 1946 to 1988.

Dependent variables: Armed conflict is a contested incompatibility that concerns government or territory or both, over which the use of armed force between the military forces of two parties results in battle-related deaths. At least one of the parties is the government of a state. (*a*) Minor armed conflict results in at least 25 deaths per year and fewer than 1000 deaths over the course of the conflict. (*b*) Intermediate armed conflict results in more than 1000 deaths during the course of the conflict, but fewer than 1000 in any given year. (*c*) War results in more than 1000 deaths in any given year.

Availability: URL <http://www.pcr.uu.se/research/data.htm> provides access to a data set, code book, definitions, summary table and contact information. The data set covering the extended period of 1946–2001 will be available in late 2002 on the Internet site of the International Peace Research Institute, Oslo (PRIO) at URL <http://www.prio.no>.

Conflict Simulation Model (Konflikt-Simulations-Modell, KOSIMO)

Location: Heidelberg Institute of International Conflict, University of Heidelberg, Heidelberg, Germany.

Principal investigator: Frank R. Pfetsch.

Purpose: To provide a searchable database of political conflicts including crises, wars, insurrections, negotiation, mediation and peace settlements.

Current coverage: 1945–99.

Dependent variables: Political conflict is defined as the clashing of overlapping interests around national values and issues between at least two parties, at least one of which is the organized state. The conflict has to be of 'some duration' and 'magnitude'. The intensity ranges from 'latent conflict' to 'non-violent crisis' to 'violent crisis' to 'war'. Possible instruments used in the course of a conflict are negotiations, authoritative decisions, threat, pressure, passive or active withdrawals, or the use of physical violence.

Availability: URL <http://www.hiik.de/en/kosimo/kosimo.htm> provides access to a searchable database, summary graphs, publications and contact information.

Major Episodes of Political Violence (MEPV)

Location: Center for Systemic Peace, University of Maryland, College Park, Maryland, USA.

Principal investigator: Monty G. Marshall.

Purpose: To list all episodes of major political violence of any type. Categories include all forms of interstate, intra-state and inter-communal warfare. This data set is one of six data sets that comprise the Armed Conflict and Intervention Project at the Center for Systemic Peace and the Center for International Development and Conflict Management (CIDCM), University of Maryland. The larger project attempts to capture the deeper qualities and complexities of violent social conflict. It collects global information on the security context, membership of international organizations, displaced populations, direct military interventions, political interactions and bilateral trade flows.

Current coverage: 1946–2000.

Dependent variables: Major episodes of political violence involve the systematic use of lethal violence and terror by organized groups and/or states that substantially affect the society or societies that directly experience the armed conflict (resulting in at least 500 directly related fatalities, substantial destruction of infrastructure and population displacements). Episodes may involve states, a state and non-state group, or non-state groups only, including interstate and independence war, ethnic and revolutionary (civil) war, inter-communal warfare, genocide and communal massacres. Each episode is rated on a 10-point scale according to its total impact on the society or societies that are directly affected by the violence.

Availability: URL <http://members.aol.com/CSPmgm/warlist.htm> provides access to a data set, code book, summary table and contact information. In addition to the MEPV data set, the Armed Conflict and Intervention Project provides access to data sets on membership of international organizations and displaced populations. Data sets on military interventions, political interactions and bilateral trade flows are scheduled to be made publicly available in the future.

Causes and processes of conflict

International Crisis Behavior Project

Location: University of Maryland, College Park, Maryland, USA and McGill University, Montreal, Quebec, Canada.

Principal investigators: Michael Brecher, McGill University, and Jonathan Wilkenfeld, University of Maryland.

Purpose: To investigate 20th century interstate crises and the behaviour of states under externally generated stress. The data describe the sources, processes and outcomes of all military–security crises involving states.

Current coverage: 22 December 1917 to 31 December 1994.

Dependent variables: Part 1: All international crises occurring during the coverage period, characterized by: (*a*) a distortion in the type and an increase in the intensity of disruptive interactions between two or more adversaries, with an accompanying high probability of military hostilities or, during a war, an adverse change in the military balance; and (*b*) a challenge to the existing structure of an international system—global, dominant or sub-system—posed by the higher-than-normal conflictual interactions. Part 2: All foreign policy crises experienced by states as a result of their involvement in the international crises defined above. A foreign policy crisis is defined as a situation in which three conditions, deriving from a change in a state's external or internal environment, are perceived by the highest-level decision makers of the state: (*a*) a threat to basic values, (*b*) an awareness of finite time for response to the external threat to basic values, and (*c*) a high probability of involvement in military hostilities.

Availability: URL <http://www.missouri.edu/~polsjjh/ICB> provides access to data sets, code books, summary tables, papers and contact information.

Correlates of War–Militarized Interstate Disputes (MID 3)

Location: Pennsylvania State University, State College, Pennsylvania, USA.

Principal investigators: Stuart Bremer, Jim Ray, Dan Geller, Paul Diehl, Doug Gibler, Paul Hensel, Chuck Gochman, Glenn Palmer, Brian Pollins, Ric Stoll, Pat Regan and Zeev Maoz.

Purpose: To identify for all militarized interstate disputes the participants, start and end dates, fatality totals, hostility levels, revision sought, outcome and method of settlement.

Current coverage: 1816–1992. Currently being updated to 2001.

Dependent variables: A militarized interstate dispute involves the threat, display or use of force short of war by one member state, explicitly directed towards the government, official representatives, official forces, property or territory of another state. The outcome variable is recorded on a five-point ordinal scale ranging from non-reciprocated action, to the threat, display or use of force, to interstate war.

Availability: Data for 1816–1992 are available at URL <http://pss.la.psu.edu/MID_DATA.HTM>. Update to 2001 available from late 2002 at URL <http://mid3.la.psu.edu/>; this site provides access to papers, operational definitions and coding procedures (in the paper by Daniel M. Jones, Stuart A. Bremer and J. David Singer), progress reports, a discussion forum and contact information.

Behavioral Correlates of War (BCOW)

Location: Middlebury College, Middlebury, Vermont, USA.

Principal investigator: Russell Leng.

Purpose: To analyse the behaviour of states engaged in potential pre-war disputes. Data on 47 cases focus on crisis dynamics by generating descriptive data on the actions and interactions of states. The cases are selected as a representative sample of crises since 1816 and they permit the testing of a number of theories of interstate crisis behaviour.

Current coverage: 1838–1980.

Dependent variables: Interstate crises in which the principal protagonist on each side is a member of the interstate system. A crisis is an event on the continuum of belligerence that extends from a simple dispute, to a militarized dispute, to a crisis, to war. A crisis is a militarized dispute that requires protracted bargaining, defined as when there are at least 50 exchanges between the two major participants.

Availability: URL <http://community.middlebury.edu/~leng> provides access to the project description, data sets, software needed for working with the data, a users' manual and contact information.

Issue Correlates of War (ICOW)

Location: Florida State University, Tallahassee, Florida, USA.
Principal investigators: Paul Hensel and Sara McLaughlin Mitchell.
Purpose: To collect systematic data on contentious issues between states, with a focus on identifying the issues regardless of any particular action that may or may not have been taken to resolve them. Presently covering territorial, riverine and maritime claims.
Current coverage: 1816–2000.
Dependent variables: Contentious issues which involve explicit statements of disagreement by official governmental representatives of at least two states. ICOW identifies each issue with reference to the involved states, the object of the claim (such as the specific river or territory) and the time frame over which it endures. ICOW then collects data on the salience of each issue and on attempts to settle it through peaceful bilateral negotiations, binding or non-binding third-party activity or militarized conflict.
Availability: URL <http://www.icow.org> provides access to the project description, data sets, code books, publications and contact information.

Rivalry Data Set

Location: Department of Political Science, University of Illinois, Urbana, Illinois, USA.
Principal investigators: Paul Diehl and Gary Goertz.
Purpose: This data set provides a comprehensive overview of 1166 rivalries, 63 of which are enduring. The data set is designed to provide the basis for the analysis of the initiation, dynamics and termination of international rivalries.
Current coverage: 1816–1992.
Dependent variables: Rivalry is defined by the frequency of militarized interstate disputes between the same pair of states. The existence of a militarized rivalry is indicated by the occurrence of militarized disputes as defined by the COW–MID data set. Disputes which occur within 10–15 years of each other are considered to be part of the same rivalry. A dispute is considered part of the same rivalry if it involves the same two states and occurs within 11 years of the first dispute of the sequence, within 12 years of the second dispute of the sequence, and within up to 15 years of the fifth dispute in the sequence.
Availability: URL <http://www.pol.uiuc.edu/faculty/Diehl/diehl3lnk.htm> provides access to data sets, explanatory information and contact information.

Internal Wars and Failures of Governance: State Failure Data Set

Location: Center for International Development and Conflict Management (CIDCM), University of Maryland, College Park, Maryland, USA.

Principal investigators: Monty G. Marshall, Ted Robert Gurr, Jack A. Goldstone and Barbara Harff.

Purpose: To provide the dependent variable (state failure) in the US Government-sponsored State Failure Task Force quantitative analyses of structural indicators of failure, the purpose of which is to create 'early warning' (two-year) models of state failure situations. Independent variables used in published Task Force analyses are being prepared for public release and should be available soon to complement the dependent variable data on state failures.

Current coverage: 1955–2000.

Dependent variables: The data set includes all cases of internal wars (ethnic war, revolutionary war, or genocide–politicide) and failures of governance (substantial reversion to more autocratic rule or collapse of central authority) that began between 1955 and 2000 in independent countries with populations greater than 500 000. A war is defined as an armed conflict involving state authorities and a challenger group that results in at least 1000 directly related deaths over the course of the episode and at least one year during which there were more than 100 directly related deaths. The case begins with the first year during which the 100-death threshold is reached and ends when deaths fall below that threshold for at least five years. An episode of genocide or politicide (politically motivated mass murder) is defined by the merits of the case (for instance, an established intent to eliminate non-combatant group members). A failure of governance is defined generally as a six-point decrease in the state's Polity IV regime score (that is, towards greater institutional autocracy) or a Polity IV 'interregnum' (a collapse of central regime authority through failure, revolution or involuntary state disintegration).

Availability: URL <http://www.bsos.umd.edu/cidcm/inscr/> provides access to data sets, code books, summary tables and contact information.

Minorities at Risk

Location: Center for International Development and Conflict Management (CIDCM), University of Maryland, College Park, Maryland, USA.

Principal investigators: Ted Robert Gurr, Monty G. Marshall and Christian Davenport.

Purpose: To monitor and analyse the status and conflicts of politically active ethno-political groups in countries with a population of at least 500 000. Coverage of 275 contemporary and 65 historical groups.

Current coverage: 1945–2000.

Dependent variables: Ethno-political groups, defined as communal groups that: (*a*) are disadvantaged by comparison with other groups in their societies, usually because of discriminatory practices, or (*b*) have organized politically to promote or defend their collective interests. Only ethno-political groups with populations greater than 100 000 or 1 per cent of the population are included.

Availability: URL <http://www.bsos.umd.edu/cidcm/mar/> provides access to data sets (registration required), a code book, publications, project history and contact information.

Violent Intrastate Nationalist Conflicts (VINC)

Location: University of Indianapolis, Indianapolis, Indiana, USA.
Principal investigator: Bill Ayres.
Purpose: To measure and study conflict outcomes in violent, intra-state nationalist conflicts—those involving ethnic and other forms of secessionism—and their antecedents and correlates. It seeks to mark starting and ending points of these conflicts, to measure the characteristics and behaviour of the actors in them, and to answer questions about how and why conflicts end the way they do.
Current coverage: 1945–96.
Dependent variables: Each conflict episode is coded for the highest level of violence or rebellion reached during that episode, using a seven-point ordinal scale adapted from the Minorities at Risk data set project. Each conflict episode is coded for estimated number of deaths caused by that conflict (rounded to the nearest 100 for estimates less than 10 000, rounded to the nearest 1000 for estimates greater than 10 000). Conflict ends when both sides are no longer either fighting or talking with each other about what the solution to the conflict should be. Different episodes of the same conflict must be separated by a 12-month lull in both fighting and negotiating. The data also specify four types of ending and four types of agreement.
Availability: URL <http://facstaff.uindy.edu/~bayres/vinc.htm> provides access to data sets, a code book, summary tables, papers and contact information.

Third Party Interventions in Intrastate Conflict

Location: University of Binghamton, Binghamton, New York, USA.
Principal investigator: Patrick M. Regan.
Purpose: To identify all military, economic and diplomatic interventions in civil conflicts, primarily to help determine the relationship between military, economic and diplomatic interventions and the duration of conflict. Current coverage includes only military and economic interventions.
Current coverage: 1945–99.
Dependent variables: Third-party interventions in intra-state conflicts are convention-breaking military and/or economic activities in the internal affairs of a foreign country targeted at the authority structures of the government with the aim of affecting the balance of power between the government and opposition forces. Intrastate conflicts are organized military hostilities between two groups in conflict in which there were at least 200 fatalities over the course of the conflict.
Availability: URL <http://bingweb.binghamton.edu/~pregan/> provides access to a data set, user's manual and contact information.

Costs of conflict and conflict early warning

Protocol for the Assessment of Nonviolent Direct Action (PANDA)

Location: Program on Nonviolent Sanctions and Cultural Survival (PONSACS), Harvard University, Cambridge, Massachusetts, USA.
Principal investigators: Doug Bond, Joe Bond, J. Craig Jenkins, Churl Oh and Charles Louis Taylor.

Purpose: PANDA is an automated early warning system that is combined with on-the-ground research of conflict regions provided by anthropological insights. These two strands of research at PONSACS work to identify conflict regions before they erupt into violence and to actively promote non-violent alternatives to armed conflict. The project's premise is that by monitoring and examining interaction events with a 'data lens' that is sensitive to non-violent direct action, it can track and compare the evolution of conflict manifest in both violent and non-violent behaviour. The project also seeks to help make the costs of conflict transparent by providing a longitudinal series of social, political and economic events gleaned from news reports, and to facilitate independent testing and peer review. The PANDA software protocol for parsing news stories has been superseded by the Integrated Data for Events Analysis (IDEA) protocol.

Current coverage: 1991–2000.

Dependent variables: The basic parameters of the data include the source actor and target actor of social, political and economic events, the events themselves, as well as their date, location and a selection of attributes of the same. In more common terms, each data record represents the 'who does what to/with whom, when, where, why and how' of an event reported in the news. Any of the events data variables may be treated variously as an independent or dependent variable, depending upon the specific research questions being asked.

Availability: The code book, data files, protocol files, publications and contact information are available at URL <http://www.wcfia.harvard.edu/ponsacs/panda.htm>.

The Global Event-Data System (GEDS)

Location: Center for International Development and Conflict Management (CIDCM), University of Maryland, College Park, Maryland, USA.

Principal investigator: John Davies.

Purpose: To allow computer-assisted identification, narrative description and analytical coding of daily international and intra-national events, describing the day-to-day actions of all states and the major non-state communities and international organizations. GEDS includes and expands on Edward Azar's Conflict and Peace Data Bank (COPDAB, covering 1948–78), updating it for selected countries. Near-real-time tracking, as needed for early warning, can be generated on request using COPDAB scales or more specific 'accelerator' models for anticipating ethno-political conflicts, genocides or politicides.

Current coverage: 1948 to the mid-1990s.

Dependent variables: Events are operationally defined as reports from reputable sources which specify who did or said what to whom, when and where. Conflict and cooperation are operationally defined and coded using Azar's 15-point COPDAB scales either as categorical variables, as ordinal variables, or as ratio variables.

Availability: URL <http://geds.umd.edu/geds/> provides access to data sets, code book and contact information. The software is currently off-line. Updated data for some countries can be generated on request (at cost). This can include near-real-time tracking.

The Kansas Events Data System (KEDS)

Location: University of Kansas, Kansas City, Kansas, USA.
Principal investigator: Philip A. Schrodt.
Purpose: To generate political event data through automated coding of English-language news reports. These data are used in statistical early-warning models to predict political change. Building on the World Event/Interaction Survey (WEIS) Project, the KEDS project has three major research concentrations: software development for the machine-coding of political event data, production of events data sets and development of early-warning methods.
Current coverage: 1979–99.
Dependent variables: Event data are nominal or ordinal codes recording the reported interactions between international actors at specific points in time.
Availability: URL <http://www.ku.edu/~keds/> provides access to software, data sets, papers and contact information.

IV. Conclusions

Caveat emptor—let the user beware. In an ironic twist on the presumption of objectivity that underlies these quantitative research projects, the diversity of systematic data collection appears to support the constructivist argument that reality lies in the eye of the beholder.[51] To what extent do the data problems reviewed here affect researchers' ability to do good work? The core issue is the balance between reliability and validity, that is, between accuracy in recording information and appropriateness of the information for addressing theoretical concepts of interest. The balance confronts both quantitative and qualitative attempts to simplify the world in order to understand it and elicits different types of solutions from different types of researcher. Quantitative researchers place primary importance on reliability. To fulfil the requirement of systematically recording a series of events in a consistent manner (reliability), conflict data projects need to delimit complex phenomena through definitions and coding rules. In the process, they limit the range of their validity.

The problem of limited validity is partially resolved by the wide variety of data-collection projects that now exist. The primary purpose of this appendix is to point researchers towards the data sets that are appropriate for the questions they seek to answer. Some projects provide data that are appropriate for studies of the patterns of conflict occurrence and the structural features of the international system and its members that make violent conflict more or less likely. Other projects operate at a different level of analysis by focusing on the issues at stake and the comparative characteristics of the antagonists. The data produced allow for the development and testing of theories on the processes of conflict initiation, sustainment and resolution. A third type of project allows for analysis of foreign policy interactions by providing detailed information on events over time. These data-collection projects offer researchers a vast array of good data with which to develop academic theories and policy-related arguments.

---

[51] Eberwein and Chojnacki (note 21).

2. Conflict prevention

RENATA DWAN*

I. Introduction

The prevention of violent conflict, as an issue of international concern, is a relatively new item on the agenda of multilateral forums. Discussion of the concept of conflict prevention since the mid-1990s has focused on the desirability and feasibility of international preventive action. In 2001, however, the United Nations (UN) and the European Union (EU) attempted to move conflict prevention from concept to practice by charting new directions. In somewhat similar processes both the UN and the EU set out frameworks for the principles of conflict prevention of their member states, reviewed existing preventive tools within their organizations, recommended institutional changes to improve and broaden the scope of these instruments, and proposed strategies for intra- and inter-organizational coordination to facilitate the effective implementation of prevention. The comprehensiveness of these reports, the high level at which they were considered and the policies they can potentially lead to mark a coming of age for conflict prevention as a norm in international politics.

Section II of this chapter traces the rise of conflict prevention as an international priority issue. Section III examines the UN and EU reports on prevention and the substance of their recommendations. In section IV two cases are examined—West Africa and Zimbabwe—in which the UN and the EU have attempted to adopt and implement some of the proposals for preventive action contained in their respective documents. Section V addresses the effect of the international response to the terrorist attacks of 11 September 2001 on the future development of conflict prevention.

Appendix 2A reviews the multilateral peace missions of 2001 and contains a comprehensive table of data on these missions.

II. The rise of conflict prevention

The release of major reports on conflict prevention by the UN and the EU in 2001 marked the culmination of efforts under way since the mid-1990s to establish conflict prevention as a priority issue within the international community of states and in multilateral and non-state organizations.[1] This has come about for a number of related reasons.

---

[1] United Nations, Prevention of armed conflict, Report of the Secretary-General, UN document A/55/985, S/2001/574, 7 June 2001.

* Sharon Wiharta assisted in the research for this chapter.

First, the end of the cold war changed perceptions about the incidence of conflict. The termination of superpower rivalry increased awareness of the prevalence of intra-state conflict around the world. Some such conflicts had been seen as consequences of East–West confrontation, while others had been 'contained' by a mixture of superpower sponsorship and coercion.

Second, the reduction of East–West tension has provided a more conducive climate for consideration of the causes of conflict. Within academic and policy-making circles there has been a shift away from the study of the dynamics of war to the wider cycle of conflict: the structural and short-term causes of conflict, the conditions in which disputes become violent, the effects of conflict on individuals and societies, the processes of conflict resolution and the substance of peace-building. This new discourse was reflected at the UN Millennium Summit in September 2000 when, for the first time, the Security Council formally acknowledged economic, social, cultural and humanitarian grievances as root causes of armed conflict.[2]

Conflict prevention can be defined as political, economic or military actions taken by third parties to keep inter- or intra-state tensions and disputes from escalating into violence. This definition includes action taken in post-conflict situations to avoid a recurrence of violence, strengthen capabilities for peaceful resolution of disputes and alleviate the underlying problems producing them.[3] The breadth of this definition makes it an issue of wide potential engagement: multilateral forums that previously had little engagement in peace and security issues, such as the Group of Eight (G8), have begun to address conflict prevention as a policy concern.[4]

Third, influential state and non-state actors have exerted concerted pressure to put the issue on the international agenda. A range of peace- and development-focused non-governmental organizations (NGOs) have lobbied governments in Europe and North America to adopt more comprehensive approaches to conflict and its prevention, the most influential of which was the Carnegie Commission on Preventing Deadly Conflict.[5] Its study on the causes of conflict and the requirements of a functional system for prevention, led by an eminent international board, provided a widely quoted assessment of the costs of conflict. It estimated that the international community spent approximately $200 billion on seven major military interventions in the 1990s and calculated that preventive action in each case would have saved the inter-

[2] United Nations Security Council Resolution 1318, 7 Sep. 2000; see also Dwan, R., 'Armed conflict prevention, management and resolution', *SIPRI Yearbook 2001: Armaments, Disarmament and International Security* (Oxford University Press: Oxford, 2001), pp. 70–77.

[3] For a discussion of the concept of conflict prevention see Lund, M. S., *Preventing Violent Conflicts: A Strategy for Preventive Diplomacy* (United States Institute of Peace Press: Washington, DC, 1996); and Lund, M. and Rasamoelina, G. (eds), *The Impact of Conflict Prevention Policy, Conflict Prevention Network (CPN) Yearbook 1999/2000* (Nomos Verlag: Baden-Baden, 2000).

[4] 'G8 Roma initiatives on conflict prevention', Conclusions of the meeting of the G8 Foreign Ministers, Rome, 18–19 July 2001, Attachment 2. For the members of the G-8 see the glossary in this volume.

[5] Other leading NGOs active in promoting conflict prevention as a policy include the European Platform for Conflict Prevention and Transformation, the Forum on Early Warning and Early Response, the International Crisis Group, International Alert and Saferworld.

national community almost $130 billion.⁶ The commitment of the UN Secretariat to the cause of conflict prevention has helped raise the profile of such reports. UN Secretary-General Kofi Annan assumed his position with a commitment to move the UN from a 'culture of reaction to a culture of prevention'. Under his leadership the UN Secretariat has reformed its internal structures to incorporate conflict prevention as an area of interdepartmental concern and taken external initiatives to develop partnerships with regional organizations, the private sector and civil society. This new approach has encouraged critical examination of the UN's own weaknesses, most notably in the publication of two reports on the consequences of the failure to adopt a preventive approach in the conflicts in Srebrenica (Bosnia and Herzegovina) and Rwanda.⁷

Annan's success in orienting the UN towards prevention has hinged, ultimately, on the support of the member states. States such as Canada, Denmark, Netherlands, Norway, Sweden and, increasingly, Japan and the UK have undertaken a comprehensive review of the relationship between development, governance and conflict. As a result, they have revised their national development aid policies to give greater emphasis to the political context of aid delivery.⁸ At the multilateral level, these states have driven the preventive agenda in the forums in which they participate and have provided crucial back-up to the initiatives of NGOs and international organizations to establish conflict prevention as a norm in international relations.⁹ Sweden, for example, made conflict prevention a key priority of its presidency of the EU in the spring of 2001.¹⁰ Non-European regional organizations with a peace and security mandate, such as the Organization of American States (OAS), the Organization of African Unity (OAU)¹¹ and the Economic Community of West African States (ECOWAS), have also taken steps to develop institutional capacity for conflict prevention, with the active encouragement of this international constituency. The EU, for example, has provided funds for the establishment in 1999 of the ECOWAS Mechanism for Conflict Prevention,

---

⁶ These figures are from the study commissioned by the Carnegie Commission on Preventing Deadly Conflict: Brown, M. and Rosecrance, R. (eds), *The Costs of Conflict: Prevention and Cure in the Global Arena* (Rowman & Littlefield: Lanham, Md., 1999).

⁷ United Nations, Report of the Secretary-General pursuant to General Assembly Resolution 53/55 (1998), Srebrenica report, UN document A/54/549, 15 Nov. 1999; and United Nations, Report of the independent inquiry into the actions of the United Nations during the 1994 genocide in Rwanda, UN document S/1999/1257, 15 Dec. 1999.

⁸ Most national development policies had hitherto focused largely on economic development and poverty reduction. The British Government has gone furthest in reorienting a substantial degree of its development assistance activities around conflict prevention. For more information see the Foreign Office Internet site, URL <http://www.fco.gov.uk/news/keythemepage.asp?pageid=254>.

⁹ Björkdahl, A., 'Conflict prevention from a Nordic perspective: putting prevention into practice', *International Peacekeeping*, vol. 6, no. 3 (autumn 1999), pp. 54–72; and Dwan, R., 'Institutionalising mainstreaming—a paradox?', eds L. van de Groor and M. Huber, *Mainstreaming Conflict Prevention: Concept and Practice, Conflict Prevention Network Yearbook 2000/01* (Nomos Verlag: Baden-Baden, 2002).

¹⁰ European Union, 'Programme of the Swedish Presidency of the European Union, 1 January to 30 June 2001', EU document SN5613/00, 14 Dec. 2000.

¹¹ The OAU member states adopted the Constitutive Act of the African Union on 11 July 2000; it entered into force on 26 May 2001, formally establishing the African Union (AU), with headquarters in Addis Ababa. The AU will replace the OAU after a transitional period.

Management, Resolution, Peacekeeping and Security, which includes regional early-warning offices as well as intergovernmental bodies.[12]

The combined effect of these efforts has been to encourage the detailing of the concept and practice of conflict prevention to facilitate a more concrete, forceful set of policies and activities. If prevention was in the past seen as an inoffensive concept to be used primarily as an interstate confidence-building measure, today's concept heralds a comprehensive new way of engaging in international politics. In practice, this makes it a far more controversial idea than its definition would suggest.

III. The UN and EU reports on conflict prevention in 2001

The active efforts of the conflict prevention constituency described in section II help explain the fact that the UN Secretary-General's report of 7 June 2001 was released just days before the European Council's adoption of the Programme for the Prevention of Violent Conflicts at its meeting in Gothenburg, Sweden, on 15–16 June. The UN Secretary-General's report, the EU Programme and the EU Commission's Communication of 11 April 2001 all differentiate between long-term (structural) prevention, which addresses the root causes of conflict, and short-term measures, which aim to prevent existing disputes from becoming violent. Moreover, all three embrace a comprehensive approach to the root causes of conflict and advocate that the bulk of international preventive effort be focused on long-term conflict prevention with an emphasis on sustainable development. They share a similar perspective: that a wide range of political, economic, social and military tools exist to address violent conflict and that both the UN and the EU are already engaged in a wide range of either indirect or deliberate preventive activities. The reports place emphasis on the need for greater coherence of current policies and better coordination between instruments and actors at both the intra- and the inter-institutional level. They represent, in essence, an attempt to create a new conflict prevention framework within which all current and future activities should take place.

The UN Secretary-General's report

The UN Secretary-General's report of 7 June 2001 was the result of the second Security Council discussion of conflict prevention, in July 2000, and of the decision to invite Kofi Annan to submit an analysis of and recommendations on the role of the UN system in prevention.[13] The length of time taken to

---

[12] Dwan, R., 'Armed conflict prevention, management and resolution', *SIPRI Yearbook 2000: Armaments, Disarmament and International Security* (Oxford University Press: Oxford, 2000), pp. 77–134; Dwan (note 2), pp. 82, 102–103; and Final Communiqué, EU–ECOWAS Ministerial Meeting, 16 Oct. 2000, EU Press Release 12309/00 (Presse 390).

[13] United Nations (note 1). Conflict prevention was formally addressed in a Security Council meeting for the first time in a 2-day debate on 29–30 Nov. 1999.

prepare the report and the fact that it was addressed to both the Security Council and the General Assembly illustrated the UN Secretariat's desire to avoid the negative reaction provoked in the General Assembly by the 2000 Report of the Panel on United Nations Peace Operations, known as the Brahimi Report after Lakhdar Brahimi, chair of the high-level panel established by the Secretary-General in March 2000.[14] Many states felt that the Assembly had not been adequately consulted in the preparation of the Secretary-General's report and that the finished product reflected primarily the views of Western states. The report went out of its way not to offend states that, in some cases, may be the most obvious candidates for international preventive action. The cautious tone demonstrates that consensus on conflict prevention remains far from universal.[15] The document, in its essence, is a claim for the legitimacy of conflict prevention as a focus of international concern.[16]

The Secretary-General's report builds its case on the premise that the primary responsibility for conflict prevention rests with national governments and that preventive action assists, rather than undermines, the national sovereignty of member states. The report argues that the UN Charter (Chapter 1, Article 2, paragraph 3) makes prevention an obligation for member states and that placing it at the centre of the international system is to bring the UN back to its core mandate. It is the potential threat of conflict prevention to the principle of non-interference in national affairs that makes the implementation of the idea so controversial for many states and the greatest obstacle to overcome among UN member states.

A key theme of the report is the link between conflict prevention and development. It stresses that an investment in national and international efforts for conflict prevention must be seen as a simultaneous investment in sustainable development since the latter can best take place in an environment of sustainable peace. This emphasis is intended to meet the concerns of some developing states that fear that a Western-led focus on prevention may result in diminished and/or more conditional development assistance.[17] It is also an argument to convince sceptical donor countries that conflict prevention is a core component of sustainable development assistance, even if the benefits of investing in conflict prevention may not be immediately evident.

The key word in the Secretary-General's review of the preventive role of the principal bodies in the UN system is 'coordination'. This not only is a consequence of the multidimensional nature of conflict but also reflects the wide range of existing preventive efforts already under way. The challenge is one of mobilization of collective potential rather than of introducing new instruments

---

[14] United Nations, Identical letters dated 21 Aug. 2000 from the Secretary-General to the President of the General Assembly and the President of the Security Council, UN document A/55/305, S/2000/809, 21 Aug. 2000.

[15] Dwan, R., 'Consensus: a challenge for conflict prevention?', SIPRI, *Preventing Violent Conflict: The Search for Political Will, Strategies and Effective Tools*, Report of the Krusenberg Seminar, Sep. 2000.

[16] This point is also noted by Griffin, M., 'A stitch in time: making the case for conflict prevention', *Security Dialogue*, vol. 32, no. 4 (2001), pp. 481–96.

[17] See the comments made during the General Assembly review of the Secretary-General's report, UN Press Release GA/9890, 12 July 2001.

and resources. This perspective underscores the reality facing the Secretary-General: that UN member states are unwilling to make a substantial financial outlay in support of reinforced prevention.[18]

Despite these broad approaches, the report does make a number of specific recommendations to improve UN preventive capacity. The Secretary-General commits himself to submitting periodic regional and sub-regional reports on conflict prevention to the Security Council and suggests that it establish an informal working group to discuss prevention cases on a continuing basis. The report also advocates increased dialogue on prevention between the Security Council and the General Assembly and recommends the creation of an open-ended group of states within the Assembly to facilitate this dialogue. It lauds the increase in the number of Security Council fact-finding missions and recommends their use for short-term preventive purposes. The UN Secretariat commits itself to more inter-agency technical assessment and confidence-building missions as well as to the creation of prevention task forces for specific regions. Another undertaking is the establishment of an informal group of eminent persons to advise the Secretary-General on prevention.

Internal institutional changes have been under way since 1998 to incorporate a preventive focus into decision making and to increase coordination within the UN system.[19] The Department of Political Affairs (DPA) has been designated the focal point for UN prevention and has established an internal conflict prevention team. It convenes the Executive Committee on Peace and Security, which addresses system-wide preventive action. The Interdepartmental Framework for Coordination was reoriented towards early-warning and preventive action in 1998. Its 14 members (which include UN agencies and the World Bank) meet monthly to exchange information and assess potential conflict and complex emergencies. The UN staff also participate in training workshops on early warning and prevention in a course administered by the International Labour Organization (ILO) International Training Centre in Turin, Italy. Measures which have been recommended include the establishment of a UN-wide unit for policy and analysis to improve the DPA's capacity to analyse and follow up on information received from desk officers as well as a staff dedicated to conflict prevention within the Secretariat.[20] The Secretary-General's report also recommends that conflict prevention activities be funded from the regular UN budget rather than from the Trust Fund for Preventive Action, which suffers from perennial financial shortfalls. This proposal reinforces the Secretary-General's argument that, although the Secretariat and agencies can do much to improve the UN's preventive capacity, the UN will fall far short of effective preventive action without the political will and active commitment of the member states.

---

[18] See the comments made during the Security Council's consideration of the Secretary-General's report, UN Press Release SC/7081, 21 June 2001.

[19] For more on conflict prevention mainstreaming see Björkdahl (note 9) and Dwan (note 9).

[20] The creation of this unit was first proposed in the Brahimi Report but this proposal has so far not received General Assembly support.

Security Council and General Assembly reactions

The Security Council held a one-day open debate on 21 June 2001 to discuss the Secretary-General's report.[21] Although the report was broadly welcomed, the lack of discussion of its substantive proposals illustrated the continued ambivalence of many member states towards conflict prevention. In such contexts, the breadth of the concept becomes a weakness, as the unfocused discussion demonstrated. It was not until 30 August that the Security Council adopted a resolution on conflict prevention, cast in the most general of terms.[22] Although the Security Council recognized the 10 principles proposed by the Secretary-General to place prevention at the core of the UN system, it did not adopt the substantive suggestions for an informal working group on prevention or for increased cooperation with the General Assembly. Rather, the Council expressed its willingness to give 'prompt consideration' to cases brought to its attention by the Secretary-General or a member state and its commitment to take 'early and effective action'. To the extent that the tone and content of the resolution almost mirrored the July 2000 statement of the President of the Security Council,[23] it must be seen as a disappointing outcome.

The General Assembly's reaction was even more limited and testified to the persistently controversial nature of active conflict prevention. Although member states were mollified by the emphasis on the Assembly's role in creating a culture of prevention, the final resolution, adopted on 13 August 2001, did nothing more than draw the attention of states, regional organizations and civil society to the report and request them to consider its recommendations. It also called on UN bodies to undertake a similar process and report their views to the Assembly during its next (56th) session.[24] The General Assembly's handling of the report on conflict prevention demonstrated that, for all its expressed desire to play a more significant role, the Assembly is unable to provide the coherence required for it to serve as a forum for substantive discussion or for collective action. This was amply illustrated in the debate on the report on 12–13 July 2001, which degenerated into a spat between Israel, Lebanon and Syria.[25]

The EU documents

The UN Secretary-General's view that regional organizations can contribute to conflict prevention in a number of specific ways would seem to be borne out by the EU Programme for the Prevention of Violent Conflicts.[26] When considered along with the European Commission Communication on Conflict

---

[21] United Nations Press Release (note 17).
[22] United Nations Security Council Resolution 1366, 30 Aug. 2001.
[23] United Nations, Statement by the President of the Security Council, UN document S/PRST/2000/25, 20 July 2000.
[24] United Nations, General Assembly Resolution 55/281, 13 Aug. 2001.
[25] United Nations Press Release GA/9893, 13 July 2001.
[26] European Union, EU Programme for the Prevention of Violent Conflicts, endorsed by the Gothenburg European Council, June 2001, available at URL <http://www.eu2001.se/static/eng/pdf/violent.PDF>; and European Commission, Communication from the Commission on Conflict Prevention, EU document COM (2001) 211, 11 Apr. 2001.

Prevention, the EU programme represents a ground-breaking step in the collective implementation of preventive action at the regional level and contrasts markedly with the general, basic level of the UN debate.

The Programme for the Prevention of Violent Conflicts

The elaboration of an EU agenda for conflict prevention was an innovation of the Swedish Government when it held the EU presidency from January to June 2001. Sweden's declaration of this goal well before assuming the presidency and its commitment to steering the draft through the European Council were important elements in the successful conclusion of an EU agreement at the summit meeting in Gothenburg in June.[27] The political ground had been set by the European Council's decision at its meeting in Cologne, Germany, in 1999 to develop the EU's capacity to take decisions 'on the full range of conflict prevention and crisis management tasks defined in the Treaty on European Union' (the Petersberg tasks).[28] Member states claimed that the international community has a political and moral responsibility to act to avoid violent conflicts, which the EU, itself a successful example of conflict prevention, cannot ignore.

The Programme for the Prevention of Violent Conflicts represents a commitment by the EU heads of state to establish conflict prevention as a priority for EU external action. The specific elements include a broad consideration of potential conflict issues at the outset of every presidency (i.e., every six months) to identify priority areas and regions for EU preventive action. This is intended to help set coherent preventive objectives and strategies for the EU, the implementation of which is to be monitored by the Council of the European Union. The EU heads of state, in marked contrast to the UN, emphasized early warning and analysis, and assigned specific EU bodies responsibility for the provision of regular information on potential conflict situations through standardized formats and reporting methods. The programme also commits the EU member states and the European Commission to enhancing the EU's short- and long-term conflict prevention tools and stresses the need for 'partnerships for prevention' with the UN, regional organizations and civil society. In this light it declares its intent to intensify information exchange with other institutions and suggests that joint training programmes in conflict prevention be developed for field and headquarters personnel of the EU, the UN and the Organization for Security and Co-operation in Europe (OSCE). Funding for training support would come from the Commission: the programme makes no commitment to increase funding for prevention policies. Member states are encouraged to develop national action plans to increase their conflict prevention capacity and to assist in bringing prevention into all the relevant EU institutions. The 15 EU member

[27] See, e.g., Swedish Ministry for Foreign Affairs, 'Preventing violent conflict: Swedish policy for the 21st century', Government Communication 2000/01:2, May 2001.
[28] European Council, Declaration of the European Council on strengthening the common European policy on security and defence, Cologne, 3–4 June 1999; see also Rotfeld, A. D., 'Europe: the new transatlantic agenda', *SIPRI Yearbook 2000* (note 12), pp. 196–98.

states agreed that the first progress report on the implementation of the programme should be submitted to the European Council in June 2002.

Although the programme is not a detailed document, it does provide a road map for a comprehensive EU preventive approach. This was immediately evident when the subsequent Belgian presidency fulfilled the new commitment to hold a broad discussion of priorities at the outset of each presidency. At the meeting on 16 July 2001 the Council noted the intention of both the Commission and the Council Secretariat, specifically the Policy Planning and Early Warning Unit (Policy Unit), to present more detailed regional/sub-regional reports on ongoing or emerging conflict issues to the EU's Political and Security Committee (PSC).[29] The fact that the PSC meets on a weekly basis makes it a more suitable forum for monitoring prevention policies than the monthly meetings of foreign ministers in the General Affairs Council (GAC).

The Commission Communication on Conflict Prevention

The most comprehensive review of European Community policies and tools related to prevention, together with substantive recommendations for future EU conflict prevention policy, is contained in the Commission's Communication on Conflict Prevention, presented in April 2001.[30] This was a follow-up to the joint paper on conflict prevention presented by the Commission and Javier Solana, the Secretary General of the Council of the European Union and High Representative for the Common Foreign and Security Policy (CFSP), to the Nice meeting of the European Council in December 2000 and represented the Commission's contribution to the EU Programme.[31] The Commission addressed the wide range of external policies that fall under Community competency, arguing that the main target of EU efforts should be long-term prevention (defined as 'projecting stability').

As the world's largest aid donor, the EU uses its development policy and cooperation programmes as its most powerful preventive tools. The incorporation of preventive perspectives and the systematic coordination of Community instruments to implement them are to be achieved through the elaboration of Country Strategy Papers (CSPs) for each recipient of European Community aid. These papers will use agreed indicators to analyse potential conflict situations and, where conflict risk factors are identified, prevention measures will be integrated into Community programmes. The Commission also noted its intention to address more comprehensively in its support programmes democratic governance issues such as electoral processes, parliamentary activities, the rule of law and security sector reform.[32] The increased use of political dialogue, in which the EU engages with all partner countries, should also be considered for more short-term preventive action. This, the Commis-

---

[29] EU Press Release 10609/01 (Presse 282), 16 July 2001.

[30] Communication from the Commission on Conflict Prevention (note 26).

[31] European Union, Council of the European Union, Report by the Secretary General/High Representative and the Commission containing practical recommendations for improving the coherence and effectiveness of EU action in the field of conflict prevention, Nice, 7–9 Dec. 2000; and EU Press Release IP/01/560, 11 Apr. 2001.

[32] For a discussion of security sector reform see chapter 4 in this volume.

sion underscored, requires that member states take a common political line on the situation in question. A procedure for political dialogue with partner states on contentious issues of concern to the EU, established in the 2000 Cotonou Agreement, has already been put in place within the EU framework for development assistance and trade relations with African, Caribbean and Pacific (ACP) countries.[33]

Coordination among actors was a second key theme of the Commission Communication, a sensitive issue within the complex governance structure of the EU. At the intra-EU level, the Commission emphasized the need for increased coordination with member states and its intention to exchange CSPs with corresponding national documents (e.g., on bilateral development aid programmes). A pilot system for information exchange has also been set up between the Commission, the Council Policy Unit and member state desk officers for two areas—the Balkans and the African Great Lakes Region. Increased cooperation on short-term preventive action between the Commission and the CFSP structures in the Council Secretariat, especially the High Representative, was also stressed. These include measures such as regular reviews of potential conflict zones, including the establishment of early-warning mechanisms, a coordinated approach to the use of political dialogue and more use of EU Special Representatives. Coordination with other international actors included proposals for integrating discussion on early-warning and monitoring systems into the political dialogue with partner countries, structured dialogue with the UN on conflict prevention, the exchange of documents, common staff training programmes in prevention with the OSCE and the Council of Europe, and co-financing of the funding instruments of the World Bank and the International Monetary Fund (IMF).

Within the Commission, structures have been established to help implement the new approach. The Directorate-General for External Relations created a unit for conflict prevention, crisis management and African, Caribbean and Pacific issues in late 2000; it is responsible for coordinating the Commission's prevention activities and liaising with the Council Secretariat on early-warning and crisis prevention strategies. An Inter-service Quality Support Group has also been established to review CSPs in order to ensure that cross-cutting issues, such as prevention, are incorporated in them. Guides are also being developed (e.g., a Conflict Prevention Handbook) to help desk officers in identifying and developing projects with preventive measures. A Rapid Reaction Mechanism has been set up to enable the release of funds for short-term crisis reaction activities.[34]

The EU Programme for Prevention and the Commission's Communication point to the role a regional organization can play in activating conflict prevention. The smaller membership size and common cultural/political perspectives

---

[33] The Cotonou Agreement is a partnership agreement between the members of the African, Caribbean and Pacific Group of States of the one part and the European Community and its member states of the other part, signed in Cotonou, Benin, on 23 June 2000. The Cotonou Agreement is also called the ACP–EC Partnership Agreement. EU document ACP/CE/EN, available at URL <http://europa.eu.int/comm/development/cotonou/agreement_en.htm>.

[34] European Commission Press Release IP/01/255, 26 Feb. 2001.

of regional organizations facilitate consensus, which in turn can help states take a common position on issues of preventive action. A regional organization's capacity for early warning and monitoring may be enriched by greater knowledge of and contacts in its own neighbourhood. The high degree of integration in the EU may further facilitate information exchange and intelligence sharing between member states on vulnerable situations. However, the EU is an atypical regional organization in its wealth and other resources, degree of integration and extra-regional reach. EU conflict prevention, therefore, is unlikely to provide an easily adaptable model for other regional actors.

The EU's new conflict prevention goals pose significant challenges to the organization, whose external profile has often been less than the sum of its parts. Its complex structure and multiple components provide serious obstacles to either swift or effective action on a cross-cutting issue such as prevention. Successful implementation depends on political leadership from the Council and follow-up from successive EU presidencies, as well as the development of active information and coordination frameworks between member states, the Commission and the Council, to develop real policy coherence.[35] It also requires the commitment of financial resources. This latter issue was noted by the European Parliament in its review of the Commission Communication, which called for a wider political detbate in Europe and an increase in the budget for external actions.[36]

IV. Prevention in practice: West Africa and Zimbabwe

The UN and EU documents were intended as guidance to move conflict prevention from a rhetorical expression to a practical policy. In 2001 both organizations made concrete efforts to implement some of the commitments outlined in their respective reports. This section explores UN and EU preventive efforts with regard to two areas of identified vulnerability: the threat of all-out regional conflict in West Africa and the reality of domestic conflict in Zimbabwe.[37] In so doing, it illustrates some of the potential as well as the limitations of international preventive action.

Regional conflict in West Africa

West Africa has experienced some of the most severe and sustained conflicts in the post-cold war period, with devastating effects on national states and populations. The current regional instability began in 1989 with civil war in Liberia, a seven-year conflict that displaced 80 per cent of the 2.5 million population.[38] The conflict had spread to Sierra Leone by 1991, when rebels of

---

[35] Dwan, R., 'Conflict prevention and CFSP coherence', ed. A. Missiroli, *Coherence for European Security Policy*, Occasional Paper 27 (Western European Union, Institute for Security Studies: Brussels, May 2001).
[36] 'News in brief', *European Security Review*, no. 10 (Jan. 2002), p. 5.
[37] For a discussion of these conflicts see chapter 1 in this volume.
[38] United Nations Office for the Coordination of Humanitarian Affairs (OCHA), UN Consolidated Inter-Agency Appeal for West Africa 2001, 23 Mar. 2001, available at URL <http://www.reliefweb.int>.

the Revolutionary United Front (RUF) began making incursions across the Liberian border into Sierra Leone. The catastrophic consequences of the war in Sierra Leone included the displacement of at least 70 per cent of the 4.5 million population. This war was declared ended only in January 2002.[39] The influx of refugees into bordering Guinea led to serious political instability, violence against refugees and rebel attacks in Guinea. Tensions between the three countries of the Mano River Union (MRU)—Guinea, Liberia and Sierra Leone—increased dramatically in the second half of 2000 with fighting along the Guinean–Liberian and Guinean–Sierra Leonean borders.[40] By early 2001 Guinean and Liberian troops had been deployed to their joint border and fighting had spread across the border into Guinea.[41] Liberia also faces growing conflict within its borders as attacks by anti-government forces in northern Lofa County mount. Economic growth in the MRU states has been non-existent: Liberia and Sierra Leone lie, respectively, in 174th and 175th place on the UN Human Development Index for 2000, and Guinea is ranked the 8th least developed country in the world.[42]

Neighbouring Côte d'Ivoire, currently the most stable and economically developed country in the West African region, has undergone a period of domestic turbulence and increasing economic woes. Political tensions are also high in Guinea-Bissau, while fighting between separatist rebels and the government in the Cassamance Province of nearby Senegal threatens a peace agreement signed in March 2001.[43] With the possibility of complete regional implosion looking increasingly likely, the UN and the EU put new emphasis on prevention in West Africa in 2001. The sub-region (as it is described by the UN) presents a real challenge for preventive efforts.

First, it has multiple, interconnected conflicts at a variety of different stages, none of which can be addressed independently. Second, a wide variety of international strategies are required simultaneously—pre- and post-conflict prevention as well as conflict mediation, peacekeeping, conflict resolution and peace-building. Third, there are many different regional and extra-regional actors already involved in West Africa, making coordination a real challenge. Fourth, the UN and the EU have both been engaged in West Africa for some

---

[39] 'The president's speech', BBC News Online, 18 Jan. 2002, available at URL <http://news.bbc.co.uk/hi/english/world/>; see also Reno, W., 'War and the failure of peacekeeping in Sierra Leone', *SIPRI Yearbook 2001* (note 2), pp. 149–61.

[40] 'Guinea rounds up refugees', BBC News Online, 11 Sep. 2000, available at URL <http://news.bbc.co.uk/hi/english/world/>; and Seyni, B., 'After Liberia, Sierra Leone, is Guinea next?', Panafrican News Agency (PANA), 20 Oct. 2000, available at URL <http://allafrica.com/ stories/20010200417.html>.

[41] Agence France-Presse (AFP), 13 Jan. 2001, in 'Guinea and Liberia reinforce troops near common border', Foreign Broadcasting Information Service–*Daily Report, Sub-Saharan Africa (FBIS-AFR)*, FBIS-AFR-2001-0113, 16 Jan. 2001.

[42] UN OCHA (note 38); and UN Development Programme, *Human Development Report 2001: Making New Technologies Work for Human Development* (Oxford University Press: Oxford, 2001).

[43] Agence France-Presse (AFP), 23 Mar. 2001, in 'Senegalese government, rebels sign second pact in week', FBIS-AFR-2001-0323, 26 Mar. 2001; and 'Senegal deploys troops to Casamance', BBC News Online, 23 May 2001, available at URL <http://news.bbc.co.uk/hi/english/world>.

UN efforts

The need for an interlocking approach to the cycle of instability in West Africa was clear to UN peacekeepers deployed in Sierra Leone and to humanitarian aid officials working in the region long before tensions among Guinea, Liberia and Sierra Leone had reached a critical point.[44] The UN Security Council acknowledged the regional nature of the crisis following its mission to Sierra Leone in October 2000.[45] In December 2000 the Secretary-General established an Inter-Agency Task Force on West Africa under the coordination of the DPA. This task force, heralding the approach emphasized in the Secretary-General's report, was instructed to undertake a mission to take stock of sub-regional priority needs in security, humanitarian affairs and development, to consult with governments and ECOWAS on enhancing cooperation with the UN, to make recommendations for a sub-regional strategy to help address identified challenges and to propose how international support for such a strategy could be mobilized. The mission included representatives from the main UN agencies as well as a representative of ECOWAS. By the time the task force undertook its mission to the region, on 7–27 March 2001, the situation had deteriorated sufficiently to prompt it to warn of a possible 'domino effect' in the entire West African sub-region unless urgent political, economic and social progress was made.[46]

Concern for regional stability was the basis of the imposition of Security Council sanctions on Liberia on 7 March, effective from May 2001. Resolution 1343 bans the export of diamonds from Liberia and imposes an arms embargo on the country as well as travel restrictions on senior officials of the Liberian Government until Liberia ceases financial and military support to the RUF and expels the RUF from its territory.[47] The decision to impose sanctions on the regime of President Charles Taylor was, in turn, the result of a report on the violation of the ban on diamond exports from Sierra Leone, an inquiry that convincingly demonstrated how Liberian involvement had prolonged and intensified Sierra Leone's war.[48] A panel of experts was appointed by the Secretary-General on 29 March to monitor violations of the sanctions as well as Liberia's compliance with Resolution 1343. Its first report recommended the expansion of the arms embargo to state and non-state actors in all three

---

[44] The Mano River Union was established among the 3 countries in 1973 with the aim of increasing sub-regional economic integration.
[45] United Nations, Report of the Security Council Mission to Sierra Leone, UN document S/2001/992, 16 Oct. 2000.
[46] United Nations, 'Towards a comprehensive approach to durable and sustainable solutions to priority needs and challenges in West Africa', Report of the inter-agency mission to West Africa, UN document S/2001/434, 2 May 2001.
[47] UN Security Council Resolution 1343, 7 Mar. 2001.
[48] United Nations, Report of the Panel of Experts appointed pursuant to UN Security Council Resolution 1306, UN document S/2000/1195, 20 Dec. 2000. For more on sanctions see chapter 5 in this volume.

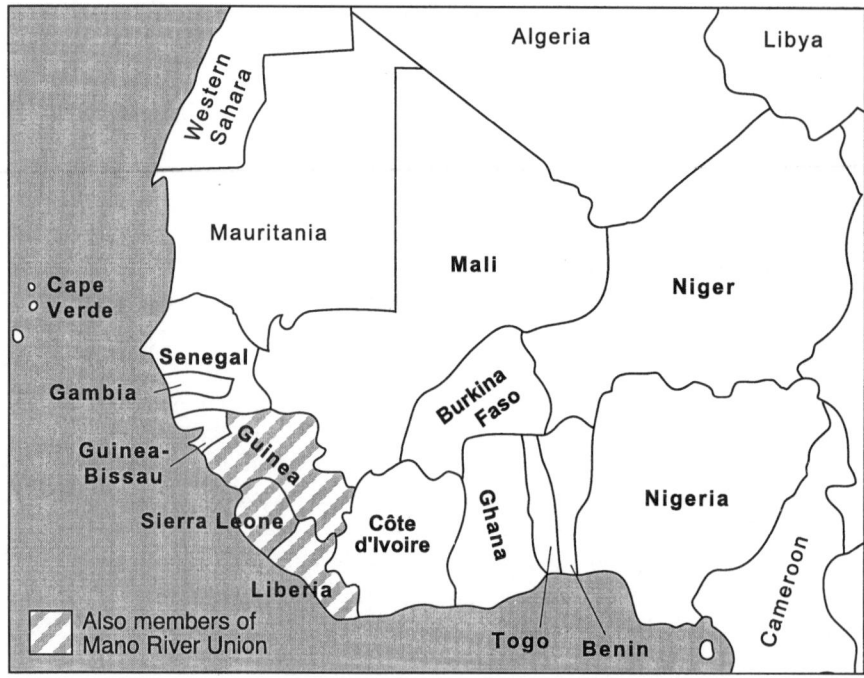

Figure 2.1. Map of West Africa: the 15 member states of ECOWAS

countries of the MRU.[49] ECOWAS, which had objected to the imposition of immediate sanctions and won the reprieve of two months for Liberia, also established a monitoring mechanism to assess Liberia's compliance with UN sanctions. In April 2001 the mission travelled to Liberia with delegates of the UN Sanctions Committee.[50]

West Africa came under Security Council consideration again on 14 May, when the report of the March Inter-Agency Task Force mission to West Africa was discussed.[51] The mission proposed a number of concrete steps to implement a regional strategy for peace and security issues, the most controversial of which was the expansion of the United Nations Mission in Sierra Leone (UNAMSIL) to cover all three MRU countries. Such a step would require a new mandate for UNAMSIL and was immediately rejected by the Security Council. The report made additional recommendations that mirrored the strategy marked out in the Secretary-General's report. These include the establishment of a UN Office for West Africa, to be headed by a Special Represen-

[49] United Nations, 'Expert Panel on Liberia presents report to Security Council with proposals for furthering peace in Mano River region', UN Press Release SC/7196, 5 Nov. 2001.

[50] 'ECOWAS reiterates stance against immediate sanctions on Liberia', ECOWAS Press Release, 18/2001, 16 Feb. 2001; and Final communiqué of the Extraordinary Summit of Heads of State and Government of ECOWAS, Abuja, 11 Apr. 2001, URL <http://www.ecowas.int/sitecedeao/english/final-com-11042001.htm>.

[51] United Nations, Letter dated 30 Apr. 2001 from the Secretary-General addressed to the President of the Security Council, UN document S/2001/434, 2 May 2001; and UN Press Release SC/7059, 14 May 2001.

tative of the Secretary-General, the strengthening of the presence of the UN Office for the Coordination of Humanitarian Affairs (OCHA) in the region, proposals for integrated UN agency programmes that address all aspects of the conflict as well as international donor conferences to mobilize financial support for two vulnerable West African countries, Guinea and Guinea-Bissau. On 26 November 2001 the Secretary-General informed the Security Council of his intention to establish the proposed regional office in Dakar, Senegal, from January 2002.[52] Its mandate is to enhance the coherence of UN activities in the region, to liaise with ECOWAS and other international actors, and to carry out good-offices roles on behalf of the Secretary-General, especially in the area of conflict prevention and peace-building.

The UN Security Council's lack of enthusiasm for extending UNAMSIL's duties to monitoring the Sierra Leonean–Liberian–Guinean borders may be comprehensible in the light of UNAMSIL's past problems and ongoing difficulties. However, it illustrates the general absence of will in the UN to undertake preventive deployment and, more specifically, new peace operations in Africa. It also reinforces the significance of ECOWAS as the UN's main interlocutor for peace and security issues in West Africa. The UN takeover of the former ECOWAS-led peace operation in Sierra Leone brought the two organizations into close, if sometimes difficult, contact, although their cooperation in Sierra Leone improved with the creation of the Coordinating Mechanism for Sierra Leone in 2000 between ECOWAS, UNAMSIL and the Sierra Leone Government.[53] In the case of border tensions among the MRU countries, ECOWAS continues to play the dominant mediating role. As early as November 2000 the organization dispatched a technical mission to investigate the border crises and, at its Bamako summit meeting one month later, approved the deployment of a 1700-strong ECOWAS Monitoring Group (ECOMOG) force along the common Guinean, Sierra Leonean and Liberian borders.[54] A mediation committee comprising the presidents of Mali, Nigeria and Togo was subsequently set up to facilitate conflict resolution.[55] Although the three parties to the conflict approved the creation of the ECOMOG force and although Niger, Nigeria, Mali and Senegal pledged troops, by the end of 2001 the peace operation had not yet been deployed. This is in part because Guinea and Liberia refuse to sign the Status of Forces Agreement enabling ECOMOG forces to be deployed on their territory, preferring to deal unilaterally with what they describe as insurgencies.[56] Another factor, emphasized by ECOWAS, is UN Security Council authorization and assistance for

---

[52] United Nations, Letter dated 26 Nov. 2001 from the Secretary-General addressed to the President of the Security Council, UN document S/2001/1128, 29 Nov. 2001.

[53] Dwan, *SIPRI Yearbook 2001* (note 2), pp. 82, 102–103; and Reno (note 39).

[54] Panafrican News Agency (PANA), 'ECOWAS moves to end border crises', 15 Nov. 2000, URL <http://allafrica.com/stories/printable/200011150254.html>; Final communiqué of the 24th Session of the Authority of Heads of State and Government, Bamako, 15–16 Dec. 2000; and ECOWAS Extraordinary Summit, Final communiqué (note 50).

[55] 'Committee of three presidents to mediate crisis in Mano River Union', ECOWAS Press Release, 37/2001, 12 Apr. 2001.

[56] Adeyemi, S., 'Summit to tackle Mano River Union crises', Panapress, 14 June 2001, URL <http://62.210.150.98/lusaka/newslat.asp?code=eng005881&dte=14/06/01>.

ECOMOG's deployment.[57] The UN Inter-Agency Task Force mission to West Africa concluded its report with a recommendation for a West African integration framework, with ECOWAS at its centre. If ECOWAS is to play the central role that the UN would like to see it play, especially in mediation, preventive deployment and peacekeeping, then it will require substantial new resources from and active partnership with the UN.

EU efforts

Since 1997 the EU has adopted a series of Common Positions on the conflicts in Africa, including the 1998 Common Position on human rights, democratic principles, the rule of law and good governance in Africa and, most recently, the 14 May 2001 Common Position concerning conflict prevention, management and resolution in Africa.[58] These policy documents emphasize the role of the EU as a supporter of African regional organizations' peace efforts. However, the 2001 Common Position also commits the EU to develop a proactive, integrated approach to conflict prevention and crisis management and to take new steps to promote coordination with other actors in this field. The implication of this Common Position would be to give the EU an enhanced political profile in the African continent beyond that of primarily a provider of substantial humanitarian and development aid.

In West Africa, where the EU is the region's leading development cooperation and trade partner, a regional aid approach has been in place alongside bilateral support programmes.[59] This was given a new political framework in June 2001, when the EU foreign ministers, in line with their new prevention programme, identified West Africa as one of the regions where the EU would increase its attention and seek priority cooperation with the UN in conflict prevention, management and resolution issues.[60] The outgoing Swedish presidency offered to assist its Belgian successor in developing an EU policy on the political and humanitarian crises in West Africa. In a step reflecting a recommendation of the European Commission's Communication, a Swedish diplomat, Hans Dahlgren, was appointed the Special Representative of the Presidency of the European Union to the countries of the MRU. His mandate is to follow developments in the three countries with a view to proposing appropriate EU action, pursuing dialogue with the UN and ECOWAS to identify coordinated measures to deal with the crises, encouraging dialogue between the three MRU states, supporting the disarmament and demobilization process in Sierra Leone, and maintaining contact with Liberia regarding its conformity with UN sanctions.[61] Dahlgren travelled to the MRU countries in October and

---

[57] ECOWAS Extraordinary Summit, Final communiqué (note 50).
[58] European Council, Common Positions of 2 June 1997 (97/356/CFSP), 25 May 1998 (98/350/CFSP) and 14 May 2001 (2001/374/CFSP).
[59] The 8th European Development Fund for the period ending in 2001 provided €228 million to West Africa.
[60] European Union, General Affairs Council (GAC) Press Release 9398/01 (Presse 226), 12 June 2001.
[61] Swedish Ministry for Foreign Affairs, Press Release, 5 Oct. 2001; and Swedish Ministry for Foreign Affairs, Email communication with Africa Desk, 30 Oct. 2001.

December 2001 to pursue these tasks and will report to the EU Council in June 2002.

In his discussions with UN and ECOWAS officials in the region, the Special Representative of the Presidency of the European Union noted the EU's willingness to provide more financial support for prevention and peace-building initiatives.[62] This has so far been borne out. In July 2001 the EU agreed a new aid package to Liberia that included a €25 million programme for the resettlement of refugees and internally displaced persons (IDPs). This was supplemented in October 2001 by a €5.1 million programme for refugees and IDPs in Sierra Leone and Guinea.[63] In line with the new commitment to political dialogue with partner countries, the EU began consultations with Liberia in November 2001 to discuss human rights, democratic principles and the rule of law, making its position clear that the Liberian Government would have to take concrete initiatives to comply with the terms of the Cotonou Agreement. These negotiations are to continue and will shape the EU's decisions on its future aid to Liberia.[64] During the year similar negotiations took place with Côte d'Ivoire authorities, following the suspension of EU cooperation in response to the government's handling of the national elections held in 2000. After three months of discussions the EU agreed to the gradual resumption of aid to the country, with full cooperation dependent on a six-monthly review of the situation.[65] These actions illustrate that, although development aid will remain the EU's main tool for external relations, the Commission is determined to give it a new political–security dimension and make it more effective in achieving EU policy goals.

Although the EU emphasizes coordination with the UN in conflict prevention, it has made efforts to deepen relations with African regional organizations, particularly the OAU, its main interlocutor for issues relating to peace and security. ECOWAS has become increasingly important to the EU, however, since relations between the two organizations were established in 1998. The EU–ECOWAS ministerial meetings in October 2000 and 2001 symbolized this new recognition.[66] During these meetings, the EU foreign ministers reiterated their commitment to continue providing assistance for the development of the ECOWAS Mechanism for Conflict Prevention, Management, Resolution, Peacekeeping and Security. However, the 2001 ministerial meeting was somewhat strained by the suspension of EU aid to Togo as a result of government efforts to amend constitutional electoral procedures and to clamp down on political opposition in that country. This illustrates one of the barriers

---

[62] UN OCHA (note 38); and Integrated Regional Information Network for West Africa (IRIN-WA), 'Sierra Leone: uphill struggle for disarmament in the east', IRIN-WA weekly round-up 101, 1–7 Dec. 2001.

[63] IRIN-WA, 'Guinea–Sierra Leone: EC allocates euro 5.1 million in humanitarian aid', 10 Oct. 2001, available at URL <http://www.reliefweb.int>.

[64] EU Press Release, 13789/01 (Presse 412), 9 Nov. 2001.

[65] EU Press Release 10245/01 (Presse 267), 25 June 2001; and IRIN-WA, 'Côte d'Ivoire: EU agrees to gradual resumption of aid', IRIN-WA weekly round-up 78, 23–29 June 2001.

[66] Final Communiqué, EU–ECOWAS Ministerial Meeting, Abuja, 16 Oct. 2000; and Final Statement, Second EU–ECOWAS Ministerial Meeting, Brussels, 12 Oct. 2001, EU Press Releases 12309/00 (Presse 390), 16 Oct. 2000 and 12884/01 (Presse 370), 16 Oct. 2001.

The impact of EU and UN preventive activity

It is impossible to draw causal connections between the activities of the UN and the EU in West Africa and events on the ground during the course of 2001. Nevertheless, if conflict prevention is to become a policy reality for international and regional organizations, then some assessment of the impact of UN and EU activities in the region is desirable. By the end of 2001 West Africa was still a region in turmoil, with one of the world's most serious humanitarian crises (in the MRU).[67] Fighting between Liberian government forces and armed dissidents in the north-west of the country surged in December 2001, causing further flights of refugees and more strain on relations with Guinea and Sierra Leone.[68] Nevertheless, war has not broken out between the three MRU countries and the parties have been persuaded to back down from some more inflammatory actions (e.g., Liberia's expulsion of the Guinean and Sierra Leonean ambassadors).[69] The situation in Sierra Leone, on the other hand, has stabilized, in large part because Liberia terminated its active support to the RUF, as demanded by the UN Security Council. This has permitted UNAMSIL to gradually expand its presence to almost all parts of the country, including diamond-producing centres formerly under RUF control.[70] UNAMSIL completed the disarmament of over 37 000 former combatants by the end of 2001 and is providing assistance for the presidential and parliamentary elections scheduled to take place in May 2002.[71]

UN and EU preventive actions in West Africa in 2001 were short-term and reactive, coming late in the day to prevent a complete conflagration of the region. The difficulty of swift reaction was particularly marked in the UN: three months passed between the establishment of an Inter-Agency Task Force and its dispatch to the region, and it took a further six weeks after the return of the task force for its report to be considered in the Security Council. Subsequently, seven months passed before a UN Office for West Africa was established. Moreover, UN Security Council discussions demonstrated how reluctant the UN member states are to consider preventive deployment or further peacekeeping commitments—at least as regards Africa. Nonetheless, the fact that the UN and the EU maintained and intensified their focus on the region was a significant development. UN sanctions against Liberia and Secretary-General updates on UNAMSIL brought West Africa to the Security Council's attention on a regular basis.

[67] 'UNHCR get figures wrong in Guinea', BBC News Online, 4 June 2001, available at URL <http://news.bbc.co.uk/hi/english/world>.
[68] UN Press Release SG/SM/8085, AFR/367, 21 Dec. 2001.
[69] IRIN-WA, 'Liberia: Government expels ambassadors, closes borders', IRIN-WA weekly round-up 64, 17–23 Mar. 2001, available at URL <http://www.reliefweb.int>; and ECOWAS Extraordinary Summit, Final communiqué (note 50).
[70] See appendix 2A.
[71] United Nations, Twelfth Report of the Secretary-General on the United Nations Mission in Sierra Leone, UN document S/2001/1195, 13 Dec. 2001.

In contrast, the formal structure of the EU's relations with external actors provides for regular policy reviews at the ministerial and working levels. At the same time, the way in which the EU took up West Africa in the second half of 2001 demonstrates the significance of the EU presidency in setting priorities for the Council's agenda. Whether the Special Representative of the Presidency to the region will play an active role in shaping EU policy on West Africa depends on successive presidencies—and the EU Representative's own government—giving him the wherewithal to do so. Without some form of political leadership from the Council, the substance of EU external relations in Africa will remain development aid administered by the European Commission.

Despite their last-minute nature, the UN and EU actions demonstrated the organizations' appreciation of the need to address the structural causes of West Africa's crises. The UN Inter-Agency Task Force mission provided a comprehensive road map of the wide range of social, economic, humanitarian and security problems that need to be tackled and demonstrated that its specialized agencies, already on the ground, are the best placed actors to deal with massive humanitarian emergencies. The design and provision of EU aid illustrate that its comparative advantage lies in its regularity and its focus on long-term capacity development. Another strength of EU development assistance is its inclusion of mechanisms for political dialogue with the states concerned. It enables the EU to traverse sovereignty-sensitive issues of internal governance while keeping channels open with the particular government. The UN's coercive instruments—primarily sanctions—may be stronger, but the EU has a wider and more scaled range of tools from which to choose.[72]

Ultimately, however, coordination has been the key to efforts to stem the rising tide of conflict in West Africa. In 2001 the UN and the EU prioritized relations with the region's primary institutional actor, ECOWAS, and as a result established a more coherent international voice than has previously been heard in West Africa. Regular communication, formalized meetings, joint missions in the region and financial support were important elements of this approach and helped secure coordination with regard to the peace process in Sierra Leone, sanctions on Liberia and mediation in the MRU crisis. The renewed focus on ECOWAS as a partner is, undoubtedly, a reflection of the lack of alternative partners in the region as well as of the Western reluctance to become embroiled in its conflicts. However, it also reflects a heightened awareness that regional strategies for prevention and crisis management are needed and that sustainable peace is, ultimately, dependent on local actors. ECOWAS remains a troubled regional organization, not just because of its lack of an institutional capacity but also because the conflicts it is called on to manage are those between and within its own members. All these countries face significant social, political and economic domestic problems, not least Nigeria, the leading power in the organization. Therefore, if ECOWAS is to play the ambitious role envisaged for it, it will need a sustained, active

[72] For a comprehensive discussion of sanctions see chapter 5 in this volume.

partnership with, and assistance from, the UN and the EU. Regional preventive strategies must be international in substance.

Civil conflict in Zimbabwe

Zimbabwe represents a contrasting, although no less challenging, case of international conflict prevention—an intra-state crisis that has not yet become a violent conflict, in a country formerly seen as an African 'success story'. Zimbabwe has had a relatively good rate of development (117th on the UN Human Development Index for 2000)[73] and is acknowledged as a powerful actor in sub-regional and African politics. Its current demise illustrates how development is not exclusively progressive and points to the significance of bad governance as a cause of conflict.

Zimbabwe has been in a political crisis since February 2000, when the government was defeated in a referendum on changing the constitution so as to permit President Robert Mugabe to remain in office for an additional 10 years with increased powers.[74] This opened the possibility that Mugabe and his Zimbabwe African National Union–Patriotic Front (ZANU–PF) Party might lose power in the 2002 parliamentary elections for the first time since 1980, when Zimbabwe gained independence. Mugabe seized on the long-standing grievance of land reform to secure voter support. His regime, through inflammatory speeches, the overriding of Zimbabwean High Court rulings and the passivity of the police authorities, facilitated the violent invasion of white-owned farms by ZANU–PF supporters and war veterans. The fact that a new political party, the Movement for Democratic Change (MDC), garnered enough votes in the June 2001 elections to become the first serious opposition to Mugabe's regime, with power to block constitutional amendments, has continued to keep the redistribution of land ownership a central issue. Although there is an agreed basis for resolving the issue between Zimbabwe and international donors, the most important of which is the UK, the former colonial power in the country, Mugabe's government has continued to champion rapid land seizures and resettlement. The effect of this violence on the agriculturally based economy has been catastrophic: Zimbabwe has been described as one of the world's fastest shrinking economies.[75] By late 2001 this former food exporter was importing grain from neighbouring South Africa, while a World Food Programme (WFP) emergency intervention was under way to provide food for an estimated 550 000 people in need.[76] Over 70 000 people were internally displaced during 2001 and an estimated 500 Zimbabwean refugees crossed the South African border daily.[77]

---

[73] UN Development Programme (note 42).
[74] International Crisis Group (ICG), *Zimbabwe in Crisis: Finding a Way Forward*, Africa Report no. 32 (ICG: Harare/Brussels, 13 July 2001).
[75] Economist Intelligence Unit, *Zimbabwe Country Report*, June 2001.
[76] Integrated Regional Information Network for Southern Africa (IRIN-SA), 'Government to be sole distributor of food aid', IRIN-SA, 13 Nov. 2001.
[77] Lamont, J., 'Mugabe's policies attacked by Mbeki', *Financial Times*, 29 Nov. 2001, p. 9.

As the March 2002 presidential election approached, the government's disregard for the rule of law turned into overt oppression. Zimbabwe's only independent daily newspaper was regularly charged for spurious criminal offences and was twice bombed, and foreign news organizations faced increasing restrictions.[78] Mounting laws curbing freedom of speech and association accompanied violence and political intimidation against lawyers, journalists, trade unionists and political opposition figures. Zimbabwe displays every classic sign of a country disintegrating into violence.

UN activities

The fact that the UN has remained almost mute on Zimbabwe's political crisis cogently demonstrates its limitations as an early-warning or preventive actor. The sanctity of the sovereignty principle among member states rules out UN engagement in domestic political affairs until a conflict has or is about to spread beyond national borders. The only likely exception, as the intervention in Kosovo partly illustrated, occurs in cases in which UN Security Council members are sufficiently interested and united to take up a particular conflict. UN aid agencies, including the WFP and the United Nations Development Programme (UNDP), undertook missions to Zimbabwe in late 2001 to assess the food and land reform situation in the country.[79] In January 2002, for the first time in over a year, Secretary-General Annan addressed the situation in the country, noting his concern at the imposition of restrictive laws and supporting the efforts of the sub-regional organization, the Southern African Development Community (SADC), to facilitate free and fair elections in Zimbabwe.[80]

EU activities

The EU, by contrast, has been engaged to an unprecedented degree in an attempt to halt Zimbabwe's international and domestic demise. This has been influenced by its role as an aid donor—the EU provides €11.9 million annually to Zimbabwe—and the UK's efforts to put Zimbabwe on the Council agenda, particularly as bilateral relations between the two countries deteriorated. In February 2001 the EU warned Zimbabwe that its human rights record threatened the continued provision of EU aid and one month later, during a visit to Europe by President Mugabe, it initiated the political dialogue procedure of the Cotonou Agreement. This effort to effect change in Mugabe's policies through negotiation yielded little result, forcing the Council to issue a warning in May 2001 that it would review its approach to the crisis. On 25 June 2001 the EU foreign ministers spelled out the progress the EU

[78] Human Rights Watch (HRW), *Zimbabwe: Submission to the Commonwealth Ministerial Action Group*, 30 Jan. 2002, available at URL <http://www.hrw.org/africa/zimbabwe.php>.
[79] 'UN land team in Harare', BBC News Online, 16 Nov. 2001, available at URL <http://news.bbc.co.uk/hi/english/world>.
[80] UN Press Release, SG/SM/8100 AFR/369, 15 Jan. 2002; and Barrow, G., 'UN plays waiting game with Zimbabwe', BBC News Online, 29 Jan. 2002, available at URL <http://news.bbc.co.uk/hi/english/world>.

expected to take place in the next 60 days, including an end to the official encouragement of political violence, an invitation to the EU to observe the 2002 elections, concrete action to protect media freedom, compliance with judiciary decisions and an end to illegal farm occupations. If Zimbabwe failed to comply, the Council warned, 'appropriate measures' would be taken.[81]

Despite the absence of any sign of progress over the summer, the EU did not impose any punitive policies against Zimbabwe after its 60-day deadline had expired. Although there was speculation that economic sanctions would be imposed, political dialogue consultations dragged on until the end of October 2001, when the EU warned Mugabe that he had one last chance to take steps towards the restoration of the rule of law and free and fair elections. The threat was explicit this time: failure would result in the imposition of sanctions at the end of January 2002.[82] On 11 January 2002 a meeting was held in Brussels between a high-level Zimbabwean delegation and the EU at which another stiff warning of EU action was given and another round of assurances provided by the Zimbabwean representatives.[83] Although the EU acknowledged the futility of pursuing the political dialogue procedure further, the EU foreign ministers pulled back from imposing any kind of sanction at their meeting on 28 January 2002. Instead, they pressed Mugabe to permit EU observers into the country ahead of the March elections and noted their concern at the 9 January 2002 threat by the head of the Zimbabwean armed forces that the military would not accept the outcome of the presidential election if it did not agree with the result.[84] Meanwhile, according to the influential NGO Human Rights Watch, 'the atmosphere of intimidation has been so intense that the presidential elections . . . cannot be free and fair',[85] while Amnesty International reported that, between late December 2001 and early January 2002, 10 people were killed by 'state-sponsored militia'.[86]

The evident lack of impact of EU actions on the Zimbabwean Government's activities throughout 2001 is a consequence of a number of factors central to conflict prevention policies. First and foremost, the EU has been internally divided on how to approach Zimbabwe's political crisis and what combination of policies should be applied. Some EU states, notably the UK, have cut bilateral aid substantially and others, such as Denmark and Sweden, have frozen all assistance to the country. France, in contrast, significantly increased its aid to Zimbabwe in 2001.[87] As a result, repeated EU threats have had little coercive power because the likelihood of actual punitive action has been weak.

[81] European Union, General Affairs Council, GAC Conclusions, EU Press Release 10228/01 (Presse 250), 25 June 2001.
[82] European Union, General Affairs Council, GAC Conclusions, EU Press Release 1329/01 (Presse 390), 29 Oct. 2001.
[83] Dempsey, J., 'Scant hopes for EU talks on Zimbabwe', *Financial Times*, 11 Jan. 2002, p. 6; and ICG, *All Bark and No Bite? The International Response to Zimbabwe's Crisis*, Africa Report no. 40 (ICG: Harare/Brussels, 25 Jan. 2002).
[84] European Union, General Affairs Council, GAC Conclusions, EU Press Release 5636/02 (Presse 16), 28 Jan. 2002.
[85] HRW (note 78).
[86] Amnesty International, Memorandum to SADC Heads of State, 11 Jan. 2002.
[87] Cook, R., Statements made on the Commonwealth Monitoring Action Group Mission to Zimbabwe, House of Commons, 1 May 2001; Afrol News, 'EU considers sanctions against Zimbabwe',

One of the main reasons behind this lack of unity among the 15 EU member states is the link between Zimbabwe and the ongoing conflict in the Democratic Republic of the Congo (DRC).[88] Since its intervention on the side of the DRC Government in 1998, Zimbabwe has been a key actor in international negotiations to end that conflict. The interest of Belgium and France in securing peace in the African Great Lakes Region has made them keen to forge a cooperative, rather than a combative, relationship with Mugabe. The Zimbabwean President's visit to both countries in March 2001, in parallel with the start of the EU's tough line in the political dialogue, demonstrated how much the credibility of an international threat depends on the unity with which it is delivered. Such unity is much harder to achieve when the target of the threat is a significant international actor.

Another main reason behind the EU's wavering line lies in the problematic nature of sanctions.[89] Almost all EU member states agree that a halt to development aid and the imposition of economic sanctions would adversely affect Zimbabwe's citizens rather than the political elite responsible for the crisis. Zimbabwe's opposition party, the MDC, as well as the country's neighbours signalled that economic sanctions would be dangerous for the situation in the country.[90] However, Mugabe's domestic opponents, along with human rights groups, have called for targeted sanctions against the regime in the form of travel bans and a freeze of the assets of senior officials.[91] The EU foreign ministers concluded that even the imposition of limited sanctions could prompt Mugabe to prevent foreign observation of the March 2002 presidential election and exploit the sanctions for his own political gain.

The lack of unity and coordination between the international and regional organizations involved with Zimbabwe—the EU, the Commonwealth of Nations, the OAU and the SADC—is the third main reason for the failure of EU action to prevent the deepening crisis. One element of this is the general sensitivity of African–European relations. President Mugabe has proved adept at manoeuvring European warnings to his own advantage by playing on colonial legacies and making the most of Zimbabwe's standing in African politics. At the July 2001 OAU summit meeting,[92] he sought and gained the support of other African leaders for his land resettlement programme. The OAU Council of Ministers stressed the UK's responsibility, as the former colonial power, and rebuked it for allegedly trying to mobilize the countries of Europe and North America against Zimbabwe.[93] The sensitivity of land issues among

15 May 2001, available at URL <http://www.afrol.com/News2001/zim022_eu_sanctions.htm>; and European Parliament Joint Resolution B5-0549, 0554, 0571, 0581, 0582 and 0592.2001, 6 Sep. 2001.

[88] Seybolt, T., 'The war in the Democratic Republic of Congo', *SIPRI Yearbook 2000* (note 12), pp. 59–75.

[89] For a more comprehensive discussion of sanctions see chapter 5 in this volume.

[90] Dempsey, J. et al., 'EU to hold back on Zimbabwe sanctions', *Financial Times*, 28 Jan. 2002, p. 5.

[91] IRIN-SA, 'Zimbabwe: EU to consider economic sanctions', IRIN-SA Country stories: Zimbabwe, 15 May 2001, available at URL <http://www.irinnews.org>.

[92] This was the last summit meeting of the OAU. See note 11.

[93] Declaration on the Resolution of the Land Question in Zimbabwe, OAU Summit, Lusaka, Zambia, 9 July 2001, AHG/Decl.2 (XXXVII); and OAU Decisions adopted by the 74th Ordinary Session of the Council of Ministers, 5–8 July, 2001, CM/Dec.625 (LXXIV).

many African states, along with the OAU's emphasis on the sovereignty principle, rules out the OAU as a partner for EU preventive action in Zimbabwe. Belatedly, the EU turned to the Commonwealth of Nations and the SADC in an attempt to coordinate international efforts.[94]

The *Commonwealth of Nations*,[95] under British pressure, has attempted to engage in the crisis and coordinate its actions with those of the EU. It is within this forum that negotiations between the UK and Zimbabwe over the funding and administration of land reform have taken place. Its mixed geographical membership has helped avoid a north–south cleavage on the issue. On 7 September 2001 a special Commonwealth meeting in Abuja, Nigeria, adopted the Abuja Agreement, by which the UK would provide financial assistance (£36 million) in exchange for Zimbabwe's implementation of land reform. In return, Zimbabwe agreed to end illegal occupations of land, to restore the rule of law and to work with the UNDP in the implementation of land reform.[96] However, even before questions could be raised as to how the Abuja Agreement would be implemented, there were reports of new farm seizures.

The Commonwealth's weakness as a preventive actor lies in its lack of political and economic tools. For example, the deal it brokered between the UK and Zimbabwe did not include any provisions as to actions to be taken if Mugabe failed to abide by the agreement. The postponement of the Commonwealth summit meeting in October 2001 merely cemented this lack of follow-up power.[97] A Commonwealth Committee delegation visited Zimbabwe in October 2001 to review progress on the agreement but was itself divided on whether the fundamental issue was land reform or the rule of law.[98] Mugabe continued to prevaricate over the Commonwealth Secretary-General's request to send a mission to the country. He was comfortable in the knowledge that the next summit meeting, scheduled for March 2002, would take place too late to put insurmountable pressure on him before the election. Zimbabwe's suspension from the organization, the Commonwealth's most powerful coercive weapon, had no consensus among the member states.[99]

The *SADC*,[100] the EU's potentially most valuable partner in the region, has been the most reluctant of all organizations to become involved in the problems of one of its members. Despite international exhortations that it should play an active regional role, South Africa, the organization's leading power,

[94] European Union, General Affairs Council, 'EU foreign ministers declared their support for SADC and Commonwealth efforts in Zimbabwe in October 2001', GAC Press Release 13291/01 (Presse 390), 29–30 Oct. 2001.

[95] For the members of the Commonwealth of Nations see the glossary in this volume.

[96] 'Text of Zimbabwe agreement', BBC News Online, 7 Sep. 2001, available at URL <http://news.bbc.co.uk/hi/english/world/>; and Holman, M. and Hawkins, T., 'Africa's deal', *Financial Times*, 8–9 Sep. 2001, p. 7.

[97] 'Zimbabwe: the pressure builds', *The Economist*, 15 Sep. 2001, pp. 41–42. The Commonwealth Summit was postponed following the 11 Sep. attacks in the USA.

[98] ICG (note 83).

[99] Commonwealth Press Release on the 18th meeting of the Commonwealth Ministerial Action Group, 30 Jan. 2002.

[100] For the members of the Southern African Development Community (SADC) see the glossary in this volume.

maintained a low-key diplomatic approach towards neighbouring Zimbabwe until August 2001, when the SADC summit meeting agreed to establish a ministerial task force to address Zimbabwe's land crisis.[101] SADC heads of state travelled to Harare, Zimbabwe, in early September and held an extraordinary summit meeting there to underscore their support for the Abuja Agreement. A committee was set up to monitor developments and is scheduled to meet every few weeks to assess progress.[102] Concern over regional and domestic stability has shaped the SADC states' actions towards Zimbabwe. South Africa, in particular, was reluctant to become involved in the debate about Zimbabwean land reform precisely because it faces its own land resettlement problems. Nor were southern African states keen to provoke active opposition movements that could influence their own domestic politics. Finally, the countries neighbouring on Zimbabwe, such as Mozambique, are heavily dependent on it for food exports, markets and even aid. The fact that continued repression and instability in Zimbabwe now threaten the economic and political climate of the entire region is the motivation for SADC action since September 2001.[103] Although the tone of the states neighbouring on Zimbabwe has become considerably more critical, by the start of 2002 regional and international actors were still some distance from each other in terms of coordinating a united approach to the country. In the absence of international coordination and regional political leadership, external preventive efforts have little effect.

V. Prevention after 11 September 2001

The international environment in which the UN and EU conflict prevention reports were presented in June 2001 was fundamentally changed by the terrorist attacks on the USA three months later. The shock generated by the attacks and the responses that have followed them carry serious repercussions for the international adoption and practice of conflict prevention.

The most immediate consequence is distraction. By virtue of their broad nature, the UN and EU Council programmes were intended to start, rather than to conclude, greater consideration of conflict prevention at the international, regional and national levels. Their release just prior to the start of the summer vacation period in Europe and North America gave little opportunity for this process to get under way before September. The attacks of 11 September catapulted international attention onto the threat of global terrorism and this attention has been glued fast by the subsequent US-led intervention in Afghanistan. Every UN-recognized international and regional organization issued a declaration against terrorism in the wake of the attacks, and the subject has been on every multilateral agenda, for example, the agendas of subsequent EU–

---

[101] Summit of Heads of State and Government of SADC, Blantyre, Malawi, 6–4 Aug. 2001.
[102] SADC Press Release, 12 Sep. 2001; and IRIN, 'Zimbabwe: All heads turn to SADC indaba in Harare', IRIN Country Stories: Zimbabwe, 7 Sep. 2001, available at URL <http://www.irinnews.org>.
[103] Lamont (note 77).

ECOWAS and EU–OAU meetings.[104] Conflict prevention, as a new subject on the international agenda and one that is non-time-specific and broad in scope, has little chance of maintaining significant political attention. This is particularly true in cases in which it has not yet been institutionalized in the structures and systems of an international organization or national government. Political leadership, in such situations, is central to successful follow-up. By the end of 2001 the prospect of such leadership looked doubtful in both the UN and, to a lesser extent, the EU.

Approaches to the threat of terrorism have the potential to incorporate much of the central tenets of conflict prevention. Issues such as the root causes of terrorism, structural and short-term approaches to its prevention, the broad range of state and non-state actors involved, and the multiple tools required to address terrorist threats are precisely the issues with which conflict prevention research and policy making have grappled over the past decade. Initially, it seemed that international organizations and states might incorporate the preventive framework into their approach to terrorism. For instance, the European Council, at its extraordinary meeting on 21 September 2001, declared that the fight against terrorism required the EU to 'play a greater part in the efforts of the international community to prevent and stabilize regional conflicts'. Efforts to address the threat would be 'all the more effective' if they were based on 'an in-depth political dialogue with those countries and regions of the world in which terrorism comes into being'.[105] The UN Security Council's third resolution in the aftermath of the attacks similarly acknowledged that the fight against international terrorism required a sustained, comprehensive approach that addressed regional conflicts and 'the full range of global issues, including development issues'.[106] US Secretary of State Colin Powell, speaking in support of this resolution, promised that the war against terrorism would be fought 'with increased support for democracy programmes, judicial reform, conflict resolution, poverty alleviation, economic reform and health and education programmes'.[107]

However, the subsequent global effort against terrorism has moved away from a preventive focus and is now characterized as a 'war against terrorism'.[108] In this narrower approach, the preventive concept is severely circumscribed. Prevention of terrorism, as currently practised, consists of measures taken to stop international terrorism, cut off the financial, political and military sources of terrorist support and, where possible, apprehend terrorists before

[104] See, e.g., Africa–Europe Ministerial Conference, Joint Declaration on Terrorism, Brussels, 11 Oct. 2001; OSCE Bucharest Plan of Action for Combating Terrorism, MC(9).DEC/1, 4 Dec. 2001; and NATO's Response to Terrorism, Statement issued at the Ministerial Meeting of the North Atlantic Council held at NATO Headquarters, Brussels, 6 Dec. 2001.

[105] European Council, Conclusions and plan of action of the extraordinary European Council meeting on 21 Sep. 2001, EU document SN 140/01

[106] United Nations Security Council Resolution 1377, 12 Nov. 2001.

[107] United Nations Security Council 4413th meeting, 12 Nov. 2001, UN document S/PV.4413, 12 Nov. 2001.

[108] For a discussion of the significance of terminology see Howard, M., 'What's in a name? How to fight terrorism', *Foreign Affairs*, vol. 81, no. 1 (Jan./Feb. 2002), pp. 8–13.

they commit acts of terror.[109] Although this approach employs a broad range of instruments, it is coercive and short-term in character. It is in origin and practice distinct from the concept of conflict prevention that was elaborated over the past decade and reflected in the UN and EU documents of 2001. Indeed, the current approach to the prevention of terrorism risks undermining the entire notion of conflict prevention.

It does this in a number of direct and indirect ways. First, although international cooperation against terrorism embraces a wide range of military and non-military instruments, the attacks on Afghanistan have inevitably focused attention on the military elements and given vent to the idea of a global military engagement against terrorism. Important as this may be, there is some risk that the prioritization of military relations between states will undermine the important progress forged in the post-cold war world in broadening international affairs so as to take greater account of non-military issues and the legitimate engagement of non-state actors. Second, the war against terrorism has led to the forging of new relationships between states that were formerly at odds with each other.[110] In many cases, these differences centred on the domestic policies of a state. Improved regional and international cooperation on shared threats may indeed contribute to stability and peace. However, the extent to which states such as Pakistan, Sudan or Tajikistan are called upon to assist in the fight against terrorism may constrain the international community's willingness to engage with these countries on such sensitive questions as governance and human rights. Indeed, for a number of states, the discourse on the war against terrorism is providing a means for legitimating a more aggressive approach to domestic and regional dissent.[111] The global effort against international terrorism marks the appearance of a new paradigm in international politics. It is important that it does not undermine the norms that have so recently been established.

---

[109] A similar point is noted by Powers, T., 'The trouble with the CIA', *New York Review of Books*, vol. 49, no. 1 (17 Jan. 2002), pp. 28–32.

[110] The renewal of US–Pakistani relations and the resumption of US aid blocked since Pakistan tested its first nuclear weapon, in 1998, constitute the most obvious examples.

[111] Examples are Israel's depiction of its conflict with Palestine after the assassination of Tourism Minister Rehavam Zeevi on 17 Oct. (e.g., Morris, H., 'Sharon's call', *Financial Times*, 18 Oct. 2001, p. 14) and India's characterization of its actions towards Pakistan after the 13 Dec. suicide attack on the Indian Parliament (e.g., Chandrasekaran, R., 'Pakistan and India spar with sanctions', *International Herald Tribune*, 28 Dec. 2001, p. 1).

Appendix 2A. Multilateral peace missions, 2001

RENATA DWAN, THOMAS PAPWORTH and
SHARON WIHARTA*

I. Peace missions in 2001

There were no new United Nations peace missions in 2001, for the first time since 1996. However, five new multilateral missions were initiated—one of them, the International Security Assistance Force (ISAF) in Afghanistan, on the authority of the UN Security Council. The remaining four missions—two led by the North Atlantic Treaty Organization (NATO) in the Former Yugoslav Republic of Macedonia (FYROM), one led by the Organization for Security and Cooperation in Europe (OSCE) in the Federal Republic of Yugoslavia (FRY), and an international protection and support force in Burundi, under the leadership of South Africa—all received UN Security Council endorsement.[1] The range of actors involved in these missions illustrates the diversity of contemporary multilateral peace missions. Despite differences in mandate and function, the new missions in 2001, with the exception of the OSCE mission in the FRY, share one common feature: they are limited in size, scope and mandate. This is illustrative of an increased reluctance on the part of regional organizations and states to commit resources to open-ended peace missions. Where large peace missions exist, as in East Timor or Sierra Leone, they are under UN control. The peacemaking and peace-building challenges they continue to present have increased UN Security Council reluctance to embark on new peacekeeping initiatives.

Three small peace missions were terminated in 2001. The International Civilian Support Mission in Haiti, under the responsibility of the UN Department of Political Affairs (DPA), was terminated once its mandate to assist preparations for national elections had expired. This ended the successive multilateral peace missions present in Haiti since 1993.[2] The incorporation of the Western European Union (WEU) into the European Union (EU) brought an end to its two missions—the Multinational Advisory Police Element for Albania (MAPE) and the WEU Demining Assistance Mission (WEUDAM) in Croatia.[3] These changes bring to 51 the total number of multilateral peace missions under way in 2001.

---

[1] UN Security Council Resolution 1345, 21 Mar. 2001; UN Security Council Resolution 1371, 26 Sep. 2001; and UN Security Council Resolution 1375, 29 Oct. 2001.

[2] See Dwan, R. et al., 'Multilateral peace missions, 1999', *SIPRI Yearbook 2000: Armaments, Disarmament and International Security* (Oxford University Press: Oxford, 2000), pp. 135–56; and Papworth, T., 'Multilateral peace operations, 2000', *SIPRI Yearbook 2001: Armaments, Disarmament and International Security* (Oxford University Press: Oxford, 2001), pp. 128–48.

[3] A European Commission follow-up training and institution-building assistance programme to the police mission in Albania was launched on 1 June 2001. European Union, General Affairs Council (GAC), 2342nd Meeting, Luxembourg, 9 Apr. 2001.

*SIPRI intern Hannes Baumann contributed to the preparation of this appendix.

The International Security Assistance Force in Afghanistan

On 20 December 2001 the UN Security Council authorized the establishment of the International Security Assistance Force, organized and led by the United Kingdom, to be deployed in Afghanistan for a period of six months.[4] The mission was the result of the UN-sponsored conference in Bonn, Germany, to establish a political framework to bring to an end the civil war that has raged in the country since 1994. The initiation of this process was a consequence of the launch of US attacks on the ruling Taliban regime on 7 October 2001. The US-led campaign helped restore the fortunes of the loose coalition of opposition forces, the United Islamic Front for the Salvation of Afghanistan (Northern Alliance), and brought about the Taliban's defeat.[5] The Agreement on Provisional Arrangements in Afghanistan Pending the Re-establishment of Permanent Government Institutions (Bonn Agreement) of 5 December provides for the creation of an Afghanistan Interim Authority (AIA) made up of representatives of Afghanistan's ethnic groups as well as an international force to assist it in the 'maintenance of security in Kabul and its surrounding areas'.[6] The Military Technical Agreement subsequently signed between the ISAF and the AIA notes that the international force may also assist the AIA in reconstruction and in developing future security structures.[7]

The establishment of the ISAF is modelled on the 1999 Australian-led International Force for East Timor (INTERFET).[8] As in that case, the Security Council authorized the ISAF under Chapter VII of the UN Charter,[9] thereby enabling it to legitimately 'take all necessary measures to fulfil its mandate'.[10] It called on the UN member states to contribute to the force but noted that expenses would be borne by the participating states, not by the UN peacekeeping budget. The advantage of this ad hoc coalition approach to peacekeeping is the relative speed with which such a force can be deployed and the potential for coherent, effective action under a lead nation. An advance party of British troops arrived in the Afghan capital, Kabul, on 22 December 2001.[11]

However, the ISAF's collaboration has been controversial in almost every aspect. British leadership of the force initially met with opposition from some Afghan leaders, conscious of the legacy of British imperial involvement in their country.[12] Domestic dissension within the UK itself on the extent and length of any commitment to a peace operation in Afghanistan led the government to underscore the limited nature of its task and its duration. The UK agreed to lead the ISAF only for the first three months of the operation while British Defence Secretary Geoff Hoon emphasized that the ISAF was not a peacekeeping mission but rather an assistance mission

---

[4] UN Security Council Resolution 1386, 20 Dec. 2001.

[5] For more on the Afghanistan civil war and the US-led attack, see chapter 1 in this volume.

[6] Annex 1, para. 3 of the Bonn Agreement, available at URL <http://www.uno.de/frieden/afghanistan/talks/agreement.htm>; and UN Security Council Resolution 1386 (note 4).

[7] Military Technical Agreement between the International Security Assistance Force (ISAF) and the Interim Administration of Afghanistan ('Interim Administration'), 30 Dec. 2001.

[8] For more on INTERFET see Dwan, R., 'Armed conflict prevention, management and resolution', *SIPRI Yearbook 2000* (note 2), pp. 116–18.

[9] Article 42 of Chapter VII grants the Security Council the authority to 'take such action by air, sea, or land forces as may be necessary to maintain or restore international peace and security'. The text of Chapter VII of the UN Charter is available at URL <http://www.un.org/aboutun/charter/chapter7.htm>.

[10] UN Security Council Resolution 1386 (note 4).

[11] Thornhill, J. and Williamson, H., 'British troops spearhead stabilisation force', *Financial Times*, 22 Dec. 2001, p. 2.

[12] McGregor, R., 'Britain "should not lead peacekeepers"', *Financial Times*, 19 Dec. 2001, p. 11.

to the AIA.[13] Germany, a key player in the negotiations on Afghanistan, objected to the proposal that the USA would have overall command of the mission. US command was seen as necessary to ensure that ISAF activities did not 'interfere' with the ongoing US military operation against suspected Taliban and al-Qaeda targets ('Operation Enduring Freedom').[14] Germany insisted on a clear separation between the ISAF mission and the operation. This forced the UK to modify its description of its relationship to one of 'essential enabling support' for the USA to support for the ISAF, a task declared as 'vital and considerable'.[15] The US relationship to the ISAF remains a sensitive question: while the US administration has ruled out US participation in any peace operation, there is recognition that the credibility of ISAF depends on the extent to which it is linked to US military power.[16] The AIA, meanwhile, illustrated the ambivalence with which the deployment of any international force is regarded among its leadership in its initial call for a restricted use-of-force mandate (Chapter VI of the UN Charter) and in its insistence that the forces number 3000 instead of the 5000 first proposed.[17]

The lack of agreement on the nature and scope of an international peace operation force in Afghanistan reflects the enormity of the challenge of peace establishment and reconstruction in that country. Nor are there strong indications that Afghans are united in their commitment to the political settlement worked out in Bonn. Renewed fighting in January 2002 between rival factions and warlords as well as widespread lawlessness testify to the dangerous situation throughout the country. This instability hampers efforts to deliver humanitarian assistance to the estimated 6 million people in need.[18] The authority of the AIA remains entirely dependent on international military, economic and financial support.[19]

By early 2002, however, UN officials and Afghan leaders had begun to campaign for an expanded international force to be deployed throughout the country.[20] External actors, already baulking at the extent of their potential engagement in Afghanistan, are unwilling to commit themselves to comprehensive action until they have some indication that a stable peace is viable. However incapable the ISAF is of undertaking peace-enforcement and -maintenance tasks, the degree of its success in fulfilling its mandate to stabilize Kabul will be a key determinant for a future UN peace operation.

---

[13] 'Blair outlines Afghan force options', BBC News Online, 17 Dec. 2001, available at URL <http://news.bbc.co.uk/hi/english/uk_politics>; 'Just the job for us', *The Economist*, 22 Dec. 2001, p. 39; and Nicoll, A., 'Force's role "to assist Afghans"', *Financial Times*, 11 Jan. 2002, p. 8.

[14] Hoyos, C. *et al.*, 'Split threatens anti-terror pact', *Financial Times*, 20 Dec. 2001, p. 1.

[15] Hoge, W., 'UK presses plan for Afghan force', *International Herald Tribune*, 20 Dec. 2001, p. 1; and Speech by Defence Secretary Geoff Hoon to parliament, 19 Dec. 2001, available at URL <http://news.mod.uk/news/press/news_press_notice_asp?newsItem_id=1298>.

[16] Leyne, J., 'US wary of peacekeeping', BBC News Online, 21 Nov. 2001, available at URL <http://news.bbc.co.uk/hi/english/world>; and Robinson, G., 'US agrees to support Afghan peacekeepers', *Financial Times*, 29 Jan. 2002, p. 4.

[17] 'Limits urged on Kabul force', BBC News Online, 15 Dec. 2001, available at URL <http://news.bbc.co.uk/hi/english/world/>; and Robinson, G., 'Afghan leaders agree on foreign troop numbers', *Financial Times*, 31 Dec. 2001, p. 3. The text of Chapter VI of the UN Charter is available at URL <http://www.un.org/Overview/Charter/chapter6.html>.

[18] Annan, K., United Nations, The situation in Afghanistan and its implications for international peace and security: report of the Secretary-General, UN document A/56/681-S/2001/1157, 6 Dec. 2001.

[19] The World Bank and the United Nations Development Programme (UNDP) estimate initial reconstruction costs over the next 2.5 years at $5 billion. Agence France-Presse (AFP), 'Afghanistan may need 30 000 peacekeepers: UN official', 23 Jan. 2002, available at ReliefWeb, URL <http://www.reliefweb.int/>.

[20] Hoyos, C., 'Troops in Kabul urged to expand mandate', *Financial Times*, 24 Jan. 2002, p. 4.

Peace operations in the Balkans

The two NATO peace operations launched in the FYROM in August and September 2001, Task Force Harvest (TFH) and Task Force Fox (TFF), respectively, were similarly limited in scope and duration. NATO became involved in the FYROM's conflict between government forces and armed ethnic Albanian rebels in June, at the request of the FYROM Government. NATO insistence on a ceasefire and peace agreement between the two parties before it would provide any security assistance helped bring about the signing of a peace accord on 13 August 2001. This accord included provisions for a NATO weapon collection operation.[21] The TFH was activated on 22 August with a 30-day mandate to disarm rebel forces of the Albanian National Liberation Army (NLA) and destroy collected weapons. The disarmament process was to be voluntary, with a figure of approximately 3000 weapons agreed in advance.[22] In return, the FYROM Government committed itself to constitutional reforms to provide the country's Albanian minority with greater rights. The swift deployment of 4500 NATO troops helped maintain the fragile peace and was successfully concluded on 27 September 2001.[23]

NATO recognition that continued engagement would be required to ensure that the peace process, slow though it was, moved forward lay behind its agreement to deploy an immediate follow-up operation, the TFF. The task of this 700-strong force is to provide security to the 150 OSCE and 30 EU civilian monitors deployed to oversee, *inter alia*, the return of refugees and police redeployment in former rebel-held villages.[24] A notable feature of these NATO operations is their European character. The USA provided only equipment support to the TFH and is not involved in the TFF.[25]

Peace operations elsewhere in the Balkans are following a similar pattern. NATO defence ministers agreed in December to cut the Stabilization Force (SFOR), the NATO-led operation in Bosnia and Herzegovina, from 18 000 to around 12 000. The size of the US force is scheduled to be reduced from 3100 to 1100.[26] The scheduled termination of the UN Mission in Bosnia and Herzegovina (UNMIBH) in December 2002 led the EU to commit its fledgling civilian crisis management capability to take over the UNMIBH's police mission component, the International Police Task Force (IPTF), from January 2003.[27] The OSCE, the other regional organization active in peace operations in the Balkans, is also expected to increase its engagement in Bosnia and Herzegovina after the UN's departure. It is the lead external actor in law reform, democratization and institution-building activities in Croatia, the FRY and the FYROM.

---

[21] Richburg, K., 'NATO planners wary of Macedonia mission', *International Herald Tribune*, 14 Aug. 2001, p. 4; and NATO, 'Task Force Harvest background information', available at URL <http://www.nato.int>.

[22] 'NATO accepts weapons count in Macedonia', *International Herald Tribune*, 25–26 Aug. 2001, p. 2.

[23] Transcript of NATO press briefing, 30 Aug. 2001, available at URL <http://www.nato.int>; and 'NATO/Macedonia', *Atlantic News*, no. 3324 (28 Sep. 2001), p. 1.

[24] Ripley, T., 'Macedonia monitoring group plan', *Jane's Defence Weekly*, 12 Sep. 2001, p. 5; and NATO, 'Operation Amber Fox', NATO Press release 133, 27 Sep. 2001.

[25] The TFH was under British command, while the TFF is led by Germany.

[26] Loeb, V., 'Rumsfeld asks NATO for Bosnia troop cuts', *International Herald Tribune*, 19 Dec. 2001, p. 1; and Hill, L., 'NATO nears decision on force reduction', *Jane's Defence Weekly*, 2 Jan. 2002, p. 4.

[27] European Union, GAC, 2406th meeting, Brussels, 28 Jan. 2002; and European Union, GAC, 2409th meeting, Brussels, 18–19 Feb. 2002.

Peace operations in Africa

The size and nature of the current UN peace operations in Africa contrast starkly with the multilateral missions described above. The UN Mission in Sierra Leone (UNAMSIL) is the largest current UN peace operation and for much of 2000 one of its most troubled.[28] UNAMSIL and the UN Organization Mission in the Democratic Republic of the Congo (MONUC) were initiated in response to peace agreements to bring an end to two catastrophic civil wars that have spread beyond national borders. The problems besetting UNAMSIL and MONUC included lack of financial and human resources, the slow deployment of troops from contributing states and disagreement between international, regional and local actors on the conduct of the peace operation.[29] The fundamental issue, however, was the continuation of the conflicts whose termination UNAMSIL and MONUC were charged with monitoring. The fact that neither mission was authorized to use force except in self-defence when threatened compounded their inability to carry out their respective mandates. In UNAMSIL's case, this led to the near-collapse of the mission after the kidnapping of 500 peacekeepers by the rebel Revolutionary United Front (RUF) in May 2000. British forces, who intervened to rescue the UN peacekeepers, continue to guarantee the security of UNAMSIL and of Sierra Leone's President, Ahmed Tejan Kabbah, in the capital, Freetown.[30]

UNAMSIL and MONUC fared better in 2001, expanding in size and in presence outside the national capitals. UNAMSIL was able to carry out its mandate to disarm and demobilize RUF rebels and assist in the start of a political process between the RUF and the government.[31] Planning for a national reconciliation process, which includes a special court to try war crimes, is under way.[32] MONUC benefited from the ceasefire that followed Joseph Kabila's assumption of the presidency in the Democratic Republic of the Congo (DRC) to expand its operational strategy and begin verification and monitoring of the disengagement of warring parties and their withdrawal to designated defensive positions.[33] By October MONUC was able to initiate a disarmament, demobilization and reintegration process.[34]

The relative success of UNAMSIL and MONUC is due to the renewed political engagement of the UN and its member states in negotiating conflict settlements in Sierra Leone and the DRC. Increased coordination between the UN and the relevant regional organizations was an important element in this. Finally, increased financial and personnel contributions enabled each operation to reach full strength and to deploy throughout the country. The extent of the commitment required and the

---

[28] Reno, W., 'War and the failure of peacekeeping in Sierra Leone', *SIPRI Yearbook 2001* (note 2), pp. 149–61.

[29] Dwan, R., 'Armed conflict prevention, management and resolution', *SIPRI Yearbook 2001* (note 2), pp. 69–127.

[30] 'Sierra Leone: The spreading battle ground', *The Economist*, 7 Apr. 2001, p. 46; and Hoon, G. (MP, Defence Secretary), House of Commons, Hansard Written Answers for 14 Jan. 2002 (pt 13).

[31] United Nations, Twelfth Report of the Secretary-General on the United Nations Mission in Sierra Leone, UN document S/2001/1195, 13 Dec. 2001.

[32] 'The president's speech', BBC News Online, 18 Jan. 2002, available at URL <http://news.bbc.co.uk/hi/english/world>.

[33] United Nations, Sixth report of the Secretary-General on the United Nations Organization Mission in the Democratic Republic of the Congo, UN document S/2001/128, 12 Feb. 2001; for more on the conflict in the DRC, see chapter 1 in this volume.

[34] United Nations, Ninth report of the Secretary-General on the United Nations Organization Mission in the Democratic Republic of the Congo, UN document S/2001/970, 16 Oct. 2001.

lessons that were demonstrated have deepened UN member states' caution with regard to new operations in Africa. This partially explains the Security Council's welcoming of South Africa's agreement to establish and lead a mission to support the fragile peace process in Burundi.[35] UNAMSIL's and MONUC's records also explain the mandate of the South African protection mission: the force is only intended to provide protection for politicians returning from exile until such time as an all-Burundi protection unit can be trained.[36]

II. Table of multilateral peace missions

Table 2A lists 51 multilateral peace missions (observer, peacekeeping, peace-building and combined peacekeeping and peace-enforcement operations) initiated, ongoing or terminated in 2001. The missions are grouped by organization, either sole or lead, and listed chronologically within these groups.

The first group, covering UN missions, is divided into three sections: 15 operations run by the Department of Peacekeeping Operations; 3 missions not properly defined as peacekeeping (under Chapters VI and VII of the UN Charter) and coordinated by the Department of Political Affairs; and 1 mission initiated by UN authority but carried out at UN request by an ad hoc coalition of member states. The next five groups cover missions conducted or led by regional organizations: 13 by the OSCE; 4 by NATO; 3 by the EU/WEU; 3 by the Commonwealth of Independent States (CIS), including 1 mission carried out by Russia under bilateral arrangements; and 3 by the Organization of African Unity (OAU).[37] A final group lists 6 missions led by other organizations or ad hoc coalitions of states recognized by the UN. Peace missions comprising non-resident individuals or teams of negotiators or operations not sanctioned by the UN are not included.

Missions initiated in 2001, or new participating states in an existing mission, are listed in bold text; operations or individual participation ending in 2001 are in italics. Legal instruments underlying the establishment of an operation—UN Security Council resolutions or formal decisions by regional organizations—are cited in the first column. Personnel numbers include civilian observers or civilian staff only where indicated. The main exception is for observers in OSCE missions, who are usually civilian. Mission fatalities are recorded from the beginning of the mission until the last reported date for 2001 and as a total for 2001. Unless otherwise stated all figures are as of 31 December 2001. Budget figures are given in millions of US dollars. Conversion from budgets set in other currencies are based on 31 December 2001 conversion rates.

---

[35] UN Security Council Resolution 1375, 29 Oct. 2001. The peace process is facilitated by former South African President Nelson Mandela. Dwan (note 8), pp. 78–129.
[36] Joint Communiqué of the 15th Summit of the Regional Peace Initiative on Burundi, 23 July 2001; and United Nations, Interim report of the Secretary-General to the Security Council on the situation in Burundi, UN document S/2001/1076, 14 Nov. 2001.
[37] The OAU will be replaced by the African Union (AU) after a transitional period. The OAU member states adopted the Constitutive Act of the African Union in 2000; it entered into force on 26 May 2001, formally establishing the AU, with headquarters in Addis Ababa. The AU will take over the OAU operations.

Table 2A. Multilateral peace missions

Acronym/ (Legal instrument[a])	Name	Location	Start date	Countries contributing troops, military observers (mil. obs) and/or civilian police (CivPol) in 2001	Troops/ Mil. obs/ CivPol	Deaths: To date/ In 2001	Cost ($m): 2001/ Unpaid
United Nations (UN) peacekeeping operations (15 operations) (UN Charter, Chapters VI and VII)					37 665 1 801 7 642[1]	1 734[2] 64	1 931.5[3] 1 979.1[4]
UNTSO (SCR 50)[5]	UN Truce Supervision Organization	Egypt/Israel/ Lebanon/ Syria	June 1948	Argentina, Australia, Austria, Belgium, Canada, Chile, China, Denmark, Estonia, Finland, France, Ireland, Italy, Netherlands, New Zealand, Norway, Russia, Slovakia, Slovenia, Sweden, Switzerland, USA[6]	– 152[7] –	38 –[8]	22.8[9] –
UNMOGIP (SCR 91)[10]	UN Military Observer Group in India and Pakistan	India/Pakistan (Kashmir)	Jan. 1949	**Austria**, Belgium, Chile, Denmark, Finland, Italy, South Korea, Sweden, Uruguay[11]	45[12] –	9 –[13]	7.3[14] –
UNFICYP (SCR 186)[15]	UN Peacekeeping Force in Cyprus	Cyprus	Mar. 1964	Argentina, Australia, Austria, Bolivia, *Brazil*, Canada, Finland, Hungary, Ireland, *Nepal, Netherlands, Paraguay*, **Slovakia**, *Slovenia*, **Sweden**, UK, *Uruguay*[16]	1 196 – 35[17]	170 –[18]	42.4[19] 21.9[20]
UNDOF (SCR 350)[21]	UN Disengagement Observer Force	Syria (Golan Heights)	Mar. 1974	Austria, Canada, Japan, Poland, Slovakia, Sweden[22]	1 036[23] (80)[24] –	40 –[25]	36.0 19.6[26]
UNIFIL (SCR 425 & 426)[27]	UN Interim Force in Lebanon	Lebanon	Mar. 1978	Fiji, *Finland*, France, Ghana, India, Ireland, Italy, Nepal, Poland, *Sweden*, Ukraine[28]	3 639[29] (50)[30] –	244 5[31]	106.2[32] 166.5[33]
UNIKOM (SCR 689)[34]	UN Iraq/Kuwait Observation Mission	Iraq/Kuwait (Khawr 'Abd Allah waterway and UN DMZ)	Apr. 1991	Argentina, Austria, Bangladesh, *Canada*, China, Denmark, Fiji, Finland, France, Germany, Ghana, Greece, Hungary, India, Indonesia, Ireland, Italy, Kenya, Malaysia, Nigeria, Pakistan, Poland, Romania, Russia, Senegal, Singapore, Sweden, Thailand, Turkey, UK, Uruguay, USA, Venezuela[35]	906 193[36] –	16 3[37]	52.8[38] 17.8[39]

CONFLICT PREVENTION 131

MINURSO (SCR 690)[40]	UN Mission for the Referendum in Western Sahara	Western Sahara	Sep. 1991	Argentina, Austria, Bangladesh, Belgium, China, Egypt, El Salvador, France, Ghana, Greece, Guinea, Honduras, Hungary, India, Ireland, Italy, Jordan, Kenya, Malaysia, Nigeria, Norway, Pakistan, Poland, Portugal, Russia, Senegal, South Korea, Sweden, Uruguay, USA[41]	27 204 22[42]	10 –[43]	48.8[44] 83.8[45]
UNOMIG (SCR 849 & 858)[46]	UN Observer Mission in Georgia	Georgia (Abkhazia)	Aug. 1993	Albania, Austria, Bangladesh, Czech Rep., Denmark, Egypt, France, Germany, Greece, Hungary, Indonesia, Jordan, Pakistan, Poland, Russia, South Korea, Sweden, Switzerland, Turkey, UK, **Ukraine**, Uruguay, USA[47]	– 106[48] –	7 4[49]	27.9[50] 15.0[51]
UNMIBH (SCR 1035)[52]	UN Mission in Bosnia and Herzegovina	Bosnia and Herzegovina	Dec. 1995	Argentina, Austria, Bangladesh, Bulgaria, Canada, Chile, **China**, Czech Rep., Denmark, Egypt, *Estonia*, Fiji, Finland, France, Germany, Ghana, Greece, Hungary, Iceland, India, Indonesia, Ireland, Italy, Jordan, Kenya, Malaysia, Nepal, Netherlands, Nigeria, Norway, Pakistan, Poland, Portugal, Romania, Russia, Senegal, Spain, Sweden, Switzerland, Thailand, *Tunisia*, Turkey, UK, Ukraine, USA, *Vanuatu*[53]	– 4 1 674[54]	11 3[55]	144.7 107.6[56]
UNMOP (SCR 1038)[57]	UN Mission of Observers in Preklava	Croatia	Jan. 1996	Argentina, Bangladesh, Belgium, Brazil, *Canada*, Czech Rep., Denmark, Egypt, Finland, Ghana, Indonesia, Ireland, Jordan, Kenya, Nepal, New Zealand, Nigeria, Norway, Pakistan, Poland, *Portugal*, Russia, Sweden, Switzerland, Ukraine[58]	– 27[59] –	– –	See UNMIBH[60]
UNMIK (SCR 1244)[61]	UN Interim Administration in Kosovo	Federal Republic of Yugoslavia (Kosovo)	June 1999	Argentina, Austria, Bangladesh, Belgium, Benin, Bolivia, Bulgaria, Cameroon, Canada, Chile, Côte d'Ivoire, Czech Rep., Denmark, *Dominican Rep.*, Egypt, *Estonia*, Fiji, Finland, France, *Gambia*, Germany, Ghana, Greece, Hungary, Iceland, India, Ireland, Italy, Jordan, Kenya, Kyrgyzstan, Lithuania, Malawi, Malaysia, Nepal, *Netherlands*, New Zealand, Niger, Nigeria, Norway, Pakistan, Philippines, Poland, Portugal, Romania, Russia, Senegal, Slovenia, Spain, Sweden, Switzerland, Tunisia, Turkey, UK, Ukraine, USA, Zambia, Zimbabwe[62]	– 37 4 519[63]	14 2[64]	227.3[65] –

Acronym/ (Legal instrument[a])	Name	Location	Start date	Countries contributing troops, military observers (mil. obs) and/or civilian police (CivPol) in 2001	Troops/ Mil. obs/ CivPol	Deaths: To date/ In 2001	Cost ($m): 2001/ Unpaid
UNAMSIL (SCR 1270)[66]	UN Mission in Sierra Leone	Sierra Leone	Oct. 1999	Bangladesh, Bolivia, Canada, China, Croatia, Czech Rep., Denmark, Egypt, France, Gambia, Ghana, Guinea, **India**, Indonesia, **Jordan**, Kenya, Kyrgyzstan, Malaysia, Mali, *Namibia*, Nepal, New Zealand, **Niger**, Nigeria, Norway, Pakistan, Russia, Senegal, Slovakia, **Sri Lanka**, Sweden, Tanzania, Thailand, UK, Ukraine, Uruguay, Zambia, Zimbabwe[67]	17 105 261 54[68]	59 (36)[69]	3 722.1[70] 3 317.1[71]
UNTAET (SCR 1272)[72]	UN Transitional Administration in East Timor	East Timor	Oct. 1999	Argentina, Australia, Austria, Bangladesh, **Benin**, Bolivia, Bosnia and Herzegovina, Brazil, Canada, *Cape Verde*, Chile, China, Denmark, Egypt, Fiji, *France*, Gambia, Ghana, Ireland, Jordan, Kenya, Malaysia, Mozambique, Namibia, Nepal, New Zealand, Niger, Nigeria, Norway, Pakistan, *Peru*, Philippines, Portugal, Russia, **Samoa**, Senegal, Singapore, **Slovakia**, Slovenia, South Korea, Spain, Sri Lanka, Sweden, Thailand, Turkey, UK, Ukraine, Uruguay, USA, Vanuatu, Zambia, Zimbabwe[73]	7 110 102 1 316[74]	17 9[75]	300.8[76] 211.8[77]
MONUC (SCR 1279)[78]	UN Observer Mission in the Democratic Republic of the Congo	Democratic Republic of the Congo	Nov. 1999	Algeria, **Argentina**, Bangladesh, Belgium, Benin, Bolivia, Burkina Faso, **Cameroon**, Canada, **China**, Czech Rep., Denmark, Egypt, France, Ghana, India, **Indonesia**, **Ireland**, **Italy**, Jordan, Kenya, *Libya*, **Malawi**, Malaysia, Mali, Morocco, **Mozambique**, Nepal, Niger, Nigeria, **Norway**, Pakistan, **Paraguay**, Peru, Poland, **Portugal**, Romania, Russia, Senegal, South Africa, **Spain**, **Sweden**, Switzerland, Tunisia, *Tanzania*, UK, Ukraine, Uruguay, Zambia[79]	2 924 449 13[80]	4[81] –	209.1[82] 246.9[83]

CONFLICT PREVENTION   133

Acronym	Name	Location	Start date	Contributing countries						
UNMEE (SCR 1312)[84]	United Nations Mission in Ethiopia and Eritrea	Ethiopia, Eritrea	July 2000	Algeria, Argentina, Austria, Bangladesh, Benin, **Bosnia and Herzegovina, Bulgaria,** Canada, China, **Croatia, Czech Rep.,** Denmark, Finland, France, **Gambia,** Ghana, **Greece,** India, **Ireland,** Italy, Jordan, Kenya, Malaysia, **Namibia,** Nepal, Netherlands, Nigeria, Norway, **Paraguay,** Peru, Poland, Romania, Russia, **Singapore,** Slovakia, South Africa, Spain, Sweden, Switzerland, Tanzania, Tunisia, Ukraine, Uruguay, USA, Zambia[85]	3 722	217[86]	—	2	2[87]	208.9[88]	128.4[89]
Other UN operations[90] (3 operations)										
UNSMA (A/RES/ 47/20B)[91]	UN Special Mission in Afghanistan	Afghanistan/ Pakistan	Mar. 1994	Denmark, *France,* Germany, Japan, Sweden, UK, Ukraine[92]	—	24[93]	—	1	—[94]	..[95]	—
MINUGUA (A/RES/ 48/267)[96]	UN Verification Mission in Guatemala	Guatemala	Oct. 1994	Argentina, **Austria, Barbados, Belgium,** Bolivia, Brazil, Canada, **Chile,** Colombia, **Costa Rica, Dominican Rep., Ecuador, Egypt,** El Salvador, France, **Germany, Greece, Guatemala, Haiti, Honduras, Ireland,** Italy, **Japan,** Mexico, **Nicaragua,** Norway, **Panama, Peru,** *Portugal,* **Russia,** Spain, *Sweden,* **UK, Ukraine,** Uruguay, USA, Venezuela[97]	—	281[98]	9[99]	1	—[100]	16.2[101]	—
MICAH (A/RES/ 54/193)[102]	International Civilian Support Mission in Haiti	Haiti	Mar. 2000	Barbados, Benin, **Cameroon,** Canada, Chile, Colombia, ***Côte d'Ivoire,*** DR Congo, **Croatia, El Salvador,** Ethiopia, France, Germany, Ghana, Greece, Guinea-Bissau, **India,** Ireland, Israel, Italy, Kenya, Mali, Mauritius, Mexico, Netherlands, Russia, Rwanda, Senegal, Spain, Sweden, Tajikistan, Trinidad and Tobago, UK, Uruguay, USA, Yugoslavia, Zimbabwe[103]	—	47[104]	—	1	—	23.5[105]	—
Multinational operations tasked and authorized by the UN (1 operation)[106]										
ISAF (SCR 1386)[107]	International Security Assistance Force	Afghanistan	Dec. 2001	UK[108]	200[109]	—	—	—	—[110]	..[111]	—

Acronym/ (Legal instrument*)	Name	Location	Start date	Countries contributing troops, military observers (mil. obs) and/or civilian police (CivPol) in 2001	Troops/ Mil. obs/ CivPol	Deaths: To date/ In 2001	Cost ($m): 2001/ Unpaid
OSCE operations (13 operations)[112]							
– (CSO 18 Sep. 1992)[113]	OSCE Spillover Mission to Skopje	Former Yugoslav Rep. of Macedonia	Sep. 1992	**Armenia, Austria, Azerbaijan, Belarus, Belgium, Canada, Croatia,** Czech Rep., **Denmark, Finland, France, Georgia,** Germany, **Ireland,** *Italy,* **Latvia, Luxembourg, Moldova, Netherlands,** Norway, **Poland, Romania, Russia, Slovakia, Slovenia, Spain,** Sweden, **Switzerland, Tajikistan, Turkey,** UK, **Ukraine,** USA[114]	– 133 77[115]	– _116	6.5[117] –
– (CSO 6 Nov. 1992)[118]	OSCE Mission to Georgia	Georgia	Dec. 1992	Austria, Azerbaijan, Belarus, Bulgaria, Croatia, **Czech Rep.,** Denmark, Estonia, France, Germany, Hungary, **Latvia,** Lithuania, Moldova, Norway, Poland, Romania, Russia, Slovakia, Spain, Sweden, **Switzerland,** UK, Ukraine, USA[119]	– 50[120] –	– _121	8.7[122] –
– (CSO 13 Dec. 1992)[123]	OSCE Mission to Estonia	Estonia	Feb. 1993	Armenia, Austria, *Canada, Denmark,* Finland, Germany[124]	– 5[125] –	– _126	0.6[127] –
– (CSO 4 Feb. 1993)[128]	OSCE Mission to Moldova	Moldova	Feb. 1993	Finland, Germany, *Lithuania,* Netherlands, Poland, Slovakia, UK, USA[129]	– 9[130] –	– _131	0.8[132] –
– (CSO 23 Sep. 1993)[133]	OSCE Mission to Latvia	Latvia	Nov. 1993	Bulgaria, Canada, Germany, Norway, Sweden[134]	– 6[135] –	– _136	0.6[137] –
– (Ministerial Council, 1 Dec. 1993)[138]	OSCE Mission to Tajikistan	Tajikistan	Feb. 1994	*Austria,* **Denmark,** France, Germany, Norway, Poland, *Romania,* Russia, **Switzerland, Ukraine,** USA[139]	– 15[140] –	– _141	1.7[142] –

CONFLICT PREVENTION 135

– (PC 11 Apr. 1995)[143]	OSCE Assistance Group in Chechnya	Chechnya	Apr. 1995	*Austria, Czech Rep.,* **Denmark, Germany,** *Moldova, Poland,* **Romania**[144]	– 6[145] –	–[146] –	1.4[147] –
– (10 Aug. 1995)[148]	Personal Representative of the Chairman-in-Office on the Conflict Dealt with by the OSCE Minsk Conference	Azerbaijan (Nagorno-Karabakh)	Aug. 1995	Czech Rep., *Germany,* Hungary, Poland, **Romania,** UK, Ukraine[149]	– 6[150] –	–[151] –	0.8[152] –
– (MC/5/DEC/ 18 Dec. 1995)[153]	OSCE Mission to Bosnia and Herzegovina	Bosnia and Herzegovina	Dec. 1995	Albania, **Armenia,** Austria, *Belarus,* **Belgium,** Bulgaria, Canada, Czech Rep., *Denmark,* Estonia, Finland, France, Germany, Greece, Hungary, **Iceland,** Ireland, Italy, **Latvia,** Moldova, Netherlands, Norway, Poland, *Portugal,* Romania, Russia, **Slovakia, Slovenia,** Spain, Sweden, Switzerland, Tajikistan, UK, *Ukraine,* USA[154]	– 180[155] –	–[156] –	22.0[157] –
– (PC/DEC 112, 18 Apr. 1996)[158]	OSCE Mission to Croatia	Croatia	July 1996	Armenia, Austria, Belarus, Belgium, Bulgaria, Canada, **Croatia,** Czech Rep., Finland, France, Georgia, Germany, *Greece,* Ireland, Italy, *Japan,* **Kyrgyzstan,** *Lithuania,* Macedonia, Moldova, Netherlands, Norway, Poland, Portugal, Romania, Russia, **Slovakia,** Spain, Sweden, Switzerland, Turkey, UK, Ukraine, USA[159]	– 100[160] –	–[161] –	11.8[162] –
– (PC/DEC 160, 27 Mar. 1997)[163]	OSCE Presence in Albania	Albania	Apr. 1997	Austria, Belarus, **Canada, Croatia,** *Czech Rep.,* France, Germany, Ireland, Italy, **Hungary,** *Kazakhstan, Luxembourg,* **Moldova,** Netherlands, **Poland, Romania,** *Russia, Spain,* Switzerland, *Tajikistan,* UK, USA[164]	– 45[165] –	–[166] –	3.7[167] –
OMIK (PC/DEC 305, 1 July 1999)[168]	OSCE Mission in Kosovo	Federal Republic of Yugoslavia (Kosovo)	July 1999	Austria, Azerbaijan, *Belarus,* Belgium, *Bosnia and Herzegovina,* Bulgaria, Canada, Croatia, Czech Rep., Denmark, *Estonia,* Finland, France, Georgia, Germany, Greece, Hungary, Iceland, Ireland, Italy, Kyrgyzstan, Latvia, Lithuania, Moldova, Netherlands, Norway, Poland, *Portugal,* Romania, Russia, *Slovakia,* Slovenia, Spain, Sweden, Switzerland, *Tajikistan,* Turkey, UK, Ukraine, USA[169]	– 831[170] –	3 –[171]	83.6[172] –

Acronym/ (Legal instrument[a])	Name	Location	Start date	Countries contributing troops, military observers (mil. obs) and/or civilian police (CivPol) in 2001	Troops/ Mil. obs/ CivPol	Deaths: To date/ In 2001	Cost ($m): 2001/ Unpaid
– (PC/DEC 401, 11 Jan. 2001)[173]	OSCE Mission to the Federal Republic of Yugoslavia	Federal Republic of Yugoslavia	Mar. 2001	Austria, Belgium, Canada, Estonia, Finland, Germany, Greece, Hungary, Ireland, Italy, Netherlands, Norway, Poland, Portugal, Romania, Russia, Slovakia, Spain, Sweden, Turkey, UK, Ukraine, USA[174]	– 51[175] –	–[176] –	4.5[177] –

North Atlantic Treaty Organization (NATO) and NATO-led operations (4 operations)

Acronym/ (Legal instrument[a])	Name	Location	Start date	Countries contributing troops, military observers (mil. obs) and/or civilian police (CivPol) in 2001	Troops/ Mil. obs/ CivPol	Deaths: To date/ In 2001	Cost ($m): 2001/ Unpaid
SFOR (SCR 1088)[178]	NATO Stabilization Force	Bosnia and Herzegovina	Dec. 1996	Albania, Argentina, Australia, Austria, Belgium, Bulgaria, Canada, Czech Rep., Denmark, Estonia, Finland, France, Germany, Greece, Hungary, Iceland, Ireland, Italy, Latvia, Lithuania, **Luxembourg**, Morocco, Netherlands, New Zealand, Norway, Poland, Portugal, Romania, Russia, Slovakia, Slovenia, Spain, Sweden, Turkey, UK, USA[179]	18 853[180] – –	804 6[181]	23.2[182] –
KFOR (SCR 1244)[183]	NATO Kosovo Force	Federal Republic of Yugoslavia (Kosovo)	June 1999	Argentina, Austria, Azerbaijan, Belgium, Bulgaria, Canada, Czech Rep., Denmark, Estonia, Finland, France, Georgia, Germany, Greece, Hungary, Iceland, Ireland, Italy, Jordan, **Latvia**, Lithuania, Luxembourg, Morocco, Netherlands, Norway, Poland, Portugal, **Romania**, Russia, Slovakia, Slovenia, Spain, Sweden, Switzerland, Turkey, UAE, UK, Ukraine, USA[184]	39 000[185] – –	65 22[186]	24.6[187] –
TFH[188]	*Task Force Harvest*	*Former Yugoslav Republic of Macedonia*	*Aug. 2001*	*Belgium, Canada, Czech Rep., France, Germany, Greece, Italy, Netherlands, Spain, Turkey, UK*[189]	*4 675*[190] – –	1 1	–[191] –
TFF (SCR 1371)[192]	Task Force Fox	Former Yugoslav Republic of Macedonia	Sep. 2001	Belgium, Denmark, France, Germany, Greece, Italy, Norway, Poland, Portugal, Spain, Turkey, UK, USA[193]	965[194] – –	– –	–[195] –

CONFLICT PREVENTION 137

European Union/Western European Union operations (3 operations)

EUMM (Brioni Agreement)[196]	European Union Monitoring Mission	Albania, Former Yugoslavia[197]	July 1991	Austria, Belgium, *Czech Rep.*, Denmark, Finland, France, Germany, Greece, Ireland, Italy, Luxembourg, Netherlands, Norway, *Poland*, Portugal, Slovakia, Spain, Sweden, UK[198]	– 145[199] –	10 3[200] –	4.2[201] –
MAPE (WEU Council, 2 May 1997)[202]	Multinational Advisory Police Element for Albania	*Albania*	May 1997	Czech Rep., Denmark, Finland, France, Germany, Greece, Italy, Netherlands, Norway, Portugal, Sweden, Turkey, UK[203]	– 33[204] –	– – –	1.3[205] –
WEUDAM (10 May 1999)[206]	Western European Union Demining Assistance Mission in Croatia	Croatia	May 1999	Finland, France, Italy, Sweden[207]	– 4[208] –	– – –	0.01[209] –

Russian and Commonwealth of Independent States (CIS) operations (3 operations)

– (Bilateral, 24 June 1992)	South Ossetia Joint Force	Georgia (South Ossetia)	July 1992	Georgia, Russia, (South Ossetia)[210]	1 200[211] 54[212] –	.. ..[213]	.. ..
– (Bilateral, 21 July 1992)	Joint Control Commission Peacekeeping Force	Moldova (Trans-dniester)	July 1992	Moldova, Russia, (Trans-Dniester), Ukraine[214]	1 413 40[215] –	.. ..	.. ..
– (CIS, 15 Apr. 1994)	CIS Peacekeeping Forces in Georgia	Georgia (Abkhazia)	June 1994	Russia[216]	1 870[217] – –	83 6[218]	.. ..

Acronym/ Legal instrument[f]/ Name	Location	Start date	Countries contributing troops, military observers (mil. obs) and/or civilian police (CivPol) in 2001	Troops/ Mil. obs/ CivPol	Deaths: To date/ In 2001	Cost ($m): 2001/ Unpaid
Organization of African Unity (3 operations)						
OMIB (OAU, 7 Dec. 1999)[219] OAU Mission in Burundi	Burundi	Dec. 1993	Rep. of Congo, Guinea, *Rwanda*	2[220] –	– [221]	0.5[222] –
OMIC (OAU, 6 Nov. 1997)[223] OAU Observer Mission in the Comoros	Comoros	Nov. 1997	Niger, Senegal, **Togo**, *Tunisia*	3[224] –	. . [225]	0.2[226] –
JMC (OAU, 3 Sep. 1999)[227] Joint Military Commission	Democratic Republic of Congo	Sep. 1999	**Angola, DR Congo,** Rep. of Congo, **Kenya, Namibia, Rwanda, Tanzania, Uganda, Zimbabwe**[228]	57[229] –	– –	0.9[230] –
Other operations (6 operations)						
NNSC (Armistice Agreement)[231] Neutral Nations Supervisory Commission	North Korea/ South Korea	July 1953	Poland, Sweden, Switzerland[232]	9[233] –	– [234]	1.3[235] –
MFO (Protocol to Treaty of Peace)[236] Multinational Force and Observers	Egypt (Sinai)	Apr. 1982	Australia, Canada, Colombia, Fiji, France, Hungary, Italy, New Zealand, Norway, Uruguay, USA	1 836[237] –	44 1[238]	51.0[239] –

CONFLICT PREVENTION

TIPH 2 (Hebron Protocol)[240]	Temporary International Presence in Hebron	Hebron	Jan. 1997	Denmark, Italy, Norway, Sweden, Switzerland, Turkey[241]	– 91[242]	–[243] –	2.0[244] –
PMG (Lincoln Agreement 1998)[245]	Bougainville Peace Monitoring Group	Papua New Guinea	May 1998	Australia, Fiji, New Zealand, Vanuatu[246]	– 75[247]	1 –	5.1[248] –
IPMT (Townsville Peace Agreement)[249]	International Peace Monitoring Team for the Solomon Islands	Solomon Islands	Nov. 2000	Australia, **Barbados, Cook Islands**, New Zealand, **Nigeria, Tonga, Vanuatu**[250]	– 49[251]	–[252] –	2.1[253] –
SAPSD (Regional Peace Initiative on Burundi)[254]	South African Protection and Support Detachment[255]	Burundi	**Nov. 2001**	South Africa[256]	1 500[257] – –	1 1[258]	35.0[259] ..

^a *Acronyms in the table and notes*: A/RES = UN General Assembly Resolution; ; CSO = OSCE Committee of Senior Officials (now the Senior Council); DMZ = Demilitarized Zone; DPKO = UN Department of Peacekeeping Operations; GA = UN General Assembly; MC = Ministerial Council; MOU = Memorandum of Understanding; SC = UN Security Council; SCR = UN Security Council Resolution; PC.DEC = OSCE Permanent Council Decision.

¹ United Nations, DPKO, Monthly summary of military and CivPol personnel deployed in current United Nations operations as of 31 Dec. 2001, 15 Jan. 2002.
² Figure as of 31 Dec. 2001, including military, observer, police, international civilian staff, local staff and 'other' UN employees. Note that this figure represents the total mission fatalities for all UN missions since 1948, not only those listed below. United Nations (note 1).
³ Total of figures listed below. Does not include UNMIK.
⁴ As of 15 Dec. 2001. United Nations, Interim report of the Secretary-General on the situation concerning Western Sahara, UN document S/2002/41, 10 Jan. 2002, para. 34.
⁵ UNTSO was established in May 1948 to assist the Mediator and the Truce Commission in supervising the observance of the truce in Palestine after the Arab–Israeli War that followed the creation of the state of Israel. The mandate was maintained during 2001.
⁶ United Nations (note 1).
⁷ United Nations (note 1).
⁸ Includes 3 locally recruited staff. United Nations, 'Fatalities by mission and appointment type—as of December 31 2001', UN Internet site, URL <http://www.un.org/Depts/dpko/fatalities/fatal1.htm>.

[9] Budget for 2001. United Nations, 'Middle East–UNTSO: Facts and figures', UN Internet site, URL <http://www.un.org/Depts/DPKO/Missions/untso/untsoF.htm>. UNTSO is funded through the UN's regular budget and consequently should not suffer arrears.

[10] UNMOGIP was established in Mar. 1951 to replace the United Nations Commission for India and Pakistan (SCR 91, 30 Mar. 1951). Its task is to supervise the ceasefire in Kashmir under the July 1949 Karachi Agreement. UNMOGIP Internet site, URL <www.un.org/Depts/DPKO/Missions/unmogip.htm>.

[11] United Nations (note 1).

[12] United Nations (note 1).

[13] Includes 2 locally recruited staff. United Nations (note 1).

[14] United Nations, 'India and Pakistan–UNMOGIP: Facts and figures', UN Internet site, URL <http://www.un.org/Depts/DPKO/Missions/unmogip/unmogipF.htm>. UNMOGIP is funded through the UN's regular budget and consequently should not suffer arrears.

[15] UNFICYP was established by SCR 186 (4 Mar. 1964) to prevent fighting between the Greek Cypriot and Turkish Cypriot communities and to contribute to the maintenance and restoration of law and order. Since 1974 UNFICYP's mandate has included monitoring the ceasefire and maintaining a buffer zone between the 2 sides. The mandate was extended until 15 June 2002 by SCR 1384 (14 Dec. 2001).

[16] United Nations (note 1).

[17] United Nations (note 1).

[18] United Nations (note 8).

[19] Includes a voluntary contribution of $13 565 715 from the Government of Cyprus and $6.5 million from the Government of Greece. United Nations, Report of the Secretary-General on the United Nations operation in Cyprus, UN document S/2001/1122, 30 Nov. 2001, para. 15.

[20] As of 31 Oct. 2001. United Nations (note 19), para. 17.

[21] UNDOF was established after the 1973 Middle East War under the Agreement on Disengagement and SCR 350 (31 May 1974), to maintain the ceasefire between Israel and Syria and to supervise the disengagement of Israeli and Syrian forces. The mandate was extended until 31 May 2002 by SCR 1381 (27 Nov. 2001).

[22] United Nations (note 1).

[23] United Nations (note 1).

[24] The military observers are seconded from UNTSO's Observer Group Golan. United Nations, 'Syrian Golan Heights–UNDOF: Facts and Figures', UN Internet site, URL <http://www.un.org/Depts/DPKO/Missions/undof/undofF.htm>.

[25] United Nations (note 8).

[26] As of 31 Oct. 2001. United Nations, Report of the Secretary-General on the United Nations Disengagement Observer Force, UN document S/2001/1079, 15 Nov. 2001, paras 8–9.

[27] UNIFIL was established by SCR 425 (19 Mar. 1978), to confirm the withdrawal of Israeli forces from southern Lebanon and to assist the Government of Lebanon in ensuring the return of its effective authority in the area. The mandate was renewed until 31 Jan. 2002 by SCR 1365 (31 July 2001).

[28] United Nations (note 1).

[29] United Nations (note 1).

[30] The military observers are seconded from UNTSO. United Nations, 'Lebanon–UNIFIL: Facts and figures', UN Internet site, URL <http://www.un.org/Depts/DPKO/Missions/unifil/unifilF.htm>.

[31] Includes 1 locally recruited staff member. United Nations (note 8).

[32] For the period July–Dec. 2001. United Nations (note 30).

[33] As of 15 Dec. 2001. UN, Report of the Secretary-General on the United Nations Interim Force in Lebanon, UN document S/2002/55, 16 Jan. 2002, para 21.

CONFLICT PREVENTION 141

34 UNIKOM was established by SCR 689 (9 Apr. 1991) as an unarmed observation mission with the mandate to monitor the Khawr 'Abd Allah and the demilitarized zone and to observe any hostile actions between the 2 states. In Feb. 1993 the mandate was expanded with the addition of an infantry battalion by SCR 806 (1993) to prevent small-scale violations of the DMZ and the borders.
35 United Nations (note 1).
36 United Nations (note 1).
37 Includes 1 locally recruited staff member. United Nations (note 8).
38 Two-thirds of this amount is paid by Kuwait. United Nations, Report of the Secretary-General on the United Nations Iraq–Kuwait Observation Mission, UN document S/2001/913, 26 Sep. 2001, para. 15.
39 As of 31 Aug. 2001. United Nations (note 38), para 16.
40 MINURSO was established by SCR 690 (29 Apr. 1991) to monitor the ceasefire between the Frente Polisario and the Moroccan Government, verify the reduction of Moroccan troops in Western Sahara and organize a free and fair referendum. The mandate was renewed until 28 Feb. 2002 by SCR 1380 (27 Nov. 2001).
41 United Nations (note 1).
42 United Nations (note 1).
43 United Nations (note 8).
44 United Nations, Interim report of the Secretary-General on the situation concerning Western Sahara, UN document S/2002/41, 10 Jan. 2002, para. 33.
45 As of 15 Dec. 2001. United Nations (note 44), para. 34.
46 UNOMIG was established by SCR 858 (24 Aug. 1993). The mission's original mandate of verifying the ceasefire between the Georgian Government and the Abkhaz authorities was invalidated by resumed fighting in Abkhazia in Sep. 1993, and UNOMIG was given an interim mandate to maintain contacts with both sides to the conflict and with Russian military contingents and to monitor and report on the situation. The mandate was renewed until 31 Jan. 2002 by SCR 1364 (31 July 2001).
47 United Nations (note 1).
48 United Nations (note 1).
49 United Nations (note 8).
50 United Nations, 'Georgia–UNOMIG: Facts and figures', UN Internet site, URL <http://www.un.org/Depts/DPKO/Missions/unomig/unomigF.htm>.
51 As of 15 Dec. 2001. United Nations, Report of the Secretary-General concerning the situation in Abkhazia, Georgia, UN document S/2002/88, 18 Jan. 2002, para. 23.
52 The International Police Task Force (IPTF) was authorized in accordance with Annex 11 of the 1995 General Framework Agreement for Peace in Bosnia and Herzegovina, the Dayton Agreement (SCR 1035, 21 Dec. 1995), together with a civilian mission proposed by the Secretary-General. United Nations, Report of the Secretary-General on Former Yugoslavia, UN document S/1995/1031, 13 Dec. 1995. The mission was later given the name UNMIBH. The mandate was extended until 21 June 2002 by SCR 1357 (21 June 2001).
53 United Nations (note 1).
54 United Nations (note 1).
55 Includes 2 locally recruited staff. United Nations (note 8).
56 Sum outstanding as of 31 Oct. 2001. United Nations, Report of the Secretary-General on the United Nations Mission in Bosnia and Herzegovina, UN document S/2001/1132, 29 Nov. 2001, para. 31.
57 UNMOP was established by SCR 1038 (15 Jan. 1996) to monitor the demilitarization of the Prevlaka peninsula, hitherto carried out by UNPROFOR and UNCRO. Its mandate was extended until 15 Jan. 2002 by SCR 1362 (11 July 2001).
58 United Nations (note 1).
59 United Nations (note 1).

60 For administrative and budgetary purposes UNMOP is treated as part of UNMIBH. United Nations, Report of the Secretary-General on the United Nations Mission of Observers in Prevlaka, UN document S/2000/1251, 29 Dec. 2000, para. 14.

61 UNMIK was established by SCR 1244 (10 June 1999). Its main tasks are: promoting the establishment of substantial autonomy and self-government in Kosovo; civilian administrative functions; maintaining law and order; promoting human rights; and assuring the safe return of all refugees and displaced persons. A positive decision by the Security Council is required to terminate the mission. SCR 1244 (10 June 1999), Article 19.

62 United Nations (note 1).

63 United Nations (note 1).

64 'Other' UN employees. United Nations (note 8).

65 Budget set at 505 million DM and is as of 30 Sep. 2001. 1 DM = $0.45 (SEBanken, Sweden). UN, Report of the Secretary-General on the United Nations Interim Administration in Kosovo, UN document S/2002/62, 15 Jan. 2002, Annex III, table A.

66 UNAMSIL was established by SCR 1270 (22 Oct 1999) following the signature of the Lomé Peace Agreement between the Sierra Leone Government and the Revolutionary United Front (RUF) on 7 July 1999. The tasks of the mission were to include *inter alia* assisting in the implementation of the Lomé Agreement, monitoring adherence to the ceasefire, encouraging the parties to create confidence-building mechanisms, supporting the anticipated elections, and ensuring the security and freedom of movement of UN personnel. In 2000 its mandate was strengthened and its numbers increased to 13 000. The mandate was extended until 31 Mar. 2002 by SCR 1370 (18 Sep. 2001).

67 United Nations (note 1).

68 United Nations (note 1).

69 United Nations (note 8).

70 Proposed budget. United Nations, Twelfth report of the Secretary-General on the United Nations Mission in Sierra Leone, UN document S/2001/1195, 13 Dec. 2001, para. 83.

71 As of 15 Nov. 2001. United Nations (note 70).

72 UNTAET was established by SCR 1272 (25 Oct. 1999). The mission was endowed with overall responsibility for the administration of East Timor and empowered to exercise all legislative and executive authority, including the administration of justice. The military component of UNTAET replaced INTERFET on 23 Feb. 2000. Its mandate was extended until 31 Jan. 2002 by SCR 1338 (31 Jan. 2001).

73 United Nations (note 1).

74 United Nations (note 1).

75 United Nations (note 8).

76 Budget for the period July–Dec. 2001. The revised budget for the period July 2001–June 2002 is $458 million. United Nations, 'East Timor–UNTAET Facts and figures', UN Internet site, URL <http://www.un.org/peace/etimor/UntaetF.htm>.

77 As of 15 Dec. 2001. United Nations, Report of the Secretary-General on the United Nations Transitional Administration in East Timor, UN document S/2002/80, 17 Jan. 2002, para. 57.

78 SCR 1279 (30 Nov. 1999) established the United Nations Organisation Mission in the Democratic Republic of the Congo (MONUC). It is mandated to liaise with the Joint Military Commission (JMC), plan for the observation of the ceasefire and the disengagement of forces, and provide humanitarian assistance. UN document S/1999/1279, 30 Nov. 1999. On 24 Feb. 2000 its mandate was extended and also expanded to include the deployment of around 5000 troops to protect UN and JMC personnel, and civilians under imminent threat of violence. UN document S/2000/1291, 24 Feb. 2000. SCR 1355 (15 June 2001) extended its mandate until 15 June 2002.

79 United Nations (note 1).

80 United Nations (note 1).

81 United Nations (note 8).

CONFLICT PREVENTION   143

82 Budget for period July–Dec. 2001. A revised budget for the period July 2001 to June 2002 is in preparation. United Nations, 'Democratic Republic of the Congo–MONUC Facts and figures', UN Internet site, URL <http://www.un.org/Depts/dpko/monuc/monucF.html>.
83 As of 30 Sep. 2001. United Nations, Ninth report of the Secretary-General on the United Nations Organization Mission in the Democratic Republic of the Congo, UN document S/2001/970, 16 Oct. 2001, para. 91.
84 UNMEE was established by SCR 1312 (31 June 2000). The mission was mandated to prepare a mechanism for verifying the cessation of hostilities, the establishment of the Military Co-ordination Commission provided for in the ceasefire and a peacekeeping deployment. The mission was later expanded with the allocation of 4200 troops and 220 military observers and tasked to monitor the ceasefire, repatriate Ethiopian and Eritrean troops and monitor the positions of Ethiopian and Eritrean troops outside a 25-km temporary security zone, to chair the Military Co-ordination Commission of the UN and the OAU, and to assist in mine clearance. SCR 1320 (15 Sep. 2000). SCR 1369 (14 Sep. 2001) extended its mandate until 15 Mar. 2002.
85 United Nations (note 1).
86 United Nations (note 1).
87 United Nations (note 8).
88 Proposed budget. United Nations, Progress report of the Secretary-General on Ethiopia and Eritrea, UN document S/2001/1194, 13 Dec. 2001, para. 60.
89 As of 31 Oct. 2001. United Nations (note 88), para. 60.
90 UN peace operations not deployed under Chapter VI or VII of the UN Charter and administered by the UN Department of Political Affairs (DPA). This list does not include UN peace-building offices.
91 In Apr. 1999, UNSMA military advisers returned to Kabul for the first time since Aug. 1998, when all UN staff were withdrawn from Afghanistan after the killing of 2 local UN staff and a military adviser. United Nations, Report of the Secretary-General on the situation in Afghanistan, UN document S/1999/698, 20 June 1999. UNSMA keeps a rotational presence in Kabul.
92 Countries listed are those whose participation has been verified; other information is not available from New York. Telephone conversations with Thomas Winkler, Minister Counsellor, Embassy of Denmark in Sweden, 1 Feb. 2002; Col. Cotte Brune, Defence Attaché, Embassy of France in Sweden, 31 Jan. 2002; Lt-Col Christian Hettfleich, Defence Attaché, Embassy of the Federal Republic of Germany in Sweden, 7 Feb. 2002; Stina Götbrink, Swedish Ministry of Foreign Affairs, 24 Jan. 2002; Barnaby Willitts-King, programme officer for Afghanistan, British Department for International Development (DFID), 4 Feb. 2002; and Andrii Semenchuk, First Secretary, Embassy of Ukraine in Sweden, 20 Feb. 2002; and Email from Col Yunosuke Kawazu, Defence Attaché, Embassy of Japan in Sweden, 1 Feb. 2002.
93 Supported by 4 locally recruited staff members. Telephone conversation with Mikhail Seliankin, Information Officer, UN Department of Public Information, 15 Jan. 2002.
94 Telephone conversation with Seliankin (note 93).
95 Funded from the UN regular budget. Telephone conversation with Seliankin (note 93).
96 MINUGUA (Misión de Verificación de las Naciones Unidas en Guatemala) had until 1997 been limited to verifying the Comprehensive Agreement on Human Rights and the human rights aspects of the Agreement on Identity and Rights of Indigenous Peoples. In 1997 the parties to the agreement requested that MINUGUA expand its functions to verify all the signed agreements, and that the mission's functions should also comprise good offices, advisory and support services and public information. The mandate was extended until 31 Dec. 2001. UN General Asssembly Resolution 55/177, 1 Mar. 2001.
97 Email from Mercedes Zalaya, MINUGUA, 5 Dec. 2001.
98 Includes 4 military observers and 277 international civilian staff. Email from Zalaya (note 96). The *SIPRI Yearbook 2001* listed only military and police observers.
99 United Nations (note 1).
100 Email from Zalaya (note 97).
101 $16.2 million. Email from Zalaya (note 97).

102 MICAH was established on 18 Feb. 2000 and took over from MICIVIH and MIPONUH on 16 Mar. 2000. It was mandated until 6 Feb. 2001 to support democratization, assist in judicial reform, help professionalize the police, assist in human rights and prepare for elections. UN General Assembly Resolution 54/193, 18 Feb. 2000). It withdrew once its mandate expired.
103 Email from Andrei Barac, Finance Department, DPKO, 2 Oct. 2001.
104 Supported by 80 locally employed staff. Email from Barac (note 103).
105 The budget for the period 16 Mar. 2000 to 6 Feb. 2001 included $8 753 900 from the regular budget and $14 734 000 from extra-budgetary resources (Trust Fund). Email from Barac (note 103).
106 Operations led by a coalition of states.
107 On 20 Dec. 2001 the SC, acting under Chapter VII, authorized a multinational force to help the Afghan Interim Authority maintain security in and around Kabul, as envisaged in Annex I of the Bonn Agreement. UN document SC/7248, 20 Dec. 2001.
108 The final force will consist of forces from 10–20 countries, but as of 31 Dec. 2001 only the UK (as the lead nation) had forces deployed in Afghanistan specifically assigned to ISAF. Telephone conversation with Andrew Smith, Pol/Mil Unit, British Foreign and Commonwealth Office, 7 Jan. 2002.
109 Approximate figure as of 31 Dec. 2001. This includes various advance elements and troops already in Afghanistan who will be transferred to ISAF. The final total has not yet been decided. Telephone conversation with Smith (note 108).
110 Telephone conversation with Smith (note 108).
111 Budget figures will not be available until the composition of the force is finalized. Telephone conversation with Smith (note 108).
112 Includes OSCE long-term missions and other field activities with a peacemaking or peace-building mandate, but not human rights offices, election monitoring groups or liaison offices.
113 Decision to establish the mission taken at 16th Committee of Senior Officials (CSO) meeting, 18 Sep. 1992, *Journal*, no. 3, Annex 1. The mission was authorized by the Government of the Former Yugoslav Republic of Macedonia (FYROM) through Articles of Understanding agreed by exchange of letters, 7 Nov. 1992. The mission's tasks include assessing the level of stability and the possibility of conflict and unrest. 25 international monitors were added until 31 Dec. 2001 to enhance the mission. PC.DEC/437/Corr.1, 6 Sep. 2001. Similarly, an additional 72 confidence-building monitors, 60 police advisers, 17 police trainers and 10 other staff were added until 31 Dec. 2001. PC.DEC/439, 29 Sep. 2001. Both deployments were under the terms of the existing mandate.
114 Email from Vasil Krpac, OSCE Spillover Mission to Skopje, 11 Dec. 2001.
115 Supported by 310 nationally employed staff. Email from Chantal Maille, OSCE Spillover Mission to Skopje, 10 Dec. 2001.
116 Email from Maille (note 115).
117 €7 424 300. Further Enhancement of the OSCE Spillover Monitor Mission to Skopje and the Deployment of Police Advisers and Police Trainers, Permanent Council Decision no. 439, OSCE Document PC.DEC/439, 28 Sep. 2001. €1 = $0.88 (SEBanken, Sweden).
118 Decision to establish the mission taken at the 17th CSO meeting, 6 Nov. 1992, *Journal*, no. 2, Annex 2. The mission was authorized by the Government of Georgia through an MOU on 23 Jan. 1993, and by South Ossetia's leaders by an exchange of letters on 1 Mar. 1993. Initially, the objective of the mission was to promote negotiations between the conflicting parties. The mandate was expanded on 29 Mar. 1994 to include *inter alia* monitoring of the Joint Peacekeeping Forces in South Ossetia. On 15 Dec. 1999 the mission's tasks were further expanded to include monitoring Georgia's border with Chechnya. OSCE Permanent Council Decision no. 334, PC.Jour/267, 15 Dec. 1999. The mandate was extended until 31 Dec. 2001. PC.DEC/442/Corr.1, 2 Nov. 2001. On 13 Dec. 2001 the mission's tasks were further expanded to include monitoring Georgia's border with Ingushetia. OSCE Permanent Council Decision no. 450, PC.DEC/450, 13 Dec. 2001.
119 Email from Anna Westerholm, Democratization Officer, OSCE Mission to Georgia, 19 Nov. 2001.
120 In the winter (from 15 Nov.) there are 20 core mission staff and 30 border monitors; in the summer the number of monitors is increased to 42. Email from Westerholm (note 119).

121 Email from Westerholm (note 119).
122 €9 257 800 from EU and €650 000 in voluntary contributions. Email from Westerholm (note 119).
123 Decision to establish the mission taken at the 18th CSO meeting, 13 Dec. 1992, *Journal*, no. 3, Annex 2. Authorized by the Estonian Government through MOU, 15 Feb. 1993. The mission's tasks include assisting in the recreation of civil society and collecting information relating to the status and rights of the communities in Estonia. In Nov. 2000, the mission received a set of guidelines from the Chairman-in-Office to direct the mission's work towards closure. Email from Neil Brennan, OSCE Mission to Latvia, 28 Nov. 2001.
124 Email from Doris Hertamph, Head of OSCE Mission to Estonia, 24 Sep. 2001.
125 Email from Hertamph (note 124).
126 Email from Hertamph (note 124).
127 €637 400. Email from Hertamph (note 124).
128 Decision to establish the mission taken at the 19th CSO meeting, 4 Feb. 1993, *Journal*, no. 3, Annex 3. Authorized by the Government of Moldova through MOU, 7 May 1993. The mission's tasks include assisting the parties in pursuing negotiations on a lasting political settlement to the conflict as well as gathering and providing information on the situation.
129 Fax from Lt-Col Jozef Gric, OSCE Mission to Moldova, 19 Oct. 2001.
130 Fax from Gric (note 129).
131 Fax from Gric (note 129).
132 $755 400. Fax from Gric (note 129).
133 Decision to establish the mission taken at the 23rd CSO meeting, 23 Sep. 1993, *Journal*, no. 3, Annex 3. Authorized by the Government of Latvia through MOU, 13 Dec. 1993. The tasks of the mission include addressing citizenship issues, providing information, advice on these issues and reporting on the implementation of OSCE norms. In Nov. 2000, the mission received a set of guidelines from the Chairman-in-Office to direct the mission's work towards closure. Email from Brennan (note 123).
134 Email from Brennan (note 123).
135 The mission is supported by 5 locally employed staff members. Email from Brennan (note 123).
136 Email from Brennan (note 123).
137 €703 000. Email from Brennan (note 123).
138 Decision to establish the mission taken at 4th meeting of the Ministerial Council, Rome (CSCE/4-C/Dec. 1), Decision I.4, 1 Dec. 1993. No bilateral MOU was signed. The tasks of the mission include facilitating dialogue, promoting human rights and informing the OSCE about further developments. The mandate was extended until 31 Dec. 2001. PC.DEC/422, 28 June 2001.
139 As of 12 Nov. 2001. Information from Ambassador Marc Gilbert, Head of Mission, 12 Nov. 2001.
140 There are also 55 locally employed staff members. Information from Gilbert (note 139).
141 Information from Gilbert (note 139).
142 €1 918 400. Telephone conversation with Maria Naydenova, Budget Department, OSCE Secretariat, 18 Dec. 2001.
143 Decision to establish the mission taken at 16th meeting of the Permanent Council, 11 Apr. 1995, Decision (a). No bilateral MOU was signed. The mission's tasks include promoting respect for human rights and a peaceful resolution to the crisis, facilitating delivery of humanitarian aid and ensuring the return of refugees and displaced persons. All international mission staff withdrew from Chechnya in Dec. 1998, but the mission continues to operate from offices in Moscow.
144 Email from OSCE Assistance Mission in Chechnya, 20 Sep. 2001.
145 OSCE Assistance Mission in Chechnya (note 144).
146 OSCE Assistance Mission in Chechnya (note 144).

147 €1 625 400. OSCE Assistance Mission in Chechnya (note 144).
148 In Aug. 1995 the OSCE Chairman-in-Office appointed a Personal Representative (PR) on the Conflict Dealt with by the OSCE Minsk Conference, which seeks a peaceful settlement to the Nagorno-Karabakh conflict. The PR's mandate consists of assisting the Minsk Group in planning possible peacekeeping operations, assisting the parties in confidence-building measures and in humanitarian matters, and monitoring the ceasefire between the parties. *Annual Report 2000 on OSCE Activities (1 Nov. 1999–31 Oct. 2000)*, 24 Nov. 2000.
149 Email from Nino Dekonozishvili, Administrative Officer/Assistant to the Personal Representative, 8 Nov. 2001.
150 The Personal Representative is assisted by 5 field assistants. However, only 4 posts are currently filled. Email from Dekonozishvili (note 149).
151 Email from Dekonozishvili (note 149).
152 € 909 600. Email from Dekonozishvili (note 149).
153 Decision to establish the mission taken at 5th meeting, Ministerial Council, Budapest, 8 Dec. 1995 (MC(5).DEC/1) in accordance with Annex 6 of the Dayton Agreement. The tasks of the mission include assisting the parties in regional stabilization measures and democracy building. The mandate was extended until 31 Dec. 2000. 260th PC meeting, PC.DEC/319, 2 Dec. 1999.
154 Fax from Maja Soldo, Personal Assistant to the Chief of Staff and Planning, OSCE Mission to Bosnia and Herzegovina, 19 Oct. 2001.
155 Fax from Soldo (note 154).
156 Fax from Soldo (note 154).
157 €25 056 600. Fax from Soldo (note 154).
158 Decision to establish the mission taken by the PC, 18 Apr. 1996, *Journal*, no. 65 (PC.DEC/112). Adjustment of the mandate by the Permanent Council, 26 June 1997, *Journal*, no. 121, PC.DEC/176, and 25 June 1998, *Journal*, no. 174, PC/DEC/239. The mission's tasks include assisting and monitoring the return of refugees and displaced persons as well as the protection of national minorities. The mandate was extended until 31 Dec. 2001. PC.DEC/396, 14 Dec. 2000.
159 Email from Alessandro Fracassetti, OSCE Mission to Croatia, 8 Nov. 2001.
160 There are also 238 locally recruited staff. Email from Fracassetti (note 159).
161 Email from Fracassetti (note 159).
162 €13 357 700. Email from Fracassetti (note 159).
163 Decision to establish the mission taken at the 198th meeting of the Permanent Council. The current mandate was set on 11 Dec. 1997, *Journal* no. 193, PC DEC/206.
164 Email from Caterina Artelli, Spokesperson, OSCE Presence in Albania, 12 Dec. 2001.
165 Email from Artelli (note 164).
166 Email from Artelli (note 164).
167 €4 253 200. Email from Artelli (note 164).
168 On 1 July 1999 the PC established the OSCE Mission in Kosovo for an initial period until 10 June 2000 to replace the transitional OSCE Kosovo Task Force, which had been established on 8 June 1999 (PC.DEC/296). The tasks of OSCE Mission to Kosovo include training police, judicial personnel and civil administrators, and monitoring and promoting human rights. On 10 Dec. 2001 the mandate was extended until 31 Dec. 2001. PC.DEC/382, 10 Dec. 2001.
169 Emails from Dominique LeDantec, Personnel Assistant, OSCE Secretariat, Vienna, 18 Dec. 2001; and Alexandra Gusarova, Personnel Officer, OSCE Secretariat, 24 Jan. 2002.
170 This figure includes 130 international police trainers and support staff attached to the OSCE-run Kosovo Police Service School. The internationally recruited personnel are assisted by 2033 locally recruited staff members. Email from Chris Cycmanick, Information Officer, OSCE Mission in Kosovo, 29 Nov. 2001.
171 Email from Cycmanick (note 170).
172 €95 056 300. Email from Cycmanick (note 170).

173 On 11 Jan. 2001 the PC established the OSCE Mission to the Federal Republic of Yugoslavia with an initial mandate of 1 year. Its mandate is to provide expert assistance to the Yugoslav authorities and civil society groups in the areas of democratization and human and minority rights, assist with the restructuring and training of law enforcement agencies and the judiciary, provide media support, and facilitate the return of refugees. PC.DEC/401, 11 Jan. 2001. The mission opened in Mar. On 15 Nov. 2001 the Permanent Council directed the mission to open an office in Podgorica, Montenegro. PC.DEC/444, 15 Nov. 2001.
174 Email from Nina Hartley, OSCE Mission to the Federal Republic of Yugoslavia, 11 Dec. 2001.
175 Supported by 87 locally employed staff members. Email from Hartley (note 174).
176 Email from Hartley (note 174).
177 €5 165 000. Email from Hartley (note 174).
178 SFOR was established in Dec. 1996 to replace the NATO Implementation Force (IFOR), created to implement the military aspects of the Dayton Agreement. SCR 1088 (12 Dec. 1996). On 21 June 2001 its mandate was extended for an additional 12 months by SCR 1357.
179 Email from Lt Phil Coope, Liaison Officer, SFOR Public Information Office, Sarajevo, 7 Dec. 2001.
180 Figure as of 5 Dec. 2001. Email from Coope (note 179).
181 Email from Coope (note 179).
182 51.6 million DM. Telephone conversation with Lt-Col Stephan Piat, Budget Financial Division, Supreme Headquarters Allied Powers Europe (SHAPE) Operations Centre, 16 Jan. 2002. This figure covers only the budget for the NATO HQs (civilian personnel and operations and maintenance costs), not investments in infrastructure necessary to support the operation. Contributing countries provide separate finances for their contingents.
183 KFOR received its mandate from the SC on 10 June 1999. Its tasks include deterring renewed hostilities, ensuring the withdrawal and preventing the return of the FRY military and police forces, demilitarizing the KLA, establishing a secure environment, supporting UNMIK and monitoring borders. SCR 1244, 10 June 1999.
184 United Nations, 'Nations contributing to KFOR', KFOR Internet site, URL <http://www.nato.int/kfor/kfor/nations/default.htm>, 18 Jan. 2002.
185 As of 31 Dec. 2001. Telephone conversation with Capt. Michael Coleman, Media Officer, SHAPE Public Information Office, 22 Jan. 2002.
186 Figures given are from Mar. 2001 to 7 Feb. 2002. Telephone conversation with Wing Commander Daz Slaven, KFOR, 7 Feb. 2002.
187 54.7 million DM. Telephone conversation with Piat (note 182). This figure covers only the budget for the NATO HQs (civilian personnel and operations & maintenance costs) and not investments in infrastructure necessary to support the operation. Contributing countries provide separate finances for their contingents.
188 The North Atlantic Council (NAC) approved a draft plan for the collection of weapons from armed insurgents on 29 June 2001. Deployment was authorized 2 days after the 13 Aug. peace agreement was signed, and the first troops began to arrive on 17 Aug. NATO Press Release (2001)112, 15 Aug. 2001. On 26 Sep. NATO announced that the collection was complete, and TFH began to withdraw. 'Task Force Harvest background information', NATO Internet site, URL <http://www.afsouth.nato.int/operations/skopje/harvest.htm#background>.
189 'Task Force Harvest composition, Operation Essential Harvest', 5 Nov. 2001, NATO Internet site, URL <http://www.afsouth.nato.int/operations/skopje/harvest.htm#composition>.
190 On 31 Aug. 2001 a TFH spokesman said that c. 4300 personnel, 92% of the force, were in theatre. '(FYRO) Macedonia: Op Essential Harvest', *Defence News Analysis*, Issue 01/33 (3 Sep. 2001), p. 1.
191 Funds for both TFH and TFF are drawn from the KFOR budget. The total amount for both operations was estimated at 300 000 DM. Telephone conversation with Piat (note 182).
192 The NAC authorized a follow-on mission to succeed TFH and 'contribute to the protection of international monitors who will oversee the implementation of the peace plan in the former Yugoslav Republic of Macedonia'. NATO Press Release (2001)133, 27 Sep. 2001. The UN Security Council authorized the establishment of a multinational security presence the same day. UN document 1371, 27 Sep. 2001. Deployment was immediate.
193 Email from Martin Klein, NATO AFSouth, 11 Jan. 2002.

194 Email from Klein (note 193).
195 Funds for both TFH and TFF are drawn from the KFOR budget. The total amount for both operations was estimated at 300,000 DM. Telephone conversation with Piat (note 182).
196 Mission established by the Brioni Agreement, signed at Brioni, Croatia, on 7 July 1991 by representatives of the European Community (EC) and the 6 republics of the former Yugoslavia. MOUs were signed with the government of Albania in 1997 and Croatia in 1998. The ECMM became the EUMM upon becoming an instrument of the EU's CFSP, and was mandated to monitor political and security developments, borders, inter-ethnic issues and refugee returns, to contribute to the early warning of the European Council, and to contribute to confidence building and stabilization in the region. Council Joint Action of 22 December 2000 on the European Union Monitoring Mission, EU document 2000/811/CFSP, 23 Dec. 2000, Intro., para. 6 and Article 1, para. 2.
197 The EUMM operates in Albania and in Bosnia and Herzegovina, Croatia, Macedonia (FYROM) and the FRY (Serbia, Montenegro, Kosovo and Presevo). Fax from Stephan Muller, Policy Unit of the General Secretariat, Council of European Union, 22 Jan. 2001.
198 Email from Stephan Muller, Policy Unit of the General Secretariat, Council of the European Union, 26 Sep. 2001.
199 The mandate calls for 120 monitors, though only 110 were assigned. At the beginning of Sep. 2001 it was decided to temporarily reinforce EUMM representation by a further 25 monitors in Macedonia (FYROM) until after the 27 Jan. 2002 parliamentary elections. This did not require a change of mandate. Email from Muller (note 198).
200 On 19 July 2001 a Slovak, a Norwegian and their Albanian translator were killed when their vehicle struck a mine in FYROM. Email from Muller (note 198).
201 €4 820 404. Council Joint Action of 22 December 2000 on the European Union Monitoring Mission, 23 Dec. 2000, Article 5, para 1.
202 Established under the authority of the Western European Union Council at the request of the EU, 2 May 1997. On 24 June 1997 an MOU between the Government of Albania and the WEU was signed. MAPE's mission was to rebuild and gradually hand over training responsibilities to the Albanian police, and was expanded to include training and advice throughout the country down to police unit level. On 1 June 2001 the EU assumed direct responsibility for a programme of advice, training and institution-building for the Albanian police, administered by the European Commission. Email from Isabelle MacDonald, Head of Council Section, 17 Oct. 2001.
203 As at close of mission, 31 May 2001. Email from MacDonald (note 202).
204 As at close of mission, 31 May 2001. Email from MacDonald (note 202).
205 €1 480 000. Email from MacDonald (note 202).
206 The Western European Union Demining Assistance Mission (WEUDAM) became operational on 10 May 1999, following a request by the EU, on the basis of Article J 4.2 of the Treaty on European Union. The mission provides advice, technical expertise and training support to the Croatian Mine Action Centre (CROMAC). WEUDAM website, URL <http://www.weu.int/eng/info/weudam.htm>. WEUDAM's mandate was extended until Dec. 2001, when it was withdrawn. WEUDAM Final Report, 30 Nov. 2001, p. 2.
207 WEUDAM (note 206), p. 3.
208 WEUDAM (note 206), p. 2.
209 €111 782. Second email from Isabelle MacDonald, 5 Feb. 2002.
210 Fax from Dieter Boden, Special Representative of the Secretary-General for Georgia, United Nations, 11 Dec. 2001.
211 Includes 500 in a Russian battalion, 250 in a Georgian battalion, 350 in a South Ossetian battalion and 100 logistical support troops. The mandated maximum is 1500. Fax from Boden (note 210).
212 18 military observers each from Russia, Georgia and South Ossetia. Email from Anna Westerholm, Democratization Officer, OSCE Mission to Georgia, 14 Jan. 2002.
213 According to official reports there have been no combat casualties, but some through accident and/or ill health. Email from Westerholm (note 212).
214 Email from Lt-Col Jozef Gric, OSCE Mission to Moldova, 24 Oct. 2001.
215 Each party provides 10 military observers. In addition there are 323 Russian, 331 Moldovan and 759 Trans-Dniestrian (including 308 not stationed in the security zone) peacekeeping troops. Email from Gric (note 214).

CONFLICT PREVENTION 149

216 Fax from Boden (note 210).
217 Fax from Boden (note 210).
218 Figure as of 11 Dec. 2001. Fax from Boden (note 210).
219 OMIB (or MIOB, Mission de l'OUA au Burundi) was established on 7 Dec. 1993 by the Central Organ of the OAU Mechanism for Conflict Prevention, Resolution and Management. The mission's mandate, to promote dialogue between military and government leaders, was endorsed by a treaty between the OAU and Burundi, 8 Apr. 1994. Ognimba, E., 'Connaissance de la Mission de l'OUA au Burundi' [Briefing on the OAU Mission in Burundi], *Resolving Conflicts*, Feb.–Mar. 1996, p. 10. OMIB was effectively withdrawn in July 1996. Fax from Said Djinnit, Assistant Secretary General, Political Affairs Department, 29 Oct. 2001.
220 The mission consists of a Special Representative of the Secretary General of the OAU and a Political Officer. Fax from Djinnit (note 219).
221 Fax from Djinnit (note 219).
222 $480 000. Fax from Djinnit (note 219).
223 OMIC (Mission d'Observation Militaire aux Comores) was established by decisions of the OAU at its 39th and 40th ordinary sessions at ambassadorial level in Adis Abeba, Ethiopia, 24 Oct. and 6 Nov. 1997. The tasks of the force include monitoring the situation on the Comoros and creating a climate of trust. De Matha, J. (Lt-Col, Logistics Officer, OMIC), 'La Mission d'Observation Militaire aux Comores', *Resolving Conflicts*, May–June 1998, pp. 25–26.
224 The mission consists of a Chief of Liaison Office/Special Representative of the Secretary General of the OAU, a Finance Officer and a Secretary. Fax from Djinnit (note 219).
225 Fax from Djinnit (note 219).
226 $209 000. Fax from Djinnit (note 219).
227 The JMC was formally established on 3 Sep. 1999 with a mandate to monitor compliance with the provisions of the July Lusaka Ceasefire Agreement and to investigate violations. OAU, Report of the Secretary-General on the DRC Peace Process, OAU Central organ/MEC/AMB/3, 23 Sep. 1999.
228 The JMC consists of personnel from Kenya, the Republic of Congo (Brazzaville) and Tanzania, as well as the 9 parties to the conflict. In addition to the states listed, these parties include the Rassemblement Congolais pour la Démocratie, the Rassemblement Congolais pour la Démocratie–Goma, the Rassemblement Congolais pour la Démocratie–Kisangi, and the Mouvement de la Libération Congolais. Fax from Djinnit (note 219); and Fax from Jean Mfasoni, Ag. Director, Political Affairs Department, 30 Nov. 2001.
229 Includes the JMC Chairman, 18 military officers attached to the main JMC and 36 to regional JMCs, as well as a Finance Officer and an OAU liaison to the JMC Chairman. Fax from Djinnit (note 219).
230 Budget for the period 1 Feb.–31 Aug. 2001. Proposed budget for Jan.–June 2002, based on 2001 total expenditure, is $2.5 million. Fax from Djinnit (note 219); and Email from BG Njuki Mwaniki, Military Chairman, JMC, 1 Feb. 2002.
231 Agreement concerning a military armistice in Korea, signed at Panmunjom on 27 July 1953 by the Commander-in-Chief, UN Command; the Supreme Commander of the Korean People's Army; and the Commander of the Chinese People's Volunteers. Entered into force on 27 July 1953.
232 Fax from Birgitta Delorme, Office of the Defence Attaché, Embassy of Switzerland in Sweden, 1 Oct. 2001.
233 The Swedish contingent was cut to 4 officers after a request from the Swedish Ministry for Foreign Affairs, which cited financial reasons. The Polish delegation of 2 personnel has no permanent presence in North Korea. Fax from Delorme (note 232).
234 Fax from Delorme (note 232).
235 Fax from Delorme (note 232).
236 The Multinational Force and Observers (MFO) was established on 3 Aug. 1981 by the Protocol to the Treaty of Peace between Egypt and Israel, signed on 26 Mar. 1979. Deployment began on 20 Mar. 1982, following the withdrawal of Israeli forces from Sinai. MFO, 'Multinational Force and Observers', Annual Report of the Director General, Jan. 2002.
237 Figure as of 1 Nov. 2001. *Annual Report* (note 236), p. 5.

238 Email from Mary Cordis, Chief of Personnel, MFO HQ, Rome, 8 Jan. 2002.
239 *Annual Report* (note 236), p. 42.
240 Protocol Concerning the Redeployment in Hebron, signed on 15 Jan. 1997.
241 Email from Lars Tore Kjerland, Senior Press and Information Officer/Official Spokesperson, TIPH, 27 Sep. 2001.
242 Email from Kjerland (note 241).
243 Email from Kjerland (note 241).
244 This figure does not include salaries, which are paid by the contributing countries. Email from Kjerland (note 241).
245 The PMG was set up in 1998 to monitor the ceasefire and to assist in democratic resolution of the conflict in Bougainville. Information provided by Capt. Lorraine Mulholland, Public Relations Officer, PMG Bougainville, by email on 14 Dec. 2000.
246 As of 20 Dec. 2001. Telephone conversation with Darren Brown, Australian Embassy in Stockholm, 14 Jan. 2002.
247 As of 20 Dec. 2001. Telephone conversation with Brown (note 246).
248 Budget set at AUD 10 million and is for the period 1 July 2001–30 June 2002. 1 AUD = $0.51 (SEBanken, Sweden). Email from Andrew Barnes, First Secretary, Embassy of Australia in Sweden, 2 Mar. 2002.
249 Annex II of the Townsville Peace Agreement between the Solomon Islands Government and the Guadalcanal and Malaitan militias, 15 Oct. 2000, agreed to the establishment of an international mission mandated to assist in confidence-building, receive and catalogue surrendered weapons and monitor treaty violations. It reports to the Peace Monitoring Council. Fax from Jemal Sharah, Executive Officer (Solomon Islands), Pacific Affairs Branch, Department Foreign Affairs and Trade, Australia, 17 Nov. 2000.
250 Email from Peter McCready, Department of Foreign Affairs and Trade, Australia, 27 Sep. 2001.
251 Email from McCready (note 249).
252 Email from McCready (note 249).
253 Between Nov. 2000 and Nov. 2001. AUD 4.1 million was spent. Email from McCready (note 249).
254 The Special Protection Force was established at the 15th Summit of the Regional Peace Initiative on Burundi on 23 July 2001 with a mandate to protect state institutions and Burundian political leaders returning from exile, and to act as a confidence-building measure. The force would be contributed to by Ghana, Nigeria, Senegal and South Africa. On 23 Oct. 2001, South Africa stated to the Security Council its intention to deploy an interim protection force on 1 Nov. 2001. The Security Council endorsed the establishment of the interim security presence with SCR 1375 (29 Oct. 2001). The OAU also expressed its support for the establishment of the SAPSD. Joint Communiqué of the 15th Summit of the Regional Peace Initiative on Burundi, 23 July 2001; and Fax from Ki Doulaye Corentin, Ag. Head, Conflict Management Centre, OAU, 23 Jan. 2002.
255 Email from S/Sgt D. Thathana, South African Department of Defence Information Centre, 14 Jan. 2002.
256 Other countries have expressed an intention to contribute troops, but only South Africa had deployed personnel by 31 Dec. 2001. Email from Thathana (note 255).
257 Fax from Chargé d'Affaires, South African Embassy, Stockholm, 12 Nov. 2001. According to Thathana, the maximum number of South African troops mandated is 701. Email from Thathana (note 255).
258 Email from Thathana (note 255).
259 Estimated official budget for the mission, provided mainly by the EU. Fax from Chargé d'Affaires (note 257).

3. The military dimension of the European Union

ZDZISLAW LACHOWSKI

I. Introduction

The modest attempts of the European Union (EU) to effectively develop its own military security policy were not immediately successful after the major changes which occurred in the late 1980s and early 1990s. In the face of the new challenges and threats that were emerging in Europe and elsewhere, the post-cold war transatlantic 'division of labour' as regards security could no longer be predicated on the traditional division into military and non-military areas. Almost a decade passed before the EU states made this realization and developed the concept of the European Security and Defence Policy (ESDP) under the EU's 'second pillar' (common foreign and security policy, CFSP).

Since the 1999 European Council meeting in Helsinki its 'Headline Goal'—to be able by 2003 to rapidly deploy a sizeable force, up to corps level, for crisis management tasks—has been pursued. Efforts have been made to better meet security threats by implementing the full range of crisis management missions: the Petersberg tasks.[1] Events in 2001 served as a mid-course test for the success of these efforts.

The EU is confronted with several major issues: the ultimate goal and shape of the ESDP; how best to pursue the Headline Goal in both institutional and capability terms; and the challenge of politico-military integration.

This chapter analyses developments in 2001 in the run-up to the deadline of 2003 and assesses the progress of the ESDP. Section II reviews West European efforts until the end of 2000. Section III addresses the steps which have been taken to transform the political commitments made by the EU states into structures and military capabilities and examines the relationship between the EU and other actors. Section IV discusses the EU–US relationship and the challenges to military security cooperation between the EU member states.[2]

---

[1] Presidency Conclusions: European Council, Helsinki, 10–11 Dec. 1999, reproduced in 'From St-Malo to Nice, European defence: core documents', compiled by M. Rutten, *Chaillot Papers*, no. 47 (May 2001). The Petersberg tasks include humanitarian intervention and evacuation operations; peacekeeping; and crisis management, including peacemaking. They are an aspect of the ESDP as reformulated in the 1997 Treaty of Amsterdam amending the Treaty on European Union, the treaties establishing the European Communities and certain related acts. Excerpts of the treaty concerning the CFSP and ESDP are reproduced in *SIPRI Yearbook 1998: Armaments, Disarmament and International Security* (Oxford University Press: Oxford, 1998), pp. 177–81. The Petersberg tasks are discussed in Rotfeld, A. D., 'Europe: an emerging power', *SIPRI Yearbook 2001: Armaments, Disarmament and International Security* (Oxford University Press: Oxford, 2001), pp. 190, 193–195. All EU documentation can be found at the EU Internet site, URL <http://europa.eu.int/>.

[2] Goals have also been set that are intended to improve non-military crisis management and conflict prevention. This chapter does not address those aspects of the EU efforts.

II. From 1998 to 2000

The conflicts in the Balkans made the EU states conscious of the urgent need to reassess and alter the past approach: the European Community and the EU providing aid, trade and cooperation, while the North Atlantic Treaty Organization (NATO) served as the military defender of Europe. Events in the 1990s also demonstrated the growing US reluctance to remain involved in European affairs on the scale of the cold war and the US preference for increased European involvement in situations which affect Europe more than the United States. NATO's intervention in Kosovo in 1999 illustrated the gap between Europe's limited military capability and US military strength. The EU's heavy reliance on the US military capacity during the Kosovo conflict moved the debate on the military security dimension forward and led to the creation of the Headline Goal and related 'collective capability goals' (deployability, sustainability, interoperability, flexibility, mobility, survivability in the areas of command and control, intelligence, logistics, and strategic transport).[3]

The 1992 Treaty on European Union (Maastricht Treaty) transformed the European Community's European Political Cooperation into a Common Foreign and Security Policy, which aims to 'include all questions related to security of the Union, including the eventual framing of a common defence policy, which might in time lead to a common defence'.[4] The EU requested that the Western European Union (WEU) develop and implement decisions and actions. In June 1992 the Petersberg WEU Council agreed on new tasks to strengthen the operational role of the WEU.[5] The 1996 Berlin NATO meeting gave practical meaning to the Petersberg tasks by envisaging the creation of a Combined Joint Task Force to 'facilitate the use of separable but not separate military capabilities in operations led by the WEU'.[6] The Berlin decisions thus paved the way from what was considered a military–technical arrangement for borrowing assets from NATO in order to carry out NATO-authorized missions to an EU political security undertaking. In 1997 the WEU's Petersberg tasks were incorporated into the 1997 Treaty of Amsterdam.[7]

A new approach was introduced in the 1998 British–French Joint Declaration on European Defence (Saint Malo Declaration), which noted that 'the [European] Union must have the capacity for autonomous action, backed up by credible military forces, the means to decide to use them, and a readiness to

---

[3] For discussion of the EU's military security dimension since 1998 see Rotfeld, A. D., 'Europe: the institutionalized security process', *SIPRI Yearbook 1999: Armaments, Disarmament and International Security* (Oxford University Press: Oxford, 1999), pp. 250–54; Rotfeld, A. D., 'Europe: the new transatlantic agenda', *SIPRI Yearbook 2000: Armaments, Disarmament and International Security* (Oxford University Press: Oxford, 2000), pp. 195–200; and Rotfeld, *SIPRI Yearbook 2001* (note 1), pp. 186–99.

[4] European Communities, *Treaty on European Union* (Office for Official Publications of the European Communities: Luxembourg, 1992). Excerpts are reproduced in *SIPRI Yearbook 1994* (Oxford University Press: Oxford, 1994), pp. 251–57.

[5] WEU Ministerial Council, 19 June 1992 in Petersberg (Bonn), *Atlantic News*, no. 2436 (23 June 1992).

[6] NATO, 'Final Communiqué, Ministerial Meeting of the North Atlantic Council, Berlin', Press communiqué, M-NAC-(96)63, 3 June 1996, para 6, URL <http://www.nato.int/docu/pr/1996/p96-063e.htm>.

[7] *SIPRI Yearbook 1998* (note 1).

do so, in order to respond to international crises'.⁸ These steps were a preparation for incorporating the WEU into the EU. In 1999 the post of Secretary General of the Council of the European Union/High Representative for the Common Foreign and Security Policy was created. The Secretary General of the Council of the European Union, Javier Solana, was appointed as the first High Representative for the CFSP.

The 1999 European Council meeting in Cologne had significant implications for the ESDP. The meeting resolved that 'the European Union shall play its full role on the international stage' and that it should be given 'the necessary means and capabilities to assume its responsibilities regarding a common European policy on security and defence'. The EU was to take over the functions of the WEU by the end of 2000. The German presidency conclusions repeated the relevant text of the Saint Malo Declaration regarding the EU's capacity for autonomous action, but the mandate for the EU forces was to be limited to the Petersberg tasks.⁹ The 1999 Finnish presidency decisions envisaged that by 2003 the EU must be able to deploy within 60 days and sustain for at least one year military forces of up to 50 000–60 000 troops (with appropriate air and naval support) capable of carrying out the full range of Petersberg tasks. Collective capability goals in the areas of command, control, communications and intelligence (C^3I) and strategic transport were also adopted. In 2000 the EU legislation underwent another change: the Treaty of Nice, which has not yet entered into force, strengthened the links between the foreign, security and defence policy of the EU states and the EU framework.¹⁰ The amendments in the Treaty of Nice reflected the operative development of the ESDP as an EU project.

The 2000 Capabilities Commitment Conference made it possible to combine the national commitments that correspond to the military capability goals set by the European Council meeting in Helsinki.¹¹ The conference identified numerous areas where efforts will be made to improve assets, investment, development and coordination in order to gradually acquire or enhance the capabilities required for autonomous EU action. Denmark opted out of all aspects of EU cooperation with defence implications, but the other EU states

---

⁸ The British–French Joint Declaration on European Defence is reproduced in *SIPRI Yearbook 1999* (note 3), p. 265. The Apr. 1999 NATO Washington summit meeting agreed on the 'Berlin-plus' deal which assured EU access to NATO planning capabilities, the presumption of availability to the EU of pre-identified NATO capabilities and common assets, identification of a range of European command options, and the further adaptation of NATO's defence planning to incorporate more comprehensively the availability of forces for EU-led operations. NATO, 'Washington Summit Communiqué issued by the Heads of State and Government participating in the meeting of the North Atlantic Council in Washington, DC on 24th April 1999: An Alliance for the 21st Century', NAC-S(99)64, 24 Apr. 1999, para. 10, URL <http://www.nato.int/docu/pr/1999/p99-064e.htm>.

⁹ It added that such responses to international crises should be made 'without prejudice to actions by NATO'. Cologne European Council, Presidency Conclusions, 3–4 June 1999, reproduced in 'From St-Malo to Nice, European defence: core documents' (note 1), pp. 41–45.

¹⁰ Treaty of Nice amending the Treaty on European Union, the treaties establishing the European Communities and certain related acts, *Official Journal of the European Communities*, C 80/1, 10 Mar. 2001, URL <http://europa.eu.int/eur-lex/en/treaties/dat/nice_treaty_en.pdf>. The treaty in its definitive form was signed on 26 Feb. 2001 in Nice after legal and linguistic editing.

¹¹ The Capabilities Commitment Conference was held in Brussels on 20 Nov. 2000. See Sköns, E. *et al.*, 'Military expenditure and arms production', *SIPRI Yearbook 2001* (note 1), p. 245–46.

have committed themselves to making the national contributions, set out in the Force Catalogue (a pool of military assets and capabilities), corresponding to the rapid reaction capabilities. The EU and its partner states pledged over 100 000 troops, 400 aircraft and 100 ships to the European rapid reaction force (ERRF) pool.[12]

The Nice European Council meeting in 2000 decided to make the ESDP operationally capable as quickly as possible, at the latest by the end of 2001.

A credible security actor

Currently, the ESDP is confined to a limited security policy for the EU members; a defence policy has not yet been elaborated. British–French cooperation, as epitomized by the Saint Malo Declaration, represents the new approach to security in Europe. Under Prime Minister Tony Blair the UK has become more 'European', and in the second half of the 1990s France became more positive towards NATO. Thus both countries moved closer towards the centre of Europe's political spectrum from the two extremes—'Atlanticist' (the UK) and 'Europeanist' (France)—around which the other EU countries were formerly grouped.

All EU members now agree that the European Union must become a credible security actor, although views differ, particularly as regards various aspects of the European–US relationship. Thus the question is not whether a European force should be created but how and to what extent the force is to be developed, what degree of 'autonomy' is to be pursued and what roles it is to play. Although there is consensus on the need for an autonomous military security role for the EU, there is a spectrum of views: ranging from those of sceptics, who point to the allegedly insurmountable complexities of the scheme, to those of the enthusiasts, who assert that without a workable security component the CFSP is doomed to fail.

It is difficult to differentiate the various views within the EU with regard to the building of a 'European force'. However, with the two major European powers—France and the UK—placed in the middle, there remain states that are either cautious about the evolution of the ESDP and opt for retaining a strong transatlantic link (e.g., Denmark, the Netherlands and Portugal) or more pro-integrationist (e.g., Greece, Luxembourg and Spain). Unlike those states which support ambitious military missions for the ERRF (e.g., France), countries such as Finland and Sweden emphasize the non-military aspects of crisis management and envisage the ERRF concentrating on peacekeeping. However, the dividing lines between these groups are not clearly definable.[13]

---

[12] 'From St-Malo to Nice, European defence: core documents' (note 1), pp. 158–63. Analysts claim, however, that because of the need to rotate military personnel 180 000 troops would be required.

[13] E.g., unlike most of the Italian political community, Italy's Foreign Minister Renato Ruggiero took a strongly pro-European stance (one indication of which was Italy's participation in the A400 military Airbus project). After Ruggiero's resignation foreign policy under Prime Minister Silvio Berlusconi shifted towards a more Atlanticist position.

The varying interpretations of the strategic vision by the major European countries are important. While most of the EU countries support the short-term pragmatic approach to building the ERRF, France champions its long-term goal of a more ambitious '*puissance* Europe', going beyond 2003. Various statements by French officials have suggested that France's goal is not merely a modest crisis management capability for the EU, but the more ambitious project of a standing force.[14]

For the UK a credible European security capability means strengthening the European pillar of NATO. The UK is also determined to avert the possibility of more radical plans being pushed through by other EU states. The UK does not want to deplete the modest European resources in order to implement a scheme that might undermine the transatlantic links.[15] It therefore advocates effective use of NATO's assets, avoiding duplication and thus cutting the cost of future EU military operations.

Germany's stand on the ESDP is cautious and points both to increasing 'Europe's ability to act in accordance with its responsibilities, its resources and with international expectations' and developing a 'real strategic partnership' with the USA. German Defence Minister Rudolf Scharping reaffirms that 'in the military field, our goal is not to create a European army' and emphasizes the ESDP's close cooperation with NATO. Germany has expressed the view that both Euro-Atlantic security institutions will benefit from 'synergy and integration'.[16]

III. Towards operational capabilities

In order to bring the common security policy into being, as laid out in the Maastricht Treaty and other legally and politically binding documents, the ESDP is now focused on the preparation for and fulfilment of the Petersberg tasks. In 2001 work on the EU's new military capabilities was conducted under the EU presidencies of Sweden and Belgium.

The main EU goal in 2001 was to achieve an 'initial operational capability' by the end of the year. In the first half of 2001 Sweden's priorities in the security field emphasized the civilian aspects of crisis management and conflict prevention. Consequently, the Swedish presidency report on the ESDP to the Gothenburg European Council meeting was predominantly oriented to

---

[14] The suggestion by French armed forces chief of staff Gen. Jean-Pierre Kelche in early 2001 that the ERRF should have its own planning staff independent of NATO was negatively received by other EU states, fuelling fears that France is pursuing an autonomous European defence at the cost of NATO's role in Europe. 'EU: controversy characterizes debate on rapid reaction force', Radio Free Europe/Radio Liberty, *RFE/RL Newsline*, 28 Mar. 2001, URL<http://www.rferl.org/nca/features/2001/03/2803200 11104323.asp>.

[15] Terriff, T. et al., Royal Institute of International Affairs (RIIA), *European Security and Defence Policy after Nice*, Briefing Paper, New Series, no. 20 (Chatham House: London, Apr. 2001), URL <http://www.riia.org>.

[16] Scharping, R., 'European security policy and global stability', *U.S.I. Journal*, vol. 131 (Jan./Mar. 2001), pp. 18–19.

the non-military aspects of European security.[17] Nevertheless, a number of steps were taken to improve the EU's ability to conduct military crisis management operations. Major progress was made in consolidating the necessary internal structures and procedures, but in other areas EU members' efforts encountered obstacles. The Belgian presidency actively turned towards the military security field.[18] As agreed at the Gothenburg European Council meeting, the civilian aspect of the ESDP was also pursued in the latter half of 2001, although the development of the EU policing capability lagged behind military crisis management efforts.[19]

Both presidencies pursued the following security-related goals: (*a*) to enable the EU to respond more rapidly in a crisis (at the latest by the December 2001 European Council meeting in Laeken, Belgium), including taking the necessary measures for implementation and validation of the crisis management mechanisms and further discussions with NATO on mutual arrangements; (*b*) to ensure the follow-up of the military capability objectives (including defining the details of the evaluation mechanism and organizing the Capabilities Improvement Conference, CIC) in order to realize the Headline Goal commitments; (*c*) to pursue permanent arrangements with the 15 non-EU European countries; (*d*) to establish similar arrangements with other potential partners; (*e*) to set up a satellite centre and an institute for security studies within the EU; (*f*) to identify possible areas and modalities of cooperation with international organizations; and (*g*) to further enhance the cohesion and effectiveness of the EU's conflict prevention capability.[20]

The ESDP structures and procedures

The EU has set up structures and procedures which enable it to analyse, plan, decide on, launch and carry out military crisis management operations when NATO 'as a whole is not involved'. In accordance with the 1999 Helsinki guidelines, the 2000 Nice European Council meeting endorsed the creation of

[17] Presidency Report to the Gothenburg European Council on the European Security and Defence Policy, Press Release, Brussels, no. 9526/1/01, 11 June 2001, URL <http://ue.eu.int/Newsroom/loadDoc.asp?max=1&bid=75&did=66829&grp=3577&lang=1>. Both the Nice and the Gothenburg mandates also included recommendations concerning civilian crisis management measures. Of its 5 annexes only the 1 on EU exercise policy is devoted to military-related issues. The other annexes deal with the policing and civilian aspects of crisis management. The appointment of Finnish General Gustav Hägglund (a military representative of a neutral country) to head a new EU Military Committee was interpreted as a possible upgrading of non-military issues at the cost of the ERRF.

[18] According to Belgian Defence Minister André Flahaut the presidency priorities were: military capabilities, a European White Paper on defence, closer contacts with public opinion and national and European parliamentary assemblies, health issues (in the context of the depleted uranium issue), and cooperation between the armed forces of the EU members and in terms of armaments. *Atlantic News*, no. 3308 (13 July 2001), p. 3.

[19] For discussion of EU policing activities see Dwan, R. (ed.), *Executive Policing: Enforcing the Law in Peace Operations*, SIPRI Research Report no. 16 (Oxford University Press: Oxford, forthcoming); and Dwan, R., *International Policing in Peace Operations: The Role of Regional Organizations*, SIPRI Research Report no. 19 (Oxford University Press: Oxford, forthcoming).

[20] European Council, Nice, 7–9 Dec. 2000, Presidency Report on the European Security and Defence Policy, VIII, reproduced in 'From St-Malo to Nice, European defence: core documents' (note 1), pp. 175–76; and Presidency Report to the Gothenburg European Council (note 17).

three new bodies—the Political and Security Committee (PSC), the European Union Military Committee (EUMC) and the European Union Military Staff (EUMS)—for the oversight of ERRF policy and strategy.[21] All decisions within these bodies require consensus, which ensures that the ESDP remains a 'common' rather than a 'single' policy.[22]

The politico-military structures of the Secretariat of the European Council were strengthened in 2001. The EU's Situation Centre monitors both civil and military occurrences and provides early warning and crisis monitoring. In order to strengthen the ESDP Solana increased the number of staff and facilitated a process whereby some of the WEU functions are to be assumed by the EU.[23]

The Political and Security Committee

On 22 January 2001 the Political and Security Committee, which deals with all aspects of the CFSP, replaced the Political Committee for the CFSP. The PSC is the 'linchpin' of the ESDP and the CFSP and exercises 'political control and strategic direction' of the EU's military response to crises.[24] It also plays a major role in coordinating consultation with NATO and third-party states that are involved in a crisis situation. The PSC meets at the ambassadorial or equivalent level in Brussels (usually twice a week). Ten tasks have been assigned to the PSC, including: drawing up 'opinions' for the Council of the European Union (hereafter Council), providing guidelines for other committees which address CFSP issues, sending guidelines to the EUMC and taking responsibility for the political direction of the development of military capabilities. The PSC examines political, diplomatic and civil measures, as well as military options. It is not yet certain how the PSC will cooperate with other bodies, such as the Committee of Permanent Representatives of the Member States at the European Union (COREPER).

As a rule, the chairmanship of the PSC rotates with the presidency. However, the Secretary General of the Council of the European Union/High Representative for the CFSP can act as chairman in a crisis.[25] Formal consultations between the EU and NATO started in early 2001 at the level of the PSC and the North Atlantic Council (NAC).

---

[21] Presidency Conclusions (note 1), pp. 171–72.
[22] Belgium, Germany and the Netherlands had opted to give the ESDP structures a more *communautaire* character but were effectively opposed by other states including France, the UK and some others.
[23] The European Union Institute for Security Studies and the European Union Satellite Centre, which are intended to support the CFSP and ESDP, formally began operation on 1 Jan. 2002.
[24] Nice European Council Meeting Presidency Conclusions, Nice, 8 Dec. 2000, Annex III, reproduced in 'From St-Malo to Nice, European defence: core documents' (note 1), pp. 191–93.
[25] The creation of the post of the Secretary General of the European Council/High Representative for the CFSP was intended to enhance the European political security identity. It is the hope of some EU states that the post will evolve into that of a 'secretary of state' for the EU.

The European Union Military Committee

There are two additional military-related bodies: the European Union Military Committee and the European Union Military Staff, which is discussed below. The EUMC was made permanent on 9 April 2001, and its first formal meeting was held on 23 May. It is composed of the national defence ministers represented by their military representatives (most of whom are also NATO military representatives) and is the highest military body within the Council. It gives advice and makes recommendations to the PSC on all military matters and has the right to initiate proposals and activities. It also provides military direction to the EUMS and acts as a liaison between it and the PSC. In crisis management situations it acts on the request of the PSC.[26] The EUMC is responsible for maintaining an official military relationship with non-EU European NATO members and organizations. The first meeting of the EUMC and NATO's Military Committee was held on 12 June at NATO Headquarters. Information was exchanged on existing assets and capabilities and the ongoing work of both bodies. It was decided that the two bodies will meet as required and at least once during each EU presidency.

Some meetings of the General Affairs Council, a group made up of foreign ministers from member states, have also included defence ministers. Discussions have begun on the establishment of a separate Defence Ministers Council to better handle military capability requirements and overall military coordination. In the past the idea had been rejected by the EU foreign ministers. The EU defence ministers held their first informal meeting on 6 April 2001 to discuss the issues of operationalization of crisis management capabilities and the agenda for the November 2001 CIC. A second informal meeting was held in October. The establishment of a Defence Ministers Council was also discussed at the CIC, but no decision was taken. The Belgian presidency report on ESDP invited the Spanish presidency to further examine the proposal. When the ERRF has developed further, the role of the defence ministers is bound to increase.

The European Union Military Staff

The third body, the European Union Military Staff, was made permanent on 11 June 2001. Its 135 staff members provide military expertise and early-warning capability and support the EUMC in situation assessment and the military aspects of strategic planning, including the 'identification of European national and international forces'.[27] However, the EUMS does not have the capacity to plan operations and there is no plan to create a separate EU planning headquarters like that of NATO. The EUMS links the EUMC with national and multinational military headquarters. These arrangements are voluntary for EU members and independent from the control or scrutiny of the European Commission or the European Parliament. The EUMS is part of the

[26] 'From St-Malo to Nice, European defence: core documents' (note 1), pp. 193–96.
[27] 'From St-Malo to Nice, European defence: core documents' (note 1), pp. 196–98.

Council Secretariat, headed by the High Representative for the CFSP. In addition to his CFSP competences, his role as an integrating force of foreign and security policy has potentially been significantly enhanced.[28]

Procedures and exercises

Deployment of the ERRF can be initiated independently or in conjunction with other international organizations such as the Organization for Security and Co-operation in Europe (OSCE) or the United Nations. Any military response to a crisis that is carried out under an EU joint action is to remain under the political and strategic control of the EU, even when the assets of NATO or another organization are used.[29]

If a deployment of the ERRF is to be made, the PSC requests, and the Military Committee issues, an 'initiating directive' to the EUMS Director General. The EUMS draws up strategic options which are sent back to the EUMC. The EUMC may add comments and returns the directive to the PSC. Once it has been given PSC approval the 'initial planning directive' provides guidelines for military action. The host country (i.e., the country to which the troops will be sent) is then asked to accept the action. Gaining the approval of all the EU states will present difficulties. There will be a need for political unity and cooperation in sharing resources and carrying out such missions.

In June 2001 crisis management procedures and measures to facilitate decision making and adequate coordination of all the EU instruments were further elaborated and tested at a PSC crisis management workshop. As a result, the EU Exercise Policy and an EU Exercise Programme for 2001–2006 were agreed.[30] The PSC has overall responsibility for all EU exercises. The EU Exercise Policy sets out the requirements for and categories of international crisis management exercises (e.g., chain of command and procedures and arrangements with NATO and other non-EU European partners), including 'the most demanding' ones. The EU has invited NATO to observe all of the EU exercises as long as such invitations are reciprocated; it can also invite non-EU European NATO members and EU candidates to take part in exercises as well as other states, organizations and non-governmental organizations (NGOs). A second workshop, a meeting of the PSC and the EUMC, was held in October 2001.[31] It examined the issues of financing crisis management operations, improving public and parliamentary knowledge of the ESDP, crisis management, health issues related to military operations and the like.

An exercise to test and validate the structures and procedures of the European crisis management mechanisms was carried out in May 2002. Joint EU–NATO command post exercises are scheduled for 2003, as soon as a formal

---

[28] Cornish, P. and Edwards, G., 'Beyond the EU/NATO dichotomy: the beginning of a European strategic culture', *International Affairs*, vol. 77, no. 3 (2001), p. 595.
[29] 'From St-Malo to Nice, European defence: core documents' (note 1), Annex III, pp. 192–93.
[30] Presidency Report to the Gothenburg European Council (note 17), Annex IV to the Annex, 'Exercise Policy of the European Union'.
[31] 'Developing procedures and preparing exercises', Presidency Report to the Laeken European Council on European Security and Defence Policy, p. 7.

agreement between the EU and NATO has been reached. No aspect of the EU Exercise Policy can be carried out until capability deficiencies have been addressed; its implementation will therefore remain limited in scope for some time to come.

Military capabilities

Although structure and procedure are important, military capability is crucial.[32] A large number of forces are to be made available for EU missions, but their quality varies. A pool of forces and capabilities exists from which forces can be rapidly assembled for particular operations on a case-by-case basis once the endorsement of the relevant national governments has been given. The units have to meet specific criteria as regards availability, deployability, sustainability and interoperability. Following the Capabilities Commitment Conference in November 2000 the EU Secretariat identified deficiencies in the force contributions made by the member states, including: insufficient long-range heavy air and sea lift capacity to rapidly deploy a substantial force; ineffective command and control systems at various levels; and problems associated with intelligence collection, interpretation and dissemination capability.

Consequently, the EU states were requested to review their contributions and take steps to remedy the shortfalls before the 2001 CIC under the Belgian presidency. Work was undertaken to further develop and refine operational and strategic capability requirements for: interoperability, rotation and readiness; C^3I; ISTAR (intelligence, surveillance, target acquisition and reconnaissance); and strategic mobility and logistics. The offers of capabilities by 15 non-EU European NATO members and states which are applying for EU membership were reviewed and clarified. The member states also worked on the details of the follow-up and evaluation mechanism for military capabilities. The focus was on reviewing the Headline Goal to ensure its compatibility with the pledges undertaken in NATO's defence planning process (Defence Capabilities Initiative, DCI) and the review process of the Partnership for Peace (PFP).

At the CIC in Brussels the EU member states made 'significant' quantitative and qualitative improvements to address the existing deficiencies. The participants adopted a statement on Improving European Military Capabilities and the European Capability Action Plan (ECAP) to gradually advance national and international solutions.[33] EU states claimed to have fulfilled some two-thirds of the 144 required capabilities. Of the remainder, 20 are considered serious and unresolved, of which 15 are addressed by NATO's DCI.[34] It was

---

[32] After the meeting with the EU defence ministers in Apr. 2001, Solana reportedly stated: 'We can have committees and procedures, but if we don't have capabilities we have nothing'. *The Guardian*, 7 Apr. 2001, p. 7.

[33] 'Statement on improving European military capabilities, European Capability Action Plan', Presidency Report to the Laeken European Council on European Security and Defence Policy, Annex I.

[34] 'EU resolves two-thirds of gap in capabilities', *Defense News*, 26 Nov.–2 Dec. 2001, p. 8; and Monaco, A., 'The rapid reaction force: the EU takes stock', *European Security Review*, no. 9 (Dec. 2001), pp. 1–2. In Jan. 2002 the Chairman of the Military Committee, Gen. Gustav Hägglund, stated that

stated that the EU ought to be able to carry out the whole range of Petersberg tasks by 2003. This was an indirect admission that work on the development of capabilities might not be complete and that some shortfalls will not be rectified by that date.

Force capabilities

The EU states confirmed the existence of a pool of more than 100 000 troops, some 400 combat aircraft and 100 ships. With regard to land-based forces, progress has been made in the areas of multiple rocket launchers, transmission, electronic warfare, armoured infantry and bridging engineering. Improvements were sought with regard to protecting deployed forces and commitment capability and logistics. The need to improve the quantity of available ground elements and the operational mobility and flexibility of deployed forces was also expressed.

Progress was reported regarding naval and aviation resources. However, improvements are required as regards naval aviation resources, maritime medical evacuation and other problems (including combat search and rescue tasks and precision guided weapons).

Strategic capabilities

There are a sufficient number of C^3I headquarters at the operation, force and component levels as well as deployable communication units. The member states have also offered to provide additional intelligence and surveillance resources. However, a qualitative analysis of these contributions has yet to be made and may reveal deficiencies. Assistance for strategic decision making is inadequate, and additional efforts are needed because the ISTAR capability remains limited.

Air and maritime strategic mobility has improved, but there is still a lack of wide-body aircraft and 'roll-on roll-off' (ro-ro) ships. No progress is being made as regards the plan to build a fleet of Airbus A400M transport aircraft, and some EU states have expressed a preference for the use of leased Ukrainian Antonov-24 aircraft or for optimizing the use of existing resources by coordinated or joint use, and the like.[35]

The EU member states are also attempting to improve the quality of their forces in eight areas: force structure; budget; staff; multinational cooperation;

---

the EU possessed 90% of the capabilities defined in the Force Catalogue. *Atlantic News*, no. 3355 (25 Jan. 2002), p. 3. See also chapter 6 in this volume.

[35] Funding and fielding the Airbus is seen as the litmus test of how serious the EU is about the Headline Goal. Italy has decided that it will not take part in the multinational project to build the A400M Airbus heavy-lift aircraft (it had committed itself to purchase 16 aircraft). 'Rome puts off decision on role in Airbus project', *Financial Times*, 9 Nov. 2001, p. 8. In Mar. 2002 Germany presented a two-stage purchase plan for the 73 A400M aircraft which it is committed to buy. With Germany's full order, the 8 EU countries (Belgium, France, Germany, Italy, Luxembourg, Spain, Turkey and the UK) would buy 196 aircraft. If the total dropped below 180, the contract would have to be renegotiated. 'Berlin clears way to A400 project funding', *Financial Times*, 20 Mar. 2002, p. 2. See also chapter 6 in this volume.

logistics; training; research, technology, industrial cooperation and public procurement; and civilian–military cooperation.

Non-EU European contributions

The offers of additional capabilities by non-EU European states (NATO states and the states which are applying for EU membership) were welcomed.[36] The offers made by these states at the November 2000 ministerial meeting were included in a supplement to the Force Catalogue. The non-EU European states were also requested to make similar offers at the CIC meeting in November 2001.

The European Capability Action Plan

The European Capability Action Plan was initially proposed by the Netherlands and subsequently addressed by the Belgian presidency. It aims to address capability deficiencies. It is not a timetable for action and goals but a set of principles and mechanisms for monitoring and encouraging gradual progress towards achieving the Headline Goal. ECAP is based on national decisions and aims at rationalizing the defence efforts of the members and increasing synergy between national and multinational projects. Its underlying principles are:

1. There is a need for enhanced effectiveness and efficiency of European military capability efforts. The plan calls for increased cooperation between member states and groups of member states in order to achieve rationalization gains.
2. The plan proposes a 'bottom–up approach' to European defence cooperation. The commitments of member states will be voluntary, rather than subject to a European-level scheme.
3. Coordination between EU member states and cooperation with NATO are targeted at removing specific shortcomings, avoiding wasteful duplication and ensuring transparency and consistency with NATO.
4. Broad public support is important and the public should be provided with a 'clear picture' of the CFSP/ESDP so that political action can be made more effective and political will strengthened.

The plan calls on member states to conclude their current projects and initiatives and make the new capabilities available to the EU. This requires: (*a*) making additional national forces and capabilities available and including them in future projects and initiatives; (*b*) making existing capabilities more effective and efficient and seeking 'creative responses' that go beyond traditional military procurement programmes; and (*c*) applying multinational solu-

---

[36] The EU has envisaged two formats for cooperation with the non-EU European countries: '15+6' (non-EU European NATO states: Czech Republic, Hungary, Iceland, Norway, Poland and Turkey) and '15 + 15' (non-EU European NATO states plus candidate states: Bulgaria, Cyprus, Estonia, Latvia, Lithuania, Malta, Romania, Slovakia and Slovenia).

tions (co-production, financing and acquisition of capabilities) and coordinating the management and use of equipment.

A meeting of senior national experts will analyse and evaluate the plan, which will continue to be developed by the EUMC. A Headline Task Force consisting of panels of experts on specific types of capability will be established to analyse remaining problems and identify feasible national and multinational solutions. The results of the analysis will be reported by the PSC to the Council at regular intervals.[37]

The Headline Goal after 11 September 2001

The CIC was held after the terrorist attacks on the USA. The hope was expressed that the changed strategic situation would positively affect the military approach of the EU and its financial status. The Afghanistan campaign re-emphasized the urgency of acquiring military airlift and airborne refuelling capabilities and the means to destroy air defences. British Defence Minister Geoff Hoon proposed that the Headline Goal be updated and refined as necessary to ensure its relevance.[38]

The EU High Representative for the CFSP pointed to new ESDP responsibilities which include putting additional emphasis on preparation for operational readiness, taking full account of the terrorist threat to European forces and civilian populations and further improving the early-warning process. In this context, Solana called for the strengthening of the Secretariat and the EU Situation Centre, in particular. He also warned against overburdening national budgets in response to suggestions by some EU countries that additional capabilities should be earmarked for combating terrorism.[39]

Although there was hope that the Petersberg tasks might be modified to address the new challenges, the goal of proclaiming the EU military structures as operational was given precedence, while the new elements were not mentioned in the December 2001 presidency conclusions.[40]

EU–NATO cooperation

Official EU documents stress that a permanent and effective relationship with NATO is a crucial element of the ESDP. The ESDP is intended to lead to an EU–NATO 'genuine strategic partnership' for the management of crises, and the decision-making autonomy of both organizations is to be retained.[41]

---

[37] European Capability Action Plan (note 33). See also de Grave, F., 'European Security and Defence Policy as a framework for defence co-operation', *RUSI Journal*, Feb. 2002, pp. 13–15.

[38] 'EU resolves two-thirds of gap capabilities' (note 34).

[39] Summary of the intervention by Javier Solana, EU High Representative for Common Foreign and Security Policy, at the informal meeting of defence ministers, Brussels, 12 Oct. 2001.

[40] In Oct. the EU defence ministers considered drawing up a White Paper on security and defence. However, this was postponed so that progress towards the Headline Goal would not be delayed. *Atlantic News*, no. 3332 (26 Oct. 2001), p. 3.

[41] The Feira European Council identified 4 areas for developing the EU's relationship with NATO: security issues, capability goals, the modalities for EU access to NATO assets and the definition of permanent consultation arrangements. Presidency Conclusions, Santa Maria da Feira, 19–20 June 2000,

During the Swedish presidency letters were exchanged with the Secretary General of NATO confirming the permanent arrangements for consultation and cooperation between the EU and NATO.[42] In February 2001 the PSC and the NAC met for the first time under the new permanent EU–NATO consultation arrangement. It has been agreed that the two bodies will meet formally at least three times during every EU presidency and that there will also be at least one EU–NATO ministerial meeting per presidency.

The first such ministerial meeting was held on 30 May 2001 in Budapest. NATO Secretary General Lord Robertson noted the progress that had been made, especially on the technical side, and urged that additional efforts be made by EU and NATO members. He stressed four aspects of EU–NATO cooperation: the equal status of both organizations; the need for coherence in defence planning to avoid unnecessary duplication; the participation of non-EU European NATO members; and the need to focus on capabilities.[43]

In addition to the NAC–PSC meeting, the first sessions of the EUMC and the NATO Military Committee were held. The EU–NATO ad hoc group on capabilities exchanged information on relevant aspects of EU and NATO work in this area, and the NATO experts were thanked for their contribution to the development of the Headline Goal and the EU Exercise Programme.

The EU and NATO have conducted joint crisis management activities in the western Balkans (in southern Serbia and Macedonia). This cooperation took the form of ministerial and political NAC–PSC consultations and joint activities of the Secretary General of the Council of the European Union/High Representative for the CFSP and the NATO Secretary General, as well as of their representatives.

The discussions between the EU and NATO on the arrangements which permit EU access to NATO's assets and capabilities (guaranteed permanent access to NATO's planning capabilities, presumption of availability of pre-identified assets and capabilities, and identification of a series of command options), as envisaged at the Nice European Council meeting, were not finalized in 2001. Work on a security agreement (exchange of classified CFSP/ESDP documents between the EU and NATO) was also not completed.[44] Because the Belgian presidency was unable to conclude the EU–NATO agreement on access to NATO's assets and capabilities, the Laeken Presidency Conclusions were limited to a reiteration of the intention to finalize the security arrangements with NATO and to conclude the relevant agreements.

---

appendix 2, reproduced in 'From St-Malo to Nice, European defence: core documents' (note 1), pp. 130–32.

[42] Because Sweden is a non-NATO EU member it may have been more constrained in attempting to develop EU–NATO cooperation than the succeeding Belgian presidency.

[43] Opening statement by the Secretary General, NATO–EU Ministerial Meeting, Budapest, 30 May 2001, URL <http://www.nato.int/docu/speech/2001/s010530a.htm>.

[44] Informal meetings between Solana and Lord Robertson had led to an interim security agreement for the exchange of documents, which laid the groundwork for the arrangements envisaged in the treaty.

Cooperation with non-members

The Nice European Council meeting established the modalities for the involvement of non-EU nations in EU-led operations. It proposed that regular (non-crisis) 'dialogue, cooperation and consultation' be carried out between the 15 EU states and the 15 non-EU European nations. Each non-EU European state is to appoint a representative from its mission to the EU and a military liaison officer as a contact to the EUMS. In the event of an EU-led operation using NATO's assets, the partners will be consulted and the non-EU European NATO countries will have the automatic right to take part. For an EU-led operation which does not use NATO's assets, the partners will be consulted in advance and may be invited to participate.

However, it was not possible to achieve a cooperation agreement between the EU and NATO. In 2001 Turkey, and later Greece, obstructed the talks on permanent arrangements (see section IV). Nevertheless, issues such as NATO's assets and capabilities and solutions to the European command problem were discussed.[45] After the 11 September terrorist attacks Solana reiterated that the EU 'should make the best possible use of the contributions of candidate countries and NATO allies'.[46]

Both the Swedish and the Belgian presidencies sought to implement the arrangements approved at the Nice European Council meeting. The EU foreign and defence ministers met their counterparts of the non-EU European NATO members and the states which are applying for EU membership ('the fifteen') and the non-EU European NATO states ('the six') to inform them of the outcome of the November 2000 Capabilities Commitment Conference and the November 2001 CIC. Civilian aspects of crisis management, the implementation of arrangements for consultation and participation, EU–NATO relations, crisis-related topics and so on were also discussed. The non-EU European countries have appointed their representatives to the PSC and the EUMS. The first meetings at the EUMC forum were held during the Swedish presidency.

The European Council meeting in Laeken stressed the need to implement the Nice arrangements, the additional contribution by 'the fifteen' to the civilian and military capabilities and participation in crisis management operations by 'the six' (in particular, by setting up a Committee of Contributors to function in the event of an operation). Altogether, the declared non-EU European members' contributions amount to some 15 000 troops plus equipment.[47]

Arrangements have also been developed with other partners such as Canada (EU–Canada summit meeting, 19 September), Russia[48] (EU–Russia summit

---

[45] According to a NATO concept, the natural choice would be the Deputy Supreme Allied Commander, Europe (SACEUR); the role of the Deputy Supreme Allied Commander, Atlantic (SACLANT) in a European operation was also discussed. *Atlantic News*, no. 3286 (25 Apr. 2001), p. 2.

[46] *Atlantic News*, no. 3329 (17 Oct. 2001), p. 3.

[47] For the breakdown of the contributions see *Jane's Defence Weekly*, vol. 36, no. 24 (12 Dec. 2001), pp. 25–27.

[48] The EU agreed that the PSC will hold monthly meetings with Russian officials. The decision reportedly overruled the position of its representatives to the PSC who signalled that such meetings

meeting, 3 October 2001) and Ukraine (EU–Ukraine summit meeting, 11 September).

Cooperation with international organizations

The evolution of European crisis management capabilities calls for enhanced, mutually reinforcing cooperation between the EU and other international organizations (e.g., the UN and the OSCE). Themes and areas for EU–UN cooperation have been endorsed by the Council. They emphasize not only civilian aspects but also the potential of developing the military aspects of crisis management and their contribution to peacekeeping operations, especially in the western Balkans, the Middle East and Africa.

In early 2002 the prospect emerged of a first test for an ESDP crisis management deployment substituting for NATO's Operation Amber Fox (Task Force Fox, TFF) in the Former Yugoslav Republic of Macedonia (FYROM).[49]

IV. Challenges to politico-military integration

The EU–US relationship

The USA has supported the idea of a stronger European role in security matters since the European Security and Defence Identity (ESDI) concept was launched within NATO in 1994. Since the 1998 Saint Malo Declaration, however, the evolving EU defence capability has led to US concern that it might eventually result in a challenge to NATO's importance in European security affairs and a transatlantic rift. In 2001 the tension between the EU and the George W. Bush Administration gradually decreased owing to several factors: the USA's growing interest in Asian security problems, which resulted in a reduced interest in Europe (a process which accelerated after the change in Russia's policy in the autumn of 2001); the problem of international terrorism which became the matter of utmost concern to the USA; and the exposure of Europe's lack of military capacity in the campaign in Afghanistan.

In the first months of 2001 there was disagreement between the European states and the USA over a range of issues, most notably the US missile defence plans and the creation of the ERRF. The disputes were magnified by misunderstandings and a considerable 'Euro-scepticism' in the new US administration and were not the result of irreconcilable differences.[50] In a

---

would be too constraining for the Committee. Zecchini, L., 'Notre ami Vladimir Poutine . . .' [Our friend Vladimir Putin . . .], *Le Monde*, 11 Oct. 2001, p. 17.

[49] In early 2002 the Spanish presidency announced the goal of taking responsibility for peacekeeping in Macedonia from NATO by the summer of 2002. It is hoped that by that time the crisis caused by Greece will have been resolved because final agreement must be reached with NATO on EU use of NATO capabilities.

[50] *The Economist* commented on this in mid-2001: 'The Europeans will swallow their reservations about missile defences, if Mr. Bush can persuade them that the pursuit is practical (and practicable), not ideological. . . . For their part, the Americans will swallow their reservations about Europe's defence initiative, if it does not draw resources away from NATO'. 'Mr. Bush goes to Europe', *The Economist*, 7 June 2001.

move to stave off a transatlantic rift, the British, French and German defence ministers sought to reassure the USA that any crisis would be considered by NATO before the EU decided to handle it. French approval for giving NATO a de facto right of first refusal was welcomed, since it was the first time that France had expressed this view.[51]

Another key issue was defence planning, with France calling for a larger European staff to ensure that the ERRF could operate on its own.[52] However, this was opposed by the UK and other EU states. During his visit to the USA, on 23–24 February, Blair sought to assure the Bush Administration that the ERRF would not compete with NATO, but would rely on NATO's planning staff and add to NATO's resources. The Blair–Bush joint statement underscored that NATO will remain the essential foundation of transatlantic security. Both states supported the goal of non-EU European NATO members and other partners assuming greater responsibility for crisis management by strengthening NATO's capabilities when NATO as a whole chooses not to engage.[53] In March German Chancellor Gerhard Schröder travelled to Washington with a 'strong and reassuring signal' that the ESDP is not intended to weaken or undermine NATO.[54]

Although reassured, the US administration still expressed concern.[55] The informal EU defence ministers meeting in April confirmed that there is no intention to set up a planning system similar to that of NATO.[56] NATO Supreme Allied Commander, Europe (SACEUR), General Joseph W. Ralston, acknowledged in May that NATO–EU relations were 'moving in the right direction, although slower'.[57]

The attacks of 11 September shifted US attention towards terrorism and the Central Asian region. However, pressure increased for the EU to play a greater role in conflict prevention and crisis management, whether or not NATO is involved. In this context, the NATO Secretary General and the USA began to demand more strongly that a greater financial contribution be made by Europe.

EU access to NATO capabilities: Turkish opposition

Following the 1999 Cologne and the June 2000 Santa Maria da Feira European Council meeting decisions, the EU and NATO worked on permanent arrangements enabling EU access to NATO's military assets and planning capabil-

---

[51] Fitchett, J., 'US and EU ponder defense trade-off', *International Herald Tribune*, 8 Feb. 2001, pp. 1, 5.
[52] In the spring of 2001 French Defence Minister Alain Richard was to have downplayed earlier calls for an ERRF planning capability independent of NATO. 'EU told to buy strategic capability', *The Guardian*, 7 Apr. 2001, p. 7.
[53] *Atlantic News*, no. 3271 (28 Feb. 2001), p. 4.
[54] Background briefing by senior administration official on the President's meeting with Chancellor Schroeder of Germany, White House Office of the Press Secretary, 29 Mar. 2001, URL <http://usinfo.state.gov/products/pdq/pdq.htm>.
[55] 'Rumsfeld hesitates to OK defense plan', *Washington Times* (Internet edn), 19 Mar. 2001, URL <http://www.washingtontimes.com/world/default-2001319221547.htm>.
[56] *Atlantic News*, no. 3284 (11 Apr. 2001), p. 3.
[57] *Atlantic News*, no. 3291 (17 May 2001), pp. 1–2.

ities. The European Union and NATO agreed that the EU's incipient military crisis management capability should be autonomous. However, different interpretations of 'autonomy' soon emerged. The controversy over access extended to the scope of guaranteed permanent access. NATO members (chiefly the USA) are willing to allow automatic access to planning when the EU leads operations. In order to maintain NATO's primary role, the USA proposes that the EU rely on NATO planning and close coordination with it. The risk is that lack of progress on assured access to NATO capabilities may lead EU countries to establish their own operational planning cell within the ERRF, thus leading to an eventual weakening of the transatlantic link.

It was Turkey, a non-EU NATO member, which blocked this proposal, insisting on allowing EU access on a case-by-case basis. The NAC ministerial meeting in December 2000 failed to reach agreement on the issue.[58] Turkey's opposition blocked progress in the talks on EU access to NATO capabilities for almost the whole of 2001. Its main concern was that any arrangements in this area which benefit the EU will potentially harm Turkish security interests.[59] Turkey insisted on being given either full say within the ESDP or participation at least comparable to the level it had as a WEU Associate Member. This was unacceptable to the EU as it requires maintaining absolute CFSP/ESDP decision-making autonomy and unanimity. Nevertheless, Turkish Foreign Minister Ismail Cem insisted that 'Turkey is not prepared to allow the use of these capabilities and assets it shares unless it has a right to participate *reasonably* in their use'.[60]

In June 2001 British and US diplomats put forward a compromise formula. In May the UK had presented a paper which proposed that non-EU European NATO members would have 'interlocutors', who would meet periodically and in the event of crises with the PSC. These countries would also have military liaison officers permanently attached to the EUMS. The British paper proposed expanding considerably the role of the Committee of Contributors, a body envisaged by the Nice European Council meeting for day-to-day management of crisis operations. The paper noted that Turkey's participation would be of 'particular benefit' in the cases of EU-led operations, although it stopped short of guaranteeing that Turkey would be included in the decision making. The compromise also included an indirect assurance that the ERRF would not intervene in Greek–Turkish disputes.[61] On this basis, British, Turkish and US diplomats drafted an accord, but it was promptly rejected by

---

[58] In addition, the EU is interested in obtaining automatic access to some NATO common assets such as command, control and communications (C^3) capabilities including the use of airborne warning and control system (AWACS) aircraft and so on, which is unlikely to be accepted by the USA. Nassauer, O. and Gourlay, C., 'Controversy over EU access to NATO capabilities', *European Security Review*, no. 4 (Mar. 2001), pp. 3–4.

[59] Turkey has also been suspected of trying to manipulate the NATO issue to achieve a 'backdoor entry into the EU'. Of 16 possible conflict scenarios envisaged by NATO, 13 would involve Turkey. The EU also envisages that several operations would take place in the area. Fitchett, J., 'Turkey puts roadblocks in EU force negotiations', *International Herald Tribune*, 26 Jan. 2001, p. 4.

[60] Cem, I., 'A necessary role in defence', *Financial Times*, 29 May 2001, p. 15.

[61] Gordon, M., 'Pact could end Turk objection to EU force', *International Herald Tribune*, 5 June 2001, p. 4.

the Turkish National Security Council, and especially by the Turkish General Staff.[62] This development was badly received by the EU.[63] The EU–Turkey talks continued, but the deadlock persisted. The 11 September terrorist attack on the USA apparently strengthened Turkey's hand, and Turkey stressed its position as a country located in a volatile region. Ankara submitted a number of conditions to the EU countries which it asked them to meet.[64]

It was not until 26 November that partial progress was made by Turkish, British and US officials. Agreement was reached on two points: the ERRF will not be used in potential crises in Cyprus or the Aegean region, and the EU will not intervene in any potential crisis between Greece and Turkey. In response to Turkey's demand for an automatic invitation to autonomous operations (when NATO's capabilities are not used) either affecting its security interest or close to Turkey geographically, the EU proposed that in each case Turkey's request would be evaluated and decided upon through a consultative mechanism.[65] The EU argued that to accept Turkey's demand would require a change in the Treaty on European Union under which the veto of a single EU state could prevent Turkey from participating in autonomous operations. On 28 November, during the discussion of the 'case-by-case' formula, Greek Foreign Minister George A. Papandreou signalled that Greece, which had not been directly involved in the negotiations, might use its right of veto within the EU to block Turkey's full participation in the ERRF.

On 2 December, British, Turkish and US diplomats announced that they had broken the two-year deadlock over EU access to NATO capabilities. They agreed on a carefully drafted 'Istanbul Document'.[66] Greece, however,

---

[62] Anatolia Agency (Ankara), 4 June 2001, in 'Turkish military hands Rumsfeld note on ESDI issues', Foreign Broadcast Information Service, *Daily Report–West Europe* (*FBIS-WEU*), FBIS-WEU-2001-0604, 4 June 2001.

[63] Foreign Minister of Belgium Louis Michel stated soon after the start of the Belgian EU presidency: 'We will not yield to Turkey's pressures. The EDSP would not be formed by non-EU member countries' approval', Anatolia Agency (Ankara), 18 July 2001, in 'Belgian envoy asks Turkey to soften stand on ESDP', FBIS-WEU-2001-0718, 18 July 2001.

[64] The conditions were reportedly: guarantees that the EU corps will not be used to resolve conflicts between allied countries; inclusion of Turkey in the EU decision-making mechanisms whenever operations affect its national interests and are close to its geographic location; participation of the non-EU NATO '6' in EU military manoeuvres; allowing Turkish military officers to maintain offices in the EU military headquarters; and strengthening the role of the Committee of Contributors. 'The European army is locked on Turkey', *Hurriyet* (Istanbul), 12 Oct. 2001, in 'Turkey's conditions for participation in the new European defense doctrine reported', FBIS-WEU-2001-1014, 12 Oct. 2001.

[65] 'Progress on European army', *Milliyet* (Istanbul), 27 Nov. 2001, in 'Turkish sources report progress in ESDP dispute with EU', FBIS-WEU-2001-1127, 27 Nov. 2001.

[66] The 'Istanbul Document' envisaged, among other things, that: (*a*) the ESDP will 'in no case and in no form of crisis' be used against any ally; (*b*) there will be expanded deliberations with the non-EU European NATO states, which will be 'associated' to decisions and actions; (*c*) the '15 + 6' consultations will be more frequent and facilitated by the appointment of 'permanent interlocutors' of the PSC and 'representatives' to the EUMC from non-EU European NATO states; (*d*) in cases of an EU planned autonomous operation in the 'geographic proximity' of a non-EU European ally or such affecting that ally's national security interests, the Council will consult it and, taking into account the outcome of these deliberations, shall decide on whether that ally should participate; and (*e*) the NATO '6' will take the role of observers for the operations in which they do not take part if coordinated by Supreme Headquarters Allied Powers Europe (SHAPE). For the text and discussion of the agreement see 'The Istanbul document', *I Kathimerini* (Athens), 11 Dec. 2001, in 'Greek paper carries "text" of "Istanbul Document" on ESDP', FBIS-WEU-2001-1212, 11 Dec. 2001.

promptly declared that it would not accept a text that is not completely clear and which contains too many ambiguities.

On 10 December Greece decided to de facto block the proposed agreement. The Greek objections referred to its national interest and guarantees that the deal with Turkey would not undermine the EU's decision-making autonomy, the main reason apparently being that the ERRF should not constrain Greece in its disputes with Turkey.[67]

Against this background, the Laeken European Council meeting declared that 'the EU is now able to conduct *some* crisis-management operations' and 'will be in a position to take on progressively more demanding operations, as the assets and capabilities at its disposal continue to develop'.[68] Putting the best face on the failure to conclude the agreement on EU access to NATO's assets, Solana insisted that the operational capability of the ERRF is a separate issue.[69]

The financial hurdle

The EU states are faced with the necessity of buying new equipment and modernizing their armed forces, which will be an expensive process. As most of the resources will have to be found within existing military budgets, there is a need to improve and accommodate defence cooperation within the EU, enhance the interoperability of forces, increase effectiveness and efficiency in building capabilities, and redistribute resources more rationally at national and EU levels.[70] The gap between the development of the new ESDP bureaucracy and the lack of progress in finding ways to finance future EU military operations were criticized by some EU members, the West European public and NATO.

In response, Solana suggested in May that certain strategic capabilities could be developed collectively, resulting in savings. Additionally, Sweden presented a paper on funding common elements of crisis management operations (*matériel* supplied by states).[71]

In addition to advancing the work done by the Swedish presidency in the first half of 2001, the Belgian presidency was requested to develop a plan for

---

[67] Greece has asked for 'counter-concessions'. E.g., it presented a demand that Turkey be denied a veto regarding the intervention of the ERRF in a hypothetical confrontation between Greece and Turkey and that Turkey should not be allowed to block an ESDP operation in the Balkans. Greece also demands that when Cyprus joins the EU it be accorded full participation rights in the ERRF. 'Greece blocks accord with Turkey', *Financial Times*, 17 Dec. 2001, p. 2.

[68] Declaration on the Operational Capability of the Common European Security and Defence Policy, Annex II (A), Laeken, 14–15 Dec. 2001, URL <http://www.eurunion.org/legislat/Defense/Laeken%20ESDP.htm>.

[69] Lobjakas, A., 'EU: Laeken summit mired in controversy over defense policy', Radio Free Europe/Radio Liberty, *RFE/RL Newsline*, 14 Dec. 2001, URL <http://www.rferl.org/nca/features/2001/12/14122001092130.asp>. The Belgian Foreign Minister went further and suggested that the EU would send a 'multilateral' EU force to participate in an Afghanistan peacekeeping operation. This declaration was disavowed by other EU states.

[70] de Grave (note 37).

[71] *Atlantic News*, no. 3292 (22 May 2001), p. 4; and *Atlantic News*, no. 3294 (29 May 2001), p. 3.

financing the implementation of crisis management operations.[72] The gap between the political declarations and desires of the EU states, on the one hand, and the financial commitments and resources, on the other, was glaring in 2001. There were sharp disagreements over how the ERRF should be financed (e.g., shared cost versus investment by individual countries).[73]

Europe has not perceived much need for increased defence spending, and, consequently, military spending had been declining.[74] The EU member governments were slow to increase their defence spending to fill the gaps in hardware identified by the two capabilities conferences held in 2000 and 2001. So far, the ESDP implementation has not led to the expected European arms industry consolidation.[75]

Solana has tried to strike a positive note in the debates on defence spending, claiming that 10 of the 15 EU member states will spend more in 2001 than in 2000.[76] Other estimates present a different analysis of the situation.[77]

At the end of the year the dispute over finances continued. In order to break the deadlock, the EUMS, headed by Solana, submitted three options to the member states. The first, supported by Greece, Luxembourg and Portugal, envisages charging the states for the entirety of the operational expenditure based on gross national product—'all costs are common'. Austria, Germany, Ireland, Spain and the UK opposed it, advocating a second proposal—the use of an existing NATO system ('costs lie where they fall'). The system could reduce the administrative burden at the 'top' of the EU, since member states would be responsible for the management of the operational costs. The third, 'intermediate' or compromise option—backed by France, Italy and the Netherlands—would increase common costs (e.g., renting buildings or the cost of temporary staff) but charge operational spending to member states participating in any operations, much like the second option.[78]

---

[72] According to a report commissioned by the British Government, the member states will have to restructure their defence budgets to meet the additional spending estimated at $25 billion dollars over the next 10–15 years, based on acquisition and initial running costs. The amount is much higher if the cost of supporting the additional systems is taken into account. Dempsey, J. and White, D., 'Not rapid enough', *Financial Times*, 19 Nov. 2001, p. 15.

[73] The European Parliament was also reluctant to approve additional funds for the ERRF, fearing that it would set a dangerous precedent because the ERRF would be outside the Commission's purview. The European Parliament has more control over the activities of the European Commission than of the Council.

[74] See also chapter 6 in this volume.

[75] Taylor, S., 'Europe's defence industry frustrated at government reluctance to boost arms spending', *European Security Review*, no. 7 (July 2001), pp. 5–6. However, apart from increased spending, restructuring of military expenditure is needed as many current programmes are of cold war origin.

[76] Solana, J., 'European defence: the task ahead', *European Voice* (Internet edn), URL <http://ue.eu.int/solana/print.asp?docID=68380&BID=108>.

[77] International Institute for Strategic Studies, *The Military Balance 2001/2002* (Oxford University Press: Oxford, 2001), pp. 34–35, noted that the spending of European NATO members fell on average by 6.7% in real terms between 1999 and 2000. Germany's spending will continue to decrease, while British and French budgets will experience modest increases in 2002. Other European NATO states will experience downward trends. See also 'Financial and economic data relating to NATO defence', NATO Press Release M-DPC-2 (2001), 18 Dec. 2001. URL <http://www.nato.int/docu/pr/2001/p01-156e.htm>.

[78] 'Wrangling over finances delays EU defence policy', *Financial Times*, 2 Nov. 2001, p. 8.

V. Conclusions

Building up the ESDP should allow it to shoulder a larger share of the burden of European security, thus rebalancing the transatlantic security relationship in order to make it work more effectively and efficiently. The military capabilities provided by the ESDP have the potential to help redefine that relationship.

In the aftermath of the terrorist acts of 11 September there was a change in the general political outlook, which led to the transformation of transatlantic relations and the improvement of NATO–Russian relations. These developments will influence the evolution of the ESDP, although exactly how and to what degree was not clearly demonstrated by the end of 2001.[79] With US interest shifting elsewhere, the potential for a crisis over the EU–NATO relationship has abated. The post-11 September developments brought home to the EU the reality of its role and potential in the transatlantic relationship. This will influence the division of labour and complementarity between Europe and the USA and will increase pressure on Europe to improve its military capabilities both in the EU and NATO.

The military-related bodies of the ESDP were established and began operation in 2001. EU–NATO institutional cooperation was also strengthened, as evidenced in the western Balkans. Some EU capability shortcomings were addressed wholly or in part in 2001, but the EU plans concerning the most critical aspects of its ERRF (intelligence, logistics, communications and strategic transport) either are still encountering political and financial obstacles or will need a much longer implementation period than the target date of 2003. Although the ESDP has been declared operational and able to perform the less demanding Petersberg tasks, the crucial issue of EU access to NATO's assets and capabilities remained unresolved. The two-year stalemate caused by Turkey's intransigence regarding this issue was broken, but Greek opposition created new problems.

The reasons why the Headline Goal schedule has not been met are complex. While the EU has avoided falling into the trap of Europeanism-versus-Atlanticism, the scope of the ERRF has not yet been clearly defined. The issue of unavoidable but rational duplication of efforts by the EU and NATO has not been sufficiently addressed. This will need to be done in order to agree how and where to allocate EU resources. Nonetheless, financial considerations are bound to constrain excessive EU duplication.

Defining the ESDP and building public support for increased spending will be challenging issues in the years ahead. Before the 11 September terrorist attacks the EU governments did not perceive an urgent need for military-related spending increases. Now their taxpayers must be persuaded of the need to spend more. The European states are slow to increase their military budgets,

---

[79] However, this does not imply that the EU has been passive as regards terrorism; both the Council and the European Commission are implementing measures to address terrorism. See European Union, '11 September attacks: the European Union's broad response', URL <http://europa.eu.int/news/110901/>.

demonstrate flexibility and inventiveness in rationalizing procurement policies (standardization, 'shopping around', leasing of equipment, possible cooperation with the USA, etc.), and embark on regulation and restructuring of the defence industry. There is a need for a synergistic and rational approach to defence spending, and the creation of a single arms-procurement organization would make a positive contribution in this respect.

The lack of leadership within the EU, its cumbersome decision-making bodies and the propensity of the major EU governments to act alone in a crisis (as was demonstrated in November 2001 during the campaign in Afghanistan) illustrate the difficulty of forging a common foreign, security and defence policy. The 'security culture' that is evolving within the EU (cooperation between its pillars and institutions, dialogue between society and its leaders, harmonization of the civil and military aspects of the ESDP, etc.) must become more deeply rooted and develop more rapidly. Bureaucratic conflicts within the EU, such as the dispute over the distribution of tasks between the ESDP bodies and the Community institutions, illustrate the challenge of harmonizing the collaboration of the EU's two pillars. The negative outcome of the Irish referendum on the Nice Treaty in June 2001 underscored the gap and the need for dialogue between the public and government. The future enlargement of the EU and NATO also poses challenges which may temporarily weaken the ESDP.

4. The challenges of security sector reform

DYLAN HENDRICKSON and ANDRZEJ KARKOSZKA*

I. Introduction

The end of the cold war gave new impetus to pressures for political and economic liberalization around the globe. States aspiring to democratic governance and strong economies require capable administrative and political structures. A key element is a well-governed security sector, which comprises the civil, political and security institutions responsible for protecting the state and the communities within it. Reform or transformation of the security sector is now seen as an integral part of the transition from one-party to pluralist political systems, from centrally planned to market economies, and from armed conflict to peace, and is a growing focus of international assistance.[1]

International interventions under the auspices of the United Nations, the North Atlantic Treaty Organization or powerful individual states carried out since the early 1990s to resolve violent conflicts and assist these transitions have shown immense limitations. External forces have often supplanted the local security apparatus or, in some cases, explicitly sought to dismantle it where it was considered to be a part of the security problem. However, without adequate efforts to restore a viable national capacity in the security domain, external interventions offer at best temporary solutions to security problems and may, in some cases, aggravate the situation.

Security sector reform aims to help states enhance the security of their citizens. The shift from state- and military-centric notions of security to a greater emphasis on human security has underscored the importance of governance issues and civilian input into policy making. The kinds of security policies that governments adopt, the instruments used to implement these policies and the interests served by these policies are critical factors.

The security sector reform agenda therefore encompasses—but is far broader than—the traditional civil–military relations approach to addressing security problems. Security sector reform has potentially wide-ranging implications for how state security establishments are organized and, by extension, for how international security and development assistance is delivered.

---

[1] The terms 'reform' and 'transformation' are used interchangeably in this chapter, although 'reform' is the term of choice because it is most commonly used by those working in the field. 'Transformation' implies a more fundamental change than reform and is emerging as the preferred term in some circles involved in security sector work. For arguments in favour of the use of the term 'transformation' see, e.g., Williams, R., 'African armed forces and the challenges of security sector transformation', *Strategic Review for Southern Africa*, vol. 23, no. 2 (Nov. 2001), pp. 1–34.

* The authors are indebted to Nicole Ball for comments on an early draft of this chapter.

These implications are only just starting to be understood and translated into policy and are eliciting mixed reactions from both the international actors that provide security assistance and the recipients of aid.

Developing countries have been cautious about embracing security sector reform. They are wary of the conditions attached to external assistance and the promotion of 'one size fits all' solutions to their problems, such as the structural adjustment programmes of the 1980s. Past security assistance programmes were often ill-conceived and poorly implemented, and resulted in outcomes that were not supportive of either citizen security or development goals.[2] The Central and East European (CEE) states[3] have responded more favourably to the reform agenda, which is seen as complementing the wider economic and political reforms in which many of them are engaged. Crucially, the prospect of integration into NATO and 'the West' has provided a powerful, additional incentive for CEE states to reform their security sectors.

Despite the fact that security sector reform is moving up on the international agenda, it remains a new area of activity. There is still no consensus on how to define the concept of security sector reform or on what the objectives and the priorities for international assistance should be.[4] Most actors are just starting to grapple with the political sensitivities of security sector work, and few have developed the policy instruments required to work in an integrated way with their partners.[5] As a consequence, there are different levels of acceptance among international actors, many of whom remain wary of how security sector reform will impinge on traditional institutional mandates or foreign policy objectives.

While the general principles that underpin security sector reform have relevance for all countries, this chapter is principally concerned with how the agenda has been conceptualized and implemented by international actors in the context of developing countries and the CEE states. Section II outlines the background to this policy agenda and some of its key features. Section III examines the relevance of security sector reform to international security, particularly in light of the new 'war on terrorism'. Section IV then looks in more detail at the context for security sector reform in Africa, Asia, Latin America and the CEE states, where the level and nature of international involvement vary significantly. Finally, drawing on recent lessons, section V highlights a number of key challenges for external assistance.

---

[2] Washington Office on Latin America (WOLA), *Demilitarizing Public Order: The International Community, Police Reform and Human Rights in Central America and Haiti*, Nov. 1995, available from WOLA, URL <http://www.wola.org/pubs.html>.

[3] These states are Albania, Bosnia and Herzegovina, Bulgaria, Croatia, the Czech Republic, Estonia, Hungary, Latvia, Lithuania, the Former Yugoslav Republic of Macedonia, Poland, Romania, Slovakia, Slovenia and Yugoslavia.

[4] Hendrickson, D., *A Review of Security-Sector Reform*, Working Paper no. 1 (Conflict, Security and Development Group, Centre for Defence Studies, King's College London: London, Sep. 1999).

[5] Organisation for Economic Co-operation and Development (OECD), Development Assistance Committee (DAC), 'Security issues and development co-operation: a conceptual framework for enhancing policy coherence', *DAC Journal*, vol. 2, no. 3 (2001), section II, pp. 31–71.

II. The policy agenda

The end of the cold war set in motion a profound rethinking of the notion of security and of strategies for international assistance in this domain. The militarized notions of security that emerged during the cold war gave rise to a narrow stress on territorial integrity and security through armaments that has been difficult to change.[6]

Before 1989, aid to the Third World—including development, humanitarian and security assistance—was closely linked to the dynamics of the cold war. Security became synonymous with the stability of the international system and regime stability—the protection of client regimes from external and internal threats. Assistance programmes paid little attention to democratic civil–military relations, to effective legislative and executive oversight over the various security branches, or to the creation of a professional ethos within security services that was consistent with the dictates of a modern democracy. No real attempt was made to include important civilian sectors (e.g., the foreign policy and finance sectors) in the formulation of security policy.

In many developing countries and CEE states, the provision of basic services such as security, employment and social welfare has sharply eroded since the end of the cold war. These problems have focused critical attention on how state security establishments shape and condition the processes of economic and political change.

In this environment, organizations involved in development assistance have been cautious about entering the arena of security sector reform but they have gradually realized that they cannot avoid it. International financial institutions (IFIs)[7] play a key role in setting the economic framework in which the major donors engage in developing countries and in CEE countries. The IFIs have a clear impact on the outcome of security sector reforms by virtue of their involvement in macroeconomic adjustment and stabilization programmes, though their direct involvement has to date been limited to a concern with the issue of military expenditure.[8] Both the International Monetary Fund (IMF) and the World Bank have traditionally been cautious about becoming involved in security-related matters because of the differing views of their board members on this issue as well as the ingrained conservatism of these institutions. Nevertheless, there is growing recognition that security sector reform should be a concern.[9]

The World Bank, in particular, is increasingly recognizing the need to set its support for demobilization programmes and the strengthening of public expenditure management systems within a broader framework of security sector reform. This is forcing the organization to reconsider the role of the tra-

---

[6] Baldwin, D. A., 'Security studies and the end of the cold war', *World Politics,* vol. 18 (Oct. 1995), pp. 117–41.

[7] See, e.g., URL <http://www.wellesley.edu/Economics/IFI>.

[8] Ball, N., 'Transforming security sectors: the IMF and World Bank approaches', *Journal of Conflict, Security and Development*, vol. 1, no. 1 (2001), pp. 45–66.

[9] World Bank, *Security, Poverty Reduction and Sustainable Development: Challenges for the New Millennium* (World Bank: Washington, DC, Sep. 1999).

ditional instruments of economic conditionality that it has often wielded, together with the IMF, in an attempt to obtain the adherence of the borrowing countries to military expenditure limits.

For similar reasons, recognition of the need for a broader approach to security has emerged from the debates on civil–military relations, particularly in relation to the CEE states where Western defence establishments have been active.[10] In Africa, Asia and Latin America, a parallel process of rethinking security concepts has also been under way and has influenced the security sector reform agenda.[11] Many countries were engaged in security sector reform activities long before this concept gained international prominence.[12]

The new security thinking is set apart from past approaches because it recognizes that: (*a*) ensuring the safety of citizens should rank alongside national defence as the primary goal of state security policy; (*b*) greater emphasis needs to be placed on the role of civilian actors in both formulating and managing security policy (the critical role of governance was largely overlooked by cold war security assistance programmes, and development actors avoided for the most part engagement in activities related to the security sector); and (*c*) different means of achieving security objectives must be acknowledged. The traditional reliance on primarily military instruments of force should be complemented more effectively with diplomatic, economic, legal, political and social mechanisms, and greater preventive action.

The need for a broad approach to security is underscored by the experiences of developing countries and the CEE states, where political and state-building processes are now seen as the foundation for efforts to enhance the security of states and their citizens.[13] In these contexts, state and regime legitimacy is constantly being challenged, and demands for economic redistribution and political participation are creating major overloads on weak administrative and political systems. Unmet social and political needs run the risk of provoking popular unrest and opposition to governments, ultimately making them more vulnerable to internal and external threats.

Defining the security sector

Because the actors involved in delivering security services and the relationships between them vary from country to country, there is not a universally applicable definition of the security sector. A narrow focus on the conventional Western security actors such as armed forces, police and intelligence services, for instance, does not capture the diversity of security actors in other

---

[10] Edmunds, T., 'Defining security sector reform', *Civil–Military Relations in Central and Eastern Europe Network Newsletter*, no. 3 (Oct. 2001), pp. 3–6, available at URL <http://civil-military.dsd.kcl.ac.uk>.

[11] Cawthra, G., *Securing South Africa's Democracy: Defence, Development and Security in Transition* (Macmillan: London, 1997), pp. 7–26.

[12] Ball, N. et al., *Governance in the Security Sector*, Occasional Paper (Centre for Democracy and Development: London, forthcoming).

[13] Ayoob, M., *The Third World Security Predicament: State Making, Regional Conflict, and the International System* (Lynne Rienner: London, 1995).

countries. In Africa, formations such as presidential guards and militia forces are common, while a whole range of 'private' security actors are emerging because of the collapse of state security structures.[14] Similarly, in the CEE states there is a wide range of internal security forces, often linked to interior ministries, which rival the military in terms of numbers and influence.

In addition, it is also clear that the management of security policy in all countries, including the industrialized states, is influenced by a range of informal norms and practices that are closely shaped by national political, cultural and social circumstances. This is a reason for the complex array of institutions and interactions that affect the relationship between the organizations authorized by states to use force and those mandated to regulate these organizations and formulate security policy. The security sector consists of the following elements.

Forces authorized to use force: armed forces; police; paramilitary forces; presidential guards; intelligence services (including both military and civilian agencies); secret services; coast guards; border guards; customs authorities; and reserve and local security units (civil defence forces, national guards, militias, etc.).

Security management and oversight bodies: presidential and prime ministerial offices; national security advisory bodies; legislature and legislative select committees; ministries of defence, internal affairs, foreign affairs; customary and traditional authorities; financial management bodies (finance ministries, budget offices, financial audit and planning units); and civil society organizations (civilian review boards, public complaints commissions, etc.).

Justice and law enforcement institutions: judiciary; justice ministries; prisons; criminal investigation and prosecution services; human rights commissions and ombudsmen; correctional services; and customary and traditional justice systems. (Unwritten, informal norms, stemming out of the local, tribal and clan traditions, culture and beliefs, are often more powerful or obligatory than the written, formal rules and norms established by central state authorities.)

Non-statutory security forces: liberation armies; guerrilla armies; private bodyguard units; private security companies; and political party militias.

Strictly speaking, the security sector can be seen to comprise the first three categories, which are part of the state machinery for providing security. However, non-statutory security forces can have a significant influence on economic and political governance and need to be taken into account. In countries emerging from war, for instance, liberation or guerrilla armies will often need to be demobilized or integrated into a new national army as part of peace settlements. Similarly, private security companies and bodyguard units

[14] Williams, R., 'Africa and the challenges of security sector reform', eds J. Cilliers and A. Hilding-Norberg, *Building Stability in Africa: Challenges for the New Millennium*, Monograph 46 (Institute for Security Studies: Pretoria, Feb. 2000), available at URL <http://www.iss.co.za/Pubs/Monographs/No46/Contents.html>.

may also have important roles to play where state capacity in the security domain is weak, and these need to be appropriately regulated.

The level of involvement by civil society and private sector actors in security sector governance differs widely from country to country. Their direct role is usually limited, although there is increased acceptance that these actors can be important agents for change when they apply political pressure and inform reform agendas. Relevant civil society actors include professional groups (lawyers and accountants), advocacy groups (human rights bodies), research and policy think tanks, religious groups and the media. Non-state groups have a particularly important role to play in conflict-torn societies, where statutory security sector capacity is usually weak.

While the concept of the security sector provides a framework for targeting international assistance, the challenges of the transformation of this sector cannot be understood in isolation from the wider institutional, societal and political context. The security sector cannot function effectively if the administrative and legal framework is fundamentally weak or corrupt. The security sector is also crucial to political power, both in the 'macro' sense of regime stability and in the 'micro' sense of exercising day-to-day political control and generating revenue. Security sector reform is therefore closely tied to domestic processes of political and social change.

Defining security sector reform

There is an increased recognition that the security sector, like any other part of the public sector, must be subject to the principles of civil oversight, accountability and transparency. How these principles are implemented and the specific ways in which the security sector is organized will depend on the circumstances.

Strengthening the institutional framework for managing the security sector involves three broad challenges: (*a*) to ensure the proper location of security activities within a constitutional framework defined by law and to develop security policies and instruments to implement them; (*b*) to build the capacity of policy makers to effectively assess the nature of security threats and to design strategic responses supportive of wider development goals; and (*c*) to strengthen mechanisms for ensuring security sector accountability by enabling the state and non-state actors responsible for monitoring security policy and enforcing the law to fulfil their functions effectively.

Within this broad framework more specific, short-term objectives may include improving the management of security expenditure, negotiating the withdrawal of the military from a formal political role, dissociating the military from an internal security role, strengthening the effectiveness of the security forces, and demobilizing and reintegrating surplus security personnel. The growing number of issues that are becoming entwined in the security sector reform agenda include conflict prevention, democratization, human rights protection and development. The wide range of governance objectives to which international actors are giving priority can be grouped in the following seven categories.[15]

Professional security forces. Professionalization encompasses doctrinal and skill development, technical modernization and an understanding of the importance of accountability and the rule of law.

Capable and responsible civil authorities. The relevant civil authorities in the executive and legislative branches of government need to have the capacity to develop security policy and to manage and oversee the security sector.

High priority to human rights protection. Respect for human rights must exist among civilians as well as members of the security forces.

Capable and responsible civil society. Civil society should have the capacity to monitor the security sector, promote change and provide input to government on security matters.

Transparency. Although some security matters require confidentiality, basic information about security policies, planning and resourcing should be accessible both to the civil authorities and to members of the public.

Conformity with international and internal law. The security sector should operate in accordance with international law and domestic constitutional law.

Regional approaches. Many security problems are shared by countries within a region, and the security of individual countries and individuals within those countries will benefit from regional approaches.

A specific focus on the security forces is now accepted as essential in order to build the human capacity and institutional instruments that they require to fulfil their legitimate functions.[16] Security forces are in a powerful position vis-à-vis other branches of government and citizens to influence governance processes. While central to preserving state sovereignty and authority, the armed forces in particular are one of the few institutions able to endanger states from the inside.[17] This makes it essential for appropriate incentives to be designed to win their support for reforms. Some reforms may focus on improving technical proficiency, but there is an increasing emphasis on organizational restructuring within the security sector in order to ensure adequate provision for civil oversight and direction of the security forces.

---

[15] See, e.g., Ball, N., 'Democratic governance in the security sector', Paper presented at the United Nations Development Programme Workshop on Learning from Experience for Afghanistan, New York, 5 Feb. 2002, available at URL <http://www.undp.org/eo/afghanistan/index.html>.

[16] Williams, R., 'Development agencies and security-sector restructuring', *Journal of Conflict, Security and Development*, vol. 2, issue 1 (2002).

[17] This is borne out by the frequency of military *coups d'état* in many parts of Africa, Asia and Latin America. See, e.g., Hanneman, R. A. and Steinback, R. L., 'Military involvement and political instability: an event history analysis 1940–1980', *Journal of Political and Military Sociology*, vol. 18 (summer 1990), pp. 1–23.

Approaches to security sector reform

There are different philosophies on how best to achieve reform objectives. Security sector reform is underpinned by a number of normative assumptions about the desirability of democratization, civilian control of the armed forces, a clear division between internal and external security functions, the independence of the judiciary and a strong civil society role.[18] These are 'ideal-type' situations that no country has fully succeeded in implementing. In practice, such institutional arrangements are difficult to achieve and not always consistent with the immediate needs or priorities of reforming countries. Instead, these are now seen as goals that countries can work towards from their different starting points.

Only a limited number of countries in which international actors are engaged today are able to undertake fundamental institutional reforms. In the past, international security assistance programmes relied excessively on external templates for reform, with little regard for the social, political and institutional context in which they were being applied. This has resulted in unrealistic assumptions about how states and their security sectors function and undue sensitivity to issues of national ownership. A key concern of governments is that reforms will undermine their power base and compromise their own efforts to address security problems. There is now increased recognition that the greatest potential for security sector reform exists where it is supported from outside but driven by strong internal dynamics. In the most successful examples, there will be a clear national vision for reform and political will at the highest levels of the government.

In countries where these conditions do not exist, particularly in conflict-torn societies, the first priority is generally to restore political stability and basic capacity in the security sector before fundamental institutional problems can be tackled. Political support for reform has to be built up. The bureaucracy and the economy are generally weak. Key security sector institutions, including civilian bodies and the various branches of the security forces, tend to lack clearly defined roles and adequate skills. Consequently, it is not possible to develop a clear national vision for reform.

In these conditions, attempting to promote security sector reform may simply mobilize opposition to change. A broader focus on building basic capacity first may itself not go beyond developing skills and confidence building among security sector personnel. The fact that security sector reform is expensive means that progress will be closely tied to improvements in the economy and living conditions. This makes security sector reform a long-term endeavour. A sustained commitment to security sector reform will depend on development assistance rather than on short-term conflict resolution.

---

[18] Chalmers, M., 'Structural impediments to security-sector reform', Paper presented at the International Institute for Strategic Studies–Centre for the Democratic Control of Armed Forces (IISS–DCAF) Conference on Security-Sector Reform, Geneva, 23 Apr. 2001.

III. Relevance to international security

The fragile security structures in developing countries and the CEE states have diverse historical roots that can be traced to the nature of state building as well as to more recent international development policies. Efforts to develop properly accountable security forces were hampered during the 1980s and 1990s by the immense pressures placed on countries to reduce public spending as a consequence of external pressures for economic liberalization. With the security forces often seen as a barrier to economic and political development, attempts were made to reduce their size and influence, and insufficient attention was paid to how the security void would be filled.

Security sector reform aims to improve governance, thereby reducing the risk of state weakness or state failure. It is often in weak or failed states that conflicts arise. Such states have contributed to a range of destabilizing transnational security problems such as population movements and trafficking in drugs, people and arms, as well as stimulating the widespread incidence of violence and disorder, including groups that carry out terrorist acts. The majority of these problems have important regional dimensions because of weakened state capacity to police borders and regulate economic activity. Insurgent groups that have traditionally relied on neighbouring countries for support and shelter are increasingly exploiting commercial opportunities linked to the expansion of the global economy to sustain their activities.

At a time when weak states facing endemic insecurity and violence have become increasingly unable to rely on the international community for assistance, their internal problems are having greater spillover effects at both the regional and the global level. The sheer scale of the crises afflicting many parts of the developing world and the CEE states has meant that there has simply not been enough international capacity to address all the problems. There has also been a reluctance on the part of Western governments to intervene in countries no longer deemed to be of strategic interest.

Consequently, the international community has a strong self-interest in integrating security sector reform into wider conflict prevention and state-building strategies that combine developmental, legal, military and political instruments. These strategies may include the peaceful resolution of non-violent disputes, peacekeeping, post-conflict peace-building and reconstruction, political participation, reforming the criminal justice system, and strengthening governance across the public sector, specifically in the security sector.

Instruments of security sector reform

This list of the instruments for promoting and implementing security sector reform is inevitably selective.

Donor countries

The main sponsors of security sector reform have been the aid donor countries, including Canada, Germany, the Netherlands, Norway, Sweden, Switzerland, the United Kingdom and the United States. Each of these countries is at a different stage in developing its policies and operationalizing programmes of assistance, and there tends to be great variation in approaches from country to country.

The British Department for International Development (DFID) has taken the lead, in cooperation with other British government departments, in developing a comprehensive security sector reform policy.[19] The DFID, the Foreign and Commonwealth Office, the Home Office and the Ministry of Defence have developed joint programmes of assistance for security sector reform in a number of countries in Africa and Asia. The Ministry of Defence's cooperation programmes in the CEE states, known as Defence Diplomacy, have been broadened to make them more supportive of security sector reform objectives. The UK has also actively pushed the security sector reform agenda at the multilateral level by seeking to encourage the further engagement of UN agencies and IFIs in this area. In the USA, security assistance is delivered by a number of government departments that focus separately on the military, the police and civilian security sector actors, with a limited coordination of activities.[20]

Donor countries have also become increasingly reliant on a wide range of non-state actors, including non-governmental organizations (NGOs), academic institutions and private security companies, to address the gaps in their expertise and capacity. These actors are playing an increasingly important role in the delivery of security assistance, although the growing number of players has also made it more difficult to achieve policy coherence and ensure accountability.

Multilateral development actors

UN work on security sector reform is spread over its specialized agencies and missions, which are engaged in a range of relevant activities, including police and justice reform, regulation of small arms transfers, and the demobilization and reintegration of ex-combatants. The UN Development Programme (UNDP) has gone furthest in defining a comprehensive framework for its involvement in security sector reform but is still developing the capacity to operationalize it. Both the Department of Political Affairs (DPA), which is the focal point within the UN for peace-building activities, and the Department of Peacekeeping Operations (DPKO), responsible for peacekeeping operations,

---

[19] This policy is broken down into 2 components, administered by different departments. The security sector reform policy focuses on the defence sector and cross-cutting governance issues. The policy on safety, security and access to justice covers personal security and justice systems.

[20] Welch, C. and Forman, J. M., *Civil–Military Relations: USAID's Role*, Technical Publication Series (Center for Democracy and Governance, US Agency for International Development: Washington, DC, July 1998).

have a clear interest and comparative advantage in other aspects of this agenda.[21]

The European Union (EU) external assistance programmes have two dimensions that are relevant to security sector reform. One is the EU assistance provided to the African, Caribbean and Pacific countries under the framework of the Cotonou Agreement of June 2000, which emphasizes the importance of good governance and entails periodic performance assessments to measure progress towards implementing political and institutional reform.[22] The other consists of the EU Common Foreign and Security Policy (CFSP) programmes, which do not mention security sector reform specifically but do require that all applicant states introduce democratic oversight of the military.[23]

Addressing security sector reform is a priority of the Organisation for Economic Co-operation and Development (OECD) Development Assistance Committee (DAC), whose Network on Conflict, Peace, and Development Cooperation carries out a range of research and policy-related activities designed to harmonize the work of its members in the conflict and security domain.[24] In April 2001 the OECD development ministers endorsed a supplement to the DAC Guidelines on Conflict, Peace and Development Cooperation which includes measures to bring conflict prevention into the mainstream of policy formulation, to take account of the relationship between security and development, including security sector reform, and to strengthen international support for peace processes.[25]

Regional security organizations

NATO adopted the Partnership for Peace (PFP) programme in 1994.[26] PFP programmes have elaborated norms and guidelines for the oversight of military institutions and the internal state security apparatus as well as the specific civil–military relations characteristic of a stable democracy. This comprehensive framework for reforming the management of the armed forces is available to nearly all of the post-communist and post-Soviet CEE states. The Membership Action Plan[27] and the 1995 NATO Study on Enlargement[28] made it clear that the application of a set of basic principles of 'democratic control over the military' is a precondition for NATO to consider any application for membership.

---

[21] For a discussion of UN conflict prevention activities see chapter 2 in this volume.
[22] The text of the Cotonou Agreement is available at URL <http://europa.eu.int/comm/development/cotonou/index_en.htm>.
[23] For a discussion of EU conflict prevention activities see chapter 2 in this volume.
[24] OECD, DAC (note 5).
[25] OECD, DAC, *The DAC Guidelines on Preventing Violent Conflict*, 2001, available at URL <http://www.oecd.org/EN/document/0,,EN-document-notheme-2-no-24-5782-0,00.htm>.
[26] An account of the PFP initiative is given in Rotfeld, A. D., 'Europe: the multilateral security process', *SIPRI Yearbook 1995: Armaments, Disarmament and International Security* (Oxford University Press: Oxford, 1995), pp. 275–81.
[27] URL <http://www.nato.int/docu/facts/2000/nato-map.htm>.
[28] North Atlantic Treaty Organization, *Study on NATO Enlargement* (NATO: Brussels, Sep. 1995).

The role of the Organization for Security and Co-operation in Europe (OSCE) in security sector reform consists mainly in setting models and norms for the individual member states and the region as a whole. In 1994 the principles guiding the role of armed forces in democratic societies were further elaborated and 'operationalized' in the OSCE Code of Conduct on Politico-Military Aspects of Security (sections VII and VIII).[29] The implementation of 'democratic oversight over the military' became a political obligation for all members of this organization, thus mandating their implementation in internal legal norms, regulations and procedures.

Compliance with these guidelines is assessed at periodic review conferences of the OSCE states. Equally important is the experience of the OSCE in confidence- and security-building measures, which have led to an improvement of interstate relations on the European continent since the mid-1970s. Among these measures are several which relate to building regional transparency in such areas as weapons procurement, budgets and restructuring of armed forces. These transparency measures, however, remain focused on interstate relations rather than on the objective of full transparency within the security sectors of the countries concerned. Sub-regional arrangements in Europe include the Process of Good Neighborliness, Stability, Security and Cooperation of the Countries of Southeastern Europe (South East European Cooperation Process, SEECP), which provides a kind of sub-regional code of conduct for relations in the region.[30] The SEECP defence ministries have worked on cooperative security reform since 1997.

Outside Europe, regional and sub-regional organizations, including the Association of South-East Asian Nations (ASEAN), the Organization of American States (OAS), the Organization of African Unity (OAU),[31] the Economic Community of West African States (ECOWAS) and the Southern African Development Community (SADC),[32] have initiated various programmes linked to the transformation or better management of the security sector in their member states. However, regional and sub-regional mechanisms are not always well coordinated and their objectives may differ, even within the same state. In Africa, for instance, there are a number of conflict prevention mechanisms, including the Conference on Stability, Security, Development and Co-operation in Africa (CSSDCA) adopted by OAU leaders

---

[29] URL <http://www.vbs.admin.ch/internet/GST/KVR/e/e-Codeofconduct.htm>. See also Organization for Security and Co-operation in Europe (OSCE,) 'Towards a genuine partnership in a new era', Budapest Document 1994, available at URL <http://www1.umn.edu/humanrts/osce/new/budapest-summit-declaration.html>; and *SIPRI Yearbook 1995* (note 26), pp. 309–13. According to the guidelines, the armed forces should be placed under, and used by, states' institutions having democratic legitimacy and abiding by legal norms, democratic values, neutrality in national political life, and human and civil rights, as well as by a rule of individuals' responsibility for possible orders and deeds inconsistent with the norms of domestic and international law.

[30] Information on this process is available at URL <http://www.seecp.gov.mk/general_info.htm>.

[31] The OAU member states adopted the Constitutive Act of the African Union on 11 July 2000; it entered into force on 26 May 2001, formally establishing the African Union (AU), with headquarters in Addis Ababa. The AU will replace the OAU after a transitional period.

[32] For the members of these organizations see the glossary in this volume.

in 1989 and the New Partnership for Africa's Development (NEPAD), adopted in 2001, which have not yet been harmonized.[33]

The implications of 11 September 2001

The 11 September 2001 attacks on the World Trade Center and the Pentagon, engineered and carried out by the al-Qaeda network, have underscored the link between state failure and international security.

While the 'war on terrorism' being led by the USA is being fought on many fronts, a central element of the strategy is to strengthen transnational intelligence and law-enforcement cooperation and military action. The less developed states that have joined the 'coalition against terrorism', and that are seen to harbour political elements that may be a threat to the USA and its allies, will probably receive increased support to bolster their intelligence and internal security capacity. These reforms may not be consistent with meaningful security sector reform since significant trade-offs can be expected between the initial primary focus on strengthening effectiveness and the longer-term goal of improving transparency and accountability in the security sector.[34]

Some of the regimes that will be in the front line in the anti-terrorism campaign are authoritarian and have security institutions that operate in a manner that is far from open and accountable. These security services enjoy substantial political influence and institutional autonomy, making them resistant to change. Moreover, it is their appreciable counter-terrorism capabilities, including powers of arrest and surveillance authority, which reform would curtail.[35] It is highly likely that, despite the potential costs to human rights and civil liberties, encouraging serious reforms will be given less priority than persuading political leaders that it is in their interest to use their intelligence and law-enforcement capacities to help the USA and its allies.[36]

From the perspective of developing and transition countries that are being strongly encouraged to support the US-led campaign, there is a clear conflict between security sector reform objectives and means. Many of these countries are aid-dependent and face significant external constraints on how they budget and manage resources, particularly in the security sector. Even as they come under persistent pressure from their key donors to reduce security spending, they are being urged to bolster their internal security and intelligence capacities. A number of leaders have also cynically used the pretext of the war

---

[33] Fayemi, K. and Hendrickson, D., 'NEPAD: the security dimension', Democracy and Development Review, *Journal of West African Affairs*, vol. 3, no. 1 (2002), p. 85.

[34] Stevenson, J., 'Counter-terrorism and the role of the international financial institutions', *Journal of Conflict, Security and Development*, vol. 1, no. 3 (2001), pp. 153–59.

[35] Stevenson (note 34).

[36] This is a particular issue of concern in Afghanistan, where the desire to limit the exposure of international forces to combat with Taliban forces has resulted in the hasty training and arming of factional forces by the USA and its allies. However, limited attention has thus far been devoted to the question of how to integrate these forces into a national army with appropriate management and oversight structures. See, e.g., Graham, B. and Loeb, V., 'US special forces to train recruits for Afghan army', *Washington Post*, 26 Mar. 2002.

against terrorism to clamp down on internal opposition figures who are deemed a threat to national security interests.

These developments raise the spectre of a return to cold war security thinking, which revolved around regime security. A growing number of states are finding it necessary to curtail individual rights, including the right to privacy in the areas of communications and personal data. Cross-border traffic has become more tightly controlled, with new restrictions pending in a number of countries. Even as international cooperation in intelligence gathering and joint action against terrorist cells are increasing, there are corresponding demands for less scrutiny by elected officials over the plans, budgets and operations of states' security organs. Increasingly 'centralized' and strengthened security sectors cannot help but exert commensurably greater influence on states' security policy and budgetary decisions.

The problems are already apparent not only in a number of developing countries but also in the USA itself. In the wake of 11 September, the US Government has tried to evade congressional oversight on defence spending related to the war on terrorism. Requests made for $10 billion to cover unspecified Department of Defense 'anti-terrorism efforts', as part of a $48 billion overall increase in the defence budget, would, if granted, mean the loss of some of Congress' 'power of the purse'.[37]

IV. Security sector reform in a regional context

The countries of Africa, Asia, Central and Eastern Europe, and Latin America face diverse security challenges. Among the countries undertaking security sector reforms are those that are (*a*) emerging from war, (*b*) shifting from communist to pluralist systems, (*c*) authoritarian regimes and (*d*) functioning democracies. While the nature and potential of reforms under way in these countries depend largely on their specific circumstances, there are also regional variations that influence reform processes. This is the case most notably with regard to the nature of external involvement and the forms of assistance and incentives on offer from region to region.

Africa

Of the 24 major conflicts that took place worldwide in 2001, seven were in Africa.[38] These were driven by a complex interplay of internal, regional and global factors and meant that about one-fifth of the people living in Africa were directly affected by armed conflict. Such a state of affairs has contributed to the rapid militarization of the continent. After decreasing during the early 1990s, military spending in Africa has been rising steadily since 1996.[39]

---

[37] Laurenz, R., 'Top legislator wary of contingency fund', *Defense Week*, 11 Feb. 2002, p. 1, available at URL <http://ebird.dtic.mil/Feb2002/e20020211top.htm>. See also appendix 6E in this volume.
[38] See chapter 1 and appendix 1A in this volume.
[39] See chapter 6 in this volume.

The issue of military spending has been the primary entry point for international actors interested in security sector reform in Sub-Saharan Africa, but there is increasing recognition that military spending issues are only a symptom of more fundamental governance problems.[40] The problems of security in some African countries arise as much from an under-investment in the security sector, both in financial and in human resource terms, as from weak state structures and mechanisms of civil oversight. The persistence of an environment of political instability and weak rule of law has placed immense demands on African security forces, which are already undergoing a profound institutional crisis.

The armed opposition faced by some governments in Africa is in part a consequence of the severe reduction in public services that occurred throughout the 1980s as a result of economic stagnation and externally imposed economic structural adjustment and stabilization measures.[41] It also stems from the rejection of non-performing African states by citizens who have resorted to a reliance on non-state sources of protection, including ethnic and religious affiliations. In the context of economic and political uncertainty and dwindling salaries, state security forces themselves have increasingly been less able to resist their manipulation by political elites.

The elementary demand for security by citizens, states and corporations has increased the trend towards the 'privatization' of security. While meeting the security needs of the privileged classes, some governments have invested less in law-and-order measures that benefit the general population. This breakdown in confidence in the rule of law has led commercial security companies and traditional militias to take on roles once fulfilled by the public sector. At the individual level, populations in many countries have resorted to arming themselves. Criminals have sometimes been lynched. Political leaders have exploited militia groups that have arisen in places such as in Nigeria's Niger Delta for their own purposes.

Military influence over the political process, either direct or indirect, is still a reality in many African states. In some countries in which the military has formally withdrawn from politics, it remains well placed to influence power owing in large part to the fact that some civilian rulers no longer enjoy a popular mandate to remain in power and thus rely on the state security apparatus. Commercial involvement by the military in many countries has further reinforced its autonomy and expanded its influence over vital issues of national governance.

International donor agencies have seen Africa as the testing ground for the security sector reform policy agenda. There are currently externally supported initiatives in Botswana, Ghana, Lesotho, Mozambique, Nigeria, Rwanda, Sierra Leone, Tanzania and Uganda. With the exception of Sierra Leone, where international actors, led by the UK, have embarked on a comprehensive

---

[40] Ball *et al.* (note 12).
[41] Hutchful, E., Draft report of the African Leadership Forum Expert Group on Demilitarisation, Security and Development, African Security Dialogue and Research, Accra, Ghana, Jan. 2002, information available at URL <http://www.africansecurity.org/publications.html>.

programme to rebuild the security sector, the initial international engagement in Africa has been partial and specific.

With the exception of South Africa after the transition to majority rule in 1994, there have been few examples of comprehensive and sustained, internally driven reform processes led by states. Nevertheless, African states are pursuing a range of initiatives to address security sector reform, which the international community is increasingly coming to understand and support. The most prominent of these is the NEPAD, launched in October 2001, which is a wide-ranging vision for promoting better government and ending Africa's wars.[42] The security dimension of NEPAD is currently being developed under the direction of South Africa.

Asia

Nine of the 24 major conflicts in 2001 were in Asia.[43] There has been limited external support for security sector reform in Asia, with the exception of a few high-profile cases including Cambodia,[44] East Timor[45] and, currently, Afghanistan. The relative neglect of Asia is notable given that the region's security sectors are, to a greater or lesser degree, afflicted with the same problems that security sector reform seeks to remedy in other parts of the developing world. The donor countries and the multilateral development actors have less influence in this region than in Africa. As a result, there have been fewer external incentives and pressures for reform, including international assistance and economic conditionality.

Nevertheless, domestic economic, social and political change, resulting in the growth of civil society and democratization, has driven significant restructuring of states' security sectors in parts of Asia. The results in South-East Asia, for instance, are compatible with the reform agendas of international actors,[46] although progress has been limited. While only Myanmar (Burma) remains under direct military rule, in Cambodia, Indonesia, Laos, Malaysia, Thailand and Viet Nam the military and secret services continue to exert strong influence over governance despite significant democratic advances.[47]

In Indonesia, the former dual function of the military, which provided it with a role in security and politics, has nominally been replaced by a single function, focused on national defence. However, in reality the disengagement from politics and internal security has been superficial. While officially the

---

[42] The New Partnership for Africa's Development (NEPAD), Framework document issued in Abuja, Nigeria, Oct. 2001, available at URL <http://www.dfa.gov.za/events/nepad.pdf>.

[43] See chapter 1 and appendix 1A in this volume.

[44] See, e.g., Hendrickson, D., 'Cambodia's security sector reforms: limits of a down-sizing strategy', *Journal of Conflict, Security and Development*, vol. 1, no. 1 (2001), pp. 67–82.

[45] Rees, E., 'Security-sector reform and transitional administrations', *Journal of Conflict, Security and Development*, vol. 2, no. 1 (2002), pp. 151–55.

[46] Huxley, T., *Reforming Southeast Asia's Security Sectors*, Working Paper no. 4 (Conflict, Security and Development Group, Centre for Defence Studies, King's College London: London, Apr. 2001).

[47] In Thailand alone, e.g., there have been 16 coups in the past 70 years. Richardson, M., 'In jittery Southeast Asia, fears of military backlash', *International Herald Tribune*, 2 Jan. 2001.

police now have sole responsibility for internal security and maintaining law and order, in practice they lack the appropriate training and resources and continue to rely on the military for back-up.[48] The military continues to enjoy most of its prerogatives, including extensive involvement in commercial activity that provides the bulk of the funding for its statutory activities.

Across South-East Asia civil institutions dealing with security issues and the political structures supporting such institutions remain weak[49] and permeated by representatives from the military and other security organs. Where economic conditions have worsened in the region, widespread social unrest and economic hardship have made it all the more difficult to introduce systematic and purposeful security sector reforms.

The opportunity for external support to reverse this trend has presented itself most clearly in the context of countries emerging from war, such as Cambodia. Here, the international focus has been on reducing the size of the army and military spending and primarily on instruments of economic conditionality wielded by the IMF and the World Bank. The failure to set demobilization within a wider framework for security sector reform has made it difficult for international actors to specifically target the problems that give rise to high military expenditure and weak military accountability.[50]

In the aftermath of 11 September 2001, the USA is bolstering its military presence across Asia, seeking levels of military cooperation not seen since the end of the cold war.[51] A central thrust of US policy is on conducting intelligence work with its national counterparts and tackling transnational security problems, such as drug and human trafficking and money laundering, which are considered to be linked with terrorist groups. While US engagement is welcomed by many of its regional allies, the narrow focus on 'anti-terrorism' activities and on stabilizing flashpoints like the India–Pakistan conflict may divert attention from broader security sector reform objectives.

In the former Soviet republics of Central Asia, where the military presence of the USA has increased the most, enhancing the capacity of civilian oversight mechanisms is a central challenge. The security sector retains key characteristics of the Soviet-era security apparatus: heavily armed forces that are entirely inadequate to meet new security needs, powerful secret services modelled on the former Committee of State Security (Komitet gosudarstvennoy bezopasnosti, KGB), police forces that are closely linked to the state political authority, and a very weak civil society.[52]

---

[48] McCulloch, L., 'Police reform in Indonesia', *Conflict, Security and Development Group Bulletin*, no. 10 (Mar./Apr. 2001).

[49] Huxley (note 46).

[50] Hendrickson (note 44).

[51] Hiebert, M. and Lawrence, S., 'Hands across the ocean', *Far Eastern Economic Review*, vol. 165 (14 Mar. 2002).

[52] 'Where Stalin lives on', *International Herald Tribune*, 20 Aug. 2001; and 'Trouble in Central Asia', *International Herald Tribune*, 17 Aug. 2001.

Eastern Europe

There have been significant Western attempts to promote the development of civil–military relations in the East European states over the past decade. Progress has been rapid in terms of building political support for the goals of democratic and civilian control over the armed forces. Concerns about regional political stability and integration into both NATO and the EU have provided powerful incentives for East European states to reform their security sector. Nevertheless, the accompanying institutional and attitudinal changes required to entrench change have been slow and the central thrust of reform efforts has been on the armed forces.

In contrast to those of Central Europe, the majority of the post-communist states of Eastern Europe remain heavily militarized and only partially integrated into the global economy. The armed forces are heavily oriented to offensive tasks and, in most countries, are complemented by equal numbers of internal police and secret security services which look to different ministries, chains of command and mechanisms for civilian control. In Russia, for instance, there are 10 different state security services with a combined workforce of several hundred thousand.[53]

Despite growing impetus for modernization as part of wider processes of economic liberalization, reform has largely been superficial. State elites often retain a strong, albeit unofficial, affiliation with the old political system as well as strong links to the secret security services. This creates a powerful conservative bloc, ill-disposed to far-reaching reforms.

With the vast bulk of the population in Eastern Europe preoccupied by low living standards and unemployment, civil society for its part remains poorly organized or motivated to pressure for security sector reform. The legacy of repressive behaviour by the security institutions continues to instil fear in the population, with the result that there is limited public debate on security reform issues. However, this legacy also means that the task of building political support for security sector reform has been relatively easier in the East European states than in Africa, Asia and Latin America.

Latin America

Two of the major armed conflicts in 2001 took place in Latin America.[54] The countries of this region can be broken down into three categories with regard to their readiness for security-related reforms: (*a*) the Central American countries engaged in transitions from cold war-era armed conflicts to peace; (*b*) the Andean countries, including Colombia, Ecuador, Peru and Venezuela, which are currently involved in armed conflicts or face the very real prospect of an outbreak of violence or military intervention in politics; and (*c*) the countries

---

[53] ['Russian Special Services'], *Biuletin Osrodka Studiow Wschodnich* [*Bulletin of the East Studies Center*] no. 5 (20), (June 1997) (in Polish).

[54] See chapter 1 and appendix 1A in this volume.

of the Southern Cone, including Argentina, Brazil, Chile and Uruguay, which have a long legacy of military regimes and only recently embarked on processes of democratization.

In the changing political and strategic landscape facing the Latin American countries, there is an urgent need for technical as well as doctrinal modernization, in particular to reform national security doctrines that mandate the military to protect the state from both internal and external enemies. However, despite the constant exhortation on these societies and their militaries to modernize their security sectors,[55] there has been limited external assistance for reforms of a substantive nature. Internally driven reforms have also been limited, tending to focus on bringing the security forces to account for human rights abuses and on strategies to enhance public or 'citizen' security.

Comprehensive and integrated reform of civil and security institutions has still not caught on in the region.[56] The few countries in which security sector reform has been attempted include Panama, following the US intervention in 1989, and El Salvador, Guatemala and Haiti, during the 1990s under the auspices of UN missions supporting post-war reconstruction processes. For the most part, the UN initiatives have placed greatest emphasis on the rule of law and judicial assistance as well as on facilitating war-to-peace transitions with support for the demobilization and reintegration of ex-combatants.

On the military side, multilateral involvement has largely been around efforts to control military expenditure and to extend civilian control over the military. However, there have been some mixed messages when attempts to limit military roles have been contradicted by exhortations from the US Department of Defense for the military to find new roles, including policing, staunching migration and tackling drug trafficking.

The US anti-narcotics agenda has generally held sway over other agendas supported by civil society, the IFIs and European countries owing to generous US funding and US prominence in the region. The drive to stamp out narcotics in the Andean region has resulted in packages of military assistance that have caused the Colombian military to be strengthened as they take on more of a law enforcement role.[57] Paramilitary groups have also benefited from US assistance. At the same time, they are not placed within any framework of

---

[55] Nunn, F. M., 'Latin American military–civilian relations between World War II and the New World Order', *University of South Africa (UNISA) Latin American Report*, vol. 13, no. 1 (Jan./June 1997); and the Internet site of the Partnership for Democratic Governance and Security, URL <http://www.pdgs.org>.

[56] Neild, R., Washington Office on Latin America, Washington, DC, personal communication with the authors.

[57] Human Rights Watch (HRW), 'The ties that bind: Colombia and military–paramilitary links', *HRW Report*, vol. 12, no. 1 (Feb. 2000), available at URL <http://www.hrw.org/reports/2000/colombia>; and HRW, 'Colombia: paramilitary groups closely tied to army, police: US funding military unit implicated in serious abuses', *HRW World Report 2001: Colombia*, available at URL <http://www.hrw.org/press/2001/10/sixthdivision.htm>.

accountability. This has come at a time when the USA has been vocal about the importance of restoring civilian authority over the military.[58]

V. Challenges to security sector reform

The lack of a shared definition of security sector reform makes it difficult to make a clear overall statement on current progress and remaining challenges. International support for security sector reform has to date been relatively limited and of an ad hoc nature. Apart from the CEE states, where the focus has been predominantly on issues relating to military reform and border security, the most notable programmes have been in developing countries emerging from war. At this more specific programme level there is only a cursory understanding of what international assistance has achieved. In part, this is because international actors have been slow to develop tools for assessing the effectiveness of their policies.

Nevertheless, it is increasingly apparent that the receptivity of different societies to the security sector reform agenda varies greatly depending on their internal circumstances and the external incentives for reform. In cases in which the domestic constituencies, institutional capacity and incentives for reform are weak, a sharp reduction of the impact of external assistance should be expected. This underscores the limits of current international efforts to support security sector reform, which have to date focused primarily on spreading Western norms and practices to inform how the security sector of aid recipients should operate.

Substantive progress in building consensus around standards of security sector governance across Europe is apparent, although the extent to which these goals have been institutionalized in the working of the security sector has been variable.[59] At one level, 'first generation' institutional issues such as the creation of constitutional frameworks and mechanisms for civil oversight have been successful. However, a 'second generation' of issues that relate to the acquisition of shared norms and values by civilians and the military has not yet made a significant impact.

The African, Asian and Latin American experiences are much less clear-cut. For the most part, the conditions for reform have not been as favourable as in Europe owing to the institutional fragility of states, political instability, resource constraints and the limited nature of external incentives on offer. The lack of strategic significance to the Western countries of those countries most in need has also played a considerable role. The larger cultural gap between

---

[58] This, e.g., is the focus of the Center for Hemispheric Defense Studies at the National Defense University in Washington, DC. Information on its programme is available at URL <http://www.ciponline.org/facts/chds.htm>.

[59] Forster, A., 'Civil–military relations and security sector reform: West looking East', Paper presented at the 4th International Security Forum Conference on Civil–Military Relations and Democratic Control of Armed Forces, Geneva, 15–17 Nov. 2000, available at URL <http://www.dcaf.ch/publications/papers/ISF_WS_III-4_Forster.pdf>.

these societies and the West has also underscored the need for international actors to reflect more carefully on what aspects of their national experiences have relevance to developing countries and on how to more effectively facilitate the development of a national vision and domestic constituencies to sustain reform processes.

Operationalizing concepts

International actors have been slow to develop a holistic and long-term approach to providing international assistance. Efforts to ensure that different national and international programmes fit together effectively on the ground have not been successful. This has led to a tendency on the part of many actors to rebrand long-standing activities as security sector reform without evaluating the needs of aid recipients or adapting policies to make them relevant to new circumstances.[60] Thus there continues to be a narrow focus, in many cases, on direct military and police training and on efforts to address the proliferation of light weapons, to demobilize and reintegrate ex-combatants, or to provide human rights training to members of the security forces. While all of these are important aspects of security sector reform, they will be of limited long-term utility unless they are carried out in such a way as to support the wider agenda of strengthening the institutional framework for managing the security sector.

The question of the sequencing of international assistance has also come to the fore as members of the development and security communities have begun to work together more closely in the context of multifunctional international assistance programmes. While the broad objective of strengthening management and oversight of the security sector is generally shared, within this framework international actors in the development and security communities often prioritize different goals that may not be compatible. For example, military assistance provided to foreign armies to increase their effectiveness may undermine efforts by other external actors to limit the political influence of the military and strengthen civilian capacities.

There are differences in national approaches to security sector reform. While there is an increased recognition that reforms cannot and should not be imposed from the outside, international actors have been constrained from helping to build local ownership by short programming cycles, poor understanding of the countries in which they work and the sensitivities of engaging with governments that are not seen as committed to reform. The prescriptive approach of the US Department of Defense contrasts with the greater British emphasis on facilitating reform—though there is evidence that the United States is changing its approach.[61]

---

[60] Forster (note 59).
[61] The US African Center for Strategic Studies (ACSS) provides senior African military and civilian leaders with academic training in civil–military relations, national security strategy and defence economics. The initial emphasis of the ACSS programme when it was launched in 1998 was on the wholesale transfer of the US model of civil–military relations into the African socio-cultural context. The initial opposition to this approach from African participants at the first regional seminars hosted by

Unfavourable environments

Implementing security sector reform in conflict-torn societies presents the greatest challenges. The lack in most cases of a strong national vision and capacity coupled with the urgency of reform results in an overwhelming emphasis on an external timetable and model. This is despite the fact that international actors rarely have a clear understanding of the situation on the ground, of what preceded a war or of how the new power dynamics are arranged. Persisting tensions, along with the enhanced role of security forces in political matters, constitute major barriers to reform.

The value placed on institutional and political stability by post-war governments is often not fully appreciated by international actors. Government reluctance to embark on a reform process tends to be confused with a weak commitment to a peace process or to democratization rather than with a lack of the instruments, resources and support needed to push through difficult changes. Civilian oversight mechanisms such as legislative select committees and financial auditing bodes, if they exist at all, are difficult to reactivate because of the centralization of security policy making by the executive branches of government.

Overcoming these barriers poses significant challenges for external actors seeking to support reforms. In most cases, there is a huge gap between the stated objectives of reform processes and the starting point, which is very hard to bridge owing to the inadequacy of local and external resources. In these contexts, critical issues such as national ownership, civilian capacity building and strategic planning in the security sector are given lower priority than other aspects of post-war reconstruction in the social and economic domains. For external actors concerned with the restoration of a national capacity in the security domain, this has required rethinking strategies of engagement.

The top priority in most conflict-torn societies is to prepare the political terrain for more fundamental institutional reforms. Greater priority should be given to small, strategic interventions designed to build relationships and trust and to setting out policy options for countries undertaking national strategic reforms in order to facilitate these efforts. This will often require international actors to become engaged in helping to create a 'comfort zone' in which disparate groups that have never spoken with each other before can begin to shape a mutually acceptable reform vision.

Building a national vision for reform is also a priority in the CEE states although, with the exception of post-war Bosnia and Herzegovina and Kosovo (Federal Republic of Yugoslavia), a relatively strong institutional framework for debate and policy planning is already in place. This is not the case in countries such as Sierra Leone and Uganda, which require more 'root and branch' reforms and where the international community has helped to organize seminars which have served to stimulate dialogue between the military, the

the ACSS has resulted in a greater effort to engage Africans in shaping the programme. See URL <http://www.africacenter.org>.

police, politicians, members of parliament, civilian policy sectors and civil society groups. This move away from a narrow reliance on technical assistance is positive, although it has also brought international actors into a sensitive domestic arena, which has traditionally, both in their own countries and in aid recipients, been out of bounds to foreigners.

Conflicting objectives

As international actors from diverse policy communities have become involved in joint assistance programmes, it has become readily apparent that security sector reforms involve conflicting objectives. Even where public investments in the security sector absorb the lion's share of state resources, they may be insufficient to meet national needs. A number of countries, including Rwanda and Uganda, have come under intense pressure from aid donors to reduce military spending at a time when they face significant external threats to national peace and stability.

Unsustainably high levels of military spending are a legitimate cause for concern in view of the impact on macroeconomic stability and poverty reduction objectives. However, the failure of international actors to anchor efforts to manage military expenditure within a broader reform programme designed to enhance the security of states and their citizens can result in a number of unintended consequences.

Two specific problems have become apparent where donors and the IFIs have relied on economic conditionality to encourage countries to reduce military spending rapidly without reference to the quality of governance in the security sector.[62] First, this strategy avoids addressing the underlying political conflicts and institutional and human-resource weaknesses, of which high levels of military spending are only one manifestation. Second, it creates a perverse incentive for governments to resort to creative accounting in order to conceal portions of their expenditure.

While the off-budget problem is difficult to detect, there are good reasons to suspect that it is relatively common where security sector governance is weak. Addressing the problem involves creating incentives for both militaries and governments to keep military spending on budget as well as to strengthen fiscal management and the management of the defence sector. The binding constraints are often political in nature and require fundamental changes in civil–military relations that cannot be achieved fully until the civilian sectors, including defence and finance ministries and parliaments, can fulfil their mandatory oversight roles effectively.

Security sector reforms can also have other unintended consequences. The relationship between security sector downsizing and the enhancement of polit-

---

[62] Hendrickson, D. and Ball, N., *Off-budget Military Spending and Revenue: Issues and Policy Perspectives for Donors*, Occasional Paper no. 1 (Conflict, Security and Development Group, King's College London: London, Jan. 2002).

ical stability or public investments in the social and economic sectors is far from straightforward. Recent examples of military restructuring in the CEE states, for instance, illustrate how reforms can increase instability. The attempts made by some countries to quickly demilitarize their economies and diminish the burden of defence budgets by drastic reductions in the size of their security forces have produced a number of undesirable outcomes. These include large numbers of untrained security personnel entering the labour market, adding to already high unemployment levels; visible disenchantment among demobilized personnel, especially in the officer corps, which has created anti-reform sentiments; and serious wastage of resources as ill-conceived reforms have had to be revoked. The decline of morale within the armed forces has also undermined combat-readiness and military discipline, resulting in the illegal transfer of weapons into the hands of criminals.

Recent experiences also suggest that reductions in the size of the armed forces will not automatically lead to increased spending in other public sectors. African cases have clearly demonstrated that the processes of downsizing and restructuring military forces themselves require ample resources and will not save money in the short run because released personnel must be re-educated and assimilated into the economy or pensioned.[63] Furthermore, the re-allocation of public spending from the security sector to the social sectors will only come about if there is a change in spending priorities, which usually requires tackling vested political and military interests.

However, there is increased recognition that defence cooperation arrangements that bind many developing countries and CEE states with the industrialized countries can also impede other reform processes. In the case of a number of the CEE states, including the Czech Republic, Hungary and Poland, the resources gained from military reductions and restructuring, which these countries had to commit themselves to undertake in order to prepare themselves for membership of NATO, have been channelled into modernization programmes for the armed forces. This has meant that the long awaited 'peace dividend', which a reduction in the size of the armed forces might bring about, can only be achieved after a longer period of sustained reforms which increase efficiency in the armed forces, usually concomitant with an overall transformation of the economy.

Integrated programmes

In practice, few countries will undertake to reform the security sector as a whole, even though there is recognition of the need for a holistic view of the process. The first test case for a comprehensive international programme to rebuild and reform the security sector was Sierra Leone. This initiative, led by the British Government from early 1999, involved inputs from ministries responsible for defence, development, foreign relations and home affairs.

---

[63] Marley, A. D., 'Military downsizing in the developing world: process, problems, and possibilities', *Parameters*, vol. 27, no. 4 (winter 1997/98), pp. 137–44.

Initial activities supported by the UK were designed to strengthen and civilianize the defence ministry, produce a new national security policy, reform the police, and train and equip 2500 soldiers for a new national army.

The resumption of hostilities between the Sierra Leonean Government and the rebel Revolutionary United Front in mid-1999 led to a pronounced shift in the focus of the British programme from strengthening the civilian components of the security sector responsible for oversight and management to winning the war. Military training provided to the national army, including the support of the UN peacekeeping mission, paid immense dividends in terms of restoring security and government control over the national territory. However, the longer-term governance agenda, including the strengthening of key regulatory mechanisms such as the finance ministry, took a back seat during this period.

Sierra Leone's case has underscored the immense challenges facing external assistance in a context in which the security sector has been weakened by years of mismanagement at the same time as there is a need to approach security sector reform as a part of a wider reconstruction programme.[64] In the urgency to rehabilitate the national army, the task of integrating the defence budget into the wider public expenditure management framework was given a back seat. Such a framework is essential if security spending is to be subject to the standard fiscal controls of the finance ministry and the appropriate legislative scrutiny, neither of which have seen their capacity to fulfil this role strengthened.

International actors have tended to overlook the development of the capacities required to make a sector-wide assessment of needs, including a clear understanding of the security threats a country faces and of the options available to the state to meet these threats.[65]

There has been a tendency to underplay the extent to which security sector problems are exacerbated by external factors, including regional conflicts, interstate rivalry and global economic forces.[66] The easy availability of arms on international markets and the emergence of lucrative 'war economies' with regional and international dimensions have received the most attention from the international community.

Assistance has not been separated from the agendas of specific donors.[67] There has been a tendency for donor countries to concentrate on geographical areas or states where they have historical connections or strategic interests.

---

[64] Ero, C., *Sierra Leone's Security Complex*, Working Paper no. 3 (Conflict, Security and Development Group, Centre for Defence Studies, King's College London: London, 2000).

[65] As the case of Cambodia has demonstrated, this has made it difficult to develop a logical and sustainable reform plan, to effectively assess what level of security spending is affordable in relation to other public sectors, or to buy in the support of government and the security forces themselves for reform. See Hendrickson (note 44).

[66] OECD, DAC (note 5); and Cooper, N. and Pugh, M., *Security-Sector Transformation in Post-Conflict Societies*, Working Paper no. 5 (Conflict, Security and Development Group, Centre for Defence Studies, King's College London: London, 2002).

[67] E.g., the British Government approved the sale of a radar system to Tanzania in spite of opposition from a cross section of groups in Tanzania and the international community. This is important to note, in view of the leading role that the UK is playing in setting the security sector reform agenda and of the fact

Different perspectives and voices also need to be integrated into the reform process. In many of the countries that are most in need of security sector reform, non-state actors offer a strategic entry point for international actors. In Africa, donors such as Denmark, Norway and the UK have actively supported networks of NGOs working on security issues. The development of non-governmental networks, in which the atmosphere is more informal and sensitive political issues can be put aside, is particularly valuable in terms of promoting security sector reform.

Recent round table discussions on security sector reform have begun to build linkages between states (e.g., within sub-regions of Africa and on an Africa-wide basis) as well as to establish cross-regional linkages (for example, through participation by representatives from Asia and Latin America in African meetings).

VI. Conclusions

The internal conditions in states not only establish the security environment within them but also have an impact on regional and international security. Where states are unable to manage developments within their borders successfully, the conditions are created for disorder and violence that may spill over onto the territory of other states and ultimately perhaps require an international intervention.

Security sector reform is part of an attempt to develop a more coherent concept for reducing the risk that state weakness or state failure will lead to disorder and violence. The set of norms, laws and institutions by which the security sector is governed represents an important element of the overall effort to improve governance.

Some governments are now working with civil society and the security sector as one part of this attempt to develop a reform programme. Although security sector reform has become established on the international security agenda, there is still not a shared understanding at the international level of what this term means. This has limited the debate on the subject. Assisting in the development of such a shared understanding should be a priority objective for the research community.

Security sector reform is a new activity and neither its scope nor the manner of its organization has been fully established. A broad set of needs have been identified and a large number of mechanisms are likely to be required to address them. Many diverse actors will have to coordinate their activities to meet these needs.

Sustainable reform depends on the cooperation and participation of stakeholders in the countries concerned. However, reform processes may be generated by different factors. In some regions and countries, reforms have

that Tanzania is dependent on aid for nearly 50% of its annual budget. See 'World Bank could bar $40 million Tanzania air traffic deal', *World Bank Development News*, 21 Dec. 2001, available at URL <http://www.worldbank.org/developmentnews>.

been initiated in response to pressure from local or domestic actors. In other cases external forces—either states or international organizations—have actively pressed the case for reform.

The prospect of participation in European integration has provided a significant positive incentive for reform among Central European states and in certain East European states. This cannot be matched by regional and sub-regional organizations in Africa, Asia or Latin America. In these regions the primary incentive for reform has been based largely on persuasion and the use of economic assistance to encourage countries to undertake reform.

There have been few cases of sustained externally driven reforms in African states in the past. As a consequence, several new, internally generated initiatives are now under way in the framework of the New Partnership for African Development.

The response of states to the 11 September 2001 terrorist attacks on the United States may slow down the development of a security sector reform agenda. Increased importance is being placed on developing cooperation with the armed forces, intelligence services and law-enforcement services of other states to identify and eliminate groups and individuals engaged in terrorist acts. There is a risk that security sector reform will become subordinate to anti-terrorism activities in countries where the development of this cooperation is seen as particularly important.

5. Sanctions applied by the European Union and the United Nations

IAN ANTHONY

I. Introduction

During 2001 sanctions continued to play an important role in the efforts to manage a range of security problems, while the reform of sanctions witnessed towards the end of the 1990s continued. Sanctions are now not only used to target states, but also applied to non-state entities and, increasingly, to individuals. After the terrorist attacks on the United States on 11 September 2001, the United Nations Security Council agreed on extensive measures against groups and individuals that have carried out acts of terrorism.

As part of the wider effort to improve the effectiveness of UN sanctions an informal working group was established in April 2000 to develop general recommendations on how to improve the effectiveness of sanctions. This working group completed a draft report in 2001. In addition, in 2001 the use of sanctions related to particular countries continued to be a focus of attention. In the 1990s very extensive sanctions were applied to Iraq and to the former Yugoslavia. On 29 November 2001 the UN Security Council approved Resolution 1382, which made important modifications to the sanctions regime against Iraq. By November 2001 all of the multilateral sanctions against Yugoslavia had been removed when the European Union (EU) lifted the remaining restrictive measures.[1]

The European Union has been in the process of developing a Common Foreign and Security Policy (CFSP). The leaders of the EU have decided that the EU must take a comprehensive approach to crisis management. Sanctions are one instrument available to implement that approach.[2] The European Union has established sanctions against states although the UN Security Council has not taken a similar decision. In some cases the EU has maintained its sanctions after the Security Council has decided to end UN measures. These decisions reflect the emergence of a political actor with an identity separate from the identity of its member states, since those states would not themselves have taken these decisions outside the EU context.

---

[1] Certain sanctions remain in place against individuals, principally financial sanctions against former Yugoslav President Slobodan Milosevic and his immediate family.
[2] The EU High Representative for the CFSP, Javier Solana, has written that 'crucial to success is the ability to provide the full range of instruments—economic and technical assistance, civilian police and institutional-building tools, trade incentives or sanctions, etc., in order to force parties to a conflict into a negotiated settlement and to rebuild the economy and restore the societies of a country or a region after a conflict'. Solana, J., 'Decisions to ensure a more responsible Europe', *International Herald Tribune*, 14 Jan. 2000.

Although the word 'sanctions' is frequently used, it does not have an agreed definition. The UN Charter does not contain the word at all but refers to measures that may be adopted in response to identified threats to the peace, breaches of the peace and acts of aggression.

Within the legal codes of states, sanctions are that part of a law that inflicts a penalty for its violation. In common usage international sanctions can be defined as any restriction or condition established for reasons of foreign policy or national security applied to a foreign country or entity by a group of states using substantially equivalent measures.

The implications of using sanctions against states are similar to a military action as their intent is always to inflict damage on the target.[3] For this reason, the legitimacy of sanctions used without a decision by the UN Security Council has been questioned.

Section II of this chapter examines United Nations sanctions and recent efforts to develop a more systematic approach to their use. Section III examines the use of sanctions by the European Union. Subsequent sections examine the use of sanctions in Iraq and Afghanistan and against terrorism.

II. The United Nations sanctions experience

While the use of force by the United Nations is envisaged in the Charter, because no armed forces are under its command sanctions are the only means of coercion available to the UN.[4]

In total, the Security Council has imposed sanctions on 16 countries: Afghanistan, Angola, Eritrea, Ethiopia, Haiti, Iraq, Liberia, Libya, Rwanda, Sierra Leone, Somalia, South Africa, Southern Rhodesia (now Zimbabwe), Sudan, the Federal Republic of Yugoslavia (FRY) and the Socialist Federal Republic of Yugoslavia.[5] Fourteen of these cases reflect decisions taken after the end of the cold war. The use of sanctions after 1990 stimulated debate about their effectiveness as an instrument for helping to manage security.

Although a decision of the Security Council establishes sanctions, the particular conditions in which sanctions have been used and the specific objectives established for them have differed widely from case to case. From the use of sanctions during the 1990s it is possible to perceive the emergence of some general principles. The sanctions imposed in the 1990s can be sorted into four general categories: (*a*) cases of cross-border conflict, (*b*) civil wars likely to have 'spillover' effects that will destabilize the region as a whole, (*c*) cases

---

[3] Space in this chapter does not permit a general discussion of sanctions and their use. For a recent overview see Brzoska, M. (ed.), *Smart Sanctions: The Next Steps* (Nomos Verlagsgesellschaft: Baden-Baden, 2001). The non-governmental organization The Fourth Freedom Forum maintains a bibliography of literature related to sanctions, available at URL <http://www.fourthfreedom.org>.

[4] Under Article 41 of the UN Charter the Security Council 'may decide what measures not involving the use of armed force are to be employed to give effect to its decisions, and it may call upon the members of the United Nations to apply such measures. These may include complete or partial interruption of economic relations and of rail, sea, air, postal, telegraphic, radio, and other means of communication, and the severance of diplomatic relations.'

[5] Several of these sanctions regimes are discussed in chapters 2 and 8 in this volume.

where internal repression by the government of one country is likely to generate conflicts that destabilize the region as a whole, and (*d*) cases where the target supports international terrorism.

Sanctions have a declaratory element through which a group of states indicates its preferred outcome in a particular crisis situation. Although they are intended to inflict damage on the target, sanctions are not intended to be punitive. Their objective is to bring about a change in the policies and behaviour of the target. Sanctions can be of different kinds.

Given the responsibilities of the UN Security Council, arms embargoes are a sanction that has been used fairly frequently in response to a threat to the peace, a breach of peace or a case of aggression.

The objective of an arms embargo was historically to signal that the particular dispute should be settled by peaceful means and that the international community will not assist either party in seeking a military solution. However, over an extended period an arms embargo could have an impact on the military capacity of warring parties, depending on the effectiveness of its enforcement.

The measures chosen could be economic—for example, the target of sanctions may be prohibited from buying or selling particular goods. They might be financial—for example, refusing to permit bank deposits held in the name of the target to be drawn upon. They might be travel-related—for example, states may deny ships or aircraft registered in the name of the target access to their territory or prohibit ships or aircraft on their national registers from visiting the target state. Diplomatic sanctions could include withdrawing support for, suspending or expelling the target state from international organizations or the drawing down of diplomatic contacts of various kinds.

In the past, sanctions were applied to the territories of states. However, from the mid-1990s they have also been imposed on particular parties to a conflict rather than on all citizens of a state and on parts of the territory of a state rather than on its entire territory.

These changes have reflected a different approach by the UN Security Council towards parties to a conflict. Increasingly, even-handedness and a general call for peaceful resolution have been replaced by a determination of responsibility. Sanctions have been applied only to the party identified as responsible for actions that represent a threat to the peace, a breach of the peace or an act of aggression.

Sanctions have been applied to non-state entities. The first such case was those applied to the União Nacional Para a Independência Total de Angola (UNITA, National Union for the Total Independence of Angola). With the development of targeted sanctions there are cases of sanctions being applied to individuals within a government.

As yet there are no cases of arms embargoes applying to the armed forces under government control while not applying to the armed forces of a non-state entity fighting that government. The UN has stopped short of endorsing military action to overthrow a government, but there are cases where an embargo has been applied to armed forces in de facto control of the territory of a state. The arms embargoes applied to the Revolutionary United Front (RUF) in Sierra Leone and the Taliban in Afghanistan are two examples where armed forces of the deposed government were exempt from embargoes. The UN may legitimize the use of military means to restore governments that have lost power in a *coup d'état*.

As such measures are mentioned in the UN Charter, relatively few concerns were raised regarding the legitimacy of UN sanctions before the mid-1990s, after which their legitimacy was increasingly questioned for three reasons.

First, the application of sanctions to some but not all parties to a dispute made the use of sanctions more politically sensitive and opened them to the charge that they have become an instrument of political coercion rather than conflict resolution. A second factor that raised questions about legitimacy was evidence that extensive sanctions could have disproportionate effects on the society of the target. A third factor was evidence that limited sanctions were not being implemented and so could not be effective.

Some states began to argue that in these conditions compliance with decisions by the Security Council was not mandatory since those decisions did not accurately reflect the broader will of the UN membership.

The status of the role of sanctions can be summarized in three points.

1. In the 1990s the use of sanctions became progressively more widespread.
2. There has been an attempt to design sanctions that can apply to particular targets (i.e., decision makers held to be responsible for threats to the peace, breaches of the peace or acts of aggression).
3. As a result, sanctions have become more complicated to design and implement.

Although sanctions proved to be complex and controversial the Security Council did not abandon this instrument but tried to improve it in cooperation with the Office of the Secretary-General and other UN organs.[6] An informal working group was established in April 2000 to develop general recommendations on how to improve the effectiveness of sanctions.[7]

The UN Working Group on the General Issues on Sanctions

The Working Group on the General Issues on Sanctions was established on 17 April 2000 with representatives of all of the states then sitting on the UN Security Council.[8] The objective of the working group was to institutionalize a

---

[6] For a more detailed discussion of how this process of improvement has been approached, see the documents available at URL <http://www.smartsanctions.ch/> for information related to financial sanctions and at URL <http://www.smartsanctions.de> for information related to travel sanctions and arms embargoes.

[7] Note by the President of the Security Council, UN document S/2000/319, 17 Apr. 2000, URL <http://www.un.org/sc/committees/sanctions/s00319.htm>.

[8] Although the report has not been published, a text of the Chairman's Proposed Outcome of the Working Group on Sanctions, Feb. 2001, is available at URL <http://www.cam.ac.uk/societies/casi/info/scwgs140201.html>. This is believed to be an authentic version of the text.

systematic, general approach to defining and implementing sanctions within the United Nations. The working group could create the basis for a legal regime for sanctions to establish norms based on international consensus.[9] While the group was intended to complete its work by 30 November 2000, consensus could not be achieved within the group by that date. The chairman of the group (Ambassador Anwarul Karim Chowdhury from Bangladesh) presented a proposed outcome of the group to the members of the Security Council at an informal meeting on 14–15 February 2001 with a view to securing support for a final report. The members decided to defer consideration of the report in the Security Council indefinitely.[10]

One reason for seeking changes was the growing view that the procedures for decision making in the UN Security Council could undermine the effectiveness of sanctions once they were agreed. The time taken for discussions of the scope, framing of language and translating the decisions into national laws could be used by a sanctions target to put in place defensive measures. This was particularly the case for financial sanctions, given the relative ease with which a sanctions target could 'hide' assets.[11]

The recommendations contained in the draft report are apparently divided into three sections: sanctions administration, sanctions design and sanctions implementation. While the contents of the sections on sanctions administration and implementation have been agreed, differences remain on two specific points contained in the section on sanctions design.

The section on administration includes measures to improve the effectiveness of the sanctions committees that are established by the Security Council each time a new set of sanctions is introduced. In particular, a system enhancing the role of the chairs of sanctions committees and for improved communication between the committees is recommended. In addition, an enhanced role for the UN Secretariat is proposed.

While making clear that implementation of sanctions is primarily a matter for states, the section on implementation includes measures for enhanced assessment, evaluation, monitoring and enforcement of sanctions. These recommendations would build on experience gained in recent investigations carried out under the auspices of the UN.

---

[9] This was the final objective of the Canadian Government, which took the initiative to establish the Working Group during its period as President of the Security Council. 'Sanctions have little, or only controversial, standing in international law. They fall into a grey zone between humanitarian law and the rules of warfare.' Remarks of Canadian representative at the 4128th meeting of the UN Security Council, reproduced in 'General issues related to sanctions', UN document S/PV.4128, 17 Apr. 2000.

[10] Report of the Secretary-General, Implementation of the provisions of the Charter of the United Nations related to assistance to third states affected by the application of sanctions, UN document A/56/303, 17 Aug. 2001.

[11] Albert Cluckers, Senior Manager of the Internal Audit Department, Bank Brussels Lambert, has noted that 'As the entering into force of sanctions is subject to a very slow procedure, the countries or the individuals to be sanctioned can easily set up evasive systems'. 'How can financial assets and financial transactions be controlled? Comments from a banker's point of view', Expert Seminar on Targeting UN Financial Sanctions organized by the Swiss Federal Office for Foreign Economic Affairs, Department of Economy, 17–18 Mar. 1998, p. 81, available at URL <http://www.smartsanctions.ch/interlaken1.htm>.

An International Commission of Inquiry (known as UNICOI) was established by UN Security Council Resolution 1013 of 7 September 1995 to conduct investigations 'relating to the sale or supply of arms and related matériel to former Rwandan government forces in the Great Lakes region'. Reports by the commission included information that pointed to violations of the UN arms embargo on Rwanda that was in effect between May 1994 and August 1995.[12] UNICOI was reconstituted in 1998 with a new mandate to collect information and investigate reports relating to these violations.[13]

Other panels of experts have investigated violations of sanctions imposed on UNITA, on rebel groups in Sierra Leone and on Liberia. The panels of experts reported on UNITA in March 2000, on Sierra Leone in December 2000 and on Liberia in October 2001.[14] In addition, a UN monitoring mechanism on Angola sanctions was established by UN Security Council Resolution 1295.[15] The reports published by the various panels and the monitoring mechanism are widely recognized as having had a significant effect on the actions of the Security Council and of states. The recommendations of the Chowdhury Report would have made UN monitoring a routine element of sanctions implementation.

Most of the recommendations on sanctions design were supported by all of the states represented in the working group. The recommendations were intended to minimize the risk that sanctions would have a serious negative impact on the humanitarian situation of people living in the target state and to focus pressure on decision makers by tailoring sanctions to the specific conditions of the target state.

It can be argued that sanctions would be most effective if applied to parties before a breach of the peace or an act of aggression has taken place. However, most members of the working group opposed the use of sanctions during the early phases of a crisis.[16] The shortage of relevant capacities both at the United Nations and in the regions where the targets of sanctions are located has been a barrier to effective enforcement. In these conditions the use of sanctions was

---

[12] The initial UNICOI reports can be found in UN documents S/1996/67, S/1996/195, S/1997/1010 and S/1998/63. The arms embargo was established in UN Security Council Resolution 918, 17 May 1994.

[13] UN Security Council Resolution 1161, 9 Apr. 1998. UNICOI published its findings in Interim report of the International Commission of Inquiry (Rwanda), UN document S/1998/777, 19 Aug. 1998.

[14] Report of the Panel of Experts on Violations of Security Council Sanctions against UNITA, UN document S/2000/203, 10 Mar. 2000 (the Fowler Report); Report of the Panel of Experts appointed pursuant to UN Security Council Resolution 1306 (2000), paragraph 19 in Relation to Sierra Leone, UN document S/2000/1195, 20 Dec. 2000; and Report of the Panel of Experts pursuant to Security Council Resolution 1343 (2001), paragraph 19, concerning Liberia, UN document S/2001/1015, 26 Oct. 2001.

[15] UN Security Council Resolution 1295, 18 Apr. 2000.

[16] 'Application of sanctions is justifiable only when all other peaceful means of dispute settlement or international peace keeping, including temporary measures provided for in Article 40 of the UN Charter, have failed and the UN Security Council determined a threat to peace, violation of peace or an act of aggression.' Zvedre, E. K., 'UN Security Council arms embargoes: implementation of the Security Council resolutions in Russia', Paper delivered at the Seminar on Arms Embargoes and Sanctions, organized jointly by the Ministry of Foreign Affairs, Hungary and the Department of Foreign Affairs and International Trade, Canada, Budapest, 26–27 Apr. 2001.

supported only in exceptional cases, at least until additional capacities were in place.[17]

Sanctions are therefore seen as a measure of last resort in two senses: they are to be applied when all possibilities for reconciliation through dialogue have been exhausted and only in cases where a reasonable probability of effective enforcement exists.

The draft report included a recommendation that, before sanctions are imposed, the UN Secretariat should prepare a report indicating the impact sanctions could be expected to have in the specific conditions of the target country, including the impact on states in the immediate vicinity. The establishment of such a general rule was considered by some to dilute the authority of the Security Council to take decisions.

The main disagreement apparently related to the conditions for suspending and lifting sanctions.[18] In particular, the proposal to establish a general rule that the resolution establishing sanctions should always have a 'sunset clause' by which the sanctions were time limited. At the end of the agreed period a new decision would be required to continue sanctions.

Adopting general rules on sanctions would take away the discretion of the Security Council to establish sanctions of indefinite duration. Lifting sanctions then requires agreement by all five permanent members of the Security Council. Moreover, each member of the Security Council determines when the conditions that will lead to the lifting of sanctions have been met according to its own interpretation of the relevant resolutions.[19]

The disagreement over the establishment of general rules reflected the influence of the ongoing discussion of sanctions imposed against Iraq during 2001. The proposed rules are intended to contribute to establishing a 'road map' that makes it more clear what a target state needs to do in order to have sanctions lifted. The United States in particular has been reluctant to accept general rules if they could increase the pressure to lift sanctions on Iraq in conditions where Iraq has not complied with all of the conditions established in existing UN Security Council resolutions.

The Security Council has adopted into its practice many of the recommendations of the UN working group, including the establishment of time-limited sanctions. Some of the recommendations of the draft report may be accepted as separate items, rather than as part of a general sanctions reform.[20] More-

---

[17] 'Cases in which the imposition of sanctions is feasible must be strictly interpreted.... We should avoid increasing the number of sanctions regimes. The United Nations already has the greatest difficulty in securing compliance with those currently in force.' Unofficial translation of remarks by the representative of France at the 4128th meeting of the UN Security Council, reproduced in General issues related to sanctions, UN document S/PV.4128, 17 Apr. 2000.

[18] Scott, D., 'Improving UN sanctions: Security Council debates time limits for future sanctions', Markland Group Report, reproduced in *Compliance Matters*, issue 10 (2001).

[19] The working group discussed the idea that the Security Council should agree in advance on guidelines for how the target of sanctions could be judged to have met the conditions contained in the sanctions resolution.

[20] In 2002 France and the United Kingdom were working to create a permanent sanctions monitoring unit within the United Nations. Proposed elements of an independent expert unit, British–French Non-paper circulated to UN Security Council sanctions experts on 25 Jan. 2002.

over, sanctions reform efforts will continue outside the framework of the UN. Between 1998 and 2001 the governments of Switzerland and Germany organized reform processes that led to the development of sanctions policy handbooks that have subsequently been used extensively by the Security Council when designing and drafting sanctions resolutions.[21] In October 2001 the Government of Sweden initiated a follow-on process to focus on the implementation and monitoring of targeted sanctions.[22]

III. The EU sanctions experience

The European Union has two objectives in using sanctions. First, the EU has acted to implement UN sanctions more effectively. Second, the EU has used sanctions as an instrument of its common foreign policy.

After 1970 the European Community discussed foreign policy as part of its European Political Cooperation (EPC) and sanctions issues reflected the need for collective implementation of UN measures against South Africa and Southern Rhodesia (now Zimbabwe). The common commercial policy of the European Community (which applied to trade in civilian goods and commodities) meant that economic sanctions required implementation by common institutions as well as by the member states.[23]

While international law—including decisions of the United Nations—provides legitimacy to EU sanctions, there is not a direct link to a decision by the UN under Article 41 or Chapter 7 of the UN Charter in every case.

In 1982 the European Community adopted sanctions against the Soviet Union in response to political developments in Poland and against Argentina following the invasion of the Falklands Islands.[24] Subsequently, the EU adopted sanctions outside the framework of UN decisions against Belarus, China, Indonesia, Kazakhstan, Libya, Myanmar (Burma) and Zimbabwe.[25]

In cases where the EU has used sanctions outside the framework of UN decisions it has usually been to promote human rights and democratization objectives in external relations. The link between sanctions and human rights has been made explicit in that sanctions are mentioned as one instrument with

---

[21] Swiss Federal Office for Foreign Economic Affairs, Department of the Economy, in cooperation with the United Nations Secretariat, *Second Interlaken Seminar on Targeting United Nations Financial Sanctions: Final Report*, 29–31 Mar. 1999, available at URL <http://www.smartsanctions.ch/int2_papers.htm>; and Brzoska, M. (ed.), *Design and Implementation of Arms Embargoes and Travel and Aviation Related Sanctions: Results of the 'Bonn–Berlin Process'*, Bonn International Center for Conversion in cooperation with the Auswärtiges Amt (German Foreign Ministry) and the United Nations Secretariat, 2001. It is available at URL <http://www.bicc.de/general/events/unsanc/2000/booklet.html>.

[22] Statement of Mr Hans Dahlgren, Sweden, to the UN Security Council, 22 Oct. 2001, reproduced in UN document S/PV.4394, 22 Oct. 2001.

[23] Bohr, S., 'Sanctions by the United Nations Security Council and the European Community', *European Journal of International Law*, vol. 4, no. 2 (1993).

[24] Kuyper, P. J., 'Community sanctions against Argentina: lawfulness under Community and international law', in eds D. O'Keefe and H. G. Schermers, *Essays in European Law and Integration* (Kluwer Law and Taxation Publishers: Boston, Mass., 1982).

[25] The case of Zimbabwe is discussed in chapter 2 in this volume.

which the Charter of Fundamental Rights of the European Union (proclaimed at the Nice summit meeting in December 2000) will be implemented.[26]

The Commission of the European Communities (hereafter referred to as the Commission), which is fully associated with the CFSP, has recommended that the Council of the European Union (hereafter referred to as the Council) debate ways of enabling the EU to devise and implement 'preventive sanctions'. The Commission has recommended establishing a system for monitoring potential conflict areas with a view to identifying parties liable to start a conflict and analysing their existing or potential power base. This information and analysis would allow the Council to introduce sanctions at a time when they could be expected to have the greatest effect.[27]

The greater use and expanded scope of sanctions in the 1990s took place against the background of continuous institutional change. Changes included the free movement of goods and services within the EC required by the 1987 Single European Act and the creation of the European Union—with a Common Foreign and Security Policy and closer cooperation in the area of justice and home affairs.

The legal basis for EU sanctions depends on the particular measure adopted. In each case the Council, using powers conferred in the 1992 Treaty on European Union (Maastricht Treaty), unanimously adopts a common position or a joint action identifying the objectives of measures to be undertaken.[28] From this point there is divergence in the legal form.

A two-stage procedure was established for economic sanctions. The 1957 Treaty Establishing the European Community (Treaty of Rome) provides the authority for implementing economic and financial sanctions through common institutions. Article 60 contains measures related to the movement of capital and payments while Article 301 provides the legal basis for trade sanctions. On this basis the Commission prepares a regulation containing specific measures that give effect to the political decision. The Council adopts this regulation through a qualified majority vote. The regulation, which can be modified only through a unanimous decision of the Council, becomes Community law, binding throughout the EC.

The use of Community law in the form of regulations whose implementation is monitored by the Commission is intended to ensure uniform application of sanctions measures. However, the use of arms embargoes by the EU requires a different legal basis because arms and military goods remain outside the scope

---

[26] The European Union's role in promoting human rights and democratisation in third countries, Communication from the Commission to the Council and the European Parliament, Document COM(2001) 252 final, Brussels, 8 May 2001, p. 6. The charter is part of the Treaty of Nice amending the Treaty on European Union, the treaties establishing the European Communities and certain related acts. *Official Journal of the European Communities*, C 80/1, 10 Mar. 2001, URL <http://europa.eu.int/eur-lex/en/treaties/dat/nice_treaty_en.pdf>.

[27] Commission of the European Communities, Communication from the Commission on conflict prevention, COM(2001)211 final, 11 Apr. 2001, p. 24.

[28] The Council of the European Union is composed of 1 representative at ministerial level from each member state, who is empowered to commit his or her government. Council members are politically accountable to their national parliaments.

of the common commercial policy.²⁹ There has been a need to reduce the risk that uneven implementation of agreed measures will diminish the effectiveness of EU arms embargoes and perhaps undermine the trust between member states.

The member states have sought greater uniformity through a dialogue that has led to political agreement on how arms embargoes should be applied. When an arms embargo is applied to a particular country, the states decide at the same time whether it should be interpreted as a 'full scope' or less than full scope embargo. If the embargo is to be full scope, then it is defined as being on 'arms, munitions and military equipment'. In that case, it will apply to all the goods on a common embargo list. If an embargo is less than full scope, it will be defined as 'an embargo on arms and munitions' and the member states then specify in the common list the categories that it will cover.³⁰

In addition, the EU has a different legal basis for travel and diplomatic sanctions since these also rest on measures that are still within the competence of member states rather than Community institutions. Travel sanctions have included bans on entry visas for specified individuals (usually senior political and military officials) and the suspension of high-level visits by officials. Diplomatic sanctions have included the expulsion of diplomatic and military personnel attached to the diplomatic representations in member states and, conversely, the withdrawal of personnel attached to diplomatic representations of member states in the target country.

Sanctions as an instrument of EU policy: the targeted sanctions against Yugoslavia after 1998

Between 1992 and 1996 the member states of the European Community implemented restrictions on trade and financial relations with the Federal Republic of Yugoslavia as well those parts of Croatia and Bosnia and Herzegovina under the control of Serb forces.³¹ These decisions were taken to implement decisions by the UN Security Council. These comprehensive sanctions were lifted in October 1996 in the context of the political settlement of the conflicts in the former Yugoslavia reached in November 1995.³²

---

²⁹ Under Article 296 of the Treaty of Rome, member states may take such measures as they consider necessary connected with the production of or trade in arms, munitions and war *matériel*.

³⁰ An unofficial copy of the European Union common embargo list is available at URL <http://projects.sipri.se/expcon/euframe/eu_commonlist.htm>.

³¹ The measures were first introduced by Council Regulations (EEC) no. 1432/92 and (EEC) no. 2656/92. More comprehensive sanctions were imposed by Council Regulation (EEC) no. 990/93 of 26 April 1993 concerning trade between the European Economic Community and the Federal Republic of Yugoslavia (Serbia and Montenegro), *Official Journal of the European Communities*, L 102, 28 Apr. 1993, pp. 14–16.

³² Council Regulation (EC) no. 2382/96 of 9 December 1996 repealing Regulations (EEC) 990/93 and (EC) no. 2471/94 and concerning the termination of restrictions on economic and financial relations with the Federal Republic of Yugoslavia (Serbia and Montenegro), the United Nations Protected Areas in the Republic of Croatia and those areas of the Republic of Bosnia and Herzegovina under the control of Bosnian Serb forces, *Official Journal of the European Communities*, L 328, 18 Dec. 1996, pp. 1–2.

When an armed conflict in the Kosovo province of the republic of Serbia of the FRY escalated in early 1998, and in particular after employment of the military against the Kosovar Albanian community, the European Union imposed restrictive measures as part of its response. However, in contrast to the comprehensive sanctions employed after 1992, after 1998 the EU applied what has been called 'a sophisticated mix of smart sanctions and incentives, regularly adjusted and fine-tuned to changing circumstances'.[33]

The EU established a moratorium on government-financed export credit support for trade and investment in the FRY in April 1998.[34] The EU had established an arms embargo against the FRY in late 1996. This embargo applied to the direct supply of arms. On 19 March 1998 the EU amended its embargo to include the provision of training as well as equipment that might be used for internal repression or for terrorism.[35] The EU decided not to issue visas to any senior representatives of the FRY and the government of the republic of Serbia. The scope of this measure was extended for the first time in December 1998 and was subsequently revised regularly following reviews of the list of individuals subject to the ban.[36] For example, in May 1999 the EU extended the visa ban to include President Slobodan Milosevic, his family, all ministers and senior officials of the governments of both the FRY and Serbia, and immediate associates of Milosevic.[37]

In May 1998 the EU introduced an asset freeze on funds held abroad by the FRY and Serbian governments as well as the assets of individuals associated with Milosevic and companies controlled by or acting on behalf of the Government of the FRY.[38] The financial sanctions were extended in May 1999 to cover additional individuals and to prohibit the provision of export finan-

---

[33] Cortright, D. and Lopez, G., 'Introduction', to Cortright, D. and Lopez, G. A., *Assessing Smart Sanctions: Lessons from the 1990s* (Rowman and Littlefield: Lanham, Md., 2002). The EU sanctions are discussed in detail in de Vries, A., 'European Union sanctions against the Federal Republic of Yugoslavia from 1998 to 2000: a special exercise in targeting' in the same volume.

[34] Council Regulation (EC) 926/98 of 27 April 1998 concerning the reduction of certain economic relations with the Federal Republic of Yugoslavia, *Official Journal of the European Communities*, L 130, 1 May 1998, pp. 1–4.

[35] Common Position of 19 March 1998 defined by the Council on the basis of Article J.2 of the Treaty on European Union on restrictive measures against the Federal Republic of Yugoslavia, *Official Journal of the European Communities*, L 95, 27 Mar. 1998 pp. 1–3.

[36] Council Common Position of 19 March 1998 (note 35); and Council Common Position of 14 December 1998 defined by the Council on the basis of Article J.2 of the Treaty on European Union on restrictive measures against the Federal Republic of Yugoslavia (98/725/CFSP), *Official Journal of the European Communities*, L 345, 19 Dec. 1998, pp. 1–2.

[37] Council Common Position of 10 May 1999 adopted by the Council on the basis of Article 15 of the Treaty on European Union concerning additional restrictive measures against the Federal Republic of Yugoslavia (1999/318/CFSP), *Official Journal of the European Communities*, L 123, 13 May 1999, pp. 1–2; and Council Decision of 10 May 1999 implementing Common Position 1999/318/CFSP concerning additional restrictive measures against the Federal Republic of Yugoslavia (1999/319/CFSP), *Official Journal of the European Communities*, L 123, 13 May 1999, pp. 3–11.

[38] Council Common Position of 7 May 1998 defined by the Council on the basis of Article J.2 of the Treaty on European Union concerning the freezing of funds held abroad by the Federal Republic of Yugoslavia (FRY) and Serbian Governments (98/326/CFSP), *Official Journal of the European Communities*, L 143, 14 May 1998, pp. 1–2.

cing by the private sector for transactions with the governments of the FRY and Serbia.[39]

In April 1999 the EU banned the sale and supply of petroleum and petroleum products to the FRY.[40] In September 1999 the ban was first modified to permit sales and supplies to Kosovo and Montenegro (but not Serbia) and subsequently to permit sales to Serb municipalities that supported the democratic transition in the FRY (the so-called 'energy for democracy' programme).[41]

During 2000 the sanctions measures were modified in the light of the changing conditions in the FRY, and in particular following the fall of the Milosevic Government and the election of President Vojislav Kostunica in September 2000. While the arms embargo and the restrictive measures applied to President Milosevic and his immediate associates were maintained, other sanctions were progressively lifted in 2000 and 2001. In October 2001 the EU lifted the arms embargo on the FRY.[42] In November 2001 the last of the sanctions on the FRY were lifted.[43]

The policy of targeted sanctions was combined with a series of inducements designed to encourage democracy and respect for human rights in the FRY. Moreover, they were implemented alongside the development of a regional initiative of the EU—the 1999 Stability Pact for South Eastern Europe—in which more than 40 countries undertook to strengthen the efforts of countries in South-Eastern Europe to foster peace, democracy, respect for human rights and economic prosperity.[44] The EU targeted sanctions were supported by Stability Pact partners, which include the United States and states bordering the FRY. Participation in the Stability Pact, which also promised to be a preparatory phase prior to integration of states into West European institutions, was not possible for the FRY until it undertook a democratic transition. Cortright and Lopez have noted that 'multilateral sanctions and incentives played into the hands of the democratic opposition while isolating the regime, they were a successful example of the application of smart sanctions'.[45]

---

[39] Council Common Position of 10 May 1999 (note 37), pp. 1–2; and Council Decision of 10 May 1999 implementing Common Position 1999/318/CFSP (note 37), pp. 3–11.

[40] Council Common Position of 23 April 1999 defined by the Council on the basis of Article J.2 of the Treaty on European Union concerning a ban on the supply of petroleum and petroleum products to the Federal Republic of Yugoslavia (FRY) (1999/273/CFSP), *Official Journal of the European Communities*, L 108, 27 Apr. 1999, pp. 1–2.

[41] Council Common Position of 3 September 1999 amending Common Position of 23 April 1999 defined by the Council on the basis of Article J.2 of the Treaty on European Union concerning a ban on the supply of petroleum and petroleum products to the Federal Republic of Yugoslavia (FRY) (1999/604/CFSP), *Official Journal of the European Communities*, L 236, 7 Sep. 1999, p. 1.

[42] Council Common Position of 8 October 2001 amending Common Position 96/184/CFSP concerning arms exports to the former Yugoslavia and Common Position 98/240/CFSP on restrictive measures against the Federal Republic of Yugoslavia (2001/719/CFSP), *Official Journal of the European Communities*, L 268, 10 Oct. 2001, p. 49.

[43] Council Regulation (EC) no. 2156/2001 of 5 November 2001 repealing Regulation (EC) no. 926/98 concerning the reduction of certain economic relations with the Federal Republic of Yugoslavia, *Official Journal of the European Communities*, L 289, 6 Nov. 2001, p. 5.

[44] The Stability Pact on South Eastern Europe is reproduced in *SIPRI Yearbook 2000: Armaments, Disarmament and International Security* (Oxford University Press: Oxford, 2000), pp. 214–20, and is available at URL <http://www.stabilitypact.org/official%20Texts/PACT.HTM>. A brief summary of the pact and the partners are listed in the glossary in this volume.

[45] Cortright and Lopez (note 33).

IV. Sanctions against Iraq

The sanctions imposed on Iraq by the United Nations were path-breaking in their scope and in the central role played by the UN in their implementation.[46] The immediate response by the UN included full trade and financial sanctions against Iraq (established in UN Security Council Resolution 661), which have been likened in their impact to a full blockade.[47]

The sanctions consist of three elements: an embargo on arms sales to Iraq, an embargo on oil purchases from Iraq and a wider embargo on economic contacts with Iraq (including trade, financial contacts of various kinds and certain kinds of travel). With the exception of the arms embargo (to which no exceptions are permitted) none of the other sanctions is implemented in a rigid and absolute manner.

The Security Council established the 'oil-for-food' programme to permit Iraq to import products for humanitarian purposes. Initially including food and health-care products, after 1995 UN Security Council Resolution 986 authorized further exemptions to provide relief for the Iraqi population. The sanctions regime has been modified to permit the reconstruction of infrastructure for humanitarian reasons—notably, the housing stock and the water, sanitation and electrical power systems. In order to finance these imports UN member states were permitted to buy a certain amount of oil from Iraq and deposit the payment for this oil into an account managed by the United Nations. This account is used to pay for Iraq's imports. In December 1999 UN Security Council Resolution 1284 lifted the ceiling on the value of oil that Iraq may sell and by the start of 2001 Iraq had accumulated $5 billion in this UN-administered account. The increased oil price and increases in Iraqi oil production meant that Iraq was projected to earn over $16 billion in 2001, of which 70 per cent would be available for purchases of humanitarian supplies.[48] As of August 2001 about $3.5 billion remain unspent in UN oil-for-food accounts while Iraq had not implemented about $1 billion of the contracts approved by the Sanctions Committee.[49]

In spite of this inbuilt flexibility, in 2001 the continued erosion of the sanctions regime against Iraq was said to be occurring in different ways. First, while civilian air travel to Iraq has never been prohibited, some air traffic to Baghdad in this period did not conform to the UN notification procedures established to reduce the risk that air traffic could undermine the sanctions. Second, Iraq was increasingly able to import non-military items outside the framework of the oil-for-food programme—that is, imports were financed by oil revenues paid directly to Iraq by the buyers rather than using money man-

---

[46] Urquhart, B., 'The role of the United Nations in the Iraq–Kuwait conflict in 1990', *SIPRI Yearbook 1991: World Armaments and Disarmament* (Oxford University Press: Oxford, 1991), pp. 617–38.

[47] Burri, J., 'Introductory paper', Expert Seminar on Targeting United Nations Sanctions, Interlaken, 17–19 Mar. 1998; and UN Security Council Resolution 661, 6 Aug. 1990.

[48] Hain, P., Minister of State, British Foreign and Commonwealth Office, 'Britain and the Gulf 2000', Speech at the Royal Institute of International Affairs, London, 7 Nov. 2000. It is available at URL <http://www.fco.gov.uk>.

[49] British Foreign and Commonwealth Office, 'Fact sheet: UN controls on Iraq', available at URL <http://www.fco.gov.uk>.

aged by the UN. Third, trade delegations visiting Baghdad were alleged to be discussing military–technical cooperation with Iraq in anticipation of the arms embargo that is part of the sanctions regime being lifted.

The approach to the sanctions against Iraq was new in part because of the way in which they were linked with a previously untried approach to arms control. UN Security Council Resolution 687 stated that sanctions would remain in place until Iraq demonstrated in a fully transparent way that its nuclear, biological and chemical (NBC) weapon programmes along with programmes to acquire long-range missile delivery systems had been dismantled permanently. The UN Special Commission on Iraq (UNSCOM) was created with the purpose of helping to verify Iraqi compliance with this obligation.[50]

After UNSCOM withdrew its personnel from Iraq in December 1998 the Iraqi authorities linked the return of a UN presence to monitor disarmament with the lifting of sanctions. Reconciling the need for the Security Council to meet its responsibilities as contained in the various resolutions passed after 1990 and the practical obstacle created by the non-cooperation of Iraq dominated the discussion of how to reform the UN approach to Iraq in 1999. In December 1999 UN Security Council Resolution 1284 disbanded UNSCOM and replaced it with a new body, the UN Monitoring, Verification and Inspection Commission (UNMOVIC), in the hope that this new arrangement would restore the minimal cooperation from Iraq required for the UN to operate inside Iraq.[51]

The debate in the Security Council in 2001

In the first half of 2001 issues that were not resolved during the debate on Resolution 1284 were raised again. More than 12 months after the creation of UNMOVIC there was no sign that Iraq had any intention of accepting inspections by the United Nations to verify its compliance with the terms of Resolution 687.[52] Therefore, it was still necessary to consider the relationship between the sanctions and the implementation of arms control in Iraq. In addition, the issue of the humanitarian impact of sanctions was unresolved. Some members of the Security Council (prompted in part by a study conducted by Norway, then chair of the Iraq Sanctions Committee) believed more firmly that sanctions were not having their intended effect and should be adjusted.[53]

---

[50] Ekéus, R., 'The United Nations Special Commission on Iraq: activities in 1992', *SIPRI Yearbook 1993: World Armaments and Disarmament* (Oxford University Press: Oxford, 1993), pp. 691–704.

[51] This phase of UN involvement in Iraq is described in Wahlberg, M., Leitenberg, M. and Zanders, J. P., 'The future of chemical and biological disarmament in Iraq: from UNSCOM to UNMOVIC', *SIPRI Yearbook 2000* (note 44), pp. 560–75.

[52] For an overview of the problems confronting UNMOVIC see 'Preventing the further proliferation of weapons of mass destruction: the importance of on-site inspection in Iraq', Lecture by Dr Hans Blix, Executive Chairman of UNMOVIC, at the 3rd training course of UNMOVIC, Vienna, 19 Feb. 2001, URL <http://www.un.org/Depts/unmovic/ExecChair/BlixVienna.htm>.

[53] For a summary of the findings of the Norwegian study see the speech by Norwegian Foreign Minister Thorbjørn Jagland, 'Norway and the UN Security Council: our experience so far', Oslo, 16 May 2001, URL <http://www.norway.org.uk/cgi-bin/wbch3.exe?d=4882&p=1790>.

The United States has always claimed that exemptions made for humanitarian reasons were abused by the Iraqi Government, which diverted materials intended for rebuilding infrastructure for its own use. In 2001 these allegations continued. It was alleged that Iraq had used telecommunications equipment imported from China to improve the command and control system for its air defence network. The USA raised this issue bilaterally with the Chinese Government and, according to US Secretary of State Colin Powell, 'China has now said that they have told the companies that were in the area doing fiber optics work to cease and desist. We are still examining whether or not it was a specific violation of the sanctions policy, and if it was, we will call that to the attention of the sanctions committee so that they can take appropriate action with respect to China'.[54]

In the Security Council meeting of 26 June 2001 draft resolutions that would modify the sanctions regime against Iraq were discussed extensively. In the discussion the Permanent Representative of the United Kingdom stated that there was a need for the UN to reconcile two principles: (*a*) to ensure that Iraq did not have and could not acquire weapons that would allow it to pose a threat to its region; and (*b*) 'to alleviate the suffering of the Iraqi people and take whatever steps we can from the outside to ensure that their needs are met. We agree to this extent with the Russian Federation that the status quo is not acceptable'. The second principle was said to be 'as important and more immediate than the first'.[55]

This Security Council meeting took place in the context of UN Security Council Resolution 1352 of 1 June 2001, in which the Security Council expressed its intention 'to consider new arrangements for the sale or supply of commodities and products to Iraq and for the facilitation of civilian trade and economic cooperation with Iraq in civilian sectors'.[56] After considering the modifications to the sanctions regime put forward, the Security Council adopted Resolution 1382 on 29 November 2001.[57] The resolution included two technical annexes that were adopted for implementation of the sanctions against Iraq. These annexes made important changes in the procedures for sanctions implementation.

In the United States the incoming administration of President George W. Bush initiated a review of the US policy on Iraq, with a particular focus on UN sanctions, as one of its first foreign policy priorities. The review reflected the understanding in the US administration of the need to restore international cooperation (in particular among the members of the Security Council) without undermining the arms control objectives of the sanctions. Secretary of State Powell underlined that the main purpose of the policy review was 'to

---

[54] Secretary of State Colin Powell, *Overview of Foreign Policy Issues and Budget*, S. HRG. 107–41, Hearing before the Committee on Foreign Relations, US Senate (US Government Printing Office: Washington, DC, 8 Mar. 2001), p. 27.

[55] Statement by the Permanent Representative of the UK to the UN Sir Jeremy Greenstock to the UN Security Council, 26 June 2001, URL <http://www.ukun.org/xq/asp/SarticleType.17/Article_ID.284/qx/articles_show.htm>.

[56] UN Security Council Resolution 1352, 1 June 2001.

[57] UN Security Council Resolution 1382 is available at URL <http://www.un.org/Docs/scres/2001/>.

rescue the sanctions policy' by bringing the coalition behind the full implementation of UN resolutions back together.[58]

Describing the situation facing the Bush Administration, former US Assistant Secretary of State for Nonproliferation Robert Einhorn noted:

Iraq won the propaganda battle and the result was that there was, in January 2001, widespread support internationally for getting rid of the UN sanctions regime. And it wasn't just Russian, French or Arab governments. We are talking about Western European governments, as well. We're even talking about populations in the West and in the United States, as non-governmental groups were calling for removal of sanctions. Members of the U.S. Congress were writing to the president to ask for relief for the Iraqi people.[59]

The changes proposed in the Security Council in 2001 reinforced the view (held in particular by certain US-based analysts) that in practice the Bush Administration was not reversing the process by which the standard of compliance required from Iraq in exchange for modification of the sanctions had been progressively lowered after 1991.[60]

Revising the sanctions regime

The sanctions against Iraq are administered by a UN Sanctions Committee (established under UN Security Council Resolution 661) that consists of representatives of all 15 members currently sitting on the Security Council. The 5 permanent members of the Security Council are also always represented on the Sanctions Committee while 10 members are replaced in line with changes in Security Council membership.

UN Security Council Resolution 1051 of 27 March 1996 established an export/import monitoring mechanism to evaluate trade with Iraq to ensure that it is consistent with the purposes of the sanctions. Under the mechanism states were required to transmit data from potential exporters on the intended sale or supply from their territories of any items or technologies to a joint unit constituted by UNSCOM (a task later taken over by UNMOVIC) and the International Atomic Energy Agency (IAEA). The items for which notification is required were identified in a technical annexe to the resolution (referred to as the '1051 list').

The items on the 1051 list are not military items. As noted above, Iraq is subject to a complete arms embargo; military items should not be traded to it. Moreover, UN Security Council Resolution 687 not only confirms the general arms embargo but also establishes an embargo on the supply to Iraq of items

---

[58] The outcome of the review is described by Powell in *Overview of Foreign Policy Issues and Budget* (note 54), pp. 5–7.

[59] Einhorn, R., 'The emerging Bush Administration approach to addressing Iraq's WMD and missiles programs', Keynote Address to the conference Understanding the Lessons of Nuclear Inspections and Monitoring in Iraq: A Ten-Year Review, Institute for Science and International Security, Washington, DC, 14–15 June 2001.

[60] E.g., Perlez, J., 'Capitol hawks seek tougher line on Iraq', *New York Times*, 7 Mar. 2001; and Crossette, B., 'UN sanctions didn't stop Iraq from buying weapons', *New York Times*, 18 June 2001.

and technology used in arms production of all kinds or for the utilization or stockpiling of NBC weapons as well as long-range missile delivery systems for such weapons—so-called 'dual-use items'. The 1051 list therefore contains civilian items that, in theory, could be applied for end-uses that are not compatible with the sanctions regime. This list is extensive.

The export/import monitoring mechanism was not established as a licensing regime since neither the joint unit nor the Sanctions Committee denies particular transactions. The joint unit assessed the technical aspects of a notified transaction to ensure that it was 'sanctions compliant'. If the notified trade activity was considered to be contradictory to the sanctions regime, the information and a technical assessment would be passed to the Sanctions Committee (whose members are the same as those of the Security Council) for a decision on how to proceed. The Sanctions Committee would inform national authorities of the state concerned and expect them to take the necessary measures to prevent the transaction from taking place.

Transactions with Iraq are not denied by the UN sanctions implementation system but they are in effect placed on hold while the contract is evaluated. As of August 2001 approximately $3.3 billion worth of contracts were on hold.[61] Most of these decisions were taken at the request of the United States and, to a lesser extent, the United Kingdom.

There are four sets of reasons why contracts are placed on hold: (*a*) because the contracts (which may include thousands of items) are too complicated to evaluate quickly; (*b*) because of a lack of information on which to make an assessment of the end-uses of the goods; (*c*) because the contract contains items controlled under Resolution 1051 that will only be approved on the condition that the equipment is monitored in Iraq by the UN Office of the Iraq Programme (OIP); and (*d*) because contracts may include items that are sufficiently sensitive that the contract will only be approved if these items are removed.

As noted above, the regime of inspections was intended to raise the level of confidence that trade with Iraq would not contribute to illegal programmes. With the cooperation of Iraq in allowing inspections, the UN could rapidly assess the end-user and end-use in a given case and reduce the number of contract holds. However, cooperation from Iraq has not been forthcoming.

As a result of the history of Iraqi procurement of NBC weapons prior to 1990 along with the development of long-range missile delivery systems and subsequent efforts to mislead UN inspection teams, there is no trust in the good faith of the Iraqi authorities.

The system created under UN Security Council Resolution 1382

Under the conditions noted above, in particular in the absence of inspections, it was difficult for the joint unit to be certain that a particular item would not

---

[61] Report of the Security Council Committee established by Resolution 661 (1990) concerning the situation between Iraq and Kuwait on the implementation of the arrangements in paragraphs 1, 2, 6, 8, 9 and 10 of Resolution 986 (1995), UN document S/2001/842, 5 Sep. 2001.

be diverted to an unauthorized end-user or for an illegal end-use. There has been a tendency to request additional information to satisfy any residual doubts leading to delays in considering particular transactions—the delays became indefinite when the information requested was not forthcoming or was insufficient. The system established under UN Security Council Resolution 1382 is intended to create procedures that minimize delays in trade with civilian items to Iraq.

Resolution 1382 includes two annexes. Annex 1 is a Goods Review List that contains three schedules of items.[62] The first schedule is the 1051 list in its entirety. The second schedule is the list of high-technology, dual-use items contained in an annexe to UN Security Council document S/2001/1120. The third is a list of individual items not contained in other lists but with potential military applications. Annex 2 contains procedures for the application of the Goods Review List.

Under the system, applications for each export of commodities and products should in future be submitted to the OIP by the responsible authorities of the exporting states. The application should include the contract, the technical specifications of items to be transferred and information about the end-user in Iraq. Once received, the information is evaluated and treated in one of three ways.

If the OIP experts conclude that the item is subject to the embargo established by Resolution 687 it is returned to the authorities who submitted it, who will be expected to apply a strong presumption of denial in considering the transaction.

If the OIP experts conclude that the application contains any item on the Goods Review List approved by Resolution 1382, this information is forwarded to the Iraq Sanctions Committee along with an assessment of the humanitarian, economic and security implications of approval or denial of the export.

If the OIP experts conclude that the application does not contain any item either subject to the Resolution 687 embargo or on the Goods Review List then both the exporting state and the Government of Iraq are informed that there is no obstacle to completing the transaction. Once verification is received that these items have arrived as contracted in Iraq, payment to the exporter is authorized from UN-managed funds.

The procedures are time constrained. Once a technically complete submission has been received it must be processed within four working days by the OIP. The procedures also include a right of appeal on technical grounds against decisions by the OIP in cases where an export application is referred to the Iraq Sanctions Committee. The responsible authorities of the exporting state may contest a decision by OIP experts that an item is contained on the Goods Review List.

After Resolution 1382 was passed, British Foreign Secretary Jack Straw stated that 'Iraq holds the key to its reintegration into the international com-

[62] UN Security Council Resolution 1382 (note 57).

munity—compliance with UN Security Council Resolution 1284. There must be independently verified compliance with the international community's insistence that Iraq give up its Weapons of Mass Destruction'. However, Straw also pointed out that 'the UN decision will soon mean no sanctions on ordinary imports into Iraq, only controls on military and weapons-related goods. Iraq will be free to meet all its civilian needs'.[63]

While calling for end-user delivery verification prior to payment, the procedures do not contain provisions for post-shipment inspection of items on the Goods Review List. To this extent the link between the sanctions and the system of in-country controls envisaged for UNMOVIC has been broken. The implementation of the new system will therefore require a detailed evaluation of requests to export prior to the shipment of goods. Resolution 1382 stresses the strong self-interest for exporters to submit 'technically complete applications' if the system is going to expedite civilian trade.

The changes do not include a timetable for lifting sanctions on Iraq—which will occur after a long period under any conditions—but they do stress the link between suspending most sanctions and the implementation of existing UN resolutions.

V. Sanctions against Afghanistan

The UN has been active in seeking a resolution to the conflicts that have taken place on the territory of Afghanistan for more than 20 years, including maintaining a Special Mission in that country. In October 1996 UN Security Council Resolution 1076 called on states to refrain from military engagement in Afghanistan, including the supply of personnel, arms or ammunition.[64] This led several states to establish national arms embargoes against Afghanistan or to introduce changes to their export licensing procedures.

In 1996 UN reports made reference to growing evidence that Afghanistan was a location where terrorist groups actively sought refuge. However, resolutions and statements did not identify any particular party to the conflict as blameworthy or subject any particular party to sanctions.

By 1998 the Security Council, on the basis of information provided in reports by the Secretary-General, had identified the Taliban as the party primarily responsible for the escalation of hostilities. In August 1998 UN Security Council Resolution 1193 expressed concern that escalation was 'due to the Taliban forces offensive in the northern parts of the country'.[65] This resolution,

---

[63] Iraq: Statement by the Foreign Secretary, 30 Nov. 2001, reproduced at URL <http://www.fco.gov.uk>.

[64] UN Security Council Resolution 1076, 22 Oct. 1996, paras 3 and 4. The trigger for this action was the growing evidence that parties to the conflict were not respecting either the status of the UN or the safety of UN personnel in the country: in particular, the forced entry to UN premises by Afghan irregular soldiers in Sep. 1996 to capture former Afghan President Najibullah Khan, who was subsequently executed.

[65] This resolution was also triggered by a particular event—the forced entry by the Taliban into the Iranian Consulate General in Mazar-e-Sharif and the kidnapping of personnel, 9 of whom were subsequently murdered.

adopted shortly after the bombing of the US embassies in Kenya and Tanzania, also referred to the presence of terrorist groups in Afghanistan. However, the resolution did not make any direct link between the terrorists and these particular bombings.

In December 1998 the Security Council pointed to the 'failure of the leadership of the Taliban . . . to comply with the demands made in its previous resolutions' and expressed its readiness to consider mandatory sanctions.[66]

UN sanctions against Afghanistan

The United Nations approved sanctions related to Afghanistan prior to the attacks on the United States on 11 September 2001.

The Security Council introduced mandatory travel and financial sanctions on the Taliban for the first time in UN Security Council Resolution 1267, adopted on 15 October 1999.[67] Resolution 1267 required states to deny permission for aircraft owned, leased or operated by the Taliban to land or take off in their territory and to freeze funds and other financial resources owned or controlled by the Taliban.[68] Resolution 1267 made much more explicit links between the Taliban and specific terrorists and acts of terrorism. It made particular reference to the attacks on US embassies in Africa and named Usama bin Laden and 'others associated with him' and included a demand that the Taliban hand over bin Laden to the appropriate authorities in the United States, where he was indicted for conspiring to kill US nationals.

In December 2000 the Security Council strengthened the sanctions against the Taliban.[69] UN Security Council Resolution 1333 introduced a mandatory arms embargo that prohibited not only arms transfers but military assistance of any kind to the Taliban. In addition, states were required to close all Taliban offices on their territories and to reduce their diplomatic presence in Afghanistan. Travel sanctions were extended to include a ban on overflight by aircraft owned or controlled by the Taliban as well as the closure of all offices of Ariana Afghana Airlines. Non-military trade sanctions were also introduced for the first time with a ban on exports of the chemical acetic anhydride (used in the production of narcotics) to any part of Afghanistan under Taliban control. Financial sanctions were extended to include the freezing of funds owned or controlled by Usama bin Laden and the al-Qaeda organization (the first time al-Qaeda was named in a resolution).

Resolution 1333 introduced a set of sanctions that were extremely complex from the point of view of implementation. First, the sanctions applied only to a part of Afghanistan, not the full territory. Second, sanctions applied only to some entities and individuals in Afghanistan, not all of them. Effective implementation required a detailed knowledge of activities and actors inside the country—information that neither the UN nor many member states possess.

[66] UN Security Council Resolution 1214, 8 Dec. 1998, para. 15.
[67] UN Security Council Resolution 1267, 15 Oct. 1999.
[68] UN Security Council Resolution 1267 (note 67), para. 4.
[69] UN Security Council Resolution 1333, 19 Dec. 2000.

This sanctions design complicated implementation by national authorities. The customs authority plays a key role in sanctions enforcement in most states. This design is difficult to translate into border control procedures used at the point of exit from a customs area. Effective implementation required a system for licensing trade and economic contacts with Afghanistan based on detailed end-user and end-use information.

Resolution 1333 in effect created a requirement for the Sanctions Committee to establish a comprehensive information system for monitoring sanctions-related developments in Afghanistan. This system would be required both to meet the needs of the Security Council and to assist many states with implementation. The basis for such an information system was authorized in July 2001 in UN Security Council Resolution 1363 in the form of a monitoring mechanism.[70] The mechanism would include an Office for Sanctions Monitoring and Coordination: Afghanistan, in New York, and cooperation with the so-called 'Six-plus-Two' states (China, Iran, Pakistan, Tajikistan, Turkmenistan and Uzbekistan, plus Russia and the USA) to establish sanctions support teams on the borders of Afghanistan. Support teams (one located in each of the six countries neighbouring Afghanistan) would be tasked with investigating allegations of sanctions busting.[71]

The Six-plus-Two states are an informal group formed in 1997 to discuss the conflict in Afghanistan under the leadership of the UN Secretary-General. The group did not play a central role in elaborating practical measures because it included states that supported different Afghan factions and have different interests in Afghanistan. The fact that this group was prepared to cooperate in the implementation of sanctions is one indication of the growing seriousness with which neighbouring states, as well as Russia and the USA, were addressing the risks of terrorism prior to the attacks on the United States.

The immediate tasks for the Office for Sanctions Monitoring and Coordination: Afghanistan included the development of detailed lists of actors (individuals and entities) to which sanctions applied. By December 2001 this list (which is continuously updated) included over 200 individuals and approximately 75 entities.[72]

European Union sanctions against Afghanistan

In October 1996 the UN called on states to stop providing military assistance to the parties to the conflict in Afghanistan. The European Union implemented

---

[70] UN Security Council Resolution 1363, 30 July 2001.

[71] Report of the Committee of Experts on Afghanistan, Statement in the Security Council of Ambassador Alfonso Valdivieso, Permanent Representative of Colombia, Chairman of the Sanctions Committee on Afghanistan, 5 June 2001, available at URL <http://www.un.int/colombia/english/consejo_seguridad/staafghanistanchairmanJune%205.htm>.

[72] The consolidated list is published in UN document S/7222, 26 Nov. 2001, and a first addendum was published as UN document S/7252 on 26 Dec. 2001. The mandate of the sanctions monitoring group was extended by 12 months by UN Security Council Resolution 1390, 28 Jan. 2002.

this request through a Common Position requiring all member states to impose an embargo on arms supplies to Afghanistan.[73]

In November 1999 the EU extended restrictive measures in the form of travel and financial sanctions that were applied specifically to the Taliban.[74]

Although the legal form that established sanctions—a Council Common Position—was the same, the modifications to the sanctions in 1999 made their implementation much more complicated. First, the new measures could not be implemented by EU member states without reference to common institutions. Second, applying the measures to the Taliban rather than to all Afghan citizens required cooperation between different authorities within the EU member states, within and between different EU institutions, and between member states and the EU institutions.

In 2001 the sanctions against Afghanistan were modified five times to reflect decisions taken at the UN Security Council.[75] In addition, the EU arms embargo was brought into conformity with UN resolutions in November 2001. Whereas previously the UN decisions related to those parts of Afghanistan under the control of the Taliban and the forces of the Taliban, the EU embargo covered the whole territory of Afghanistan. The decisions of November facilitated the participation of British forces in the military operations in Afghanistan by permitting them to receive and supply arms and other equipment inside the country without any risk of violating EU law.[76]

---

[73] Common Position of 17 Dec. 1996 defined by the Council on the basis of Article J.2 of the Treaty on European Union concerning the imposition of an embargo on arms, munitions and military equipment on Afghanistan (96/746/CFSP), *Official Journal of the European Communities*, L 342, 31 Dec. 1996, p. 1.

[74] Council Common Position of 15 Nov. 1999 concerning restrictive measures against the Taliban (1999/727/CFSP), *Official Journal of the European Communities*, L 294, 16 Nov. 1999 pp. 1–2.

[75] Council Common Position of 26 Feb. 2001 concerning additional restrictive measures against the Taliban and amending Common Position 96/746/CFSP (2001/154/CFSP), *Official Journal of the European Communities*, L 57, 27 Feb. 2001, pp. 1–2; Council Regulation (EC) no. 467/2001 of 6 Mar. 2001 prohibiting the export of certain goods and services to Afghanistan, strengthening the flight ban and extending the freeze of funds and other financial resources in respect of the Taliban of Afghanistan and repealing Regulation (EC) no. 337/2000, *Official Journal of the European Communities*, L 67, 9 Mar. 2001, pp. 1–23; Commission Regulation (EC) no. 2062/2001 of 19 Oct. 2001 amending, for the third time, Council Regulation (EC) 467/2001 prohibiting the export of certain goods and services to Afghanistan, strengthening the flight ban and extending the freeze of funds and other financial resources in respect of the Taliban of Afghanistan and repealing Regulation (EC) no. 337/2000, *Official Journal of the European Communities*, L 277, 20 Oct. 2001, pp. 25–26; Commission Regulation (EC) no. 2199/2001 of 12 Nov. 2001 amending, for the fourth time, Council Regulation (EC) 467/2001 prohibiting the export of certain goods and services to Afghanistan, strengthening the flight ban and extending the freeze of funds and other financial resources in respect of the Taliban of Afghanistan and repealing Regulation (EC) no. 337/2000 *Official Journal of the European Communities*, L 295, 13 Nov. 2001, pp. 16–18; and Commission Regulation (EC) no. 2373/2001 of 4 Dec. 2001 amending, for the fifth time, Council Regulation (EC) 467/2001 prohibiting the export of certain goods and services to Afghanistan, strengthening the flight ban and extending the freeze of funds and other financial resources in respect of the Taliban of Afghanistan and repealing Regulation (EC) no. 337/2000, *Official Journal of the European Communities*, L 320, 5 Dec. 2001, p. 11.

[76] Council Common Position of 5 Nov. 2001 concerning restrictive measures against the Taliban and amending Common Positions 1996/746/CFSP, 2001/56/CFSP and 2001/154/CFSP (2001/771/CFSP), *Official Journal of the European Communities*, L 289, 6 Nov. 2001, p. 36.

VI. Sanctions against terrorism

In September 2001, following the terrorist attacks on the United States, the UN Security Council adopted Resolution 1373.[77] The resolution included a range of different measures with steps and strategies to combat terrorism. Paragraph 1(c) of the resolution decided that states shall: 'freeze without delay funds and other financial assets or economic resources of persons who commit, or attempt to commit, terrorist acts or participate in or facilitate the commission of terrorist acts; of entities owned or controlled directly or indirectly by such persons; and of persons and entities acting on behalf of, or at the discretion of such persons and associated persons and entities'.

These measures seek to eliminate, rather than change the behaviour of, terrorists and as such there is a question about whether they are sanctions. However, sanctions have been applied by the Security Council in response to acts of terrorism in the past.

In 1992 UN Security Council Resolution 748 introduced travel and diplomatic sanctions as well as an arms embargo on Libya.[78] These sanctions were introduced after Libya had not responded to requests by France, the UK and the USA for assistance in establishing responsibility for the bombing of Pan Am flight 103 over the Scottish town of Lockerbie in October 1988. The travel sanctions were subsequently strengthened in 1993 through the addition of financial sanctions and trade sanctions focused on the oil industry.[79]

In 1996 UN Security Council Resolution 1054 introduced diplomatic and travel sanctions against Sudan.[80] This resolution was intended to bring about Sudanese compliance with a request for the extradition to Ethiopia of three individuals suspected of carrying out an assassination attempt on Egyptian President Hosni Mubarak during a visit to Adis Abeba on 26 June 1995. The travel sanctions were strengthened in August 1996.[81]

Like those contained in Resolution 1267 on Afghanistan, these measures were applied to specific targets linked to acts of terrorism designated as such by the Security Council. However, although the measures adopted in September 2001 were a response to the attacks on the United States, they are to be applied globally in an effort to prevent all terrorist acts. This was the first use by the UN of sanctions to address a threat to the peace outside the context of a specific location.

In spite of the recommendations contained in the draft report of the Working Group on the General Issue of Sanctions, the UN resolutions do not provide specific directions about the scope of application of sanctions. No time limit is established for the sanctions. Moreover, there is no UN list of individuals and entities that have carried out terrorist acts.

---

[77] UN Security Council Resolution 1373, 28 Sep. 2001.
[78] UN Security Council Resolution 748, 31 Mar. 1992.
[79] UN Security Council Resolution 883, 11 Nov. 1993.
[80] UN Security Council Resolution 1054, 26 Apr. 1996.
[81] UN Security Council Resolution 1070, 16 Aug. 1996.

Recognizing that many states would require assistance in implementing Resolution 1373, the Security Council established a Counter-Terrorism Committee consisting of all the Security Council members. The terms of reference of the committee have been laid down by the Security Council.[82] The committee is tasked with providing UN member states with the appropriate expertise needed to implement Resolution 1373, including the preparation of model anti-terrorism laws and identifying technical, financial, regulatory and legislative resources that might facilitate implementation.

In practice, states are likely to receive more specific guidance from the information related to terrorism published by states (in particular the United States) and regional organizations (in particular the European Union).

On 23 September President Bush published an executive order that included both a definition of terrorist acts and an annexe that included a list of individuals and designated global terrorist entities.[83] Under separate legislation the USA publishes a list of countries that, in the view of the US Government, support international terrorism. The governments of these countries are not necessarily subject to financial sanctions. However, no US legal person (either an individual or a company) may engage in financial transactions with the governments of these countries without authorization by the Secretary of the Treasury, who decides in consultation with the Secretary of State.

EU sanctions related to terrorism

Prior to the 11 September attacks the European Union had used sanctions as part of its effort to combat international terrorism. EU measures were adopted in this regard before similar actions were taken in the United Nations. Member states adopted sanctions against Libya after a series of terrorist attacks in Europe in the mid-1980s.[84] An EPC declaration was used to underline the importance of a joint response including cases where terrorist acts were assisted by abuses of diplomatic immunity. In April 1986 the member states identified Libya as a country that deliberately encouraged recourse to acts of violence that were a threat to Europe. An arms embargo and diplomatic sanctions were adopted against Libya.[85] A limited trade embargo and travel sanctions were adopted in 1993 to facilitate the implementation of the sanctions adopted by the UN Security Council referred to above.[86]

[82] UN Security Council Resolution 1377, 12 Nov. 2001.

[83] Executive Order 13224 Blocking Property and Prohibiting Transactions with Persons who Commit, Threaten to Commit or Support Terrorism, The White House, 23 Sep. 2001. Documents on US sanctions against terrorism are available at URL <http://www.treas.gov/ofac/>.

[84] In Dec. 1985 the US and Israeli check-in desks at airports in Rome and Vienna were attacked simultaneously. The attacks killed 20 people, including the 4 perpetrators. In Apr. 1986 a bomb exploded on board a TWA aircraft as it made its descent to Athens airport, killing 4 passengers. A few days later, a bomb blast in a West Berlin discotheque frequented by US service personnel killed two civilians—1 US citizen and 1 German citizen.

[85] Statement by the Foreign Ministers of the Twelve on International Terrorism and the Crisis in the Mediterranean, The Hague, 14 Apr. 1986, URL <http://projects.sipri.se/expcon/euframe/eu_libya86.htm>.

[86] Council Decision of 22 Nov. 1993 on the Common Position defined on the basis of Article J.2 of the Treaty on European Union with regard to the reduction of economic relations with Libya, *Official*

Following the attacks on the USA on 11 September 2001 the European Union rapidly adopted a series of diplomatic, economic, financial, political and security-related measures in response. Sanctions are one element in this overall response, although the package includes many other measures. The adopted sanctions have two purposes. The first purpose is to implement elements of UN Security Council Resolution 1373 which, as described above, decided that all states shall freeze funds and other economic assets and financial resources owned or controlled by legal persons who commit, support or plan terrorist attacks.[87] Second, the sanctions decisions are intended to support the EU programme of action against terrorism.

In December 2001 the Council agreed on key definitions that are essential to implementing sanctions. The member states defined terrorist acts and agreed a specific consolidated list of persons, groups and entities that have committed terrorist acts.[88] In addition, the Council agreed on the scope of the terms 'funds', 'other financial assets', 'economic resources' and 'financial services' as well as a list of persons, groups and entities whose funds were to be frozen.[89]

These measures were adopted as a part of the CFSP. However, financial sanctions will probably be applied in consultation and cooperation with officials and agencies already engaged in programmes to combat organized crime. In particular, the EU will work closely with the UN Security Council Counter-Terrorism Committee and with the intergovernmental Financial Action Task Force on Money Laundering (FATF) that was established jointly by the Group of Seven (G7) leading industrialized nations and the European Commission in 1989.[90]

VII. Conclusions

During the 1990s the increased use of sanctions in the changed international environment sparked a debate about when and how sanctions could be used in a legitimate and an effective manner. Sanctions have not been effective in bringing about a change in the behaviour of their targets when applied as a 'stand-alone' measure. At the same time, sanctions are recognized as a

---

Journal of the European Communities, L 295, 30 Nov. 1993, p. 7; and Council Regulation 3274/93 of Nov. 1993 preventing the supply of certain goods and services to Libya, *Official Journal of the European Communities*, L 295, 30 Nov. 1993, pp. 1–3.

[87] As noted above, the EU had already implemented measures against the assets and resources of specific persons and entities in Afghanistan.

[88] Council Common Position of 27 Dec. 2001 on the application of specific measures to combat terrorism (2001/931/CFSP), *Official Journal of the European Communities*, L 344, 28 Dec. 2001, pp. 93–96.

[89] Council Regulation (EC) no. 2580/2001 of 27 Dec. 2001 on specific restrictive measures directed against certain persons and entities with a view to combating terrorism, *Official Journal of the European Communities*, L 344, 28 Dec. 2001, pp. 70–75; and Council Decision of 27 Dec. 2001 establishing the list provided for in Article 2(3) of Council Regulation 2580/2001 on specific restrictive measures directed against certain persons and entities with a view to combating terrorism, *Official Journal of the European Communities*, L 344, 28 Dec. 2001, pp. 83–84.

[90] The activities of the FATF are described at URL <http://www1.oecd.org/fatf/index.htm>. The members of the G7 are listed in the glossary in this volume.

necessary and an important instrument for conflict resolution and the efforts of states have been directed to improving their design and implementation as well as considering how sanctions might be used as a part of a wider strategy. This discussion has been carried on in different international organizations, intergovernmental discussions, official (but informal and ad hoc) processes and outside government.

Although there have been some calls for the elaboration of an integrated and comprehensive legal framework for sanctions, in practice the elaboration of international law in this area has taken place through an operational approach in reaction to a particular event.

The United Nations has established a working group on sanctions. The results of the work of this group, which seem to have a 'lessons learned' character, may go some way towards providing general rules based on the experiences of the UN and of states during the 'sanctions decade'. However, the main results of the group seem more likely to be aimed at enhancing the capacity of the UN to take and implement decisions.

Those general rules that were proposed in the report of the group were not applied in the decisions taken in late 2001 imposing sanctions against entities engaged in terrorism. The use of sanctions against terrorism—a general and global threat rather than a threat to the peace, breach of the peace or act of aggression in a specific location—is unprecedented. However, it is not currently proposed to apply similar measures to other general threats identified by the Security Council.

The European Union has also elaborated its approach to the use of sanctions as part of its Common Foreign and Security Policy, mainly in response to specific events rather than through a more 'top–down' approach.

The member states have increasingly used the EU to give effective expression to decisions of the United Nations. However, the EU has also developed a distinctive approach to the use of sanctions in foreign policy areas where the UN has not provided direction, notably to support the parts of the CFSP aimed at improving human rights. A recommendation by the Commission in 2001 that the EU should think in a broader manner about how sanctions should be decided and implemented may lead to further development in this area.

Part II. Military spending and armaments, 2001

Chapter 6. Military expenditure

Chapter 7. Arms production

Chapter 8. International arms transfers

6. Military expenditure

ELISABETH SKÖNS, EVAMARIA LOOSE-WEINTRAUB,
WUYI OMITOOGUN and PETTER STÅLENHEIM

I. Introduction

World military expenditure in 2001 is estimated at $839 billion (in current dollars), accounting for 2.6 per cent of world gross domestic product (GDP) and a world average of $137 per capita. This estimate, which is based on adopted defence budgets, is likely to be revised upwards when supplementary expenditures resulting from the 11 September 2001 attacks on the United States and the ensuing 'war on terrorism' have been taken fully into account.

A few countries account for the major part of the world total. Five countries account for over half and the 15 major spenders account for over three-quarters of the world total. The high-income countries—the industrial countries and those in the Middle East—have the highest per capita spending. The developing countries—particularly those in Africa and the Middle East—have the heaviest economic burden of military expenditure in terms of its share of GDP.

After the decline from 1987 to 1998, military expenditure began to rise again, both globally and in most regions of the world. Over the three-year period 1998–2001, it increased by around 7 per cent in real terms. The increase of 2 per cent in 2001 is smaller than the increases in 1999 and 2000, but world military expenditure is likely to rise much faster in the coming years, owing primarily to a substantial increase in US military spending.

Apart from the spending increases that were planned in 2001, the attacks of 11 September are also likely to have an impact on future trends in military expenditure, not only in the USA but in several other countries as well.

Section II of this chapter provides an overview of the global trends in military expenditure. Section III describes the impact of the war on terrorism on military expenditure on the basis of information available at the end of 2001. Regional developments in military expenditure are summarized in section IV. Section V presents the main findings of the chapter.

Appendix 6A presents SIPRI data on military expenditure for 158 countries for the 10-year period 1992–2001. Country data are provided in three formats: in their original form, in local currency and current prices (table 6A.2); in constant US dollars, to establish changes in military expenditure in real terms, that is, after adjusting for inflation (table 6A.3); and as a share of GDP, which provides a rough measure of the economic burden of military expenditure (table 6A.4). Appendix 6B presents data on NATO countries' expenditure on personnel and military equipment.

The SIPRI military expenditure data are based on official statistics, in most cases allocations for the ministry of defence or for a broader functional category such as national defence. It is important to note that official statistics on military and military-related expenditure have a range of limitations and that their interpretation requires some caution. The main limitations of official data are described in appendix 6C, which presents the sources and methods for SIPRI's military expenditure data. Appendix 6D describes the responses to requests for data by SIPRI and the United Nations as well as other initiatives to improve transparency in military expenditure. Appendix 6E analyses the long-term trends in US military expenditure and the US defence budget for fiscal year (FY) 2003.

II. World military expenditure: trends and pattern

The level of world military expenditure in 2001, on the basis of adopted defence budgets, is estimated to be $772 billion, at 1998 prices and exchange rates (appendix 6A, table 6A.1). This figure corresponds to roughly $839 billion in current dollars.[1] After an 11-year period of decline, world military expenditure has risen since 1998. The total increase was of the order of 7 per cent in real terms—1.3 per cent in 1999, 4.0 per cent in 2000 and about 2 per cent in 2001. Actual expenditure for 2001 is likely to be considerably higher owing to supplementary expenditure after 11 September, primarily in the United States, to finance the war on terrorism and counter-terrorism measures.

On a regional basis, the increase since 1998 seems to be a general one. Military expenditure has increased in all regions except Oceania, although the increase in Western Europe was very small (table 6.1). The regions with the highest rates of growth in military expenditure over the period 1998–2001 are Africa (31 per cent), Central and Eastern Europe (28 per cent), South Asia (26 per cent) and the Middle East (25 per cent) (table 6.1). The regions that have contributed most to the global increase in terms of volume of expenditure over this period are the Middle East (an increase of $15 billion), Central and Eastern Europe ($13 billion), North America and East Asia ($7 billion each).

Among the country income groups, military expenditure increased most rapidly in the poorest countries, while the high-income countries had the smallest rate of growth between 1998 and 2001 (table 6.1, column 3). In terms of absolute increase (table 6.1, column 6), the high-income group nonetheless represents a significant increase in spending—$9 billion (at constant 1998 prices and exchange rates) between 1998 and 2001. The volume increases in the upper–middle-income and low-income countries were only slightly higher ($11 billion and $14 billion, respectively), although their growth rates were much higher. The group with a high increase in both relative and absolute terms is the lower–middle-income countries. The combined military expendi-

---

[1] This estimate in current dollars is derived by applying the US inflation rate between 1998 and 2001 (8.7% over 3 years) to the world figure of $772 billion at constant (1998) prices and exchange rates.

ture of this group increased by 29 per cent in real terms and by $20 billion in absolute terms. This is to a great extent explained by the fact that Russia is included in this group; excluding Russia, the military expenditure of this group increased by 18 per cent, or $7 billion.

Further disaggregation of the world total shows that the global increase in military expenditure is due primarily to the volume increases of a few major spenders. Of the total increase of $53 billion (at constant 1998 prices and exchange rates) in total world military expenditure over the period 1998–2001, 10 countries—Russia, China, the USA, Saudi Arabia, Iran, India, Brazil, Italy, Oman and Nigeria—contributed almost $50 billion.

Two of the major spenders have announced large military expenditure increases. The US budget proposal for FY 2003 includes a total increase in outlays for national defence of $54 billion, or 16.6 per cent in real terms, between FY 2001 and FY 2003 (see appendix 6E). China has announced an increase of 17.6 per cent in its 2002 budget for national defence.

The global pattern

The global pattern of military expenditure corresponds by and large to the global economic and political structure. The greater part of military spending takes place in the rich regions. In 2001, the 32 high-income countries accounted for 70 per cent of the world total, while the 51 low-income countries accounted for 8 per cent (table 6.1). The 63 countries in Africa and Latin America together accounted for only 5 per cent of the world total.

The major spenders

World military expenditure is concentrated in a few countries. The 15 major spenders accounted for over three-quarters of the world total and the 5 major spenders for more than half of the total in 2001 (table 6.2). The USA is by far the largest spender, accounting for 36 per cent of world military expenditure. The next four in size are Russia, France, Japan and the UK. Their level of military expenditure is significantly lower, each accounting for 5–6 per cent of the world total, and their combined military expenditure is roughly half that of the USA. The next layer of major spenders consists of Germany, China, Saudi Arabia, Italy and Brazil, each with 2–4 per cent of the world total.

The rank order of military spenders, as with all cross-country economic comparisons, is highly dependent on the method used to convert local currencies into dollars. If Russian military expenditure is converted by use of the market exchange rate instead of the purchasing power parity (PPP) rate,[2] Russia ranks number 12.

Many of the major spenders listed in table 6.2 are regional powers that dominate the regional military expenditure totals presented in appendix 6A, table 6A.1. India accounted for 72 per cent of South Asian military expendi-

[2] See appendix 6C.

Table 6.1. Changes in world and regional military expenditure estimates, 1992–2001

Figures are calculated and expressed in US $b., at constant (1998) prices and exchange rates. Figures in italics are percentages.

Country groups[a]/ (no. of countries)	Relative change (%)			Absolute change ($b.)			World share (%)
	(1) 1992–01	(2) 1992–98	(3) 1998–01	(4) 1992–01	(5) 1992–98	(6) 1998–01	(7) 2001
Africa (44)	*+ 32*	*+ 1*	*+ 31*	+ 3	± 0	+ 3	*2*
North (3)	..	*+ 33*	..	..	+ 1	..	..
Sub-Saharan (41)	..	*– 14*	..	..	– 1	..	..
Americas (21)	*– 17*	*– 20*	*+ 3*	– 66	– 75	+9	*41*
North (2)	*– 21*	*– 23*	*+ 3*	– 75	– 82	+ 7	*37*
Central (8)	*+ 21*	*+ 17*	*+ 4*	+ 1	± 0	± 0	–
South (11)	*+ 46*	*+ 36*	*+ 8*	+ 8	+ 6	+ 2	*3*
Asia and Oceania (30)	*+ 23*	*+ 11*	*+ 10*	+ 24	+ 12	+ 12	*17*
Central Asia (5)	..	..	..	..	..	..	–
East Asia (16)	*+ 19*	*+ 11*	*+ 8*	+ 16	+ 9	+ 7	*13*
South Asia (5)	*+ 54*	*+ 22*	*+ 26*	+ 6	+ 2	+ 4	*2*
Oceania (4)	*± 0*	*+ 4*	*– 4*	± 0	± 0	± 0	*1*
Europe (41)	*– 18*	*– 23*	*+ 7*	– 54	– 69	+ 15	*31*
Central & Eastern (21)	*– 37*	*– 50*	*+ 28*	– 35	– 48	+ 13	*8*
Western Europe (20)	*– 10*	*– 11*	*+ 1*	– 20	– 22	+ 2	*23*
Middle East (13)	*[+ 38]*	*[+ 11]*	*+ 25*	[+ 20]	[+ 5]	+ 15	*9*
World total (149)	*– 9*	*– 15*	*+ 7*	**– 75**	**– 128**	**+ 53**	*100*
High income (32)	*– 15*	*– 16*	*+ 2*	– 93	– 102	+ 9	*70*
Upper middle income (23)	*+ 44*	*+ 25*	*+ 15*	+ 25	+ 14	+ 11	*11*
Lower middle income (43)	*– 21*	*– 39*	*+ 29*	– 24	– 44	+ 20	*11*
Low income (51)	*+ 44*	*+ 10*	*+ 31*	+ 18	+ 4	+ 14	*8*

[a] For the country coverage of the regional and income groups, see appendix 6A, table 6A.1. Some countries are excluded because of the lack of consistent time-series data. Africa excludes Angola, Benin, Congo (Republic of), Congo (Democratic Republic of the), Libya and Somalia; Asia excludes Afghanistan; Europe excludes Yugoslavia; and the Middle East excludes Iraq. The world total excludes all these countries.

Source: Calculated on the basis of data in appendix 6A, table 6A.1.

ture, Russia for 71 per cent of Central and East European (CEE), Brazil for 44 per cent of South American, Japan for 39 per cent and China for 24 per cent of East Asian, and Saudi Arabia for 33 per cent of Middle Eastern military expenditure.

Economic burden

World military expenditure of $839 billion (in current dollars) in 2001 corresponds to an average of $137 per capita[3] and 2.6 per cent of world

---

[3] The estimate of world military expenditure per capita is based on a world population of 6.134 billion in 2001. UN Department of Economic and Social Affairs, Population Division, *World Urbanization Prospects: The 2001 Revision* (United Nations: New York, 2002), table A1: 'Population of urban and

Table 6.2. The 15 major spenders, military expenditure in 2001[a]

Figures are in US $b., at constant (1998) prices and exchange rates. Figures in italics are percentages.

Rank[a]	Country	Size ($b.)	World share (%)
1	USA	281.4	*36*
2	Russia (PPP)[b]	[43.9]	*[6]*
3	France	40.0	*5*
4	Japan	38.5	*5*
5	UK	37.0	*5*
Sub-total top 5		**440.8**	*57*
6	Germany	32.4	*4*
7	China	[27.0]	*[3]*
8	Saudi Arabia	26.6	*3*
9	Italy	24.7	*3*
10	Brazil	14.1	*2*
Sub-total top 10		**565.6**	*72*
11	India	12.9	*2*
12	South Korea	10.2	*1*
13	Israel	9.1	*1*
14	Turkey	8.9	*1*
15	Spain	8.0	*1*
Sub-total top 15		**614.7**	*78*
World total		**772**	*100*

[a] The rank order of countries differs with the base year and the method of conversion to dollars. The base year should ideally be the same as the year of comparison, while this table is based on military expenditure figures in constant (1998) prices and exchange rates because of the lack of PPP data for Russia for 2001.

[b] Conversion to dollars is made by use of the market exchange rate for most countries. The main exception in this table is Russian military expenditure, which is converted by use of the PPP conversion factor (see appendix 6C). If the market exchange rate is used for Russia, its military expenditure in 2001 amounts to $12.7 billion at constant (1998) prices and exchange rates.

Sources: Appendix 6A (tables 6A.1 and 6A.3) and the SIPRI database on military expenditure.

GDP.[4] This is a substantial level of resource consumption for military purposes and thus an economic burden on the global economy. However, the burden varies significantly between regions.

While the major part of military expenditure is spent by the Western industrial countries, the heaviest economic burden—the defence burden—is in developing countries with a high level of poverty. Around 1.2 billion people— one-sixth of the world population and one-fourth of the population of the

rural areas at mid-year and percentage urban, 2001', 20 Mar. 2002, URL <http://www.un.org/esa/population/publications/wup2001/wup2001dh.pdf>.

[4] This share is based on an estimate for world GDP in 2001 of $32 150 billion, calculated by applying the 2.6% annual growth rate during the 1990s to the GDP figure for 2000 of $31 337 billion. World Bank, *World Development Report 2002: Building Institutions for Markets* (Oxford University Press: New York, 2002), table 3, p. 237.

Table 6.3. Global pattern of military expenditure, population and gross national income, 2000

Figures are percentages. They may not add up to totals due to the conventions of rounding.

Country groups[a]	Shares of world total		
	Mil. expenditure[b]	Population	Gross national income[c]
World total	**100**	**100**	**100**
High income	75	15	80
Upper middle income	10	11	10
Lower middle income	8	34	7
Low income	7	41	3
Low- and middle-income countries			
Africa, Sub-Saharan	1	11	1
America, Latin	4	9	6
Asia, East	6	31	6
Asia, South	2	22	2
CEE[d] and Central Asia	4	8	3
Middle East and North Africa	7	5	2
Total low and middle income	**25**	**85**	**20**

[a] The income and regional groups in this table differ from those of the SIPRI tables on military expenditure in appendix 6A. In order to enable a comparison between SIPRI and World Bank data, this table is based on the World Bank country classification. For the coverage of country groups for SIPRI tables, see appendix 6A. The main difference in this table is that the geographical regions exclude high-income countries: East Asia excludes Brunei, Japan, Singapore and Taiwan; and the Middle East excludes Israel, Kuwait and the United Arab Emirates.

[b] The shares are calculated on military expenditure figures at constant (1998) prices and exchange rates.

[c] The shares are calculated on GNI figures at current prices and exchange rates. Gross national income (GNI)—a measure of the income side of the national economy—is replacing gross national product (GNP)—a measure of the output of the national economy—as the standard measure in national accounts. They are different ways of measuring national economic activity.

[d] CEE = Central and Eastern Europe.

Sources: Military expenditure: appendix 6A; Population and GDP: World Bank, 'Key indicators of development', *World Development Report 2002: Building Institutions for Markets* (Oxford University Press: New York, 2002), table 1, pp. 232–33.

developing world—subsist on less than $1 per day.[5] Under these circumstances, even a world average defence burden of 2.6 per cent constitutes a serious diversion of resources from the fulfilment of basic needs. The burden is even higher in many of these countries.

[5] United Nations Development Programme, *Human Development Report 2001: Making New Technologies Work for Human Development* (Oxford University Press: New York, 2001), p. 9.

Table 6.3 shows the global distribution of military expenditure, population and gross national income (GNI).[6] The global pattern of military expenditure is fairly similar to the distribution of national income but very different from the global pattern of population. High-income countries account for only 15 per cent of the world population but for as much as 75 per cent of world military expenditure, which means that average per capita military expenditure in these countries is much higher than the world average. The countries with the highest per capita spending are located in the Middle East, followed closely by high-income countries in North America and Western Europe. In some of these countries, per capita military expenditure is between $400 and $1500. Among the low- and middle-income countries (a group which corresponds roughly to the 'developing countries'), the African and Asian countries account for a much larger share of world population than of world military spending. This means that per capita military expenditure is very low in these regions, although many countries of the regions have a high defence burden because of their relatively lower national incomes.

On a regional basis, the share of national income devoted to military expenditure is relatively even. However, this does not apply to the low-income countries, which account for a larger share of military expenditure (7 per cent) than of world income (3 per cent). This indicates that military expenditure imposes a much greater burden on the economies of low-income countries than those of countries in the other groups. The burden is more than twice as high as the world average, even on the basis of official data. In reality, the burden is even higher because official data often understate the actual military expenditure levels of these countries. Furthermore, the fact that these countries are poor means that the surplus available for resource allocation is very small after basic needs are provided for. Thus, in low-income countries, military expenditure imposes a severe burden on the economy.

Among the low- and middle-income regions, those with the highest defence burden are the Middle East and North Africa, as shown by their much higher share of global military expenditure than their share of world GNI. In contrast, Latin America has a lower defence burden than the world average.

Countries with the highest defence burden

Among the countries with the highest defence burden—where military expenditure is higher than 4 per cent of GDP—seven are low-income countries that are or have recently been involved in armed conflict: Burundi, Eritrea, Ethiopia and Rwanda in Africa; and Bosnia and Herzegovina, Croatia and Yugoslavia in Europe (table 6.4). In reality, the number of countries in this group is probably much higher because official expenditure data are severely under-reported in most of the countries involved in armed conflict.

In poor countries, a high defence burden can be expected to seriously impinge on public expenditure for education and health. The official data pre-

---

[6] The World Bank used GNI instead of GNP for the first time in the 'Selected world development indicators', *World Development Report 2002* (note 4), p. 231.

Table 6.4. Countries with the highest defence burden in 1999[a]: social and military expenditure as a share of gross domestic product, 1995–2000

Figures are percentages.

Country[b]	Public expenditure on Education 1995–97	Public expenditure on Health 1998	Military expenditure 1995	1996	1997	1998	1999	2000
Eritrea	1.8	..	19.9	22.8	13.5	29.0	22.9	..
Saudi Arabia	7.5	..	10.3	9.5	12.0	16.2	12.3	11.6
Oman	4.5	2.9	14.6	12.5	11.5	[11.4]	[10.4]	[9.7]
Ethiopia	4.0	1.7	2.0	1.9	3.2	5.1	9.4	..
Jordan	7.9	5.3	[9.4]	8.6	8.8	8.9	9.2	9.5
Kuwait	5.0	..	13.9	10.4	8.2	9.0	8.1	8.2
Israel	7.6	6.0	8.3	8.6	8.4	8.4	8.0	8.0
Burundi	4.0	1.6	4.2	5.8	6.0	6.5	6.2	5.4
Syria	4.2	0.8	7.1	5.9	5.7	5.8	[5.6]	[5.5]
Yemen	7.0	..	7.3	6.4	6.5	6.7	5.6	[5.2]
Singapore	3.0	1.2	4.4	4.5	4.7	5.4	5.4	4.8
Turkey	2.2	..	3.9	4.1	4.1	4.3	5.0	4.9
Bahrain	4.4	2.6	4.7	4.7	4.6	4.8	4.9	4.0
Greece	3.1	4.7	4.3	4.5	4.6	4.8	4.8	4.9
Bosnia and Herzegovina	..	..	[14.3]	[3.8]	5.3	4.7	4.6	4.2
Pakistan	2.7	0.9	5.3	5.1	4.9	4.8	4.6	4.5
Rwanda	..	2.0	4.4	5.3	4.1	4.4	4.6	[3.0]
Yugoslavia	..	..	..	..	[4.8]	4.4	4.4	5.9
Croatia	5.3	..	9.4	7.2	5.7	5.5	4.1	3.0
Morocco	5.3	1.2	4.6	4.0	3.9	3.7	4.1	4.2

[a] Countries with a known military expenditure share of GDP of 4% or more in 1999.

[b] Countries are ranked by their share of military expenditure in GDP in 1999.

Sources: Military expenditure as a share of GDP: appendix 6A, table 6A.4; Public expenditure on education and health: United Nations Development Programme (UNDP), *Human Development Report 2001: Making New Technologies Work for Human Development* (Oxford University Press: Oxford, 2001), appendix table 16, 'Priorities in public spending', pp. 195–97.

sented in table 6.4 verify this to some extent. However some poor countries with a high defence burden do not fit this pattern. Burundi, Ethiopia and Yemen are all low-income countries with a high defence burden that are still able to spend a relatively high share of their GDP on public education. Jordan, Morocco and Syria are lower–middle-income countries that also have relatively high budgetary allocations for education. This could be due to their high prioritization of public-sector spending in general, but it could also be the result of other factors, such as high volumes of foreign economic assistance.

Many of the countries with a high defence burden are, however, among the more wealthy Middle Eastern countries, classified by the World Bank as high-income and upper–middle-income economies—Bahrain, Israel, Kuwait, Oman and Saudi Arabia. These countries are less seriously affected by their high defence burden and can therefore afford rather high allocations for social expenditure.

III. The war on terrorism

The war on terrorism launched by the USA after the terrorist attacks of 11 September 2001 is likely to have a significant impact on international relations and security for at least a decade. The extent to which and the way in which the war will also have an impact on military expenditure were not clear by the end of 2001. Before 11 September, there were few examples of military operations against terrorism. Anti-terrorist operations were largely regarded as an internal security matter, conducted by governments within their own territories. With few exceptions, international terrorism was combated through reliance on international law enforcement and international cooperation between national intelligence agencies. The US-led war in Afghanistan, initiated on 7 October 2001, opened a new era for combating terrorism because of its heavy reliance on military force. This will have an impact not only on budgets for internal security, including police forces, intelligence, customs and other non-military counter-terrorist measures, but also on allocations for military forces and military intelligence.

The actual and potential costs of the war on terrorism can be divided into two main components: the costs of the war in Afghanistan, including those for the conduct of the war and the reconstruction of Afghanistan; and the longer-term costs of future military and internal security measures associated with the threat of terrorism.

The costs of the war in Afghanistan

Expenditures to cover the costs of the war in Afghanistan are significant. The costs include not only those for the war itself but also for buying support from neighbouring countries and for reconstruction after the war. Seen from an economic perspective, there are losers and, ironically, also winners of wars. While the USA and Afghanistan can be seen as the major losers in economic terms, some of the countries bordering on Afghanistan have made economic gains, in particular Pakistan. The losses for the USA include: (a) lives and property, from the terrorist attacks of 11 September; (b) the economic consequences of the attacks; and (c) the direct and the indirect costs of the war in Afghanistan. The size of these costs is indicated both by figures on funding (approved appropriations) and by calculated cost estimates. Immediately after the 11 September attacks, the US Congress authorized an emergency appropriation of $40 billion for anti-terrorism activities, about half of which was for the Department of Defense (DOD) and the rest for other departments involved in these activities (see appendix 6E). The actual amount appropriated by the

DOD based on these authorizations was $17.2 billion.[7] In March 2002 the Administration of President George W. Bush requested an additional $14 billion in emergency appropriations for FY 2002 (1 October 2001– 30 September 2002) to continue the 'global war on terrorism'.[8] Thus, the total additional allocations for the war on terrorism for the period from 11 September 2001 to 30 September 2002 was $31 billion for the DOD and about $20 billion for non-DOD activities. These allocations have been made as emergency appropriations outside the ordinary budget, but they are likely to have an impact on other spending categories in future budgets. A trade-off between military and social expenditures can be expected.

The US Congressional Budget Office (CBO) has estimated the cost of the war in Afghanistan as $10.2 billion for FY 2002.[9] This estimate is limited to the incremental direct costs to the DOD of conducting military operations in and around Afghanistan. It does not include the costs of humanitarian and economic assistance to Afghanistan and neighbouring states, the costs of homeland defence, the costs of support to federal, state and local agencies in the USA or counter-terrorism operations in other parts of the world.

Other NATO countries with a smaller involvement in the war in Afghanistan have also incurred significant expenditures. The cost to the UK for the British forces involved in the war will be covered by an extra allocation of £100 million ($150 million) to the ministry of defence from treasury contingency funds.

Australia's additional military expenditure for domestic anti-terrorist measures, primarily for border protection, in or near Afghanistan was estimated at A$362 million (US$186 million) in early 2002. To cover these additional costs, the Australian Government decided to generate savings by cancelling or curtailing a number of training and maintenance programmes.[10] Japan, which has a constitutional barrier to the external use of military force, has contributed financial and other support to the US troops in Afghanistan. In October 2001 the Japanese Parliament (Diet) passed a new anti-terrorism bill that would allow the Japanese Self Defense Forces to perform a range of military support functions for the USA in the war in Afghanistan, including the delivery of fuel and supplies, repair work, communications and surveillance.[11]

Afghanistan has suffered major physical destruction in the war. At a donor conference held in Tokyo in January 2002, about $4 billion in aid was pledged for reconstruction of the country over the next five years. In addition to about

---

[7] US Department of Defense, *DOD FY2002 Supplemental to Continue the Global War on Terrorism* (DOD: Washington, DC, Mar. 2002), URL <http://www.dtic.mil/comptroller/fy2003budget/fy2002_supp.pdf>.

[8] US Department of Defense (note 7). For the general background see 'DoD seeks extra $10–20B: request comes on top of $379B defense budget proposal for 2003', *Defense News*, 4–10 Feb. 2002, pp. 1, 4.

[9] 'CBO estimates war costs', *Forecast International/DMS*, 10 Apr. 2002, URL <http://www.forecast1.com>.

[10] 'War on terror forces budget cuts', *Defense News*, 4–10 Mar. 2002.

[11] 'The Japanese military: new rules of defence', *Far Eastern Economic Review*, 1 Nov. 2001, pp. 20–21; and 'Chinese advice has Japan up in arms', Stratfor.com, 17 Dec. 2001, URL <http://www.stratfor.com/premium/analysis_print.php?ID=201315>.

$1.3 billion pledged by Western donors, several countries in the region pledged major amounts, including Iran ($560 million), Saudi Arabia ($220 million), and India and Pakistan ($100 million each).[12] The total cost of the reconstruction of Afghanistan, however, is likely to be much higher than these pledged contributions.[13]

The costs for maintaining stability in Afghanistan cannot be estimated. According to the 2001 Bonn Agreement,[14] an International Security Assistance Force (ISAF) was established to help the Afghan Interim Authority to establish and train new Afghan security and armed forces. By March 2002, 18 countries were contributing personnel, equipment and other resources to this force.[15]

The Pakistani economy has improved as a result of the war, after having deteriorated for 10 years, aggravated by the sanctions imposed after the 1998 nuclear tests and a heavy debt burden (debt servicing accounted for more than half of public expenditure). As a result, the number of people living in absolute poverty doubled during the decade.[16] By February 2002 the economic situation in Pakistan was much brighter: exports of textiles, garments and leather had risen, and both manufacturing growth and aid inflows had increased.[17] A number of political decisions related to the war in Afghanistan contributed to this growth. The USA lifted its economic sanctions on Pakistan on 22 September and its ban on military aid on 4 October.[18] It also made trade concessions for textile imports from Pakistan. The Pakistani economy benefited from a total of about $1 billion in economic aid, primarily from the USA but also from Japan and the European Union (EU). There was a new loan from the International Monetary Fund (IMF), a loan extension from China, and other debt rescheduling and write-offs from *inter alia* the Paris Club of government creditors, the USA and Canada.[19]

Other countries in the area, notably Georgia and Uzbekistan, have also received increased or new economic or military assistance in return for their contributions to the war. Some of them, including Indonesia and the Philip-

---

[12] 'Donor nations vow billions for Kabul', *International Herald Tribune*, 22 Jan. 2002, p. 1.
[13] Recovery efforts will cost $10–15 billion over the next decade. URL <http://www.cfrterrorism.org/policy/refugees.html>.
[14] The Agreement on Provisional Arrangements in Afghanistan Pending the Re-establishment of Permanent Government Institutions, Bonn, 5 Dec. 2001, URL <http://www.uno.de/frieden/afghanistan/talks/agreement.htm>. See also 'Stability in view', *Jane's Defence Weekly*, 19 Dec. 2001.
[15] UN Security Council Resolution 1386, 20 Dec. 2001, URL <http://www.un.org/Docs/scres/2001/res1386e.pdf>. On 10 Jan. 2002, 14 European countries, New Zealand and Turkey signed a joint Memorandum of Understanding (MOU), formalizing their contributions to the force. Subsequently, Belgium and the Czech Republic signed the MOU, and Bulgaria is contributing personnel. British Ministry of Defence, 'International Security Assistance Force (Operation Fingal)', URL <http://www.operations.mod.uk/fingal/index.hytm>. See also the introduction in this volume.
[16] 'War good for economy—even in Pakistan', Stratfor.com, 12 Oct. 2001, URL <http://www.stratfor.com/standard/analysis_print.php?ID=200953>.
[17] 'Shaukat expects growth rate at 3.5 pc', *The News* (Islamabad), (Internet edn, in English), 19 Feb. 2002, in 'Pakistani Finance Minister foresees better economic growth in 2002–2003', Foreign Broadcast Information Service, *Daily Report–Near East and South Asia (FBIS-NES)*, FBIS-NES-2002-0219, 20 Feb. 2002.
[18] 'Pakistan emphasizes financial over military ties', *Defense News*, 8–14 Oct. 2001, p. 8.
[19] *The News* (note 17).

pines,[20] have also received or been promised military assistance to fight terrorism on their own territories.

The impact on future budgets

After the 11 September attacks, most Western countries initiated some form of review of their long-term requirements to counter the threat of terrorism. Measures to build preparedness against threats in the areas of both military defence and internal security, cooperation between these two sectors and international cooperation were considered. However, other than the United States, only a small number of countries announced such measures in 2001. Nevertheless, it is very likely that measures will be implemented in more countries and will have a gradual impact on future military expenditure.

In the United States, the attacks had a strong impact on future military expenditure. The defence budget for FY 2003 was proposed to increase by $48 billion, or 14 per cent (appendix 6E). The bulk of the increase in US military expenditure is not devoted to new activities directly related to the threat of terrorism but to general improvements for military personnel and in weapon programmes developed for the cold war security environment. However, this sharp increase in US military expenditure would not have been politically feasible if the attacks had not occurred. It can therefore be argued that the main determinant of the change in trend in US military expenditure is the change in US public opinion in favour of a stronger defence.

The part of the US budget specifically dedicated to counter-terrorism is the Defense Emergency Reserve Fund (DERF), for which $20.1 billion was requested for FY 2003. This includes $9.4 billion to cover known costs for the war on terrorism and $10 billion as a contingency reserve fund for future wars on terrorism in other countries. The $9.4 billion allocation was for force protection ($2.7 billion), air patrols over the continental USA ($1.2 billion), and procurement of precision-guided and other munitions ($812 million), tanker and transport aircraft (c. $800 million) and unmanned aerial vehicles (UAVs) ($189 million), while $2.6 billion was for classified programmes.[21]

In addition to the defence budget, the FY 2003 budget request included a large allocation ($37.7 billion) under the heading 'homeland security', of which only 22 per cent was for the DOD and the rest divided between a number of government agencies within the departments of justice and transportation. A large part of this allocation was devoted to counter-terrorism activities (appendix 6E).

The war will also have an impact on the US research and development (R&D) budget, with a shift in priorities from civil to military R&D. While under the previous administration the goal was to reduce military R&D to less

---

[20] 'Aid for Jakarta's war on terrorism', *Financial Times*, 29 Jan. 2002, p. 3; and 'Philippines to see boost in US military financing', *Defense News*, 5–11 Nov. 2001, p. 16.
[21] 'Pentagon details extra money for war on terror', *Defense News*, 25 Feb.–3 Mar. 2002, p. 4.

than half of the total R&D budget,[22] in the budget for FY 2003 there is a disproportionate increase in military R&D. Of the total R&D increase of 8 per cent, military R&D is proposed to increase by 11 per cent, to $54.5 billion, while civilian R&D will increase by only 6 per cent, to $57.2 billion.[23] The greatest increases, and those most closely related to the threat of terrorism, were the 700 per cent increase for R&D on measures to combat bio-terrorism ($2.4 billion), 19 per cent for space technology ($3.4 billion) and 17 per cent for nanotechnology ($679 million),[24] while reductions were proposed for some environment and energy research programmes.

Two other countries decided in 2001 to reorient their future budgets in response to the threat of terrorism. Canada announced on 10 December that its budget for 2002 would include a major allocation for counter-terrorism activities, although most of the increase was for internal security. Of the total increase of C$7.7 billion ($4.8 billion) in security expenditure, C$6.5 billion was allocated to homeland defence against terrorist activities—border guards, police, customs, airport security and security intelligence services—while C$1.2 billion was to cover costs for the deployment of two ships in the Indian Ocean and the Persian Gulf as part of the war effort.[25] The German Government decided in September 2001 to raise taxes in order to generate an annual surplus of DM 3 billion ($1.4 billion) to spend on institutions dealing with domestic and foreign security. Of the total, the Bundeswehr will receive DM 1.5 billion, the ministry of interior affairs DM 500 million for increasing the readiness of border patrols, the ministry for foreign affairs DM 200 million for increased security at German embassies, and the ministry of development DM 200 million to support crisis management overseas. About DM 500 million will be kept in reserve.[26] The exact content of the DM 1.5 billion allocation to the Bundeswehr was not announced, but it reportedly includes DM 1.2 billion for the procurement of new equipment required for closing capability gaps in critical areas, such as mobility, protection, and reconnaissance and surveillance.[27] The German defence expenditure plan for the period 2003–2006 was also revised upwards by DM 1.5 billion, to DM 47.7 billion.[28]

---

[22] Sköns, E. et al., 'Military expenditure and arms production', *SIPRI Yearbook 2001: Armaments, Disarmament and International Security* (Oxford University Press: Oxford, 2001), p. 231.

[23] 'War effort shapes US budget, with some program casualties', *Science*, 8 Feb. 2002, pp. 952–54.

[24] R&D in nanotechnology—nano-scale science, engineering and technology—deals with the manipulation of matter at the molecular level. Priority funding includes innovative nanotechnology solutions to biological–chemical–radiological–explosive detection and protection. Roco, M. C., 'National nanotechnology investment in the fy 2003 budget request', *AAAS Report XXVII: Research & Development FY 2003*, American Association for the Advancement of Science, 25 Mar. 2002, URL <http://www.aaas.org/spp/dspp/rd/03pch24.htm>.

[25] 'Security needs shape Ottawa's plans', *Financial Times*, 12 Dec. 2001, p. 4; 'Canada boosts budget for counterterrorism', *Defense News*, 17–23 Dec. 2002, p. 20; and 'Canadian defence budget gets short shrift', *Jane's Defence Weekly*, 19 Dec. 2001, p. 5.

[26] 'In Germany, more money for counterterrorism' *Defense News*, 15–21 Oct. 2001, p. 76.

[27] 'Terrorism, threats drive spending for Bundeswehr', *Defense News*, 3–9 Dec. 2001, pp. 32, 36.

[28] 'Interview: Rudolf Scharping, Minister of Defence', Germany, *Defense News*, 17–23 Dec. 2001, p. 54.

In its defence budget for FY 2003/2004 the United Kingdom made smaller extra allocations of £20 million ($28 million) for anti-terrorism measures, in addition to a supplement of £30 million ($42 million) for policing.[29] Longer-term planning is under way for new roles for the British armed forces in the light of 11 September. Work was initiated in 2001 on a new chapter to supplement the 1998 Strategic Defence Review, the blueprint for current British military strategy. It is expected to result in substantial additional future funding for anti-terrorism activities.[30]

Other countries may have made similar budget allocations for counter-terrorism but not in the form of separate allocations. Allocations may reflect different perceptions of the threat of terrorism—whether it requires primarily a military or a non-military response. France has taken a negative stand on the use of military means for combating terrorism, as illustrated by the statement of Defence Minister Alain Richard. He argued that the tools to fight terrorism were in the hands of the courts and the police rather than the military.[31] However, France has urged the defence ministers of the European Union to increase spending on the European Security and Defence Policy, with reference to the war on terrorism. While this proposal was rejected at an informal meeting of the EU defence ministers on 12 October 2001, the ministers declared their intention to review preparations for defence against terrorist threats to EU forces deployed in the future.[32]

The 18 December meeting of NATO defence ministers reiterated the alliance's resolve for zero tolerance of terrorism.[33] The NATO defence plan review manifested a consensus to increase the proportion of forces that can be deployed and sustained in operations beyond alliance territory. According to NATO Secretary General Lord Robertson, additional resources were required: 'The simple message from NATO defence ministers is this—you can't get defence on the cheap'.[34] In early 2002 he again criticized the low level of European spending in comparison with that of the USA: 'Too many governments spend too little on defence. And too many governments waste what they spend on capabilities that contribute nothing to their own security, the security of Europe or our wider collective interests. Smart investment is the only way to share the transatlantic burden'. He called for better homeland defence, better intelligence, more deployable civil police and more effective monitoring of money laundering.[35] The NATO stance on combating terrorism is likely to have a great impact on the defence debate in the NATO countries during 2002.

---

[29] 'Chancellor: pre-budget statement', *Defense News Analysis*, 3 Dec. 2001.
[30] British Ministry of Defence, '11 September—a new chapter for the Strategic Defence Review', 14 Feb. 2002, URL <http://moddev.dera.gov.uk/news/press/news_press_notice.asp?newsitem_id=1247>.
[31] 'Across Europe, defense spending falls', *Defense News*, 17–23 Dec. 2001, p. 44.
[32] 'EU rejects proposal for more spending on security, defence', *Jane's Defence Weekly*, 24 Oct. 2001, p. 3.
[33] NATO, 'Statement on combating terrorism: adapting the alliance's defence capabilities', Press release (2001) 173, 18 Dec. 2001, URL <http://www.nato.int/docu/pr/2001/p01-173e.htm>.
[34] 'NATO/bi-annual meeting: response to terrorism', *Atlantic News*, 20 Dec. 2001, p. 1.
[35] 'The transatlantic link', Speech by the NATO Secretary General at the Annual Conference of Defence and Society, Sälen, Sweden, 21 Jan. 2002, URL <http://www.nato.int/docu/speech/2002/s020121a.htm>.

IV. Regional trends in military expenditure[36]

Africa

Military expenditure in Africa has been rising steadily since 1996, after a period of fluctuation. The rise can be attributed almost entirely to domestic factors—internal armed conflicts, threats of new conflicts and continuing modernization programmes. However, some countries deviate from this pattern. In Ghana, the 40 per cent rise in military expenditure in 2000 was due primarily to salary increases for the members of the armed forces, which follows the trend of a general increase in public sector wages. In 2001, however, Ghana's defence allocation dropped by 37 per cent, in real terms, owing to a cut in the investment and administration budget of the military.[37]

Conflicts still account for the high and rapidly rising military expenditure in Burundi, the Democratic Republic of the Congo (DRC), Sierra Leone, Sudan and other states that are directly or indirectly involved in the conflicts, including Guinea, Rwanda and Uganda.

The economic and human burden of military activities in Africa is very high. The greatest burden is due to the armed conflicts in the region but is not fully reflected in official figures on military expenditure. Official figures tend to underestimate the level of military activities in the region for several reasons.

First, they reflect only those activities that are conducted and financed by states. They do not include the military activities of non-state actors, foreign military assistance from other states or foreign military presence in the form of peacekeeping forces. Thus, for example, the moderate rise in Sierra Leone, seen against the intensity of the war and rebel activities there, is due to the massive external assistance to its security sector from the UK and to the presence of UN peacekeepers.

Second, in several African states there is both non-deliberate and deliberate under-reporting of real spending on military activities. With the exception of flagrant cases, it is difficult to investigate or even estimate real spending in these countries, but a UN panel found that the official figures of Rwanda and Uganda have greatly under-reported their actual military expenditure.[38]

Third, many African countries have significant extra-budgetary revenues and expenditures.

Countries providing development assistance to African states involved in conflict try to impose ceilings on their military expenditure levels. Some recipient countries try to avoid the imposition of ceilings. Thus, in a letter to the British Government in 2001, Ugandan President Yoweri Museveni asked for

---

[36] For the countries included in the data for each region, see the notes to table 6A.1 in appendix 6A. For the armed conflicts mentioned in this section, see chapter 1 and appendix 1A in this volume.

[37] Republic of Ghana, 'Consideration of annual estimates', in *Parliamentary Debates: Official Report* (Accra), Fourth Series, vol. 28, no. 41 (28 Mar. 2001), pp. 2949–52.

[38] United Nations, Report of the Panel of Experts on the Illegal Exploitation of Natural Resources and Other Forms of Wealth in the Democratic Republic of the Congo, Press conference, United Nations, 16 Apr. 2001, URL <http://www.un.org/News/briefings/docs/2001/DRCPressCfc.doc.htm>.

its support to convince other donor countries to allow Uganda to raise its military expenditure above the officially agreed level of less than 2 per cent of GDP.[39]

A few countries—for example, Namibia and Zimbabwe—have reduced their defence budgets for 2001. This probably reflects their intention to end or reduce their military presence in the DRC. However, supplementary allocations, primarily to cover the costs of withdrawing their forces, especially in the case of Namibia, may result in a continued increase in these countries.[40]

In other countries—Botswana, Nigeria, Senegal and South Africa—high and rising military expenditure reflects large arms procurement programmes for the modernization of their armed forces. In Botswana, the arms procurement programme has resulted in a rising trend in military expenditure over the past decade and a 14.6 per cent increase in FY 2000/2001. The Nigerian Government continues to modernize its armed forces through training, arms purchases and refurbishment of plants for the production of small arms. The armed forces have also requested additional resources to finance the deployment of military forces for the prevention of internal violence, which has become increasingly rampant in 2000–2001. In this respect, the 42 per cent increase in the FY 2001 defence budget over the previous year reflects some of the challenges faced by the military. The explosion in 2001 in an ammunition dump in Lagos became a catalyst for increased military expenditure in Nigeria to finance the reconstruction of a large number of similar military facilities and general repair of military installations throughout the country. The Senate Committee on Defence has indicated its willingness to provide funding for this purpose in addition to other funds for the general repair of military installations. Senegal, under the government of President Abdoulaye Wade, is restructuring and increasing the salaries of its armed forces, reflected in its 14 per cent increase in military spending for 2001. The 15 per cent increase in South Africa's defence budget for FY 2001/2002 reflects its large arms procurement programme, which is expected to continue for the next 10–15 years.

In North Africa, there was a strong increase in military spending by Algeria and Morocco.[41] The major reasons for this rise are the civil war in Algeria and, in the case of Morocco, the Western Sahara crisis. The programme for modernization of the Algerian armed forces is also a factor in the rise of Algerian military expenditure over the decade.

The Middle East

Military expenditure in the Middle East has been rising since 1996, after reaching a low point in 1995.[42] The gradual but steady rise was maintained

[39] 'UK moves to prevent Uganda, Rwanda clash', *East African*, 22 Oct. 2001, URL <http://www.nationaudio.com/news/eastAfrican/29102001/Regional/Regional19.html>.

[40] Bank of Namibia, *Quarterly Bulletin*, Dec. 2001, p. 16, URL <http://www.bon.com.na>.

[41] Libya, one of the major spenders in North Africa, does not release figures on its military expenditure.

[42] Iraq, one of the major spenders in the Middle East, does not release figures on its military expenditure.

until 2000 and then accelerated in 2001, according to preliminary estimates for the year. The increase can be interpreted as the result of several factors. First, the Arab–Israeli conflict has motivated continued increases in the military expenditure of Israel and other states involved in the conflict, particularly Egypt and Syria. Second, Iran continues its armaments programme, which is reflected in rapidly rising military expenditure since 2000. Third, in 2001 the defence budgets of most of the Persian Gulf states have also begun to increase, after having been more or less stagnant since 1997. The rise in 2001 coincided with high oil prices throughout 2000 and the first three quarters of 2001. High oil revenue has facilitated increases in government spending, since it constitutes the major source of income for most of these states.

For the second consecutive year, in 2001 the Government of Israel introduced a supplementary budget for defence because of the increase in violence in the conflict with the Palestinians. The initial defence budget was increased to meet the rising military requirements brought about by the conflict. The 2001 allocation is 7 per cent above expenditure in 1999, the last year in which the country kept within the originally budgeted allocation to defence. Israel has one of the world's highest ratios of military expenditure to GDP. The defence budget for 2002, endorsed by the government in November 2001, shows an increase to 41.4 billion shekels ($9.8 billion). More than 20 per cent of the defence budget is financed by annual US military grant assistance of about $2.06 billion.[43]

All of the major oil-producing states in the Middle East have planned significant increases in their military expenditure for 2001, including Iran, Kuwait, Oman, Saudi Arabia and the United Arab Emirates (UAE).

Iran continued the steep rise in military expenditure that began in 2000 and in 2001 increased its military expenditure by about 26 per cent in real terms within the larger total government budget made possible by increased oil revenue. Iran continues to build an indigenous capacity for arms production and is forging stronger ties with Russia for the supply of conventional weapons.[44]

Although the defence minister of Kuwait, appointed in 2001, promised a reduction in Kuwait's arms procurement—and more transparency in military spending—the country still renewed its long-term agreement for security and military supplies with the United States in February 2001.[45] The implication of this is a continuous rise in military spending.

Oman increased its military spending in 2001 by 39 per cent in real terms, well above the increase in overall government expenditure of 15 per cent. It has resumed the previously suspended programme to modernize its armed forces. The air force is the greatest beneficiary of the new programme.[46]

---

[43] 'Limited war forces Israel to boost defense budget: extra funds earmarked for readiness, modernization', *Defense News*, 26 Nov.–5 Dec. 2001, p. 6, available at URL <http://www.defensenews.com>.

[44] See chapter 8 in this volume for information on transfers of major conventional weapons.

[45] Economist Intelligence Unit, *Country Report: Kuwait*, Mar. 2001.

[46] Deen, T., 'War threatens to trigger arms race', *Dawn* (Internet edn), 14 Oct. 2001, URL <http://www.dawn.com/2001/10/14/int12.htm>; and 'US DOD approves Omani F-16 sale', *Jane's Defence Weekly*, 17 Oct. 2001, p. 15.

The exact size of the increase in Saudi Arabia's military expenditure is not known because figures for defence are not provided in the government budget. However, judging on the basis of the category 'other expenditure', which includes military allocations,[47] the increase is likely to have been almost 30 per cent in real terms. This steep rise reflects the major rise in oil revenue, which accounts for about 70 per cent of total government revenue. Although Saudi Arabia is believed to prioritize the settling of its domestic debt and providing jobs for its increasing number of unemployed,[48] talks have also resumed on the replacement of its ageing fleet of F-5 military aircraft and increasing its fleet of F-15s.[49]

In 2001 the UAE continued its practice of presenting a constant figure for military expenditure: the figure for 2001 is identical to the figure it has presented for the past seven years. In reality, however, Abu Dhabi, the richest emirate in the union, has a large arms procurement programme that has not been included in the figure for total union expenditure.[50] The UAE has ordered new military aircraft estimated at a value of $11 billion, including an order for 80 F-16 combat aircraft from the USA due for delivery from 2004.

Central Asia

Official figures for the Central Asian states indicate a significant increase in the military expenditure of the region over the period 1998–2001. It is difficult to obtain precise information on the rate of increase and overall level of their military spending because it is not known what is included or how reliable the official figures are. While there has been a clear improvement in the availability of official data from some of the states,[51] transparency is still low, which makes it difficult to assess the extent to which the official figures reflect the real cost of military activities in these countries.

Except in Turkmenistan, official military expenditure accounts for about 1 per cent of GDP in the Central Asian states, which is a comparatively low defence burden. This could be explained by two factors: (*a*) the reliance in several of these countries on paramilitary forces, the cost of which is not included in official military expenditure (e.g., Kazakhstan and Uzbekistan); and (*b*) their cooperation with and military assistance from Russia and to a lesser extent from some Western countries, including France, Germany, Turkey and the USA. Against the background that at least the two major powers in the region—Kazakhstan and Uzbekistan—are engaged in major military reform programmes to build up modern and professional armed

---

[47] Economist Intelligence Unit, *Country Report: Saudi Arabia*, Feb. 2001, p. 16.
[48] 'Surge in Saudi oil revenue will not flow to arms buys', *Defense News*, 7 May 2001, pp. 1, 28.
[49] Economist Intelligence Unit, *Country Profile 2001: Saudi Arabia*, p. 13.
[50] Economist Intelligence Unit, *Country Profile 2001: United Arab Emirates*, p. 14.
[51] Kazakhstan and Kyrgyzstan responded to the SIPRI request for data in 2001 with completed questionnaires on their disaggregated military expenditure. Uzbekistan reported to the United Nations in 2000. See appendix 6D.

forces, the defence burden is probably significantly greater than official figures indicate.[52]

Military expenditure figures for 2001 are available for only two states, Kazakhstan and Kyrgyzstan. The extraordinary rise of 42 per cent in Kazakh military spending reflects a shift in its threat perceptions, military doctrine and procurement pattern since 2000. The focus of its threat perceptions has changed from a large-scale war with China to low-intensity conflict with non-state Islamic groups operating in Afghanistan, Kyrgyzstan and Uzbekistan. Therefore, under the Kazakh military doctrine adopted in February 2000, emphasis is placed on force mobility, border protection, and command, control, communications and intelligence (C^3I) equipment. The structure of the defence budget has changed accordingly: a significantly increasing share of Kazakh expenditure is for arms acquisitions, both in the adopted defence budget for 2001 and even more so in the planned budget for 2002.[53]

South Asia

Military expenditure in South Asia, with some of the poorest countries of the world, continued to rise in 2001. With a 54 per cent increase in real terms over 10 years and a 26 per cent increase over the period 1998–2001, South Asia is one of the regions with the most rapidly increasing military expenditure. This can be interpreted as the effect of conflicts and the tense security situation in the region, including the conflicts between India and Pakistan over Kashmir and Jammu, the civil war in Sri Lanka, and border problems between Bangladesh and neighbouring India and Myanmar.

The defence budgets for 2001, adopted in late 2000, increased in India and Nepal, while the budgets of Bangladesh and Pakistan were comparatively flat in real terms. India's adopted defence budget for FY 2001/2002 was 14 per cent higher than actual expenditure for FY 2000/2001. The Nepalese defence budget for 2001 shows an increase of 10 per cent in real terms over 2000.

East Asia

Military expenditure in East Asia increased by 19 per cent over the 10-year period 1992–2001. The adopted defence budgets for 2001 show a combined increase of 5 per cent in real terms for the region. The major spenders in the region are China and Japan, with 38 and 27 per cent, respectively, of the regional total in 2001, followed by South Korea and Taiwan, with 10 and 7 per cent, respectively.

---

[52] For more detail on the coverage and reliability of official military expenditure for each of the Central Asian states see Eaton, M., 'Major trends in military expenditure and arms acquisitions by the states of the Caspian region', ed. G. Chufrin, SIPRI, *The Security of the Caspian Sea Region* (Oxford University Press: Oxford, 2001), chapter 5.

[53] Pukhov, R., 'Transfers of Russian arms into Central Asia and the regional military balance', Unpublished paper presented at a SIPRI/Partnership for Peace workshop, 29–30 Nov. 2001.

The level of Japan's military expenditure continued to remain virtually flat—the defence budget for FY 2001/2002 represented a growth of less than 1 per cent in real terms—at less than 1 per cent of GDP. It nonetheless funds a major arms procurement programme, as outlined in the Mid-Term Defense Program for FYs 2001–2005, adopted in December 2000.[54]

In South Korea the military build-up continues. The proposed defence budget for 2002 was set to increase by 6.3 per cent to 16.4 trillion won ($12.5 billion).[55] The mid-term defence plan for 2002–2006 includes 91.9 trillion won ($70 billion) in combined military expenditure over the five-year period, of which 34.5 trillion won ($26.5 billion) is for modernization projects.[56] The planned acquisitions include an airborne warning and control system (AWACS), Aegis Class destroyers and indigenous combat aircraft. The share of 'force improvement programmes'—the official South Korean term for arms procurement—in total military expenditure is scheduled to grow from 34 per cent in 2001 to 40 per cent in 2006.

China

The rate of increase in China's official military expenditure has been sustained at a high level since 1995. Over the period 1995–2000 it increased at an annual average rate of almost 11 per cent in real terms. In 2001 the increase was 16.2 per cent (to 141.04 billion yuan, or renminbi). The defence budget for 2002 involves an increase of 17.6 per cent (to 166 billion yuan/renminbi).[57] Since the inflation rate is less than 1 per cent, the real increase is virtually the same. Since 1998, when the government banned the business activities of the People's Liberation Army (PLA), part of the increase has been in compensation for lost revenues from these activities. Thus, the actual increase in China's military resource consumption is somewhat lower than these figures suggest. The revenues that the PLA previously obtained from its commercial activities have been estimated as roughly 10 per cent of the official defence budget, or about 11 billion yuan in 1998.[58] Thus, if the shift from commercial revenues to state financing was completed by 2002, as reported,[59] the PLA lost 11 billion yuan in revenues over the four years 1999–2002.

The official reason for the increase in the 2002 national defence budget, announced on 6 March 2002, was four-fold: (*a*) to safeguard national sov-

[54] Japan Defense Agency, *Defense of Japan 2001* (Japan Defense Agency: Tokyo, 2001), described in Sköns *et al.* (note 22), pp. 254–55.
[55] *Yonhap* (Seoul), (in English), 25 Sep. 2001, in 'ROK's Yonhap: S. Korea to kick off AWACS project next year', in Foreign Broadcast Information Service, *Daily Report–East Asia (FBIS-EAS)*, FBIS-EAS-2001-0925, 26 Sep. 2001.
[56] *Korea Herald* (Seoul), (Internet edn, in English), 29 June 2001, in 'Further on ROKG defense budget for next 5 years', in FBIS-EAS-2001-0628, 29 June 2001.
[57] 'Military & armed forces', *China Today* (Internet edn), URL <http://www.chinatoday.com/arm/>.
[58] Wang, S., 'The military expenditure of China, 1989–98', *SIPRI Yearbook 1999: Armaments, Disarmament and International Security* (Oxford University Press: Oxford, 1999), pp. 334–49.
[59] 'China threat theory collapses of itself, as military spending of both US and Japan has far exceeded that of China', *Wen Wei Po* (Hong Kong), (in Chinese), 6 Mar. 2002, p. A6, in Foreign Broadcast Information Service, *Daily Report–China (FBIS-CHI)*, FBIS-CHI-2002-0306, 11 Mar. 2002.

ereignty; (b) to adapt to changes in the international situation; (c) to raise the technological level of the armed forces; and (d) to increase military pay, allowances and pensions.[60] Reference was also made to the increase in the US military budget and in particular to the US national missile defence system.[61] There is a continued effort to address the problem of the salaries of Chinese military personnel, which have been lagging behind the general salary level.[62] The effort to raise the technological level of military equipment for the Chinese armed forces is a long-term programme, part of the 'four modernizations' launched in 1988, but in the 2001 and 2002 defence budgets the emphasis was on the need to enhance combat capability in the 'context of high technology'. An additional factor influencing the growth in Chinese military expenditure is affordability—military expenditure takes a low share of GDP, both officially (1.4 per cent in 2001) and according to most Western estimates (2.1 per cent according to SIPRI).

Western estimates of Chinese military expenditure are higher than the official Chinese figures, partly because they include estimates of a number of military-related expenditure items that are known or believed not to be included in the official figures. The major items usually added in Western estimates of the official Chinese defence budget are arms imports, additional military R&D, and extra-budgetary expenditures financed by revenues from the commercial activities of the PLA.

Some of the lack of detail and transparency in the Chinese defence budget may be the result of a weak budgeting system. Efforts are under way to improve this system. In 2001 the Central Military Commission adopted a new budgeting system, introducing a 'zero-base' budgeting system for the defence budget, which centralizes the allocations of funds and revenues.[63] The purpose of the new 'Scheme for Reforming the Drawing up of the Armed Forces Budget' is to modernize the budgeting process and the management of military expenditure by introducing more transparency and control or, as expressed by the director of the PLA finance department, to 'gradually build a new military budget model with concentrated finance and financial powers, scientific allocation of military funds, specifically transparent budget items, and tight supervision and constraint, and make efforts to explore a new road of relatively high returns for relatively low input'.[64]

---

[60] 'Increasing military expenditure is normal and necessary', *Ta Kung Pao* (Hong Kong), (Internet edn, in Chinese), 7 Mar. 2002, in 'PLA NPC deputies explain defense budget increase', FBIS-CHI-2002-0307, 8 Mar. 2002.

[61] *Wen Wei Po* (note 59).

[62] 'Sun Zhiqiang discloses most of China's added defense budget to be used mainly to increase soldiers' wages, benefits', *Zhongguo Xinwen She* (Beijing), (in Chinese), 7 Mar. 2002, in 'PLA senior official says added defense budget to be used on soldier's welfare', FBIS-CHI-2002-0307, 8 Mar. 2002.

[63] 'China reforms budget system', *Jane's Defence Weekly*, 3 Oct. 2001, p. 17; 'PRC plans reform of army's purchasing system', Xinhua Domestic Service (Beijing), (in Chinese), 9 Jan. 2002, in FBIS-CHI-2002-0109, 11 Jan. 2002; and 'Bearing overall situation in mind, making concerted efforts to promote reform', *Jiefangjun Bao* (Beijing), (Internet edn, in Chinese), 10 Jan. 2002, in 'Jiefangjun Bao commentator on reform of PLA procurement system', FBIS-CHI-2002-0110, 11 Jan. 2002.

[64] Interview with Ding Jiye, Director of the Finance Department of the General Logistics Department in 'PLA logistics officer on reform of drawing up military budget', *Jiefangjun Bao* (Beijing), (Internet edn, in Chinese), 30 Apr. 2001, in FBIS-CHI-2001-0430, 2 May 2001.

South America

Data on military expenditure in South America are difficult to interpret and analyse for several reasons. Some countries are relatively transparent and report detailed expenditure data in the SIPRI questionnaire, while others do not respond, and official data are impossible to obtain (see appendix 6D). With these reservations, the level of military expenditure in this region can be described as relatively flat and comparatively low in terms of the share of GDP over the period 1997–2001. Only Chile's share of GDP is higher than the world average of 2.6 per cent (appendix 6A, table 6A.4).

Brazil's defence budget for 2001, which accounts for roughly half of the regional total at 19.5 billion reais ($14 billion), shows a considerable increase over actual expenditure of 14 billion reais ($11 billion) in 2000. However, this increase is most likely to be revised into a roughly constant trend in actual military expenditure, as was the case in 1999 and 2000.

Countries in the region increasingly face common threats, particularly internal unrest, drug trafficking and organized crime, while the risk of conflict between them is low. This is also reflected in the composition of US military aid to the region. Of a total of $2130 million in US military aid over the period FY 2000–2002, aid for International Narcotics Control (INC) amounted to $2.1 billion, primarily for Bolivia, Colombia, Ecuador, Mexico and Peru. The purpose of INC aid is to prevent drugs from entering the USA, dismantle drug cartels, and reduce and ultimately eliminate drug crop cultivation.

Brazil and Chile are embarking on programmes for the acquisition of major conventional weapons. To the extent that payments will be reflected in the official military budgets, they will result in increased military expenditure in the future. Chile decided in February 2002 to purchase 10 F-16 combat aircraft from the USA in a deal worth $660 million. This is the first major US sale to South America since the 1970s, when the USA initiated a restrictive policy on arms sales to the region. There is a risk that this will have an impact on military expenditure in neighbouring countries, setting off a regional arms race.[65] Chile's expenditure on arms procurement is not included in its official military expenditure. According to the 'reserved copper law' of 1958, the state-owned copper company CODELCO sets aside 10 per cent of its annual sales revenues for the financing of military purchases for the armed forces. When President Ricardo Lagos assumed office in March 2000, he promised to review this arrangement of off-budget allocations for military purchases. With the purchase of combat aircraft, the government may have locked this arrangement in place until 2009, when the payment schedule ends for these aircraft.[66]

---

[65] 'Chile says will buy US fighters, denies arms race', *Air Letter*, 1 Feb. 2002, p. 5.

[66] For a brief description of this method of financing see 'Chile's defence policy copper-bottomed; a very strange way to pay for an army, a navy or an air force', *The Economist*, 8 Feb. 2002, p. 47. See also Rojas Aravena, F., 'Chile', ed. Singh, R. P., SIPRI, *Arms Procurement Decision Making, Volume II: Chile, Greece, Malaysia, Poland, South Africa and Taiwan* (Oxford University Press: Oxford, 2000), pp. 9–38.

Western Europe

Military expenditure in Western Europe has been roughly flat since the mid-1990s, when the post-war decline in spending ended. In 2001 there was a small reduction again, after two years of slight real growth (table 6.5). The reduction in 2001 is due primarily to the declining defence budgets of Germany and Italy, while those of France and the UK were flat. However, actual expenditure for 2001, in contrast to budgeted spending, may show an increase in real terms—as was the case in 2000, when military budgets were reduced[67] but actual spending increased in real terms. Furthermore, the terrorist attacks of 11 September and the subsequent war on terrorism have led to supplemental allocations in some European countries, which will also result in an upward shift in actual military expenditure. However, neither of these factors is sufficiently strong to involve a significant increase in actual West European military expenditure in 2001.

West European governments are under great pressure to raise their military expenditure, but there is strong public resistance to tax increases in these countries. While the terrorist attacks on the United States brought a major change in US public opinion on military defence, this has not been the case in Europe.

The dynamics of the restructuring of military forces in many European countries are changing. During the early post-cold war years, restructuring was driven by the need to downsize and adapt to the more benign security environment and by the emergence of new tasks associated with peace operations. While adaptation to new tasks, in particular those of peace operations, continues to be an important objective, the rationale and aim of current restructuring programmes for European military forces are increasingly framed in terms of shortfalls in their military capabilities, experienced in the operations in the Balkans. To a great extent they involve the same issues that were previously on the NATO agenda—harmonization of forces and standardization of military equipment. Current restructuring programmes are focused on three areas: (a) restructuring of the armed forces to increase their mobility; (b) a change in recruitment system, from conscript to professional forces; and (c) modernization of military equipment. The key word is interoperability, which to a great extent means adaptation of European forces to US military technology and equipment.

While the transatlantic capability gap is often illustrated by the gap in military expenditures, in particular spending on equipment and military R&D, the major issue concerns the lack of interoperability, in particular in advanced C^3I technology. This has been the conclusion primarily from experience in joint operations in the Balkans.[68] In the US-led war in Afghanistan, 'Operation Enduring Freedom', the USA therefore had less interest in joint operations

---

[67] Sköns et al. (note 22), table 4.7, p. 244.
[68] As described in a number of reports, including US General Accounting Office (GAO), *European Security: US and European Contributions to Foster Stability and Security in Europe*, GAO-02-174 (GAO: Washington, DC, Nov. 2001).

Table 6.5. West European and NATO military expenditure, 1992–2001[a]

Figures are in US $b., at constant (1998) prices and exchange rates. Figures may not add up to totals because of the conventions of rounding.

	1992	1993	1994	1995	1996	1997	1998	1999	2000	2001
Total W. Europe	201	194	189	179	180	179	179	182	183	181
NATO Europe	187	180	175	167	168	166	166	174	175	174
NATO W. Europe	187	180	175	167	168	166	166	169	170	168
New members	–	–	–	–	–	–	–	5	5	6
Non-NATO W. Europe	14	13	13	12	12	13	13	13	13	13
European Union	184	178	172	172	173	172	173	175	177	175
France	44	44	44	42	41	41	40	40	40	40
Germany	42	38	35	35	34	33	33	34	33	32
Italy	22	22	22	20	22	23	23	24	26	25
UK	45	44	42	39	39	37	37	37	37	37
Total NATO	557	533	508	481	466	462	457	467	478	472
USA	355	336	317	298	282	291	274	275	286	281

[a] There are 2 breaks in the NATO series. The first break is in 1999, when 3 states joined NATO. Therefore, a separate series is shown for the new members: the Czech Republic, Hungary and Poland. The second break is in 2001, when the UK changed its accounting system for military expenditure from 'cash basis' to 'resource basis', but this is not likely to have a significant impact on the data.

Source: Appendix 6A, tables 6A.1 and 6A.3. SIPRI data for NATO member states are based on NATO data. Military expenditure, as defined by NATO, may diverge significantly from nationally defined defence budgets.

with its NATO allies. This has evoked two types of reaction among European states: (*a*) a fear of US unilateralism; and, as emphasized by the arms industry, (*b*) concern about the gap in military industrial capability.

It has been argued that one reason for the lack of resolve on the issue of the transatlantic capability gap is that both sides also benefit from the status quo.[69] It provides the USA with an argument for taking the lead without strong European interference in decision making, and it allows European states to set other budget priorities because the extra expenditure required to match the US effort would be enormous and thus seriously jeopardize their commitments. On the other hand, if Europe is to be able to conduct military operations without US commitments, it must have an autonomous technological capability.

The defence debate in Western Europe during 2001 continued to focus on the question of the appropriate level of military expenditure. The answers differed widely, primarily depending on two basic and sometimes interrelated views on: (*a*) the extent to which European security and defence policy should rest on military strength; and (*b*) the extent to which Europe should have an autonomous military capability. The pressure for increased military expendi-

[69] 'Nato's welcome imbalance in military might', *Financial Times*, 7 Feb. 2002, p. 11.

ture in West European countries originates in these questions and is expressed primarily in terms of: (*a*) the sharing of the defence burden within NATO; (*b*) the interoperability of US and European forces in joint NATO military activities; and (*c*) the build-up of a European rapid reaction force. In response to these requirements, European governments have made commitments in terms of future military capabilities which in some countries go beyond what can realistically be expected to be funded by future defence budgets. As a result, many European governments are trying to introduce new and innovative financing mechanisms for military activities.

The transatlantic gap in military capability

Burden sharing within NATO has been an issue since the foundation of the alliance. While the context of the debate changed after the cold war, the issue of burden sharing came into focus again after the NATO involvement in the wars in the former Yugoslavia and the adoption of the NATO Defence Capabilities Initiative (DCI) in April 1999. The 11 September terrorist attacks in the United States raised the issue of NATO burden sharing to a level of urgency.

The distribution of the defence burden can be perceived and measured in several ways, depending on the weight assigned to military capability in the defence concept. Six different measures were used by the US Congressional Budget Office in a 2001 report on NATO burden sharing.[70] They included three standard measures of military expenditure—as a share of GDP, per capita spending and the proportion of the workforce in the military—and three new measures that were meant to be more applicable to the post-cold war environment—contributions to NATO's rapid reaction forces, contributions to the peacekeeping forces in Bosnia and Herzegovina and Kosovo, and economic assistance to Central and Eastern Europe. The results of the CBO study are summarized in table 6.6. While US military expenditure is higher in terms of the ratios of GDP and population, the gap has narrowed. Moreover, the gap reflects the United States' global security interests in addition to its contributions to NATO. As regards specific contributions to NATO peacekeeping operations and to economic aid, the European allies are taking on more than a proportional share of the burden.

Another study, conducted by the US General Accounting Office (GAO) in 2001, concluded that the post-cold war security environment has brought new requirements for fostering security—smaller military forces, heavier reliance on non-military contributions such as development assistance, and increased reliance on multilateral organizations for the provision of security. In these areas, the study concluded that the European countries had made significant contributions. In spite of the reductions in their forces, they have been actively engaged in peacekeeping and other security-enhancing activities in Europe,

---

[70] US Congressional Budget Office (CBO), *NATO Burden Sharing After Enlargement* (CBO: Washington, DC, Aug. 2001), available at URL <http://www.cbo.gov/showdoc.cfm?index=2976&sequence=1>.

Table 6.6. Burden sharing within NATO

Measure	Result
Military expenditure as a share of GDP	With the exception of Greece and Turkey, the USA spends a larger share of its GDP on defence than its European allies, but the gap has narrowed significantly since 1985.
Military expenditure per capita	The USA spends more per capita than any of its allies. This gap reflects the global nature of US security interests.
Proportion of the labour force in the military	Several European allies make a larger contribution to the common defence than the USA does. In 1999, 1.6% of the US workforce was employed in the military sector, compared with 1.7% for the European allies.
Contributions to NATO's reaction forces	The UK has been in the forefront of these efforts. France has also developed forces that augment NATO's capability. Germany plans to develop a significant force.
Peacekeeping missions	Many of the European allies bear more than their proportional share of the burden in Bosnia and Herzegovina and Kosovo. They also took on a larger share of the burden of the Kosovo Force (KFOR) peacekeeping mission. The USA maintained (in June 2000) the largest national presence in Kosovo, but the majority of the ground troops were European.
Economic aid to Central and East European countries	Many of the European allies shoulder at least a proportional share of these costs. Germany has provided more aid to the region than any other country.

Source: US Congressional Budget Office (CBO), *NATO Burden Sharing After Enlargement* (CBO: Washington, DC, 2001), Summary, URL <http://www.cbo.gov/showdoc.cfm?index=2976&sequence=1>.

and they have provided about $47 billion of a total of $71 billion in disbursed assistance to CEE countries over the period 1990–99.[71] While total US military expenditure is higher than European expenditure, the US cost of supporting its military presence in European NATO countries in 2000 was estimated at $11.2 billion, a 50 per cent reduction since 1990. The shortcomings of European countries were in specific military capabilities, such as mobility of forces and the technological level of their equipment.

These two studies show the difference in the emphasis of the role of military strength in security building between Europe and the USA.

The European rapid reaction force

The creation of a European rapid reaction force (ERRF) within the development of the EU European Security and Defence Policy will require additional economic resources. Some progress was made in 2001 in efforts to allocate the required resources, as stipulated in the Headline Goal set in Helsinki in

---

[71] US General Accounting Office (note 68), p. 5.

December 1999. At the EU Capabilities Improvement Conference (CIC) in November 2001, it became clear that about two-thirds of the required capabilities for the ERRF were secured. They include a pool of more than 100 000 troops, around 400 combat aircraft and 100 ships ready to be deployed for the ERRF.[72] Out of the remaining 20 shortfalls, 15 are reportedly addressed also in the NATO Defence Capabilities Initiative. This was subsequently confirmed at the European Council summit meeting in Laeken, Belgium, on 14–15 December, when the ERRF was declared operational.[73]

While the DCI and ERRF capabilities have been seen as overlapping, a perception seems to be increasing that Europe needs to have the necessary autonomous capacity in the event the USA is not involved. This also requires greater financial commitments. The purpose of the European Capability Action Plan (ECAP), adopted at the November CIC meeting, was to optimize the use of available resources by way of cooperation. Its first option for improving existing resources for the development of European military capabilities is to make available more national forces and capabilities to the ERRF. Its second option is to employ multinational solutions for the production, financing, acquisition and management of military capabilities.

The cooperative procurement of A400M military transport aircraft

The main multilateral European procurement project is the A400M military transport aircraft, to be negotiated and managed by the joint procurement agency Organisme Conjoint de Coopération en Matière d'Armement (OCCAR). The project is seen as important to Europe's plans to create a rapid reaction force that can deploy without relying on US air-lift capacity. Furthermore, if agreed, this programme, at a total value of $16 billion, would be the most important joint European procurement programme administered by OCCAR.

Originally, eight countries—Belgium, France, Germany, Italy, Luxembourg, Spain, Turkey and the UK—agreed to procure a total of 212 A400M aircraft. However, in June 2001 Italy refrained from signing the first Memorandum of Understanding for 16 aircraft, and in November Germany announced that it had problems in financing its purchase of 73 aircraft.[74] The German defence budget for 2002, as approved by the Bundestag (parliament), included only DM 10 billion ($4.5 billion) for the A400M programme, which would finance the procurement of only 40–50 aircraft.

Germany's approval is decisive for the entire programme, because its order for 73 aircraft is the largest of those of the seven nations involved, and Airbus

---

[72] European Union, *Statement on Improving European Capabilities*, EU's General Affairs Council with the participation of the ministers for defence of the EU, Brussels, 19 Nov. 2001, URL <http://www.defense-aerospace.com/data/verbatim/data/ve236/index.htm>. See also 'EU resolves two-thirds of gap in capabilities', *Defense News*, 26 Nov.–2 Dec. 2001, p. 8; and 'EU goals mean increased defense spending', *Defense News*, 17–23 Dec. 2001.

[73] Declaration of the Operational Capability of the Common European Security and Defence Policy, Laeken, Belgium, 14–15 Dec. 2001, available at URL <http://europa.eu.int/futurum/documents/offtext/doc151201_en.htm>. See also chapter 3 in this volume.

[74] 'Germany, European allies compromise on A400M', *Defense News*, 17–23 Dec. 2002, p. 10.

has declared that the minimum total order is 180 aircraft. Thus, there is a risk that the project could fail if Germany procures only 40–50 aircraft.

Pressured by the governments of France and the UK, the German Government was able to reach an agreement with leaders of parliament that Germany would sign the contract for 73 aircraft, under the condition that parliament would approve additional funding in the future.[75] On 18 December German Defence Minister Rudolf Scharping signed the order for 73 aircraft with this reservation. On the same day OCCAR signed an order for 196 A400M military transport aircraft to be built by the EADS subsidiary Airbus Military Company, but with the reservation that the start of production is contingent on approval by the Bundestag. With the German defence budget frozen until 2006, Scharping has been requested by the parliamentary opposition to explain how he would cover preliminary costs through private funding mechanisms. This is likely to be a focus of the German and European defence debate in 2002.

Private financing of military capabilities

Other ways to resolve the current mismatch in Europe between planned military capabilities and planned funding for defence include various forms of partnerships between government and private industry. The concept of private financing of military activities originates in the UK as an arms procurement technique under the heading of Smart Acquisition. Currently, Private Finance Initiative (PFI) techniques are being explored and employed in other European countries. In PFI arrangements, a private company is contracted to fund and supply equipment for the defence ministry, while the defence ministry buys only the associated services, not the equipment itself. The idea behind the concept is that the private partner company is driven by its commercial need to service the capital it has raised privately, and at the same time the company can benefit from the ability to create and exploit non-military capacity as part of the project.[76] The perceived benefit for the defence ministry is that the cost is shifted from the design, development and production phase, as in ordinary arms procurement programmes, to a later stage, when the service is provided.

The British PFI programme includes contracts for accommodation, housing, information systems, utilities, training facilities and equipment. By October 2001, a total of 37 PFI deals had been signed, resulting in some £32 billion in private sector investment in military projects.[77]

A major project due for a PFI contract decision in 2002 is the Future Strategic Tanker Aircraft (FSTA) for the British Royal Air Force, planned to be operational from 2006. With a contract value of £13 billion over a 25-year life

---

[75] Nicholl, A., 'Private funding's biggest test is to come', *Financial Times*, 13 Dec. 2001.
[76] British Ministry of Defence, 'Public private partnerships in the MoD: MoD's approach to the Private Finance Initiative', 20 Dec. 2001, URL <http://www.mod.uk/business/pfi/intro.htm>.
[77] Nicholl (note 75).

cycle, this programme will become the first major aerospace procurement programme to be run under PFI auspices. The RAF will not own the aircraft it operates; they will be owned and managed by the 'service provider', that is, the contracted private company. Under the arrangement, the contractor will provide the aircraft, fully maintained and 'fuelled ready to fly', and on completion of a mission the RAF will hand it back to the company. At the same time, it allows the service provider to use the spare capacity of the aircraft to generate revenue by leasing it to third parties, thus offsetting the cost of the capability to the British Ministry of Defence.[78]

The arms industry sees PFI contracts with the defence ministry as an important emerging area of future business. This is reflected in their creation of specialized divisions and even companies for the management of PFI contracts. In early January 2002, BAE Systems announced the establishment of a new subsidiary, BAE Systems Capital Limited, to provide 'innovative financing options'. The company will concentrate on military contracts for the procurement of services under privately financed long-term supply contracts. According to company estimates, the annual value of such contracts may amount to £4 billion in the UK alone.[79]

Russia

Russian military expenditure fell sharply in 1992 and continued to decline until 1998. With the improvement of the national economy and fiscal discipline and with the reorientation of defence and defence industrial policy, Russian military expenditure has been increasing in real terms since 1998. The sharpest increase was in 1999, the first year in which the adopted budget for national defence was nearly fully implemented. Since then, the growth in spending has slowed down somewhat and the budget for 2002 shows negligible growth.

Over the period 1998–2001, official Russian expenditure on national defence, excluding pensions, increased from 56.7 billion roubles to 246.7 billion roubles (table 6.7), an increase of about 60 per cent in real terms. In reality, the increase may have been smaller, about 52 per cent, because of the change in the definition of 'national defence' in 2001. The federal budget for 2002, as signed into law by President Vladimir Putin on 31 December 2001, allocated 284.158 billion roubles for national defence, an increase of less than 1 per cent in real terms compared with actual expenditure of 246.7 billion roubles in 2001.[80]

---

[78] British Ministry of Defence, 'MOD celebrates smart acquisition successes', 1 May 2001, URL <http://www.defense-aerospace.com/data/communiques/data/2001May5486/index.htm>.
[79] BAE Systems, 'Establishment of BAE Systems Capital announced', Press release, 9 Jan. 2002, URL <http://www.defense-aerospace.com/data/communiques/archives/2002Jan/data/2002Jan8465/index.htm>.
[80] Its nominal increase of 15.2% is only slightly greater than the forecast inflation of 14.5% (as measured by the GDP deflator).

Table 6.7. The Russian Federation: military expenditure, 1992–2002[a]
Figures in italics are percentages.

	National defence[b] (m. current roubles)	Total military expenditure[c] (m. current roubles)	Total military expenditure[c] (constant 1998 US $b.)[d]	GDP (b. current roubles)	National defence as % of GDP	Total military expenditure as % of GDP
1992	855	1 049	80.4	19.1	*4.5*	*5.5*
1993	7 213	9 037	70.9	171.5	*4.2*	*5.3*
1994	28 500	35 890	68.6	610.7	*4.7*	*5.9*
1995	49 600	63 220	43.4	1 540.5	*3.2*	*4.1*
1996	63 891	82 485	39.5	2 145.7	*3.0*	*3.8*
1997	79 692	105 034	42.2	2 478.6	*3.2*	*4.2*
1998[e]	56 704	. .	. .	2 741.1	*2.1*	. .
1998[f]	68 004	85 574	30.6	2 741.1	*2.5*	*3.1*
1999[e]	115 594	. .	. .	4 766.8	*2.4*	. .
1999[f]	134 412	165 477	35.9	4 766.8	*2.8*	*3.5*
2000[e]	191 728	. .	. .	7 302.2	*2.6*	. .
2000[f]	213 488	[262 800]	[40.3]	7 302.2	*2.9*	*[3.6]*
2001[e]	246 700	. .	. .	9 040.8	*2.7*	. .
2001[f]	273 500	[335 600]	[43.9]	9 040.8	*3.0*	*[3.7]*
2002B[e]	[284 158]	. .	. .	10 950.0	*[2.6]*	. .
2002B[f]	[324 432]	[439 500]	[50.3]	10 950.0	*[3.0]*	*[4.0]*

[a] Figures show actual expenditure if not otherwise indicated. B = budget as first adopted and signed into law. There is a series break in 2001 because of the change in the definition of 'national defence' in that year. If the figure for 2001 is adjusted to the 2000 definition, the figure for 2001 is lower by about 22.6 billion roubles. This difference does not have an impact on the figure for military expenditure, since it covers all items.

[b] Military pensions were included in the budget chapter 'national defence' before 1998. From 1998 this table also provides a figure for national defence including pensions.

[c] Total military expenditure (the SIPRI figure) includes military pensions and military-related items under other budget chapters, such as expenditures for paramilitary forces and military research and development.

[d] Constant dollar figures are in PPP terms with 1998 as the base year.

[e] Excluding military pensions.

[f] Including military pensions.

Sources: Julian Cooper, Centre for Russian and East European Studies, University of Birmingham, using: **1992–96:** Cooper, J., 'The military expenditure of the USSR and the Russian Federation, 1987–97', *SIPRI Yearbook 1998: Armaments, Disarmament and International Security* (Oxford University Press: Oxford, 1998), appendix 6D; **1997–1999:** 'On the execution of the Federal Budget to 1 January', Russian Ministry of Finance (URL <http://www.minfin.ru>); **2000:** 'On the execution of the Federal Budget to 1 January 2001', Russian Ministry of Finance (URL <www.minfin.ru>); actual expenditure on military pensions is calculated on the basis of the pensions share of the budget item 'Social policy'; **2001:** (URL <http://www.minfin.ru>); **2002B:** Table 6.8, and Federal Law 'On the Federal Budget for 2002' as adopted by the president on 1 Feb. 2002, Russian Ministry of Finance Internet site URL <http://www.minfin.ru>, also available on the Internet site of *Rossiyskaya Gazeta* URL <http://www.rg.ru>; and PPP rate for **1998:** World Bank, *World Development Indicators 2000* (World Bank: Washington, DC, 2000), pp. 280–82.

Table 6.8. The Russian defence budget for 2002, as adopted on 30 December 2001
Figures are in million roubles and current prices. Figures in italics are percentages.

Budget item	Allocation (m. roubles)
National defence	
Ministry of Defence	263 864
Current maintenance and training	164 605
Procurement, R&D construction	99 259
Minatom (nuclear weapons)	13 994
Mobilization and training (outside the forces)	3 270
Collective security/peacekeeping	2 728
National security provisions by branches of economy	303
Total national defence	**284 158**
Military pensions	40 274
National defence including pensions	**324 432**
Other military expenditure	
Fund for support of military reform	16 545
Paramilitary forces	
Interior troops	13 571
Border troops	17 558
Security services	31 813
Total paramilitary forces	62 943
Military 'science' (est. 40% of total)	12 127
Liquidation of weapons, including fulfilment of international arms agreements	10 315
Mobilization preparation of economy	500
Subsidies and subventions to closed cities	10 544
Baikonur	674
Total subsidies	11 218
Realization of international obligations in military–technical cooperation	1 418
Total other military expenditure	**115 066**
Total military expenditure	**439 498**
Total federal budget expenditure	1 947 386
National defence (excl. pensions) as % of total expenditure	*14.6*
Total military exp. as % of total expenditure	*22.6*
GDP	10 950 000
National defence as % of GDP	*2.6*
Total military exp. as % of GDP	*4.0*
Memoranda items	
Science: total	30 318

Sources: Julian Cooper, Centre for Russian and East European Studies, University of Birmingham, using: URL <http://www.rg.ru/official/doc> (the Internet site of *Rossiyskaya Gazeta*), 8 Jan. 2002 (as published in *Rossiyskaya Gazeta*, 31 Dec. 2001); and 'On the Federal Budget for 2002', as adopted by the president on 1 Feb. 2002, Russian Ministry of Finance Internet site URL <http://www.minfin.ru>, also available on the Internet site of *Rossiyskaya Gazeta*, URL <http://www.rg.ru>.

Total Russian military expenditure, as calculated for SIPRI,[81] is significantly higher. This includes not only expenditure for pensions, military reform and paramilitary forces, but also an estimate for a military share of the general science budget and for other military-related allocations outside the ministry of defence and national defence function (table 6.8). These figures for total military expenditure show a total increase over the period 1998–2001 of 43 per cent in real terms, from $30.6 billion in 1998 to $43.9 billion in 2001.[82] The adopted budget for 2002 includes allocations for total military expenditure of 439.5 billion roubles (table 6.8). This implies a much higher growth rate in total military expenditure than in 'national defence'. Military-related items outside the official defence budget that have shown above-average growth include the paramilitary forces, military pensions and the estimates for non-ministry of defence military R&D. Actual expenditure figures for these items were not available by the end of 2001, which means that the SIPRI estimates for Russian military expenditure in 2001 may be slightly lower than actual expenditure.

The Russian defence budget for 2002 has a strong focus on arms procurement, which includes both weapon acquisitions and military research and development (R&D). This was the result of a major change in the 10-year State Armaments Programme for 2001–2010. Until October 2001, the focus of the armaments programme was on R&D for future weapon systems and on the modernization of existing systems, while any significant acquisitions of new weapon systems were deferred for another seven to eight years. This programme was fundamentally revised, according to a new concept for the development of the Russian arms industry to 2010, approved by the Russian National Security Council on 30 October 2001.[83] The new concept involves a 'step-by-step' increase in expenditure on military equipment, with the first increase, of 27 billion roubles ($900 million at the market exchange rate), in 2002 for so-called 'state-of-the-art weapons'.[84] According to President Putin, this meant that from now on the arms industry would have top priority. He warned that the US-led anti-terrorist campaign in Afghanistan, with its 'significant military element', had forced Russia to upgrade its priorities.[85]

---

[81] The military expenditure estimates for Russia were calculated by Julian Cooper, Centre for Russian and East European Studies (CREES), University of Birmingham.

[82] This figure is calculated by use of the purchasing power parity (PPP) conversion factor rather than the market exchange rate, in order to reflect international differences in relative prices. If Russian military expenditure is converted into dollars at the market exchange rate, its military expenditure in 2001 is $12.7 billion (at constant 1998 prices and exchcange rates). One PPP dollar (PPP$) has the same purchasing power in Russia as it has in the USA, reflecting the fact that the rouble has a higher purchasing power on the domestic market than suggested by the market exchange rate.

[83] Pronina, L., Interview with Ilya Klebanov, Russian Deputy Prime Minister, in 'For Russian military, research spending rises', *Defense News*, 3–9 Dec. 2001, p. 40.

[84] 'Russia to increase defence procurement in 2002', ITAR-TASS (Moscow), 19 Oct. 2001, in Foreign Broadcast Information Service, *Daily Report–Central Eurasia (FBIS-SOV)*, FBIS-SOV-2001-1020, 22 Oct. 2001.

[85] 'Putin orders increase in Russian military spending', *Air Letter*, 19 Oct. 2001, p. 5.

The state defence order for 2002, approved on 17 January 2002, incorporated the planned increase from 52 billion roubles in 2001 to 79 billion roubles ($2.5 billion at the market exchanage rate) in 2002 for R&D, new procurement, and modernization of military equipment for the military and paramilitary forces.[86] The new armaments programme was formalized by the approval of the Russian Federation Presidential Edict of 20 January 2002 of the State Ordnance Programme for the period to 2010.[87] While most of the 43 volumes of this document are classified, some information has become available. Priorities for new development of weapon systems include two new types of helicopters, a fifth-generation multi-role fighter aircraft, a new generation of submarines and surface ships, new combat vehicles and a tank.[88] Planned allocations for the strategic nuclear forces constitute 16 per cent of the armaments programme and 18 per cent of the state defence order for 2002.[89] A shift is planned in the composition of the procurement budget from research, development and engineering (RD&E) to new production. While the 2001 state defence order included 41 per cent for the development of new weapons, this share was reduced to 37 per cent in the 2002 order. The share for series production and purchase of new weapons increased from 48 per cent to 51 per cent. The potential for modernization of existing equipment is nevertheless emphasized, and the share for repair and servicing of existing equipment will remain stable at 11–12 per cent. According to Deputy Defence Minister and Chief of Procurement Colonel-General Aleksey Moskovskiy, the introduction of new computer technology will increase the efficiency of missile control systems by 20 per cent and increase the combat capability of the air force by 15 per cent and of navy ships by 30 per cent.

The impact of the Russian armaments programme on the arms industry is also emphasized by defence ministry officials.[90] The planned new development programmes are distributed among the major arms-producing companies to preserve capability. In the future, a gradual shift is planned from RD&E to the industry, with the aim of allocating 65–70 per cent of the total value of the state defence order to the production of new equipment, while the share of RD&E will decrease to 15 per cent.[91]

---

[86] 'State OKs $2.5 billion arms budget', *Moscow Times*, 18 Jan. 2002, p. 5.

[87] Talov, B., 'A military secret in rubles and missiles: RF Deputy Minister Of Defence Colonel-General Aleksey Moskovskiy has stated more precisely the details of the new state ordnance program', *Rossiyskaya Gazeta* (Moscow), (in Russian), 1 Feb. 2002, in FBIS-SOV-2002-0201, 11 Feb. 2002.

[88] 'Russian deputy PM outlines scope of the new arms program', Interfax (Moscow), (in English), 24 Jan. 2001, in FBIS-SOV-2002-0124, 25 Jan. 2002.

[89] 'Defence ministry official outlines Russia's arms program up to 2010', ITAR-TASS (Moscow), (in English), 29 Jan. 2002, in FBIS-SOV-2002-0129, 30 Jan. 2002.

[90] Litovkin, V., 'The defense order: the government has sent money to heal the VPK [military–industrial complex] and the army', *Obschaya Gazeta* (Moscow), (in Russian), 24 Jan. 2002, in 'Newly approved state defence order seen as inadequate', FBIS-SOV-2002-0124, 28 Jan. 2002; and 'Vasilchenko, Y., 'Military secret', *Rossiyskaya Gazeta* (Moscow), (in Russian), 18 Jan. 2002, in 'RF Government discusses 2002 state military order behind closed doors', FBIS-SOV-2002-0118, 22 Jan. 2002.

[91] Talov (note 87); and 'Russia puts money into research, development', *Jane's Defence Weekly*, 20 Feb. 2002, p. 20, URL <http://www.jdw.janes.com>.

V. Conclusions

The change from a decline in world military expenditure to an increase since 1998 is the result of several factors. On a geographical basis, the increase has taken place primarily in those countries that made the greatest reductions in military spending over the period 1987–98, that is, the states members of the two military alliances in the Euro-Atlantic area, NATO and the Warsaw Pact, foremost among them Russia. In the developing regions, with the exception of Sub-Saharan Africa, there was no corresponding trend of post-cold war reductions.

The increase in military spending since 1998 is the result primarily of the change in trend in Europe, North America and Africa, although there have also been higher rates of growth in Asia and the Middle East. The most marked change in trend has taken place in Russia, where the rapid reduction of military spending changed into growth in 1999 and stabilized in 2001 at a level comparable to that of some of the major West European countries. In Western Europe, military expenditure has increased only slightly. In the United States the increase has been greater—probably significantly higher than shown by the figures reported to NATO, which do not fully take into account supplementary expenditures after 11 September 2001.

There are different reasons for the change in trend. Military expenditure can be seen as a function of driving forces within prevailing economic and political constraints. Determinants of military expenditure are of four broad types: (*a*) security-related; (*b*) technological; (*c*) economic and industrial; and (*d*) more broadly political. One of the factors behind the change into growth in Europe and North America is the assumption of new military tasks in the form of peace support operations while at the same time the inertia in existing procurement programmes means that they continue to absorb large-scale funding. In Russia, the main explanation for the change in trend is economic: the earlier economic constraints that constituted the primary reason for the reduction in Russian military expenditure have eased since the late 1990s. In East Asia, economic feasibility also seems to be a determinant factor for the trend in military spending. In addition, there is a strong security-related element, in particular for the trend in military expenditure in China and the Korean peninsula. External security factors play a major role also in South Asia and the Middle East, while in Africa the acceleration in military expenditure is due primarily to domestic armed conflict and restructuring of armed forces.

The attacks of 11 September have already had a strong impact on US military expenditure, because of both the costs for the war in Afghanistan and the change in public opinion on military spending that they have brought about. The long-term impact is likely to be even greater as a result of the big boost in budget authority for future defence outlays. The attacks will also have an impact on the military expenditure trends of several other countries that have incurred costs for the war on terrorism and are planning to supplement their military forces with capabilities for anti-terrorist activities. The US allies in Europe and elsewhere are being encouraged by the US Government to contribute more to defence, while a reactive pattern cannot be excluded in parts of Asia and the Middle East. Therefore, if the current approach of relying heavily on military capability to combat terrorism continues, it is most likely that there will be a strong rise in military expenditure in the coming years. At the same time, the new security environment after 11 September has reinforced the emerging trend of a blurred distinction between military security and other types of security. Budgets for internal security are also likely to be affected as countries reinforce domestic counter-terrorism measures.

Appendix 6A. Tables of military expenditure

ELISABETH SKÖNS, EVAMARIA LOOSE-WEINTRAUB, WUYI OMITOOGUN and PETTER STÅLENHEIM[1]

Sources and methods are explained in appendix 6C. Notes and explanations of the conventions used appear below table 6A.4. Data in this appendix should not be combined with those in previous SIPRI Yearbooks because of revision.[2]

Table 6A.1. Military expenditure by region, in constant US dollars, 1992–2001
Figures are in US $b., at constant 1998 prices and exchange rates.[3] Figures do not always add up to totals because of the conventions of rounding.

	1992	1993	1994	1995	1996	1997	1998	1999	2000	2001
World total	847	814	793	741	722	732	719	728	757	772
Geographical regions										
Africa	9.3	8.8	9.3	8.9	8.5	8.8	9.3	10.9	11.3	12.2
North Africa	2.7	2.8	3.3	3.1	3.3	3.5	3.6	3.8	4.2	. .
Sub-Saharan Africa	6.6	6.0	6.1	5.8	5.3	5.3	5.7	7.0	7.1	. .
Americas	383	367	348	333	314	315	308	308	319	317
North America	364	345	326	307	290	288	282	283	294	289
Central America	2.4	2.5	3.0	2.7	2.7	2.8	2.8	2.9	2.9	2.9
South America	16.9	19.4	19.2	23.0	21.2	24.0	22.9	22.1	22.4	24.7
Asia and Oceania	105	108	109	112	115	117	117	119	123	129
Central Asia	. .	. .	1.8	1.8	1.9	2.2	. .	2.4	. .	. .
East Asia	84.6	85.6	87.5	90.2	92.9	93.9	93.6	94.4	96.4	101
South Asia	11.3	12.3	12.3	12.9	13.1	13.7	13.8	15.1	16.4	17.4
Oceania	7.5	7.8	7.7	7.5	7.4	7.5	7.8	7.5	7.4	7.5

MILITARY EXPENDITURE 267

Europe	296	278	275	239	235	238	227	233	241	242
Central and Eastern	95.4	84.6	86.5	60.4	55.4	59.0	47.3	51.6	57.4	60.5
Western	201	194	189	179	180	179	179	182	183	181
Middle East	52.3	51.0	50.9	47.9	48.9	53.5	57.8	56.1	63.1	72.4
Organizations										
ASEAN	7.9	8.3	8.7	10.2	10.5	18.2	16.6	16.8	16.6	17.0
CIS	87.0	78.2	79.9	53.3	49.0	52.7	40.7	45.6	51.6	54.7
CIS Europe	84.9	75.9	78.0	51.5	47.0	50.5	38.7	43.2	49.0	51.8
CIS Asia	. .	. .	1.8	1.8	1.9	2.2	. .	2.4	. .	. .
EU	184	178	172	172	173	172	173	175	177	175
NATO	557	533	508	481	466	462	457	467	478	472
NATO Europe	187	180	175	167	168	166	166	174	175	174
OECD	614	590	566	541	539	537	532	537	549	543
OPEC	31.2	31.2	30.5	27.6	27.2	32.0	35.6	33.8	39.8	. .
OSCE	669	633	610	555	535	536	520	529	546	542
Income group (GNP/capita 1998)										
Low (≤ $670)	41.1	41.1	40.6	41.7	43.6	44.7	45.3	50.3	53.5	59.3
Lower middle ($761–3030)	113	104	108	80.6	76.6	81.5	69.0	73.7	83.5	89.1
Upper middle ($3031–9360)	57.4	62.0	60.2	62.9	61.6	68.5	71.1	68.3	71.8	81.6
High (≥ $9361)	635	607	584	556	540	537	533	535	548	542

Notes:

The world total and the totals for regions, organizations and income groups in table 6A.1 are estimates, based on data in table 6A.3. When military expenditure data for a country are missing for a few years, estimates are made, most often on the assumption that the rate of change in that country's military expenditure is the same as that for the region to which it belongs. When no estimates can be made, countries are excluded from the totals. The countries excluded from all totals in table 6A.1 are: Afghanistan, Angola, Benin, the Republic of Congo, the Democratic Republic of the Congo (DRC), Iraq, Libya, Somalia and Yugoslavia.

Regional totals are presented only when based on country data accounting for at least 90% of the regional total.

Totals for geographical regions add up to the world total and subregional totals add up to regional totals.

Totals for regions and income groups cover the same groups of countries for all years, while totals for organizations cover only the member countries in the year given.

The country coverage of income groups is based on figures of 1998 GNP per capita as calculated by the World Bank and presented in its *World Development Report 1999/2000* (International Bank for Reconstruction and Development and Oxford University Press: Washington, DC, Sep. 1999).

Africa: Algeria, Angola, Benin, Botswana, Burkina Faso, Burundi, Cameroon, Cape Verde, Central African Republic, Chad, Congo (Republic of), Congo (Democratic Republic of the, DRC), Côte d'Ivoire, Djibouti, Equatorial Guinea, Eritrea, Ethiopia, Gabon, Gambia, Ghana, Guinea, Guinea-Bissau, Kenya, Lesotho, Liberia, Libya, Madagascar, Malawi, Mali, Mauritania, Mauritius, Morocco, Mozambique, Namibia, Niger, Nigeria, Rwanda, Senegal, Seychelles, Sierra Leone, Somalia, South Africa, Sudan, Swaziland, Tanzania, Togo, Tunisia, Uganda, Zambia, Zimbabwe.

North Africa: Algeria, Libya, Morocco, Tunisia.

Sub-Saharan Africa: Angola, Benin, Botswana, Burkina Faso, Burundi, Cameroon, Cape Verde, Central African Republic, Chad, Congo (Republic of), Congo (Democratic Republif of the, DRC), Côte d'Ivoire, Djibouti, Equatorial Guinea, Eritrea, Ethiopia, Gabon, Gambia, Ghana, Guinea, Guinea-Bissau, Kenya, Lesotho, Liberia, Madagascar, Malawi, Mali, Mauritania, Mauritius, Mozambique, Namibia, Niger, Nigeria, Rwanda, Senegal, Seychelles, Sierra Leone, Somalia, South Africa, Sudan, Swaziland, Tanzania, Togo, Uganda, Zambia, Zimbabwe.

Americas: Argentina, Belize, Bolivia, Brazil, Canada, Chile, Colombia, Costa Rica, Ecuador, El Salvador, Guatemala, Guyana, Honduras, Mexico, Nicaragua, Panama, Paraguay, Peru, Uruguay, USA, Venezuela.

North America: Canada, USA.

Central America: Belize, Costa Rica, El Salvador, Guatemala, Honduras, Mexico, Nicaragua, Panama.

South America: Argentina, Bolivia, Brazil, Chile, Colombia, Ecuador, Guyana, Paraguay, Peru, Uruguay, Venezuela.

Asia and Oceania: Afghanistan, Australia, Bangladesh, Brunei, Cambodia, China, Fiji, India, Indonesia, Japan, Kazakhstan (1992–), New Zealand, North Korea, South Korea, Kyrgyzstan (1992–), Laos, Malaysia, Mongolia, Myanmar (Burma), Nepal, Pakistan, Papua New Guinea, Philippines, Singapore, Sri Lanka, Taiwan, Tajikistan (1992–), Thailand, Turkmenistan (1992–), Uzbekistan (1992–), Viet Nam.

Central Asia: Kazakhstan (1992–), Kyrgyzstan (1992–), Tajikistan (1992–), Turkmenistan (1992–), Uzbekistan (1992–).

East Asia: Brunei, Cambodia, China, Indonesia, Japan, North Korea, South Korea, Laos, Malaysia, Mongolia, Myanmar (Burma), Philippines, Singapore, Taiwan, Thailand, Viet Nam.

South Asia: Afghanistan, Bangladesh, India, Nepal, Pakistan, Sri Lanka.

Oceania: Australia, Fiji, New Zealand, Papua New Guinea.

Europe: Albania, Armenia (1992–), Austria, Azerbaijan (1992–), Belarus (1992–), Belgium, Bosnia and Herzegovina (1992–), Bulgaria, Croatia (1992–), Cyprus, Czechoslovakia (–1992), Czech Republic (1993–), Denmark, Estonia, Finland, France, Georgia (1992–), Germany, Greece, Hungary, Iceland, Ireland, Italy, Latvia, Lithuania, Luxembourg, Macedonia (Former Yugoslav Republic of, FYROM) (1992–), Malta, Moldova (1992–), Netherlands, Norway, Poland, Portugal, Romania, Russia (1992–), Slovakia (1993–), Slovenia (1992–), Spain, Sweden, Switzerland, UK, Ukraine (1992–), Yugoslavia (Serbia and Montenegro, 1992–).

Central and Eastern Europe: Albania, Armenia (1992–), Azerbaijan (1992–), Belarus (1992–), Bosnia and Herzegovina (1992–), Bulgaria, Croatia (1992–), Czechoslovakia (– 1992), Czech Republic (1993–), Estonia, Georgia, Hungary, Latvia, Lithuania, Macedonia (Former Yugoslav Republic of, FYROM) (1992–), Moldova (1992–), Poland, Romania, Russia (1992–), Slovakia (1993–), Slovenia (1992–), Ukraine (1992–), Yugoslavia (1992–).

Western Europe: Austria, Belgium, Cyprus, Denmark, Finland, France, Germany, Greece, Iceland, Ireland, Italy, Luxembourg, Malta, Netherlands, Norway, Portugal, Spain, Sweden, Switzerland, UK.

Middle East: Bahrain, Egypt, Iran, Iraq, Israel, Jordan, Kuwait, Lebanon, Oman, Saudi Arabia, Syria, Turkey, United Arab Emirates, Yemen.

Association of South-East Asian Nations (ASEAN): Brunei, Cambodia (1999–), Indonesia, Laos (1997–), Malaysia, Myanmar (Burma) (1997–), Philippines, Singapore, Thailand, Viet Nam (1995–).

Organization for Security and Co-operation in Europe (OSCE): Albania, Armenia (1992–), Austria, Azerbaijan (1992–), Belarus (1992–), Belgium, Bosnia and Herzegovina (1992–), Bulgaria, Canada, Croatia (1992–), Cyprus, Czechoslovakia (–1992), Czech Republic (1993–), Denmark, Estonia, Finland, France, Georgia (1992–), Germany, Greece, Hungary, Iceland, Ireland, Italy, Kazakhstan (1992–), Kyrgyzstan (1992–), Latvia, Lithuania, Luxembourg, Macedonia (1995–), Malta, Moldova (1992–), Netherlands, Norway, Poland, Portugal, Romania, Russia (1992–), Slovakia (1993–), Slovenia (1992–), Spain, Sweden, Switzerland, Tajikistan (1992–), Turkey, Turkmenistan (1992–), UK, Ukraine (1992–), USA, Uzbekistan (1992–), Yugoslavia (2000–).

Commonwealth of Independent States (CIS): Armenia, Azerbaijan, Belarus, Georgia (1993–), Kazakhstan, Kyrgyzstan, Moldova, Russia, Tajikistan, Turkmenistan, Ukraine, Uzbekistan.

CIS Europe: Armenia, Azerbaijan, Belarus, Georgia (1993–), Moldova, Russia, Ukraine.

CIS Asia: Kazakhstan (1992–), Kyrgyzstan (1992–), Tajikistan (1992–), Turkmenistan (1992–), Uzbekistan (1992–).

European Union (EU): Austria (1995–), Belgium, Denmark, Finland (1995–), France, Germany, Greece, Ireland, Italy, Luxembourg, Netherlands, Portugal, Spain, Sweden (1995–), UK.

North Atlantic Treaty Organization (NATO): Belgium, Canada, Czech Republic (1999–), Denmark, France, Germany, Greece, Hungary (1999–), Iceland, Italy, Luxembourg, Netherlands, Norway, Poland (1999–), Portugal, Spain, Turkey, UK, USA.

NATO Europe: Belgium, Czech Republic (1999–), Denmark, France, Germany, Greece, Hungary (1999–), Iceland, Italy, Luxembourg, Netherlands, Norway, Poland (1999–), Portugal, Spain, UK,

Organization of Petroleum-Exporting Countries (OPEC): Algeria, Ecuador (–1992), Gabon (–1995), Indonesia, Iran, Iraq, Kuwait, Libya, Nigeria, Qatar, Saudi Arabia, United Arab Emirates and Venezuela.

Organisation for Economic Co-operation and Development (OECD): Australia, Austria, Belgium, Canada, Czech Republic (1995–), Denmark, Finland, France, Germany, Greece, Hungary (1996–), Iceland, Ireland, Italy, Japan, South Korea (1996–), Luxembourg, Mexico (1994–), Netherlands, New Zealand, Norway, Poland (1996–), Portugal, Slovak Rep. (2000–), Spain, Sweden, Switzerland, Turkey, UK, USA.

Table 6A.2. Military expenditure by region and country, in local currency, 1992–2001

Figures are in local currency, current prices, and are for calendar years.

State	Currency	1992	1993	1994	1995	1996	1997	1998	1999	2000	2001
Africa											
North Africa											
Algeria[4]	m. dinars	[23 000]	29 810	46 800	58 847	79 519	101 126	112 248	121 600	141 600	. .
Libya	m. dinars	. .	. .	. .	. .	. .	. .	. .	. .	. .	. .
Morocco	m. dirhams	10 488	11 071	13 557	12 957	12 890	12 476	12 666	13 921	14 639	15 686
Tunisia	m. dinars	256	277	301	324	387	396	417	424	442	[459]
Sub-Saharan											
Angola[5]	th./m. kwanzas	0.4	3.5	130	2 469	163	(391)	288	3 670	. .	. .
Benin	m. francs	. .	. .	. .	. .	. .	. .	. .	. .	. .	. .
Botswana	m. pulas	376	450	458	460	467	586	808	855	974	1 141
Burkina Faso	m. francs	18 824	17 139	16 800	18 400	19 000	22 500	23 300	25 700	25 200	28 400
Burundi	m. francs	8 121	8 805	10 589	10 517	15 408	20 800	26 400	28 700	31 100	35 000
Cameroon	m. francs	[48 650]	47 621	52 477	56 691	59 819	69 288	80 969	89 095	87 598	89 118
Cape Verde	m. escudos	. .	220	281	477	352	382	443	518	814	. .
Centr. Afr. Rep.[6]	m. francs	6 137	5 421	5 935	6 496	6 239	. .	. .	. .	. .	. .
Chad	m. francs	. .	11 085	12 333	10 000	12 681	9 700	9 500	. .	. .	. .
Congo, Rep. of	m. francs	. .	. .	. .	. .	. .	. .	. .	. .	. .	. .
Congo, DRC[7]	m. francs	. .	. .	. .	. .	52 516	54 588	. .	. .	. .	. .
Côte d'Ivoire	m. francs	41 503	42 088	46 677	4 481	3 712	4 019	4 013	. .	. .	. .
Djibouti	m. francs	5 089	4 702	4 648	1 721	. .	. .	. .	. .	. .	. .
Equatorial Guinea	m. francs	. .	. .	1 321	771	968	634	1 459	1 335	. .	. .
Eritrea[8]	m. birr	. .	539	439	754	803	1 462	2 481	4 836	5 500	. .
Ethiopia[9]	m. birr	[716]	[819]	813	. .	. .	. .	9 000	. .	. .	. .
Gabon	m. francs	. .	. .	. .	27.6	38.5	42.6	43.1	40.1	(57.2)	. .
Gambia[10]	m. dalasis	31.2	23.3	22.2	58 823	72 644	93 148	133 000	158 000	277 269	231 740
Ghana	m. cedis	18 201	26 600	36 147							

MILITARY EXPENDITURE 271

Guinea	m. francs	50 200	42 000	44 800	:	:	48 600	67 700	:	(80 000)	100 000
Guinea-Bissau[11]	m. francs	:	:	400	615	770	1 061	1 711	:	:	:
Kenya	m. shillings	5 027	6 131	6 577	7 668	9 756	10 302	10 610	11 824	(14 566)	(17 363)
Lesotho	m. maloti	60.1	62.4	81.9	95.1	107	135	156	171	:	:
Liberia[12]	m. dollars	23.6	37.3	41.3	:	:	:	:	330	:	:
Madagascar[13]	b. francs	68.9	72.4	84.6	116	201	267	287	[316]	319	:
Malawi	m. kwachas	90.9	113	149	169	309	434	450	634	774	894
Mali	b. francs	:	[16.8]	22.2	26.9	27.1	31.3	32.2	36.0	[45.3]	:
Mauritania[14]	m. ouguiyas	3 427	3 640	3 640	3 640	3 680	3 660	:	:	:	:
Mauritius	m. rupees	178	190	213	234	233	206	203	217	228	271
Mozambique[15]	b. meticais	259	399	762	522	704	840	1 013	1 251	1 442	1 825
Namibia[16]	m. dollars	355	229	202	248	286	383	435	660	786	736
Niger	m. francs	:	:	9 700	9 200	8 900	9 600	12 700	14 500	:	:
Nigeria[17]	m. naira	3 004	3 500	7 032	14 000	15 350	17 920	23 100	45 400	34 181	59 000
Rwanda[18]	m. francs	11 863	12 900	5 700	14 700	22 600	23 300	27 340	29 000	[20 300]	25 100
Senegal[19]	m. francs	29 056	33 962	36 725	40 389	40 809	41 324	44 300	48 200	44 400	51 500
Seychelles	m. rupees	105	67.1	60.1	55.2	52.4	51.1	55.5	60.8	62.5	:
Sierra Leone	m. leones	10 081	13 244	15 546	18 898	17 119	9 315	7 800	19 800	20 002	:
Somalia	shillings	:	:	:	:	:	:	:	:	:	:
South Africa	m. rand	10 724	10 713	12 352	11 942	11 143	11 131	10 716	10 678	13 031	15 303
Sudan[11]	m. dinars	1 576	3 527	5 939	8 060	9 520	16 300	42 800	62 200	84 100	:
Swaziland	m. emalangeni	58.1	73.6	85.5	99.7	108	117	142	152	151	158
Tanzania	b. shillings	25.8	21.3	26.7	44.0	52.8	61.2	(70.0)	(86.5)	:	:
Togo	m. francs	13 000	14 200	14 100	15 400	:	:	:	:	:	:
Uganda[20]	m./b. shillings	56 904	72 174	[81 050]	[87.7]	116	134	181	204	185	204
Zambia[21]	b. kwachas	16.8	22.0	39.0	65.8	56.9	90.8	114	73.7	[58.0]	:
Zimbabwe[22]	m. dollars	1 269	1 439	1 826	2 214	2 742	3 441	3 710	7 200	15 100	13 300
America											
Central America											
Belize	th. dollars	10 584	12 261	15 799	16 106	15 932	18 790	:	:	:	:
Costa Rica	m. colones	:	:	:	:	:	:	:	:	:	:

State	Currency	1992	1993	1994	1995	1996	1997	1998	1999	2000	2001
El Salvador	m. colones	975	888	829	849	843	853	843	873	839	955
Guatemala	m. quetzals	[677]	693	806	843	784	801	894	914	1 225	1 306
Honduras	m. lempiras	. .	. .	(385)	. .	. .	. .	. .	(485)	. .	. .
Mexico	m. new pesos	[5 430]	[6 514]	[8 694]	10 368	14 637	18 306	20 950	25 825	28 335	29 920
Nicaragua[11]	m. gold córdobas	211	224	232	235	240	245	265	290	[324]	. .
Panama[23]	m. balboas	78.8	95.2	101	96.6	101	118	104	112	. .	. .
North America											
Canada[24]	m. dollars	13 111	13 293	13 008	12 457	11 511	10 831	11 716	12 360	12 314	12 174
USA[24]	m. dollars	305 141	297 637	288 059	278 856	271 417	276 324	274 278	280 969	301 698	305 886
South America											
Argentina[25]	m. pesos	[3 280]	[3 830]	4 021	4 361	4 136	4 016	3 962	4 143	3 739	3 726
Bolivia	m. bolivianos	473	537	569	612	682	857	1 128	864	792	858
Brazil[26]	th./m. reais	7 018	188	4 108	10 008	9 994	13 104	12 743	12 328	13 988	19 521
Chile[27]	b. pesos	[498]	575	730	800	910	980	1 180	1 070	[1 243]	[953]
Colombia	b. pesos	[882]	1 104	1 296	1 775	2 500	3 376	3 109	3 785	3 881	. .
Ecuador[28]	b. sucres/m. dollars	532	841	982	893	1 260	. .	. .	. .	. .	. .
Guyana	m. dollars	453	562	759	808	780	. .	. .	. .	. .	. .
Paraguay	m. guaranies	154	167	. .	. .	. .	275	284	258	280	. .
Peru[11]	m. new soles	1 001	. .	2 083	1 816	2 228	2 638	2 847	3 115	2 734	. .
Uruguay	m. pesos	813	974	111	197	241	473	(685)	(805)	(949)	. .
Venezuela	b. bolivares	55.0	95.0	(991)	(3 355)	(6 900)	[13 700]	. .	. .	. .	. .
Asia and Oceania											
Central Asia											
Kazakhstan[29]	b. tenge	. .	(0.3)	(3.8)	10.8	16.3	17.9	18.9	17.2	17.2	26.6
Kyrgyzstan[29]	m. soms	5.5	38.8	105	267	334	501	515	839	1 173	[1 270]
Tajikistan[29]	m. roubles	2.6	243	347	[713]	3 977	10 713	13 562	18 723	21 907	. .
Turkmenistan[30]	b. manats	. .	. .	1.5	15.1	158	440	436	582	[850]	. .
Uzbekistan[29]	m. soms	. .	. .	. .	. .	. .	[13 700]	. .	[34 860]	. .	. .

MILITARY EXPENDITURE 273

Country	Unit										
East Asia											
Brunei[31]	m. dollars	[430]	[398]	[420]	[425]	474	555	614	[520]	[485]	[548]
Cambodia	b. riels	[90.0]	[165]	[302]	302	298	305	312	336	311	[301]
China, P. R.[32]	b. yuan	[69.2]	[73.1]	[87.2]	[105]	[124]	[139]	[157]	[172]	[189]	[223]
Indonesia	b. rupiahs	[3 504]	[3 689]	[4 424]	[5 096]	[5 980]	[6 877]	[8 969]	11 399	13 900	. .
Japan	b. yen	4 510	4 618	4 673	4 714	4 815	4 922	4 942	4 934	4 935	4 950
Korea, North[33]	b. won	(4.6)	(4.7)	(4.8)	. .	. .	. .	2.9	2.9	3.0	3.1
Korea, South	b. won	8 410	9 215	10 075	11 074	12 243	13 102	13 594	13 337	14 477	15 388
Laos	b. kip	. .	. .	. .	. .	49.2	53.5	. .	. .	. .	. .
Malaysia	m. ringgits	4 500	4 951	5 565	6 121	6 091	5 877	4 545	6 321	6 414	7 332
Mongolia	m. tugriks	1 184	4 147	6 766	9 547	11 850	14 830	16 749	18 416	26 126	25 300
Myanmar	b. kyats	8.4	12.7	16.7	22.3	27.7	29.8	37.3	43.7	[42.9]	. .
Philippines	m. pesos	17 462	21 132	24 401	30 510	32 269	37 405	38 412	[36 520]	[38 370]	. .
Singapore	m. dollars	3 799	4 010	4 273	5 206	5 782	6 618	7 475	7 616	7 675	7 790
Taiwan	m. dollars	233	246	288	296	277	288	302	265	244	245
Thailand	m. baht	64 961	73 708	78 300	88 983	93 959	97 783	100 328	85 513	[76 183]	[74 141]
Viet Nam	b. dong	3 730	3 168	4 730	. .	. .	. .	. .	. .	. .	. .
South Asia											
Afghanistan	m. afghanis	. .	. .	. .	. .	. .	. .	. .	. .	. .	. .
Bangladesh	m. taka	14 396	16 105	19 021	21 582	22 065	24 546	27 390	30 255	33 045	34 630
India	b. rupees	174	209	230	260	291	339	387	464	530	601
Nepal[34]	m. rupees	1 607	1 801	1 939	2 064	2 242	2 471	2 789	3 240	3 678	4 200
Pakistan	m. rupees	81 606	89 619	98 144	112 085	123 550	131 803	139 818	146 931	153 945	160 150
Sri Lanka[35]	b. rupees	12.9	15.4	19.4	35.2	38.1	37.1	42.5	40.1	56.9	57.0
Oceania											
Australia	m. dollars	9 584	10 201	10 326	10 472	10 608	10 761	11 298	[11 083]	[11 360]	[12 030]
Fiji	m. dollars	45.9	49.4	49.3	48.8	51.2	48.0	47.7	53.9	58.0	. .
New Zealand	m. dollars	1 097	1 050	1 016	1 004	976	1 002	1 057	1 081	1 108	1 135
Pap. New Guinea	m. kina	56.5	67.1	84.2	72.2	104	114	. .	91.7	85.0	. .
Europe											
Albania	b. leks	[2.5]	4.0	4.7	4.7	4.8	4.4	5.1	6.0	6.6	6.1

MILITARY SPENDING AND ARMAMENTS, 2001

State	Currency	1992	1993	1994	1995	1996	1997	1998	1999	2000	2001
Armenia[29]	m. dram	6.5	89.5	. .	21 200	21 700	30 500	33 300	[36 000]	[45 000]	. .
Austria	m. shillings	19 600	20 500	21 200	21 500	21 690	22 012	22 272	22 339	22 727	22 986
Azerbaijan[29]	b. manats	0.8	7.9	85.6	248	305	353	376	435	494	. .
Belarus[36]	m. roubles	1.4	25.8	604	1 933	2 266	6 079	9 834	38 740	115 250	[240 000]
Belgium	m. francs	132 819	129 602	131 955	131 156	131 334	131 796	133 007	136 252	139 711	138 564
Bosnia & Herz.[37]	m. marks	. .	. .	[278]	[383]	[156]	327	346	381	386	377
Bulgaria	m. leva	6.5	8.5	14.3	22.8	41.4	460	580	681	[754]	[772]
Croatia[38]	m. kuna	200	3 422	7 149	9 282	7 760	7 000	7 500	5 798	4 784	4 266
Cyprus	m. pounds	191	90.0	99.0	91.0	141	[158]	[143]	[162]	[177]	. .
Czech Rep.[39]	m. korunas	. .	[23 627]	24 375	25 070	26 817	27 582	33 570	37 211	39 200	44 615
Czechoslovakia[40]	m. korunas	48 503	. .	. .	. .	. .	. .	. .	. .	. .	. .
Denmark	m. kroner	17 129	17 390	17 293	17 468	17 896	18 521	19 071	19 428	19 339	20 455
Estonia[41]	m. kroons	68.0	174	327	417	499	736	843	1 083	1 332	1 651
Finland	m. markkaa	9 298	9 225	9 175	8 594	9 776	9 246	10 194	8 885	9 791	9 368
France	m. francs	238 874	241 199	246 469	238 432	237 375	241 103	236 226	239 488	240 752	245 537
Georgia[42]	th./m. lari	[3.5]	[200]	[40.0]	[55.0]	[76.0]	[95.0]	[69.0]	[68.0]	[54.0]	[40.0]
Germany	m. marks	65 536	61 529	58 957	58 986	58 671	57 602	58 327	59 854	59 758	59 858
Greece	b. drachmas	835	933	1 053	1 171	1 343	1 511	1 725	1 853	2 018	2 129
Hungary	m. forints	61 216	67 492	67 996	76 937	85 954	96 814	132 602	166 685	189 400	220 812
Iceland	m. kronur	—	—	—	. .	. .	. .	. .	. .	. .	. .
Ireland	m. pounds	376	385	412	426	457	491	507	533	606	715
Italy	b. lire	30 813	32 364	32 835	31 561	36 170	38 701	40 763	43 062	47 100	46 009
Latvia[43]	m. lats	. .	12.0	19.0	23.0	21.0	22.1	24.8	33.1	42.4	48.4
Lithuania[44]	m. litai	. .	85.4	79.3	115	169	302	553	479	796	921
Luxembourg	m. francs	3 963	3 740	4 214	4 194	4 380	4 797	5 197	5 330	5 613	6 657
Mac. (FYROM)[45]	m. denars	. .	. .	. .	. .	5 223	4 163	4 301	4 746	4 564	4 419
Malta	th. liri	8 513	9 419	10 533	10 996	12 002	12 020	11 297	11 164	13 009	12 661
Moldova[29]	m. lei	. .	9.7	36.7	60.0	70.7	80.5	57.0	63.0	63.6	76.8
Netherlands	m. guilders	13 900	13 103	12 990	12 864	13 199	13 345	13 561	14 534	14 284	15 582

MILITARY EXPENDITURE 275

Country	Unit										
Norway	m. kroner	23 638	22 528	24 019	22 224	22 813	23 010	25 087	25 809	25 722	26 853
Poland	m. zlotys	2 624	3 980	5 117	6 595	8 313	10 075	11 687	12 242	13 239	15 091
Portugal	m. escudos	341 904	352 504	360 811	403 478	401 165	418 772	420 654	452 843	479 663	504 480
Romania	b. lei	196	420	1 170	1 525	2 058	5 784	7 342	10 309	17 033	22 463
Russia[46]	m. roubles	[1 049]	[9 037]	[35 890]	[63 220]	[82 485]	[105 034]	[85 574]	[165 477]	[262 800]	[335 600]
Slovak Rep.[47]	m. korunas	. .	8 211	9 614	13 588	13 412	13 901	14 009	13 532	15 761	18 464
Slovenia[48]	m. tolars	[22 875]	[26 081]	[30 651]	39 664	44 666	46 434	50 030	50 013	49 518	66 726
Spain[49]	b. pesetas	928	1 055	995	1 079	1 091	1 123	1 124	1 180	1 264	1 303
Sweden	m. kronor	[35 769]	36 992	37 182	33 194	28 847	38 825	40 034	41 980	43 652	44 246
Switzerland	m. francs	6 014	5 524	5 723	5 011	4 782	4 634	4 532	4 416	4 503	4 420
UK[50]	m. pounds	22 850	22 686	22 490	21 439	22 330	21 612	22 477	22 548	23 532	23 772
Ukraine[51]	m. hryvnias	. .	8.0	337	1 665	2 833	3 428	3 712	3 908	6 200	5 800
Yugoslavia[52]	m. dinars	. .	. .	. .	. .	. .	[5 450]	6 441	8 517	21 292	31 501

Middle East

Country	Unit										
Bahrain	m. dinars	94.6	94.4	96.3	103	109	109	111	123	121	. .
Egypt[53]	m. pounds	5 211	5 723	6 142	6 682	7 164	7 573	7 986	8 154	8 312	8 708
Iran[54]	b. rials	1 482	2 255	4 023	4 457	6 499	8 540	10 624	12 933	21 936	32 163
Iraq	m. dinars	. .	. .	. .	. .	. .	. .	. .	. .	. .	. .
Israel[55]	m. new shekels	16 919	17 539	19 836	22 519	26 979	29 581	32 449	34 021	36 430	37 265
Jordan[56]	m. dinars	[316]	[347]	[403]	[448]	429	458	502	529	563	[561]
Kuwait	m. dinars	1 852	900	979	1 102	971	745	695	740	947	. .
Lebanon	b. pounds	499	518	704	795	760	686	692	864	871	1 394
Oman[57]	m. riyals	778	738	779	776	737	698	[617]	[626]	[740]	[1 017]
Saudi Arabia[58]	m. riyals	54 000	61 636	53 549	49 501	50 025	66 000	78 000	66 000	75 000	[97 000]
Syria	m. pounds	33 412	29 948	37 270	40 500	40 746	42 842	46 064	[46 064]	[49 600]	[52 000]
Turkey	tr. liras	42.3	77.7	157	303	612	1 183	2 289	4 168	6 248	9 030
UAE[59]	m. dirhams	5 827	5 827	5 827	5 827	[5 827]	6 027	6 027	6 027	6 027	6 027
Yemen	m. rials	16 812	19 752	30 273	35 897	44 964	55 104	53 824	58 311	[65 000]	[77 300]

Table 6A.3. Military expenditure by region and country, in constant US dollars, 1992–2001

Figures are in US $m., at constant 1998 prices and exchange rates, and are for calendar years.[3]

State	1992	1993	1994	1995	1996	1997	1998	1999	2000	2001
Africa										
North Africa										
Algeria[4]	[1 041]	1 119	1 362	1 319	1 502	1 761	1 911	2 017	2 341	. .
Libya	. .	. .	. .	. .	. .	. .	. .	. .	. .	. .
Morocco	1 370	1 375	1 602	1 442	1 393	1 335	1 319	1 440	1 486	1 554
Tunisia	289	300	311	316	363	359	366	362	367	[371]
Sub-Saharan										
Angola[5]	1 215	656	2 323	1 592	2 476	(1 860)	733	2 419	. .	. .
Benin	. .	. .	. .	. .	. .	. .	. .	. .	. .	. .
Botswana	158	166	153	139	128	148	191	188	197	217
Burkina Faso	49.2	44.6	34.9	35.6	34.6	40.1	39.5	44.0	43.3	46.4
Burundi	50.8	50.3	52.6	43.8	50.8	52.3	59.0	62.0	54.1	55.6
Cameroon	[132]	134	109	108	110	121	137	148	144	143
Cape Verde	. .	2.9	3.7	5.8	4.1	4.0	4.5	5.1	8.1	. .
Central African Rep.[6]	15.5	14.1	12.4	11.4	10.5	. .	. .	. .	. .	. .
Chad	. .	38.3	30.3	22.6	25.5	18.4	16.1	. .	. .	. .
Congo, Rep. of	. .	. .	. .	. .	. .	. .	. .	. .	. .	. .
Congo (DRC)[7]	116	115	101	. .	96.9	96.9	. .	. .	. .	. .
Côte d'Ivoire	35.8	31.6	29.4	27.0	21.4	22.6	22.6	. .	. .	. .
Djibouti	. .	. .	2.8	3.2	. .	. .	. .	. .	. .	. .
Equatorial Guinea	. .	95.3	74.1	118	137	86.8	197	167	. .	. .
Eritrea[8]	[123]	[136]	125	106	119	211	349	642	730	. .
Ethiopia[9]	. .	. .	. .	. .	. .	. .	15.3	. .	. .	. .
Gabon	3.6	2.5	2.3	2.7	3.8	4.0	4.0	3.6	(5.1)	. .
Gambia[10]	42.0	49.2	53.5	54.6	46.0	46.1	57.5	60.7	85.1	53.5
Ghana										

MILITARY EXPENDITURE

Guinea	52.7	41.2	42.2	. .	. .	41.3	54.7	. .	(57.9)	67.6
Guinea-Bissau[11]	. .	. .	2.4	2.5	2.1	1.9	2.9	. .	. .	. .
Kenya	204	170	142	164	192	181	176	191	(222)	(265)
Lesotho	18.6	17.1	20.7	22.0	22.6	26.3	28.2	29.2	. .	. .
Liberia[12]	. .	. .	. .	. .	. .	. .	. .	7.8	. .	. .
Madagascar[13]	38.3	36.6	30.8	28.3	41.0	52.2	52.8	[52.8]	47.6	12.4
Malawi	17.3	17.5	17.1	10.6	14.1	18.1	14.5	14.1	13.3	. .
Mali	. .	[44.1]	47.2	50.5	47.6	55.2	54.6	61.7	[78.3]	. .
Mauritania[14]	26.1	25.4	24.3	22.8	22.1	21.0	. .	. .	. .	. .
Mauritius	11.3	11.0	11.4	11.8	11.1	9.2	8.5	8.4	8.5	9.6
Mozambique[15]	120	130	153	67.7	62.2	69.8	83.6	101	104	125
Namibia[16]	106	63.2	50.4	56.0	59.9	73.8	78.7	110	120	103
Niger	. .	. .	20.6	17.7	16.2	17.0	21.5	25.2	. .	. .
Nigeria[17]	903	670	857	987	837	903	1 055	1 945	1 369	1 981
Rwanda[18]	108	105	28.4	60.1	86.1	79.2	87.5	95.1	[64.1]	76.6
Senegal[19]	73.8	86.7	70.9	72.3	71.1	70.9	75.1	81.0	74.1	84.5
Seychelles	21.0	13.2	11.6	10.7	10.3	10.0	10.5	10.9	10.5	. .
Sierra Leone	23.6	25.4	24.0	23.2	17.1	8.1	5.0	9.4	9.6	. .
Somalia	3 140	. .	. .	. .	. .	. .	. .	. .	. .	. .
South Africa	229	2 859	3 026	2 692	2 340	2 152	1 938	1 836	2 128	2 361
Sudan[11]	18.7	255	199	160	81.4	95.1	213	267	334	. .
Swaziland	132	21.1	21.6	22.4	22.8	23.0	26.0	26.1	23.2	. .
Tanzania	39.4	86.8	81.5	105	104	104	(105)	(121)	. .	. .
Togo	66.5	43.1	31.6	29.9	. .	. .	. .	. .	. .	144
Uganda[20]	122	79.5	[81.3]	[81.1]	100	108	146	154	136	. .
Zambia[21]	215	55.3	63.9	80.3	47.5	60.7	61.1	31.2	[19.5]	228
Zimbabwe[22]		191	199	197	200	212	173	213	401	

America
Central America

Belize	6.0	6.9	8.7	8.6	8.0	9.3	. .	. .	. .	. .
Costa Rica	. .	. .	. .	. .	. .	. .	. .	. .	. .	. .

State	1992	1993	1994	1995	1996	1997	1998	1999	2000	2001
El Salvador	189	145	123	114	103	99.9	96.3	99.2	93.3	102
Guatemala	[185]	169	177	171	143	134	140	136	172	173
Honduras	. .	. .	(63.0)	. .	. .	. .	. .	(32.6)	. .	. .
Mexico	[1 770]	[1 935]	[2 414]	2 133	2 240	2 323	2 293	2 425	2 430	2 412
Nicaragua[11]	39.2	34.5	33.5	30.6	28.0	26.2	25.0	24.6	[24.7]	. .
Panama[23]	83.5	100	105	99.7	103	119	104	110	. .	. .
North America										
Canada[24]	9 600	9 557	9 333	8 750	7 958	7 372	7 898	8 190	7 942	7 635
USA[24]	354 507	335 940	316 776	298 376	282 231	280 785	274 278	275 057	285 679	281 426
South America										
Argentina[25]	[3 972]	[4 193]	4 226	4 433	4 198	4 055	3 964	4 194	3 821	3 834
Bolivia	140	147	144	141	140	167	205	153	134	143
Brazil[26]	5 656	7 471	7 503	11 011	9 499	11 648	10 976	10 126	10 734	14 111
Chile[27]	[1 762]	1 805	2 056	2 082	2 205	2 239	2 564	2 250	[2 518]	[1 862]
Colombia	[1 954]	1 995	1 891	2 141	2 508	2 849	2 180	2 387	2 235	. .
Ecuador[28]	490	535	490	363	411	. .	. .	. .	. .	. .
Guyana	4.9	5.5	6.6	6.2	5.6	. .	. .	. .	. .	. .
Paraguay	119	109	. .	. .	. .	113	104	88.5	88.1	. .
Peru[11]	907	. .	. .	. .	. .	. .	. .	. .	. .	. .
Uruguay	420	326	482	296	283	279	272	282	236	. .
Venezuela	1 453	1 817	1 319	1 464	895	1 174	(1 251)	(1 190)	(1 207)	. .
Asia and Oceania										
Central Asia										
Kazakhstan[29]	. .	(1 140)	(729)	752	812	759	750	632	556	796
Kyrgyzstan[29]	105	85.3	79.9	146	138	168	156	187	220	[223]
Tajikistan[29]	163	151	86.0	[61.8]	66.5	95.3	84.2	91.2	80.3	. .
Turkmenistan[30]	. .	. .	423	377	363	550	466	504	[681]	. .
Uzbekistan[29]	. .	. .	(495)	(414)	(553)	[642]	. .	[982]	. .	. .

MILITARY EXPENDITURE

East Asia										
Brunei[31]	[292]	[260]	[267]	[255]	279	322	356	..	..	..
Cambodia	[70.7]	[60.5]	[106]	105	94.3	93.4	83.3	86.3	80.5	[78.6]
China, P. R.[32]	[15 400]	[14 200]	[13 600]	[13 900]	[15 300]	[16 600]	[19 000]	[21 100]	[23 100]	[27 000]
Indonesia	[828]	[795]	[878]	[925]	[1 005]	[1 083]	[896]	945	1 111	..
Japan	35 989	36 384	36 554	36 920	37 664	37 845	37 748	37 819	38 080	38 468
Korea, North[33]	(2 083)	(2 133)	(2 190)	..	..	..	1 327	1 327	1 364	1 409
Korea, South	8 220	8 597	8 849	9 309	9 811	10 048	9 700	9 437	10 016	10 201
Laos	..	..	..	..	36.3	31.0	..	..	..	..
Malaysia	1 425	1 514	1 641	1 744	1 677	1 576	1 158	1 568	1 567	1 766
Mongolia	34.1	32.3	28.1	25.3	21.1	19.3	19.9	20.4	25.9	23.1
Myanmar	6 179	7 113	7 559	8 036	8 581	7 136	5 873	5 824	[5 727]	..
Philippines	676	766	816	945	916	1 004	939	[837]	[843]	..
Singapore	2 512	2 591	2 678	3 208	3 514	3 944	4 466	4 550	4 523	4 542
Taiwan	8 189	8 411	9 428	9 352	8 507	8 765	9 030	7 896	7 194	7 207
Thailand	2 179	2 392	2 419	2 598	2 593	2 555	2 426	2 061	[1 808]	[1 731]
Viet Nam	456	358	488	..	..	..	..	..	..	..
South Asia										
Afghanistan	..	..	..	..	..	..	..	..	..	..
Bangladesh	428	465	521	545	536	566	584	607	647	666
India	7 211	8 139	8 111	8 342	8 567	9 310	9 390	10 733	11 793	12 879
Nepal[34]	38.1	39.8	39.5	39.1	38.9	41.2	42.3	45.4	50.8	55.8
Pakistan	3 292	3 287	3 204	3 257	3 252	3 115	3 111	3 139	3 151	3 159
Sri Lanka[35]	361	387	450	757	707	628	658	593	793	700
Oceania										
Australia	6 796	7 104	7 058	6 835	6 751	6 831	7 111	[6 876]	[6 748]	[6 859]
Fiji	28.1	28.8	28.5	27.7	28.1	25.5	24.0	26.6	27.9	..
New Zealand	665	627	593	565	537	544	567	581	580	580
Papua New Guinea	45.8	51.8	63.2	46.2	59.6	63.1	..	38.8	31.1	..
Europe										
Albania	[72.8]	63.7	60.9	56.7	51.0	35.6	33.6	40.0	43.7	39.1

280 MILITARY SPENDING AND ARMAMENTS, 2001

State	1992	1993	1994	1995	1996	1997	1998	1999	2000	2001
Armenia[29]	270	151	. .	256	220	272	273	[293]	[370]	. .
Austria	1 798	1 815	1 823	1 808	1 792	1 794	1 799	1 794	1 783	1 759
Azerbaijan[29]	1 181	949	583	330	338	379	406	514	573	. .
Belarus[36]	1 435	2 049	2 058	817	627	1 027	961	961	1 065	[1 377]
Belgium	4 088	3 883	3 861	3 783	3 709	3 665	3 664	3 710	3 711	3 592
Bosnia and Herzegovina[37]	. .	. .	[136]	[196]	[106]	196	197	217	214	201
Bulgaria	618	467	401	395	323	310	330	377	[378]	[361]
Croatia[38]	1 253	1 340	1 351	1 687	1 352	1 171	1 179	879	688	587
Cyprus	454	204	214	192	289	[312]	[277]	[309]	[324]	. .
Czech Rep.[39]	. .	[1 148]	1 077	1 015	998	946	1 040	1 129	1 144	1 244
Czechoslovakia[40]	2 702	. .	. .	. .	. .	. .	. .	. .	. .	. .
Denmark	2 865	2 872	2 800	2 771	2 780	2 815	2 846	2 829	2 736	2 826
Estonia[41]	25.7	34.6	44.0	43.6	42.4	56.6	59.9	74.5	88.1	103
Finland	1 871	1 819	1 789	1 660	1 877	1 754	1 908	1 643	1 751	1 631
France	44 457	43 964	44 191	42 003	40 993	41 143	40 042	40 379	39 914	40 013
Georgia[42]	[180]	[319]	[407]	[213]	[211]	[246]	[173]	[143]	[109]	[76.5]
Germany	42 407	38 121	35 546	34 962	34 289	33 037	33 146	33 816	33 117	32 371
Greece	4 675	4 564	4 642	4 742	5 025	5 355	5 836	6 110	6 449	6 577
Hungary	889	800	678	598	541	515	618	707	731	781
Iceland	—	—	—	—	—	—	—	—	—	—
Ireland	603	609	636	642	677	717	723	748	805	913
Italy	21 958	22 075	21 529	19 663	21 675	22 727	23 478	24 397	26 025	24 731
Latvia[43]	. .	46.1	53.7	52.0	40.4	39.2	42.0	54.8	68.4	76.2
Lithuania[44]	. .	73.2	39.5	41.1	48.3	79.3	138	119	195	223
Luxembourg	122	111	123	120	123	133	143	145	148	171
Macedonia (FYROM)[45]	. .	. .	. .	. .	97.5	76.9	79.0	88.3	80.3	73.7
Malta	26.7	28.4	30.5	30.6	32.6	31.7	29.1	28.1	32.0	30.4
Moldova[29]	. .	88.4	57.2	83.5	81.4	85.8	57.0	43.2	33.2	35.5
Netherlands	8 005	7 356	7 094	6 892	6 932	6 861	6 836	7 168	6 871	7 172

MILITARY EXPENDITURE 281

Norway	3 536	3 295	3 464	3 129	3 172	3 118	3 325	3 343	3 232	3 276
Poland	2 717	3 011	2 905	2 923	3 075	3 239	3 363	3 283	3 223	3 484
Portugal	2 398	2 315	2 259	2 426	2 339	2 390	2 336	2 457	2 530	2 553
Romania	1 382	834	981	967	940	1 037	827	797	904	888
Russia[46]	[80 400]	[70 900]	[68 600]	[43 400]	[39 500]	[42 200]	[30 600]	[35 900]	[40 300]	[43 900]
Russia (MER)[47]	[23 200]	[20 500]	[19 800]	[12 500]	[11 400]	[12 200]	[8 800]	[10 400]	[11 600]	[12 700]
Slovak Rep.[48]	. .	348	359	462	431	421	398	347	361	394
Slovenia[48]	[318]	[275]	[270]	310	318	303	301	282	252	311
Spain[49]	7 655	8 323	7 494	7 765	7 586	7 655	7 524	7 720	7 997	7 954
Sweden	[4 978]	4 921	4 840	4 213	3 643	4 879	5 036	5 260	5 416	5 358
Switzerland	4 462	3 969	4 077	3 506	3 319	3 200	3 126	3 024	3 035	2 951
UK[50]	44 532	43 528	42 108	38 815	39 442	37 019	37 232	36 778	37 307	36 975
Ukraine[51]	. .	1 445	6 184	6 410	6 053	6 315	6 187	5 305	6 565	5 489
Yugoslavia[52]	. .	. .	. .	. .	. .	[1 179]	1 074	997	1 450	1 122
Middle East										
Bahrain	271	264	267	277	295	289	296	332	328	
Egypt[53]	2 521	2 470	2 451	2 304	2 304	2 329	2 357	2 335	2 316	2 371
Iran[54]	3 596	4 516	6 129	4 537	5 131	5 745	6 064	6 148	9 110	11 515
Iraq	. .	. .	. .	. .	. .	. .	. .	. .	. .	. .
Israel[55]	7 808	7 296	7 346	7 578	8 159	8 207	8 539	8 511	9 012	9 107
Jordan[56]	[552]	[587]	[658]	[715]	642	666	708	742	784	[769]
Kuwait	5 880	2 845	3 019	3 311	2 815	2 146	2 000	2 066	2 598	. .
Lebanon	600	501	630	644	565	473	456	568	575	. .
Oman[57]	1 995	1 876	1 988	2 005	1 902	1 800	[1 605]	[1 620]	[1 938]	[2 689]
Saudi Arabia[58]	15 508	17 516	15 133	13 339	13 318	17 561	20 828	17 904	20 512	26 645
Syria	4 592	3 635	3 923	3 948	3 669	3 786	4 104	[4 184]	[4 526]	[4 737]
Turkey	6 470	7 153	7 006	7 184	8 044	8 380	8 781	9 696	9 383	8 885
UAE[59]	1 992	1 893	1 790	1 715	[1 665]	1 674	1 641	1 607	1 585	. .
Yemen	559	490	515	391	377	438	396	393	[406]	[443]

Table 6A.4. Military expenditure by region and country, as a percentage of gross domestic product, 1992–2000

State	1992	1993	1994	1995	1996	1997	1998	1999	2000
Africa									
North Africa									
Algeria[4]	[2.2]	2.6	3.1	2.9	3.1	2.7	4.0	3.8	3.5
Libya	. .	. .	. .	. .	. .	. .	. .	. .	. .
Morocco	4.3	4.4	4.9	4.6	4.0	3.9	3.7	4.1	4.2
Tunisia	1.9	1.9	1.9	1.9	2.0	1.9	1.8	1.7	1.7
Sub-Saharan									
Angola[5]	12.0	12.5	19.8	17.6	19.5	(22.3)	11.4	21.2	. .
Benin	. .	. .	. .	. .	. .	. .	. .	. .	. .
Botswana	4.3	4.5	3.9	3.5	2.9	3.1	3.9	3.7	3.7
Burkina Faso	2.3	2.1	1.7	1.7	1.6	1.7	1.6	1.7	1.6
Burundi	3.6	3.7	3.9	4.2	5.8	6.0	6.5	6.2	5.4
Cameroon	[1.5]	1.3	1.2	1.2	1.1	1.2	1.4	1.5	1.3
Cape Verde	. .	0.8	0.8	1.3	0.9	0.8	0.9	0.9	1.3
Central African Rep.[6]	1.6	1.5	1.3	1.2	1.2	. .	. .	. .	. .
Chad	. .	2.7	1.9	1.4	1.5	1.1	1.0	. .	. .
Congo, Rep. of	. .	. .	. .	. .	. .	. .	. .	. .	. .
Congo (DRC)[7]	. .	. .	. .	. .	. .	. .	. .	. .	. .
Côte d'Ivoire	1.4	1.4	1.1	. .	0.9	0.9	. .	. .	. .
Djibouti	6.1	5.6	5.4	5.1	4.2	4.5	4.4	. .	. .
Equatorial Guinea	. .	. .	2.3	2.2	. .	. .	. .	. .	. .
Eritrea[8]	. .	21.4	13.0	19.9	22.8	13.5	29.0	22.9	. .
Ethiopia[9]	[2.7]	[2.9]	2.4	2.0	1.9	3.2	5.1	9.4	. .
Gabon	. .	. .	. .	. .	. .	. .	0.3	. .	. .
Gambia[10]	1.0	0.7	0.6	0.8	1.0	1.9	0.9	0.8	(1.1)
Ghana	0.6	0.7	0.7	0.8	0.6	0.7	0.8	0.8	1.0
Guinea	1.9	1.4	1.4	. .	. .	1.2	1.5	. .	(1.5)

MILITARY EXPENDITURE 283

Guinea-Bissau[11]	:	:	0.3	0.5	0.6	0.7	1.3	:
Kenya	1.9	1.8	1.6	1.6	1.8	1.7	1.5	1.6
Lesotho	2.6	2.3	2.8	2.8	2.6	2.9	3.2	3.1
Liberia[12]	10.6	23.3	31.2	:	:	:	:	1.8
Madagascar[13]	1.2	1.1	0.9	0.9	1.2	1.5	1.4	[1.4]
Malawi	1.4	1.3	1.5	0.8	0.8	1.0	0.8	0.8
Mali	:	[2.4]	2.3	2.3	2.1	2.2	2.0	2.2
Mauritania[14]	3.5	3.2	2.9	2.7	2.5	2.3	:	:
Mauritius	0.4	0.3	0.3	0.3	0.3	0.2	0.2	0.2
Mozambique[15]	5.1	5.0	5.7	2.5	2.2	2.1	2.2	2.4
Namibia[16]	4.3	2.6	1.9	1.9	1.9	2.3	2.3	3.1
Niger	:	:	1.2	1.1	1.0	1.0	1.2	1.4
Nigeria[17]	0.5	0.4	0.6	0.7	0.5	0.6	0.8	1.3
Rwanda[18]	4.4	4.6	3.4	4.4	5.3	4.1	4.4	4.6
Senegal[19]	1.8	2.2	1.8	1.8	1.7	1.6	1.6	1.6
Seychelles	4.7	2.8	2.4	2.3	2.1	1.8	1.8	1.9
Sierra Leone	2.5	2.6	2.5	2.4	2.0	1.0	0.8	1.6
Somalia	:	:	:	:	:	:	:	:
South Africa	2.9	2.5	2.6	2.2	1.8	1.6	1.5	1.3
Sudan[11]	2.5	2.8	2.5	1.7	0.9	1.0	2.2	2.6
Swaziland	1.9	2.1	2.0	2.0	1.9	1.8	2.0	2.0
Tanzania	1.9	1.2	1.2	1.5	1.4	1.3	(1.3)	(1.3)
Togo	2.9	4.0	2.6	2.4	:	:	:	:
Uganda[20]	1.5	1.8	[1.6]	[1.5]	1.8	1.9	2.1	2.1
Zambia[21]	3.0	1.5	1.7	2.2	1.4	1.8	1.9	1.0
Zimbabwe[22]	3.7	3.4	3.3	3.6	3.2	3.4	2.6	3.4
America								
Central America								
Belize	1.1	1.2	1.4	1.4	1.3	1.5	:	:
Costa Rica	:	:	:	:	:	:	:	:
El Salvador	2.0	1.5	1.2	1.0	0.9	0.9	0.8	0.8

(1.8)
:
:
1.2
0.8
[2.5]
:
0.2
2.5
3.3
:
0.9
[3.0]
1.4
1.8
1.4
:
1.5
3.0
1.6
:
:
1.8
[0.6]
4.8
:
:
0.7

State	1992	1993	1994	1995	1996	1997	1998	1999	2000
Guatemala	[1.3]	1.1	1.1	1.0	0.8	0.7	0.7	0.7	0.8
Honduras	. .	. .	(1.3)	. .	. .	. .	. .	(0.6)	. .
Mexico	[0.5]	[0.5]	[0.6]	0.6	0.6	0.6	0.5	0.6	0.5
Nicaragua[11]	2.4	2.1	1.9	1.7	1.5	1.3	1.2	1.1	[1.1]
Panama[23]	1.2	1.3	1.3	1.2	1.2	1.4	1.1	1.2	. .
North America									
Canada[24]	1.9	1.8	1.7	1.5	1.4	1.2	1.3	1.3	1.2
USA[24]	4.8	4.5	4.1	3.8	3.5	3.3	3.1	3.0	3.1
South America									
Argentina[25]	[1.4]	[1.6]	1.6	1.7	1.5	1.4	1.3	1.5	1.3
Bolivia	2.1	2.2	2.1	1.9	1.8	2.1	2.4	1.8	1.5
Brazil[26]	1.1	1.3	1.2	1.5	1.3	1.5	1.4	1.3	1.3
Chile[27]	[3.3]	3.2	3.4	3.1	3.2	3.1	3.5	3.1	[3.3]
Colombia	[2.6]	2.5	1.9	2.1	2.5	2.8	2.2	2.5	2.3
Ecuador[28]	2.7	3.1	2.7	1.9	2.1	. .	. .	. .	. .
Guyana	1.0	1.0	1.0	0.9	0.8	. .	. .	. .	. .
Paraguay	1.6	1.4	. .	. .	. .	1.3	1.2	1.1	1.0
Peru[11]	2.2	. .	. .	. .	. .	. .	. .	. .	. .
Uruguay	2.1	1.6	2.4	1.5	1.4	1.3	1.2	1.3	1.1
Venezuela	1.3	1.7	1.3	1.4	0.8	1.1	(1.3)	(1.3)	(1.2)
Asia and Oceania									
Central Asia									
Kazakhstan[29]	. .	(1.0)	(0.9)	1.1	1.1	1.1	1.1	0.9	0.7
Kyrgyzstan[29]	0.7	0.7	0.9	1.7	1.5	1.6	1.5	1.7	1.9
Tajikistan[29]	0.4	3.9	1.7	[1.1]	1.3	1.7	1.3	1.4	1.2
Turkmenistan[30]	. .	. .	1.8	2.3	2.0	4.0	3.1	3.4	[3.8]
Uzbekistan[29]	. .	. .	(1.5)	(1.1)	(1.2)	[1.4]	. .	[1.7]	. .
East Asia									

MILITARY EXPENDITURE

Brunei[31]	[6.5]	[6.0]	[6.3]	[5.7]	6.2	6.9	7.6	:	2.4
Cambodia	[3.6]	[3.0]	[4.9]	4.2	3.6	3.3	2.9	2.8	[2.1]
China, P. R.[32]	[2.7]	[2.1]	[1.9]	[1.8]	[1.8]	[1.9]	[2.0]	[2.1]	[2.1]
Indonesia	[1.2]	[1.1]	[1.2]	[1.1]	[1.1]	[1.1]	[0.9]	1.0	1.1
Japan	0.9	0.9	1.0	0.9	0.9	0.9	1.0	1.0	1.0
Korea, North[33]	:	:	:	:	:	:	:	:	:
Korea, South	3.4	3.3	3.1	2.9	2.9	2.9	3.1	2.8	2.8
Laos	:	:	:	:	:	:	:	:	:
Malaysia	3.0	2.9	2.8	2.8	2.4	2.1	1.6	2.1	1.9
Mongolia	2.5	2.5	2.4	1.7	1.8	1.8	2.0	2.0	2.5
Myanmar	3.4	3.5	3.5	3.7	3.5	2.7	2.3	2.0	[1.7]
Philippines	1.3	1.4	1.4	1.6	1.5	1.5	1.4	[1.2]	[1.2]
Singapore	4.8	4.3	4.0	4.4	4.5	4.7	5.4	5.4	4.8
Taiwan	4.4	4.2	4.5	4.2	3.6	3.5	3.4	2.8	2.5
Thailand	2.3	2.3	2.2	2.1	2.0	2.1	2.2	1.9	[1.6]
Viet Nam	3.4	2.3	2.6	:	:	:	:	:	:
South Asia									
Afghanistan	:	:	:	:	:	:	:	:	:
Bangladesh	1.1	1.2	1.2	1.3	1.2	1.2	1.2	1.3	1.3
India	2.3	2.4	2.3	2.2	2.1	2.2	2.2	2.4	2.4
Nepal[34]	0.9	0.9	0.9	0.8	0.8	0.8	0.8	0.9	0.9
Pakistan	6.1	5.7	5.3	5.3	5.1	4.9	4.8	4.6	4.5
Sri Lanka[35]	3.0	3.1	3.4	5.3	5.0	4.2	4.2	3.6	4.5
Oceania									
Australia	2.3	2.3	2.2	2.1	2.0	1.9	2.0	[1.8]	[1.7]
Fiji	2.0	1.9	1.8	1.7	1.7	1.6	1.5	1.5	:
New Zealand	1.5	1.4	1.2	1.1	1.0	1.0	1.1	1.1	1.0
Papua New Guinea	1.3	1.4	1.6	1.2	1.5	1.6	:	1.0	0.8
Europe									
Albania	[4.6]	3.2	2.5	2.1	1.7	1.3	1.1	1.2	1.2
Armenia[29]	2.2	2.3	:	4.1	3.3	3.8	3.5	[3.6]	[4.4]

State	1992	1993	1994	1995	1996	1997	1998	1999	2000
Austria	1.0	1.0	0.9	0.9	0.9	0.9	0.9	0.8	0.8
Azerbaijan[29]	3.3	5.0	4.6	2.3	2.2	2.2	2.3	2.6	2.7
Belarus[36]	1.5	2.6	3.4	1.6	1.2	1.7	1.4	1.3	1.3
Belgium	1.8	1.7	1.7	1.6	1.4	1.4	1.5	1.4	1.4
Bosnia and Herzegovina[37]	. .	. .	[13.7]	[14.3]	[3.8]	5.3	4.7	4.6	4.2
Bulgaria	3.2	2.8	2.7	2.6	2.4	2.7	2.7	3.0	[3.0]
Croatia[38]	7.6	8.2	8.2	9.4	7.2	5.7	5.5	4.1	3.0
Cyprus	6.2	2.7	2.7	2.3	3.4	[3.6]	[3.0]	[3.2]	[3.2]
Czech Rep.[39]	. .	[2.3]	2.1	1.8	1.7	1.6	1.8	2.0	2.0
Czechoslovakia[40]									
Denmark	1.9	1.9	1.8	1.7	1.7	1.7	1.6	1.6	1.5
Estonia[41]	0.5	0.8	1.1	1.0	1.0	1.1	1.2	1.4	1.6
Finland	1.9	1.9	1.8	1.5	1.7	1.5	1.5	1.2	1.3
France	3.4	3.3	3.3	3.1	3.0	2.9	2.8	2.7	2.6
Georgia[42]	. .	[1.9]	[4.4]	[2.3]	[2.0]	[2.1]	[1.4]	[1.2]	[0.9]
Germany	2.1	1.9	1.7	1.7	1.6	1.6	1.5	1.5	1.5
Greece	4.5	4.4	4.4	4.3	4.5	4.6	4.8	4.8	4.9
Hungary	2.1	1.9	1.6	1.4	1.2	1.1	1.3	1.5	1.5
Iceland	—	—	—	—	—	—	—	—	—
Ireland	1.2	1.1	1.1	1.0	1.0	0.9	0.8	0.8	0.7
Italy	2.0	2.1	2.0	1.8	1.9	1.9	2.0	2.0	2.1
Latvia[43]	. .	0.8	0.9	1.0	0.7	0.7	0.7	0.8	1.0
Lithuania	. .	0.7	0.5	0.5	0.5	0.8	1.3	1.1	1.8
Luxembourg	0.9	0.8	0.8	0.8	0.8	0.8	0.8	0.7	0.7
Macedonia (FYROM)[44]	. .	. .	. .	. .	3.0	2.3	2.3	2.4	2.1
Malta	1.0	1.0	1.0	1.0	1.0	0.9	0.8	0.8	0.8
Moldova[29]	. .	0.5	0.8	0.9	0.9	0.9	0.6	0.5	0.4
Netherlands	2.4	2.2	2.0	1.9	1.9	1.8	1.7	1.8	1.6
Norway	3.0	2.7	2.8	2.4	2.2	2.1	2.3	2.2	1.8

Poland	2.3	2.6	2.3	2.1	2.1	2.1	2.1	2.0	1.9
Portugal	2.7	2.6	2.5	2.6	2.4	2.3	2.2	2.1	2.1
Romania	3.3	2.1	2.4	2.1	1.9	2.3	2.0	1.9	2.1
Russia[45]	[5.5]	[5.3]	[5.9]	[4.1]	[3.8]	[4.2]	[3.1]	[3.5]	[3.6]
Slovak Rep.[46]	:	2.1	2.1	2.5	2.2	2.0	1.9	1.7	1.8
Slovenia[47]	[2.2]	[1.8]	[1.7]	1.8	1.7	1.6	1.5	1.4	1.2
Spain[48]	1.6	1.7	1.5	1.5	1.5	1.4	1.4	1.3	1.3
Sweden	[2.5]	2.6	2.3	1.9	1.6	2.1	2.1	2.1	2.1
Switzerland	1.8	1.6	1.6	1.4	1.3	1.2	1.2	1.1	1.1
UK[49]	3.8	3.5	3.3	3.0	2.9	2.7	2.6	2.5	2.5
Ukraine[50]	:	0.5	2.8	3.1	3.5	3.7	3.6	3.0	3.6
Yugoslavia[51]	:	:	:	:	:	[4.8]	4.4	4.4	5.9
Middle East									
Bahrain	5.2	5.0	4.6	4.7	4.7	4.6	4.8	4.9	4.0
Egypt[52]	3.3	3.3	3.0	2.9	2.8	2.7	2.6	2.4	2.3
Iran[53]	1.9	2.5	2.9	2.4	2.6	3.0	3.1	2.7	3.8
Iraq	:	:	:	:	:	:	:	:	:
Israel[54]	10.5	9.4	8.8	8.3	8.6	8.4	8.4	8.0	8.0
Jordan[55]	[8.7]	[8.8]	[9.2]	[9.4]	8.6	8.8	8.9	9.2	9.5
Kuwait	31.8	12.4	13.3	13.9	10.4	8.2	9.0	8.1	8.2
Lebanon	5.2	4.0	4.6	4.4	3.7	3.0	2.8	3.6	3.6
Oman[56]	16.2	15.4	15.7	14.6	12.5	11.5	[11.4]	[10.4]	[9.7]
Saudi Arabia[57]	11.7	13.9	11.9	10.3	9.5	12.0	16.2	12.3	11.6
Syria	9.0	7.2	7.4	7.1	5.9	5.7	5.8	[5.6]	[5.5]
Turkey	3.7	3.8	4.1	3.9	4.1	4.1	4.3	5.0	4.9
UAE[58]	4.5	4.5	4.3	4.0	[3.6]	3.3	3.5	3.2	2.6
Yemen	9.1	8.7	10.4	7.3	6.4	6.5	6.7	5.6	[5.2]

Conventions:
( ) Uncertain figure
[ ] SIPRI estimate
| Change of multiple or change of currency

Notes:

[1] Contributions of military expenditure data, estimates and advice are gratefully acknowledged from Emmanuel Kwesi Aning (African Security Research and Dialogue, Accra), Valentyn Badrak (Center for Army Conversion and Disarnament Studies, Kiev), Mesfin Binega (Organization of African Unity, Adis Abeba), Julian Cooper (Centre for Russian and East European Studies, University of Birmingham), David Darchiashvili (Center for Civil–Military Relations and Security Studies, Tbilisi), Dimitar Dimitrov (University of National and World Economy, Sofia), Paul Dunne (Middlesex University, London), Guy Lamb (Centre for Conflict Resolution, Cape Town), Reuven Pedatzur (Tel Aviv University), Thomas Scheetz (Buenos Aires), Ron Smith (Birkbeck College, London), Shaoguang Wang (Chinese University of Hong Kong) and Ozren Zunec (University of Zagreb).

[2] Military expenditure data from different volumes of the SIPRI Yearbook should not be combined because of data revision between volumes. Revision can be significant; e.g., when a better time series becomes available, the entire SIPRI series is revised accordingly. When data are available in local currency but not in constant US$ or as a share of GDP, this is due to lack of economic data. Revisions in constant dollar series can also originate in significant revisions in the economic statistics of IMF that are used for these calculations.

[3] Figures in constant dollars are converted using the market exchange rate for all countries except Armenia, Azerbaijan, Belarus, Georgia, Kazakhstan, North Korea, Kyrgyzstan, Moldova, Russia, Tajikistan, Turkmenistan, Ukraine and Uzbekistan. For these countries, conversion to dollars has been done using the purchasing power parity (PPP) rates from *World Development Indicators 2000* (International Bank for Reconstruction and Development: Washington, DC, Mar. 2000).

[4] The figures for Algeria are budget figures for recurrent expenditure only.

[5] Figures for Angola should be seen in the context of highly uncertain economic statistics because of the impact of war on the Angolan economy.

[6] The figures for the Central African Republic are for current expenditure only.

[7] Formerly Zaire.

[8] Eritrea became independent from Ethiopia in May 1993. Figures for 1995 include expenditure for demobilization.

[9] The figure for Ethiopia in 1999 includes an allocation of 1 billion birr in addition to the original defence budget.

[10] Figures for the Gambia are for current expenditures only.

[11] This country has changed currency during the period. All figures have been converted to the most recent currency.

[12] The figure for Liberian military expenditure in 1999 is for security, which represents 13% of total expenditure of $64 million.

[13] The figure for Madagascar include expenditure for the gendarmerie and the National Police.

[14] Figures for Mauritania are for operating expenditures only.

[15] Figures for Mozambique include expenditure for the demobilization of government and RENAMO soldiers and the formation of a new unified army from 1994 onwards. Figures are for defence and security.

[16] Namibia became independent on 21 Mar. 1990. During the period 1990/91–1992/93 military construction accounted for more than half of Namibian military expenditure. Figures for 1999 refer to the budget of the ministry of defence only. In addition, the 1999 budget of the ministry of finance includes a contingency provision of 104 million ND for the Namibian military presence in the Democratic Republic of the Congo (DRC).

[17] Figures for Nigeria before 1999 are understated because of the use by the military of a favourable specific dollar exchange rate.

[18] Figures for Rwanda in 1997 do not include a demobilization allowance of 1.0 billion francs. The figure for 1998 is the official defence budget. According to the International Monetary Fund (IMF) there are additional sources of funding for military activities, both within budget and extra-budgetary. Alternative estimates put Rwanda's military expenditure at twice the official figure.

[19] Figures for Senegal do not include expenditure for paramilitary forces, which in 1998 amounted to 21 100 million francs.

[20] Figures for Uganda are for current expenditure only.

[21] Figures for Zambia are uncertain, especially in constant dollars and shares of GDP, because of very rapid inflation and several changes in the currency.

[22] The figure for Zimbabwe in 1999 includes a supplementary allocation of 1800 million ZD.

[23] The Panamanian defence forces were disbanded in 1990 and replaced by the national guard, consisting of the national police and the air and maritime services.

[24] Figures are for fiscal year rather than for calendar year.

[25] Figures for Argentina are uncertain because of very rapid inflation and a change in the currency. All figures have been converted to the most recent currency.

[26] The figure for Brazil in 2001 is likely to be significantly revised downwards because in Brazil actual defence expenditure are usually much lower than budgeted expenditure.

[27] Figures for Chile are based on estimates by the IMF of military expenditures including military pensions and direct transfers from the Corporación Nacional del Cobre (CODELCO, National Copper Corporation) for military purchases. Source: IMF Staff Country Report no. 00/104, Aug. 2000, p. 151.

[28] Ecuador changed its currency from the sucre to the US dollar on 13 Mar. 2000, at a rate of one dollar to 25 000 sucres.

[29] Figures are converted to dollars using the PPP.

[30] Figures are converted to dollars using the PPP. The coverage of this series varies over time due to classification changes in the Turkmenistan system of public accounts.

[31] Figures for Brunei are current expenditure on the Royal Brunei Armed Forces.

[32] Figures for China are for estimated total military expenditures. On the estimates in local currency and share of GDP, see Shaoguang Wang, 'The military expenditure of China, 1989–98', *SIPRI Yearbook 1999: Armaments, Disarmament and International Security* (Oxford University Press: Oxford, 1999), pp. 334–49. Dollar figures are converted using the market exchange rate.

[33] Dollar figures for North Korea are in current dollars. Figures are converted to dollars using the PPP.

[34] Figures for Nepal do not include expenditure on paramilitary forces, which in fiscal year 1998/99 amounted to 3315 million taka.

[35] Figures for Sri Lanka are for current expenditure only. The special allocation in 2000 of Rs 28 billion for war-related expenditure is therefore not reflected in the official figure.

[36] Figures are for central and local expenditures on defence. Figures are converted to dollars using the PPP.

[37] Bosnia and Herzegovina declared its independence from the former Yugoslavia in Mar. 1992 and was recognized by the European Community and the USA in Apr. 1992. The local currency since Jan. 1998 is the convertible mark, set at 1 convertible mark = 1 Deutsche Mark. Figures for Bosnia and Herzegovina include expenditure for both the Army of the Federation of Bosnia and Herzegovina and the Army of the Republika Srpska. The Army of the Federation of Bosnia and Herzegovina is divided into 2 components—a Croat and a Bosniak. The figures are based on estimates by the IMF of military expenditure for both entities, excluding off-budget assistance from other countries for the Republika Srpska. Source: IMF Staff Country Report no. 01/106, July 2001, pp. 27–28.

[38] Croatia declared its independence from the former Yugoslavia in June 1991 and was recognized by the European Community in Jan. 1992 and by the United Nations in May 1992.

[39] The Czech Republic was formed on 1 Jan. 1993 after the break-up of Czechoslovakia.

[40] The Czech Republic and Slovakia were formed on 1 Jan. 1993 after the break-up of Czechoslovakia. Figures in the table for constant dollars are at current prices and 1990 exchange rates.

[41] Figures do not include expenditures for paramilitary forces.

[42] Figures are converted to dollars using the PPP. Figures probably do not include the significant amounts of military aid received from Turkey.

[43] Figures do not include: (*a*) allocations for military pensions paid by Russia, which averaged 27 million lats per year over the 3 years 1996–98; or (*b*) expenditure on paramilitary forces, which amounted to 98.5 million lats in 1999.

[44] The Republic of Macedonia declared its independence from the former Yugoslavia in Nov. 1992 and was admitted as the Former Yugoslav Republic of Macedonia (FYROM) to the United Nations in Apr. 1993.

[45] For sources and methods of the military expenditure figures for the USSR and Russia, see Cooper, J., 'The military expenditure of the USSR and the Russian Federation, 1987–97', *SIPRI Yearbook 1998: Armaments, Disarmament and International Security* (Oxford University Press: Oxford, 1998), appendix 6D, pp. 243–59. Dollar figures are converted using the PPP. If the market exchange rate is used, the level of Russian military expenditure in constant (1998) US dollars is less than 30% of the level in the SIPRI tables. This series is shown in italics.

[46] Slovakia was formed on 1 Jan. 1993 after the break-up of Czechoslovakia. Figures do not include expenditure on pensions or paramilitary forces. Expenditure on paramilitary forces amounted to 400 million korunas in 1998 and 458 million in 1999.

[47] Slovenia declared its independence from the former Yugoslavia in June 1991 and was recognized by the European Community in Jan. 1992 and by the United Nations in May 1992.

[48] Figures for Spain do not include a major part of government military R&D expenditure of *c.* 111.7 billion pesetas ($700 million) in 1998, 163.1 billion pesetas ($1000 million) in 1999, and 159.4 billion pesetas ($900 million) in 2000, financed by the ministry of industry.

[49] The series for the UK has a break between 2000 and 2001, because in 2001 the UK changed its accounting system for defence expenditure from 'cash basis' to 'resource basis'. Figures are for fiscal year rather than for calendar year.

⁵⁰ Figures are converted to dollars using the PPP. Figures for Ukraine are for the adopted budget for 'National Defence' and some other defence items. Actual expenditure was reportedly 95–99% of budgeted expenditure for the years 1996–99 and about 80–90% of budgeted expenditure for 1994–95.
⁵¹ The Federal Republic of Yugoslavia was created in Apr. 1992, comprising the republics of Serbia and Montenegro. Figures for Yugoslavia include military pensions and arms imports.
⁵² Figures for Egypt include military aid from the USA of approximately $1.3 billion annually.
⁵³ Figures for Iran include expenditures for public order and safety.
⁵⁴ Figures for Israel include military aid from the USA of approximately $2 billion annually.
⁵⁵ Figures for Jordan are expenditure for defence and security.
⁵⁶ Figures for Oman are for recurrent expenditure on defence and national security.
⁵⁷ Figures for Saudi Arabia are for defence and security.
⁵⁸ Figures for the UAE exclude the local military expenditure of each of the 7 emirates that form the United Arab Emirates.

Source: SIPRI military expenditure database.

Appendix 6B. Tables of NATO military expenditure by category

Table 6B.1. NATO military expenditure on personnel and equipment, 1992–2001

Figures are in US $m. at 1998 prices and exchange rates. Figures in italics are percentage changes from previous year.

Country	Item	1992	1993	1994	1995	1996	1997	1998	1999	2000	2001
North America											
Canada	Personnel	4 790	4 559	4 797	4 178	3 652	3 132	3 412	3 446	3 484	3 290
	Person. change	*1.6*	*−4.8*	*5.2*	*−12.9*	*−12.6*	*−14.3*	*9.0*	*1.0*	*1.1*	*−5.6*
	Equipment	1 786	1 835	1 624	1 621	1 239	950	869	669	985	1 038
	Equip. change	*3.4*	*2.8*	*−11.5*	*−0.2*	*−23.6*	*−23.3*	*−8.6*	*−23.0*	*47.2*	*5.4*
USA	Personnel	139 321	130 345	123 543	118 811	109 514	109 855	106 907	104 710	107 778	101 986
	Person. change	*−3.9*	*−6.4*	*−5.2*	*−3.8*	*−7.8*	*0.3*	*−2.7*	*−2.1*	*2.9*	*−5.4*
	Equipment	81 182	73 907	92 499	82 661	75 781	72 939	70 143	68 530	62 610	61 792
	Equip. change	*−11.4*	*−9.0*	*25.2*	*−10.6*	*−8.3*	*−3.8*	*−3.8*	*−2.3*	*−8.6*	*−1.3*
Europe											
Belgium	Personnel	2 669	2 703	2 676	2 689	2 561	2 541	2 509	2 539	2 442	2 430
	Person. change	*−20.8*	*1.2*	*−1.0*	*0.5*	*−4.8*	*−0.8*	*−1.3*	*1.2*	*−3.8*	*−0.5*
	Equipment	335	272	301	203	196	228	216	243	215	195
	Equip. change	*−16.4*	*−18.9*	*10.8*	*−32.7*	*−3.3*	*16.1*	*−5.3*	*12.6*	*−11.4*	*−9.1*
Czech Rep.	Personnel								529	490	572
	Person. change									*−7.5*	*16.9*
	Equipment									257	253
	Equip. change								184	*39.9*	*−1.5*
Denmark	Personnel	1 624	1 631	1 644	1 677	1 660	1 656	1 709	1 698	1 493	1 545
	Person. change	*−2.6*	*0.4*	*0.8*	*2.0*	*−1.0*	*−0.3*	*3.2*	*−0.7*	*−12.1*	*3.5*
	Equipment	510	419	445	346	348	386	394	323	331	323
	Equip. change	*10.7*	*−17.8*	*6.2*	*−22.3*	*0.5*	*11.2*	*1.9*	*−18.1*	*2.5*	*−2.2*

MILITARY EXPENDITURE 293

France	Personnel	24 851	22 644	21 612	21 552	21 282	23 503	24 273	24 364	24 104	24 087
	Person. change	−1.7	−8.9	−4.6	−0.3	−1.3		3.3	0.4	−1.1	−0.1
	Equipment	5 640	4 231	3 875	3 980	3 802	9 057	7 752	7 833	7 527	7 952
	Equip. change	−19.0	−25.0	−8.4	2.7	−4.5		−14.4	1.1	−3.9	5.7
Germany	Personnel						20 714	20 271	20 221	20 091	19 822
	Person. change						−2.7	−2.1	−0.3	−0.6	−1.3
	Equipment						3 557	4 206	4 464	4 475	4 215
	Equip. change						−6.5	18.3	6.1	0.2	−5.8
Greece	Personnel	2 870	2 839	2 924	3 003	3 074	3 333	3 526	3 751	4 029	4 218
	Person. change	−0.9	−1.1	3.0	2.7	2.4	8.4	5.8	6.4	7.4	4.7
	Equipment	1 094	1 127	1 133	940	1 062	1 040	1 204	1 183	1 149	992
	Equip. change	19.8	3.1	0.5	−17.0	13.0	−2.1	15.9	−1.7	−2.9	−13.7
Hungary	Personnel								330	356	374
	Person. change									7.8	5.2
	Equipment								149	90	82
	Equip. change									−39.2	−9.2
Italy	Personnel	13 987	13 885	14 123	13 249	15 009	17 128	17 219	18 060	18 574	17 803
	Person. change	−3.5	−0.7	1.7	−6.2	13.3	14.1	0.5	4.9	2.9	−4.2
	Equipment	3 294	3 797	3 337	2 947	3 108	2 580	2 919	2 852	3 732	3 076
	Equip. change	−10.7	15.3	−12.1	−11.7	5.5	−17.0	13.2	−2.3	−30.8	−17.6
Luxembourg	Personnel	92	86	96	97	101	105	110	110	112	115
	Person. change	12.0	−7.2	12.1	0.9	4.6	3.2	5.3	0.1	1.9	2.5
	Equipment	6	3	3	3	5	5	9	7	7	28
	Equip. change	−11.2	−44.6	−16.9	9.9	78.3	−8.8	101	−21.5	−5.7	304
Netherlands	Personnel	4 603	4 369	4 136	4 134	3 865	3 840	3 573	3 570	3 493	3 329
	Person. change	3.5	−5.1	−5.4	0.0	−6.5	−0.7	−6.9	−0.1	−2.1	−4.7
	Equipment	1 137	1 030	1 192	1 072	1 297	1 080	1 049	1 211	1 171	1 265
	Equip. change	−9.5	9.4	15.7	−10.0	21.0	−16.7	−2.9	15.4	−3.3	8.0

Country	Item	1992	1993	1994	1995	1996	1997	1998	1999	2000	2001
Norway	Personnel	1 549	1 186	1 209	1 168	1 189	1 200	1 254	1 303	1 319	1 348
	Person. change	2.5	−23.4	1.9	−3.4	1.8	0.9	4.5	3.9	1.2	2.3
	Equipment	863	909	987	794	800	768	831	755	628	765
	Equip. change	20.1	5.4	8.6	−19.6	0.7	−4.0	8.3	−9.1	−16.8	21.8
Poland	Personnel								2 049	2 007	2 263
	Person. change									−2.0	12.8
	Equipment								365	282	312
	Equip. change									−22.6	10.7
Portugal	Personnel	1 930	1 847	1 778	1 886	1 887	1 911	1 934	2 044	2 070	2 067
	Person. change	10.4	−4.3	−3.8	6.1	0.0	1.3	1.2	5.7	1.3	−0.2
	Equipment	53	167	95	143	147	197	89	103	162	160
	Equip. change	−73.4	216	−43.1	50.5	2.9	33.8	−54.6	15.2	57.9	−1.2
Spain	Personnel	5 320	5 185	4 961	5 103	5 108	5 063	5 076	5 092	5 112	5 165
	Person. change	−0.6	−2.5	−4.3	2.9	0.1	−0.9	0.3	0.3	0.4	1.0
	Equipment	834	1 124	914	1 057	1 015	1 041	902	890	1 035	1 051
	Equip. change	−21.8	34.7	−18.6	15.7	−4.0	2.6	−13.4	−1.3	16.3	1.5
UK	Personnel	19 505	18 935	17 433	16 237	15 878	14 569	14 145	13 950	14 256	17 781
	Person. change	−5.1	−2.9	−7.9	−6.9	−2.2	−8.2	−2.9	−1.4	2.2	3.7
	Equipment	8 060	11 317	10 485	8 550	9 429	9 232	9 848	9 875	9 601	9 160
	Equip. change	−15.7	40.4	−7.4	−18.5	10.3	−2.1	6.7	0.3	−2.8	−4.6
Middle East											
Turkey	Personnel	3 151	3 898	3 573	3 658	3 717	4 059	4 255	4 559	4 230	3 454
	Person. change	5.6	23.7	−8.4	2.4	1.6	9.2	4.8	7.2	−7.2	−18.4
	Equipment	1 605	1 638	2 053	2 132	2 482	2 264	1 813	2 473	2 655	3 261
	Equip. change	14.9	2.1	25.3	3.9	16.4	−8.8	−19.9	36.4	7.4	22.8

MILITARY EXPENDITURE 295

	1992	1993	1994	1995	1996	1997	1998	1999	2000	2001
NATO Western Europe										
Personnel	79 002	75 310	72 591	70 796	71 614	72 061	71 327	72 337	72 992	72 622
Person. change	–2.9	–4.7	–3.6	–2.5	1.2	0.6	–1.0	1.4	0.9	–0.5
Equipment	21 826	24 397	22 766	20 035	21 208	20 112	21 667	21 906	22 506	21 230
Equip. change	–13.5	11.8	–6.7	–12.0	5.9	–5.2	7.7	1.1	2.7	–5.7
NATO total										
Personnel	226 264	214 112	204 504	197 443	188 497	189 107	185 901	187 961	191 338	184 561
Person. change	–3.3	–5.4	–4.5	–3.5	–4.5	0.3	–1.7	1.1	1.8	–3.5
Equipment	106 398	101 776	118 941	106 448	100 709	96 265	94 492	94 275	89 385	87 968
Equip. change	–11.3	–4.3	16.9	–10.5	–5.4	–4.4	–1.8	–0.2	–5.2	–1.6

Note: The figures in this table were calculated on the basis of NATO statistics on the distribution of total military expenditure by category. The shares for personnel and equipment are applied to the figures for total military expenditure in constant 1998 US dollars, as presented in table 6A.3. Prior to 2001, France was not included in the NATO statistics on military expenditure by category. In table 6B.2, SIPRI therefore provides its own calculation of French expenditure on military equipment, as presented in its official defence budget. The revision of figures in constant dollars from previous SIPRI Yearbooks is due almost entirely to revisions in the economic statistics (consumer prices indices), as provided by the IMF, and not to changes in the original figures for military expenditure and the shares thereof.

Sources: NATO, *Financial and Economic Data Relating to NATO Defence–Defence Expenditures of NATO Countries (1980–2001)*, Press release M-DPC-2(2001)156, 18 Dec. 2001, URL <http://www.nato.int/docu/pr/2001/p01-156e.htm>; and NATO Press releases M-DPC-2 (2000)107 (5 Dec. 2000); M-DPC-2 (1999)152 (2 Dec. 1999); M-DPC-2(97)147 (2 Dec. 1997); M-DPC-2(96)168 (17 Dec. 1996); and M-DPC-2(95)115 (29 Nov. 1995).

Table 6B.2. Military equipment expenditure of France, 1992–2001

Figures are in US $m. at 1998 prices and exchange rates. Figures in italics are percentage changes from previous year.

Item	1992	1993	1994	1995	1996	1997	1998	1999	2000	2001
Equipment	17 473	16 161	15 863	13 165	13 470	12 968	11 685	11 634	11 624	13 570
Equipment change	–1.8	–7.5	–1.8	–17.0	2.3	–3.7	–9.9	–0.4	–0.1	16.7

Note: The figures in this table were calculated based on data as presented in the French defence budget. They refer to actual expenditures on military equipment and include all items covered by titles V and VI of the French defence budget (i.e., research and development, prototype construction, procurement of finished equipment, infrastructure and technical and industrial investment, and investment subsidies). These figures are not comparable to the equipment expenditure as defined by NATO and presented in table 6B.1.

Sources: French Assemblée Nationale, 'Avis présenté au nom de la Commission de la défense nationale et des forces armées sur le projet de loi de finances pour 2001, tome VIII, Défense: crédits d'équipement', Report 2627, Paris, 11 Oct. 2000, p. 11; French Assemblée Nationale, 'Avis présenté au nom de la Commission de la défense nationale et des forces armées sur le projet de loi de finances rectificative pour 2000', Report 2764, Paris, 29 Nov. 2000; and French Assemblée Nationale, 'Avis présenté au nom de la Commission de la défense nationale et des forces armées sur le projet de loi de finances pour 2002, tome VIII, Défense: crédits d'équipement', Report 3262, Paris, 11 Oct. 2001, pp. 3–4.

Appendix 6C. Sources and methods for military expenditure data

This appendix describes the sources and methods for the SIPRI military expenditure data provided in the tables in chapter 6, appendix 6A and on the SIPRI Internet site, URL <http://projects.sipri.se/milex.html>. For a more comprehensive overview of the conceptual problems and sources of uncertainty involved in all sets of military expenditure data, the reader is referred to other sources.[1] A major revision of the SIPRI military expenditure series has been made during recent years to improve its consistency over time. Thus the revised series, for the period beginning in 1988, cannot always be combined with the SIPRI series for earlier years, 1950–87. There is also a continuous revision and updating of the data, in particular for the most recent years, as data for budget allocations are replaced by data for actual expenditures. The base year for the constant dollar series was changed from 1995 to 1998 in the *SIPRI Yearbook 2001*.

I. Purpose of the data

The main purpose of the data on military expenditure is to provide an easily identifiable measure of the scale of resources absorbed by the military. Military expenditure is an input measure which is not directly related to the 'output', of military activities, such as military capability or military security. Long-term trends in military expenditure and sudden changes in trend may be signs of a change in military output, but such interpretations should be made with caution.

Military expenditure data as measured in constant dollars (table 6A.3) are an indicator of the trend in the volume of resources used for military activities with the purpose of allowing comparisons over time for individual countries and comparisons between countries. The share of gross domestic product (GDP, see table 6A.4) is an indicator of the proportion of national resources used for military activities, and therefore of the economic burden imposed on the national economy.

II. Coverage of the data

The military expenditure tables in appendix 6A cover 158 countries, including most countries with a population exceeding 1 million. The time coverage in this Yearbook is the 10-year period 1992–2001. Consistent SIPRI data are available from 1988 onwards for all countries. These are not always consistent with the SIPRI series for the period 1950–87.

---

[1] Such overviews include: Brzoska, M., 'World military expenditures', eds K. Hartley and T. Sandler, *Handbook of Defense Economics*, vol. 1 (Elsevier: Amsterdam, 1995); Herrera, R., *Statistics on Military Expenditure in Developing Countries: Concepts, Methodological Problems and Sources* (OECD Development Centre: Paris, 1994); and Ball, N., 'Measuring third world security expenditure: a research note', *World Development*, vol. 12, no. 2 (1984), pp. 157–64.

Definition of military expenditure

The definition of military expenditure adopted by SIPRI, based on the NATO definition, is used as a guideline. Where possible, SIPRI military expenditure data include all current and capital expenditure on: (*a*) the armed forces, including peace-keeping forces; (*b*) defence ministries and other government agencies engaged in defence projects; (*c*) paramilitary forces, when judged to be trained and equipped for military operations; and (*d*) military space activities. Such expenditures should include: (*a*) military and civil personnel, including retirement pensions of military personnel and social services for personnel; (*b*) operations and maintenance; (*c*) procurement; (*d*) military research and development; and (*e*) military aid (in the military expenditure of the donor country). Excluded are civil defence and current expenditures for previous military activities, such as for veterans' benefits, demobilization, conversion and weapon destruction.

In practice it is not possible to apply this definition for all countries, since this would require much more detailed information than is available about what is included in military budgets and off-budget military expenditure items. In many cases SIPRI is confined to using the national data provided, regardless of definition. Priority is then given to the choice of a uniform time series for each country to achieve consistency over time, rather than to adjusting the figures for single years according to a common definition. In cases where it is impossible to use the same source and definition for all years, the percentage change between years in the deviant source is applied to the existing series in order to make the trend as correct as possible. Such figures are shown in square brackets. In the light of these difficulties, military expenditure data are not suitable for close comparison between individual countries and are more appropriately used for comparisons over time.

III. Methods

Estimation

SIPRI data reflect the official data reported by governments. As a general rule, SIPRI assumes national data to be accurate until there is evidence to the contrary. Estimates are made primarily when the coverage of official data does not correspond to the SIPRI definition or when there is no consistent time series available. In the first case, estimates are made on the basis of an analysis of official government budget and expenditure accounts. The most comprehensive estimates, those for China and Russia, have been presented in detail in previous Yearbooks.[2] In the second case, differing time series are linked together. In order not to introduce assumptions into the military expenditure statistics, estimates are always based on empirical evidence and never based on assumptions or extrapolations. Thus, no estimates are made for countries which do not release any official data, and these countries are displayed without figures. SIPRI estimates are presented in square brackets in the tables (these are most often used when two different series are linked together). Round brackets are

---

[2] Cooper, J., 'The military expenditure of the USSR and the Russian Federation, 1987–97', *SIPRI Yearbook 1998: Armaments, Disarmament and International Security* (Oxford University Press: Oxford, 1998), pp. 243–59; and Wang, S., 'The military expenditure of China, 1989–98', *SIPRI Yearbook 1999: Armaments, Disarmament and International Security* (Oxford University Press: Oxford, 1999), pp. 334–49.

used when data are uncertain for other reasons, such as the reliability of the source or the economic context.

Data for the most recent years include two types of estimate which apply to all countries: (*a*) figures for the most recent year(s) are for adopted budget, budget estimates or revised estimates, and are thus more often than not revised in subsequent years; and (*b*) the deflator used for the last year in the series is an estimate based on a limited number of months or as provided by the International Monetary Fund (IMF). Unless exceptional uncertainty is involved in these estimates, they are not bracketed.

The world total and the totals for regions, organizations and income groups in table 6A.1 are estimates because data are not always available for all countries in all years. These estimates are most often made on the assumption that the rate of change in an individual country for which data are missing is the same as for the average in the region to which it belongs. When no estimate can be made, countries are excluded from the totals.

Calculations

The SIPRI military expenditure figures are presented on a *calendar-year* basis, with a few exceptions. The exceptions are Canada, the UK and the USA, for which NATO statistics report data on a fiscal-year basis. Calendar-year data are calculated on the assumption of an even rate of expenditure throughout the fiscal year. The ratio of military expenditure to GDP is calculated in domestic currency at current prices and for calendar years.

The original data are provided in local currency at current prices (as presented in table 6A.2). In order to enable comparisons between countries and over time, these are converted to US dollars at constant prices (table 6A.3). *The deflator* used for conversion from current to constant prices is the consumer price index (CPI) of the country concerned. This choice of deflator is connected to the purpose of the SIPRI data—that they should be an indicator of resource use on an opportunity cost basis.[3]

Conversion to dollars is for most countries done using the average market exchange rate (MER). However, for some countries purchasing power parity (PPP) rates are used. The PPP dollar rate of a country's currency is defined as the number of units of the country's currency required to buy the same amount of goods and services in the domestic market as $1 would buy in the United States.[4] While MERs are based on price ratios in foreign transactions only, the PPPs are based on price comparisons for the entire economy. For economies with a low degree of foreign exposure, PPP rates thus reflect the price ratios of the entire economy more accurately than MERs. SIPRI uses PPP rates for most countries in transition and for North Korea (as indicated in the footnotes to appendix 6A). Also for many developing countries, The use of PPP rates would be more appropriate also for many developing countries. However, the lack of good PPP data imposes the use of MERs for conversion to constant dollars for developing countries. For a discussion of the advantages and disadvantages of the use of PPP rates and the impact of using PPP rates instead of MERs, see the *SIPRI Yearbook 1999*.[5]

---

[3] A military-specific deflator would be the more appropriate choice if the objective were to measure the purchasing power in terms of the amount of military personnel, goods and services that could be bought for the monetary allocations for military purposes.

[4] *World Bank Indicators 2000* (International Bank for Reconstruction and Development/World Bank: Washington, DC, Mar. 2000), p. 283.

[5] 'Sources and methods for military expenditure data', *SIPRI Yearbook 1999* (note 2), pp. 330–33.

In the *SIPRI Yearbook 2001* the data in constant US dollars are presented to base year 1998. The choice of base year has a significant impact on the comparison between countries because different national currencies move against the dollar in different ways. Therefore, the base year has a significant impact also on regional shares in the world total. Thus, while the share of Asia in world military expenditure in 2000 is 17.9 per cent when expressed at constant 1995 prices and exchange rates (as in the *SIPRI Yearbook 2000*), it is 15.3 per cent with 1998 as the base year.

Total military expenditure figures are calculated for three country groupings—geographical region, membership in international organizations and income per capita. The coverage of these groupings is provided in the notes to table 6A.1.

IV. Limitations of data

Data on military expenditure are associated with a number of limitations. The limitations are of three main types: reliability, validity and comparability.

The main reliability problems are due to the limited and varying inclusiveness of expenditure items. The coverage of official defence expenditure varies significantly between countries and over time for the same country. In many countries, the official data cover only part of actual military expenditure. Important items can be hidden under non-military budget headings or even be financed entirely outside the government budget. A multitude of such off-budget mechanisms are employed in practice.[6] Furthermore, in some countries, actual expenditure may be very different from budgeted expenditure—it is most often higher but in some cases it may be significantly lower. These factors limit the utility of military expenditure data for reasons of reliability.

Another reason for the limited utility is the very nature of expenditure data. The fact that expenditure data are merely input measures makes them rather useless as an indicator of military strength or capability. They are nonetheless widely used for that purpose. In reality, the composition of military expenditure has a major impact on the military capability it provides, as does the technological level of military equipment, the status of maintenance and repair, and so on. Therefore, military expenditure data, even when reliably measured and reported, provide only an indicator of the economic resources consumed for military purposes.

For the purpose of international comparison, a third complicating factor is the method for conversion into a common currency, usually the US dollar. As illustrated by the case of Russia (chapter 6, table 6.2), the choice of conversion factor makes a great difference in the cross-country comparisons of military expenditure. In the most extreme cases, the choice of a purchasing power parity (PPP) conversion factor instead of the market exchange rate can result in a ten-fold increase in the dollar value of a country's military expenditure.[7] This is a general problem in international comparisons of economic data which is not specific to military expenditure. Still, it

---

[6] For an overview of such mechanisms, see Hendrickson, D. and Ball, N., 'Off-Budget Military Expenditure and Revenue: Issues and Policy Perpsectives for Donors', *CSDG Occasional Papers* (Conflict, Security and Development Group (CSDG), King's College: London), no. 1 (Jan. 2002).

[7] The problems involved in dollar conversion, are described in the appendix on sources and methods in *SIPRI Yearbook 2000: Armaments, Disarmament and International Security* (Oxford University Press: Oxford, 2000), pp. 288–90. The appendix also presents a table showing the impact of using PPP rates rather than market exchange rates.

does pose a major limitation, which should be borne in mind when using military expenditure data.

V. Sources

The sources for military expenditure data are, in order of priority: (*a*) primary sources, that is, official data provided by national governments, either in their official publications or in response to questionnaires (see appendix 6D); (*b*) secondary sources which quote primary data; and (*c*) other secondary sources.

The first category consists of national budget documents, defence white papers and public finance statistics as well as responses to a SIPRI questionnaire which is sent out annually to ministries of finance and of defence, central banks and national statistical offices of the countries in the SIPRI database. It also includes government responses to questionnaires about military expenditure sent out by the United Nations and the Organization for Security and Co-operation in Europe (OSCE).

The second category includes international statistics, such as those of NATO and the IMF. Data for NATO countries are taken from NATO defence expenditure statistics published in a number of NATO sources. Data for many developing countries are taken from the IMF's *Government Financial Statistics Yearbook*, which provides a defence line for most of its member countries. This category also includes publications of other organizations which provide proper references to the primary sources used. The three main sources in this category are the *Europa Yearbook* (Europa Publications Ltd, London), the *Country Reports* of the Economist Intelligence Unit (London), and the *Country Reports* by IMF staff.

The third category of sources consists of specialist journals and newspapers.

The main sources for economic data are the publications of the IMF: *International Financial Statistics*, *World Economic Outlook* and *Staff Country Reports*. The source for most PPP rates is *World Development Indicators* (International Bank for Reconstruction and Development).

Appendix 6D. Official data on military expenditure

EVAMARIA LOOSE-WEINTRAUB

I. Introduction

Military expenditure data are used for a wide range of purposes in academic analysis and international politics and are essential for the determination of the economic burden of military activities on society. Data on military spending can be used to measure and compare resource allocation between core government functions—such as defence, law and order, education, health and social security—as well as between military budget headings—such as operating cost, procurement, research and development, and construction expenditure. The defence budget can be the central instrument for transparency and accountability: the public can hold parliaments accountable, and parliaments can hold defence planners in the ministries of defence and the armed forces accountable. Military expenditure data can also serve as a confidence- and security-building measure (CSBM). Disclosure of military expenditure data within a transparent framework of public expenditure is therefore an important element of an open, democratic society.

In reality, however, public access to military expenditure data is generally poor. The reporting of military expenditure information by governments is far less comprehensive, detailed and standardized than is the case for general economic statistics. While most industrialized countries provide disaggregated data on their military expenditure, information on military budgets is especially poor in many developing and transition countries. A few countries regard military expenditure as confidential and refuse to provide any information at all, while others hide large portions of their military budget in different parts of the government budget. In addition, because there is no international standardized definition of military spending on which to base a on set of budget methods, individual governments can basically define 'military expenditure' to suit their purposes. The choice of definition is reflected in the reported size of the military budget. Countries also often change their accounting procedures, which makes it extremely difficult to compare different time series. The quantity of information has increased over the years, but the availability and standard of information are still poor and differ substantially between countries and regions.

SIPRI has compiled data on military expenditure from open sources on a global scale since 1969. This appendix describes the responses to requests for military expenditure data by SIPRI and the United Nations and reviews initiatives introduced in 2001 to enhance transparency and reduce military expenditures.

II. International reporting of military expenditure

When SIPRI began to compile information on military expenditure there was no international system for reporting data, with the exception of that of the North Atlantic Treaty Organization (NATO). Today, there are a number of such international reporting systems. They are of five different types and serve different purposes: (*a*) obligatory reporting within NATO as a collective defence organization; (*b*) voluntary reporting within the United Nations, as a general transparency measure; (*c*) exchange of information within regional organizations or initiatives, as part of a broader set of CSBMs; (*d*) the collection of national data by international statistical organizations, such as the International Monetary Fund (IMF);[1] and (*e*) the collection of data by research institutes and other organizations with an interest in issues related to armament and disarmament.[2] The first three of these mechanisms, all intergovernmental reporting systems, are described in this section.

The 19 NATO member states are obliged to report each year to the NATO Economics Directorate. The purpose of the reporting is for assessment of burden-sharing within the alliance. NATO provides some of these data in an annual press release. The data include figures in local currency, at current and constant prices, as a share of gross domestic product (GDP), on expenditure by category, and per capita.[3]

The 189 UN member states are requested to report annually, by 30 April, to the Secretary-General, their military expenditure for the most recent fiscal year for which data are available, using the standardized international reporting instrument for military expenditure adopted by the UN General Assembly in 1980.[4] The replies are reproduced in annual reports to the General Assembly.[5] The level of participation in this system continues to be low since it is a voluntary reporting system and there are few incentives to participate.[6]

Several initiatives have been made in the third type of reporting system, for CSBM purposes. In addition to the system of the Organization for Security and Co-operation in Europe (OSCE), similar systems have recently been initiated in South-Eastern Europe, Bosnia and Herzegovina, Latin America and Africa.

Since 1991, the 55 OSCE member states have exchanged data on military expenditures as part of a broader system of OSCE CSBMs, based on the Vienna Document 1994.[7] The member states are obliged to report their military expenditure to the Conflict Prevention Centre in Vienna. The circulation of the reports is restricted and they can be released only with the permission of the originating government. The exchange of information within the OSCE is better than the submission of reports to

---

[1] The main current example in this category is the IMF, which presents its data in *Government Finance Statistics Yearbook* (IMF: Washington, DC).

[2] Apart from SIPRI, the main examples in this category are the International Institute for Strategic Studies (IISS), which presents its data in *The Military Balance* (Oxford University Press: London), and the US Department of State's Bureau of Arms Control (formerly ACDA), which presents its data in *World Military Expenditures and Arms Transfers* (US Government Printing Office: Washington, DC).

[3] SIPRI uses these data as the basis for its military expenditure tables. For 2001: 'Financial and economic data relating to NATO defence: defence expenditures of NATO countries 1980–2001', Press release M-DC-2 (2001) 156, 18 Dec. 2001, URL <http://www.nato.int/docu/pr/2001p01-156.htm>.

[4] United Nations, 'Objective information on military matters, including transparency and military expenditures', UN General Assembly Resolution 35/142, 12 Dec. 1980.

[5] United Nations, Objective information on military matters, including transparency of military expenditures, Report by the Secretary-General, UN document A/56/267, 3 Aug. 2001.

[6] United Nations, 'Objective information on military matters, including transparency of military expenditures', UN General Assembly Resolution A/53/218, 4 Aug. 1998.

[7] Conference on Security and Co-operation in Europe, Vienna Document 1994 of the Negotiations on Confidence- and Security-Building Measures, CCs document 113/94, 28 Nov. 1994, reproduced in *SIPRI Yearbook 1995: Armaments, Disarmament and International Security* (Oxford University Press: Oxford, 1995), pp. 799–820. Paragraph 15 of the document includes the rules for reporting of military expenditures.

the UN. One of the reasons is probably that OSCE member states are obliged to report within the CSBM system but not to the public.

Several new initiatives to enhance transparency and accountability and/or to reduce excessive military expenditure were introduced in 2001. Under the auspices of the Special Co-ordinator of the Stability Pact for South Eastern Europe, the Initiative for Transparency of Defence Budgeting was launched at a meeting in Vienna in March 2001.[8] The mission of this project is twofold: (*a*) to promote domestic and international transparency of military expenditure budgets, and the defence budgeting process, throughout South-Eastern Europe; and (*b*) to encourage good practice in defence decision making (policy making, planning, programming and budgeting), with particular reference to accountability.

In another initiative, launched in October 2001,[9] the OSCE Mission to Bosnia and Herzegovina aims to discourage what it considers excessive military expenditure by the country, tries to raise public awareness of spending levels, and encourages citizens to demand transparency and accountability in the budget process. According to the OSCE, Bosnia and Herzegovina, a country with two armies that include three military components[10] with different national bases, does not face an immediate security threat. The reallocation of funds away from military expenditure would significantly improve the economic situation in Bosnia and Herzegovina. Preliminary evidence indicates that the military budget of the Republika Srpska is also economically unsustainable. Therefore, a similar OSCE meeting will be held with the authorities of the Federation of Bosnia and Herzegovina in the Republika Srpska when a review of the entity's military audit has been concluded.[11]

Actions to improve transparency have also been initiated by the Organization of American States (OAS). A regional conference, held in El Salvador in February 1998, adopted the Declaration of San Salvador on Confidence- and Security-Building Measures. The declaration recommended, among other measures, 'studies for establishing a common methodology in order to facilitate the comparison of military expenditure in the region'.[12] As the first result of this initiative, in November 2001 Argentina and Chile introduced a common method for registering their military expenditures,[13] Paraguay and Peru expressed their interest in establishing such an

---

[8] 'Initiative for Transparency of Defence Budgeting', Joint meeting of the Multinational Steering Group (MSG) and the Academic Working Group (WAG) under the Stability Pact Working Table III, was held on 15–17 Mar. 2001, Vienna. URL <http://www.stabilitypact.org/stabilitypactcgi/catalog/view_file.cgi?prod_id=5117&prop_unit=file&prop_type=en>.

[9] 'Lower military expenditures–higher standard of living', Military expenditure reduction initiative launched by the Head of the OSCE Mission to Bosnia and Herzegovina, Robert M. Beecroft, Sarajevo, 9 Oct. 2001, available at URL <http://www.oscebih.org/military/eng/military.htm>.

[10] There are 2 armed forces in Bosnia and Herzegovina, as stipulated by the 1995 General Framework Agreement for Peace in Bosnia and Herzegovina—the Army of the Federation of Bosnia and Herzegovina and the Army of the Republika Srpska. The former is divided into 2 components, a Croat and a Bosniak.

[11] OSCE, Mission to Bosnia and Herzegovina, 'OSCE urges cuts in defence spending by Bosnia and Herzegovina' , OSCE press release, 28 Jan. 2002, URL <http://www.osce.org./news/generate.php3-news_id=2260>.

[12] Organization of American States (OAS), Conference on Confidence- and Security-Building Measures (CSBMs), Declaration of San Salvador on Confidence- and Security-Building, San Salvador, 28 Feb. 1998.

[13] 'A common standardized methodology for the measurement of defence spending' presented at the Public Presentation and Intergovernmental Meeting on a Standardized Methodology for Comparing Defence Spending and on its Application in Argentina and Chile. Report submitted to the Governments of Argentina and Chile by the UN Economic Commission for Latin America and the Caribbean (ECLAC), Santiago, Chile, 29–30 Nov. 2001, pp. 1–60.

Table 6D. SIPRI and UN requests for military expenditure data, 2001

Figures are numbers of countries.[a]

Regions	SIPRI coverage 1	SIPRI request 2	SIPRI replies 3[b]	UN coverage 4	UN replies 5[c]	Total replies 6[d]
Africa	50	50	3	52	1	4
America, North	2	0	0	2	2	2
America, Central[e]	8	8	0	13	3	3
America, South	11	11	4	12	4	6
Asia, Central	5	5	2	5	2	3
Asia, East	16	16	6	16	3	8
Asia, South	6	6	1	6	1	2
Oceania	4	4	3	6	1	3
Europe, West	20	21	12	20	18	19
Europe, Central/East[g]	15	15	12	15	11	13
Europe, CIS	7	7	4	7	5	6
Middle East	14	13	1	15	2	2
Small states[h]	–	–	–	20	2	2
Total	**158**	**156**	**48**	**189**	**55**	**73**

[a] The number of replies actually received by SIPRI is higher than the number of countries because more than 1 reply was received from some countries.

[b] Albania, Australia, Austria, Azerbaijan, Belgium, Belarus, Bolivia, Brazil, Brunei, Bulgaria, Chile, Colombia, Croatia, Czech Republic, Denmark, Estonia, Fiji, Finland, Georgia, Greece, India, Italy, Japan, Jordan, Kazakhstan, North Korea, South Korea, Kyrgyzstan, Latvia, Lithuania, Luxembourg, Macedonia (FYROM), Malta, Mauritius, Moldova, Mongolia, New Zealand, Norway, Poland, Romania, Slovenia, South Africa, Switzerland, Taiwan, Tunisia, Turkey, United Kingdom and Yugoslavia.

[c] Austria, Belarus, Belgium, Brazil, Bulgaria, Burkina Faso, Canada, Chile, Costa Rica, Croatia, Cyprus, Czech Republic, Denmark, El Salvador, Estonia, Finland, France, Georgia, Germany, Greece, Hungary, Italy, Japan, Jordan, Kazakhstan, Kiribati, Latvia, Lebanon, Lithuania, Luxembourg, Macedonia (FYROM), Malta, Mexico, Moldova, Nepal, Netherlands, New Zealand, Peru, Philippines, Poland, Portugal, Russia, Slovak Republic, Slovenia, Spain, Sweden, Switzerland, Thailand, Turkey, Ukraine, United Kingdom, United States, Uruguay, Uzbekistan and Vanuatu.

[d] Totals may be smaller than the sums of column 3 and 5 because the same country may appear in 2 columns.

[e] Excludes the Caribbean states.

[g] Excludes the Commonwealth of Independent States (CIS) member states.

[h] At least 20 UN member states are too small to have a defence force but are included here for the sake of completeness.

Sources: SIPRI questionnaires for 2001; and United Nations, Report of the Secretary-General on objective information on military matters, including transparency on military expenditure, UN document A/56/267, 3 Aug. 2001.

instrument. However, Brazil, the largest country in South America, has rejected the idea that the reporting of military spending by countries in the region should follow a common structure. Brazilian Defence Minister Geraldo Quintao emphasized the need to preserve the difference in each country's military budget and financial management methods because of their geopolitical and strategic peculiarities.[14]

In Africa, the Conference on Security, Stability, Development and Co-operation in Africa (CSSDCA), an organization similar to the OSCE, has on its agenda, as part of its security 'basket', the reduction of military expenditure. The organization is still in its infancy; its first meeting took place in Pretoria in December 2001.[15]

III. Reporting of military expenditure data to SIPRI and the UN

Each year SIPRI sends out a questionnaire to most countries of the world—except for the very small countries assumed not to have any sizeable armed forces—asking them to provide official data on their military expenditure for the preceding five years. The request is sent to their embassies in Stockholm or another nearby embassy as well as to relevant ministries, central banks and national statistical offices. The SIPRI questionnaire is much less detailed than the UN reporting instrument. SIPRI disaggregates military expenditure into six categories: (*a*) military and civil personnel, including retirement pensions and military personnel and social services for personnel; (*b*) operations and maintenance; (*c*) procurement; (*d*) military construction; (*e*) military research and development; and (*f*) paramilitary forces, when judged to be trained and equipped for military operations. Table 6D shows the rates of response to requests by SIPRI and the UN for military expenditure data for 2001.

The rates of response to both SIPRI and the UN continued to be low in 2001. While SIPRI received 48 replies (column 3), roughly one-third of the 158 countries covered in the SIPRI database, the UN received 55 replies (column 5), about 29 per cent of all 189 member states. A total of 73 countries (column 6) provided data in 2001. This was an increase compared to the reporting in 2000,[16] although the results are not entirely comparable because SIPRI did not send out requests to NATO countries in previous years. Responses to the UN increased from 32 in 2000 to 55 in 2001. Responses to SIPRI increased from 33 in 2000 to 48 in 2001, including five NATO countries which were not asked to report in 2000. The combined number of responses increased from 55 in 2000 to 73 in 2001. (The total increase differs from the sum increases of SIPRI and the UN because of overlaps.) Whether this increase in reporting in 2001 is the beginning of a new trend remains to be seen.

Of the regions listed in table 6D, only three countries in *Africa*—Mauritius, South Africa and Tunisia—replied to SIPRI's questionnaire. The UN received one reply, from Burkina Faso. Because of the low level of information available for Africa, SIPRI is conducting a study to assess the availability of military expenditure data in the African region.[17]

Latin America is a region with little transparency in military expenditure. A reporting lag of two years or more for the majority of countries makes accurate estimates of

---

[14] 'Brazil rejects Chilean proposal on military ' *O Estado de Sao Paulo* (Internet edn WWW), 14 Nov. 2001, in Foreign Broadcast Information Service, *Daily Report–Latin America (FBIS-LAT)*, FBIS-LAT-2001-1114, 14 Nov. 2001, pp. 1–2.

[15] First Experts' Meeting of the Conference on Security, Stability, Development and Co-operation in Africa (CSSDCA), CSSDCA/Expt/RPT1, Pretoria, South Africa, 9–14 Dec. 2001.

[16] 'Sources and methods for military expenditure data', *SIPRI Yearbook 2001: Armaments, Disarmament and International Security* (Oxford University Press: Oxford, 2002) p. 301, table 4C.

[17] Omitoogun, W., *Military Expenditure of African States,* SIPRI Research Report no. 17 (Oxford University Press Oxford, forthcoming 2002).

military expenditure difficult. Of the 11 *South American* countries from which SIPRI has requested information, only Bolivia, Brazil, Chile and Colombia have responded during recent years, while Argentina has never replied to SIPRI's questionnaire; none of the Central American countries replied in 2001. The UN received returns from Costa Rica, El Salvador and Mexico in Central America and from Brazil, Chile, Peru and Uruguay in South America.

Two of the five states of *Central Asia* replied to both SIPRI and the UN. Military expenditure data in this region are particularly uncertain since their coverage is not known and economic statistics are in general unreliable. SIPRI provides data based on the official data available, although these constitute only very rough indicators of actual military expenditure.

The countries in *East Asia* which replied to SIPRI were Brunei, Japan, Mongolia, North Korea, South Korea and Taiwan; Japan, the Philippines and Thailand reported back to the UN. The largest country in the region, China, did not reply to SIPRI or the UN; the official Chinese military budget that is publicly available substantially understates its total expenditure on national defence.[18]

In *South Asia* only India reported back to SIPRI and only Nepal to the UN. In *Oceania* three of the four countries—Australia, Fiji and New Zealand—responded to the SIPRI request in 2001.

Most, although not all, governments in Europe provide aggregate data on their military expenditure. In *Western Europe* 12 of the 21 countries covered by SIPRI responded to the questionnaire; 18 countries responded to the UN reporting system. For *Central and Eastern Europe* 12 countries—Albania, Bulgaria, Croatia, the Czech Republic, Estonia, Latvia, Lithuania, Macedonia, Poland, Romania, Slovenia and Yugoslavia—out of 15 reported back to SIPRI. Of the members of the Commonwealth of Independent States (CIS) in Europe, Azerbaijan, Belarus, Georgia and Moldova reported to SIPRI, each reporting for the first time. The largest country, Russia, did not reply to SIPRI but did report to the UN for the year 2000.

For the *Middle East*, the response rate is extremely low: only one country, Jordan, replied to SIPRI, and only three countries—Jordan, Lebanon and Turkey—reported to the UN.

IV. Conclusions

There is a clear need for better access to public information on military budgets. This is a fundamental precondition for strengthening the institutions of accountability and control. In spite of the fact that sensitive military expenditure information is more widely available in some parts of the world as a by-product of the global information revolution, not all societies place equal importance on the collection and dissemination of military expenditure data. It is usually difficult to obtain military data from states with autocratic systems, as well as from countries at war or in local armed conflict, for example, in Africa, Central America, Central Asia and the Middle East. Other possible reasons for the low response rate are that there is often insufficient basic skill and expertise in the appropriate government departments, and that the demand for accountability of financial resources allocated to the military from the public is low. In some countries, providing military expenditure information to

---

[18] Wang, S., 'The military expenditure of China, 1989–98', *SIPRI Yearbook 1999: World Armaments and Disarmament* (Oxford University Press: Oxford, 1999), appendix 7D, p. 335.

potential adversaries runs counter to traditional thinking about the protection of national security.

Despite these reservations, transparency is accelerating in other regions, for example, Central and Eastern Europe. One reason may be that many of these countries aspire to NATO membership and have been supplied with information about and training in NATO standards in military expenditure reporting. More information on military expenditure is also provided in some parts of the CIS in Europe. However, while some material on the Russian military budget is made available on the Internet, at the national level deputies still regard the budgetary procedure as highly unsatisfactory and demand greater transparency and parliamentary control over the military budget.[19]

If widely adopted as policy, transparency measures can further reduce the risks of destabilizing military activities and achieve mutual confidence and security at lower costs.

---

[19] Cooper, J., 'Russian military expenditure and arms production', especially 'Transparency and parliamentary control', *SIPRI Yearbook 2001* (note 16), pp. 316–17.

Appendix 6E. US military expenditure and the 2001 Quadrennial Defense Review

DAVID GOLD

I. Introduction

US defence budgets have been increasing since 1998, after almost a decade in which levels of real US military expenditure declined every year, with the exception of 1992.[1] The initial increases were modest. Outlays by the Department of Defense (DOD) increased by $15.6 billion between fiscal years (FYs) 1998 and 2001 (at constant, FY 2003, prices[2]) or by 3 per cent in real terms over the three years.[3]

The Administration of President George W. Bush requested and received an additional $5.5 billion for FY 2001, almost entirely for operations and maintenance (O&M), and an additional $18.4 billion for FY 2002. The latter increase was mostly for personnel—to improve pay, housing and other incentives—for O&M, and for research, development, testing and evaluation (RDT&E), including a substantial increase for missile defence. The resulting budget request for FY 2002 was 7 per cent higher, in real terms, than the FY 2001 budget request of the Administration of President Bill Clinton.[4]

No increases were requested for weapons procurement in the FY 2001 budget. The Bush Administration had promised a far-reaching revision of US military strategy in the Quadrennial Defense Review (QDR), scheduled for release at the end of FY 2001 (30 September). This was followed, in early February 2002, by the administration's budget request for FY 2003 and its projections for future years. Defense Secretary Donald Rumsfeld had indicated that a large increase in military spending was unlikely and that the QDR was likely to result in scaled back US military requirements and possibly in the postponement or cancellation of some major weapon systems in order to free resources for a transformation and modernization of the military. During the 2000 presidential election campaign, Bush had indicated that his administration would consider skipping a generation of weapons in order to free funding for a major transformation of the US military.

The prospect for holding down weapons expenditure was reinforced by the release of the mid-year reports of the Office of Management and Budget (OMB), an agency of the executive branch, and the Congressional Budget Office (CBO), a research arm of the Congress—both of which projected a sharp decline in the expected budget

---

[1] The data described and analysed in this appendix are national defence data provided by the US Department of Defense. These data differ significantly from the data provided by NATO and used for the SIPRI tables on military expenditure in chapter 6 and appendix 6A. These figures cannot be compared because they are calculated according to different definitions and thus do not have the same coverage.

[2] All budget figures cited in this appendix are expressed at base year FY 2003, as presented by the US Department of Defense in its budget for FY 2003. For sources see table 6E.

[3] US Department of Defense, 'National defense budget estimates for FY2003', Office of the Under Secretary of Defense (Comptroller), Mar. 2002, table 6-11, URL <http://www.defenselink.mil>.

[4] Belasco, A. and Daggett, S., *Appropriations and Authorization for FY2002: The Defense Budget*, RL31005 (Congressional Research Service, Library of Congress: Washington, DC, updated 14 Dec. 2001).

Table 6E. US military expenditure, FYs 1955–2003[a]

Figures are actual expenditures, in US $b, at constant (FY 2003) prices. Figures in italics are percentages.

	1955	1960	1965	1970	1975	1980	1985	1990	1995	2000	2001	2002[b]	2003[c]
Department of Defense outlays													
Personnel[d]	135.0	116.8	128.3	155.2	120.0	117.1	123.6	119.5	95.2	85.4	80.9	83.9	92.8
O&M[e]	75.4	72.4	75.4	102.6	84.6	92.1	119.0	124.0	110.7	115.2	117.2	133.9	143.5
Procurement	86.6	78.5	69.7	99.6	54.4	65.1	106.4	103.8	61.6	54.2	56.7	59.7	62.0
RDT&E[f]	16.4	28.9	32.5	31.3	27.7	26.8	41.7	48.6	39.5	39.7	41.8	45.5	50.8
Construction	10.3	8.6	5.2	5.0	4.5	5.0	6.7	6.6	7.7	5.4	5.2	5.8	6.0
Family housing	–[g]	0.1	3.2	2.7	3.5	3.3	4.0	4.5	4.0	3.6	3.6	3.8	3.9
Other[h]	–5.1	–0.8	–4.6	–4.0	–0.3	–1.9	1.2	–1.6	–2.3	1.7	1.1	0.8	1.7
Total DOD	**318.7**	**304.5**	**309.6**	**392.5**	**294.4**	**307.5**	**402.5**	**405.5**	**316.4**	**305.2**	**306.5**	**333.5**	**360.7**
National Defense outlays													
Total National Defense[i]	. .	. .	. .	. .	. .	. .	. .	. .	. .	319.5	324.9	355.2	379.0
National Defense as a share of GDP (%)	*10.8*	*9.3*	*7.4*	*8.1*	*5.5*	*4.9*	*6.1*	*5.2*	*3.7*	*3.0*	*3.0*	*3.2*	*3.3*

[a] The US fiscal year runs from 1 Oct. of the previous year to 30 Sep. of the named year. These data are national defence data as provided by the US Department of Defense. They differ significantly from data provided by NATO and used for the SIPRI tables on military expenditure in chapter 6 and appendix 6A. [b] Budgeted for FY 2002. [c] Requested for FY 2003. [d] Includes pensions. [e] Operations and maintenance. [f] Research, development, testing and evaluation. [g] Less than $0.045 billion. [h] Includes revolving and management funds, revenues and intra-government receipts, changes in foreign currency valuations and contingency funds. [i] Defence function, including DOD, nuclear weapon programmes in the Department of Energy, military activities of the Coast Guard in the Department of Transportation, and other military activities.

Sources: US Department of Defense, 'National defense budget estimates for FY2003', Office of the Under Secretary of Defense (Comptroller), Mar. 2002, URL <http://www.dtic.mil/comptroller/fy2003budget>: **DOD outlays:** Table 6-11, 'Department of defense outlays by title'; **Total National Defense outlays:** Table 1-1, 'National defense budget summary'; **National Defense as a share of GDP:** Table 7-7, 'Defense shares of economic and budgetary aggregates'.

surplus as a result of the large tax cut and the weakening US economy.[5] With the tightening of US federal finances, a large increase in military spending was thought to be even less likely. Indeed, Secretary Rumsfeld was rumoured to be encountering severe conflicts within the DOD as he sought to impose stricter ceilings on budget requests.

The 11 September 2001 terrorist attacks in the United States changed this outlook. With the near-unanimous consent of the Congress, President Bush declared a 'war on terrorism' that was expected to be long-lasting and near-global in scope. Congress authorized a supplementary appropriation of $40 billion to be applied immediately to anti-terrorism activities, half in FY 2001 and half in FY 2002. About $20 billion was expected to be allocated to the DOD, although the exact disposition of the entire supplemental package was to be determined in future appropriations legislation. Congress quickly accepted the earlier Bush add-ons to the FY 2002 budget, as the Democratic Party-controlled Senate shelved any objections. Rather than postponing any large projects, administration officials indicated that the existing procurement projects would be retained, along with improvements in readiness and personnel expenditures and further commitments to transformative technologies, in its budget request for FY 2003 and in its programme for future years. Indeed, the administration had in August given its approval for continued development of the F-22, a supersonic, stealth fighter, and in October selected Lockheed Martin to be the prime contractor for the Joint Strike Fighter (JSF) programme. Both of these 'legacy' programmes had been widely reported as candidates for drastic reduction or even cancellation. Thus, the United States was poised to begin a major expansion of its military spending.

Section II of this appendix describes the post-World War II trends in US military spending and the US defence budget. Section III summarizes the 2001 Quadrennial Defense Review and analyses its relevance for future US military spending. Section IV comments on the US defence budget for FY 2003 and on future trends in spending, and section V offers the main findings of this appendix.

II. The US defence budget in the long-term perspective

Since the end of World War II, US military spending has undergone a series of cycles associated with actual or perceived major shifts in the security environment, but it has not exhibited any sustained long-term growth. When measured in constant, FY 2003, prices, the DOD outlays, that is, expenditure, of $333.5 billion in FY 2002 are considerably below the peaks reached in 1953 during the Korean War ($389 billion), in 1968 during the Viet Nam War ($434 billion), and in 1987 and 1989 at the end of the build-up of the Administration of President Ronald Reagan ($425 billion). Since the US economy has grown considerably over the past half-century, the defence burden—military spending as a share of gross domestic product (GDP)—has dropped. With the exception of the peaks reached during build-ups, the defence burden declined from about 10 per cent of GDP in the second half of the 1950s to about 3 per cent by the end of the 1990s (table 6E). To illustrate this shift, even those arguing for a substantial growth in military spending envisaged an increase to 4 per cent of GDP, which was only slightly higher than the low point reached after the Viet Nam War.

[5] Executive Office of the President, Office of Management and Budget, 'Mid-session review, budget of the United States Government', 22 Aug. 2001, URL <http://www.whitehouse.gov/omb/budget/fy2002/index.html>; and US Congressional Budget Office, 'The budget and economic outlook: an update', Congressional Budget Office, Washington, DC, Aug. 2001, URL <http://www.cbo.gov>.

The decline in US military spending after the end of the cold war, while substantial in real terms, was consistent with the pattern observed throughout the post-World War II period. US military expenditure in real terms dropped by a quarter between FYs 1989 and 1998, with DOD spending on procurement and on RDT&E declining by 42 per cent.

The composition of the DOD budget has also undergone cyclical shifts but has been remarkably steady over time (see table 6E for the period 1955–2003). Personnel costs were at a steady level until the end of the cold war and declined only with the post-cold war downsizing of the active-duty military. O&M costs increased to a higher level in the 1970s as the USA expanded its global military presence in response to rising tensions in the Middle East, but these outlays have not shown a general tendency to rise since the mid-1980s. Procurement expenditures have fluctuated sharply but when combined with RDT&E—together they comprise what is usually referred to as the 'investment' accounts—the fluctuations have been less marked. With the exception of the build-up in the 1980s, fluctuations in procurement and RDT&E expenditures have tended to offset each other, and the size of revenue flows to the arms industry has been reasonably stable. (If military exports are included, the stability trend is reinforced since exports also tend to fluctuate in a manner that offsets the fluctuations in DOD procurement.)

The decline in the defence burden

The decline in the defence share of GDP reflects an important constraint on the future growth of US military spending. The US defence budget is determined by a complex process. It involves the military services; the Secretary of Defense, who must adjudicate the spending priorities of the various services; the president, working primarily via the OMB, which must adjudicate military and non-military spending requests; both houses of Congress and relevant congressional armed services and budget committees; the voting public; and various interest groups that seek to influence decision makers. Over time, this set of actors and activities has yielded a result whereby defence budgets have not grown, and the defence burden has declined. The long-term decline in the defence burden is the result of two sets of factors.

The first factor is the growing importance of civilian priorities. One of these priorities is an increase in spending for both public and private civilian consumption. A prominent example is US expenditure on health care, which grew from 5.1 per cent of GDP in 1960 to 13.6 per cent of GDP in 1997. Social security, a public-sector expenditure, has also grown. While social security is a transfer payment and not a direct claim on GDP, it is a claim on tax revenues and it competes with national defence for public sector resources. The widespread preference for tax cuts, or at least limits on tax increases, also reflects a desire to protect and expand private consumption and investment.

Another civilian priority is a sound macroeconomic environment. At several points in the post-World War II years—in the late 1950s, the 1970s and the 1990s—limitations on the growth of military spending were a component of a broader macroeconomic policy. In the 1980s and 1990s, the national objective to eliminate the federal budget deficit led to a hardening of the ceilings on military spending. Congress adopted a more centralized procedure for determining spending levels in the 1980s.

The amounts allocated for each of the discretionary spending categories,[6] including defence, were established at the beginning of the budget cycle and could not be altered except in the event of a national security emergency. In the absence of such an emergency, changes in military spending could be accomplished only if an increase in one defence account were accompanied by a decrease in another. This procedure effectively removed significant power for determining spending levels from the congressional committees responsible for defence authorizations and appropriations and gave greater power to the budget committees.[7]

The second set of factors that explains the declining defence burden is a consequence of both the successes and the failures of military spending. The successes are reflected in the fact that the defence budget has purchased a large, powerful and technologically dominant military apparatus. This has been clear in the four major military confrontations in which the USA has been involved since the end of the cold war: the Persian Gulf War in 1991, the conflicts in the former Yugoslavia in 1995 and 1999, and the 2001–2002 conflict in Afghanistan as part of the 'war on terrorism'. The USA has had clear military superiority in the application of technology and mobility in relation to its opponents and in comparison with its allies.[8] Even during the cold war, many observers believed that the USA and its allies held overall superiority in weaponry, mobility and skills in comparison with the Soviet Union and its allies.

The failures of US military spending are reflected in the fact that the DOD has had difficulty in efficiently managing the funds it has received. There have been numerous examples of excess profits and bloated procurement programmes. Newspaper accounts of the high prices of standard items in the 1980s contributed to congressional and public dissatisfaction with important aspects of the Reagan Administration's build-up. The DOD has long sought, not entirely successfully, to reduce its infrastructure of unneeded military bases. Despite ongoing efforts to reform procurement and management, the DOD has not been able to make sufficient progress.[9] In this context, many among the electorate and policy makers are reluctant to add to DOD budgets, except during major shifts in the security environment.

Pressures to increase the defence budget

These pressures to constrain military spending often conflict with the pressures to expand the defence budget. Changes in national security objectives are one source of

---

[6] In US budget terminology there are 2 types of spending: 'discretionary', or budgetary resources provided in appropriations acts; and 'mandatory', sometimes called 'direct', or spending controlled by laws other than the appropriation acts. These laws, such as those governing social security and Medicare, are changed infrequently, and spending for these items is determined by pre-existing formulae and eligibility criteria. Thus, it is only discretionary spending that is determined in the annual budgetary process.

[7] Gold, D., 'Could we have done better? A retrospective on the peace dividend of the 1990s', ed. A. Markusen, *America's Peace Dividend* (Council on Foreign Relations (CFR) and Columbia International Affairs Online (CIAO): New York, 2000), available only online, at URL <http://wwwc.cc.Columbia.edu/sec/dlc/ciao/book/markusen/>.

[8] Alexander, M. and Garden, T., 'The arithmetic of defence policy', *International Affairs*, July 2001.

[9] US General Accounting Office, *Best Practices: Better Matching of Needs and Resources Will Lead to Better Weapon System Outcomes*, GAO-01-288 (US Government Printing Office: Washington, DC, Mar. 2001); and Walker, D., Comptroller General of the United States, *DOD Financial Management: Integrated Approach, Accountability, Transparency, and Incentives Are Keys to Effective Reform*, GAO-02-497T (US Government Printing Office: Washington, DC, Mar. 2002), available at URL <www.gao.gov>.

pressure to increase the defence budget, and they occur either because national security needs change, often rapidly, or because policies change, as, for example, with the election of a new administration that seeks to alter the policies of its predecessor.

A second source of pressure emanates from demands from different constituencies within the national security establishment. The separate services push for weapons and forces that they believe are required for them to carry out their missions, and in the competition for budgetary resources this sometimes conflicts with what other services perceive as their needs. Congressional, business, labour and regional interests advocate expenditures for weapon programmes and infrastructure largely to preserve or provide a flow of resources. In addition, an administration often has to cater to the interests of constituents within the electorate and among elites, especially because of the need to raise substantial sums of money to conduct an election campaign in the United States.

During the recent period of shrinking overall budgets, the DOD has maintained its R&D activities at a high level and kept its entire plan for future procurement on the drawing board. Thus, major weapon systems—including those for both national (NMD) and theatre missile defence (TMD); the F-18E/F, F-22 and JSF tactical aircraft; the V-22 Osprey rotor-lift aircraft for the Marine Corps; the Army's Crusader mobile artillery system; new attack submarines and aircraft carriers; and so on—remained in the R&D and procurement pipeline. Personnel and O&M costs were often squeezed. There was some success in reducing infrastructure and lowering future infrastructure maintenance costs with the closure of military bases, and procurement reforms and the adjustment of internal management practices also produced savings. However, the fundamental conflict between weapons and readiness costs in the context of strict budget ceilings has continued to dominate defence budget planning. By the end of the Clinton Administration in 2001, there was growing agreement that there was a mismatch between stated force requirements and budgetary resources, but there was less agreement as to whether it was the force structure or the budget that should bear most of the burden of adjustment. This mismatch, along with the expectation that substantial additional resources would be needed for new technologies, contributed to the view that some weapon systems that were legacies of the cold war could be dropped.

III. The Quadrennial Defense Review and US military strategy

The QDR is prepared by the DOD in response to a legislative mandate of the Congress. The first QDR was issued in 1997, when a majority in Congress, supported by many within the DOD and in the larger defence community, were dissatisfied with the pace and quality of the transformation of the military after the cold war. A number of official analyses and annual reports of the Secretary of Defense had failed to yield an overall vision that was thought to be sufficiently transformative in the light of a rapidly evolving global security situation. In the context of authorizing the defence budget for FY 1997, Congress required the preparation of the first QDR and made this mandate permanent in 1999, with the second QDR due for delivery in 2001 and subsequent QDRs to be issued every four years.[10] Congress also mandated that the

---

[10] Brake, J. D., *Quadrennial Defense Review (QDR): Background, Process, and Issues*, CRS Report for Congress RS20771 (Congressional Research Service, Library of Congress: Washington, DC, 21 June 2001).

initial QDR be followed by an evaluation by a panel of outside experts, the National Defense Panel. This evaluation process was not included in the mandate for the second QDR.

The first major post-cold war review resulted in the 'Base Force structure' formulated by Chairman of the Joint Chiefs of Staff Colin Powell and Secretary of Defense Cheney in the 1989–93 Administration of President George Bush. The Base Force was not a single document but was contained in a series of planning efforts and public statements. It established the ability to fight and win two major theatre wars simultaneously as the primary military objective of the United States. The model was clearly the Gulf War, and the two most likely regional contingencies were thought to be the Persian Gulf and North-East Asia.[11] Defense Secretary Les Aspin led the Bottom–Up Review (BUR) in 1993,[12] the first year of the Clinton Administration. The BUR essentially re-affirmed the ability to fight and win two major theatre wars as the primary objective of the US military, although it indicated that such an objective could be met with a smaller force and lower spending levels than those advocated by the Base Force.

Budget ceilings dominated the first QDR. At the beginning of the process, General John Shalikashvili, then Chairman of the Joint Chiefs of Staff, the body with responsibility for drafting the report, ordered his staff to prepare the document taking into account the strict budgetary ceilings.[13] The 1997 QDR re-affirmed the basic elements of the two-theatre war strategy and contained detailed descriptions of the force structure needed to implement the strategy. This force structure essentially ratified the structure that was in place and the existing procurement programme. The National Defense Panel evaluation criticized the strategy and force structure as being unrealistic, in the light of both the budget constraints and the changing global security situation, but these criticisms had little effect on explicit DOD strategy or on budget requests.

During the 2000 election campaign, presidential candidate Bush criticized the Clinton Administration for failing to transform the military and indicated that his administration might 'skip a generation' of weapons in order to free resources for the necessary transformation in technology and personnel.

The 2001 QDR, released less than three weeks after the 11 September attacks in the USA and the declaration of a 'war on terrorism', articulated a shift in US military strategy.[14] Instead of basing force structure on the ability to meet specific and explicit threats, the focus would be on developing forces with a range of capabilities to meet both unforeseen and predictable threats. This, it is argued, represents a shift from a

---

[11] Jaffe, L. S., *The Development of the Base Force 1989–1992* (Joint History Office, Office of the Chairman of the Joint Chiefs of Staff: Washington, DC, 1993). General Powell began formulating the Base Force in 1989, fully a year before Iraq invaded Kuwait. Powell, C., *My American Journey* (Random House: New York, 1995).

[12] On the BUR see Aspin, L., *Report on the Bottom–Up Review* (US Department of Defense: Washington, DC, Oct. 1993); and Gunzinger, M., 'Beyond the Bottom–Up Review', ed. M. A. Sommerville, *Essays on Strategy XIV* (Institute for National Strategic Studies, National Defense University Press: Washington, DC, 1996), available at URL <http://www.ndu.edu/inss/books/essa/essabtbu.html>. The Base Force, the BUR and the 1997 QDR are analysed in Larson, E. V., Orletsky, D. T. and Leuschner, K., *Defense Planning in a Decade of Change: Lessons From the Base Force, Bottom–Up Review, and Quadrennial Defense Review* (Rand Corporation: Santa Monica, Calif., 2001), available at URL <http://www.rand.org/publications/MR/MR1387>.

[13] Wilson, G. C., *This War Really Matters: Inside the Fight for Defense Dollars* (Congressional Quarterly Press: Washington, DC, 2000), chapters 2 and 3.

[14] US Department of Defense, 'Quadrennial Defense Review report', 30 Sep. 2001, URL <http://www.defenselink.mil/pubs/qdr2001.pdf>.

'threat-based' strategy to a 'capabilities-based' strategy, with a wide range of capabilities called for to address the range of possible and unforeseen threats. This emphasis on a capabilities-based strategy represents a strengthening of a number of trends, including the expanded use of information technology in the Revolution in Military Affairs, the development of defences against asymmetric threats and the renewal of attention to homeland defence. The QDR states that the ability to protect the US homeland from a variety of threats will become one of the two main pillars of the new force structure.

The second pillar is the ability to wage and win two theatre wars in overlapping time frames. Thus, the new force-sizing strategy does not make a sharp break with the past. The QDR states that 'U.S. forces will remain capable of swiftly defeating attacks against U.S. allies and friends in any two theaters of operation in overlapping timeframes'. In addition, 'U.S. forces will be capable of decisively defeating an adversary in one of the two theaters . . . This capability will include the ability to occupy territory or set the conditions for a regime change if so directed'.[15]

Thus, while making the military objectives and requirements more flexible, the QDR retains a two-theatre war requirement. One of these wars could be a global war against terrorism.

The main thrust of the 2001 QDR is to give primacy to the development, introduction and deployment of the transformative technologies and strategies that will allow the USA to maintain its military supremacy. However, it contains very little guidance as to the new force structure or budget requirements. It eschews the type of detailed descriptions of forces and spending levels that dominated the 1997 QDR. Despite Secretary Rumsfeld's numerous statements that budgetary resources would be shifted to accommodate and accelerate the necessary transformation, the 2001 QDR supports the existing force structure and suggests that expansions and improvements in existing forces will be needed to meet the new requirements, along with technological transformation. This apparent contradiction soon brought forward criticisms from the DOD that the QDR did not provide the guidance needed to move forward.[16] It also meant that the difficult decisions regarding spending levels and the allocation of funds would have to await the FY 2003 budget submissions and congressional debates.

IV. US military spending in FY 2002 and beyond

President Bush formally introduced his administration's budget for FY 2003 (1 October 2002 to 30 September 2003) on 4 February 2002. The budget request for the defence function emphasized a substantial increase over the budget approved for FY 2002, largely in the context of prosecuting the war in Afghanistan and the global war against terrorism. The administration had used most of its first year in office to conduct a major review of US defence policy, which was released in the 2001 QDR, and had postponed major changes in the Clinton Administration's defence programme until the FY 2003 budget.

Following accepted practice, the US federal budget, which includes detailed expenditure plans for each of the government's functions and for each government

[15] US Department of Defense (note 14), p. 21.
[16] Fulgham, D. A., 'QDR became "pabulum" as decisions slid', *Aviation Week & Space Technology*, 8 Oct. 2001.

department and agency, is introduced eight months before the start of the fiscal year to allow sufficient time for Congress to exercise its legislative oversight function.[17] As a first step, following the budgetary reforms enacted in the 1980s, the House of Representatives and the Senate establish spending ceilings for the discretionary portion[18] of the budget and its major components, including the defence function. Once established, the ceilings on discretionary spending can be exceeded only in an emergency. This occurred after the 11 September terrorist attacks when the president requested and received $20 billion in supplementary appropriations for FY 2001 and $20 billion for FY 2002. In the absence of an emergency, a breach of the ceiling for the defence function, for example, because of unanticipated cost growth in a procurement programme, leads to a sequestration process whereby the DOD must find cuts in other programmes of equal amounts before the larger expenditure to cover the cost overrun can be approved. This may be the most important way in which the budget process has transferred budgetary power away from the military-oriented committees of Congress.

The defence spending request is then debated in the relevant committees of the House of Representatives and the Senate. The budget for the DOD itself represents most of the funding for defence, but the defence function also includes spending for nuclear warhead research and production in the Department of Energy, military functions of the Coast Guard in the Department of Transportation, and other, smaller defence activities outside the DOD. In addition, some activities of the DOD, most prominently the work of the Army Corps of Engineers in maintaining civilian rivers and harbours, are not part of the defence function. Spending legislation is split between budget authority, which establishes the legal basis for the government to make contractual commitments, and budget appropriations, which represent the spending commitments for a given year. Changes in budget authority tend to lead to changes in appropriations, since some contractual commitments involve appropriations that are spread out over several years. Outlays are what is actually spent and may differ from appropriations because of differences in the timing of planned and actual spending. Thus, there is often confusion regarding the different totals; press accounts and many scholarly analyses rarely make distinctions between the different types of defence expenditure figures, as provided by the DOD, nor state which type of figure they are citing.

The Bush Administration budget request for FY 2003 included $396.8 billion in budget authority for National Defense, an increase of $48 billion over FY 2002. DOD outlays are expected to reach $360.7 billion in FY 2003 (table 6E). At the time of writing, congressional opposition to the president's budget request is minimal. If it is approved, the FY 2003 budget will represent an increase over expected FY 2002 DOD outlays of 8.1 per cent, in real terms. Taking DOD outlays in FY 2001 as a base, and assuming that Congress approves the FY 2003 request, the Bush Administration will have added $54.2 billion to defence outlays, in FY 2003 prices (table 6E). This represents an increase of 17.7 per cent in real terms in only two years, the largest two-year increase since the Viet Nam War. The key question is what this increase will buy.

---

[17] For explanations of US budget categories and a chronology of the US budgetary processes see Executive Office of the President of the United States, Office of Management and Budget, *Budget System and Concepts and Glossary* (US Government Printing Office: Washington, DC, 2002), available at URL <www.whitehouse.gov/omb/budget/index.html>.

[18] For explanations of the terms 'discretionary' and 'mandatory' see note 6.

The investment accounts

DOD procurement and RDT&E are projected to jump by 14.4 per cent in real terms from FY 2001, the last budget of the Clinton Administration, to FY 2003, reaching a combined total of $112.8 billion (table 6E). In the ongoing debate on defence planning, much has been made of the so-called 'procurement holiday' of the 1990s, the period of the post-cold war 'peace dividend'. Supporting this are the data showing a 70 per cent drop in budget authority for arms procurement from FY 1985 to FY 1997. However, outlays for procurement fell by a significantly smaller amount, 50 per cent in real terms over the period FY 1985–97. Moreover, RDT&E funds were used to support some activities that had previously been funded out of procurement accounts. Overall, actual spending on the investment accounts fell by 42 per cent, much less than the drawdowns after the build-ups for the wars in Korea and Viet Nam.

The Bush programme includes funding for weapon systems already in the stage of RDT&E or production. There are no new weapon initiatives, although some programmes, such as those for missile defence, are being expanded. Other programmes, such as the V-22 Osprey, are being reorganized because of their serious testing and management problems, and missile defence has already been reorganized with the aim of avoiding duplication and reducing overhead expenses. Thus, various TMD programmes have been combined with the NMD programme to create a single programme under the Missile Defense Agency of the DOD. Most procurement and RDT&E programmes are continuing along much the same trajectory as they did under the Clinton Administration. The DOD announced that it was cancelling 18 army procurement programmes, along with the US Navy's DD-21 destroyer and Area Missile Defense programme.[19] However, both of the Navy systems continue to receive RDT&E funding: the DD-21—now renamed the DD-X—as a test bed for new naval technologies, and the Area Missile Defense programme under the account of missile defence rather than under the account of Navy systems.

Many of the procurement and RDT&E programmes are controversial. Missile defence is scheduled to absorb 18.4 per cent of the RDT&E funding for FYs 2003 and 2004, and the administration maintains FY 2004 as the target date for initial deployment. However, many questions remain regarding missile defence technical feasibility, military utility and system costs.[20] The F-22 has entered low-rate production, but problems that have led to delays in the past have not been solved and it is considered by outside observers to be a high-risk programme.[21] All three of the tactical combat aircraft are projected to be produced at considerably higher unit costs than those of the systems they are replacing. These costs could rise in the future, putting the status of the programmes in doubt. Even if the overall budget is approved fairly easily, these and other systems are likely to be the subject of considerable congressional debate.

---

[19] US Department of Defense, 'Secretary Rumsfeld briefs the fiscal 2003 DoD budget', 4 Feb. 2002, URL <http://www.defenselink.mil/news/Feb2002/t02042002_t0204sd.html>. The DOD also announced that it was retiring the Peacekeeper missile, initially known as the MX, which had been a centrepiece of the Reagan Administration's build-up. This weapon programme has not received procurement funding for some time, so the budgetary implications of retiring the missile are quite small.

[20] US Congress, Congressional Budget Office, *Estimated Costs and Technical Characteristics of Selected National Missile Defense Systems* (US Congressional Budget Office: Washington, DC, Feb. 2002), available at URL <www.cbo.gov>.

[21] US General Accounting Office, *Tactical Aircraft: F-22 Delays Indicate Initial Production Rates Should Be Lower to Reduce Risks*, GA0-02-298 (US Government Printing Office: Washington, DC, Mar. 2002), available at URL <www.gao.gov>.

Personnel

Personnel outlays are projected to rise from $80.8 billion in FY 2001 to $92.8 billion in FY 2003, and then to $103.6 billion in FY 2005, all figures in FY 2003 prices. President Bush has made improvement of military pay and benefits a major objective, as a means of offsetting shortfalls in retention and improving morale. However, many of the complaints from enlisted personnel and officers concern issues such as the length of overseas postings, separation from families, and uncertainties as to the nature of missions—not remuneration issues. Some problems of retention surfaced during the strong economic expansion of the late 1990s, but they are not long-term in nature. The quality of US military personnel has risen since the early 1980s, as measured by years of education and performance on standardized tests, and remuneration levels are consistent with civilian occupations with similar skill levels.[22] Some inequities, especially at the lower pay scales, need to be rectified, and the Bush programme is seeking to address them.

One of the major budgetary increases is for health care, especially for military retirees. The DOD added $8 billion in budget authority to fund health care improvements, in response to a congressional mandate. This money is in the O&M portion of the budget, but it is clearly an item designed to benefit personnel. As health care costs are expected to continue to rise for the economy as a whole, this item is likely to absorb significant budgetary resources in the future.

Improving morale is less a budgetary item and more a question of organizational issues. Some of the lessons learned in the early months of the war in Afghanistan, in terms of the importance of mobility and the real-time use of information, may help to generate organizational innovations that make better use of individual skills. Paradoxically, with a clear enemy and continued combat success, the war against terrorism is likely to yield a significant boost to military morale.

Operations and maintenance

O&M funding has been another contentious issue in the debates on military readiness and in the continual attempts by the DOD to reduce its infrastructure expenses by closing excess military bases, outsourcing some activities to private companies and increasing the proportion of supplies purchased through commercial channels in an 'off the shelf' manner. The Clinton Administration's Revolution in Business Affairs did lead to significant savings, although the actual amounts appear to have been lower than projected. The prospects for future savings appear limited, partly because the best opportunities—sometimes called the 'low hanging fruit'—have already been taken.[23]

Since O&M takes such a large proportion of total DOD resources—39.8 per cent of requested outlays for FY 2003—there will be continued attempts to generate greater economies in these accounts. One area is the closing of redundant military bases, depots and other facilities. The Reagan Administration and Congress established a Base Realignment and Closure (BRAC) process. Previously, it had been nearly impossible to close facilities because congressional representatives of districts with targeted facilities, in coalition with other representatives who feared they would lose

---

[22] Williams, C., 'Our GIs earn enough', *Washington Post,* 12 Jan. 2000.
[23] Williams, C., 'Holding the line on infrastructure spending', ed. C. Williams, *Holding the Line: US Defense Alternatives for the Early 21st Century* (MIT Press: Cambridge, Mass., 2001).

their facilities, acted to stymie the administration's attempts to close bases. The BRAC process established a commission with members appointed by both Congress and the administration. The commission compiled a list of facilities to be closed, and Congress and the administration were required to accept or reject the entire list without amendment. This de-politicizing of the process was successful and three rounds of base closure were completed. However, when the Clinton Administration sought to keep two facilities open by privatizing their functions, Congress refused to continue the process. The Bush Administration and Congress have agreed to come back to the BRAC process, but not before FY 2005. If that schedule were to be maintained, actual savings would not be available to ease possible budget pressures for some years.

At the same time, readiness needs may increase as the war on terrorism proceeds, given the greater need to maintain troops, equipment and supplies in forward locations. However, the success of the operation in Afghanistan, with resources inherited from the previous administration, suggests that readiness problems may have been overstated and may be resolved more easily than expected.

Homeland security

Homeland security is treated as a separate category in the budget documentation. Although there is an Assistant to the President for Homeland Security in the Office of Homeland Security, established in October 2001 within the Executive Office of the President, the budget for that office is not large. Instead, funding for homeland security is spread over a number of departments and allocated to a number of budgetary categories. The budget request includes $37.7 billion for homeland security, up from $18 billion in FY 2002, with 22 per cent of the total earmarked for the DOD, 20 per cent for the Department of Transportation and 19 per cent for the Department of Justice.[24]

Defence of the US homeland has traditionally been a function shared among a large number of government agencies, including the DOD; the customs service in the Treasury Department; the Federal Bureau of Investigation in the Department of Justice; the Immigration and Naturalization Service in the State Department; the Coast Guard in the Department of Transportation; various federal, state and local agencies dealing with public health issues; and fire, police and emergency medical services run by state, county and city governments throughout the country. The homeland defence activities of the DOD and the Coast Guard have been regarded as a component of the defence function, but the other agencies and activities have not. However, homeland defence also includes a number of civilian activities, such as the support of 'first responders' to attacks on the homeland, which involves local police, fire and emergency medical personnel; defence against biological warfare, which involves both public and private medical personnel; and enhanced security at domestic airports. Thus, a number of civilian activities become defined as components of national defence. While such activities remain under civilian authority, the continuation of the war against terrorism may well blur the distinction in future budgetary and policy discussions.

---

[24] Kosiak, S., 'FY 2003 budget request for homeland security', Center for Strategic and Budgetary Assessments, 8 Feb. 2002, URL <www.csbaonline.org>.

Transformation

The president and the defence secretary have, since entering office, placed increased emphasis on the organizational and technological transformation of the US military, although the details of this transformation have not been spelled out. Most observers have emphasized the need for enhanced applications of information technology designed to transmit real-time information on battlefield conditions in order to accelerate weapons deployment and support tactical adjustments. In discussions of these issues, the emphasis on information technology was often accompanied by the argument that a number of cold war-era weapon programmes should be cancelled, since they were either redundant, having been designed to meet threats that no longer exist, or not suited to the new defence environment. Moreover, it would be necessary to cancel such programmes to find the budgetary resources for transformation. Both the president and the defence secretary made comments to this effect in the months prior to 11 September.

However, the military services, contractors and other interests opposed programme cuts and, with the increase in defence resources made available after the terrorist attacks, these legacy programmes have survived. It was reported that Secretary Rumsfeld had accepted the continuation of legacy programmes as the price for avoiding clashes within the DOD.[25] There are no explicit initiatives in the FY 2003 budget request, although substantial technological transformation may occur within existing programmes. In some instances, altering the missions and organization of existing forces can introduce the flexibility required in the new environment. Moreover, the administration has blurred the distinction between legacy and transformation systems by claiming that programmes such as missile defence are part of its transformation initiatives. The absence of more explicit attention to transformation and the continuation of expensive legacy systems suggest that there will continue to be a conflict between the old and the new.[26]

V. Conclusions

There are many uncertainties in the short-term outlook for US defence spending, the course of the war against terrorism being the most important. The programme put forward by the Bush Administration for the US armed forces, both in its budget requests and in its other statements, most prominently the QDR, also raises uncertainties in terms of its ultimate affordability and its conformity to an overall vision of the role of the US military in the emerging global environment. Issues involving the investment accounts, procurement and RDT&E illustrate these latter uncertainties.

The investment accounts are 'back-loaded', that is, a large part of the funding for the completion of current programmes is concentrated in the later years of the five-year projections. Such back-loading assumes that funding will be available when it is needed, but a future budget squeeze would force the DOD to make the choices between programmes that have so far been postponed. The Reagan programme ran into similar difficulties when large federal budget deficits, political opposition to the planned nuclear weapon build-up in the USA and Western Europe, and other economic and political changes forced a slowdown in the procurement programme. DOD

---

[25] 'Old hawk learns new tricks', *The Economist*, 11 Oct. 2001.
[26] Keller, B., 'The fighting next time', *New York Times*, 10 Mar. 2002; and 'Transformation postponed', *The Economist*, 14 Feb. 2002.

officials have stated that the funding for existing programmes has absorbed the lion's share of available funding. However, they have the authority to make choices among programmes, reducing or cancelling some to shift resources to activities that are deemed more urgent. By continuing the existing set of procurement and RDT&E programmes, they are effectively choosing to make less sharp a break with the past than they stated they would.

In this regard, the 2001 QDR did not provide appropriate guidance. The shift to a capabilities-based emphasis is too open-ended, since it supports any set of capabilities that the USA can produce. The F-22 may provide an example. It is a stealthy, supersonic fighter plane, potentially a technological marvel, but it was designed during the cold war to counter an expected new generation of Soviet aircraft and air defences that never materialized. The F-15, which the F-22 will replace, gives the USA air superiority over any conceivable enemy well into the future. Thus, the F-22 may be a system without a threat to combat; if so, it will absorb resources that could be better used elsewhere.

The 2001 QDR has provoked relatively little discussion, especially when compared with the 1997 QDR. This may be due to the environment that has emerged after the terrorist attacks in the United States, with its emphasis on the steps to be taken in the war against terrorism, specifically in Afghanistan. The failure of the 2001 QDR to articulate a more specific vision of US military policy, and the emphasis in the budget on continuity rather than change, suggests that a major opportunity has been lost. It also suggests that providing the military with substantially more funding, however justifiable in terms of short-term security perceptions, may, over time, prevent the very reforms that leaders claim are needed.

7. Arms production

ELISABETH SKÖNS and REINHILDE WEIDACHER

I. Introduction

The arms production sector has undergone profound change in many countries, both quantitatively and qualitatively. The catalyst for this change was the end of the cold war, which brought a sharp reduction in Russian arms procurement and arms exports and a smaller although still significant decline in arms procurement and arms exports in the industrial countries in the West. The change in security policies and force structures since the end of the cold war, long-term developments in military technology and the continuous increases in the cost of advanced weapon systems have also brought about a change in the type of weapon systems and services demanded. The 11 September 2001 terrorist attacks in the USA are likely to result in yet another shift in military–industrial developments because of their impact on security requirements and international relations, including the military–industrial sphere.

Most of the decline in the volume of arms production took place in the first half of the 1990s, when the main trends in the arms industry were downsizing, rationalization and diversification into civilian production.[1] Since the mid-1990s the decline in the volume of output has levelled out. The main trends in the development of the arms industry in this period were associated with the strategies developed by the major surviving companies to enable them to remain competitive on a global scale. Section II of this chapter analyses the results of these developments for the concentration, internationalization and privatization of arms production. It assesses how far these processes had developed by the early 2000s, both in the main centres of arms production in the United States and Western Europe and in some of the minor arms-producing countries in Europe and other regions as a result of the sales and marketing strategies of large supplier companies. While the dearth of data in the field of arms production makes it difficult to obtain a comprehensive empirical picture of these trends, the chapter draws on the SIPRI database on arms-producing companies and data on national arms production to identify the broad characteristics of the arms industry currently. Section III discusses the specific features of developments in the Russian arms industry, although the data are still too weak to allow a comparative analysis. Appendix 7A presents financial and employment data on the 100 largest arms-producing companies in the Organisation for Economic Co-operation and Development (OECD) and developing countries. For the first time it has been possible to

---

[1] See previous editions of the SIPRI Yearbook.

Table 7.1. National arms sales[a] and arms exports,[b] Western Europe, 1990–2000
Figures are in US $m. at 2000 prices and exchange rates. Figures in italics are percentages.

Country		1990	1995	1996	1997	1998	1999	2000	Change (%) 1990–2000
UK	Arms sales	24 470	19 940	21 910	22 730	22 530	19 410	..	*–21*
	Arms exports	9 140	8 170	10 430	10 940	9 550	6 630	6 680	*–27*
France	Arms sales	20 740	13 940	14 280	15 170	14 810	12 370	11 060	*–47*
	Arms exports	6 430	2 840	4 300	6 260	5 920	3 560	2 490	*–61*
Germany	Arms sales	12 930	5 790	..	..	..	..	..	..
	Arms exports	890	1000	500	680	650	1 370	630	*–29*
Italy	Arms sales	..	3 220	..	..	..	..	..	..
	Arms exports	980	660	620	750	960	840	560	*–63*
Nether- lands	Arms sales	1 920	1 440	1 190	1 520	1 490	..	..	..
	Arms exports	850	420	730	850	..	..	..	..
Sweden	Arms sales	1 020[c]	1 040	1 110	1 300	1 490	1 230	1 210	*+19*
	Arms exports[d]	460	370	340	340	390	400	480	*+4*
Spain	Arms sales	2 770	2 100	2 170	2 310	2 350	2 600	..	*–6*
	Arms exports[e]	360	..	..	500	450	..	..	..

[a] Data on arms sales are for total arms sales of the country (i.e., for domestic arms procurement and for export). For some countries, such data are provided by the government or a defence industry association, for other countries they are estimated by SIPRI, by adding arms exports and subtracting arms imports from figures for national arms procurement.

[b] Data on arms exports are provided separately from arms sales and are seldom comparable to arms sales because of differences in the definition of military equipment. Therefore, the table does not show arms exports as a percentage of arms sales. The use of such shares in a cross-country comparison would be somewhat misleading.

[c] Data are for 1991.

[d] The Swedish definition of arms export changed in 1993.

[e] Data on arms exports are derived from Ministry of Defence, *La industria de defensa en España, 1999* [The defence industry in Spain, 1999]. These data differ significantly from arms exports statistics provided by the Spanish Ministry of Economy (see appendix 8E).

Sources: Appendix 7B.

compile similar data, although still tentative, on Russian companies. These data are also provided in appendix 7A together with the sources and methods used in the data compilation. Appendix 7B describes the availability of data on national arms production by governments and defence industry associations and their limitations.

The reduction in the demand for military equipment during the 1990s was significant, both in the aggregate and for some individual countries. NATO statistics show that the combined military equipment expenditures of all NATO countries dropped by 40 per cent in real terms from the peak level in 1987 to 2001—by 43 per cent in the United States and by 35 per cent in NATO Europe, although with great variation between countries. In Europe the reductions took place during the first half of the 1990s. Since 1997 equipment expenditure in NATO Europe has increased by 6 per cent in real terms. According to the NATO statistics, the decline in total NATO equipment expenditure since 1997 is due to the continuing reduction in US expenditure on equipment.[2]

Estimates of national arms sales—used as an approximation of arms production—for the seven largest arms-producing countries in Western Europe show a sharp decline between 1990 and 1995 in most countries, and a slower decline thereafter (table 7.1).[3] The decline has presumably also been sharp in Germany and Italy, although no statistics are available for these countries. Among the seven countries listed in table 7.1, arms production has increased only in Sweden, a reflection of the JAS-39 Gripen combat aircraft programme. In recent years the decline in arms exports has been sharper than in arms production. Attempts to compensate for decreased domestic arms procurement by increased arms exports thus may not have been successful.

II. Concentration, internationalization and privatization

In the early post-cold war period the decline in the demand for military equipment and the continuously rising research and development (R&D) requirement for major weapon systems reinforced economic pressure to concentrate and internationalize arms production activities and led to increased acceptance of foreign ownership of arms-producing facilities. In countries where much of the arms industry was still under state ownership there was also pressure for privatization in order to enable or facilitate mergers with other companies, particularly in cases of cross-border acquisitions.

The international system for the production and trade in weapons has evolved in line with developments in the mode of production, military technology and the security environment and with military requirements.[4] Changes in the organization and control of the production and sale of weapons have created major changes in the rate of concentration, internationalization and privatization.

The mass production of weapons began in the 19th century as a result of the industrial and technological revolution. A separate military–industrial sector emerged, which was privately owned and produced weapons largely for profit. It was dominated by a few large companies with a markedly international orientation. Exports and production were relatively unregulated, because they were perceived to facilitate the maintenance of innovative capabilities and the productive base. In the late 19th century, one of the largest companies, Krupp (Germany) exported 86 per cent of its arms production. In the period between

---

[2] See appendix 6B in this volume; and NATO, *Financial and Economic Data Relating to NATO Defence–Defence Expenditures of NATO Countries (1980–2001)*, Press Release M-DPC-2(2001)156, 18 Dec. 2001, URL <http://www.nato.int/docu/pr/2001/p01-156e.htm>.

[3] Similar data for the US arms industry are not available.

[4] Krause, K., *Arms and the State: Patterns of Military Production and Trade* (Cambridge University Press: Cambridge, 1992), summarized in Held, D. et al., *Global Transformations* (Polity Press: Cambridge and Oxford, 1999). See also Dunne, P., 'The defense industrial base', eds K. Hartley and T. Sandler, *Handbook of Defense Economics*, vol. 1 (Elsevier Science: Amsterdam, 1995).

the two world wars, when the demand for weapons declined, the companies formed cartels to divide the reduced international market among them.

Beginning in the 1930s and continuing through World War II the major powers built up indigenous arms industries under national control. The development and production of weapons were focused solely on war. World War II involved a dramatic increase in arms production, a profound reconfiguration of the pattern of arms production and a sharp reduction in the international arms trade. Massive state intervention in the arms trade and production system was required to ensure the supply of arms and continuous military innovation.

During the cold war, arms production and the arms trade were dominated by the two superpowers. Arms production and trade reacquired a global dimension as arms transfers were used by the superpowers in their East–West rivalry. The cold war arms trade system appears to have been an historical exception in that the system of arms production and arms trade was so strongly dominated by governments and conditioned by the bipolar structure of world politics.

Since the end of the cold war, the global system of arms production and arms trade has again undergone a transformation. During this period of shrinking demand and rapid developments in military technology, the arms-producing companies which have emerged as large suppliers have used strategies of growth (primarily through mergers and acquisitions) to remain competitive. They have had to cope with rising R&D costs, access foreign markets and gain leverage with their main customers: national governments. In this restructuring the USA has taken the lead, aided by the size of the arms procurement and R&D budgets of the US Government. This has resulted in changes in the global structure of arms production, its industrial structure, state–industry relations and company characteristics.[5]

Concentration

In modern times, the arms industry has been much less concentrated than comparable civilian high-technology industries. A probable factor behind this difference is the preference for procurement from national production in countries which had the industrial capability and economic resources to do so. However, since the end of the cold war, economic pressure for concentration has resulted in a major shift in government perceptions and policies towards the domestic defence industrial base. There has been an increased acceptance of concentration and monopolistic tendencies in the arms industry.

In the post-cold war period, the rate of concentration has increased significantly among the 100 largest arms-producing companies on the SIPRI Top 100 list. Table 7.2 presents the change in concentration ratios for different groups of the largest of these companies, both in their total markets (total sales) and in their military markets (arms sales). In 1990 the concentration

[5] For an analysis of the consolidation process and its implications for the 1990s see Markusen, A. R. and Costigan, S. S. (eds), *Arming the Future: A Defense Industry for the 21st Century* (Council on Foreign Relations Press: New York, 1999).

Table 7.2. Change in concentration ratios, SIPRI Top 100 companies, 1990–98
Figures are percentages.

| | Concentration ratios (% of combined total of Top 100) ||||||
| | Arms sales ||| Total sales |||
Company section	1990	1995	2000	1990	1995	2000
5 largest companies	22	28	42	33	34	40
10 largest companies	37	42	58	51	53	57
15 largest companies	48	53	66	61	65	68
20 largest companies	57	61	72	69	73	76

Source: The SIPRI arms industry database.

ratios for arms sales were very low—much lower than for their total sales. This reflects the fact that commercial markets were more concentrated than military markets. By 2000 concentration ratios were considerably higher in the military markets. The increase in concentration in the 1990s was most marked for the group of the 5 and 10 largest companies, but there were also significant increases in other groups. The process of concentration was relatively slow during the first half of the decade. In the second half, the rate of concentration increased considerably.

By 1995 the rate of concentration in arms production was still not as high as in comparable non-military production, but by 2000 the difference had tended to disappear. For the 10 largest companies the concentration was higher in arms sales than in total sales in 2000. This indicates that during the 1990s the dilemma of the choice between the benefits of economies of scale and the benefits of competition, the central defence industrial policy dilemma for the past 40 years, had gradually been resolved in favour of scale.[6] Economic forces have been given freer play in military sales, which has resulted in a concentration rate almost similar to that of the non-military markets—at least for the largest arms-producing companies. This will probably raise difficult political issues. In their arms procurement processes countries will confront large international arms-producing companies with strong market power.[7]

Increased concentration has resulted in an increase in the size of the largest arms-producing companies, both in relation to other companies and in relation to the total procurement budgets of domestic governments. The few very large companies at the very top of the SIPRI Top 100 list in 2000 (appendix 7A) are significantly larger in terms of arms sales than their counterparts in 1990.

---

[6] For more on the economic dynamics of this policy dilemma see Smith, R., 'Defence procurement and industrial structure in the UK', *International Journal of Industrial Organisation*, vol. 8 (1990), pp. 185–205.

[7] Dunne, J. P. and Smith, R. P., 'The evolution of the international arms industry', Paper presented to the Fifth Annual Middlesex Conference on Economics and Security, Middlesex University, London, 15–16 June 2001, URL <http://bobbins.mdx.ac.uk/~john6/conf2001/paper522001.pdf>. This paper presents a quantitative analysis of the changes in the structure of the market and the degree of concentration in the arms industry based on the SIPRI data on arms-producing companies for the period 1990–98.

These are the companies that emerged as the largest defence contractors from the series of mergers and acquisitions (M&A) in the late 1990s. Growth is the result primarily of acquisitions, not internal growth. The problems associated with large-scale acquisitions—increased debts and difficulties with integrating acquired activities—have led several of the largest companies (e.g., Lockheed Martin, Boeing and Raytheon) to begin divesting some of their non-core activities with the aim of focusing their business activities and reducing their debts.[8] Other companies, such as Northrop Grumman and General Dynamics, continue their expansion in military activities, in particular in information technology (IT) services to the US Department of Defense (DOD).[9]

The high rate of concentration can also be seen in the dominance of a few contractors in government arms procurement. In the USA, the share of the five largest recipients of prime contract awards has increased from 22 per cent of total US DOD prime contract awards in 1990 to 31 per cent in 2000. Two companies, Lockheed Martin and Boeing, received 11 and 9 per cent, respectively, of the total value of US DOD prime contract awards in fiscal year (FY) 2000.[10] Similarly, the domestic arms sales of the British BAE Systems and French Thales—two of the largest European arms-producing companies—accounted for 15–20 per cent of total arms procurement from domestic production in 2000.[11]

Concentration takes place primarily through mergers and acquisitions and through joint ventures. Table 7.3 lists the major acquisitions during 2001 in the Euro-Atlantic area. It shows that although the US M&A peaked in the period 1994–97, the process continued to 2001, albeit on a less intensive scale.

The main development in US concentration in 2001 was the $2.1 billion acquisition of Newport News Shipbuilding, owner of one of six major shipyards in the USA. General Dynamics, owner of three of the other shipyards, was blocked from the acquisition, based on the conclusions of the DOD that it 'would eliminate competition for nuclear submarines' and 'harm competition for surface combatants and for the development of emerging technologies for both nuclear submarines and surface ships'.[12] Instead, the DOD decided to allow the bid by Northrop Grumman, owner of the other two shipyards.

---

[8] Lockheed Martin had a debt of $10 billion at end-2000. Lockheed Martin, *Annual Report 2000*, p. 56.

[9] 'Flat DOD budgets force contractors to diversify into booming sector', *Defense News*, 26 Feb. 2001, p. 34.

[10] US Department of Defense, *100 Companies Receiving the Largest Dollar Volume of Prime Contract Awards*, fiscal years 1990 and 2000.

[11] Estimates are based on the share of the company's domestic arms sales in total domestic armaments spending. For BAE Systems: £1.5–2 billion in £10 billion; for Thales: €1.5 billion in €10 billion. Company annual reports; British and French official data on national arms sales; and appendix 7B in this volume.

[12] 'Northrop Grumman expects Newport News buyout soon', *Jane's Defence Weekly*, 31 Oct. 2001, p. 34.

Table 7.3. Major acquisitions of arms-producing companies in North America and Western Europe, 2001

Figures are in US $m.

Buyer company (country)	Acquired company	Seller country	Sector*a*	Price	Deal status
Intra-US					
AlliantTech Systems	Thiokol Propulsion	USA	. .	685	Completed
AlliantTech Systems	Unit of Blount Int.	USA	SA/A	250	Agreed
BF Goodrich	Unit of Raytheon	USA	El	. .	Completed
DRS Technologies	Unit of Boeing	USA	El	84	Agreed
EDO	Dynamic Systems	USA	IT	. .	Agreed
General Dynamics	Primex Technologies	USA	SA/A	. .	Completed
L-3 Communications	Unit of AlliantTech	USA	El	. .	Agreed
L-3 Communications	Gov't Service Group	USA	El/IT	38	Agreed
Lockheed Martin	OAO Corp.	USA	IT	. .	Agreed
Northrop Grumman	Newport News	USA	Sh	2 600	Cleared
Northrop Grumman	Unit of Aerojet (Gencorp)	USA	El	. .	Completed
Northrop Grumman	Litton	USA	Sh	2 600	Completed
Titan	BTG	USA	IT	. .	Agreed
Veritas Capital	Raytheon Aerospace	USA	Ac service	270	Agreed
Intra-European					
EADS (FRG/FRA/SPA)	Patria Industries	FIN	El MV SA/A	42	Completed
EADS (FRG/FRA/SPA)	CAC Systèmes	FRA	Ac	5	Cleared
HDW/Ferrostaal (FRG)	Hellenic Shipyard	GRE	Sh	6	Agreed
Transatlantic by USA/Canada					
CAE (CAN)	Unit of BAE Systems	UK	El	80	Agreed
Carlyle Group (USA)	Unit of BAE Systems	UK	El	200	Agreed
FLIR Systems (USA)	Unit of Saab Tech Elecs	SWE	El	. .	Agreed
General Dynamics (USA)	Santa Barbara	SPA	MV SA/A	5	Completed
ONCAP (CAN)	BAE Systems Canada	UK	El	200	Agreed
Transatlantic by Europe					
EADS (FRG/FRA/SPA)	Cogent	CAN	El	. .	Agreed
Thales (FRA)	Magellan Corp and Navigation S.	USA	El	70	Completed
ASML (NET)	Silicon Valley Group	USA	Oth	. .	Agreed
GKN (UK)	Unit of Boeing	USA	Ac	. .	Agreed

USA = United States; FRG = Germany; FRA = France, SPA = Spain, FIN = Finland; GRE = Greece; CAN = Canada, UK = United Kingdom; SWE = Sweden; NET = Netherlands

a For sector codes, see appendix 7A.

Sources: The SIPRI arms industry files on mergers and acquisitions.

The decision in favour of Northrop Grumman could be interpreted as a sign that the US Government prefers a degree of competition in its weapon acquisitions. Whether this will be successful is uncertain. In an examination of the implications of this acquisition, the Congressional Research Service (CRS) concluded that it could reduce competition in the construction of aircraft carriers and amphibious assault ships as well as in naval radar and combat systems.[13] The deal is an illustration of the dilemma for governments of balancing the promotion of rationalization and economies of scale to achieve cost reduction against preserving competition in an already highly oligopolistic market.

In Europe, the consolidation of national arms industries took place earlier, because of the smaller domestic markets. Further concentration efforts have involved international joint ventures and mergers, resulting in integration on the European level. European integration and transatlantic industrial linkages are described in the section on internationalization.

Impact on company specialization

Reduced budgets for arms procurement after the end of the cold war left arms-producing companies with three options: (*a*) leaving the military market, (*b*) reducing their dependence on military sales through diversification into civilian products, or (*c*) strengthening their position within the military market, primarily through acquisitions. The outcome of industrial adjustment strategies is likely to have a significant influence on government–industry relations. Companies specializing in specific weapon programmes often hold a dominant position in the market and thereby gain strong leverage over governments to favour their systems in the weapon acquisitions selection process.

A comparison between the 10 largest arms-producing companies in 1990 and 2000 illustrates that there has been no reduction in their specialization on military sales between 1990 and 2000. Growth and consolidation through acquisitions does not seem to have led to reduced dependence on arms sales for these companies. The same finding was arrived at by a study of defence specialization in a number of US firms.[14]

The restructuring process has not led to a further specialization of single companies on one specific sector of military production. Companies have maintained, if not increased, their diversification within the military sector largely as a result of acquisitions. None of the world's largest producers of military aircraft—Lockheed Martin, Boeing, and BAE Systems—derived more than 20 per cent of their total sales from sales of military aircraft in 2000. However, several of these companies have focused on a few major weapon programmes. For example, the F-16 and the F-35 (Joint Strike Fighter, JSF) combat aircraft programmes accounted for more than 15 and 25 per cent, respectively, of Lockheed Martin's order backlog in 2001.[15] The contract for the JSF, which is expected to be the only new major combat aircraft development programme for several decades, was awarded to Lockheed Martin as single prime contractor in October 2001, under a 'winner-takes-all'

---

[13] *Navy Shipbuilding: Proposed Mergers Involving Newport News Shipbuilding: Issues for Congress*, Report by the Congressional Research Service, May 2001, summarized in 'DOD signals relaxed attitude towards mergers', *Defense News*, 8–14 Oct. 2002, p. 6.

[14] Markusen, A., 'The post-cold war persistence of defense specialized firms', ed. G. Susman, *Defense Diversification in the Post-Cold War Era: Corporate Strategies and Public Policy Perspectives* (Elsevier: London, 1998).

[15] See chapter 8 in this volume.

procurement strategy. The contract award was accompanied by an intensive debate on the possibilities of maintaining competing technological capabilities in fighter aircraft production in the USA within the framework of this strategy.[16] Lockheed Martin's third major aircraft programme, the F-22, entered a low-rate initial production phase in 2001. Similarly, for Boeing, its C-17 transport aircraft and F/A-18E/F combat aircraft programmes account for a high share of the company's Military Aircraft and Missiles division sales. Both aircraft are, however, unlikely to be produced in large numbers and Boeing's failure to win the JSF prime contract award may significantly reduce its role as a military aircraft producer.[17] Therefore, government intervention has been discussed to support the company's military activities through other programmes, such as the C-17 and 767 tanker, based on industry considerations rather than on military requirements.[18] Contracts for the US missile defence system are another major source of revenue for Boeing—$1200 million in 2000. These activities accounted for roughly 15 per cent of its Space and Communications sales in 2000.[19]

Internationalization

The arms-producing activities of companies are generally much less internationalized than their commercial activities, because of national security considerations. While there has been a higher degree of internationalization since the end of the cold war, it is still relatively limited. The main forms of internationalization in the arms industry are: international trade, foreign investment, sub-contracting, licensing, mergers and acquisitions, joint ventures and looser forms of inter-firm agreements, including co-production, management consortia and teaming arrangements.[20]

The types and drivers of internationalization vary with the geographical context and between the tiers of producers. The largest Western arms-producing companies derive a significant portion of their sales from exports (10–40 per cent in 2000).[21] For individual weapon programmes, exports can account for an even larger share of total production. Thus, Lockheed Martin sales of F-16 combat aircraft since 1975 included slightly over 2000 aircraft to the US armed forces and almost the same amount (slightly over 1800) for export.[22] New orders in 2000 for the F-16 included 220 for export and only 14 for the

---

[16] RAND, 'Assessing competitive strategies for the Joint Strike Fighter: opportunities and options', 2001, URL <http://rand.org/publications/MR/MR1362/>, p. 80.

[17] De Briganti, G., 'After JSF, Boeing glides in military market', defense-aerospace.com, 5 Dec. 2001, URL <http://www.defense-aerospace.com/data/features/data/fe215/index.htm>.

[18] Project on Government Oversight, 'The Pentagon attempts to quietly push two sweetheart deals for Boeing through Congress', 26 Nov. 2001, URL <http://www.pogo.org/mici/c17/c17alert.htm>.

[19] Boeing, *Annual Report 2000*, p. 58.

[20] Sköns, E., 'Western Europe: internationalization of the arms industry', ed. H. Wulf, SIPRI, *Arms Industry Limited* (Oxford University Press: Oxford, 1993), p. 190.

[21] Estimates based on available data for the 10 largest arms-producing companies in the OECD and developing countries in 2000.

[22] Lockheed Martin, URL <http://www.lockheedmartin.com/factsheet/product2.html>.

USA.[23] Similarly, Boeing has very high export shares for some of its programmes, including its Apache and Chinook military helicopters, all of which were exported in recent years.

As a result of international mergers and acquisitions many of the major Western arms-producing companies are increasingly expanding their access to foreign markets through the establishment of foreign subsidiaries rather than direct exports alone. Thales has called this strategy a 'multi-domestic' industrial presence.[24] The term epitomizes the acknowledgement by an international company of the preference of national governments to procure domestically produced arms.

The largest West European companies are more internationalized than US companies in terms of foreign subsidiaries. This is true, in particular, for the three companies among the Top 10 in 2000—BAE Systems, EADS and Thales. These companies have pursued very active strategies for investment in foreign companies, not only within Europe but also in the USA, Australia and South Africa. BAE Systems has a strong presence in North America, where its foreign subsidiaries had combined sales of $3.7 billion in 2001 and employed 22 000.[25] Its sales from foreign subsidiaries worldwide accounted for 45 per cent of total sales in 2000.[26] The number of foreign subsidiaries owned by Thales increased from 55 in 1998 to 213 in 2000, largely as a result of its acquisition of Racal (UK). Thales has an extremely high dependence on foreign markets. Foreign sales (exports and sales from foreign subsidiaries) accounted for about 75 per cent of its total sales in 2000.[27]

Internationalization occurs in three different geographical contexts and layers of producers: (a) among the major arms-producing companies in Europe, a process which is currently focused on the signatories of the 2000 six-nation Framework Agreement;[28] (b) on the transatlantic level; and (c) on acquisitions by major Western arms-producing companies in minor producer countries in the context of major arms export deals.

These developments are also linked to different political and institutional processes. Since governments are the arms industry's main customers, their procurement plans and defence policies play a major role in shaping the industry. The future of European defence industrial integration is intertwined with the process of European integration in general and with the development of a European Security and Defence Policy (ESDP) in particular.[29] The tension between the goal of an autonomous military capability for Europe and that of maintaining and developing the transatlantic partnership within NATO is also reflected in the defence industry policies and the developments in industry.

---

[23] Lockheed Martin, URL <http://www.lockheedmartin.com/spotlight/newslines/newsline18 7.html>.
[24] Thales, *Annual Report 2000*, p. 38
[25] BAE Systems North America, URL <http://www.na.baesystems.com/aboutus.htm>.
[26] BAE Systems *Annual Report 2000*, p. 43.
[27] Thales, *Annual Report 2000*, pp. 8, 71.
[28] Framework Agreement between the French Republic, the Republic of Germany, the Italian Republic, the Kingdom of Spain, the Kingdom of Sweden, and the United Kingdom of Great Britain and Northern Ireland Concerning Measures to Facilitate the Restructuring and Operation of the European Defence Industry, 27 July 2000, URL <http://projects.sipri.se/expcon/loi/indrest02.htm>.
[29] See chapter 3 in this volume.

Integration among the major arms-producing companies in Europe

European integration of arms production capabilities has been limited to cross-border joint ventures because of the difficulties that confront international mergers and acquisitions, most notably the lack of a legal and political framework for transnational companies in otherwise fragmented military markets. However, many of the joint ventures that were formed during the 1990s represent rather broad cooperative structures that may be forerunners to more integrated structures in the future. A turning point was reached in 2000 with the creation of EADS through a cross-border merger of three major aerospace companies—following the creation of BAE Systems through a national merger in 1999. While it is still not clear to what extent EADS represents genuine integration at the management level,[30] and despite the fact that roughly 80 per cent of the company's sales are for the non-military market, its creation marked a milestone in European military–industrial integration. Intra-European integration continued in 2001, with the continued concentration of arms production within large transnational joint ventures (table 7.4). The major events in 2001 included the creation of the British–Italian helicopter joint venture AgustaWestland and the expansion of the British–Italian avionics joint venture Alenia Marconi Systems.

The major European joint venture created in 2001 was the British–French–German–Italian MBDA for the design and production of missiles.[31] Owned by BAE Systems, EADS (37.5 per cent each) and Finmeccanica (25 per cent), it will include the missile activities of Anglo/French Matra BAe Dynamics, of former Daimler-Chrysler Aerospace, of Italian Finmeccanica, and some additional French and British activities with a combined turnover of €2.3 billion ($2.1 billion) and total orders worth €13 billion ($12 billion). The merger was subject to review by national monopoly commissions but national governments invoked an EU clause exempting the merger from a review by EU Commission competition authorities for national security reasons, since arms sales account for 99 per cent of MBDA sales. The EU Commission competition authorities were reported to be considering a challenge of the legality to use this clause and an investigation.[32]

While a number of other joint ventures were created and old ones expanded in the European arms industry in 2001, a development in early 2002 illustrated the difficulties that remain for the consolidation of the European arms industry. This was the collapse of the plans by EADS and Finmeccanica to create a joint venture company—the European Military Aircraft Company (EMAC). In effect, EMAC would have resulted in a merger between the military aircraft capabilities of Germany, France, Italy and Spain with a combined employment

---

[30] Betts, P., 'Take-off delayed by squabbles in the cockpit', *Financial Times,* 16 Nov. 2001, p. 12.
[31] MBDA is an abbreviation for Matra, BAe Dynamics, Alenia Marconi Systems and Aérospatiale Matra.
[32] 'Merger wins go-ahead for missile venture', *Air Letter*, 30 Dec. 2001, p. 5.

Table 7.4. International, West European and transatlantic joint ventures and mergers among arms-producing companies, established in 2000–2001

Company name	Owner companies, parent company (country)	Sector[a]
West European		
Aero Propulsion Alliance (APA)	24.8% MTU (Germany); 24.8% Rolls-Royce (UK); 24.8% Snecma (France); 13.6% ITP (Spain); 8% Fiat Avio (Italy),[b] 4% Techspace Aero (Belgium)	Aircraft engines (A400M)
AgustaWestland	50% GKN (UK); 50% Finmeccanica (Italy)	Helicopters
Astrium	50% DaimlerChrysler Aerospace (Germany); 50% Matra Marconi Space (France/UK)	Space
Diehl Avionik Systeme	51% Diehl (Germany); 49% Thomson-CSF Sextant, Thomson-CSF (France)	Avionics
European Aeronautic Defence and Space Company (EADS)	30% DaimlerChrysler (Germany); 15% French State; 15% Lagardère (France); 5.5% SEPI (Spain)	Aircraft, electronics, missiles
ET Marinesysteme	50% EADS (Germany/France); 50% Thales Nederland, Thales (France)	Naval electronics
Eurofighter Simulation Systems	26% Thales (France); 26% Indra (Spain); 24% CAE Elektronik, CAE (Canada) and STN Atlas Elektronik (Germany); 24% Finmeccanica (Italy)	Simulation
MBDA[c]	37.5% EADS (Germany/France); 37.5% BAE Systems (UK); 25% Finmeccanica (Italy)	Missiles
Nordic Support and Service Centre (NSCC)	AerotechTelub (Sweden); Danish Aerotech (Denmark); Astec Helicopter Serv. (Norway)	Helicopter logistics
Rolls-Royce Snecma	50% Rolls-Royce (UK); 50% Snecma (France)	Aircraft engine development
Stand-Off Surveillance and Target Acquisition Radar (SOSTAR)	28% EADS (Germany/France); 28% Thales (France); 28% FIAR (Italy); 11% Indra (Spain); 5% Fokker Space (Netherlands)	Radars
Turboprop International	33% Snecma (France); 33% MTU (Germany); 22% Fiat Avio (Italy); 12% IPT (Spain)	Aircraft engines (A400M)
Transatlantic		
Aviation Communication & Surveillance Systems (ACSS)	70% L–3 Communications (USA); 30% Thales (France)	Electronics
Performance Diesels Company	MTU, DaimlerChrysler (Germany); General Dynamics (USA)	Engines for military vehicles
Rotorism	50% AgustaWestland (Italy/UK); 50% CAE (Canada)	Helicopter simulation
Thales Raytheon Systems	50% Raytheon (USA); 50% Thales (France)	Radars

[a] For sector codes, see appendix 7A.

[b] FiatAvio was excluded from the joint venture in early 2002 following the Italian withdrawal from the A400M programme.

[c] MBDA is an abbreviation for Matra, BAe Dynamics, Alenia Marconi Systems and Aérospatiale Matra.

Sources: The SIPRI arms industry files on joint ventures.

of 17 000 and annual revenues of €2.5 billion.³³ The official reason for the failure was that the planned mix of commercial and military aircraft, which the Alenia contribution had planned to include in order to bring its participation up to rough parity with the others, would not 'fit with the situation post September 11'.³⁴ Another, perhaps more crucial, complication was the Italian Government's decision to participate in the US F-35 JSF project, in which BAE Systems is a partner. The failed integration of Alenia Aerospazio (of Finmeccanica) into EADS prolongs the competition between EADS and BAE Systems for a dominant position in the European military aerospace sector. However, EADS and Finmeccanica continued negotiations to link their military aircraft activities in a looser structure of cooperation.

Continued industrial integration is supported by government policy initiatives aimed at harmonizing armament requirements. The defence ministers of six European countries—France, Germany, Italy, Spain, Sweden and the United Kingdom—signed a declaration in late 2001 that commits them to: (*a*) cooperate on advanced technologies that will develop Europe's future capabilities for combat air systems towards the end of the next decade (2020); (*b*) to launch, in cooperation with industry, a programme known as the European Technology Acquisition Programme (ETAP) for this purpose; and (*c*) to encourage European industry to make a suitable financial contribution to the effort.³⁵ The declaration is an indication of the determination of the major European governments to create a framework for the continued integration of arms-producing activities in Europe.³⁶ It also aims to achieve a balance between European and transatlantic integration. ETAP allows individual partner countries to cooperate only on selected parts of the programme. This made it possible for the UK, which shares the same kind of sensitive military technology with the USA within the JSF project,³⁷ to participate in ETAP. While this represented yet another step in efforts to promote and support the integration of the European arms industry at the government level, future European cooperation in the development and production of advanced weapon systems is still only developing slowly.

Transatlantic military–industrial links

The establishment of military–industrial links between the USA and Western Europe has been subject to even stronger political and regulatory challenges than the efforts to integrate the arms industry in Western Europe. As a result,

---

³³ 'EMAC to allow Europe to take on US fighter industry', *Jane's Defence Weekly*, 18 Apr. 2001, pp. 26–27.

³⁴ Nicoll, A., 'EADS withdraws from Italian deal', *Financial Times*, 24 Jan. 2002, p. 1.

³⁵ French Ministry of Defence, 'European governments and industry to cooperate on future capabilities and technologies for combat air systems', Press notice on behalf of the defence ministries of France, Germany, Italy, Spain, Sweden and the United Kingdom, Paris, 19 Nov. 2001, URL <http://www.defense.gouv.fr/english/news/shortnews/b201101/201101.htm>.

³⁶ It is a continuation of previous policy measures by the same countries: the 1997 statement of their joint interest in an efficient and globally competitive European aerospace and defence electronics industry, and the 2000 Framework Agreement on Measures to Facilitate the Restructuring and Operation of the European Defence Industry.

³⁷ Barrie, D., 'ETAP partners look to seal stealth deal', *Defense News*, 8–14 Oct. 2001, p. 4.

transatlantic mergers and acquisitions among the largest arms-producing companies on both sides of the Atlantic have been rare.

The most important transatlantic acquisitions since 1995 were the acquisition of the US military electronics company Tracor by the British company GEC in 1998 and the acquisition of two Lockheed Martin military electronics units by BAE Systems in 2000. The acquisitions were made possible because of the particular military–political relation between the UK and the USA. Companies from other West European countries have not been able to establish a similar strong foothold in the USA.

The large West European arms-producing companies have sought acquisitions in the USA in order to gain access to its vast budget for military equipment. US arms-producing companies, on the other hand, seem to be less interested in large-scale acquisitions in Europe, mainly because of the significantly smaller market for military equipment and its continuing fragmentation into national markets.[38] Despite this, minor US acquisitions in Europe have taken place in recent years. Among the most significant were the acquisitions of the Swedish Bofors Defence by United Defense in 2000 and of the Spanish Santa Barbara by General Dynamics in 2001.

While direct mergers and acquisitions have been rare, transatlantic military–industrial links have developed in more flexible forms, particularly in joint ventures. This trend is likely to continue. A number of industrial links have been established among major arms-producing companies on both sides of the Atlantic as an outgrowth of government-to-government programmes. These include the partnerships between Boeing and BAE Systems, between Lockheed Martin and EADS, between Northrop Grumman and EADS, and between Raytheon and Thales.[39]

The establishment of the large transatlantic joint venture Thales Raytheon Systems was approved in June 2001 by the French and US governments. Thales Raytheon Systems includes the air defence activities of Thales and Raytheon and represents the first US–European partnership across an entire product sector. Having solved a range of regulatory issues, in particular in the fields of security and export licensing, it may serve as a model for future transatlantic strategic partnerships.[40]

The most significant decision in 2001 with regard to its impact on future transatlantic cooperation was the decision by the British Government to participate in the JSF project. The project will clearly strengthen existing British–US armaments cooperation and result in a number of new military–industrial links between the two countries.

---

[38] US General Accounting Office (GAO), 'Defence trade, contractors engage in varied international alliances', GAO/NSIAD-00-231, Sep. 2000, URL <http://www.gao.gov>.

[39] The reasons for European efforts to establish themselves in the USA are the subject of a study by Andrew James. James, A. D., 'The prospects for a transatlantic defence industry', in ed. B. Schmitt, *Between Cooperation and Competition: The Transatlantic Defence Market*, Chaillot Paper 44 (Western European Union Institute for Security Studies: Paris, Jan 2001).

[40] 'Thales–Raytheon teaming raises regulatory issues', *Defense News*, 10–16 Sep. 2001.

Internationalization in minor arms-producing countries

In their search for access to markets, major Western arms-producing companies are establishing close industrial links with companies in some of the smaller arms-producing countries. These are often countries that are embarking on large military procurement programmes. The arms industries and governments in these minor arms-producing countries perceive a close relationship with a large Western arms-producing company as a means of gaining access to advanced military technology and financial resources essential to restructuring, and thereby maintaining parts of their domestic defence industrial bases.

These industrial links take different forms ranging from collaboration agreements to acquisitions. The acquisition of shares in companies that are in need of not only technological and marketing support but also financial investment is increasingly becoming a central part of the offers to supply armaments to countries which have failed to carry through long-standing arms industry restructuring and privatization plans. Such offers are often made voluntarily—on the initiative of the supplier company—but they can also be part of formal offset requirements on the side of the recipient country. In both forms they have become a central marketing strategy of foreign supplier companies. Investment in the arms industry of recipient countries may also be a means for a company that is transferring military technology through licence agreements to maintain control over the technology.

There are various implications of this kind of linkage between arms exports and direct participation in arms-producing companies in the recipient country. Such offset requirements may lead supplier companies to compete in areas outside of their core activities and result in job losses in the supplier country and in technology transfers that decrease the competitive advantage of the supplier.[41] On the recipient side, military offsets may divert the focus of decisions on weapon procurement from military requirements to industry considerations. The close link between industry offsets and armament imports may also increase the pressure on governments to implement military procurement plans. Moreover, buyer companies often agree to broad investment and restructuring schemes. Therefore, direct military offsets in the form of the acquisition of arms-producing companies in the recipient country by the supplier company may lead to the maintenance or strengthening of arms production in companies that would otherwise have left the military market, either through diversification into civilian production or by closing down their facilities for arms production. The majority of the companies that were subject to foreign acquisitions in connection with arms deals in 2000 and 2001 faced severe financial problems and might have gone bankrupt in the absence of foreign investment.

---

[41] The role of offsets in international trade and their possible adverse effects on the US industry, economy and national security are the subject of a report by the US Presidential Commission on Offsets in International Trade. *Status Report of the Presidential Commission on Offsets in International Trade*, 18 Jan. 2001, URL <http://www.offsets.brtrc.net/statusreport/statusreport.htm>. See also chapter 8 in this volume.

Table 7.5. Company acquisitions linked to arms deals with minor producer countries, 2000–2001

Figures are in US $m. Figures in italics are percentages.

Arms procurement deal			Related company acquisition			
Buyer country	No. and weapon system	Value	Acquired company	Buyer company	Share (%)	Price
Agreed						
Brazil	(48) Combat Ac upgrade	230	Aeroeletronica	Elbit (Israel)	*60*	2.3
Greece	3 Submarines	1 300	Hellenic Shipyard	HDW (Germany)	*100*	30
Finland	20 Helicopters	480	Patria Industries	EADS (France/ Germany/Spain)	*27*	39
Poland	8 Transport Ac	212	PZL Warszawa–Okecie	EADS (France/ Germany/Spain)	*51*[a]	7
Poland	.. Combat Ac engines	..	PZL Rzeszow	UTC (USA)	*85*	70
Planned						
Czech Rep.	24 Combat Ac	1 300	Aero Vodochody	BAE Systems/ Saab (UK/Sweden)	*64*	..
Greece	60+ Combat Ac	≈ 4 600	Hellenic Aerosp.	..	*49*	..
Poland	10–12 Helicopters	..	Swidnik	..	*37*	..

Ac = aircraft

[a] EADS agreed to increase its share in the company to 85% within 2 years.

Source: SIPRI arms industry files.

During the 1990s new arms markets for Western suppliers have emerged in South Africa and Central and Eastern Europe (CEE). The increase in the procurement requirements of these countries has coincided with a need for foreign investment and technology input for the restructuring of their domestic industries. Major West European armaments suppliers are negotiating a series of agreements on direct investments in the South African arms industry in connection with its large procurement programme, which was initiated in 1999.[42] During 2000 and 2001, similar investment agreements were made with or offered to arms-producing companies in Poland and the Czech Republic (table 7.5), stimulated by the competition between major aerospace producers to gain control over the markets in these countries.

The arms industry restructuring plan adopted by the Polish Government in 1999 linked the purchase of foreign military equipment to the sale of the state-owned domestic arms-producing companies. Under the scheme foreign suppliers were not only required to place offset contracts with the Polish arms industry but also to participate in the privatization of the Polish company

---

[42] Acquisitions of South African companies by Western arms producers in 1999 are discussed in Sköns, E. and Weidacher, R., 'Arms production', *SIPRI Yearbook 2000: Armaments, Disarmament and International Security* (Oxford University Press: Oxford, 2000), pp. 311–14.

involved in these offsets.⁴³ The aim is to achieve a successful restructuring and privatization of the Polish arms industry by linking it closely to the implementation of arms import programmes.

In 2001 the first privatization of a Polish arms-producing company in direct connection with the purchase of foreign military equipment was agreed. As part of a deal involving the procurement of eight CASA (EADS) military transport aircraft, the supplier companies EADS and Spanish AVIA Systems Group acquired a 51 per cent stake in the Polish aerospace company PZL Warszawa-Okecie and agreed on an investment plan for the company.⁴⁴ The dependence on military sales of the Polish aircraft company PZL Warszawa-Okecie is likely to increase if the offset agreement related to its acquisition is implemented. The company currently produces the PZL-130 Orlik military trainer aircraft, but its main production is civilian aircraft—mostly for agricultural use. The 2001 offset agreement stipulated the participation of the Polish company in the production of the military transport aircraft and the transformation of the company into the primary in-service support centre for the aircraft in Poland.⁴⁵ Similar offers have been made by foreign suppliers in connection with the Polish procurement plans for combat helicopters.⁴⁶ These included offers to acquire shares in a helicopter company (PZL Swidnik). The postponement of the procurement programmes has resulted in withdrawals of these offers. In another example, a US engine company (Pratt & Whitney) acquired a majority share in a Polish company producing aircraft components (PZL Rzeszow) with the purpose of increasing its chances of gaining involvement in a procurement contract for combat aircraft.⁴⁷ In late 2001 the Czech Government set a requirement for direct military offsets similar to the Polish one. The Czech Government required the acquisition of the state-owned majority share in the domestic aerospace company Aero Vodochody in connection with its combat aircraft procurement tender.⁴⁸ Aero Vodochody is facing severe financial difficulties, largely as a result of the low demand for one of its main military products, the L-159 combat aircraft. The team winning the fighter contract, BAE Systems (UK) and Saab (Sweden), offered to acquire a 64 per cent stake in the company.⁴⁹

---

⁴³ Krason, M. A., 'An offer to investors: arms industry closer to NATO', *Rynki Zagraniczne* (Warsaw), 6–7 May 1999, p. 5, in 'Polish military industry discussed', Foreign Broadcast Information Service, Daily Report–East Europe (FBIS-EEU), FBIS-EEU-1999-0524, 25 May 1999; and Piskorski, M., 'Arms industry more attractive', *Rzeczpospolita* (Warsaw), 24 June 1999, in 'Polish Cabinet drafts arms industry privatization law', FBIS-EEU-1999-0624, 25 June 1999.

⁴⁴ Ratajczyk, A., 'Privatization takes off', *Warsaw Voice*, vol. 672, no. 36 (9 Sep. 2001), URL <http://www.warsawvoice.pl/>.

⁴⁵ Holdanowicz, G., 'Warsaw buys C–295s with offsets to revamp PZL', *Jane's Defence Weekly*, 5 Sep. 2001, p. 19.

⁴⁶ 'Helicopter producer for sale', *Warsaw Voice*, 28 Nov. 1999, p. 10.

⁴⁷ Lockheed Martin F–16 fighter aircraft are powered by Pratt & Whitney engines. See Holdanowicz, G., 'Country briefing: Poland, an uphill task', *Jane's Defence Weekly*, 26 Sep. 2001, p. 24; and 'Pratt & Whitney buys Polish aero firm for $70m', *Air Letter*, 25 Sep. 2001, p. 4.

⁴⁸ Elch, J., 'Nya krav kan sinka affär' [New demands can sink deal], *Svenska Dagbladet* (Stockholm), 21 Nov. 2001.

⁴⁹ 'Report: Czech Army grounds L–159s', *Air Letter*, 19 Nov. 2001, p. 1.

Similar, although less clear, cases of foreign ownership participation linked to procurement programmes occurred in two other minor arms-producing countries during 2000 and 2001—Finland and Greece. The partial privatization of the largest Finnish arms-producing company Patria Industry in 2001 took place in close connection with a major procurement decision by the Nordic countries. In June 2001 Finland, Sweden and Norway decided to procure NH-90 helicopters, produced by a joint venture in which the EADS helicopter subsidiary Eurocopter maintains a majority stake. In February 2001 EADS had agreed to acquire a minority stake in state-owned Patria Industries.

The Greek Government's decision in 2000 to sell off a 49 per cent share of the loss-making domestic aerospace company Hellenic Aerospace Industries (HAI) was strongly linked to a procurement plan for combat aircraft.[50] The Eurofighter companies—EADS, BAE Systems and Alenia Aerospazio—competed with a French–Greek bid led by Dassault Aviation, producer of the Rafale combat aircraft, for the share in Hellenic Aerospace (HAI).[51] In late 2000, however, the fighter procurement plan was suspended and with it the related privatization plan for HAI.[52]

These are examples of cases where direct investment in the arms industry of minor arms-producing countries has been part of the market access strategies of supplier companies. Supplier companies may also see the acquisition of shares of arms-producing companies in the recipient country as a means to secure control over their weapon technology. This can be important in cases where the importing country has an offset policy for arms imports which includes requirements for technology transfers from the supplier. It is exemplified by the acquisition of Hellenic Shipyards (Greece) in 2001 by the German consortium Howaldtswerke Deutsche Werft (HDW) and Ferrostaal. In 2000 the German consortium signed a contract for the construction of three Type 214 submarines, one to be built in Germany and two in Greece by Hellenic Shipyards.[53] The acquisition of the Greek company responsible for the construction of the German-designed submarines allows HDW/Ferrostaal to maintain control over the technology transferred through the deal.[54]

---

[50] In 1999 the Greek Government decided to acquire 50 F–16 and 15 Mirage 2000. In early 2000 negotiations were still continuing about the possible acquisition of 60–90 Eurofighters. 'KYSEA briefed on course of armaments program', Athens News Agency, 21 Jan. 2000, in 'KISEA discusses Greek five-year arms programme', Foreign Broadcast Information Service, *Daily Report–West Europe (FBIS-WEU)*, FBIS-WEU-2000-0121, 24 Jan. 2000.

[51] Bombeau, B., 'Mirage 2000–5 pour la Grèce', *Air & Cosmos*, no. 1776 (22 Dec. 2000), p. 31.

[52] Hellenic Aerospace was awarded a significant share of the subcontract work as part of the offset agreement for the Greek purchase of French Mirage 2000 (5 aircraft, or 65 billion drachmas). Valmas, T. L., 'Greece signs US$1.4 billion contract for 32 Mirage 2000s', *Jane's International Defense Review*, Nov. 2000, p. 24.

[53] HDW/Ferrostaal acquired Hellenic Shipyards for €6 million and agreed to invest in the company: €33 million immediately and another €15 million within 5 years. HDW Press Release, 12 Oct. 2001, on Defense-aerospace.com, URL http://www.defense-aerospace.com/data/communiqués.

[54] Agüera Büchenbeuren, M., 'German buy continues European shipyard mergers', *Defense News*, 29 Oct.–4 Nov. 2001, p. 12.

Privatization

The national arms-producing facilities which were built up in the 1930s to provide states with effective control over military production have gradually been replaced by, or transformed into, private commercial companies that produce weapon systems for the state on contract. The privatization of arms production continued throughout the 1990s. By the early 2000s a large part of the arms industry was privately owned in most major arms-producing countries.

Three major drivers of privatization in the post-cold war period can be distinguished. Table 7.6, shows the major events of privatization since 1990: (*a*) the privatization of the major remaining state-owned arms-producing companies in Western Europe (France, Italy and Spain) and Australia in the first stage of their participation in measures of concentration, often also involving their internationalization; (*b*) the transition of the formerly centrally planned CEE economies to a capitalist system with private ownership, which also involved the arms industry; and (*c*) the privatization in other minor arms-producing countries as a result of industrial offsets in major arms import programmes.

In France, where the state controlled most of the development and production of military equipment as late as 1998, the aim of European integration brought about a series of privatizations of its main military aerospace and electronics companies in 1998–99. However, significant assets remain under state management or ownership, including the shipbuilding company DCN, the aircraft maintenance company SMA, the military vehicle company GIAT Industries and the aeronautics engine company Snecma. There are plans to transform DCN from a state-managed into a state-owned company by 2003,[55] while the partial privatization of Snecma, planned for late 2001, was postponed when the aeronautics market declined after 11 September.[56] In Italy—where throughout the 1990s almost all arms-producing enterprises (except FIAT) belonged to large state holding companies—the major aerospace company (Finmeccanica) was privatized and privatization of the shipyard Fincantieri was initiated in 2000.

Spain, another country with state ownership as the dominant mode in the arms industry, has initiated a series of privatizations since 1999 in order to be able to join in the internationalization of the European arms industry. Thus, by the early 2000s large private companies were the dominant mode of ownership in the arms industry in all major arms-producing countries in the West, similar to the situation before World War II.

---

[55] Mackenzie, C., 'France's DCN approaches privatization with task list', *Defense News*, 12–18 Nov. 2001, p. 32.
[56] Lewis, J. A., 'Snecma privatization plan is put on hold until markets recover', *Jane's Defence Weekly*, 26 Sep. 2001, p. 14.

Table 7.6. Major cases of company privatization, 1990–2001

Year	Country	Company	Share privatized (%)	Form of privatization (sales of shares)	Buyer type	Nationality
1990	Norway	Raufoss	47	Public offering	IS	–
1993	Netherlands	Fokker	51	Private sales	C	F (FRG)
1993	Norway	NFT	49	Public offering	IS	–
1993	Sweden	Celsius	75	Public offering	IS	–
1994	Brazil	Embraer	55	Private sales	IS	D/F (USA)
1994	Germany	IABG	45	Private sales	C	F (USA)
1995	Germany	IABG	23	Employee buyout	–	D
1995	Argentina	AMC	–	Leasing	C	F (USA)
1995	Australia	ASTA	Majority	Private sales	C	D/F (USA)
1997	Greece	Elefsis Shipyards	..	Private sales	C	D
1998	Czech Rep.	Aero Vodochody	34	Private sales	C	F (USA)
1998	France	Thomson-CSF	33	Public offering	IS	–
1999	Australia	ADI	100	Private sales	C	D/F (FRA)
1999	Bulgaria	Arsenal	51	Employee buyout	–	D
1999	France	Aérospatiale	–	Merger	–	–
1999	Norway	Norsk Jetmotor	33	Private sales	C	F (SWE)
1999	Spain	Indra	66	Public offering	IS	–
1999	Sweden	Celsius	25	Private sales	C	D
2000	Bulgaria	Trema	50	Employee buyout	–	D
2000	Greece	Hellenic Vehicle Ind.	43	Private sales	C	D
2000	Italy	Finmeccanica	38	Public offering	IS	–
2000	Spain	CASA	–	Merger	–	–
2001	Czech Rep.	Tatra	91.6	Private sales	C	F (USA)
2001	Finland	Patria Industries	26.8	Private sales	C	F (EUR)
2001	Greece	Hellenic Shipyards	51	Private sales	C	D
2001	Italy	Fincantieri	17	Public offering	IS	–
2001	Poland	PZL Warszawa-Okecie	51	Private sales	C	F (EUR)
2001	Poland	WSK PZL Rzeszow	85	Private sales	C	F (USA)
2001	Spain	Santa Barbara	100	Private sales	C	F (USA)

IS = Individual share holders; C = Company; F = Foreign; FRG = Germany; D = Domestic; USA = United States; FRA = France; SWE = Sweden; EUR = Europe.

Sources: SIPRI arms industry files.

Outsourcing of military services and functions

In recent years not only military hardware but also the provision of services has become subject to contracting to private industry (outsourcing). Outsourcing includes a range of services (support services for military equipment, military facilities and military operations), which until recently were the prerogative of government organizations such as units of the armed forces or departments of ministries of defence. Private companies are thus assuming an important role also within the field of military support services.

This practice is the result of increasing budgetary constraints and the view that there is potential for greater efficiency with increased participation by private industry in the provision of government functions. The services supplied by private companies vary with regard to their proximity to war fighting capabilities and range from non-military specific services such as management of housing, to equipment support services and the provision of a variety of military support functions.

The key distinction between public and private, or outsourced, provision is whether the provider is acting as a private entity on contract, subject to profit-making discipline, or is operating within the public sector and subject to direct democratic and civil service accountability systems. A study of the implications of privatization and outsourcing in the USA found that 'the enormity of the difference in behavior and motivation of agents operating under these two very different systems is not well understood or acknowledged by most analysts.'[57]

The study concluded that there are no clear benefits from the privatization of military purchases but there are significant risks. While the main argument in favour of privatization is its positive impact on cost through increased competition, there is broad acknowledgement that it is not the private ownership per se, but competition that can induce better quality services at more reasonable cost. The study found that this is the case only under certain conditions: that there are more than three competitors; that competition persists over time; that the task and performance requirements are clear; and that there is active monitoring by the government customer and sustained capacity to do so. The risks associated with privatization included the potential for corruption and the capture of political decision making by politicians. In the USA privatization is associated with the potential for heightened influence over military policy by private contractors to the DOD—through lobbying and financial campaign support for presidential and congressional candidates, domination of DOD advisory committees and growing monopolization of the expertise needed to design, build and operate modern weapons.[58]

The outsourcing of support services and functions is considered most advanced in the UK.[59] The process has been supported over the past decade by a number of government initiatives, such as the Competing for Quality (CFQ), Private Finance Initiative (PFI) and Public Private Partnership (PPP) programmes. In the USA outsourcing of military support activities was stalled during the first half of the 1990s as a result of strict competition requirements but gained increased importance during the Bill Clinton Administration.[60] In

---

[57] Markusen, A., 'The case against privatizing national security', *Governance,* vol. 16, no. 4 (forthcoming 2003), available at URL <http://www.hhh.umn.edu/people/amarkusen/writings.htm>.

[58] Sapolsky, H., Gholz, E. and Kaufman, A., 'Security lessons from the cold war', *Foreign Affairs,* vol. 78, no. 4 (1999), pp. 77–89, cited in Markusen (note 57).

[59] RAND, Public–Private Partnerships: Proceedings of the US–UK Conference on Military Installation Assets, Operations, and Services, 14–16 Apr. 2000, URL <http://www.rand.org/publications/CF/CF164/>.

[60] A RAND publication that summarizes US and British efforts in the field found that the US Congress has placed a variety of restrictions on outsourcing and privatization, in particular through Circular A–76. RAND (note 59). See also Bailey Grasso, V., *Defense Outsourcing: The OBM Circular*

Germany the Gesellschaft für Entwicklung Beschaffung und Betrieb, (GEBB) was established in late 2000 with the purpose of freeing the armed forces from service functions that were not part of core military capabilities by finding private industry solutions for them.[61] Similar developments are under way in other countries. As a result, services account for an increasing share of private industry revenues from military customers. In the USA the share of the value of services in the total value of prime contract awards to US companies increased from 12 per cent in 1988 to 29 per cent in 1999.[62] BAE Systems expected the market for outsourcing of defence services to grow by 5 per cent as compared to military procurement by 2.4 per cent.[63] According to Serco, 'In the UK alone the market for defence services is expected to reach £15.1 billion by 2009'.[64]

Equipment support (i.e., life-cycle support of military equipment) is accounting for an increasing share of system costs. The system itself often accounts for less than half of total revenues, the rest being different kinds of services associated with the programme. According to Boeing:

the design, development and production of a military aircraft system make up only 30 per cent of a government's investment in total ownership cost. The overwhelming 70 per cent of that total cost is in sustainment and support—from program planning and management, through training, technical manuals and support equipment, to maintenance, modifications, upgrades and other ageing-aircraft sustainment activities.[65]

Equipment support services are provided primarily by large prime contractors, which supply services that cover the entire life-cycle of the weapon they produce. Aircraft maintenance and repair services contribute significantly to the arms sales of a large number of major military aerospace companies such as BAE Systems, Boeing and Lockheed Martin (table 7.7). Roughly one-half of Bombardier's arms sales in 2000 were derived from support services to the military. Bombardier not only provides pilot training services but also maintains ownership over the training aircraft ('power by the hour') within the NATO Flying Training in Canada (NFTC) programme.[66]

Services related to command, control, communication and information systems (C^3IS) equipment are assuming particular importance within the broader field of equipment support services. The rapid advance in information

---

A–76 Policy, Congressional Research Service (CRS) Report to Congress (Library of Congress: Washington, DC, 21 Feb. 2002).

[61] The GEBB Internet site can be accessed at Bundesministerium der Verteidigung, URL <http://wirtschaft.bundeswehr.de/index_.html>.

[62] Department of Defence, Directorate for Information Operations and Reports (DIOR), *Prime Contract Awards,* annual.

[63] BAE Systems, *Annual Report 2000,* p. 6.

[64] Serco, *Annual Report 2000,* p. 9.

[65] Boeing, 'Military aerospace support', URL<http://www.boeing.com/defense-space/military/as.htm>.

[66] See NFTC Internet site, URL <http://www.nftc.net/introduction/ExecutiveSummary.html>.

Table 7.7. Selected companies providing services to the military, 2000
Figures are in US $m.

Company, unit, country	Main military services	Sales to MODs 2000
Anteon, USA	IT services for US Navy	410
BAE Systems, Customer Solutions & Support, UK	Aircraft training and maintenance	2 500
Boeing, Military Aircraft Support Unit, USA	Aircraft training and maintenance	
Bombardier, Defence Services, Canada	Fleet management, aircraft training	[160]
Computer Sciences Corp., USA	IT services	1 610
Dyncorp, USA	IT services; fleet management; policy support	800
EDS, USA	IT services	950
Lockheed Martin, Technology Services, USA	Space operations support, aircraft support, management of nuclear weapon programme	2 280
MPRI, USA	Policy support; armed forces training	
Science Applications Int., USA	IT services	1 950
Serco, UK	Management of facilities	[300]
Silicon Graphics, USA	IT services	370
Titan, USA	IT services	780
Veridian, USA	IT services; R&D, test and evaluation of aircraft and spacecraft	[590]
Vinnell, USA	Management of facilities, armed forces training	..

Sources: SIPRI arms industry database and SIPRI arms industry files.

technologies is considered to have changed the conduct of warfare and lead to a shift in military requirements from single platforms to integrated networks, so-called Network Centric Warfare (NCW). This is an evolving concept based on the idea that linking various systems together will generate greater military benefits than could be derived from individual weapon platforms.[67] A broad range of companies, from major traditional arms-producing companies to small and fast-growing military IT specialized companies, provide services related to the integration of individual surveillance, information management and combat platforms. Large prime contractors for weapon platforms perceive diversification into federal IT products and services as a way to expand in a growth sector and apply technologies and knowledge they have accumulated through weapon systems integration.[68] IT-specialized companies, such as the large US companies Computer Science Corporation and Science Applications International, play an equally strong role in this market. The largest contract in

---

[67] Holzer, R., 'Center brings together pieces of Network Centric Warfare puzzle', *Defense News*, 27 Aug.–2 Sep. 2001, p. 26.
[68] Ratnam, G., 'Information technology market draws US firms: flat DoD budgets force contractors to diversify into booming sector', *Defense News*, 26 Feb. 2001, p. 34.

the field was awarded in 2000 to EDS—a contract worth $6.9 billion over eight years to upgrade the US Navy–Marine Corps Intranet.[69]

Military support functions that have been privatized (sold and/or outsourced to private contractors) include: (*a*) management of base facilities and related services, (*b*) logistics (military supply chain), and (*c*) military advice (planning and intelligence) and training services. Examples of companies specializing in this field—often referred to as Private Military Companies—are Serco (UK), which provides facilities management and ground maintenance work for the British Ministry of Defence (MOD), and Dyncorp (USA), which provides a wide range of services to the US military—from policy support to operating and maintaining ships for the US Military Sealift Command (MSC) and providing support services to US forces deployed in peacekeeping operations.[70] MPRI, a US company specializing in the provision of military training services and policy consulting to armed forces, was acquired in July 2001 by the US military electronics and IT company L-3 Communications.

Although small in terms of financial importance in comparison with equipment support services, the provision of military support functions by private companies has raised concerns as regards government control, in particular as a significant share of these support functions are exported from major Western countries to areas of conflict. In a recent initiative—the Green Paper on Private Military Companies—the British Government has started to discuss the necessity and possibilities for regulation of this relatively new sector.[71]

III. Russia

The efforts to transform the Russian arms industry are slowly beginning to produce changes in its structure, ownership and dynamics. During the period since the disintegration of the Soviet Union in 1991, and the subsequent dramatic cuts in Russian arms procurement and arms production, the Russian arms industry has been subject to a range of different policies and strategies. In the first half of the 1990s these policies were aimed primarily at conversion of productive capabilities from military to civil products, but in recent years the overall aim has been to secure a level and capability of Russian arms production comparable to that of major West European countries. Current restructuring efforts are focused on rationalization and consolidation by reduction of excess capacity and concentration of production. In 2001 there were several mergers into larger structures and the government launched new initiatives to reinforce this process, aimed at downsizing, concentration and promoting the technological level of the arms industry.

[69] Wakeman, N., 'Companies ride the e-gov tidal wave', *Washington Technology*, URL <http://www.washingtontechnology.com/top-100-2000/top-100-20002.html>.
[70] See the Dyncorp Internet site, URL <http://www.dyncorp.com/companies/index.htm>. The company also maintains the US State Department's aerial fleet in the Andes. Vest, J., 'State outsources secret war', *The Nation*, 23 May 2001, URL <http://www.thenation.com>.
[71] UK Foreign and Commonwealth Office, Private Military Companies: Options for Regulation, Feb. 2002, HC 577, Stationery Office, London, accessible at URL <http://www.fco.gov.uk>.

Quantitative trends

Russian arms production has been increasing since 1998. Official statistics on output trends in the Russian defence industrial complex (DIC) for the period 1991–2000 show a reduction in the index numbers for military output by more than 90 per cent in 1991–97.[72] This was the result of the dramatic drop in domestic procurement of new weapon systems and loss of major arms export markets in the early 1990s. This was followed by an almost doubling of the level in 1997–2000, a very rapid growth but from a low starting point, resulting in a level of Russian military production in 2000 that was only 18 per cent of the Soviet level in 1991 and one-third of the Russian level in 1992.[73] The expectations of continued growth, although more moderate, do not seem to have been fulfilled in 2001. While official data for DIC output trends no longer include data for military output, it can be derived from these data that there was a slight decline—by 1–2 per cent—in the military output of the DIC in 2001.[74] The official value of foreign currency earnings through arms exports increased from $3.7 billion in 2000 to $4.4 billion in 2001.[75]

However, it is still likely that moderate growth in Russian military output will be resumed in 2002, primarily because of the decision in October 2001 to begin a step-by-step increase in Russian procurement of new weapon systems.[76] This assessment is supported by the government decision on 17 January 2002 to approve a 40 per cent increase in the defence order for 2002, which is a significant increase also in real terms. The prioritized items were R&D on new generations of military hardware, aircraft manufacturing, communications, spacecraft, strategic forces and weapon systems for the land force and navy.[77] Altogether, this suggests that Russian arms production has passed its low point and is set for at least a moderate increase.[78]

---

[72] The only available official statistics on Russian arms production are the index numbers on the output of the Russian defence industrial complex. They refer to the defence industrial complex (DIC). It was officially designated the *voyenno-promyshlenny kompleks* (VPK)—now *oboronno-promyshlenny kompleks* (OPK). There are no official data on the amount, in monetary terms, of total national arms production in Russia.

[73] Cooper, J., 'Russian military expenditure and arms production', *SIPRI Yearbook 2001: Armaments, Disarmament and International Security* (Oxford University Press: Oxford, 2001), pp. 311–22.

[74] DIC data for 2001 only include figures for its total output and its civilian output. These show a 5% growth in total DIC output for the period Jan.–Nov. 2001 and a growth of 12% for civilian production. Data provided by Alexander Kabanov, deputy head of the Joint department of defence industries of the Russian Ministry of Industry, Science and Technology; 'The Russia's DIC production volume has grown', *Daily Express* (Internet newsletter of the Center for Army Conversion and Disarmament), 27 Dec. 2001, URL <http://www.defense-ua.com/eng/news/?id=1488>. The trend in military output can be derived from these data on the basis that they constitute roughly one-half of total DIC output.

[75] Boyle, J., 'Putin announces boom in arms sales', *St Petersburg Times*, 28 Dec. 2001, p. 3.

[76] See chapter 6 in this volume.

[77] 'Minister: Russian Government defense orders to almost double in 2002', Interfax (Moscow), 17 Jan. 2002, in Foreign Broadcast Information Service, *Daily Report–Central Eurasia* (*FBIS-SOV*), FBIS-SOV-2002-0117, 18 Jan. 2002.

[78] A new set of figures was presented in late Dec. 2001 according to which total DIC output was forecast to grow by 16% in 2002, with civilian and military output rising at roughly the same rate. However, this was in comparison with new figures for 2001 (showing an increase of 7.6% in DIC total output, of which 16.5% is civilian output), which makes it difficult to compare with previous statistics. Statement by First Deputy Minister of Science, Industry and Technology, Alexander Brindikov. 'Russian minister expects defence industry production to grow 16 per cent in 2002', Interfax (Moscow), 29 Dec. 2001, in FBIS-SOV-2001-1229, 31 Dec. 2001.

Privatization of the Russian arms industry[79]

In the 1990s there was a process of privatization of the Russian arms industry. While most of this process took place during the period 1992–95, it has implications for the structure and dynamics of the Russian arms industry today. The arms-producing enterprises and institutions were privatized according to the same pattern as applied in Russian industry at large, yet, there were great differences in the dynamics and stages of privatization in this sector. It has also been associated with more resistance from actors with vested interests in the status quo—enterprise managers and regional and local governments—as well as with major rivalries between potential private buyers/financiers, which suggests the magnitude of benefits to be gained from acquiring these enterprises.

Privatization of defence enterprises has proceeded according to two different methods in two major stages of relatively short duration. The first stage was concentrated to the second half of 1992 under the decrees of the Yegor Gaidar Government. It included the partial privatization of a considerable number of enterprises, including the Sukhoi design bureau, although the state kept a controlling share. The second stage was based on the system of voucher privatization (large-scale privatization for checks), initiated by Chairman of the State Property Committee Anatoliy Chubais. The vouchers were distributed in the autumn of 1992 and used for privatization in 1993. Decisions on major defence industry privatization using vouchers were taken from the autumn of 1992 and implemented from 1993. This stage lasted until 1995 when the privatization of arms-producing plants was frozen. This was when the bulk of arms industry privatization took place. It included the privatization of two major military industrial plants—the Krasnoe Sormovo shipbuilding plant in Nizhny Novgorod and the Mil helicopter plant in Moscow. During this time two banks, ONEXIMbank and Incombank, became the largest non-state actors in the process of defence industry restructuring and began an intensive battle to gain control over the Sukhoi design bureau and two major shipyards in St Petersburg—Severnaya Verf and Baltiyskiy Zavod.

After the suspension of privatization of the arms industry in 1995 the large banking–industrial empires began a redivision of military–industrial property. This led to an intense tug-of-war in 1997 between, on the one hand, the financial and industrial groups owned by the Russian oligarchs and, on the other hand, the emerging hegemony of the ONEXIMbank, in particular for access to the enterprises with large export orders or in anticipation of such orders. ONEXIMbank won control over several major companies—including the

---

[79] This section is based partly on an unpublished background paper by Makienko, K. and Pukhov, R., 'Privatization in the Russian defence industry: the situation in 2001', Centre for Analysis of Strategies and Technologies (CAST), Moscow, 24 Dec. 2001. See also Pukhov, R., 'MiG design and production system: post-Soviet transformations', *Eksport Vooruzheniy*, no. 2(24), (Mar./Apr. 2001), pp. 25–33; and Pukhov, R., 'Sukhoi Group: post-Soviet transformation', *Eksport Vooruzheniy*, no. 3(25), (May/June 2001), pp. 21–27.

Table 7.8. Ownership in the Russian arms industry, as of 31 December 2000
Figures in italics are percentages.

Type of enterprise	Number of enterprises	Share	Share of DIC production	Share of DIC employment
State-owned	701	*43.0*	*46.0*	*49.7*
JSC (S)	470	*28.8*	*34.0*	*34.7*
JSC (P)	460	*28.2*	*20.0*	*15.6*
Total	**1 631**	***100.0***	***100.0***	***100.0***

JSC (S) = State-owned joint stock company; JSC = privatized joint stock company.
Sources: Number and share of enterprises: URL <http://i.vpk.ru/rest/vlast>; share of production: URL <http://i.vpk.ru/vpkrus>; and share of employment: URL <http://i.vpk.ru/rest/vpkrus/kadri>.

Severnaya Verf, which some months later received a major contract worth $1 billion for the construction of two Sovremenny Class destroyers for China—and over Baltiyskiy Zavod, another shipyard that had a defence order worth up to $1 billion for the construction of three frigates for India.

The current structure of the private arms industry in Russia has emerged as the result of this tough rivalry between oligarchic empires to redistribute privatized property and from the financial crisis of 1998 that disrupted the financial–industrial groups. In aerospace, the major non-state owner is the Kaskol Group, which has a controlling share in the Gidromash plant, the sole Russian designer and manufacturer of aircraft undercarriages. It also has a minority stake (38 per cent) in the Sokol aircraft manufacturing plant in Nizhny Novgorod, which builds MiG combat aircraft and trainer aircraft. In shipbuilding, the largest non-state owners are the New Programs and Concepts (NPC) holding company (which inherited the arms-producing assets of ONEXIMbank) and the IST Group, which has a controlling share in Baltiyskiy Zavod.

However, the largest increasingly active, and most successful players in the non-state sector of the Russian arms industry are the company managers, who in the process of privatization became owners of the facilities of which they were the managers. The best examples of this are the owners of the IAIA and the Rybinsk Motors plant, who initiated the mergers, until then a prerogative of the government, when they acquired the Russian Avionics and the Lyulka–Saturn design bureau in 2001. A third category of potential non-state owners was introduced in 2001, when President Vladimir Putin announced the possibility of foreign investment in the Russian defence industry.[80]

Of the 1 631 enterprises of the Russian DIC at the end of 2000, 28 per cent were completely privatized joint stock companies and 29 per cent were partly privatized joint stock companies (table 7.8). The remaining 43 per cent were

---

[80] 'Foreigners in our defense sector?', *Moscow News*, 7–13 Nov. 2001, p. 7.

Table 7.9. Major arms-producing companies in Russia, 2000

This is a tentative list of the 20 largest arms-producing companies in Russia. Data on arms sales are to a large extent based on estimated ranges of the share of arms sales in total sales. Figures are in million roubles and percentages. Companies are ranked in order of their estimated arms sales in 2000.

Company (parent)	Ownership type[a]	Sector[b]	Arms sales	Arms share (%)	Export share (%)	Employ-ment
1. PK Antey	State	El Mi	12 770	99	99	50 000
2. KnAAPO	State	Ac	> 11 400	> 90	95	18 850
3. Severnaya Verf	Private	Sh	10 500	70	68	3 500
4. Baltiyskiy Zavod (IST)	Private	Sh	6 500	99	96	6 100
5. IAIA	Semi-private	Ac	4 950	96	96	13 814
6. Avitek	State	Ac Mi	> 4 200	> 90	..	7 200
7. Progress	Semi-state	Ac Mi	3 920	99	96	4 294
8. UMPO	State	Eng	> 3 160	> 70	..	20 172
9. RAC MiG	State	Ac	2 900	*100*	99	18 000
10. Defense Systems Group	State	El Mi	[> 2 650]	*[> 90]*	..	15 181
11. Zvezda-Strela	State	Mi	> 2 600]	*[> 90]*	..	1 800
12. Izhmash Group	Semi-state	SA/A	< 2 400]	*[< 50]*	26	25 400
13. Pirometr (K'AO')	Private	El	>1 800]	*[> 70]*	..	600
14. Kazan Helicopters[c]	Private	Ac	[1 500]	*[100]*	72	7 288
15. Rostvertol (Kaskol)	Private	Ac	[> 1 300]	*[> 90]*	..	7 058
16. Sokol (Kaskol)	Semi-private	Ac	[> 1 200]	*[> 90]*	..	10 000
17. Uralvaonzavod	State	MV	[1 200]	*[30]*	..	28 993
18. Degtyarev (MDM)	Private	Mi SA/A	[> 1 170]	*[> 70]*	..	15 368
19. ALMAZ(OMZ)	Private	Sh	870	97	99	1 076
20. Zvezdochka	State	Sh	[> 750]	*[> 90]*	..	8 735

[a] Semi-private = open joint stock company for which the state has a share package but not a control share; semi-state = open joint stock company with state capital control share.

[b] For sector codes, see appendix 7A.

[c] All helicopters produced by Kazan helicopters are dual use, here categorized as military.

Sources: Data provided by the Centre for Analysis of Strategies and Technologies (CAST), URL<http://www.cast.ru/english/index.html>, Moscow, processed in cooperation with SIPRI with comments by Julian Cooper, Center for Russian and East European Studies (CREES), University of Birmingham. See also appendix 7A, table 7A.3.

still state-owned enterprises, about two-thirds of which were prohibited from privatization.[81] The state-owned companies account not only for a proportionately higher share of production and employment in the DIC, but they also manage the prioritized domestic procurement programmes, such as the Topol-M missiles and the nuclear submarines.

Among the 20 largest companies, the degree of privatization was slightly higher: 7 of these were privately controlled companies in 2000; 9 were under continued state control; and the remaining 4 were semi-state or semi-private

[81] 'Russia approves plan to trim defense industry', *Defense News*, 5–11 Nov. 2001, p. 18.

companies (table 7.9). The dependence on arms sales is in general very high. In 14 of the top 20 companies arms sales accounted for 90 per cent or more of their total sales in 2000 and only 1 of these had an arms sales share lower than 50 per cent (table 7.9). Privately owned arms producing companies tend to differ from state-owned enterprises in three ways: first, they appear to be more inclined to diversify into civilian products; second, they display a higher degree of transparency of financial information, since this is a requirement for their access to financial instruments; and third, they are more successful at corporate restructuring and the formation of international alliances.

Industrial structure

The Russian arms industry is still characterized by an extreme degree of over-capacity and a strong dependence on exports, both of which are primarily the result of the collapse of domestic arms procurement that began in 1992. The DIC employs a total of almost 2 million people, and reportedly operates at only 20 per cent of manufacturing capacity.[82]

While reliable data are difficult to obtain, a comparison which illustrates the magnitude of the problem of excess capacity was provided in 2001, according to which the Russian aerospace (including missile and space) sector employs over 800 000 people and has an annual production volume of roughly $2 billion. The comparable sector in Western Europe has a combined employment of 98 000 persons and annual production of $22 billion.[83]

There have been several attempts to speed up the restructuring of the Russian arms industry. The process has been slow, however, and met by resistance from actors with a vested interest in the status quo, including company managers and regional governments. A new plan for the development of the DIC for the period to 2010 was adopted by the government on 30 October 2001.[84] Its aim to give top priority to arms production, and it includes measures to concentrate the industry into new, more competitive structures in order to develop new technology and joint marketing. According to the plan, approximately one-half of the defence enterprises will be merged into not more than 50 holding companies and the rest left without support and thus expected to close down.

A number of company mergers took place in 2001, most of which are associated with the production of the Su-27 family of combat aircraft. These include the merger of 20 aviation-related companies into one avionics company (Aerospace Equipment), which reportedly accounts for 60 per cent of the domestic military avionics market;[85] the decision to merge seven design and production companies into another major aviation holding company (Scientific Produc-

[82] 'Russia approves plan to trim defense industry' (note 81).
[83] 'Getting to grips with Russia's reforms', *Interavia*, Nov. 2001, p. 12.
[84] 'The fundamentals of RF policy on the development of the Defence Industrial Complex in the period to 2010 and beyond', was adopted on 30 Oct. 2001 in a joint session of the State Council Presidium and the Security Council of the Russian Federation. 'Foreigners in our defense sector?' (note 80).
[85] Kozyrev, M., 'Rynok avioniki peredelyat' [Market of avionics will be redivided], *Vedenosti Daily*, 20 July 2001.

tion Center Tekhnokompleks);[86] the merger of two major aero-engine companies into what became the first Russian integrated research and production association (NPO Saturn); and the decision in October 2001 to merge the major parts of the Sukhoi design bureau and two aircraft manufacturing companies (KnAAPO and NAPA) into one holding company (Sukhoi Aviation).[87]

Russian defence industrial policy continues to be strongly focused on arms exports because export revenues are seen as the only feasible option for obtaining the investment finance required for the necessary restructuring of the arms industry. Exports accounted for over 60 per cent of total military output in 2000.[88] Among the top 20 Russian arms-producing companies the export share is significantly higher. Of the 10 companies which provide export data, seven companies had a share of exports in total sales higher than 90 per cent (table 7.9).

In sum, the Russian arms industry is still characterized by an extremely high degree of over-capacity, a low degree of civilian production and a very high dependence on arms exports. The level of military output has been increasing since 1998 and continued, although moderate, growth is probable as the result of plans for increasing domestic procurement and a relatively large stock of export orders. Thus, it appears that the free fall of Russian arms production has been halted and that this will lead to an increased rate of consolidation of the industry, as was reflected in the concentration of the arms industry during 2001. There is an increasing rate of private ownership and control in the industry, primarily by banks but also by company directors, and there may even be some opening up towards foreign investment in the Russian arms industry. These trends will probably change the dynamics of Russian arms production in the direction of increased commercialization, and private sector interests, similar to those prevailing in the West.

IV. Conclusions

The arms industry entered a state of profound restructuring after the cold war. In the first half of the 1990s the arms industry experienced a significant cutback in orders, both domestically and from foreign governments. The level of arms production declined significantly in all major arms-producing countries.

Since the mid-1990s the main goal of the large arms-producing companies that emerged from the process of concentration has been to grow in size and to improve their capability to acquire arms procurement contracts, through takeovers, mergers, joint ventures and other forms of company-to-company cooperation, both nationally and internationally. These developments combined with the processes of commercialization and privatization are resulting in fundamental changes in the global system of arms production and arms trade.

The increasing commercialization of arms production is a result of changes in technology but also of the privatization of the ownership of the arms indus-

[86] Kozyrev (note 85).
[87] 'Russia approves plan to trim defense industry' (note 81).
[88] Cooper (note 73), table 4E.3, p. 319.

try and outsourcing of an increasing range and amount of military services and functions.

The process of concentration of ownership—and to some extent also of production—within the arms industry has continued from the national to the international level, driven by the largest companies in their search for access to military markets. A limited number of extraordinarily large companies—Boeing, General Dynamics, Lockheed Martin, Northrop Grumman, and Raytheon in the USA; and BAE Systems, EADS and Thales in Western Europe—have emerged, each producing military goods and services for an annual value ranging from $5 billion to $19 billion. All are searching for a strong international 'identity', while also increasing their leverage on their 'home' governments by virtue of their dominant role in major current weapon programmes.

The internationalization efforts in Europe are aimed to achieve further concentration, in order for European companies to become larger, which is seen as a prerequisite for becoming competitive with the USA, but also for establishing military industrial partnerships with US companies. However, European industrial integration is proceeding slowly, and there has been renewed interest in the establishment of transatlantic industrial links, largely within the context of government-to-government programmes for the development and production of specific weapon systems.

Market access is also the predominant motive for European and US acquisitions of arms-producing companies in minor producer countries that constitute potential markets. The increased acceptance of foreign ownership in the arms industry by governments in these countries primarily reflects their search for access to advanced technology and to some extent also to foreign markets. Requirements for direct offsets, including foreign investment in the domestic arms industry, are often used as a means to gain such access and sustain some domestic capability.

Both the commercialization and the internationalization of arms production are driven by companies in search of higher profit margins, but carried out within an international political and economic framework marked by (*a*) the end of the cold war and the related change in military requirements and shift in budget priorities, (*b*) a general shift towards the privatization of government functions, (*c*) the rapid development of information technologies, and (*d*) an increasing international interdependence in economic relations. Governments have maintained their role as key supporters of arms producing activities within their countries—through R&D funding, procurement and export support—while at the same time the sustainability of private arms-producing companies assumes an important role in defence industrial policy decisions.

This raises the question of the extent to which the role of national governments is diminishing with regard to the control and regulation of the supply of armaments to national and foreign armed forces. It also raises the issue of transparency in the development of military technology and the production of equipment and services that increasingly take place in privately owned large and powerful companies.

Appendix 7A. Arms industry data

REINHILDE WEIDACHER and the SIPRI ARMS INDUSTRY NETWORK*

I. Selection criteria and sources of data

Table 7A.2 lists the 100 largest arms-producing companies in the Organisation for Economic Co-operation and Development (OECD) and the developing countries, ranked by their arms sales in 2000.[1] The table contains information on their arms sales in 1999 and 2000, and their total sales, profit and employment in 2000. Table 7A.3 provides a tentative list of the 20 largest arms-producing companies in Russia, ranked by their arms sales in 2000. The table contains information on their arms sales in 1999 and 2000, and their total sales, export shares and employment in 2000.

Limited information about the world arms industry is publicly available. Tables 7A.2 and 7A.3 present data gathered from the following sources: company reports, a questionnaire sent to over 200 companies and corporation news published in the business sections of newspapers, military journals and by Internet news services specializing in military matters. Company archives, marketing reports, government publication of prime contracts and country surveys were also consulted. In the absence of data from these sources, estimates have been made by SIPRI. The scope of the data and the geographical coverage are largely determined by the availability of data and the he available data are not standardized across publications.

II. Definitions

Arms sales: The data on arms sales are not standardized because there is no generally agreed definition of arms sales. The comparability of the company arms sales is therefore limited.

SIPRI defines arms sales as sales of military goods and related services to military customers. Data for arms sales include only sales of goods and services which are

---

[1] For the membership of the OECD, see the glossary in this volume. The category of developing countries covers all countries other than the OECD and the former and current centrally planned economies.

* Participants in the network are: Sibylle Bauer (Institute d'Études Européennes, Université libre de Bruxelles); Konstantin Makienko and Ruslan Pukhov (Centre for Analysis of Strategies and Technologies, CAST, Moscow); Paul Dunne (Middlesex University, London); Ken Epps (Project Ploughshares Canada, Ontario); Gülay Günlük–Senesen (Istanbul University); Richard Haines (University of Port Elizabeth); Jean-Paul Hébert (Centre Interdisciplinaire de Recherches sur la Paix et d'Études Stratégiques, CIRPES, Paris); Christos Kollias (School of Business and Economics, Larissa); Shinichi Kono (Mitsubishi Research Institute, Tokyo); Luc Mampaey (Groupe de Recherche et d'Information sur la Paix et la Sécurité, GRIP, Brussels); Michele Nones and Giovanni Gasparini (Istituto Affari Internazionali, IAI, Rome); Arcadi Oliveres and Manuel Manonelles (Centre d'Estudis sobre la Pau i el Desarmament, Barcelona); Ton van Oosterhout, (Netherlands Organization for Applied Scientific Research, TNO, The Hague); and Reuven Pedatzur (Tel Aviv University).

designed specifically for military purposes. Excluded are sales of general-purpose goods (i.e., oil, electricity, office computers, cleaning services, uniforms and boots, but not surveillance and other electronic technology related to weapon systems). Whenever possible, data for arms sales include all revenues related to the sale of military equipment, that is, not only for their manufacture but also for their research and development (R&D), maintenance, servicing and repair.

Data for arms sales include sales both for domestic procurement and for export. The data for arms sales are used as an approximation of the annual value of arms production. An exception is made for shipbuilding companies. For these companies there exists a significant discrepancy between the value of annual production and annual sales because of the long lead (production) time of ships and the low production run (number). Annual production values therefore have to be estimated. Most shipbuilding companies are able to provide such estimates of the value of their military production on an annual basis.

Estimates: Estimates of arms sales are made for companies that do not report their arms sales, if there is sufficient basis for making an estimate. SIPRI estimates are of different types. In some cases, SIPRI uses the figure for the total sales of a 'defence' division, although the division may also have some, often unspecified, civil sales. In other cases, when the company reports neither a figure for arms sales nor a figure for a defence division, SIPRI makes its own estimates, based on figures for contract awards, information on the company's current armament production programmes and figures provided by company officials in media or other reports.

Total sales, profits and employment: Data on total sales, profits and employment are for the entire company, not for the arms-producing sector alone. Profit data are after taxes. Employment data are a year-end figure, except for those companies which publish only a yearly average. All data are presented on the fiscal year basis reported by the company in its annual report.

Conversion rate: The period average of market exchange rates of the International Monetary Fund's *International Financial Statistics* is used for conversion from local currencies to US dollars.

III. Coverage

Country coverage: Table 7A.2 covers arms-producing companies in the OECD and the developing countries except China. Data on arms sales are not available for companies in all these countries. This is the case for South Korea and Taiwan. No comparable data at the enterprise level are available for the former and current centrally planned economies. Table 7A.3 presents a tentative list of the 20 largest arms-producing companies in Russia.

Types of company: Tables 7A.2 and 7A.3 include public and private companies engaged in arms production. Manufacturing or maintenance units of the armed services are not included.

Table 7A.1. Regional/national shares of arms sales[a] for the top 100 arms-producing companies in the OECD and developing countries[b] in 2000

Figures do not always add up because of the conventions of rounding.

Number of companies	Region/ country	Arms sales (US $ b.)	Share of total arms sales (%)
43	USA	94.6	60.0
33	*West European OECD*	*47.8*	*30.5*
13	UK	22.4	14.2
7	France	11.0	7.0
3	Italy	3.4	2.2
5	FRG	3.4	2.2
1	Sweden	1.2	0.8
2	Spain	0.6	0.4
1	Switzerland	0.5	0.3
1	Other[c]	5.3	3.4
14	*Other OECD*	*8.7*	*5.5*
10	Japan[d]	7.4	4.7
2	Canada	0.6	0.4
1	Australia	0.4	0.2
1	Turkey	0.3	0.2
10	*Developing countries*	*6.5*	*4.1*
5	Israel	3.5	2.2
3	India	1.9	1.2
1	Singapore	0.8	0.5
1	South Africa	0.4	0.2
100	**Total**	**157.6**	**100.0**

[a] Arms sales include both domestic procurement and arms exports.

[b] For a list of member countries in the OECD, see appendix 6A.1. The category of developing countries covers all countries other than the OECD and the former and current centrally planned economies.

[c] Other (European) refers to EADS, a merger of French, German and Spanish companies.

[d] For Japanese companies data are for new military contracts rather than for arms sales.

Source: Table 7A.2.

IV. The 100 largest arms-producing companies in the OECD and developing countries, 2000

The 100 largest arms-producing companies in the OECD and developing countries had combined arms sales of roughly $157 billion in 2000. The US dominance is clear. Among the total 100 companies, there were 43 in the USA, compared to a combined total of 33 for all of Western Europe, 14 in other OECD countries and 10 in developing countries, excluding China. US companies accounted for an over proportional share of arms sales, 60 per cent, reflecting the larger size of their companies.

ARMS PRODUCTION

Table 7A.2. The 100 largest arms-producing companies in the OECD and developing countries, 2000

Figures in columns 6, 7, 8 and 10 are in US$ million. Figures in italics are percentages.

1	2	3	4	5	6	7	8	9	10	11
Rank[a]					Arms sales					
2000	1999	Company (parent company)	Country	Sector[b]	2000	1999	Total sales 2000	Col. 6 as % of col. 8	Profit 2000	Employment 2000
1	1	Lockheed Martin	USA	Ac El Mi	18 610	19 790	25 329	*73*	−519	130 000
2	2	Boeing	USA	Ac El Mi	16 900	16 000	51 521	*33*	2 128	198 000
3	3	BAE SYSTEMS	UK	A Ac El Mi SA/A Sh	14 400	15 470	18 473	*78*	1 440	84 900
4	4	Raytheon	USA	El Mi	10 100	11 530	16 895	*60*	141	93 700
5	5	Northrop Grumman	USA	Ac El Mi SA/A	6 660	7 070	7 618	*87*	608	39 300
6	6	General Dynamics	USA	El MV Sh	6 520	5 550	10 356	*63*	901	43 300
7	–	EADS[c]	France/FRG/Spain	Ac El Mi	5 340	0	22 303	*24*	−832	88 880
8	7	Thales[d]	France	El Mi SA/A	5 160	4 450	8 476	*61*	..	57 230
9	8	Litton	USA	El Sh	3 950	3 910	5 588	*71*	218	40 300
10	13	TRW	USA	El Oth	3 370	2 990	17 231	*20*	438	102 880
11	9	United Technologies	USA	El Eng	2 880	3 480	26 583	*11*	1 808	153 800
12	14	Mitsubishi Heavy Industries[e]	Japan	Ac MV Mi Sh	2 850	2 460	28 255	*10*	−189	..
13	S	Finmeccanica[f]	Italy	A Ac El MV Mi SA/A	2 440	2 790	5 540	*44*	312	39 370
14	15	Rolls Royce	UK	Eng MV	2 130	2 410	8 890	*24*	126	46 600
15	17	Newport News	USA	Sh	2 030	1 830	2 072	*98*	90	17 000
16	18	Science Applications[g]	USA	Oth	1 950	1 740	5 896	*33*	2 059	41 500
17	16	GKN	UK	Ac MV	1 740	1 860	7 726	*23*	491	42 970

Rank[a] 2000	Rank[a] 1999	Company (parent company)	Country	Sector[b]	Arms sales 2000	Arms sales 1999	Total sales 2000	Col. 6 as % of col. 8	Profit 2000	Employment 2000
S	S	Pratt & Whitney (UTC)	USA	Eng	1 620	2 120	7 366	22	. .	. .
18	21	Computer Sciences Corp.[g]	USA	Oth	1 610	1 470	10 524	15	233	68 000
19	19	DCN	France	Sh	1 600	1 700	1 603	100	-85	16 000
20	20	General Electric	USA	Eng	1 600	1 600	129 853	1	12 735	313 000
21	23	Honeywell International	USA	El	1 550	1 420	25 023	6	1 659	125 000
22	22	Rheinmetall	FRG	A El MV SA/A	1 460	1 420	4 137	35	76	29 880
S	S	Rheinmetall DeTec (Rheinm.)	FRG	A El MV SA/A	1 460	1 420	1 464	100	21	9 180
23	27	Israel Aircraft Industries	Israel	Ac El Mi	1 350	1 200	2 180	62	84	14 520
24	31	L-3 Communications	USA	El	1 340	1 000	1 910	70	83	14 000
25	24	ITT Industries	USA	El	1 330	1 410	4 829	28	265	41 550
26	40	Saab	Sweden	Ac El Mi	1 210	740	1 947	62	113	15 450
27	25	Textron	USA	Ac El Eng MV	1 200	1 300	13 090	9	218	68 000
28	26	United Defense[h]	USA	MV	1 180	1 210	1 184	100	22	5 350
29	30	Ordnance Factories	India	A SA/A	1 130	1 150	1 247	91	. .	146 070
30	33	Mitsubishi Electric[e]	Japan	El Mi	1 120	980	38 318	3	1 158	116 720
31	28	CEA	France	Oth	1 050	1 190	6 329	17	83	15 990
S	S	Alenia Marconi Systems (Finmeccanica, Italy/BAE Systems, UK)	Italy/UK	El	990	1 040	1 163	85	. .	8 000
32	39	SNECMA Groupe[i]	France	Eng	970	780	5 989	16	. .	36 650
33	44	EDS[g]	USA	Oth	950	630	19 227	5	1 143	122 000
34	32	Dassault Aviation Groupe	France	Ac	930	990	3 211	29	218	11 420
35	29	Kawasaki Heavy Industries[e]	Japan	Ac Eng Mi Sh	920	1 160	9 840	9	-96	. .
36	38	Alliant Tech Systems	USA	SA/A	900	800	1 142	79	68	6 020
S	S	Eurocopter Group (EADS)	France	Ac	850	890	1 884	45	37	9 910
S	S	Thales Systèmes Aéroportés	France	El	830	970	830	100	. .	5 100

Rank 2000	Rank 1999	Company	Country	Sector	Arms sales 2000	Arms sales 1999	Total sales	%	Profit	Empl.
37	37	(Thales)	France	El	820	810	3 934	21	140	15 900
38	52	SAGEM Groupe	France	El	800	540	1 809	44	10	19 400
39	–	Dyncorp[g]	USA	Oth	780	220	1 033	75	:	:
40	47	Titan[g]	USA	Oth	770	590	:	:	:	:
S	S	Singapore Technologies, ST	Singap.	Ac El MV SA/A Sh	770	600	3 854	20	2 287	14 060
S	S	SAGEM (SAGEM Groupe)	France	El	770	590	1 310	59	167	:
41	66	ST Engineering (ST)	Singap.	Ac El MV SA/A Sh	700	440	700	100	17	4 000
42	43	Elbit Systems[j]	Israel	El	700	700	7 151	10	636	41 200
43	57	Rockwell International[k]	USA	El Mi	670	500	674	100	–2	4 300
44	42	Rafael	Israel	SA/A Oth	670	710	53 027	1	612	223 950
45	41	FIAT	Italy	Eng MV SA/A	660	720	660	100	:	2 100
46	63	Krauss-Maffei Wegmann	FRG	MV	640	460	10 525	6	666	56 000
47	46	Marconi	UK	El	610	610	1 807	34	18	10 000
S	S	Harris	USA	El	610	290	610	100	:	2 000
[48]	51	EDS Defence (EDS, USA)	UK	El	:	550	650	90	:	:
49	54	Veridian[g]	USA	Oth	540	520	184 632	:	4 452	:
S	S	General Motors, GM	USA	El Eng MV	530	510	28 281	2	:	:
50	55	GM Canada (GM, USA)	Canada	Eng MV	530	510	2 219	24	268	16 500
51	48	Smiths Industries	UK	El	520	570	1 491	35	63	11 190
52	35	Diehl	FRG	Mi SA/A	510	890	510	100	–261	7 500
53	36	GIAT Industries	France	A MV SA/A	510	870	1 844	28	31	7 730
54	59	Hunting	UK	SA/A Oth	500	480	549	90	:	4 100
55	61	Israel Military Industries	Israel	A MV SA/A	500	470	10 344	5	85	:
56	53	Ishikawajima-Harima[e]	Japan	Eng Sh	480	520	1 047	46	129	7 900
S	S	Gencorp[j]	USA	El Eng	470	470	697	67	57	5 010
57	56	Agusta (Finmeccanica)	Italy	Ac	460	500	547	83	54	35 000
58	49	Hindustan Aeronautics	India	Ac Mi	450	550	557	81	40	3 800
[59]	67	RUAG SUISSE	Switzerl.	A Ac Eng SA/A	:	440	:	:	:	:
60	76	Devonport Management[m]	UK	Sh	430	370	50 197	1	525	4 200
61	65	NEC	Japan	El	420	440	:	:	:	149 930
		Babcock Borsig	FRG	Sh						

360 MILITARY SPENDING AND ARMAMENTS, 2001

Rank[a] 2000	Rank[a] 1999	Company (parent company)	Country	Sector[b]	Arms sales 2000	Arms sales 1999	Total sales 2000	Col. 6 as % of col. 8	Profit 2000	Employment 2000
62	68	Primex Technologies	USA	SA/A	420	430	530	78	25	2 850
S	S	HDW (Babcock Borsig)	FRG	Sh	420	440	529	79	28	3 330
63	87	Anteon[g]	USA	Oth	410	300	543	75	-3	5 000
64	62	Toshiba[e]	Japan	El Mi	400	470	55 223	1	892	188 040
65	72	Cobham	UK	Comp (Ac El)	400	390	862	47	95	6 960
[66]	88	Vosper Thornycroft	UK	Sh	..	290	574	69	33	7 500
67	79	DRS Technologies	USA	El	390	360	428	90	12	2 200
S	S	Marconi Mobile (Marconi, UK)	Italy	El	390	410	..	..	..	..
S	S	Thales Nederl. (Thales, France)	Netherl.	El	390	310	391	100	-16	2 890
S	S	The Aerospace Corp.[n]	USA	Oth	390	350	434	91	..	..
68	70	Tenix	Australia	Sh	380	390	576	65	..	5 000
69	69	Mitre[n]	USA	Oth	380	410	604	63	..	..
[70]	71	ThyssenKrupp, TK	FRG	Sh	..	390	34 281	1	486	193 320
S	S	TK Werften (TK)	FRG	Sh	..	..	718	51	..	2 890
71	34	SEPI	Spain	A MV Sh	370	970	..	..	..	..
72	58	Silicon Graphics[g]	USA	Oth	370	500	2 300	16	..	..
73	90	BF Goodrich	USA	Comp (Ac)	350	280	4 364	8	326	26 320
74	–	Mitsui Shipbuilding[e]	Japan	Sh	340	..	4 033	8	20	..
[75]	78	TI Group	UK	Comp (Ac)	..	370	4 791	7	..	..
S	S	IZAR[o]	Spain	Sh	340	390	723	47	-154	10 650
76	99	Bombardier	Canada	El Mi	330	230	10 841	3	657	79 000
77	81	Komatsu[e]	Japan	MV SA/A	330	330	10 173	3	64	32 000
[78]	73	Denel	S. Africa	A Ac El MV Mi SA/A	..	380	541	61	4	10 380
79	–	Alcoa[p]	USA	Comp (Eng)	330	0	22 936	1	1 484	142 000
S	82	Cordant Technologies[p] (Alcoa)	USA	Comp (Eng)	330	330	2 513	13	..	17 200

ARMS PRODUCTION 361

		Company	Country	Sector					
S	S	IVECO (FIAT)	Italy	MV	330	310	..	..	..
80	12	IRI*f*	Italy	Sh	320	3 000	..	..	..
S	S	Fincantieri Gruppo (IRI)	Italy	Sh	320	210	1 769	*18*	9 410
81	74	Alvis	UK	MV Oth	310	380	310	*100*	1 460
82	89	AM General Corporation*g*	USA	MV	310	290	448	*69*	1 360
83	–	Engineered Support Systems	USA	El	310	130	361	*85*	2 400
S	–	Thales Optronics (Thales, France)	UK	El	310	0	387	*80*	..
84	93	Bharat Electronics	India	El	300	260	382	*79*	14 180
85	75	EG&G*h*	USA	Comp (El Oth)	300	380	488	*62*	..
86	S	Stewart & Stevenson	USA	MV	300	150	1 153	*26*	4 100
S	S	Shin Maywa*e*	Japan	Ac	300	90	1 247	*24*	2 000
87	60	MKEK	Turkey	SA/A	290	480	371	*78*	8 940
88	86	Babcock International Group	UK	Sh	290	300	669	*44*	6 310
89	91	Motorola*r*	USA	El	290	280	37 580	*1*	147 000
S	S	ADI (Transfield Holding/Thomson-CSF, France)	Australia	El SA/A Sh	290	290	351	*83*	2 850
S	S	FIAT Aviazione (FIAT)	Italy	Eng	290	350	1 374	*21*	5 360
S	S	Singapore Aerospace (ST Engineering)	Singap.	Ac Eng	290	260	556	*53*	..
90	92	CAE	Canada	El	280	260	802	*34*	6 000
91	83	Koor Industries	Israel	El	280	320	2 031	*14*	8 700
S	S	Elisra (Tadiran)	Israel	El	280	..	309	*91*	1 550
92	106	Oshkosh Truck	USA	MV	280	220	1 324	*21*	4 600
S	S	Thales Avionics (Thales)	France	El	280	240	921	*30*	6 600
93	–	Cubic Corporation	USA	Comp (El Oth)	270	220	532	*51*	3 800
94	–	Japan Electronic Computer*e*	Japan	El	260	220	..	..	..
95	100	Ultra Electronics	UK	El	260	230	344	*75*	2 300
96	–	CACI International	USA	El	250	220	491	*51*	..
97	96	Teledyne Technologies	USA	El	250	250	795	*31*	..
[98]	97	United States Marine Repair	USA	Comp (Sh)	..	250	..	..	..
99	98	Nissan Motor*e s*	Japan	A MV	240	240	..	..	..

Rank[a]		Company (parent company)	Country	Sector[b]	Arms sales		Total sales	Col. 6 as	Profit	Employment
2000	1999				2000	1999	2000	% of col. 8	2000	2000
100	–	Indra	Spain	El	240	220	629	38	35	5 310

Notes:

[a] Companies are ranked according to the value of their arms sales in 2000. Companies with a designation in square brackets [] are ranked according to a rough estimate of their arms sales, in which case there is a (..) in the column for arms sales. Companies with the designation S in the column for rank are subsidiaries. A dash (–) in the column for rank designations indicates either that the company did not produce arms in 1999, or that it did not exist in 1999 as it was structured in 2000, in which case there is a zero (0) in column 7, or that it did not rank among the 100 largest companies in 1999. Rank designations in the column for 1999 may not correspond to those given in table 4D.2 in the *SIPRI Yearbook 2001* because of subsequent revision.

[b] Key to abbreviations: A = artillery, Ac = aircraft, El = electronics, Eng = engines, Mi = missiles, MV = military vehicles, SA/A = small arms/ammunition, Sh = ships, and Oth = other. Comp () = components of the product within the parentheses. It is used only for companies which do not produce any final systems.

[c] The European Aeronautic Defence and Space Company (EADS) was created in July 2000 through the merger of DaimlerChrysler Aerospace (with the exception of MTU), Aérospatiale Matra, and Construcciones Aeronáuticas (CASA). The company is owned by 33% by DaimlerChrysler, (Germany), 31% by private investors, 15% by Lagardère (France), 15% by the French state (through direct and indirect share holdings), and 6% by SEPI (Spain). EADS is registered in the Netherlands.

[d] Data for Thales are pro–forma sales as if the acquisition of Racal (UK) had occurred on 1 Jan. 2000. Racal was acquired by Thales in Aug. 2000.

[e] For Japanese companies figures in the arms sales column represent new military contracts rather than arms sales.

[f] The former arms–producing subsidiary of IRI (state) was partly privatized in June 2000. The share of Finmeccanica owned by IRI was reduced from more than 50% to 5%. An additional 32% of Finmeccanica is directly owned by the Italian Treasury.

[g] This company is a provider of IT services and products to defence ministries. Figures are for total sales to defence ministries, an unknown share of which is for military applications.

[h] The company is owned by the private investment group The Carlyle Group.

[i] Data for Snecma are pro–forma sales as if the acquisition of Labinal had occurred on 1 Jan. 2000. Labinal was acquired by Snecma in July 2000.

[j] Data for Elbit Systems are pro–forma sales as if the acquisition of El–Op had occurred on 1 Jan. 2000. El–Op was acquired by Elbit Systems in July 2000.

[k] In July 2001 Rockwell International spun off its avionics and communications business in Rockwell Collins.
[l] In Oct. 2001 Gencorp sold part of the arms-producing activities within its Aerojet division to Northrop Grumman.
[m] Development Management is owned by the US company Halliburton since 1996.
[n] This company operates Federally Funded Research and Development Centers (FFRDC) for the US Department of Defense.
[o] IZAR was established in July 2000 through the merger of the SEPI (state) owned naval shipyard Bazan and the private civilian shipyard Astilleros Españoles. The company is fully owned by SEPI.
[p] Alcoa acquired Cordant Technologies in May 2000. Arms sales for Cordant Technologies are for 1999. In Apr. 2001 Alcoa sold Thiokol Propulsion (part of Cordant Technologies) to Alliant Tech Systems.
[q] AM General is owned by Renco Group.
[r] In Aug. 2001 Motorola sold its arms-producing subsidiary Integrated Information Systems Group to General Dynamics.
[s] In July 2001 Nissan Motor sold its aerospace and defence division to Ishikawajima-Harima.

364 MILITARY EXPENDITURE, PRODUCTION AND TRADE

Table 7A.3. A tentative list of the 20 largest arms-producing companies in Russia, 2000

Figures in columns 6, 7, and 8 are in million current roubles. Figures in italics are percentages.

1	2	3	4	5	6	7	8	9	10	11
Rank[a] 2000	Company (parent company)	Ownership type[b]	Owner	Sector[c]	Arms sales 2000	Arms sales 1999	Total sales 2000	Arms share of total sales	Export share of total sales	Employment 2000
1	PK Antey	State OJSC	RCSA[d]	El Mi	12 770	5 640	12 900	99	99	50 000
2	KnAAPO[e]	State	RASA[f]	Ac	>11 400	>1 800	12 650	>90	95	18 850
3	Severnaya Verf	Private OJSC	NPC Ltd (53.5%), Interros Group (46.5%)	Sh	10 500	8 500	14 950	70	68	3 500
4	Baltiyskiy Zavod (IST)	Private OJSC	IST Group (54%), Incombank (25.4%), MIB bank (17%)	Sh	6 500	2 790	6 590	99	96	6 100
5	Irkutsk Aviation Industrial Ass., IAIA[e]	Semi-private OJSC	Management, State (14.7%)	Ac	4 950	5 250	5 160	96	96	13 810
6	Avitek	State		Ac Mi	4 190	..	4 660	>90	..	7 200
7	Progress	Semi-state OJSC	State (50% plus one share)	Ac Mi	3 920	210	3 960	99	96	4 290
8	UMPO	Semi-private OJSC	..	Eng	3 160	..	4 510	>70	..	20 172
9	Russian Aircraft Corporation MiG	State		Ac	2 900	2 940	2 900	100	99	18 000
10	Defense Systems Group	State OJSC		El Mi	(>2 650)	..	2 950	(>90)	..	15 180
11	Zvezda–Strela	State		Mi	(>2 600)	..	2 880	(>90)	..	1 800
12	Izhmash Group	Semi-state OJSC	State (76.8%)	SA/A	(<2 400)	..	4 800	(<50)	26	25 400
13	Pirometr (K'AO')	Private JSC	Management	El	(>1 800)	..	2 600	(>70)	..	600
14	Kazan Helicopters	Private JSC	Management (>50%)	Ac	1 500	..	1 500	100	72	7 290
15	Rostvertol (KASKOL)	Private OJSC	Management	Ac	(>1 300)	..	1 460	(>90)	..	7 060
16	Sokol (KASKOL)[g]	Semi-private OJSC	KASKOL Group	Ac	(>1 200)	..	1 380	(>90)	..	10 000
17	Uralvaonzavod	State		MV	(≈1 200)	..	4 010	(30)	..	28 990

18	Degtyarev (MDM)[h]	Private OJSC	MDM Group	Mi SA/A(>1 170)	..	1 680	(>70)	..	15 370
19	ALMAZ (OMZ)	Private OJSC	Obedinennye Mashino-stroitelnye Zavody (OMZ)	Sh	870	890	97	99	1 080
20	Zvezdochka	State		Sh	(>750)	830	(>90)	..	8 740

Notes:
[a] Companies are ranked according to the value of their arms sales in 2000.
[b] Ownership type designations: OJSC = Open Joint Stock Company (OAO, Oktrytoye Aktsionernoye Obshchestvo). State = Completely state–owned enterprises in which there are no privatization procedures. State OJSC = Completely state–owned OJSC. Semi-state OJSC = OJSC with state capital control share. Semi–private OJSC = OJSC for which the state has a share package but not a control share. Private OJSC = OJSC without state capital.
[c] Key to abbreviations: A = artillery, Ac = aircraft, El = electronics, Eng = engines, Mi = missiles, MV = military vehicles, SA/A = small arms/ammunition, Sh = ships, and Oth = other.
[d] RCSA = Russian Control Systems Agency [Rossiyskoye Agentstvo po Sistemam Upravleniya (RASU)].
[e] All companies manufacturing Sukhoi designed aircraft, KnAAPO, IAIA and NAPA, and the Sukhoi Design Bureau are to be merged into one new holding, Aktsionernaya Kholdingovaya Kompaniya Sukhoi (AKhK Sukhoi, Joint Stock Holding Company Sukhoi) according to 2001.
[f] RASA = Russian Aviation and Space Agency [Rossiyskoye Aviakosmicheskoye Agentstvo (Raka; Rosaviakosmos)].
[g] Nizhniy Novgorod Aviatsyiomy Zavod 'Sokol'.
[h] OJSC 'V. A. Degtyarev Plant'.

Sources: Arms sales share and exports share: The Centre for Analysis of Strategies and Technologies (CAST), URL<http://www.cast.ru/english/index.html>, Moscow; ( ) = estimates. Total sales and employment: 'The 200 largest companies in Russia', *Ekspert*, no. 35 (295), 24 Sep. 2001; 'The largest machine-building companies in Russia', *Ekspert*, no. 42 (302), 12 Nov. 2001.

Appendix 7B. Available government and industry data on the arms industry

REINHILDE WEIDACHER

I. Introduction

The arms industry is the supplier of the military means of power. While similar to other industrial activities in its pursuit of economic wealth, goods produced by this industry play a central role for peace and war. It is therefore essential that arms production be subject to rules of public accountability. Accountability in arms production should facilitate oversight and evaluation of the activities of arms-producing companies and of government decisions related to the arms industry.

However, information on world arms production is patchy and generally poor, although the amount of information made publicly available has increased significantly in the past decade. The intense post-cold war effort to reduce overcapacities in arms production led to a profound restructuring of the world arms industry. For all the actors involved in the process—industrial, political, and military leaders as well as individual shareholders—it is essential to gain insight into the size and structure of the arms industry. Available information has also become more easily accessible to the general public through the rapid advances in information technology, which have vastly expanded the possibilities for swift dissemination of information.

Yet, while the sheer amount of available information has grown, open, valid and reliable information on the arms industry (made available through regulated procedures and on a regular basis) has increased only in a few countries. Public understanding of and research into the dynamics of arms production therefore largely rely on information disclosed voluntarily and on an ad hoc basis.

In a large number of countries national security and commercial confidentiality motives severely limit the collection and disclosure of information on the arms industry. While efforts towards an international standardization of reporting rules and practices on related military matters, such as military expenditures[1] and arms transfers,[2] have resulted in some improvements, no similar efforts have been initiated in the area of arms production.

This appendix therefore reviews information on the size of the arms industry that was made available by industry associations and government organizations in the 20 largest arms-producing countries in the late 1990s.[3] The focus is on quantitative information that provides an indication of the economic importance of the arms industry, such as the value of arms sales or arms production and the number of employees in arms production.[4]

---

[1] See appendix 6D in this volume.
[2] See appendix 8E in this volume.
[3] According to data and estimates of the value of national arms sales in US$ for the most recent year available, in the period 1996 through 2000 these countries are (in alphabetical order): Australia, China, Canada, France, Germany, India, Italy, Israel, Japan, South Korea, the Netherlands, Russia, Singapore, South Africa, Spain, Sweden, Taiwan, Ukraine, the United Kingdom and the United States.
[4] A possible alternative approach could focus on the number or volume of weapons produced.

In the past decade SIPRI has compiled statistical information on the largest arms-producing companies in the Organisation for Economic Co-operation and Development (OECD) and developing countries. This has been an important tool for analysing the adjustment strategies of arms-producing companies to changes in the military market and in the broader industrial environment.

Company data, however, are inadequate to assess the overall weight of the arms industry in national and international economic and political dynamics for two reasons: the fundamental lack of comparable data for all arms-producing companies, and the difficulty of deriving the overall size of and trends in arms production from company data. As a result of the restructuring of the arms industry in the 1990s the top layer of the world arms industry is made up of large private corporations with a broad shareholder base. These corporations publish a large amount of information on their organization and financial performance. However, few arms-producing companies provide open and clear information on their sales related to the production of arms. While the format of general financial reporting by companies is highly standardized across countries in other areas, there is no obligation for companies to report specifically on the production of military equipment.[5] Information on the number of individuals employed in the arms-producing parts of companies is provided even more seldom than information on the value of output in arms production.

Available company data do not suffice to derive information on the national level or on trends in arms production. Owing to the complex structure of the world arms industry, with large prime contractors with production facilities in more than one country and a large number of subcontractors, a simple addition of company sales or production values does not produce national totals.

For these reasons, and facilitated by the increased availability of data in the late 1990s, SIPRI has begun to complement the collection of data on arms-producing companies with a collection of available national statistics. This appendix presents the results of that process; it addresses only the availability of information on arms production at the national, not the company, level.

II. Reporting practices

Government organizations or arms industry associations in more than one-half of the 20 largest arms-producing countries in the late 1990s have made available certain information on the size of the national arms industry at some point in time in the 1990s. However, the lack of standardized rules and practices for gathering and disclosing information on the arms industry severely limits the value of what is available.

Accountability demands that the disclosed data be valid, reliable and provided in a format that facilitates comparison not only across time, but also with production for civilian purposes and across countries. In order to be valid economic statistics on the arms industry have to provide a sound indicator of the economic importance of arms production within the wider economy. Such indicators are the financial value of arms

---

[5] Within the industry there are various initiatives that are intended to improve and harmonize non-financial reporting by private corporations. The primary focus of these initiatives is on environmental and employment aspects of business activities. See, e.g., Global Reporting Initiative, 'Sustainability Guidelines on Economic, Environmental and Social Performance', June 2000, URL <http://www.globalreporting.org/>.

sales for single companies and on a value-added basis on the national level,[6] including a breakdown into domestic and foreign sales as well as by product type. An additional indicator of the economic importance of the arms industry is the number of people employed in the production of arms.

However, the arms industry does not form a distinct industrial sector according to generally accepted industrial codes, and there is no agreement on the definition of 'arms production' and the methodology required for its measurement. The use of other terms, such as 'defence' or 'national security' without any clear definition of their content contributes to obfuscate knowledge about the arms industry. In the absence of a standard definition of 'arms' or 'military goods', published data need be supported by detailed information on the methodology applied for producing them.

External rules for implementing data collection and dissemination are a necessary precondition for guaranteeing reliability. In the absence of such standards and rules vested interests, such as industries that seek government support, may determine the content of the information disclosed.

No government or industry organization in the 20 largest arms-producing countries in the late 1990s provided comprehensive, valid and reliable information on the national arms industry in a single, official document that included employment and financial statistics, as well as information on defence industrial policy issues and developments in the wider economic and political framework in which arms-producing companies operate.

Available information on the arms industry is generally provided in publications with a broad coverage, such as military statistical publications or defence White Papers. Exceptions to this are the arms industry surveys published by the Spanish Government and the arms industry associations of Canada and Sweden.

Detailed statistics on the size of the national arms industry, including financial values and employment statistics, are available for only two countries (France and the United Kingdom), while less detailed statistics are available for Canada, the Netherlands, Spain and Sweden (table 7B). Because of the lack of international standardization the available statistics are not comparable even for these countries.

No regular data on the national arms industry are provided for Australia, China, Germany, Israel, Italy, South Africa and Ukraine. For Australia, Italy and South Africa comprehensive surveys of the state of the national arms industry and its perspective for the future were published in the late 1990s to provide the basis for a review of defence industrial strategy and policy options. The surveys include some statistics on the value of and employment in arms production.

While no regular statistics on the value of the arms sales of the Russian industry are available, the Teleinformation Network of the military–industrial complex, (Teleinformatsionnaya Set, TS-VPK) has provided detailed information on quantitative trends with regard to output and employment in arms production since 1991. These were published only for the months of January to November in 2001 but not for December and not for the full year 2001, and it is unclear whether they have been discontinued.

---

[6] Because of the difficulty of covering all arms-producing activities within a country and obtaining information on the value-added share of their arms sales, estimates of national arms sales are often based on data on domestic arms procurement expenditure plus arms exports.

ARMS PRODUCTION 369

Table 7B. Open and regular sources for national data on the arms industry in the 20 largest arms-producing countries

Country[a]	Source	Statistics Output[b]	Employ-ment[c]	Defini-tion[d]	Background information[e]	First year of data
USA	Government[f]	–	+	–	–	
	Government[g]	(+)	–	–	–	
	Government[h]	–	–	–	+	1996
France	Government[i]	+	+	+	–	1999
UK	Government[j]	+	+	+	–	
Germany	–	–	–	–	–	
Japan	Government[k]	+	–	–	–	
Russia	Government[l]	Index	Index	–	–	
Canada	Industry[m]	+	+	–	–	1997
China	–	–	–	–	–	
Israel	–	–	–	–	–	
Italy	Government[n]	–	–	–	–	
South Korea	Government[o]	+	–	–	–	
Australia	Government[p]	–	–	–	–	
India	Government[q]	+	–	–	+	
Netherlands	Government[r]	(+)	–	(+)	(+)	1997
Singapore	–	–	–	–	–	
South Africa	Government[s]					
Spain	Government[t]	+	+	–	+	1998
Sweden	Government[u]	+	–	–	–	1999
	Industry[v]	(+)	(+)			1987
Taiwan	Government[w]	(+)	–	–	–	
Ukraine	–	–	–	–	–	

+ = Valid data and information; (+) = invalid data (i.e., company data rather than total national values or numbers, and contract values rather than sales/production values); – = no data or information.

Notes:
[a] Countries are grouped within broad ranges according to the estimated value of their arms sales. Within these groups countries are listed in alphabetic order.
[b] Values of total arms production or arms sales.
[c] Total, direct, or indirect employment in arms production on a national level.
[d] Clear definition of what is measured.
[e] Qualitative assessment of the development of the arms industry and reference to relevance policy issues.

Sources (most recent edition):
[f] US Department of Defense, Office of the Under Secretary of Defense (Comptroller), *National Defense Budget Estimates for FY 2002*, Aug. 2001, URL <http://www.dtic.mil/comptroller/fy2001budget>.
[g] US Department of Defense, Directorate for Information Operations and Reports, *Prime Contract Awards* (annual); URL <http://web1.whs.osd.mil/peidhome/procstat/procstat.htm>.
[h] US Secretary of Defense, Deputy Under Secretary of Defense (Industrial Affairs), *Annual Industrial Capabilities Report to Congress*, Mar. 2002, URL <http://www.acq.osd.mil/ip/ip_products.html >.

i Ministry of Defence, Observatoire Economique de la Défense, *Annuaire statistique de la defense 2001* [Defence statistical yearbook 2001], June 2001.

j Ministry of Defence, UK, Defence Analytical Service Agency, *UK Defence Statistics 2001*, 2001, URL <http://www.mod.uk>.

k Amount of defence production: Japan Defense Agency, *Defense of Japan 2001*; and contract awards: Japan Defense Agency, annual list of 20 largest contractors.

l Teleinformatsionnaya Set, TS-VPK, URL < http://www.vpk.ru/eng/index.htm>.

m Grover, B., *Canadian Defence Industry 1999: A Statistical Overview of the Canadian Defence Industry*, Dec. 1999, URL <http://www.cdia.ca/fullreport.htm>.

n Ministry of Defence, Italy, Defence–Industry Committee, *Lineamenti di Politica Industriale per la Difesa* [Defence industry policy outlines], Oct. 1996, p. 41.

o Ministry of National Defense, South Korea, *Defence White Paper 2000*, 2001, pp. 159 ff.

p Department of Defence, Australia, *Defence and Industry Strategic Policy Statement,* June 1998, URL <http://www.dmo.defence.gov.au/id/di_policy/policy.pdf>.

q Ministry of Defence, India, *Annual Report 2000–2001*, pp. 52 ff., 2001, URL <http://www.mod.nic.in/reports/report01.htm>.

r National Conventional Arms Control Committee, *White Paper on the South African Defence Related Industries*, 1999, URL <http://www.polity.org.za/govdocs/white_papers/defence/defenceprocure1.htm>.

s Domestic orders for military equipment: Ministry of Defence, Directorate General for Armament, *Jaaroverzicht Materieelbeleid* [Procurement policy] (annual); arms export licences: Ministry of Economic Affairs, *Nederlandse Wapenexportbeleid 2000* [The Netherlands arms export policy in 2000], July 2001, URL < http://www.ez.nl/beleid/ext_frame.asp?site=/beleid/home_ond/handelspolitiek/hpinx01.htm>.

t Ministry of Defence, Spain, *La industria de defensa en España* [The defence industry in Spain], 2000, URL <http://www.mde.es/mde/infoes/indus3/>.

u Swedish Ministry for Foreign Affairs, *Swedish Arms Exports 2000*, 2001, URL <http://www.utrikes.regeringen.se/propositionermm/skrivelser/pdf/s20002001_114.pdf>. The data are compiled by Inspektionen för strategiska produkter [National Inspectorate of Strategic Products], URL <http://www.isp.se>.

v Association of Swedish Defence Industries, *Facts about the Swedish Defence Industry2001–2002,* Aug. 2001; and *Statistics 2001*, URL http://www.defind.se/pdf/statistik.htm, Mar. 2002.

w Ministry of National Defence, Republic of China, *National Defence Report 2000*, 2000, pp. 83, ff.

Industry associations

Only few of the national arms industry associations provide valid economic statistics on their member companies to the public at large. This is probably a result of the fact that they are established primarily for political lobbying on behalf of their member companies.

Two national arms industry associations, the Canadian and Swedish, published statistics on national output and employment in arms production in the late 1990s. The Canadian association commissioned an ambitious statistical survey in 1997 as the first of a series of biannual publications. Data on total national arms sales compiled by the Swedish association since 1987 are of limited value as they are derived from an aggregation of total company arms sales, rather than value-added sales. The aggregate values therefore include sales of weapon components and parts by subcontractors, which are also included in the value of sales of final weapons by prime contractors. Until 2000 the industry association used a different definition than the

Swedish Government (Swedish National Inspectorate of Strategic Products, ISP) and published the statistics in the government arms export report to the Swedish Parliament since 1999.[7]

Government organizations

In the 1990s two countries initiated efforts to enhance the dissemination of statistics on the military sector in order to improve public understanding of its role in the wider economy. The *UK Defence Statistics* has been published since 1992. The publication was initiated in order to 'improve the availability and presentation of statistical information on defence', traditionally compiled in the *Statement on the Defence Estimates*,[8] which is presented annually to parliament by the Secretary of State of Defence.[9] The information on the UK's arms industry that is provided includes arms industry employment statistics, statistics on domestic equipment expenditure and export sales, as well as Ministry of Defence payments to contractors.

The French Ministry of Defence first published its *Annuaire statistique de la defense* in 1999. The publication aims to provide a comprehensive overview of the role of the military sector in the national and international economy. The yearbook therefore contains extensive comparative notes.[10] It also includes a presentation of major arms-producing companies, arms industry employment statistics by sector and geographical region, and arms sales statistics by markets.[11]

No single comprehensive statistical report is provided by the government in the country with the world largest arms industry: the United States. While the US President is empowered to collect data on the national defence technology and industrial base, there are severe limitations on the disclosure of the information.[12] Statistics on prime contract awards, arms exports and employment in arms production are available from separate sources. An assessment of the state of the arms industry and an overview of relevant policy developments has been presented since 1996 by the Secretary of Defense to the Committees on Armed Services of the Senate and the House of Representatives in the *Annual Industrial Capabilities Report to Congress*.

---

[7] The definition of arms sales used by the ISP is more restrictive than the one used by the arms industry association.

[8] The British *Statement on the Defence Estimates* was published until 1996; it was not produced in 1997 because of the Strategic Defence Review. In 1998 the *Statement on the Defence Estimates* was replaced by the *Strategic Defence Review Report: Strategic Defence Review*; in 1999 it was renamed the *Defence White Paper*.

[9] The decision followed a 'detailed review of the dissemination of Defence statistics'. Ministry of Defence, *Defence Statistics, 1992 Edition* (Government Statistical Service: London, 1992).

[10] The publication of the yearbook represents a clear improvement in the amount and quality of information on arms production provided by the French Government. Hébert, J.-P., 'L'information économique du ministère de la Défense, Le Débat Stratégique' [Economic information from the Ministry of Defence: the strategic debate], no. 44 (May 1999), URL <http://www.ehess.fr/centres/cirpes/ds/ds44/infoeco.html>.

[11] Data on arms sales are based on government expenditure data rather than on industry sales data. Ministry of Defence, *Annuaire statistique de la defense 2001* [Defence statistical yearbook 2001] (Ministry of Defence: Paris, June 2000), p. 101.

[12] All information that is deemed 'confidential or with reference to which a request for confidential treatment is made by the person furnishing such information shall not be published or disclosed unless the President determines that the withholding thereof is contrary to the interest of the national defense'. United States Code, Title 10, chapter 148: National Defense Technology and Industrial Base, Defense Reinvestment, and Defense Conversion, section 2507: Data Collection Authority of President. It can be accessed via US House of Representatives, Office of the Law Revision Counsel, URL <http://uscode.house.gov/uscode.htm>.

Since 1998 the Spanish Ministry of Defence has published a comprehensive assessment of the national arms industry and its role in the world arms market. The assessment is supported by some statistical information, but it lacks adequate methodological notes.[13] The Dutch Office of Military Production, which is part of the Directorate General for Industry of the Ministry of Economic Affairs, compiles annual statistics on the size of the national arms industry based on a questionnaire which is distributed to companies. Selected results of the survey have been published in recent editions of the government's arms export report to parliament.[14]

The defence white papers of India, Japan, South Korea, and Taiwan contain some information on the size of the national arms industries. However, with the exception of India, the information provided is very limited.[15]

III. Conclusions

There is a clear need for increased public information on the arms industry. In the 1990s few governments undertook efforts to improve insight into the dynamics of arms production. Significant progress in public transparency was made throughout the 1990s as regards arms transfers. A considerable number of countries report to the United Nations Register of Conventional Arms on transfers of specific categories of major conventional weapons. In Western Europe transparency regarding arms transfers improved significantly at the end of the 1990s. In 1999 the European Union (EU) published the aggregate values of arms exports submitted by its members within the framework of the 1998 EU Code of Conduct for Arms Exports.[16] A significant number of governments of major arms-producing countries have also submitted comprehensive arms reports on export their respective parliaments.[17] The debate on transparency in arms production could effectively build on these efforts.

Commercial confidentiality is among the primary obstacles to increased transparency and accountability as regards arms exports and production. In order to avoid putting companies in a disadvantageous position with respect to competitors, the issue needs to be addressed within the framework of the debate on international harmonization of government regulation.

An open debate on transparency in arms production could lead to the establishment of common reporting rules and procedures. In the current situation corporate self-regulation with regard to public transparency prevails over government regulation.

---

[13] Data on the arms exports share in total arms sales for 1998 have been reduced substantially in the 2001 edition, as compared to the 2000 edition, without explanatory notes.

[14] See also Ministry of Economic Affairs, Office of Military Production, *De Nederlandse defensiegerelateerde industrie, Een inventarisatie van omvang en kenmerken* [The Dutch defence industry: an inventory of its size and characteristics], 18 Mar. 1999, is available at URL <http://www.ez.nl/cmp/doc/industrie.pdf>.

[15] Since 2000 the Department of Defence Production and Supply of the Indian Ministry of Defence has also made a considerable amount of information available to a larger public on its Internet site, including financial information on arms-producing companies. Ministry of Defence, Government of India, 'Defence production & supply', URL <http://www.mod.nic.in>.

[16] The 1st and 2nd editions of the report are available on the SIPRI Export Control Project Internet site at URL <http://projects.sipri.se/expcon/eu_documents.html>.

[17] A comprehensive review of reporting practices in Europe is provided by Mariani, B. and Urquhart, A., *Transparency and Accountability in European Arms Export Controls: Towards Common Standards and Best Practice*, Saferworld, Dec. 2000, URL <http://www.saferworld.co.uk/pubtrans.htm>.

8. International arms transfers

BJÖRN HAGELIN, PIETER D. WEZEMAN,
SIEMON T. WEZEMAN and NICHOLAS CHIPPERFIELD

I. Introduction

The SIPRI Arms Transfers Project identifies trends in international *transfers* of major conventional weapons using the SIPRI trend indicator.[1] The trend-indicator value represents the volume of international transfers of both major conventional weapons and military technology for the foreign licensed production of these weapons. As shown by the five-year moving averages presented in figure 8.1, global arms transfers for the period 1997–2001 continued to decline. This is explained mainly by a reduction in the deliveries by the United States.[2] Section II presents the dominant trends of individual suppliers and recipients of major weapons in 1997–2001.

Section III gives examples of transfers of all types of weapons to regions of conflict and discusses the effects of arms transfers and planned acquisitions to India, Pakistan and countries in West Africa. Certain countries are prohibited from receiving arms, some because they are involved in armed conflicts. Information on multilateral arms embargoes in force in the period 1997–2001 is also presented.

Section IV provides, first, an account of SIPRI's estimate of the value of the global *arms trade* in 2000 and a discussion of the major suppliers based on their own national reporting.[3] Second, it presents the factors which influence the international arms trade in the short- and long-term perspective. The future supply of advanced major weapons is affected by the uncertainty concerning the organization of transatlantic production and trade. Section V examines the Joint Strike Fighter (JSF) project as a case study of transatlantic cooperation and the effects it may have on transfers of military technology and combat aircraft.

---

[1] SIPRI data on arms *transfers* refer to actual deliveries of major conventional weapons. To permit comparison between the data on such deliveries of different weapons and identification of general trends, SIPRI uses a *trend-indicator value*. The SIPRI values are therefore only an indicator of the volume of international arms transfers and not of the actual financial values of such transfers. Thus they are not comparable to economic statistics such as gross domestic product or export/import figures. The method used in calculating the trend-indicator value is described in appendix 8D. A more extensive description of the methodology used, including a list of sources, is available on the SIPRI Internet site, URL <http://www.sipri.se/projects/armstrade/atmethods.html>. The figures may differ from those given in previous SIPRI Yearbooks; the SIPRI arms transfers database is constantly updated as new data become available, and the trend-indicator values are revised each year.

[2] Five-year moving averages are calculated as a more stable measure of the trend in arms transfers than the often erratic year-to-year figures.

[3] The value of the arms *trade* refers to the financial values of arms transfers.

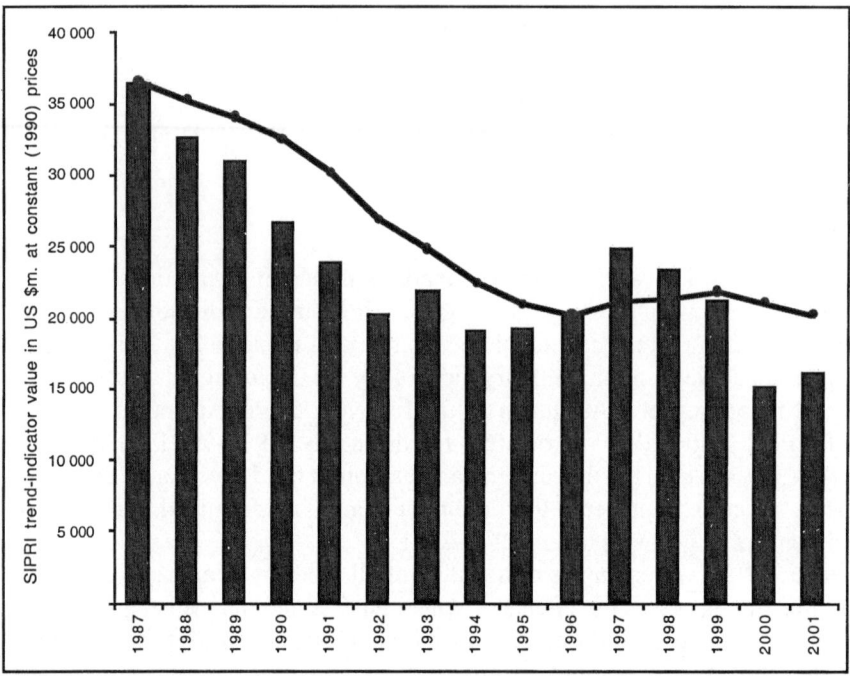

Figure 8.1. The trend in international transfers of major conventional weapons, 1987–2001
Note: The histogram shows annual totals and the curve denotes the five-year moving average. Five-year averages are plotted at the last year of each five-year period.

Section VI reports on national and international transparency in arms transfers in 2001. Section VII summarizes the main findings of the chapter.

II. The suppliers and recipients

The major suppliers

The United States was the largest supplier (44.5 per cent) in the period 1997–2001 (see table 8A.2) with deliveries to a large number of recipients. However, the USA experienced its third consecutive year of reduced deliveries. In 2001 the USA accounted for 28 per cent of global arms transfers. From 1998 to 2001 its deliveries fell by 65 per cent, caused mainly by a drop in deliveries of combat aircraft to major recipients. The major recipients of US weapons in 1997–2001 were Saudi Arabia and Taiwan. Known foreign orders will result in an increased volume of US deliveries. Such orders include expensive weapon systems that have a strong impact on the trend-indicator value, such as approximately 350 F-16 combat aircraft as well as helicopters and early-warning and transport aircraft.[4]

---

[4] After 11 Sep. 2001, the US Department of Defense's Defense Security Cooperation Agency, which handles all government-to-government arms transfers, established an Enduring Freedom Response Cell. Its purpose is to put on a fast track weapon requests from allies and friends in the fight against terrorism. 'New war, old weapons', *The Nation*, 29 Oct. 2001, URL <http://www.thenation.com>. See also

The US arms embargo on India was lifted after September 2001, and in 2001 there were indications that India's acquisition policy may change in a way that could strengthen the US position among the major suppliers.[5] India's long-standing arms transfer relationship with Russia makes it unlikely that the USA, at least in the short term, will threaten Russia's strong position as India's main supplier. Instead, a change in India's policy is likely to involve the USA as a supplier of military sub-systems and components for Indian indigenous projects, many of which have experienced technological difficulties and delays.[6] The new US policy will also permit previously embargoed US sub-systems and components to be included in weapons delivered to India by other Western suppliers (see below).[7]

Russia was the second-largest supplier in 1997–2001, accounting for 17 per cent of total arms transfers for the period and 30.7 per cent for 2001. After a 24 per cent increase in arms transfers from 2000 to 2001 Russian deliveries exceeded that of the USA—Russia therefore became the largest supplier in 2001.[8] (The USA ranked second.) The increase is explained by the greater volume of weapons delivered by Russia to China and India, in particular.

Russian military–industrial reorganization could lead to the more efficient use of available resources for arms production. This, in turn, could lead to the availability of more advanced weapons for sale, supported also by the December 2001 Russian (Rosoboronexport)–Ukrainian (Ukrspetsexport) agreement to promote joint military projects and regulate their relationship as arms exporters.[9] In addition, better after-delivery support could result in more satisfied customers, who would continue to order Russian weapons. However, the small number of major recipients of weapons from Russia—basically China and India—and Russia's limited investment in military research and development (R&D) are major drawbacks for its long-term competitiveness with regard to the most advanced weapons.[10] The known outstanding orders cannot sustain the level of Russian arms transfers beyond 2001.

India's role as a major Russian recipient may also change. In the period 1997–2001 it accounted for 37 per cent of Russia's combined deliveries to China and India. Despite a June 2001 Indian–Russian agreement that listed a number of possible joint projects and future Russian deliveries, the lifting of

---

Federation of American Scientists, 'America's war on terrorism', URL <http://www.fas.org/terrorism/at/index.html>.

[5] Powell, C., 'US looks to its allies for stability in Asia and the Pacific', *International Herald Tribune*, 27–28 Jan. 2001, p. 8.

[6] In early 2002 the US administration accepted sales of jet engines, advanced avionics, weapon-locating radar, ground sensors and other military items to India. Bedi, R., 'Bush clears sale of 20 military items to India', *Jane's Defence Weekly*, 13 Feb. 2002, p. 3.

[7] Luce, E., 'India to ask US for speedy military equipment sales', *Financial Times*, 17 Jan. 2002, p. 5.

[8] Makienko, K., 'Preliminary estimates of Russian performance in military–technical cooperation with foreign states in 2000', *Eksport Vooruzhenii*, no. 1 (Jan./Feb. 2001), p. 5.

[9] Svyatko, S., 'Ukrainian–Russian military cooperation: from rivalry to partnership', *Military Parade*, Jan. 2002, URL <http://www.milparade.com/cgi-milparade/reader.cgi?>; and Lake, D., 'Russia reports strong arms exports in 2001', *Jane's Defence Weekly*, 9 Jan. 2002, p. 20.

[10] Saradzhyan, S., 'Experts: Russian firms must break out of current "client ghetto"', *Air Defence*, 28 May–3 June 2001, p. 38.

Table 8.1. Transfers of major conventional weapons from the 10 leading suppliers to the 38 leading recipients, 1997–2001

Figures are trend-indicator values expressed in US $m. at constant (1990) prices. Figures may not add up because of the conventions of rounding.

Recipients	Suppliers											
	USA	Russia	France	UK	FRG	Ukraine	Netherlands	Italy	China	Belarus	Other	Total
Africa	**243**	**1 296**	**3**	**29**	**33**	**192**	**–**	**50**	**144**	**1 056**	**1 044**	**4 090**
Algeria	46	549	–	16	–	93	–	–	16	362	22	1 106
Angola	–	192	–	–	–	–	–	–	–	435	275	902
Other	197	555	3	13	33	99	–	50	128	259	745	2 082
Americas	**1 670**	**663**	**641**	**1 316**	**489**	**133**	**340**	**425**	**–**	**333**	**1 254**	**7 264**
Argentina	351	3	60	1	143	–	6	108	–	–	19	688
Brazil	280	–	415	341	128	–	1	25	–	–	264	1 454
Canada	136	–	39	731	64	–	–	50	–	–	26	1 045
Mexico	342	223	–	–	–	83	–	8	–	–	9	664
Peru	16	292	4	–	–	–	7	10	–	–	15	671
USA	..	–	–	221	25	50	–	201	–	333	492	988
Other	547	147	123	23	130	–	325	24	–	–	135	1 754
Asia	**15 438**	**12 013**	**5 221**	**1 329**	**600**	**1 881**	**422**	**631**	**1 253**	**78**	**2 387**	**41 253**
China	31	6 515	81	26	–	365	–	11	n.a.	–	90	7 117
India	–	3 771	78	107	40	123	270	20	–	–	300	4 710
Japan	3 137	–	21	18	3	–	–	5	–	–	18	3 203
Malaysia	350	16	8	530	–	–	19	426	–	–	106	1 454
Pakistan	147	–	667	–	–	1 280	38	39	624	78	63	2 931
South Korea	2 549	301	392	195	373	–	51	10	–	–	60	3 931
Singapore	1 168	28	7	7	–	–	–	–	–	–	459	1 669
Taiwan	7 482	–	3 883	–	–	–	–	–	–	–	30	11 397
Thailand	421	–	4	12	132	–	–	109	25	–	604	1 304
Other	152	1 382	83	433	50	112	44	14	566	–	701	3 537
Europe	**12 695**	**1 142**	**962**	**703**	**1 739**	**293**	**513**	**365**	**2**	**–**	**1 632**	**20 046**
Finland	2 224	–	24	3	–	–	1	–	–	–	26	2 273
Greece	2 220	639	267	60	632	120	325	–	–	–	174	4 436

INTERNATIONAL ARMS TRANSFERS

Italy	1 009	—	—	40	184	6	—	—	n.a.	—	1	1 237
Netherlands	835	—	—	59	—	100	—	—	—	—	43	1 036
Norway	476	—	—	67	131	11	—	32	—	—	257	973
Spain	607	—	—	100	119	—	—	10	66	—	45	947
Sweden	128	—	—	110	—	538	—	—	—	—	6	782
Switzerland	1 333	—	—	58	—	—	—	—	—	—	5	1 395
UK	2 359	—	—	34	n.a.	46	—	—	200	—	41	2 680
Other	1 506	503	—	206	208	409	175	145	102	2	1 031	4 287
Middle East	**14 000**	**2 237**	—	**2 886**	**2 967**	**1 819**	**127**	**588**	**133**	**156**	**760**	**25 719**
Egypt	3 023	182	—	4	—	16	—	4	—	17	3	3 250
Iran	—	1 225	—	—	—	—	45	—	—	96	>0.5	1 367
Israel	2 046	—	—	24	—	765	—	—	—	—	—	2 835
Jordan	362	—	—	—	316	—	46	—	33	43	59	782
Kuwait	315	—	—	—	163	—	—	—	—	—	3	919
Qatar	1	—	—	—	270	—	—	21	—	—	—	942
Saudi Arabia	4 490	—	—	117	1 784	—	—	—	30	—	296	6 717
Turkey	3 271	—	—	383	44	1 035	—	43	69	—	184	5 028
UAE	59	303	—	1 307	139	5	36	507	—	—	57	2 412
Other	432	527	—	41	251	—	—	—	—	—	170	1 467
Oceania	**769**	—	—	**3**	**354**	**90**	—	—	**65**	—	**918**	**2 207**
Australia	670	—	—	—	354	90	—	—	65	—	292	1 471
New Zealand	100	—	—	3	—	—	—	—	—	—	610	713
Other	—	—	—	—	—	—	—	—	—	8	15	23
Other[a]	3	—	—	5	—	49	—	—	—	—	91	148
Total	**44 8 210**	**17 354**	**9 808**	**6 699**	**4 821**	**2 627**	**1 862**	**1 671**	**1 555**	**1 518**	**7 998**	**100 732**

[a] Includes the UN and NATO (as non-state actors, not as combinations of all member states) and unknown recipients.

Note: The SIPRI data on arms transfers refer to actual deliveries of major conventional weapons. To permit comparison between the data on such deliveries of different weapons and identification of general trends, SIPRI uses a *trend-indicator value*. The SIPRI values are an indicator only of the volume of international arms transfers, not of the actual financial values of such transfers. Thus they are not comparable to economic statistics such as gross domestic product or export/import figures.

Source: SIPRI arms transfers database.

the US embargo makes it easier for India to acquire military supplies from France and Israel—both important suppliers to India—and from other European suppliers and the USA.[11] An April 2001 Indian parliamentary report noted that Russia no longer grants India 'friendship prices' based on a fixed exchange rate, an arrangement that also allowed barter deals. This change, in addition to the high Russian commission fees, is part of the reason for the proposal that India should request global bids for all its defence procurements.[12] If this report results in a change of India's acquisition policy, in the long term it could permit India to play off Russia against other suppliers in an attempt to get better deals with regard to price, content and delivery schedules. For India this illustrates the difficulty of having one major supplier that defines the conditions. For Russia it illustrates the problems connected with having few major recipients in a situation where friendship prices no longer provide an incentive to purchase Russian weapons.[13]

Both France and the United Kingdom are major suppliers of combat aircraft. The distribution of new combat aircraft orders between France, the UK, the USA and possibly Russia will be important for their future positions among the major suppliers.[14] Apart from Russia only *France* showed a marked increase in arms deliveries in 2001, thereby reversing the decline since 1998. France accounted for 10 per cent of global arms transfers in 1997–2001, ranking as the third-largest supplier. The main recipients were Taiwan and the United Arab Emirates.

Ranking fourth, in 1997–2001 *the UK* accounted for 7 per cent of international arms transfers, showing only a slight increase in 2001, when it ranked fourth. Saudi Arabia was the UK's only major recipient. *German*y, the fifth-largest supplier in the period 1997–2001 as well as for 2001, showed a similar dependence on deliveries to Turkey. Germany experienced a drop in arms deliveries by 46 per cent from 2000 to 2001 and accounted for 5 per cent of international arms transfers in 1997–2001. Total arms transfers from *Ukraine*, the sixth-largest supplier for the period 1997–2001 and the eighth-largest supplier for 2001, were only 54 per cent of German arms transfers in 1997–2001. However, deliveries to Ukraine's main recipient, Pakistan, were roughly equivalent to German and British deliveries to their main

[11] One example is India's co-development of a light attack helicopter with France that is equipped with a French engine. Taverna, M. A., 'New Turbomeca engine to power India's LAH', *Aviation Week & Space Technology*, 7 Jan. 2002, p. 30; Raghuvanshi, V., 'Absent sanctions, India may shift weapon buys', *Defense News*, 1–7 Oct. 2001, p. 3; *Interavia*, Sep. 2001, p. 12; Luce, E., 'US seeks to strengthen military links with India', *Financial Times*, 6 Nov. 2001, p. 2; and Rajagopalan, R. P., 'Indo-US relations in the Bush White House', *Strategic Analysis*, vol. 25, no. 4 (2001), pp. 545–56.

[12] Raghuvanshi, V., 'Report urges India to widen contracting process', *Defense News*, 30 Apr. 2001, p. 6.

[13] A bilateral agreement on military–technological cooperation signed in Oct. 2001 might make Iran the third most important Russian recipient. Pronina, L., 'Tehran turns to Moscow to fulfill weapon needs', *Defense News*, 22–28 Oct. 2001, p. 18. US–Iranian relations, which seemed to improve in 2001, might worsen as a result of the alleged linkage to Iran of the arms shipment seized by Israel in Jan. 2002. Hockstader, L., 'Freighter captain implicates Iran in weapons shipment', *International Herald Tribune*, 10 Jan. 2002, p. 6.

[14] 'France isolated in military aircraft programs' (translation from *Le Monde*), 24 Feb. 2002, in Foreign Broadcast Information Service, *Daily Report–West Europe (FBIS-WEU)*, FBIS-WEU-2002-0225, 25 Feb. 2002.

recipients. Ukraine accounted for 2.6 per cent of global arms transfers in 1997–2001, as well as in 2001.

With increasing cooperation being the main European industrial strategy, the global share of international arms transfers of individual European countries as well as the *European Union* (EU) as a whole will remain low. Cooperation in the development and production of major weapons for common European procurement reduces the number of weapons being produced and the number of countries in Europe from which transfers are made. Taken together, EU transfers to non-European countries (in effect, accounted for by France, Germany, Italy, Spain, Sweden and the UK, the six countries of the Framework Agreement Concerning Measures to Facilitate the Restructuring and Operation of the European Defence Industry) added up to 23 per cent of global arms transfers in 1997–2001 (appendix 8B).[15]

The major recipients

The long-term relative stability of the positions of the major suppliers is reflected among the major recipients. The nine largest recipients in the period 1996–2000 were also the largest recipients in 1997–2001 (appendix 8A), although their internal ranking altered. As a result of the continuous increase in China's arms imports since 1999 and a 44 per cent increase in 2001—including Russian combat aircraft and the second Sovremenny destroyer—China was the second-largest recipient in the period 1997–2001 and by far the largest recipient in 2001. Imports by India increased by 50 per cent but from a much lower level than for China, making India the third-largest recipient in 2001 after the UK. The reduction in South Korean imports in 2000 continued in 2001. As a result, the ranking order of the five major importers in 1997–2001 was Taiwan, China, Saudi Arabia, Turkey and India. These five recipients accounted for 35 per cent of global arms imports in 1997–2001.

Taiwan remained the largest recipient in 1997–2001 partly because of deliveries of 150 F-16 combat aircraft from the USA in 1997–99; 66 per cent of Taiwan's imports in this period were supplied by the USA and the rest were mainly from France. Another factor explaining Taiwan's high ranking is the decrease in mainly Saudi Arabia's imports.

Taiwan's 'wish list' is long, but the US Government has been restrictive in its deliveries of certain weapons in order not to complicate relations with China. However, the George W. Bush Administration has indicated a less restrictive position on arms transfers to Taiwan. In 2001 it declared its support for Taiwan and proposed that Taiwan should receive the same benefits as important non-NATO allies, a controversial proposal in Congress.[16] The administration's proposal was further complicated by the US offer of conventional submarines that are not produced in the USA and which would require

---

[15] The Framework Agreement is available at URL <http://projects.sipri.se/expcon/loi/indrest02.htm>.
[16] Sherman, J., 'Proposal gives Taiwan US defense privileges', *Defense News*, 21 May 2001, p. 4.

the involvement of European producers and possibly Australia.[17] A European–US combined effort to provide Taiwan with such submarines was discussed in late 2001, but no agreement was reported.[18]

There is international and regional concern over the effects of the nuclear and conventional weapon programmes in India and Pakistan—Pakistan was the 10th-largest importer in the period 1997–2001. Acquisitions and plans by India and Pakistan to acquire major weapon systems, some capable of delivering nuclear weapons, and their threat perceptions are described below.

III. Arms transfers to areas of conflict

Arms deliveries may strengthen military capabilities and influence bilateral or regional stability. They are of particular concern when made to recipients in regions where there is military–political tension, conflict or war. While India and Pakistan have been in the focus of concern largely because of their nuclear programmes they are also examples of countries which receive major conventional weapons despite a long-lasting military conflict. Moreover, in some cases the line between nuclear and conventional weapons is not clear in that acquisitions of conventional weapons can also support nuclear ambitions.

In other types of conflict, such as armed conflicts involving less well-armed actors, a focus on only major conventional weapons is insufficient for an understanding of how weapons may influence the development of a conflict.[19] This is illustrated below by the discussion of the transfer of all types of arms to actors in intra-state wars in Guinea, Liberia and Sierra Leone.

Governments and groups in countries located in areas of conflict receive arms—sometimes legally and sometimes illegally and in breach of international arms embargoes. While suppliers have different reasons for supplying weapons, the arms suppliers cannot control whether arms deliveries will stabilize or destabilize a particular relationship. Sometimes the weapons help to end a war; in other situations the acquisition of new weapons increases insecurity and could thereby reduce the likelihood of a peaceful solution.

Arms imports by India and Pakistan

Arms transfers to India and Pakistan are probably the most disturbing examples of the failure to restrict the transfer of weapons to areas of tension and war. For most of the period since their independence in 1947 India and Pakistan have been fighting a low-intensity war. In 1965 and 1971 they fought intensive wars, and on several other occasions they have been very close to

---

[17] Pomfret, J. and Mufson, S., 'Submarines for Taipei opposed by Europeans', *International Herald Tribune*, 26 Apr. 2001, p. 4.

[18] Sherman, J., 'Pentagon seeks European sub designs for Taiwan', *Defense News*, 22–23 Oct. 2001, p. 6.

[19] See also chapter 5 in this volume. For the texts of the UN embargoes see URL <http://projects.sipri.se/expcon/un_d3.htm>; and the EU embargoes see URL <http://projects.sipri.se/expcon/euframe/euembargo.htm>. See also table 8.2 in this chapter.

full-scale war. In recent years the threat of all-out war has again emerged. In 1999 the intensity of the fighting increased when Kashmiri rebels, supported by and operating from Pakistan, occupied a small area in the Kargil region of Kashmir. In December 2001 the situation became even more volatile after an attack on the Indian Parliament by terrorists. India accused Pakistan of having organized the attack and vowed to combat terrorists and their supporters everywhere, if necessary attacking Pakistan, quoting the US-led 'war on terrorism' as just as valid for India and its war against Kashmiri rebels as it is for the US attacks on Afghanistan and the Taliban.[20]

Despite the half-century of tension and fighting between India and Pakistan, arms exporters have largely been willing to supply India and Pakistan with weapons, even weapons that could be deemed offensive. While all of the major arms suppliers have national guidelines cautioning against arms exports to regions of tension and/or countries at war or have signed international policy declarations to that effect, many have not perceived the existing tension between India and Pakistan as a strong reason to forbid exports. One exception is the USA, which imposed a national embargo on Pakistan in 1992, when it refused to stop its nuclear weapon programme, and imposed sanctions in 1998 on India after India tested nuclear weapons. The embargo on Pakistan was not supported by any other country and, although the sanctions on India were supported by most EU countries, they were limited to equipment directly linked to nuclear weapon development and production.

India and Pakistan have for many years been large importers of major weapons, usually ranking among the highest 15 importers. Both countries are dependent on imports to a high degree. With the exception of ballistic missiles, almost none of the weapons in service in India is of Indian design despite India's long-standing policy to establish an indigenous arms industry capable of developing and producing even the most advanced major weapons. Pakistan has a more modest arms industrial policy, largely focused on the assembly, maintenance and modification of major weapons and the production of ammunition and foreign-designed light weapons. Even Pakistan's ballistic missiles are dependent on technology from China and North Korea.

In the period 1997–2001 India acquired 80 per cent of its imports from Russia. Russia is willing to supply the most advanced weapons and to transfer technology under licence-production or cooperation programmes. However, India has other options.[21] France, Germany and the UK are other major suppliers, for example, of combat aircraft and submarines as well as components and technologies for several Indian indigenous programmes (e.g., LCA combat aircraft and the Arjun tank) and for several weapons bought from Russia (e.g., Su-30MKI combat aircraft and Kilo Class submarines). In recent years Israel has become an important supplier, mainly of electronics.

[20] 'Indien ser sig som terroroffer' [India sees itself as a victim of terror], *Svenska Dagbladet* (Stockholm), 20 Dec. 2001, p. 20; 'India feels rival will not resort to nuclear weapons', *Financial Times*, 29–30 Dec. 2001, p. 3; and Slater, J. and Rashid, A., 'Dangerous manoeuvres', *Far Eastern Economic Review*, 10 Jan. 2002, pp. 14–18.

[21] See also the discussion below of changes in US policy and possible influences.

Pakistan is in a more difficult position. Its lack of financial resources and Indian pressure on possible suppliers have left it with few options. Its three main suppliers are China, France and the USA. The USA embargoed all sales of military equipment to Pakistan in 1992 and has shown little interest in supplying large amounts of weapons even after Pakistan's assistance in the US fight against the Taliban. French weapons are expensive, and Pakistan considers Chinese weapons as second rate.[22]

Tactical or strategic systems

Because India and Pakistan are neighbours, large parts of both countries are within easy reach of combat aircraft and medium-range ground- and sea-launched missiles. While the effectiveness of such conventional systems in the past was rather limited, new technology has increased their potential to destroy critical strategic targets such as central military and government headquarters or bases for weapons of mass destruction. In addition, both countries have developed and tested nuclear weapons that are or can be used to arm combat aircraft, air-, sea- and land-based missiles.[23] For these reasons, such weapons may be considered 'strategic weapons'.

India and Pakistan have made clear that they want to acquire additional means of delivering nuclear weapons and more long-range precision weapons, most of which must be imported. India has advanced and ambitious plans. In recent years it has ordered 180 Su-30MKI combat aircraft from Russia and has ordered or is evaluating several types of precision-guided weapons for these aircraft. The Su-30MKI has a range which enables it to reach all of Pakistan, and with the support of four recently ordered Il-78 tanker aircraft (from Uzbekistan) the Su-30MKI is able to reach deep into the Middle East, possibly deterring an attack on India by the Arab states that support Pakistan. Older submarines that are being modernized and new submarines (including nuclear-powered submarines from Russia) are planned to be equipped with Russian and indigenous land-attack missiles. Nuclear warheads are planned to be added to some of these missiles.[24]

The Pakistani Navy has countered with a plan for submarine-launched missiles with nuclear warheads, possibly modifying SM-39 Exocet missiles supplied by France or UGM-84 Harpoon missiles supplied by the USA.[25] The navy is also still seeking to acquire advanced combat aircraft. Plans for 44 Mirage 2000-5s from France were cancelled for financial reasons in 1994,

---

[22] 'Pakistan Air Force said planning to buy 52 planes, arms, surveillance radars', *The News* (Islamabad), 5 Sep. 2001, in Foreign Broadcast Information Service, *Daily Report–China (FBIS-CHI)*, FBIS-CHI-2001-0905, 7 Sep. 2001.

[23] See also appendix 10A in this volume.

[24] *NAVINT*, vol. 13, no. 6 (15 Mar. 2001), p. 5; and Siddiqu-Agha, A., 'Nuclear navies?', *Bulletin of the Atomic Scientists*, Sep./Oct. 2000, pp. 12–14.

[25] 'Pakistan intends to equip its three armed services with nuclear weapons', *Pravda*, 11 June 2001, URL <http://english.pravda.ru/world/2001/06/11/7498.html>; and 'Pakistan's nuclear navy', *Bulletin of the Atomic Scientists*, 22 Feb. 2002, URL <http://www.bullatomsci.org/bulletinwirearchive/BulletinWire 010222.html>. An option to modify US-supplied RGM-84 Harpoon missiles has also been mentioned. Siddiqu-Agha (note 24).

but the plan may be revived.[26] There is, however, no evidence that Pakistan plans to arm its aircraft with long-range conventional missiles as India plans to do.

Air defence systems

In addition to building up an offensive capability, India has sought to acquire an anti-ballistic missile defence system and to improve its air defences against aircraft.[27]

India has an ambitious air defence and anti-ballistic missile (ABM) programme. In recent years Indian air defences have been strengthened with new surface-to-air missile (SAM) systems acquired mainly from Russia, and other SAM systems in Indian service have been modernized.[28] Reports claim that more SAM systems are planned.[29] Plans for an air-defence system with a secondary ABM role—based on the Russian S-300/SA-10 SAM system—were drawn up in the early 1990s, primarily to counter Pakistani aircraft and short-range surface-to-surface missiles (SSMs). Pakistan's development of both nuclear weapons and medium-range SSMs in the late 1990s has been closely matched by Indian plans, which now include several different SAM systems to be bought from Russia and integrated with locally developed Akash and Trishul SAM systems and Green Pine early-warning radars bought from Israel.[30] The Russian systems include the latest version of the S-300/SA-10 with a limited ABM capability, the S-300V/SA-12 with a more pronounced ABM capability and another system, possibly the new S-400, which has even more ABM capabilities.[31]

If the argument that defences against ballistic missiles (armed with nuclear warheads) undermine deterrence is true for the Russian–US relationship, it should also be valid for India and Pakistan. An Indian defence against ballistic missiles and aircraft could, on the one hand, be perceived as purely defensive and guarantee an Indian second-strike capability. It could also, on the other hand, be perceived by Pakistan as part of an offensive policy to defeat any possible Pakistani retaliation after an Indian first strike.

---

[26] 'Pakistan Air Force said planning to buy 52 planes, arms, surveillance radars' (note 22).

[27] Pakistan has no stated requirements for ABMs, and its air defences are mainly based on combat aircraft and limited numbers of short-range missiles.

[28] *Asian Defence Journal*, Nov. 2000, p. 52; and 'India plans to update air defenses', *Defense News*, 9–15 July 2001, p. 20.

[29] Sengupta, P. K., 'China, India in new arms buying spree', *Asian Defence Journal*, Aug. 2001, p. 16.

[30] The Green Pine radar was developed by Israel for its Arrow ABM system. While there is little debate in Israel about the deal, it is unclear if the USA will allow Israel to export the Green Pine radar since it includes US technology. *Jane's Defence Weekly*, 12 July 2000, p. 3; and *Aviation Week & Space Technology*, 25 Sep. 2000, p. 19.

[31] Sengupta, P. K., 'More military hardware flows into South, Southeast Asia', *Asian Defence Journal*, Nov. 2001, p. 12; 'India may acquire Russian "missile shield"', *Jane's Defence Weekly*, 14 Nov. 2001, p. 5; and Sengupta (note 29).

The significance of small transfers of arms

In many current armed conflicts there are poorly armed actors, including both non-state and government actors. Even limited transfers of conventional arms (e.g., small quantities of arms or of relatively simple or old arms) can be an important addition to the military capabilities of such actors. Small transfers can significantly affect how conflicts develop. This significance is poorly reflected, if at all, in the SIPRI arms transfers statistics because the arms involved either contribute to a low SIPRI arms transfer trend-indicator value or are not included in the trend-indicator value, as in the case of small arms.[32]

Arms transfers to conflict areas are common but often small. As a result few non-state actors in current conflicts appear in SIPRI's list of arms recipients, while government actors in current conflicts usually rank low. The latter is illustrated with the following examples of governments involved in armed conflicts and their rank in the SIPRI list of recipients of major conventional weapons in the period 1997–2001: Algeria (23), Angola (31), Colombia (40), Indonesia (42), Sri Lanka (51), Macedonia (61), the Democratic Republic of the Congo, DRC (63), the Philippines (67), Sudan (70) and Rwanda (87).

How small transfers of arms are perceived to be significant as they either fuel conflicts or contribute to the resolution of conflicts is illustrated below with examples of recent arms transfers to actors in the conflicts in Guinea, Liberia and Sierra Leone. For the period 1997–2001 the government actors in these conflicts rank 96, 110 and 115, respectively, in the SIPRI list of recipients of major conventional weapons. The non-state actors in these conflicts do not appear in the SIPRI list.

Arms transfers to Guinea, Liberia and Sierra Leone

In the UN the constant flow of small arms from abroad to the Revolutionary United Front (RUF) was considered as fuelling the conflict in *Sierra Leone* and led to a UN embargo on arms supplies to the RUF.[33] *Liberia* was accused of acquiring small arms from Eastern Europe by circumventing normal controls and then supplying them to the RUF to sustain its war effort. In order to cut off these supplies to the RUF, in March 2001 the UN placed an arms embargo on Liberia.[34]

While the arms supplies to the RUF were generally considered as fuelling the conflict in West Africa, opinions were mixed about the effect of arms supplies to *Guinea*, another actor in the conflict. When the conflict spread from

---

[32] For a discussion of the UN Conference on the Illicit Trade in Small Arms and Light Weapons see appendix 13A in this volume. Of the weapons generally considered to fall in the category of small arms and light weapons the SIPRI arms transfers trend-indicator value includes only man-portable SAMs and anti-tank missiles.

[33] UN Security Council Resolution 1171, 5 June 1998.

[34] United Nations, Report of the Panel of Experts appointed pursuant to Security Council Resolution 1306 (2000), paragraph 19 in relation to Sierra Leone, contained in 'Note by the President of the Security Council', UN Document S/2000/1195, 20 Dec. 2000; and United Nations, Report of the Panel of Experts pursuant to Security Council Resolution 1343 (2001), paragraph 19, concerning Liberia, 26 Oct. 2001.

Sierra Leone to Guinea the Guinean Government acquired small batches of weapons, including mortars from Croatia in 2000, artillery from Moldova and Romania in 2000, ammunition and weapon upgrade packages from Russia in 2001, and four second-hand combat helicopters from Ukraine in 2001.[35] On the one hand, these arms supplies caused concern because their use in indiscriminate attacks on the RUF caused civilian casualties and because, in reaction to the acquisition of combat helicopters by Guinea, Liberia and Côte d'Ivoire sought to procure similar combat helicopters.[36] On the other hand, the newly acquired weapons were an important element of the successful Guinean Government offensive against the RUF. The resulting defeat of the RUF in Guinea contributed to the weakening of the RUF in Sierra Leone to such an extent that it was prepared to agree to a peace agreement and disarmament.[37]

Many of the weapons supplied to the actors in the conflicts in Guinea, Liberia and Sierra Leone came from East European countries and former Soviet republics. They were sold via arms brokers and dealers, often operating through companies with unclear ownership, and were shipped through both willing and unknowing transit countries.[38] The financial turnover of such arms is, at most, in the range of millions of dollar. Nevertheless, profit has been a sufficient motive for individuals and cash-strapped governments to engage in this type of trade, regardless of the possible detrimental effect on conflicts. While the supplies of weapons to Guinea may have helped to end a conflict, this was not the specific objective of the suppliers.

In other cases, however, governments have intentionally supplied weapons as one element of the efforts to resolve a conflict. In the case of the Sierra Leone conflict the British Government has adopted such a policy. It donated small arms to the Sierra Leone Government in 2000 and continued training the Sierra Leone armed forces in 2001.[39] Military assistance to Sierra Leone was considered part of a strategy to 'repel the RUF, to restore the peace process and to rebuild Sierra Leone' and a 'key factor in helping transform the security situation in the country'.[40]

British support was only possible when, in 1998, the UN limited its 1997 arms embargo on the whole of Sierra Leone to an arms embargo on the RUF

---

[35] SIPRI database; and Agence France-Presse, 'Guinean president inks military deal with Russia', 27 July 2001, URL <http://www.iansa.org/news/2001/jul_01/index.htm>.

[36] Human Rights Watch, 'Guinean forces kill, wound civilians in Sierra Leone', 28 Feb. 2001, URL <http://www.hrw.org/press/2001/02/guinea0227.htm>; and Report of the Panel of Experts pursuant to Security Council Resolution 1343 (note 34), pp. 38–39.

[37] See chapter 1 in this volume.

[38] Report of the Panel of Experts pursuant to Security Council Resolution 1343 (note 34).

[39] Wezeman, P. D., Wezeman, S. T. and Chipperfield, N., 'Transfers of small arms and other weapons to armed conflicts', *SIPRI Yearbook 2001: Armaments, Disarmament and International Security* (Oxford University Press: Oxford, 2001), p. 417; and British Ministry of Defence, 'Sierra Leone Army to receive more British military training', 26 Jan. 2001, URL <http://www.mod.uk/index.php3?page=2&nid=1059&view=870&cat=0#news1059>.

[40] 'Statement by the Foreign Secretary Robin Cook, House of Commons, London, Tuesday, 6 June 2000', UKonline, URL <http://www.fco.gov.uk/news/newstext.asp?3753>; and 'UK military assistance in Sierra Leone', 10 Downing Street Newsroom, 3 Sep. 2001, URL<http://www.pm.gov.uk/news.asp?NewsId=2497>.

alone. The UN made the change because the Sierra Leone Government was considered to have legitimate military needs.[41]

In general, the challenge in such cases is to determine whether and under which circumstances arms supplies can be a responsible policy instrument as part of a broader set of conflict resolution instruments.

The ECOWAS moratorium on small arms and light weapons

On the recipient side, the governments of West Africa have recognized the need to control the proliferation of arms in the region. The focus of the ensuing effort has been on small arms and light weapons. In order to stem the spread of such weapons in West Africa, on 31 October 1998 the member states of the Economic Community of Western African States (ECOWAS) declared a three-year Moratorium on the Importation, Exportation and Manufacture of Small Arms and Light Weapons.[42] The moratorium was extended for another three years on 5 July 2001. A code of conduct for the implementation of the moratorium was adopted on 10 December 1999. It calls for the ECOWAS member states to: exchange information on procurement of small arms and develop a regional register on small arms stocks, improve and harmonize the control of transfers of small arms, cooperate with other members in border controls, destroy all small arms surplus to national requirements and small arms confiscated or collected in the context of peace accords, and establish national commissions to implement the moratorium.[43]

Even though several ECOWAS states have made efforts to fulfil their commitments, the implementation of the moratorium has proved difficult.[44] The plans for a regional arms register have not been successful, and only three national commissions had been set up by early 2001. If the implementation of the moratorium were to improve it would have an important function as a first step in regional arms control. The exchange of information on military matters between governments, in particular, is a new development in the region.[45]

---

[41] UN Security Council Resolution 1171, 5 June 1998. For a broader discussion of how even-handedness and a general call for peaceful resolution have increasingly been replaced by a determination of responsibility see chapter 5 in this volume.

[42] The Declaration of a Moratorium on the Importation, Exportation and Manufacture of Small Arms and Light Weapons in West Africa is contained in 'Letter dated 17 December 1998 from the Permanent Representative of Togo to the United Nations addressed to the Secretary-General', UN document A/53/763, S/1998/1194, 18 Dec. 1998.

[43] Haddad, M. N., 'West Africa and the moratorium on small arms', *The Perspective*, URL <http://www.theperspective.org/moratorium.html>. For a full description of the moratorium see the UN Regional Centre for Peace and Disarmament in Africa, in cooperation with the Norwegian Institute of International Affairs (NUPI) and the Norwegian Initiative on Small Arms Transfers (NISAT), *The Making of a Moratorium on Light Weapons* (NUPI: Oslo, 2000), URL <http://www.nisat.org/publications/moratorium/default.htm>. The Code of Conduct for the Implementation of the Moratorium is reproduced on pp. 49–54.

[44] Small Arms Survey, 'Tackling the small arms problem: multilateral measures and initiatives', *Small Arms Survey 2001: Profiling the Problem* (Oxford University Press: Oxford, 2001), p. 260; and Edwards, S., 'UN's African gun control program firing blanks', *National Post*, Apr. 14 2001, URL <http://www.nisat.org/west%20africa/news%20from%20the%20region/Critique_of_PCASED.htm>.

[45] United Nations Office for the Coordination of Humanitarian Affairs (OCHA), Integrated Regional Information Network for Western Africa (IRIN-WA), 'West Africa: IRIN focus on renewal of small

However, in practice the moratorium is limited in scope. It does not stop all imports of weapons by ECOWAS countries because it provides for exemptions which allow countries to import small arms to meet legitimate national security needs.[46] Because the moratorium does not define 'legitimacy' each government can interpret it as it sees fit and procure the weapons it wants. While the moratorium appears to address small arms proliferation to all actors in the region, in reality it is mainly an instrument for cooperation between governments to keep weapons under state control.

Another limitation of the moratorium is the focus on small arms and light weapons. Major conventional weapons played a role in the build-up of tension in the region, and the UN report on Liberia therefore suggested that, for reasons of transparency and confidence building, the moratorium should be broadened to an information exchange for all weapon types procured by ECOWAS states.[47]

International arms embargoes

There were 36 partial or complete international embargoes (on 21 countries, 1 territory and 3 rebel groups) on arms transfers, military services or other military-related transfers in the period 1997–2001 (table 8.2).[48] At the end of 2001, 12 countries, 1 territory and 3 rebel groups were under international arms embargoes. Of these, 7 were under mandatory UN Security Council embargoes—legally binding for all UN members.

A peace agreement was signed between *Eritrea* and *Ethiopia* in December 2000, and the mandatory one-year arms embargo (imposed on 17 May 2000 by the UN on both countries) came to an end in May 2001. The EU embargoes on both countries were also lifted at that time.[49]

The UN arms embargo imposed in 1998 on *Yugoslavia* was lifted on 10 September 2001 when the Security Council announced that it was satisfied that Yugoslav actions against Albanians in Kosovo had come to an end. The EU lifted its sanctions on Yugoslavia in October 2001.

The terrorist attacks of 11 September 2001 and the US attacks on Afghanistan increased support for the mandatory embargo on arms transfers to *Afghan territory held by the Taliban* imposed by the UN Security Council on 19 December 2000. The purpose was to force the Taliban to stop supporting and training 'international terrorists' and to cease harbouring Usama bin Laden. By the end of 2001, however, the embargo had become almost meaningless since the Taliban held little or no Afghan territory. The voluntary UN

---

arms moratorium', IRIN-WA, 11 June 2001, URL <http://www.reliefweb.int/w/rwb.nsf/6686f15dbc852 567ae00530132/311c56442257003585256a690077a1ca?OpenDocument>.

[46] *The Making of a Moratorium on Light Weapons* (note 43), p. 17.

[47] Report of the Panel of Experts pursuant to Security Council Resolution 1343 (note 34), p. 12.

[48] See chapter 1, appendix 1A in this volume for descriptions of some of the relevant conflicts. For the texts of UN embargoes and relevant documents see URL <http://www.un.org/Docs/scinfo.htm#SANCTIONS>. See also the texts of EU and UN embargoes (note 19).

[49] Arms deliveries to Eritrea from Russia were resumed as soon as the embargo was lifted.

Table 8.2. International arms embargoes, 1997–2001

Target	Entry into force	Lifted	Legal basis
Mandatory UN embargoes			
Afghanistan (Taliban)	19 Dec. 2000	–	UNSCR 1333
Angola (UNITA)	15 Sep. 1993	–	UNSCR 864
Eritrea	17 May 2000	15 May 2001	UNSCR 1298
Ethiopia	17 May 2000	15 May 2001	UNSCR 1298
Iraq	6 Aug. 1990	–	UNSCR 661
Liberia[a]	19 Nov. 1992	7 Mar. 2001[b]	UNSCR 788
	7 Mar. 2001[b]	–	UNSCR 1343
Libya	31 Mar. 1992	5 Apr. 1999[c]	UNSCR 748
Rwanda (rebels)[d]	16 Aug. 1995	–	UNSCR 1011
Sierra Leone	8 Oct. 1997	5 June 1998	UNSCR 1132
Sierra Leone (rebels)[e]	5 June 1998	–	UNSCR 1171
Somalia[f]	23 Jan. 1992	–	UNSCR 733
Yugoslavia (FRY)	31 Mar. 1998	10 Sep. 2001	UNSCR 1160
Non-mandatory UN embargoes[g]			
Afghanistan	22 Oct. 1996	(Oct. 2001)[h]	UNSCR 1076
Armenia	29 July 1993	–	UNSCR 853
Azerbaijan	29 July 1993	–	UNSCR 853
Eritrea	12 Feb. 1999	17 May 2000[i]	UNSCR 1227
Ethiopia	12 Feb. 1999	17 May 2000[i]	UNSCR 1227
EU embargoes (mandatory)			
Afghanistan[j]	17 Dec. 1996	(5 Nov. 2001)[k]	96/746/CFSP
Bosnia and Herzegovina[l,m]	5 July 1991	–	–
China	27 June 1989	–	–
Croatia[l]	5 July 1991	20 Nov. 2000	–
DRC[j]	7 Apr. 1993	–	–
Eritrea[j,l]	15 Mar. 1999	31 May 2001	1999/206/CFSP
Ethiopia[j,l]	15 Mar. 1999	31 May 2001	1999/206/CFSP
Indonesia[l]	17 Sep. 1999	17 Jan. 2000	1999/624/CFSP
Iraq	4 Aug. 1990	–	–
Libya	27 Jan. 1986	–	–
Myanmar (Burma)[j]	29 July 1991[n]	–	–
Nigeria[j]	20 Nov. 1995	1 June 1999	95/515/CFSP
Sierra Leone (rebels)[e,l]	8 Dec. 1997	–	98/409/CFSP
Slovenia	5 July 1991	26 Feb. 1996[o]	–
Sudan[j]	15 Mar. 1994	–	94/165/CFSP
Yugoslavia[l]	5 July 1991	8 Oct. 2001	–
Other international embargoes (non-mandatory)			
Nagorno-Karabakh (Azerbaijan)[p]	28 Feb. 1992	–	–
Burundi[q]	6 Aug. 1996	23 Jan. 1999	–
Nigeria[r]	24 Apr. 1996	Nov. 1999	–

Acronyms: CFSP = Common Foreign and Security Policy; DRC = Democratic Republic of the Congo; EU = European Union; FRY = Federal Republic of Yugoslavia; UNITA = União Nacional Para a Independência Total de Angola (National Union for the Total Independence of Angola); UNSCR = UN Security Council Resolution.

a Does not apply to deliveries to Economic Community of West African States Monitoring Group (ECOMOG) forces in Liberia.

b The arms embargo imposed by UNSCR 788 was terminated by UNSCR 1343 and replaced by a new embargo imposed for different reasons.

c The embargo was suspended on this date, not lifted.

d Does not apply to deliveries to government forces in Rwanda. The embargo is also on equipment for persons in neighbouring states if the equipment is for use in Rwanda.

e Does not apply to deliveries to government or ECOMOG forces in Sierra Leone.

f Modified in June 2001 to allow certain non-lethal equipment for UN, humanitarian and media workers in Somalia.

g UN voluntary embargoes are in the form of a non-binding 'call' or 'urge' not to supply weapons. The dates when UN voluntary embargoes end are difficult to assess since there is generally no formal time limit or announcement of the end. The embargoes mentioned here are those deemed by the authors to still be in effect in the period 1997–2001 since the original grounds for the resolution have not been resolved.

h The voluntary UN embargo on Afghanistan was not officially lifted but ceased to have any effect around Oct. 2001 when several countries began supplying the Northern Alliance as part of the war on terrorism.

i On 17 May 2000 the UN Security Council implemented a mandatory embargo.

j Does not apply to deliveries under existing contracts.

k The embargo was modified on 5 Nov. 2001 by Council decision 2001/771/CFSP to include only deliveries to Taliban-held territory, in line with the UN mandatory embargo.

l The EU associate members, the candidate country Cyprus and the European Free Trade Association (EFTA) countries (Iceland, Liechtenstein, Norway and Switzerland), members of the European Economic Area, have declared that they share the objectives of these embargoes.

m The embargo was modified on 17 July 1999 (99/481/CFSP) to exclude small arms for the police and demining equipment.

n A 'decision to refuse the sale of any military equipment' was made by the EU General Affairs Council on 29 July 1991. On 28 Oct. 1996 a decision confirming the embargo (96/635/CFSP) was made by the EU Council of Ministers for Foreign Affairs.

o On this date the embargo was changed to a case-by-case evaluation governed by the EU common criteria on arms exports adopted in 1991. The embargo was officially lifted on 10 Aug. 1998.

p Organization for Security and Co-operation in Europe embargo only on deliveries to forces engaged in combat in Nagorno-Karabakh (i.e., the local forces of Nagorno-Karabakh and those of Armenia and Azerbaijan in Nagorno-Karabakh).

q Embargo by the Democratic Republic of the Congo, Eritrea, Ethiopia, Kenya, Rwanda, Tanzania, Uganda and Zambia.

r Commonwealth embargo.

Source: SIPRI arms transfers archives.

embargo on all parties in the conflict in Afghanistan since 1996 was not implemented when the USA and its allies joined the Northern Alliance in the fight against the Taliban.[50]

On 7 March 2001, UN Security Council Resolution 1343 lifted the 1992 embargo against Liberia after the civil war ended. The UN imposed a new embargo to force Liberia to cease its support of the RUF in Sierra Leone.[51]

---

[50] The 1996 embargo was meant as a signal of concern regarding the war in Afghanistan. The war was not mentioned as a reason for the 2000 mandatory embargo.

[51] UN Security Council Resolution 1343, 7 Mar. 2001.

Several media reports have claimed that there were continuing breaches of the arms embargoes, including the mandatory UN embargoes. The reports focused on continued deliveries mainly by East European countries and former Soviet republics to UNITA (União Nacional Para a Independência Total de Angola, National Union for the Total Independence of Angola) in *Angola* and to the RUF in *Sierra Leone*.[52] These claims were supported by the United Nations' 2001 report investigating its embargo on *Liberia*.[53]

IV. Arms trade and competition

The value of the arms trade

The SIPRI trend indicator cannot be used to assess the economic scale of the global arms market or national arms markets. For this purpose, data are needed on the financial value of international sales of weapons, here called the arms *trade*.

Most of the major supplier governments release data on the value of their arms trade (appendix 8E). By adding these together, it is possible to arrive at a rough estimate of the total financial value of the global arms trade. The value of the global arms trade in 2000 (the latest available year) is estimated at $27–33 billion.[54] This is a rough estimate because the available data are not entirely reliable or comparable, as explained in appendix 8E. Since 1999, when SIPRI began to estimate this global financial value from the national financial values, it has accounted for less than 1 per cent of total world trade.[55]

The national reports differ with regard to the definition of arms and the sources used. Some nations publish reports using different definitions (see appendix 8E for a presentation of these problems). It is therefore not possible to make a completely reliable comparison. Table 8.3 provides the best comparison that can be made of the value of arms deliveries in 1996–2000 as reported by the major suppliers and converted to constant 1998 prices.[56] The lower US value of 'arms deliveries' is compiled by the Congressional Research Service (CRS), while the higher values of 'arms transfer deliveries' are compiled by the Department of State. Both British values for deliveries of 'defence equipment' are compiled by the Ministry of Defence (MOD), but the

---

[52] For earlier reports on Angola and Sierra Leone see Hagelin, B., Wezeman, P. D., Wezeman, S. T. and Chipperfield, N., 'Transfers of major conventional weapons', *SIPRI Yearbook 2001* (note 39), pp. 323–52; and Human Rights Watch, *Neglected Arms Embargo on Sierra Leone Rebels* (Human Rights Watch: Washington, DC, 15 May 2000).

[53] Report of the Panel of Experts pursuant to Security Council Resolution 1343 (note 34).

[54] The lower estimate is the aggregation of reported minimum values; the higher estimate is the aggregation of reported maximum values of delivered arms. The US Department of State figure for 2000 is not available but the Department of State always reports a higher value than the one reported by the CRS. For some smaller countries, only data on arms licences are available. When this is the case, these values have been used. For the 1998 values see Hagelin, Wezeman, Wezeman and Chipperfield (note 52), p. 350.

[55] Total world trade in 2000 amounted to $6310.1 billion. International Monetary Fund (IMF), *International Financial Statistics Yearbook* (IMF: Washington, DC, 2001), p. 131.

[56] The conversion to constant prices was made according to the methodology used by the SIPRI military expenditure project. See appendix 6C in this volume.

higher values include items which are not distinguished as either military or civil aerospace equipment by the official commodity classifications. The definitions also differ between the other suppliers: the Russian values represent exports of 'military equipment', while France reports on deliveries of 'arms and associated services' and Germany on 'weapons of war'.

The USA was by far the largest supplier in 1996–2000 even when using the low values (table 8.3). For Russia, however, these values create a different position than that usually reported. Russia ranks fourth, before Germany, although there is a substantial gap between the total values for Russia and Germany. Consequently, this also affects the other major suppliers. The high values for the UK made it the second-largest supplier. If the lower values are used for the UK, France becomes the second-largest supplier.

Table 8.3 also shows arms exports as a share of total national exports. Russia, the UK and the USA show significant arms export shares. Using the higher arms export values, the shares for both Russia and the USA were over 4 per cent for individual years, while the share for the UK—even with the high values—has been on the decline from close to 4 per cent in 1996 and 1997 to less than 3 per cent by 1999.

With the exception of the USA (even using the low figures), the UK in 1996 and 1997, and France in 1997 and 1998 the shares are less than 2 per cent of total exports. German arms exports accounted for less than 1 per cent of total exports for each year. The conclusion is that arms exports did not contribute significantly to the national economy.

This conclusion is supported by a study of the economic costs and benefits from a 50 per cent reduction in British defence exports. It concludes that the economic costs are limited and that the discussion of defence exports should focus mainly on non-economic considerations.[57]

Competition

For the buyer, competition between two or more suppliers is important in order to obtain the lowest price. Success in international competition for larger contracts has become more important for supplier company survival because such contracts have become fewer and infrequent. The existence of several suppliers that are eager to sell gives the buyer a strong bargaining position. The buyer may receive benefits which would not otherwise be available, such as a lower price or access to military technology. The examples given below show how compensation arrangements affect the trade in arms and military technology.

---

[57] Chalmers, M. et al., *The Economic Costs and Benefits of UK Defence Exports* (Centre for Defence Economics, University of York: York, Nov. 2001). Nationally available data on the number of weapons exported can also be compared with data reported by other sources. Such a comparison of Russian data with data presented by the CRS was made by Makienko, K., 'US Congressional Research Service report on Russia's place on arms market', *Eksport Vooruzhenii*, no. 5 (Sep./Oct. 2001), pp. 2–6. The conclusion was that the CRS report was not a reliable source of information, nor did it present an accurate evaluation of Russia's position in the international arms market.

Table 8.3. Arms deliveries according to national reporting, 1996–2000
Figures are in US $m. at constant (1998) prices.

	1996	1997	1998	1999	2000	1996–2000
USA: High values (1)	23 708	32 212	27 000	32 306	. .	115 226
USA: Low values (2)	15 410	16 537	16 482	17 558	13 434	79 421
% of total US exports (1)	3.7	4.6	4.0	4.7	. .	. .
% of total US exports (2)	2.4	2.4	2.4	2.6	1.8	2.3
UK: High values (1)	10 911	11 449	9 988	6 932	6 985	46 265
UK: Low values (2)	6 009	5 754	3 260	1 598	2 728	19 349
% of total UK exports (1)	3.9	3.9	3.7	2.6	2.6	3.3
% of total UK exports (2)	2.2	2.0	1.2	0.6	1.0	1.4
France	5 621	8 176	7 728	4 633	3 298	29 457
% of total French exports	1.9	2.8	2.5	1.6	1.1	2.0
Russia	4 055	3 658	2 700	3 328	3 485	17 227
% of total Russian exports	4.4	4.1	3.6	4.5	3.5	4.0
Germany	588	794	760	1 607	737	4 486
% of total UK exports	0.1	0.2	0.1	0.3	0.1	0.2

Sources: Nationally reported arms export values as listed in appendix 8E converted to constant prices and International Monetary Fund (IMF) total national export figures. IMF, *International Financial Statistics Yearbook* (IMF: Washington, DC, 2001), p. 131.

Inventive approaches by the recipient may influence the price and terms of a deal. The main contenders to sell new combat helicopters to South Korea are companies from the USA and Russia. Usually, US weapons are purchased under the Foreign Military Sales (FMS) programme, involving US Government guarantees, or they are bought directly from the producing company (commercial sale). US policy has been that a potential buyer may request an FMS programme or a commercial sale, but not both. However, South Korea has asked for three separate offers from each of two different US companies: one under the FMS programme, one for a commercial sale, and one for Korean licence manufacture.[58]

Compensation and arms transfers

While competition can result in the best price for the buyer, the purpose of compensation arrangements is for the supplier to help 'offset' the buyer's acquisition costs. Such offset deals are often long-term arrangements involving the purchasing country in a mix of trade and industrial arrangements.[59]

Offsets can be used as a competitive tool to win an export contract. The specific offset arrangement may involve the transfer of military equipment and technologies. First, it is not uncommon today to include military technology

---

[58] Sherman, J., 'South Korea driving hard bargain for choppers, analysts say', *Defense News*, 2–8 July 2001, p. 8.

[59] For more information see Sköns, E. and Weidacher, R., 'Arms production', *SIPRI Yearbook 2000: Armaments, Disarmament and International Security* (Oxford University Press: Oxford, 2000), pp. 311–14; and Hagelin, Wezeman, Wezeman and Chipperfield (note 52), pp. 334–38.

transfers to the buyer in addition to the weapon system itself. For instance, in South Korea, French Dassault in cooperation with other French companies offered the Rafale combat aircraft with 70–100 per cent offsets, including the transfer of technologies to build air-to-air and air-to-surface missiles. Its US competitor, Boeing, offered the F-15K with 70 per cent offsets, including technology transfers that would enable South Korea to produce its own aircraft by 2015.[60] Similarly, Italian Agusta transferred some production of A-109 helicopters to South Africa as part of a South African order of 30 A-109s.[61]

Second, some arrangements involve transfers of military equipment from the buyer. As part of their offer to sell 24 JAS-39 Gripen aircraft to the Czech Republic, in December 2001 BAE Systems agreed to attempt to find buyers for 36 L-159 light combat/trainer aircraft being produced for but no longer wanted by the Czech Republic.[62] Third, more common than such direct military export support are military transfers from the recipient country that result from co-production arrangements. As a result of the selection of NH-90 transport helicopters by Finland, Norway and Sweden under the Nordic Standard Helicopter Programme (NSHP), industry in all three countries will be involved in NH-90 production not only for NSHP countries but also for other customers.[63] The Dassault and Eurofighter International offers to South Korea also included local production of components for aircraft not ordered by South Korea.[64] Parts of the order of Italian A-109 helicopters ordered by Sweden are to be supplied from the South African production line.[65]

International ownership of companies in the most advanced arms-producing countries is part of defence industrial strategy, but foreign investment may also be a particular form of offset. As such, it is mainly found in the less advanced arms-producing countries and in countries that want to privatize state-owned companies.[66] For the buyer's company it brings work and preserves skills, for the government it sustains military–industrial capacity, and for both it secures employment.

---

[60] 'France to provide missile technology for fighter contract', *Korean Herald* (Internet edn), 17 Sep. 2001; 'Dogfight over Seoul', *Interavia*, Oct. 2001, pp. 43–44; and *Jane's Defence Weekly*, 24 Oct. 2001, p. 22. In addition, the French Government has agreed to sell SCALP 300-km range air-to-surface missiles for the Rafale, while the US Government has agreed to sell the similar SLAM for the F-15K.

[61] 'Sweden orders A109s', *AirForces Monthly*, Aug. 2001, p. 23; and *Aviation Week & Space Technology*, 25 Jan. 2002, p. 21.

[62] Anderson, R., 'Czechs give jet fighter deal to UK–Swedish consortium', *Financial Times*, 11 Dec. 2001, p. 6. The additional demand to participate in the L-159 exports was mentioned before the government's decision. 'Spänd väntan på Saab' [Tense wait at Saab], *Svenska Dagbladet* (Stockholm), 6 Dec. 2001, p. 16; and 'Spänd väntan på Gripenbesked' [Tense wait for Gripen decision], *Dagens Nyheter* (Stockholm), 6 Dec. 2001, p. C2.

[63] *Defence Industry*, Nov. 2001, p. 13; and 'NSHP offset will boost Saab's earnings', *Countertrade & Offsets*, vol. 19, no. 20 (22 Oct. 2001), pp. 5–6.

[64] *Air Letter*, 11 June 2001; and Taverna, M. A., 'United order lifts Dassault', *Aviation Week & Space Technology*, 5 Nov. 2001, p. 61.

[65] 'Sweden orders A109s', *AirForces Monthly*, Aug. 2001, p. 23, URL <www.airforcesmonthly.com>.

[66] See Sköns and Weidacher (note 59), pp. 311–14; and Hagelin, Wezeman, Wezeman and Chipperfield (note 52), pp. 335–38.

Table 8.4. Examples of reportedly offered or demanded offsets

Weapon system	Customer	Offset share (%)
Demanded in combat aircraft deal by	Brazil	100
Demanded in combat aircraft deal by	Slovakia	100
Gripen combat aircraft offer to	Hungary	100
Gripen combat aircraft offer to	Poland	100
C-235/295 transport aircraft deal with	Poland	100
Gripen combat aircraft offer to	Czech Republic	150
Eurofighter combat aircraft offer to	South Korea	70
NH-90 helicopter deal with	Sweden	100
Zuzana howitzer deal with	Greece	65
PzH2000 howitzer deal with	Greece	120
US tank offer with	Greece	100

Source: SIPRI files.

However, arrangements such as these may be controversial from a competitive point of view. Dassault is a minority owner in Brazil's main aircraft company Embraer. Dassault is also one of the competitors for a Brazilian combat aircraft order involving offsets in the form of co-production. It has been questioned whether Dassault's minority share gives it an unfair advantage over its competitors.[67]

The future of offsets

In the light of the increasing use of offsets since the 1980s the question of what demands are 'reasonable' has been raised. Paradoxically, increasing recipient demands could lead to less competition in some cases. While offsets covering 100 per cent of the contract value seem to be acceptable to major arms suppliers (table 8.4), higher offsets might be controversial. The 150 per cent of value demanded by the Czech Republic in 2001 caused the French and US companies to withdraw their offers before a contract had been negotiated, claiming that the conditions were unfair. The one remaining alternative was the Saab–BAE Systems Gripen aircraft. In such a situation the buyer cannot play off different supplier offers against each other in an attempt to reduce the price or to get other benefits. In other words, Saab–BAE Systems could have revised their offer. However, they accepted the demands and in addition reportedly offered a 100 per cent financing package spread over 15 years.[68]

---

[67] *Countertrade & Offset*, vol. 19, no. 20 (22 Oct. 2001), p. 6.
[68] Fiorenza, N., 'Integration challenges: Czech Republic stretches its defense budget to meet NATO standards', *Armed Forces Journal*, Aug. 2001, pp. 18–19. An Irish spokesman is quoted as saying that countertrade is not allowable under EU procurement regulations. *Countertrade & Offset*, vol. 19, no. 20 (22 Oct. 2001), p. 5. If the plan presented by the Belgian Prime Minister in 2001 not to link Belgian procurements directly to industrial offset agreements becomes policy it will be the exception to the rule. *Aviation Week & Space Technology*, 2 July 2001, p. 21. In early 2001 Turkey changed its offset policy. *Air Letter*, 22 Jan. 2001, p. 4. However, as a result of the focus on terrorism, after Sep. Turkey sought speedy deliveries rather than maximum industrial benefits. Bekdil, B. E. and Enginsoy, U., 'Turkish pol-

INTERNATIONAL ARMS TRANSFERS 395

While there has been criticism by supplier companies of the increasing use of offsets as a distortion of the competitive market—companies are not equally able to offer offsets—it has also been regarded as necessary to accept offsets in order to compete for contracts. The 2001 Status Report by the US Presidential Commission on Offsets in International Trade downgraded the overall negative effect of offsets in military trade when valued against the benefits.[69] The conclusions are likely to reduce the impact of the general criticism of offsets in the USA and other countries and focus debate on policies to reduce the most negative consequences of offsets.

A more specific discussion of offsets is found in a 2001 European Defence Industries Group (EDIG) policy paper. The EDIG arguments reflect the criticism of offsets as a market-distorting mechanism.[70] Basically, only when exporting to developing nations is it acknowledged that offsets must be accepted for the benefits they bring to the buyer. If there were a single defence market in Europe, offsets would no longer be used. This would also apply to the Central and East European countries, especially with regard to direct (military) offsets, when these countries are fully engaged in European multinational projects.

With regard to the USA, European offset demands are presented as a compensation—or as a punitive response—for the lack of European access to the US defence market. EDIG suggests a policy similar to the 'Buy American' policy. Europe should implement policies which exclude non-European companies from bidding for work (i.e., European preference) unless: (*a*) the technology or goods/services are not available at an affordable price within Europe; and (*b*) comparable and effective reciprocal access to markets and other methods of control over trade have been agreed. Until these conditions are met, according to EDIG, European nations should not abolish offset requirements when buying from the USA.

V. New weapons and transatlantic cooperation: the JSF

Transatlantic cooperation in the development of major military platforms is rare—especially European involvement in the development of US military equipment to be acquired by US forces. The Joint Strike Fighter (JSF) combat aircraft is an exception in that it attempts to meet the need of three US military services and foreign customers for a common platform. It is a continuation of the US JAST (Joint Advanced Strike Technology) project of the early 1990s, which aspired to develop avionics systems, components and propulsion tech-

---

icy shift aims to meet wartime needs', *Defense News*, 15–21 Oct. 2001, p. 10; and Anderson, R., 'BAE–Saab in move for stake in Czech group', *Financial Times*, 22 Oct. 2001, p. 16.

[69] Presidential Commission on Offsets in International Trade, Status Report, Washington, DC, 2001, approved by the Commission, 18 Jan. 2001. Because of the change of the US administration, and the fact that the acting Commission Director left in Oct. 2001, the Final Report was not published in Oct. 2001 as planned. Shumskas, J., Presidential Commission on Offsets, Private communication with the authors; and Barrie, D. and Svitak, A., 'Global look at offsets could alter future trade', *Defense News*, 3–9 Dec. 2001 (Special report), pp. 27–28.

[70] *EDIG Policy Paper on Offsets*, EPP/00/18, Brussels, 26 June 2001.

nologies to be used in future joint service combat aircraft designs.[71] In November 1996 Lockheed Martin and Boeing were selected as the prime competing contractors to develop prototypes by 2000.[72] In June 1997 BAe (which became BAE Systems in December 1999) joined the new Lockheed Martin–Northrop Grumman team,[73] but there is also significant British participation in the competing team.[74] In October 2001 the JSF entered the engineering and manufacturing development (EMD) phase as the F-35 after the US Government chose the Lockheed Martin version of the aircraft.

If, as has been argued, the project is a model for the future organization of transatlantic cooperation, it appears that the transatlantic market will remain unbalanced with regard to the involvement of governments and industries on both sides of the Atlantic, that the US and European markets will remain unequally accessible and that future European aircraft projects may even suffer from the close British–US cooperation.

The arguments for the JSF

There appear to be four main reasons why the JSF combat aircraft is considered an important project. First, the initial competitive development phase (CDP) involved a specified and difficult balance between advanced design (including new manufacturing procedures), high performance and fixed cost. An important point is that foreign industrial participation has been accepted only according to 'best value' based on competition. There is no *juste retour* principle, common in earlier international projects (guaranteed industrial involvement in relation to the national financial contribution), and no offset arrangement.

---

[71] In 1995 the design concepts were so promising that the project refocused on producing the next-generation strike fighter through competitive development. It thus became the JSF, where 'joint' refers to joint US services use.

[72] *Naval Forces*, no. 1 (1996), p. 62. BAe joined the JSF shortly after it entered an agreement with French Dassault on future fighter technology. In 1995 it also created the JAS Gripen joint venture together with Saab, giving BAe a strong position in both the US and the European combat aircraft programmes. The long-term cooperation between the UK and the USA—including the AV-8B Harrier, the T-45 Hawk programmes and BAe participation in the ARPA (Advanced Research Project Agency) Common Affordable Lightweight Fighter programme in 1994—led the Pentagon in 1996 to 'commit' Lockheed Martin and Boeing to restructure their work to include BAe.

[73] 'McDonnell Douglas out of JSF programme', *NAVINT*, vol. 8, no. 24 (29 Nov. 1996), p. 7. The Pentagon choice of prime contractors in Nov. 1996 resulted in the elimination of BAe's partner McDonnell Douglas–Northrop. Lockheed Martin is or has been involved with the UK in several projects, such as the Royal Navy Merlin helicopter (see Cook, N., 'US know-how exported to Europe', *Interavia*, May 1999, pp. 41–42), the Tomahawk submarine integration, and the Tracer/Future Scout and Cavalry System programme. In the summer of 1999 it created Lockheed Martin UK Ltd, a new company based in London, combining all of its British defence and commercial business interests under a single UK-registered company. 'Lockheed Martin sets up new UK company', *World Aerospace & Defense Intelligence*, 9 July 1999, p. 11. In early 2000 Sanders selected BAE Systems as an electronic warfare and countermeasures subcontractor. 'BAE SYSTEMS joins Sanders Litton JSF team', *Defence Systems Daily*, 1 Feb. 2000.

[74] The British participation includes Flight Refuelling Ltd (fuel system), BAe (vehicle management system, cockpit display and flight-control systems for the Boeing team), Messier-Dowty Ltd (main and nose landing gear system), and Rolls-Royce plc (vertical lift propulsion system, attitude control system). 'Boeing establishes JSF industry team', *Defence Systems Daily*, 1 Oct. 1998; and 'Small firms also gain', *Defense News*, 23–29 Nov. 1999, p. 8.

Second, as noted above, the JSF is the first US combat aircraft project that attempts to meet the need of three US military services and foreign customers for a common platform. Designs for the US Air Force, Navy and Marine Corps as well as the British Air Force and Navy are to be produced from the same production line. There is therefore an initial market of over 3000 aircraft because the JSF is planned to replace the A-6, A-10, AV-8B, F-16 and, to some extent, the F/A-18 combat aircraft in the USA and the Harrier and Tornado combat aircraft in the UK.[75]

Third, foreign countries are given the opportunity to be involved in the programme at different levels of involvement and costs. For the EMD phase—which could last for more than 10 years—the types of association are defined as levels 1–3 plus a fourth, 'major participant', level with basically the same costs as in the four types of association during the CDP phase (table 8.5).[76]

1. Full partner (10 per cent of the cost) participation permitted direct influence on requirements. Only the UK signed as a non-US CDP full partner for the Navy in 1995 and, in 1999, for the Air Force.

2. The associate partners (2–5 per cent of the cost) were Denmark, the Netherlands and Norway. Associate partner status enabled them to influence the requirements for the conventional take-off and landing (CTOL) variant as long as the results were perceived to be mutually beneficial. Each financial contribution was matched by an equal US contribution.

3. The informed partners (1–2 per cent of the cost) were Canada and Italy, but they were not permitted to influence the requirements. The USA contributed $50 million to joint activities related to the Canadian CTOL version.

4. The 'major participants'—Israel, Singapore and Turkey—were countries that want to receive extensive unclassified and non-proprietary information in order to evaluate the JSF as a possible future acquisition.

Fourth, for the UK in general and BAE Systems in particular (as the major foreign participant) the JSF is the only combat aircraft under development that involves transatlantic technology exchange and the potential for future British–US cooperation. The UK was the only full foreign partner in the initial CDP, and it has signed up for full participation in the EMD phase. For the British Government the JSF may also be seen as a means to insure against unsuccessful European combat aircraft projects in the future.

The JSF as a model for transatlantic cooperation

The JSF project is regarded as a blueprint for future transatlantic cooperation because it has included foreign government and industry participation from the beginning and because acquisitions by European nations are planned. The

---

[75] Garamone, J., 'Lockheed-Martin team wins Joint Strike Fighter competition', *American Forces Information Service*, 26 Oct. 2001, URL <http://www.defenselink.mil/news/Oct2001/n10262001_200110266.html>.

[76] Holzer, R., 'Joint Strike Fighter draws international interest', *Defense News*, 10–16 June 1996, p. 4.

Table 8.5. International participation in the Joint Strike Fighter competitive development phase, 1995–2001
Figures are in US $m.

Country	Date joined	Status	Foreign financial contribution	US contribution
UK	Dec. 1995	Full partner	200	–
Netherlands	Apr. 1997	Associate partner	10	10
Norway	Apr. 1997	Associate partner	10	10
Denmark	Sep. 1997	Associate partner	10	10
Canada	Jan. 1998	Informed partner	10	50
Italy	Dec. 1998	Informed partner	10	–
Singapore	Mar. 1999	Major participant	3.6	–
Turkey	Jun. 1999	Major participant	6.2	–
Israel	Sep. 1999	Major participant	0.75	–

Source: Birkler, J. et al., *Assessing Competitive Strategies for the Joint Strike Fighter: Opportunities and Options*, RAND report MR-1362.0 (RAND Corporation: Washington, DC, 2001), p. 14.

EMD phase is likely to include most of the foreign CDP participants and possibly some additional participants. When analysing the JSF project as a model, a number of issues that provide financial, technological, military and political lessons should be considered.

The cost paid by foreign partners, and thus the financial risk taken, is expressed as a percentage of the total cost of the specific project phase. The higher the level of participation, the higher the financial involvement and financial risk. The major industrial argument in Europe for involvement in the project is the potential technology gain, but only the British Government has agreed to pay the price for full foreign partnership in both the CDP and the EMD phase. BAE Systems has invested so much in the JSF that it announced in 1999 that it would remain involved even if the British Government does not order any aircraft.[77]

Since there is no *juste retour* all foreign companies compete for JSF contracts. Specific technological benefits may be important for participating companies but the benefits to states may be more difficult to calculate. One possible exception is British aerospace competence.[78] In 2001 a RAND Corporation study concluded that British industry is likely to gain more from its JSF involvement than the proportional value of the British financial contribution.[79] A 2001 British–US agreement allegedly defines a set of principles that provide a framework for long-term British involvement in the programme: safeguarding the UK's 'national interests' and ensuring that the military capabilities of the aircraft are properly managed and maintained throughout the lifetime of

[77] 'BAe–Lockheed link to stay regardless of UK orders', *Financial Times*, 23 July 1999, p. 10.
[78] James, B., 'Lockheed fighter award also cheered in Britain', *International Herald Tribune*, 29 Oct. 2001, p. 13; and Jones, S., 'Something for everyone in the industry', *Financial Times*, 29 Oct. 2001, p. 2.
[79] Birkler, J. et al., *Assessing Competitive Strategies for the Joint Strike Fighter: Opportunities and Options*, RAND report MR-1362.0 (RAND Corporation: Washington, DC, 2001), p. 75.

the aircraft.⁸⁰ Officials and industry executives have stated that technical data would include stealth technology, software codes and the ability to integrate British weapons on the JSF.⁸¹

Even when the benefits of participating in the JSF are acknowledged by foreign governments, they have been reluctant to allocate funds from hard-pressed defence budgets to a project with no guaranteed national industrial involvement and for which there may not be a military requirement and thus no acquisition. Although there are different association alternatives, there is a fairly high cost for becoming a foreign partner with the right to influence the technical requirements. This is especially true in the EMD phase, when the costs are substantially higher than in the CDP. For example, had Turkey become an EMD associate partner the estimated cost would have been between $500 million and $1.2 billion.⁸² Turkey therefore decided to become an informed partner, and at the end of 2001 there were no associate partners.

Although there is transfer of US technology to foreign JSF partners, the major beneficiary is US industry. It is the USA that defines which foreign companies are to be allowed to participate and bring their skills into the project. As the JSF is a US project, European or other industrial participation is defined as foreign participation, with the exception of that of BAE Systems. Technology transfer as part of the project is a crucial political issue. Because of US re-export restrictions the British MOD cannot allow British companies to share stealth technology with European companies. In the light of the future US export of the JSF aircraft the issue of technology transfer has surfaced, including the transfer of stealth technology in particular, and also with regard to technologies for receiving intelligence and sensor data from aircraft and satellites.⁸³

Long-term considerations

The limited acquisition budgets in European countries indicate that only a few new development projects can be selected. There is therefore competition in Europe between the JSF and current as well as future aircraft projects. If the JSF Joint Program Office is successful in its attempt to bring more countries into the project this could reduce the European Typhoon/Eurofighter market and British funding of future European aircraft. This as well as the restrictions in sharing stealth technology could complicate cooperation in the European Technology Acquisition Plan (ETAP), the 2001 agreement by the six major European aerospace producers to develop future air-platform technologies.⁸⁴

---

⁸⁰ The Right Honourable Baroness Symons of Vernham Dean (Minister of State for Defence Procurement, UK), US Department of Defense News Briefing, 17 Jan. 2001, URL <http://www.defense-aerospace.com/data/verbatim/data/ve156/index.htm>.

⁸¹ Nicoll, A., 'Fighter deal will allow UK access to stealth secrets', *Financial Times*, 18 Jan. 2001, p. 8.

⁸² Bekdil, B. E., 'Turkey remains committed to JSF', *Defense News*, 30 Apr. 2001, p. 12.

⁸³ Sweetman, B., *Interavia*, Feb. 1999, pp. 42–43; and Fulghum, D. A., 'Tech gap affects JSF exports', *Aviation Week & Space Technology*, 14 June 1999, p. 205.

⁸⁴ French Ministry of Defence, 'Press notice on behalf of the defence ministries of France, Germany, Italy, Spain, Sweden and the UK', 19 Nov. 2001, URL <http://www.defense.gouv.fr/english/news/

The transatlantic market is still characterized by political and industrial mistrust. As with other US exports, compromises will be made concerning JSF transfers as regards the sharing and transfer of technology. This will be balanced against military interoperability and the political risks of transatlantic discord. It is therefore too early to conclude that the JSF is the best model for a new, more efficient and open transatlantic arms market.

VI. Arms transfer reporting and transparency

International transparency

The UN Register of Conventional Arms

The year 2000 (the latest reported year) was a success for the UN Register of Conventional Arms with regard to participation and timely reporting. The UN Secretary-General's report included responses from 105 countries.[85] In the period 1992–99 only about 80–90 countries responded to the request for information. However, many of the countries reporting for the first time in 2000 were of little or no significance for the purpose of the register—to provide early warning against a possible destabilizing build-up of weapons—nor did the quality of data provided improve.

EU transparency

The EU published aggregate values of arms exports as submitted by its members in the third annual review of the implementation of the 1998 EU Code of Conduct for Arms Exports.[86] There were few changes compared with the report for 2000. Denmark supplied data for the first time, and a number of countries followed the Swedish example and provided data on both arms export deliveries and licences issued. The report noted that, in order to make national reports on arms exports more comparable and to improve transparency, a matrix containing statistical data from the national reports had been compiled.[87] However, as the matrix was not published, apparently only intergovernmental transparency is intended.

shortnews/620110/201101.htm>; and Cook, N., 'UK's JSF MoU deal with USA infuriates Europe', *Jane's Defence Weekly*, 24 Jan. 2001, p. 2. See also Morocco, J. D., 'ETAP to harvest Europe's technological expertise', *Aviation Week & Space Technology*, 2 July 2001, p. 63. The systems now studied include unmanned combat aircraft (UCAV) that are restricted under the Missile Technology Control Regime and conventional air-launched cruise missiles (CALCMs). 'Europe's future fighter quandary', *Interavia*, May 2001, pp. 40–41.

[85] The 9th UN Secretary-General's report containing information received from governments on their arms export and/or imports was released on 31 July 2001. United Nations Register of Conventional Arms: Report of the Secretary-General. UN document A/56/257, 31 July 2001. The document and its addenda and corrigenda are at URL <http://www.un.org/Depts/dda/CAB/register.htm>. A database containing all import and export data from reports between 1993 and 2001 is at URL <http://domino.un.org/UN_REGISTER.nsf>.

[86] Council of the European Union, *The Third Annual Report according to Operative Provision 8 of the European Union Code of Conduct on Arms Exports*, Brussels, 7 Nov. 2001. The Code of Conduct is reproduced in *SIPRI Yearbook 1999: Armaments, Disarmament and International Security* (Oxford University Press: Oxford, 1999), pp. 503–505, and at URL <http://projects.sipri.se/expcon/eucode.htm>.

[87] Council of the European Union (note 86), p. 4.

The court case concerning a 1997 report on arms exports prepared for the EU Council of Ministers demonstrated the division within the EU on the extent of transparency. The Council was supported by Spain in an appeal against a judgement from the EU Court of First Instance ruling that the Council could not refuse a member of the European Parliament (MEP) access to the report. The Council argued that disclosure of sensitive information in the report would harm EU relations with other countries. The MEP was supported in her request for access by Denmark, Finland and the UK. In 2001 the European Court of Justice upheld the decision of the Court of First Instance and ruled that partial disclosure of the document must be considered.[88]

National transparency

Government and industry statistics on annual values of national arms exports are presented in appendix 8E. Compared to the major increase in arms export transparency in previous years there were few new developments in 2001, although the increased openness which has developed in Western Europe is spreading to Central Europe. By late 2001 the Polish Government had prepared a report on arms exports but its publication was delayed.[89] It was also announced that the Czech Republic intends to work towards compliance with EU requirements concerning transparency in arms exports.[90]

France, Germany and the UK provided details of the transfer of small arms and light weapons. In late 2001 the Ministry of Foreign Affairs of the Czech Republic also published a report detailing the quantities and types of imports and exports of small arms and light weapons.

In a report published in 2001 the US State Department failed to comply with the mandate of the US Security Assistance Act of 2000, namely, to provide details on arms delivered directly from the producer to the recipient foreign company.

As noted in the *SIPRI Yearbook 2001*, with more arms projects becoming multinational rather than national, especially in Europe, there is a risk that national transparency in transfers of arms and arms-related equipment will be reduced.[91] No government reports in detail about its involvement in multinational programmes.

---

[88] European Court of Justice, Judgement of the Court of Justice in Case C-353/99 P. The Court of Justice upholds the judgement of the Court of First Instance annulling the Council's decision to refuse Ms Hautala access to a report on arms exports, 6 Dec. 2001, URL <http://europa.eu.int/cj/en/cp/aff/cp0163en.htm>.

[89] Wyganowski, P., Polish Ministry of Foreign Affairs, Personal communication with the authors.

[90] Czech Ministry of Foreign Affairs, 'Report on the Czech Republic's approach to international negotiations concerning small arms and light weapons', 2001, URL<http://www.mzv.cz/_dokumenty/rucnialehkezbraneeng.pdf>.

[91] Hagelin, Wezeman, Wezeman and Chipperfield (note 52), p. 350.

VII. Conclusions

The five-year moving average level of global arms transfers fell in the period 1997–2001. The USA was the largest supplier in 1997–2001 despite a 65 per cent reduction in its arms deliveries since 1998. Russia was the second-largest supplier in the period. A 24 per cent increase in arms transfers from 2000 to 2001 made Russia the largest supplier in 2001.

China was by far the largest recipient in 2001 after an increase by 44 per cent from 2000. Imports by India increased by 50 per cent, making it the third-largest recipient in 2001. The other major recipients in the period 1997–2001 were Saudi Arabia, Taiwan and Turkey.

It is impossible for the arms supplier to control whether arms deliveries will stabilize or destabilize a particular bilateral relationship. The possibly destabilizing effect of arms transfers or acquisition plans is illustrated by India and Pakistan. Even relatively minor acquisitions, as illustrated by three countries in West Africa, may influence war-fighting and affect the acquisition behaviour of neighbouring countries. The United Nations continues to criticize the efficiency of arms embargoes.

Competition on the global arms market has strengthened new forms of marketing and transfer arrangements. Offset arrangements granted to the buyer may include military technology transfers in addition to the weapon system itself. Some arrangements involve transfers of military equipment from the buyer. In both cases offsets stimulate international military transfers.

If the JSF project is treated as an example of the future organization of transatlantic cooperation, the transatlantic market will remain unbalanced with regard to the involvement of government and industry on both sides of the Atlantic. Only the UK has been willing to participate fully and pay the cost of influencing JSF requirements. The cost of the highest form of participation in JSF development will remain too high for most European countries to afford.

Appendix 8A. The volume of transfers of major conventional weapons: by recipients and suppliers, 1997–2001

BJÖRN HAGELIN, PIETER D. WEZEMAN, SIEMON T. WEZEMAN and NICHOLAS CHIPPERFIELD

Table 8A.1. The recipients of major conventional weapons

The list includes all countries/non-state actors with imports of major conventional weapons in the period 1997–2001. Ranking is according to the 1997–2001 aggregate imports. Figures are trend-indicator values expressed in US $m. at constant (1990) prices. Figures may not add up because of the conventions of rounding.

Rank order								
1997–2001	1996–2000[a]	Recipient	1997	1998	1999	2000	2001	1997–2001
1	1	Taiwan	4 863	4 026	1 641	492	375	11 397
2	5	China	541	230	1 500	1 746	3 100	7 117
3	2	Saudi Arabia	2 783	2 507	1 215	69	143	6 717
4	3	Turkey	955	1 767	1 180	684	442	5 028
5	6	India	1 502	551	1 062	531	1 064	4 710
6	7	Greece	820	1 461	573	685	897	4 436
7	4	South Korea	718	941	1 131	740	401	3 931
8	8	Egypt	905	511	530	818	486	3 250
9	9	Japan	575	1 206	1 035	181	206	3 203
10	12	Pakistan	624	588	797	163	759	2 931
11	10	Israel	42	1 296	1 169	283	45	2 835
12	17	UK	74	379	98	882	1 247	2 680
13	13	UAE	678	748	420	278	288	2 412
14	11	Finland	393	558	799	513	10	2 273
15	15	Singapore	159	655	194	520	141	1 669
16	29	Australia	14	109	337	324	687	1 471
17	20	Malaysia	527	37	783	87	20	1 454
18	21	Brazil	449	163	205	40	597	1 454
19	19	Switzerland	388	448	499	27	33	1 395
20	18	Iran	232	287	234	279	335	1 367
21	16	Thailand	830	61	168	83	162	1 304
22	26	Italy	556	10	8	235	428	1 237
23	30	Algeria	35	103	428	175	365	1 106
24	34	Canada	103	21	40	411	470	1 045
25	25	Netherlands	137	259	299	188	153	1 036
26	23	USA	546	119	103	106	114	988
27	24	Norway	198	187	193	286	109	973
28	22	Spain	204	76	318	259	90	947
29	28	Qatar	491	338	97	8	8	942
30	14	Kuwait	438	204	110	133	34	919
31	36	Angola	3	183	350	111	255	902
32	32	Sweden	147	278	165	99	93	782

Table 8A.1 contd

Rank order 1997–2001	1996–2000[a]	Recipient	1997	1998	1999	2000	2001	1997–2001
33	45	Jordan	108	202	47	145	280	782
34	35	New Zealand	322	14	317	–	60	713
35	38	Argentina	88	120	199	184	97	688
36	39	Peru	351	28	114	–	178	671
37	33	Mexico	184	256	33	178	13	664
38	52	Bangladesh	26	–	193	222	180	621
39	44	Denmark	98	171	144	85	116	614
40	54	Colombia	164	99	37	71	222	593
41	31	Chile	122	90	194	169	16	591
42	27	Indonesia	97	95	193	164	38	587
43	40	Myanmar (Burma)	220	203	119	–	–	542
44	37	Viet Nam	101	171	152	8	74	506
45	50	Austria	192	206	48	25	15	486
46	49	Syria	–	20	20	420	–	460
47	48	Germany (FRG)	5	118	136	121	80	460
48	51	France	161	137	94	44	–	436
49	53	Ethiopia	62	194	75	95	–	426
50	42	Bahrain	74	8	–	314	30	426
51	46	Sri Lanka	2	69	44	254	40	409
52	43	Cyprus	113	21	242	2	15	393
53	47	Kazakhstan	163	–	62	113	31	369
54	55	Morocco	135	–	60	124	–	319
55	41	Oman	175	22	–	88	30	315
56	61	Venezuela	32	12	59	82	116	301
57	57	Eritrea	36	180	14	–	60	290
58	60	Belgium	60	67	60	30	33	250
59	62	Yemen	–	–	54	158	33	245
60	64	Romania	25	57	33	17	110	242
61	74	Macedonia	–	7	72	6	126	211
62	59	Poland	–	–	1	135	63	199
63	58	Congo (DRC)	18	–	72	108	–	198
64	63	North Korea	4	2	172	11	2	191
65	56	Hungary	70	31	50	14	14	179
66	82	Georgia	8	18	22	6	80	134
67	69	Philippines	45	51	–	–	13	109
68	71	Uganda	–	62	39	6	–	107
69	73	Botswana	67	4	2	–	32	105
70	67	Sudan	66	–	36	–	–	102
71	89	Croatia	41	–	–	–	59	100
72	65	Bosnia-Herzegovina	71	3	25	–	–	99
73	70	Ecuador	52	14	20	5	–	91
74	107	Afghanistan/NA[b]	–	–	–	14	68	82
75	75	Belarus	–	41	41	–	–	82
76	66	Tunisia	38	16	9	–	18	81
77	90	Slovenia	8	4	16	–	53	81
78	95	Ireland	–	2	30	–	46	78
79	84	NATO[c]	–	49	–	–	22	71

Rank order								
1997–2001	1996–2000[a]	Recipient	1997	1998	1999	2000	2001	1997–2001
80	76	Uruguay	19	30	9	4	–	62
81	78	Kenya	61	–	–	–	–	61
82	79	United Nations[c]	–	1	8	51	–	60
83	83	Bahamas	–	–	–	54	–	54
84	86	Czech Republic	5	–	2	18	27	52
85	110	Portugal	–	7	2	2	38	49
86	85	Lithuania	–	18	5	4	19	46
87	87	Rwanda	2	2	26	14	–	44
88	80	Brunei	30	7	5	–	1	43
89	91	Laos	14	20	–	7	–	41
90	77	South Africa	9	–	14	–	17	40
91	93	Estonia	14	2	1	22	–	39
92	97	Zimbabwe	–	1	24	4	7	36
93	117	Namibia	2	7	–	–	25	34
94	88	Nigeria	1	–	–	26	1	28
95	96	Zambia	–	–	–	27	–	27
96	112	Guinea	–	4	–	8	15	27
97	128	Latvia	–	–	4	–	22	26
98	81	Cambodia	22	–	–	–	–	22
99	103	Ghana	4	–	7	1	9	21
100	100	Congo	18	1	2	–	–	21
101	101	Papua New Guinea	20	–	–	–	–	20
102	99	Mauritania	8	3	–	9	–	20
103	92	Lebanon	6	5	4	4	1	20
104	102	Chad	–	7	–	13	–	20
105	113	Unknown[d]	–	3	5	–	9	17
106	105	Cameroon	6	9	–	–	1	16
107	106	Mali	7	3	–	5	–	15
108	104	Dominican Republic	–	–	2	13	–	15
109	108	Trinidad & Tobago	–	2	–	11	1	14
110	109	Sierra Leone	8	–	5	–	–	13
111	111	Panama	12	–	–	–	–	12
112	143	Nepal	–	–	–	–	10	10
113	114	Jamaica	–	5	5	–	–	10
114	115	Cape Verde	–	9	–	1	–	10
115	118	Liberia	–	–	1	8	–	9
116	119	Yugoslavia	8	–	–	–	–	8
117	120	Suriname	–	–	8	–	–	8
118	121	Togo	7	–	–	–	–	7
119	122	Malta	6	–	–	–	–	6
120	68	Bulgaria	–	–	6	–	–	6
121	141	Uzbekistan	–	–	–	–	5	5
122	127	Micronesia	4	–	–	–	–	4
123	124	El Salvador	–	3	–	–	–	3
124	129	Niger	3	–	–	–	–	3
125	130	Azerbaijan	–	–	–	3	–	3
126	131	Albania	–	3	–	–	–	3

Table 8A.1 contd

Rank order 1997–2001	Rank order 1996–2000[a]	Recipient	1997	1998	1999	2000	2001	1997–2001
127	146	Lesotho	–	–	–	–	2	2
128	72	Armenia	–	–	2	–	–	2
129	132	Lebanon/Hizbollah	1	–	–	–	–	1
130	134	Swaziland	–	–	–	1	–	1
131	135	Northern Cyprus	–	1	–	–	–	1
132	94	Mauritius	1	–	–	–	–	1
133	125	Luxembourg	–	–	1	–	–	1
134	136	Guatemala	–	1	–	–	–	1
135	148	Djibouti	–	–	–	–	1	1
136	137	Bolivia	–	–	–	1	–	1
137	138	Turkey/PKK[b]	–	–	–	–	–	–
138	139	Sri Lanka/LTTE[b]	–	–	–	–	–	–
139	140	Macedonia/NLA[b]	–	–	–	–	–	–
140	123	Lebanon/SLA[b]	–	–	–	–	–	–
141	142	Tonga	–	–	–	–	–	–
142	98	Slovakia	–	–	–	–	–	–
143	133	Paraguay	–	–	–	–	–	–
144	126	Palau	–	–	–	–	–	–
145	116	Palestinian AA[e]	–	–	–	–	–	–
146	144	Mozambique	–	–	–	–	–	–
147	145	Maldives	–	–	–	–	–	–
148	147	Cote d'Ivoire	–	–	–	–	–	–
149	149	Belize	–	–	–	–	–	–
		Total	24 832	23 325	21 179	15 165	16 231	100 732

[a] The rank order for recipients in 1996–2000 differs from that published in the *SIPRI Yearbook 2001* (pp. 353–56) because of the subsequent revision of figures for these years.

[b] Non-state actor: rebel group. SLA = South Lebanese Army; NA = Northern Alliance (UIFSA = United Islamic Front for the Salvation of Afghanistan); NLA = National Liberation Army; PKK = Kurdish Workers' Party; LTTE = Liberation Tigers of Tamil Eelam.

[c] Non-state actor: international organization.

[d] One or more unknown recipient(s).

[e] Non-state actor: Palestine Autonomous Authority.

Notes:
 – = between 0 and 0.5.

The SIPRI data on arms transfers refer to actual deliveries of major conventional weapons. To permit comparison between the data on such deliveries of different weapons and identification of general trends, SIPRI uses a *trend-indicator value*. The SIPRI values are only an indicator of the volume of international arms transfers and not of the actual financial values of such transfers. Thus they are not comparable to economic statistics such as gross domestic product or export/import figures.

Source: SIPRI arms transfers database.

Table 8A.2. The suppliers of major conventional weapons

The list includes all countries/non-state actors with exports of major conventional weapons in the period 1997–2001. Ranking is according to the 1997–2001 aggregate exports. Figures are trend-indicator values expressed in US $m. at constant (1990) prices. Figures may not add up because of the conventions of rounding.

Rank order								
1997–2001	1996–2000[a]	Supplier	1997	1998	1999	2000	2001	1997–2001
1	1	USA	11 277	12 930	9 957	6 095	4 562	44 821
2	2	Russia	2 837	1 885	3 874	3 779	4 979	17 354
3	3	France	2 963	3 340	1474	743	1 288	9 808
4	4	UK	2 441	1 040	990	1 103	1 125	6 699
5	5	Germany (FRG)	542	1 147	1 261	1 196	675	4 821
6	6	Ukraine	671	765	568	193	430	2 627
7	7	Netherlands	548	545	340	204	225	1 862
8	8	Italy	368	345	404	196	358	1 671
9	9	China	323	292	192	160	588	1 555
10	10	Belarus	401	57	474	253	333	1 518
11	13	Sweden	83	112	146	296	486	1 123
12	11	Israel	247	168	98	259	203	975
13	12	Spain	619	167	29	51	4	870
14	14	Canada	163	131	130	68	152	644
15	15	Australia	317	3	298	–	–	618
16	16	Slovakia	81	10	141	83	21	336
17	18	Moldova	316	–	–	3	5	324
18	17	Czech Republic	31	23	65	81	95	295
19	26	Norway	58	2	9	45	156	270
20	21	Bulgaria	4	48	163	4	4	223
21	19	Switzerland	66	31	41	44	36	218
22	30	South Korea	29	31	–	6	150	216
23	20	Belgium	89	23	28	2	72	214
24	22	Kazakhstan	–	2	180	16	9	207
25	23	Georgia	–	–	72	108	–	180
26	24	Poland	20	1	67	26	44	158
27	27	Singapore	78	42	–	1	–	121
28	32	Brazil	28	15	–	–	55	98
29	28	South Africa	9	28	17	22	20	96
30	33	Indonesia	10	–	60	–	20	90
31	29	Greece	52	21	1	–	11	85
32	31	Kuwait	–	82	–	–	–	82
33	41	Austria	5	12	2	2	61	82
34	25	Unknown[b]	16	1	2	52	8	79
35	34	Turkey	–	3	43	21	2	69
36	35	Qatar	37	–	9	–	–	46
37	63	Lebanon	–	–	–	–	45	45
38	42	Finland	1	8	13	9	3	34
39	37	UAE	33	–	–	–	–	33
40	39	Romania	8	2	19	3	–	32
41	43	Hungary	24	–	–	–	–	24
42	44	Denmark	–	–	–	18	–	18
43	45	India	–	–	–	16	1	17

Table 8A.2 contd

Rank order 1997–2001	1996–2000[a]	Supplier	1997	1998	1999	2000	2001	1997–2001
44	48	New Zealand	13	–	–	–	–	13
45	40	North Korea	–	13	–	–	–	13
46	50	Malaysia	–	–	8	–	–	8
47	51	Taiwan	5	–	–	–	–	5
48	52	Jordan	5	–	–	–	–	5
49	49	Egypt	5	–	–	–	–	5
50	59	Argentina	–	–	–	2	3	5
51	38	Chile	–	2	1	1	–	4
52	53	Pakistan	–	–	–	3	–	3
53	54	Libya	3	–	–	–	–	3
54	46	Japan	3	–	–	–	–	3
55	55	Yugoslavia	2	–	–	–	–	2
56	58	Croatia	–	–	–	2	–	2
57	65	Bahrain	–	–	–	–	2	2
58	60	Uruguay	–	–	–	1	–	1
59	61	Malawi	–	–	–	1	–	1
60	57	Iran	1	–	–	–	–	1
61	62	Syria	–	–	–	–	–	–
62	56	Saudi Arabia	–	–	–	–	–	–
63	64	Ireland	–	–	–	–	–	–
64	47	Estonia	–	–	–	–	–	–
65	36	Cyprus	–	–	–	–	–	–
		Total	24 832	23 325	21 179	15 165	16 231	100 732

[a] The rank order for suppliers in 1996–2000 differs from that published in the *SIPRI Yearbook 2001* (pp. 357–58) because of the subsequent revision of figures for these years.

[b] One or more unknown supplier(s).

Notes:

– = between 0 and 0.5.

The SIPRI data on arms transfers refer to actual deliveries of major conventional weapons. To permit comparison between the data on such deliveries of different weapons and identification of general trends, SIPRI uses a *trend-indicator value*. The SIPRI values are only an indicator of the volume of international arms transfers and not of the actual financial values of such transfers. Thus they are not comparable to economic statistics such as gross domestic product or export/import figures.

Source: SIPRI arms transfers database.

Appendix 8B. The volume of transfers of major conventional weapons: by regions and other groups of recipients and suppliers, 1992–2001

BJÖRN HAGELIN, PIETER D. WEZEMAN, SIEMON T. WEZEMAN and NICHOLAS CHIPPERFIELD

Table 8B.1. Volume of imports of major conventional weapons
Figures are SIPRI trend-indicator values expressed in US $m. at constant (1990) prices. Regional and group figures include transfers between countries/non-state actors in the same region or organization, unless otherwise noted. Figures may not add up because of the conventions of rounding.

	1992	1993	1994	1995	1996	1997	1998	1999	2000	2001
World total	20 216	21 868	19 045	19 272	20 291	24 832	23 325	21 179	15 165	16 231
Intl organizations	–	10	29	11	–	–	50	8	51	22
Africa	386	322	565	553	468	596	787	1 164	736	807
Sub-Saharan	310	196	259	122	256	387	669	668	437	425
Americas	1 082	1 242	1 867	1 401	1 524	2 121	962	1 027	1 330	1 824
North	537	721	1 031	514	473	649	139	143	517	584
Central	17	158	121	50	81	195	266	39	257	14
South	529	363	715	837	969	1 277	557	844	557	1 226
Asia	5 251	5 678	5 403	8 049	8 047	11 064	8 912	9 252	5 336	6 689
Central	–	–	24	99	170	163	–	62	113	36
North-East	3 809	3 602	2 811	4 855	5 261	7 632	6 637	5 799	3 261	4 320
South-East	271	780	1 266	1 720	1 153	1 114	1 068	1 294	778	213
South	1 172	1 294	1 304	1 376	1 462	2 154	1 208	2 096	1 270	2 053
Europe	6 325	5 175	4 462	3 013	3 409	3 802	4 570	3 988	3 710	3 976
Middle East	6 843	9 031	6 426	6 109	6 699	6 888	7 916	5 079	3 680	2 156
Oceania	316	392	291	129	141	359	123	654	324	747
Rebel groups	1	1	148	2	7	1	–	–	14	68
ASEAN	614	669	1 919	2 368	1 865	2 024	1 299	1 613	869	449
CIS	106	62	360	154	273	171	59	127	122	115
CIS Europe	106	62	337	54	103	8	59	65	9	80
EU	4 976	2 989	3 184	2 183	2 155	2 847	3 728	2 773	3 169	3 246
EU from non-EU	4 119	2 068	1 976	1 346	1 508	1 988	2 603	2 005	2 605	2 525
NATO	7 170	5 822	5 535	3 834	3 347	3 917	4 778	3 247	4 019	4 218
NATO Europe	6 634	5 102	4 506	3 319	2 874	3 268	4 639	5 348	3 159	3 737
GCC	2 423	3 844	1 765	2 268	3 975	4 639	3 827	1 842	890	533
OECD	9 566	8 178	7 078	5 148	6 541	6 925	8 827	7 693	6 139	5 822
OSCE/CSCE	8 202	7 844	6 791	4 948	5 091	5 498	6 472	5 348	5 024	5 038
P5	2 948	1 538	882	960	1 663	1 322	865	1 795	2 778	4 461
Wassenaar	10 352	9 921	7 938	7 115	6 827	6 854	8 749	7 900	6 297	6 080

Note: Tables 8B.1 and 8B.2 show the volume of arms transfers for different geographical regions and subregions, selected groups of countries, rebel groups and international organizations. Countries/rebel groups can belong to only one region. Since many countries are included in more than one group or organization, totals cannot be derived from these figures. Countries are included in the values for the different international organizations from the year of joining. The following countries/rebel groups are included in each region or group.

Table 8B.2. Volume of exports of major conventional weapons
Figures are SIPRI trend-indicator values expressed in US $m. at constant (1990) prices. Regional and group figures include transfers between countries/non-state actors in the same region or organization, unless otherwise noted. Figures may not add up because of the conventions of rounding.

	1992	1993	1994	1995	1996	1997	1998	1999	2000	2001
World total	20 216	21 868	19 045	19 272	20 291	24 832	23 325	21 179	15 165	16 231
Intl organizations	–	–	–	–	–	–	–	–	–	–
Africa	103	63	19	18	36	12	28	17	23	20
Sub-Saharan	103	63	11	18	36	9	28	17	23	20
Americas	12 363	11 891	9 726	9 634	9 425	11 468	13 077	10 088	6 168	4 771
North	12 215	11 827	9 684	9 586	9 364	11 440	13 061	10 087	6 163	4 714
Central	86	23	–	5	–	–	–	–	–	–
South	62	41	43	43	61	28	17	1	4	58
Asia	861	1 395	835	1 042	719	449	380	440	201	768
Central	–	–	–	85	12	..	2	..	16	9
North-East	852	1 370	801	925	707	360	336	192	166	738
South-East	8	22	31	30	–	88	42	68	1	20
South	..	3	3	2	–	–	–	–	19	1
Europe	6 764	8 275	8 160	8 373	9 784	12 227	9 583	10 181	8 445	10 411
Middle East	116	216	223	164	232	328	252	150	279	252
Oceania	8	28	24	20	14	330	3	298	–	–
Rebel groups	1	–	–	–	–	–	–	–	–	–
ASEAN	8	22	31	30	–	88	42	68	1	20
EU	3 430	4 270	5 584	4 381	5 627	7 709	6 759	4 688	3 822	4 310
EU to non-EU	2 455	3 192	4 342	3 548	4 981	6 851	5 635	3 922	3 258	3 599
CIS	2 530	3 176	1 569	3 477	3 678	4 225	2 709	5 168	4 351	5 755
CIS Europe	2 530	3 176	1 569	3 392	3 667	4 225	2 707	4 988	4 336	5 747
Framework	3 313	3 926	4 944	3 717	5 053	7 015	6 157	4 308	3 589	3 942
Framework to non-Framework	3 142	3 810	4 599	3 381	4 845	6 648	5 699	3 967	3 274	3 770
NATO	15 650	16 195	15 454	13 811	14 867	19 119	19 693	14 666	9 744	8 631
NATO Europe	3 436	4 370	5 770	4 224	5 502	7 680	6 633	4 711	3 686	4 055
OECD	16 145	16 492	15 706	14 131	15 314	19 690	19 913	15 231	10 182	9 462
OSCE/CSCE	18 979	20 102	17 844	18 043	19 160	23 668	22 649	20 491	14 645	15 136
P5	16 493	17 719	13 227	14 930	16 222	29 841	19 487	16 487	11 880	12 542
Wassenaar	18 746	20 174	17 783	17 927	19 024	23 311	22 623	20 062	14 271	14 943

International organizations: NATO and the United Nations as non-state actors—not as combinations of all member states.

Africa: Algeria, Angola, Benin, Botswana, Burkina Faso, Burundi, Cameroon, Cape Verde, Central African Republic, Chad, Comoros, Congo (Rep. of), Congo (DRC), Côte d'Ivoire, Djibouti, Equatorial Guinea, Eritrea, Ethiopia, Gabon, Gambia, Ghana, Guinea, Guinea-Bissau, Kenya, Lesotho, Liberia, Libya, Madagascar, Malawi, Mali, Mauritania, Mauritius, Morocco, Mozambique, Namibia, Niger, Nigeria, Rwanda, Sao Tomé and Principe, Senegal, Seychelles, Sierra Leone, Somalia, South Africa, Sudan, Swaziland, Tanzania, Togo, Tunisia, Uganda, Zambia, Zimbabwe

Sub-Saharan Africa: Angola, Benin, Botswana, Burkina Faso, Burundi, Cameroon, Cape Verde, Central African Republic, Chad, Comoros, Congo (Rep. of), Congo (DRC), Côte d'Ivoire, Djibouti, Equatorial Guinea, Eritrea, Ethiopia, Gabon, Gambia, Ghana, Guinea, Guinea-Bissau, Kenya, Lesotho, Liberia, Madagascar, Malawi, Mali, Mauritania, Mauritius, Mozambique, Namibia, Niger, Nigeria, Rwanda, Sao Tomé and Principe, Senegal, Seychelles, Sierra Leone, Somalia, South Africa, Sudan, Swaziland, Tanzania, Togo, Uganda, Zambia, Zimbabwe

Americas: Argentina, Bahamas, Barbados, Belize, Bolivia, Brazil, Canada, Chile, Colombia, Costa Rica, Cuba, Dominica, Dominican Republic, Ecuador, El Salvador, Grenada, Guatemala, Guyana, Haiti,

Honduras, Jamaica, Mexico, Nicaragua, Panama, Paraguay, Peru, St Vincent and the Grenadines, Suriname, Trinidad and Tobago, Uruguay, USA, Venezuela

North America: Canada, USA

Central America: Bahamas, Barbados, Belize, Costa Rica, Cuba, Dominica, Dominican Republic, El Salvador, Grenada, Guatemala, Haiti, Honduras, Jamaica, Mexico, Nicaragua, Panama, St Vincent and the Grenadines, Trinidad and Tobago

South America: Argentina, Bolivia, Brazil, Chile, Colombia, Ecuador, Guyana, Paraguay, Peru, Suriname, Uruguay, Venezuela

Asia: Afghanistan, Bangladesh, Bhutan, Brunei, Cambodia, China, India, Indonesia, Japan, Kazakhstan, North Korea, South Korea, Kyrgyzstan, Laos, Malaysia, Maldives, Mongolia, Myanmar (Burma), Nepal, Pakistan, Philippines, Singapore, Sri Lanka, Taiwan, Tajikistan, Thailand, Turkmenistan, Uzbekistan, Viet Nam, Khmer Rouge (Cambodia), Liberation Tigers of Tamil Eelam (LTTE, Sri Lanka), Mujahideen (Afghanistan), United Islamic Front for the Salvation of Afghanistan (UIFSA, Afghanistan)

Central Asia: Kazakhstan, Kyrgyzstan, Tajikistan, Turkmenistan, Uzbekistan

North-East Asia: China, Japan, North Korea, South Korea, Taiwan

South-East Asia: Brunei, Cambodia, Indonesia, Laos, Malaysia, Myanmar, Philippines, Singapore, Thailand, Viet Nam

South Asia: Bangladesh, Bhutan, India, Maldives, Nepal, Pakistan, Sri Lanka, Liberation Tigers of Tamil Eelam (LTTE, Sri Lanka)

Europe: Albania, Armenia, Austria, Azerbaijan, Belarus, Belgium, Bosnia and Herzegovina, Bulgaria, Croatia, Cyprus, Czechoslovakia (1992–), Czech Republic (1993–), Denmark, Estonia, Finland, France, Georgia, Germany, Greece, Hungary, Iceland, Ireland, Italy, Latvia, Liechtenstein, Lithuania, Luxembourg, Macedonia, Malta, Moldova, Monaco, Netherlands, Norway, Poland, Portugal, Romania, Russia, Slovakia (1993–), Slovenia, Spain, Sweden, Switzerland, UK, Ukraine, Yugoslavia (FRY)

Middle East: Bahrain, Egypt, Iran, Iraq, Israel, Jordan, Kuwait, Lebanon, Oman, Palestinian Autonomous Authority), Qatar, Saudi Arabia, Syria, United Arab Emirates, Turkey, Yemen, Hizbollah (Lebanon), Kurdish Workers' Party (PKK, Turkey), Lebanese Forces (LF, Lebanon), South Lebanese Army (SLA, Lebanon), Southern Rebels (Yemen)

Oceania: Australia, Fiji, Kiribati, Marshall Islands, Micronesia, New Zealand, Palau, Papua New Guinea, Samoa, Solomon Islands, Tonga, Tuvalu, Vanuatu

Rebel groups (only those rebel groups which had imports/exports in the period 1992–2001 are listed): Hizbollah (Lebanon), Kurdish Workers' Party (PKK, Turkey), Lebanese Forces (LF, Lebanon), Liberation Tigers of Tamil Eelam (LTTE, Sri Lanka), Northern Alliance (Afghanistan), National Liberation Army (NLA, Macedonia), United Islamic Front for the Salvation of Afghanistan (UIFSA, Afghanistan), South Lebanese Army (SLA, Lebanon), Southern Rebels (Yemen)

Association of South-East Asian Nations (ASEAN): Brunei, Indonesia, Laos (1997–), Malaysia, Myanmar (Burma, 1997–), Philippines, Singapore, Thailand, Viet Nam (1995–)

European Union (EU): Austria (1995–), Belgium, Denmark, Finland (1995–), France, Germany, Greece, Ireland, Italy, Luxembourg, Netherlands, Portugal, Spain, Sweden (1995–), UK

Commonwealth of Independent States (CIS): Armenia, Azerbaijan, Belarus, Georgia (1993–), Kazakhstan, Kyrgyzstan, Moldova, Russia, Tajikistan, Turkmenistan, Ukraine, Uzbekistan

Commonwealth of Independent States (CIS) Europe: Armenia, Azerbaijan, Belarus, Georgia (1993–), Moldova, Russia, Ukraine

Framework (Framework Agreement Concerning Measures to Facilitate the Restructuring and Operation of the European Defence Industry): France, Germany, Italy, Spain, Sweden, UK

GCC (Gulf Co-operation Council): Bahrain, Kuwait, Oman, Qatar, Saudi Arabia, United Arab Emirates

NATO: Belgium, Canada, Czech Republic (1999–), Denmark, France, Germany, Greece, Hungary (1999–), Iceland, Italy, Luxembourg, Netherlands, Norway, Poland (1999–), Portugal, Spain, Turkey, UK, USA

412 MILITARY SPENDING AND ARMAMENTS, 2001

NATO Europe: Belgium, Czech Republic (1999–), Denmark, France, Germany, Greece, Hungary (1999–), Iceland, Italy, Luxembourg, Netherlands, Norway, Poland (1999–), Portugal, Spain, Turkey, UK

Organization for Security and Co-operation in Europe (OSCE)/Conference on Security and Co-operation in Europe (CSCE): Albania, Andorra, Armenia, Austria, Azerbaijan, Belarus, Belgium, Bosnia and Herzegovina, Bulgaria, Canada, Croatia, Cyprus, Czechoslovakia (1992–), Czech Republic (1993–), Denmark, Estonia, Finland, France, Georgia, Germany, Greece, Holy See, Hungary, Iceland, Ireland, Italy, Kazakhstan, Kyrgyzstan, Latvia, Liechtenstein, Lithuania, Luxembourg, Macedonia (1995–), Malta, Moldova, Monaco, Netherlands, Norway, Poland, Portugal, Romania, Russia, San Marino, Slovakia (1993–), Slovenia, Spain, Sweden, Switzerland, Tajikistan, Turkey, Turkmenistan, UK, Ukraine, USA, Uzbekistan, Yugoslavia (FRY)

Organisation for Economic Co-operation and Development (OECD): Australia, Austria, Belgium, Canada, Czech Rep. (1995–), Denmark, Finland, France, Germany, Greece, Hungary (1996–), Iceland, Ireland, Italy, Japan, South Korea (1996–), Luxembourg, Mexico (1994–), Netherlands, New Zealand, Norway, Poland (1996–), Portugal, Spain, Sweden, Switzerland, Turkey, UK, USA

P5 (5 Permanent members of the UN Security Council): China, France, Russia, UK, USA

Wassenaar Arrangement: Argentina, Australia, Austria, Belgium, Bulgaria, Canada, Czech Republic, Denmark, Finland, France, Germany, Greece, Hungary, Iceland, Ireland, Italy, Japan, Luxembourg, Netherlands, New Zealand, Norway, Poland, Portugal, Romania, Russia, Slovakia, South Korea, Spain, Sweden, Switzerland, Turkey, UK, USA, Ukraine

Appendix 8C. Register of the transfers and licensed production of major conventional weapons, 2001

BJÖRN HAGELIN, PIETER D. WEZEMAN, SIEMON T. WEZEMAN and NICHOLAS CHIPPERFIELD

The register in table 8C.1 lists major weapons on order or under delivery, or for which the licence was bought and production was under way or completed during 2001. Sources and methods for the data collection are explained in appendix 8D. Entries in table 8C.1 are alphabetical, by recipient, supplier and licenser. 'Year(s) of deliveries' includes aggregates of all deliveries and licensed production since the beginning of the contract. 'Deal worth' values in the comments refer to real monetary values as reported in sources and not to SIPRI trend-indicator values. Conventions, abbreviations and acronyms are explained below the table. For cross reference, an index of recipients and licensees for each supplier can be found in table 8C.2.

Table 8C.1. Register of transfers and licensed production of major conventional weapons, 2001, by recipients

Recipient/ supplier (S) or licenser (L)	No. ordered	Weapon designation	Weapon description	Year of order/ licence	Year(s) of deliveries	No. delivered/ produced	Comments
Afghanistan/Northern Alliance							
S: Russia	(10)	Mi-17/Hip-H	Helicopter	2000	2000–2001	(10)	Probably second-hand; possibly financed by India
	(10)	BM-21	MRL	(2001)	2001	(10)	Ex-Russian; aid
	(27)	BMP-1	IFV	(2001)	2001	(27)	Ex-Russian; aid
	(26)	BMP-2	IFV	(2001)	2001	(26)	Ex-Russian; aid
	(30)	BTR-60PB	APC	(2001)	2001	(30)	Ex-Russian; aid
	(20)	T-55M	Main battle tank	(2001)	2001	(20)	Ex-Russian; aid; no. delivered could be 60
	(20)	T-62	Main battle tank	(2001)	2001	(20)	Ex-Russian; aid; no. delivered could be 40
	(270)	AT-3 Sagger/9M14M	Anti-tank missile	(2001)	2001	(270)	Ex-Russian; aid; for BMP-1 IFVs
	(260)	AT-4 Spigot/9M111	Anti-tank missile	(2001)	2001	(260)	Ex-Russian; aid; for BMP-2 IFVs
Algeria							
S: Belarus	(28)	MiG-29S/Fulcrum-C	FGA aircraft	(1998)	1999–2001	(28)	Ex-Belarussian; part of deal incl 8 MiG-29UB trainer version delivered from Russia via Belarus
China	(24)	C-802/CSS-N-8 Saccade	Anti-ship missile	(1999)	2000–2001	(16)	For 3 Djebel Chinoise Class FAC
Czech Republic	17	L-39Z Albatros	Jet trainer aircraft	2001		. .	Deal worth $30 m; L-39ZA version; originally produced for Nigeria but never delivered for financial reasons; status uncertain
Russia	8	MiG-29/Fulcrum-A	Fighter aircraft	(1999)	2000–2001	(8)	Ex-Russian; MiG-29UB trainer version; ordered and delivered via Belarus; part of deal incl 28 MiG-29 delivered from Belarus

Recipient/ supplier (S) or licenser (L)	No. ordered	Weapon designation	Weapon description	Year of order/ licence	Year(s) of deliveries	No. delivered/ produced	Comments
	(3)	Su-24MK/Fencer-D	Bomber aircraft	1999	2001	(3)	Ex-Russian; probably modernized before delivery
	22	Su-24MK/Fencer-D	Bomber aircraft	2000	2001	10	Ex-Russian; possibly modernized before delivery; delivery 2001–2002
	(24)	TEST-71	AS/ASW torpedo	(1997)	2000–2001	(8)	For 3 modernized Koni Class frigates; designation uncertain
	3	Drum Tilt	Fire control radar	(1997)	2000	1	For modernization of 3 Koni (Mourad Raïs) Class frigates
	6	Pozitiv-ME1.2	Surveillance radar	(1997)	2000	3	For modernization of 3 Koni (Mourad Raïs) Class frigates and 3 Nanuchka (Hamidou) Class corvettes
South Africa	3	Plank Shave	Surveillance radar	(1997)	2000	1	For modernization of 3 Nanuchka (Hamidou) Class corvettes
USA	(660)	Ingwe	Anti-tank missile	(1998)	2000–2001	(660)	For 33 Mi-24 helicopters modernized in South Africa
	6	Beech-1900/C-12J	Light transport ac	2000	2001	(6)	
	6	Beech-1900D HISAR	AGS aircraft	2000	2001	(2)	

Angola

S: Belarus	20	Su-24MK/Fencer-D	Bomber aircraft	1997	2000–2001	(20)	Ex-Belarussian; may incl some from Russia
	(15)	Su-27S/Flanker-B	Fighter aircraft	(1999)		..	Probably ex-Belarussian; supplier could be Ukraine; status uncertain
Russia	(3)	Mi-24P/Mi-35P/Hind-F	Combat helicopter	(2000)	2000–2001	(3)	Probably ex-Russian; possibly modernized before delivery

Argentina

S: Italy	(16)	RAT-31S/L	Surveillance radar	1999	2000–2001	(8)	Part of $185 m deal with US company for civilian/military 'NRP' air surveillance system; delivery probably 2000–2002
Netherlands	6	DA-05	Surveillance radar	(1979)	1985–2000	(5)	For 6 MEKO-140 Type (Espora Class) frigates
	6	WM-28	Fire control radar	(1979)	1985–2000	(5)	For 6 MEKO-140 Type (Espora Class) frigates
USA	16	Bell-205/UH-1H	Helicopter	1996	1997–2001	(16)	Ex-US; aid
	12	Bell-209/AH-1F	Combat helicopter	(2000)	2001	(2)	Ex-US; delivery 2001–2002
	2	Hughes-300/TH-55	Light helicopter	(2001)	2001	2	Schweizer-300C version; for Coast Guard
	1	SA-315B Lama	Light helicopter	2001	2001	1	Second-hand; incl for SAR
	(425)	BGM-71 TOW	Anti-tank missile	2000	2001	(425)	Probably BGM-71E TOW-2A version
	(44)	Mk-48 533mm	AS/ASW torpedo	(1999)	2001	(11)	For 2 modernized Santa Cruz Class (TR-1700 Type) submarines
L: Germany (FRG)	6	MEKO-140 Type	Frigate	1979	1985–2000	5	Argentine designation Espora Class; delivery of last 2 delayed some 10 years for lack of funding until 2000–2002

Australia

S: Canada	68	Piranha 8x8	APC	1998	2000–2001	(22)	Deal worth $180–210 m incl licensed production of 82 IFV version; incl 5 ambulance, 5 ARV, 16 CP, 18 radar reconnaissance and 11 repair version;

INTERNATIONAL ARMS TRANSFERS 415

Supplier	No. ordered	Weapon designation	Weapon description	Year of order	Year of delivery	Comments
Israel	(18)	EL/M-2022	MP aircraft radar	1995	2001	Australian designations ASLAV-PC/R/A/C/S/F; assembled in Australia; delivery 2000–2005
	2					Part of 'Project Air-5276' worth $372 m for modernization of 18 P-3C ASW/MP aircraft to AP-3C Sea Sentinal version by US company; delivery of aircraft 2001–2004
Multiple sellers	22	AS-665 Tiger	Combat helicopter	2001		'Project Air-87' worth $670 m (offsets incl production of components and production of EC-120 helicopter for Asian market); bought from France and FRG; Aussie Tiger version; incl 21 assembled in Australia; delivery 2004–2007
Norway	(60)	Penguin Mk-2-7	Anti-ship missile	1998	2001	'Project Sea-1414 Phase-1' worth $40–49 m (offsets incl production of warheads in Australia); for SH-2G helicopters; delivery 2001–2002
	(60)	Penguin Mk-2-7	Anti-ship missile	1999		'Project Sea-1414 Phase-2' worth $36–49 m; for SH-2G helicopters; delivery 2002–2003
Sweden	8	9LV	Fire control radar	(1991)	1996–2001	(30) 9LV453 version for 8 MEKO-200ANZ Type (Anzac Class) frigates; incl for use with Seasparrow SAM
	8	Sea Giraffe-150	Surveillance radar	1991	1996–2001	(3) For 8 MEKO-200ANZ Type (Anzac Class) frigates
UK	12	Hawk-100	FGA/trainer aircraft	1997	2000–2001	(3) Deal worth $494 m incl 21 licensed production; Hawk-127 version
	6	MSTAR	Battlefield radar	1999		(12) 'NINOX' programme worth $32 m incl 55 licensed production; Australian designation Amstar; delivery from 2002
	..	Rapier	SAM system	2001		Ex-UK; 'Project Land-140' worth $8 m; modernized before delivery; Rapier B1M version
USA	(400)	ASRAAM	AAM	1998	2000–2001	Deal worth $62 m; for F/A-18 FGA aircraft; delivery 2000–2002
	4	Boeing-737 AEW	AEW&C aircraft	2000		(219) 'Wedgetail' programme worth $1.8 b; Australian designation A-30; option on 2 more; delivery from 2006–2007
	11	SH-2G Super Seasprite	AS/ASW helicopter	1997	2001	3 Ex-US SH-2F rebuilt to SH-2G(A) version; 'Project Sea 1411' worth $421 m (incl $67 m for 10-year support); incl some assembly in Australia; delivery 2001–2002
	8	Mk-45 127mm/54	Naval gun	(1989)	1996–2001	(3) Mk-45 Mod-2 version; for 8 MEKO-200ANZ Type (Anzac Class) frigates
	8	AN/SPS-49	Surveillance radar	1993	1996–2001	(3) AN/SPS-49V(8) version; for 8 MEKO-200ANZ Type (Anzac Class) frigates
	(64)	RGM-84 Harpoon	Anti-ship missile	(2001)		For 8 MEKO-200ANZ Type (Anzac Class) frigates; contract not yet signed
	(672)	RIM-7PTC ESSM	SAM	(2001)		For 6 new and 2 modernized MEKO-200ANZ Type (Anzac Class) and for 6 modernized Adelaide (Perry) Class frigates; delivery probably from 2003; contract probably not yet signed
	71	AN/APG-73	Combat ac radar	2000		For modernization of 71 F/A-18 FGA aircraft; delivery 2002–2005
	4	AN/TPS-117	Surveillance radar	1998	2000–2001	(4) Deal worth $68–90 m; assembled in Australia
	(400)	AIM-120B AMRAAM	AAM	(2000)	2001	(100) 'Project Air-5400'
	51	Popeye-1	ASM	1998		'Project Air-5398' worth $90 m; for F-111C/G bomber aircraft; status uncertain
L: Canada	82	Piranha/LAV-25	IFV	1998		Deal worth $180–210 m incl direct delivery of 68 APC version; Australian designation ASLAV-25; delivery 2002–2005

416 MILITARY SPENDING AND ARMAMENTS, 2001

Recipient/ supplier (S) or licenser (L)	No. ordered	Weapon designation	Weapon description	Year of order/ licence	Year(s) of deliveries	No. delivered/ produced	Comments
Germany (FRG)	8	MEKO-200ANZ Type	Frigate	1989	1996–2001	3	Australian designation Anzac Class; delivery 1996–2006; part of deal incl 2 produced in Australia for New Zealand
Italy	6	Gaeta Class	MCM ship	1994	1999–2001	4	Deal worth $523–636 m; Australian designation Huon Class; delivery 1999–2002
Sweden	6	Type-471	Submarine	1987	1996–2001	6	Deal worth $2.8 b; Australian designation Collins Class
UK	21	Hawk-100	FGA/trainer aircraft	1997	2001	21	Deal worth $494 m incl 12 delivered direct; Hawk-127 version
	55	MSTAR	Battlefield radar	1999		. .	'NINOX' programme worth $32 m incl 6 delivered direct; Australian designation Amstar; delivery from 2002
Austria							
S: Sweden	(1 700)	RBS-56 Bill-2	Anti-tank missile	1996	1998–2001	(1 700)	Austrian designation PAL-2000
USA	9	S-70A/UH-60L	Helicopter	2000		. .	Deal worth $183 m (incl $47 import duties; offsets worth $394 m); incl for SAR; S-70A-42 version; option on 3 more
Bahrain							
S: UK	1	BAe-146	Transport aircraft	2001	2001	1	Deal worth $25 m; Avro RJ-100 version
USA	(10)	AGM-65D Maverick	ASM	1999	2001	(10)	
	26	AIM-120B AMRAAM	AAM	1999		. .	Deal worth $110 m; delivery probably 2002/2003
	(153)	BGM-71F TOW-2B	Anti-tank missile	2000	2001	(153)	
	30	MGM-140A ATACMS	SSM	2000		. .	Deal worth $20 m; delivery 2002
Bangladesh							
S: Italy	(8)	Otomat Mk-2	Anti-ship missile	(1998)	2001	(8)	For 1 DW-2000H Type (Bangabandhu Class) frigate
Netherlands	1	DA-08	Surveillance radar	(1999)	2001	1	For 1 DW-2000H Type (Bangabandhu Class) frigate delivered from South Korea
	1	LIROD	Fire control radar	(1999)	2001	1	For 1 DW-2000H Type (Bangabandhu Class) frigate delivered from South Korea
	1	Variant	Surveillance radar	(1999)	2001	1	For 1 DW-2000H Type (Bangabandhu Class) frigate delivered from South Korea
South Korea	1	DW-2000H Type	Frigate	1998	2001	1	Deal worth $100 m; Bangladeshi designation Bangabandhu Class
Belgium							
S: France	60	Box Mortar	Mortar	(2000)	2000–2001	(60)	
Israel	18	B-Hunter UAV	UAV	1998	2001	(6)	Deal worth $72.7 m incl 3 control systems; delivery 2001–2002

INTERNATIONAL ARMS TRANSFERS 417

USA	...	surveillance radar	(1999)	2001	1	Ex-UK
	(92)	Combat ac radar	1993	1996–2001	(62)	For 'Mid-Life Update' (MLU) modernization of 92 F-16A/B FGA aircraft to F-16AM/BM (F-16C/D) version
	AN/APG-66					
	BGM-71 TOW	Anti-tank missile	2001			BGM-71E TOW-2A version; delivery 2002–2003
	(562)					
Bosnia and Herzegovina						
S: Unknown	5	Mi-24V/Mi-35/Hind-E	1998		..	Second-hand; aid; possibly in storage in Turkey
Botswana						
S: Austria	2	4K-7FA-G-127	2000	2001	(2)	Deal worth ASH500 m incl 20 SK-105A1 tank destroyers; CP version
	20	SK-105A1 Kurassier	1997	2001	(20)	Deal worth ASH500 m incl 2 4K-7FA APC/CPs; option on 30 more not used
France	(80)	Eryx	2000		..	Designation uncertain
USA	1	C-130B Hercules	(1999)		..	Ex-US; modernized before delivery
Brazil						
S: Austria	1	4K-4FA-SB20 Greif	2000	2001	1	
	17	SK-105A1 Kurassier	2000	2001	17	Ex-Belgian; modernized before delivery
Belgium	37	M-109A3 155mm	1999	2001	(37)	
Canada	3	IRIS	1997	2001	(2)	Deal worth $143 m; part of 'SIVAM' air-surveillance network; for 3 ERJ-145RS/R-99B AEW aircraft produced in Brazil; delivery of aircraft 2001–2002
France	8	AS-532U2/AS-332L2	2001		..	Deal worth $160 m; assembled in Brazil
	8	AS-532U2/AS-332L2	2000	2001	8	Deal worth $91 m
	(8)	MM-40 Exocet	(1995)		..	For 1 Barroso Class frigate
	1	Clemenceau Class	2000	2001	1	Ex-French; deal worth $10.5 m; Brazilian designation Sao Paulo Class
Italy	54	Grifo	(2000)		..	For modernization of AMX (A-1A) FGA aircraft
	48	Grifo	2000		..	For modernization of 48 F-5E/F FGA aircraft to F-5BR version by Israeli company
	7	RAN-20S	1995	2001	(1)	Deal worth $111.5 m incl 13 RTN-30X radars and 6 Albatros SAM systems; for 1 Barroso Class frigate produced in Brazil and modernization of 6 Niteroi Class frigates; delivery of ships 2001–2006
	13	RTN-30X	1995	2001	(2)	Deal worth $111.5 m incl 7 RAN-20S radars and 6 Albatros SAM systems; for 1 Barroso Class frigate produced in Brazil and modernization of 6 Niteroi Class frigates; delivery of ships 2001–2006
	(144)	Aspide Mk-1	1996	2001	(36)	Deal worth $48.5 m; for 6 modernized Niteroi Class frigates; delivery of ships 2001–2004
Sweden	5	PS-890 Erieye	1997	2001	(2)	Deal worth $143 m; part of 'SIVAM' programme; for 5 ERJ-145AEW&C/R-99A AEW aircraft produced in Brazil; delivery of aircraft 2001–2002

Recipient/ supplier (S) or licenser (L)	No. ordered	Weapon designation	Weapon description	Year of order/ licence	Year(s) of deliveries	No. delivered/ produced	Comments
		Torpedo-2000	AS/ASW torpedo	1999		. .	Deal worth $59.7 m; for Type-209/1400 (Tikuna and possibly Tupi Class) submarines
UK	(12)	L-118 105mm	Towed gun	(2000)	2001	(12)	
USA	10	C-130H Hercules	Transport aircraft	2001	2001	6	Ex-Italian aircraft sold back to US producer; 'C-X' programme worth $66–70 m; delivery 2001–2002
	2	Metro-3	Transport aircraft	(1999)	2001	(2)	Ex-US; modernized before delivery
	6	AN/TPS-117	Surveillance radar	1997	1999–2001	(6)	For 'SIVAM' air-surveilance network
	20	RGM-84 Harpoon	Anti-ship missile	(1999)		. .	Deal worth $39 m; could be AGM-84 version
L: Germany (FRG)	1	Type-209/1400	Submarine	1995		. .	Brazilian designation Tikuna Class; 1 more planned but cancelled; delivery 2006

Brunei

Recipient/ supplier (S) or licenser (L)	No. ordered	Weapon designation	Weapon description	Year of order/ licence	Year(s) of deliveries	No. delivered/ produced	Comments
S: France	(36)	MM-40 Exocet	Anti-ship missile	2000		. .	MM-40 Block-2 version; for 3 Yarrow-95m Type frigates
	8	Mistral	Portable SAM	2000	2001	(8)	
Indonesia	3	CN-235MPA	MP aircraft	(1995)		. .	Status uncertain
UK	3	Yarrow-95m Type	Frigate	1998		. .	Bruneian designation Nakhada Ragam Class; delivery from 2002/2003
	(96)	Seawolf VL	SAM	(2001)		. .	For 3 Yarrow-95m Type frigates

Cameroon

Recipient/ supplier (S) or licenser (L)	No. ordered	Weapon designation	Weapon description	Year of order/ licence	Year(s) of deliveries	No. delivered/ produced	Comments
S: France	6	Tetras	Light aircraft	(2000)	2001	6	

Canada

Recipient/ supplier (S) or licenser (L)	No. ordered	Weapon designation	Weapon description	Year of order/ licence	Year(s) of deliveries	No. delivered/ produced	Comments
S: Germany (FRG)	121	Leopard-1A5 turret	Turret	1996	1999–2001	(121)	Ex-FRG; deal worth $89–105 m; modernized before delivery; for modernization of 114 Canadian Leopard-1 (Leopard C1) tanks to Leopard C2 version; incl 5 for training and 2 as spare turret; 2 more delivered for testing only
Italy	15	EH-101-500	Helicopter	1998	2001	(6)	Deal worth $404–500 m (offsets 110%); for SAR; Canadian designation CH-149 or AW-520 Cormorant; delivery 2001–2002
UK	18	Hawk-100	FGA/trainer aircraft	1997	2000–2001	(18)	Deal worth $574 m; for civilian company for training of pilots from Canadian and other air forces under 'NATO Flying Training in Canada' (NFTC) programme; Hawk Mk-115 version; Canadian designation CT-155

INTERNATIONAL ARMS TRANSFERS 419

Supplier/Recipient	No. ordered	Weapon designation	Weapon description	Year of order	Year(s) of delivery	Comments
			FGA/trainer aircraft	2000		For civilian company for training of pilots from Canadian and other air forces under 'NATO Flying Training in Canada' (NFTC) programme; option on 2 more; Hawk Mk-115 version; Canadian designation CT-155
	4	Upholder Class	Submarine	1998	2000–2001	Ex-UK; lease worth $504 m; in exchange for 8-year UK use of Canadian bases for training; Canadian designation Victoria Class; delivery 2000–2002
USA	47	MSTAR	Battlefield radar	2001		Deal worth $13 m; for 47 Piranha-3 APC (artillery fire conrol version); delivery 2002–2003
	..	AGM-65G Maverick	ASM	(1999)		
	10	RIM-66M Standard-2	SAM	1998	2000–2001	
	..	RIM-7PTC ESSM	SAM	2001		For 12 modernized Halifax Class frigates; delivery 2003–2010
L: Switzerland	120	Piranha/LAV-25	IFV	1999	2001	Deal worth $169 m; Canadian designation Kodiak
	171	Piranha/LAV-25	IFV	(1999)		Incl 71 tank-destroyer, 39 AEV and 47 artillery fire control version; Canadian designation Kodiak; delivery 2002–2003
Chile						
S: Bahrain	1	Bell-412	Helicopter	2000	2001	Second-hand
France	1	Scorpene Class	Submarine	1997		'Neptune' programme worth $400 m incl 1 from Spain; Chilean designation Hyatt Class; delivery 2004
Italy	(114)	M-113A2	APC	(1996)		Ex-Italian; deal incl also 14 M-548A1 vehicles
Netherlands	6	BrPz-1 Biber	ABL	(2001)		Ex-Dutch; modernized before delivery; incl 3 bridge-carrying version
	2	Leopard-1 chassis	Tank chassis	2001		Ex-Dutch; Leopard-1 tanks modified to mine-clearer/AEV before delivery
	3	PiPz-1	AEV	2001		Ex-Dutch; modernized before delivery
Spain	2	Bell-412	Helicopter	2000	2001	Second-hand
	1	Scorpene Class	Submarine	1997		'Neptune' programme worth $400 m incl 1 from France; Chilean designation Hyatt Class; delivery 2006
USA	(1)	Bell-412	Helicopter	(2000)	2001	Second-hand
	3	Cessna-525 Citation	Light transport ac	(2000)	2001	For training
	10	F-16C	FGA aircraft	(2001)		'Caza-2000' or 'F-2000' programme worth $637 m; F-16C/D Block-50/52 version; contract not yet signed; delivery from 2005–2006
L: Germany (FRG)	4	MEKO-200 Type	Frigate	2000		'Tridente' programme; MEKO-200ACH version; delivery 2006–2009
China						
S: France	(14)	Castor-2	Fire control radar	(1986)	1994–2000	For 2 Luhu Class (Type-052), 2 Luhai Class and modernization of 2 Luda-1 Class (Type-051) destroyers and for 8 Jiangwei-2 Class frigates; probably assembled in China; for use with Crotale EDIR (Chinese designation HQ-7) SAM system

Recipient/ supplier (S) or licenser (L)	No. ordered	Weapon designation	Weapon description	Year of order/ licence	Year(s) of deliveries	No. delivered/ produced	Comments
	(336)	R-440N Crotale	SAM	1986	1990–2001	(270)	For 2 Luhu Class (Type-052), 2 Luhai Class and 2 modernized Luda-1 Class (Type-051) destroyers and for 8 Jiangwei-2 Class frigates; possibly assembled or produced in China; US/NATO designation CSA-4
Italy	(18)	RTN-20S	Fire control radar	(1985)	1991–99	(15)	For 2 Luhu, 1 Luda-3 and 2 Luhai Class destroyers and 7 or 8 Houjian Class FAC; Chinese designation Type-347G; more produced for export
Russia	(4)	A-50U Mainstay	AEW&C aircraft	(2001)			No. ordered could be up to 5; contract not yet signed; delivery possibly from 2002
	28	Su-27SK/Flanker-B	FGA aircraft	1999	2000–2001	18	Deal worth $1 b; payment for Russian debt to China; Su-27UBK trainer version; delivery 2000–2002
	50	Su-27SK/Flanker-B	FGA aircraft	1996	1998–2001	(20)	Assembled in China; prior to licensed production; deal worth $1.46–2.5 b incl 150 licensed production; Chinese designation J-11; delivery 1998–2007/2008 incl 150 licensed production
	38	Su-30MK/Flanker	FGA aircraft	1999	2000–2001	38	Deal worth $1.8–$2 b; Su-30MKK version
	38	Su-30MK/Flanker	FGA aircraft	2001			Deal worth $2 b; Su-30MKK version; delivery 2002–2003
	(3 720)	AA-11 Archer/R-73	AAM	(1995)	1996–2001	(1 200)	For some 310 Su-27SK and Su-30MKK FGA aircraft
	(100)	AA-12 Adder/R-77	AAM	(2000)			For Su-30MKK and Su-27UBK FGA aircraft; Chinese designation R-129
	(4)	SA-10e/S-300PMU-2	SAM system	2001			
		SA-10 Grumble/48N6	SAM	2001			
	..	Sokol	Combat ac radar	(2001)	2001	(20)	Probably for F-10 FGA aircraft
	100	Zhuk	Combat ac radar	2001	2001	1	For F-8-II fighter aircraft; delivery 2001–2003
	..	AS-17/Kh-31A1	Anti-ship missile	(1997)		..	For J-8MIIM and/or JH-7 FGA aircraft; may incl ARM version; may incl assembly or licensed production in China; status uncertain
	..	AS-17/Kh-31P1	Anti-radar missile	(1998)	2001	(10)	Incl for Su-30MKK FGA aircraft
	2	Sovremenny Class	Destroyer	1996	1999–2001	2	Type-956E version; originally ordered for Soviet Union/Russia but cancelled before completion and sold to China
	2	Sovremenny Class	Destroyer	(2001)			Type-956EM version; deal worth over $1 b; option on 2 more; delivery 2005/2006
	(132)	SA-17 Grizzly/9M38M2	SAM	(2001)		..	SA-N-12/9M38M2 version for 2 Sovremenny Class destroyers; designation uncertain
UK	(6)	Searchwater	AEW radar	1996	1999	1	Deal worth $62–66 m; for Y-8 AEW and MP aircraft or possibly SA-341/Z-8 helicopters
Ukraine	(1 860)	AA-10a/b Alamo/R-27	AAM	(1995)	1996–2001	(600)	For some 310 Su-27SK and Su-30MKK FGA aircraft; supplier could be Russia
	(1 860)	AA-10c/d Alamo/R-27E	AAM	(1995)	1996–2001	(600)	For some 310 Su-27SK and Su-30MKK FGA aircraft; supplier could be Russia
L: France	..	AS-365/AS-565	Helicopter	1988	1992–2001	(19)	Chinese designation Z-9A-100 Haitun and Z-9B/C/G; no. produced could be much higher; incl ASW version; more produced for civilian customers

INTERNATIONAL ARMS TRANSFERS 421

	Weapon designation	Weapon description	Year of order	Year(s) of deliveries	No. delivered	Comments
	Su-27SK/Flanker-B	FGA aircraft	1996		(150)	Chinese designation PL-8; no. delivered could be much lower or higher Deal worth $1.46 b to $2.5 b incl 50 assembled in China; Chinese designation J-11; delivery 1998–2007/2008 incl 50 assembled in China
Colombia						
S: Russia	Mi-17/Hip-H	Helicopter	2001	2001	6	Deal worth $36 m; Mi-17-1V version
USA	Bell-205/UH-1H	Helicopter	(2000)		(42)	Ex-US; deal worth $323 m; part of $1.3 b 'Plan Colombia' anti-narcotics aid; modernized to Huey-II before delivery; delivery from 2002
	S-70A/UH-60L	Helicopter	2000	2001	14	Deal worth $106 m
	S-70A/UH-60L	Helicopter	2000	2001	16	Deal worth $116 m; part of $1.3 b 'Plan Colombia' anti-narcotics aid; incl 2 for Police
	SA-2-37A	Reconnaissance ac	2000	2001	(5)	Part of $1.3 b 'Plan Colombia' anti-narcotics aid
	AN/APG-66	Combat ac radar	(2000)	2001	2	Part of $1.3 b 'Plan Colombia' anti-narcotics aid; for modification of 2 Metro-3 transport aircraft to AEW aircraft
	AN/TPS-70	Surveillance radar	2000	2001	1	
Croatia						
S: USA	AN/FPS-117	Surveillance radar	1999	2001	5	Deal worth $94 m
Cyprus						
S: France	MILAN	Anti-tank missile	2000	2000–2001	784	MILAN-3 version
Greece	Bell-205/UH-1H	Helicopter	(2001)	2001	(8)	Ex-Greek; for SAR; loan
	M-48A5 Patton	Main battle tank	(1999)	2001	(10)	Ex-Greek
Czech Republic						
S: Italy	Grifo	Combat ac radar	(1998)	2000–2001	72	For 72 L-159A FGA/trainer aircraft; Grifo-L version
Russia	Mi-24V/Mi-35/Hind-E	Combat helicopter	2001	2001	6	Ex-Russian; payment for Russian debts to Czech Republic
Sweden	JAS-39 Gripen	FGA aircraft	(2001)		24	Deal worth $1.3 b (offsets 150%); incl 4 JAS-39B version; delivery 2004–2008; contract not yet signed
Denmark						
S: Canada	Challenger-604	Transport aircraft	2000	2001	2	Deal worth $44 m; for MP, medical evacuation and VIP transport
Finland	XA-185	APC	2001	2001	11	Ex-Finnish: ambulance version; lease for use with Danish peacekeeping forces in Bosnia and Kosovo, delivery 2001–2002
Germany (FRG)	MU-90 Impact	ASW torpedo	1999	2001	(16)	For 4 Flyvefisken FAC
Norway	Arthur	Artillery radar	1997	1999–2001	8	Deal worth $40 m
Sweden	Näcken Class	Submarine	2001	2001	1	Ex-Swedish; deal worth $35 m

Recipient/ supplier (S) or licenser (L)	No. ordered	Weapon designation	Weapon description	Year of order/ licence	Year(s) of deliveries	No. delivered/ produced	Comments
UK	14	EH-101-500	Helicopter	2001		. .	Deal worth $329 m (offsets incl production of EH-101 components in Denmark); incl for SAR; delivery 2004–2006
USA	3	C-130J-30 Hercules	Transport aircraft	2000		. .	Delivery from 2003
	(70)	AN/APG-66	Combat ac radar	1993	1996–2001	(70)	For 'Mid-Life Update' (MLU) modernization of 70 F-16A/B FGA aircraft to F-16AM/BM (F-16C/D) version
	(180)	RIM-7M Seasparrow	SAM	(1994)	1998–2001	(120)	For 4 Flyvefisken Class FAC/MCM ships and 3 modernized Niels Juel Class corvettes
Djibouti							
S: South Africa	9	Casspir	APC	2000	2001	9	Ex-South African; modernized before delivery
Dominican Republic							
S: Brazil	10	EMB-314 Super Tucano	Trainer aircraft	2001		. .	Incl for anti-narcotics operations
Egypt							
S: Austria	170	M-60A1 Patton-2	Main battle tank	(2001)		. .	Ex-Austrian; deal worth $27 m
China	35	K-8 Karakorum-8	Jet trainer aircraft	2000	2001	(5)	Deal worth $345 m incl 45 licensed production; K-8E version; incl 25 assembled in Egypt
Finland	2	155-GH-52	Towed gun	(1999)	2000–2001	2	Deal worth $17–21 m incl licensed production; incl 1 assembled in Egypt
Germany (FRG)	74	G-115D	Trainer aircraft	2000	2000–2001	(48)	G-115EG version; delivery 2000–2002
USA	1	E-2C Hawkeye	AEW&C aircraft	2001		. .	Ex-US; aid; modernized to Hawkeye-2000 before delivery under deal worth $25 m
	24	F-16C	FGA aircraft	1999	2001	8	'Peace Vector-6' deal worth $1.2 b; incl 12 F-16D version; delivery 2001–2002
	201	M-109A2 155mm	Self-propelled gun	(2001)		. .	Ex-US; deal worth $77 m; contract not yet signed
	26	M-270 MLRS 227mm	MRL	2001		. .	Deal $354 m incl rockets, 3 M-88A2 ARVs and 30 M-577A2 APC/CPs; 'FMF' aid; delivery probably 2003
	30	M-113A2	APC	2001		. .	M-577A2 CP version; deal worth $354 m 26 M-270 MLRS MRL, rockets and 3 M-88A2 ARVs; 'FMF' aid
	50	M-88A2 Hercules	ARV	1998	2000–2001	(30)	Deal worth $197.9 m; assembled in Egypt; delivery 2000–2002
	13	M-88A2 Hercules	ARV	2000		. .	Deal worth $354 m 26 M-270 MLRS MRL, rockets and 30 M-577A2 APC/CPs; 'FMF' aid; contract not yet signed
	3	M-88A2 Hercules	ARV	2001		. .	
	5	AN/APS-145	AEW radar	1999		. .	Deal worth $138 m; for modernization of 5 E-2C AEW&C aircraft to Hawkeye-2000 version; delivery from 2002

INTERNATIONAL ARMS TRANSFERS 423

	No. ordered	Weapon designation	Weapon description	Year of order	Year(s) of delivery	No. delivered	Comments
	8	I-HAWK	SAM system	(1996)	1998–2001	(8)	Deal worth $143 m
	4	Phalanx Mk-15	CIWS	2001			Ex-US; deal worth $206 m; modernized before delivery
							Ex-US; deal worth $0.83 m; modernized before delivery incl 1 to Block-1B version
	(2 372)	BGM-71 TOW	Anti-tank missile	1996	1998–2000	(2 250)	Deal worth $59 m; BGM-71E TOW-2A version
	1 058	FIM-92A Stinger	Portable SAM	1998	2000–2001	(600)	For 50 Avenger air-defence systems
	..	FIM-92A Stinger	Portable SAM	2000			For 24 Avenger air-defence systems
	180	MIM-23B HAWK	SAM	1996	1998–2000	(100)	Ex-US
	(42)	RGM-84 Harpoon	Anti-ship missile	1998	2000–2001	(42)	Deal worth $51 m; AGM-84 version for F-16 FGA aircraft
	(32)	RGM-84 Harpoon	Anti-ship missile	1997	2000–2001	(32)	
	4	Ambassador Type	FAC(M)	2000			Deal worth $400 m; Ambassador Mk-3 version; delivery 2004–2005
L: China	45	K-8 Karakorum-8	Trainer aircraft	2000			Deal worth $345 m incl 35 direct delivered; K-8E version
Finland	..	155-GH-52	Towed gun	(1999)			Deal worth $17–21 m incl 2 direct delivered
Germany (FRG)	..	Fahd	APC	1978	1986–2000	(680)	Developed for production in Egypt; more produced for export
USA	100	M-1A1 Abrams	Main battle tank	1999			Deal worth $564 m; delivery 2001–2003
	100	M-1A1 Abrams	Main battle tank	2001			Deal worth $590 m; contract not yet signed
Eritrea							
S: Russia	(8)	MiG-29S/Fulcrum-C	FGA aircraft	(1999)	2001	(2)	
Estonia							
S: USA	4	R-44	Helicopter	2001			Aid; delivery 2002
	1	AN/TPS-117	Surveillance radar	2001			Deal worth $12 m; part of 'BALTNET' air-surveillance network
Ethiopia							
S: Bulgaria	140	T-55	Main battle tank	(1998)	1999	100	Ex-Bulgarian; may incl some from Ukraine and/or Romania sold via Bulgaria; delivery of last 40 suspended 2000–2001 because of UN embargo
Finland							
S: France	(540)	Mistral	Portable SAM	(1989)	1990–2001	(500)	For Sako (modified SADRAL) SAM system on 2 Hamina and 2 Hameenma minelayers, and on 4 modernized Helsinki and 4 Rauma Class FAC, and 1 modernized Pohjanmaa Class minelayers
Israel	(6)	Ranger	UAV	1999			Deal worth $30 m; assembled or produced in Germany; Spike-2.5 version
	..	Spike/NT-G Gill	Anti-tank missile	2000	2001	(6)	Deal worth $350 m (€520 m incl support) incl licensed production of 18; supplied by France and Germany; delivery 2004
Multiple sellers	2	NH-90 TTH	Helicopter	2001			

Recipient/ supplier (S) or licenser (L)	No. ordered	Weapon designation	Weapon description	Year of order/ licence	Year(s) of deliveries	No. delivered/ produced	Comments
Netherlands	2	Scout	Surveillance radar	1998	1998	(1)	For 2 Hamina Class FAC produced in Finland
Sweden	57	CV-9030	IFV	(2000)		. .	Deal worth $176 m (offsets incl production of components in Finland); 'TA-2000' project; delivery 2002–2005
	2	9LV	Fire control radar	(1997)	1998	1	9LV225 version for 2 Hamina Class FAC produced in Finland
	(16)	RBS-15SF	Anti-ship missile	(1997)	1998–2001	(13)	For 2 Hamina Class FAC
	. .	RBS-15 Mk-3	Anti-ship missile	2001		. .	Finnish designation RBS-15 SF-3
L: Multiple sellers	18	NH-90 TTH	Helicopter	2001		. .	Deal worth $350 m (€520 m incl support) incl direct delivery of 2; supplied by France and Germany; delivery from 2004
France							
S: Israel	(4)	Eagle	UAV	2001		. .	Option on 3; delivery 2002
Spain	2	CN-235	Transport aircraft	(2001)		. .	Delivery from 2003
Sweden	4	Giraffe AMB-3D	Surveillance radar	2001		. .	Deal worth $894 m incl 2 ordered in 1995 (offsets worth $440 m incl French production of components); delivery 2003
USA	1	E-2C Hawkeye	AEW&C aircraft	1999			
Georgia							
S: Bulgaria	2	Vydra Class	Landing craft	2001	2001	2	Ex-Bulgarian; aid
Czech Republic	6	D-30	Towed gun	(2001)	2001	6	Ex-Czech
	120	T-55AM-2	Main battle tank	1998	2000–2001	(120)	Ex-Czech; incl some T-54 tanks
Turkey	2	Bell-205/UH-1H	Helicopter	(2001)	2001	2	Ex-Turkish; aid
USA	6	Bell-205/UH-1H	Helicopter	1999	2001	6	Ex-US; aid; 4 more delivered for spares only
Germany (FRG)							
S: Netherlands	3	APAR	Surveillance radar	(1997)		. .	For 3 Sachsen Class (F-124 or Type-124) frigates produced in FRG; delivery 2002–2005
	3	SMART-L	Surveillance radar	(1997)		. .	For 3 Sachsen Class (F-124 or Type-124) frigates produced in FRG; delivery 2002–2005
Sweden	(31)	Bv-206S	APC	(2001)		. .	Deal worth $12 m; incl CP and ambulance version; delivery 2002–2004; contract not yet signed
	10	HARD	Surveillance radar	1998	2000–2001	(4)	For ASRAD SAM systems; delivery 2000–2003
	. .	RBS-15 Mk-3	Anti-ship missile	(2001)		. .	For K-130 Class corvettes; incl assembly in FRG; contract not yet signed
UK	17	Seaspray-3000	MP aircraft radar	(1999)	2001	(2)	Part of $125 m modernization of 17 Lynx helicopters to Super Lynx version

INTERNATIONAL ARMS TRANSFERS 425

L: USA	..	RIM-66M Standard-2	SAM	2001	(..)	Deal worth $170 m; for F-4F FGA aircraft
	(1 400)	FIM-92C Stinger	Portable SAM	1986	1998–2001	(1 300)	FRG designation Fliegerfaust-2; part of 'European Stinger Production Programme' involving production of components in FRG, Greece, Netherlands and Turkey and final assembly in FRG; delivery 1998–2001/2002

Ghana							
S: UK	20	Tactica	APC	2000	2000–2001	(20)	Deal worth $9 m
USA	2	Balsam Class	Depot ship	2000	2001	(2)	Ex-US; aid

Greece							
S: Brazil	4	EMB-145AEW&C	AEW&C aircraft	1999		..	Deal worth $476–500 m incl 4 PS-890 Erieye radars from Sweden; delivery from Sweden to Greece from 2002 after fitting of PS-890 radar; option on 2 more
Canada	2	CL-415	Transport aircraft	1998		..	For combat SAR; delivery 2002
France	4	AS-532UL/AS-332L1	Helicopter	2000		..	Deal worth $90 m (offsets 100%); for combat SAR; option on 2 more transport version; delivery 2003
	15	Mirage-2000-5	FGA aircraft	2000		..	Deal worth $1.4 b incl 200 MICA and 56 SCALP-EG missiles and modernization of 10 Greek Mirage-2000EG to Mirage-2000-5; delivery 2003–2004
	200	MICA	AAM	2000		..	Deal worth $1.4 b incl 15 Mirage-2000-5 FGA aircraft, 56 SCALP-EG missiles and modernization of 10 Greek Mirage-2000EG to Mirage-2000-5; MICA-EM and MICA-IR version; option on 50 more
	56	Storm Shadow/SCALP	ASM	2001		..	Deal worth $1.4 b incl 15 Mirage-2000-5 FGA aircraft, 200 MICA missiles and modernization of 10 Greek Mirage-2000EG to Mirage-2000-5; SCALP-EG version
Germany (FRG)	11	Crotale-NG	SAM system	1999	1999–2001	(11)	Deal worth $266 m (incl offsets) incl VT-1 missiles; probably incl 4 ex-French
	60	Eurofighter/Typhoon	FGA aircraft	(2001)		..	Deal worth $4.5–4.9 b; contract not yet signed
	28	PZH-2000 ALV	ALV	2001		..	Deal worth $164 m or $228 m (offsets 120%) incl 24 PzH-2000 155mm self-propelled guns; incl 24 ammunition carriers and 4 ammunition loader version; delivery 2003–2004
	24	PzH-2000 155mm	Self-propelled gun	2001		..	Deal worth $164 m or $228 m (offsets 120%) incl 28 ALVs; delivery 2003–2004
	175	Leopard-1A5	Main battle tank	1997	1998–2001	(175)	Ex-FRG; offsets for Greek order for modernization of F-4E FGA aircraft in FRG
	38	AN/APG-65	Combat ac radar	1997	2001	(25)	Part of $315 m 'Peace Icarus-2000' modernization programme of 38 F-4E FGA aircraft; delivery 2001–2002

Recipient/ supplier (S) or licenser (L)	No. ordered	Weapon designation	Weapon description	Year of order/ licence	Year(s) of deliveries	No. delivered/ produced	Comments
	(350)	FIM-92C Stinger	Portable SAM	1986	1998–2001	(330)	Part of 'European Stinger Production Programme' involving production of components in FRG, Greece, Netherlands and Turkey and final assembly in FRG; delivery 1998–2001/2002
	(432)	FIM-92C Stinger	Portable SAM	2000		. .	Deal worth $134 m incl ASRAD launchers; delivery from 2002
	(126)	RIM-116A RAM	SAM	2000		. .	Deal worth $25 m incl 3 launchers; for 3 Super Vita Class FAC
	1	Type-214	Submarine	2000		. .	Deal worth $0.92–1.26 b incl 2 licensed production (offsets 115%); Greek designation Katsonis Class; delivery 2005
Netherlands	10	Bergepanzer-1	ARV	2001	2001	10	Ex-Dutch; deal worth $6 m; modernized in Germany before delivery
	(7)	LIROD	Fire control radar	2000		. .	For 3 Super Vita Class and 4 Osprey-55 Type (Pirpolitis Class) FAC
	3	MW-08	Surveillance radar	2000		. .	For 3 Super Vita Class FAC
	3	STING	Fire control radar	2000		. .	For 3 Super Vita Class FAC
	3	Scout	Surveillance radar	2000		. .	For 3 Super Vita Class FAC
	(7)	Variant	Surveillance radar	2000		. .	For 3 Super Vita Class and 4 Osprey-55 Type (Pirpolitis Class) FAC
	1	Kortenaer Class	Frigate	2001	2001	1	Ex-Dutch; deal worth $37 m; Greek designation Elli Class
Russia	19	SA-15/9A331 Tor-M1	Mobile SAM system	(2001)		. .	Deal worth $400 m or $700 m incl missiles; status uncertain; delivery possibly from 2002
	(323)	SA-15 Gauntlet/9M338	SAM	(2001)		. .	Deal worth $400 m or $700 m incl 19 SA-15/9A331 Tor-M1 SAM systems; status uncertain
	10	SA-15/9A331 Tor-M1	Mobile SAM system	2000	2000–2001	8	Deal worth $300 m; replacing 6 of 1999 order for 21 transferred to Cyprus; delivery 2000–2002
	(170)	SA-15 Gauntlet/9M338	SAM	(2000)	2000–2001	(150)	For 10 SA-15/9A331 Tor-M1 SAM systems
	1 100	AT-14/Kornet	Anti-tank missile	2001		. .	Deal worth $95 m
Slovakia	1	Pomornik Class	ACV/landing craft	2000	2001	1	Deal worth $102 m incl 1 ex-Russian; Greek designation Kefallania Class
	12	Zuzana 155mm	Self-propelled gun	2000	2001	12	Option on 6 more
Sweden	2	Saab-340AEW	AEW&C aircraft	1999	2001	2	Orginally produced for Sweden but on loan until 2002 delivery of EMB-145SA AEW&C aircraft
	4	PS-890 Erieye	AEW radar	1999		. .	For 4 EMB-145AEW&C AEW aircraft; deal worth $476–500 m incl aircraft; option on 2 more; delivery from 2002
UK	2	Hunt Class	MCM ship	1999	2000–2001	2	Ex-UK; part of order for 3 Super Vita Class FAC
USA	(6)	A-7E Corsair-2	FGA aircraft	1998	2000–2001	(6)	Ex-US; aid
	7	CH-47D Chinook	Helicopter	1999	2001	7	Deal worth $376 m
	50	F-16C	FGA aircraft	2000		. .	'Peace Xenia' deal worth $2.1 b; incl 16 F-16D version; delivery 2003–2004
	10	F-16C	FGA aircraft	2001		. .	'Peace Xenia' deal worth $183 m; incl 4 F-16D version; delivery 2004
	45	PC-9/T-6A Texan-2	Trainer aircraft	1999	2000–2001	(15)	Deal worth $223 m (offsets 120% incl production in Greece of parts for 300 PC-9/T-6A); option on 5 more; delivery 2000–2003
	2	S-70B/SH-60B Seahawk	AS/ASW helicopter	2000		. .	S-70B-6 Aegean Hawk version; option on 2 more

INTERNATIONAL ARMS TRANSFERS 427

No.	Weapon designation	Weapon description	Year of order	Year of delivery	No. delivered	Comments
	... ...	Light transport ac	2000	2001	2	Deal worth $11.4 m; for photographic reconnaissance and survey; C-12R/AP version
10	T-2E Buckeye	Jet trainer aircraft	1998	2001	5	Ex-US; aid; T-2C version; delivery 2001–2002
7	M-728	AEV	(2000)		..	Ex-US; aid
4	Patriot	SAM system	1999	2001	(1)	Deal worth $887 m ($1.13 b incl option on 2 more; offsets 120%); delivery 2001–2002
..	MIM-104 PAC-2	SAM	(1999)		..	For 4 Patriot SAM systems
248	AGM-114K Hellfire	Anti-tank missile	1998		..	Deal worth $24 m; for AH-64A helicopters
100	AIM-7M Sparrow	AAM	(2000)		..	Ex-US; aid; designation uncertain
1 322	FIM-92A Stinger	Portable SAM	2001		..	Deal worth $150 m; deal also incl 188 launchers
(30)	MGM-140A ATACMS	SSM	1998	2000–2001	(30)	Deal worth $245 m incl 18 M-270 MRLS MRLs, 11 M-577 APC/CPs and 146 rockets
(51)	MGM-140A ATACMS	SSM	1999	2001	(20)	
208	RIM-66B Standard-1MR	SAM	1999		..	Ex-US; aid; status uncertain
(176)	VT-1	SAM	1998	1999–2001	(176)	Deal worth $266 m (incl offsets) incl 11 Crotale NG SAM systems from France; ordered and delivered via France
Ukraine						
2	Pomornik Class	ACV/landing craft	2000	2001	2	Deal worth $98.5 m; Greek designation Kefallania Class
L: Denmark						
4	Osprey-55 Type	Patrol craft	1999		..	Greek designation Pirpolitis Class or Hellenic-56 Type; delivery from 2002
Germany (FRG)						
2	Type-214	Submarine	2000		..	Deal worth $0.92–1.26 b incl 1 direct delivered (offsets 115%); Greek designation Katsonis Class; option on 1 more; delivery 2007–2008
Italy						
1	Etna Class	Support ship	1999		..	Deal worth $128 m; delivery 2004
UK						
3	Super Vita type	FAC(M)	1999		..	Deal worth $324 m; option on 4 more; delivery 2003–2004
Guinea						
S: Ukraine						
(4)	Mi-24V/Mi-35/Hind-E	Combat helicopter	(2000)	2001	(4)	Ex-Ukrainian
Hungary						
S: Italy						
9	SHORAR-2D	Surveillance radar	1997	1999–2001	(9)	Deal worth $100 m incl 180 Mistral missiles, 45 ATLAS launchers and 54 UNIMOG trucks; sold via France
Sweden						
14	JAS-39 Gripen	FGA aircraft	2001		..	Ex-Swedish; modernized before delivery; 10-year lease worth $500 m (offsets 100%); option to extend lease or buy; delivery from 2004
India						
S: France						
10	Mirage-2000E	FGA aircraft	2000		..	Deal worth $325–353 m; Mirage-2000H version; incl 6 Mirage-2000TH trainer version; Indian designation Vajra; delivery 2003
Germany (FRG)						
2	Type-209/1500	Submarine	(2001)		..	Indian designation Project-75 and Shishumar Class; delivery 2006–2008; contract not yet signed

Recipient/ supplier (S) or licenser (L)	No. ordered	Weapon designation	Weapon description	Year of order/ licence	Year(s) of deliveries	No. delivered/ produced	Comments
	(56)	SUT	AS/ASW torpedo	(1999)	. .	. .	For 2 Type-209 (Shishumar Class) submarines; status uncertain
Israel	(6)	Heron-2	UAV	2001			
	32	Searcher UAV	UAV	(2000)	2001	(16)	
	. .	EL/M-2022	MP aircraft radar	(2000)		. .	For Do-228 MP aircraft
	10	EL/M-2032	Combat ac radar	1999	2000–2001	(10)	For modernization of 10 Jaguar-M FGA aircraft
	3	EL/M-2080	Surveillance radar	(2001)		. .	For modification of 3 Il-76 transport aircraft delivered from Uzbekistan via Russia to Il-76/A-50EI AEW&C aircraft, contract not yet signed
	(56)	EL/M-2129	Artillery radar	1999	2001	(5)	
	(200)	EL/M-2140	Battlefield radar	1999	2001	(25)	
	(7)	EL/M-2221	Fire control radar	(2001)		. .	Deal worth $280 m; for modernization of 1 Viraat Class aircraft carrier, possibly 3 Rajput Class destroyers and 3 Brahmaputra Class frigates and/or 3 Delhi Class destroyers produced in India; for use with Barak SAM
	. .	Barak	SAM	(2001)		. .	Deal worth $280 m; for 1 modernized Viraat Class aircraft carrier and possibly 3 modernized Rajput Class destroyers and 3 Brahmaputra Class frigates and/or 3 Delhi Class destroyers produced in India, contract not yet signed
Italy	9	Seaguard TMX	Fire control radar	1993	2000–2001	(6)	For 3 Brahmaputra Class (Project-16A) frigates produced in India
	(36)	A244/S 324mm	ASW torpedo	(1993)	2000–2001	(12)	For 3 Brahmaputra Class (Project-16A) frigates; possibly licensed produced in India as NST-58
Netherlands	4	DA-05	Surveillance radar	(1989)	1995–2001	(3)	For modernization of 1 Viraat Class aircraft carrier and for 3 Brahmaputra Class (Project-16A) frigates produced in India; incl assembly in India; Indian designation RAWS, RAWS-03 or PFN-513
	7	LW-08	Surveillance radar	(1989)	1997–2001	(6)	For modernization of 1 Viraat Class aircraft carrier and for 3 Delhi Class (Project-15) destroyers and 3 Brahmaputra Class (Project-16A) frigates produced in India; incl assembly in India; Indian designation RALW, RAWL-2 or PLN-517
	(10)	ZW-06	Surveillance radar	(1989)	1997–2001	(8)	For modernization of 1 Viraat Class aircraft carrier and for 3 Delhi Class (Project-15) destroyers and 3 Brahmaputra Class (Project-16A) frigates; incl assembly in India; Indian designation Rashmi
Poland	44	WZT-3	ARV	1999		. .	Deal worth $31.1 m
Russia	4	Ka-31/Helix	AEW helicopter	1999	2001	(1)	Deal worth $92 m; delivery 2001–2002
	5	Ka-31/Helix	AEW helicopter	2001		. .	Deal worth $100–108 m; delivery from 2002
	40	Mi-17/Hip-H	Helicopter	2000	2000–2001	(40)	Deal worth $170 m; Mi-17-1V version; modified in India to combat helicopter
	(24)	MiG-29K/Fulcrum-D	FGA aircraft	(2001)		. .	For use on Gorshkov Class aircraft carrier, no. ordered could be up to 66; incl some MiG-29KUB trainer version

INTERNATIONAL ARMS TRANSFERS   429

No.	Designation	Type	(Year of order) Year(s) of delivery	(Delivered)	Comments
10	Su-30MK/Flanker	FGA aircraft	1998		Su-30MKI after delivery; delivery 1997–2003/2004
(1 140)	AA-10c/d Alamo/R-27E	AAM	1996 1997–2001	(250)	Su-30MKI version; ordered while still being developed; delivery 2003/2004
(1 520)	AA-11 Archer/R-73	AAM	(1996) 1997–2001	(500)	For 190 Su-30MKI FGA aircraft
. .	AA-12 Adder/R-77	AAM	(2000)	. .	For 190 Su-30MK FGA aircraft
					For Su-30MK FGA aircraft and 125 MiG-21-93 (modernized MiG-21bis) fighter aircraft; possibly also for MiG-29 fighter aircraft
(4)	Tu-22M3/Backfire-C	Bomber aircraft	2001	. .	Ex-Russian; lease; contract not yet signed
(4)	140mm RL	Naval MRL	(1992) 1997	(2)	For 2 Magar Class landing ships; designation uncertain
(14)	BM-9A52/BM-23	MRL	(2001)	. .	Deal worth Rs 12 b; delivery probably 2002; contract not yet signed
(14)	2S6M Tunguska	AAV(G/M)	(2001)	. .	Deal worth Rs 6 b; delivery probably 2002; contract not yet signed
(720)	SA-19 Grisom/9M111	SAM	(1998)	. .	For 45 2S6 AAV(G/M)s; status uncertain
310	T-90S	Main battle tank	2001 2001	(80)	Deal worth $600–700 m (incl 55% advance payment); ordered as reaction to Pakistani acquisition of 320 T-80UB tanks; incl 186 assembled in India; delivery 2001–2004
(6)	CADS-N-1 Kashtan	AD/SAM system	(2000)	. .	For 3 Nilgiri Class (Project-17) frigates produced in India; status uncertain
4	Cross Dome	Surveillance radar	(1992) 1998–2001	(3)	For 4 Kora Class (Project-25A) corvettes produced in India
18	Front Dome	Fire control radar	(1986) 1997–2001	(18)	For 3 Delhi Class (Project-15) destroyers produced in India; for use with SA-N-7 SAM; ordered from Soviet Union and delivered from Russia after breakup of Soviet Union
3	Half Plate	Surveillance radar	(1983) 1997–2001	(3)	For 3 Delhi Class (Project-15) destroyers produced in India; ordered from Soviet Union and delivered from Russia after breakup of Soviet Union
3	Kite Screech	Fire control radar	(1986) 1997–2001	(3)	For 3 Delhi Class (Project-15) destroyers produced in India; ordered from Soviet Union and delivered from Russia after breakup of Soviet Union
7	Plank Shave	Surveillance radar	(1993) 1997–2001	(6)	For 3 Delhi Class (Project-15) destroyers and 4 Kora Class (Project-25A) corvettes produced in India; for use with SS-N-25 missiles
10	Bass Tilt	Fire control radar	(1992) 1997–2001	(9)	For 3 Delhi Class (Project-15) destroyers and 4 Kora Class (Project-25A) corvettes produced in India
3	AK-100	Naval gun	(1986) 1997–2001	(3)	For 3 Delhi Class (Project-15) destroyers produced in India; ordered from Soviet Union and delivered from Russia after breakup of Soviet Union
(216)	SA-11 Gadfly/9M38M1	SAM	(1993) 1997–2001	(216)	SA-N-7/9M38M1 version for 3 Delhi Class (Project-15) destroyers
(15)	SET-65E	ASW torpedo	(1993) 1997–2001	(15)	For 3 Delhi Class (Project-15) destroyers
(15)	53-65K 533mm	AS torpedo	(1993) 1997–2001	(15)	For 3 Delhi Class (Project-15) destroyers
(416)	SS-N-25/Kh-35 Uran	Anti-ship missile	(1992) 1997–2001	(260)	For 3 Delhi Class (Project-15) and 2 modernized Kashin-2 (Rajput) Class destroyers, 3 Brahmaputra Class (Poject-16A) frigates, 4 Kora Class (Project-25A) corvettes and 1 Tarantul-1 (Vibhuti) Class FAC
3	Krivak-4 Class	Frigate	1997	. .	Deal worth $0.82–1 b; Indian designation Talwar Class; Russian designation Type-1135.6; ordered due to problems with indigenous production of major

Recipient/ supplier (S) or licenser (L)	No. ordered	Weapon designation	Weapon description	Year of order/ licence	Year(s) of deliveries	No. delivered/ produced	Comments
	(108)	SA-11 Gadfly/9M38M1	SAM	(2000)		..	SA-N-7/9M38M1 version for 3 Krivak-4 Class frigates; warships; delivery delayed from 2001–2002 to 2002–2003 due to financial problems of producer
	(84)	SS-N-27/3M54E1	Anti-ship missile	(1998)		..	For 4 Kilo Class submarines and 3 Krivak-4 Class frigates
	(360)	SA-19 Grisom/9M111	SAM	(1997)		..	SA-N-11/9M111 version for 6 Kashtan AD/SAM systems on 3 Krivak-4 (Talwar) Class frigates
	..	SA-15 Gauntlet/9M338	SAM	(1997)		..	SA-N-9 version for 3 Krivak-4 Class frigates; status uncertain
	(125)	Kopyo	Combat ac radar	1996	2001	(2)	Part of deal worth $428–626 m for modernization of up to 125 MiG-21bis fighter aircraft to MiG-21bis UPG/MiG-21I version; option on 50 more; delivery 2001–2004/2005
	(4)	SA-12/S-300V	SAM system	(2001)		..	Deal worth $1.5 or $2.5 b incl 150 missiles; contract probably not yet signed
	150	SA-12b Giant/9M82	SAM	(2001)		..	Deal worth $1.5 or $2.5 b incl 4 SA-12 SAM systems; contract probably not yet signed
	8	Zmei/Sea Dragon	Airborne MP radar	(2000)		..	For modernization of 8 Tu-142 ASW/MP aircraft to Tu-142J
	..	AT-16/9M120	Anti-tank missile	(2000)		..	For 40 Mi-17 helicopters
	(2 500)	SA-18 Grouse/Igla	Portable SAM	2001	2001	(500)	Deal worth $32–50 m
	1	Gorshkov Class	Aircraft carrier	(2001)		..	Ex-Russian; modernized and modified to conventional take off/landing (CTOL) carrier before delivery; contract not yet signed; delivery 2004/2005
	(360)	SA-19 Grisom/9M111	SAM	(2001)		..	SA-N-11/9M111 version for 12 Kashtan AD/SAM systems on 1 Gorshkov Class aircraft carrier and 3 Nilgiri Class (Project-17) frigates; status uncertain
	1	Victor-3 Class	Nuclear submarine	2001		..	Ex-Russian; 3-year lease worth $75 m; delivery 2002
Slovakia	42	VT-72B	ARV	1999		..	Deal worth $30.4 m
South Africa	75	Casspir	APC	2000	2000–2001	75	Ex-South African; modernized before delivery
UK	(47)	Super Marec	MP aircraft radar	(1983)	1988–97	(20)	For Do-228-200M MP aircraft
Ukraine	(1 140)	AA-10a/b Alamo/R-27	AAM	(1996)	1997–2001	(304)	For 190 Su-30MKI FGA aircraft; designation uncertain
Uzbekistan	6	Il-78M/Midas	Tanker aircraft	2001		..	Deal worth $150 m; Il-78MK version; no. ordered could be 4; delivery 2002/2003
L: France							
Germany (FRG)	..	MILAN	Anti-tank missile	(1981)	1984–2001	(48 500)	MILAN-2 version; incl for BMP-2 IFVs
	33	Do-228-200MP	MP aircraft	1983	1988–97	(15)	For Coast Guard
	14	Do-228-200MP	MP aircraft	(1989)	1994–96	(4)	
Netherlands	(20)	Reporter	Surveillance radar	(1997)	1998–2001	(20)	
Russia	140	Su-30MK/Flanker	FGA aircraft	2000		..	Deal worth $3 b; Su-30MKI version; for delivery 2005–2017
	..	AT-5 Spandrel/9M113	Anti-tank missile	(1988)	1992–2001	(4 300)	For BMP-2 IFVs; ordered from Soviet Union and produced under Russian licence after breakup of Soviet Union

INTERNATIONAL ARMS TRANSFERS 431

			Weapon designation	Weapon description	Year of order	Year(s) of deliveries	No. delivered	Comments
	UK	17	Jaguar International	FGA aircraft	1999	2001	7	Indian designation Vibhuti Class; ordered from Soviet Union and produced under Russian licence after breakup of Soviet Union
		20	Jaguar International	FGA aircraft	2000	2001	(5)	Indian designation Shamsher; Jaguar-B version
		1	Magar Class	Landing ship	(1996)			
Indonesia								
S:	France	(13)	AS-332B Super Puma	Helicopter	1997	2001	(5)	Incl 6 for SAR; 3 more ordered for VIP transport
		(3)	EC-120B Colibri	Light helicopter	(2000)			
		3	Ocean Master	MP aircraft radar	2001			For 3 CN-235MPA aircraft produced in Indonesia; deal worth $47 m; delivery of aircraft 2002–2003
		9	Ocean Master	MP aircraft radar	1996	2000–2001	(9)	For 6 C-212MP MP aircraft and 3 Bo-105 helicopters
	Singapore	(15)	SF-260M	Trainer aircraft	(2001)			Ex-Singapore; status uncertain
	South Korea	7	KT-1 Woong Bee	Trainer aircraft	2001			Deal worth $60 m (offsets for Korean order for CN-235 transport aircraft); delivery from 2003
	USA	16	AN/APG-66	Combat ac radar	1996	1999	10	For 16 Hawk-200 FGA aircraft delivered from UK; last 6 embargoed by USA
L:	Germany (FRG)	4	PB-57 Type	Patrol craft	1993	2000	2	Deal worth $260 m; Indonesian designation Singa Class; delivery 2000–2004
	Spain	6	C-212MP Aviocar	MP aircraft	1996	2000–2001	(6)	C-212-200MP version
Iran								
S:	China	14	Y-7	Transport aircraft	1996	1998	(2)	Delivery 1998–2006
	Russia	21	Mi-17/Hip-H	Helicopter	1999	2000–2001	(21)	Incl some for SAR; Mi-171Sh version
		(30)	Mi-17/Hip-H	Helicopter	2001			Deal worth $150 m; Mi-171Sh version; delivery 2002–2003
		(540)	AT-6 Spiral/9M114	Anti-tank missile	(1999)	2000–2001	(275)	For Mi-171Sh helicopters; possibly incl AT-9 version
	Ukraine	(12)	An-74/Coaler-B	Transport aircraft	(1997)	1998–2001	(3)	Deal worth $133 m; An-74T-200 version
L:	Russia	(1 420)	BMP-2	IFV	(1991)	1994–2001	(370)	Deal worth $2.2 b for 1500 BMP-2 IFVs and 1000 T-72 tanks probably incl some delivered direct
		(874)	T-72S1	Main battle tank	(1996)	1997–2001	(350)	Deal worth $2.2 b for 1500 BMP-2 IFVs and 1000 T-72 tanks probably incl some delivered direct
		..	AT-3 Sagger/9M14M	Anti-tank missile	(1995)	1996–2001	(1 500)	Iranian designation RAAD; incl I-RAAD version
		..	AT-5 Spandrel/9M113	Anti-tank missile	(1998)	1999–2001	(300)	Iranian designation Towsan-1
	Ukraine	..	An-140	Transport aircraft	1997		..	Iranian designation Iran-140; status uncertain
Ireland								
S:	France	1	AS-365/AS-565 Panther	Helicopter	1999	2001	1	

Recipient/ supplier (S) or licenser (L)	No. ordered	Weapon designation	Weapon description	Year of order/ licence	Year(s) of deliveries	No. delivered/ produced	Comments
Switzerland	40	Piranha-3 8x8	APC	1999	2001	(35)	Deal worth $50.8 m; incl 4 CP, 1 ARV and 1 ambulance version; Piranha-3H version; delivery 2001–2002
UK	1	Mod. Guardian Class	OPV	2000	2001	1	Deal worth $21 m; Irish designation Roisin Class
USA	3	S-92C Helibus	Helicopter	(2001)		..	Deal worth $55–62 m; incl for SAR; option on 2 more; contract not yet signed; delivery 2002
Israel							
S: USA	8	AH-64D Apache	Combat helicopter	2001		..	Deal worth $509 m incl modernization of 10 Israeli AH-64A to AH-64D; option on 12 more; delivery from 2004
	(480)	AGM-114K Hellfire	Anti-tank missile	(2000)		..	For AH-64D helicopters
	1	Boeing-707-320C	Transport aircraft	2000	2001	1	Second-hand; modernized and modified to tanker aircraft after delivery
	50	F-16I	FGA aircraft	1999		..	'Peace Marble-5' deal worth $2.5 b (offsets 25%); financed by USA; Israeli designation Suefa; delivery 2003–2005
	52	F-16I	FGA aircraft	2001		..	'Peace Marble-5' deal worth $2 b (incl $1.3 b for aircraft and $300 m for engines); Israeli designation Suefa; delivery 2006–2009
	3	Gulfstream-5	Transport aircraft	2001		..	Deal worth $174 m; modified in Israel to ELINT aircraft
	24	S-70A/UH-60L	Helicopter	2001		..	Deal worth $217 m; delivery 2002
	(8)	Super King Air-200	Light transport ac	1999	1999–2001	(8)	Israeli designation Zufut; incl for EW and ELINT
	(10)	AN/APG-78 Longbow	Combat ac radar	2000		..	Aid worth $509 m; for modernization of 10 Israeli AH-64A helicopters to AH-64D; delivery 2003; status uncertain
	(64)	AIM-120B AMRAAM	AAM	(1998)	1998–2001	(64)	Deal worth $28 m
	48	AIM-120B AMRAAM	AAM	2001		..	Deal worth $31 m
	40	Popeye-1	ASM	2001		..	
Italy							
S: France	18	Iguane/Varan	MP aircraft radar	(1994)	1997–2001	(14)	For modernization of 18 Atlantic-1 ASW/MP aircraft
Germany (FRG)	2	PzH-2000 155mm	Self-propelled gun	2000	2001	(2)	Deal worth $455 m incl 68 licensed production; bought to start training
Sweden	58	Bv-206S	APC	(2001)		..	Contract probably not yet signed
UK	(200)	Storm Shadow/SCALP	ASM	1999		..	Delivery from 2003
USA	12	C-130J Hercules-2	Transport aircraft	1997	2000–2001	(12)	Incl some tanker/transport version
	4	C-130J-30 Hercules	Transport aircraft	2000		..	Delivery 2002
	6	C-130J-30 Hercules	Transport aircraft	1997	2001	(3)	Delivery 2001–2002
	34	F-16A	FGA aircraft	2001		..	Ex-US; $760 m lease until Eurofighter enters service in 2010; F-16A Block-15ADF version; incl 4 F-16B version; modernized before delivery; possibly 4 more for spares only; delivery 2003/2004

INTERNATIONAL ARMS TRANSFERS 433

							Comments
	4	RQ-1A Predator	UAV	2001			Deal worth $619 m (offsets worth up to $1.1 b); option on 2 more; delivery 2004–2006
	9	LVTP-7A1/AAV-7A1	APC	1999	2001	(9)	Deal worth $55 m; option on 2 more; delivery 2002 Ex-US; deal worth $90 m incl modernization with US-supplied equipment after delivery and modernization of 25 Italian LVTP-7A1
	233	AIM-120B AMRAAM	AAM	1997	2000–2001	(133)	Deal worth $116 m; for AV-8B+ FGA aircraft
	(735)	FIM-92A Stinger	Portable SAM	1998	2001	(500)	Deal worth $110 m
	..	FIM-92A Stinger	Portable SAM	2000		..	Delivery 2002–2004
L: Germany (FRG)	68	PzH-2000	Self-propelled gun	2000		..	Deal worth $455 m incl 2 direct delivered; delivery from 2004
	2	Type-212	Submarine	1997		..	Option on 2 more; delivery 2004–2005

Japan

							Comments
S: France	2	Ocean Master	MP aircraft radar	(2001)		..	For 2 Gulfstream-5 MP aircraft delivered from USA
Italy	5	127mm/54	Naval gun	(1999)		..	For 5 Improved Murasame Class frigates produced in Japan
USA	(20)	AH-64D Apache	Combat helicopter	2001		..	'AH-X' programme; delivery from 2005
	4	B-767	Tanker aircraft	(2001)		..	'KC-X' programme; contract not yet signed
	(21)	BAe-125/RH-800	MP aircraft	1995	1997–2001	(13)	'H-X' programme; for SAR; Japanese designation U-125A
	(9)	Gulfstream-4	Transport aircraft	1994	1996–2000	(5)	'U-X' programme; Japanese designation U-4
	2	Gulfstream-5	Transport aircraft	2001		..	Deal worth $100 m; for Coast Guard long-range MP incl against piracy in South-East Asia
	(20)	Super King Air-350	Light transport ac	1997	1999–2001	(3)	Incl for reconnaissance; Japanese designation LR-2
	(72)	M-270 MLRS 227mm	MRL	1993	1995–2001	(63)	Assembled in Japan
	(13)	AN/APS-145	AEW radar	2000		..	For modernization of E-2C AEW&C aircraft; delivery from 2004
	30	Phalanx Mk-15	CIWS	(1993)	1996–2001	(12)	For 12 Murasame Class frigates and 3 Oosumi Class AALS produced in Japan; incl some Block-1B version
	40	AIM-120B AMRAAM	AAM	(1998)	2001	40	Deal worth $22 m
	(16)	RIM-66M Standard-2	SAM	1999	2001	(16)	
	16	RIM-66M Standard-2	SAM	(2001)		..	Deal worth $27 m; delivery 2003
	(336)	RIM-7M Seasparrow	SAM	1993	1996–2001	(167)	For 9 Murasame Class and 5 Improved Murasame Class frigates
L: France	..	MO-120-RT-61 120mm	Mortar	1992	1993–2001	(331)	Incl for use with Type-96 mortar carrier produced in Japan
USA	(40)	AH-64D Apache	Combat helicopter	2001		..	'AH-X' programme; delivery from 2005
	79	CH-47D Chinook	Helicopter	1986	1988–2001	(57)	CH-47J and CH-47JA versions
	(203)	F-15C Eagle	FGA/trainer aircraft	1978	1982–99	(191)	F-15J version; incl 38 F-15DJ version; delivery 1982–2006
	210	Hughes-500M/OH-6D	Light helicopter	1977	1978–2001	(210)	OH-6D and OH-6DA version; incl for training
	(64)	S-70/UH-60J Blackhawk	Helicopter	1988	1991–2001	(41)	
	(80)	S-70/UH-60J Blackhawk	Helicopter	1995	1998–2001	(18)	UH-60JA version; deal worth $2.67 b
	101	S-70B/SH-60J Seahawk	ASW helicopter	1988	1991–2001	(86)	

434 MILITARY SPENDING AND ARMAMENTS, 2001

Recipient/ supplier (S) or licenser (L)	No. ordered	Weapon designation	Weapon description	Year of order/ licence	Year(s) of deliveries	No. delivered/ produced	Comments
S: Belgium	(27)	Sea Vue	MP aircraft radar	1992	1995–2001	19	For 27 BAe-125-800/RH-800 (U-125A) SAR aircraft delivered from UK and USA
Jordan							
S: UK	(110)	Scorpion	Light tank	(1999)	2001	(110)	Ex-Belgian
	16	T-67 Firefly	Trainer aircraft	2001		. .	T-67M260 version; delivery 2002
	(4)	Aigis	APC	2000	2000–2001	(4)	Jordanian designation AB-2
	(288)	Challenger	Main battle tank	1999	1999–2001	(242)	Ex-UK; Jordanian designation Al Hussein
USA	(9)	Bell-209/AH-1F	Combat helicopter	1998	2000–2001	(9)	Ex-US; aid
	(23)	M-901 ITV	Tank destroyer (M)	1999	2000–2001	(23)	Ex-US; aid
	(425)	BGM-71 TOW	Anti-tank missile	2000	2001	(425)	Probably BGM-71E TOW-2A version
	(562)	BGM-71 TOW	Anti-tank missile	2001		. .	BGM-71E TOW-2A version; delivery 2002–2003
	(110)	Javelin	Anti-tank missile	2001		. .	Deal worth $12 m incl 30 launchers
Ukraine	50	BTR-94	IFV	1999	2000–2001	(50)	
Kazakhstan							
S: Russia	(38)	Su-27S/Flanker-B	Fighter aircraft	(1995)	1996–2001	(18)	Ex-Russian; payment for Russian debt to Kazakhstan
Kuwait							
S: China	(27)	PZL-45 155mm	Self-propelled gun	1998	2000–2001	(27)	Deal worth $186.5 m
	(24)	PZL-45 155mm	Self-propelled gun	1998	2000–2001	(27)	
Egypt	2	Skyguard	SAM system	2001		. .	
Germany (FRG)	11	Tpz-1 Fuchs	APC	2000		. .	NBC reconnaissance version
UK	(80)	Sea Skua	Anti-ship missile	1997	2000–2001	(80)	Deal worth $89 m; Sea Skua SL version; for 8 PB-37BRL Type FAC
Latvia							
S: Czech Republic	(5)	T-55AM-2	Main battle tank	(1999)	2001	(5)	Ex-Czech; aid
Norway	2	Storm Class	FAC(M/T)	2001	2001	2	Ex-Norwegian; modernized before delivery; aid; 1 more delivered for spares only
Sweden	(1)	Giraffe-40	Surveillance radar	(2000)	2001	(1)	Ex-Swedish; aid
Unknown	1	AN/TPS-117	Surveillance radar	2001		. .	Part of 'BALTNET' air-surveillance network; delivery 2003

INTERNATIONAL ARMS TRANSFERS 435

S: USA	M-113A2	APC		(19)	
Lesotho					
S: Germany (FRG)	Bo-105L	Light helicopter	(2000)	2	Bo-105LSA-3 version
Lithuania					
S: Norway	Storm Class	FAC(M/T)	(2000)	2	Ex-Norwegian; modernized before delivery; aid; 1 more delivered for spares only; Lithuanian designation Dzukas Class
Sweden	Giraffe-40	Surveillance radar	(2000)	(1)	Ex-Swedish; aid
USA	Javelin	Anti-tank missile	2001	(75)	Deal worth $10 m; delivery 2002/2004
Macedonia					
S: Greece	Bell-205/UH-1H	Helicopter	2001	2	Ex-Greek; aid; for use against NLA rebels; 2 more cancelled
USA	Cessna-337/O-2	Light aircraft	2001	1	Leased from and operated by US civilian company for surveillance
Ukraine	Mi-17/Hip-H	Helicopter	(2000)	(8)	Ex-Ukrainian; for use against NLA rebels
	Mi-24V/Mi-35/Hind-E	Combat helicopter	(2000)	12	Ex-Ukrainian; incl 2 Mi-24K reconnaissance version; for use against NLA rebels
	Su-25/Frogfoot-A	Ground attack ac	2001	4	Ex-Ukrainian; incl 1 Su-25UB trainer version; mainly for use against NLA rebels
	BM-21	MRL	2001	(6)	Ex-Ukrainian
	T-72	Main battle tank	(2001)	(31)	Ex-Ukrainian
Malaysia					
S: Brazil	Astros-2	MRL	(2000)	. .	Delivery 2002
France	AS-555UN Fennec	Light helicopter	2001	. .	Deal worth $38 m; delivery 2003
	AS-555UN Fennec	Light helicopter	(2001)	. .	Contract not yet signed
Germany (FRG)	TRS-3D	Surveillance radar	(2000)	. .	For 6 MEKO-A100 Type frigates
	MEKO-A100 Type	Frigate	1999	. .	'New Generation Patrol Vessel' (NGPV) programme worth $1.42 b incl 4 licensed production; MEKO-100RMN version; assembled in Malaysia; delivery 2004
Italy	RAT-31S	Surveillance radar	2001	. .	Deal worth $54 m; RAT-31DL version
	Skyguard	Fire control radar	(2000)	. .	For 6 MEKO-A100 Type frigates; designation uncertain
Pakistan	HN-5A/Anza-1	Portable SAM	2001	. .	Deal worth $12.8 m
	Red Arrow-8	Anti-tank missile	2001	. .	Deal worth $8.1 m
Russia	Mi-17/Hip-H	Helicopter	(2001)	. .	Mi-17-1V version
	N-019ME Topaz	Combat ac radar	(1999)	. .	For modernization of 17 MiG-29N FGA aircraft

Recipient/ supplier (S) or licenser (L)	No. ordered	Weapon designation	Weapon description	Year of order/ licence	Year(s) of deliveries	No. delivered/ produced	Comments
	(204)	AA-12 Adder/R-77	AAM	(1997)			For 17 MiG-29N FGA aircraft; status uncertain
	..	AT-13 Saxhorn/9M131	Anti-tank missile	2001	2001	(100)	Deal worth $30 m
South Africa	22	G-5 155mm	Towed gun	2000	2001	(3)	Deal worth $50 m; G-5 Mk-3 version; delivery 2001–2002
Switzerland	9	PC-7-2	Trainer aircraft	2000	2001	(9)	Deal worth $28 m; order delayed from 1997 to 2000 due to financial crisis
Turkey	44	AIFV	IFV	2000		..	Deal worth $300 m incl 167 other versions; incl assembly in Malaysia; delivery from 2002
	167	AIFV-APC	APC	2000		..	Deal worth $300 m incl 44 IFV version; incl ambulance, ALV and CP version; incl assembly in Malaysia, delivery from 2002
UK	6	Super Lynx	AS/ASW helicopter	1999		..	Deal worth $158 m; delivery 2002/2003
	(48)	Sea Skua	Anti-ship missile	2001		..	For 6 Super Lynx helicopters; delivery 2002
L: Germany (FRG)	4	MEKO-A100 Type	Frigate	1999		..	'New Generation Patrol Vessel' (NGPV) programme worth $1.42 b incl 2 delivered direct; MEKO-100RMN version; delivery from 2005
Malta							
S: UK	1	Bulldog	Trainer aircraft	2001	2001	1	Ex-UK
Mexico							
S: Brazil	1	EMB-145AEW&C	AEW&C aircraft	2001		..	Deal worth $230–250 m incl 2 MP version; mainly for anti-narcotics operations; delivery 2004
	2	EMB-145MP	MP aircraft	2001		..	Deal worth $230–250 m incl 1 AEW version; mainly for anti-narcotics operations; delivery around 2004
Canada	1	DHC-8 Dash-8	Transport aircraft	2000	2001	1	DHC-8-200 version; option on 1 more
Italy	30	SF-260M	Trainer aircraft	1999	2000–2001	(30)	SF-260E version
Sweden	1	PS-890 Erieye	AEW radar	2001		..	For EMB-145 AEW aircraft delivered from Brazil; delivery 2004
USA	2	Sea Vue	MP aircraft radar	2001		..	For 2 EMB-145 MP aircraft delivered from Brazil
Morocco							
S: France	2	Floreal Class	Frigate	1998		..	Deal worth $130–140 m; delivery delayed from 2001–2002 to 2003–2004
Myanmar (Burma)							
S: Russia	10	MiG-29S/Fulcrum-C	FGA aircraft	2001		..	Deal worth $130 m; incl 2 MiG-29UB trainer version
	(60)	AA-10c/d Alamo/R-27E	AAM	2001		..	For MiG-29 FGA aircraft
	(60)	AA-11 Archer/R-73	AAM	2001		..	For MiG-29 FGA aircraft

INTERNATIONAL ARMS TRANSFERS 437

	No.	Weapon designation	Weapon description	Year of order	Year(s) of deliveries	No. delivered	Comments
S: Israel	4	A-4N Skyhawk-2	FGA aircraft	2000	2001	(4)	Ex-Israeli; modernized and modified for FRG before delivery; leased and operated by civilian company in FRG for target towing for NATO
Namibia							
S: China	4	K-8 Karakorum-8	Jet trainer aircraft	2000	2001	4	
Moldova	(2)	Mi-8T/Hip-C	Helicopter	2001	2001	(2)	Second-hand; lease
Unknown	2	Mi-24D/Mi-25/Hind-D	Combat helicopter	2001	2001	(2)	Second-hand; supplier could be Libya; possibly leased
Nepal							
S: India	2	SA-315B Lama	Light helicopter	2001	2001	2	Probably ex-Indian; aid
Kazakhstan	2	Mi-17/Hip-H	Helicopter	(1999)	2001	2	Deal worth $5.8–6.97 m; second-hand; modernized before delivery; for use against Maoist rebels; supplier could be Kyrgyzstan
Netherlands							
S: Germany (FRG)	60	PzH-2000 155mm	Self-propelled gun	(2001)		..	Offsets 100%; contract not yet signed; delivery from 2003/2004
	874	FIM-92C Stinger	Portable SAM	(1992)	1998–2001	(800)	Part of 'European Stinger Production Programme' involving production of components in FRG, Greece, Netherlands and Turkey and final assembly in FRG; delivery 1998–2001/2002
Israel	2 400	Spike/NT-G Gill	Anti-tank missile	2001		..	Deal worth $150–225 m; delivery from 2003
Italy	4	127mm/54	Naval gun	1996		..	Ex-Canadian guns sold back to producer and modernized before delivery; for 2 De Zeven Provincien Class destroyers produced in Netherlands
USA	30	AH-64D Apache	Combat helicopter	1995	1998–2001	(24)	Deal worth $686 m (offsets $873 m); delivery 1998–2002
	(139)	AN/APG-66	Combat ac radar	1993	1996–2001	(139)	For 'Mid-Life Update' (MLU) modernization of 139 F-16A/B FGA aircraft to F-16AM/BM (F-16C/D) version
	605	AGM-114K Hellfire	Anti-tank missile	1995	1996–2001	(495)	Deal worth $127 m; for AH-64D helicopters
	16	RIM-66M Standard-2	SAM	(1998)		..	Deal worth $24 m incl 8 training missiles; for De Zeven Provincien Class destroyers
	(240)	RIM-7PTC ESSM	SAM	(2001)		..	For 4 De Zeven Provincien Class destroyers; contract not yet signed
New Zealand							
S: Canada	105	Piranha/LAV-25	IFV	2001		..	Deal worth $263 m (offsets worth $3 m); delivery 2003–2004
USA	4	SH-2G Super Seasprite	AS/ASW helicopter	1997	2001	4	Deal worth $185 m (offsets 36%); SH-2G(NZ) version
	1	SH-2G Super Seasprite	AS/ASW helicopter	1999		..	Deal worth $23 m; SH-2G(NZ) version; delivery 2002

Recipient/ supplier (S) or licenser (L)	No. ordered	Weapon designation	Weapon description	Year of order/ licence	Year(s) of deliveries	No. delivered/ produced	Comments
Nigeria							
S: Russia	8	Mi-34/Hermit	Light helicopter	1999	2000–2001	3	
North Korea							
L: Russia	..	AT-4 Spigot/9M111	Anti-tank missile	(1987)	1992–2001	(1 000)	Ordered from Soviet Union and produced under Russian licence after breakup of Soviet Union
	..	SA-16 Gimlet/Igla-1	Portable SAM	(1989)	1992–2001	(500)	More possibly produced for export; probably ordered from Soviet Union and produced under Russian licence after breakup of Soviet Union
Norway							
S: France	1 225	Eryx	Anti-tank missile	1998	2001	(1 225)	
Multiple sellers	6	NH-90 NFH	AS/ASW helicopter	2001		..	Deal worth $400 m incl 8 NH-90 TTH version; for Coast Guard; ordered from France and Germany and produced in Finland; delivery 2005–2008
	8	NH-90 TTH	Helicopter	2001		..	Deal worth $400 m incl 6 NH-90 NFH version; for Coast Guard; for SAR; ordered from France and Germany and produced in Finland; option on 10 more; delivery 2005–2008
Netherlands	52	Leopard-2A4	Main battle tank	2000	2001	(20)	Ex-Dutch; deal worth $168 m incl ammunition; option on 4 more; delivery 2001–2002
Spain	5	Mod. F-100 Class	Frigate	2000		..	Deal worth $1.5 b incl radars from USA (offsets 100% in 10 years incl NASAMS SAM system and Penguin missiles for Spain); incl 3 assembled or licensed production in Norway; Norwegian designation Nansen Class; delivery 2005–2009
Sweden	12	Arthur	Artillery radar	1997	1999–2001	(11)	Deal worth $85 m; delivery 1999–2002
USA	(58)	AN/APG-66	Combat ac radar	1993	1996–2001	(58)	For 'Mid-Life Update' (MLU) modernization of 58 F-16A/B FGA aircraft to F-16AM/BM (F-16C/D) version
	500	AIM-120A AMRAAM	AAM	1996	1998–2000	(384)	Deal worth $150 m; assembled in Norway; for F-16AM/BM FGA aircraft
	5	AN/SPY-1F	Surveillance radar	(2000)		..	Deal worth $500 m incl AEGIS combat system and Mk-41 SAM launchers; for 5 Nansen Class frigates
	(240)	RIM-7PTC ESSM	SAM	(2000)		..	For 5 F-100 (Nansen) Class frigates
Oman							
S: France	(10)	VBL	Recce vehicle	2000	2001	(10)	Fitted with ALBI Mistral SAM launcher
	(100)	Mistral	Portable SAM	2000	2001	(10)	For use with ALBI launcher on VBL armoured vehicle
UK	16	Super Lynx-300	Helicopter	2001		..	

INTERNATIONAL ARMS TRANSFERS 439

Supplier	No. ordered	Weapon designation	Weapon description	Year of order	Year(s) of deliveries	No. delivered	Comments
USA	80	Piranha 8x8	APC	...	2000	20	Deal worth $172 m; Challenger-2 (Oman) Phase-2 version
	...				2001	(30)	Incl ARV, CP, 81mm mortar carrier, ambulance, artillery observation and 1 other version; delivery 2001–2002
	...	Martello S-743D	Surveillance radar	1999			Delivery 2002
	12	F-16C	FGA aircraft	(2001)			Deal worth $1.12 b incl armament and LANTIRN radars; F-16C/D Block-50/52 version; delivery 2004–2005; contract not yet signed
	14	LANTIRN	Aircraft radar	(2001)			For F-16C/D FGA aircraft; contract not yet signed
	80	AGM-65D Maverick	ASM	(2001)			For F-16C/D FGA aircraft; AGM-65D/E version; contract not yet signed
	50	AIM-120B AMRAAM	AAM	(2001)			For F-16C/D FGA aircraft; AIM-120C version; contract not yet signed
	100	AIM-9M Sidewinder	AAM	(2001)			For F-16C/D FGA aircraft; AIM-9M-8/9 version; contract not yet signed
	20	RGM-84 Harpoon	Anti-ship missile	(2001)			For F-16C FGA aircraft; AGM-84D version; contract not yet signed
	(562)	BGM-71 TOW	Anti-tank missile	2001			BGM-71E TOW-2A version; delivery 2002–2003

Pakistan

S:

Supplier	No. ordered	Weapon designation	Weapon description	Year of order	Year(s) of deliveries	No. delivered	Comments
China	(40)	F-7MG	Fighter aircraft	(2001)	2001	40	Pakistani designation F-7PG; no. ordered could be up to 80
	(3)	Type-347G	Fire control radar	(1996)	1997–2001	(3)	For 3 Jalalat-2 Class FAC produced in Pakistan
	(24)	C-802/CSS-N-8 Saccade	Anti-ship missile	(1996)	1997–2001	(24)	For 3 Jalalat-2 Class FAC
France	40	Mirage-5	FGA aircraft	1996	1998–2001	40	Ex-French; 'Blue Flash-6' deal worth $120 m; modernized before delivery; incl 6 Mirage-3D trainer version
	2	Agosta-90B Type	Submarine	1994	1999	1	Deal worth $750 m (+ $200 m for modernization of Pakistan Naval Dockyard to build submarines) incl 1 licensed production; incl 1 assembled in Pakistan; Pakistani designation Khalid Class; delivery 1999–2002
	(96)	F-17P	AS torpedo	(1996)	1999–2001	(72)	F-17P Mod-2 version; for 3 Agosta-90B Type (Khalid Class) submarines
	(24)	SM-39 Exocet	Anti-ship missile	1994	1999–2001	(20)	Deal worth $100 m; for 3 Agosta-90B Type (Khalid Class) submarines
Indonesia	4	CN-235MPA	MP aircraft	2001	2001	...	Deal worth $49 m
Lebanon	(10)	Mirage-3E	FGA aircraft	(2000)		10	Ex-Lebanese; deal worth $4.7 m; Mirage-3EL version; incl 1 Mirage-BL trainer version
Sweden	...	RBS-70	Portable SAM	(1985)	1988–2001	(350)	Assembled in Pakistan
	(24)	Type-43	ASW torpedo	1994	1999–2001	(12)	Type-43X2 version; for 6 modernized Tariq (Amazon) Class frigates
USA	(6)	S-70/UH-60 Blackhawk	Helicopter	(2001)		...	Designation uncertain; probably ex-US; part of $73 m aid for Afghan border patrol

L:

Supplier	No. ordered	Weapon designation	Weapon description	Year of order	Year(s) of deliveries	No. delivered	Comments
China	(150)	FC-1	FGA aircraft	1999		...	Developed for Pakistan; delivery possibly from 2003
	...	QW-1 Vanguard	Portable SAM	(1993)	1994–2001	(550)	Pakistani designation Anza-2
	...	Red Arrow-8	Anti-tank missile	1989	1990–2001	(8 600)	Pakistani designation Baktar Shikan
France	1	Agosta-90B Type	Submarine	1994			Deal worth $750 m (+ $200 m for modernization of Pakistan Naval Dockyard to build submarines) incl 2 delivered direct; Pakistani designation Khalid Class; delivery 2002

440   MILITARY SPENDING AND ARMAMENTS, 2001

Recipient/ supplier (S) or licenser (L)	No. ordered	Weapon designation	Weapon description	Year of order/ licence	Year(s) of deliveries	No. delivered/ produced	Comments
Italy	(215)	Grifo-7	Combat ac radar	1995	2000–2001	(51)	For modernization of some 35 Mirage-3, 80 F-7MG and 100 F-7MP fighter aircraft
Peru							
S: Poland	2	An-28TD Bryza-1TD	Light transport ac	(2001)	2001	2	Possibly second-hand; supplier uncertain
Russia	150	T-72M1	Main battle tank	(1999)	2001	(150)	Probably ex-Russian
USA	5	K-1200 K-MAX	Helicopter	2000	2001	5	Deal worth $21 m; aid; for Police anti-narcotics operations
Unknown	1	L-410UVP Turbolet	Transport aircraft	2001	2001	1	Second-hand
Philippines							
S: Israel	2	Blue Horizon	UAV	2001	2001	2	Deal worth $1–1.2 m or $2–12 m; for use against Abu Sayyaf and other Muslim rebels; possibly delivered from Singapore
USA	(5)	Bell-205/UH-1H	Helicopter	(2001)	2001	5	Ex-US; aid
	1	C-130B Hercules	Transport aircraft	2000	2001	1	Ex-US; aid
	4	C-130K Hercules	Transport aircraft	(2001)		..	Ex-UK aircraft sold back to US producer, deal worth $41 m; modernized before delivery; incl for MP
	(1)	Cyclone Class	Patrol craft	(2001)		..	Ex-US; aid
Poland							
S: Germany (FRG)	23	MiG-29S/Fulcrum-C	FGA aircraft	(2001)		..	Ex-FRG; aid; incl 4 MiG-29UB trainer version; contract signed 2002; delivery 2002–2004
	(100)	AA-10a/b Alamo/R-27	AAM	(2001)		..	Ex-FRG; aid; part of deal for 23 MiG-29 fighter aircraft
	(400)	AA-11 Archer/R-73	AAM	(2001)		..	Ex-FRG; aid; part of deal for 23 MiG-29 fighter aircraft
	(150)	AA-8 Aphid/R-60	AAM	(2001)		..	Ex-FRG; aid; part of deal for 23 MiG-29 fighter aircraft
	128	Leopard-2A4	Main battle tank	2001		..	Ex-FRG
Multiple sellers	..	MU-90 Impact	ASW torpedo	(2001)		..	Incl for Mi-14PL helicopters and MEKO-A100 Type and Perry (Pulaski) Class frigates
Netherlands	3	STING	Fire control radar	2001		..	For modernization of 3 Orkan Class corvettes
Spain	8	CASA C-295M	Transport aircraft	2001		..	Deal worth $212 m (offsets 100%); delivery 2003–2005
Sweden	3	Giraffe AMB-3D	Surveillance radar	2001		..	For modernization of 3 Orkan Class corvettes
	(60)	RBS-15 Mk-3	Anti-ship missile	(2001)		..	For 3 Orkan Class corvettes and probably for 2 MEKO-A100 Type frigates; contract not yet signed
	(24)	RBS-15SF	Anti-ship missile	2000	2001	(24)	Ex-Swedish; deal worth $10 m; ordered for temporary use until RBS-15 Mk-3 delivered

USA	4	SH-2G Super Seasprite	AS/ASW helicopter	(2001)			Ex-US; licensed production; for use on Polish chassis; Polish designation Chrobry or Krab
	9	MSTAR	Battlefield radar	2000			Ex-US; deal worth $20 m; delivery 2002
	2	Perry Class	Frigate	1999			Deal worth $4.1 m; Polish designation uncertain
	(108)	RIM-66B Standard-1MR	SAM	(1999)	2000–2002/2003		Ex-US; aid; Polish designation Pulaski Class; delivery 2000–2002/2003
	..	RIM-7PTC ESSM	SAM	(2001)	2000–2001		Ex-US; for 2 Perry (Pulaski) Class frigates
L: Germany (FRG)	2	MEKO-A100 Type	Frigate	2001			For 2 MEKO-A100 Type frigates; contract not yet signed
Russia	(9)	An-28/M-28B Bryza-1R	MP aircraft	(1992)	1993–2001		Polish designation Project-621 or Gawron-2 Type; option on 4–5 more; delivery 2003–2005
UK	(72)	AS-90 turret	Turret	1999			For use on Polish chassis; Polish designation Chrobry or Krab
Portugal							
S: Germany (FRG)	9	EC-135/EC-635	Helicopter	1999	2001		Deal worth $35–38 m
Multiple sellers	(12)	EH-101-500	Helicopter	(2001)			Deal worth $287 m; for SAR; suppliers Italy and UK; delivery 2004–2006; contract not yet signed
USA	20	F-16A	FGA aircraft	1998	2001	(2)	Ex-US; 'Peace Atlantis-2' deal worth $268 m; modernized to F-16AM/BM (F-16C/D) in Portugal after delivery; incl 4 F-16B version; 5 more ordered for spares only; delivery 2001–2003
	20	AN/APG-66	Combat ac radar	(2000)	2001	(2)	For 'Mid-Life Update' (MLU) modernization of 20 F-16A/B FGA aircraft to F-16AM/BM (F-16C/D) 2001–2003
Qatar							
S: France	(48)	Apache	ASM	1994	1999–2001	(30)	For Mirage-2000-5 FGA aircraft; Black Pearl version
Romania							
S: France	(660)	R-550 Magic-2	AAM	(1996)	2001	(100)	For 110 MiG-21 Lancer (modernized MiG-21MF/UM) and possibly for MiG-29 fighter aircraft; may incl assembly or licensed production in Romania
Germany (FRG)	(32)	Gepard	AAV(G)	(1997)	1999–2001	(32)	Ex-FRG; aid worth $37 m; probably modernized before delivery; 11 more delivered for spares only
Israel	(960)	Spike/NT-G Gill	Anti-tank missile	(1998)	1999–2001	(250)	
USA	5	Shadow-600	UAV	2000	2001	(5)	For 24 modernized SA-330 (IAR-330) helicopters; designation uncertain Deal worth $7.5 m
Saudi Arabia							
S: Canada	(8)	Bell-412	Helicopter	1999	2001	(7)	No. ordered could be up to 36

442 MILITARY SPENDING AND ARMAMENTS, 2001

Recipient/ supplier (S) or licenser (L)	No. ordered	Weapon designation	Weapon description	Year of order/ licence	Year(s) of deliveries	No. delivered/ produced	Comments
	492	Piranha 8x8	APC	1990	1994–2001	(492)	Ordered via USA as FMS deal worth $700 m incl 625 other version; incl 71 ambulance, 18 ALV, 182 CP, 67 ARV and 34 AEV version and 73 AFSVs fitted with UK AMS 120mm mortar; for National Guard
	130	Piranha/LAV-90	AFSV	2000	2000–2001	(66)	Deal worth $416 m incl 1827 BGM-71 missiles; ordered via USA as part of 1990 FMS deal worth $700 m incl 987 other version; for National Guard
France	12	AS-532U2/AS-332L2	Helicopter	1996	1998–2001	12	Deal worth $508 m; for combat SAR; armed with 20mm gun
	3	La Fayette Class	Frigate	1994		..	'Sawari-2' deal worth $3.42 b incl other weapons and construction of a naval base (offsets 35%); also designated F-3000S Type; Saudi designation Al Ryadh Class; delivery from 2002
	48	ASTER-15	SAM	(1997)		..	For 3 La Fayette Class frigates
	(48)	MM-40 Exocet	Anti-ship missile	1994	2000–2001	(20)	For 3 La Fayette Class frigates; MM-40 Block-2 version
Italy	(8)	Bell-412/AB-412	Helicopter	1999	2001	(7)	Delivery 2001–2002
USA	105	AGM-65D Maverick	ASM	2001		..	Deal worth $21 m; including 7 AGM-65G version
	500	AIM-120B AMRAAM	AAM	2000		..	For F-15 FGA aircraft; delivery 2003
	1 827	BGM-71 TOW	Anti-tank missile	2000	2001	(1 000)	Deal worth $416 m incl 132 Piranha IFV from Canada; BGM-71E TOW-2A version; for National Guard
	(562)	BGM-71 TOW	Anti-tank missile	2001		..	BGM-71E TOW-2A version; delivery 2002–2003
	(108)	VT-1	SAM	(1994)		..	For use with Crotale SAM systems on 3 La Fayette Class frigates; ordered and delivered via France

Singapore

Recipient/ supplier (S) or licenser (L)	No. ordered	Weapon designation	Weapon description	Year of order/ licence	Year(s) of deliveries	No. delivered/ produced	Comments
S: France	(671)	MILAN	Anti-tank missile	(1996)	1997–2001	(671)	Designation uncertain
	6	La Fayette Class	Frigate	2000		..	Incl 5 assembled in Singapore; delivery 2005–2009
Israel	(600)	Python-4	AAM	(1997)	1997–2001	(360)	For F-5S and F-16 FGA aircraft
	..	Spike/NT-G Gill	Anti-tank missile	1999	2001	(100)	Spike version; assembled in Singapore
Sweden	(32)	Type-43	ASW torpedo	(1997)	2000–2001	(32)	Type-431 version for 4 Sjöormen (Challenger) Class submarines
	(80)	Type-613	AS torpedo	(1997)	2000–2001	(80)	For 4 Sjöormen (Challenger) Class submarines
	3	Sjöormen Class	Submarine	1997	2000–2001	3	Ex-Swedish; modernized before delivery; incl 1 based in Sweden until 2003 for training under 'Riken' programme; Singaporean designation Challenger Class; 1 more delivered for spares only
USA	8	AH-64D Apache	Combat helicopter	1999		..	Deal worth $647 m (incl $25.9 m for Longbow radars) incl 192 AGM-114K missiles; delivery from 2002
	12	AH-64D Apache	Combat helicopter	2001		..	Deal worth $617 m; contract not yet signed; delivery from 2005
	(192)	AGM-114K Hellfire	Anti-tank missile	(2001)		..	For AH-64D helicopters; contract not yet signed
	(6)	CH-47D Chinook	Helicopter	(1997)	2000–2001	(6)	
			Helicopter	1999		..	

100	AIM-120B AMRAAM	AAM	2001		. .	Deal worth $100 m; for F-16 and possibly F-5S FGA aircraft
L: Russia						
(3 000)	SA-16 Gimlet/Igla-1	Portable SAM	(2000)		. .	Status uncertain
Slovenia						
S: France						
2	AS-532UL/AS-332L1	Helicopter	2001		. .	Deal worth $24 m; AS-532AL version; delivery 2003
Germany (FRG)						
6	Roland-2	SAM system	2000	2001	6	Ex-FRG; aid worth $12 m incl missiles
(120)	Roland-2	SAM	2000	2001	(120)	Ex-FRG; aid worth $12 m incl 6 Roland-2 SAM systems
USA						
1	M-1114 ECV	APC	(2000)	2001	(1)	
L: Austria						
(20)	Pandur	APC	1998	1999–2001	(15)	Slovenian designation Valuk; option on 50 more
South Africa						
S: France						
(64)	MM-40 Exocet	Anti-ship missile	2000		. .	For 4 MEKO-A200 Type frigates; delivery 2002–2004
. .	Mistral	Portable SAM	(2001)		. .	For Rooivalk combat helicopter; contract not yet signed
4	Lindau Class	Minesweeper	2000	2001	4	Ex-FRG; aid; 2 more delivered for spares only
4	MEKO-A200 Type	Frigate	(2000)		. .	Deal worth $0.8–1.115 b (offsets $3.2 b incl $403 m for arms industry); for delivery 2004–2005
Germany (FRG)						
3	Type-209/1400	Submarine	1999		. .	Deal worth $600–795 m (offsets 375–430%); Type-209/1400MOD version; delivery 2005–2007
Italy						
30	A-109 Hirundo	Light helicopter	1999		. .	Deal worth $240 m (offsets $977 m incl $191 m for arms industry; incl transfer of production-line of A-109 and A-119 to South Africa); option on 10 more; incl 25 assembled in South Africa; delivery 2002–2005
Sweden						
9	JAS-39 Gripen	FGA aircraft	1999		. .	Deal worth $1.16 b incl 12 Hawk-100 from UK (offsets $8.7 b incl $1.5 b for arms industry); JAS-39B version; for delivery 2006/2007–2009; option on 19 more
UK						
(12)	Hawk-100	FGA/trainer aircraft	1999		. .	Deal worth $1.16 b incl 9 JAS-39 from Sweden (offsets $8.7 b incl $1.5 b for arms industry); assembled in South Africa; Hawk-120 version; delivery 2005; option on 12 more
South Korea						
S: France						
(48)	Crotale-NG	SAM system	(1999)	1999–2001	(12)	Korean designation Pegasus; for use with Korean developed SAM; mounted on Korean K-200 APC
Indonesia						
8	CN-235	Transport aircraft	1997	2001	2	Deal worth $120–143 m (offsets incl Indonesian order for KT-1 trainer aircraft and other military equipment from Korea); CN-235-220 version; delivery 2001–2002

444 MILITARY SPENDING AND ARMAMENTS, 2001

Recipient/ supplier (S) or licenser (L)	No. ordered	Weapon designation	Weapon description	Year of order/ licence	Year(s) of deliveries	No. delivered/ produced	Comments
Netherlands	3	Goalkeeper	CIWS	1999		..	For 3 KDX-2 Type frigates; delivery from 2002
UK	13	Super Lynx	AS/ASW helicopter	1997	1999–2001	(13)	Deal worth $328 m incl Sea Skua missiles and modernization of 11 South Korean Super Lynx helicopters; Super Lynx Mk-99A version
USA	8	RH-800XP	Reconnaissance ac	1996	2000–2001	(8)	'Peace Pioneer' deal worth $461 m; incl 4 RH-800RA ground-surveillance and RH-800SIG SIGINT version; deal temporarily suspended in 1998 after corruption charges
	(44)	AN/APG-67	Combat ac radar	(2001)		..	For 44 A-50 FGA aircraft produced in South Korea
	7	AN/FPS-117	Surveillance radar	2000		..	Delivery 2002–2004
	3	AN/SPS-49	Surveillance radar	(1999)		..	For 3 KDX-2 Type frigates produced in South Korea
	110	RIM-66M Standard-2	SAM	2000		..	Deal worth $159 m; for 3 KDX-2 Type frigates
	64	RIM-116A RAM	SAM	(2001)		..	For 3 KDX-2 Type frigates; delivery probably 2003
	(425)	BGM-71 TOW	Anti-tank missile	2000	2001	(425)	Probably BGM-71E TOW-2A version
	(111)	MGM-140A1 ATACMS	SSM	2001		..	Deal worth $80.71 m
	100	Popeye-1	ASM	(1997)	2000–2001	(50)	Deal worth $125 m incl modernization of 30 F-4E FGA aircraft; US designation AGM-142; delivery 2000–2003
	(72)	RGM-84 Harpoon	Anti-ship missile	(1994)	1998–2001	(32)	UGM-84 version for Type-209 (Chang Bogo Class) submarines
L: Germany (FRG)	3	Type-209/1200	Submarine	1994	1999–2001	3	Deal worth $510 m; Korean designation Chang Bogo Class
	3	Type-214	Submarine	2000		..	'KSS-2' programme worth $1.1 b (incl $711 m import from FRG); delivery 2007–2009
Netherlands	3	MW-08	Surveillance radar	1999		..	For 3 KDX-2 Type frigates produced in South Korea
	6	STIR	Fire control radar	1999		..	For 3 KDX-2 Type frigates produced in South Korea
USA	20	F-16C	FGA aircraft	2000		..	Deal worth $663 m; delivery from 2003
	3	Mk-45 127mm/54	Naval gun	1999		..	Deal worth $22 m; Mk-45 Mod-4 version for 3 KDX-2 Type frigates produced in South Korea
	200	K-1A1/Type-88	Main battle tank	(1994)	1996–2001	(14)	Deal worth $781 m; incl 2 or 3 prototypes
	57	LVTP-7A1/AAV-7A1	APC	1995	1997–2001	(57)	Deal worth $91 m; incl 5 ARV and 4 CP version; Korean designation Korean Armoured Amphibious Vehicle (KAAV)
	67	LVTP-7A1/AAV-7A1	APC	2000	2001	(10)	Deal worth $99–120 m; Korean designation Korean Armoured Amphibious Vehicle (KAAV); delivery 2001–2006

Spain

S: Argentina	1	CH-47C Chinook	Helicopter	2001	2001	1	Ex-Argentine
Canada	4	AN/APS-504(V)	MP aircraft radar	(1999)		..	For 4 CN-235MP MP aircraft produced in Spain
	4	AN/DS-2 Modified	Fire control radar	1994	1998	2	For 4 Meroka CIWS on 2 Galicia Class AALS

INTERNATIONAL ARMS TRANSFERS 445

	15	EC-120B Colibri	Light helicopter	1999	1999–2001	(12)	Deal worth $205 m (offsets 100%); delivery 1998–2002
Germany (FRG)	4	Buffel	ARV	1998	2000–2001	(15)	Deal worth $15 m; for training
	30	Leopard-2A5	Main battle tank	1998		. .	Deal worth $2.2 b (offsets 80%) incl 12 licensed production and 219 Leopard-2A5+ tanks; delivery from 2002/2003
						. .	Deal worth $2.2 b (offsets 80%) incl 189 licensed production and 16 Buffel ARVs; Leopard-2A5E version; delivery from 2002
Italy	22	B-1 Centauro	Armoured car	1999	2000–2001	(22)	Deal worth $70 m (offsets 100%)
	62	B-1 Centauro	Armoured car	(2001)		. .	Deal worth $185 m; delivery 2004–2006; contract signed 2002
	4	RAN-12L/X	Surveillance radar	(1996)		. .	For 4 De Bazán (F-100) Class frigates produced in Spain 2002–2006
Netherlands	2	RAN-30X	Surveillance radar	(1993)	1998	(1)	For use with Meroka CIWS on 2 Galicia Class AALS produced in Spain
	2	DA-08	Surveillance radar	(1994)	1998	(1)	For 2 Galicia Class AALS produced in Spain
Norway	(20)	Penguin Mk-2-7	Anti-ship missile	(2000)		. .	Deal worth $26.3 m (offsets for Norwegian order for 5 frigates); option on more; for S-70/SH-60B helicopters
Sweden	20	Bv-206S	APC	2000		. .	Incl 10 ordered in 2001 for SEK43 m; option on up to 30 more
Switzerland	18	Piranha-3 8x8	APC	2001		. .	Delivery 2003–2004
USA	6	S-70B/SH-60B Seahawk	AS/ASW helicopter	2000		. .	Deal worth $77 m incl modernization of 6 Spanish S-70B; Spanish designation HS-23; delivery by 2004
	8	AN/SPG-62	Fire control radar	(1996)		. .	For use with Standard and ESSM SAMs on 4 De Bazán (F-100) Class frigates produced in Spain 2002–2006
	4	AN/SPY-1F	Surveillance radar	1996		. .	Deal worth $750 m; part of AEGIS combat system for De Bazán (F-100) Class frigates produced in Spain 2002–2006
	4	Mk-45 127mm/54	Naval gun	1999		. .	Ex-US ; modernized before delivery; for 4 De Bazán (F-100) Class frigates produced in Spain
	(80)	RIM-66M Standard-2	SAM	(2001)		. .	Deal worth $105 m incl 32 training missiles; for 4 F-100 (De Bazán) Class frigates; contract not yet signed
	(384)	RIM-7PTC ESSM	SAM	(2001)		. .	For 4 F-100 (De Bazán) Class frigates; contract not yet signed
	4	AN/MPQ-64 Sentinel	Surveillance radar	(2000)		. .	Ordered via Norway as part of 4 NASAMS SAM systems
	44	AGM-65F Maverick	Anti-ship missile	1999	2001	(44)	For AV-8B FGA aircraft
	. .	AIM-120A AMRAAM	SAM	(2000)		. .	For 4 NASAMS SAM systems ordered from Norway
	100	AIM-120B AMRAAM	AAM	(1998)	2000–2001	(100)	Deal worth $52 m
	226	Javelin	Anti-tank missile	(2001)		. .	Deal worth $25 m incl 12 launchers; contract not yet signed
L: Germany (FRG)	12	Buffel	ARV	1998		. .	Deal worth $2.2 b (offsets 80%) incl 4 direct delivered and 219 Leopard-2A5+ tanks
	189	Leopard-2A5+	Main battle tank	1998		. .	Deal worth $2.2 b (offsets 80%) incl 30 delivered direct and 16 Buffel ARVs; Leopard-2A5E version; delivery from 2003

Sri Lanka
S: China   6   K-8 Karakorum-8   Jet trainer aircraft   (2000)   2001   6

Recipient/ supplier (S) or licenser (L)	No. ordered	Weapon designation	Weapon description	Year of order/ licence	Year(s) of deliveries	No. delivered/ produced	Comments
Czech Republic	(16)	RM-70 122mm	MRL	2000	2000–2001	(16)	Ex-Czech; possibly aid; for use against LTTE rebels
	(41)	T-55AM-2	Main battle tank	2000	2000–2001	(41)	Ex-Czech; possibly aid; for use against LTTE rebels
USA	2	AN/TPQ-36 Firefinder	Artillery radar	2000		. .	
Sweden							
S: Finland	104	XA-200	APC	2000	2001	(8)	Deal worth $67 m; for peacekeeping forces; XA-203S version; including XA-202S CP version; option on 60 more; delivery 2001–2002
Germany (FRG)	(360)	BMP-1	IFV	1994	1998–2001	(360)	Former GDR equipment; modernized in Poland for $25.5 m before delivery; Swedish designation Pbv-501; 66 more delivered for spares only
Italy	10	Buffel	ARV	1999		. .	Deal worth $125 m; option on 4 more; delivery 2002–2003
	20	A-109M	Light helicopter	2001		. .	Deal worth $113 m; Swedish designation Hkp-15; for training; incl components made in South Africa as offsets for South African order for JAS-39 combat aircraft; delivery from 2002
Multiple sellers	18	NH-90 TTH	Helicopter	2001		. .	Deal worth $600 m (offsets incl production of parts for 200 NH-90 worth $200 m); incl 5 for SAR; Swedish designation Hkp-14; ordered from France and Germany and produced in Finland; delivery 2005–2009
Switzerland	5	Piranha-3 10x10	APC	2000		. .	Delivery 2002
L: Germany (FRG)	91	Leopard-2A5+	Main battle tank	1994	1998–2001	(91)	Deal worth $770 m incl 160 ex-FRG Leopard-2 tanks (offsets 120%); option on 90 more; Swedish designation Strv-122
Switzerland							
S: France	12	AS-532UC/AS-332	Helicopter	1998	2000–2001	(7)	Deal worth $208 m; incl 10 assembled in Switzerland; delivery 2000–2002
	2	Master-A	Surveillance radar	1998	2000	(1)	Part of Florako air surveillance network
Germany (FRG)	25	Buffel	ARV	2001		. .	Deal worth $63 m; delivery 2004–2005
UK	(2 000)	Rapier Mk-2	SAM	(2001)		. .	Assembled in Switzerland; contract not yet signed
L: Sweden	185	CV-9030CH	IFV	2000		. .	'Schutzenpanzer-2000' programme worth $424 m incl 1 direct delivered; incl 32 CV-9030CH-COM CP version; delivery 2002–2005; option on 124 more
Syria							
S: Russia	(14)	Su-27SK/Flanker-B	FGA aircraft	(1999)			Status uncertain

INTERNATIONAL ARMS TRANSFERS 447

	No. ordered	Weapon designation	Weapon description	Year of order	Year(s) of deliveries	No. delivered/ produced	Comments
S: USA	13	Bell-206/OH-58D(I)	Combat helicopter	1999	1999–2001	(8)	Deal worth $172 m incl ammunition; assembled in Taiwan
	(21)	Bell-209/AH-1W	Combat helicopter	1997	2000–2001	(21)	Deal worth $479 m
	9	CH-47D Chinook	Helicopter	1999	2001	(1)	Deal worth $300–486 m; delivery 2001–2002
	2	E-2C Hawkeye	AEW&C aircraft	1999			Deal worth $400 m; E-2T version; delivery 2003–2004/2005
	11	S-70B/SH-60B Seahawk	AS/ASW helicopter	1997	2000–2001	(11)	S-70C(M)-2 Thunderhawk version
	146	M-109A5 155mm	Self-propelled gun	(2000)			Deal worth $405 m; status uncertain
	300	M-60A3 Patton-2	Main battle tank	1996	1998–2001	(300)	Ex-US; deal worth $223 m
	20	AN/PPQ-2 PSTAR	Surveillance radar	2000	2001	(20)	Deal worth $18 m; for use with Stinger SAM
	240	AGM-114K Hellfire	Anti-tank missile	1999	2001	(240)	Deal worth $23 m
	40	AGM-65G Maverick	ASM	2001			Deal worth $18 m; for F-16 FGA aircraft; contract not yet signed
	200	AIM-120B AMRAAM	AAM	2000		(50)	Deal worth $150 m; AIM-120C version; for F-16 aircraft; stored in USA until China introduces similar missiles
	1 786	BGM-71 TOW	Anti-tank missile	1997	1999–2001	(1 786)	Deal worth $80 m; BGM-71E TOW-2A version
	(1 299)	FIM-92A Stinger	Portable SAM	1997	2000–2001	(600)	Deal worth $200 m incl 79 Avenger air-defence systems and 50 portable launchers; no. ordered could be up to 1599
	728	FIM-92A Stinger	Portable SAM	1998			Deal worth $180 m incl 61 launchers
	360	Javelin	Anti-tank missile	(2001)			Deal worth $51 m incl 40 launchers; contract not yet signed
	(77)	RGM-84 Harpoon	Anti-ship missile	2001			Delivery before 2004
	(9)	RGM-84 Harpoon	Anti-ship missile	2001			
	(383)	RIM-66B Standard-1MR	SAM	(1994)	1994–2001	(303)	For Perry (Cheng Kung) Class frigates
	4	Kidd Class	Destroyer	2001			Ex-US; deal worth $750–783 m; contract not yet signed
L: Ireland	..	CM-31	APC	(1997)	1999–2001	(30)	
USA	8	Perry Class	Frigate	1989	1993 98	7	Taiwanese designation Cheng Kung Class; 'Kwang Hua-1' project; order for last 1 delayed from 1997 to 2001 for financial reasons; delivery 1993–2003/2004

Thailand

	No. ordered	Weapon designation	Weapon description	Year of order	Year(s) of deliveries	No. delivered/ produced	Comments
S: Germany (FRG)	20	Alpha Jet	FGA/trainer aircraft	1999	2000–2001	20	Ex-FRG; deal worth $34.5 m; modernized before delivery; 5 more delivered for spares only; for ground-attack role
Israel	4	Searcher UAV	UAV	(2000)		..	Deal worth $12 m
UK	2	Super Lynx	AS/ASW helicopter	2001		..	Deal worth $25 m (offsets 50%)
USA	30	Bell-205/UH-1H	Helicopter	2001		..	Ex-US; aid; modernized before delivery for $37 m; delivery 2002–2003/2004
	6	Bell-209/AH-1F	Combat helicopter	2001		..	Ex-US; aid; modernized before delivery for $7 m; delivery 2002–2004
	16	F-16A	FGA aircraft	2000	2001	(5)	Ex-US; deal worth $130–157 m incl 2 more for spares only
	8	AIM-120B AMRAAM	AAM	(2001)		..	For F-16 FGA aircraft; to be stored in USA; reaction to Myanmarese order for MiG-29 FGA aircraft

Recipient/ supplier (S) or licenser (L)	No. ordered	Weapon designation	Weapon description	Year of order/ licence	Year(s) of deliveries	No. delivered/ produced	Comments
	2	S-70A/UH-60L	Helicopter	2001	2001	2	Deal worth $20 m; for patrol and anti-narcotics operations along border with Myanmar
	(2)	S-70A/UH-60L	Helicopter	2001		. .	Delivery 2002
L: UK	3	Khamronsin Class	OPV	1997	2000–2001	(3)	Thai designation Hua Hin Class
Trinidad & Tobago							
S: USA	2	PA-31-300 Navajo	Light transport ac	(2000)	2000–2001	(2)	Probably aid
Tunisia							
S: France	(20)	MILAN	Anti-tank missile	1997		. .	Designation uncertain
Italy	2	G-222	Transport aircraft	2000	2001	2	Ex-Italian; aid
USA	1	C-130B Hercules	Transport aircraft	(2001)	2001	1	Ex-US
Turkey							
S: France	9	Ocean Master	MP aircraft radar	2001		. .	For 9 CN-235MP produced under licence in Turkey
	19 200	Eryx	Anti-tank missile	1998	1998–2001	(1 512)	
	(24)	MM-38 Exocet	Anti-ship missile	(2000)	2001	(8)	Second-hand; for 6 D'Orves Class corvettes
	6	D'Orves Class	Corvette	2000	2001	3	Ex-French; deal worth $210 m (incl $150 m for reactivation of 5); delivery 2001–2002
Germany (FRG)	(1 500)	FIM-92C Stinger	Portable SAM	1986	1998–2001	(1 300)	Part of 'European Stinger Production Programme' involving production of components in FRG, Greece, Netherlands and Turkey and final assembly in FRG; delivery 1998–2001/2002
	1	Frankenthal Class	MCM ship	1999		. .	Deal worth $625 m incl 5 licensed production; delivery 2003/2004
	1	Kiliç Class	FAC(M)	2000		. .	Delivery 2002
Israel	54	EL/M-2032	Combat ac radar	1997	2001	(10)	For 54 F-4E FGA aircraft modernized in deal worth $600–700 m in Israel to F-4E-2020
	(46)	Popeye-1	ASM	(1998)	2000–2001	(15)	For F-4E-2020 FGA aircraft
Italy	5	Bell-412EP/AB-412EP	Helicopter	1998	2001	(3)	Deal worth $52 m; for SAR; AB-412EP version; for Coast Guard
	4	Bell-412EP/AB-412EP	Helicopter	1999		. .	Deal worth $35 m; for SAR; AB-412EP version; for Coast Guard; delivery from 2002
Netherlands	4	LIROD	Fire control radar	2001		. .	For 4 Kiliç Class FAC
	4	MW-08	Surveillance radar	2001		. .	For 4 Kiliç Class FAC
	4	STING	Fire control radar	2001		. .	For 4 Kiliç Class FAC

INTERNATIONAL ARMS TRANSFERS 449

Supplier	No.	Weapon designation	Weapon description	Year of order	Year(s) of delivery	No. delivered	Comments
Norway	16	Penguin Mk-2-7	Anti-ship missile	2000	2001	(8)	For 4 Kiliç Class FAC; Deal worth $34–40 m; for S-70/SH-60B helicopters; originally refused by Norway because of human rights violations in Turkey but allowed after Turkey became EU candidate
South Korea	(20)	K-9 155mm	Self-propelled gun	2001			Deal worth $1 b incl licensed production of 280 in Turkey; assembled in Turkey
USA	(4)	Boeing-737 AEW	AEW&C aircraft	(2001)			Contract not yet signed; delivery from 2005
	50	S-70A/UH-60L	Helicopter	1999	1999–2001	(50)	Originally ordered 1992, but deal suspended 1994–1999 for financial reasons and as reaction to US policy towards Turkish actions against Kurds; deal worth $561 m (offsets worth $110 m); S-70A-28 and S-70A-28D versions; Turkish designation Karaku
	8	S-70B/SH-60B Seahawk	AS/ASW helicopter	1998	2001	(2)	S-70B-28 version; orginally 4 ordered 1997 but changed to 8 in 1998; delivery 2001–2002
	84	AGM-114B Hellfire-2	ASM	1999	2000–2001	(84)	Deal worth $6.7 m; AGM-114M version; for S-70B/SH-60B helicopters; bought after Norwegian refusal to sell Penguin
	140	Dragoon	APC	(1998)			Deal worth $45 m; for Police
	(48)	RGM-84 Harpoon	Anti-ship missile	(1993)	1997–2001	(48)	UGM-84 version for 4 Type-209/1400 (Preveze Class) submarines
	16	RGM-84 Harpoon	Anti-ship missile	1995			Deal worth $15.3 m; for 1 MEKO-200T-2 Type (Barbaros Class) frigate; delivery before 2003
	1	Perry Class	Frigate	2001			Ex-US; lease worth $28 m; Turkish designation Gaziantep Class
L: France	30	AS-532UL/AS-332L1	Helicopter	1997	2000–2001	16	'Phoenix-2' deal worth $430 m incl 2 delivered direct; incl 18 for SAR and 5 for combat SAR; delivery 2000–2003
Germany (FRG)	5	Frankenthal Class	MCM ship	1999			Deal worth $625 m incl 1 direct delivered; delivery 2004–2007
	3	Kilic Class	FAC(M)	2000			Delivery from 2002/2003
	4	Type-209/1400	Submarine	1998			Deal worth $556 m; Turkish designation Gur Class; delivery 2004–2006/2007
South Korea	280	K-9 155MM	Self-propelled gun	(2001)			Deal worth $1 b incl 20 direct delivered
Spain	9	CN-235MP	MP aircraft	1998			'Meltem' programme worth $103–120 m; incl 3 MSA version for Coast Guard; delivery probably from 2002
UK	840	Rapier Mk-2	SAM	1999			Deal worth $130–150 m; for use with Rapier SAM systems modernized to Rapier B1X (Rapier-2) version; delivery 2002–2010
USA	50	Bell-209/AH-1Z	Combat helicopter	(2001)			Deal worth $4 b incl option on 95 more; contract not yet signed
	551	AIFV	IFV	2000	2001	(100)	Deal worth $338 m; incl APC and mortar carrier versions; delivery 2001–2004

UAE

Supplier	No.	Weapon designation	Weapon description	Year of order	Year(s) of delivery	No. delivered	Comments
S: France	14	AS-350/550 Fennec	Light helicopter	1999	2001	(14)	Deal worth $27 m; for Dubai; incl for training; AS-350B version
	(4)	AS-365F/565SA Panther	ASW helicopter	1997	2001	(4)	For Dubai
	(32)	AS-15TT	Anti-ship missile	(1997)	2001	(32)	For 4 AS-565SB helicopters; for Dubai

Recipient/ supplier (S) or licenser (L)	No. ordered	Weapon designation	Weapon description	Year of order/ licence	Year(s) of deliveries	No. delivered/ produced	Comments
	30	Mirage-2000-5 Mk-2	FGA aircraft	1998	. .	. .	Deal worth $3.4 b incl modernization of 33 UAE Mirage-2000 to Mirage-2000-5 Mk-2 version; incl 11 Mirage-2000DAD version; incl 12 ex-French Mirage-2000 rebuilt to Mirage-2000-5 Mk-2 version; delivery 2002–2005
	(756)	MICA	AAM	1998	. .	. .	For Mirage-2000-5 Mk-2 FGA aircraft; MICA-EM version; delivery 2003
	500	R-550 Magic-2	AAM	(1998)	. .	. .	For Mirage 2000-5 Mk-2 FGA aircraft
	390	Leclerc	Main battle tank	1993	1994–2001	(349)	Deal worth $3.4 b incl 46 Leclerc ARVs (offsets 60%); incl 2 Leclerc Driver Training Tank version; delivery 1994–2002
	46	Leclerc DNG	ARV	1993	1997–2001	(37)	Deal worth $4.6 b incl 390 Leclerc tanks (offsets 60%); delivery 1997–2003
	4	Ocean Master	MP aircraft radar	2000	. .	. .	For 4 CN-295MPA MP aircraft delivered from Spain
	1	P-37BRL Type	FAC(M)	2001	. .	. .	'Baynunah' programme; prior to licensed production; for Abu Dhabi; also designated Combattante-1 Type
Germany (FRG)	(64)	Tpz-1 Fuchs	APC	2001	. .	. .	NBC reconnaissance version
Netherlands	10	Scout	Surveillance radar	1996	1997–2001	(10)	For modernization of 2 Kortenaer Class frigates, 6 TNC-45 Type FAC and 2 other ships
Romania	10	SA-330 Puma	Helicopter	2001	. .	. .	Deal worth $125 m incl modernization of 15 UAE SA-330; for Abu Dhabi; delivery probably 2002–2005
Russia	50	96K9 Pantzyr-S1	AAV(G/M)	2000	. .	. .	Deal worth $734 m; delivery probably 2002–2004
	(1 200)	SA-19 Grisom/9M111	SAM	2000	. .	. .	For 96K9 Pantzyr-S1 SAM AAV(G/M)s
Spain	4	C-295MPA	ASW/MP aircraft	2001	. .	. .	'Shaheen-1' programme worth $114 m; for Abu Dhabi
UK	. .	Black Shahine	ASM	1998	. .	. .	For Mirage-2000-5 Mk-2 FGA aircraft
USA	1	C-130/L-100-30	Transport aircraft	2001	2001	1	Second-hand; sold via UK; modernized in UK before delivery
	80	F-16C Block-60	FGA aircraft	2000	. .	. .	Deal worth $5 b (incl $400 m for engines); incl 40 F-16D version; delivery 2004–2007
	24	RGM-84 Harpoon	Anti-ship missile	1998	1998–2001	(24)	RGM-84G-4 version; for 2 Kortenaer (Abu Dhabi) Class frigates
Ukraine	(80)	BTR-3U	IFV	2000	2001	(40)	No. ordered could be 90
L: France	5	P-37BRL Type	FAC(M)	2001	. .	. .	'Baynunah' programme; for Abu Dhabi; also designated Combattante-1 Type

UK

Recipient/ supplier (S) or licenser (L)	No. ordered	Weapon designation	Weapon description	Year of order/ licence	Year(s) of deliveries	No. delivered/ produced	Comments
S: Canada	5	BD-700 Global Express	Transport aircraft	1999	. .	. .	Deal worth $1.3 b (offsets 100%) incl ASTOR radars; for modification to AGS aircraft in USA and UK with ASTOR radars; delivery around 2005
Germany (FRG)	99	G-115D	Trainer aircraft	1998	1999–2001	(99)	Deal worth $28 m; for civilian company for training of UK pilots; G-115E version; UK designation Tutor T-1
	4	Ro-Ro 20000-ton Type	Cargo ship	2000	. .	. .	Deal worth $1.8 b incl 2 licensed production and 25 years operation by civilian company for UK Joint Rapid Reaction Force; delivery 2002–2004

INTERNATIONAL ARMS TRANSFERS 451

Norway	4	FBRV	ARV		(1999)	..	For 2 Albion Class AALS produced in UK
Sweden	108	BvS-10	APC	2000	2001	2	Deal worth $12.4 m; option on 1 more; delivery 2002–2003 Deal worth $90 m (armoured parts produced in UK); incl CP and ARV version; delivery 2001–2005
USA	10	C-130J Hercules-2	Transport aircraft	1994	2000–2001	(10)	Deal worth $1.56 b (offsets 100%) incl 10 C-130J-30 version; UK designation Hercules C-5
	4	C-17A Globemaster-3	Transport aircraft	2000	2001	4	'STSA' programme; 7-year lease worth $750–825 m (and $300 m for training and support); C-17 Block-12 version
	8	MH-47E Chinook	Helicopter	1995	2000–2001	(8)	Deal worth $365 m incl 6 CH-47D version; delivery delayed from 1998 to 2001 because of technical problems; UK designation Chinook HC-Mk-3
	8	Cougar	APC	2001		..	Deal worth $3.6 m; UK designation Tempest; delivery 2003
	980	AGM-114 Longbow	Anti-tank missile	1995	2000–2001	(350)	Deal worth $3.95 b (offsets 100%) incl 67 AH-64D helicopters; assembled in UK; delivery 2000–2003
	(1 600)	AGM-114K Hellfire	Anti-tank missile	(1996)	2000–2001	(400)	For AH-64D helicopters; assembled in UK
	(200)	AGM-65G Maverick	ASM	2000	2000–2001	(200)	Deal worth $60 m; for Harrier GR-7 FGA aircraft; AGM-65G2 version; ordered as result of experience in 1999 Kosovo war
	(50)	AIM-120A AMRAAM	AAM	(1997)	2000–2001	(50)	For Tornado F-3 fighter aircraft
	..	AIM-120B AMRAAM	AAM	2000		..	Deal worth $285 m; delivery from 2002
	65	BGM-109 Tomahawk	SLCM	1995	1998–2001	(65)	Deal worth $316 m; BGM-109 T-LAM Block-III version; for Swiftsure and Trafalgar Class submarines
	20	BGM-109 Tomahawk	SLCM	1999	2000–2001	(20)	Deal worth $50 m; BGM-109 T-LAM Block-IIIC version; for Swiftsure and Trafalgar Class submarines
	(48)	BGM-109 Tomahawk	SLCM	(2001)		..	Deal worth $87 m; BGM-109 T-LAM Block-IIIC version
L: Germany (FRG)	2	Ro-Ro 20000-ton Type	Cargo ship	2000		..	Deal worth $1.8 b incl 4 delivered direct and 25 years operation by civilian company for UK Joint Rapid Reaction Force; delivery 2002–2004
USA	59	AH-64D Apache	Combat helicopter	1996	2000–2001	(21)	Deal worth $3.95 b (offsets 100%) incl 8 delivered direct and 980 AGM-114 missiles; WAH-64D Apache Mk-1 version; delivery 2000–2003
	5	ASTOR	AGS radar	1999		..	Deal worth $1.3 b (offsets 100%) incl 5 BD-700 aircraft; for modification of 5 BD-700 transport aircraft delivered from Canada to AGS aircraft
USA							
S: Canada	(160)	Bell-206 JetRanger	Light helicopter	1993	1993–2001	(151)	For training; Bell-206B-3 version; US designation TH-67A Creek
	366	Piranha/LAV-25	IFV	2000		..	Incl APC and armoured car version; assembled in USA; option on 1765 more; delivery from 2002
Israel	13	A-4E Skyhawk	FGA aircraft	2000	2001	13	Ex-Israeli; operated by civilian company in USA for US and other military customers for training; incl 3 TA-4J trainer version
Italy	8	A-109K	Light helicopter	2000	2000–2001	8	For Coast Guard 'Airborne Use of Force' anti-narcotics operations; A-109E Power version; US designation MH-68A or MK-68 Mako

Recipient/ supplier (S) or licenser (L)	No. ordered	Weapon designation	Weapon description	Year of order/ licence	Year(s) of deliveries	No. delivered/ produced	Comments
Norway	6	Penguin Mk-2-7	Anti-ship missile	1997	2000–2001	(6)	Deal worth $6 m
UK	8	UFH 155mm	Towed gun	1997	2000–2001	(8)	US designation XM-777; prior to licensed production
L: Austria	(50)	Pandur	APC	1999	2000–2001	(20)	Deal worth $51 m; US designation AGMS (Armored Ground Mobility System); delivery 2000–2003; more produced for export
Israel	..	Popeye-1/AGM-142	ASM	1998			US designation AGM-142A Have Nap or Raptor
Switzerland	(710)	PC-9/T-6A Texan-2	Trainer aircraft	1996	1999–2001	(44)	'JPATS' programme worth $7 b (incl $4.7 b for aircraft only); delivery 1999–2014
UK	187	Hawk/T-45A Goshawk	Jet trainer aircraft	1981	1990–2001	(141)	'VTXTS' or 'T-45TS' programme; T-45A and T-45C Goshawk versions; delivery 1990–2005
	(686)	UFH 155mm	Towed gun	(2000)		..	US designation M-777; delivery 2003–2010

Unknown

S: South Africa	(10)	RG-12 Trojan	APC	2001	2001	(10)	Recipient in Middle East
	(10)	RG-12 Trojan	APC	2001	2001	(10)	Recipient in Middle East
Sweden	(1)	Giraffe AMB-3D	Surveillance radar	1998	2001	(1)	Recipient possibly Singapore

Uzbekistan

S: Russia	50	BTR-80	APC	2001	2001	23	Deal worth $30 m incl other weapons

Venezuela

S: Brazil	8	AMX-T	FGA/trainer aircraft	1999	2001	(4)	Deal worth $150 m; AMX-ATA version
Israel	(6)	EL/M-2238	Surveillance radar	1999	2001	(1)	For modernization of 6 Sucre (Lupo) Class frigates
	(54)	Barak	SAM	1999	2001	(27)	Deal worth $20 m incl 3 ADAMS launchers; option on more; ordered via Dutch company; delivery 2001–2002
Italy	12	SF-260M	Trainer aircraft	1998	2000–2001	(12)	Deal worth $12 m; SF-260EU version
Netherlands	3	Flycatcher Mk-2	Fire control radar	1999	2001	(1)	For use with 3 ADAMS SAM launchers delivered from Israel; option on 3 more; delivery 2001–2002
Poland	12	M-28 Skytruck	Light transport ac	1997	1999–2001	(12)	Deal worth $20 m; for National Guard
	(18)	M-28 Skytruck	Light transport ac	(1999)	2000–2001	(12)	
South Korea	1	Endeavour Class	Support ship	1999		..	Designation uncertain; delivery 2002
Sweden	4	Giraffe-AD	Surveillance radar	1998		..	For use with RBS-70 SAMs
	(200)	RBS-70	Portable SAM	1999	2000–2001	(200)	Deal worth $45 m incl other weapons

S: Russia	(144)	SA-16 Gimlet/Igla-1	Portable SAM	(1996)	2001	(24)	SA-N-10 version for 4 BPS-500 Type (Ho-A Class) FAC; designation uncertain
	(64)	SS-N-25/Kh-35 Uran	Anti-ship missile	(1996)	2001	(16)	For 4 BPS-500 Type (Ho-A Class) FAC
	2	Svetlyak Class	Patrol craft	2001		. .	Delivery probably 2002
L: Russia	(4)	BPS-500/Type-1241A	FAC(M)	1996	2001	(1)	Vietnamese designation Ho-A Class

Yemen

S: Czech Republic	(106)	T-55AM-2	Main battle tank	1999	2000–2001	(106)	Ex-Czech; incl some T-54 tanks; possibly modernized before delivery
Poland	3	Deba Class	Landing craft	1999	2001	3	Deal worth $50 m incl 1 Lublin Class landing ship
	1	Lublin Class	Landing ship	1999		. .	Deal worth $50 m incl 3 Deba Class landing craft; Yemeni designation Bilquis Class; delivery 2002
Russia	(14)	MiG-29S/Fulcrum-C	FGA aircraft	(2001)		. .	Deal worth $437 m; no. ordered could be up to 24
	(39)	T-72B	Main battle tank	(1999)	2000–2001	(39)	Ex-Russian; modernized before delivery

Zimbabwe

S: Ukraine	(1)	An-12/Cub-A	Transport aircraft	(2000)	2001	(1)	Second-hand

Abbreviations and acronyms

ac	Aircraft	NBC	Nuclear, chemical and biological
AAA	Anti-aircraft artillery	no.	Number
AALS	Amphibious assault landing ship	OPV	Offshore patrol vessel
AAV	Anti-aircraft vehicle	Recce	Reconnaissance
ABL	Armoured bridge-layer	SAM	Surface-to-air missile
ACV	Air-cushion vessel (hovercraft)	SAR	Search and rescue
AEV	Armoured engineer vehicle	ShAM	Ship-to-air missile
AEW	Airborne early-warning	ShShM	Ship-to-ship missile
AEW&C	Airborne early-warning and control	SIGINT	Signals intelligence
AGS	Airborne ground-surveillance	SLCM	Submarine-launched cruise missile
ALV	Armoured logistic vehicle	SP	Self-propelled
AMV	Anti-mine vehicle	SSM	Surface-to-surface missile
APC	Armoured personnel carrier	SuShM	Submarine-to-ship missile
APC/CP	Armoured personnel carrier/command post	UAE	United Arab Emirates
		UAV	Unmanned aerial vehicle (drone)
		UIFSA	United Islamic Front for the Salvation of Afghanistan (Northern Alliance)
ARV	Armoured recovery vehicle		
AS	Anti-ship	VIP	Very important person
ASM	Air-to-surface missile		
AShM	Air-to-ship missile		
ASW	Anti-submarine warfare	VTOL	Vertical take-off and landing (aircraft)
CDM	Coast defence missile		
CIWS	Close-in weapon system		**Conventions**
CP	Command post	. .	Data not available or not applicable
ELINT	Electronic intelligence	( )	Uncertain data or SIPRI estimate
EW	Electronic warfare	m	million (10⁶)
FAC	Fast attack craft	b	billion (10⁹)
FGA	Fighter/ground attack		
FMF	Foreign Military Funding (US)		'Contract not yet signed' is used in the comments field when sources indicate that preliminary agreements have been signed, but when the final contract has not yet been signed.
FMS	Foreign Military Sales (US)		
FRG	Federal Republic of Germany		
(G)	Gun-armed		
GDR	German Democratic Republic		
IFV	Infantry fighting vehicle		'Status uncertain' is used in the comments field when sources are contradictory about the (continued) existence of the reported deal.
incl	Including/include(s)		
LTTE	Liberation Tigers of Tamil Eelam		
(M)	Missile-armed		'Unknown' is used in cases where it has not been possible to identify a supplier or recipient with an acceptable degree of certainty.
MCM	Mine countermeasures		
MP	Maritime patrol		
MRL	Multiple rocket launcher		

Table 8C.2. Index of suppliers of major conventional weapons and their recipients and licensees, 2001

This index lists recipients and licensees by suppliers of major weapons on order or under delivery, or for which the licence was bought and production was under way or completed during 2001. The types of weapon involved in the transfers can be found by cross-referencing with the register of the transfers and licensed production of major conventional weapons in 2001 in table 8C.1. Entries are alphabetical, by supplier, recipient and licensee. 'Unknown' is used in cases where it has not been possible to identify a supplier or recipient with an acceptable degree of certainty.

Supplier	Recipients (**R**) and licensees (**L**)
Argentina	R: Spain
Austria	R: Botswana, Brazil, Egypt
	L: Slovenia, USA
Bahrain	R: Chile
Belarus	R: Algeria, Angola
Belgium	R: Brazil, Jordan
Brazil	R: Dominican Republic, Greece, Malaysia, Mexico, Venezuela
Bulgaria	R: Ethiopia, Georgia
Canada	R: Australia, Brazil, Denmark, Greece, Mexico, New Zealand, Saudi Arabia, Spain, UK, USA
	L: Australia
China	R: Algeria, Egypt, Iran, Kuwait, Namibia, Pakistan, Sri Lanka
	L: Egypt, Pakistan
Czech Republic	R: Algeria, Georgia, Latvia, Sri Lanka, Yemen
Denmark	L: Greece
Egypt	R: Kuwait
Finland	R: Denmark, Egypt, Sweden
	L: Egypt
France	R: Belgium, Botswana, Brazil, Brunei, Cameroon, Chile, China, Cyprus, Finland, Greece, India, Indonesia, Ireland, Italy, Japan, Malaysia, Morocco, Norway, Oman, Pakistan, Qatar, Romania, Saudi Arabia, Singapore, Slovenia, South Africa, South Korea, Spain, Switzerland, Tunisia, Turkey, UAE
	L: China, India, Japan, Pakistan, Turkey, UAE
Germany	R: Canada, Denmark, Egypt, Greece, India, Italy, Kuwait, Lesotho, Malaysia, Netherlands, Poland, Portugal, Romania, Slovenia, South Africa, Spain, Sweden, Switzerland, Thailand, Turkey, UAE, UK
	L: Argentina, Australia, Brazil, Chile, Egypt, Greece, India, Indonesia, Italy, Malaysia, Poland, South Korea, Spain, Sweden, Turkey, UK
Greece	R: Cyprus, Macedonia
India	R: Nepal
Indonesia	R: Brunei, Pakistan, South Korea
Ireland	L: Taiwan
Israel	R: Australia, Belgium, Finland, France, India, NATO, Netherlands, Philippines, Romania, Singapore, Thailand, Turkey, USA, Venezuela
	L: China, USA
Italy	R: Argentina, Bangladesh, Brazil, Canada, Chile, China, Czech Republic, Hungary, India, Japan, Malaysia, Mexico, Netherlands, Saudi Arabia, South Africa, Spain, Sweden, Tunisia, Turkey, USA, Venezuela

TRANSFERS OF MAJOR CONVENTIONAL WEAPONS   455

Kazakhstan	R: Nepal
Lebanon	R: Pakistan
Moldova	R: Namibia
Netherlands	R: Argentina, Bangladesh, Chile, Finland, Germany, Greece, India, Norway, Poland, South Korea, Spain, Turkey, UAE, UK, Venezuela
	L: India, South Korea
Norway	R: Australia, Denmark, Latvia, Lithuania, Spain, Turkey, UK, USA
Pakistan	R: Malaysia
Poland	R: India, Peru, Venezuela, Yemen
Romania	R: UAE
Russia	R: Afghanistan/Northern Alliance, Algeria, Angola, China, Colombia, Czech Republic, Eritrea, Greece, India, Iran, Kazakhstan, Malaysia, Myanmar, Nigeria, Peru, Syria, UAE, Uzbekistan, Viet Nam, Yemen
	L: China, India, Iran, North Korea, Poland, Singapore, Viet Nam
Singapore	R: Indonesia
Slovakia	R: Greece, India
South Africa	R: Algeria, Djibouti, India, Malaysia, unknown
South Korea	R: Bangladesh, Indonesia, Turkey, Venezuela
	L: Turkey
Spain	R: Chile, France, Norway, Poland, UAE
	L: Indonesia, Turkey
Sweden	R: Australia, Austria, Brazil, Czech Republic, Denmark, Finland, France, Germany, Greece, Hungary, Italy, Latvia, Lithuania, Mexico, Norway, Pakistan, Poland, Singapore, South Africa, Spain, UK, unknown, Venezuela
	L: Australia, Switzerland
Switzerland	R: Ireland, Malaysia, Spain, Sweden
	L: Canada, USA
Turkey	R: Georgia, Malaysia
UK	R: Australia, Bahrain, Belgium, Brazil, Brunei, Canada, China, Denmark, Germany, Ghana, Greece, India, Ireland, Italy, Jordan, Kuwait, Malaysia, Malta, Oman, Poland, South Africa, South Korea, Switzerland, Thailand, UAE, USA
	L: Australia, Greece, India, Poland, Thailand, Turkey, USA
Ukraine	R: China, Greece, Guinea, India, Iran, Jordan, Macedonia, UAE, Zimbabwe
	L: Iran
USA	R: Algeria, Argentina, Australia, Austria, Bahrain, Belgium, Botswana, Brazil, Canada, Chile, Colombia, Croatia, Denmark, Egypt, Estonia, France, Georgia, Germany, Ghana, Greece, Indonesia, Ireland, Israel, Italy, Japan, Jordan, Lebanon, Lithuania, Macedonia, Mexico, Netherlands, New Zealand, Norway, Oman, Pakistan, Peru, Philippines, Poland, Portugal, Romania, Saudi Arabia, Singapore, Slovenia, South Korea, Spain, Sri Lanka, Taiwan, Thailand, Trinidad and Tobago, Tunisia, Turkey, UAE, UK
	L: Egypt, Germany, Japan, South Korea, Taiwan, Turkey, UK
Uzbekistan	R: India
Multiple sellers	R: Australia, Finland, Norway, Poland, Portugal, Sweden
	L: Finland
Unknown	R: Bosnia and Herzegovina, Latvia, Namibia, Peru

Appendix 8D. Sources and methods for arms transfers data[1]

The SIPRI Arms Transfers Project reports on international flows of conventional weapons. Since publicly available information is inadequate for the tracking of all weapons and other military equipment, SIPRI covers only what it terms *major conventional weapons*.

Data are collected from open sources in the SIPRI arms transfers database and presented in a register that identifies the suppliers, recipients and weapon deliveries; and in tables that provide a measure of the trends in the total flow of major weapons and its geographical pattern. SIPRI has developed a unique trend-indicator value system. This system is not comparable to official economic statistics such as gross domestic product, public expenditure and export/import figures.

The database covers the period from 1950. Data collection and analysis are a continuous process. As new data become available the database is updated for all years included in the database.[2]

I. Selection criteria and coverage

Selection criteria

SIPRI uses the term 'transfer' rather than 'trade' since the latter is usually associated with 'sale'. SIPRI covers not only sales of weapons, including manufacturing licences, but also other forms of weapon supply, including aid and gifts.

The transferred weapons must be destined for the armed forces, paramilitary forces or intelligence agencies of another country. Weapons supplied to or from rebel forces in an armed conflict are included as deliveries to or from the individual rebel forces, identified under separate 'recipient' or 'supplier' headings. Supplies to or from international organizations are also included and categorized in the same fashion. In cases where deliveries are identified, but where it is not possible to identify either the supplier or the recipient with an acceptable degree of certainty, transfers are registered as coming from 'unknown' suppliers or going to 'unknown' recipients. Suppliers are termed 'multiple' only if there is a transfer agreement for weapons that are produced by two or more cooperating countries and if it is not clear which country will make the delivery.

Weapons must be transferred voluntarily by the supplier. This includes weapons delivered illegally—without proper authorization by the government of the supplier or recipient country—but excludes captured weapons and weapons obtained from defectors.

---

[1] A complete description of the SIPRI Arms Transfers Project methodology, including a list of the sources used, is available on the SIPRI Internet site at URL <http://www.sipri.se/projects/armstrade/atmethods.html>.

[2] Thus data from several SIPRI Yearbooks or other SIPRI publications cannot be combined. Readers who require time-series trend-indicator value data for periods before the years covered in this Yearbook and readers who require updated registers should contact SIPRI, preferably via the Internet site at URL <http://projects.sipri.se/armstrade/atrequest.html>.

The weapons must have a military purpose. Systems such as aircraft used mainly for other government branches but registered with and operated by the armed forces are excluded. Weapons supplied for technical or arms procurement evaluation purposes only are not included.

Major conventional weapons: the coverage

SIPRI covers only what it terms *major conventional weapons*, defined as:

1. *Aircraft:* all fixed-wing aircraft and helicopters, including unmanned reconnaissance/surveillance aircraft, with the exception of micro-light aircraft, powered and unpowered gliders and target drones.
2. *Armoured vehicles:* all vehicles with integral armour protection, including all types of tank, tank destroyer, armoured car, armoured personnel carrier, armoured support vehicle and infantry fighting vehicle.
3. *Artillery:* naval, fixed, self-propelled and towed guns, howitzers, multiple rocket launchers and mortars, with a calibre equal to or above 100-mm.
4. *Radar systems:* all land-, aircraft- and ship-based surveillance and fire-control radars, with the exception of navigation, weather and range-only radars. In cases where the radar is fitted on a platform (vehicle, aircraft or ship), the register only notes those radars that come from a different supplier than the supplier of the platform.
5. *Missiles:* all powered, guided missiles and torpedoes with conventional warheads. Unguided rockets, guided but unpowered shells and bombs, free-fall aerial munitions, anti-submarine rockets and target drones are excluded.
6. *Ships:* all ships with a standard tonnage of 100 tonnes or more, and all ships armed with artillery of 100-mm calibre or more, torpedoes or guided missiles, with the exception of most survey ships, tugs and some transport ships.

The statistics presented refer to transfers of weapons in these six categories only. Transfers of other military equipment—such as small arms/light weapons, trucks, artillery under 100-mm calibre, ammunition, support equipment and components, as well as services or technology transfers—are not included.

II. The SIPRI trend indicator

The SIPRI system for the valuation of arms transfers is designed as a *trend-measuring device*. It permits the measurement of changes in the total flow of major weapons and its geographical pattern. The trends presented in the tables of SIPRI trend-indicator values are based only on *actual deliveries* during the year/years covered in the relevant tables and figures, not on orders signed in a year.

The trend-indicator value system, in which similar weapons have similar values, reflects both the quantity and the quality of the weapons transferred. The value reflects the transfer of *military resources*.

Arms transfers can be measured with several objectives in mind. The two most common objectives are to gain knowledge about the economic factor and about the military factor of arms transfers. However, different goals require different statistical approaches.

The SIPRI values do not reflect the financial value (payments for) of weapons transferred. This is impossible for three reasons. First, in many cases no reliable data on the value of the transfer are available. Second, even if the value of a transfer is known, it is in almost every case the total value of a deal, which may include not only the weapons entered in the SIPRI database but also other items related to these weapons (e.g., spare parts, armament or ammunition) as well as support systems (e.g., specialized vehicles) and items related to the integration of the weapon in the armed forces (e.g., software changes to existing systems or training). Third, even if the value of a transfer is known, there remains the problem that important details about the financial arrangements of the transfer (e.g., credit/loan conditions and discounts) are usually not known.[3]

Measuring the military factor would require a concentration on the value of the weapons as a military resource. Again, this could be done from the actual financial values of the weapons transferred, assuming that these values generally reflect the military capability of the weapon. However, the problems enumerated above would still be valid (e.g., a very expensive weapon may be transferred as aid at a 'zero' price and therefore not show up in financial statistics, but it would still be a significant transfer of military resources). The SIPRI solution is a system in which military resources are measured by including an evaluation of the technical parameters of the weapons. The tasks and performance of the weapons are evaluated and the weapons are assigned a value in an index. These values reflect the military resource value of the weapon in relation to other weapons. This can be done under the condition that a number of benchmarks or reference points are established by assigning some weapons a fixed place in the index. These are the core of the index, and all other weapons are compared to these *core weapons*.

In short, the process of calculating the SIPRI trend-indicator value for individual weapons is as follows.

For a number of weapon types (noted in the register as the 'weapon designation') it is possible to find the actual average unit acquisition price in open sources. It is assumed that such real prices roughly reflect the military resource value of a system. For example, a combat aircraft bought for $10 million may be assumed to be twice the resource as one bought for $5 million, and a submarine bought for $100 million may be assumed 10 times the resource as the $10 million combat aircraft. Those weapons with a real price are used as the core weapons of the valuation.

Weapons for which a price is not known are compared with core weapons. This comparison is made in the following steps.

1. The description of a weapon is compared with the description of the core weapon. In cases where no core weapon exactly matches the description of the weapon for which a price is to be found, the closest match is sought.

2. Standard characteristics of size and performance (weight, speed, range and payload) are compared with those of a core weapon of a similar description. For example, a 15 000-kg combat aircraft would be compared with a combat aircraft of similar size.

3. Other characteristics, such as the type of electronics, loading/unloading arrangements, engine, tracks or wheels, armament and materials, are compared.

4. Weapons are compared with a core weapon from the same period.

---

[3] It is possible to present a very rough idea of the economic factors from the financial statistics now available from most arms-exporting countries. However, most of these statistics lack sufficient detail. See appendix 8E.

Production under licence is included in the arms transfer statistics to reflect the average percentage of licenser-imported components embodied in the weapon (in reality this import share may fluctuate, often gradually decreasing over time). Supplies of subsystems from other sources than the licenser registered in the database are not included (unless these subsystems are weapons as defined by SIPRI for the database, in which case a separate record is included in the database with details for these systems).

Weapons delivered in 'second-hand' condition are given a standard value of 40 per cent of the value assigned to the new weapon; second-hand weapons that have been significantly refurbished or modified by the supplier before delivery (and have thereby become a greater military resource) are given a value of 66 per cent of the new value. In reality there may be huge differences in the military resource value of a second-hand weapon depending on its condition after use and the modifications during the years of use.

The SIPRI trend indicator does not measure military value or effectiveness. It does not take into account the conditions under which a weapon is operated (e.g., an F-16 combat aircraft operated by well-balanced, well-trained and well-integrated armed forces has a much greater military value than the same aircraft operated by a developing country; the resource is the same but the effect is very different). It also assumes that the real prices of the core weapons are genuinely real and do not include costs that, even if officially part of the programme, are actually not exclusively related to the weapon itself—for example, funds that seem to be part of a programme could actually be related to optional add-ons and armament or to the development of basic technology that will also be included (free of cost) in other programmes but have for the sake of convenience been put under one programme, and hidden government subsidies to keep industry in being by paying more than the weapon is worth.

III. Sources

The sources for the data presented in the arms transfers register are of a wide variety: newspapers; periodicals and journals; books, monographs and annual reference works; and official national and international documents. The common criterion for all these sources is that they are open—published and available to the general public.

Such open information cannot, however, provide a comprehensive picture of world arms transfers. Published reports often provide only partial information, and substantial disagreement among reports is common. Order and delivery dates and exact numbers, or even types, of weapons ordered and delivered, or the identity of suppliers or recipients, may not always be clear from the sources. Therefore, the exercise of judgement and the making of estimates are important elements in compiling the SIPRI arms transfers database. Estimates are kept at conservatively low levels (and may very well be underestimates).

All sources of data as well as calculations of estimates, while not published by SIPRI, are documented in the SIPRI database.

Appendix 8E. Government and industry data on national arms exports

PIETER D. WEZEMAN

Publicly available government and industry statistics on the value of national arms exports are listed in table 8E. These data are included here for four reasons: (*a*) to make them more accessible; (*b*) to illustrate the current state of government transparency in arms export data; (*c*) to underline the fact that arms export data from different countries are only partially comparable; and (*d*) to provide a rough indication of the financial scale of arms exports.

Caution should be exercised when using the data in table 8E for detailed analysis. Only some of the statistics are fully explained, definitions are not consistent from country to country, and the reports give different definitions for what is included in the category 'arms'. Some countries release figures only on arms exports, while others aggregate exports of arms and dual-use equipment. Some release data on the value of items *delivered*, others on the value of items *approved* for export, some on both. In some countries different reports present different national arms export data. To underline this last type of inconsistency, all relevant data are included in the table. No attempt has been made to compensate for any of these comparability problems or possible lack of reliability.

Despite these methodological reservations, in the absence of good alternatives the values are considered useful as a rough indication of the financial scale of arms exports. Such an indication cannot be derived from the SIPRI arms transfer trend-indicator values, as these indicate the volume of international arms transfers and not actual financial values (see appendix 8D).

The table is not comprehensive and there are other countries, such as China, whose exports would be larger than those of some of the countries listed in the table. However, SIPRI estimates that the countries in the table together account for over 90 per cent of the total volume of deliveries of major conventional weapons in 2001, and it can be assumed that these countries together account for a similarly high percentage of total arms exports in financial terms.

Table 8E is based on publications of governments and arms industry associations, well-documented reports on government statements, and government and arms industry association replies to SIPRI's requests for information. Comments are worded as closely as possible to the details given in the documents cited. If the comment does not specify whether the values refer to permits or deliveries, it is because this distinction is not specified in the original source. Sources refer to the last year reported for each country. Sources for previous years are given in previous editions of the SIPRI Yearbook. The 2000 US dollar series is calculated on the basis of the exchange rates on 31 December 2000.[1] SIPRI collects hyperlinks to Internet sites containing official arms export data, which in several cases include data disaggregated by recipients and category.[2]

---

[1] As compiled by OANDA Corporation, available at URL <http://www.oanda.com>.
[2] These links are listed at URL <http://www.sipri.se/projects/armstrade/atlinks.html>.

Table 8E. Government and industry data on national arms exports, 1996–2000

Country	Currency unit (current prices)	1996	1997	1998	1999	2000	2000 (US $m.)	Explanation of data
Australia	m. A. dollars[a]	435.2	19	14	516	..	..	Value of shipments of military goods (fiscal years)
Austria	m. euros[b]	..	..	..	395	563	497	Value of licences for arms exports
Belgium	m. B. francs[c]	8 180	7 460	12 537	9 536	7 778	182	Value of exports of arms and ammunition
	m. euros[d]	..	..	649.7	622	779	734	Value of licences for arms exports
Brazil	m. US dollars[e]	8.7	26	70	98	..	..	Value of arms exports
Canada	m. C. dollars[f]	464.8	304.3	421	434	478	319	Shipments of military goods, excluding exports to the USA
Industry	m. C. dollars[g]	798	..	851	..	..	..	Defence revenues from markets outside Canada and the USA
Industry	m. C. dollars[g]	996	..	1 010	..	..	..	Defence revenues from US market
Czech Rep.	m. US dollars[h]	117	182	104	101	98	98	Value of arms exports
Denmark	m. euros[d]	..	..	..	..	(31)	..	Value of military equipment exported between 1 July and 31 Dec. 2000
Finland	m. F. markkaa[i]	69	81.5	184	239	140	22	Value of exports of defence *matériel*
France	m. francs[j]	18 593	27 683	28 767	17 466	..	..	Value of exports of military *matériel*
	m. francs/euros[k]	..	..	..	150 (franc)	27 (euro)	25	Value of paid transfers of war *matériel* by the Ministry of Defence
	m. euros[k]	3 102	4 751	7 771	4 724	6 955	6 552	Value of export orders for arms and associated services (constant prices)
	m. euros[k]	4 666	6 787	6 415	3 846	2 738	2 579	Value of deliveries of arms and associated services (constant prices)
Germany	m. D. marks[l]	1 006	1 384	1 338	2 844	1 330	633	Value of exports of weapons of war
	m. euros[d]	..	..	2 829	3 026	2 843	2 678	Value of licences for arms exports
Greece	m. euros[d]	..	..	..	43	21	20	Value of arms exports
India	m. rupees[m]	1 430	1 860	1372	..	..	..	Value of exports by the ordnance factories and of surplus armed forces stores
Ireland	m. euros[d]	..	..	20	60	31	29	Value of arms export licences

INTERNATIONAL ARMS TRANSFERS 461

462   MILITARY SPENDING AND ARMAMENTS, 2001

Country	Currency unit (current prices)	1996	1997	1998	1999	2000	2000 (US $m.)	Explanation of data
Israel	m. US dollars[n]	1 466	1 654	1 879	1 606	1 764	1 764	Value of the shipment of military goods
	m. US dollars[o]	1 331	1 684	1 890	1 618	1 759	1 759	Value of defence exports
Italy	b. lire[p]	1 196	2 065	1 944	1 715	1 169	561	Value of deliveries of military equipment
	b. lire[p]	2 165	1 726	1 838	2 596	1 658	796	Value of export licences for military equipment
Kazakhstan	m. tenge[q]	. .	. .	. .	. .	900	6	Value of arms exports
Korea, South	m. US dollars[r]	31.9	69.4	. .	. .	. .	. .	Value of defence industrial products export
	m. US dollars[s]	45	58	147	197	55	55	Value of arms exports
Netherlands	m. guilders[t]	922	2 438	952	807	918	392	Value of export licences for military goods
Industry	m. guilders[u]	1 600	1 900	. .	. .	. .	. .	Value of exports as reported by defence industry
Norway	m. kroner[v]	985	1 060	1 135	1 300	1 169	133	Value of actual deliveries of defence matériel
Pakistan	m. US dollars[w]	. .	. .	. .	30	. .	. .	Value of arms exports
Portugal	m. escudos[x]	4 157	3 205	3 806	2 133	2 552	11	Value of exports of defence matériel, equipment, technology
	m. euros[d]	. .	. .	. .	. .	22	21	Value of arms export licences
Romania	m. US dollars[y]	77	56	56	67	38	38	Value of arms exports
Russia	m. US dollars[z]	3 900	3 600	2 700	3 400	3 680	3 680	Value of exports of military equipment
	m. US dollars[z]	. .	. .	. .	. .	2 840	2 840	Value of currency received for exported military equipment
Slovakia	m. koruny[aa]	2 214	1 277	1 272	2 300	2 041	44	Value of exports of military production
Slovenia	m. tolars[bb]	1 568	2 290	726	0.7	. .	. .	Value of exports of defence equipment
	m. US dollars[cc]	. .	. .	5.3	. .	. .	. .	Value of arms exports
South Africa	m. rand[dd]	517	1 324.9	646	1 096	. .	. .	Value of export permits issued
Spain	m. pesetas[ee]	19 473	95 128	27 262	23 524	23 007	130	Value of exports of defence matériel (excl. dual-use equipment)
	m. pesetas[ff]	73 995	84 185	76 955	. .	. .	. .	Value of exports of defence production
Sweden	m. kronor[gg]	3 087	3 101	3 514	3 654	4 371	464	Value of actual deliveries of military equipment
	m. kronor[gg]	2 859	5 061	3 273	7 153	4 640	493	Value of export permits granted for sales of military equipment
Industry	m. kronor[hh]	4 289	3 667	4 434	3 974	5 715	608	Foreign sales of military and civil products to military customers
Switzerland	m. S. francs[ii]	232.9	294.3	212.7	231.5	214	133	Value of exports of war matériel

INTERNATIONAL ARMS TRANSFERS   463

Country	Unit							Description
Taiwan	m. NT dollars[jj]	..	..	(25 500)	(27 400)	..	..	Value of military sales in 2-year periods
Turkey	m. US dollars[kk]	..	139	80	84	..	..	Value of arms exports
Industry	m. US dollars[ll]	236	..	..	..	..	..	Export turnover of defence industries
UK	m. pounds[mm]	..	3 359.6	1 968.3	980.5	1 720.5	2 570	Value of deliveries of military equipment
	m. pounds[nn]	3 402	3 359	1 968	980	1 721	2 570	Value of deliveries of defence equipment
	m. pounds[nn]	6 177	6 684	6 030	4 250	4 406	6 580	Value of deliveries of defence equipment and items where the official commodity classifications do not distinguish between military and civil aerospace equipment
	m. pounds[nn]	5 080	5 540	6 049	5 044	4 737	7 075	Value of export orders for defence equipment
Ukraine	m. US dollars[oo]	..	600	..	..	500	500	Value of arms exports
USA	m. dollars[pp]	11 389	15 448	12 618	16 802	11 421	11 421	Value of deliveries of defence articles and services through the US Government (foreign military sales) in fiscal years
	m. dollars[pp]	1 563	1 818	2 045	654	478	478	Value of deliveries of munitions-controlled items directly from US manufacturers
	m. dollars[pp]	9 517	8 509	8 793	11 874	11 851	11 851	Value of agreements on sales of defence articles and services through US Government (foreign military sales) in fiscal years
	m. dollars[qq]	22 800	31 700	27 000	33 000	..	..	Value of arms transfer deliveries
	m. dollars[qq]	38 300	34 400	35 600	59 200	..	..	Value of arms transfer agreements
	m. dollars[rr]	14 820	16 274	16 482	17 935	14 187	14 187	Value of arms deliveries
	m. dollars[rr]	10 956	7 324	10 030	12 379	18 562	18 562	Value of arms transfer agreements

.. = No data available or received.

[a] Australian Department of Defence, Industry Division, *Annual Report: Exports of Defence and Strategic Goods from Australia, 1999/2000*, May 2001.

[b] *Report on the Export of Arms by the Republic of Austria for the Year 2000*, received from the Embassy of Austria.

[c] Belgian Ministry of Foreign Affairs, *Verslag van de regering aan het parlement over de toepassing van de wet van 5 augustus 1991 betreffende de in-, de uit-, en de doorvoer van wapens, munitie, en speciaal voor militair gebruik dienstig materiaal en de daaraan verbonden technologie, 1 januari 2000 tot 31 december 2000* [Report from the government to the parliament on the implementation of the law of 5 August 1991 regarding the import, export and transit of weapons, ammunition, and material for military use and related technologies, 1 January 2000 until 31 December 2000], 5 July 2001.

[d] Council of the European Union, *Third Annual Report according to Operative Provision 8 of the European Union Code of Conduct on Arms Exports*, Brussels, 7 Nov. 2001, p. 19.

[e] Information received from the Brazilian Embassy.
[f] Canadian Department of Foreign Affairs and International Trade, Export Controls Division, Export and Import Controls Bureau, *Annual Report: Export of Military Goods from Canada, 2000*, Feb. 2002.
[g] Grover, B., Canadian Defence Industries Association, 'Canadian defence industry 1999: a statistical overview of the Canadian defence industry', Dec. 1999, URL <http://www.cdia.ca/fullreport.htm>.
[h] 'Czech arms deals burgeoning even after transformation, claims Czech Weekly', CTK (Prague), 18 June 2001, in Foreign Broadcast Information Service, *Daily Report–East Europe* (FBIS-EEU), FBIS-EEU-2001-0618, 18 June 2001.
[i] Finnish Ministry of Defence, *Export of Defence Materiel*, 12 Apr. 2001, URL <http://www.vn.fi/plm/ekvas.htm#puolu>.
[j] Ministère de la défense, *Annuaire statistique de la défense* [Yearbook on defence statistics], June 2001, p. 99.
[k] Ministère de la défense, *Rapport au Parlement sur les exportations d'armement de la France en 2000* [Report to Parliament on French arms exports in 2000], Dec. 2001.
[l] *Bericht der Bundesregierung über ihre Exportpolitik für konventionelle Rüstungsgüter im Jahr 2000* (Rüstungsexportbericht 2000) [The German report of the Federal Government on its export policy for conventional armaments in 2000], Berlin, 21 Nov. 2001.
[m] Indian Ministry of Defence, Department of Defence Production and Supply, *Annual Report 2000/2001*.
[n] Information received from the Foreign Defence Assistance and Defence Exports organization (SIBAT), Ministry of Defence, Israel.
[o] As reported by government-owned Israeli Aircraft Industries Ltd, URL <http://www.iai.co.il/dows/dows/Serve/level/English/1.1.1.html>.
[p] Camera dei Deputati, *Relazione sulle operazioni autorizzate e svolte per il controllo dell'esportazione, importazione e transito dei materiali di armamento nonchè dell'esportazione e del transito dei prodotti ad alta technologia (anno 2000)* [Chamber of Deputies, Report on operations authorized and carried out concerning the control of export, import and transit of weapons *matériel* as well as the export and transit of high-technology products (2000)], 31 Mar. 2001.
[q] Centre for Analysis of Strategies and Technologies (CAST), *Eksport Vooruzhenii* (Moscow), Mar./Apr. 2001, p. 44.
[r] Ministry of National Defense, *South Korean Defence White Paper 1998*, 1999, p. 370.
[s] The South Korean Defense Ministry, quoted in 'Seoul to export $200 mil. worth of defense goods' *Korea Herald*, 26 Mar. 2001, URL <http://www.koreaherald.co.kr/SITE/data/html_dir/2001/03/26/200103260036.asp>.
[t] Netherlands Ministry of Economic Affairs, *Jaarrapport Nederlands wapenexportbeleid 2000* [The Netherlands arms export policy in 2000], URL <http://www.ez.nl/Beleid/home_ond/handelspolitiek/strateg/KST54902.pdf>.
[u] Information received from the Commissariat for Military Production of the Netherlands Ministry of Economic Affairs.
[v] Norwegian Ministry of Foreign Affairs, *Eksport av forsvarsmateriell fra Norge i 2000* [Arms exports from Norway in 2000], St meld nr 45 (2000–2001), 11 May 2001.
[w] 'Pakistani Minister Haider: arms, ammunition worth US dollar 30 million exported in 1999', *The Nation* (Islamabad), in Foreign Broadcast Information Service, *Daily Report–Near East and South Asia* (FBIS-NES), FBIS-NES-2000-0909, 9 Sep. 2000.
[x] Portuguese Ministry of Defence, *Anuario Estatístico da Defesa Nacional 2000* [Annual national defence statistics, 2000], Oct. 2001, p. 113.
[y] Information received from the Embassy of Romania.
[z] Makienko, K., 'The official results of Russia's arms exports in 2000', Centre for Analysis of Strategies and Technologies (CAST), *Eksport Vooruzhenii* (Moscow), Mar./Apr. 2001, p. 42.

aa Information received from the Ministry of Economy, Slovak Republic; and 'Slovak ministry publishes arms production, arms trade for 2000', TASR (Bratislava), in Foreign Broadcast Information Service, *Daily Report–East Europe* (*FBIS-EEU*), FBIS-EEU-2001-0521, 21 May 2001.

bb Information received from the Ministry of Foreign Affairs, Republic of Slovenia.

cc Information from the Ministry of Defence, Republic of Slovenia, received through the Slovenian Embassy in Stockholm.

dd Internet site of the National Conventional Arms Control Committee, Republic of South Africa, URL <http://www.mil.za/SANDF/DRO/NCACC/ncacc.htm>.

ee Spanish Ministry of Economy, *Estadísticas españolas de exportación de material de defensa y de doble uso realizadas en 2000* [Spanish statistics on exports of defence and dual use *matériel* as achieved in 2000], 31 July 2001.

ff Figures are calculated on the basis of relevant data on the Internet site of the Spanish Ministry of Defence, URL <http://www.mde.es/mde/infoes/indus2/pec.htm>.

gg Swedish Ministry for Foreign Affairs, *Redogörelse för den svenska exportkontrollpolitiken och exporten av krigsmateriel år 2000* [Swedish export control policy and exports of military equipment in 2000], Regeringens skrivelse 2000/01:114, 5 Apr. 2001.

hh Information received from the Association of Swedish Defence Industries [Försvarsindustriföreningen, FIF], 13 Nov. 2001.

ii Staatssekretariat für Wirtschaft, 'Ausfuhr von Kriegsmaterial im Jahr 2000 [Exports of war *matériel* 2000], Pressemitteilung, 6 Feb. 2001.

jj Ministry of National Defense, Republic of China, *2000 National Defense Report*, Taipei, 2000.

kk Information received from the Turkish Ministry of National Defence, Ankara.

ll Information received from the Defense Industry Manufacturers Association, Turkey.

mm British Foreign and Commonwealth Office, Ministry of Defence and Department of Trade and Industry, *Annual Report on Strategic Export Controls 2000*, 20 July 2001, URL <http://www.fco.gov.uk/news/newstext.asp?5176>.

nn British Ministry of Defence, Defence Analytical Services Agency, *UK Defence Statistics 2001*, URL <http://www.dasa.mod.uk/products/ukds/2001/ukds.html>.

oo A Ukrainian government official stated that the overall value of Ukrainian arms trade amounted to nearly $500 m in 2000. 'Ukraine denies report of drop in arms sales', ITAR-TASS (Moscow), 17 Aug. 2001, in Foreign Broadcast Information Service, *Daily Report–Central Eurasia* (FBIS-SOV), FBIS-SOV-2001-0817, 17 Aug. 2001.

pp Foreign Military Sales, Foreign Military Construction Sales and Military Assistance Facts, as of 30 Sep. 2000, Deputy for Financial Management Comptroller, Department of Defense Security Cooperation Agency, Washington, DC, URL <http://web.deskbook.osd.mil/>.

qq US Department of State, Bureau of Verification and Compliance, *World Military Expenditures and Arms Transfers 1999–2000* (US Department of State: Washington, DC, 2001), URL <http://www.state.gov/t/vc/rls/rpt/wmeat/99_00/ >.

rr Grimmett, R. F., *Conventional Arms Transfers to Developing Nations, 1993–2000*, CRS Report for Congress (Library of Congress, Congressional Research Service: Washington, DC, 16 Aug. 2001), pp. 71, 76.

Part III. Non-proliferation, arms control and disarmament, 2001

Chapter 9. Arms control after the attacks of 11 September 2001

Chapter 10. Ballistic missile defence and nuclear arms control

Chapter 11. The military uses of outer space

Chapter 12. Chemical and biological weapon developments and arms control

Chapter 13. Conventional arms control

Chapter 14. Multilateral export controls

9. Arms control after the attacks of 11 September 2001

IAN ANTHONY

I. Introduction

A number of events in 2001 led both practitioners and observers to question the usefulness of arms control as an instrument for managing security problems under the present conditions. This was prompted by problems in implementing existing arms control agreements as well as an identified lack of momentum in discussions about new agreements. Apart from the events of 2001, in recent years there has been a more general tendency to argue that arms control is, if not in crisis, then at least failing to play its role in the management of international security.[1]

Reviewing the current multilateral arms control agenda, the group of senior experts who advise the United Nations Secretary-General went as far as to report 'a crisis of multilateral disarmament diplomacy' in September 2001.[2]

This chapter does not describe developments in specific arms control processes, many of which are examined in detail in other chapters of this Yearbook.[3] This chapter focuses on the impact on the wider arms control agenda of two developments in 2001: the change in the US administration and the terrorist attacks that took place in the United States on 11 September 2001.

As the international system changes, international law cannot be immune to the consequences of those changes. The most important recent change in the international system has been the emergence of one dominant power, the United States. In 2001 the new administration subjected a range of arms control processes to an unaccustomed level of critical scrutiny. Although there were elements of discontinuity in US arms control policy during the first year of the George W. Bush Administration, the approach also reflected some positions that had been evolving in Washington over several years.

Two questions are at the root of US concerns about the role of arms control. The first question is how to respond when parties violate an agreement to which they are a party. The second is whether arms control processes and agreements can modify the behaviour of at least the key states.

---

[1] This view was reflected in many of the contributions published in Anthony, I. and Rotfeld, A. D. (eds), SIPRI, *A Future Arms Control Agenda* (Oxford University Press: Oxford, 2001).

[2] United Nations, Review of the implementation of the recommendations and decisions adopted by the General Assembly at its tenth special session: Advisory Board on Disarmament Matters, UN document A/56/418, 27 Sep. 2001, p. 2.

[3] Chapter 10 examines nuclear arms control issues, chapter 11 examines proposals to control the military use of space, chapter 12 examines biological and chemical weapon disarmament treaties, chapter 13 examines developments in conventional arms control, and chapter 14 examines multilateral efforts to control missiles.

While these questions of compliance and effectiveness—which cut across all the forms of arms control—are not new, the Bush Administration demonstrated a high degree of clarity in its public statements and a greater assertiveness in decision making than had previously been the case.

The policies adopted by the USA stimulated the wider discussion of how arms control could contribute to international security. The discussions took on an added dimension after the terrorist attacks against the USA on 11 September 2001. These attacks reinforced the view in the USA that there is a close correlation between the states that sponsor and carry out terrorist acts and those that actively seek to acquire nuclear, biological and chemical (NBC) weapons through clandestine programmes. Moreover, the same states are seeking to acquire ballistic missiles and other means that could be used to deliver one or more of these types of weapon.

II. Salient characteristics of arms control

Although there is no precise agreed definition of arms control, in its usage in English it can best be described as a cooperative, purposive approach to armaments policy. The primary aim is to produce effects on the actions of participants that would not otherwise have occurred.[4] Unilateral measures may fall within this general definition if they are undertaken in order to bring about reciprocal actions. Export control cooperation falls within this general definition, although it is undertaken in order to produce effects on the armaments policy of actors other than those that are cooperating.

Arms control is only one of the factors that influence the volume and distribution of arms. Constraints include the limits of current technology, the resources (human and financial) available to develop, produce, acquire and use arms as well as political decisions about force levels and force structure by responsible authorities in the context of existing threats.

The level of armaments can also be determined through restrictions adopted as part of a post-conflict settlement. These measures may result from external pressure rather than being voluntary acts by the parties.[5]

Other elements that can form part of a definition of arms control are flexible. The scope of coverage can include restrictions on structure (i.e., the levels and types of arms) or restrictions on operations (i.e., the deployment and use of arms). Where restrictions apply to structure, these may include total bans on particular categories of weapons. In the past, the parties to arms control agreements have usually been states. However, sub-state entities are parties to recent arms control agreements.[6] Arms control agreements can be bilateral,

---

[4] A recent discussion of how to define arms control is contained in Rotfeld, A. D., 'The future of arms control', *Polish Quarterly of International Affairs*, spring 2001, p. 10.

[5] The warring parties in Bosnia and Herzegovina adopted such measures following the 1995 Dayton Agreement. The forced disarmament of Iraq after 1990 was subsequent to United Nations Security Council resolutions.

[6] E.g., the Agreement on Sub-Regional Arms Control established subsequent to the 1995 Dayton Agreement includes the Federation of Bosnia and Herzegovina (consisting of Croat and Bosnian entities)

involving two parties to an agreement, or multilateral, involving more than two parties.

While most arms control agreements aim to restrict capacities for purposes related to military stability and security, there are others that aim to restrict capacities for humanitarian purposes.[7] Although the purposes for seeking an agreement can differ, in most cases the aims of arms control have been associated with military security and stability.[8] While relatively few military capabilities were made subject to controls, in the cold war period the main value of arms control was to act as a channel of communication between adversaries in conditions where few such channels existed. Communication was believed to reduce the risks that particular weapon programmes, deployments or actions would be misperceived by adversaries or potential adversaries. In this way arms control could help make behaviour more predictable.[9]

Arms control arrangements take different forms. Legally binding agreements are undertaken on behalf of the state. They must be signed by representatives authorized to bind the state under international law to the commitments contained in them. These agreements bind not only the current government but also its successors to take steps and allocate resources needed to implement the commitments they contain for as long as the agreement is in force. Such agreements are likely to include remedies that parties could expect to be applied in cases of non-compliance. Politically binding measures are also a commitment by the parties to a particular course of action, and parties could expect criticism in cases of non-compliance. A government would still commit itself to allocate the resources and modify legislation, practices and policies in ways that implement the arrangement.[10] However, political measures would not contain judicial remedies and parties would not expect to be subject to sanctions in cases of non-compliance.[11]

Compliance and effectiveness in arms control

In analyses of international agreements and political arrangements a distinction has been drawn between compliance and effectiveness.[12] An agreement

---

and the Republika Srpska as distinct parties, separate from the state of Bosnia and Herzegovina. For further discussion of this agreement see chapter 13 in this volume.

[7] E.g., the primary objectives of arms control agreements related to anti-personnel landmines are humanitarian rather than strategic.

[8] In what is often quoted as a classic statement of objectives, Thomas Schelling and Morton Halperin saw arms control as helping to reduce the likelihood of war, reduce its scope and violence should it occur, and reduce the political and economic costs of preparing for it. Schelling, T. C. and Halperin M. H., *Strategy and Arms Control* (Twentieth Century Fund: New York, 1961), p. 2.

[9] It should be noted that confidence-building measures were and are also intended to contribute to stability, in part through increases in transparency. Recent confidence-building measures are examined in chapter 13 in this volume.

[10] Ahlström, C., *The Status of Multilateral Export Control Regimes: An Examination of Legal and Non-Legal Agreements in International Cooperation* (Iustus Förlag: Uppsala, 1999).

[11] Roberts, G., 'International agreements and arms control', *Treaty Times*, Feb. 2001, pp. 5–6.

[12] Raustiala, K., 'Compliance and effectiveness in international regulatory cooperation', *Case Western Reserve Journal of International Law*, vol. 23, no. 453 (2000).

enjoys compliance if the parties to it act in accordance with its provisions. An agreement is effective if it leads parties to change their behaviour in an attempt to become and remain compliant with it.

Accordingly, it is possible for there to be compliance with an agreement without there being effectiveness. This would be the case if an agreement codifies the current practices of parties. Such an agreement would be 'self-implementing' because parties are already fully compliant without any changes in behaviour.

There is a distinction between non-compliance and the violation of an agreement stemming from the intent of parties. Non-compliance might be inadvertent—as a result of a technical failure or a failure in communication, for example. There may also be different views among parties about what constitutes compliance. By contrast, a violation of an agreement is a deliberate decision not to take the actions required or to take actions that are prohibited by that agreement.

While potentially serious, inadvertent or accidental non-compliance need not threaten the continuation of an agreement. In such cases the parties can work together to improve compliance.[13] If an agreement has been violated, on the other hand, other parties are likely to question whether it still has value unless remedies are applied in an effort to address the violation.

An agreement can be effective without enjoying full compliance. For example, if a state modifies its national laws or regulations or changes its policies as the result of an international agreement, then it has been effective even if the modifications do not bring about full compliance.

Arms control incorporates a wide spectrum of agreements, measures and processes.[14] No single general statement can be made about the extent of compliance and effectiveness with these various regimes, which can be sorted by categories.

One category of agreements is intended to produce effects on the force levels of parties. Bilateral agreements between the USA and Russia place limits on strategic and intermediate-range weapons, as well as limiting missile defences prior to the termination of the 1972 ABM Treaty[15] that was announced in 2001 and will take effect in 2002. Regional agreements limit certain force levels. In Europe, the 1990 CFE Treaty[16] and the 1996 Florence Agreement[17] limit certain conventional equipment. Nuclear weapons are prohibited in the Pacific, by the 1985 Treaty of Rarotonga;[18] in South

---

[13] E.g., the Open Skies Consultative Commission (OSCC) was established by the 1992 Open Skies Treaty to resolve questions of compliance with the treaty. For other such bodies see the glossary in this volume.

[14] For a comprehensive overview see annex A in this volume; and Goldblat, J., International Peace Research Institute, Oslo (PRIO) and SIPRI, *Arms Control: The New Guide to Negotiations and Agreements* (SAGE Publications: London, forthcoming 2002).

[15] The Treaty on the Limitation of Anti-Ballistic Missile Systems.

[16] The Treaty on Conventional Armed Forces in Europe.

[17] The Agreement on Sub-Regional Arms Control.

[18] The South Pacific Nuclear Free Zone Treaty.

America, by the 1967 Treaty of Tlatelolco;[19] in South-East Asia, by the 1995 Treaty of Bangkok;[20] and in Africa, by the 1996 Treaty of Pelindaba.[21]

In addition, a number of treaties aim at eliminating particular categories of weapon on a global basis. The 1968 Non-Proliferation Treaty (NPT)[22] reflects the desire of states parties to cease the manufacture of nuclear weapons, liquidate existing nuclear weapon stockpiles and eliminate from national arsenals nuclear weapons and the means of their delivery. The 1972 Biological and Toxin Weapons Convention (BTWC)[23] represents a commitment by states parties to eliminate biological weapons. The 1993 Chemical Weapons Convention (CWC)[24] represents a commitment by states parties to eliminate chemical weapons. The 1997 APM Convention[25] represents a commitment by states parties to eliminate anti-personnel mines.

The second category of agreements are intended to restrict the further development and deployment of weapons without prohibiting their acquisition and possession. This category includes bilateral agreements such as the 1976 US–Soviet Peaceful Nuclear Explosions Treaty.[26] Other such agreements include the 1959 Antarctic Treaty, the 1963 Partial Test Ban Treaty, the 1967 Outer Space Treaty, the 1971 Seabed Treaty,[27] the 1974 Threshold Test Ban Treaty,[28] the 1981 Certain Conventional Weapons Convention[29] and the 1996 Comprehensive Nuclear Test-Ban Treaty.

The third category of measures are intended to establish conditions under which states may transfer agreed items to others without restricting either the possession or the use of these items by the exporting state.

Three export control regimes—the Australia Group, the Missile Technology Control Regime (MTCR) and the Nuclear Suppliers Group (NSG)—have developed guidelines that participating states are committed to implement through their national export control systems.[30] While the decisions about whether to authorize the export of a particular controlled item are taken nationally, in these three regimes participating states have accepted a so-called 'no undercut' obligation. If a participating state denies authorization to export an item that is controlled for reasons relevant to the purposes of the regime concerned, it informs other participating states of that decision. Regime part-

---

[19] The Treaty for the Prohibition of Nuclear Weapons in Latin America and the Caribbean.
[20] The Treaty on the Southeast Asia Nuclear Weapon-Free Zone.
[21] The African Nuclear-Weapon-Free Zone Treaty.
[22] The Treaty on the Non-Proliferation of Nuclear Weapons.
[23] The Convention on the Prohibition of the Development, Production and Stockpiling of Bacteriological (Biological) and Toxin Weapons and on their Destruction.
[24] The Convention on the Prohibition of the Development, Production, Stockpiling and Use of Chemical Weapons and on their Destruction.
[25] The Convention on the Prohibition of the Use, Stockpiling, Production and Transfer of Anti-Personnel Mines and on their Destruction.
[26] The Treaty on Underground Nuclear Explosions for Peaceful Purposes.
[27] The Treaty on the Prohibition of the Emplacement of Nuclear Weapons and other Weapons of Mass Destruction on the Seabed and the Ocean Floor and in the Subsoil thereof.
[28] The Treaty on the Limitation of Underground Nuclear Weapon Tests.
[29] The Convention on Prohibitions or Restrictions on the Use of Certain Conventional Weapons which may be Deemed to be Excessively Injurious or to have Indiscriminate Effects (also known as the 'Inhumane Weapons' Convention).
[30] For more on these regimes see chapter 14 in this volume.

ners have committed themselves not to authorize an essentially identical export without prior consultation. A fourth regime, the Wassenaar Arrangement on Export Controls for Conventional Arms and Dual-Use Goods and Technologies, has not agreed either guidelines or a no-undercut principle. However, it has elaborated a set of factors that will be taken into account when a decision to authorize a particular export is taken. All four regimes have elaborated one or more lists of items that should not be exported without authorization by the national authorities of participating states.

The fourth category consists of practical disarmament measures. Through these measures, states provide financial and technical assistance to regulate or safeguard military capacities that are located in other states and considered surplus to requirements and outside the existing force structure of those states. The procedure of safe and secure storage is temporary, pending the destruction of weapons or the liquidation of weapon-related materials and technology. The USA and Russia have carried out a range of activities in the framework of the US-sponsored Cooperative Threat Reduction (CTR) programme, other non-proliferation efforts, and initiatives to enhance the protection, control and accounting of NBC weapon-related materials.[31] Other examples include the collection and destruction of weapons considered surplus to requirements in a range of post-conflict locations.

Compliance and effectiveness in existing arms control regimes

The various agreements noted above have had different levels of compliance and effectiveness. Those that had a direct impact on the military capabilities of the major powers during the period of the cold war combined a high level of compliance with a high degree of effectiveness. Bilateral US–Soviet agreements on nuclear arms control as well as agreements on conventional weapons between European states could be named in this context. As a result of these agreements, the parties were required to modify domestic procedures and regulations significantly. The requirement for verification led to the creation of new institutions.[32] While cases of non-compliance with these agreements have come to light, the agreements have not been violated.

In certain cases, compliance has been achieved without requiring agreements to be effective. For example, many of the states that are parties to multilateral nuclear arms control agreements have neither a military nuclear programme nor a civilian nuclear industry and infrastructure. For these states, commitments on nuclear weapon-free zones and nuclear non-proliferation are self-implementing. Similarly, many of the parties to the Outer Space Treaty do not

---

[31] The current status of these programmes in the nuclear sphere is examined in chapter 10 and that of the programmes in the chemical and biological sphere in chapter 12 in this volume.

[32] E.g., the Joint Compliance and Inspection Commission (JCIC) was established as the forum to resolve questions of compliance, clarify ambiguities and discuss ways to improve implementation of the 1991 Treaty on the Reduction and Limitation of Strategic Offensive Arms (START I Treaty) and the 1993 Treaty on Further Reduction and Limitation of Strategic Offensive Arms (START II Treaty), and the Joint Consultative Group (JCG) was established by the CFE Treaty for the parties to reconcile ambiguities of interpretation and implementation of the treaty.

have and are never likely to acquire the capacity to take actions prohibited under the treaty—such as establishing military bases and installations or conducting military manoeuvres in outer space.

States that participate in multilateral processes in spite of the fact that the resulting agreements have little relevance to their actions do so for more than one reason. In part they demonstrate political support for the purposes of the agreement in the belief that widespread (ideally universal) support will increase the likelihood of the norm underpinning the regime being implemented. States may participate in order to further other political objectives.[33] Participation is also a means of gaining information that would not be available to non-state parties. While these are reasonable and important objectives, they make only an indirect contribution to the effectiveness of regimes. Effectiveness depends mainly on participation by states whose actions can have a material bearing on the issues under discussion and that are subject to an agreement. To illustrate, if the approximately 70 states parties that do not carry out military or civilian nuclear activities left the NPT while India, Israel and Pakistan (three non-parties with military nuclear programmes) joined it, the treaty would be more effective.

Some regimes have been effective although levels of compliance have been questioned. For example, within the MTCR and the NSG information is periodically exchanged, suggesting that one or more participating states have authorized exports of controlled items that are inconsistent with the guidelines agreed by the regime.[34] However, all the participating states have modified their domestic laws and regulations as a consequence of agreements reached in the framework of the regimes. In many cases these revisions to domestic laws have been far-reaching, introducing new primary legislation and completely restructuring the administrative and bureaucratic apparatus used to implement export controls. In addition, these regimes have begun to establish common institutions to help manage processes that, while informal, have come to involve frequent meetings of officials and technical experts to discuss a wide range of issues as well as a high volume of exchanged documents.

While conceptually distinct, the issues of effectiveness and compliance are interrelated. Moreover, the regimes within these four categories can be mutually reinforcing. For example, the effectiveness of informal mechanisms can be enhanced by the existence of multilateral treaties and conventions, even though they carry on their work outside the framework of a treaty. As a specific example, decisions of the Nuclear Suppliers Group have led states with civilian nuclear industries and infrastructure to modify their national export control laws and regulations partly because they are keen to demonstrate their compliance with the NPT. Similarly, decisions by the Australia Group have been facilitated by the desire of its participants to be seen to be in compliance with the commitments related to biological weapon disarmament established

---

[33] As one example, some European states have participated in multilateral arms control in the expectation that it will accelerate the process of European integration.

[34] In 2001 a series of analyses were published criticizing the multilateral export control regimes. The criticisms related in part to questionable records of compliance.

in the BTWC and the commitments related to chemical weapon disarmament established in the CWC.

In summary, there is significant evidence that most arms control regimes enjoy a high degree of compliance. Moreover, many regimes are effective even if full compliance is not achieved. This effectiveness is evidence that states parties are working in good faith to make their national practices compliant. These observations tend to support the view that arms control does have a useful role to play in managing security problems. It also tends to support the view that the main problem in arms control is how to respond in cases where regimes are violated—that is, what actions to take when regimes enjoy neither compliance nor effectiveness.

III. The Bush Administration and arms control

The Bush Administration has injected a sense of urgency into discussions of arms control. As Rose Gottemoeller has expressed it, 'President Bush is prepared to engage partners in arms control and nonproliferation efforts, but will also move rapidly and unilaterally if they are not willing to join him'.[35]

President Bush has set ambitious objectives for US foreign and security policy. The reluctance to be drawn into extended or indefinite arms control negotiations is partly a function of domestic politics. Given the nature of his election, for most of the year 2001 he could not be confident about securing a second term in office. This approach also removes the use of extended deliberations as a tactic by which states interested to maintain existing conditions can influence US decisions.

The approach of the Bush Administration to arms control has been characterized as 'a marked disdain for multilateralism' and 'a penchant for go-it-alone policies'.[36] Senior officials reject this characterization. The US State Department Special Representative for Nuclear Nonproliferation has stated that 'multilateral regimes are important . . . global non-proliferation and arms control regimes will continue to be an important and valuable part of US strategy'.[37] In addition to supporting some existing arms control regimes the Bush Administration has advocated the creation of certain new regimes. For example, it supports the negotiation of a multilateral treaty to end the production of unsafeguarded fissile material.[38] In June 2001, following a review of US policy towards North Korea, President Bush instructed the State Depart-

[35] Gottemoeller, R., 'The current US agenda for the nonproliferation regime', Paper delivered at 'How to Harmonize Peaceful Uses of Nuclear Energy and Non-Proliferation', International Symposium of the Study Group on Peaceful Uses of Nuclear Energy and Non-Proliferation Policy, Tokyo, 7–8 Mar. 2001, available at URL <http://www.jaif.or.jp/english/npsympo/sympo_2nd.html>.

[36] Korb, L. and Tiersky, A., 'The end of unilateralism? arms control after September 11', *Arms Control Today*, Oct. 2001.

[37] Norman Wulf, Remarks at the panel discussion 'Do NPT review conferences really matter?' at 'New Leaders, New Directions', 2001 Carnegie International Non-Proliferation Conference, Washington, DC, 18–19 June 2001. For additional information see URL <http://www.ceip.org/files/projects/npp/resources/Conference%202001/panels/nptreview.htm>.

[38] Statement by Ambassador Eric M. Javits, US Representative to the Conference on Disarmament, Geneva, 7 Feb. 2002.

ment to discuss with North Korea how to create a less threatening conventional military posture on the Korean peninsula.[39] In a third example, the administration supports the development of a Draft International Code of Conduct (ICOC) against Ballistic Missile Proliferation.[40]

While denying a predisposition against arms control agreements, senior officials have made it clear that continued US participation in any given arms control process (whether multilateral or bilateral) is conditional on its perceived utility in helping to solve current US security problems. For example, Under Secretary of State for Arms Control and International Security John Bolton has stated that 'arms control can be an important part of American foreign policy, but I think the real question is what advances our national interest. And in those cases where, for example, arms control treaties are ineffective or counterproductive or obsolete, they shouldn't be allowed to stand in the way of the development of our foreign policy'.[41] Similarly, the US representative to the Conference on Disarmament stated that:

although maintaining international peace and security is our primary goal and overarching purpose, in the final analysis preserving national security is likewise necessary and essential. Mutual advantage is one key factor, for any arms control treaty must enhance the security of all States Parties. Basic obligations need to be well-focused, clear, and practical, so States will have a rational basis for committing themselves to the future treaty.[42]

President Bush did not reverse a policy of strong commitment to arms control in general, still less a commitment to multilateral arms control. The Bush Administration has placed arms control much closer to the heart of US foreign and security policy than its predecessor did.[43]

In the period 1993–2001, the Department of Defense showed sustained interest in providing technical and financial assistance to Russia and some other states that emerged on the territory of the former Soviet Union to help them secure and then safely dismantle weapons and weapon-related know-how.[44] However, President Bill Clinton, Vice-President Al Gore, and succes-

---

[39] The outcome of the review was described by Assistant Secretary of State for East Asian and Pacific Affairs James E. Kelly in 'United States policy in East Asia and the Pacific: challenges and priorities', Testimony before the Subcommittee on East Asia and the Pacific, House Committee on International Relations, 12 June 2001.

[40] US Department of State, 'US supports universal code against missile proliferation', Press Release, Washington, DC, 11 Feb. 2002. The ICOC is available at URL <http://projects.sipri.se/expcon/drafticoc.htm> and is discussed further in chapter 14 in this volume.

[41] 'Bolton: missile defense may help prevent proliferation', Interview with Under Secretary John Bolton on 14 Aug. 2001, available at URL <http://www.usinfo.state.gov/topical/pol/arms/stories/bolt 0814.htm>.

[42] Statement by Ambassador Eric M. Javits (note 38).

[43] The most dynamic period of arms control ever seen began in 1987 with the signing of the Treaty on the Elimination of Intermediate-Range and Shorter-Range Missiles (INF Treaty) and ended in 1993 with the signing of the Chemical Weapons Convention. Allan Krass has called this the 'arms control revolution'. Krass, A. S., *The United States and Arms Control: The Challenge of Leadership* (Praeger: Westport, Conn., 1997), p. 29.

[44] These CTR measures were described by Secretary of Defense William Perry as 'defence by other means'. Perry, W. J., 'Defense by other means', Remarks to US/Russian Business Council, Washington, DC, *Defense Issues*, vol. 10, no. 43 (29 Mar. 1995).

sive secretaries of state and national security advisers showed only limited and sporadic interest in arms control—reflecting a tendency to reduce the attention paid to politico-military aspects of foreign affairs.

The main foreign policy priority of successive Clinton administrations was to bring about a transformation of the political and economic systems of countries in Asia, Africa, Europe and the Middle East. Issues related to politico-military stability, including bilateral and multilateral arms control, were allocated a lower priority.[45]

Senior members of the Bush Administration have long experience and personal interest in strategic and politico-military issues. Writing in early 2000 Condoleezza Rice, subsequently appointed as National Security Advisor to the President, observed that in setting priorities for the USA 'peace is the first and most important condition for continued prosperity and freedom. America's military power must be secure because the United States is the only guarantor of global peace and stability'.[46] Moreover, the leadership believes that the global military security environment contains threats to the vital interests of the USA for which arms control can be one (although not the most important) element in an overall response.

Central to this view of arms control is the belief among the senior decision makers of the Bush Administration that some arms control agreements have been and continue to be violated. In other words, some parties to agreements are not inadvertently failing to comply with agreements but are deliberately cheating. These agreements can only continue to be supported by the USA on the condition that violations cease and the capacities created through violations are eliminated, thereby bringing parties back into compliance.

Under Secretary of State Bolton has characterized the main emphasis of current US arms control policy as 'the determination to enforce existing treaties, and to seek treaties and arrangements that meet today's threats to peace and stability, not yesterday's. Fundamental to the Bush Administration's policy is the commitment to honor our arms control agreements, and to insist that other nations live up to them as well'.[47]

The logic underpinning the approach towards enforcing existing treaties was trailed in a presentation by Richard Perle in which he observed that:

many treaty constraints that would be desirable if honoured are not desirable where there can be no assurance of compliance. . . . We must recognize that the world will be a safer place when countries that respect the rights of others are more powerful than those that do not. . . . agreements that weaken the Western democracies relative

---

[45] Talbott, S., 'US support for reform in Russia and the other new independent states', Statement before the Subcommittee on European Affairs, Senate Foreign Relations Committee, 7 Sep. 1993, *US Department of State Dispatch*, vol. 4, no. 38 (20 Sep. 1993), pp. 633–37; and Lake, A., Assistant to the President for National Security Affairs, 'From containment to enlargement', Address at the School of Advanced International Studies, Johns Hopkins University, Washington, DC, 21 Sep. 1993, *US Department of State Dispatch*, vol. 4, no. 39 (27 Sep. 1993), pp. 658–64.

[46] Rice, C., 'Promoting the national interest', *Foreign Affairs*, vol. 79, no. 1 (Jan./Feb. 2000), p. 50.

[47] Statement of the Honorable John R. Bolton, Under Secretary of State for Arms Control and International Security, US Department of State, to the Conference on Disarmament, 24 Jan. 2002.

to states that support terror or launch wars of aggression are foolish excursions that allow statesmen to feel good while they are actually doing bad.[48]

The identification of 'backlash states' occurred in 1994, when National Security Advisor Anthony Lake described Iran, Iraq, Libya, North Korea and Cuba in these terms.[49] However, the logic underpinning this listing was different from that put forward by the Bush Administration. Outlaw states were identified as states that do not accept certain values, including the pursuit of democratic institutions, free markets, peaceful conflict resolution and collective security. The Clinton Administration advocated a strategy of containment and isolation to pressure the governments of these countries to change their behaviour.[50] In line with this approach the Clinton Administration raised the issue of supplies of military and dual-use technologies to these states in export control regimes, in bilateral discussions with supplier states and as part of the transatlantic dialogue with the European Union (EU).

The Bush Administration has not characterized arms control as an instrument to be applied to bring about domestic political changes. It has tried to develop a comprehensive strategy based on the understanding that NBC weapons along with missile delivery systems for them pose 'a direct and serious threat to the national security of the United States, our friends, forces and allies'.[51] In this light, arms control has been evaluated first and foremost against the role it might play in reducing what are perceived to be military threats to the USA and its interests.

Within this overall strategy three separate elements have been identified. These elements are to provide protection in conditions where NBC weapon capabilities and missile delivery systems already exist, to adapt the approach to deterrence to take into account contemporary threats, and to prevent or slow the spread of NBC weapons and missile delivery systems. Arms control is expected to play the primary role with regard to prevention and to consist of the following four elements: (*a*) efforts to persuade or induce governments engaged in proliferation to change their behaviour; (*b*) efforts to deny proliferators the supply of equipment, material and technology from foreign suppliers; (*c*) the provision of technical and financial assistance to secure or eliminate surplus capabilities; and (*d*) efforts to strengthen existing treaties, promote new ones that meet US interests and upgrade the means of verifying implementation.

In line with this approach, US responses to weapon programmes of concern have sometimes but not always taken place within the framework of existing treaties. The issue was put in sharp focus in November 2001, when John

---

[48] Perle, R., 'Good guys, bad guys and arms control', eds Anthony and Rotfeld (note 1), p. 50.
[49] Lake, A., 'Confronting backlash states', *Foreign Affairs*, vol. 73, no. 2 (Mar./Apr. 1994). The concept itself is not new. Robert Litwak has traced a long history of references to 'outlaw', 'pariah', 'rogue' and 'renegade' states in US foreign policy. Litwak, R., *Rogue States and US Foreign Policy: Containment after the Cold War* (Johns Hopkins University Press: Washington, DC, 2000).
[50] The logic of the approach was explained in Lake (note 49), pp. 45–55.
[51] 'US nonproliferation efforts', US Department of State Fact Sheet, Washington, DC, 7 Sep. 2001, available at URL <http://usembassy.state.gov/japan/wwwhse0305.html>.

Bolton named Iraq, Libya, North Korea and Syria as countries that had violated the BTWC and Iran as a state that had 'probably' violated the convention.[52] In his statement Bolton noted the belief in the United States that other states parties to the BTWC that were not named were conducting offensive biological weapon programmes. Bolton mentioned one state, Sudan, that is not a party to the BTWC but which the USA believes to have an active biological weapon programme. While arms control agreements will sometimes play a role in the US response to perceived military threats, in other cases different instruments will be brought to bear.

In some cases where US responses occur outside the framework of an agreement this could be because the state in which a programme is located is not a party to the agreement establishing controls over programmes of that type. In other cases it could be because there is no agreement establishing controls over the type of programme causing concern.

In one case—Iraq—international efforts to eliminate NBC weapon programmes are primarily connected to the obligations contained in UN Security Council resolutions. The USA has played a leading role in the discussion of how to implement these resolutions.[53] At the time Iraq was conducting a clandestine nuclear weapon development programme it was a party to the NPT. However, although a strengthened NPT review process was created in 1995, efforts to address violations through the mechanisms of the treaty have been less consequential.[54]

There is also likely to be a differentiated response from the USA to weapon programmes with similar technical characteristics. This differentiation may reflect the different political relations between the USA and the particular states where programmes are located.

Sanctions and other measures could be applied where the US interests at stake are considered sufficiently important and where this approach is considered likely to yield results. Although the Bush Administration, in common with its predecessor, would prefer not to be required by domestic legislation to use sanctions, in given cases it would keep this instrument available for use at the discretion of the executive.

In September 2001 President Bush determined that institutes and enterprises in China and Pakistan had engaged in missile technology proliferation activities. This determination triggered the imposition of sanctions on both countries under the conditions established by US domestic legislation.[55]

---

[52] John R. Bolton, Under Secretary of State for Arms Control and International Security, Remarks to the 5th Biological Weapons Convention RevCon Meeting, Geneva, Switzerland, 19 Nov. 2001, available at URL <http://www.state.gov/t/us/rm/janjuly/6231.htm>.

[53] Recent developments in the UN efforts to implement the relevant resolutions are discussed in chapter 5 in this volume.

[54] Described in the Final Document of the 2000 Review Conference of the Parties to the Treaty of the Non-Proliferation of Nuclear Weapons, available at URL <http://www.un.org/Depts/dda/WMD/finaldoc.html>. It can be predicted that the issue of NPT non-compliance and responses to it is very likely to be raised by the United States at forthcoming review conferences.

[55] 'Bureau of Nonproliferation: imposition of missile proliferation sanctions against a Chinese entity and a Pakistani entity', *Federal Register*, vol. 66, no. 176 (11 Sep. 2001), p. 47256. This and other cases are discussed further in chapter 14 in this volume.

In other cases coercive measures are less likely to be applied. During 2001 the administration continued the policy initiated by President Clinton of seeking support in Congress to lift the sanctions imposed on India and Pakistan following the nuclear tests carried out by those two countries in 1998. The imposition of sanctions was required by US domestic legislation and did not reflect the preferred approach of the administration.[56] In August 2001 the administration was reported to have prepared a request to Congress to remove from law the requirement that sanctions be imposed in response to particular acts.[57] In the meantime, on 22 September President Bush determined that maintaining sanctions was not in the national security interests of the USA. In October the policy of denying authorization for all exports and re-exports of items controlled for nuclear proliferation and missile technology reasons to India and Pakistan was ended.[58]

At the same time the USA has continued to pursue a regular and high-level dialogue with all three countries—China, India and Pakistan—about arms control, non-proliferation and export control. As these three countries have in recent years developed their own industries producing strategic goods, export control questions have come to play a more significant role in the bilateral discussions between each of them and the United States.

Arms control and missile defence

During 2001 the second element of the Bush Administration's approach to arms control, 'seeking agreements that meet today's threats to peace and stability, not yesterday's', was reflected in the need to reconcile the changes in security policy considered necessary by the new administration with bilateral arms control discussions with Russia.[59]

Of particular importance from the US perspective was the need to modify the existing arms control framework in ways that did not prevent progress in the development of a ballistic missile defence system able to defend the USA from limited attacks.

During the 1980s and 1990s ballistic missiles with progressively longer ranges were acquired by a number of states that had never previously had the capacity to project power. The list of states known to be acquiring (in some cases through indigenous programmes and in some cases through foreign assistance) long-range missiles correlates closely with the group of states alleged to be developing nuclear and/or biological weapons.

---

[56] Neither India nor Pakistan is a party to either the NPT or the 1996 Comprehensive Nuclear Test-Ban Treaty. For a discussion of international responses to the testing see Anthony, I. and French, E. M., 'Non-cooperative responses to proliferation: multilateral dimensions', *SIPRI Yearbook 1999: Armaments, Disarmament and International Security* (Oxford University Press: Oxford 1999), pp. 682–83.

[57] Perlez, J., 'US ready to end sanctions on India to build an alliance', *New York Times* (Internet edn), 27 Aug. 2001.

[58] 'India and Pakistan: lifting of sanctions, removal of Indian and Pakistani entities and revision in license review policy', *Federal Register*, vol. 66, no. 190 (1 Oct. 2001), pp. 50090–93.

[59] This is discussed further in chapters 10 and 11 in this volume.

A December 2001 national intelligence estimate finds that 'before 2015 the USA most likely will face [intercontinental ballistic missile] threats from North Korea and Iran, and possibly from Iraq—barring significant changes in their political orientations—in addition to the longstanding missile forces of Russia and China'.[60] The architecture of a national missile defence system has not yet been agreed in detail. However, the Ballistic Missile Defense Organization anticipates a multi-tiered system able to apply countermeasures throughout the entire trajectory of a missile launched against the USA, from its launch and boost phase to terminal stages of an attack. The USA will not have a capacity for countering existing and emerging medium- to long-range ballistic missile threats in the near term. Responsible officials have therefore stated that 'there is some urgency behind our missile defense development and test efforts. The deployment of missile defenses requires commitment and focus in our programs over many years'.[61]

In the view of President Bush, the required testing programme could not be accomplished within the framework of the 1972 ABM Treaty, and on 13 December 2001 Bush announced the United States' plans to withdraw from that treaty and formally notified the governments of Russia, Belarus, Kazakhstan and Ukraine. According to Bush, the ABM Treaty was no longer required since the hostility that once led the USA and the former Soviet Union 'to keep thousands of nuclear weapons on hair-trigger alert, pointed at each other' had been replaced by 'a new, much more hopeful and constructive relationship. We are moving to replace mutually assured destruction with mutual cooperation'. This would include a new strategic relationship between the United States and Russia, acknowledging that 'the greatest threats to both our countries come not from each other, or other big powers in the world, but from terrorists who strike without warning, or rogue states who seek weapons of mass destruction'.[62]

IV. International effects of the Bush Administration approach

Decisions by the Bush Administration in 2001 were seen internationally as evidence that the United States might pursue its military security interests without due consideration for the concerns of other states. Domestic critics of the administration took up this point during the year.[63] Senate Majority Leader Tom Daschle criticized the Bush Administration for a 'willingness to walk away from agreements that were embraced by many of our closest friends and

---

[60] US National Intelligence Council, *Foreign Missile Developments and the Ballistic Missile Threat through 2015*, Unclassified summary of a National Intelligence Estimate, Dec. 2001.

[61] Remarks of Lieutenant General Ronald T. Kadish, Director, Ballistic Missile Defense Organization, at the Hearing on Ballistic Missile Defense Testing, before the Military Research and Development Subcommittee, House Armed Services Committee, 14 June 2001.

[62] The White House, 'Remarks by the president on national missile defense', Press Release, 13 Dec. 2001.

[63] E.g., leading figures in the Democratic Party raised the issue of foreign perceptions of the Bush Administration. Gephardt, R., 'The future of trans-Atlantic relations: collaboration or confrontation?', Speech at the Carnegie Endowment for International Peace, Washington DC, 2 Aug. 2001, available at URL <http://usinfo.state.gov/topical/pol/arms/stories/01080311.htm>.

allies, and broadly supported by the international community'. Four of the six cases cited in support of this argument were related to arms control.[64]

Subsequently, the administration began to address this criticism directly. During its first year the Bush Administration was not averse to consultation on arms control-related issues. Reflecting the relatively high priority accorded to military security questions by the administration, US officials undertook visits to many countries to take up such matters. However, the administration also made clear that, while US positions may be modified or reviewed in the light of countervailing arguments and evidence, modifications would not be undertaken solely out of deference to other states.

In testimony before Congress Secretary of State Colin Powell noted:

I think we have demonstrated that we are anxious to reach out to the world. We are not unilateralists pulling back . . . But where there is a matter of principle, where we believe strongly about something and we have to stick by our principles, we will do that, and lead, and try to convince others to go with us. This isn't unilateralism; this is leadership. And our friends, I think, are increasingly coming to the understanding that this is principled leadership, the kind that they should respect, follow where they think it is appropriate to follow, and where they think it is not appropriate to follow, let them make their own individual sovereign choice.[65]

The evidence from the first year of the Bush Administration tends to support the interpretation put forward by Powell. Most participants in arms control regimes have paid close attention to ensuring that they comply with provisions of agreements and other measures while paying relatively less attention to compliance by others. However, the approach taken by the USA has raised the issue of whether a more collective approach to compliance is necessary for regimes to succeed.

There are differences between states about the most important priorities in arms control. For example, many European countries would have wished to see more rapid progress in the development of a treaty-based framework for anticipated reductions in the nuclear weapon arsenals of the United States and Russia. With the demise of the START framework, these countries would still prefer to see the reductions in the numbers of nuclear weapons codified in a treaty. Similarly, it remains the view of many European states that it is feasible to develop adequate responses to cases of treaty violations within the framework of multilateral treaties.[66]

At the same time, European Union representatives best placed to evaluate the impact of EU policies in the United States have pointed to a need for an adjustment in thinking about how to implement these preferences.

---

[64] The cases were the CTBT, the BTWC protocol, the ABM Treaty and the July 2001 UN Conference on the Illicit Trade in Small Arms and Light Weapons in All Its Aspects. Daschle, T., 'A new century of American leadership', Remarks at the Woodrow Wilson International Center for Scholars, 9 Aug. 2001, available at URL <http://usinfo.state.gov/topical/pol/arms/stories/01080903.htm>.

[65] Secretary of State Colin L. Powell, Testimony at Budget Hearing Before the Senate Budget Committee, Senate Budget Committee, Washington, DC, 12 Feb. 2002.

[66] See, e.g., the discussion of European Union approaches to the BTWC in chapter 12 in this volume.

Ambassador Günter Burghardt, Head of the European Commission Delegation to the United States, has noted that 'it will be up to the EU to demonstrate that we have practical solutions to difficult problems and that these solutions are sometimes best achieved by international cooperation. . . . It will clearly be fruitless and counter-productive to preach the virtues of international cooperation and multilateralism as ends in themselves'.[67]

European initiatives in arms control have been relatively few in recent years. There has been a tendency among states in Western Europe to emphasize the need to enhance their autonomous military capabilities in the context of the Defence Capabilities Initiative within NATO and the European Security and Defence Policy (ESDP) within the EU.[68] Against that background European participation has tended to emphasize the steps taken to comply with existing multilateral treaty commitments as well as measures largely aimed at constraining powers in regions other than Europe. Finally, there has been a reluctance to risk open disagreement with the United States within existing 'West–West' forums, none of which has so far addressed arms control in a regular and systematic manner.

There is an understanding in Europe of the need for the arms control agenda to emphasize progress in areas of common interest in the present political environment. German Defence Minister Rudolf Scharping has pointed to the need to emphasize 'the potential use of biological, chemical, radiological and nuclear weapons' as well as 'the evolution in ballistic missile delivery means and the fact that we are confronted with forms of asymmetric warfare'.[69]

It is not clear that this view is shared in other states of critical importance to arms control. In China, Russia and the Middle East there has been a tendency to try to link arms control processes. Rather than treating the issues involved in each process on their merits, this approach assumes that concessions can be traded across processes in search of a 'grand bargain'.

While there are many recent examples of this approach, three can be used for purposes of illustration.

For a considerable time Russian leaders emphasized that a US withdrawal from the ABM Treaty would jeopardize other nuclear arms control agreements.[70] Russia is moving away from this approach under the leadership of President Vladimir Putin. The US decision to give the development of the ballistic missile defence programmes higher priority than the preservation of the ABM Treaty required the Russian Government to formulate a response.

[67] 'Remarks to a renewed transatlantic security partnership in the aftermath of 11 September 2001 terror attacks', Joint Hearing, Committee on Foreign Affairs, Human Rights, Common Security and Defence Policy and Committee on Industry, External Trade, Research and Energy, European Parliament, Brussels, 19 Feb. 2002, available at URL <http://www.eurunion.org/news/speeches/2002/020219gb.htm>.

[68] Chapter 3 examines the development of the ESDP. The link to arms control is further developed in Bayles, A., 'Arms control: an endangered species in the new security environment?', eds Anthony and Rotfeld (note 1), pp. 17–21.

[69] Speech of Rudolf Scharping at Global Security: New Challenges, New Strategies, 38th Munich Conference on Security Policy, 3 Feb. 2002, available at URL <http://www.securityconference.de>.

[70] Speech of the Russian Minister of Defence Sergey Ivanov at 'Global and Regional Security at the Beginning of the 21st Century', 37th Munich Conference on Security Policy, 4 Feb. 2001, available at URL <http://www.securityconference.de>.

President Putin, while regretting the decision, made clear that it represented no threat to Russian security. Moreover, he reiterated his previous commitment to further deep reductions in nuclear forces.[71]

Leading Chinese officials have taken the view that progress in processes aimed at non-proliferation of weapons depends not only on the merits of the particular cases but also on developments in other areas of arms control and strategic affairs.[72] Chinese officials continue to emphasize the existence of an 'international arms control and disarmament system' that should be addressed as an integrated whole, not in a piecemeal manner, rather than as a series of discrete attempts to solve particular problems.[73]

The refusal by some Arab states to sign the CWC or ratify the BTWC without the simultaneous elimination of nuclear weapons—and in particular those of Israel—is a third such example.

The Bush Administration also brought several issues under discussion to a point where decisions would be required from other countries.

The fact that the USA chose a conference of states parties as the forum in which to name particular states as violators of the BTWC forced other parties to make a determination about whether they share the US finding. In addition, states parties have been forced to begin thinking in a less abstract way about how to respond to detected violations.[74]

President Bush decided that there was convincing evidence that China had engaged in missile proliferation, knowing that the USA would be obliged to impose sanctions in response to this determination. This in turn increased the pressure on the Chinese Government to respond to questions about what steps it was taking to implement undertakings related to its national missile-related export control system given to the USA in November 2000.[75]

The US approach will require other states to consider their approach to two factors in particular: the degree to which existing arms control regimes enjoy compliance; and the degree to which these regimes are effective instruments with which to manage security problems.

V. The impact of the attacks of 11 September

In the period immediately after the terrorist attacks against the United States on 11 September 2001, it is natural for states to pay considerable attention to the risks posed by the acquisition of military capacities of different kinds. Two issues in particular appear to have become the main focus of attention:

---

[71] For further discussion see chapters 10 (in particular) and 11 in this volume. The position may also reflect a judgement that the character of Russian statements about missile defence will influence the US approach to the form in which reciprocal deep cuts in nuclear arsenals are codified.

[72] E.g., Sha Zukang, 'Non-proliferation at a crossroads,' Address to the Wilton Park Conference on Non-Proliferation, 14 Dec. 1999, available on the Internet site of the Ministry of Foreign Affairs of China, URL <http://fmprc.gov.cn>.

[73] Statement by Hu Xiaodi at the Plenary of the 2002 Session of the Conference on Disarmament, Geneva, 7 Feb. 2002, available at URL <http://www.fmprc.gov.cn/eng/c464.html>.

[74] For further discussion see chapter 12 in this volume.

[75] For further discussion see chapter 14 in this volume.

(a) whether states that sponsor and support actors that carry out terrorist acts can deter military attacks carried out against them; and (b) how to prevent those carrying out terrorist acts from acquiring military capacities and in particular capacities that would allow them to carry out acts of 'catastrophic terrorism'.

Military force was applied directly against armed groups in Afghanistan that were identified as being complicit in the attacks of 11 September. At the same time, what is expected to be an extended campaign against terrorism was initiated by a coalition of states led by the USA. As part of that wider campaign the immediate focus was on putting in place counter-terrorist measures based on law enforcement and political, diplomatic, financial and intelligence-sharing activities.

Apart from the actions in Afghanistan, US Secretary of State Powell made clear that cabinet ministers had 'not made any recommendation to the President about the major use of military force and the President has made no decision as yet with respect to such use of force'.[76] At the same time, the future use of force was not excluded as part of the wider campaign against terrorism.[77]

In this context those arms control processes intended to prevent the acquisition of nuclear and biological weapons along with missile delivery systems for them can be expected to receive greater attention. This focus would be logical because these weapons and delivery systems could provide states with military capacities that might deter the use of force against them by the USA and, conceivably, other states.

The second issue that has been addressed more urgently is the risk that terrorist groups might gain access to nuclear, biological or chemical weapons.

Arms control agreements have historically been concluded between states and establish rules that are binding on states. The 11 September attacks have increased the attention being given to the idea that agreements might establish commitments with regard to non-state actors. Existing agreements contain general obligations in line with this idea. For example, the CWC includes a general obligation not 'to develop, produce, otherwise acquire, stockpile or retain chemical weapons, or transfer, directly or indirectly, chemical weapons to anyone' under any circumstances. Similarly, the BTWC includes a commitment not to transfer biological weapons to 'any recipient whatsoever'.[78]

In this context the role of arms control in preventing the acquisition of NBC weapons by terrorist groups is related to three issues: first, ensuring compliance with existing non-proliferation agreements; second, achieving participation in those agreements by all states with capacities that could contribute to NBC weapon programmes of concern, including those that are not currently

---

[76] Powell, C. L., Testimony before the House Appropriations Subcommittee on Foreign Operations, Export Financing, and Related Programs, 13 Feb. 2002.

[77] President Bush has stated that 'every nation now knows that we cannot accept—and we will not accept—states that harbour, finance, train, or equip the agents of terror. Those nations that violate this principle will be regarded as hostile regimes. They have been warned, they are being watched, and they will be held to account'. Remarks by the President at the Citadel, Charleston, S.C., 11 Dec. 2001 available at URL <http://www.whitehouse.gov/news/releases/2001/12/20011211-6.html>.

[78] For further information see chapter 12 in this volume.

parties; and third, ensuring that states that choose to remain outside existing agreements act in a responsible manner in regard to any entity that may sponsor or carry out terrorist acts.

A general rule that non-state actors should not have access to any military capacities has not found universal support. The USA has been a particularly strong opponent of a general rule of this kind. For example, during discussions held in the framework of the Organization for Security and Co-operation in Europe (OSCE) in 2000 and in the United Nations in 2001 the USA made clear its opposition to such a rule if applied to small arms and light weapons.

The USA described the limitation of trade in small arms and light weapons only to governments as conceptually flawed since it would 'preclude assistance to an oppressed non-state group defending itself from a genocidal government. Distinctions between governments and non-governments are irrelevant in determining responsible and irresponsible end-users of arms'.[79]

VI. Conclusions

Developments in 2001 have been seen in some quarters as evidence of a loss of confidence by key actors—in particular the USA—in the capacity of arms control to manage security problems. The evidence suggests that recent developments reflect an adaptation of arms control, which is in essence a framework in which structured dialogue can be organized around armaments policy, rather than abandonment.

As part of this process of adaptation there may be a certain loss of coherence in the position of particular states with regard to arms control. For example, a state may agree measures in the framework of one regional process based on principles that would not be acceptable if applied in a different location or on a global basis. This may be a transitory phenomenon as new norms and principles develop in a changing security environment.

A rigid distinction between law and politics is difficult to draw where legal changes are attempted in parallel with efforts to modify approaches to the underlying issues subject to legislation. Where discussions of law are used as an instrument of political change, frictions are inevitable at the places where legal and political processes meet. At present, because of its dominant position, the views and policies put forward by the USA generate the most influential pressures shaping the attitudes on which future agreements will be based.

In 2001 this friction was felt in the discussions of the ABM Treaty, of a protocol to verify the BTWC and of whether a general rule should be adopted prohibiting military assistance to non-state actors. Each of these discussions dealt with an important but contested underlying issue of principle. In helping to frame the issues and by providing a context for structured discussion, arms control was fulfilling one of its most important functions.

[79] Plenary Address by John R. Bolton, US Under Secretary of State for Arms Control and International Security Affairs, at the UN Conference on the Illicit Trade in Small Arms and Light Weapons in All Its Aspects, New York, 9 July 2001, available at URL <http://www.state.gov/t/us/rm/janjuly/4038.htm>.

A change in approach to arms control in the USA under the Bush Administration has brought about a need for adaptation. While restoring the traditional linkage between arms control and military security, the administration also sought to adapt agreements, processes and arrangements to its view of the contemporary strategic environment.

During 2001 the need for arms control dialogue to be conducted in a cooperative security framework was underlined. However, it is not clear whether other states share the US view of the contemporary strategic environment or, to the extent that they do, draw the same lessons with regard to the need to adapt arms control regimes. It is increasingly understood that existing institutions are not adequate to allow those states that share values to organize their activities in pursuit of common purposes. The main challenge will be to engage the United States in this common framework on the basis of responsible leadership and with the willing acceptance of its partners.

The proposals put forward by the Bush Administration built on positions whose evolution can be traced over the previous decade. While the issues were not new, the manner in which the Bush Administration addressed them changed the character of the discussion. Many governments have stated that they regard arms control as a valuable and necessary activity. However, while existing and putative bilateral and multilateral arms control processes had been discussed among officials for several years, in the absence of sustained involvement by senior political decision makers these discussions produced few results. In particular, discussions rarely engaged with the critical choices that would have to underpin any new agreements.

Senior ministers in the Bush Administration have created a political environment in which these critical choices will be required by governments within a relatively short time frame to establish the basic frame of reference for arms control processes.

While the Bush Administration had already changed the context of the debate about arms control prior to the terrorist attacks on the USA on 11 September, these attacks are likely to contribute to further changes. In particular, there is likely to be greater focus on the issue of how to ensure compliance with agreements relating to NBC weapons along with delivery systems for them. However, while arms control is likely to play an indirect and supporting role as one of the instruments applied in the wider campaign against terrorism, its main emphasis and impact will continue to be in other areas of international security.

10. Ballistic missile defence and nuclear arms control

SHANNON N. KILE

I. Introduction

In 2001 the international controversy over the United States' missile defence plans and the future of the 1972 Treaty on the Limitation of Anti-Ballistic Missile Systems (ABM Treaty) came to a head. On 13 December, President George W. Bush announced that the USA would withdraw from the ABM Treaty. Bush's announcement was widely expected and did not undermine commitments made by Russia and the USA the previous month to further reduce their nuclear arsenals. Against the background of improving political relations, Bush and Russian President Vladimir Putin had pledged to make significant new cuts in US and Russian strategic nuclear forces. As the year ended, however, there was disagreement between Russia and the USA over whether these reductions would be made within the framework of an arms control treaty or as parallel, non-legally binding initiatives.

This chapter reviews the principal developments in missile defence and nuclear arms control in 2001. Section II describes the US administration's decision to withdraw from the ABM Treaty and assesses the reaction of Russia and other states. It also examines changes in the US programme to develop and deploy a missile defence system designed to protect the United States and its allies from a limited ballistic missile attack. Section III examines the Russian and US commitments to make further nuclear force reductions. It also notes the completion of the reductions in strategic nuclear delivery vehicles (SNDVs) and accountable warheads mandated by the 1991 Treaty on the Reduction and Limitation of Strategic Offensive Arms (START I Treaty). Section IV summarizes developments related to the international cooperative programmes to dismantle nuclear weapons and enhance the safety and custodial security of nuclear materials in the former Soviet Union. Section V surveys the status of efforts to bring the 1996 Comprehensive Nuclear Test-Ban Treaty (CTBT) into force. Section VI presents the conclusions.

Appendix 10A provides data on the nuclear forces of the five legally recognized nuclear weapon states and on the nuclear arsenals of India, Israel and Pakistan. Appendix 10B analyses the arms control challenges posed by non-strategic (or tactical) nuclear weapons and describes proposals for controlling and eventually eliminating these weapons. Appendix 10C provides an overview of changes under way in the US and Russian nuclear weapon production complexes. Appendix 10D examines recent international efforts to strengthen the physical protection of nuclear facilities.

SIPRI Yearbook 2002: Armaments, Disarmament and International Security

II. Ballistic missile defence and the future of the ABM Treaty

Ballistic missile defence (BMD) and the future of the ABM Treaty have generated controversy both within the USA and between the USA and Russia.[1] This controversy came to a head in 2001 with the change of US administration. The incoming Bush Administration pledged that one of its immediate policy priorities would be to pursue the deployment of a more extensive missile defence system than that envisaged by its predecessor; it submitted an amended fiscal year (FY) 2002 defence budget authorization bill that significantly increased funding for missile defence research and development (R&D) programmes. The change of administration also led to a shift in the US position on the ABM Treaty. President Bush announced that the USA would withdraw from the treaty rather than seek to amend it to permit the deployment of a limited national missile defence (NMD) system. The announcement elicited a restrained reaction from President Putin, who signalled that the decision would not derail improving Russian–US relations.

The US missile defence debate

Missile defence has been a source of recurring partisan dispute in the USA. In the late 1990s a consensus gradually emerged in Washington that a BMD system was needed to protect the USA against an attack by a small number of long-range missiles—possibly armed with nuclear, chemical or biological weapons—launched by 'rogue states' such as North Korea or Iraq.[2] This consensus was reflected in the US Congress' overwhelming approval of the National Missile Defense Act of 1999, which committed the USA 'to deploy as soon as is technologically possible an effective National Missile Defense system capable of defending the territory of the United States against limited ballistic missile attack (whether accidental, unauthorized, or deliberate)'.[3]

However, missile defence has remained a controversial issue in Congress. There has been a spirited debate over how limited in scope and scale a future BMD system should be and over the pace of its development. There has also

---

[1] The ABM Treaty was signed by the USA and the USSR in May 1972 and entered into force in Oct. 1972. In Sep. 1997, Belarus, Kazakhstan, Russia, Ukraine and the USA signed a Memorandum of Understanding on Succession (MOUS) pursuant to which the 4 former Soviet republics collectively assumed the rights and obligations of the USSR under the ABM Treaty. The MOUS was not subsequently ratified by the US Senate. However, representatives of all 5 states participated in the Standing Consultative Commission, the forum for the parties to discuss ABM Treaty-related questions. For a summary of the main provisions of the treaty and the MOUS see annex A in this volume. The text of the ABM Treaty, the Agreed Statements, Common Understandings and Unilateral Statements, and the 1974 Protocol are presented in Stützle, W., Jasani, B. and Cowen, R., SIPRI, *The ABM Treaty: To Defend or Not to Defend?* (Oxford University Press: Oxford, 1987), pp. 207–13.

[2] For a summary of the US missile defence debate prior to 2001 see Kile, S., 'Nuclear arms control and ballistic missile defence', *SIPRI Yearbook 2001: Armaments, Disarmament and International Security* (Oxford University Press: Oxford, 2001), pp. 424–26.

[3] National Missile Defense Act of 1999, Public Law 106–38, 22 July 1999.

been considerable disagreement over the degree to which, if any, this system should be constrained by the ABM Treaty.

In 2001 there were important changes in the tone and substance of this debate. Paradoxically, both advocates and critics of BMD claimed that the 11 September 2001 terrorist attacks on the USA reinforced their views about the relative priority that should be accorded to missile defence in countering new threats to US security.[4] However, in the wake of the attacks, the partisan conflict largely disappeared as Republicans and Democrats moved to present a united front on defence and security issues. Popular support for a large increase in US defence spending also effectively removed the budgetary constraint that critics had hoped would derail, or at least delay, the new administration's ambitious missile defence plans. In addition, Bush's announcement that the USA would withdraw from the ABM Treaty removed one of the main points of contention from the debate.

The Bush Administration's arguments for missile defence

The Bush Administration entered office in January 2001 committed to the development of a robust missile defence system to protect the USA and its allies. Conservatives in the Republican Party have supported the idea of building a strategic missile defence shield for some time.[5] Support has widened in the light of growing scepticism about the adequacy of the existing framework of arms control treaties and multilateral supplier arrangements designed to prevent the spread of non-conventional weapons and the means to deliver them.[6] The Bush Administration's approach to missile defence was part of a broader shift in emphasis from attempting to halt proliferation at its source to a greater focus on responding to—and managing the consequences of—proliferation. It also reflected an inclination to favour unilateral responses to proliferation challenges.

The new administration lost little time in urging Congress to push ahead with missile defence as an urgent priority. One argument put forward by senior administration officials was that a nationwide missile defence system would usefully supplement nuclear deterrence; this supplement was increasingly needed in the light of the emergence of states armed with long-range ballistic missiles which might not be deterred by threats of devastating retaliation.[7] Other officials downplayed the risk posed by potentially 'undeterrable' states, focusing instead on the prospect that a state might initiate a regional conflict involving US allies and important national interests in the mistaken

---

[4] See, e.g., Cirincione, J. and Payne, K., 'Debate: in the wake of 11 September where does missile defence fit in security spending priorities?', *NATO Review*, vol. 49 (winter 2001/2002), pp. 26–30.

[5] See FitzGerald, F., *Way Out There in the Blue: Reagan, Star Wars and the End of the Cold War* (Simon & Schuster: New York, 2000), pp. 114–46.

[6] See, e.g., Perle, R., 'Good guys, bad guys and arms control', eds I. Anthony and A. D. Rotfeld, SIPRI, *A Future Arms Control Agenda: Proceedings of Nobel Symposium 118, 1999* (Oxford University Press: Oxford, 2001), pp. 43–49. See also chapter 11 in this volume.

[7] Testimony of Secretary of Defense Donald H. Rumsfeld before the Armed Services Committee, US Senate, 12 July 2001, URL <http://www.defenselink.mil/speeches/2001/s20010621-secdef2.htm>.

belief that the USA would be deterred, by their missiles, from intervening in the conflict. In their view, the deployment of a nationwide missile defence system—even one using unproven technologies—would force potential adversaries to reassess the risks they would face by confronting the USA, thereby enhancing US freedom of action when responding to regional crises.[8] In addition, it was argued that the deployment of missile defences by the United States would discourage aspiring proliferators from developing or otherwise acquiring long-range ballistic missiles and weapons of mass destruction.[9]

New planning guidelines

During the 2000 presidential campaign, Bush had called for extensive missile defences to protect both the USA and its allies.[10] Upon taking office he ordered a re-evaluation of the scale and scope of the NMD system architecture put in place during the Administration of President Bill Clinton. That architecture relied exclusively on ground-based interceptor missiles guided by external sensors, to collide with incoming missile warheads in the mid-course phase of their flight trajectories (i.e., after they have separated from their booster rockets outside the earth's atmosphere). This mid-course-intercept approach had been criticized by both supporters and opponents of missile defence as providing an inherently fragile defence. Particular concern had been expressed about the ability of this approach to overcome the range of countermeasures (e.g., various types of decoys) that an attacker could be expected to employ.[11]

In May 2001, Bush indicated that he favoured building a more robust system that would eventually consist of several layers of defences.[12] While acknowledging that significant 'technological difficulties' would have to be overcome, he expressed confidence that 'complementary and innovative approaches' to missile defence would eventually succeed. Bush said that the Department of Defense (DOD) was examining options for deploying an initial defence capability against limited missile threats; this capability could be supplemented

---

[8] Prepared Statement on Ballistic Missile Defense, Deputy Secretary of Defense Paul Wolfowitz, US House of Representatives, Armed Services Committee, 18 July 2001, URL <http://www.defenselink.mil/speeches/2001/s20010719-depsecdef1.htm>. The argument that a national missile defence system would help to prevent the USA from being deterred by rogue state missile threats had been put forward by senior Pentagon officials during the Clinton Administration. See the testimony of Walter B. Slocombe, Under Secretary of Defense for Policy, to the Armed Services Committee, US House of Representatives, 13 Oct. 1999, in US Information Service (USIS), *Washington File*, available at the Federation of American Scientists (FAS) Space Policy Project site at URL <http://www.fas.org/spp/starwars/program/news99/991013-missile-usia.htm>.

[9] Wolfowitz (note 8).

[10] 'New leadership on national security', Address by Governor George W. Bush to the National Press Club, Washington, DC, 23 May 2000.

[11] See, e.g., Heritage Foundation Commission on Missile Defence, *Defending America: A Plan to Meet the Urgent Missile Threat* (Heritage Foundation: Washington, DC, 1999); and Lewis, G., Gronlund, L. and Wright, D., 'National missile defense: an indefensible system', *Foreign Policy*, no. 117 (winter 1999/2000), pp. 120–31.

[12] 'Remarks by the president to students and faculty at National Defense University', Washington DC, The White House, Office of the Press Secretary, 1 May 2001, URL <http://www.whitehouse.gov/news/releases/2001/05/20010501-10.html>; and Knowlton, B., 'Bush calls for missile shield, saying ABM pact is outdated', *International Herald Tribune*, 2 May 2001, pp. 1, 10.

later by sea- and land-based sensors and interceptors. He noted in particular that he saw 'substantial advantages' in systems capable of intercepting missiles in the boost phase (i.e., the powered ascent phase) of their flight trajectories.[13] He also expressed interest in proposals to deploy advanced sensors and interceptors in space as part of an integrated, multi-layer missile defence system.[14] Following the address, administration officials emphasized that no final decision on the system architecture had been taken.[15]

In January 2002 US Secretary of Defense Donald Rumsfeld issued a memorandum outlining the future direction of the DOD Missile Defense Program.[16] He identified four main missile defence priorities: (*a*) 'to defend the USA, deployed forces, allies and friends'; (*b*) 'to employ a Ballistic Missile Defense System (BMDS) that layers defences to intercept missiles in all phases of their flight'; (*c*) 'to enable the services to field elements of the overall BMDS as soon as practicable'; and (*d*) to develop and test technologies and 'improve the effectiveness of deployed capability by inserting new technologies as they become available or when the threat warrants an accelerated capability'.[17] Rumsfeld's memorandum directed the DOD to develop for deployment an integrated BMD system capable of addressing 'all ranges of threats'. This set out a clear planning requirement for a multi-layer missile defence system capable of countering larger, more technically sophisticated missile threats than the limited NMD system envisaged by the Clinton Administration.

CBO cost estimates

In a report released in January 2002 the US Congressional Budget Office (CBO) presented estimates of the potential cost (in constant 2001 dollars) of several different types of national missile defence systems.[18] It examined three

---

[13] Transcript of 'Remarks by the President' (note 12). A number of prominent missile defence proponents have advocated the development of a sea-based system, based on current US Navy theatre missile defense programmes, to intercept missiles during the boost phase of their trajectories. Since such a system would intercept ascending missiles before they could deploy warheads and decoys, it would not face the discrimination problem inherent in the mid-course intercept approach. Deutch, J., Brown, H. and White, J., 'National missile defense: is there another way?', *Foreign Policy*, no. 119 (summer 2000), pp. 91–104; and Garwin, R., 'A defense that will not defend', *Washington Quarterly*, vol. 23, no. 3 (summer 2000), pp. 109–23.

[14] The development of space-based assets for use in missile defence and other military roles and missions has been accorded a high priority by US Defense Secretary Donald Rumsfeld, who headed a congressionally mandated commission charged with reviewing US space activities in 2000. See *Report of the Commission to Assess United States National Security Space Management and Organization*, 11 Jan. 2001, Executive Summary, available at URL <http://www.space.gov/commission/report.htm>.

[15] Gordon, M., 'Bush describes his brave new world but not how to get there', *International Herald Tribune*, 3 May 2001, p. 6; and Suro, R., 'Plan for missile defense not clear', *Washington Post*, 9 May 2001, p. A8.

[16] Secretary of Defense Donald Rumsfeld, 'Missile Defense Program direction', Memorandum, Office of Secretary of Defense, 2 Jan. 2002, available at URL <http//www.defenselink.mil/news/Jan2002/b01042002_bt008-02.html>.

[17] 'DOD establishes Missile Defense Agency', US DOD News release no. 008-02, 4 Jan. 2002, URL <http://www.defenselink.mil/news/Jan2002/b01042002_bt008-02.html>.

[18] Congressional Budget Office (CBO), *Estimated Costs and Technical Characteristics of Selected National Missile Defense Systems*, Jan. 2002, available at the CBO Internet site, URL <http://www.cbo.gov>.

architectures under consideration by the DOD: a ground-based mid-course intercept system; a stand-alone sea-based mid-course intercept system (i.e., a sea-based system not seen as a complement to a ground-based one); and a constellation of satellite-based lasers and interceptors.

The CBO looked first at the missile defence system architecture proposed during the Clinton Administration.[19] It estimated that a system consisting of 100 ground-based interceptors deployed at a single site in Alaska (the so-called Expanded Capability 1 system) would cost $23–25 billion to develop, deploy and operate to 2015. If this system were expanded to include a second site with 150 additional interceptor missiles, along with satellite-based sensors and additional X-band (very high resolution) radar, the total cost would rise to $51–58 billion.[20]

The CBO report cautioned that the costs of sea- and satellite-based systems were more difficult to estimate since these systems were in the early phases of technology demonstration or the concepts for them were under development. It estimated that a stand-alone sea-based mid-course intercept system would cost $43–55 billion to develop, deploy and operate to 2015. It did not provide an estimate for a sea-based boost-phase system because the DOD 'had not released a description, however preliminary, of what might compose such a system'.[21] The report estimated that the cost of a space-based laser system, consisting of a constellation of lasers deployed in low earth orbit, would range from $56 billion to $68 billion to 2025. It did not provide an estimate of the cost of the 'Brilliant Pebbles' satellite-based interceptor system because of a lack of relevant technical and operational documentation.[22]

The CBO report concluded that 'the total cost of national missile defense cannot be determined definitively at this time' because of numerous uncertainties about the scale and configuration of the missile defence system to be deployed. [23] Nevertheless, some independent analysts used the CBO estimates as the basis for calculating the total cost of NMD, which was projected to be as much as $238 billion over the next 15–25 years. [24] While this would make NMD one of the most expensive DOD weapon procurement programmes, it would be similar in scale to the cost of other major US procurement programmes, such as the Joint Strike Fighter.[25]

---

[19] For a description of the components and operational concept of this system see Kile (note 2), pp. 426–29.
[20] CBO (note 18), p. 9.
[21] CBO (note 18), p. 2.
[22] CBO (note 18), pp. 29–30. The Brilliant Pebbles system was part of the missile defence architecture of the Administration of President George Bush (1989–93), known as Global Protection Against Limited Strikes (GPALS). It was to consist of 500–1000 hit-to-kill interceptors. Each interceptor would be housed in an orbiting satellite which would provide communications with ground stations.
[23] CBO (note 18), pp. 2–3.
[24] Center for Arms Control & Non-Proliferation (formerly the Council for a Livable World Education Fund), 'CBO report indicates missile defense could cost $238 billion', 31 Jan. 2002, available at the Council for a Livable World Internet site, URL <http://www.clw.org>.
[25] See chapter 8 in this volume.

Missile defence funding and programme changes

In December 2001 Congress approved a $317.4 billion amended defence appropriations bill for FY 2002.[26] The bill included the largest appropriation yet—$7.78 billion—for missile defence. This was $525 million less than requested by the Bush Administration in June, but it represented an increase of $2.5 billion over the FY 2001 appropriation for missile defence. Coupled with increases for counter-terrorism programmes that were added after the 11 September terrorist attacks, Congress approved a total of $8.24 billion for BMD and increased counter-terrorism activities.[27] The administration's FY 2003 defence budget request kept overall funding for missile defence programmes essentially unchanged from the final FY 2002 appropriation, allocating $7.76 billion.[28]

Reorganization of US missile defence programmes

The amended FY 2002 defence budget called for a major reorganization of the Ballistic Missile Defense Organization (BMDO), the DOD office with primary responsibility for administering BMD programmes.[29] These programmes were reorganized into six main areas, with the aim of facilitating the development and deployment of an integrated, multi-layer missile defence system employing complementary sensors and weapons[30] (see table 10.1). Among other changes, this involved dropping the distinction between theatre missile defence (TMD) and NMD systems.[31] These systems are now considered to be programme elements in a single BMD architecture and are grouped according to the stages of the flight trajectory—boost, mid-course or terminal—in which incoming targets are to be intercepted. Congress rejected the administration's request to transfer funding responsibility for three 'lower tier' missile defence programmes—Patriot PAC-3, Medium Extended Air Defense System (MEADS) and Navy Area Defense (NAD)—from the BMDO to army and

---

[26] Garamone, J., 'Bush signs defense bill into law during Pentagon ceremony', American Forces Information Service, 10 Jan. 2002, URL <http://www.defenselink.mil/news/Jan2002/n01102002_200201104.html>. For more detail on the FY 2002 defence budget as well as on trends in US military spending see appendix 6E in this volume.

[27] The bill allows the president to use, at his discretion, up to $1.3 billion of the appropriated amount either for missile defence R&D programmes or for DOD activities to combat terrorism. US House Armed Services Committee, 'Conferees reach bipartisan accord on Fiscal Year 2002 Defense Authorization Bill', Press release, 12 Dec. 2001, URL <http://www.house.gov/hasc/pressreleases/2001/01-12-12confsummary.html>.

[28] 'Details of Fiscal 2003 Department of Defense (DOD) budget request', Press release no. 049-02, Office of Secretary of Defense (Public Affairs), 4 Feb. 2002, URL <http://www.defenselink.mil/news/Feb2002/b02042002_bt049-02.html>.

[29] In Jan. 2002 the BMDO was designated a DOD agency and renamed the Missile Defense Agency (MDA). 'DOD establishes Missile Defense Agency' (note 17).

[30] Statement of Lt Gen. Ronald T. Kadish, Director, BMDO, to the House Armed Services Committee, 19 July 2001, URL <http://www.acq.osd.mil/bmdo/bmdolink/html/kadish19jul01.html>.

[31] For a description of these programmes see chapter 11 in this volume.

Table 10.1. Funding of US ballistic missile defence programmes, FY 2002[a]
Figures are for budget authority, in US $m. at current (FY 2002) prices.

Programme	Description	Amount
Systems Engineering	Battle management, command and control (BMC2) system; communications; integration of multi-layered defences into interoperable BMD system	808.0
Terminal Segment	Ground- and sea-based systems designed to intercept target missile or warhead inside the earth's atmosphere, in the final phase of its flight trajectory	200.1
Midcourse Defense Segment (MDS)	Ground- and sea-based systems designed to intercept a target missile or warhead above the earth's atmosphere, in the mid-course phase of its flight trajectory[c]	3 762.3
Boost Segment	Air, sea- and space-based systems, including directed energy weapons such as the Airborne Laser, designed to intercept target missile during the powered ascent phase of its flight trajectory	599.8
Sensor Segment	Satellite-based sensors and other systems to detect ballistic missile launches and provide tracking data in all phases of the flight trajectory[b]	335.4
Technology	Components, sub-systems and new concepts for sensors and weapons for future missile defence platforms	139.3
Theater High-Altitude Area Defense (THAAD)	Truck-mounted launchers equipped with high-speed hit-to-kill interceptor missiles, mobile ground-based radar (GBR) and BMC2 system; designed for the defence of larger areas against short- to medium-range ballistic missiles inside and outside the earth's atmosphere	866.5
Patriot PAC-3	Land-based, mobile launcher equipped with high-speed hit-to-kill interceptor missiles and associated engagement radar; designed for the defence of point targets/limited areas against short- to medium-range missiles inside the earth's atmosphere	898.7
Navy Area Defense[d] (NAD)	Navy cruisers and destroyers equipped with reconfigured Aegis radar and upgraded Standard SM-2 interceptor missiles designed for defence of point targets/limited areas against short- to medium-range missiles inside the earth's atmosphere	99.3
Other[e]		65.6
Total		**7 775.0**

[a] Figures include funding for US Air Force, Army and Navy missile defence programmes as well as for Missile Defence Agency (formerly known as the Ballistic Missile Defence Organization) programmes and related Defense Department activities.

[b] Includes funding authorization for restructured Space-Based Infrared System–Low (SBIRS–Low) satellite programme.

[c] Includes funding authorization for MDS Test Bed Facility ($786 million).

[d] The NAD programme was cancelled by the Department of Defense in Dec. 2001 because of cost overruns and technology development problems.

[e] Includes funding authorization for military construction ($8.2 million) and the Joint Air Missile Defense Organization ($26.9 million).

Sources: Statement of Lt Gen. Ronald Kadish, Director, US Missile Defense Agency, Joint Hearing before the House of Representatives Procurement and Research and Development Subcommittees, 27 Feb. 2002, URL <http://www.house.gov/hasc/openingstatementsandpress releases/107thcongress/02-02-27kadish.html>; and US Department of Defense, *Budget for Fiscal Year 2003: Program Acquisition Costs by Weapon System*, Feb. 2002, pp. 64–65, available at URL <http://www.dtic.mil/comptroller/fy2003budget/fy2003weabook.pdf>.

navy service accounts, citing concern that the services would not be able to adequately support them.[32]

Focus on RDT&E

The organizational changes were accompanied by a reorientation of BMDO programme activities towards research, development, testing and evaluation (RDT&E) activities and away from production and deployment.[33] Congress approved a large increase in RDT&E funding for BMD in the amended FY 2002 budget, authorizing $7.0 billion, compared to $4.9 billion in FY 2001 and $3.1 billion in FY 2000.[34]

The Director of the BMDO, Lieutenant General Ronald Kadish, testified before Congress that the new emphasis on RDT&E reflected a 'broader, more flexible approach' to missile defence. It did not involve defining a specific defence architecture from the outset or committing the DOD to 'arbitrary' dates for production and deployment.[35] In order to reduce the technology development risks, components would be deployed incrementally, as they are proven through testing and meet specific performance criteria and programme milestones. Kadish emphasized that the new approach involved putting a robust testing programme in place. This would more than double the number of planned tests and increase their complexity.

Continuing concerns about the development programme

These changes came against the background of the release of a DOD internal report, completed in August 2000, that re-ignited concerns about the deployment readiness and likely effectiveness of missile defences designed to protect

---

[32] US House of Representatives, *National Defense Authorization Act for Fiscal Year 2002*, Conference report to accompany S. 1438, 12 Dec. 2001, p. 595. These systems are designed to intercept incoming short- to medium-range missile warheads inside the earth's atmosphere.

[33] Associated Press, 'Bush's missile defense shifts focus to testing', *International Herald Tribune*, 10 July 2001, p. 6.

[34] US Department of Defense, *Budget for Fiscal Year 2003: Program Acquisition Costs by Weapon System*, Feb. 2002, pp. 64–65, available at URL <http://www.dtic.mil/comptroller/fy2003budget/fy2003weabook.pdf>; and BMDO, 'A budgetary history of the Ballistic Missile Defense Organization', Fact Sheet no. 408-00-11, Nov. 2000, p. 2.

[35] Kadish (note 30). The ambitious schedule for deploying an initial NMD system had been criticized as a 'rush to failure' in the 1998 Welch Report, prepared by an independent team of experts appointed by the Pentagon. 'Report of the Panel on Reducing Risk in Ballistic Missile Defense Flight Test Programs', 27 Feb. 1998, URL <http://www.acq.osd.mil/bmdo/bmdolink/pdf/welchrpt.pdf>.

the USA.[36] The unclassified report had been withheld from Congress until the end of May 2001, prompting critics there to charge that the Pentagon was keeping the report 'hidden from view' because it showed 'that there are critical flaws in the missile defense program'.[37] The report, which had been prepared during the Clinton Administration's Deployment Readiness Review for the NMD system then under consideration, concluded that missile defence technologies were too immature to be able to assess the system's operational effectiveness or predict realistic deployment dates.[38] This conclusion was underscored in 2001 by the announcement of further delays in engineering development work on the Ground-Based Interceptor (GBI) missile.

The report also identified problems with the BMDO's test and evaluation programmes. It pointed out that the integrated flight test (IFT) programme incorporated 'significant limitations on achieving realistic engagement conditions'; among other shortcomings, the BMDO was criticized for failing to schedule tests against multiple targets, even though 'multiple engagements are expected to be the norm'.[39] The report also expressed concern that the components of the system were not being tested against the range of countermeasures expected to be available to a state with the capability to deploy a long-range ballistic missile.[40] It recommended that future flight tests be made more challenging and that more consideration be given to potential countermeasures that a missile defence system could realistically be expected to face.

Flight test developments

In 2001 the BMDO announced two successful interception tests in the integrated flight test programme. The first (IFT-6) came on 14 July 2001, when a prototype interceptor missile successfully collided with a target vehicle carried by a modified Minuteman intercontinental ballistic missile (ICBM) over the central Pacific Ocean.[41] The second test (IFT-7) was carried out on 3 December 2001, when a prototype exo-atmospheric kill vehicle (EKV) successfully 'discriminated' a target warhead from a large balloon decoy and manoeuvred to collide with it.[42] The purpose of the tests was to demonstrate that it was feasible for 'hit-to-kill' technology to intercept and destroy a long-

[36] Coyle, P., Director, DOD Office of Operational Test and Evaluation, *Operational Test and Evaluation Report in Support of National Missile Defense Deployment Readiness Review* (The Coyle Report), 10 Aug. 2000, URL <http://www.dote.osd.mil/reports/FY00/index.html>. Many of these concerns had been raised in previous reports by government-appointed panels and by independent experts. See Kile (note 2), pp. 431–32.

[37] Letter from US Representative John Tierney to Secretary of Defense Donald Rumsfeld, 12 June 2001, quoted in 'August 2000 Pentagon report on NMD technology', *Arms Control Today*, vol. 31, no. 6 (July/Aug. 2001), p.32.

[38] Coyle (note 36), pp. 45–46, 49. For a description of the results of the Deployment Readiness Review see Kile (note 2), pp. 432–33.

[39] Coyle (note 36), p. 20.

[40] Coyle (note 36), p. 5.

[41] 'Missile intercept test successful', News release no. 313-01, Office of Secretary of Defense (Public Affairs), 15 July 2001.

[42] 'Missile intercept test successful', Press release, BMDO, Office of External Affairs, 3 Dec. 2001, URL <http://www.acq.osd.mil/bmdo/bmdolin k/html/ift7.htm>.

range ballistic missile target. In addition, the tests were designed to show whether prototype elements of the planned Ground-based Midcourse Defense (GMD) architecture—including the SBIRS satellite-based early-warning system, a ground-based tracking radar, and a battle management and communications system—could work together. A total of 26 tests are currently scheduled in the flight test programme to the end of FY 2006.

The successful interceptions followed two consecutive test failures in 2000 that had fuelled concern about the readiness and reliability of the technologies being developed for NMD. The BMDO's claim that the successful test interceptions had demonstrated the 'basic functionality' of the proposed GMD system was greeted with scepticism by some analysts.[43] Critics charged that the flight test programme did not realistically simulate combat engagement conditions.[44] Among other shortcomings, they pointed out that the mock warheads carried transponders which served as a radio beacon to guide kill vehicles to the vicinity of their targets in space.[45]

BMDO officials acknowledged that the flight tests did not simulate realistic engagement conditions but added that this had not been an aim of the tests. They pointed out that artificialities are inherent in the early stages of weapon development testing programmes, when the main goal is to identify basic weaknesses and acquire confidence in new technology.[46] According to BMDO Director Kadish, the initial flight tests were never intended to be 'pass–fail' tests of the system's operational effectiveness or the basis for an early deployment decision.[47] He emphasized that, over time, the test programme would employ 'more realistic scenarios and countermeasures' designed to 'demonstrate increasing capability'.[48]

A test bed facility

The amended FY 2002 defence budget approved the administration's funding request for a Midcourse Defense Segment (MDS) Test Bed Facility designed to enhance DOD missile defence testing capabilities.[49] The facility, which is scheduled to be completed by the end of 2004, is based on the Clinton Administration's plan to build an ABM interceptor site in central Alaska. It will con-

---

[43] 'Missile intercept test successful' (note 42).
[44] Dao, J, 'Missile shield experts cautious on success', *International Herald Tribune*, 17 July 2001, p. 3.
[45] For an overview of the criticism of the Integrated Flight Testing programme, including a detailed analysis of the role played by the warhead transponder in the recent tests, see Gronlund, L. *et al.*, 'An assessment of the intercept test program of the Ground-Based Midcourse national missile defense system', *Briefing Paper*, Union of Concerned Scientists, 30 Nov. 2001, URL <http://www.ucsusa.org/arms/ift7.html>.
[46] Graham, B., 'Missile defense test's value questioned', *Washington Post*, 2 Dec. 2001, p. 6.
[47] Graham (note 46). Some observers have argued that the missile defence R&D programme is, for political reasons, being prematurely pressed to justify an early deployment decision—a purpose which the initial flight tests were not intended to serve. See Graham, B., *Hit to Kill: The New Battle over Shielding America from Missile Attack* (PublicAffairs Books: New York, 2001).
[48] Kadish (note 30).
[49] US House Armed Services Committee (note 27).

sist of a set of launchers, radar, and command and control installations in Alaska and California and at Kwajalein Atoll in the central Pacific Ocean. According to BMDO officials, the facility will allow for more realistic testing of the GMD system by providing 'trajectory, sensing and interception scenarios that resemble conditions under which the system might be expected to operate'.[50]

The MDS Test Bed Facility will provide several sites from which to launch interceptor and target missiles as part of the integrated flight testing programme.[51] The plan calls for the construction of two test launch silos on Kodiak Island off the southern coast of Alaska, for both target missiles aimed towards the continental USA and interceptors that could shoot down test missiles coming towards Alaska from either California or Kwajalein Atoll.[52] Analysts have charged that there is no 'clear or convincing rationale' for the facility in terms of addressing specific testing shortcomings identified by government-appointed commissions and panels of independent experts.[53]

The facility will also include an installation to be built at Fort Greely in central Alaska that will house five silos for GBI missiles. This installation is intended to be used as a missile storage site and a command centre for launching test missiles from Kodiak Island. However, BMDO officials stated that Fort Greely could also be used to provide an 'emergency' missile defence capability if there were credible evidence of an imminent missile threat to the USA and if the technology were sufficiently mature.[54] This provoked criticism that the test bed facility was an attempt by the Bush Administration to move ahead with preparations for the early deployment of a rudimentary missile defence system under the guise of improved testing.[55]

US–Russian discussions on the future of the ABM Treaty

The shift in US position

The change in administration in 2001 resulted in a new US approach to the ABM Treaty. During the Clinton Administration, the USA had sought, unsuccessfully, to obtain Russia's agreement on a series of amendments to the

---

[50] Statement by Patricia Sanders, Deputy for Test, Simulation and Evaluation, BMDO, before the US Senate Committee on Appropriations, Subcommittee on Military Construction, 31 July 2001, URL <http://www.acq.osd.mil/bmdo/bmdolink/html/sanders html>.

[51] US missile interceptors are currently tested solely at the Reagan Test Range at Kwajalein Atoll. Flight tests involve launching target missiles from Vandenburg Air Force Base in California towards Kwajalein, a distance of c. 7500 km.

[52] These flight tests are intended to simulate the speed and trajectory of ballistic missiles launched from north-eastern Asia more realistically than current flight tests.

[53] Gronlund, L. and Wright, D., 'The Alaska test bed fallacy: missile defense deployment goes stealth', *Arms Control Today*, vol. 31, no. 7 (Sep. 2001), p. 9.

[54] Dao, J., 'Pentagon to propose an ABM site in Alaska', *International Herald Tribune*, 11 July 2001, p. 3.

[55] Dao (note 54); and Council for a Livable World, 'The Bush Administration's national missile defense proposal', *Backgrounder*, 1 Aug. 2001, URL <http://www.clw.org/coalition/nmdbkground0801.htm>.

ABM Treaty that would permit the USA to deploy a limited missile defence system but not interfere with the basic purpose of the treaty. US officials insisted that only modest amendments were needed to accommodate a system consisting of a single site with 100 missile interceptors based in Alaska.[56] This led to complaints from some missile defence advocates that Clinton was more concerned about preserving the ABM Treaty intact—and not upsetting China, Russia and US allies—than about considerations of operational effectiveness.

In contrast, the Bush Administration came to office deeply sceptical about the desirability of preserving the ABM Treaty. Senior officials identified two main problems with the treaty. The first had to do with its restrictions on the testing of anti-missile systems.[57] These restrictions were criticized for limiting the ability of the DOD 'to explore fully' promising new BMD technologies.[58] At a NATO ministerial meeting in June 2001, Rumsfeld warned the allies that US plans to test various anti-missile technologies would begin 'bumping up' against the ABM Treaty. While declining to specify what planned testing would violate the treaty or predict when this would occur, he declared that the Bush Administration would not be deterred from conducting tests that might violate the treaty.[59] In October 2001, however, it was reported that Rumsfeld had ordered the BMDO to postpone three anti-missile tracking tests that would have violated the ABM Treaty.[60]

The administration's second main criticism was that the ABM Treaty is an outdated agreement that does not reflect the fundamental transformation of the security environment that has taken place since it was signed 30 years ago. In his May 2001 address on missile defence, President Bush described the accord as an anachronism that enshrined the 'grim premise' of mutual assured destruction.[61] He stressed that new concepts of deterrence were needed that rely on defensive as well as offensive forces—among other benefits, missile

---

[56] Kile (note 2), pp. 435–36. This primarily would have involved amending the treaty to permit the USA to change the location of its designated ABM site. It would also involve amending the treaty's restrictions on early-warning and ABM engagement radars and its prohibition on the use of satellite-based sensors.

[57] The ABM Treaty imposes strict limitations on the testing of permitted ABM interceptors and components. In addition, it prohibits the development, testing or deployment of sea-, air-, space- or mobile land-based ABM systems or components.

[58] Transcript of press briefing by National Security Advisor Condoleeza Rice, 8 Nov. 2001, The White House, Office of Press Secretary, in US Information Service (USIS), *European Washington File* (US Embassy: Stockholm, 8 Nov. 1999).

[59] Dao, J., 'Rumsfeld outlines to NATO fast track for missile shield', *New York Times* (Internet edn), 8 June 2001, URL <http://www.nytimes.com/2001/06/08/world/08Nato.html>.

[60] One test would have involved using a ship-based Aegis radar system to track a missile interceptor while a separate tracking radar located at Vandenburg AFB, Calif. tracked a strategic target missile. The ABM Treaty bans the tracking of strategic missiles and anti-missile interceptors by sea-based radars or by other radar systems not initially designed for this purpose. The treaty also prohibits a radar like the one at Vandenburg AFB from being used to track strategic missiles unless it is located at a designated ABM test range. Shanker, T. and Sanger, D., 'US, awaiting Putin, delays missile defense tests', *New York Times* (Internet edn), 26 Oct. 2001, URL <http://www.nytimes.com/2001/10/26/international/26MISS.html>; and Gertz, B., 'Rumsfeld orders tests limited to comply with ABM Treaty', *Washington Times* (Internet edn), 26 Oct. 2001, URL <www.washingtontimes.com/national/20011026-27648144.htm>.

[61] 'Remarks by the president' (note 12); and Knowlton, B., 'Bush calls for missile shield, saying ABM pact is outdated', *International Herald Tribune*, 2 May 2001, pp. 1, 10.

defence 'can strengthen deterrence by reducing the incentives for proliferation'. Bush declared that Russia was 'no longer an enemy' and urged it to work together with the USA to forge a new framework for their strategic relations. This framework would supplant the ABM Treaty's cold war-era constraints and allow both countries to build missile defences to counter new threats emerging in a less predictable world.[62]

Bush did not follow up on his call for a new US–Russian strategic framework by spelling out what its main elements should be. Statements made by senior administration officials offered a rationale for abandoning the ABM Treaty but were similarly vague on what should follow in its place. This suggested to some analysts that the administration had yet to formulate a coherent new framework beyond the fact that it would require the abandonment of the ABM Treaty.[63]

The administration's insistence on pushing forward with missile defences led to warnings from Democrat congressional leaders that they would block any move to unilaterally abrogate the ABM Treaty.[64] They complained that an abrogation of the treaty would maximize the political costs of developing as yet unproven technologies by damaging US relations with Russia, China and key US allies. In the Senate, where the Republicans had lost their majority position during 2001, the new Democrat chairman of the Appropriations Committee added a provision to the amended FY 2002 defence authorization bill that prohibited the DOD from conducting missile defence tests that would violate the ABM Treaty without congressional approval. However, in the wake of the 11 September terrorist attacks, Democrats shelved their objections to the administration's missile defence test plans in order to present a united front on national security issues.[65]

Russian concerns

In Russia, Bush's call to replace the 'anachronistic' ABM Treaty with a new framework featuring a mixture of deterrence and strategic defence was greeted with considerable scepticism. Many Russian officials and analysts continued to view the missile defence issue primarily in terms of its impact on nuclear deterrence and the US–Russian strategic balance inherited from the cold war. Concern was expressed in some conservative quarters that the USA was seeking to dismantle the ABM Treaty in order to proceed with the development of a large-scale missile defence system capable of neutralizing Russia's nuclear deterrent; the USA would thereby achieve 'a multifold military superiority' that would allow it to unilaterally shape the global order according to its

[62] 'Remarks by the president' (note 12).
[63] Miller, S. E., 'The flawed case for missile defence', *Survival*, vol. 43, no. 3 (autumn 2001), pp. 95–110.
[64] Loeb, V. and Morgan, D., 'Democrats pare missile defense funds', *Washington Post*, 6 Sep. 2001, p. A5.
[65] Loeb, V., 'Levin agrees to cut missile test curbs from defense bill', *Washington Post*, 19 Sep. 2001, p. A3.

liking.⁶⁶ Underlying this concern was the fear of a continuous expansion of the US missile defence system juxtaposed to the continuous decline, imposed by financial exigencies, in Russia's strategic nuclear forces.

Senior Russian officials also continued to express concern about the consequences of a US abandonment of the ABM Treaty. Defence Minister Sergey Ivanov emphasized that the treaty 'constituted a single whole with an entire series of other interrelated agreements in the overall arms control and disarmament system'.⁶⁷ He warned that a unilateral US withdrawal from the accord would lead to a collapse of that system and usher in a 'phase of complete unpredictability in the sphere of global security'.

In addition, Russian officials argued that the problem of ballistic missile proliferation must be considered within the broader framework of international legal and political non-proliferation arrangements; these could be supplemented by the creation of a new Global Control System for the Non-Proliferation of Missiles and Missile Technologies, as proposed by Foreign Minister Igor Ivanov at the 2000 Non-Proliferation Treaty (NPT) Review Conference.⁶⁸ The Bush Administration's missile defence plans were widely condemned in Russia as an inappropriate response to the problem of missile proliferation as well as a worrying sign that the USA was unwilling to engage in the patient, multilateral diplomacy needed to address proliferation incentives.

Improved political climate for US–Russian talks

During 2001 there was an important change in the tone of the missile defence dispute. As part of a broader rapprochement in US–Russian relations, Bush and Putin moved to defuse the political tensions between their countries arising from the dispute. At a series of meetings held in the summer and autumn, the two presidents struck a conciliatory note in discussing the differences in their assessments of ballistic missile threats and in their approaches to addressing them. After the 11 September terrorist attacks, they expressed their determination not to allow these differences to stand in the way of fostering better bilateral relations or creating a climate conducive to pragmatic cooperation.

Bush and Putin met for the first time at a summit meeting held in Ljubljana, Slovenia, on 16 June 2001. Following discussions characterized by an unexpectedly positive tone, they agreed to initiate a 'constructive dialogue' between their countries on enhancing strategic stability.⁶⁹ This would consist of a series of regular, expert-level bilateral consultations to discuss potential

---

⁶⁶ See, e.g., Ladygin, F., 'Major plan: will they not proceed?', *Tribuna* (Moscow), 22 Aug. 2001, in 'Col-Gen (ret.) Fedor Ladygin, former General Staff Main Intelligence Directorate chief, on US decision to deploy a missile defence, differing US/Russian threat perceptions, and viability of ABM Treaty', Foreign Broadcast Information Service, *Daily Report–Central Eurasia (FBIS-SOV)*, FBIS-SOV-2001-0822, 22 Aug. 2001.

⁶⁷ Interfax (Moscow), 5 June 2001, in 'Russian defense minister: US withdrawal from ABM Treaty may destroy strategic stability', FBIS-SOV-2001-0605, 5 June 2001.

⁶⁸ See Kile (note 2), pp. 439–40; and chapter 14 in this volume.

⁶⁹ Tyler, P., 'Bush and Putin: new era of trust?', *International Herald Tribune*, 18 June 2001, pp. 1, 4.

'new threats' posed by ballistic missile proliferation as well as means of countering them.[70] Despite the cordial atmosphere, there was no sign of a convergence of views: Bush insisted that the ABM Treaty had been rendered obsolete by the transformation of the international security system; and Putin reiterated Russia's view that the ABM Treaty remained the 'cornerstone of the modern architecture of international security' that must be preserved.

The two presidents met for a second time on 22 July 2001 at the meeting of the Group of Eight (G8) industrialized countries in Genoa, Italy. In a Joint Statement, they announced that they had agreed to begin 'intensive consultations on the interrelated subjects of offensive and defensive systems'.[71] This meant that discussions on modifying or scrapping the ABM Treaty would be linked to talks on making further reductions in strategic offensive nuclear forces.

The Joint Statement issued at Genoa fuelled speculation that a Russian–US arms control deal was taking shape.[72] This would involve Russia's agreement to amend the ABM Treaty to permit the USA to proceed with the development of a limited strategic missile defence system. In return, the USA would agree with Russia to make further cuts in their respective strategic offensive nuclear forces (see section III below).

The prospects for reaching an agreement appeared to improve when the Russian Government indicated in the early autumn that it would be willing to consider adjustments to the 'present-day system of agreements on strategic stability', including the ABM Treaty.[73] According to Defence Minister Ivanov, 'changes which do not weaken the main part of the document—the ban on the deployment of a national missile defence system—may be introduced'.[74] However, Foreign Minister Ivanov declared that Russia would not make any 'swaps or bargains' involving mutual reductions in strategic offensive arms in exchange for a joint withdrawal from the ABM Treaty. Ivanov noted that progress towards reductions in nuclear arsenals 'is possible only in the context of strategic stability . . . and it is the ABM Treaty and other related agreements that give this stability'.[75]

---

[70] Transcript of press conference remarks by President Bush and President Putin, Brdo Castle, Slovenia, The White House, Office of the Press Secretary, 16 June 2001.

[71] Joint Statement by President Bush and President Putin on Upcoming Consultations on Strategic Issues, The White House, Office of the Press Secretary, 22 July 2001.

[72] There had been similar speculation about a possible 'grand bargain' on strategic defensive and offensive forces prior to a June 2000 summit meeting between Putin and Clinton. Gordon, M., 'Moscow talks fail to forge the big breakthrough', *International Herald Tribune*, 5 June 2000, pp. 1, 4.

[73] Quoted by Tyler, P., 'Kremlin willing to review missile accords, aide says', *New York Times* (Internet edn), 7 Sep. 2001, URL <http://www.nytimes.com/2001/09/07/international/europe/07MISS.html>.

[74] Quoted by Interfax (Moscow), in 'Russian Defense Minister says changes to ABM Treaty possible', 10 Sep. 2001, FBIS-SOV-2001-0910, 10 Sep. 2001; and Associated Press, 'New Russian declaration on ABM pact', *International Herald Tribune*, 11 Sep. 2001.

[75] Quoted by ITAR-TASS (Moscow), 1 Nov. 2001, in 'Ivanov says Russia still sees ABM Treaty as "cornerstone" of strategic stability', FBIS-SOV-2001-1101, 1 Nov. 2001.

Bilateral discussions on strategic stability

Informal discussions held during the summer and autumn of 2001 under the auspices of a bilateral working group on strategic stability yielded few results. Senior Russian officials participating in the discussions complained repeatedly that the USA had not provided any details about the basing modes and technical capabilities of its planned missile defence system.[76] They also complained that the US side's professed interest in forging a new framework of strategic stability appeared to have little substantive content beyond the idea of jointly withdrawing from the ABM Treaty in the near future. This idea was firmly ruled out by Russia, which also cautioned the Bush Administration against moving with undue haste to abandon the treaty. Russian officials stressed that extensive consultations were needed to 'clarify each other's positions on security matters in the twenty-first century' before work could begin on the joint drafting of proposals for a new framework.[77] These talks might last for at least one year and probably longer.[78] In addition, the discussions would eventually have to be widened to take into account the views of the other nuclear weapon states—China, France and the UK.[79]

Bush Administration officials grew increasingly impatient with what they saw as Russia's deliberate go-slow approach. There was speculation in the USA that Russia was essentially playing for time in the hope that the administration's ambitious missile defence plans would have to be scaled down or abandoned in the face of budget concerns and negative public opinion in the USA and abroad.[80] During a visit to Moscow in August 2001, the Under-Secretary of State for Arms Control and International Security, John Bolton, reportedly told Russian interlocutors that the administration had an informal deadline of November to convince Russia to join the USA in withdrawing from the ABM Treaty and agreeing to a new strategic framework. In the event of Russia's refusal to withdraw from the treaty, the USA would proceed to do so unilaterally.[81] The White House subsequently denied that Bolton's comments were tantamount to an ultimatum. At the same time, however, Bush announced that he intended to give notice of a US withdrawal from the ABM Treaty 'at a time convenient to America'.[82]

---

[76] Interfax (Moscow), 6 Aug. 2001, in 'Defense minister hopes for more details on US national missile defense', FBIS-SOV-2001-0806, 6 Aug. 2001; and Vasilyev, Ye., 'Rumsfeld imagines enemies in Moscow', *Vremya MN* (Moscow), 18 Aug. 2001, in 'Response to US Defense Secretary Rumsfeld's negative characterization of Russia's NMD stance', FBIS-SOV-2001-0817, 18 Aug. 2001.

[77] Interfax (Moscow), 6 Sep. 2001, in 'Adjustments in strategic stability agreements possible if ABM Treaty preserved', FBIS-SOV-2001-0906, 6 Sep. 2001.

[78] Interfax (Moscow), 6 Sep. 2001, in 'Russian official rules out quick deal on ABM issue with US', FBIS-SOV-2001-0906, 6 Sep. 2001.

[79] Interfax (Moscow), 27 July 2001, in 'Russia says more countries should take part in ABM Treaty discussions', FBIS-SOV-2001-0731, 31 July 2001.

[80] Baker, P., 'Kremlin rules out quick missile deal with US', *International Herald Tribune*, 7 Sep. 2001, p. 7.

[81] Wines, M., 'US disclaims deadline for Russia on missiles', *International Herald Tribune*, 24 Aug. 2001, p. 3.

[82] Sanger, D., 'Bush vows to quit ABM pact', *International Herald Tribune*, 25–26 Aug. 2001, pp. 1, 5.

Disagreement over missile defence testing

Against the background of US–Russian strategic cooperation following the 11 September terrorist attacks, Bush and Putin held a summit meeting in Washington DC and Crawford, Texas on 11–13 November 2001. Prior to the meeting, there had been renewed media speculation that a US–Russian deal on missile defence and the future of the ABM Treaty might be imminent.[83] It was widely noted that the White House had ordered the Pentagon to postpone a series of missile defence tests scheduled for mid-November that had raised a number of ABM Treaty compliance questions.

However, high-level talks aimed at reaching a compromise solution reportedly broke down over the issue of missile defence testing.[84] Russia refused to agree to changes in the ABM Treaty that would open the door to unrestricted US testing. For its part, the Bush Administration was unwilling to engage in detailed discussion of each element of the BMDO's missile-defence testing programme, as insisted upon by Russia. It feared that doing so would effectively give Russia a veto over the US testing programme whenever Moscow deemed that a particular test would violate the ABM Treaty.[85] By the end of the summit meeting, according to one senior US administration official, both sides concluded that there was no way to accommodate an ambitious testing programme for a nationwide BMD system within the framework of a treaty designed to prevent the development of such a system.[86]

The US decision to withdraw from the ABM Treaty

On 13 December 2001 the United States gave formal notice that it would withdraw from the ABM Treaty in six months.[87] In explaining the decision, President Bush stressed that the USA wanted to 'move beyond' the constraints of the ABM Treaty and forge a new strategic relationship with Russia that

---

[83] Donovan, J., 'Bush–Putin summit could mark historic shift in relations', Radio Free Europe/Radio Liberty (RFE/RL), *Weekday Magazine*, 13 Nov. 2001, URL <http://www.rferl.org/nca/features/2001/11/12112001075633.asp>; and Sanger, D. and Shanker, T., 'US gains on shield tests', *International Herald Tribune*, 29 Oct. 2001, p. 4.

[84] Mufson, S. and LaFraniere, S., 'ABM withdrawal: a turning point in arms control', *Washington Post*, 13 Dec. 2001, pp. A1, 13.

[85] Bogdanov, V., 'Treaty and provisos; from missile test site to courtroom', *Rossiyskaya Gazeta*, 8 Dec 2001, p. 7, in 'US NMD tests seen as possible ABM Treaty "breach"', FBIS-SOV-2001-1210, 12 Dec. 2001; and Tyler, P., 'US and Russia to complete talks on arms control', *New York Times* (Internet edn), 11 Dec. 2001, URL <http://www.nytimes.com/2001/12/11/international/europe/11DIPL.html>.

[86] Sanger, D. and Tyler, P., 'Bush pulls out of ABM Treaty: aides recount road to deadlock', *New York Times* (Internet edn), 13 Dec. 2001, URL <http://www.nytimes.com/2001/12/13/international/13CND-MISS.html>.

[87] Text of diplomatic note sent to Russia, Belarus, Kazakhstan and Ukraine (see note 1). Press Statement, Office of Spokesman, US Department of State, 14 Dec. 2001, URL <http://www.state.gov/r/pa/prs/ps/2001/6859.htm>. Article XV, paragraph 2 of the ABM Treaty gives each party the right to withdraw from the agreement if it decides that 'extraordinary events related to the subject matter of the treaty have jeopardized its supreme interests'. The party must give notice of its decision 6 months prior to withdrawing from the treaty as well as a statement of the extraordinary events that prompted its decision.

would 'replace mutual assured destruction with mutual cooperation'. He argued that, 'as the events of September 11 made all too clear', the greatest threats to the USA and Russia 'come not from each other, or other big powers in the world, but from terrorists who strike without warning, or rogue states who seek weapons of mass destruction'. Bush insisted that since terrorist groups—and some of the states which support them—were known to be seeking 'the ability to deliver death and destruction to our doorstep via missiles', the USA must 'have the freedom and the flexibility' to develop effective missile defences. He had therefore concluded that the USA could not remain party to a treaty that 'hindered our ability to develop ways to protect our people from future terrorist or rogue state missile attacks'.[88]

The Russian response

The Kremlin's reaction to Bush's announcement was a restrained one. President Putin expressed regret over the US decision, which he described as 'mistaken', but said that it had not come as a surprise to the Russian Government.[89] He characterized the unilateral move by the USA as a difference between friends that should not, if properly handled, disrupt 'the spirit of partnership and even alliance' between Russia and the USA.[90]

Putin and his senior ministers emphasized that Bush's decision did not pose a military threat to Russia. The country would continue to possess for the foreseeable future robust offensive forces capable of overcoming anti-missile defences.[91] They rejected calls to build up the Strategic Rocket Forces, in particular the widely mentioned idea of deploying multiple warheads on the single-warhead Topol-M (SS-27) ICBM. Defence Minister Ivanov declared that it would be 'senseless' to 'waste lots of money on an arms race' given that the US national missile defence system was a 'myth'.[92]

At the same time, however, Russian officials predicted that the US decision would be likely to have a negative impact on global non-proliferation efforts and international stability. There was particular concern that it would lead China, India and Pakistan to build up their nuclear arsenals and spur other countries to pursue nuclear and other non-conventional weapon programmes.[93] Russian officials also renewed their complaints that US missile defence plans

[88] Remarks by the president on National Missile Defense, The White House, Office of the Press Secretary, 13 Dec. 2001, URL <http://www.whitehouse.gov/news/releases/2001/12/20011213-4.html>.
[89] 'Statement made by Russian President Vladimir Putin on December 13, 2001, regarding the decision of the administration of the United States of America to withdraw from the Anti-Ballistic Missile Treaty of 1972', available at the Russian Federation Ministry for Foreign Affairs Internet site, URL <http://www.ln.mid.ru/website/brp_4.nsf/english>.
[90] Gowers, A. et al., 'Interview with Vladimir Putin', *Financial Times* (Internet edn), 15 Dec. 2001, URL <http://news.ft.com>.
[91] 'Statement made by Russian President Vladimir Putin' (note 89).
[92] Quoted by Interfax (Moscow), 18 Dec. 2001, in 'Russian DM rules out arms race after US exit from ABM Treaty', FBIS-SOV-2001-1218, 18 Dec. 2001; and ITAR-TASS, 2 Feb. 2002, in 'Russian official says end of ABM Treaty "very destructive" for Asia', FBIS-SOV-2002-0202, 2 Feb. 2002.
[93] Interfax (Moscow), 13 Dec. 2001, in 'Russian MP: US withdrawal from ABM Treaty may trigger nuclear arms race', FBIS-SOV-2001-1213, 13 Dec. 2001.

relied heavily on the use of space-based assets, which could lead to a destabilizing arms race in outer space.[94]

Reactions to the unilateral US decision from some commentators and analysts were more critical. According to one Russian defence analyst, the considerable anger and resentment felt throughout Russia's military–political elite were due in part to a perceived loss of prestige: the country's inability to respond in military terms meant that Russians had 'lost the last opportunity to pretend that we are equal with the USA'.[95] There was also a widespread view that the move, coming in the midst of a potentially historic East–West rapprochement, had been calculated to humiliate Russia.[96]

In the State Duma, there were warnings about a nationalist backlash in Russia that could put at risk the recent improvement in US–Russian relations.[97] According to the deputy chairman of the Defence Committee, Alexei Arbatov, the decision would be likely to strengthen groups in Russia which argue that the USA cannot be trusted; these groups could be expected to 'exert strong pressure on President Putin to slow down or even freeze the cooperation with the United States in Afghanistan and elsewhere'.[98]

Agreeing to disagree

Despite the failure to reach a deal on the ABM Treaty, Bush and Putin continued to accentuate the positive development of US–Russian relations. Both leaders appeared determined to prevent an acrimonious falling out over missile defence that might jeopardize the recent warming in relations between their countries and, more specifically, their unprecedented intelligence and logistics cooperation in the war in Afghanistan. They also reaffirmed their pledge to make deep cuts in their countries' strategic nuclear forces. This was an especially important consideration for Putin, since the size of Russia's nuclear forces was set to fall sharply over the decade owing to chronic underfunding. Aware that there was little that Russia could do to slow down or derail US missile defence plans, Putin may have expected that a pay-off for his muted reaction to the US move would come in the form of a treaty mandating mutual reductions in strategic offensive forces to an equal ceiling.

The presidents also had other motivations for playing down the impasse. Bush sought to allay the concerns of US allies, particularly those in Europe, and other foreign governments that his missile defence plans would lead to renewed rivalry with Russia and an unrestrained arms race. This was largely a tactical, 'damage limitation' consideration: it was an attempt by the White

---

[94] See chapter 11 in this volume.
[95] Alexander Golts, quoted by Traynor, I., 'Russia puts brave face on the inevitable', *The Guardian* (Internet edn), 14 Dec. 2001, URL <http://www.guardian.co.uk/Archive/Article/0,4273,4319813,00.html>.
[96] Wines, M., 'Putin calls US withdrawal from ABM pact a mistake', *International Herald Tribune*, 14 Dec. 2001, p. 3.
[97] Interfax (Moscow), 13 Dec. 2001, in 'Parliamentary leaders unified in criticism of US exit from ABM Treaty', FBIS-SOV-2001-1213, 13 Dec. 2001.
[98] Quoted by Wines (note 96).

House to reduce the short-term political and diplomatic costs of moving forward with the development of an expansive missile defence system.

By contrast, Putin's insistence that Bush's 'mistaken' decision would not harm US–Russian relations reflected an underlying shift in the Russian Government's strategic priorities. As one analyst has argued, Putin's foreign policy 'serves his domestic economic goals: to stabilize, regularize and restructure the economy to support a twenty-first century Russian society and cultivate a newly confident Russian state'.[99] The promotion of economic growth and integration into the global economy requires, above all, substantially improved relations with the United States. At the same time, for domestic political reasons Putin was seeking reassurance that the USA was not looking for confrontation or for unilateral strategic advantage at a time when Russia faced serious internal problems.

Putin's muted reaction also reflected a re-ordering of Russia's security policy priorities. The issue of missile defence and the future of the ABM Treaty, while important symbolically in terms of Russia's status as an equal partner with the USA, has been eclipsed on the security policy agenda by more pressing concerns about Russia's relations with NATO and the growing instability along the southern rim of Russia. Implicit in this shift is an underlying judgement that the USA does not pose a military threat to Russia. By playing down the impact of Bush's decision on US–Russian relations, Putin appears to be putting himself in a better position to extract tangible 'rewards' from the USA. In particular, this might involve gaining US backing for efforts to give Russia a more influential role in European security arrangements, including a greater voice in NATO's decision-making process.[100]

The Chinese response

The US decision to withdraw from the ABM Treaty also drew a subdued reaction from China. The Chinese Government expressed concern about the 'negative impact of the US retreat' from the treaty, emphasizing that it 'is of crucial importance to maintain the international disarmament control regime and global strategic stability'.[101] Officials noted that the UN General Assembly had, in November 2001, overwhelmingly adopted a resolution sponsored by China (along with Belarus and Russia) calling for the parties to the ABM Treaty to 'preserve and strengthen the treaty through full and strict com-

---

[99] Wallander, C., 'Russia's strategic priorities', *Arms Control Today*, vol. 32, no. 1 (Jan./Feb. 2002), pp. 4–6.
[100] Wallander (note 99); and Litovkin, V., 'I and the last seven days: the love of peace is settling in the General Staff', *Obshchaya Gazeta* (Moscow), 13 Dec. 2001, in 'Deputy Chief of Staff Baluyevskiy on strategic arms reductions, missile defense, US presence in Central Asia', FBIS-SOV-2001-1212, 17 Dec. 2001.
[101] Zhuqing, J., 'State criticizes US for abandoning treaty', 14 Dec. 2001, *China Daily* (Internet edn), URL <http://www1.chinadaily.com.cn/cndy/2001-12-14/47905.html>.

pliance'.[102] They also reiterated their warnings that the USA's missile defence plans could spark an arms race in outer space.[103]

The restrained reaction from Beijing was part of a broader trend in 2001 in which Chinese officials toned down their criticism of US missile defence plans.[104] This was in part the result of the Bush Administration's consultations aimed at assuring China that it was not the intended 'target' of a US strategic missile defence system. Many Chinese officials and analysts had maintained that the real purpose of the USA's missile defence shield was to neutralize the deterrent value of China's small force of ICBMs rather than to defend against attacks from states such as Iraq and North Korea, which do not have missiles capable of reaching US territory. For its part, the Bush Administration displayed relatively little public concern about China's ongoing programme to modernize and expand its strategic nuclear forces, which is likely to result in a significant increase in the number of Chinese ICBMs capable of reaching the USA.[105] Another factor contributing to China's restrained response was the US decision, in April 2001, to defer the sale of advanced-capability theatre missile defences to Taiwan. This proposed sale had aroused considerable unease in Beijing.[106]

The European response

In Europe, there was a muted reaction to Bush's announcement both from US allies and other states. The circumspect tone of European responses reflected in part a recognition that the US decision to move ahead with strategic missile defence was a foregone conclusion.[107] More important, however, was the unexpected equanimity with which Russia and China accepted the US move to withdraw from the ABM Treaty. Many European leaders, especially in France and Germany, had previously voiced serious concern that the abandonment of the ABM Treaty would complicate relations with Russia and China, sound the death knell for nuclear disarmament and possibly reverse the progress made to date. At the same time, however, the restrained reactions of Russia and China did not assuage European misgivings about what was seen as a worrying tendency in US foreign policy to eschew international agreements—the promotion of which has traditionally been considered an important US national interest—in favour of unilateral undertakings.

[102] United Nations General Assembly Resolution 56/24, 29 Nov. 2001, available at URL <http://www.un.org/Depts/dhl/resguide/r56c1.htm>.

[103] Xia, Z., 'ABM withdrawal a dangerous sign', *China Daily* (Internet edn), 21 Dec. 2001, URL <http://www.1chinadaily.com.cn/cndy/2001-12-21/48910.html>. For a description of China's efforts to open negotiations in the Conference on Disarmament on an international treaty to prohibit the militarization of outer space see chapter 11 in this volume.

[104] Gill, B., 'Can China's tolerance last?', *Arms Control Today*, vol. 32, no. 1 (Jan./Feb. 2002), pp. 7–9.

[105] Sanger, D., 'Bush won't oppose China missile buildup', *International Herald Tribune*, 3 Sep. 2001, pp. 1, 5.

[106] Gill (note 104).

[107] Erlanger, S., 'Bush's move on ABM pact gives pause to Europeans', *New York Times* (Internet edn), 12 Dec. 2001, URL <http://www.nytimes.com/2001/12/13/international/13Euro.html>.

Bush's explanation of his decision to withdraw from the ABM Treaty underscored the gap in threat perceptions that separates the USA from many of its European allies on the missile defence issue. In Europe, US claims about the emerging ballistic missile threat posed by states such as Iran, Iraq and North Korea tend to be dismissed as exaggerated. At a meeting of NATO foreign ministers in May 2001, the alliance refused to endorse a US call to take urgent measures to cope with the 'common threat' posed by emerging long-range ballistic missile capabilities in potentially hostile states.[108] In addition, the events of 11 September were seen by many Europeans as lending credence to those who argued that the real threat to security came from terrorists with no access to missile technologies.[109]

III. US–Russian strategic nuclear arms control

Implementation of the START I Treaty

On 5 December 2001 Russia and the USA marked the completion of the third and final phase of reductions in deployed strategic offensive arms mandated by the START I Treaty.[110] Under START I, Russia and the USA undertook to make phased reductions in their strategic offensive nuclear forces over a seven-year period, starting from the treaty's entry into force on 5 December 1994, to no more than 1600 strategic nuclear delivery vehicles and 6000 treaty-accountable nuclear warheads. Interim limits on SNDVs and accountable warheads were to be reached within three and five years, respectively, after the treaty's entry into force. START I also placed limits on inventories of mobile and heavy ICBMs and aggregate ballistic missile throw-weight (or lifting capacity).

The START I Treaty has a 15-year duration, which may be extended by agreement of the parties for successive five-year periods. The verification and inspection arrangements will continue for as long as the treaty remains in force. These include 12 types of on-site inspections as well as data exchanges and notifications regarding the parties' strategic nuclear forces and facilities. The START verification and inspection arrangements are likely to be used, in streamlined form, to monitor compliance with the pledges made by Bush and Putin in November 2001 to further reduce their countries' strategic nuclear forces. The parties will continue to meet as necessary in the Joint Compliance and Inspection Commission (JCIC), which START I established as the forum for resolving compliance questions and discussing ways to facilitate implementation.

---

[108] Drozdiak, W., 'NATO divided on missile defense', *Washington Post*, 30 May 2001, p. A15.

[109] Norton-Taylor, R., 'Europe resigned while Britain clicks its heels', *The Guardian* (Internet edn), 14 Dec. 2001, URL <http://www.guardian.co.uk/Archive/Article/0,4273,4319724,00.html>.

[110] US Department of State, Bureau of Arms Control, 'Fact Sheet: START Treaty final reductions', 5 Dec. 2001, URL <http://www.state.gov/t/ac/rls/fs/index.cfm?docid=6669>.

START I accomplishments

The START I Treaty was signed by the USSR and the USA on 31 July 1991, following over a decade of negotiation. It remains the only in force, legally binding agreement regulating the size and composition of the US and Russian nuclear arsenals. The treaty's ceilings on deployed strategic nuclear forces have brought about significant reductions in the US and Russian nuclear arsenals, albeit to levels that many arms control advocates find, more than a decade after the end of the cold war, disappointingly high. Between 1990 and 2001, the number of deployed treaty-accountable nuclear warheads declined by 44 per cent on the US side and by 46 per cent on the Russian side.

The START I Treaty proved instrumental in settling the fate of the former Soviet strategic nuclear arsenal in Belarus, Kazakhstan and Ukraine. With the dissolution of the USSR these new states had inherited over 3400 strategic nuclear warheads based on their territories, although operational control over the weapons remained in Russian hands. A key concern in the international community, particularly in the United States, was to preserve a centralized command and control system for the post-Soviet strategic nuclear forces and to ensure their security and custodial safety. At a meeting of foreign ministers in Lisbon, Portugal, in May 1992, Belarus, Kazakhstan and Ukraine signed the Lisbon Protocol with Russia and the USA, making these three countries also parties to START I; the three non-Russian former Soviet republics committed themselves in the protocol to meet the USSR's nuclear arms reduction obligations and pledged to accede to the NPT as non-nuclear weapon states.[111] The START I Treaty thereby provided the basis for consolidating Soviet nuclear warheads in Russia and for eliminating the delivery vehicles and associated infrastructure in Belarus, Kazakhstan and Ukraine.

Towards deeper reductions in strategic nuclear arms

In 2001 there was a breakthrough in the US–Russian strategic arms reduction process. Progress towards making deeper negotiated cuts in strategic nuclear arsenals had been blocked by an impasse in bringing into force the 1993 Treaty on Further Reduction and Limitation of Strategic Offensive Arms (START II Treaty).[112] The impasse had arisen in April 2000 when the Russian Parliament passed a ratification law which, *inter alia*, stipulated that Russia would ratify the START II Treaty only after the US Senate ratified a package of legally binding Agreed Statements from 1997, relating to the ABM

---

[111] Excerpts from the text of the Lisbon Protocol are reproduced in *SIPRI Yearbook 1993: World Armaments and Disarmament* (Oxford University Press: Oxford, 1993), pp. 574–575.

[112] The START II Treaty was ratified by the US Senate in Jan. 1996 and, in amended form, by both houses of the Russian Federal Assembly in Apr. 2000. For a description of the provisions of the START II Treaty see Lockwood, D., 'Nuclear arms control', *SIPRI Yearbook 1993* (note 111), pp. 554–59.

Treaty.¹¹³ This led to a situation in which the START II Treaty had been ratified by both parties but could enter into force, since the so-called ABM Treaty demarcation agreement established by the Agreed Statements was unacceptable to the Bush Administration—and was explicitly identified as being so in the 2001 Nuclear Posture Review (NPR). For its part, the Putin Administration showed no interest in asking parliament to amend the ratification law. This linkage was set aside when Bush and Putin agreed at their November 2001 summit meeting to supersede, or 'leap over', the START II Treaty and undertake a new round of deeper arms reductions.¹¹⁴ In doing so, they effectively rendered the long-stalled treaty a dead letter. At the same time, however, they paved the way for progress towards further cuts in strategic nuclear forces where none had appeared possible.

Interest in further strategic force reductions

The idea of negotiating deeper reductions has become a particularly attractive one in Russia, since it holds out the prospect of requiring the USA to reduce its forces to levels that Russia could afford to match as it eliminates missiles and submarines reaching the end of their service lives. In November 2000, President Putin proposed that Russia and the USA should reduce their strategic nuclear arsenals to below the 2500-warhead limit envisaged in the proposed START III accord. While not specifying a new limit, he called for 'radically reduced ceilings for nuclear warheads that could be reached either jointly or in parallel moves.¹¹⁵ Russian officials subsequently proposed a ceiling of 1500 nuclear warheads for each side. They emphasized, however, that any deeper cuts in nuclear forces would depend on progress in preserving and strengthening the ABM Treaty.¹¹⁶

In the USA there has been renewed political interest in adjusting US nuclear targeting doctrine and nuclear force levels to reflect a strategic environment in which Russia is no longer seen as an enemy. One argument made by supporters of deeper cuts was that the USA was forcing Russia to retain nuclear forces

---

¹¹³ The Agreed Statements established a set of criteria for distinguishing between theatre (or non-strategic) missile defence systems, which are permitted by the ABM Treaty, and strategic missile defence systems, which are not. For a description of the Agreed Statements and related documents see Kile, S., 'Nuclear arms control', *SIPRI Yearbook 1998: Armaments, Disarmament and International Security* (Oxford University Press: Oxford, 1998), pp. 420–23.

¹¹⁴ In order for Bush to be able to proceed, the US Congress had to approve language contained in the FY 2002 amended defense authorization bill that lifted a 1998 restriction, effectively barring the president from unilaterally reducing US strategic nuclear forces below START I levels. Bleek, P., 'Bush, Putin pledge nuclear cuts; implementation uncertain', *Arms Control Today*, vol. 31, no. 10 (Dec. 2001), pp. 19, 24.

¹¹⁵ 'Statement by the President of the Russian Federation Vladimir V. Putin', 13 Nov. 2000, Press release no. 48, Embassy of the Russian Federation in the United Kingdom, 14 Nov. 2000, URL <http://www.great-britain.mid.ru/GreatBritain/pr_rel/pr48.htm>

¹¹⁶ ITAR-TASS, 1 Feb. 2001, in 'Foreign Minister says Russia ready for talks with US on START III', FBIS-SOV-2001-0201, 1 Feb. 2001; Interfax (Moscow), 28 June 2001, in 'Sergeyev calls for immediate talks with US on arms cuts', FBIS-SOV-2001-0628, 28 June 2001; and Interfax (Moscow), 26 July 2001, in 'Russia "repeated" proposal to reduce nuclear warheads to 1500', FBIS-SOV-2001-0726, 26 July 2001.

Table 10.2. Soviet/Russian and US strategic offensive nuclear forces, by delivery vehicles and START-accountable warheads, September 1990 and December 2001[a]

Category	1990 USSR	1990 USA	2001 Russia[b]	2001 USA	START I final limit
Strategic nuclear delivery vehicles (SNDVs)[c]	2 338	1 672	1 136	1 237	1 600
Warheads attributed to ICBMs and SLBMs	9 416	8 210	4 894	4 821	4 900
Total treaty-accountable warheads	10 271	10 563	5 518	5 948	6 000

ICBM = intercontinental ballistic missile; SLBM = submarine-launched ballistic missile

[a] The numbers given in this table are in accordance with the START I Treaty counting rules and include delivery vehicles which have been deactivated in preparation for elimination or conversion but which remain treaty-accountable.

[b] The USSR's obligations under the START I Treaty were assumed by Russia as its legal successor state and later by Belarus, Kazakhstan and Ukraine. Only Russia retained SNDVs and nuclear warheads at the end of the implementation period.

[c] Deployed ICBMs and their associated launchers, deployed SLBMs and their associated launchers, and deployed heavy bombers.

Sources: START I Treaty Memorandum of Understanding, 1 Sep. 1990; US Department of State, Bureau of Arms Control, 'Fact Sheet: START I aggregate numbers of strategic offensive arms', 1 Apr. 2002, URL <http://www.state.gov/t/ac/rls/fs/9075.htm>.

beyond a level which it can afford to maintain safely.[117] Arms control advocates also argued that the current US nuclear posture has changed little from the cold war, which means that 'friends' are now targeting one another.[118] However, the US military has been noticeably unenthusiastic about embracing reductions below the 2000- to 2500-warhead level in the absence of new presidential targeting guidance.[119] Analysts note that cuts below this level would require the removal of targets from the US strategic war plan or reductions in the level of damage to targets believed necessary for deterrence.[120] In addition, reductions below this level would be likely to require the DOD to restructure its 'triad' (heavy bombers, submarines and land-based missiles) of stra-tegic nuclear forces.

---

[117] Pincus, W. and Dewar, H., 'Approved nuclear measure unlikely to affect Clinton', *Washington Post*, 8 June 2000, p. A12.

[118] Kimball, D., 'Fuzzy nuclear math', *Arms Control Today*, vol. 31, no. 10, (Dec. 2001), p. 2.

[119] Statement on Command Posture by Admiral Richard Mies, Commander-in-Chief, US Strategic Command, before the Strategic Subcommittee, US Senate Armed Services Committee, 11 July 2001, available at URL <http://www.senate.gov/~armed_services/hearings/2001/f010711.htm>.

[120] Despite the end of the cold war, there are currently 2230 'vital' Russian targets on the US strategic war plan (the Single Integrated Operating Plan, SIOP); targets in China were reintroduced into the SIOP in 1998–99 after an absence of nearly 20 years. US strategic planners have traditionally set the required level of damage against vital targets at 80%. With current targeting guidance (which was last modified in 1997 by a Presidential Decision Directive), c. 2500 deployed strategic nuclear warheads are considered to be the minimum necessary to execute the SIOP. Blair, B., 'Background paper on the strategic war plan and START reductions', Center for Defense Information, 18 May 2000, URL <http://www.cdi.org/issues/proliferation/blairbckReduc.html>.

During the 2000 presidential campaign, Bush had vowed to pursue deep cuts in warheads and missiles based on a new strategic doctrine and approach to arms control.[121] In his May 2001 address on missile defence, Bush stated that he would consider reducing US strategic nuclear forces—possibly in a unilateral step—to 'the lowest possible number consistent with [US] national security'.[122] He also said that he would consider reducing the alert status of US ICBMs, which remain primed for rapid launch. Some observers interpreted these statements as an attempt to overcome concern among US allies that the administration's missile defence plans would reverse the post-cold war trend towards lower nuclear force levels.[123]

The Bush–Putin understanding on deeper reductions

During their November 2001 summit meeting in Washington, Bush and Putin agreed to move ahead with making deeper reductions in strategic nuclear forces. At a joint White House news conference, Bush announced that the United States would, over the next decade, unilaterally reduce the number of its operationally deployed strategic nuclear warheads to 1700–2200. This would involve a two-thirds cut in the current number of deployed nuclear warheads; it would also entail cuts substantially below the 3500-warhead ceiling mandated by the START II Treaty. Putin promptly pledged that his government would respond in kind by making reductions to 1500 warheads, although he gave no timetable for doing so.[124]

A key question left unanswered by the summit meeting was in what form, if any, the unilateral reductions promised by Bush and Putin would be codified. Shortly after the meeting, the White House issued a statement pledging 'to work with Russia to formalize this arrangement on offensive forces, including appropriate verification and transparency measures'.[125] While welcoming the US offer, the Russian Government insisted that this had to be done in the form of a treaty.[126] It emphasized that a legally binding agreement, containing streamlined verification arrangements based on those in the START treaty regime, was essential to ensure predictability in US and Russian nuclear poli-

---

[121] Myers, S., 'Bush plans nuclear review, clearing way for unilateral reductions', *International Herald Tribune*, 9 Feb. 2001, p. 3; and Pincus, W., 'US considers shift in nuclear targets', *Washington Post*, 29 Apr. 2001, p. A23.

[122] Transcript of 'remarks by the president' (note 12).

[123] Fitchett, J. 'Europeans receptive to a broad strategy', *International Herald Tribune*, 2 May 2001, pp. 1, 10. Others point out that the govts of US allies and others have traditionally seen arms control agreements, and not reductions alone, as an integral part of the process to reverse the arms race and curb the spread of nuclear weapons.

[124] 'Statement made by Russian President Vladimir Putin' (note 89).

[125] 'Response to Russian statement on US ABM Treaty withdrawal', The White House, Office of the Press Secretary, 13 Dec. 2001, URL <http://www.whitehouse.gov/news/releases/2001/12/2001/20011213-8.html>.

[126] Interfax (Moscow), 14 Nov. 2001, in 'Defense Ministry sharply criticizes US idea of unilateral arms cuts', FBIS-SOV-2001-1114, 14 Nov. 2001; Fidler, S., 'Bush, Putin strive for arms accord', *Financial Times*, 15 Nov. 2001, p. 5; and Sipress, A., 'US seeks deal on arms cuts by summer', *Washington Post*, 11 Dec. 2001, p. A28.

cies.[127] In Russia's view, this was a necessary precondition for the preservation of stability in US–Russian relations. It was also seen as important to provide assurance to other states, particularly China, about the future size and structure of their nuclear arsenals. The absence of formal commitments could encourage a build-up of nuclear forces by China and possibly by other states.[128]

Bush Administration officials made clear that they opposed Russia's call for codifying parallel but unilateral undertakings in the form of a legally binding arms control agreement.[129] This opposition reflected a deep-rooted scepticism, shared by key national security policy makers in the administration, about the relevance of treaty-based approaches to strategic nuclear arms control.[130] The Bush team had come to office with little interest in engaging in cumbersome, time-consuming negotiations leading to complex arms reduction agreements that mandated precisely equilibrial force limits accompanied by detailed verification provisions.[131] White House advisers argued that the USA could, through unilateral reductions, move to much lower force levels and still accomplish any conceivable military mission.[132] In this view, Russia will follow the USA's lead out of its own national interest, since it can no longer afford to maintain current nuclear force levels.

Furthermore, Bush Administration officials maintained that, with the end of the cold war, there was no need to begin another protracted arms control negotiation with a Russia that was no longer viewed as an enemy. For the same reason they also showed no interest in rescuing the START II Treaty, even though this meant abandoning the ban on land-based missiles carrying multiple warheads, which had been a key US objective in negotiating the treaty.[133] Senior administration officials argued that the USA was facing an increasingly uncertain world. As a matter of prudence, it should seek to preserve its flexibility and freedom of action in responding to new or unforeseen threats.[134] It should therefore not lock itself into a new set of binding treaty limits. Rather, the USA should decide how many nuclear warheads it needs,

---

[127] ITAR-TASS, 18 Jan. 2002, in 'Russian official says Moscow, Washington continue to disagree on arms reductions', FBIS-SOV-2002-0118, 18 Jan. 2002.

[128] 'Commit to paper', *Washington Post*, 16 Nov. 2001, p. A46, available at URL <http://www.washingtonpost.com/ac2/wp-dyn?pagename=article&node=contentId=A38010-2001Nov15>. It was also argued that a formal treaty was essential in part because so much of the recent warming in US–Russian relations seemed to hinge on the personal chemistry between Bush and Putin and might not survive a change in administration.

[129] Baker, P. 'A familiar Bush strategy on disarmament', *Washington Post*, 14 Nov. 2001, p. A6.

[130] For a discussion of the Bush Administration's approach to arms control and non-proliferation see chapter 9 in this volume.

[131] For an influential study that provides a guide to the Bush Administration's thinking about arms control and nuclear doctrine see Payne, K. *et al.*, *Rationale and Requirements for US Nuclear Forces and Arms Control* (National Institute for Public Policy: Washington, DC, Jan. 2001), available at URL <http://www.nipp.org/publications.php>. Several of the individuals who contributed to the study now occupy high-level national security policy-making positions in the Bush Admin-istration.

[132] According to William Odom, former director of the National Security Agency, 'if we spent 10 years in arms control forums, we'd never get it done'. Myers (note 121).

[133] Gordon, M, 'US arsenal: treaties vs. nontreaties', *New York Times* (Internet edn), 14 Nov. 2001, URL <http://www.nytimes.com/2001/11/14/international/14nuke.html>.

[134] See Payne *et al.* (note 131), pp. 12–15.

based on a thorough review of nuclear strategy, and then reduce or restructure its nuclear arsenal accordingly.

The White House's reluctance to enter into legally binding arms reduction commitments drew criticism from inside the USA and abroad. In February 2002, the administration indicated that it would not rule out the possibility of reaching a legally binding agreement with Russia to reduce nuclear arsenals. US Secretary of State Colin Powell suggested the possibility of a treaty or an 'executive agreement' that Congress could debate and approve as a joint resolution. Such a document would state US intentions, as in the preamble of many treaty documents, and set out in general terms the verification procedures to be applied, but it would not specify undertakings and commitments in detail.[135]

Irreversibility of nuclear reductions

A second question left unanswered at the November 2001 summit meeting was whether the two sides would require the verified elimination of surplus nuclear warheads identified for removal from operational deployment. Bush's statement announcing the unilateral reductions did not specify whether the warheads to be removed from operational deployment would be dismantled or held in reserve as a 'hedge' against unforeseen future threats, as the Clinton Administration had done with surplus warheads under START I.[136]

The Bush Administration subsequently informed Congress that many of the nuclear warheads removed from delivery vehicles would be placed in reserve stockpiles and not be dismantled. This gave rise to a new dispute with Russia. Some Russian analysts complained that the US refusal to physically destroy warheads made the agreement on the reduction of strategic arms 'absolutely pointless'.[137] The Bush Administration's position meant that 'Russia and the USA would not have equal rights in the sphere of strategic arms'.[138]

The idea of requiring surplus warheads to be dismantled has gained support in Russia as a mechanism for addressing concerns about asymmetries in the 'reconstitution potential' of the US and Russian strategic nuclear forces. These concerns were first raised during the debate in Russia over whether to ratify the START II Treaty.[139] Analysts there point out that major reductions in

[135] Slevin, P. and Pincus, W., 'US now seeking binding deal with Russia on nuclear arms', *Washington Post*, 6 Feb. 2002, p. A15.

[136] START I and START II do not require the dismantlement of the warheads removed from delivery vehicles as scheduled for elimination or conversion. At a summit meeting held in Helsinki in 1997, Clinton and Yeltsin agreed that a future START III Treaty should contain 'measures relating to the transparency of strategic nuclear warhead inventories and the destruction of strategic nuclear warheads'. The goal of these measures was to make permanent US–Russian reductions in their strategic nuclear forces. Joint Statement on Parameters on Future Reductions in Nuclear Forces, The White House, Office of the Press Secretary, 21 Mar. 1997, available at URL <http://www.ceip.org/files/projects/npp/resources/summits6.htm#parameters>.

[137] Pikayev, A., quoted by Interfax (Moscow), 10 Jan. 2002, in 'US refusal to destroy warheads makes strategic arms control accord pointless', FBIS-SOV-2002-0110, 10 Jan. 2002.

[138] Pikayev (note 137).

[139] Kile (note 113), pp. 415–16.

Russia's Strategic Rocket Forces are inevitable over the next decade as ageing ICBMs reach the end of their service lives and are not replaced. In contrast, the USA plans to move to a lower number of deployed strategic nuclear warheads primarily by 'downloading' (that is, by removing one or more warheads from a missile carrying multiple warheads) and retaining most of its Minuteman III ICBMs and highly accurate, long-range Trident II SLBMs.[140] The USA also plans to continue to maintain reserve stockpiles consisting of thousands of nuclear weapons in various stages of readiness. Russian analysts argue that this has the effect of leaving the USA in a better position than Russia to rapidly reconstitute its strategic forces by 'uploading' stored nuclear warheads back onto its land- and sea-based ballistic missiles and thereby achieve a significant strategic advantage over Russia.[141]

Russian concerns were fuelled in January 2002 by the release of the results of the DOD Nuclear Posture Review, a comprehensive 10-month review of the US strategic and tactical nuclear force posture.[142] The NPR set out a three-phase schedule for reducing the number of 'operationally deployed strategic warheads' to between 1700 and 2000 by the year 2012. According to Assistant Secretary of Defense J. D. Crouch, the USA would maintain a substantial number of nuclear warheads in reserve as a 'responsive capability'. He noted, however, that 'there have been no final decisions made at this point on what should be the size' of this capability or about the overall size of the US nuclear stockpile.[143]

IV. Cooperative nuclear security initiatives

Since 1991 the USA has funded an expanding range of cooperative initiatives to assist with the dismantlement or conversion of the former Soviet Union's vast non-conventional weapon complexes and help to safeguard nuclear and other hazardous materials.[144] These initiatives have played a central, albeit sometimes controversial, role in the international community's efforts to manage proliferation risks in the former USSR and to address the challenges aris-

---

[140] Many Russian defence analysts had argued during the START II Treaty ratification debate that a requirement in a future START III Treaty to dismantle warheads removed from ballistic missiles would help to compensate for the absence of a rule in START II requiring that a downloaded missile must be fitted with an entirely new 'bus', or front-end platform, able to hold only the smaller number of warheads.

[141] Frolov, V., 'A new start on the banks of the Potomac', *Vremya MN* (Moscow), 15 Jan. 2002, in '"Reconstitution potential": major issue in US–RF strategic arms consultations', FBIS-SOV-2002-0115, 15 Jan 2002.

[142] US Department of Defense, Transcript of special briefing on the Nuclear Posture Review, 9 Jan. 2002, URL <http://www.defenselink.mil/news/Jan2002/t01092002_t0109npr.html> (news briefing slides available at URL <http://www.defenselink.mil/news/Jan2002/t01092002_t0109npr.html>). For more detail on the NPR see appendix 6E and appendix 10A in this volume.

[143] US Department of Defense (note 142).

[144] These initiatives have grown from the original Cooperative Threat Reduction programme (also called the Nunn–Lugar programme after the senators who co-sponsored the original authorizing legislation), which began in 1991 with funding from the US Department of Defense. The programme has since evolved to encompass a wide range of non-proliferation and demilitarization activities under the auspices of the Department of Energy and the Department of State as well as the DOD.

ing from the Soviet nuclear legacy. An important focus of US-funded cooperative initiatives in recent years has been to prevent former Soviet scientists working on nuclear, chemical or biological weapon programmes from selling their skills to unfriendly regimes or terrorist groups.

With regard to nuclear-related dangers, considerable progress has been made in eliminating former Soviet strategic nuclear weapons and enhancing the safety and custodial security of nuclear weapons remaining in Russia. However, because of the scale and scope of the former Soviet nuclear weapon complex, international efforts to prevent the 'leakage' or misappropriation of fissile and other weapon-usable material will face formidable challenges for years to come. It is estimated that there are approximately 650 tonnes of weapon-usable nuclear material in the former Soviet Union, not including the contents of nuclear warheads.[145] This material is currently held at 66 sites, of which 56 are located in Russia.[146] These include nuclear weapon R&D facilities, nuclear fuel production and fabrication plants, civilian research institutes and naval fuel facilities.

The security shortcomings identified at many of these facilities have raised concern about the possible theft or unauthorized diversion of highly enriched uranium (HEU), plutonium and other weapon-usable nuclear material.[147] This has inspired, since 1995, the launch of a variety of urgent measures aimed at creating an effective fissile material physical control and accounting (MPC&A) regime. In January 2001 a bipartisan panel report commissioned by the Department of Energy (DOE) had stressed the seriousness of the national security threat to the USA posed by the possibility that terrorist groups or hostile states could acquire weapons of mass destruction or weapon-usable material from the former Soviet Union. The report advocated a ten-fold increase in funding for US threat reduction programmes over the next decade.[148]

In March 2001 the Bush Administration announced that it would undertake a comprehensive review of over 30 US-funded non-proliferation and nuclear security programmes in the former Soviet Union.[149] The purpose was to examine the 'cost–benefit ratio' of each programme and to assess whether they focused on 'priority threat reduction and non-proliferation goals'.[150] It would

---

[145] Carnegie Endowment for International Peace and the Monterey Institute of International Studies, *The Nuclear Successor States of the Soviet Union: Nuclear Weapon and Sensitive Exports Status Report*, no. 6 (June 2001), p. 75.

[146] Of the 10 facilities outside Russia, 1 is in Belarus, 3 are in Kazakhstan, 1 in Latvia, 3 in Ukraine and 2 in Uzbekistan. Carnegie Endowment for International Peace and the Monterey Insti-tute of International Studies (note 145).

[147] For an analysis of incidents since 1991 involving illicit trafficking in nuclear and other radiological material see appendix 10D in this volume.

[148] Pincus, W., 'Panel urges $30 billion to secure Russian arms', *Washington Post*, 11 Jan. 2001, p. A21.

[149] Miller, J., 'US will review its aid to Russia for stopping the spread of weapons', *International Herald Tribune*, 30 Mar. 2001, p. 3.

[150] Fact Sheet: 'Administration review of nonproliferation and threat reduction assistance to the Russian Federation', The White House, Office of the Press Secretary, 27 Dec. 2001, URL <http://www.whitehouse.gov/news/releases/2001/12/print/20011217.html>.

Table 10.3. Summary of funding for principal DOD and DOE non-proliferation programmes in the former Soviet Union, February 2002

Figures are for appropriated funds, in US $m. at current prices.

Programme activity	FY 2001[a]	FY 2002
Cooperative Threat Reduction Programme		
Strategic nuclear arms elimination (Russia and Ukraine)	206.9	184.9
WMD infrastructure elimination (Kazakhstan and Ukraine)	–	12.0
Nuclear weapon transportation & storage security (Russia)	103.7	65.5
Fissile material storage facility (Russia)	57.4	–
Weapon-grade plutonium elimination (Russia)	32.1	41.7
Warhead dismantlement processing (Russia)	9.3	–
Chemical weapons destruction	–	50.0
Biological weapons proliferation prevention	12.0	17.0
Military-to-military contacts	9.0	18.7
Management and support	13.0	13.2
Department of Defense programme total	**443.4**	**403.0**
Material Protection, Control and Accounting (MPC&A)	169.7	293.0
Arms control and non-proliferation[b]	148.5	75.7
Russian Transition Initiative[c]	–	57.0
HEU Agreement[d] transparency	14.5	14.0
Fissile materials disposition[e]	226.5	252.0
Non-proliferation and verification R&D	244.5	322.3
International nuclear safety	19.3	20.0
Programme direction[f]	51.4	–
Department of Energy programmes total	**874.4**	**1 034.0**[g]

[a] Figures include $223 million emergency supplemental appropriation for non-proliferation and nuclear security programmes in the former Soviet Union.

[b] Includes funding for the Nuclear Cities Initiative (NCI) in FY 2001.

[c] Created in FY 2002 by the merger of the Nuclear Cities Initiative and the Initiative for Proliferation Prevention (IPP) programmes.

[d] The 1993 Highly Enriched Uranium Agreement, available at URL <http://www.nti.org/>.

[e] Conducts activities in Russia and the USA to eliminate surplus weapons-usable fissile material, including programmes to dispose of 68 tons of excess Russian and US military plutonium.

[f] Programme direction was transferred in FY 2002 from the DOE to the National Nuclear Security Administration.

[g] Less use of $7.5 million of prior year unobligated balances.

Sources: Center for Arms Control and Non-proliferation, 'Summary of major US non-proliferation programs—FY 2002', 24 Jan. 2002, available at the Council for a Livable World Internet site, URL <http://www.clw.org/control/proliferation.html>; and US House of Representatives, *National Defense Authorization Act for Fiscal Year 2002*, Conference report to accompany S. 1438, 12 Dec. 2001.

also examine ways to improve the coordination of these programmes and consider possible new initiatives. The review was announced at a time when some senior administration officials were expressing doubts about the effectiveness of these programmes in reducing nuclear-related threats in the former Soviet

Union.[151] The announcement came against the background of a Russian–US dispute over access rights to their respective nuclear weapon facilities. It also coincided with mounting concern in the USA that Russia's sharing of nuclear and other sensitive technologies with Iran was undermining wider US non-proliferation goals.[152]

The Bush Administration's FY 2002 defence budget called for modest reductions in funding levels for nuclear security initiatives in the former Soviet Union.[153] The proposed reduction came primarily at the expense of nuclear material security, disposition and safety programmes administered by the DOE. However, these programmes enjoyed considerable bipartisan support in Congress, which subsequently restored most of the funding for them.[154]

The events of 11 September heightened concern in the USA about the danger of nuclear weapons from the former Soviet Union—or of fissile or other hazardous material—falling into the hands of terrorist groups. Congress approved a $223 million emergency appropriation to expand non-proliferation and nuclear security activities in the former Soviet Union. This included $120 million for the MPC&A programme and $10 million to improve the safety of Soviet-era nuclear power reactors and facilities.[155] Congress also approved an additional $15 million for the Russian Transition Initiative, which consolidated two programmes aimed at preventing a 'brain drain' of experts from the former Soviet nuclear, chemical and biological weapon complexes by creating new, non-defence-sector jobs for them.[156]

In December 2001 the Bush Administration announced the results of its review of non-proliferation and threat reduction assistance programmes. The report concluded that most programmes 'work well, are focused on priority tasks and are well managed'.[157] It identified four programme areas for expansion: MPC&A activities, including cooperation with Russia to install nuclear detection equipment at border posts; the DOE's Warhead and Fissile Material Transparency programme; the State Department's International Science and Technology Centers (ISTC); and the Redirection of Biotechnical Scientists programme. The review also recommended accelerating the Cooperative Threat Reduction (CTR) project to construct a pilot chemical weapons destruction facility at Shchuch'ye.[158]

---

[151] Pincus, W., 'Bush targets Russia nuclear programs for cuts', *Washington Post*, 18 Mar. 2001, p. A23.
[152] Luongo, K., 'Improving US–Russian nuclear cooperation', *Nonproliferation Review*, vol. 8, no. 3 (fall 2001), pp. 85–91.
[153] Bleek, P., 'Bush seeks cuts in Pentagon threat reduction programs', *Arms Control Today* vol. 31, no. 7, (Sep. 2001), p. 28.
[154] Johnson, J., 'Securing the nuclear threat', *Chemical and Engineering News*, vol. 75, no. 51, (17 Dec. 2001), pp. 43–44.
[155] Russian–American Nuclear Security Advisory Council (RANSAC), 'Anticipated FY 2003 budget request for the Department of Energy cooperative nuclear security programs in Russia', 9 Jan. 2002, URL <http://www.ransac.org/new-web-site/whatsnew/fy03budget.html>.
[156] RANSAC (note 155).
[157] Fact Sheet (note 150).
[158] For more detail see chapter 12 in this volume.

For FY 2003, the Bush Administration has announced that it will seek an increase in funding for non-proliferation and threat reduction activities. The administration has requested $416 million for the Defense Department's CTR programme.[159] It has also asked for a record $1.11 billion for the DOE's defence nuclear non-proliferation programmes.[160] The largest increases in the budget request, compared to FY 2002 appropriations, are earmarked for the DOE's MPC&A and Fissile Material Disposition programmes.[161]

V. The Comprehensive Nuclear Test-Ban Treaty

During 2001 five states signed the CTBT and 19 ratified it. As of 1 January 2002, the CTBT had been ratified by 90 states and signed by a further 76 states. Of the 44 states whose ratification is required for the treaty to enter into force, 31 had ratified the treaty and an additional 10 states had signed but not ratified the treaty by the end of 2001.[162] The USA has signed the treaty but later voted not to ratify it.[163] There are three states among the 44—India, North Korea and Pakistan—which have not signed the accord.

On 11–13 November 2001 the Conference on Facilitating the Entry into Force of the Comprehensive Nuclear Test-Ban Treaty was held at United Nations Headquarters in New York. The meeting was attended by the delegates of 109 states.[164] The USA did not take part. The conference issued a Final Declaration that *inter alia* reaffirmed the importance of universal adherence to the CTBT for nuclear non-proliferation and disarmament efforts and called on all states to maintain a moratorium on nuclear weapon test explosions or any other nuclear explosions.[165]

During the year the Provisional Technical Secretariat for the Comprehensive Nuclear Test-Ban Treaty Organization (CTBTO) continued to make progress towards implementing the global verification regime to monitor compliance with the test ban. The Secretariat is responsible for supervising the construc-

[159] Council for a Livable World, 'Quick analysis of Fiscal 2003 budget request US nonproliferation programs', Press release, 5 Feb. 2002, URL <http://www.clw.org/control/03proliferation.html>.

[160] Council for a Livable World (note 159); and US Department of Energy, *FY 2003 Congressional Budget Request: Budget Highlights*, DOE/ME-0008, Feb. 2002, p. 7.

[161] Council for a Livable World (note 159).

[162] The treaty will enter into force 180 days after it has been ratified by the 44 members of the Conference on Disarmament with nuclear power or research reactors on their territories, as listed in annex 2 of the treaty. For the parties and signatories of the CTBT see annex A in this volume.

[163] President Clinton signed the CTBT in Sep. 1996, but the US Senate voted narrowly in Oct. 1999 not to ratify the treaty. See Kile, S., 'Nuclear arms control and non-proliferation', *SIPRI Yearbook 2000: Armaments, Disarmament and International Security* (Oxford University Press: Oxford, 2000), pp. 464–66.

[164] Article XIV of the CTBT provides for the convening of an annual conference by the states which have deposited their instruments of ratification (other states may participate as observers) to consider 'what measures consistent with international law may be undertaken to accelerate the ratification process in order to facilitate the early entry into force of the treaty'. The text of the CTBT is reproduced in *SIPRI Yearbook 1997: Armaments, Disarmament and International Security* (Oxford University Press: Oxford, 1997), pp. 414–31.

[165] Report of the Conference on Facilitating the Entry into Force of the CTBT, CTBT-ART.XIV/2001/6, Public Information Section, Preparatory Commission for the CTBTO, Vienna, 15 Nov. 2001, URL <http://www.ctbto.org>.

tion and certification of an International Monitoring System (IMS), which will consist of 321 monitoring stations and 16 laboratories located in 90 countries.[166] By the end of February 2002, installations had been completed at 122 stations.[167] Work also continued to connect the IMS stations through a satellite communication network to an International Data Centre (IDC) in Vienna, Austria. The IDC is responsible for receiving, processing and distributing raw data received from the IMS stations to member states.

During 2001, the continuing uncertainty about the timing of the treaty's entry into force contributed to some erosion of international support for the CTBTO. This stemmed largely from concern about the rising cost of the organization. Brazil and Argentina took the lead in questioning the sizeable annual increases in the CTBTO's budget for building the IMS when it was unclear when—or if—the treaty might enter into force.[168] In addition, China and Iran delayed or halted the transmission of data to the IDC from a number of monitoring stations on their territories. There was speculation that this may have been in reaction to the USA's announcement in August 2001 that it would contribute to the costs associated with the monitoring system but not to the other functions of the CTBTO.[169]

VI. Conclusions

In December 2001 the long-running controversy over the United States' missile defence plans and the future of the 1972 ABM Treaty came to a head when President Bush announced that the USA intended to withdraw from the treaty in order to proceed with the development of a large-scale ballistic missile defence system. At its core, the missile defence controversy had involved a doctrinal dispute over the relationship between deterrence and strategic defence in the post-cold war world and the continued relevance of the ABM Treaty as the 'cornerstone of strategic stability'. Bush's announcement, which drew a notably restrained response from China and Russia, effectively brought the debate to a close and heralded the collapse of one of the main pillars of the nuclear arms control framework inherited from the cold war.

The US decision to withdraw from the ABM Treaty came as part of its rejection of the relevance of traditional nuclear arms control treaties to the US national security strategy. The Bush Administration brought to office an ideological aversion to the ABM Treaty's constraints on strategic defence and its

---

[166] These stations use 4 verification technologies—seismic, infrasound, hydro-acoustic and radionuclide—to monitor the earth for evidence of a nuclear explosion. Preparatory Commission for the CTBTO. *The Global Verification Regime and the International Monitoring System,* Basic Facts: Booklet 3, 2001, available at URL <http://www.ctbto.org>.

[167] Preparatory Commission for the CTBTO, 'Provisional Technical Secretariat—five years old', Press release 303/06/Ann.5/02, 15 Mar. 2002.

[168] Johnson, R., 'Boycotts and blandishments: making the CTBT visible', *Disarmament Diplomacy,* no. 61 (Oct./Nov. 2001), URL<http://www.acronym.org.uk/dd/dd61/61ctbt.htm>. The CTBTO Preparatory Commission's budget for 2001 was $83.5 million, compared to $79.9 million in 2000 and $74.7 million in 1999.

[169] Giacomo, C., 'China, Iran said balking at test ban pact cooperation', Reuters, 7 Mar. 2002,

codification of the cold war-era logic of mutual assured destruction. The administration also rejected as outdated the complex and painstakingly balanced arms limitation agreements developed as a means for regulating the superpower nuclear arms competition. This type of agreement was criticized by the administration as inhibiting the United States' flexibility to adapt to a new and changing security environment. Although Bush joined Putin in November 2001 in pledging to cut the US and Russian strategic nuclear forces, US officials insisted that these reductions should be carried out as unilateral initiatives rather than in the form of a treaty, as insisted on by Russia.

The future of control and disarmament agreements is uncertain. The value of these agreements has come under increasingly critical scrutiny in the United States in recent years as a result of allegations that, or clear-cut cases in which, states have violated their legal commitments. Underlying the Bush Administration's disinterest in arms control is a deep-rooted scepticism about the efficacy of the existing framework of restraint agreements and multilateral supplier arrangements designed to prevent the spread of weapons of mass destruction and their means of delivery. Its conclusion that formal arms control is neither necessary nor desirable is a significant development, suggesting that a new strategic environment is emerging which is likely to be very different from that which existed in recent decades. In turn, the clear disinterest in multilateral treaties by the USA, the dominant state in the international system, is raising concern in many countries about the prospects for building an international security system based on stability, restraint and deep cuts in armaments.

Appendix 10A. World nuclear forces

HANS M. KRISTENSEN and JOSHUA HANDLER

I. Introduction

The world's eight nuclear weapon states maintain a total of about 17 150 deployed nuclear warheads, of which the United States and Russia together hold 93 per cent. Of the smaller nuclear weapon states, China has slightly over 400 warheads, France 348, and Israel and the United Kingdom each about 200. The two new nuclear weapon states, Pakistan and India, have 24–48 warheads each, although it is thought that not all of them are fully deployed.

In addition to these deployed warheads, thousands more are held in reserve; these are not counted in official declarations on the size of stockpiles. In recent years, an increasing proportion of the warheads have been transferred from declared categories into 'unaccountable' categories. If all warheads are counted—deployed, spares, those in both active and inactive storage, and 'pits' (plutonium cores) held in reserve—the total world stockpile consists of an estimated 36 800 warheads.

During 2001 Russia and the USA continued to reduce their deployed strategic nuclear delivery vehicles within the framework of the 1991 Treaty on the Reduction and Limitation of Strategic Offensive Arms (START I Treaty). Their reductions were completed on 1 December 2001. The USA pledged to make further reductions in its operational warheads and Russia promised to follow suit, although there was disagreement over whether these reductions would be made within the framework of a traditional arms control treaty or as parallel, non-legally binding initiatives. The latter approach was followed when Russia and the USA decided to withdraw from deployment several types of non-strategic (or tactical) nuclear weapons. Both countries continue to maintain sizeable inventories of non-strategic weapons, which remain unconstrained by any formal arms control agreement.[1]

All the nuclear weapon states have nuclear weapon modernization and maintenance programmes under way and appear to be committed to retaining nuclear weapons for the foreseeable future.

In the USA, the 2001 Nuclear Posture Review (NPR) revealed long-term plans for new ballistic missiles, strategic submarines, long-range bombers and nuclear weapons. US modernization plans call for the deployment of new Trident II missiles on older Trident nuclear-powered ballistic-missile submarines (SSBNs). Russia is modernizing its strategic forces by deploying new intercontinental ballistic missiles (ICBMs) and additional strategic bombers and is slowly constructing a new generation of SSBNs. Moreover, both countries continue to underscore the role of nuclear weapons in their security policies. Tables 10A.3 and 10A.5 show the composition of the US and Russian deployed nuclear forces.

The nuclear arsenals of the UK, France and China are considerably smaller than those of Russia and the USA, but these countries also remain committed to retaining their nuclear arsenals. Data on the British, French and Chinese delivery vehicles and nuclear warhead stockpiles are presented in tables 10A.6, 10A.7 and 10A.8, respec-

---

[1] For an overview of US and Russian tactical nuclear weapon inventories and the proliferation risks and arms control challenges arising from them see appendix 10B.

Table 10A.1. World nuclear forces, January 2002[a]

Country	Strategic warheads	Non-strategic warheads	Total warheads
USA	6 480	1 120	7 600
Russia	4 951	3 380	8 331
UK	185	–	185
France	348	–	348
China	282	120	402
India	–	–	(30–35)[a]
Pakistan	–	–	(24–48)[a]
Israel	–	–	(~200)[a]
Total			~17 150

[a] By the number of deployed warheads. The stockpiles of India, Pakistan and Israel are thought to be only partly deployed.

tively. China's strategic modernization plan has received a great deal of attention, but whether its efforts are aimed at deploying a much larger strategic force or a more modern force of the same size is still unclear. Meanwhile, France is currently engaged in developing and deploying a new generation of SSBNs, submarine-launched ballistic missiles (SLBMs) and air-launched weapons. The nuclear weapon stockpile of the UK is about the same size as that of Israel, owing to the unilateral reductions that have been made by both countries (table 10A.11).

It is particularly difficult to obtain official information about the nuclear arsenals of India and Pakistan. Tables 10A.9 and 10A.10 present estimates of the size of their stockpiles and information about their potential nuclear weapon delivery means.

The figures in the tables are estimates based on publicly available information and the authors' best estimates. The uncertainties are reflected in the text.

II. US nuclear forces

The United States maintains an estimated stockpile of about 7600 active deployed nuclear warheads, consisting of 6480 strategic and 1120 non-strategic warheads. Another 370 are spares, while about 2700 intact warheads are held in reserve. In addition to these over 10 600 intact warheads, about 5000 plutonium pits are stored at the Oak Ridge Y-12 Plant in Oakridge, Tennessee, and at the Pantex Plant in Amarillo, Texas.[2] Approximately the same number of canned assemblies (thermonuclear secondaries) are maintained as a strategic reserve. This stockpile is considerably larger than the 5949 'accountable' warheads announced by the State Department on 5 December 2001, after the completion of the final reductions mandated by the START I Treaty. The main reasons for this discrepancy are that the treaty only counts warheads on strategic launchers, attributes 'artificial' weapon loads to bombers, and ignores non-strategic weapons and warheads in the reserve. The START number is widely but often incorrectly cited by the media in reports on the size of the total US nuclear arsenal.

[2] For a description of the facilities in the US nuclear weapon complex see appendix 10C.

Table 10A.2. US nuclear warhead status and modifications, January 2002

Warhead	Number of warheads[a] and status of work
B61-3/5/10	840 deployed (150 in Europe); 450 in storage; Common Radar testing completed in 2000; life extension programme continued in 2001. Refurbishment programme to begin later this decade.
B61-7	370 deployed; 100 in storage. Refurbishment scheduled for 2006–2008.
B61-11	50 deployed; 5 in storage; introduced in 1997; fully certified in 2000. Refurbishment scheduled for 2006–2008.
W62	315 deployed; 300 in storage; all to be retired by 2009.
W76	2880 deployed; 336 in storage; completed multi-year dual-revalidation programme in 2000. Refurbishment of 800 warheads planned for 2007–12, with the remaining 2400 warheads scheduled for 2012–22, depending on a future decision.
W78	920 deployed; 600 to be transferred to 'responsive force' by 2007; life extension programme under way, including replacement of gas transfer system.
W80-0/1	900 deployed; 900 in storage; Phase 6.2/6.2A Study (feasibility, design, and cost study) completed in 2000; Phase 6.3 Study (development engineering) scheduled to start in 2001. A refurbishment programme—including new neutron generators, a gas transfer system and the redesign of the warhead electrical system—is scheduled to refurbish 600 warheads in 2006–10. If approved, the remaining warheads will be refurbished in 2011–17.
B83-0/1	420 deployed; 200 in storage; Common Radar testing completed in 2000; life extension programme continued in 2001. Candidate for new warhead modifications.
W84	400 in storage.
W87	550 deployed; 200 expected to be transferred to Minuteman IIIs (Safety Enhanced Reentry Vehicle programme) and 350 to the 'responsive force' by 2006; full-scale life extension programme under way to refurbish primary and secondary warhead components. About 375 were completed by early 2002 and the rest will be completed by late 2003.
W88	400 deployed; plutonium pit re-manufacturing under way at Los Alamos, with 3 development and 3 standard pits produced in FY 2001. Delivery of the first 'certifiable' pit is scheduled for 2003 and the first new war-reserve warheads are scheduled to enter the stockpile in 2007.

[a] The number of deployed warheads includes spares.

Sources: US Department of Energy; US Department of Defense; Norris, R. *et al.*, 'NRDC Nuclear Notebook: US nuclear forces, 2002', *Bulletin of the Atomic Scientists*, May/June 2002, URL < http://www.thebulletin.org/issues/nukenotes/mj02nukenote.pdf>; Natural Resources Defense Council, 'Faking nuclear restraint: the Bush Administration's secret plan for strengthening US nuclear forces', Press Release, Washington, DC, 13 Feb. 2002, URL <http://www.nrdc.org/media/pressreleases/020213a.asp>; and 'Alterations, modifications, refurbishments and possible new designs for the US nuclear weapons stockpile', Nuclear Watch of New Mexico, 2001, URL <http://www.nukewatch.org/weaponsfactsheet.html>.

The Administration of President George W. Bush completed a year-long Nuclear Posture Review of US nuclear forces in December 2001 and announced that the number of US 'operationally deployed strategic warheads' would be reduced to

1700–2200 over the next 10 years.[3] The NPR appears to have abandoned the 1993 Treaty on Further Reduction and Limitation of Strategic Offensive Arms (START II Treaty, not in force) and the proposed START III accord and instead sets the following new three-phase schedule for reductions.

Phase 1 (2003–2006): reduce the number of 'accountable' warheads by 1300 'as a result of the retirement of Peacekeeper, the Tridents and the like'.

Phase 2 (2006–2007): reduce the number of accountable strategic warheads further by 'taking additional operationally deployed warheads off existing ICBMs and SLBMs down to a level of about 3800' by fiscal year (FY) 2007.

Phase 3 (2008–2012): make 'the force structure decisions on how we will be bringing down the force to 1700 to 2200 operationally deployed warheads'.

The NPR did not announce any new cuts in the number of US warheads or weapons. The decisions to retire the MX/Peacekeeper ICBM, reduce the SSBN fleet to 14 submarines, dismantle the W62 warhead and denuclearize the B-1B bomber were all made in 1994, during the previous NPR. What will change is the way in which warheads are categorized and counted. The 2001 NPR preserves the existing force structure. 'The drawdown of the operationally deployed strategic nuclear warheads will preserve force structure in that ... no additional strategic delivery platforms are scheduled to be eliminated from strategic service'.

Unlike the counting rules agreed in the 1972 and 1979 Strategic Arms Limitation Treaties (SALT) and the START Treaties, warheads removed from weapon systems in overhaul, such as submarines in dry dock, are not included in the projected level of about 3800 warheads in 2007 or the goal of 1700–2200 warheads by 2012. The Bush Administration has said that it will only count *operationally deployed strategic warheads*.

Overall, the force level set by the NPR reaffirms the force structure planning developed by the US Strategic Command (STRATCOM) nearly a decade ago: the number of operationally deployed strategic warheads will remain at around 2000; the 'triad' of strategic forces will be retained; and the modernization of nuclear forces will continue.

The 'enduring nuclear stockpile'

The about 7600 deployed US strategic and non-strategic nuclear warheads consist of 13 different versions of eight basic warhead designs: the B61-3, B61-4, B61-7, B61-10, B61-11, W62, W76, W78, W80-0, W80-1, B83-1, W87 and W88. Two other warhead types are maintained in reserve: the W84, which previously armed the ground-launched cruise missiles (GLCMs) destroyed under the 1987 Treaty on the Elimination of Intermediate-Range and Shorter-Range Missiles (INF Treaty); and the B83-0 strategic bomb. Except for the W62, which will be dismantled in 2009, these basic warhead designs make up what the US Department of Defense (DOD) refers to

---

[3] US Department of Defense, Transcript of special briefing on the Nuclear Posture Review, 9 Jan. 2002, available at URL <http://www.defenselink.mil/news/Jan2002/t01092002_t0109npr.html>. The Congress-mandated NPR was completed in Dec. 2001 and submitted to Congress on 8 Jan. 2002. A copy of the classified report was leaked to the press in Mar. 2002. Arkin, W., 'Secret plan outlines the unthinkable', *Los Angeles Times*, 10 Mar. 2002, URL <http://www.latimes.com/news/opinion/la-op-arkinmar10.story>. Excerpts of the report are available at URL <http://www.globalsecurity.org/wmd/library/policy/dod/npr.htm>.

as the 'enduring nuclear stockpile' of weapons that will constitute the US nuclear arsenal over the next few decades.

The reliability of the stockpile is certified annually in a joint report to the president from the commander-in-chief of STRATCOM, the Nuclear Weapons Council and the three national nuclear weapon laboratories. Reliability is certified by randomly selecting an average of 99 warheads (11 of each of the eight basic designs and 11 of the non-deployed W84 warheads) each year and physically examining all the critical components; a few are even subjected to 'fatal' experiments that destroy the warhead. After full-scale US nuclear testing was halted in 1992, a new programme of 'sub-critical' testing was started at the Nevada Test Site (NTS) in 1997 to simulate the behaviour of nuclear weapons.[4] The first subcritical test was conducted at the U1A complex at the NTS on 2 July 1997, followed by a second test on 18 September 1997. Three more tests were conducted in 1998, two in 1999, five in 2000 and two in 2001. A new series started on 14 February 2002, when a joint US–British test code-named Vito was conducted at the NTS. The time required to return the NTS to a full-scale nuclear testing capability is 36 months, but the NPR recommended that it be shortened to no more than one year.

The tools developed to certify the reliability of the stockpile without full-scale nuclear testing bring new capabilities to modify and re-manufacture existing warheads and even to develop new ones. Between 1995 and 1997, for example, this capability enabled the development, certification and deployment of the B61-11 earth-penetrating bomb. However, the NPR identified an operational need for a more capable nuclear earth-penetrating capability than that of the B61-11, and the USA is currently studying development of the Robust Nuclear Earth Penetrator by incorporating 'an existing warhead' (probably the B61 or the B83) into a penetration weapon weighing about 5000 lbs (2273 kg). Other modifications to warheads are likely to be made in the refurbishments scheduled for all warheads in the enduring stockpile.

'Phantom arsenals'

Arms control agreements have ignored certain categories of warhead and created notional numbers for others, but the framework for the START III accord that the USA and Russia agreed to in 1997 sought to increase transparency in stockpiles, address non-strategic nuclear warheads and make reductions irreversible. The 2001 NPR reverses this effort by further limiting the warheads that are counted, ignoring non-strategic nuclear warheads and increasing the portion of the total stockpile held in reserve. As mentioned above, of the various categories, the 2012 force level of 1700–2200 warheads includes only 'operationally deployed strategic warheads'.

The NPR continues the trend of removing more warheads from the 'accountable' category and putting them into various unaccountable categories. Before the START I Treaty was signed, about 5 per cent of the total stockpile was in the 'reserve' category. The Administration of President Bill Clinton decided during the 1994 NPR that 'most weapons' removed from active status should not be destroyed but be placed in reserve to maintain a 'hedge' against unforeseeable developments. If START II, with 3500 'accountable' strategic warheads, had been implemented, it was expected that the reserve stockpile would increase to a greater than 1 : 1 ratio with the deployed

---

[4] Although subcritical tests do not produce a nuclear yield, they expose small amounts of plutonium to powerful chemical explosions in order to provide data for physical examination and computer simulation of nuclear warhead performance.

stockpile. As the NPR is implemented over the next decade, the ratio is expected to increase to 1:3. By 2012 there will be more than three times as many warheads in various 'unaccountable' categories as there are operationally deployed warheads.

The NPR creates an entirely new category of unaccountable warheads: the 'responsive force'. This will consist of active warheads that are not on deployed systems but are not considered inactive. Responsive force warheads are kept in secure storage but are available to be returned to the operationally deployed force to meet contingencies. Depending on the particular weapon system, this may take days, weeks, months, or as long as a year or more. For warheads in the inactive category, on the other hand, their limited-life components, such as tritium and neutron generators, have usually been removed. When the weapon is transferred to the active stockpile, these components are reinstalled in the weapon, which is a more time-consuming process.

Strategic bombers

The USA has two types of long-range bomber certified for nuclear missions: the B-2A Spirit and the B-52H Stratofortress. These bombers can deliver either cruise missiles or gravity bombs or a combination of both, but they are not maintained on day-to-day alert. Because the US bomber force is shrinking, only about 430 Air-Launched Cruise Missiles (ALCMs) and 430 Advanced Cruise Missiles (ACMs) are deployed, with several hundred other ALCMs in reserve. Unlike ICBMs and SSBNs, the bomber fleet was de-alerted by President George Bush in 1991, but the bombers continue to exercise their nuclear mission and can be returned to alert status within a few days if so ordered. In addition to front-line US Air Force personnel, Air Force Reserve personnel also take part in nuclear operations.

The B-2A bombers are deployed with the 509th Bombardment Wing at Whiteman Air Force Base (AFB), Missouri, where they are organized in two squadrons: the 393rd squadron, which was declared operational on 1 April 1997, and the 325th, which was activated on 8 January 1998. The first aircraft was delivered on 17 December 1993, and the 21st and final aircraft arrived in 1999. All six aircraft from the original test programme have been modified to full operational capability (FOC), bringing the total number to 21. The B-2A is scheduled to be replaced in about 2040, and a follow-on bomber programme was begun in 1998.

The B-2A is configured to carry various combinations of nuclear and conventional munitions. The nuclear weapons include the B61-7, B61-11 and B83 gravity bombs. Each B-2A can be armed with a load of either B61 or B83 bombs, but not a mix. The B-2A is designated as the 'only' carrier of the new B61-11 earth-penetrating nuclear bomb introduced in November 1997. When the B61-11 was first produced, each aircraft would initially have been forced to load with one or the other B61 version, but in late 1995 STRATCOM issued a new requirement for mixed loads of B61-7 and B61-11 bombs. All the aircraft are being upgraded to Block 30 versions, which can carry all three types of nuclear bomb and an assortment of conventional bombs, munitions and missiles. The upgrade is scheduled to be completed in 2002. Of the 21 aircraft, only 16 are considered primary mission inventory (PMI) aircraft assigned to nuclear and conventional wartime missions.

Modernization of the B-2A bomber is continuing. Like the B-52H, the B-2A is being equipped with the Air Force Mission Support System (AFMSS), a new system used for planning sorties that are part of the Single Integrated Operational Plan (SIOP). Development problems with the AFMSS in 1997 delayed full nuclear certification of the B-2A, but the problems have been resolved and the AFMSS is scheduled to support extremely high frequency (EHF) Satellite Communications (SATCOM) system integration on the B-2A from 2002.[5]

The bat-winged B-2 bomber has been plagued by technical problems, partly because of its sensitive radar-absorbing surface. In March 2002 the Air Force announced that cracks had developed in the titanium plates behind the rear exhausts of 16 of the 21 aircraft. During 2001, the average B-2 was available for combat duty only 31 per cent of the time, half of the US Air Force's goal of 60 per cent.

A total of 94 B-52H bombers are operationally deployed, of which 56 are PMI aircraft. The 1994 NPR recommended retaining only 66 B-52Hs, but the Air Force has decided to undergo a transition to a force of 76 aircraft. The current force of 94 B-52Hs has been consolidated at two bases: the 2nd Bomb Wing at Barksdale AFB, Louisiana; and the 5th Bomb Wing at Minot AFB, North Dakota. The B-52H is scheduled to remain in operation until 2040 and has recently been equipped with the AFMSS to modernize its SIOP mission planning capability. As the only carrier of cruise missiles, the B-52H is referred to by the Air Force as the 'workhorse of nuclear weapons employment'. Each B-52H can carry up to 20 ALCMs/ACMs, with ALCMs carried both internally (up to 8 missiles) and externally (up to 12 missiles); the ACM is only carried externally.

The ALCM, or AGM-86B, is equipped with the W80-1 warhead with a yield of 5–150 kt. Although only about 430 ALCMs are deployed, hundreds of others are held in reserve. Between 1982 and 1986 a total of 1739 ALCMs were produced by Boeing for the US Air Force. The Air Force states that there are approximately 1140 ALCMs in the active inventory.[6] Another 200 ALCMs are kept in inactive storage but could be fully reconstituted within about six months. Long-range Air Force planning envisages an ALCM force of 760 missiles in FYs 1999–2003, and a $134 million Service Life Extension Program (SLEP) is under way to extend the service life of the missile until 2030.

The ACM, or AGM-129A, is equipped with the 5- to 150-kt W80-1 warhead. Compared with the ALCM, the ACM has a significantly longer range (over 2000 nautical miles) and greater accuracy. It also has stealth features to increase its survivability. The ACM is designed to evade air- and ground-based defences and strike heavily defended, hardened targets at any location. There are currently about 400 missiles in the inventory. Originally, as many as 1461 ACMs were planned, but the DOD announced in January 1992 that production would stop at 640 missiles. The current design life expires in 2003–2008, but a SLEP will enter a third phase in 2002 to extend the service of the ACM until 2030.

Although the B-1B Lancer was for years described as a 'conventional-only' aircraft, the Air Force maintained the bomber in 'Nuclear Rerole' status; that is, it could be returned to nuclear missions within months if necessary.[7] Under the Nuclear Rerole Plan, spare B61 and B83 nuclear bombs were maintained in STRATCOM's Active Reserve Stockpile. According to the NPR, the B-1 will no longer be in 'Rerole' status, but the bombs will be retained. Of the original 100 B-1Bs, 92 are left.

Conceptual development of a new strategic bomber to replace the B-1B, B-2A and B-52H began in 1998.

---

[5] See also chapter 11 in this volume.
[6] 'Memo (U), ACC/LGWNA, calendar year 97 cruise missile activity/inventory', 15 June 1997, obtained under the Freedom of Information Act.
[7] See URL <http://www.nautilus.org/nukestrat/USA/bombers/b1rerole.html>.

Intercontinental ballistic missiles

As of January 2002, the USA deployed 550 operational ICBMs of two types: 500 Minuteman IIIs and 50 MX/Peacekeepers. All of the missiles are maintained at a high alert rate (over 98 per cent) and can be launched at short notice. During 2000, for example, the 341st Space Wing at Malmstrom AFB kept 200 Minuteman III missiles on alert 99.74 per cent of the year, a record for the ICBM force. ICBMs provide 'prompt' strike capability but, since the signing of the 1994 US–Russian Agreement to De-target Strategic Nuclear Missiles,[8] target coordinates have normally not been stored in the missiles' on-board guidance system; they can, however, be reloaded within a few minutes. ICBMs carry a total of 1700 warheads of three types.

The 500 operationally deployed Minuteman IIIs are located at three bases: 200 are deployed at Malmstrom AFB, Montana, in four missile squadrons (10th, 12th, 490th and 564th) of 50 missiles each as part of the 341st Space Wing; 150 are deployed at Minot AFB, North Dakota, in three missile squadrons (740th, 741st and 742nd) as part of the 91st Space Wing; and 150 are deployed at F.E. Warren AFB, Wyoming, in three missile squadrons (319th, 320th and 321st) as part of the 90th Space Wing. In addition to the 500 operational Minuteman III missiles, there are 107 missiles for spares, operational testing and evaluation, ageing and surveillance. The destruction of silos at Ellsworth AFB, South Dakota, and Whiteman AFB—two bases that once deployed Minuteman II ICBMs—was completed by December 2001, in accordance with the START I Treaty.

Total Minuteman III warhead loading dropped from 1500 to 1200 warheads in 2001. The change was needed to meet the limits on force levels set by the START I Treaty and involved downloading the 150 missiles in the 90th Space Wing from three W62 warheads to one each. The remaining 350 Minuteman missiles still carry three warheads each, but all Minuteman III missiles will be downloaded to single-warhead configuration by 2007 in order to meet the schedule set by the NPR.

An extensive modernization of the Minuteman missile force continues under a $5.5 billion five-part programme intended to improve the accuracy and reliability of the weapon and extend its service life beyond 2020.

1. The missile alert facilities (Launch Control Centers) were equipped with Rapid Execution and Combat Targeting (REACT) consoles in 1996, reducing by 50 per cent the time it takes to target the missiles. REACT is scheduled to undergo a $55 million SLEP in 2002–2005 to correct deficiencies.

2. The Guidance Replacement Program (GRP) is replacing the current NS-20 guidance system with the new NS-50 system, improving accuracy and extending service life, at a cost of $1.9 billion. The GRP was initiated in August 1993. Initial production began in March 1998 and full-rate production was achieved in December 1999. Eight annual contracts for 652 NS-50 guidance sets (a total value of $1.3 billion) are planned to be awarded by 2008. The new guidance system was expected to increase the accuracy of the Minuteman III to nearly that of the current MX ICBM—a circular error probable (CEP) of 100 metres—but it may not achieve that goal, according to DOD progress reports. Despite these setbacks, installation of the NS-50 continues. The initial operational capability (IOC) was achieved on 20 July 2000 at Malmstrom

---

[8] This agreement was contained in the Moscow Declaration of the US and Russian presidents, signed on 14 Jan. 1994, available at URL <http://www.fas.org/nuke/control/detarget/docs/940114-321186.htm>.

AFB, when the first 10 sets on operational Minuteman III missiles (plus four spares) exceeded the 30-day on-alert requirement.

3. The Propulsion Replacement Program (PRP) involves re-pouring the first and second stages and re-manufacturing the third stage of the Minuteman III missile; it incorporates the latest solid-propellant and bonding technologies and replaces obsolete or environmentally unsafe materials and components. Nine missiles were scheduled to undergo propulsion replacement in FY 2001, followed by 33, 86 and 96 missiles in the subsequent three years.

4. The Propulsion System Rocket Engine SLEP involves refurbishing the fourth, post-boost, liquid propulsion stage of the Minuteman III.

5. The Safety Enhanced Reentry Vehicle (SERV) Program, scheduled to begin in 2002, will replace all the remaining W62 and some W78 warheads on Minuteman IIIs with the newer W87 warhead from deactivated MX ICBMs. More than $250 million is earmarked for SERV to 2006 to design, develop and test the modifications needed to implement this programme. Flight testing will occur before the new Minuteman III/W87 is deployed. The present authors estimate that the 150 missiles at Warren AFB and the 50 at Malmstrom AFB will be equipped with the W87, while the other 150 Minuteman IIIs at Malmstrom and the 150 at Minot AFB will retain the W78.

The first experimental launch of a combined GRP/PRP Minuteman III ICBM took place on 13 November 1999, from Vandenberg AFB, California, to the Kwajalein Missile Range in the Pacific Ocean. Normally, three full-scale test launches are conducted each year, but four took place in 2000. One missile was launched in February 2001, and two test launches were scheduled for September but were cancelled after the terrorist attacks of 11 September 2001. Test launches for 2002 were scheduled for February, June and September.

All of the 50 MX/Peacekeeper ICBMs are currently operational. The Bush Administration plans to deactivate the weapon system in phases over a three-year period beginning on 1 October 2002. Withdrawal will occur in conjunction with the introduction of the Trident II missile into the Pacific-based submarine fleet. Current plans call for the MX silos to be retained, rather than destroyed as specified in the SALT and START treaties. MX booster stages will also be retained for potential use as space launch or target vehicles. The majority of the W87 warheads will arm Minuteman III ICBMs and the balance will be placed in the 'responsive force'.

Studies are under way to consider the acquisition of a new ICBM to be ready in 2018. Among the new ICBM capabilities which the DOD says it needs are extended range and the ability to hit re-locatable, hard and deeply buried targets.

Ballistic missile submarines

Eighteen Ohio Class (or Trident) submarines constitute the current US SSBN fleet. The US administration plans to cut the number to 14 by FY 2007 (of which two will be in overhaul at any given time and will not be counted as part of the 'operationally deployed force'). The four oldest SSBNs (*Ohio*, *Michigan*, *Florida* and *Georgia*) will be converted to carry up to 154 conventional SLCMs each and may also be used to support the Special Operations Forces, although the submarines remain accountable under the START I Treaty. To balance the future 14-submarine fleet between the Atlantic and Pacific oceans, the home port of three submarines may be moved from

Kings Bay, Georgia, to Bangor, Washington, beginning in 2002, establishing a seven-submarine force on each coast.

The US Navy has extended the Trident hull life to 44 years. The first of the 14 SSBNs that will remain in service is scheduled for retirement in 2029. The DOD is currently studying two options for a new SSBN that would be introduced in 2029. The first is a variant of the Virginia Class nuclear-powered submarine (SSN). The second is a dedicated SSBN, either a new design or a derivative of the Trident. The new project would begin in 2016.

Trident SSBNs carry two types of SLBM. Seven Pacific-based submarines carry the Trident I (C-4) missile and 10 Atlantic-based submarines carry the Trident II (D-5) missile. There is also one newly converted Trident II SSBN at Bangor, the USS *Alaska*, which completed its refit in November 2001. *Alaska* is expected to conduct its first Trident II test launch in the spring of 2002 but is already counted as a Trident II SSBN under the START I Treaty. The other three SSBNs scheduled for Trident II refit are, in order of their conversion, *Nevada* (SSBN-733), *Henry M. Jackson* (SSBN-730) and *Alabama* (SSBN-731).

Although the Trident I missile is being retired it is still being flight tested. On 9 December 2001 the *Ohio* launched a salvo of four Trident Is. A total of 570 Trident I missiles were produced between 1976 and 1986, and 222 missiles have been launched in 117 different flight test events. Each event has involved firing from one to four missiles. Of the 222 attempted launches, 188 were successful, while the remaining 34 failed for various reasons. Until the early 1990s, Trident I flight tests were carried out in both the Atlantic and Pacific oceans, but since 29 July 1993, after the last Trident I test was conducted at the Pacific Test Range, all SLBM flight tests have been at the Atlantic Test Range off the coast of Florida.

Procurement of the Trident II (D-5) continues at a rate of 12 missiles per year. A total of 384 Trident II missiles were purchased by 2001. As a result of the upgrading of four Trident I-equipped SSBNs and the extension of the service life of the submarines from 30 to 44 years, the total number of Trident II missiles to be procured will increase from 390 to 568, at an additional cost of $12.2 billion. The total cost of the programme is now $37 543.9 billion, or $66 million per missile. To arm the submarines throughout their entire life, existing Trident II missiles will be upgraded to a new variant called the D-5A. Of the 568 Trident IIs, 288 will arm 12 operational SSBNs (with another two in overhaul at any given time), 48 will be held in reserve to arm the two submarines in overhaul, while the remaining missiles will be expended in flight tests.

Four Trident II (D-5) missiles were test launched from two SSBNs in 2001. Since January 1987, 116 Trident II missiles have been expended in 72 test-launch events. Each event may launch from one to four missiles. Compared to the performance of the Trident I programme, the Trident II programme has been extraordinarily successful. Of the 116 missiles launched, only five failed or did not work, and since December 1989 the programme has accomplished a record of 94 consecutive successful launches, making the Trident II the most reliable strategic nuclear missile ever built. Despite this proven reliability, the DOD has said that the current level of flight tests, which is set by the Strategic Command, is the 'minimum acceptable to meet weapon system reliability requirements'.[9] STRATCOM's analysis suggests that it may be necessary to increase flight test requirements in the future.

---

[9] US Department of Defense, Office of the Secretary of Defense, 'Report on the D5 missile program for the Committees on Armed Services of the Senate and House of Representatives', 12 Jan. 2000, p. 7,

The US Department of State declared in December 2001 that the SSBN force carried a total of 3120 warheads, a reduction from 3456 warheads in 2000. The reduction was necessary to comply with the warhead limit set by the START I Treaty and involved downloading all Trident I (C-4) SLBMs from eight to no more than six warheads each. To meet the reductions in 'operationally deployed strategic forces' for 2012 there will be further SLBM downloading after 2007.

The SLBMs can carry two types of re-entry vehicle (RV) and warhead: either the Mk-4 with the W76 warhead or the Mk-5 with the W88 warhead. The W76/Mk-4 is by far the more numerous, with as many as 2736 warheads deployed on 16 submarines. Since its construction began in 1976, Lockheed Martin Missile and Space Operations has manufactured more than 5000 Mk-4 re-entry body assembly kits for the US and British navies. Refurbishment of the W76 is scheduled to begin in 2007 in order to ensure that the W76/Mk-4 re-entry body can support SSBN operations until 2040.

The Mk-5 RV carries the W88 warhead, the most powerful missile warhead in the US arsenal. W88 warhead production ceased after the Rocky Flats Plant closed in 1989 after 400 warheads had been completed. Although President Bush announced in February 1992 that no more W88s would be built, the 400 warheads were insufficient to support pit surveillance activities. Small-scale production of plutonium pits for the W88 therefore resumed at the TA55 facility at Los Alamos National Laboratory in 2000 to replenish the W88 pits destroyed in reliability testing. The first 'war reserve' pits are scheduled to enter the stockpile in 2007, and the current plan for the TA55 facility is to produce 20 pits per year by 2007. The facility cannot meet the increased warhead refurbishment requirement set by the NPR, however, so development of a Modern Pit Facility has begun.

In October 2003 the Navy is expected to deploy a new SLBM Retargeting System (SRS) after more than a decade in development. The SRS will enable SSBNs 'to quickly, accurately, and reliably retarget missiles to targets' and 'to allow timely and reliable processing of an increased number of targets'. The operational requirement for the SRS dates from October 1989, and the system will 'reduce overall SIOP processing' time and 'support adaptive planning'. SSBNs at sea will have a greater capability to attack fixed and mobile sites.[10]

Design of a new SLBM warhead is under way in the Navy's SLBM Warhead Protection Program (SWPP). The SWPP maintains the capability to develop replacement nuclear warheads for both the W88/Mk-5 and the W76/Mk-4. One design is described as 'near-term' and the other as 'long-term'.

Non-strategic nuclear weapons

The USA retains about 1120 non-strategic nuclear weapons, consisting of 800 B61 gravity bombs of three modifications and 320 Tomahawk Land-Attack Cruise Missiles (TLAM/Ns), a portion of which are in reserve or inactive. Although the number of non-strategic nuclear weapons has declined dramatically compared with the number in the cold war period and may change further in the future, the NPR announced no new reductions.

---

released under the Freedom of Information Act, available at URL <http://www.nautilus.org/nukestrat/USA/subs/index.html>.

[10] US Department of Defense, 'Summary explanation of significant SAR cost changes (as of December 31, 2001)', Apr. 2002, URL <http://www.defenselink.mil/news/Apr2002/d20020411changes.pdf>.

Table 10A.3. US nuclear forces, January 2002

Type	Designation	No. deployed	Year first deployed	Range (km)[a]	Warheads x yield	Warheads
Strategic forces						
Bombers[b]						
B-52H	Stratofortress	93/56	1961	16 000	ALCM 5–150 kt	430
					ACM 5–150 kt	430
B-2	Spirit	21/16	1994	11 000	Bombs	800[c]
Subtotal		*114/72*				*1 660*
ICBMs						
LGM-30G	Minuteman III					
	Mk-12	50	1970	13 000	3 x 170 kt	150
		150			1 x 170 kt[d]	150
	Mk-12A	300	1979	13 000	3 x 335 kt	900
LGM-118A	MX/Peacekeeper	50	1986	11 000	10 x 300 kt	500
Subtotal		*550*				*1 700*
SSBNs/SLBMs						
UGM-96A	Trident I (C-4)[e]	168	1979	7 400	6 x 100 kt	1 008
UGM-133A	Trident II (D-5)	264				
	Mk-4		1992	> 7 400	8 x 100 kt	1 728
	Mk-5		1990	> 7 400	8 x 475 kt	384
Subtotal		*432*				*3 120*
Strategic subtotal						*6 480*
Non-strategic forces						
B61-3, -4, -10 bombs		..	1979	..	0.3–170 kt	800[f]
Tomahawk SLCM		325	1984	2 500	1 x 5–150 kt	320[g]
Non-strategic subtotal						*1 120*
Total						**7 600**[h]

[a] Range for aircraft indicates combat radius, without in-flight refuelling.

[b] The first figure in the *No. deployed* column is the total number of B-52Hs in the inventory, including those for training, testing and back-up. The second figure is the PMI (primary mission inventory) aircraft, i.e., the number of operational aircraft assigned for nuclear and conventional wartime missions.

[c] Available for both the B-52H and the B-2A.

[d] Each of the 150 Minuteman III missiles of the 90th Space Wing at F.E. Warren AFB has been downloaded from 3 to 1 W62 warhead.

[e] The Trident I (C-4) missiles carried on 7 SSBNs based in the Pacific Ocean have been downloaded from 8 to no more than 6 warheads each to meet the START I warhead ceiling. Three of these SSBNs will be upgraded to carry the Trident II (D-5) and the remaining 4 will be converted to non-nuclear missions. By 2006, all US SSBNs will carry the Trident II.

[f] Approximately 150 of these are forward deployed in 7 European countries. Almost 500 more have been transferred to the inactive reserve.

[g] The TLAM/N is no longer deployed with the fleet but is stored on land.

[h] Another 370 warheads are spares, and 2700 additional intact warheads are kept in the reserve stockpile.

Sources: US Department of Defense, various budget reports; START I Treaty MOUs, Sep. 1990, 5 Dec. 1994, 1 July 1995, 1 Jan. 1996, 1 July 1996, 1 Jan. 1997, 1 July 1997, 1 Jan. 1998, 1 July 1998, 1 Jan. 1999, 1 July 1999, 1 Jan. 2000, 1 July 2000, 1 Jan. 2001, 1 July 2001 and 1 Apr. 2002; US Department of State; Cohen, W., Secretary of Defense, *Annual Report to the President and the Congress* (US Department of Defense: Washington, DC, Jan. 2001), pp. 89–99; International Institute for Strategic Studies, *The Military Balance 2001/2002* (Oxford University Press: Oxford, 2001); US Senate Committee on Foreign Relations, START II Treaty, Executive Report 104-10, 15 Dec. 1995; US Navy, personal communication; US Department of Defense, various documents obtained under the Freedom of Information Act; Natural Resources Defense Council, 'NRDC Nuclear Notebook', *Bulletin of the Atomic Scientists*, various issues; and Authors' estimates.

An ample supply of nearly 1300 B61 non-strategic nuclear bombs exists for various US and European NATO aircraft. Most of the bombs are stored at Kirtland AFB, New Mexico, and Nellis AFB, Nevada, for delivery by F-16C/D Fighting Falcon and F-15E Strike Eagle aircraft, with a small number deployed in Europe (see below). The F-117A Nighthawk is also considered nuclear-capable but is normally not listed in the US Air Force budget for nuclear weapons support and is maintained at a lower level of nuclear readiness than the other aircraft. In 1992 the Air Combat Command recommended de-nuclearization of the F-117A to free resources for training and on-board computer capacity, but the Air Force intervened and decided to maintain the platform in a nuclear-capable configuration. The DOD is considering whether to extend the life of the dual-capable F-16s and F-15Es or to make a block upgrade to the Joint Strike Fighter (JSF). The JSF is being designed to permit future nuclear capability after it enters service in 2012.[11]

Approximately 150 B61 bombs remain forward deployed at 10 airbases in seven European NATO states. The Weapons Storage and Security System (WS3) used to store the weapons at these locations was installed between 1990 and 1998, and plans are under way to modernize the WS3 before 2005 to maintain the system for another decade. A service-life extension study for the B61 began in 1999. The aircraft of NATO countries that are assigned nuclear missions include US-supplied F-16 aircraft and German and Italian Tornado bombers. Several NATO countries that are currently assigned strike missions with US nuclear bombs are considering purchase of the JSF.

All of the about 320 TLAM/Ns (with W80-0 warheads) were removed from their previous storage areas at Naval Air Station (NAS) North Island in San Diego, California, and Naval Weapon Station (NSW) Yorktown in Norfolk, Virginia, and are now stored at the Strategic Weapons Facilities, with strategic weapons for the SSBNs. NWS Yorktown was de-certified in August 1997 after its complement of TLAM/Ns was shipped to the Strategic Weapons Facility Atlantic at Kings Bay, which was first certified to receive the missiles in April 1997. NAS North Island's nuclear certification expired in April 1998 after all of its TLAM/Ns had been airlifted to the Strategic Weapons Facility Pacific in Bangor, Washington.

As a result of the 1994 NPR, surface vessels are no longer equipped to carry nuclear-armed Tomahawk missiles. However, the option to redeploy them on attack submarines was retained. While most US attack submarines were credited with some nuclear capability during the cold war, most SSNs today do not have nuclear missions. In the Pacific Fleet, for example, less than half of the number of attack submarines regularly undergo nuclear certification. The reduced nuclear requirement is further illustrated by the fact that SSNs which pass inspection are subsequently de-certified to save resources for more urgent non-nuclear responsibilities. If the order is given to do so, however, TLAM/Ns can be redeployed in only 30 days. To

---

[11] For more on the Joint Strike Fighter, see chapter 8 in this volume.

ensure training and force integration, TLAM/N operations are now included in STRATCOM's annual 'Global Guardian' nuclear exercises.

As directed in the 2001 NPR, the DOD will evaluate the future of the TLAM/N and decide whether to replace, retire, or retain and enhance the missile.

Nuclear command and control

The 2001 NPR calls for improved command, control, communication and intelligence (C3I) systems. The measures include expansion of the current architecture to 'a true national command and control conferencing system' that would supplement the programmes that were under way before the NPR was completed.

Currently, the EHF system on the Milstar satellites is scheduled to take over the nuclear command and control function from the Defense Satellite Communications System (DSCS) in 2003. Development is also under way of a constellation of Advanced Extremely High Frequency (AEHF) Military Satellite Communications (MILSATCOM) satellites to replenish the existing Milstar satellites and provide additional capabilities.[12] The first AEHF, called Pathfinder, is scheduled to be launched in December 2006, and three AEHF spacecraft are planned to achieve an IOC in FY 2008. According to the NPR, the purpose of these satellites is to 'provide nuclear survivable (e.g., against high altitude electromagnetic pulse), anti-jam, low and medium data rate communications to strategic and tactical users'. The NPR identifies a replacement satellite for the AEHF, the Advanced Wideband System (AWS), which is scheduled to be launched in FY 2009.

To integrate all of these command and control capabilities, the MILSATCOM Terminals programme is developing equipment, at a cost of more than $2.3 billion, that will enable users to communicate via Milstar, AEHF, Ultra High Frequency (UHF), Wideband Gapfiller System (WGS), Defense Satellite Communications System (DSCS) and other military satellites, as well as commercial satellites, to support Aerospace Expeditionary Force requirements and maintain essential strategic connectivity for nuclear forces.

In FY 2003 the DOD will also initiate an Extremely High Frequency (EHF) communications satellite programme 'primarily for national and strategic users requiring nuclear protected communications in the mid-latitude and polar regions'.[13] The first satellite is scheduled to be launched in FY 2009. The polar capability will complement the Navy Polar EHF satellites currently being deployed, which are designed to provide *inter alia* nuclear command and control in high-latitude areas. The first operational test of Navy Polar EHF was conducted with the attack submarine USS *Scranton* (SSN-756) during a deployment to the Arctic Ocean in June 2001.

Extensive modernization of nuclear command and control aircraft is also under way. A fleet of 16 E-6B TACAMO ('Take Charge And Move Out') aircraft serve as the primary relay stations for Emergency Action Messages (EAM) from the National Command Authority (NCA) to SSBNs at sea. TACAMO, which is also known as the Airborne Launch Control Center (ALCC), can—under restricted conditions—launch any missile in the Minuteman ICBM force. Additional modernization is under way to transfer the Air Force EC-135 Airborne National Command Post (ABNCP) to the E-6B, thereby consolidating command and control of all strategic forces in a single

[12] See also chapter 11 in this volume.
[13] US Department of Defense (note 3), p. 27.

airborne platform. When the modernization programme has been completed, TACAMO's improved ability to relay EAMs from the NCA to strategic forces will permit the commander-in-chief of the US Strategic Command to directly execute command and control of those forces. An IOC was achieved on 1 October 1998, with the implementation of SIOP-99, and an FOC is scheduled for October 2003, coinciding with the entry into force of the SIOP-04 war plan.

III. Russian nuclear forces

Russia was estimated to have an arsenal of 8331 operational nuclear warheads, consisting of 4951 strategic and 3380 non-strategic and air defence warheads, at the beginning of 2002. The primary changes from 2001 involved the further decrease in the number of MIRVed (equipped with multiple, independently targetable re-entry vehicles) ICBMs and SSBNs, which reduced the total number of deployed strategic warheads from 5606 in 2001 to 4951 in 2002. The number of SS-18 ICBMs and Typhoon and Delta IV SSBNs removed from service is particularly notable. The number of non-strategic nuclear weapons deemed operational declined slightly from 3590 in 2001 to 3380 in 2002.

The number of deployed strategic nuclear weapons is lower than the 5520 START I-accountable warheads attributed to Russia in December 2001, after the force reductions mandated by the START I Treaty had been completed. This is because START I counts launchers which have been deactivated or otherwise removed from service as remaining deployed with their associated warheads until several further steps have been taken to eliminate or convert them.

In 2000 President Vladimir Putin announced that Russia was interested in reductions to 1500 or fewer strategic warheads, and resources began to be shifted from nuclear to conventional forces—both of which underscored the likelihood that Russia's strategic forces will continue to decline. In June 2001 the Strategic Rocket Forces (SRF), long the leading service in the Soviet and Russian armed forces, was downgraded to a branch of the armed forces, and there were some indications that SRF troops would be subordinated to the Russian Air Force. Further decreases in the missile forces are expected; an aide to the commander-in-chief of the Russian Strategic Rocket Forces said in an interview on Russian radio that the number of ICBMs could be reduced to some 500 over the next few years. The number of missile units would be halved as well.[14]

Strategic aviation

Strategic bombers are part of the Russian Air Force's 37th Air Army. According to the 31 January 2002 START I Treaty Memorandum of Understanding (MOU), Bear bombers are deployed at the following airbases (ABs): the Bear-H16—16 at Ukrainka AB in Siberia (79th Heavy Guard Bomber Regiment), 13 at Engels AB (121st Heavy Bomber Regiment) and 2 at Ryazan AB; and the Bear H-6—25 at Ukrainka, 5 at Engels and 2 at Ryazan.

According to the 31 January 2002 START I MOU, 15 Blackjacks are based at Engels AB. Eight of these bombers were transferred from Ukraine to Russia in late

---

[14] 'Russian strategic missile forces to be cut by a third by 2006—aide to commander', BBC Worldwide Monitoring, 25 Dec. 2001.

1999 and early 2000 in exchange for partial payment of Ukrainian natural gas debts to Russia. The operational status of the bombers transferred from Ukraine is unclear, however, since they reportedly require moderate to extensive overhaul and modernization.[15] One of these is a new bomber delivered in May 2000 by the Kazan Gorbunov production plant. The Tu-160 force may increase slightly in the coming years. Although there was a lack of funding in 2001, three more Blackjacks are under construction, one of which may be delivered by late 2002 or early 2003.[16] The Tu-160s are to undergo modernization to extend their service lives and, according to Air Force Commander-in-Chief Vladimir Mikhailov, to allow them to carry 'new types of missiles with conventional and nuclear warheads'.[17]

The larger force led to the creation of a new unit for the Tu-160s, the 22nd Donbass Guard Heavy Bomber Aviation Division (Tu-160s had operated as part of the 121st Heavy Bomber Regiment).

In exercise activities, on 14 February 2001 a pair of Tu-160 Blackjack bombers flew along Norway's northern border and some four medium-range Tu-22 Backfire bombers flew near Japanese airspace. This resulted in Norway dispatching interceptor aircraft and Japan lodging a protest over possible violation of its airspace. On 16 February a Tu-95 Bear-H bomber launched a strategic cruise missile and two Tu-22M Backfire bombers launched non-strategic cruise missiles as part of the same general military exercise, which also included the ICBM and SLBM launchers listed below. A large Pacific area air exercise which was to involve Tu-160, Tu-95 and Tu-22 strategic and theatre bombers started on 10 September. Blackjack bombers were spotted at the Anadyr AB, and additional US and Canadian interceptors had been moved to the area to monitor the exercise. However, the Russian Defence Ministry cancelled the exercise after the 11 September 2001 attacks at the request of the US Government, which wanted to ensure that there would be no incidents involving aircraft flying towards the USA.

Intercontinental ballistic missiles

SS-18s. The September 1990 START I MOU states that 204 SS-18s were deployed in Russia (30 at Aleysk, 64 at Dombarovskiy, 46 at Kartaly and 64 at Uzhur). Another 104 were deployed at two basing areas in Kazakhstan. The START I Treaty called for the number of warheads on heavy ICBMs to be reduced to 1540. This meant that the number of SS-18s was to be reduced by half by the end of 2001, the date of START I final implementation.

Russia has more than fulfilled its START I obligations. The SS-18 missiles in Kazakhstan and their warheads had been shipped back to Russia by April 1995. In Russia, 60 SS-18s have been removed from service, leaving 144 (52 at Dombarovskiy, 46 at Kartaly and 46 at Uzhur; the division at Aleysk was disbanded in 2001 and its silos destroyed). The START II Treaty banned all MIRVed heavy ICBMs, although up to 90 SS-18 silos may be converted for deployment of single-warhead ICBMs. However, since the START II Treaty has not entered into force and

---

[15] Nikolayev, Y., '"Black Jacks" sent for reforging', *Izvestiya*, 19 Jan. 2002 (in Russian).
[16] AVN Military News Agency, 'Russian Air Force to commission new Tu-160 strategic bomber', 28 Mar. 2002; Interfax, 'Russian Air Force to get 3 Tu-160 bombers', 29 Mar. 2002; and AVN Military News Agency, 'Russia plans strategic aviation overhaul', 2 Jan. 2002.
[17] AVN Military News Agency, 'Russia starts modernization of Tu-160 strategic bombers', 5 Apr. 2002.

its future is in doubt, Russia may retain MIRVed SS-18s, although some may be retired by the middle of this decade, if not before, because of their age. Two variants of the SS-18 have been deployed—the older RS-20B and the newer RS-20V. Under START all SS-18s are counted as carrying 10 warheads, but the RS-20B variant can carry a single warhead and a few of these may be deployed. The range of fully loaded SS-18s is 11 000 km. Single-warhead missiles have a range of 15 000 km. Warheads on the RS-20B have yields of 500–550 kt and, on the RS-20V, 550–750 kt.

SS-19s. According to the September 1990 START I Treaty MOU, 170 SS-19s were in Russia (60 at Kozelsk and 110 at Tatishchevo). Another 130 were at two basing areas in Ukraine (the Ukrainian missiles were taken out of service by mid-1996). A November 1995 Ukrainian–Russian agreement included the sale of 32 SS-19s, which had been stored in Ukraine, back to Russia. Some SS-19s in Russia are being withdrawn from service (33 so far, at Tatishchevo) to allow space for new SS-27 missiles, which are deployed in SS-19 silos. Under START II, Russia may keep up to 105 SS-19s downloaded to a single warhead (they currently carry six warheads). Because START II has not entered into force, Russia may retain MIRVed SS-19s, although their numbers may decrease later in the decade because of ageing.

SS-24s. According to the December 1994 START I MOU, 46 SS-24s were in service in Russia—10 silo-based and 36 rail-based. Another 46 were in Ukraine (the Ukrainian missiles were taken out of service by mid-1996; the last silo for housing them was destroyed in 2001). The 10 silo-based SS-24 M2s deployed at Tatishchevo were removed from service in 2000 to accommodate deployments of new SS-27 silo-based missiles. Thirty-six rail-based SS-24M1s were at garrisons at Bershet, Kostroma and Krasnoyarsk. There are plans to remove them from service.

SS-25s. In Russia, the road-mobile, single-warhead SS-25 missile system is known as Topol. When the Soviet Union was dissolved, a number of SS-25s were left in Belarus. By 1997 the last of these missiles and their warheads had been shipped back to Russia. There are 360 SS-25s deployed at 10 basing areas in Russia. The deployment of new regiments of SS-25s (nine missiles each) ended by 1997 as Russia shifted to producing and deploying the follow-on to the SS-25—the Topol-M, or the SS-27, as it is designated by the US Government.

SS-27s. Flight testing of the SS-27 began on 20 December 1994. Two silo-based SS-27s were put on 'trial service' in December 1997 at the Tatishchevo missile base near Saratov in south-western Russia. One regiment of 10 missiles was declared operational in December 1998 and a second regiment with another 10 missiles in December 1999. A third regiment was activated in late December 2000, but with only four missiles out of the planned 10 because of a cut in the anticipated funding for 2000. Another five missiles were deployed in 2001, bringing the total deployed to 29. The SS-27s are housed in former SS-19 and SS-24 silos at Tatishchevo.

In 1998 the Strategic Rocket Forces intended to deploy 20–30 new SS-27 Topol-M missiles per year over the next three years and 30–40 per year for three years after that, but deployments have fallen far short of this schedule. Only six more missiles may be deployed in 2002 and only 50–60 by the end of 2005, considerably fewer than the 160–220 previously anticipated.

At least five ICBM launches took place in 2001: on 16 February, an SS-25 missile was launched from Plesetsk; on 27 June, an SS-19 missile was launched from Baikonur; on 3 October, a training launch of a 15-year-old SS-25 missile was conducted from Plesetsk; on 26 October, a more than 25-year-old SS-19 missile was launched from Baikonur (to test the ability of Russia to download warheads on

Table 10A.4. Russian SSBN and SSN/SSGN patrols per year, 1991–2001

Patrols	1991	1992	1993	1994	1995	1996	1997	1998	1999	2000	2001
SSBNs	37	28	19	19	14	12	13	11	7	6	1
SSNs/SSGNs	18	9	13	14	13	14	11	13	9	3	1
Total	**55**	**37**	**32**	**33**	**27**	**26**	**24**	**24**	**16**	**9**	**2**

Source: US Navy, Office of Naval Intelligence.

SS-19s under START II and to confirm the reliability of the SS-19's service life extension); and on 1 November, an SS-25 missile was test fired from Plesetsk.

Strategic submarines and SLBMs

The September 1990 START I MOU listed 62 SSBNs. At the end of 2001, only 14 were thought to remain operational: 6 Delta IIIs, 6 Delta IVs and 2 Typhoons. All Yankee, Delta I and Delta II SSBNs have been withdrawn from operational service. Of the original six Typhoon submarines, one was being scrapped in 2001, another was being prepared for scrapping and two more appear to be out of service. Unless further funding is found or a replacement for the ageing SS-N-20 missiles is developed, the remaining Typhoons are likely to be retired. Of the original 14 Delta IIIs, seven have been removed from service and one has been converted to a deep submergence rescue vehicle (DSRV) carrier. Of the original seven Delta IVs, one has been removed from service.[18] In 1999, in order to keep the remaining Delta IVs in service, it was decided to restart the SS-N-23 production line. There are reports that a new variant of this SLBM is being considered to carry 10 warheads instead of four.[19] Steps are also being taken to extend the service life of the deployed SS-N-23s. Operational SSBNs in the Northern Fleet are based on the Kola Peninsula (at Nerpichya and Yagelnaya) and in the Pacific Fleet (at Rybachiy, 15 km south-west of Petropavlovsk) on the Kamchatka Peninsula.

The keel of the first of a new Borey Class SSBN was laid in November 1996. Construction has been intermittent and was suspended altogether in 1998 while the submarine was being redesigned to accommodate a new SLBM. The Russian Navy intends to have the first boat in commission in 2005 or shortly thereafter, but it is unclear whether there will be adequate funding to finish it by then.

Combat training launches of SLBMs in 2001 included the following. On 16 February, the Delta IV Class Northern Fleet submarine K-407 *Novomoskovsk* launched an SS-N-23 from the Barents Sea; on 5 June, a Northern Fleet SSBN launched an SLBM (type not reported); on 18 September, the Delta III Class Pacific Fleet submarine K-223 *Podolsk* launched an SS-N-18 SLBM from the Sea of

---

[18] A succession of Delta IVs has been and will be in refit at the submarine shipyard in Severodvinsk, lowering at least by 1 the number available for deployment. E.g., the *Yekaterinburg* was undergoing repairs from 1996 to 2002.

[19] Krutikov, Y. and Safonov, D., 'A missile in somebody else's eye', *Izvestiya*, 19 June 2001; and Golotyuk, Y., 'Ten warheads are better than three', *Vremya Novostei*, 16 Mar. 2001.

Okhotsk; and on 18 October, the Typhoon Class Northern Fleet submarine TK-20 *Severstal* launched two SS-N-20 SLBMs from the White Sea before returning to its base at Nerpichya in November.

Economic problems, a shrinking SSBN fleet and perhaps safety concerns in the aftermath of the sinking of the *Kursk* in August 2000 have led to a large decrease in the number of SSBN patrols—along with nuclear-powered general-purpose submarine (SSN/SSGN) patrols—from 1991 to 2001, as shown in table 10A.4.[20]

The Russian nuclear stockpile and non-strategic nuclear weapons

Estimating the size of the Russian nuclear arsenal, and specifically the non-strategic nuclear weapon arsenal, is difficult. Some 30 000 nuclear weapons, plus or minus several thousand, may have been in the Soviet arsenal in 1991. Estimates of the dismantlement rates of Russian warheads vary from several hundred to 1000–2000 per year. US Defense Department and Central Intelligence Agency (CIA) estimates suggest that Russia dismantled slightly more than 1000 warheads per year during the 1990s; that is, more than 10 000 have been dismantled since 1991.[21] If so, the remaining arsenal may contain some 20 000 nuclear weapons, plus or minus several thousand. Approximately 5000 of these may be deployed on strategic nuclear weapon systems, while some 3400 may be non-strategic weapons kept for operational use by the Russian Navy and Air Force, including those for air defence. The remainder are non-strategic and strategic weapons in storage, some or all of which are destined for dismantlement.

In October 1991 and January 1992, as part of the US–Russian Presidential Nuclear Initiatives, Russia announced that it would take several unilateral steps to withdraw and eliminate some non-strategic nuclear weapons. With regard to the Russian Navy, non-strategic nuclear weapons were removed from surface ships and submarines and placed in regional or central storage sites. Nuclear weapons deployed on naval aircraft, or at front-line storage facilities servicing naval airbases, were also placed in regional or central storage sites. One-third of the Navy's non-strategic nuclear weapons were eliminated by 1996. The number of ships capable of carrying nuclear weapons has declined from about 400 in 1990 to about 100 in 2001.

With regard to the Russian Ground Forces, all nuclear weapons are thought to have been withdrawn from operational forces by 1998 and consolidated at regional or central storage sites. Although final elimination of Ground Forces nuclear weapons was expected in 2000–2001, Russia announced in April 2002 that the destruction of nuclear warheads for tactical missiles, nuclear artillery shells and nuclear mines continues. If there is sufficient funding, Russia will finish eliminating all Ground Forces nuclear weapons by 2004. With regard to the Air Force, half of its inventory of nuclear air-bombs has been eliminated. Half of the warheads for surface-to-air missiles were also destroyed. In 1992, President Yeltsin declared that production of nuclear warheads for ground-launched tactical missiles, nuclear artillery shells and nuclear mines had recently been halted. In April 2002, the Russian Government

---

[20] Some SSBNs may be ready to launch their SLBMs while in port.
[21] For further detail about Russia's nuclear warhead dismantlement activities as well as attendant changes in the size and composition of the Russian nuclear weapon complex since the end of the cold war see appendix 10C.

Table 10A.5. Russian nuclear forces, January 2002

Type	NATO designation	No. deployed	Year first deployed	Range (km)a	Warheads x yield	Warheads
Strategic offensive forces						
Bombers						
Tu-95MS6	Bear-H6	32	1984	6 500–10 500	6 x AS-15A ALCMs, bombs	192
Tu-95MS16	Bear-H16	31	1984	6 500–10 500	16 x AS-15A ALCMs, bombs	496
Tu-160	Blackjack	15	1987	10 500–13 200	12 x AS-15B ALCMs or AS-16 SRAMs, bombs	180
Subtotal		78				*868*
ICBMs						
SS-18	Satan	144	1979	11 000–15 000	10 x 500–750 kt	1 440
SS-19	Stiletto	137	1980	10 000	6 x 500–750 kt	822
SS-24 M1	Scalpel	36	1987	10 000	10 x 550 kt	360
SS-25	Sickle	360	1985	10 500	1 x 550 kt	360
SS-27	..	29	1997	10 500	1 x 550 kt	29
Subtotal		706				*3 011*
SLBMs						
SS-N-18 M1	Stingray	96	1978	6 500	3 x 200 kt (MIRV)	288
SS-N-20	Sturgeon	40	1983	8 300	10 x 100 kt (MIRV)	400
SS-N-23	Skiff	96	1986	9 000	4 x 100 kt (MIRV)	384
Subtotal		232				*1 072*
Total strategic offensive forces						**4 951**
Strategic defensive forces						
SAMs						
SA-5B Gammon, SA-10 Grumble		1 200				**1 200**
Non-strategic forces						
Land-based non-strategic						
Bombers and fighters						
Tu-22M Backfire		105			AS-4 ASM,	
Su-24 Fencer		280			AS-16 SRAM, bombs	
Subtotal		385				*1 540*
Naval non-strategic						
Attack aircraft						
Tu-22M Backfire		45			AS-4 ASM, bombs	
Su-24 Fencer		50				
Subtotal		95				*190*
SLCMs						
SS-N-9, SS-N-12, SS-N-19, SS-N-21, SS-N-22						240
ASW weapons						
SS-N-15, SS-N-16, torpedoes		..				210
Total defensive and non-strategic						**3 380**
Total						**8 331**

a Range for aircraft indicates operational range at maximum and standard payloads

Sources: START I Treaty Memoranda of Understanding (MOU), 1 Sep. 1990, 5 Dec. 1994, 1 July 1995, 1 Jan. 1996, 1 Jan. 1997, 1 July 1997, 1 Jan. 1998, 1 July 1998, 1 Jan. 1999, 1 July 1999, 1 Jan. 2000, 1 July 2000, 31 July 2001 and 31 Jan. 2002; 'NRDC Nuclear Notebook', *Bulletin of the Atomic Scientists*, various issues; International Institute of Strategic Studies, *The Military Balance 2001/2002* (Oxford University Press: Oxford, 2001); Podvig, P. (ed.), *Russian Strategic Nuclear Forces* (MIT Press: Cambridge, Mass., 2001); *Strategic Nuclear Forces*, Volume 1 of *Russia's Arms and Technologies, the XXI Century Encyclopedia* (Arms and Technologies Publishing House: Moscow, 2000); US Navy, Office of Naval Intelligence memos on 'Russian strategic and general purpose nuclear submarine patrols' covering 1991–2001, released under the Freedom of Information Act to Program on Science and Global Security, Princeton University; *Jane's Fighting Ships*, 2001–2002; and *Combat Fleets of the World 2000–2001* (Naval Institute Press: Annapolis, Md., 2000); US National Intelligence Council, *Foreign Missile Developments and the Ballistic Missile Threat Through 2015*, Dec. 2001; US National Intelligence Council, *Annual Report to Congress on the Safety and Security of Russian Nuclear Facilities and Military Forces*, Feb. 2002; US Department of Defense, *Proliferation: Threat and Response*, Jan. 2001; Ivanov, I. S., Minister of Foreign Affairs of the Russian Federation, 'Statement at the Review Conference of the Parties to the Treaty on the Non-Proliferation of Nuclear Weapons', New York, 25 Apr. 2000; 'Statement of the delegation of the Russian Federation at the First Session of the Preparatory Committee for the 2005 NPT Review Conference under Article VI of the Treaty', New York, 11 Apr. 2002; and Authors' estimates.

repeated that production of these three types of nuclear weapon system had been 'completely stopped'.[22]

IV. British nuclear forces

The UK maintains an arsenal of about 185 warheads for use by a fleet of four nuclear-powered ballistic missile submarines, consisting of 160 operational warheads and an additional 15 per cent of that number for spares. This makes the British arsenal the smallest of the five NPT-defined nuclear weapon states, and it may even be exceeded in size by Israel's nuclear arsenal.

---

[22] 'Yeltsin delivers statement on disarmament', Moscow Teleradiokompaniya Ostankino Television First Program Network (in Russian), 29 Jan. 1992, in Foreign Broadcast Information Service, *Daily Report–Central Eurasia (FBIS-SOV)*, FBIS-SOV-92-019, 29 Jan. 1992; Lobov, V., General of the Army, 'The motherland's armed forces today and tomorrow', *Krasnaya Zvezda*, 29 Nov. 1991; Yakovlev, G. (Gen.), 'Realization of reduction and limitation programs for nuclear weapons and the opportunity of an information exchange on amount of produced fissile materials and their localization', Talk prepared for the US–Russian Workshop on CTB, Fissile Material Cutoff and Plutonium Disposal, 15–17 Dec. 1993, Washington, DC, Natural Resources Defense Council, Federation of American Scientists, Moscow Physical–Technical Institute; Maslin, Y. (Col. Gen.), 'Remarks on US and Russian perspectives on the Cooperative Threat Reduction Program', made at the US Defense Special Weapons Agency Conference, 'Walking the walk: controlling arms in the 1990s', in 'Summary of the Fifth Annual International Conference on Controlling Arms', 3–6 June 1996, Norfolk, Va.; Press Conference with Lt. Gen. Igor Valynkin, Chief of the 12th Main Directorate of the Russian Ministry of Defence, regarding the nuclear security in Russian Federation armed forces, Russian Ministry of Defense, Official Kremlin International News Broadcast, 25 Sep. 1997 (Federal News Service); and 'Statement of the delegation of the Russian Federation at the First Session of the Preparatory Committee for the 2005 NPT Review Conference under Article VI of the Treaty', New York, 11 Apr. 2002, Russian Ministry of Foreign Affairs, Information and Press Department Internet site, URL <http://www.ln.mid.ru/website/bl.nsf/900b2c3ac91734634325698f002d9dcf/f8906fa2a4723ef843256ba300394eae?OpenDocument>.

The Royal Air Force (RAF) previously operated eight squadrons of dual-capable Tornado GR.1/1A aircraft. At the end of March 1998, with the withdrawal of the last remaining WE-177 bombs from operational service, the nuclear role of the Tornado was terminated. This brought to an end a four-decade-long history of RAF aircraft carrying nuclear weapons. By the end of August 1998 the remaining WE-177 bombs had been dismantled. The about 40 Tornadoes previously based at RAF Bruggen in Germany were reassigned to RAF Lossiemouth and RAF Marham in the UK by the end of 2001. The base at Bruggen will be closed.

Strategic submarines

The fourth and final Vanguard Class SSBN, *Vengeance*, was launched on 19 September 1998, commissioned on 27 November 1999 and deployed on its first patrol in February 2001. The *Vengeance* followed the HMS *Vanguard*, which sailed on its first patrol in December 1994. The second submarine, *Victorious*, entered service in December 1995, while the third, *Vigilant*, was launched in October 1995 and became operational in the autumn of 1998. The Vanguard Class submarine has a total complement of 205, providing a Ship's Company of 130 for patrols. The current estimated cost of the programme is $18.8 billion.

Each Vanguard Class SSBN carries 16 US-produced Trident II (D-5) SLBMs. Each missile carries one to three warheads, which are thought to be variations of the US W76 warhead enclosed in a US Mk-4 re-entry vehicle. The range of the missile can be extended by reducing the number of RVs. In its 'sub-strategic' configuration (see below), for example, a missile carrying a single warhead, it would have a range of more than 10 000 km.

There are no specifically British or US Trident II missiles but there is a pool of SLBMs at the Strategic Weapons Facility Atlantic at the Kings Bay Submarine Base, Georgia. The UK has title to 58 SLBMs but does not actually own them. A missile that is deployed on a US SSBN may at a later date be deployed on a British one, or vice versa. British SSBNs conduct their missile flight tests at the US Eastern Test Range off the coast of Florida. The *Vanguard* conducted two successful Demonstration and Shakedown Operations (DASOs) in May and June 1994, launching two Trident II missiles. The *Victorious* held DASOs in July and August 1995, with two missiles fired. In October 1997 the *Vigilant* launched two missiles during two DASOs, and on 21 September 2000 the *Vengeance* launched a missile during a single DASO exercise.

The current operational characteristics of the SSBN force were laid out by the Labour Government in July 1998 with the announcement of the results of the Strategic Defence Review (SDR). The decisions with regard to the British nuclear forces were the following:

1. Only one SSBN will be on patrol at any time, carrying a reduced load of 48 warheads—half the Conservative Government's announced ceiling of 96.

2. The submarine on patrol will be at a reduced alert state and will carry out a range of secondary tasks; its missiles will be detargeted and, after notice, the SSBN will be capable of firing its missiles within several days rather than several minutes, as during the cold war.

3. There will be fewer than 200 operationally available warheads, a one-third reduction from the Conservative Government's plans.

4. The number of Trident II (D-5) missiles already purchased or ordered was reduced from 65 to 58.

As a result of these decisions, the total explosive power of the operationally available weapons was reduced by almost 70 per cent compared to the force level planned under the Conservative Government. The explosive power of each Trident submarine will be one-third less than that carried on the Chevaline-armed Polaris submarines, the last of which was retired in 1996. At any given time the sole SSBN on patrol will carry about 40 warheads. The second and third SSBNs can be put to sea fairly rapidly, with similar loadings, while the fourth might take longer because of its cycle of overhaul and maintenance.

Several factors enter into the calculation of the number of warheads in the British stockpile. It is assumed that the UK will produce only enough warheads for three boatloads of missiles, a practice it followed with the Polaris missile. As stated in the SDR, there will be 'fewer than 200 operationally available warheads' in the stockpile and no more than 48 warheads per SSBN. The government also stated that it will be the practice that normally only one SSBN will be on patrol, with the other three in various states of readiness.

A further consideration is the 'sub-strategic mission'. A Ministry of Defence (MOD) official described it as follows: 'A sub-strategic strike would be the limited and highly selective use of nuclear weapons in a manner that fell demonstrably short of a strategic strike, but with a sufficient level of violence to convince an aggressor who had already miscalculated our resolve and attacked us that he should halt his aggression and withdraw or face the prospect of a devastating strategic strike'.[23] Shortly after the US 2001 Nuclear Posture Review was leaked to the press, in March 2002, British Defence Secretary Geoff Hoon stated that 'states of concern' armed with weapons of mass destruction 'can be absolutely confident that in the right conditions we would be willing to use our nuclear weapons'.[24]

The sub-strategic SSBN mission was briefed to the NATO Nuclear Planning Group in June 1995 and commenced late that year on board the *Victorious*. The mission was scheduled to 'become fully robust' when the *Vigilant* entered service in 1998, according to the 1996 Statement on the Defence Estimates.[25] If this has remained the policy, then some Trident II SLBMs already have a single warhead and are assigned targets once covered by WE-177 gravity bombs. For example, when the *Vigilant* is on patrol, 10, 12 or 14 of its SLBMs may carry up to three warheads per missile, while the other two, four or six missiles may be armed with just one warhead. There is some flexibility in the choice of yield of the Trident warhead: choosing to detonate only the unboosted primary could produce a yield of 1 kt or less, while choosing to detonate the boosted primary could produce a yield of a few kilotons. With these two missions an SSBN would have about 36–44 warheads on board during its patrol.

---

[23] Ormond, D., 'Nuclear deterrence in a changing world: the view from a UK perspective', *RUSI Journal*, June 1996, pp. 15–22.

[24] Joint Memorandum submitted by the Ministry of Defence and Foreign and Commonwealth Office, Parliamentary Defence Committee, 26 Feb. 2002.

[25] British Ministry of Defence, *Statement on the Defence Estimates 1996*, Cm 3223 (Her Majesty's Stationery Office: London, 1996), chapter 2, para. 203, available at URL <http://www.archive.official documents.co.uk/document/mod/defence/ch2.htm>.

Table 10A.6. British nuclear forces, January 2002

Type	Designation	No. deployed	Year first deployed	Range (km)	Warheads x yield	Warheads in stockpile
SLBMs						
D-5	Trident II	48	1994	> 7 400	1–3 x 100 kt	185

Sources: British Ministry of Defence (MOD), *Defence White Paper 1999*, Cm 4446 (Her Majesty's Stationery Office: London, 1999); MOD press releases and the MOD Internet site, URL <http://www.mod.uk/issues/sdr/index.htm>; British Ministry of Defence, *Strategic Defence Review* (MOD: London, July 1998); MOD, *Statement on the Defence Estimates 1996*, Cm 3223 (Her Majesty's Stationery Office: London, 1996); Ormond, D., 'Nuclear deterrence in a changing world: the view from a UK perspective', *RUSI Journal*, June 1996, pp. 15–22; Norris, R. S. et al., *Nuclear Weapons Databook, Vol. V: British, French, and Chinese Nuclear Weapons* (Westview: Boulder, Colo., 1994), p. 9; British House of Commons, *Parliamentary Debates (Hansard)*; 'NRDC Nuclear Notebook', *Bulletin of the Atomic Scientists*, various issues; and Authors' estimates.

Nuclear warhead maintenance

In 2000 the MOD awarded a contract for the operation of the Atomic Weapons Establishment (AWE) to an industrial consortium consisting of Lockheed Martin, Serco Limited and British Nuclear Fuels Limited. The 10-year contract is for £2.2 billion ($3.6 billion). On 1 April 1999 the Chief of Defence Logistics assumed overall responsibility for the routine movement of nuclear weapons within the UK. Day-to-day duties are being transferred, in phases, from RAF personnel to the MOD Police, with support from AWE civilians and the Royal Marines. The process will occur gradually and will be completed by 31 March 2002.

V. French nuclear forces

France maintains an operational arsenal of 348 nuclear warheads for delivery by land-based strike aircraft, ballistic missile submarines and naval aviation on a single aircraft carrier. The modernization of the French nuclear forces continues, including construction of the third and fourth Triomphant Class SSBNs, the M51 SLBM with a new nuclear warhead, the ASMP-A (Air-Sol Moyenne Portée) cruise missile and the Rafale nuclear-capable strike aircraft. The 2002 French defence budget increases funding for nuclear forces to FFr 17.2 billion ($2.5 billion), a 13 per cent increase over 2001.

The main lines of the current force structure were set out in February 1996 when President Jacques Chirac announced several reforms for the French armed forces, including the nuclear forces, for the period 1997–2002. This involved a combination of withdrawing several systems and modernizing others. The most significant of these changes involved a decision to eliminate land-based missiles as a component of the nuclear forces. All 18 S3D intermediate-range ballistic missiles (IRBMs) based on the Plateau d'Albion were deactivated, starting on 16 September 1996; it took two years and cost $77.5 million to fully dismantle the silos and the complex. After the land-based missiles had been deactivated, Chirac declared during his visit to Moscow in

September 1997 that 'no nuclear weapon in France's deterrent force was thenceforth targeted'.[26] Other actions included completion of the dismantlement of the South Pacific test facilities at Mururoa and Fangataufa. France stopped the production of plutonium for weapons in 1992 and of highly enriched uranium (HEU) in 1996. It has closed down and pledged to dismantle the Marcoule reprocessing plant and the Pierrelatte enrichment plant, which it started in 1998.

Nuclear strike aircraft

Three squadrons with 60 Mirage 2000Ns currently have nuclear strike roles. Two of these (*Dauphine* and *La Fayette*) are based at Luxeuil and the third (*Limousin*) at Istres. Since the 1991 Persian Gulf War, in which France was unable to use the night-attack capability of the then nuclear-only Mirage 2000N, the aircraft has been given some conventional capability to increase its utility. However, in a speech in May 1994 President François Mitterrand identified the 'N' in Mirage 2000N as standing for nuclear ('Mirage 2000N, c'est-à-dire nucleaires')[27] and Dassault, the producer of the aircraft, states on its Internet site that the 'primary assignment' of the Mirage 2000N is the nuclear strike role.[28]

The predecessor to Mirage 2000N, the Mirage IVP, was converted from its nuclear role in July 1996 and retired after 32 years of service. The Mirage IVP's ASMP missiles may have been reassigned to the Mirage 2000N. Five Mirage IVPs were retained for reconnaissance missions and are in the 1/91 Gascogne squadron at Mont-de-Marsan. The other aircraft were put into storage at Châteaudun.

The Mirage 2000N will eventually be replaced by the Rafale (B-301), which is planned to be the multi-purpose French Navy and Air Force fighter-bomber for the 21st century. Its roles include conventional ground attack, air defence, air superiority and nuclear delivery of the ASMP and/or ASMP-A. The naval version (Rafale M) entered the inventory in 2001 with Squadron 12F at Landivisiau and first entered service on board the *Charles de Gaulle* during its deployment in support of US operations in Afghanistan in early 2002. In both cases the initial role was air defence, but the Rafale M will gradually replace the nuclear strike mission of the Super Étendard on board the carrier. The Air Force's Rafale D will attain a nuclear strike role in about 2005. The Air Force still plans to buy a total of 234 Rafales, although it appears that delivery may be extended over some time.

The ASMP is equipped with the 300-kt TN-81 warhead. It is estimated that France has about 60 operational ASMPs, but additional missiles may be in inactive storage. There are conflicting reports about the inventory of missiles and warheads. A report from the French Senate stated in 1991 that France initially produced 80 warheads and 90 ASMP missiles. In May 1994, however, when 15 Mirage IVPs (plus three spares) still had nuclear roles and only 45 Mirage 2000Ns were operational, President Mitterrand identified 60 ASMP missiles for use by both air force and navy aircraft.[29] He did not disclose the number of warheads, however, and used different language

---

[26] Bourgo, N., Ambassador and Permanent Representative of France to the Conference on Disarmament, Statement on draft resolution A/C.1/53/L.16, 13 Nov. 1998, available at URL <http://www.gas.org/news/france/981113Ebis.htm>.

[27] 'Intervention de Monsieur François Mitterrand sur la politique française de dissuasion' [Statement by Mr François Mitterrand on French deterrent policy], Palais de l'Elysée, 5 May 1994, pp. 4–5.

[28] Dassault Aviation, Dassault Défense, 'Mirage 2000 N', n.d. [2002], URL <http://www.dassault-aviation.com/defense/gb/armes/M2000/presentation/m2000ND.cfm>.

[29] Bourgo (note 26).

Table 10A.7. French nuclear forces, January 2002

Type	No. deployed	Year first deployed	Range (km)^a	Warheads x yield	Warheads in stockpile
Land-based aircraft					
Mirage 2000N	60	1988	2 750	1 x 300 kt ASMP	50
Carrier-based aircraft					
Super Étendard	24	1978	650	1 x 300 kt ASMP	10
SLBMs					
M4.71^b	16	1985	6 000^c	6 x 150 kt	96
M45	32	1996	6 000^c	6 x 100 kt	192
Total					**348**

^{a} Range for aircraft assumes combat radius, without in-flight refuelling.

^{b} The M4.70 with TN-70 warheads was retired in 1996.

^{c} The range of the M4 and the M45 is listed as only 4000 km in a 2001 report from the National Defence Commission of the Assemblée Nationale.

Sources: Assemblée Nationale, Au Nom de la Commission de la Défense Nationale et des Forces Armées, sur le project de loi de finances pour 2002 (no. 3262), Tome II, Défense, 'Dissuasion Nucléaire', M. René Galy-Dejean (Député), 11 Oct. 2001, available at URL <http://assemblee-nationale.fr/budget/plf2002/a3323-02.asp>; French Ministry of Defence, 'Activities of the naval forces', Fact sheet [n.d. (2000)], URL <http://www.defense.gouv.fr/marine/anglais/present/dim2000/e_missions2.htm>; French Ministry of Defence, 'Nuclear disarmament and non-proliferation', *Arms Control, Disarmament and Non-Proliferation: French Policy* (La Documentation française: Paris, 2000), chapter 3, pp. 36–56; Address by M. Jacques Chirac, President of the Republic, at the Ecole Militaire, Paris, 23 Feb. 1996; Assemblée Nationale, *Projet de loi relatif à la programmation militaire pour les années 1997 à 2002*, no. 2766 (20 May 1996), section 2.3.4, Evolution de l'équipement des forces armées (1996–2002), p. 45; Intervention de Monsieur François Mitterrand sur la politique française de dissuasion [Statement by Mr François Mitterrand on French deterrent policy], Palais de l'Elysée, 5 May 1994, pp. 4–5; Norris, R. S. et al., *Nuclear Weapons Databook, Vol. V: British, French, and Chinese Nuclear Weapons* (Westview: Boulder, Colo., 1994), p. 10; *Air Actualités*, various issues; *Aviation Week & Space Technology*, various issues; 'NRDC Nuclear Notebook', *Bulletin of the Atomic Scientists*, various issues; and Authors' estimates.

to describe the number of missiles assigned to the different types of aircraft. For the Mirage IVP, he gave a fixed number, saying, 'we possess 15 missiles' ('nous disposons de quinze missiles'). For the Mirage 2000N/Super Etendard aircraft, however, the number was less precise, namely, 'these forces possess 45 missiles' ('les forces disposent de quarante-cinq missiles'), indicating that the exact number may be dependent on the number of operational aircraft.[30] Since then an additional 15 Mirage 2000Ns have become operational.

The ASMP is scheduled to be replaced by a longer-range (500 km as opposed to 300 km) version, sometimes called the 'ASMP Plus' (the official name is ASMP Amélioré, ASMP-A). The FFr 870 million ($117.5 million) development and production contract was awarded to EADS Aérospatiale Matra Missiles on 29 December 2000. The new missile is scheduled to enter service in 2007 on the Mirage 2000N and

[30] Bourgo (note 26).

in 2008 on the Rafale. The ASMP-A may be equipped with a modified warhead designated the TNA (tête nucléaire aéroportée).

Ballistic missile submarines

France has in operation four SSBNs of three classes: two of the new Triomphant Class SSBNs, one L'Inflexible Class SSBN and one Redoubtable Class SSBN.[31] The two Triomphant SSBNs each carry 16 M45 SLBMs with six of the new TN-75 warheads, which are thought to have been tested at the Mururoa test site in 1995. *Le Triomphant* (S616) was rolled out from its construction shed in Cherbourg on 13 July 1993 and became operational in September 1996. The second SSBN, *Le Téméraire* (S617), which was commissioned in December 1999, some six months behind schedule, successfully test launched an M45 missile in May 1999. The schedule for the third submarine, *Le Vigilant* (S618), has slipped and it will not be ready for delivery until December 2003 and for full service until December 2004. FFr 1.9 billion ($256.5 million) was allocated for the fourth SSBN (S619) in September 2000, but this boat is not scheduled to enter service until 2010. The total cost of the Triomphant Class programme is estimated to be FFr 96.3 billion ($13 billion).

France has deployed 48 SLBMs of two versions: 16 M4s with six 150-kt TN-71 warheads each; and 32 M45s equipped with six 100-kt TN-75 warheads each. Until 2000, the two older SSBNs both carried 16 M4 SLBMs. Faced with the delay of the third Triomphant SSBN, however, France converted *L'Inflexible* to carry the newer M45 SLBM in 2000–2001.[32] *L'Inflexible* test launched the M45 in April 2001. The conversion was a necessary, albeit expensive, solution to match a reduced inventory of only three sets of SLBMs (2 M45s and 1 M4). Without this refit France would, in certain situations, have been able to deploy only two SSBNs as opposed to three. The last M4-equipped SSBN, *L'Indomptable*, was refitted to carry the M4 in 1989. This missile carries six TN-71 warheads; the TN-70, which previously armed the M4, was retired in 1996.

A new SLBM, the M51, is scheduled to replace the M45 and the M4 in 2010, coinciding with the completion of the fourth Triomphant SSBN. The M51 is expected to have a range in excess of 6000 km (possibly 8000 km) and carry up to six warheads. The M51, which is a modified version of the cancelled M5 missile, was initially planned to carry an entirely new type of warhead (TNO, tête nucléaire océanique) but will instead be equipped with a more robust version of existing designs. The French Ministry of Defence credited the M51 with a 'capability of hitting several widely separated targets thanks to the spacing system incorporated in the upper compartment'. It further stated that the missile 'can adapt to changes in the interception threat' and 'will be hardened against nuclear attack'.[33] The first flight test is scheduled for 2005 and the first test launch from an SSBN in 2007.

France has undergone a transition to an operational inventory of 288 warheads for two sets of M45 SLBMs and one set of M4 SLBMs, enough to arm three of four SSBNs. There was also a lower number of missiles than launch-tubes when there were five submarines in the fleet, at which point only four sets of M4 SLBMs were

---

[31] Some sources list the L'Inflexible Class and the Redoubtable Class as the same SSBN class.

[32] French Ministry of Defence, 'FOST and the submarine force, n.d. [updated 24 Apr. 2002], URL <http://www.defense.gouv.fr/marine/anglais/present/dim/e_forces2.htm>.

[33] Ministry of Defence, 'The French armaments policy: industrial state, European strategy, operational capability', 2001, fiche 20.

procured. Of the four submarines, three are maintained in the operational cycle, although only one or two are normally 'on station' in designated patrol areas at any given time, compared with three in the early 1990s.

The SSBN force is organized under the Oceanic Strategic Task Force (Force Océanique Stratégique, FOST) and home-ported at the Île Longue base in Brest. The French Navy has recently reorganized its submarine fleet and will base all of its submarines (including SSNs formerly at Toulon) at Brest in the future. Under this reform the SSBN command centre at Houilles (Yvelines) will also be relocated to Brest, although communications facilities at Rosnay (Indre) will continue. Communication with SSBNs on patrol is also maintained with 4 C-160H Astarté communications relay aircraft.

French SSBNs are protected during their transit by nuclear attack submarines, maritime patrol aircraft (Atlantique 2), anti-submarine frigates and minesweepers. SSBN protection will also be an important mission for the planned Barracuda Class SSN. Like the SSBNs, French attack submarines each have two crews to optimize their operational availability.

Naval nuclear aviation

The nuclear-powered aircraft carrier *Charles de Gaulle* has finally entered service, after technical problems (including a propeller failure) delayed delivery for almost five years. France has spent over FFr 20 billion ($2.8 billion) on the 40 500-ton carrier, or over FFr 7 billion ($1 billion) more than its 1987 estimate of FFr 13 billion ($1.8 billion). Another FFr 50 billion ($7.1 billion) has been spent on 60 Rafale M and three E-2C Hawkeye aircraft. France is currently considering whether to include funding for a second carrier in its 2003–2008 defence spending plan, but this carrier may be built with a non-nuclear propulsion system in order to save money.

The *Charles de Gaulle*, which has a crew of 1850, can accommodate 35–40 aircraft. The first 10 Rafale Ms were deployed on the ship in 2002 when it was dispatched to the Arabian Sea for operations in support of the war in Afghanistan. The Rafale Ms were used for air defence and the *Charles de Gaulle* carried a single squadron of Super Étendards (presumably with about 10 ASMPs) for strike operations. The Navy plans to purchase a total of 60 Rafale Ms, of which the first 16 will perform an air-to-air role. However, it appears that the pace of delivery will be slow and only 19 Rafale Ms will be delivered in 2003–2008. Missions for subsequent aircraft will include the ASMP and/or the ASMP Plus.

VI. Chinese nuclear forces

China is estimated to have an arsenal of more than 400 nuclear weapons for delivery by aircraft, land-based ballistic missiles, submarine-launched ballistic missiles and possibly also non-strategic systems, including artillery.

Predictions of a Chinese build-up of nuclear forces continue to be made. In early 2002 the US CIA predicted that 'the overall size of Chinese strategic ballistic missile forces' over the next 15 years will increase to 75–100 warheads, deployed 'primarily against the United States'.[34] In addition, the CIA stated that China would have about 24 shorter-range DF-31 and DF-4 (CSS-3) missiles that could reach parts of the USA. For comparison, a 1999 CIA estimate had concluded that by 2015 China would probably have added 'a few tens' of more survivable land- and sea-based mobile missiles with smaller warheads and that 'tens of missiles' would be targeted against the USA.[35] This estimate was followed by Pentagon statements in 2000 that China does not seem to have aspirations for a large strategic force and, although modernization is under way, 'their strategic force is really quite small'.[36] Rumours that China has deployed nuclear weapons in Tibet have not been confirmed.

China's modernization of its missile force partly influenced the 1997 US presidential guidance for nuclear weapons planning to increase US targeting of China. After President Clinton signed Presidential Decision Directive-60 (PDD-60) in November 1997, STRATCOM brought China back under SIOP planning. It had been removed in 1982 following the normalization of Sino-US relations. The return of China to SIOP planning was followed by the creation of the Chinese Integrated Strategic Operations Plan (CHISOP), a hypothetical Chinese nuclear war plan created by STRATCOM planners and used to 'wargame' US nuclear strike plans against Chinese nuclear forces. The 2001 NPR reaffirmed this development.

Nuclear aviation

The Chinese bomber force is based on Chinese-produced versions of 1950s-vintage Soviet aircraft. With the retirement of the H-5, a copy of the Soviet Il-28 Beagle medium-range bomber, the main bomber is the H-6. This aircraft is based on the Soviet Tu-16 Badger medium-range bomber, which entered service with Soviet forces in 1955. China began producing the H-6 in the 1960s. It was used to drop weapons in two nuclear tests, a fission bomb in May 1965 and a multi-megaton bomb in June 1967. China attempted unsuccessfully to purchase Tu-22M Backfire aircraft from Russia in 1993 to replace the H-6. Future candidates for the nuclear strike mission may include Russian-designed Su-27 and Su-30 Flanker fighter-bombers, although there is no official Chinese confirmation of a nuclear role for these aircraft.

Land-based ballistic missiles

Chinese nuclear-capable land-based ballistic missiles consist of four different types with ranges that span from 1800 km to 13 000 km. China defines missile ranges as follows: short-range, less than 1000 km; medium-range, 1000–3000 km; long-range, 3000–8000 km; and intercontinental range, over 8000 km.

The 2800-km range DF-3 (NATO designation CSS-2) missile has been deployed for more than 25 years and is gradually being retired. Estimates on the number of operational missiles vary considerably, ranging from 40 to 150 missiles. The weapon

[34] Central Intelligence Agency, 'Foreign missile developments and the ballistic missile threat through 2015', Unclassified summary of a National Intelligence Estimate, 9 Jan. 2002, URL <http://www.cia.gov/nic/pubs/other_products/Unclassifiedballisticmissilefinal.htm>.
[35] Central Intelligence Agency, 'Foreign missile developments and the ballistic missile threat to the United States through 2015', Sep. 1999, URL <http://www.cia.gov/nic/pubs/other_products/foreign_missle_developments.htm>.
[36] Bacon, K., Assistant Secretary of Defense (Public Affairs), News Briefings, 12 Sep. 2000, URL <http://www.defenselink.mil/news/Sep2000/t09122000_t0912asd.html>, and 12 Dec. 2000, URL <http://www.defenselink.mil/news/Dec2000/t12122000_t1212asd.html>.

is deployed in Dalong, Datong, Dengshahe, Ching-yu, K'un-ming, Lianxiwang, Tonghua and Yidu, and the US Department of Defense Annual Report to Congress in 2000 described the CSS-2 as 'China's primary regional missile system'.[37] The two-stage, liquid-fuelled missile is deployed in silos and in a transportable mode but is not mentioned by the 2002 CIA report and may be phasing out with the deployment of the DF-31.

The 5500-km range DF-4 (CSS-3) is also deployed, but only about 'a dozen' missiles are listed by the 2002 CIA report. According to this report, the weapon is 'almost certainly' intended as a retaliatory deterrent against targets in Russia and Asia, although it could also reach parts of the USA.

The liquid-fuelled DF-5 (CSS-4) is China's only truly intercontinental missile. It has a range of 13 000 km and is deployed in silos. Estimates of the number of missiles vary. The 2002 CIA National Intelligence Estimate lists 'about 20' missiles, while a June 2000 US Defense Department report claimed that China had built 18 DF-5 silos.[38] The US National Air Intelligence Center stated that as of 1998 the deployed DF-5 force consisted of 'fewer than 25' missiles.[39] A CIA report leaked to *The Washington Times* shortly before President Clinton's visit to China in June 1998 estimated that 13 of the DF-5 missiles were targeted at the USA. A senior US administration official subsequently said that China does not keep its nuclear warheads mounted on its missiles but keeps them in separate storage. The DF-5 is deployed in Hsuan-hua, Lo-ning and Shuangjiang.

The DF-21 (CSS-5) is a two-stage solid-propellant missile carried in a canister on a transporter–erector–launcher (TEL). The missile has a range of 1800 km and is deployed in Ching-yu, Chuxiong, Datong, Liangkengwang and Tonghua. An improved Mod. 2 version has not been deployed.

In addition to these operational missiles, China has three new ballistic missiles in development: the road-mobile DF-31 (CSS-X-10); a longer-range version of the DF-31; and the Julang II SLBM. Development of these missiles began in the mid-1980s. The DF-31, which is thought to be in the flight-test stage of development, is a three-stage, solid-propellant, mobile ICBM with a range of about 8000 km and a CEP of 0.3–0.5 km. The missile has been flight tested three times, most recently on 4 November 2000 with several decoy warheads. The flight path was much shorter than the missile's estimated range of 8000 km. A DF-31 TEL was displayed in a 'northwestern town' in 2001, and a report cited a 'classified US intelligence report' which concluded that the DF-31 would have its first 'operational capability by the end of 2001'.[40] The 2002 CIA report predicted that garrison deployment of the DF-31 will take place by 2006, that the missile will be targeted primarily against Russia and Asia, and that 12 DF-31s could be deployed by 2015. The longer-range version of the DF-31 is expected to be deployed at some time in the latter half of the decade.

The Julang II SLBM is under development for deployment with a new Type 094 Class SSBN. Some reports suggest that the missile is a variant of the DF-31 and that it underwent tests in October 1997 that simulated launching the missile from

---

[37] US Department of Defense, 'Annual report on military power of People's Republic of China', 2000, section III: The security situation in the Taiwan Strait, URL <http://usinfo.state.gov/regional/ea/uschina/dodrpt00.htm>.

[38] Central Intelligence Agency (note 34); and Bacon (note 36).

[39] US Department of Defense, National Air Intelligence Center (NAIC), *Ballistic and Cruise Missile Threat* (NAIC: Wright-Patterson Air Force Base, Ohio, Apr. 1999).

[40] Gertz, B., 'China ready to deploy its first mobile ICBMs', *Washington Times* (Internet edn), 6 Sep. 2001, URL <http://asp.washtimes.com/printarticle.asp?action=print&ArticleID=20010906-16891927>.

submarine tubes. The 2002 CIA report does not confirm this, however, but lists the Julang II as a separate missile that may be deployed in the last half of this decade.

Allegations of Chinese theft of US nuclear warhead designs have fuelled speculation that China may soon deploy missile systems with multiple warheads. The CIA estimates that China has had the technical capability to develop and deploy MRV (multiple re-entry vehicle) payloads for many years, including a MIRV system, but has not done so.[41] If China needed an MRV capability in the short term, according to the CIA, one option might be to use a DF-31-type RV to develop and deploy a simple MRV or MIRV capability on the DF-5 in a few years. However, the CIA concluded that 'Chinese pursuit of a multiple RV capability for its *mobile* ICBM and SLBMs would encounter significant technical hurdles and would be costly'.[42] In contrast, the so-called Cox Report of 1999 concluded that China, with 'aggressive development of a MIRV system', could deploy 'upwards of 1000 thermonuclear warheads on ICBMs by 2015'.[43]

The USA's continued forward deployment of SSBNs in the Pacific, combined with its programme to develop and deploy an advanced missile defence system, may stimulate Chinese efforts to deploy a multiple-warhead system and mobile ICBMs in an attempt to reduce the vulnerability of its nuclear deterrent to pre-emptive strikes.

Ballistic missile submarines

China has had great difficulty in developing a sea-based deterrent. After decades of development, it has only managed to deploy one operational Xia Class (Project 092) SSBN. The submarine carries 12 Julang I SLBMs, each equipped with a single warhead. Although the submarine participated in a naval exercise in December 2000, it may not have achieved FOC.

The *Xia*, which was built at Huludao Naval Base and Shipyard in the northern Bohai Gulf and launched in April 1981, is equipped with the 1700-km range Julang I SLBM. The missile was initially test launched from a Golf Class diesel-powered submarine in late 1982, and a full-scale submerged launch from the *Xia* took place in 1988. The following year the *Xia* was deployed to the Jianggezhuang Submarine Base, where the nuclear warheads for its Julang I missile are believed to be stored.

The *Xia* began a major refit in 1995 and is not thought to have ever sailed beyond China's regional waters. Production of a second Xia Class submarine was started but never finished.

A new SSBN project, designated Project 094, has begun with one submarine under construction. Three to five more may be planned. The new class is expected to carry 16 three-stage Julang II SLBMs. The CIA expects the missile to be tested 'within the next decade'. Given previous difficulties with the development of a sea-based deterrent and the lack of a multiple-warhead system on land, operational deployment of the Project 094 system may be many years away. The new Julang (Giant Wave) may have

---

[41] An MRV system releases 2 or more RVs along the missile's linear flight path to a single target, which land in a relatively confined area at about the same time. The more sophisticated and flexible MIRV system can manoeuvre multiple RVs to several different release points to provide targeting flexibility against several independent targets over a much wider area and longer period of time.

[42] Central Intelligence Agency (note 34), (emphasis in original).

[43] Report of the Select Committee on US National Security and Military/Commercial Concerns with the People's Republic of China, 3 Jan. 1999 (classified) (the Cox Report), URL <http://www.fas.org/spp/starwars/congress/1999_r/cox/preface.htm>.

Table 10A.8. Chinese nuclear forces, January 2002

Type	NATO designation	No. deployed	Year first deployed	Range (km)[a]	Warheads x yield	Warheads in stockpile
Aircraft[b]						
H-6	B-6	120	1965	3 100	1–3 bombs	120
Q-5	A-5	30	1970	400	1 x bomb	30
Land-based missiles						
DF-3A	CSS-2	40	1971	2 800	1 x 3.3 Mt	40
DF-4	CSS-3	12	1980	5 500	1 x 3.3 Mt	12
DF-5A	CSS-4	20	1981	13 000	1 x 4–5 Mt	20
DF-21A	CSS-5	48	1985–86	1 800	1 x 200–300 kt	48
SLBMs						
Julang I	CSS-N-3	12	1986	1 700	1 x 200–300 kt	12
Strategic weapons							282
Non-strategic weapons						
Artillery/ADMs, Short-range missiles					Low kt	120
Total						**~ 402**

[a] Range for aircraft indicates combat radius, without in-flight refuelling.

[b] All figures for bomber aircraft are for nuclear-configured versions only. Hundreds of aircraft are also deployed in non-nuclear versions. The table assumes 150 bombs for the bomber force, with yields estimated between 10 kt and 3 Mt.

Sources: US Central Intelligence Agency, 'Foreign missile developments and the ballistic missile threat through 2015', Unclassified summary of a National Intelligence Estimate, 9 Jan. 2002, URL <http://www.cia.gov/nic/pubs/other_products/Unclassifiedballisticmissile final.htm>; US Department of Defense, Office of the Secretary of Defense, 'Proliferation: threat and response', Washington, DC, Jan. 2001, URL <http://www.defenselink.mil/pubs/ptr20010110.pdf>; Department of Defense, Report to Congress Pursuant to the FY2000 National Defense Authorization Act, 'Annual Report on the Military Power of the People's Republic of China', June 2000, URL <http://www.defenselink.mil.news/Jun2000/china 06222000.html>; Moore, F. W., *China's Military Capabilities* (Institute for Defense and Disarmament Studies: Cambridge, Mass., June 2000), URL <http:www.comw.org/cmp/fulltext/iddschina.html>; Kan, S. A. et al., *China's Foreign Conventional Arms Acquisitions: Background and Analysis,* Congressional Research Service (CRS) Report for Congress (Library of Congress: Washington, DC, 10 Oct. 2000); US State Department International Information Programs, *Pentagon Spokesman's Regular Briefing,* 12 Dec. 2000; Baker III, A. D., 'Combat fleets', *US Naval Institute Proceedings,* Dec. 2000, p. 90; US Department of Defense, National Air Intelligence Center (NAIC), *Ballistic and Cruise Missile Threat* (NAIC: Wright-Patterson Air Force Base, Ohio, Apr. 1999); Norris, R. S. et al., *Nuclear Weapons Databook, Vol. V: British, French, and Chinese Nuclear Weapons* (Westview: Boulder, Colo., 1994); US Central Intelligence Agency, various documents; 'NRDC Nuclear Notebook', *Bulletin of the Atomic Scientists,* various issues; and Authors' estimates.

a range of up to 8000 km and is estimated by the CIA to 'probably' be able to target the USA from launch areas near China.

Non-strategic weapons

Information on Chinese non-strategic nuclear weapons is limited and contradictory. There is no confirmation of their existence from official Chinese sources. Several low-yield nuclear tests conducted in the late 1970s and a large military exercise in June 1982 simulating the use of non-strategic nuclear weapons suggest that they may have been developed.

According to the US Defense Intelligence Agency (DIA), non-strategic weapons may consist of atomic demolition munitions (ADMs) (nuclear landmines), aircraft bombs and short-range ballistic missiles. The latter include the DF-15 (CSS-6), also called the M-9; and the DF-11 (CSS-X-7), also called the M-11. Both are solid-fuelled, dual-capable SRBMs and were deployed in 1995. The DF-15 has a range of 200–600 km and may carry a 10-kt neutron warhead or a 20-kt warhead. One regiment-size DF-15 unit is deployed in south-eastern China and may soon be augmented by an additional unit. The DF-11 is thought to have a range of 200–300 km. An improved version was displayed in a military parade in Beijing on 1 October 1999. Western estimates of the number of SRBMs possessed by China range from 100 to 300 DF-15s and from 40 to 100 DF-11s. In 1984 the DIA did not believe that Chinese ground forces had been equipped with artillery-fired nuclear projectiles, although this capability could have been added later.

China is reported to be developing long-range cruise missiles with ranges of 1500–2500 km. A missile programme known as X-600 appears to be based partly on Russian and US cruise missile designs. Although there is speculation about a possible nuclear capability for one of the systems, this has not been confirmed. China's other cruise missiles include the SS-N-22 missiles on two Russian-built Sovremenny Class destroyers imported from Russia. In the Russian Navy the SS-N-22 is credited with a nuclear capability, but there are no reports that China plans to equip the missile with a nuclear warhead.

VII. Indian nuclear forces

The size and composition of India's nuclear arsenal are difficult to determine. Unofficial and semi-official estimates range from a few to almost 100 nuclear warheads. An estimate is made here of a stockpile of 30–35 nuclear warheads (fewer than Pakistan has), of which some may not yet be fully assembled. This stockpile is thought to be expanding, though, and India is estimated to have produced enough fissile material for 45–95 nuclear warheads.

The Indian Atomic Energy Commission (AEC) stated that the series of five nuclear test explosions in May 1998 involved both fission and fusion weapon designs. The government claims that the first three tests, which were carried out on 11 May 1998, achieved yields of 43 kt (a 'thermonuclear' device), 12 kt (a fission device) and 200 tons (a low-yield device). If the devices actually produced the yields claimed by Indian weapon scientists, one would have expected to observe a seismic signal strength corresponding to 55 kt, or 5.76 on the Richter scale. Instead, the average recorded magnitude was 5.0, which indicates a probable yield of 12 kt, with the range possibly as low as 5 kt and as high as 25 kt. A mid-point of 12 kt is less than one-quarter of what Indian weapon scientists claimed, calling into question whether the thermonuclear milestone was achieved and whether the tests were 'completely successful' and gave India 'the capability to design and fabricate weapons ranging from low yield to around 200 kilotons', as India's AEC Chairman Rajgopal Chidambaram stated in October 2000.[44]

India established the National Security Council in April 1999 to implement its nuclear weapon policy, but its progress in setting up a nuclear command and control system is unknown. On 17 August 1999 a widely publicized draft document on Indian nuclear doctrine (prepared by a 27-member National Security Advisory Board) called for the creation of a 'credible minimum deterrent' to be based 'on a triad of aircraft, mobile land-based missiles and sea-based assets'.[45] The Board's recommendations, however, had no official standing.

While the Indian Army and Air Force have been refining their respective nuclear strategies, the government has been considering a proposal by the Tri-Services Chiefs of Staff Committee to create a strategic nuclear force. The proposal followed a report by the Group of Ministers which proposed that a Chief of Defence Staff (CDS) be established to act as a military advisor to the Prime Minister on issues related to the management and control of nuclear weapons and strategic forces. The CDS would 'exercise administrative control, as distinct from operational military control over these strategic forces'.[46]

In 2001, even during heightened tension with Pakistan, Indian government officials reaffirmed India's commitment to a nuclear no-first-use policy. However, an Indian Foreign Ministry official said in 2001 that a '"no first strike" policy does not mean India will not have a first strike capability'. He explained that India was 'working toward having a first strike capability' but added that how this option would be exercised was a political decision within the 'no first strike' policy.[47]

On 31 May 2001 the Indian Defence Ministry released a report describing its principal security concerns and detailing plans for modernization of the armed forces. Not surprisingly, Pakistan's support of terrorist groups headed the list of security concerns. After the 13 December 2001 terrorist attack on the Indian Parliament, allegedly carried out by Pakistan-backed guerrillas, the two nations mobilized their armed forces. This led to a tense situation and heated exchange of words. India reportedly moved Prithvi short-range missiles to positions near its border with Pakistan.

Strike aircraft

India has several types of aircraft that could be used to deliver a nuclear weapon, but the most likely are the MiG-27 and the Jaguar, given their range, payload and speed. The MiG-27 Flogger is a nuclear-capable Soviet aircraft produced in the 1970s and 1980s. Hindustan Aeronautics licence-assembled 165 aircraft which India calls the Bahadhur (Valiant or Brave). The single-seat aircraft weighs almost 18 000 kg when fully equipped and has a range of approximately 800 km. It can carry up to 4000 kg of bombs on external hard points. There are nine operational squadrons. It is not known which of India's bases may host nuclear-capable aircraft, but one likely candidate is Hindan, north of New Delhi. Some 50 MiG-27MLs are deployed there, less

[44] 'India can build 200kt nuclear weapons', *Jane's Defence Weekly*, 8 Nov. 2000, p. 6.
[45] 'Draft report of the National Security Advisory Board on Indian nuclear doctrine', 17 Aug. 1999, URL <http://www.meadev.gov.in/govt/indnucld.htm>.
[46] Government of India, 'Reforming the national security system', Recommendations of the Group of Ministers, 2001, p. 100, URL <http://mod.nic.in/newadditions/rcontents.htm>.
[47] Ahmedullah, M., 'Indian Air Force advocates "first strike capability"', *Defense News*, 2 Jan. 2001, p. 1.

than 640 km from Lahore, Pakistan. A few aircraft from Squadrons 2, 9 or 18 may be specially modified to carry one or more nuclear bombs.

The Jaguar IS/IB, known as the Shamsher (Sword), was nuclear-capable with the British Royal Air Force from 1975 to 1985 and with the French Air Force from 1974 to 1991. Originally a joint Anglo-French aircraft, the first 40 were supplied by British Aerospace, with the remaining 91 assembled or manufactured by Hindustan Aeronautics. The Jaguar has a gross weight of 15 450 kg and a range of 1600 km with a maximum external load of 4775 kg. There are four operational squadrons. It is not known which bases may host nuclear-capable aircraft, but one likely candidate is Ambala, 525 km from Islamabad. A few aircraft from Squadrons 5 or 14 may be specially modified to carry one or more nuclear bombs. In the Indian Air Force organization, Hindan and Ambala are part of the Western Command, located at Palam and reporting to headquarters in New Delhi.

Other aircraft, such as the Su-30K and Mirage 2000H, could be equipped to deliver nuclear bombs but are more likely to be used for air defence missions. A Mirage 2000H may have been used in May 1994 to test-drop a dummy nuclear bomb, but this has not been officially confirmed. In late 1999 India was reported to have initiated preliminary talks with France about the purchase of up to 18 Mirage 2000Ds to form part of its nuclear strike force.[48] Ten Mirage 2000H/THs were ordered in September 2000, and the Indian Air Force is said to have plans to acquire 126 Mirage 2000-5s for seven squadrons as the 'backbone' of India's airborne nuclear strike force. In December 2000 India signed a $3 billion contract with Russia for the licensed production of 140 Su-30MKI aircraft at Hindustan Aeronautics over the next 17 years. Forty Su-30K fighters procured in 1996 may also be upgraded to MKI standard. Air Chief Marshal A. Y. Tipnis said prior to the deal that the indigenous Su-30MKI will 'enable the Air Force to finalize its vision-2020 long term perspective planning',[49] which involves acquiring up to 20 squadrons of multi-role aircraft over the next 15–20 years. The first Indian-produced Su-30MKI is scheduled to roll off the production line in 2004 and will be supported by six Ilyushin-78 Midas tanker aircraft acquired from Uzbekistan.[50] During 2001, India also attempted, unsuccessfully, to lease a small number of Russian Tu-22 Backfire bombers. France has offered to sell its new Rafale aircraft to India.

Land-based missiles

India deploys one ballistic missile, the 150-km range Prithvi I SRBM. The Prithvi (Earth) is a single-stage, dual-engine, liquid-fuelled, road-mobile missile which began development in 1983 and was first tested in 1988. There have been 15 tests since 1988. The Prithvi II, an improved version with an extended range of 250 km, is under development. It was test fired on 31 March 31 2001. Of the two versions, only the Prithvi I is assessed by the CIA to have a nuclear role.

Several versions of the Agni (Fire) IRBM are under development. The initial version, which was flight tested three times between 1989 and early 1994 to a range of up to 1500 km, is thought to have been shelved. Instead, development of an improved version (Agni II) with a range of more than 2000 km is under development. The

---

[48] Bedi, R., 'India's ties with France will soar with Mirage buy', *Jane's Defence Weekly*, 1 Sep. 1999, p. 13.
[49] 'India to buy more Mirage from France', *India Today News*, 7 Oct. 2000.
[50] Thapar, V., 'IAF to get mid-air refueling aircraft', *Hindustan Times*, 6 Aug. 2001.

Table 10A.9. Indian nuclear forces, January 2002

Type/Designation	Range (km)^a	Payload (kg)	Comment
Aircraft			
MiG-27 Flogger/Bahadhur	800	3 000	At Hindan Air Base
Jaguar IS/IB/Shamsher	1 600	4 775	At Ambala Air Base
Missiles			
Prithvi I	150	1 000	Deployed, may have nuclear role
Agni I	1 500	1 000	Flight tested but status unclear.
Agni II	>2 000	1 000	Flight tested in Jan. 2001, deployment expected soon. A 700-km range version was flight tested on 25 Jan. 2002.

^{a} Range for aircraft indicates combat radius, without in-flight refuelling.

Sources: US Department of Defense, Office of the Secretary of Defense, 'Proliferation: threat and response', Washington, DC, Jan. 2001, URL <http://www.defenselink.mil/pubs/ptr20010110.pdf>; Indian Ministry of Defence; Indian Air Force; Indian Ministry of External Affairs; Albright, D., 'India's and Pakistan's fissile material and nuclear weapons inventories, end of 1999', Background paper, Institute for Science and International Security (ISIS), 11 Oct. 2000, URL <http://www.isis-online.org/publications/southasia/stocks1000.html>; US Department of Defense, National Air Intelligence Center (NAIC), *Ballistic and Cruise Missile Threat* (NAIC: Wright-Patterson Air Force Base, Ohio, Apr. 1999); 'Draft report of the National Security Advisory Board on Indian nuclear doctrine', 17 Aug. 1999, URL <http://www.meadev.gov.in/govt/indnucld.htm>; US Central Intelligence Agency, 'Foreign missile developments and the ballistic missile threat to the United States through 2015', Sep. 1999, URL <http://www.cia.gov/nic/pubs/other_products/foreign_missle_developments.htm>; Albright, D., Berkhout, F. and Walker, W., SIPRI, *Plutonium and Highly Enriched Uranium 1996: World Inventories, Capabilities and Policies* (Oxford University Press: Oxford, 1997); Burrows, W. E. and Windrem, R., *Critical Mass* (Simon & Schuster: New York, 1994); *Jane's Defence Weekly*, various issues; and Authors' estimates.

missile, which is 20 metres long, weighs about 16 tons and can carry a 1000-kg payload 2000–2500 km. The first test in April 1999 covered 2000 km in 11 minutes and may have involved a nuclear warhead assembly without the plutonium core. The second flight was on 17 January 2001 from a mobile launcher at the Chandipur missile test range off the eastern coastal state of Orissa. The missile, which was said to be in 'final operational configuration', flew 2200 km and, according to Indian officials, landed less than 100 metres from its intended target. After the test, Defence Minister Jaswant Singh informed the Indian Parliament that the 'Agni II is planned to be inducted into the armed forces during 2001–02'.[51] Both road- and rail-mobile versions of Agni II are under development and are expected to become the mainstay of India's nuclear-armed missile force.[52]

A new short-range version of the Agni II, with a range of about 700 km, was test launched from Wheeler's Island on the Indian east coast on 25 January 2002.

---

[51] 'Production of Agni missile begins: govt', *Times of India*, 25 July 2001, URL <http://timesofindia.indiatimes.com/articleshow.asp?art_id=1186900905>.

[52] Central Intelligence Agency (note 34).

Rumours of a longer-range Agni III with a range of up to 3500 km have not been confirmed.

Rumours also persist about Indian plans for an ICBM programme, referred to as the Surya (Sun). Most components needed for an ICBM are available from India's indigenous space programme. The CIA predicts that it would take one to two years for India to convert the Polar Space Launch Vehicle (PSLV) to an ICBM after a decision is made to do so.[53] The latest model, the four-stage PSLV-C3 is capable of launching a satellite weighing up to 1200 kg into polar sun-synchronous orbit (570 km) or a 3500-kg satellite into low earth orbit (400 km). Three satellites were placed in orbit on 22 October 2001. An attempt to develop a Geo-Synchronous Satellite Launch Vehicle (GSLV) made headway on 18 April 2002, when a 401-tonne, 49-metre tall GSLV launched from the Sriharikota High Altitude Range (SHAR) Centre placed a 1540-kg satellite into an elliptical geostationary transfer orbit (181-km perigee and 32 051-km apogee). Further GSLV progress to achieve full geostationary equatorial orbit (36 000 km above the same point on earth) would allow India to place permanent command and control satellites in orbit.

Naval weapons

In addition to air- and land-based nuclear-capable forces, India is working on at least two naval systems that may be equipped to carry nuclear warheads in the future. The submarine-launched Sagarika (Oceanic) missile, begun in 1991, is in an advanced stage of development. The CIA designates it as an SLBM and predicts that it will not be deployed until 2010 or later. Another potential candidate is the Dhanush (Bow) SLBM, which has been under development since 1983 for possible completion in 2003. A test firing on 11 April 2000 was only a 'partial success' and may delay the programme further. The 8.56-metre missile, which is a naval version of the army's Prithvi, is capable of carrying a 1000-kg payload to a range of 250 km. It was launched from the reinforced helicopter deck of the INS Subhadra, a modified offshore patrol vessel anchored some 20 km offshore in the Bay of Bengal. Neither the Dhanush nor the Sagarika has been declared nuclear-capable by Indian authorities.

The Advanced Technology Vessel (ATV), a nuclear-powered submarine project that has been under way in various stages since at least 1985, may be a navy nuclear-weapon launch platform. Design and operational experience was gained from operation of a Charlie I Class nuclear-powered cruise missile submarine (INS *Chakra*) leased from the Soviet Union from 1988 to 1991. Full-scale work on the ATV began in 1991, shortly after the INS Chakra was returned to Russia, and construction started in 1997. A launch date may be scheduled for 2007 at the Mazagon Dockyard in Mumbai (design work has taken place in Vishakapatnam on the east coast), but technical challenges are likely to delay the ATV further. Efforts to lease one or more nuclear submarines from Russia continue. The ATV is thought to be based partly on the INS Chakra, but the reactor is reported to be of Indian design. A land-based prototype reactor has been built and is believed to be undergoing testing at the Indira Gandhi Centre for Atomic Research at Kalpakkam in southern India. Vice Admiral R. N. Ganesh, who commanded the INS Chakra, was appointed as new director general of the ATV project in 2000 in an apparent attempt to kick-start the much delayed project.

[53] Central Intelligence Agency (note 34).

VIII. Pakistani nuclear forces

It is extremely difficult to estimate the number and types of nuclear weapons in Pakistan's arsenal. Outside experts estimate that Pakistan may possess a stockpile of 24–48 nuclear weapons, of which only some may be fully operational. The implosion design of the weapons uses a solid-core of HEU rather than plutonium, requiring an estimated 15–20 kg of HEU per warhead.

Seismic measurements of the nuclear test explosions conducted by Pakistan on 28 and 30 May 1998 suggest that the yields were of the order of 9–12 kt and 4–6 kt, respectively—lower than the yields announced by the Pakistani Government. Early Chinese tests in the 1960s used similar bomb designs, and it is suspected that Chinese experts assisted Pakistan in developing its nuclear weapon programme in the 1970s and 1980s. Over a 20-year period Pakistan pursued a gas centrifuge uranium-enrichment method to produce the material for its nuclear weapons, at the Abdul Qadeer Khan Research Laboratories in Kahuta. There is some uncertainty about how many centrifuges Pakistan has and thus how much weapon-grade uranium has been produced. By the early 1990s, some 3000 centrifuges were thought to be operating. The most cautious estimate is that Pakistan has produced enough fissile material for 30–52 nuclear weapons. A moratorium on HEU production was declared in 1991. It is unclear when production was resumed but it is thought to have started well before the 1998 nuclear tests.

Like the other nations that have developed nuclear weapons, Pakistan does not seem content with a first-generation nuclear weapon and may be pursuing advanced designs and refinements. The 40–50 Megawatt thermal (MWth) Khushab reactor constructed at Joharabad in the Khushab district of Punjab has the capability to produce weapon-grade plutonium. Loading the reactor's target materials with lithium-6 could produce tritium. Producing plutonium provides the Pakistani military with several options: making weapons with plutonium cores, mixing plutonium with HEU to make composite cores or using tritium to 'boost' a weapon's explosive yield. Separation of the plutonium is reported to take place at the 'New Labs' reprocessing plant near the Pakistan Institute of Nuclear Science and Technology (PINSTECH) at Rawalpindi. Through these efforts Pakistan seems to be positioning itself to increase and enhance its nuclear forces significantly in the coming years.

In November 2000 Pakistan placed its key nuclear institutions under the control of the National Command Authority, established in February 2000 in an apparent effort to create an effective nuclear command and control system. After the terrorist attacks of 11 September 2001, attention was focused on the security of Pakistan's nuclear arsenal. One potential danger to the arsenal is the seizure of nuclear weapon control by extremist Islamists within the intelligence service, the armed forces, the nuclear weapon programme and the population at large. President General Pervez Musharraf reportedly took several actions in the autumn of 2001 to mitigate this problem: he fired his chief of intelligence and other officers, detained several suspected retired nuclear weapon scientists, redeployed the arsenal to at least six new secret locations and appointed Lieutenant General Ghulam Mustafa as the first three-star commander of the upgraded strategic command.[54]

---

[54] Mufson, S., 'US worries about Pakistan nuclear arms', *Washington Post*, 4 Nov. 2001, p. A27; Moore, M. and Khan, K., 'Pakistan moves nuclear weapons', *Washington Post*, 11 Nov. 2001, p. A01; Albright, D., 'Securing Pakistan's nuclear weapons complex', Institute for Science and International Security, Oct. 2001, URL <http://www.isis-online.org/publications/terrorism/stanleypaper.html>; and

Strike aircraft

The aircraft of the Pakistani Air Force that is most likely to be used in the nuclear weapon delivery role is the US-manufactured F-16, although other aircraft, such as the Mirage V or the Chinese-produced A-5, could also be used. Twenty-eight F-16A (single-seat) and 12 F-16B (two-seat) trainers were delivered to the Pakistani Air Force between 1983 and 1987. At least eight of the original order are no longer in service. In December 1988 Pakistan ordered 11 additional F-16A/Bs as attrition replacements but to date they have not been delivered because of the 1984 Pressler Amendment, which forbids military aid to suspected nuclear weapon states. The US Government announced on 6 October 1990 that it had embargoed any further arms deliveries to Pakistan. The 11 embargoed aircraft are being stored in the Arizona desert near Davis-Monthan AFB. In September 1989 plans were announced for Pakistan to acquire 60 more F-16s. Of that order, 17 were built by the end of 1994, but because of the embargo they were not delivered and were also stored at Davis-Monthan AFB. In a Presidential Determination signed on 22 September 2001, President Bush waived the Pressler Amendment, but the aircraft have not been released for delivery.

The F-16s most likely to have been modified to carry nuclear weapons are deployed with Squadrons 9 and 11 at Sargodha AB, 160 km north-west of Lahore. The F-16 has a range of 1600 km, or more if drop tanks are used. It can carry up to 5450 kg externally on one under-fuselage centreline pylon and on six under-wing stations. Given the weight and size payload constraints of the F-16, if it carried a nuclear bomb it would probably weigh about 1000 kg and be attached to the centreline pylon. The assembled nuclear bombs and/or bomb components for these aircraft may be stored in an ammunition depot near Sargodha. Another possibility is that, fearing a first strike by India if war were to break out, Pakistan stores the weapons at other operational or satellite bases further to the west, near the border with Afghanistan, where the F-16s would disperse to pick up the bombs.

Ballistic missiles

The Ghauri missile is thought to be Pakistan's only operational nuclear-capable missile, although other missiles in the Pakistani arsenal could be configured to carry a nuclear warhead. The single-stage Ghauri I was first flight tested on 6 April 1998 to a distance of 1100 km, probably with a payload of up to 700 kg. The liquid-fuelled Ghauri is basically a North Korean No Dong missile, which is a Scud derivative. A two-stage Ghauri II missile was flight tested on 14 April 1999, three days after the Indian Agni II test flight. It was launched from a mobile launcher at Dina, near Jhelum, and landed in Jiwani, near the coast in the south-western Baluchistan province, after an eight-minute flight. A third version of the Ghauri, with an unconfirmed range of up to 3000 km, is under development and was test launched on 15 August 2000.

Sawant, G. C., 'India mulls options as Pak upgrades its nuclear command', *National Network*, 11 Apr. 2002.

Table 10A.10. Pakistani nuclear forces, January 2002

Type/Designation	Range (km)^a	Payload (kg)	Comment
Aircraft			
F-16A/B	1 600	5 450	At Sargodha AB
Missiles			
Shaheen I (Hatf-4)	700	1 000	Test fired 15 Apr. 1999; possible nuclear role
Ghauri I (Hatf-5)	1 300– 1 500	500– 750	Test fired on 6 Apr. 1998. Version of North Korean No Dong missile
Ghauri II (Hatf-6)	2 000– 2 300	750– 1 000	Test fired on 14 Apr. 1999

^{a} Range for aircraft indicates combat radius, without in-flight refuelling.

Sources: Islamic Republic of Pakistan, Official Internet site, URL <http://www.pak.gov.pk/>; Embassy of Pakistan, Washington, DC; US Department of Defense, Office of the Secretary of Defense, *Proliferation: Threat and Responses*, released in Jan. 2001, URL <http://www.defenselink.mil>; US Central Intelligence Agency, National Intelligence Office for Strategic and Nuclear Programs, *Foreign Missile Developments and the Ballistic Missile Threat to the United States Through 2015*, Sep. 1999, and Jan. 2002; Albright, D., 'India's and Pakistan's fissile material and nuclear weapons inventories, end of 1999', Background Paper, Institute for Science and International Security (ISIS), 11 Oct. 2000, URL <http://www.isis-online.org/publications/southasia/stocks1000.html>; Albright, D., Berkhout, F. and Walker, W., SIPRI, *Plutonium and Highly Enriched Uranium 1996: World Inventories, Capabilities and Policies* (Oxford University Press: Oxford, 1997); Burrows, W. E. and Windrem, R., *Critical Mass* (Simon & Schuster: New York, 1994); 'Three-Four-Nine: The Ultimate F-16 Site', URL <http://www.f-16.net/reference/users/f16_pk.html>; *Jane's Intelligence Review*, various issues; and Authors' estimates.

Other missiles may also be candidates for nuclear capability. The single-state solid-fueled Shaheen I (Eagle) was test fired on 15 April 1999 and is thought to have a range of 700 km and carry a payload of 1000 kg. The road-mobile Shaheen I, which is said by some to be based on either the Chinese M-9 or M-11, was said to be in serial production by late 2000,[55] and may have a nuclear role. Pakistan has obtained 30 or more complete M-11 missiles from China since 1992 and subsequently received Chinese assistance for the construction of maintenance and storage facilities. The M-11 missiles may be stored at the depot near Sargodha.

The two-stage Shaheen II medium-range missile was unveiled at the Pakistan Day parade on 23 March 2000, and the US CIA estimates that the missile has a range of 2500 km and a 1000-kg payload.[56] The solid-fuelled missile is transported on a 16-wheel mobile launcher similar to the Russian MAZ-547V that used to transport the Soviet Union's SS-20 IRBM. The Shaheen II may have a nuclear role in the future.

[55] Farooq, U., 'Pakistan starts production of Shareen 1 missile', *Jane's Defence Weekly*, 4 Oct. 2000, p. 4.
[56] Central Intelligence Agency (note 34).

IX. Israeli nuclear forces

Israel has not confirmed or denied that it has nuclear weapons. It is thought to have as many as 200 nuclear warheads, consisting of aircraft bombs, missile warheads and non-strategic weapons. Although a 2001 US DOD report omits Israel from its review of the Middle East, a 1991 US Strategic Air Command study lists Israel as a 'de facto' nuclear weapon state along with India and Pakistan.[57]

Since the late 1990s, the question of the continued validity of Israel's long-standing policy of nuclear ambiguity has been raised in several contexts: parliamentary pressure for a policy review in 1997–98; the nuclear tests by India and Pakistan in 1998; rumours in 2000 of Iran's alleged development of a nuclear weapon capability; and the terrorist attacks in the United States of 11 September 2001. However, no apparent change has resulted from these debates. Defence Minister Yitzhak Mordechai stated in July 1998 that the 30-year policy not to be the first to introduce nuclear weapons into the region was 'a formula that has served us well and, at least for the time being, should remain in place'. The policy 'has achieved a deterrent', a senior defence official was reported to have said, 'yet it hasn't broken any taboos'.[58]

Israel first began to separate plutonium at the Dimona facility in 1966, and design work on the first nuclear explosive device was successfully completed around the same time. On the eve of the Six-Day War of June 1967, Israel reportedly 'improvised' two deliverable nuclear explosive devices.[59] After the war, efforts to build an operational nuclear arsenal intensified and an explosion on 22 September 1979 high in the atmosphere over the South Indian Ocean off the coast of South Africa is believed by some to have been a clandestine Israeli test. Today, nuclear weapons are assembled at the design laboratory at Yodefat Rafael facility, outside Haifa, known as Division 20. Dimona, in the Negev desert, is the location of a plutonium–tritium production reactor and underground chemical separation and nuclear component fabrication facilities. Nuclear weapon research and design takes place at the Soreq Center near the town of Yavne and the 'Soreq Center runs the full nuclear gamut of activities . . . required for nuclear weapons design and fabrication', according to a 1987 Pentagon study.[60] The Soreq Center shares a security zone with the Palmikhim AB, from where missiles are assembled and test launched into the Mediterranean Sea. Operational nuclear weapons are thought to be stored near the Tel Nof AB, south-east of Tel Aviv, which is also adjacent to the Sdot Micha AFB for the Jericho missiles. Some sources say that these weapons may be stored at the nearby Tirosh depot, but all three facilities may form part of a larger complex (it is only 9 km from Tel Nof in the north to Sdot Micha in the south).

Strike aircraft

Over the past 30 years Israel has had many different types of aircraft capable of carrying nuclear bombs. These include the F-4 Phantom, the A-4 Skyhawk and, more

---

[57] US Department of Defense, Office of the Secretary of Defense, *Proliferation: Threat and Responses*, released in Jan. 2001, URL <http://www.defenselink.mil>; US Strategic Air Command, 'The Phoenix Study', Sep. 1991, p. 67, URL <http://www.nautilus.org/nukestrat/USA/Force/phoenix.html>.

[58] Blanche, E., 'Nuclear reactions', *Jane's Defence Weekly*, 17 June 1998, p. 22; and Rodan, S., 'Israel mulls nuke stance amid new threats', *Defense News*, 29 June–5 July 1998, p. 3.

[59] Cohen, A., *Israel and the Bomb* (Columbia University Press: New York, 1998), pp. 273–74.

[60] 'Strategic Israel', MSNBC International News, n.d. [1998], URL <http://archive.msnbc.com/modules/Israel_Strategic/>.

Table 10A.11. Israeli nuclear forces, January 2002

Type	Year first deployed	Range (km)[a]	Comment
Aircraft			
F-16A/B/C/D/I Fighting Falcon	1980	1 600	Bombs probably stored at Tel Nof
F-15I Thunder	1998	3 500	Selected for long-range strike role
Land-based missiles			
Jericho I	1972	1 200	Possibly 50 at Sdot Micha
Jericho II	1984–85	1 800	Possibly 50 at Sdot Micha, on TELs in caves
Non-strategic/battlefield [b]			
Artillery and landmines			

[a] Range for aircraft indicates combat radius, without in-flight refuelling.
[b] The status of Israeli non-strategic nuclear weapons is unknown.

Sources: Cohen, A., *Israel and the Bomb* (Columbia University Press: New York, 1998); Albright, D., Berkhout, F. and Walker, W., SIPRI, *Plutonium and Highly Enriched Uranium 1996: World Inventories, Capabilities and Policies* (Oxford: Oxford University Press, 1997); Hough, H., 'Could Israel's nuclear assets survive a first strike?', *Jane's Intelligence Review*, vol. 9, no. 9 (1997), pp. 407–10; Burrows, W. E. and Windrem, R., *Critical Mass* (Simon & Schuster: New York, 1994), pp. 275–313; 'US Strategic Air Command/XP, "Secret/ Phoenix"', 11 Sep. 1991, p. 67, released under the US Freedom of Information Act and available at URL <http://www.nautilus.org/nukestrat/USA/Force/phoenix.html>; Hersh, S. M., *The Samson Option* (Random House: New York, 1991); *Jane's Defence Weekly*, various issues; *Defense News*, various issues; and Authors' estimates.

recently, the F-16 and the F-15E. In January 2000 the Israeli Government purchased 50 F-16Is from the USA worth about $2.5 billion. Israel will begin to take delivery of the aircraft at the beginning of 2003, and the last aircraft will be supplied two years later. An additional 52 F-16Is were ordered in December 2001 and will be delivered by 2008. In four previous orders, Israel purchased or received 260 F-16s between 1980 and 1995. These included 103 F-16A, 22 F-16B, 81 F-16C and 54 F-16D models. A number of nuclear bombs may be allocated to dedicated, certified aircraft, probably at the Tel Nof AB.

Ballistic missiles

Israel's quest for a missile capability began simultaneously with its quest for nuclear weapons. In April 1963—several months before the Dimona reactor began operating—Israel signed an agreement with the French company Dassault to produce a surface-to-surface ballistic missile. Israeli specifications called for a two-stage missile capable of delivering a 750-kg warhead 235–500 km with a CEP of less than 1 km. The missile system, known as the Jericho (or MD-620), was also to take less than two hours to prepare, launch from fixed or mobile bases and fire at a rate of four to eight per hour. In early 1966 it was reported that Israel had purchased the first instalment of

30 missiles,[61] but soon after the June 1967 Six-Day War France imposed an embargo on new military equipment. Because of the French embargo, Israel began to produce the Jericho missile on its own. In 1974 the CIA cited the Jericho as evidence that Israel had made nuclear weapons—it stated that the Jericho made little sense as a conventional missile and was 'designed to accommodate nuclear warheads'.[62]

Israel subsequently developed the Jericho II, a missile similar to the US Pershing II IRBM. In May 1987 it tested an improved version of the Jericho II that flew 800 km. A document published in 1989 by the US Arms Control and Disarmament Agency gave the maximum range of the improved Jericho as 1450 km, long enough to reach the southern border of the Soviet Union.[63] Israel pursued technology in the USA and elsewhere for the missile, including a terminal guidance system using radar imaging. It is thought that the range has been increased to 1800 km. An article published in 1994 reported that about 50 Jericho II missiles were stored at the Sdot Micha base, some 45 km south-east of Tel Aviv in the Judean Hills.[64] According to an analysis of satellite images of the base, the missiles appear to be stored in caves. Upon warning they would be dispersed on their TELs so as not to be destroyed. The shorter-range Jericho I is deployed nearby in approximately equal numbers. In April 2000 Israel test launched a Jericho missile into the Mediterranean Sea. The missile landed near a US warship and the crew reportedly thought that they were under attack. Israel did not inform the USA of the test in advance.

Other weapon systems

There are also reports that Israel has developed nuclear artillery shells and possibly ADMs, stored at the Eilabun storage facility west of the Sea of Galilee. Following a report in March 2000 that Israel planned to lay neutron landmines to deter a Syrian attack following a withdrawal from the Golan Heights, Israeli Deputy Defence Minister Ephraim Sneh responded that 'this report is truly stupid. The person that wrote it not only doesn't know, but also doesn't understand anything'.[65]

In addition to the Israeli aircraft and missiles listed in table 10A.11, there are persistent rumours that Israel may also be pursuing a sea-based nuclear capability for its three new Dolphin Class submarines. Israeli officials repeatedly dismissed the rumours in 2000, and the Israeli Navy stated that the submarines 'are completely conventional' and that 'these foreign sources have fantasies'.[66] Israel has attempted to acquire long-range BGM-109 Tomahawk cruise missiles, but the US Government rejected an Israeli request for 48 missiles in March 2000.

---

[61] Finney, J. W., 'Israel said to buy French missiles', *New York Times*, 7 Jan. 1966, p. 1.
[62] Central Intelligence Agency, 'Prospects for further proliferation of nuclear weapons', DCI NIO 1945/74, 4 Sep. 1974.
[63] Arms Control and Disasrmament Agency (ACDA), 'Ballistic missile proliferation in the developing world', *World Military Expenditures and Arms Transfers 1988* (ACDA: Washington, DC, June 1989), pp. 17–20.
[64] Hough, H., 'Could Israel's nuclear assets survive a first strike?', *Jane's Intelligence Review*, vol. 9, no. 9 (1997), pp. 407–10.
[65] 'Sneh denies nuclear landmine report', *Jane's Defence Weekly*, 5 Apr. 2000, p. 6.
[66] Rodan, S., 'Israel received last Dolphin-class submarine', *Jane's Defence Weekly*, 8 Nov. 2000, p. 16; and Mahnaimi, U. and Campbell, M., 'Israel makes nuclear waves with submarine missile test', *Sunday Times*, 18 June 2000.

Appendix 10B. Tactical nuclear weapons

NICHOLAS ZARIMPAS

I. Introduction

More than a decade after the end of the cold war and successive nuclear weapon reductions, thousands of tactical nuclear weapons continue to be stockpiled in the arsenals of Russia and the United States. Far from the attention of the public and low on the political agenda, tactical nuclear weapons remain outside formal arms control agreements and—because of their small size, mobility, and decentralized command and control arrangements—pose clear security threats.

Hopes for the speedy elimination of tactical nuclear weapons were raised in the early 1990s in the aftermath of unilateral reciprocal initiatives by Russia and the USA. However, recent developments have given rise to concerns that increased reliance on, and new missions for, these weapons can be expected. Such concerns are exacerbated by the continued lack of transparency surrounding their numbers and operational status.

This appendix provides an overview of the major issues and developments related to tactical nuclear weapons, focusing on publicly available information, and reviews proposals made for controlling them. Section II discusses definitions, past deployments and the current status. Section III outlines the risks and challenges associated with tactical nuclear weapons. The prospects and means for their control are examined in section IV, and section V presents the conclusions.

II. Definitions, history and current status

Definitions

The term 'tactical nuclear weapon' emerged early in the cold war.[1] Although ambiguous and imprecise, the term is still widely used by arms control analysts, practitioners and academics to describe different and diverse types of Russian and US nuclear weapon systems.[2] 'Tactical', in the classical military sense, predominantly denotes a

---

[1] Existing bilateral nuclear arms control treaties and agreements, as well as multilateral treaties of global application, such as the 1968 Non-Proliferation Treaty and the 1996 Comprehensive Nuclear Test-Ban Treaty, do not provide definitions of the terms 'nuclear warhead' and 'nuclear weapon'. For the purposes of this appendix, a nuclear warhead is a mass-produced, reliable, predictable nuclear device capable of being carried by missiles, aircraft or other means. A nuclear weapon is a nuclear warhead mated and fully integrated with a delivery platform. Cochran, T. B., Arkin, W. M. and Hoenig, M. M., *Nuclear Weapons Databook, Vol. I: US Nuclear Forces and Capabilities* (Ballinger: Cambridge, Mass., 1984), p. 2.

[2] A distinction is sometimes made between 'theatre' (ranges of more than a few tens of kilometres) and 'tactical' nuclear weapons. However, there is no clear dividing line between these 2 categories. The term 'theatre' has had a perceived negative connotation in the European political debate. For a discussion of problems associated with attempts to define or categorize tactical nuclear weapons see Müller, H. and Schaper, A., 'Definitions, types, missions, risks and options for control: a European perspective', eds W. C. Potter *et al.*, *Tactical Nuclear Weapons: Options for Control*, UNIDIR 2000/20 (United Nations Institute for Disarmament Research: Geneva, Dec. 2000), pp. 31–35. Other terms, principally relating to the range, are frequently used: non-strategic, sub-strategic or pre-strategic (these 2 emerged from the French nuclear doctrine) and short-range.

short-range battlefield use.[3] In contrast, 'strategic' denotes weapons that, if fired from Russia or the USA, would be able to reach predetermined targets on the territory of the other country.[4]

Attempting to define tactical nuclear weapons by mission or range is challenging because some types can be used to carry out strategic tasks. This applies particularly to the long-range sea-launched cruise missiles (SLCMs), air-launched cruise missiles (ALCMs) and aircraft-delivered gravity bombs in the inventories of both Russia and the USA. For instance, gravity bombs can be delivered by tactical aircraft at relatively long ranges to hit 'strategic' targets, such as major cities or nuclear missile silos.[5] In the context of the nuclear arms reduction process, Russia, concerned with the launch capabilities of US attack submarines, viewed SLCMs as 'strategic' weapons.[6] Moreover, what the two nuclear superpowers consider 'tactical' could be seen as 'strategic' in the context of, for example, a regional conflict between one of the superpowers and a neighbouring country.[7]

A definition based on the nuclear explosive yield of the weapon is also problematic. Tactical nuclear weapons have nominally smaller yields than most strategic weapons, as low as 0.3 kt. However, some have much higher yields than those of the bombs that destroyed Hiroshima and Nagasaki (c. 15 kt) or of certain strategic weapons.[8]

Definitions based on other criteria have been put forward. One analyst has defined tactical nuclear weapons as those weapons deployed with general-purpose forces, in contrast to strategic nuclear weapons, which are normally operated by special nuclear units.[9] Others have attempted to classify them on the basis of their delivery platforms.

Any criteria chosen to differentiate tactical from strategic weapons must appropriately accommodate the purpose of the definition. In most practical cases, as in discussions about controlling them or estimating their numbers, a definition by exclusion is used, such as: 'Tactical nuclear weapons are those nuclear weapons whose deploy-

---

[3] 'Tactical nuclear weapon employment is the use of nuclear weapons by land, sea, or air forces against opposing forces, supporting installations or facilities, in support of operations that contribute to the accomplishment of a military mission of limited scope, or in support of the military commander's scheme of manoeuvre, usually limited to the area of military operations.' US Department of Defense, *Dictionary of Military and Associated Terms*, 12 Apr. 2001 (as amended through 15 Oct. 2001), p. 426, available at URL <http://www.dtic.mil/doctrine/jel/new_pubs/jpl_02.pdf>.

[4] For a detailed historical account see Müller and Schaper (note 2), pp. 22–33.

[5] 'In many cases, the distinction between a tactical nuclear weapon and a strategic nuclear weapon is an artificial one. Many tactical or non-strategic nuclear weapons can be used with strategic effect against the United States, our forward-deployed forces, and our allies.' Statement by Admiral Richard W. Mies, Commander-in-Chief, United States Strategic Command, US Strategic Nuclear Force Requirements, Hearing before the Committee on Armed Services, United States Senate, 106th Congress, 2nd session, 23 May 2000 (US Government Printing Office: Washington, DC, 2000), p. 38.

[6] Diakov, A. et al., 'NATO expansion and the nuclear reductions process', Paper presented at the conference NATO Movement to the East–Security Problems for Russia and the CIS States, St Petersburg, 28–29 Apr. 1999, available at URL <http://www.armscontrol.ru/start/publications/nato0430.htm>.

[7] In 1962 the Soviet Union stationed in Cuba several nuclear-armed SS-4 intermediate-range ballistic missiles as a response to US deployments in Europe, in particular in Turkey. These weapons, although they did not have 'strategic' ranges, were capable of reaching the USA. For a discussion of the ensuing Soviet–US crisis, which came close to a nuclear exchange, see 'The Cuban missile crisis', History and Politics Out Loud (HPOL) Archive, URL <http://www.hpol.org/jfk/cuban>; and 'The Cuban missile crisis, 1962', The National Security Archive, URL <http://www.gwu.edu/~nsarchiv/nsa/cuba_mis_cri/declass.html>.

[8] E.g., the USA's B61 Mod. 3 warhead intended for tactical missions has a yield of up to 170 kt. See section II of appendix 10A.

[9] Sokov, N., cited in Müller and Schaper (note 2), p. 35.

ment is not regulated under the START treaties (and which have not been eliminated by the INF treaty)'.[10]

The use of tactical nuclear weapons could result in enormous destruction and massive casualties in certain scenarios. Therefore, they should not be viewed as entirely different from or deserving less attention than their strategic counterparts. In this appendix, the term 'tactical nuclear weapon' is used to encompass all Russian and US battlefield and theatre nuclear weapon systems with lower than intercontinental ranges (less than 5500 km).

A brief history

Tactical nuclear weapons were first deployed by the USA in Europe, beginning in the 1950s, as a way to compensate for NATO's perceived inferiority vis-à-vis the Soviet Union in conventional military capabilities. At first they consisted only of nuclear artillery shells. These were followed by nuclear gravity bombs and, in the late 1950s and 1960s, by the introduction of the nuclear-armed Honest John and Pershing I missiles. The maximum build-up of US tactical nuclear weapons in Europe (about 7000) occurred at the beginning of the 1970s.[11] Throughout most of the 1970s and 1980s, NATO maintained a broad mix of nuclear weapon systems, including atomic demolition munitions (ADMs), nuclear artillery, air-to-surface missiles (ASMs), surface-to-air missiles (SAMs), short- and intermediate-range surface-to-surface missiles (SSMs), ground-launched cruise missiles, torpedoes and gravity bombs delivered by dual-capable aircraft.[12] Gravity bombs were also deployed by the United Kingdom in the Federal Republic of Germany.[13] Apart from those in Europe, a substantial number of nuclear weapons were deployed on board US aircraft carriers and other ships. The USA also deployed nuclear weapons in other countries, such as South Korea and Taiwan, but on a smaller scale and of a lesser diversity.[14]

---

[10] Lewis, G. and Gabbitas, A., *What Should Be Done About Tactical Nuclear Weapons?*, Occasional Paper (Atlantic Council of the United States: Washington, DC, Mar. 1999), p. 4. The treaties mentioned are the 1991 Treaty on the Reduction and Limitation of Strategic Offensive Arms (START I Treaty), the 1993 Treaty on Further Reduction and Limitation of Strategic Offensive Arms (START II Treaty, not in force), and the 1987 Treaty on the Elimination of Intermediate-Range and Shorter-Range Missiles (INF Treaty). See annex A in this volume.

[11] Shevtsov, A. et al., *Tactical Nuclear Weapons: A Perspective from Ukraine*, UNIDIR/2000/21 (United Nations Institute for Disarmament Research: Geneva, Dec. 2000), p. 1; and Wezeman, S. T., Private communication with the author. Germany hosted by far the most nuclear weapons (at the peak of NATO's deployment it stored c. 50% of them), with 21 different types of US warheads deployed on its soil in the years since 1955. Norris, R. S., Arkin, W. M. and Burr, W., 'Where they were', *Bulletin of the Atomic Scientists*, vol. 55, no. 6 (Nov./Dec. 1999), p. 29.

[12] 'NATO's nuclear forces in the new security environment', NATO Basic Fact Sheet, Apr. 1999, p. 2.

[13] For a detailed description of developments and deployments of British and French nuclear weapons see Norris, R. S., Burrows, A. S. and Fieldhouse, R. W., *Nuclear Weapons Databook, Vol. V: British, French and Chinese Nuclear Weapons* (Westview Press: Boulder, Colo., 1994); and Shevtsov et al. (note 11), pp. 2–5.

[14] In some cases, the host country was not aware of the presence of US nuclear weapons on its territory. It has been reported that non-nuclear components for nuclear warheads were stored in Japan. In 1967 the total number of tactical nuclear weapons possessed by the USA reached a peak of 20 000. For a discussion of US deployments based on a recently declassified US DOD study on the history of US tactical nuclear weapons see Norris, Arkin and Burr (note 11), pp. 26–35; and Norris, R. S., Arkin, W. M. and Burr, W., 'How much did Japan know?', *Bulletin of the Atomic Scientists*, vol. 65, no. 1 (Jan./Feb. 2000), pp. 11–13. It has been a long-standing official US policy not to comment specifically on where tactical nuclear weapons are stationed, primarily for security reasons but also for political reasons in the host countries. See Bacon, K. H., former US Assistant Secretary of Defense, US Department of Defense

The Soviet nuclear arsenal consisted of a large variety of land-based weapons (short-range SSMs, nuclear artillery, ADMs and SAMs), sea-based weapons (torpedoes and sea-launched anti-ship and land attack cruise missiles) and airborne systems (gravity bombs, short-range cruise missiles, depth charges and torpedoes).[15] Many were stationed outside Russia, in Soviet republics that were closer to prospective theatres of operation.[16] Tactical nuclear weapons were also deployed in some countries of the Warsaw Treaty Organization, probably in Bulgaria, Czechoslovakia, the German Democratic Republic, Hungary and Poland.[17]

The Presidential Nuclear Initiatives

Tactical warheads are the only types of warheads that the two nuclear superpowers agreed to eliminate, although without any transparency or verification arrangements.[18] The Presidential Nuclear Initiatives (PNIs) of 1991–92 were formulated in the context of intensive rounds of agreements in the early 1990s. On 27 September 1991, spurred by the August 1991 failed coup in the Soviet Union and in an attempt to contain proliferation risks posed by the Soviet nuclear arsenal in such precarious political conditions, US President George Bush announced, among other arms control initiatives, his intention to unilaterally and swiftly reduce US tactical nuclear weapons and called upon Russia to reciprocate. The Bush proposals included: (*a*) the complete elimination of all ground-launched, short-range nuclear weapons, including those in Europe and South Korea, and the dismantlement and destruction of all such warheads; and (*b*) the withdrawal of all tactical nuclear weapons from US ships and submarines as

---

News Briefing, 20 Oct. 1999, available at URL <http://www.defenselink.mil/news/Oct1999/t10211999_t1020asd.html>.

[15] The first Soviet tactical nuclear weapons appeared in the mid-1950s (the SS-3 missiles, bombs for the Il-28 and Tu-16 bombers, and nuclear artillery shells). They were later supplemented with, *inter alia*, intermediate-range missiles, other bombers and sea-launched missiles. Shevtsov *et al.* (note 11), p. 5; Safranchuk, I., 'Tactical nuclear weapons in the modern world and Russia's sub-strategic nuclear forces', no. 16 (PIR Center for Nonproliferation and Arms Control: Moscow, Mar. 2000), p. 6; and Wezeman, S. T., Private communication with the author.

[16] Information concerning the total number of operational Soviet tactical nuclear weapons and their deployment was never made public. According to some estimates, when the Soviet Union collapsed in late 1991, it possessed over 15 000 warheads for tactical nuclear weapons. The majority of weapons outside Russia reportedly were in Belarus, Kazakhstan and Ukraine, with perhaps less than 5% in Georgia and the Central Asian states (Kyrgyzstan, Tajikistan, Turkmenistan and Uzbekistan). Woolf, A. F., 'Nuclear weapons in Russia: safety, security, and control issues', IB98038, Congressional Research Service (CRS) Issue Brief for Congress, Washington, DC, 21 Nov. 2000, pp. 2–3, available at URL <http://www.cnie.org/nle/inter-64.html>. Other sources provide different numbers, e.g., *c.* 22 000 warheads. Arbatov, A. (Deputy Chairman of the Defence Committee of the Russian State Duma), 'Deep cuts and de-alerting: a Russian perspective', ed. H. A. Feiveson, *The Nuclear Turning Point: A Blueprint for Deep Cuts and De-Alerting of Nuclear Weapons* (Brookings Institution Press: Washington, DC, 1999), p. 320. There are also different views as to whether the Soviet Union actually deployed tactical nuclear weapons outside the territories of the 4 republics where its strategic weapons were stationed. Sokov, N., 'Tactical nuclear weapons', *Disarmament Diplomacy*, no. 21 (1998), URL <http://www.acronym.org.uk/21tactic.htm>.

[17] Shevtsov *et al.* (note 11), p. 7.

[18] The 1987 INF Treaty banned Soviet and US ground-launched cruise missiles with ranges of 500–5500 km. It thus eliminated the majority of the longer-range tactical nuclear weapons, apart from nuclear-armed sea-launched cruise missiles. The INF Treaty entered into force in June 1988 and by May 1991 the Soviet Union had completed the dismantling of all missiles covered by the treaty, a total of 1846 SS-4, SS-5, SS-20 and SS-21 ballistic missiles. Wolfsthal, J. B., Chuen, C. A. and Daughtry, E. E., *Nuclear Status Report: Nuclear Weapons, Fissile Material, and Export Controls in the Former Soviet Union* (Monterey Institute of International Studies and Carnegie Endowment for International Peace), no. 6 (June 2001), p. 32. The USA dismantled 846 missiles, including all Pershing IA and II ballistic missiles, and all land-based Tomahawk cruise missiles. Wolfsthal, Chuen and Daughtry, p. 32.

well as nuclear depth bombs for land-based naval aircraft, followed by their storage at depots in the USA (for the newer systems) or their dismantlement and the destruction of their warheads (about one-half).[19]

Soviet President Mikhail Gorbachev responded to the US initiatives on 5 October 1991, matching the US moves and going further in several respects: he not only agreed to remove all tactical nuclear weapons from Soviet ships, submarines and land-based naval aircraft bases but also suggested eliminating them altogether on a reciprocal basis instead of storing them. In addition, Gorbachev proposed that both sides remove tactical air-delivered bombs and missiles from forward-deployed units and store the warheads at separate bases.[20] Russian President Boris Yeltsin continued these initiatives by announcing on 29 January 1992 that the production of warheads for land-based tactical missiles, artillery shells and landmines had ceased and that Russia had begun to eliminate one-third of its naval tactical warheads and one-half of its nuclear SAM warheads. He stated that tactical air force weapons would be reduced by one-half and proposed removing the remaining weapons from their units and placing them in centralized storage bases on a reciprocal basis with the USA.[21] According to an authoritative estimate, some 13 700 tactical warheads were subject to elimination in accordance with the Soviet/Russian pledges.[22]

Some fragmented, largely unverifiable, official information about progress made in implementing the PNIs is available. In 2000 Russian Minister of Foreign Affairs Igor S. Ivanov announced that Russian tactical nuclear weapons had been completely removed from surface ships and multi-purpose submarines, as well as from land-based naval aircraft, and were stored at central storage facilities. He stated that one-third of all nuclear munitions for sea-based tactical missiles and naval aircraft had been eliminated, and that the destruction of nuclear warheads removed from tactical missiles, artillery shells and nuclear mines was ongoing. In addition, one-half of the number of nuclear warheads for anti-aircraft missiles and nuclear gravity bombs had been destroyed.[23] Since the early 1990s, the number of nuclear weapons earmarked for NATO in Europe has been reduced by over 85 per cent. Associated delivery systems have been reduced from 11 to 1, which is no longer maintained on alert. The USA has cancelled numerous tactical nuclear programmes and US army, marine, and navy surface and air components are no longer equipped with nuclear weapons.[24]

---

[19] Fieldhouse, R., 'Nuclear weapon developments and unilateral reduction initiatives', *SIPRI Yearbook 1992: World Armaments and Disarmament* (Oxford University Press: Oxford, 1992), p. 67. Appendix 2A (pp. 85–92) provides excerpts of the presidential announcements.

[20] Fieldhouse (note 19), p. 72.

[21] Fieldhouse (note 19), p. 73. The USA did not respond to Russia's offers to consolidate tactical aviation weapons.

[22] Arbatov (note 16), p. 319.

[23] Ivanov, I. S., Statement at the Review Conference of the Parties to the Treaty on the Non-Proliferation of Nuclear Weapons, New York, 25 Apr. 2000, unofficial translation from Russian, available at URL <http://www.ln.mid.ru/website/Brp_4.nsf/e78a48070f128a7b43256999005bcbb3/ee86737 47bab02eb4325699c00260990?OpenDocument>. See also remarks by Marshal Igor Sergeyev, former Russian Minister of Defence, Advisor to President Putin, on strategic stability, at the Carnegie International Non-Proliferation Conference New Leaders, New Directions, 18 June 2001, available at URL <http://www.ceip.org/files/projects/npp/resources/ Conference%202001/sergeyev.htm>. For further detail on reductions of tactical nuclear weapons by Russia see appendix 10A.

[24] Fact Sheet: US Commitment to NPT Article VI, Bureau of Nonproliferation, US Department of State (Washington File), 1 Apr. 2000, URL <http://usinfo.state.gov/topical/pol/arms/stories/article6. htm>. In line with the 1994 US Nuclear Posture Review the option was retained, however, to redeploy SLCMs on attack submarines in times of crisis. See Lewis and Gabbitas (note 10), p. 7.

Since the end of the cold war, the USA has reduced its tactical nuclear warheads by over 80 per cent.[25]

It should be noted, however, that the numbers of remaining Russian and US tactical nuclear weapons have not been officially disclosed. Given that initial inventories were never made public, the value of statements announcing the percentages by which the various force components have been reduced is limited.[26]

The current status[27]

Russia

All tactical nuclear weapons were withdrawn from the former Soviet territories and returned to Russia by the spring of 1992.[28] Assessing the composition and number of the remaining Russian tactical nuclear weapons is subject to considerable uncertainty. Frequently quoted statistics are mainly based on a few official statements, unclassified US intelligence reports and analyses by experts. Most published estimates agree that there are fewer than 4000 Russian operational warheads for air defence missiles, tactical aviation and naval weapons, but large variations have been reported in their composition. The majority of the tactical warheads, apart from gravity bombs, are believed to be consolidated at regional or central storage sites.[29] According to one analyst, warheads for anti-ballistic missiles and naval weapons have already been dismantled or are awaiting dismantlement.[30] Weapons still retained as spares and reserves, or stored prior to elimination, may number several thousand (numbers as high as 12 000 have been quoted).[31]

---

[25] Statement of Admiral Richard W. Mies, US Navy, Commander-in-Chief, United States Strategic Command, before the Senate Armed Services Committee Strategic Subcommittee on Command Posture, 11 July 2001, available at URL <http://www.senate.gov/~armed_services/statemnt/2001/010711mies.pdf>.

[26] The numbers of tactical nuclear weapons each side possessed in 1991 were not even disclosed at confidential meetings held in the aftermath of the parallel initiatives. Sokov (note 16). In the 1997 Founding Act, NATO and Russia established a working group on nuclear weapons in the framework of their Permanent Joint Council. The initial agenda for this group included tactical nuclear weapon transparency, safety and security issues. It is known that the 2 sides shared limited information about their tactical nuclear weapons, but such cooperation stopped following the 1999 NATO bombing of Yugoslavia. NATO continued, on many occasions, to propose exchanges of data on Russian and US tactical weapons. See NATO–Russia Establish Nuclear Weapons Working Group, NATO Briefing, 4 Dec. 1997, available at URL <http://www.basicint.org/pjcworkinggroup.htm>; *NATO Report on Options for Confidence and Security Building Measures, Verification, Non-Proliferation, Arms Control and Disarmament*, Press Communiqué M-NAC-2(2000)121, 14 Dec. 2000; and Kozaryn, L. D., 'Russians say Yeltsin's nuclear pledge fulfilled', American Forces Press Service, URL <http://www.defenselink.mil/news/May1998/n05081998_9805086.html>.

[27] This section contains only a short summary. A detailed description of the current types, delivery systems and locations of tactical nuclear weapons can be found in Müller and Schaper (note 2) and on the Internet site of the Federation of American Scientists, URL <http://www.fas.org>. See also appendix 10A.

[28] On 1 Feb. 1992 Russian President Boris Yeltsin announced that the transfer of tactical nuclear weapons from Kazakhstan had been completed in Jan. On 28 Apr. 1992 Belarussian Defence Minister Pavel Koszlevsky announced that all tactical nuclear warheads in Belarus had been transferred to Russia. On 6 May 1992 Ukrainian President Leonid Kavchuk confirmed that all tactical nuclear weapons had been transferred to Russia except those on the ships and submarines of the Black Sea Fleet. US Department of State, *Arms Control and Disarmament: The US Commitment*, Pt. 5, Intermediate- and short-range nuclear forces, available at URL <http://usinfo.state.gov/products/pubs/armsctrl/pt5.htm>.

[29] The Soviet nuclear weapon storage system once comprised more than 500 sites. It has been considerably reduced, to some 90 sites. Shevtsov *et al.* (note 11), pp. 21–22.

[30] Sokov, N., cited in Müller and Schaper (note 2), pp. 59–60.

[31] See appendix 10A.

While there is little doubt that the large number and diversity of Russian tactical nuclear weapons are of concern, it should be noted that most of them are likely to reach obsolescence within a few years.[32] Therefore, the rate of warhead re-manufacturing and the production of new weapon systems are of crucial importance.[33]

The USA and NATO

On 2 July 1992 President Bush announced that the USA had completed the withdrawal of its ground- and sea-launched tactical nuclear weapons.[34] According to published estimates the USA currently possesses fewer than 1000 operational tactical nuclear warheads, consisting of some 650 B61 gravity bombs (including those stationed in NATO countries) and 320 nuclear-armed Tomahawk SLCMs, which are stored in depots for possible redeployment on attack submarines. An additional several hundred B61s may be in storage awaiting dismantlement.[35]

Following the drastic reductions made since the early 1990s, according to various estimates, about 150–180 US-produced and -owned B61 gravity bombs of variable yields are still deployed in seven European countries and Turkey for use on US or NATO allies' dual-capable aircraft.[36] The readiness requirements of such aircraft have been reduced from minutes at the height of the cold war to weeks today.[37] At present, these are the only nuclear weapons located outside the territory of a nuclear weapon state.[38] NATO storage sites underwent a massive reduction by about 80 per cent between 1991 and 1993 as entire weapon systems were eliminated and the aggregate number of weapons was reduced. However, a new, more secure and sur-

---

[32] According to Alexei Arbatov, most Russian tactical warheads earmarked for elimination under the 1991–92 PNIs were to be dismantled anyway because their design lives would have expired (note 16), p. 319. See also Rowny, E. L., Additional Commentary, in Lewis and Gabbitas (note 10), p. 29.

[33] The re-manufacturing of warheads after the end of their design life is presumably taking place but may be slow because assembly–disassembly facilities are handling the dismantlement of a backlog of strategic and tactical warheads. According to one estimate, in 2000 actual warhead production amounted to *c.* 200–300 warheads per year, the same rate as in the USA. See Timerbaev, R., 'Dealing with cold war nuclear legacy: Russian perspective', *Yaderny Kontrol (Nuclear Control) Digest*, vol. 5, no. 3 (summer 2000), p. 29. Reportedly, Russian officials contend that they have begun to dismantle tactical warheads at a rate of *c.* 2000 per year, but this number cannot be independently confirmed. See Woolf (note 16).

[34] US Department of State (note 28).

[35] See appendix 10A.

[36] Although officially unacknowledged, these airbases reportedly are: Araxos (Greece), Aviano and Ghedi-Torre (Italy), Büchel, Stangdalem and Ramstein (Germany), Inçirlik (Turkey), Klein Brogel (Belgium), Lakenheath (United Kingdom) and Volkel (Netherlands). Butcher M. *et al.*, *Questions of Command and Control: NATO, Nuclear Sharing and the NPT*, British American Security Information Council (BASIC); and Berlin Information-center for Transatlantic Security (BITS), PENN (Project on European Nuclear Non-proliferation) Research Report 2000.1, Mar. 2000, p. 43.

[37] NATO Basic Fact Sheet (note 12), p. 6.

[38] In the mid-1960s, NATO established the system of 'nuclear sharing', which in essence remains unchanged in its basic functions. It enables non-nuclear weapon state members of NATO to participate in nuclear decision making and discussions about NATO's nuclear policy and doctrine. The European NATO allies on whose territories US tactical nuclear weapons are stationed have bilateral nuclear cooperation agreements with the USA. Although such weapons remain under full custody and control of the US forces and cannot be armed without a US Presidential Order, in time of war control can be handed over to allied forces. There is a controversy as to whether NATO nuclear sharing arrangements are in full compliance with Articles I and II of the NPT. For a discussion see Butcher *et al.* (note 35); and 'NATO nuclear sharing and the NPT—questions to be answered', PENN Research Note 97.3, BASIC–BITS–Centre for European Security and Development (CESD)–Austrian Study Center for Peace and Conflict Research (ASPR), June 1997, available at URL <http://www.basicint.org/natonpt.htm>.

vivable weapon storage system has been installed.[39] Press reports in early 2001 to the effect that tactical nuclear weapons had been removed from the Araxos airbase in Greece remain unconfirmed by the Greek Government or NATO.[40]

III. Risks and challenges

The presence of a large number of tactical nuclear weapons poses unique dangers because of their physical characteristics, prescribed missions, the absence of monitoring or control by formal agreements, and the lack of transparency regarding their status and deployment. Analysts have argued that their 'very existence in national arsenals increases the risk of proliferation and reduces the nuclear threshold'.[41] The risks associated with tactical nuclear weapons can be grouped as follows.

Safety and security. The small size and portability of some weapons (e.g., ADMs), the lack of sophisticated electronic safeguards (permissive action links) against unauthorized use among older systems, combined with their wide dispersal, forward basing and movement with conventional forces, and pre-delegation of launch authority to local commanders make them vulnerable to theft and use by terrorists or to accidental detonation. Concerns have frequently been expressed about the level of physical safety, security and control of stored Russian tactical nuclear weapons.[42] Particularly in the early years after the demise of the Soviet Union, the large number of tactical nuclear weapons redeployed in Russia from former Soviet territories and the unavailability of adequate storage facilities, combined with political turmoil and unrest in the military, raised fears about the custodial security of the weapons. However, some accounts claim that their safety is in fact satisfactory.[43] The consolidation of Russian nuclear weapons in the early 1990s to a smaller number of storage facilities has rad-

[39] *NATO Handbook* (NATO Office of Information and Press: Brussels, 2001), p. 54, available at URL <http://www.nato.int/docu/handbook/2001/hb0206.htm>; and Bacon (note 14). Photographs released by the US Air Force (USAF) show that at least some storage vaults in Europe can hold 2 weapons. A maximum of 428 US nuclear weapons can be deployed on all bases and 360 weapons on all bases currently operational. The USAF plans to modernize the nuclear weapon storage systems in Europe in order to keep them operational beyond 2005. If upgraded, the storage systems can remain in use until 2018. *PENN Newsletter*, no. 14 (May 2001), p. 8, available at URL <http://www.bits.de/main/penntoc.htm>. See also appendix 10A.

[40] *Athens News*, 'Araxos warheads removed', 18 Jan. 2001, p. A01; and ['US nuclear weapons removed'], *De Standaard Online*, 18 Jan. 2001 (in Flemish).

[41] Potter, W. C. and Sokov, N., 'The nature of the problem', eds Potter *et al.* (note 2), p. 4.

[42] Woolf (note 16); and Woolsey, R. J., 'Putin's futile warhead-rattling', *Washington Post*, 26 June 2001, p. A17. For an extensive discussion of the security of Russian ADMs (the so-called 'suitcase nuclear bombs') see Parrish, S., 'Are suitcase nukes on the loose? The story behind the controversy', CNS Reports, Center for Non-Proliferation Studies, Monterey Institute of International Studies, URL <http://cns.miis.edu/pubs/reports/lebedlg.htm>. Following the 11 Sep. 2001 terrorist attacks against the USA there have been renewed concerns about the custodial security of Russian tactical nuclear weapons and about the possibility of their acquisition by the al-Qaeda network. See, e.g., McCloud, K. and Osborne, M., 'WMD terrorism and Usama bin Laden', CNS Reports, Center for Non-Proliferation Studies, Monterey Institute of International Studies, URL <http://cns.miis.edu/pubs/reports/binladen.htm>; Dolnik, A., *America's Worst Nightmare? Osama bin Laden and Weapons of Mass Destruction* (PIR Center Nonproliferation and Arms Control Hotline: Moscow, 12 Sep. 2001), p. 3; and 'The specter of nuclear terror', Editorial/Op-Ed, *New York Times*, 19 Nov. 2001, URL <http://www.nytimes.com/2001/11/19/opinion/19MON1.html>.

[43] Habiger, E., Commander-in-Chief, United States Strategic Command, US Department of Defense News Briefing, 16 June 1998, URL <http://www.defenselink.mil/news/Jun1998/t06231998_t616hab2.html>; Moscow ITAR-TASS, 'Foreign Minister denies al-Qa'idah has Russian nuclear devices', Foreign Broadcast Information Service, *Daily Report–Central Eurasia* (*FBIS-SOV*), FBIS-SOV-2001-1112, 12 Nov. 2001; and Moscow ITAR-TASS, 'Defense Ministry to control nuclear warheads by 2003', FBIS-SOV-2001-1119, 19 Nov. 2001.

ically increased their safety and security. Despite numerous press reports, there has been no credible claim of the evidence of the theft or attempted theft of Russian nuclear warheads or their components; nevertheless, episodes of serious proliferation concern may have occurred that are unknown outside government, police or intelligence circles.[44]

Missions and proliferation. Tactical nuclear weapons are predominantly intended to be used in nuclear war-fighting. Given that their yield is smaller than that of most strategic weapons, their engagement in a conflict could be more easily contemplated because of the false perception that damage would be localized. In addition, their deployment in times of crisis increases the likelihood of pre-emptive strikes and a nuclear exchange. An escalation to general nuclear war is thus possible.

In recent years, there has been interest in both Russia and the USA in expanding the roles and missions of tactical nuclear weapons. Both countries have indicated that they would consider using tactical nuclear weapons against targets in non-nuclear weapon states with chemical or biological weapon capabilities. Giving an expanded role to nuclear weapons in military planning serves to legitimize their use and encourages de facto or aspiring nuclear weapon states to follow suit, leading to a worldwide build-up.

Arms control and disarmament. The large number of tactical nuclear weapons, asymmetries in the stockpiles of Russia and the USA, the uncertainties surrounding their operational status, the difficulties with appropriately defining them, and the absence of any meaningful controls and transparency present significant challenges for arms control and hinder disarmament. These problems will become more pronounced after deeper reductions in strategic weapons have been achieved.

Recent developments concerning tactical nuclear weapons

In the aftermath of the 1991–92 PNIs, both Russia and the USA have in general exercised restraint with regard to their tactical nuclear weapons. No new weapon systems have been produced and no credible evidence of large-scale redeployment has been recorded. However, there have been indications, principally in Russia but also in the USA, of a revived, albeit controversial, interest in such weapons in their security debates.

Russia's strategic nuclear forces are rapidly declining and funding for their modernization is scarce. Russia's conventional capabilities, as evidenced in the conflicts in the republic of Chechnya, are weak. Tactical nuclear weapons are increasingly seen by Russia as a means to offset these weaknesses. Some Russian experts and planners, recognizing NATO's conventional superiority and concerned about the eastward enlargement of the alliance, have called for the abandonment of the informal regime of the 1991–92 PNIs, the production of a new generation of tactical nuclear weapons, and their redeployment in Belarus and Kaliningrad and on ships in the Baltic Sea region.[45] Such calls were strengthened after the 1999 NATO bombing of Yugoslavia.

[44] Zarimpas, N., 'The illicit traffic in nuclear and radioactive materials', *SIPRI Yearbook 2001: Armaments, Disarmament and International Security* (Oxford University Press: Oxford, 2001), p. 506; and Albright, D., Buehler, K. and Higgins, H., 'Bin Laden and the bomb', *Bulletin of the Atomic Scientists*, vol. 58, no. 1 (Jan./Feb. 2002), p. 23.

[45] 'After 2000 . . . NATO aircraft would be capable of destroying 60% of Russia's fixed and 15% of its mobile ICBM launchers.' Arbatov (note 16), p. 322. For an analysis of Russian concerns see Lewis and Gabbitas (note 10), pp. 15–19; and Sokov, N., 'Tactical nuclear weapons elimination: next step for arms control', *Nonproliferation Review*, vol. 4, no. 2 (winter 1997), pp. 17–27.

In addition, the abandonment of the Soviet no-first-use pledge in 1993 and the revision of the Russian military doctrine in 2000 have widened the role and salience of tactical nuclear weapons in Russian military planning.[46]

Decisions taken at the 29 April 1999 Russian Security Council meeting, chaired by President Boris Yeltsin, reportedly included a resolution providing guidance for the development and use of tactical nuclear weapons.[47] Vladimir Putin, then Secretary of the Russian Security Council, commenting on the meeting, emphasized that Russia would comply with its obligations, unilateral or otherwise, to reduce its nuclear forces. The decisions of the Security Council probably pertained to the development of new low-yield nuclear munitions to be delivered by strategic launchers, including the modernization of existing weapons to reduce their yield.[48]

In the USA a small but vocal group consisting mainly of leading scientists at the Los Alamos and Sandia national laboratories, and backed by some conservative politicians and government officials, have been lobbying for the development of new, low-yield, earth-penetrating nuclear warheads (known as 'mini-nukes').[49] Coupled with high-precision delivery systems, the main purpose of such devices would be to destroy hardened or deeply buried underground targets, arguably with minimal collateral damage.[50] Congressional legislation from 1994 prohibits all research and development leading to a precision nuclear weapon with a yield of less than 5 kt, clearly recognizing that the existence of such weapons would blur the line separating nuclear and conventional warfare.[51] An analysis based on the nuclear tests conducted in the 1960s as part of the US Plowshare Program[52] concluded that 'the use of any nuclear weapon capable of destroying a buried target that is otherwise immune to conventional attack will necessarily produce enormous numbers of civilian casualties'.[53] It has recently been revealed that the US defence and energy departments have com-

---

[46] 'Russia's military doctrine', *Arms Control Today*, vol. 30, no. 4 (May 2000), pp. 29–38; and Trenin, D., 'Russia and the future of nuclear policy', ed. B. Schmitt, *Nuclear Weapons: A New Great Debate*, Chaillot Papers 48 (Institute for Security Studies–Western European Union: Paris, July 2001), p. 115.

[47] Khripunov, I., 'Russia's MINATOM struggles for survival: implications for US–Russian relations', *Security Dialogue*, vol. 31, no. 1 (Mar. 2000), p. 57.

[48] Safranchuk (note 15), pp. 4–5.

[49] Although initiatives and studies regarding earth-penetrating and low-yield nuclear weapons have a long history, recent proposals notably include: Younger, S. M. (Associate Laboratory Director for Nuclear Weapons, Los Alamos National Laboratory), *Nuclear Weapons in the Twenty-first Century*, LA-UR-00-2850 (Los Alamos National Laboratory: Los Alamos, N. Mex., 27 June 2000); and Robinson, C. P. (President and Director, Sandia National Laboratories), 'A White Paper: pursuing a new nuclear weapons policy for the 21st century', 22 Mar. 2001, available at URL <http://www.sandia.gov/media/whitepaper/2001-04-Robinson.htm>.

[50] The strategic B61 Mod. 11 earth-penetrating gravity bomb with various yields, which can also be delivered by tactical aircraft, was added to the US arsenal in 1997. It was the first new capability since 1989, despite US assurances that no new nuclear weapons were being developed. For an analysis see Mello, G., 'New bomb, no mission', *Bulletin of the Atomic Scientists*, vol. 53, no. 3 (May/June 1997), pp. 28–32. Since no changes were introduced into the nuclear explosive package, the US national laboratories do not consider the B61 Mod. 11 as a new design.

[51] US National Defense Authorization Act for Fiscal Year 1994, Public Law 103-160, sec. 3136, Prohibition on Research and Development of Low-Yield Nuclear Weapons, URL <http://www.fcnl.org/issues/arm/sup/tesban_nukfurse.htm>. Two US Republican senators, Wayne Allard of Colorado and John Warner of Virginia, included a provision to the 2001 Defense Authorization Bill to pave the way for a new generation of nuclear weapons. See URL <http://www.fcnl.org/issues/arm/sup/min_allard52400.htm>.

[52] See, e.g., 'Researching atoms for peace', URL <http:www.sandia.gov/recordsmgmt/apr2001_pg2.htm>.

[53] Nelson, R. W., 'Low-yield earth penetrating nuclear weapons', Federation of American Scientists, Public Interest Report, Jan./Feb. 2001, p. 2, URL <http://www.fas.org/faspir/2001/v54n1/weapons.htm>.

pleted an initial study on how existing nuclear weapons could be modified to attack hardened and deeply buried complexes. These departments also continue to assess the requirements for such a weapon and have formed a joint planning group to define the scope for a possible design feasibility and cost study. However, no decision has been made to go ahead with a programme to design a new or modified nuclear weapon for hardened targets.[54] In addition, given the rapid advance of conventional weapons technology, few in the US military appear interested in nuclear weapons for new missions.[55]

The production of a new generation of tactical nuclear weapons would undermine the efforts to control them and would be contrary to the spirit of the 1968 Non-Proliferation Treaty (NPT) as well as to the pledges of the nuclear weapon states to reduce their forces. Moreover, new warheads would probably require testing and would thus undermine the 1996 Comprehensive Nuclear Test-Ban Treaty (CTBT).

Arms control developments

Shortcomings of informal, non-legally binding reductions

The 1991–92 Presidential Nuclear Initiatives were timely, significant and courageous arms control steps. However, while warranted by the extraordinary geopolitical situation of the early 1990s, they are not without flaws. First, they are not legally binding and therefore easy to modify, reverse or violate, in particular since they do not contain any verification or transparency measures regarding compliance. Second, because lengthy consultations did not take place, the two parties may have interpreted and carried out their obligations in different ways. Third, there are no mechanisms to resolve disputes, ensure cooperation and provide continuity. Fourth, they do not prohibit the modernization or manufacture of new warheads. Although the initiatives undoubtedly resulted in the consolidation of—and substantial net reductions in—tactical nuclear weapon systems, they did not eliminate existing asymmetries and thereby contributed to unpredictability and instability in the nuclear relations of Russia and the USA. This unpredictability is further exacerbated by the lack of transparency.

The START framework

Following the announcement and implementation of the PNIs, tactical nuclear weapons were largely ignored on the bilateral arms control agenda.[56] Nevertheless, in the mid-1990s presidents Bill Clinton and Boris Yeltsin agreed to exchange data on warhead and material stockpiles and on a number of occasions confirmed their com-

---

[54] Pincus, W., 'Nuclear strike on bunkers assessed', *Washington Post*, 20 Dec. 2001, p. A29; and Report to Congress on the Defeat of Hard and Deeply Buried Targets, submitted by the Secretary of Defense in conjunction with the Secretary of Energy in response to Section 1044 of the Floyd D. Spence National Defense Authorization Act for Fiscal Year 2001, PL 106–398, July 2001. The US Department of Energy budget request for FY 2003 called for completion of the preliminary study on the modification of existing nuclear weapons for hard target operations. Nuclear Threat Initiative (NTI): Global Security Newswire, 11 Feb. 2002, URL <http://www.nti.org/d_newswire/issues/newswires/ 2002_2_11.html#3>; and Pincus, W., 'Nuclear plans go beyond cuts', *Washington Post*, 19 Feb. 2002, p. A13.

[55] Schwartz, S. I., 'The new-nukes chorus tunes up', *Bulletin of the Atomic Scientists*, vol. 57, no. 4 (July/Aug. 2001), p. 35; and *PENN Newsletter*, no. 15 (Nov. 2001), p. 3, available at URL <http://www.bits.de/main/penntoc.htm>.

[56] START I contains a non-legally binding commitment concerning ceilings for SLCM deployment.

mitment to transparency and irreversibility in nuclear reductions. Moreover, they set up a joint working group to explore various transparency options, including verified warhead dismantlement. However, such commitments were never implemented.[57] At their Helsinki summit meeting in March 1997, the two presidents agreed that, in the context of the START III negotiations, but as separate issues, their experts would explore, 'possible measures relating to nuclear long-range sea-launched cruise missiles and tactical nuclear systems, to include appropriate confidence-building and transparency measures'.[58] This statement clearly reflected a compromise between the Russian desire to pursue the issue of US SLCM reductions and the US goal of increasing transparency in the Russian nuclear arsenal. It marked a significant departure from the previous exclusive focus on strategic systems and raised, for the first time, the prospect of including tactical nuclear weapons in formal bilateral arms control. The 1997 agreement to start a dialogue on tactical nuclear weapons was fragile because it was reached within a complicated and controversial framework of mutual concessions and linkages involving developments related to the 1972 Treaty on the Limitation of Anti-Ballistic Missile Systems (ABM Treaty) and the 1993 START II Treaty. Progress was undermined by problems in the strategic nuclear field and bilateral consultations lasted for less than two years, that is, until early 1999.[59]

The Kaliningrad controversy

Indicative of the lack of transparency surrounding tactical nuclear weapons and of the absence of instruments to monitor compliance with the PNIs is the controversy over allegations published in the media that, for the first time since the end of the cold war, Russia had, since June 2000, been redeploying battlefield nuclear weapons in the small Russian exclave of Kaliningrad.[60] There was speculation that these redeployments were Russia's response to NATO's eastward enlargement and constituted an attempt to compensate for Russia's rapidly deteriorating conventional forces. Doubts about the accuracy of the allegations persisted, principally because of the timing of the alleged deployment.[61] Estonia, Latvia, Lithuania and Poland expressed concern, sought clarification from Russia or demanded inspections.[62] The US Administration and NATO made no public statements. Several high-level Russian officials, including President Putin, have categorically denied that tactical nuclear weapons are stationed

---

[57] An account is given in Goodby, J., 'Transparency and irreversibility in nuclear warhead dismantlement', ed. Feiveson (note 16), pp. 171–92.

[58] Joint Statement on Parameters on Future Reductions in Nuclear Forces, The White House, Office of the Press Secretary, 21 Mar. 1997, available at URL <http://www.ceip.org/files/projects/npp/resources/summits6.htm#parameters>.

[59] Pikayev, A., 'Towards increasing transparency of reduced nuclear arsenals?', Draft paper prepared for the SIPRI Workshop on Nuclear Transparency, Paris, 8–9 Feb. 2001.

[60] Gertz, B., 'Russia transfers nuclear arms to Baltics', *Washington Times*, 3 Jan. 2001, p. 1; Pincus, W., 'Russia moving warheads', *Washington Post*, 4 Jan. 2001, p. A16, available at URL <http://washingtonpost.com/wp-dyn/articles/A16331-2001Jan3.htm>; and Gertz, B., 'Satellites pinpoint Russian nuclear arms in Baltics', *Washington Times*, 15 Feb. 2001, p. 1, available at URL <http://intranet.ceip.org/library/clippings/e20010215satellites.htm>. The articles, citing anonymous US intelligence and defence officials, alleged that Russia had redeployed nuclear warheads in Kaliningrad in order to deploy them in a new, short-range missile, known as the SS-21 Tochka. They also reported that a Tochka missile was test fired in Kaliningrad in Apr. 2000. The Tochka can carry either a nuclear or a conventional warhead.

[61] See Sokov, N., 'The tactical nuclear weapons controversy', *Jane's Defence Weekly*, vol. 35, no. 5 (31 Jan. 2001), pp. 16–17.

[62] Gertz, *Washington Times*, 15 Feb. 2001 (note 60).

in Kaliningrad; nevertheless, Russia noted that there are no legal restrictions on such redeployments.[63]

IV. Prospects and means for further reductions

Influential individuals, commissions and international forums have recently voiced their deep concerns about the absence of progress towards ending the deployment of tactical nuclear weapons and further reducing the remaining arsenals.[64] Although largely ignored during the 1995 NPT Review and Extension Conference, tactical nuclear weapons received particular attention at the 2000 NPT Review Conference. Several countries, including NATO member states, raised the need to include them in arms reductions agreements. The European Union called for all tactical nuclear weapons to be brought into future reduction and disarmament arrangements 'with the objective of their reduction and eventual complete elimination'.[65] Indeed, the nuclear weapon states agreed, for the first time in this forum, to further reduce tactical nuclear weapons as practical steps towards the implementation of Article VI of the NPT.[66]

Several authoritative analyses have detailed useful plans and proposals for dealing with the problem of tactical nuclear weapons.[67] Among them, some of the most frequently discussed are formalization of the 1991–92 PNIs, transparency measures and data exchanges, negotiated and unilateral reductions, agreements on numerical limitations, the creation of nuclear weapon-free zones, a global prohibition, removal of US gravity bombs from Europe and consolidation at secure storage facilities, or a combination of several of these proposals.

Two important factors affect the prospects for a further limitation of tactical nuclear weapons in a reciprocal, cooperative and transparent manner. First, there is the disdain of the George W. Bush Administration for legally binding arms control treaties, which are seen as constraining US flexibility.[68] Second, there is an apparent lack of incentive for Russia to engage in tactical nuclear weapons transparency. Fur-

---

[63] See, e.g., Moscow Interfax, 'Russian President denies deployment of nuclear arms in Kaliningrad', FBIS-SOV-2001-0106, 6 Jan. 2001; Moscow Interfax, 'No nuclear weapons in Baltic enclave–foreign minister', FBIS-SOV-2001-0309, 9 Mar. 2001; and Agence France-Presse (North European Service), 'Russian official says Moscow entitled to deploy nuclear weapons in Kaliningrad', Foreign Broadcast Information Service, *Daily Report–Europe* (*FBIS-EUR*), FBIS-EUR-2001-0325, 25 Mar. 2001.

[64] See, e.g., 'Annan rejects n-powers' stand on proliferation', *The Hindu*, 3 Apr. 2000, available at URL <http://www.the-hindu.com/holnus/01032003.htm>; and *Report of the Canberra Commission on the Elimination of Nuclear Weapons* (National Capital Printers: Canberra, Aug. 1996), p. 54.

[65] Johnson, R., 'Implications of the outcome of the NPT review conference', International Security Information Service, Special Briefing Series on UK Nuclear Weapons Policy, no. 5 (Jan. 2001), p. 6.

[66] The text states the need for 'Steps by all the nuclear-weapon States leading to nuclear disarmament ... [including] ... the further reduction of non-strategic nuclear weapons, based on unilateral initiatives and as an integral part of the nuclear arms reduction and disarmament process', '2000 NPT Review Conference Final Document', Article VI.9, *Arms Control Today*, vol. 30, no. 5 (June 2000), p. 31.

[67] The most recent analyses are Lewis and Gabbitas (note 10); Potter, W. C., 'Unsafe at any size', *Bulletin of the Atomic Scientists*, vol. 53, no. 3 (May/June 1997), pp. 25–27; Potter and Sokov, eds Potter et al. (note 2); Arbatov (note 16); and Millar, A. and Alexander, B., 'Uncovered nukes: arms control and the challenge of tactical nuclear weapons', Fourth Freedom Forum, 16 Nov. 2001, available at URL <http://www.fourthfreedom.org>.

[68] Gottemoeller, R., 'Offense, defense, and unilateralism in strategic arms control', *Arms Control Today*, vol. 31, no. 7 (Sep. 2001), p. 11; and chapters 9 and 10 in this volume. As far as the issue of reductions of tactical nuclear weapons is concerned, the unclassified version of the congressionally mandated US Nuclear Posture Review was silent. It is noteworthy that the USA is planning to keep in storage non-operational warheads resulting from reductions in its strategic forces. US Department of Defense, 'Findings of the Nuclear Posture Review', News briefing slides, 9 Jan. 2002, URL <http://www.defenselink.mil/news/Jan2002/g020109-D-6570C.html>.

thermore, the US domestic political agenda in the aftermath of the 11 September 2001 terrorist attacks is dominated by counter-terrorism efforts. As far as Russia is concerned, the unrivalled military superiority of the USA[69] and the other NATO member states, the emerging potential of China, and the rapid deterioration of its conventional weaponry may be seen as political motivations for retaining, if not a large number, perhaps a non-transparent stockpile of tactical nuclear weapons.[70] Their possession could also form an asymmetric response to the development of US missile defences. For these reasons, the possibility of any short-term breakthrough appears remote.

Nevertheless, the 11 September terrorist attacks have brought Russia and the USA closer together and have galvanized interest in moving forward on a variety of fronts, including nuclear reductions. To make sustained progress, more trust needs to be restored, and this will probably take time. Overcoming differences over missile defences will be a critical factor in this regard. The two countries seem committed to making large cuts in their deployed strategic nuclear arsenals over the next several years, although the parameters they will adopt to this end have not yet been decided. As deployed strategic weapons are further reduced and bilateral security cooperation broadens, they could also explore a variety of confidence-building measures to address tactical nuclear weapons. First and foremost, it will be necessary, without lengthy consultations, to declare how many tactical nuclear weapons remain in their inventories. Information could be periodically updated and supported by historical data. In addition, the two countries may be willing to confirm that all battlefield nuclear weapons have been fully dismantled. Decisions to diminish the role of tactical nuclear weapons in their military strategies and to forgo any further modernization could complement such undertakings.

The USA, being more comfortable with the status quo, may lead the way. In the longer run it could well consider limiting or withdrawing its nuclear gravity bombs from Europe because these weapons no longer serve any meaningful military purpose.[71] This would be a major step forward, albeit not an easy one.[72] In return, Russia could further restrict its tactical nuclear weapons to fewer locations, provide assurances about their deployments and elimination, or even allow monitoring at its storage sites. Such exchanges could be either bilateral or take place in the framework of the NATO–Russia Permanent Joint Council.[73]

The diversion of even a single tactical nuclear weapon by a terrorist group would be a proliferation catastrophe. Urgent measures are necessary to ensure that all tactical warheads, in particular those of older designs and those that are fully detached from their delivery systems, are securely stored and guarded. Indeed, despite the souring

---

[69] According to one estimate, if Russian nuclear forces continue to deteriorate as currently projected, by 2015 the US arsenal will be 5 times the size of the combined arsenals of all the other nations. Cirincione, J., 'The assault on arms control', *Bulletin of the Atomic Scientists*, vol. 56, no. 1 (Jan./Feb. 2000), pp. 32–33.

[70] Although Russia appears to have recently embraced the notions of transparency and irreversibility of deep strategic reductions, there are no signs that it is prepared to apply them to tactical nuclear weapons. See, e.g., Statement by Ambassador Leonid A. Skotnikov, Permanent Representative of the Russian Federation to the Conference on Disarmament at the Plenary Meeting of the Conference on Disarmament, Geneva, 22 Jan. 2002, URL <http://www.ln.mid.ru>.

[71] Cotta-Ramusino, P., 'The unasked question', *Bulletin of the Atomic Scientists*, vol. 55, no. 4 (July/Aug. 1999), pp. 42–45.

[72] 'We emphasised again that nuclear forces based in Europe and committed to NATO continue to provide an essential political and military link between the European and North American members of the Alliance.' Final Communiqué of the NATO Ministerial Meeting of the Defence Planning Committee and the Nuclear Planning Group held in Brussels on 18 Dec. 2001, Press Communiqué PR/CP(2001)170.

[73] See note 26; and *NATO Handbook* (note 39).

Russian–US political relations since the mid-1990s, a distinct characteristic of that period has been their continued, unprecedented and broad nuclear security cooperation.[74] The two countries may fruitfully consider ways to further expand and strengthen these programmes and, given that the dangers are global, invite other countries to contribute.

Nuclear warhead transparency and control

Tactical nuclear weapons are essentially delivered by dual-capable means, that is, by delivery vehicles that have both nuclear and conventional capabilities. Traditional strategic nuclear arms control measures cannot therefore be applied to tactical nuclear weapons. The only meaningful way to provide assurances about the implementation of the PNIs and any additional future agreement or treaty limiting tactical nuclear weapons would be to directly apply controls on the warheads. In this respect, from a technical viewpoint, dealing with tactical warheads will not be different from dealing with strategic warheads. The implementation of warhead transparency, however, will pose formidable challenges, as evidenced by the difficulties in making progress with the agenda proposed in the mid-1990s to address this issue.

It is important for Russian–US relations to move forward in order to bring about conditions conducive to innovative technical approaches to arms control and, specifically, to deal with tactical nuclear weapons. These would require, apart from comprehensive data exchanges, a rigorous inspection and verification scheme.[75] Directly imposing controls on warheads would be an immense and ambitious technical task that would require unprecedented intrusiveness into what hitherto have been some of the most sensitive segments of the national defence establishments.

A comprehensive regime could, at least conceptually, include a full account of warheads, verification of their dismantlement and monitoring of their production facilities, broadly comprising the following main elements: (*a*) establishing declarations of warhead inventories and verifying their accuracy and, more importantly, their completeness; (*b*) providing assurances that warheads earmarked for dismantlement or elimination are not diverted or replaced by decoys; (*c*) guaranteeing that no new warheads are manufactured; and (*d*) disposing of fissile material from dismantled warheads in an irreversible way.

The uncertainties surrounding warhead inventories must be reduced to a minimum in order to establish an essential basis for deep reductions. Indeed, exchanges of stockpile information constitute a logical next step in arms control.[76] The Russian and

---

[74] The US Government, and principally the DOD and the DOE, have since the passage of the 1991 Nunn–Lugar Soviet Nuclear Threat Reduction Act, established an array of programmes to assist in the dismantlement of Russian nuclear weapons, improve their security and prevent their diversion. See Baker, H. and Cutler, L., *Russian Task Force, A Report Card on the Department of Energy's Nonproliferation Programs with Russia* (US Department of Energy: Washington, DC, 10 Jan. 2001).

[75] For a presentation and analysis of procedures and technologies for nuclear warhead transparency and verifiable dismantlement see Zarimpas, N. (ed.), SIPRI, *Transparency in Nuclear Warheads and Materials: The Political and Technical Dimensions* (Oxford University Press: Oxford, forthcoming 2002); Bukharin, O. and Luongo, K., *US–Russian Warhead Dismantlement Transparency: The Status, Problems and Proposals*, CEES Report no. 314 (Princeton University Center for Energy and Environmental Studies (CEES): Princeton, N.J., Apr. 1999), available at URL <http://www.ransac.org/new-web-site/pub/reports/transparency.html>; and UK Atomic Weapons Establishment, 'Confidence, security and verification—the challenge of global nuclear weapons arms control', Aldermaston, AWE/TR/2000/001, available at URL <http://www.awe.co.uk>.

[76] Müller, H., *The Nuclear Weapons Register: A Good Idea Whose Time Has Come*, PRIF Reports no. 51 (Peace Research Institute Frankfurt (PRIF): Frankfurt, 1998).

US presidents agreed in 1994 to develop a process for sharing classified stockpile data at regular intervals, but formal implementation was never successfully negotiated. The USA nevertheless declassified certain characteristics of its warhead stockpile (the total yield and numbers retired) from 1945 to 1994, as well as the total numbers produced from 1945 to 1961.[77]

After confidence is gained from exchanging aggregate data, more detailed accounts could be provided in a phased manner. These might include inventories by types, as well as itemized lists of warheads, including their locations. Such detailed declarations might also involve formal verification arrangements to provide assurances about their accuracy and completeness. Verification will become imperative either when current stockpiles are substantially reduced or when an agreement is reached to impose quantitative limits on them. In this regard, two tasks are of concern: (*a*) demonstrating the authenticity of a declared warhead without disclosing classified design information; and (*b*) providing guarantees about the completeness of the declarations. The first task could be accomplished either by measuring some of the 'attributes' of the warhead (at least the presence of a minimum mass of fissile material) or by making use of its detailed spontaneous and/or stimulated radiation spectrum, the so-called 'template' approach. Information-barrier systems involving both technology and procedural elements could be applied to reliably protect sensitive information. Challenge inspections and examination of facility operating records could be used to fulfil the second task.

The use of appropriate tags to show a warhead's claimed identity and seals to guarantee its presence, portal-perimeter monitoring, as well as more intrusive chain-of-custody techniques would ensure confidence in the verification of warhead dismantlement. The Russian–US Laboratory-to-Laboratory Warhead Dismantlement Transparency Programme, an unclassified technical exchange which was initiated in 1995, made major advances in many areas, such as radiation measurement, information protection, remote monitoring, disposition of non-nuclear components and chain-of-custody monitoring, including the monitoring of tags and seals. However, much remains to be done since the technology base for warhead dismantlement transparency is far from complete.

Detection of the undeclared manufacture of new warheads would of course be a much more difficult problem. Satellite imagery, remote sensing and environmental monitoring, complemented by societal verification, would all be valuable tools towards this end.[78]

Verifiable warhead elimination would have a major impact on warhead production and maintenance complexes, because such facilities were not designed to receive foreign inspectors or accommodate any other transparency measures, for example, monitoring. Consequently, warhead stewardship and re-manufacturing operations, which are probably carried out in adjoining or the same buildings where dismantlement is performed, could be seriously disrupted. Moreover, the demands on technical, support and security personnel, services and equipment would probably be significant. Physical segregation of verifiable warhead dismantlement processes or the use for this purpose of dedicated facilities or closed-down plants are options that could be

---

[77] US Department of Energy, 'Declassification of certain characteristics of the United States nuclear weapon stockpile', 1994, URL <http://www.osti.gov/html/osti/opennet/document/ press/pc26.html>.

[78] Bukharin, O. and Doyle, J., *Verification of the Shutdown or Converted Status of Excess Warhead Production Capacity: Technology Options and Policy Issues*, LA-UR-01-5001 (Los Alamos National Laboratory: Los Alamos, N. Mex., Oct. 2001).

investigated for satisfying the rigorous operational and security conditions required.[79] More serious challenges will be posed by existing asymmetries in the number, capacities, structure, functions and technical organization of warhead production and dismantlement facilities in Russia and the USA.[80] These asymmetries must be clearly identified and well understood before warhead dismantlement inspection and monitoring arrangements are formally negotiated.

Last, in order to ensure the irreversibility of the process, transparency and verification should be fully extended to fissile material made available as a result of the dismantlement process, covering both its intermediate storage (in the form of pits, components or other forms) and its final disposition. Noteworthy in this respect is the Trilateral Initiative, launched in 1996 by the IAEA, Russia and the USA to voluntarily place both classified and unclassified forms of fissile material under international verification, as well as bilateral efforts to monitor such material at the Mayak storage site in Russia.[81]

V. Conclusions

The large number of tactical nuclear weapons remaining in the Russian and US nuclear arsenals is a legacy of the cold war. Their existence is detrimental to global security and the security of Russia and the USA. These weapons pose unique challenges and dangers and therefore deserve urgent attention. Russia and the USA should take steps to address these dangers. First, they should ensure that tactical nuclear weapons are safely and securely stored. Second, they should jointly reaffirm their commitments to the 1991–92 Presidential Nuclear Initiatives, provide updates on progress made in the elimination of tactical nuclear weapons and pursue increased transparency. Third, they should reassess the perceived utility of such weapons in their military and deterrence doctrines and halt further weapon modernization. Fourth, they should unilaterally proceed with additional reductions. Fifth, they should consider ways of constructing a cooperative framework that drastically limits the numbers and locations of tactical nuclear weapons.

In the long term, effectively addressing limitations on tactical nuclear weapons can only be achieved by directly imposing controls on their warheads. To this end, it is imperative to strengthen technical arms control research, cooperation and funding.

---

[79] Dubinin, V. and Doyle, J., *Item Certification for Arms Reduction Agreements: Technological and Procedural Approaches,* LA-UR-00-2740 (Los Alamos National Laboratory: Los Alamos, N. Mex., 2000).

[80] See appendix 10C.

[81] See Shea, T., 'Potential roles for the IAEA in warhead dismantlement and fissile materials transparency', and Schaper, A., 'Monitoring and verifying the storage and disposition of fissile materials and the closure of nuclear facilities', ed. Zarimpas (note 75).

Appendix 10C. The changing Russian and US nuclear warhead production complexes

OLEG BUKHARIN

I. Introduction

The Soviet Union (and Russia, as its successor state) and the United States have developed dedicated infrastructures to design, test, mass-produce and support field deployment of tens of thousands of nuclear warheads. During the cold war, the principal task of the Soviet and US nuclear weapon laboratories was to advance nuclear weapon science, to design and test new warheads, and to provide scientific oversight of the handling of nuclear warheads throughout their life cycle. Their production facilities dismantled obsolete warheads, modernized and refurbished warheads in the stockpile, and put into production and mass-produced warheads of more advanced types. Because of the high pace of technological innovation, nuclear warheads were often replaced before the end of their design service life. Their nuclear material production complexes produced and processed highly enriched uranium (HEU), plutonium, tritium and other essential nuclear materials. Underground nuclear explosive testing was crucial for warhead development efforts as well as for stockpile safety and reliability assurances.

The direction of nuclear weapon activities in both states changed dramatically after the cold war owing to reduced funding, the end of nuclear explosive testing and nuclear warhead stockpile reductions. The end of nuclear explosive testing by the USSR (1990) and the USA (1992) has altered the nature of warhead research and development (R&D) activities. The development of more advanced warhead designs was prohibited by the US Congress and has lost priority in Russia.[1] The USA has launched an ambitious and expensive Science-Based Stockpile Stewardship (SBSS) programme, which seeks to develop supercomputing capabilities to provide for 3-D modelling of nuclear weapons, to modernize the existing facilities and build new experimental ones, to improve the understanding of the physics of nuclear weapons, and to maintain the Nevada Test Site in a state of readiness for the resumption of nuclear explosive testing.

The US SBSS programme has been criticized as unnecessary, excessively expensive, and economically and politically motivated. In particular, according to the programme's opponents, the policy of maintaining the warhead design capability, which will be enhanced by new experimental facilities and computing tools, is misguided and dangerous from the arms control and non-proliferation standpoint.[2] On the other

---

[1] As of Mar. 2002, the congressional prohibition remained in effect. However, the USA has resolved to re-establish warhead advanced concept teams in its national nuclear weapon laboratories. See Statement of US Secretary of Energy Abraham Spencer before the Committee on Armed Services, US House of Representatives, 13 Mar. 2002, URL <http://www.house.gov/hasc/openingstatements andpressreleases/107thcongress/02-03-13abraham.html>.

[2] Critics of the SBSS point out that the programme would: (*a*) undermine the 1968 Non-Proliferation Treaty and the 1996 Comprehensive Nuclear Test-Ban Treaty by advancing weapon capabilities; (*b*) describe accurate models of physics processes in nuclear weapons in unclassified publications, thereby facilitating an increase in the weapon capabilities of other nations; and possibly (*c*) facilitate development of proliferation-prone pure-fusion weapons. See, e.g., Paine, C. and McKinzie, M., *Does*

side of this debate, a number of US nuclear weapon scientists and policy makers have expressed concerns about the ability of the USA to maintain the safety and reliability of its warheads without underground nuclear explosive testing.[3]

The Russian stewardship programme is significantly less costly and less ambitious in its goals. Its R&D activities are largely limited to stockpile surveillance, maintenance of nuclear warhead design skills and prevention of surprise breakthroughs in nuclear weapon developments in other states.

Warhead stockpile reductions have also deeply affected nuclear weapon activities. As a result of the 1987 Treaty on the Elimination of Intermediate-Range and Shorter-Range Missiles (INF Treaty), the 1991 Treaty on the Reduction and Limitation of Strategic Offensive Arms (START I Treaty), the 1991–92 Bush–Gorbachev/Yeltsin Presidential Nuclear Initiatives (PNI), the decommissioning of naval ships, aircraft and other delivery systems, and the retirement of unsafe and obsolete weapons, Russia and the USA have significantly reduced their nuclear warhead stockpiles. The US active stockpile declined from 23 000 warheads of 25 types in 1985 to approximately 8000 warheads of 7 major types in 2001.[4] Nuclear warhead reductions in Russia have been even more significant, with a decline from 35 000 warheads in the mid-1980s to approximately 10 000 warheads in 2001. Reductions to approximately 5000 active warheads are projected to take place by 2010.[5]

Because of these reductions there is currently no need for new nuclear materials. The scale of new production and refurbishment of nuclear warheads has declined. The two states have undertaken a massive warhead dismantlement effort.

These changes in mission, operating environment and resources call for comprehensive restructuring and downsizing of the Russian and US nuclear warhead production complexes. Each state needs an economic, safe and secure complex consisting of a small number of compact, specialized facilities capable of providing for all critical aspects of nuclear weapon production: nuclear weapon R&D and non-nuclear explosive testing, tritium production and processing, fissile material component manufacturing, fabrication of specialized non-nuclear components and warhead assembly–disassembly. Because of their historical, economic, technical and other differences, the two states have been pursuing this objective differently and differ in the progress they have made.

---

the US Science-based Stockpile Stewardship Program Pose a Proliferation Threat? (Natural Resources Defense Council: Washington, DC, Nov. 1998).

[3] Glanz, J., 'Testing the aging stockpile in a test ban era,' *New York Times*, 28 Nov. 2000, p. 1.

[4] The US Nuclear Posture Review (NPR), completed in Dec. 2001, divides the US nuclear warhead stockpile into several categories. The active stockpile is composed of operationally deployed warheads and responsive force warheads. Responsive force warheads are stored in a ready-to-use configuration and can be deployed in a period ranging from weeks to one year or more. The responsive force stockpile is intended to ensure a flexible force that could be reconstituted in response to changing national security and threat environments. There is also an inactive stockpile, which is largely intended to replace catastrophic failures of similar warheads. US Department of Defense, Transcript of special briefing on the Nuclear Posture Review, 9 Jan. 2002, available at URL <http://www.defenselink.mil/news/Jan2002/t01092002_t0109npr.html>. For more detail on the NPR see appendix 10A in this volume. It is believed that, instead of irreversibly eliminating warheads removed from delivery systems according to the Nov. 2001 declarations by presidents Bush and Putin, the USA would shift a large fraction of the removed warheads to its responsive force or inactive stockpile. 'The Bush–Putin summit: in 2 leaders' words: cordial discord', *New York Times*, 16 Nov. 2001, p. A12; and Natural Resources Defense Council, 'Faking nuclear restraint: the Bush Administration's secret plan for strengthening US nuclear forces', Press Release, Washington, DC, 13 Feb. 2001, available at URL <http://www.nrdc.org/media/pressreleases/020213a.asp>.

[5] There is no public information on the number of warhead types in the Russian arsenal.

Sections II and III of this appendix review the efforts and progress made by Russia and the USA, respectively, in restructuring their nuclear warhead production complexes.

II. The Russian complex

At present, the Russian nuclear warhead production complex is managed by the Ministry of Atomic Energy (Minatom) and consists of 17 research institutes and manufacturing facilities (see table 10C.1).[6] The core functions of the complex—production and processing of nuclear materials, nuclear warhead R&D and assembly–disassembly of nuclear warheads—are located in 10 'closed cities'. (There are also numerous research institutes and production facilities in Moscow and other 'open cities'.) The existence of the closed nuclear cities was not officially acknowledged until 1992 and their present status is that of closed administrative–territorial units. The cities are surrounded by double fences that are patrolled by armed guards. Access is limited and thoroughly controlled. The main purpose of stringent security arrangements is to protect the secrecy of sensitive nuclear operations and to prevent terrorist attacks against them. The cities currently have a combined population of approximately 760 000 people, of whom 130 000 work at nuclear facilities (half of them on civilian projects).

Unlike the US complex (see section III), the Russian complex has a complete set of nuclear warhead production facilities. The complex remains oversized and much of its R&D and production infrastructure is redundant. The downsizing of the complex, however, is inevitable. The strategic rationale for maintaining a massive weapon production infrastructure has been relegated to the past. Moreover, the Russian economy is no longer capable of supporting the cold war complex. In fact, its technical infrastructure has already been contracting because of ageing and lack of maintenance while its pool of scientific and technical talent has been shrinking because of demographic shifts and economic conditions. Minatom's nuclear weapon workforce has been reduced from approximately 130 000 workers in the late 1980s to approximately 75 000 workers at present, and several facilities no longer perform nuclear weapon functions.

Nuclear weapon complex reductions: the early years

Cuts in the nuclear weapon programme began in the late 1980s. The initial phase of downsizing lasted approximately 10 years and can be characterized by the following three developments: the termination of HEU and plutonium production for weapons; defence conversion without complex restructuring; and spontaneous contraction of weapon capabilities.

---

[6] The weapon complex's facilities are managed by 3 Minatom departments: the Nuclear Fuel Cycle Department (formerly the Fourth Main Directorate), the Nuclear Weapons Development and Testing Department (formerly the Fifth Main Directorate), and the Department of Nuclear Weapons Production (formerly the Sixth Main Directorate). Another facility of the complex, the Novaya Zemlya Test Site, is managed by the Ministry of Defence.

Table 10C.1. The Russian Minatom nuclear warhead production complex, 2001

Facility English (Russian) name	Location (old name if applicable)	Nuclear warhead production functions
Institute of Experimental Physics (Vserossiyskiy Nauchno-Issledovatelskiy Institut Experimentalnoy Fiziki, VNIIEF)	Sarov (Arzamas-16)	Nuclear warhead design Stockpile support
Institute of Technical Physics (Vserossiyskiy Nauchno-Issledovatelskiy Institut Tekhnicheskoy Fiziki, VNIITF)	Snezhinsk (Chelyabinsk-70)	Nuclear warhead design Stockpile support
Institute of Automatics (Vserossiyskiy Nauchno-Issledovatelskiy Institut Avtomatiki, VNIIA)	Moscow	Nuclear warhead design and engineering Design of non-nuclear components Nuclear weapon maintenance instrumentation
Institute of Impulse Technologies (Vserossiyskiy Nauchno-Issledovatelskiy Institut Impulsnoy Tekhiki, VNII IT)	Moscow	Nuclear test diagnostics
Institute of Measurement Systems (Nauchno-Issledovatelskiy Institut Izmeritelnykh Sistem, NII IS)	Nizhni Novgorod	Design of non-nuclear components
Design Bureau of Road Equipment (Konstruktorskoye Buro Avtotransportnogo Oborudovaniya, KB ATO)	Mytischy, Moscow region	Nuclear warhead transportation and handling equipment
Siberian Chemical Combine (Sibirskiy Khimicheskiy Kombinat, SKhK)	Seversk (Tomsk-7)	Fabrication of HEU and plutonium weapon components
Production Association 'Mayak' (Proizvodstvennoye Obyedinenie 'Mayak)	Ozersk (Chelyabinsk-65)	Production of tritium and tritium components of nuclear warheads Fabrication of HEU and plutonium weapon components
Mining and Chemical Combine (Gorno-Khimicheskiy Kombinat, GKhK)	Zheleznogorsk (Krasnoyarsk-26)	Plutonium management
Electrokhimpribor (Kombinat Elektrochimpribor)	Lesnoy (Sverdlovsk-45)	Nuclear warhead assembly–disassembly
Electromechanical Plant 'Avangard' (Elektromechanicheskiy Zavod 'Avangard')	Sarov (Arzamas-16)	Nuclear warhead disassembly
Production Association 'Start' (Proizvodstvennoye Obyedinenie 'Start')	Zarechny (Penza-19)	Nuclear warhead disassembly
Device-Building Plant (Priboro-Storitelnyiy Zavod)	Trekhgorny (Zlatoust-36)	Nuclear warhead assembly–disassembly
Production Association 'Sever' (Proizvodstvennoye Obyedinenie 'Sever')	Novosibirsk	Production of non-nuclear weapon components

Facility English (Russian) name	Location (old name if applicable)	Nuclear warhead production functions
Production Association 'Molnia' (Proizvodstvennoye Obyedinenie 'Molnia')	Moscow	Production of non-nuclear weapon components
Urals Electromechanical Plant (Uralskiy Electromechanicheskiy Zavod)	Yekaterinburg	Production of non-nuclear weapon components
Nizhneturinsky Mechanical Plant (Nizhneturinskiy Mechanicheskiy Zavod)	Nizhnyaya Tura	Production of non-nuclear weapon components and support equipment

Sources: Podvig, P. (ed.), *Russian Strategic Nuclear Forces* (MIT Press: Cambridge, Mass., 2001); and Bukharin, O., von Hippel, F. and Weiner, S., *Conversion and Job Creation in Russia's Closed Nuclear Cities* (Program on Nuclear Policy Alternatives, Princeton University: Princeton, N.J., Nov. 2000).

The termination of defence orders for new fissile materials effectively excluded the uranium enrichment and plutonium production plants from the weapon programme.[7] As a result, no nuclear weapon activities presently take place in three of the closed nuclear cities: Novouralsk, Zelenogorsk and Zheleznogorsk. However, other activities in these cities remain critical to the mission of storing and managing hundreds of tonnes of fissile materials, some of which could be a part of Russia's strategic reserves.

In the late 1980s, the Soviet Government also developed a number of defence conversion programmes to redirect excess workers and equipment to civilian work. However, most defence conversion efforts have failed because of insufficient investment, the collapse of Russia's domestic markets, the lack of entrepreneurial and market skills, secrecy, inflexible institutional bureaucracies and high production costs. As a result, the reductions have been largely spontaneous. Although the infrastructure deterioration and personnel attrition have already made the complex much less capable, with the exception of the separation of the HEU and plutonium production facilities, it has not changed structurally.

The 1998 programme and its implementation

It was not until after the mid-1990s that Minatom and facility managers accepted the fact that weapon programme cutbacks were irreversible and that a serious restructuring and downsizing effort was needed for the complex to survive in the new environment. Such an effort was launched, and it appears that its main objective is to focus defence order funds by reducing facility duplication and by separating the defence part of the complex from the part that has become excess to defence requirements. The process of separation involves the establishment of separate lines of funding (state defence budget for defence facilities versus revenues from commercial sales for civilian parts) and management, division of the production and support infrastructure, and consolidation of remaining defence activities inside isolated buildings and/or

---

[7] Instead, they provide uranium enrichment and spent fuel management services to domestic and international customers and are involved in a variety of other nuclear and non-nuclear commercial activities. All of these facilities, with the exception of the plutonium production centre in Zheleznogorsk, are also involved in the HEU down-blending work under the 1993 US–Russian HEU Agreement, reproduced in *SIPRI Yearbook 1994* (Oxford University Press: Oxford, 1994), pp. 673–75.

technical areas. The second and companion objective is to create civilian jobs for excess personnel.

Minatom's plans were formalized in the programme 'On Restructuring and Conversion of the Nuclear Weapons Complex in 1998–2000', adopted by the Russian Government in June 1998 as a part of a broader plan to restructure Russia's defence industries.[8] The programme and other planning documents call on Minatom to: (*a*) stop warhead assembly at the serial production facilities in Sarov and Zarechny by 2000; (*b*) stop warhead dismantlement at these two facilities by 2003; (*c*) transfer the production of certain non-nuclear warhead components and assemblies to the pilot production plants of the warhead R&D institutes by 2000; (*d*) consolidate weapon work at the remaining non-nuclear component manufacturing facilities by 2000; (*e*) phase out nuclear weapon work at one of the two fissile material processing plants in 2003; (*f*) cut the number of defence programme personnel from 75 000 to 35 000 by 2005;[9] and (*g*) cut the number of defence personnel at the serial production plants from 71 000 during the cold war to approximately 11 000 within the next several years.

Downsizing is also planned for individual facilities and would involve defence personnel reductions and consolidation of weapon activities in fewer buildings and production areas. For example, the number of defence programme personnel at the warhead assembly facility in Trekhgorny was to decrease from 5766 in 1997 to 2800 in 2001.[10] At the Urals Electromechanical Plant in Yekaterinburg, which produces nuclear warhead electronic components, the plan is to split the facility into two separate entities.[11] The weapon part would be located in a single building and would retain about one-third of the equipment and infrastructure. It would be supported exclusively by defence-order funding. The remainder of the plant would have to support itself by producing and selling commercial products on the open market. The number of personnel employed in the weapon programme would decline from the cold war level of 12 000 to 1500.

Certain steps have already been taken. In April 1999 Minatom formed the Department for Conversion of the Nuclear Industry, which has the responsibility for defence conversion and complex restructuring.[12] All research institutes and production plants of the warhead complex have made progress and are working to implement facility level restructuring programmes. Essentially no weapon work is taking place at the Molnia plant in Moscow. The production association Sever in Novosibirsk, a nuclear warhead electronic components and sub-assemblies production facility, has already consolidated all weapon work in a single technical area and reduced defence pro-

---

[8] The programme was announced for the first time by Minatom's First Deputy Minister Lev Ryabev at the 7th Carnegie International Nonproliferation Conference, 11–12 Jan. 1999, Washington, DC. See Ryabev, L., 'The role of the NCI in meeting Russia's nuclear complex challenges', available at URL <http://www.ceip.org/files/events/conf99ryabevslides.asp>, where there is also a complete list of remarks and presentations from the conference.

[9] According to Minatom's First Deputy Minister Lev Ryabev, the total number of workers at Minatom's facilities in the 10 closed cities is *c*. 150 000. This presumably includes workers in transport, utilities and other support divisions. Many of these divisions have recently been transferred to municipal control. Bukharin, O. et al., *Helping Russia Downsize its Nuclear Complex: A Focus on the Closed Nuclear Cities* (Program on Nuclear Policy Alternatives, Princeton University: Princeton, N.J., June 2000).

[10] ['Trekhgorny's plans'], *Atompressa*, no. 13 (Apr. 1999), p. 3 (in Russian).

[11] Sachkova, S., ['Hopes of defence workers'], *Atompressa*, no. 46 (Dec. 1999), p. 3 (in Russian).

[12] See, e.g., ['Conversion: interview with A. Antonov'], *Atompressa*, no. 1 (Jan. 2000), pp. 1–2 (in Russian).

gramme staff.¹³ This facility is projected to lose its weapon functions some time between 2005 and 2007. Warhead assembly work was terminated at the Avangard plant in Sarov and its primary weapon function at present is warhead dismantlement.¹⁴ The Zarechny warhead production facility reportedly has very few defence orders.¹⁵ Seversk has also essentially become a civilian nuclear technology centre. Already, the bulk of the workload at the chemical and metallurgical plant in Seversk, which in the past produced HEU and plutonium components of nuclear warheads, is non-military and is related to HEU down-blending under the 1993 US–Russian HEU Agreement. Finally, the pilot plants at the nuclear warhead design institutes of the All-Russian Scientific Research Institute of Experimental Physics (Vserossiyskiy Nauchno-Issledovatelskiy Institut Eksperimentalnoy Fiziki, VNIIEF) in Sarov, the All-Russian Scientific Research Institute of Technical Physics (Vserossiyskiy Nauchno-Issledovatelskiy Institut Tekhnicheskoy Fiziki, VNIITF) in Snezhinsk, and the All-Russian Scientific Research Institute of Automatics (Vserossiyskiy Nauchno-Issledovatelskiy Institut Avtomatiki, VNIIA) in Moscow are assuming responsibility for the production of certain components and assemblies that in the past were manufactured by serial facilities.

The USA and other Western countries have launched several cooperative programmes that facilitate the downsizing of the Russian nuclear complex.¹⁶ The most significant of them is the US–Russian HEU Agreement, which provides the primary source of funding for Minatom to conduct complex conversion and downsizing activities.¹⁷ In addition, the HEU down-blending process involves thousands of workers in the closed cities. The International Science and Technology Centers (ISTC), the Cooperative Threat Reduction (CTR) programme and other cooperative efforts also provide productive work for nuclear weapon experts. Of particular note is the Nuclear Cities Initiative (NCI), which was established by the US Department of Energy (DOE) in 1998 and is intended specifically to facilitate the downsizing of the Russian nuclear weapon complex.¹⁸ Each of these cooperative programmes has made a positive contribution. However, their effect so far has been limited and it is clear that most of the job remains to be done.

Problems of downsizing and restructuring

A rational approach to optimizing Russia's warhead complex would be to consolidate weapon work at the smallest number of facilities possible. It would be based on a cost–benefit analysis of the existing infrastructure, future missions, and stockpile and funding projections. In reality, however, there are many other factors that could influ-

---

[13] Gorb, A., ['PO "Sever" in the program of restructuring and conversion of the nuclear industry'], *Atompressa*, no. 1 (Jan. 2000), p. 3 (in Russian).

[14] The assembly of the last warhead at the Avangard Plant was finished on 30 Dec. 1997. See Zavalishin, Yu., [*'Avangard' Atomic*], (Krasny Oktyabr': Saransk, 1999), p. 292 (in Russian).

[15] Saratova, L., ['How do you live, you weapons plant?'], *Gorodskoy Kuryer*, no. 3 (23 Jan. 1999) (in Russian), available at URL <http://courier.sarov.ru>.

[16] For a discussion of international assistance programmes in the closed nuclear cities see Bukharin et al. (note 9).

[17] According to the 1993 US–Russian HEU Agreement, the USA agreed to buy 500 t of excess weapon-grade HEU after it had been blended down to 4–5% enriched uranium for use in nuclear power reactors.

[18] Kile, S., 'Nuclear arms and non-proliferation', *SIPRI Yearbook 2000: Armaments, Disarmament and International Security* (Oxford University Press: Oxford, 2000), pp. 457–62; and chapter 10 in this volume.

ence Minatom's ability to plan and execute the downsizing and restructuring of the complex.

The most significant near-term problem is the redirection of excess workers. Downsizing, particularly in the closed nuclear cities, is not possible unless new jobs or other opportunities are created for displaced nuclear weapon workers. Indeed, massive lay-offs in the closed cities would threaten their social stability and increase the danger of thefts of nuclear materials or proliferation of weapon expertise. Minatom estimates that it needs to create approximately 35 000 jobs in the closed cities by 2005.

Defence conversion and job creation in the closed cities are challenging tasks. Their isolation, security restrictions, lack of modern communications and other business infrastructure and, at some locations, radioactive contamination, make the development of normal business relationships difficult. The crisis of the Russian economy and insufficient foreign investment will also continue to inhibit defence conversion at Minatom facilities and economic development in the local communities. Workforce reductions due to retirement, personnel losses to the commercial sector and minimal new hiring will hopefully relieve this pressure in five to ten years. However, massive retirement would increase the social protection and pension needs of workers in the nuclear complex.

Funding shortages constitute another critical problem. Minatom estimates that it will cost approximately $1 billion to implement the planned reductions.[19] About one-half of the cost would be for consolidation and clean-up; the rest would be for creating jobs for displaced workers. Generally, until the national economy recovers, making rapid progress on complex downsizing might prove difficult.

The pace of downsizing will also depend on domestic politics in Russia. A decision to terminate defence orders at a large production facility, especially in a closed city, would be politically unpopular (unless attractive non-military jobs were created) and would encounter opposition from facility workers, local communities, regional authorities and the Russian Duma. Pressure from these groups, compounded by anti-Western sentiment and nationalism, is likely to slow down the downsizing process.

Finally, complex downsizing could be complicated by strategic uncertainties. These include the USA's plans to keep many warheads in storage after they have been removed from strategic delivery systems and the US Senate's rejection of the 1996 Comprehensive Nuclear Test-Ban Treaty[20] (CTBT, not in force). In the absence of irreversible arms reductions, the Russian Government will be under pressure to maintain a 'national emergency' option of rebuilding its nuclear stockpile to cold war levels and of resuming massive warhead R&D efforts. There have already been proposals to initiate a weapon R&D effort aimed at countering the potential deployment of a strategic missile defence system in the USA.

These difficulties are serious. They could slow down, or even derail, the downsizing of the nuclear warhead complex unless the Russian Government and the international community provide strong political support, leadership and sufficient funding.

---

[19] Remarks by Minatom's Deputy Minister Lev Ryabev at an international conference, Helping Russia Downsize its Nuclear Complex, Princeton University, Princeton, N.J., Mar. 2000, documented in Bukharin et al. (note 9).
[20] For details of the treaty see URL <http://www.clw.org/pub/clw/coalition/ctbindex.htm>.

Beyond planned reductions

Although ambitious, the planned reductions may result in a complex that is still oversized relative to Russia's future nuclear defence missions and economic resources.[21] Further reductions of the warhead production infrastructure, therefore, could be expected in the future.[22] All warhead re-manufacturing and surveillance operations, for example, could be consolidated at one facility (most likely in Lesnoy). Deep reductions in nuclear arms by the five nuclear weapon powers recognized by the 1968 Non-Proliferation Treaty (NPT) would make it possible in a more distant future to further consolidate all warhead production and maintenance activities into the warhead design institutes in Sarov and Snezhinsk. In that case, Lesnoy would focus on warhead dismantlement and later be adapted for civilian purposes.

III. The US complex

The US nuclear warhead production complex is managed by the Office of Defense Programs of the National Nuclear Security Administration (NNSA), a semi-autonomous organization within the Department of Energy (DOE). In the mid-1980s, just prior to the end of the cold war, the complex consisted of 12 facilities and employed approximately 70 000 workers (see table 10C.2).[23] The downsizing of the complex began in the late 1980s. In 1988 and 1989, in the aftermath of the Chernobyl disaster in the Soviet Union, the DOE shut down its remaining tritium and plutonium production reactors at the Savannah River, South Carolina, and Hanford, Washington, sites. (The production of HEU for nuclear weapons in the USA stopped in 1964.) Also in 1989, the production of plutonium pits was terminated at the Rocky Flats Plant outside of Denver, Colorado, because of environmental and safety concerns.

The US industrial infrastructure for mass production of nuclear warheads has further contracted during the 1990s. In 1994 the DOE terminated the remaining weapon functions at the Rocky Flats Plant and declared the facility to be an environmental management site. In 1995 the DOE closed the Pinellas Plant near St Petersburg, Florida, and the Mound Laboratory near Miamisburg, Ohio. Many of their warhead production and management activities have been consolidated to the Kansas City Plant or transferred to the DOE's national laboratories. The weapon programme workforce has declined as of 2001 to approximately 25 000 workers and many former weapon workers have been transferred to environmental clean-up on non-proliferation work.

The US nuclear warhead complex currently consists of eight facilities.[24] The complex is projected to retain its current structure in the foreseeable future with consoli-

---

[21] Minatom is currently working on a new complex restructuring plan extending to 2010 but its contents have not been yet made public. See ['Notable dates and events in Minatom—2000'], *Atompressa*, no. 49–50 (Dec. 2000), p. 5 (in Russian).

[22] For an analysis of downsizing options for the Russian nuclear warhead complex see Bukharin, O., *Downsizing of Russia's Nuclear Warhead Production Infrastructure*, PU/CEES Report no. 323 (Princeton University, Center for Energy and Environmental Studies: Princeton, N.J., May 2000).

[23] For a general description of the US nuclear warhead complex during the cold war see Cochran, T. et al., *US Nuclear Warhead Facility Profiles, Nuclear Weapons Databook, Vol. III* (Ballinger: Cambridge, Mass., 1987). Table 10C.2 does not include facilities that were not directly involved in nuclear weapon production work.

[24] The present status of the US weapon complex is discussed in US Department of Energy, *FY 2000: Stockpile Stewardship Plan*, Sanitized Version (DOE Office of Defense Programs: Washington, DC, 15 Mar. 1999).

Table 10C.2. The US DOE nuclear warhead production complex, 2001

Facility/location	Location	Nuclear warhead production functions
Los Alamos National Laboratory (LANL)	Los Alamos, New Mexico	Basic R&D and advanced technologies development Nuclear weapon physics experiments Maintenance of capability to design/certify nuclear explosive packages (NEPs) Stockpile safety/reliability assessments Pit surveillance, modification, fabrication Production and surveillance of non-nuclear components[a]
Lawrence Livermore National Laboratory (LLNL)	Livermore, California	Basic R&D and advanced technologies development Nuclear weapon physics experiments Maintenance of capability to design/certify NEPs Stockpile safety/reliability assessments
Sandia National Laboratories (SNL)	Albuquerque, New Mexico	Non-nuclear components and systems R&D and engineering Nuclear weapon tests and experiments on weapon effects Manufacturing of neutron generators and select non-nuclear components Stockpile safety/reliability assessments
Kansas City Plant	Kansas City, Missouri	Production of non-nuclear components (electrical, mechanical materials) Surveillance, testing, repair of non-nuclear components
Pantex Plant	Amarillo, Texas	Assembly, surveillance and maintenance of nuclear warheads Dismantlement of retired warheads Production of highly explosive (HE) components Storage of plutonium pits
Oak Ridge Y-12 Plant	Oak Ridge, Tennessee	Surveillance of thermonuclear canned secondary sub-assemblies (CSAs) Maintenance of capability to produce CSAs and radiation cases Dismantlement of CSAs of retired warheads Storage of HEU and lithium materials and parts Production support to national laboratories
Savannah River Site	Aiken, South Carolina	Recycling/loading of tritium Surveillance of tritium reservoirs Support of tritium source projects Pit conversion and disposition (planned) Pit manufacturing (possible in the future)
Nevada Test Site	Las Vegas, Nevada	Maintenance of capability to conduct/ evaluate underground nuclear tests Nuclear weapon physics experiments Emergency response and radiation sensing support

Facility/location	Location	Nuclear warhead production functions
DOE warhead complex facilities shut down after 1985		
Rocky Flats Plant	Denver, Colorado	Pit manufacturing Production of beryllium and other non-nuclear components
Mound Laboratory	Miamisburg, Ohio	Fabrication/surveillance of non-nuclear warhead components
Pinellas Plant	St Petersburg, Florida	Production of neutron generators and other non-nuclear warhead components
Hanford Reservation	Hanford, Washington	Plutonium production

[a] In addition to pits, LANL is assigned responsibilities for detonator production and surveillance, neutron tube target loading, beryllium component manufacturing, non-nuclear pit parts production, mock pits production, surveillance of radioisotopic thermoelectric generators (RTGs) and certain valves.

Sources: Cochran, T. et al., *US Nuclear Warhead Facility Profiles, Nuclear Weapons Databook, Vol. III* (Ballinger: Cambridge, Mass., 1987); and US Department of Energy, *FY 2000: Stockpile Stewardship Plan*, Sanitized Version (DOE Office of Defense Programs: Washington, DC, 15 Mar. 1999).

dation of nuclear weapon activities and restructuring taking place within individual facilities. According to NNSA officials, the complex needs a major investment in order to maintain and improve the basic infrastructure of individual facilities (buildings, roads, utilities, etc.).

Although the US weapon production capability has been reduced, it remains significant. For example, the Pantex Plant has a capacity to produce approximately 1100 warheads per year, compared to the cold war production level of 2000 warheads per year.[25] Other key DOE facilities also maintain a sizeable production capacity.[26]

The USA currently lacks a source of new tritium. Tritium is an essential component of all modern thermonuclear (and presumably fission) weapons. Tritium–deuterium gas is used to boost the yield of the fission primary to make it powerful enough to ignite the thermonuclear secondary. Because it decays at a rate of approximately 5.5 per cent per year, there is a need for periodical removal of helium, the decay product, and for replenishment of tritium stocks. In this context the 1993 Treaty on Further Reduction and Limitation of Strategic Offensive Arms (START II Treaty), which is not in force, can be mentioned. According to the DOE, in order to maintain a five-year reserve of tritium without the treaty's entry into force the USA would need to resume the production of tritium in 2005. In December 1998 the DOE decided to

---

[25] Pantex's capacity is dependent on the complexity and mix of specific weapon systems and activities (dismantlement, disassembly and inspection, rebuilding, etc.), e.g., the disassembly- and inspection-only capacity is 250–350 warheads per year. See US Department of Energy (note 24).

[26] The Oak Ridge Y-12 plant maintains the capability to manufacture 300 secondaries per year, compared to 1500 secondaries per year during the cold war. The Savannah River Tritium facility is capable of tritium recycling/reloading of 2500 reservoirs per year, compared to 6000 reservoirs per year in the past. See US Department of Energy (note 24).

use the commercial nuclear power reactors of the Sequoyah and Watts Bar nuclear power plants in Tennessee for these purposes.[27]

The USA also currently lacks an industrial-scale capability to produce plutonium pits. This might not be an obstacle, however. The Los Alamos National Laboratory, the only US facility with complete plutonium-handling capabilities, is expected by 2007 to reach a manufacturing capacity of 20 pits per year. Eventually, it could be able to produce 50 (with a surge capacity of 80) pits per year. This capability is generally viewed as sufficient to maintain the US stockpile. The DOE is also developing a contingency plan that would allow the USA to have a manufacturing facility capable of producing 500 pits per year within five years of a decision to build such a facility.[28] The conviction that such a contingency plan is needed is largely based on incomplete scientific knowledge about plutonium ageing, uncertain predictions about pit longevity and the possibility of common-mode failure mechanisms that could render inoperative a large fraction of the warhead stockpile. There are also proposals within the DOE to build a new state-of-the-art uranium-processing facility to replace the ageing facilities of the Oak Ridge Y-12 plant in Tennessee.

Future deep arms reductions might lead to further contraction of the complex, including a transfer of certain production functions to national weapon laboratories and a closure of some facilities. For example, for a stockpile of a few hundred weapons, warhead maintenance and refurbishment operations eventually could be moved to the Device Assembly Facility (DAF). Located at the Nevada Test Site, DAF is a state-of-the-art safe and secure facility that was originally designed to assemble nuclear explosive devices for underground testing; it is now primarily used to support the DOE's subcritical experiments and for training.[29] It has Pantex-style warhead assembly–disassembly bays and cells, and staging areas for warhead and nuclear component storage.[30]

IV. Conclusions

The end of the cold war called for the radical downsizing and restructuring of the Russian and US nuclear warhead production complexes. The Russian complex, although smaller compared to its size during the cold war, remains oversized. The implementation of Russia's downsizing plans has been slow because of insufficient funding and difficulties in finding alternative employment for displaced nuclear weapon workers. The USA has already concluded the first phase of its infrastructure reductions. However, the US decision to rely on the SBSS programme for stockpile maintenance, the US Senate's rejection of the CTBT and the US policy of maintaining a large reserve of non-deployed warheads as well as plutonium pits could have a negative impact on non-proliferation and future arms reductions.

Future nuclear arms reductions in Russia and the USA would call for further downsizing and restructuring of their respective nuclear warhead production infra-

---

[27] 'Secretary certifies nuclear stockpile; selects tritium source, pit disassembly sites', *DOE This Month*, Jan. 1999, p. 3.

[28] See US Department of Energy (note 24), chapter 12, p. 8.

[29] 'Device assembly facility: new facilities for handling nuclear explosives', *Science and Technology Review*, May 1998, available at URL <http://www.llnl.gov/str/05.98.html>.

[30] Disassembly cells are used to conduct operations with uncased explosives and fissile material components. Should conventional explosives detonate, disassembly cells are designed to vent such an explosion and trap fissile materials. Operations with uncased insensitive high explosives and nuclear materials may be performed inside a disassembly bay.

structures. Under deep stockpile reductions, for example, most nuclear warhead operations could be consolidated at pilot-scale facilities associated with nuclear weapon R&D laboratories.

High-level policy support and technical cooperation between the two countries on complex downsizing and restructuring, and international cooperation to facilitate the contraction of the Russian complex, must become integral elements of the downsizing effort. The Russian–US nuclear relationship will be a major determinant in the reshaping of the national weapon complexes. Greater transparency in nuclear activities in both countries and expanded cooperation are thus critical for developing rational post-cold war policies.

Appendix 10D. Efforts to improve nuclear material and facility security

GEORGE BUNN and LYUDMILA ZAITSEVA

I. Introduction

International efforts to strengthen the physical protection of nuclear materials and facilities from theft and sabotage have been intensified in the aftermath of the attacks carried out in the United States on 11 September 2001. However, the magnitude of the changes that are needed to protect against terrorist attacks of that nature has not yet been widely appreciated, perhaps in part because of beliefs in some states that what happened in the USA 'can't happen here'. This appendix provides evidence that, in other countries, terrorists and thieves have already threatened or attacked nuclear facilities and tried to purchase or steal nuclear and other radioactive material. Section II of this appendix summarizes the relevant features of the 11 September attacks and the measures taken prior to that date to protect nuclear facilities against sabotage, as far as is known publicly. Section III discusses the illicit traffic in the nuclear and other radioactive materials that might be used by terrorists. Section IV addresses international efforts to improve the physical protection of such materials in both military and non-military contexts.

II. The attacks of 11 September and threats to nuclear facilities

The attacks of 11 September suggest that the threat to nuclear facilities is more complex than many states contemplated when they were built. Data published by the International Atomic Energy Agency (IAEA) show that there are at least 284 research reactors in 55 countries and 472 power reactors (operating or under construction) in 31 countries.[1] Since there is no multilateral treaty requiring physical protection of these facilities or the nuclear material used or stored for use by them, variations from state to state on how they are protected are to be expected.[2] Even in wealthy industrial countries, such as the USA, with many nuclear facilities and well-established regulatory systems, non-governmental organizations have long complained that civilian nuclear reactors are not adequately protected against truck bombs, much less against

---

[1] International Atomic Energy Agency (IAEA), *Nuclear Power Reactors in the World*, Reference Data Series no. 2 (1999), table 1; and IAEA, *Nuclear Research Reactors in the World*, Reference Data Series no. 3 (2000), table 1. Much of these data are available at URL <http://www.iaea.org/worldatom/rrdb/shtml>.

[2] The only multilateral treaty providing any standards for physical protection is the 1980 Convention on the Physical Protection of Nuclear Material (CPPNM). IAEA document INFCIRC/274/Rev 1, Add. 7, 22 Sep. 2000, available at URL <http://www.iaea.org/worldatom/Documents/Infcircs/Others/inf274r1a5.shtml>. It does so only for civilian nuclear material while in *international* transport, or in storage pending or after international transport, not for any material in *domestic* use, storage and transport. The IAEA is the depositary for the CPPNM. Article 1 (a) defines 'nuclear material' as 'plutonium except that with isotopic concentration exceeding 80% in plutonium-238; uranium-233; uranium enriched in the isotope 235 or 233; uranium containing the mixture of isotopes as occurring in nature other than in the form of ore or ore residue; any material containing one of the foregoing'.

large airliners loaded with fuel.³ A US Nuclear Regulatory Commission (NRC) technical report on a reactor to be located not far from a populated area listed possible worst-case power reactor accident scenarios such as sabotage might produce. It concluded that over 100 000 people could eventually die from the health effects caused by the radioactivity dispersed as a result of one such accident.⁴ Moreover, research reactors tend to be less well protected than power reactors but more likely to contain weapon-usable highly enriched uranium (HEU).⁵

Reports from Russia suggest that, in general, protection practices for weapon-usable materials vary and need strengthening at some installations.⁶ The IAEA experts who helped 10 smaller states, mostly in Eastern Europe, strengthen their physical protection practices said that the protections they found varied from state to state: 'Differences in culture, perceived threat, financial and technical resources and national laws are some of the reasons for variations'.⁷ A survey of physical protection practices made in 1997 by the Sandia National Laboratories (SNL) in cooperation with Stanford University showed that only 11 of 19 respondent states reported that their security was designed to deal with terrorists or saboteurs.⁸ Responses from 6 states to a Stanford University questionnaire in 2001 showed that none of them had planned protection against an attack involving truck bombs that 'spreads radioactive material over and beyond the protected area', for example, the fenced-in area around

---

³ See, e.g., Hirsch, D., 'The truck bomb and insider threats to nuclear facilities', eds P. Leventhal and Y. Alexander, *Preventing Nuclear Terrorism* (Lexington Books: Lexington, Mass., 1987), p. 207; and Bunn, G., Steinhausler, F. and Zaitseva, L., 'Strengthening nuclear security against terrorists and thieves through better training', *Nonproliferation Review* vol. 8, no. 3 (2001), pp. 139–41. For recent calls by a non-governmental organization for higher protection standards for reactors in the USA, see the Internet site of the Nuclear Control Institute, URL <http://www.nci.org>. It includes a number of links to warnings and discussions concerning possible plans by the al-Qaeda network to use airliners to dive on nuclear reactors.

⁴ US Nuclear Regulatory Commission, 'Supplement to Draft Environmental Statement Related to the Operation of San Onofre Nuclear Generating Station, Units 2 and 3', NUREG-0490, Jan. 1981, especially figure 7.1.4-4, 'Probability distribution of acute fatalities', which estimates 130 000 deaths in the event of a worst-case accident.

⁵ See Bunn, Steinhausler and Zaitseva (note 3), p. 139. Because there is no international treaty requiring protection, there is limited information available on protection practices. Cases of uranium thefts from research reactors in the Democratic Republic of the Congo and Georgia are described below. Even in an industrialized West European country after 11 Sep., individuals with false identity papers gained entry to a research reactor and were not apprehended until after they had managed to get inside it.

⁶ See, e.g., Bunn, M., 'A detailed analysis of the urgently needed new steps to control warheads and fissile material', ed. J. Cirincione, *Repairing the Regime* (Routledge: New York, 2000), pp. 74–77 (this chapter quotes Russian Minister of Atomic Energy Evgeniy Adamov as acknowledging that 'the weakening of our ability to manage nuclear material has been immeasurable'); and Orlov, V., Timerbaev, R. and Khlopkov, A., *Nuclear Nonproliferation in US–Russian Relations: Challenges and Opportunities* (PIR Center for Policy Studies: Moscow, 2002), pp. 37–49. In a television interview after 11 Sep., the head of the material protection and control department of the Russian nuclear regulatory agency (Gosatomnadzor), Yuri Volodin, acknowledged 'complaints' but said that no 'large thefts of nuclear material' had yet taken place in the Russian Federation. Lenta.Ru, ['Theft of nuclear material in Russia invented by journalists'], 13 Nov. 2001 (in Russian), available at URL <http://lenta.ru/terror/2001/11/13/volodin>. An earlier description of the protection, control and accounting of Russian nuclear weapons by an American and a Russian appears in Lepingwell, J. and Sokov, N., 'Strategic offensive arms elimination and weapons protection, control and accounting', *Nonproliferation Review*, vol. 7, no. 1 (spring 2000), p. 99.

⁷ Soo Hoo, M. et al., 'International Physical Protection Advisory Service: observations and recommendations for improvement', *Proceedings of the 40th Annual Meeting of the Institute for Nuclear Materials Management (2000)* (on CD), available from the Institute of Nuclear Materials Management, email address inmm@inmm.org.

⁸ Harrington, K., *Physical Protection of Civilian Fissile Material: National Comparisons* (Sandia National Laboratories: Livermore, Calif., 1999), p. 18.

the power reactor.⁹ In published reports of these small surveys, the particular problems at particular facilities were not generally identified with the name of the facility, or sometimes even with the name of the country where it was located, because of the fear that saboteurs or thieves would then learn where weaknesses existed. In general, facts about particular physical protection practices are kept confidential.

It is known that the al-Qaeda network and Usama bin Laden have sought weapon-usable and other radioactive materials, as well as nuclear weapons and radioactive dispersal devices.¹⁰ A question is whether they could or would also attack nuclear power reactors using, for example, aircraft or trucks carrying explosives. The magnitude of the destruction, the total disregard for life, both their victims' and their own, shown by the 11 September terrorists, as well as the enormity of the effort, coordination, organization, financial backing and sense of religious mission that probably went into their preparations all tend to suggest an answer in the affirmative. Other terrorist groups may well have similar goals.

The IAEA reported in November 2001 that its 'past efforts have focused largely on diversion of nuclear material by States for non-peaceful purposes, without the same degree of focus on malicious activities by sub-national groups'.¹¹ Thus, the IAEA's estimate of the extent of the damage to a nuclear facility from the intentional crash of a 'large, fully fuelled jetliner' was 'still a matter for analysis. Nuclear facilities vary from state to state, so studies will have to take specific plant designs into account'. According to the IAEA Director General, 'After 11 September, we realized that nuclear facilities—like dams, refineries, chemical production facilities or skyscrapers—have their vulnerabilities. There is no sanctuary anymore, no safety zone'. Moreover, IAEA experts 'are concerned that terrorists could develop a crude radiological device using radioactive sources commonly used in every day life'.¹² This could mean using radioactive materials to make 'dirty bombs' with conventional explosives to disperse the radioactivity.¹³

---

⁹ Bunn, M. and Bunn, G., 'Nuclear theft and sabotage: priorities for reducing threats'', *IAEA Bulletin*, vol. 43, no. 1 (Dec. 2001), pp. 8–9.

¹⁰ See, e.g., 'US indictment: 'detonated an explosive device', *New York Times*, 5 Nov. 1998, p. A9; Weiser, B., 'US says Bin Laden aide tried to get nuclear weapons', *New York Times*, 26 Sep. 1998, p. A3; and 'Responsibility for the terrorist atrocities in the United States', 11 Sep. 2001, *New York Times*, 5 Oct. 2001, p. B4. According to the US Central Intelligence Agency (CIA), al-Qaeda and several other terrorist organizations have expressed interest in nuclear weapons. See e.g., Zakaria, T., 'CIA: threat of weapons of mass destruction up', Reuters, 30 Jan. 2002; and 'Words of the CIA chief on terror', *New York Times*, 7 Feb. 2002, p. A10. Other recent accounts are available at the Nuclear Control Institute Internet site (note 3).

¹¹ IAEA, 'IAEA outlines measures to enhance protection against nuclear terrorism', IAEA Press Release PR 2001/26, 30 Nov. 2001, p. 1, available at URL <http://www.iaea.org/worldatom/Press/P_release/2001/prn0126.shtml>. See also IAEA, 'Summary of report on protection against nuclear terrorism, presented to the IAEA Board of Governors on 30 November 2001', IAEA Press Release PR 2001/26a, 30 Nov. 2001, available at URL <http://www.iaea.org/worldatom/Press/P_release/2001/prn0126a.shtml>. (Both of these press releases summarized a report by the IAEA Director General to the IAEA Board of Governors, a report which was 'restricted' and was not made available to the public.)

¹² IAEA, 'Calculating the new global nuclear terrorism threat', IAEA Press Release, 1 Nov. 2001, p. 3, available at URL <http://www.iaea.org/worldatom/Press/P_release/2001/nt_pressrelease.shtml>. (This is a summary of the statements made by the IAEA Director General at a press conference on the day of an IAEA Symposium on Nuclear Terrorism.)

¹³ IAEA, 'Calculating the new global nuclear terrorism threat' (note 12), p. 5. Measures being taken to address these issues are discussed in section IV.

III. Illicit trafficking in nuclear and other radioactive material

The *SIPRI Yearbook 2001* contains a summary of illicit trafficking in nuclear and other radioactive material through March 2001 based principally on the IAEA Illicit Trafficking Database.[14] All the conclusions of this summary are confirmed by Stanford University's analysis of its Database on Nuclear Smuggling, Theft and Orphan Radiation Sources (DSTO).[15] However, in addition to the state-confirmed incidents from the IAEA database, the DSTO includes open source reports and data for the years 1991 and 1992. It thereby provides a broader insight into the problem of illicit trafficking over the past 10 years.[16]

Analysis on a global level of both the state-confirmed incidents in the IAEA database and the data in the DSTO indicates that there was a noticeable increase in the number of incidents in 1998–2000, following a sharp peak in 1993 and 1994 and a subsequent decline from 1995 to 1997 (see figure 10D.1). Preliminary data show that the number of incidents declined in 2001 compared to the period 1998–2000.

The current number of incidents involving thefts and seizures of nuclear material is considerably lower than in the early 1990s. From 1991 to 1996, nuclear material was seized or stolen more frequently than other radioactive material (see figure 10D.2).[17]

---

[14] Zarimpas, N., 'The illicit traffic in nuclear and radioactive materials', *SIPRI Yearbook 2001: Armaments, Disarmament and International Security* (Oxford University Press: Oxford, 2001), pp. 503–11.

[15] The DSTO is a collection of illicit trafficking incidents (e.g., thefts and seizures of nuclear and other radioactive material) that have taken place worldwide. It includes the state-confirmed incidents presented in the IAEA Illicit Trafficking Database. See 'Comprehensive list of incidents involving illicit trafficking in nuclear materials and other radioactive sources as of 1 March 2001: confirmed by states', available from the IAEA Office of Physical Protection and Material Security; see also the abstracts on all reported instances of nuclear trafficking in and from the Newly Independent States (NIS), collected by the Monterey Institute of International Studies, Center for Nonproliferation Studies (CNS) in its NIS Nuclear Trafficking Database, available at URL <http://www.nti.org/db/nistraff>. The IAEA database is missing many incidents reported in open mass media sources because they had happened before 1993, when the IAEA started its database programme, or because the involved states failed to report them (e.g., over 200 incidents collected by the CNS researchers are not part of the IAEA database). The CNS database, in its turn, does not include c. 200 incidents from the IAEA database, because they are either unrelated to NIS countries or were not covered by the press. By combining all the incidents from these 2 databases and from additional open sources in a single format, the DSTO has achieved a more complete international picture of illicit trafficking. The unified, user-friendly computer format allows for a quick statistical analysis of the input data. It has an added advantage of corroborating the open source information with the state-confirmed IAEA data, such as the date and location of the incident and the description and exact amount of the material involved. Additional open sources used by DSTO are books and other publications, conference proceedings, international print and electronic media, and the Internet.

Orphan radiation sources are 'sources that were never subject to regulatory control; sources that were subject to regulatory control but since have been abandoned, lost or misplaced; and sources that were stolen or removed without proper authorization'. Ortiz, P. et al., 'Lost and found dangers: orphan radiation sources raise global concerns', *IAEA Bulletin*, vol. 41, no. 3 (1999), p. 18. See also US Environmental Protection Agency, 'Orphan sources initiative', available at URL <http://www.epa.gov/radiation/cleanmetals/orphan.htm>.

[16] A special parameter—reliability factor—was devised for the Stanford DSTO to define the degree of reliability of information presented in each particular case: high, medium or low. *High* denotes high credibility of data (confirmed by IAEA and/or confirmed by competent national authorities), *medium* denotes reasonable credibility of data (not confirmed to the IAEA, but confirmed by local authorities directly involved in the incidents, as referenced in mass media reports) and *low* denotes less credible or conflicting data. It should be noted that over 75% of the incidents recorded in the DSTO are in the reliability categories *high* or *medium*.

[17] For the purposes of the DSTO, nuclear material is defined as uranium, plutonium, thorium or a compound containing any of these elements, and irradiated nuclear reactor fuel. Although nuclear material is radioactive, the term 'other radioactive materials' refers primarily to ionizing radiation sources (e.g., americium, cesium, cobalt, radium, strontium, etc.). See also the definition given in note 2.

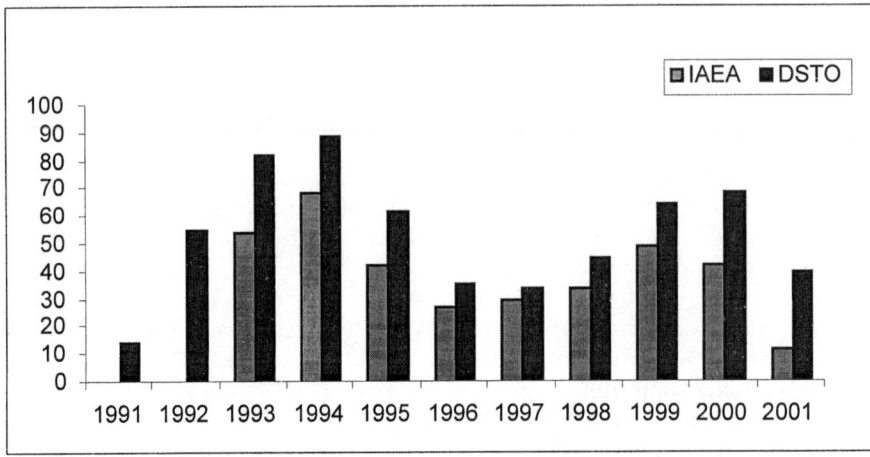

Figure 10D.1. Incidents of illicit trafficking in nuclear and other radioactive material, 1991–2001

Sources: For IAEA data: Data reported by states to the IAEA between Jan.1993 and Mar. 2001, available upon request from the IAEA Office of Physical Protection and Material Security; for DSTO data: Data include both confirmed and unconfirmed incidents with high, medium and low reliability. Database on Nuclear Smuggling, Theft and Orphan Radiation Sources (DSTO), Center for International Security and Cooperation, Stanford University, Stanford, Calif., 2002 (restricted access).

This trend started to change in 1997 and there were fewer cases of illicit trafficking involving nuclear material in 1997–2001 as compared to incidents involving other radiation sources. In the period 1998–2001, the incidents involving nuclear material have accounted for less than one-third of the total number of illicit trafficking cases.

Of 643 illicit trafficking cases recorded in the DSTO database for the period January 1991 to December 2001, almost one-half (303) involved thefts and seizures of nuclear material. Of these, 129 incidents (42 per cent) were of no proliferation concern (e.g., natural uranium, depleted uranium, 'yellow cake'), 126 (42 per cent) of low proliferation concern (e.g., low-enriched uranium, LEU, and minuscule amounts of plutonium, including those in radiation sources) and 48 (16 per cent) of high proliferation concern (e.g., HEU and plutonium).[18] However, the majority of proliferation-significant incidents took place in the period 1991–95, suggesting that the efforts to improve the physical security of weapon-usable nuclear material in Russia and other former Soviet republics have started to bear fruit.

---

[18] For comparison, the IAEA database contains 168 incidents involving nuclear material over the period Jan. 1993 to Mar. 2001. Of those, 89 incidents (53%) were of no proliferation concern, 65 (38%) of low proliferation concern and 15 (9%) of high proliferation concern. See 'Comprehensive list of incidents' (note 15).

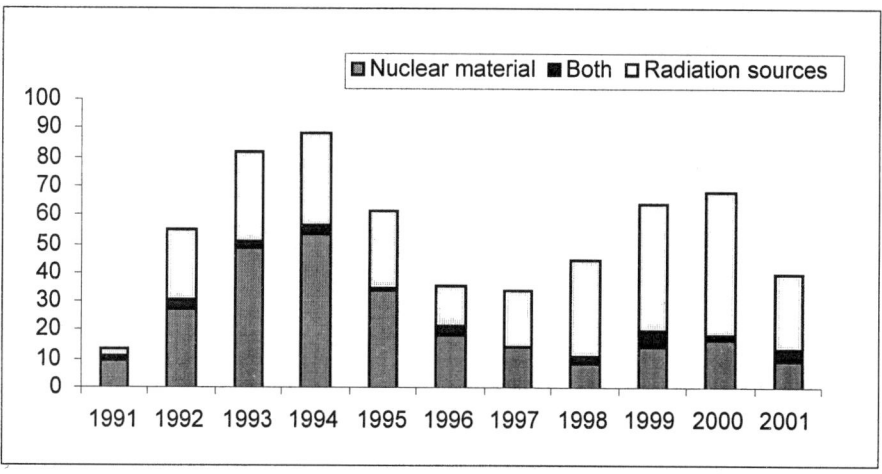

Figure 10D.2. Incidents of illicit trafficking involving nuclear material, other radioactive material and both, 1991–2001[a]

[a] Orphan radiation sources are not included because, for the purpose of the DSTO database, they are considered to be outside the illicit trafficking problem since there is no known underlying criminal intent to sell them to third parties or to use them with malicious intent.

Source: Database on Nuclear Smuggling, Theft and Orphan Radiation Sources (DSTO), Center for International Security and Cooperation, Stanford University, Stanford, Calif., 2002 (restricted access).

According to the IAEA database, the total amount of stolen and seized weapon-usable material is approximately 10.820 kg. The largest amounts were intercepted in St Petersburg, Russia, in June 1994 (2.972 kg of 90 per cent HEU), and in Prague, Czech Republic, in December 1994 (2.73 kg of 87.7 per cent HEU).[19] However, the DSTO database shows that the total could be as high as 37.158 kg if other credible proliferation-significant cases—unconfirmed by states to the IAEA—were to be included. If these 37 kg represent 10–40 per cent of the material actually smuggled, as is the case with drug trafficking in the USA, the actual situation in states with border control and law enforcement less efficient than in the USA may be a reason for serious concern.[20] Although all of this material originated in the former Soviet Union (FSU), there has been at least one known theft of enriched uranium from a state other than the FSU. In 1998, members of a smuggling ring in possession of 19.9 per cent enriched uranium (20 per cent is defined as weapon-usable) were arrested in Italy. The US-fabricated 190 g fuel rod—one of the eight reportedly missing—had been stolen from the Triga II research reactor in Kinshasa, the Democratic Republic of the Congo, where security was described as appalling.[21]

---

[19] Calculations are based on the data provided in 'Comprehensive list of incidents' (note 15).
[20] At best, law enforcement officials seize only 10–40% of the illegal drugs smuggled into the USA each year. See Williams, P. and Woessner, P., 'Nuclear material trafficking: an interim assessment', Working Paper 95-3, *Ridgway Viewpoints* (Matthew B. Ridgway Center for International Security Studies, University of Pittsburgh: Pittsburgh, Pa., 1995), p. 2, available at URL <http://www.pitt.edu/~rcss/viewpoints.htm>.
[21] Wrong, M., 'More wreck than reactor', *Financial Times*, 21 Aug. 1999, p. 8.

In general, research reactors around the world have been a reason for concern with regard to theft of weapon-usable nuclear material. Despite the ongoing US effort to convert research reactors using HEU to LEU and retrieve the HEU which it originally supplied, many states in the world, including 28 developing states, still operate on HEU.[22] Some of them are reportedly not well guarded, presenting a potential target for theft, especially in times of political crises. For example, during the Kosovo conflict in the late 1990s there was serious concern about the Vinca research reactor in Serbia, which holds some 50 kg of Soviet-produced enriched fresh HEU and 10 kg of low-irradiated HEU. This concern persists because of the questionable physical security arrangements at the facility.[23] About 2 kg of 90 per cent enriched HEU went missing from another research reactor in Sukhumi, Georgia, during the political unrest between 1992 and 1997.[24] Russia, which has supplied HEU for research reactors in these and many other states, is only at the planning stage for an effort similar to the US conversion programme.[25] The reactors at Sosny, Belarus, and Kharkiv, Ukraine, which hold between them some 445 kg of HEU, would be the prime candidates for the material retrieval.[26]

The regional trends in illicit trafficking of nuclear and other radioactive material have changed over the past 10 years. After the highest peak in 1993 and 1994, Western Europe witnessed a sharp decline in illicit trafficking in 1995–97 and the number of incidents has remained relatively low since then. By comparison, the decline recorded in the mid-1990s in Eastern Europe was less pronounced than in Western Europe and the number of incidents increased again significantly in 1999 and 2000. The improved border control and policing for radioactive materials in Eastern Europe may now serve as a barrier for the trafficking flow from the FSU, allowing less material to reach the West European frontiers.

A new peak in illicit trafficking was also observed in Russia in 1998–2001. However, during the period January 1998 to March 2001, only 3 incidents were confirmed by the Russian Government to the IAEA, whereas the Stanford DSTO database contains 37 incidents—most of them involving radiation sources—reported in open sources over the same period of time. The actual number of illicit trafficking cases may be higher still, because they can go unnoticed owing to inadequacies in the detection capabilities at many border crossings in Russia. For example, 61 events of radiation detection were recorded at the Sheremetyevo international airport in Moscow in 1999, after a radiation monitoring system had been installed, whereas in 1997, prior to its installation, only 2 such events were detected.[27] Because the

---

[22] IAEA, *Nuclear Research Reactors in the World* (note 1).

[23] In 1996 the IAEA installed an electronic surveillance system at Vinca. However, Vinca officials apparently did not consider the resulting improvement of security as sufficient and approached the USA and the IAEA with a request to remove the HEU from the country. It was considered to be at risk because of economic and political instability. So far, the material has not been removed from Serbia. See Potter, W., Miljanic, D. and Slaus, I., 'Tito's nuclear legacy', *Bulletin of the Atomic Scientists*, vol. 56, no. 2 (Mar./Apr. 2000), p. 69.

[24] Daughtry, E. and Wehling, F., 'Cooperative efforts to secure fissile material in the NIS', *Nonproliferation Review*, vol. 7, no. 1 (spring 2000), p. 100. See also the Center for Nonproliferation Studies, CNS Reports: *Confirmed Proliferation-Significant Incidents of Fissile Material Trafficking in the Newly Independent States (NIS), 1991–2001*, 30 Nov. 2001, available at URL <http://cns.miis.edu/pubs/reports/traff.htm>.

[25] Research reactors in 12 countries still use Russian/Soviet-supplied HEU. See IAEA, *Nuclear Research Reactors in the World* (note 1).

[26] Daughtry and Wehling (note 24), pp. 99, 102.

[27] Ukhlinov, L. and Bojko, V., 'Organization of customs control of fissionable and other radioactive materials', *Proceedings of the IAEA International Conference 'Security of Nuclear Material: Measures*

Sheremetyevo officials did not single out any particular seizures, it can probably be assumed that most of the detected material was of no proliferation concern. However, if someone decided to smuggle weapon-usable material on board an aircraft, as was the case with 360 g of plutonium seized in the Munich airport from passengers en route from Moscow in August 1994, he or she might have been able to do so undetected before the installation of the detection equipment.

Russia's reluctance to publicly acknowledge all the facts of continuing smuggling attempts, even if they have been successfully countered, and to report them to the IAEA may be caused by its unwillingness to be subjected again to humiliating international criticism. In addition, Russia may not want to demonstrate the weak spots in the security of its nuclear facilities in order to prevent attempts by those interested in acquiring nuclear material to do so.

Whatever the reasons are for Russia's not reporting all of its illicit trafficking cases to the IAEA, they may be shared by some other states. For example, as of March 2001, the USA has not reported a single case to the IAEA database. However, according to the US Nuclear Regulatory Commission, an average of about 200 licensed radiation sources are lost, abandoned or stolen in the USA each year and the media occasionally report thefts of radiation sources.[28] Of the other nuclear weapon states, as of March 2001 France and the UK reported three incidents each, all involving radiation sources, whereas China has not reported any. However, in July 2001, 5 grams of 80 per cent HEU—presumably a sample of a larger cache—were seized in Paris, France.[29] Given the level of enrichment, the HEU could have been stolen either from a research reactor or from a nuclear submarine depot.

Of special concern is the increased illicit trafficking from Russia and other former Soviet republics through the Southern Tier.[30] In 1992–98, the number of incidents detected in this region remained low (on the average, 4 cases per year) and then sharply increased in 1999 and 2000 (18 and 11, respectively). Six incidents were reported in 2001. Although these incidents represent only a fraction of the number of cases recorded in Eastern Europe or Russia, the quality of the material smuggled through the Southern Tier in terms of its proliferation potential is noticeably higher. This may indicate that better educated traffickers are using the southern routes. For example, of 60 trafficking incidents that took place in Eastern Europe from January 1999 to December 2001, 19 involved nuclear material, including 2 seizures of minuscule amounts of plutonium in radiation sources and 2 seizures of LEU. A total of 35 incidents were reported to have taken place in the Southern Tier over the same period, of which 18 involved nuclear material, including 3 confirmed and 2 unconfirmed seizures of HEU and plutonium and 10 seizures of LEU.

to Prevent, Intercept and Respond to Illicit Uses of Nuclear Material and Radioactive Sources', Stockholm, 7–11 May 2001, p. 80, available at URL <http://www.iaea.or.at/worldatom/Press/Focus/Stockholm/sw-papers010402.pdf>. For a review of illicit trafficking in Russia in 1989–2000 see Orlov, Timerbaev and Khlopkov (note 6).

[28] Dicus, G., 'USA perspectives: safety and security of radioactive sources', *IAEA Bulletin*, vol. 41, no. 3 (1999), p. 22. The DSTO lists 11 thefts of radiation sources in the USA.

[29] Reuters, 'French arrest 3 for nuclear trafficking', 22 July 2001, available at URL <http://www.wise-paris.org/english/intro/othersnewsarchives.html>. See also Anzelon, G., 'Improving the knowledge base on nuclear terrorism threats', Paper presented at the IAEA Special Session on Combating Nuclear Terrorism, 2 Nov. 2001, URL <http://www.iaea.org/worldatom/Press/Focus/Nuclear_Terrorism/anzelon.pdf>.

[30] For the purposes of the DSTO database, the Southern Tier includes the Caucasus (including the adjacent republics in the south of Russia—Dagestan, Karachaevo-Cherkessia, North Ossetia and Chechnya), Central Asia and Turkey.

The above evidence suggests that southern routes are used for illicit trafficking more than before. The borders of the region are still not up to the challenge. Of the 18 seizures of nuclear material that took place in the Southern Tier over the past three years, 15 were reported to have resulted from police or intelligence operations and one was intercepted at a border crossing by a US-trained official.[31] Of the 17 seizures of radioactive material, 6 took place at border crossings, 2 of which using the detection equipment provided by the US Customs Service.[32] Despite these successes of the US assistance programmes, more remains to be done to have a significant impact on curbing illicit trafficking in the region, particularly in Turkey, where only 2 of the existing 120 border posts are reportedly equipped with radiation detection systems, both donated by the USA.[33]

In all of the cases with an adequate description, the intention of the traffickers was to sell the material for profit. The scenario that appears to be most frequent is obtaining the material in Russia or Kazakhstan and transporting it through Georgia to Turkey for the final sale to end-users. As for the possible buyers in the region, the most frequently reported destinations of the smuggled nuclear material over the past 10 years have been Iraq (6 incidents), Iran (5), Libya (5) and 'a Middle Eastern country' (5).[34] Iran seems to have at least three different supply routes: from the Caucasus via Georgia and Turkey, from the Caucasus via Armenia and from Central Asia via Afghanistan. The end-users, however, are the least known link in the supply chain because of the lack of hard evidence connecting them to particular incidents. Therefore, one can only guess about the final destination of the smuggled material by the route its traffickers take.

IV. International efforts after 11 September to improve security against terrorists

Is the physical protection of nuclear power reactors against suicide attacks such as those carried out on 11 September by several large fuel-laden jet airliners beyond what is financially feasible? As seen by the IAEA, these reactors are 'industrial facilities and as such are not hardened to withstand acts of war'.[35] Better security for commercial airliners, their passengers and their airports may be the most likely way of dealing with this threat—except, perhaps, when warnings permit the use of protective fighter aircraft or when the nuclear facility is so dangerous that installation of anti-aircraft weapons is justified.[36] However, diving smaller, medium-sized rental air-

---

[31] Since 1998, there have been 8 significant seizures of nuclear material by customs or police agencies outside the USA which could be attributed to non-proliferation training carried out under the auspices of the US Customs Service. See US Customs Service, 'US Customs kicks off training to help former Soviet republics combat spread of weapons of mass destruction', Washington, DC, 21 Aug. 2001, URL <http://usinfo.state.gov/topical/pol/arms/stories/01082203.htm>.

[32] US Customs Service (note 31); and US Department of Defense, Federal Bureau of Investigation and US Customs Service Counterproliferation Program, 'Success stories', available at URL <http://www.dtra.mil/os/fbi-uscs/os_successstor.html >.

[33] Frank, D., 'Nuclear booty: more smugglers use Asia route', *New York Times*, 11 Sep. 2001, p. A1.

[34] Database on Nuclear Smuggling, Theft and Orphan Radiation Sources (DSTO), Center for International Security and Cooperation, Stanford University, Stanford, Calif., 2002 (restricted access).

[35] IAEA, 'Calculating the new global nuclear terrorism threat' (note 12), p. 3. See also IAEA, 'Summary of report on protection against nuclear terrorism' (note 11), p. 2.

[36] IAEA, 'Calculating the new global nuclear terrorism threat' (note 12). The Cogema fuel cycle complex at Cap La Hague in France has installed anti-aircraft missiles primarily to protect its large storage ponds for highly radioactive spent fuel and for the waste from its reprocessing plant. More radio-

craft flying from private airfields and loaded with high explosives and fuel onto a nuclear site could not be controlled by better public airport and airline security and might cause a major release of radioactivity. The same concern holds for trucks loaded with high explosives attacking the reactor or its spent fuel pond. In some countries, protections have been planned against an attack involving one truck bomb.[37] However, if there are two trucks and the bombs from the first truck blow up the outer protection facilities near the gate, and the second truck is then able to reach the spent fuel pond or reactor before it explodes, these protections could be inadequate.

Nuclear security in the largest nuclear weapon states

Despite efforts to improve the security of US military weapon-usable materials going back to 1970, army and navy commando teams recently demonstrated that they were able to penetrate security systems at a number of government-owned nuclear facilities and escape with significant quantities of weapon-usable nuclear materials.[38] The situation in Russia is thought to be worse.[39] The security system that Russia inherited from the Soviet Union relied on a closed society with closed borders, well-paid nuclear workers and personnel under the surveillance of the KGB (Komitet gosudarstvennoy bezopasnosti, the security services of the FSU). It began to break down with the increasing freedom and declining living standards of the past decade. As in the USA, Russian nuclear weapons, which are readily accountable, remain under what appear to be high levels of security. For Russian weapon-usable material, however, the security is generally lower and varies greatly from place to place.[40] Since 1994 the USA and a number of other industrialized countries have been providing financial assistance to help Russia install improved protection systems at many sites, both civilian and military, where nuclear materials are used and stored. However, as of late 2001, rapid security upgrades had been completed on facilities containing less than one-third of the hundreds of tons of weapon-usable nuclear materials in Russia, and programme managers in the USA estimated that completion of these upgrades would take until the end of 2007, assuming the current assistance to Russia is continued at the present levels.[41]

While Russia and the USA are believed to have some 95 per cent of the nuclear weapons in existence, the rest of the world has a much larger than 5 per cent share of weapon-usable material—including that in civilian facilities. Enough military or

---

activity is present at such a plant than at most nuclear power reactors. See Jeffries, S. and Brown, P., 'France positions missiles to protect nuclear plant', *Manchester Guardian*, 20 Oct. 2001, available at URL <http://www.guardian.co.uk/archive/article/0,4273,4281424,00.html>. An earlier report contains an analysis of the extent of the larger possible danger from the huge storage ponds. See Coeytaux, X. *et al.*, 'La Hague particularly exposed to plane crash risk', World Information Service on Energy (WISE), Paris, 26 Sep. 2001, p. 5, URL <http://www.wise-paris.org/english/oumews/news2.html>.

[37] Bunn, Steinhausler and Zaitseva (note 3), pp. 139–40.

[38] von Hippel, F., 'Recommendations for preventing nuclear terrorism', F.A.S. Public Interest Report, *Journal of the Federation of American Scientists*, vol. 54, no. 6 (Nov./Dec. 2001), p. 4, available at URL <http://www.fas.org/faspir/2001/v54n6/index.html>.

[39] See the discussion and authorities in note 6. A special report on 'Assessing US nonproliferation assistance to the NIS' appeared in *Nonproliferation Review*, vol. 7, no. 1 (spring 2000), pp. 55–125.

[40] Bunn, M. and Bunn, G., 'Reducing the threat of nuclear theft and sabotage', *Journal of the Institute of Nuclear Materials Management*, vol. 30, no. 3 (2002, forthcoming).

[41] von Hippel (note 38); and Spector, L., Testimony before the Subcommittee on Internal Security, Nonproliferation, and Federal Services, US Senate Committee on Governmental Affairs, 14 Nov. 2001. Rapid security upgrades include such measures as installing sensors and cameras to detect intruders and alarms to warn the guard forces as well as bricking up windows.

civilian plutonium for many nuclear weapons exists in Belgium, China, France, Germany, India, Israel, Japan, Switzerland and the UK. In addition, according to estimates made in 2000, more than 2 tonnes of civilian HEU exist in research reactors in 43 countries, sometimes in quantities large enough to make a bomb.[42]

International efforts to strengthen physical protection

Nuclear materials and facilities

At the centre of new international efforts to strengthen worldwide physical protection is the IAEA. Its admission that security against terrorists such as those who carried out the 11 September attacks had been neglected in the past is quoted in section II. Its November 2001 report stated that IAEA activities for the protection of nuclear material had been severely limited by lack of funds.[43] One of the most important IAEA programmes to strengthen physical protection is the International Physical Protection Advisory Service (IPPAS), which sends teams of experts to requesting states to provide advice on the adequacy of their security systems. When the experts have advised strengthening, the state has often gained financial help to do so from the European industrialized countries, just as Russia has been receiving such help from the USA and others to strengthen nuclear security for its many nuclear facilities. However, because of lack of resources, the IAEA has been able to conduct IPPAS missions in only 12 states since the initiation of the programme in 1995.[44] Other important IAEA physical protection services include training, guidance publications and information exchange. However, for example, reviews and tests of emergency responses to sabotage and terrorism have not been conducted because of the inadequate budget. The physical protection programme had received less than $1 million in the regular IAEA budget plus somewhat less in non-budgetary voluntary contributions from the USA and several other industrialized states. This was far from enough to provide adequate assistance to IAEA member states for their physical protection efforts.[45]

Lack of information on important security practices also hindered the IAEA from finding out what was necessary for improvement, emphasizing its urgent need 'to identify the most vulnerable locations and see that they get the necessary security

---

[42] Bunn and Bunn (note 40); and IAEA, *Nuclear Research Reactors of the World* (note 1). A discussion of potential problems for research reactors in former Soviet republics other than Russia appears in Daughtry and Wehling (note 24).

[43] IAEA, 'IAEA outlines measures to enhance protection against nuclear terrorism' (note 11).

[44] IAEA, 'Summary of report on protection against nuclear terrorism' (note 11); IAEA Working Group of the Informal Open-Ended Expert Meeting to Discuss Whether there is a Need to Revise the Convention on the Physical Protection of Nuclear Material, 'Final Report of the Working Group', 2 Feb. 2001, p. 5. (This final report is to be distinguished from the final report which reflects the final consensus on amending the Convention on Physical Protection of Nuclear Material, 'Final report' of 23 May 2001, which invited the IAEA Director General to convene an open-ended group of legal and technical experts to prepare a draft amendment to the convention. It was based on the 2 Feb. 'Final report of the Working Group', which contains information on many issues besides whether to amend the Convention). Two reports of the IAEA Secretariat to the Working Group describe IAEA and country assistance programmes for physical protection: IAEA Secretariat Paper no. 9, 'IAEA International Physical Protection Advisory Service (IPPAS) Programme', June 2000; and IAEA Secretariat Paper no. 15, 'Bilateral physical protection support—compilation of input from member states', Nov. 2000.

[45] IAEA, 'IAEA outlines measures to enhance protection against nuclear terrorism (note 11); and IAEA, 'Summary of report on protection against nuclear terrorism' (note 11).

upgrades'.⁴⁶ Because there were no treaty-required IAEA inspections for security measures as there were for control and accounting measures for non-nuclear weapon parties to the 1968 Non-Proliferation Treaty (NPT), the IAEA did not regularly collect confidential information on security, as it did on accounting and control. As indicated above, IAEA member states regarded this information as confidential. What information exists suggests major variations in the protection of similar facilities from state to state.⁴⁷ Thieves and terrorists desiring to steal weapon-usable material are likely to seek out the places where it is least well guarded on the basis of what they can see from the outside or learn from cooperative insiders. In order to judge the scope of the problems, the IAEA intends to make an immediate attempt to gain more information about what the actual physical protection practices of states are insofar as they are willing to report that confidentially.

The November 2001 IAEA report on protection against nuclear terrorism describes threats ranging from terrorist acquisition of nuclear material to make a nuclear weapon to attacks on facilities containing nuclear materials (reactors and fuel enrichment, fabrication, reprocessing and waste management facilities). For some time, the IAEA has assisted efforts by some members to strengthen international standards for physical protection of nuclear material. The IAEA Director General convened a group of experts in 1998 to consider strengthening the one multilateral treaty that deals with physical protection, the 1980 Convention on the Physical Protection of Nuclear Material (CPPNM). The most important reason for doing so was that the CPPNM's protection requirements apply only to nuclear material in international transport—not to that in domestic use, storage and transport.⁴⁸ In May 2001 the expert group reported a consensus on amending the treaty to apply to nuclear material within a state—material that was not in international transport.⁴⁹ Among the reasons for the amendment, according to the report, was that none of the illicit trafficking reports in the IAEA database described theft of nuclear materials in international transport; they all appeared to involve trafficking from domestic storage and use of material for which the CPPNM provided no standards.⁵⁰

While the expert group agreed to extend the CPPNM to include domestic nuclear materials, it reached no agreement on *required standards* for protection except that protection should be offered against sabotage of nuclear facilities as well as theft of nuclear material. The group did agree on some 'fundamental principles' for physical protection, which have been approved by the IAEA conference of member states and by its Board of Governors.⁵¹ On 19 March 2002 the IAEA announced pledges of funds for this effort totalling almost $3 million from several countries, the largest being from a US foundation, the Nuclear Threat Initiative.⁵² However, the IAEA reported at the same time that it required an estimated $12 million to carry out the work that was needed.

⁴⁶ IAEA, 'IAEA outlines measures to enhance protection against nuclear terrorism' (note 11).
⁴⁷ See Bunn and Bunn (note 9), pp. 8–9; and Soo Hoo *et al.* (note 7).
⁴⁸ Bunn, G., 'Raising international standards for protecting nuclear materials from theft and sabotage', *Nonproliferation Review*, vol. 7, no. 2 (summer 2000), pp. 152–53.
⁴⁹ IAEA, 'Final report', 23 May 2001 (note 44).
⁵⁰ IAEA, 'Final report of the Working Group', 2 Feb. 2001 (note 44), p. 2.
⁵¹ IAEA General Conference, *Measures to Improve the Security of Nuclear Materials and Other Radioactive Materials,* GC(45)/INF/14 (14 Sep. 2001), available at URL <http://www.iaea.or.at/worldatom/About/Policy/GC/GC45/Documents>; and IAEA, 'IAEA General Conference adopts resolution on the physical protection of nuclear material and nuclear facilities: Agency to redouble efforts to combat nuclear terrorism', IAEA Press Release PR 2001/21, 21 Sep. 2001.
⁵² See URL <http://www.nti.org/>.

The CPPNM now contains general standards for the protection of weapon-usable material when it is stored temporarily, awaiting international transport. According to the CPPNM, for example, more than 2 kg of unirradiated plutonium or more than 5 kg of HEU must be stored in a 'protected area' with access restricted to 'persons whose trustworthiness has been determined', and with surveillance of this material by guards in communication with response forces.[53] The experts' consensus report does not state whether this requirement would continue in an amended treaty. Except for the general principles, the consensus report does not suggest any specific requirements for protection of such facilities as (*a*) nuclear storage buildings or reactors, (*b*) nuclear separation, fuel fabrication or reprocessing plants, or (*c*) spent fuel or waste disposal facilities.

One of the general principles approved by the IAEA Board of Governors is that a state should base physical protection 'on a graded approach, taking into account the current evaluation of the threat, the relative attractiveness, the nature of the material and the potential consequences associated with the unauthorized removal of nuclear material and with the sabotage against nuclear facilities or nuclear material'.[54] Under this principle, if state officials decided that terrorists are not a threat to their state, they could choose to provide no protection against terrorist attacks. These states could become weak points from which terrorists could steal weapon-usable material or sabotage reactors. The consensus report on amending the CPPNM also rejects all forms of international oversight—from IAEA inspection to peer group review to periodic reports or periodic meetings of the parties to discuss practices.[55]

The IAEA publishes recommended rules for the protection of nuclear material and facilities from theft and sabotage.[56] Although many bilateral nuclear assistance agreements call for the application of these rules by the state receiving assistance, there are still major variations in physical protection by these states.[57] However, the experts' consensus opposed any requirement in the CPPNM that states follow these recommendations—or even that they be given 'due consideration'.[58]

The November 2001 IAEA report called for a revision of the IAEA recommended standards after the CPPNM amendment is agreed upon. However, that will probably not happen for a while. The amendment discussions are going slowly because, despite the 11 September terrorist attacks, most participants have not been prepared to go beyond the pre-11 September consensus described above, although they have been encouraged by the IAEA to do so.

Other radioactive material

'Other radioactive material' in IAEA regulations and policies means radioactive sources that are not uranium or plutonium but are used for industrial, medical and other uses not involving fission.[59] The November 2001 IAEA report described the

---

[53] CPPNM, Annex 1 (note 2).

[54] IAEA, 'Final report of the Working Group', 2 Feb. 2001 (note 44), Attachment 4, Secretariat Paper no. 13, 'Physical protection objectives and fundamental principles', Principle H, p. 3.

[55] IAEA, 'Final report', 23 May 2001 (note 44), p. 3.

[56] IAEA, The Physical Protection of Nuclear Material, IAEA document INFCIRC/225/Rev. 4 (Corrected) (May 1999), available at URL <http://www.iaea.org/worldatom/Documents/Infcircs/index.shtml>.

[57] Bunn and Bunn (note 9), p. 8.

[58] IAEA, 'Final report', 23 May 2001 (note 44), p. 3.

[59] See the definition in note 17.

dangers of terrorist acquisition of such radioactive material.[60] These materials are often so poorly guarded that they become 'orphaned' from control by the licensed users or those responsible for disposing of them as waste after their creation or use. They cannot be used for making nuclear weapons but can be combined with conventional high explosives to make 'dirty bombs' that scatter radioactivity over a populated area. A container in which there were radioactive materials attached to a landmine was found in Chechnya in 1998.[61] Dirty bombs of this sort are much less dangerous than nuclear weapons but easier to make[62] and could cause panic, and probably long-term injury to some people, if radioactivity were dispersed over a populated area.

The November 2001 IAEA report on protection against nuclear terrorism urged better security for this radioactive material, stating that security was lax in some places.[63] The most important of the treaties that are relevant to this problem is the 1997 Joint Convention on the Safety of Spent Fuel Management and on the Safety of Radioactive Waste Management. This convention requires parties to maintain a national regulatory framework to govern radioactive waste management.[64] The IAEA, in cooperation with other international organizations and interested states, has adopted International Basic Safety Standards for Protection against Ionizing Radiation and for the Safety of Radioactive Sources. These and related safety standards contain recommended rules for states to adopt in their national regulations or legislation to improve the security of radiation sources.[65]

The IAEA has provided some assistance to states to strengthen practices for maintaining control over radioactive material of this sort. In general, this assistance has not been designed to address the use of these materials for malicious purposes such as for

---

[60] IAEA, 'Summary of report on protection against nuclear terrorism' (note 11).

[61] 'Container with radioactive substances found in Chechnya', ITAR-TASS, 29 Dec. 1998. Other 'dirty bomb' threats are described in Orlov, V. and Khlopkov, A., 'Super-terrorism: an immediate threat to the world', PIR Center for Policy Studies, Nonproliferation and Arms Control Hotline, Moscow, 13 Sep. 2001, p. 7; and Albright, D., O'Neill, K. and Hinderstein, C., 'Nuclear terrorism: the unthinkable nightmare', Institute for Science and International Security Issue Brief, 13 Sep. 2001, available at URL <http://www.isis-online.org>.

[62] Albright, O'Neill and Hinderstein (note 61); and O'Neill, K., 'The nuclear terrorist threat', Institute for Science and International Security Issue Brief, Aug. 1997, pp. 6–8, available at URL <http://www.isis-online.org>. Following the June 1996 attack on US troops stationed in Saudi Arabia, Secretary of Defense William Perry warned that US military personnel had to prepare for a possible attack by terrorist groups using radiological weapons. US Department of Defense News Briefing, 17 July 1996.

[63] IAEA, 'Summary of report on protection against nuclear terrorism' (note 11).

[64] IAEA, The 1997 Joint Convention on the Safety of Spent Fuel Management and on the Safety of Radioactive Waste Management, IAEA document INFCIRC/546, 24 Dec. 1997, available at URL <http://www.iaea.or.at/worldatom/Documents/Legal/jointconv.shtml>. E.g., when a party proposes to establish a new spent fuel management facility or radioactive waste management facility, it must consult other parties in the vicinity of the proposed facility and provide them, upon request, with data enabling them to evaluate the likely safety impact of the facility on their territory. Articles 6.1(iv) and 13.1(iv). The party locating such a new facility must take 'appropriate steps to ensure that such facilities shall not have unacceptable effects' on other parties by locating the facility in compliance with the general safety requirements of the convention. Articles 6.2 and 13.12.

[65] IAEA, *International Basic Safety Standards for Protection against Ionizing Radiation and for the Safety of Radiation Sources*, Safety Series no. 115 (IAEA: Vienna, 1996). These standards were accepted by other international organizations having some regulatory responsibility for radioactive materials. These include, e.g., the UN Food and Agriculture Organization, the World Health Organization and several other international organizations. The standards are described in IAEA, *Legal and Governmental Infrastructure for Nuclear, Radiation, Radioactive Waste and Transport Safety*, IAEA Safety Standard Series no. GS-R-1 (IAEA: Vienna, 2000), available at URL <http://www.iaea.org/worldatom/Books/FeaturedSeries/generalsafety.shtml>.

a 'dirty bomb'. The November 2001 IAEA report proposed a new peer review programme to evaluate state regulatory structures, to assess the new threats relating to malicious acts involving radioactive waste, to find ways to help states regain control over large orphan sources, to review the existing standards and their implementation, and to consider what new norms might be needed.[66] This will also require a major increase in the IAEA budget, which members have been unwilling to provide in the past.[67]

V. Conclusions

Since 1995 there have been fewer cases of illicit trafficking of significant quantities of weapon-usable nuclear material recorded than in earlier years. This suggests that the security of such material in Russia and other former Soviet republics has been improved—probably due in large part to the collaborative efforts of Russia and other former Soviet republics with the USA and other industrialized countries. However, only about one-third of the hundreds of tons of Russian weapon-usable material outside of nuclear weapons has been secured so far as a result of the security upgrades accomplished to date. In addition, given the increased illicit trafficking through the Southern Tier over the past three years, intelligence, law enforcement and border control need to be strengthened in the Caucasus, Central Asia, Turkey and the southern regions of Russia. Moreover, security needs to be strengthened at research reactors using weapon-usable material worldwide.

The most important recommendations on strengthening security fall into three categories. First, the major existing Russian–US bilateral programmes to improve the security of Russian weapon-usable nuclear material need to be continued at the present level or a higher level. Second, multilateral efforts such as those involving the IAEA are at least as important if terrorists are to be prevented not only from acquiring weapon-usable material, but also from sabotaging reactors and causing death, illness and panic through the release of radioactivity. The most important of these efforts are: (*a*) the plans of the IAEA to determine where security assistance is needed in the many smaller countries around the world with weapon-usable material and nuclear facilities; (*b*) expansion of the bilateral financial assistance from industrialized countries to pay for such improvements; and (*c*) a major multilateral effort to amend the CPPNM and to gain as many adherents to a stronger amended version as possible. Third, the planned IAEA programme for peer review of state regulatory structures for dealing with other radioactive materials in order to prevent them from being either 'orphaned' or stolen for 'dirty bombs' should be funded by the member states, and a major multilateral effort should be made to evaluate existing international standards for these materials to consider whether new norms are needed.

---

[66] IAEA, 'Calculating the new global nuclear terrorism threat' (note 12), pp. 5–7; and IAEA, 'Summary of report' (note 11), p. 2.

[67] IAEA, 'IAEA outlines measures to enhance protection against nuclear terrorism' (note 11).

11. The military uses of outer space

JOHN PIKE

I. Introduction

Since the dawn of the space age, outer space has been regarded as the ultimate high ground from which the earth below could be controlled. Reflecting this view, the cold war space race between the superpowers was a natural corollary of the arms race. At the end of 2001, the one dominant power, the United States, had nearly 110 operational military spacecraft—well over two-thirds of all the military spacecraft orbiting the earth. Russia was a distant second, with about 40. The rest of the world had only about 20 satellites in orbit.

This chapter presents the current space programmes and provides an inventory of military spacecraft that were operational at the end of 2001. While there are various approaches to research on military space activities, an inventory provides a foundation for research on nuclear and conventional weapons. In the case of military space systems, however, an inventory is more difficult to construct and thus all the more important.

In the path-breaking chapters on military satellites published in the SIPRI Yearbook in the 1970s, the tables listed the satellites that had been launched over the course of the year.[1] At that time, when satellite launches were frequent and operational lifetimes were brief, this focus on annual launches was appropriate. Over time, however, annual launch rates have declined and operating lifetimes have been extended, so today it is meaningful to report on the spacecraft in operation.

The counting of operational military spacecraft poses greater challenges than those encountered in compiling inventories of, for example, nuclear weapons or naval forces. There is an abundance of literature on weapons, but the literature on military space activities is sparse.[2] Part of the reason for this has to do with normal secrecy and part with the relative invisibility of satellites in orbit. Fortunately, many spacecraft (even some highly classified satellites) are visible to amateur observers, a rich source of data.

---

[1] See, e.g., [Jasani, B.], 'Military satellites', *World Armaments and Disarmament: SIPRI Yearbook 1977* (Almqvist & Wiksell International: Stockholm, 1977), pp. 103–79; and [Jasani, B.], 'Military satellites', *World Armaments and Disarmament: SIPRI Yearbook 1978* (Taylor & Francis: London, 1978), pp. 69–130. See also Pike, J., Lang, S. and Stambler, E., 'Military use of outer space', *SIPRI Yearbook 1992: World Armaments and Disarmament* (Oxford University Press: Oxford, 1992), pp. 121–46.

[2] See Mehuron, T. A., '2001 Space Almanac', *Air Force Magazine*, Aug. 2001, pp. 29ff. This source provides the most readily accessible information, although it is apparently focused on nominal design constellation rather than actual operable spacecraft. One of the databases provided by Analytical Graphics for use with the Satellite Toolkit software, available at URL <http://www.stk.com>, is represented as consisting of 'operational' spacecraft, although close examination suggests that many spacecraft that are no longer operable are included in this database.

The most difficult task is to determine the lifetime of satellites. In the 1960s, the end of the operational life of a spacecraft was generally known since the satellite would burn up upon re-entering the earth's atmosphere. Today, a sizeable fraction of the 'operable' military spacecraft in orbit are not currently 'operational' but could be put into service at short notice. Thus it is not sufficient to count only the nominal constellation; it is also necessary to include in the inventory all the back-up, spare, residual, reserve and other operable spacecraft that form the total space order of battle.

Even this inventory of operable military spacecraft fails to capture the essence of military space power, which ultimately resides not in the spacecraft in orbit but in the user equipment on the ground and the integration of this equipment with terrestrial military forces. The increase in the number of US military spacecraft in orbit since the 1991 Persian Gulf War is modest compared to the revolutionary increase in the number, diversity and capabilities of terrestrial user equipment sets. Somewhat less tangible but equally important changes in doctrine, procedures and organization also contribute to translating satellites in space into military power on earth.

Sections II–IV of this chapter review the satellite programmes and military applications of the United States, Russia and all other countries. Section V discusses the companies that operate commercial satellites and their military uses. Section VI reports on the efforts to control the arms race in outer space in the Geneva-based Conference on Disarmament (CD) and section VII on the space-based systems for ballistic missile defence (BMD). Section VIII presents the main findings of this chapter, and section IX contains tables of the operational military satellites in orbit as of 31 December 2001.

II. The United States

Space operations are one of the distinctive attributes of the United States. While a few other countries conduct military space programmes of some significance, at present no other country can rival or contest US space dominance or the advantages this provides to US terrestrial military operations. Modern precision warfare is largely an artefact of the system of systems that combines intelligence, communications, navigation and other military space systems. While other countries may deploy tanks, ships and aircraft that are not individually inferior to their US counterparts, no other country can tie all these various platforms together, using military space systems, into a single, integrated precision-warfare system of systems.

US security managers are acutely aware of the fact that the advantages that accrue from this military space prowess are simultaneously a potential source of vulnerability. This awareness was reflected in a report issued at the outset of the George W. Bush Administration, produced by a commission headed by the incoming Secretary of Defense, Donald Rumsfeld:

we know from history that every medium—air, land and sea—has seen conflict. Reality indicates that space will be no different. Given this virtual certainty, the U.S. must develop the means both to deter and to defend against hostile acts in and from space. This will require superior space capabilities. . . . the U.S. has not yet taken the steps necessary to develop the needed capabilities and to maintain and ensure continuing superiority. . . . The relative dependence of the U.S. on space makes its space systems potentially attractive targets. . . . If the U.S. is to avoid a 'Space Pearl Harbor' it needs to take seriously the possibility of an attack on U.S. space systems.[3]

At the risk of over-simplification, it can be said that both proponents and critics of US space power would probably agree on a few core propositions. The USA enjoys a global preponderance of conventional military power that is unrivalled in human history. Its power-projection capabilities are uniquely enabled by military space systems. The Bush Administration is committed to ensuring this dominance for the USA and denying it to other countries. Ballistic missile defence, much of it based in space or dependent on space systems, is a critical element of 'full-spectrum dominance' to the extent that it denies adversaries the opportunity to offset US conventional supremacy through the resort to weapons of mass destruction. Of course, proponents and critics may differ as to the possibility and desirability of the realization of this vision.

Communications satellites

The USA maintains several geostationary communications satellite networks, which have been used extensively to support US military operations in the Balkans, Afghanistan and other areas.[4] With the US military increasingly focused on power projection in relatively undeveloped theatres of operation, the ability to rapidly implement a dense communications network using satellite systems has become essential.

US satellite systems, as those of other countries, operate on several different bands, each with distinct advantages. Ultra-high frequency (UHF) satellites operate on 225–400 megahertz (MHz) and provide simple, low-cost communications, although on a relatively low bandwidth. Super-high frequency (SHF) satellites, operating on the X-band at 7.25–8.4 gigahertz (GHz), are the backbone high-bandwidth fixed and transportable networks. Extremely-high-frequency (EHF) satellites, which uplink in the V-band on 43–45 GHz and downlink on the K-band at 20.2–21.2 GHz, can support both high bandwidth and highly mobile users. Although authorized for commercial rather than governmental services, some military applications have been found for C-band satellites, which downlink at 3.6–4.2 GHz.

---

[3] *Report of the Commission to Assess United States National Security, Space Management and Organization*, 11 Jan. 2001, Executive Summary, available at URL <http://www.defenselink.mil/pubs/space20010111.html>. To avoid the appearance that this report represented official policy, however, Rumsfeld resigned as chairman of the commission a few days before the report was issued.

[4] Friend, T., 'Search for bin Laden extends to earth orbit', *USA Today*, 5 Oct. 2001, p. 9A.

EHF capabilities consist primarily of the Milstar advanced communications satellites, designed during the 1980s to provide anti-jam, low probability of intercept/detection communications for mobile ground terminals on vehicles, ships, submarines and aircraft. Two Block 1 spacecraft were launched in 1994 and 1995. They carried the Low Data Rate (LDR) payload, which supported 192 channels with data rates of 75–2400 bits per second (bps).

The Milstar programme was significantly restructured after the end of the cold war to improve support to tactical users, who had greater bandwidth requirements than those originally established for strategic nuclear users. The modified Block 2 spacecraft added the Medium Data Rate (MDR) payload, which supports data rates of 4800 bps to 1.544 megabits per second (Mbps) per channel, representing a sixfold increase in aggregate data throughput capacity. The first Block 2 Milstar satellite was launched on 30 April 1999 atop a Titan IVB booster. Because of a malfunction in the Centaur upper stage of the booster, the spacecraft was placed into a very low, useless orbit (740 km by 5000 km) and the Milstar satellite was declared a complete loss on 4 May 1999. On 27 February 2001 another Milstar Block 2 was successfully launched by a Titan 4/Centaur booster from Cape Canaveral.[5] The third Block 2 Milstar spacecraft was planned for launch in January 2002. The final Milstar is planned for launch in November 2002 to replace Milstar 1.

When the Milstar programme was restructured in 1992, the requirement for Milstar to provide polar EHF coverage was dropped. In July 1995 an interim Polar Adjunct programme was initiated to fly a modified EHF payload from the Navy's UHF Follow-On (UFO) system on a classified host satellite. The first Hosted Polar Package was launched in November 1997, apparently on USA 136 (believed to be a TRUMPET signals intelligence satellite). The last two will be available in fiscal year (FY) 2003 and FY 2006.[6] EHF packages were also carried on-board FLTSAT-7 and FLTSAT-8, launched in the late 1980s and apparently still in service.

The Global Broadcast Service (GBS) is a US Air Force-led joint programme to implement a high-capacity broadcast system providing continuous, one-way, high-speed, high-volume transmission of classified and unclassified data and imagery to US forces. The GBS programme is intended to reduce the dependence of the US Department of Defense (DOD) on expensive leased commercial satellite communications. The GBS transponders are hosted on the US Navy's UFO satellites, replacing the SHF payload beginning with UFO F8, launched in 1998. The GBS package includes four 24-Mbps military Ka-band (30/20 GHz) transponders.

---

[5] Chuter, A., 'Milstar gets new capability to boost military comms', *Flight International*, 6 Mar. 2001.

[6] US Department of Defense, Defense Technical Information Center (DTIC), 0603432F, Polar MILSATCOM (Space), RDT&E Budget Item Justification Sheet (R-2 Exhibit), June 2001. The 'R-2 Exhibit' documents are the highly detailed descriptions of DOD programmes submitted to the Congress in support of the annual budget request and as such represent the most authoritative source of information on current programmes. They are available on the DTIC Internet site at URL <http://www.dtic.mil>.

Advanced EHF (AEHF) is the follow-on satellite communications system to replenish the existing Milstar 1/2 (LDR/MDR) satellite constellations.[7] The AEHF system will be compatible with existing EHF terminals and will provide a tenfold increase in communications capacity relative to Milstar 2. It will provide an increase in single-service capability from 1.5 Mbps to 8 Mbps, increase the number of coverage areas, and retain anti-jam and low-probability-of-intercept features.[8] In November 2001 a team of Lockheed Martin and TRW was awarded the contract for the AEHF programme, with an initial launch planned for early 2006.[9] This launch date represented a two-year delay from the originally projected date of 2004, and the programme's cost had risen from $2.7 billion to $3.7 billion.[10] The full four-satellite AEHF constellation is planned to be operational by 2010.

SHF satellite communications programmes include the Defense Satellite Communications System (DSCS), the DSCS Service Life Extension Program (SLEP), the Wideband Gapfiller Satellite (WGS) System and the Advanced Wideband System (AWS) satellites. The SHF satellite systems are undergoing a transition from old-technology DSCS III satellites to the more advanced DSCS SLEP and WGS satellites; this began in 1999 and will continue until 2005. The population of SHF users is growing at a rapid pace.[11]

The DSCS is the backbone of the US national security satellite communications system, providing secure voice and high data rates in the SHF band. The DSCS supports communications for global command and control, crisis management, intelligence and early-warning data relay, and diplomatic traffic for the ground mobile forces of all the military services. The constellation consists of five primary satellites (normally the most recent launches) dispersed along the equator for global coverage, as well an equal number of older 'residual' spacecraft that provide back-up coverage. When an older residual satellite is replaced at its location by a newer satellite, it is muted and sent into a super-synchronous orbit.

The DSCS SLEP will upgrade payloads on the last four DSCS satellites and provide up to five times the data throughput compared to the original DSCS III satellites. DSCS III B-8, the first SLEP satellite, was launched in January 2000. B-11, the second of four SLEP satellites, was successfully launched in October 2000 and was operating at 12° West by January 2001. DSCS III B-4 was retired in 2000, and A-2 and B-5 were scheduled for final retirement in mid-2002. DSCS III B-6 (SLEP) was planned for launch in May 2002, with A-3, the final SLEP spacecraft, scheduled for launch in May

---

[7] Satellite Communications (Space) Program Element: 0303109N Exhibit R-2, FY 2002 RDT&E, Navy Budget Item Justification, June 2001.
[8] Satellite Communications (Space) Program Element: 0303109N (note 7).
[9] Merle, R., 'Lockheed, TRW win US satellite contract', *Washington Post,* 20 Nov. 2001, p. E05; and Gildea, K., 'Boeing pulls out of advanced extremely high frequency satellite program', *C4I News,* 29 Nov. 2001.
[10] Schneider, G., 'New satellite system hit with delays', *Washington Post,* 15 Aug. 2001, p. E01.
[11] Satellite Communications (Space) Program Element: 0303109N (note 7).

2003.[12] The SLEP will extend the life of DSCS satellites until the Advanced Wideband System can be orbited.

The new WGS satellites will provide high-data-rate military satellite communications in accordance with the Joint Space Management Board's Military Satellite Communications (MILSATCOM) Architecture of August 1996. This programme was conceived to fill a short-term 'bandwidth gap' in military communications needs. The first WGS launch is scheduled for early 2004, with the remaining two launches scheduled for 2005. These dual-frequency WGS satellites will provide two-way X-band service (now provided by the DSCS), one-way Ka-band capabilities (now provided by the GBS) and a new high-capacity two-way Ka-band service.[13]

UHF communications are hosted on a wide variety of spacecraft, ranging from dedicated satellites to classified packages on objects that are supposedly 'space junk'.[14]

The US Navy, Air Force, Army and DOD share the Fleet Satellite Communication (FLTSATCOM) system, which provides reliable and secure communications for ships and submarines at sea, aircraft and military ground units throughout the world. Primarily intended to support communications between naval aircraft, ships and submarines, FLTSATCOM also supports the Strategic Command and the national command authority network. Fully operational since January 1981, the FLTSATCOM system has three satellites in geosynchronous orbit, all in reserve status.[15]

The UHF Follow-On communications satellite constellation fulfils DOD worldwide UHF communications requirements. The FLTSATCOM constellation has been largely replaced by the UFO spacecraft. The current constellation will be approaching the end of its design lifetime in 2003. One additional UFO spacecraft (F11) is planned to be launched in 2003. Even with this launch, the UFO constellation is expected to require phased replacement starting in 2007.

The Mobile User Objective System (MUOS) programme is the next-generation DOD advanced narrow-band UHF communications satellite constellation. It is intended to address the exponential growth of narrow-band communications demands.[16]

In addition to these overt military communications systems, the National Reconnaissance Office (NRO) operates a parallel network of less visible satellite communications systems to support the global collection and dissemination of intelligence data.

---

[12] Defense Satellite Communications System (DSCS), Missile Procurement, Air Force, Budget Activity 05, Other Support, Item no. 26 (June 2001).
[13] Wideband Gapfiller Satellites (Space), Missile Procurement, Air Force, Budget Activity 05, Other Support, Item no. 18 (June 2001).
[14] US Army, *The Army Satellite Communications Architecture Book* (US Army Signal Center: Fort Gordon, Ga., Apr. 2000), p. 4-2.
[15] US Army (note 14), p. 4-5.
[16] Satellite Communications (Space) Program Element: 0303109N (note 7).

The satellites of the Satellite Data System (SDS) support near-real-time communications between low-altitude imagery intelligence satellites and ground control stations, using highly elliptical semi-synchronous Molniya-type orbits, optimized for coverage of the North Pole region. The most recent launch of this programme, USA 162, was placed in orbit on 10 October 2001 on an Atlas booster launched from Cape Canaveral Air Force Station, Florida. This was apparently the second launch of a new generation of spacecraft, with the first (USA 137) having been launched in January 1998. These two launches probably replaced the second-generation spacecraft launched in 1989 and 1992 (USA 40 and USA 89). The current SDS constellation may also consist of another second-generation spacecraft, USA 125, launched in 1996, although both the mission and the status of USA 125 are somewhat unclear.

The US National Aeronautics and Space Administration (NASA) operates a constellation of six Tracking and Data Relay Satellite (TDRS) spacecraft, with TDRS-A having been withdrawn from service. The first six TDRS satellites were launched by the Space Shuttle in 1983–95. The next generation, TDRS-H, was launched on an Atlas 2A on 30 June 2000, featuring twice the capacity of previous spacecraft, with the inclusion of a Ka-band payload. TDRS supports near-real-time data transmission from the Lacrosse/Onyx low-altitude imaging intelligence satellites.[17]

In 1998 the NRO disclosed that it was 'developing a Future Communications Architecture (FCA) that will be critical to the success of these future imagery and signals intelligence systems. The FCA will consist of a network of satellites and ground communications systems that will allow us to move and process large volumes of information from operational collection systems'.[18] Little has been revealed about the FCA beyond the fact of the existence of the programme. A contract for the FCA is expected to be awarded before the end of 2003.[19]

The NRO successfully launched the Geosynchronous Lightweight Technology Experiment (GeoLITE) advanced demonstration satellite on a Boeing Delta II rocket on 18 May 2001. Built by TRW, the GeoLITE satellite has both a laser communications experiment and an operational UHF communications mission. The relationship between the GeoLITE and the FCA is unclear, although the FCA might use high-data-rate laser communications links if they are successfully demonstrated on the GeoLITE.

The Lincoln Experimental Satellites (LES 8 and LES 9) were experimental communications satellites, powered by radioisotope generators and built by the Lincoln Laboratory at the Massachusetts Institute of Technology. The Air Force Space Command continues to manage programme funding and retains

---

[17] Charles, D., 'Spy satellites: entering a new era', *Science*, 24 Mar. 1989, pp. 1541–43; and Covault, C., 'NASA, Boeing dispute major TDRS problem', *Aviation Week & Space Technology*, 8 July 2001.

[18] National Reconnaissance Office, Presentation by Keith R. Hall, Director, Assistant Secretary of the Air Force (Space), Senate Armed Services Committee, Strategic Forces Subcommittee, 11 Mar. 1998.

[19] Taylor, C., 'It's not Hughes' satellites, exactly, that Boeing was after', *Seattle Times*, 23 Jan. 2000, p. E1.

control over the LES-9 satellite. In 2000 LES-9 was still active, relaying communications with the South Pole.[20]

The civilian Iridium constellation of 66 low-earth-orbit communications satellites was fully deployed by the end of 1998, providing commercial mobile telephone, data and messaging services worldwide. However, the network attracted few customers; the Iridium owners declared bankruptcy and by March 2000 planned an immediate end to Iridium services. In November 2000 a new company, Iridium Satellite LLC, completed acquisition of the bankrupt company's satellites and control network. Iridium Satellite LLC contracted with Boeing to operate and maintain the satellite constellation. At the same time, the DOD awarded Iridium Satellite LLC a 24-month contract (with extension options until 2007) for unlimited Iridium satellite airtime for 20 000 government users.

Navigation satellites

The Navstar navigation system has fundamentally altered US military operations. Reaching full operational capability in the mid-1990s, the Navstar Global Positioning System (GPS) has profoundly improved the effectiveness of reconnaissance, weapon delivery and rapid deployment by US military forces. The most recent demonstration of this capability was in Operation Enduring Freedom in Afghanistan, where for the first time a significant fraction of all the munitions expended used GPS guidance.

The Navy Navigation Satellite System, also known as TRANSIT, was the world's first operational satellite navigation system, using satellites also known as Oscar. The last Oscar satellite was launched in August 1988, and the TRANSIT Program terminated navigation service on 31 December 1996, after Navstar reached full operational capability. Several Oscar satellites remain active in orbit, as part of the Navy Ionospheric Monitoring System (NIMS).

The Navstar GPS provides position, velocity and precise-time data for military aircraft, ships and ground personnel. The satellites broadcast high-accuracy, precisely synchronized signals that are received and processed by user equipment that computes position and velocity. This provides steering directions to target locations or navigation waypoints. The system provides location in three dimensions to a Spherical Error Probable (SEP) probability of 16 metres worldwide. GPS provides users with this worldwide three-dimensional positioning based on a constellation of 24 satellites. By the end of 2001, a total of 28 operational Navstar spacecraft were in orbit.[21]

[20] URL <http://www.phys.unsw.edu.au/~map/aasto/satellite_schedule_2000.pdf>.
[21] All sources agree that as of the end of 2001 there were a total of 28 operational Navstar spacecraft in orbit, although they diverge on precisely which spacecraft remained in service. All of the Block 1 satellites, as well as GPS 2-1, 2-3, 2-6 and 2-7, have been retired. GPS 2R-1 was destroyed in a Delta launch failure on 17 Jan. 1997. Keeping track of the Navstar constellation is complicated by the remarkable diversity of designations attached to each spacecraft. Thus, the spacecraft launched on 10 Apr. 1992 is variously known as GPS 2A-13, GPS 2A-4 and Navstar 26, each of which refers to the type of spacecraft and the sequence in which it was launched. This satellite is also known by a Space Vehicle Number (SVN), assigned to spacecraft prior to launch and independent of subsequent launch sequence, which in this case is SVN 25. There are several public sources for the current status of the Navstar constellation,

The initial purchase of 28 Block IIA satellites was awarded as a multi-year contract in September 1982. A follow-on multi-year procurement of 20 Block IIR replenishment satellites plus one optional satellite began in FY 1991, with final delivery in FY 2000. The acquisition strategy for the Block IIF satellites was a competitive multi-year contract for six satellites, with advance procurement in FY 1996, and annual purchases of three modernized satellites in FY 2005 and FY 2006. Up to 12 Block IIR satellites will be modernized to include a second civil signal and a new military signal. The first six Block IIF satellites will be modernized to include a second and third civil signal and a new military signal. GPS III satellites will incorporate full modernization with a higher-power military signal, and there will be a competitive annual purchase of three satellites in FY 2007 and three satellites from FY 2008, with advance purchase beginning in FY 2006.[22]

The Navstar spacecraft also host the Nuclear Detonation Detection System.

Weather satellites

The Defense Meteorological Satellite Program (DMSP) provides worldwide visible and infrared cloud imagery and other specialized meteorological, oceanographic, land surface and space environmental data to support US military strategic and tactical missions. It also provides real-time direct read-out of local weather to ground- and ship-based tactical terminals supporting DOD forces worldwide.[23] The primary Operational Linescan System (OLS) instrument monitors the global distribution of clouds and cloud-top temperatures, while other instruments provide more specialized data. The programme consists of two satellites in sun-synchronous polar orbits, with one satellite passing overhead in early morning and the other at noon local time. Additional spacecraft on which the primary OLS sensor has failed continue to provide data from other sensors.

As of mid-2001, four DMSP satellites were operational.[24] DMSP F10, which suffered an OLS failure on 8 February 1995, was finally retired on 14 November 1997. DMSP F11, which suffered an OLS failure on 22 April 1995, remained in service until early 1999, when it was retired. DMSP 5D-2 F12, launched on an Atlas E in 1994, remains in service. DMSP 5D-2 F13, launched on 24 March 1995, was the last flight of the Atlas E booster. As of late 2001 this spacecraft was reportedly in back-up

---

of which the SEM Almanac is the only one that includes SVN designators. Recent editions are available at URL <http://www.navcen.uscg.gov/ftp/GPS/almanacs/sem>. Other sources of Navstar constellation status reference the Pseudo Random Number (PRN) designator of the spacecraft transmitter, which is not permanently associated with a specific spacecraft. While some sources suggest that GPS 2A-04 and GPS 2A-13 are no longer in service, the SEM Almanac suggests that GPS 2-04 and GPS 2A-03 have been retired from service.

[22] Global Positioning System (GPS), Missile Procurement, Air Force, Budget Activity 05, Other Support, Item no. 22 (June 2001).

[23] Defense Meteorological Satellite Program (DMSP), Missile Procurement, Air Force, Budget Activity 05, Other Support, Item no. 24 (June 2001).

[24] Testimony of Lieutenant General Edward G. Anderson, Deputy Commander in Chief, US Space Command, House Armed Services Committee, 20 June 2001.

status.[25] DMSP 5D-2 F14 was launched from Vandenberg Air Force Base (VAFB) by a Titan 2 rocket. On 12 December 1999, DMSP 5D-3 F15, the first Block 5D-2 spacecraft, was launched from VAFB on a Titan 2 rocket. Both of the Block 5D-3s remain in service.

The new DMSP-16 (5D-3-F16) is intended to replace DMSP 5D-3 F14, launched in March 1995. Following the first launch attempt, on 18 January 2001, which was delayed by weather, the launch was delayed numerous times throughout the year because of ground equipment, fuel valve, guidance system and fuel pump problems. At the end of the year the spacecraft was scheduled to be launched in early 2002.[26]

In May 1994 the president directed the departments of defense and commerce to converge their separate polar-orbiting weather satellite programmes. The convergence into a single national resource will be completed once existing spacecraft are launched, with the first National Polar-orbiting Operational Environmental Satellite System (NPOESS) ready for launch in 2007.

The US Navy's GeoSat Follow-On (GFO) satellite was launched in February 1998 and remained in service at the end of 2001. Although initially thought of as a geodetic satellite, the system is used for real-time monitoring of the oceans. Satellite altimetry is a highly efficient method for precisely measuring the shape of the sea surface over large areas, which is directly related to the large-scale movements of water that influence the propagation of sound in the sea. GFO supports the environmental predictions that enhance US naval undersea war-fighting capabilities. All GFO data have been authorized for unconditional release and use by the civilian community.[27]

Although no longer used for navigation purposes, the Oscar navigation satellites remain in service as part of the NIMS, sponsored by the Naval Security Group (NSG). This provides three-dimensional ionospheric models to obtain high geo-location accuracies to support signals intelligence collection. The system uses computer models, similar to those used in medical resonance imaging, to process signals transmitted from Oscar navigation satellites. Also known as the Tactical Regional Area Ionospheric Tomography System (TRAITS) under NRO sponsorship, this capability was validated in the Radiant White demonstration in 1996.[28]

Early-warning satellites

The Defense Support Program (DSP) is a system of satellites in geostationary orbits, fixed and mobile ground processing stations, and a ground communica-

---

[25] URL <http://www.isciences.com/NewSite/sensors/current.html>.
[26] 'DMSP delayed again, rescheduled Dec. 20', *Space & Missile*, 8 Nov. 2001.
[27] 'GeoSat Follow-On METOC satellite—Factsheet', Space & Naval Warfare Systems Command, 11 Jan. 2001, available at URL <http://enterprise.spawar.navy.mil/spawarpublicsite/pao/fs/index.cfm?news_year=2002>.
[28] Navy Ionospheric Monitoring System (NIMS), URL <http://sgdwww.arlut.utexas.edu/projects/nims>.

tions network.[29] The DSP's primary mission is to provide strategic and tactical warning of ballistic missile attack. During the Gulf War, these satellites provided warning of Iraqi launches of Scud missiles.

During the cold war, the operational constellation consisted of three active spacecraft, but by the mid-1990s the constellation typically consisted of five operational spacecraft. DSP 20 was successfully launched on 8 May 2000, apparently replacing DSP 15, launched in 1990. Although the launch of DSP 21 was scheduled for 2001, it had not been launched by the end of the year. The programme includes subsequent launches of two additional satellites (DSP 22 and 23).[30]

The Space-Based Infrared System (SBIRS) is designed to greatly improve warning of ballistic missile launches. SBIRS incorporates new technologies to enhance launch detection and improve reporting capabilities. The SBIRS will consist of satellites in Geosynchronous Orbits (GEO), Highly Elliptical Orbits (HEO), Low Earth Orbits (LEO), and an integrated centralized ground station serving all SBIRS space elements and DSP satellites.[31]

The US Nuclear Detonation (NUDET) Detection System (USNDS) provides a highly survivable capability to detect, locate and report nuclear detonations in the earth's atmosphere or in near space. The space segment consists of NUDET detection sensors on both the GPS/Nuclear Detonation System (NDS) satellites and the DSP/NDS satellites.[32] The USNDS payload contains optical, X-ray, electromagnetic pulse (EMP/W-sensor) and dosimeter sensors.[33]

Ocean-surveillance satellites

During the cold war the White Cloud Naval Ocean Surveillance System (NOSS) was the primary ocean-surveillance satellite system.[34] Each NOSS launch placed a cluster of one primary satellite and three smaller sub-satellites (that trail along at distances of several hundred kilometres) into low polar orbit. This satellite array could determine the location of radio and radar transmitters, using triangulation, and the identity of naval units, by analysis of their operating frequencies and transmission patterns. In 1990 the White Cloud constellation apparently consisted of at least three clusters of primary and secondary satellites, launched in 1987, 1988 and 1989. There have been no subsequent launches, and the system probably ended service in the mid-1990s.

The first launch of the second-generation NOSS system took place in June 1990, with a Titan 4 booster orbiting a trio of much larger satellites. As with

[29] Richelson provides an authoritative, comprehensive treatment of the DSP in Richelson, J. T., *America's Space Sentinels: DSP Satellites and National Security* (University Press of Kansas: Lawrence, Kans., Apr. 2001).

[30] Defense Support Program (DSP), Missile Procurement, Air Force, Budget Activity 05, Other Support, Item no. 25 (June 2001).

[31] Space-Based Infra-Red System (SBIRS) High Advance Procurement, Missile Procurement, Air Force, Budget Activity 05, Other Support, Item no. 30 (June 2001).

[32] Nudet Detection System (NDS), Space Budget Item Justification (Exhibit P-40), June 2001.

[33] NUDET Detection System (NDS), Missile Procurement, Air Force, Budget Activity 05, Other Support, Item no. 23 (June 2001).

[34] Richelson, J. T., *The US Intelligence Community* (Ballinger: Cambridge, Mass., 1985), pp. 140–43.

the earlier system, these flew in close formation, apparently to facilitate the tracking of ships at sea through triangulation. By 1996 three of these triplets (a total of nine spacecraft) were in orbit, despite a launch failure on 2 August 1993 that destroyed the payload.

On 8 September 2001 an Atlas booster launched USA 160, probably the first of a third-generation ocean-surveillance satellite system. The use of the Atlas booster implies that it was a somewhat smaller spacecraft than that launched by the Titan booster. By the end of 2001 the characteristics of this new, third-generation system remained rather obscure. Observers believed that one object which had not received an official USA designation was in fact a second spacecraft, with a third spacecraft expected to be deployed from the primary USA 160 payload in early 2002.[35]

Signals intelligence satellites

The USA operates several constellations of signals intelligence (SIGINT) satellites in geostationary and highly elliptical orbits.

The geostationary SIGINT constellation probably consists of about half a dozen satellites. Although the precise number and status of these satellites are speculative, a fair approximation may be obtained by assuming that these spacecraft have an operational lifetime of about a decade, as is the case with commercial communications satellites. During the cold war a progressively larger and more capable series of spacecraft were placed into orbit, with programme names reportedly including Rhyolite, Chalet and Vortex. By the early 1990s all these spacecraft had almost certainly been withdrawn from service or relegated to back-up reserve status.

There were no new launches of signals intelligence satellites in 2001. By the end of the year, the active constellation probably consisted of a pair of MERCURY spacecraft, launched in 1994 and 1996, as well as a trio of ORION spacecraft, launched between 1990 and 1998. The nomenclature associated with these spacecraft is uncertain; the ORION spacecraft have also been referred to as MENTOR and MAGNUM.

In addition to these geostationary signals intelligence satellites, during the cold war a pair of Jumpseat satellites operated in highly elliptical Molniya-type orbits.[36] These satellites provided focused coverage of the far northern regions of the Soviet Union. Beginning in 1994 they were replaced with the much larger TRUMPET spacecraft, as many as three of which were apparently operational at the end of 2001.

In 1998 the NRO disclosed that it was 'introducing an Integrated Overhead SIGINT Architecture (IOSA) that will improve SIGINT performance and avoid costs by consolidating systems, utilizing medium lift launch vehicles wherever possible, and using new satellite and data processing technologies.

---

[35] Molczan, T., 'RE: Centaur and NOSS payload observed', 9 Sep. 2001, URL <http://www.satobs.org/seesat/Sep-2001/0121.html>.

[36] Richelson (note 34), p. 122.

At the urging of Congress, [it had] initiated the study phase for the follow-on architecture, IOSA-2'.[37]

Imagery intelligence satellites

Imagery intelligence satellites provide US military planners and political decision makers with unrivaled global situation awareness.[38] Coupled with collateral intelligence from other sources, these systems have become an integral part of US war-fighting, as demonstrated most recently in Operation Enduring Freedom in Afghanistan.[39]

The USA continued operation of three Improved Crystal (advanced KH-11) electro-optical real-time digital imagery intelligence satellites throughout the year. USA 161 was launched on 1 October 2001, apparently replacing USA 86, launched in 1992. During the cold war the KH-11 KENNAN constellation normally consisted of two spacecraft, but by the mid-1990s the constellation was expanded to as many as three spacecraft. These satellites can provide high-resolution imagery—generally reported to have a resolution of about 10 cm—which can be rapidly disseminated to users around the world.

The ONYX spacecraft (formerly known as LACROSSE) provide cloud-piercing coverage of targets using synthetic aperture imaging radar. Three spacecraft of this class, launched between 1991 and 2000, remained in service at the end of 2001.

In 1994 it was reported that Congress had funded 'Satellite 8X, an over $1 billion spacecraft that trades off the extremely high resolution of Keyhole surveillance satellites for a wider field of view that would make it easier to map theatre operations'.[40] By 1998 the NRO stated that it was 'completing the development of the Enhanced Imaging System in response to growing customer demands and large area imagery collection shortfalls'.[41] On 22 May 1999, USA 144 was launched into an orbit with a perigee of about 2600 km and an apogee of over 3100 km. The Improved Crystal spacecraft are in orbits of 300 km by 975 km, while the ONYX radar satellites are in roughly circular orbits at an altitude of about 650 km.

The NRO proposed the Future Imagery Architecture (FIA) Program in its FY 1998 budget submission to Congress in March 1997.[42] Under FIA, the NRO would specify performance requirements such as resolution and revisit rates, while the means by which the requirements are met would be specified by the contractor. Reportedly, FIA is intended to collect 8–20 times the vol-

---

[37] National Reconnaissance Office, Presentation by Keith R. Hall (note 18).
[38] The indispensable guide to this subject is Richelson, J. T., *America's Secret Eyes in Space: The US Keyhole Satellite Program* (Ballinger: Cambridge, Mass., Feb. 1990).
[39] Petrie, G., 'Imagery for surveillance and intelligence over Afghanistan', *GI News*, Oct./Nov. 2001, URL <http://www.ginews.co.uk>.
[40] 'Intelligence conferees fund Satellite 8X, transfer U-2 to DARO', *Aerospace Daily*, vol. 171, no. 60 (26 Sep. 1994), p. 477.
[41] National Reconnaissance Office, Presentation by Keith R. Hall (note 18).
[42] National Reconnaissance Office, Presentation by Keith R. Hall (note 18).

ume of imagery compared to existing systems. On 27 April 1999 Raytheon was awarded a contract by the NRO to develop the Mission Integration and Development (MIND) ground infrastructure portion of FIA. On 3 September 1999 the NRO awarded a contract to Boeing to develop, provide launch integration and operate the FIA imagery intelligence satellites.[43]

The Discoverer II imaging radar satellite system was originally planned to provide high-resolution (better than 1 metre) continuous coverage using a constellation of 24 medium-size satellites. This US Air Force programme represented something of a rival to the NRO's Future Imagery Architecture and failed to receive congressional approval. By the end of 2001 it had been recast as the Space-Based Radar, a 10-satellite surveillance and targeting constellation planned for deployment by 2008.[44]

Anti-satellite systems

During the 1960s the United States had two different nuclear-tipped anti-satellite (ASAT) systems, although both were withdrawn from operational service by the early 1970s. In the late 1970s the USA began development of an air-launched ASAT system that would destroy target satellites by the direct impact of a miniature homing warhead. This programme was cancelled after limited testing in the 1980s, although it is believed that components of this system remain in storage for potential reactivation.

The Mid-Infrared Advanced Chemical Laser (MIRACL) at White Sands, New Mexico, is believed to have a limited ASAT capability against some satellites in low earth orbit. The Miniature Sensor Technology Integration (MSTI-3) satellite was launched in 1996 by the Air Force experimental satellite. Among the experimental technologies on board were special sensors designed to detect a laser weapon attack. On 19 October 1997 the Army's MIRACL laser was tested against MSTI-3. Two shots were fired, the first of 1-second duration to trigger the sensors on the MSTI-3 satellite designed to detect the attack, followed by a 10-second laser burst that attempted to overload the sensors. The results of the test remain classified, although it is known that the satellite was not destroyed in the test.[45]

Technology development

The Space Test Program (STP) conducts space flight experiments for the military research and development community. These range from basic research to advanced development and from large free-flying spacecraft to small packages

---

[43] Dornheim, M. A., 'FIA outline takes shape', *Aviation Week & Space Technology*, 10 Dec. 2001, p. 73.
[44] Caceres, M., 'Military satellites: the next generation', *Aerospace America*, Jan. 2002, p. 20.
[45] Broad, W. J., 'Military is hoping to test-fire laser against satellite', *New York Times*, 1 Sep. 1997.

flown on other spacecraft, the Space Shuttle or the International Space Station.[46]

III. Russia

During the cold war the Soviet space programme was characterized by a high annual rate of launch of spacecraft with relatively short operating lifetimes by Western standards.[47] At the end of 1990 the USSR had a total of 75 launches, one more than its total for 1989 but in marked contrast to the 90 launches of 1988 or the peak of 101 in 1982.[48] The Russian space programme has managed to achieve only a fraction of the Soviet launch rate. However, the elimination of obsolete short-lived systems and the extension of the operating lives of the remaining systems have significantly offset the declining launch rate.

The high launch rate and short operating lives of Soviet satellites during the cold war meant that accounting for the launches and re-entry of spacecraft provided the primary indicator of Soviet military space activity. The rather different operating pattern of Russian space forces means that accounting for operational spacecraft in orbit is a more meaningful indicator. It is also a decidedly difficult analytical challenge.

By one estimate, as of late 2001 Russia had 93 operational spacecraft in orbit, compared to a peak of nearly 200 satellites during the Soviet era. These spacecraft reportedly included 43 military satellites and about 20 dual-use satellites, with the remaining 30 spacecraft performing civilian and commercial functions. Of the 93 operational spacecraft, over 80 per cent of them had exceeded their design lifetimes.[49] Other sources suggest that over 70 per cent of the operational spacecraft had exceeded their design lifetimes.[50]

In Russia, as in other countries, there is no hard and fast dividing line between military and dual-use space systems. A system-by-system survey suggests that by the end of 2001 Russia had over 40 operational military spacecraft in orbit, along with another 15 dual-use communications and navigation satellites, a finding remarkably consistent with official Russian statements.

Navigation satellites

Russia has continued operation of both Soviet-era navigation satellite networks. The Tsikada constellation consists of small satellites of modest capabilities in low earth orbits, similar to the US TRANSIT system that was

---

[46] US Department of Defense, *Space Test Program Status Report,* vol. 6, issue 2 (16 Apr. 2001), available on the STP Internet site at URL <http://www.te.plk.af.mil/stp/stp.html>.

[47] *Soviet Space History at a Glance* is a very useful resource for information on the Soviet and Russian space programmes; see URL <http://www.astronautix.com/articles/sovlance.htm>.

[48] 'Soviets finish year with 75 launches', *Defense Dail—Special Supplement,* 9 Jan. 1991, p. S-1.

[49] Feller, G. and Stein, K., 'Russian space assets not getting any better', *Space & Missile,* vol. 2, no. 23 (8 Nov. 2001).

[50] Saradzhyan, S., 'Russia's space forces work to restore fire-ravaged facility', *Space.com,* 10 May 2001.

phased out in 1995. The GLONASS (Global'naya Navigatsionnaya Sputnikovaya Sistema, Global Navigation Satellite System) network of semi-synchronous satellites provides higher accuracy fixes and is similar to the US Navstar system.

In contrast to the US TRANSIT system, which was used by both civilian and military operators, the Russian low-earth-orbit network uses similar satellites in separate military (Cosmos designation) and civilian (Tsikada) networks. The Parus system is sometimes referred to as 'military Tsikada' or 'Tsikada-M'. Orbital planes 1–6 of the military system are spaced 30 degrees apart, while planes 11–14, used by the civilian system, are spaced 45 degrees apart. There was one launch of the Parus system in 2001, Cosmos 2378, on 28 May 2001. Normally, only a single satellite is operational in each orbital plane at any given time, and this launch apparently replaced Cosmos 2279, launched in 1994 to sustain an operational constellation of six spacecraft. However, these satellite constellations are subject to frequent changes as older spacecraft are reactivated and newer spacecraft are temporarily inactivated. The total number of potentially operable spacecraft may be over a dozen.[51]

The GLONASS network has experienced major developmental problems since its introduction in 1982. The fully deployed GLONASS constellation is intended to be composed of 24 operational satellites in three orbital planes. The deployment of GLONASS peaked in 1995 with a total of 22 operating satellites. Owing to financial difficulties in Russia, only three satellites were launched over the following five years, and by June 2000 only 10 satellites of the constellation were operational. Another trio was launched in 2000, and three more were launched on 1 December 2001. Of these, at least one was an improved GLONASS-M, which has a seven- to eight-year design life versus the three-year life of the previous satellites. By the end of the year, the active constellation remained at 10 spacecraft, with four of the spacecraft launched in 1995 having been withdrawn from service.[52]

Communications satellites

The Soviet military communications network included three classes of satellites that operate in low-altitude orbits, only one of which remained in service at the end of 2001.

---

[51] The status of Parus and other spacecraft with continuous radio beacons was detailed in *SPACEWARN Bulletin*, no. 520 (25 Feb. 1997) and updated in *SPACEWARN Bulletin*, no. 545 (1 Apr. 1999), URL <http://nssdc.gsfc.nasa.gov/spacewarn/spx545-catiradiobeacon.html>. A more current appreciation of the status of these satellites is available on the HearSat Internet site (URL <http://hearsat.org>) and at Russian Navigation Satellites, URL <http://www.asahi-net.or.jp/~VQ3H-NKMR/satellite/freq-Nav.html> (apparently updated in late 1997). This is an instance in which the fidelity of the Analytical Graphics Inc (AGI) database can be independently validated, and several discrepancies are noted. AGI reports a total of 12 of these spacecraft as 'active' although the nominal active constellation consists of only 6 spacecraft. AGI reports Cosmos 2327 as active, while other sources report that it failed in orbit.

[52] Ministry of Defence of the Russian Federation, Coordination Scientific Information Center (KNITs), URL <http://www.rssi.ru/SFCSIC/english.html>.

The first-generation Strela-1M spacecraft were launched eight at a time on the SL-8 booster into a single orbital plane. Although the number of satellites active in this constellation was difficult to determine, the three most recently launched octuplets were usually thought to constitute the bulk of the nominal constellation of 24 satellites. The final launch under this programme (Cosmos 2187–2194) was conducted in 1992, and by the late 1990s this network was almost certainly inactive. The second-generation Strela-2M low-altitude communications satellites were launched one at a time on the SL-8 booster, with a constellation consisting of three satellites, each in a unique orbital plane separated by 120 degrees. Cosmos 2298, launched on 20 December 1994, was the final launch in this programme, which had almost certainly been terminated by the end of the decade.

The third-generation Strela-3 low-altitude satellites are launched in groups of six on a single SL-14 booster. The first launch under this programme was on 15 January 1985. Of the three classes of Soviet low-earth-orbit communications satellites, this is the only one to continue in Russian service. These store-dump communications satellites are generally believed to have been initially developed for the military by the Main Intelligence Administration (Glavnoye Razvedyvatelnoye Upravlenie, GRU). A commercial version was marketed as the Gonets-D1 in the 1990s. At the end of 2001, the constellation included Cosmos 2337–2339, which were launched in 1997, along with Gonets-D1 4–6. The next launch consisted entirely of Strela-3 military satellites, but all six were left in an elliptical orbit instead of the usual 1400-km circular orbit because of a booster malfunction.[53] The most recent launch, on 27 December 2001, again consisted of three Strela-3 military satellites and three Gonets D commercial spacecraft.

The Molniya satellites operate in highly inclined, highly elliptical orbits that are optimized for coverage of the northern hemisphere. The primary user of the Molniya-1 system's X-band transponders is the Russian Government and military, while the Molniya-3 system is used by both military and civilian agencies for *inter alia* television transmission.

The complete Molniya-1 constellation initially consisted of eight satellites in eight orbital planes separated by 45 degrees. This was subsequently modified to two constellations of four vehicles, with each consisting of four orbital planes spaced 90 degrees apart and with the ascending node of one constellation shifted 90 degrees from the other.[54] However, at the end of 2001 there appeared to be no more than six Molniya-1 spacecraft in operational service, with only the most recently launched pair—Molniya 1-90 and 1-91—displaying the prescribed regularity of orbital separation.[55]

---

[53] Covault, C., 'Promise and peril mark Russian launch surge', *Aviation Week & Space Technology*, 13 July 1998, p. 78.

[54] Johnson, N. and Rodvold, D., *Europe and Asia in Space 1993–1994* (Kaman Sciences Corp./ US Air Force Phillips Laboratory: Kirtland AFB, N. Mex., 1995).

[55] The puzzling irregularities of the Molniya constellation may be due to either an ongoing intentional reconfiguration of the constellation or the inability of Russia to sustain a symmetrical configuration, but the open literature is silent on this issue.

The first Soviet geostationary orbit satellites were the Raduga military and government communications satellites, first launched in 1975. By the early 1990s as many as 12 spacecraft of this series were operational in orbit, although by the end of 2001 only half this number remained operational. The oldest was Raduga 29, launched in 1993, with the most recent being Raduga 1-6, launched on 6 October 2001. The Geizer (Potok) Soviet military communications satellites operated in geostationary orbit to provide data relay support to imagery intelligence satellites, as well as fixed ground points. At the end of 2001 as many as five of these spacecraft were operational, ranging from Cosmos 2085, launched in 1990, to Cosmos 2371, launched on 4 July 2000.

Weather satellites

The Russian low-altitude weather satellite network supports both civilian and military users, in contrast to the separate systems operated by the USA. The Russian military presumably use data from the several Meteor 2 and Meteor 3 satellites, which are usually operational.

Early-warning satellites

During the cold war, the Soviet ballistic missile early-warning satellite network consisted of nine Oko satellites in Molniya-type orbits. These satellites were designed to detect launches of intercontinental ballistic missiles (ICBMs) from the continental United States but provided no coverage of sea-based missile launches. A total of six launches were conducted in 1990, indicative of the effort required to sustain this constellation.[56] By the mid-1990s Russia had evidently contented itself with maintaining only four or five operational Oko spacecraft in orbit, despite the fact that this left gaps of several hours in coverage of US land-based missile launch facilities. At the end of 2001 only four satellites—Cosmos 2340, 2342, 2351 and 2368—remained in service, with the most recent launch in December 1999. A fire damaged the Oko's Serpukhov-15 ground control facility near Moscow on 10 May 2001, although by the next day the Golitsyno-2 back-up facility had regained control of the spacecraft.[57]

The second-generation Prognoz early-warning spacecraft, known as the SPRN (Spetsializirovannim apparatom dlya obnaruzhenniya yadernix vzrivov, Special Apparatus for Observation of Nuclear Forces) is a geostationary system similar in concept to the US Defense Support Program. Following an experimental flight in 1975 (Cosmos 775), launches resumed in 1984 (Cosmos 1546), and the USSR registered a total of seven orbital slots for the Prognoz system. In practice, only three locations (24° West, 12° East and 80° East) were actually used, and by the end of the cold war the operational constella-

---

[56] 'Congress splits on milspace budget', *Military Space*, 25 Sep. 1989, p. 2.
[57] Saradzhyan (note 50).

tion apparently consisted of two spacecraft positioned at 12° East and 24° West (which was apparently the primary location).

Cosmos 2224 was launched on 17 December 1992 and was on station at 12° East until May 1999. Subsequent flights were less successful.[58] Cosmos 2282 was launched in July 1994 but drifted off station after 15 months. Cosmos 2345 was launched on 14 August 1997 and was positioned at 24° West but reportedly failed at some time between late 1997 and early 1999. Another Prognoz class early-warning satellite, Cosmos 2350, was launched on 29 April 1998, replacing Cosmos 2244 at 12° East.[59] Several weeks after the launch, it was reported that contact had been lost with Cosmos 2350 on 6 July and that the spacecraft had been written off as lost.[60] It appears that by mid-1999 Russia was without an operational geostationary early-warning spacecraft, a situation that was not remedied until the launch of Cosmos 2369 on 24 August 2001. Initially stationed at 80° East, in early December 2001 Cosmos 2379 was drifting west and by the end of the year was stabilized at the 24–25° West primary location.

Electronic intelligence satellites

The USSR apparently never launched more advanced SIGINT systems into highly elliptical or geosynchronous orbits, although in the late 1980s several launches of communications and early-warning spacecraft were initially incorrectly attributed to the electronic intelligence (ELINT) mission. By the final years of the cold war the Soviet ELINT capability consisted of two complementary low-earth-orbit systems.[61]

One of these systems consisted of a constellation of six low-altitude satellites, comprising the so-called 'third-generation' Tselina-D ('Virgin Land') ELINT system. The final launch attempt under this programme, in May 1994, ended with a booster failure. Based on the demonstrated short lifetimes of these spacecraft, this system is certainly no longer operational.[62]

The more advanced Tselina-2 ELINT satellites were initially launched in 1984 on the Proton booster, with operational flights using the Zenit-2 booster. This 6-ton spacecraft is placed into a 71° inclination orbit at an altitude of 840–860 km.[63] The absence of launches during 1989 had raised doubts about the future of this fourth-generation satellite,[64] although by the early 1990s it

---

[58] 'Missile warning system fails', *Jamestown Foundation Monitor*, 17 July 1998; and Lantratov, K., 'Cosmos-2379 in flight', *Novosti Kosmonavtiki*, no. 10 (2001), pp. 37–38.

[59] Golotyuk, Yu. and Golotyuk, S., 'Kosmos-2350—the president's new "eye"', *Defense and Security*, 5 May 1998.

[60] Bogatyrev, V., 'Russian missile defense is breaking down', *Kommersant Daily*, 15 July 1998, p. 2; and Bratersky, A., 'Russian defense satellite crippled', *Space News*, 20 July 1998, p. 7.

[61] Clark, P., 'Soviet worldwide ELINT satellites', *Jane's Soviet Intelligence Review*, July 1990, pp. 330–32.

[62] The Analytical Graphics Inc. database reports Cosmos 2221, 2228 and 2242 as being active as of the end of 2001, an assessment that is almost certainly in error.

[63] URL <http://www.russianspaceweb.com/tselina.html>.

[64] Clark, P., 'Economic changes in the Soviet space program', *Jane's Soviet Intelligence Review*, May 1990, p. 236.

appeared that this programme was intended to consist of a constellation of four spacecraft in orbital planes separated by 45 degrees.[65] In practice, these plans were marred by persistent launch vehicle failures.

The most recent flight under this programme was Cosmos 2369, launched on 3 February 2000 on a Zenit-2 from the Baikonur Cosmodrome, apparently replacing Cosmos 2263, which had been in orbit since 1993. The overall operational status of this constellation is difficult to assess in the absence of overt indicators as to the operational health of individual spacecraft and in the absence of apparent orbital alignments among recently launched spacecraft.[66] Cosmos 2369 replicated the rough alignment of Cosmos 2262 with Cosmos 2278 (launched in April 1994) and Cosmos 2297 (launched in November 1994), with at least two satellites of this trio providing simultaneous coverage of targets. There are no simple orbital relations among three more recent flights—Cosmos 2322 (launched in October 1995), Cosmos 2233 (launched in September 1996) and Cosmos 2360 (launched in July 1998). At the time of the launch of Cosmos 2360, two other spacecraft of the series were reportedly operational (presumably Cosmos 2322 and 2233).[67] By the end of 2001, probably at least three, but no more than six, of these spacecraft remained in operational service.

Ocean-surveillance satellites

During the cold war the USSR operated two classes of satellites for locating and identifying Western naval units. The nuclear-powered Radar Ocean Reconnaissance Satellites (RORSATs) used a radar with a power of several kilowatts to detect surface ships. Following the problems with the nuclear-powered Cosmos 1900 RORSAT, which malfunctioned on 12 April 1988,[68] there were no further RORSAT launches.

The Electronic Ocean Reconnaissance Satellites (EORSAT) intercept radio and radar transmissions. The first US-P (US–Upravlenniye Sputnik) and the first improved US-PM spacecraft were launched in 1974. In an apparent response to the withdrawal of the RORSAT from peacetime service, the EORSAT constellation was significantly expanded. Until the end of 1988 the EORSAT network consisted of two spacecraft flying in a single orbital plane. However, additional launches in 1989 led to a brief period during which five EORSATs were operating simultaneously in two distinct orbital planes.[69] The first of the improved US-PM spacecraft was launched in 1993, with subse-

---

[65] Tarasenko, M., 'Launch of Cosmos-2360', *Novosti Kosmonavtiki*, no. 15/16 (1998), p. 25.
[66] The AGI database reports a total of 10 spacecraft in this series (ranging back to Cosmos 2219, launched in Nov. 1992) as being operational, an assessment that is almost certainly in error.
[67] Zak, A., 'Classified Russian spy satellite launch delayed', *Space.com*, 2 Feb. 2000.
[68] 'Soviets confirm Cosmos 1900 difficulties', *Aerospace Daily*, 16 May 1988, p. 252.
[69] Covault, C., 'Soviet military space operations developing longer life satellites', *Aviation Week & Space Technology*, 9 Apr. 1990, pp. 44–49.

quent launches establishing a constellation of three spacecraft in orbital planes separated by 120 degrees.[70]

By 2001, only EORSAT was operating. Cosmos 2367 had been launched in December 1999 to replace Cosmos 2347, launched in December 1997. On 20 December 2001, Cosmos 2382 was launched, probably to replace Cosmos 2367 since the EORSAT spacecraft had typically operated for about two years before being replaced.[71]

Imagery intelligence satellites

During the cold war, launches of short-lived film-return imagery intelligence satellites constituted a significant fraction of annual Soviet launch activity. The USSR launched a total of 32 imagery intelligence satellites in 1988, 32 in 1989 and 21 in 1990. These third- and fourth-generation systems used film returned to earth in re-entry capsules and typically remained in orbit for only a few weeks.[72]

With the end of the cold war, the pace of Russian launches of imagery intelligence satellites declined precipitously, and more detailed information concerning the remaining operational systems became publicly available.[73] The sixth-generation Orlets-1 film-return spacecraft was first launched in 1986, with the most recent launch, Cosmos 2290, in August 1994. However, the absence of subsequent launches indicates that this programme has been abandoned.

The fourth-generation Yantar-4K1 Oktant is a high-resolution film-return imagery intelligence satellite. The first satellite, Cosmos 1097, was launched on 27 April 1979. The most recent flight of the derivative Yantar-4K2 Kobalt class satellite, Cosmos 2377, was launched from Plesetsk on a Soyuz booster on 29 May 2001. Cosmos 2377 re-entered on 10 October 2001 after a four-month mission.

The fifth-generation Yantar-4KS1 Terilen digital-transmission imagery intelligence satellite first flew in late 1982, with the launch of Cosmos 1426. With a mission duration ranging from six months to a year, one of these spacecraft, also called Neman, was in orbit providing almost continuous coverage. Cosmos 2320, launched in September 1995, appeared to continue this effort, although when it was de-orbited in September 1996 Russia was, for several months, apparently left without an imagery intelligence satellite in orbit. Subsequently, Cosmos 2359 was launched in 1998, remaining in orbit

---

[70] URL <http://www.astronautix.com/craft/uspm.htm>.
[71] Feller and Stein (note 49).
[72] Detailed descriptions of the fourth-generation Yantar systems are provided in Sorokin, V., 'Yantar history', *Novosti Kosmonavtiki* (Videokosmos) no. 17 (1997), p. 57, and no. 18/19 (1997), available at URL <http://www.users.wineasy.se/svengrahn/histind/Recces/fourth.htm>.
[73] Some confusion remains among various sources as to the nomenclature and classification associated with specific launches. Authoritative sources include URL <http://www.astronautix.com/articles/sovlance.htm> and URL <http://www.users.wineasy.se/svengrahn/histind/Recces/Recces.htm>.

for a year. Cosmos 2370 was launched on 3 May 2000 and continued in operation in orbit until it re-entered on 4 May 2001.

On 6 June 1997 Russia launched Cosmos 2344, an advanced real-time digital imagery intelligence satellite, subsequently identified as Arkon-1.[74] This 20-ton spacecraft was launched into a much higher orbit (roughly 1500 km by 2800 km) than that of any previous Soviet or Russian imagery intelligence spacecraft (which typically orbited at altitudes of a few hundred kilometres). The system is reportedly capable of providing imagery with a resolution of up to 2 metres.[75]

Although some reports suggested that the spacecraft had failed shortly after launch,[76] the continued commercial availability of 1- and 2-metre resolution digital imagery at the end of 2001 strongly suggests that the Arkon-1 spacecraft remains operational, since it was the only Russian imagery intelligence satellite in orbit at the end of the year.

In October 2001 Russian officials indicated that additional intelligence satellites would be launched in connection with the campaign against terrorism in Afghanistan.[77] By the end of the year, however, no additional launches had taken place.

Several Russian camera systems provide imagery that is commercially available. The KVR-3000 provides 2-metre imagery, while the DD-5 provides 1-metre imagery (it appears that this imagery is collected by the Arkon-1 spacecraft). The imagery archive dates back to 1992, and tasking is accepted for future imagery acquisition. However, 1-metre imagery of some areas may be denied, delayed or only available if re-sampled to 2-metre resolution.[78]

Anti-satellite systems

The USSR conducted the final test of the co-orbital ASAT system in 1982. According to most sources the system was deactivated around the time of the end of the cold war. Some reports suggest that an improved version had been placed on alert status, although it was not flight tested. Other reports suggest that an untested air-launched system, similar to the US Miniature Homing Vehicle system tested in the 1980s, remained under development in the early 1990s. With the end of the cold war, various other untested Soviet ASAT projects came to light.[79]

---

[74] 'Kosmos 2344 is Arkon-1', *News Bulletin of the Astronautical Society of Western Australia*, vol. 23, no 3 (Dec. 1997), pp. 27–29, URL <http://www.users.wineasy.se/svengrahn/histind/Recces/Arkon.htm>.

[75] Covault, C., 'New Russian recon keyed to area surveillance', *Aviation Week & Space Technology*, 23 Feb. 1998, p. 37. The continued commercial availability of Russian 1-metre digital imagery suggests that the best resolution of imagery from Arkon-1 is no worse than 1 metre and possibly better.

[76] Zak, A., 'Russian spy sats nearly extinct, offer little help in Afghanistan', *Space.com*, 4 Oct. 2001.

[77] United Press International, 'Russia to lift satellites over Afghanistan', 3 Oct. 2001, quoting the chief of Russia's Space Forces, Col. Gen. Anatoly Perminov.

[78] Satellite Imagery Systems, URL <http://www.cartographic.com/satellite_systems.asp>.

[79] Wade, M., 'ASAT', *Encyclopedia Astronautica*, available at URL <http://www.astronautix.com/project/asat.htm>.

IV. Other countries

Australia

The Royal Australian Navy has leased transmission capacity on PanAmSat Corporation's Leasat 5 satellite from Hughes Space and Communications Company. Leasat 5 was leased to the US Navy until early 1997. Under the terms of this unique contract, Hughes Global Services led the effort to relocate the satellite to an orbital station from which PanAmSat will monitor and control Leasat 5's payload on behalf of the Australian Defence Force. The satellite arrived on station at 156° East on 3 April 1998 to provide UHF satellite communications services to Australia.[80]

China

China became the third country, after the USA and the USSR, to launch an imagery intelligence satellite when its first FSW (Fanhui Shi Weixing, or Return Type Satellite) spacecraft was launched in 1975. A total of 14 of these spacecraft had been launched by the end of 1994. The first launch of an improved version, the FSW-2, was conducted in 1992, with another launched in 1994. The third (and most recent) FSW-2 launch took place on 20 October 1996, with the spacecraft returning to earth on 4 November 1996. By the end of 2001 China, surprisingly, had not launched an imagery intelligence satellite for over five years. It is generally believed that China has met its limited imagery intelligence requirements through the commercial purchase of imagery from Russia. Published reports suggesting that the China–Brazil Earth Resources Satellite (CBERS) is 'Beijing's first high-resolution imaging satellite ... disguised as a civilian earth monitoring system ... being used to target U.S. forces in the region' would appear to be without foundation.[81]

China launched the Zhongxing-22 (Zhongxing means 'China Star'), also known as ChinaSat-22, communications satellite on 26 January 2000. Officially characterized by China as a civilian satellite with a life expectancy of eight years, Zhongxing-22 is reportedly operated by the China Telecommunications Broadcast Satellite Corporation.[82] This launch was evidently the first of the previously announced Feng Huo network, which according to registration with the International Telecommunications Union would consist of up to five satellites (ChinaSat-21 to -25) providing mobile communications services.[83] The US Defense Intelligence Agency (DIA) was reported to have 'identified the satellite as Feng Huo-1, the first of several military communications satellites for the Qu Dian C4I system'. According to the classified DIA assessment, the FH-1 system 'will allow theater commanders to communicate

---

[80] Hughes Global Services, 'Hughes to lease defense satellite transponders to Australian defence force', News release, 11 May 1998.

[81] Gertz, B., 'Chinese "civilian" satellite a spy tool', *Washington Times*, 1 Aug. 2001, p. A1.

[82] Yee, A., 'China launches Telco Bird', *Space Daily*, 26 Jan. 2000, URL <http://www.spacedaily.com/news/china-00b.html>.

[83] Clark, P., 'Chinese communications satellites', *Via Satellite,* Feb. 2001.

with and share data with all forces under joint command' and will provide the Chinese military with 'a high-speed and real-time view of the battlefield which would allow them to direct units under joint command more effectively'. The satellite would reportedly provide the military with both C-band and UHF communications.[84]

The Beidou (Big Dipper) Navigation Test Satellite 1 (BNTS-1) was launched by a Chinese Long March 3M booster on 31 October 2000 into a geostationary orbit[85] slot at 140° East, to the east of China. It was followed by Beidou 1B on 21 December 2000, which was placed in a geostationary orbit at 80° East. The launch of this second Beidou completed the two-satellite navigational system, which will provide positional information for highway, railway and marine transportation.[86]

The precise nature of this system remains somewhat obscure, but it appears to be analogous to the Wide Area Augmentation System (WAAS) implemented in the USA to supplement the Global Positioning System. In the US WAAS, a network of precisely surveyed ground reference stations receive GPS signals and determine if errors exist and compute corrections. These corrections are then transmitted from a geostationary communications satellite on the same frequency as GPS.[87] This could enable China to continue to use the US GPS system, even in the face of US efforts to deny GPS to adversaries in wartime.

Europe

The NATO IV Satellite Communications (SATCOM) System provides strategic and tactical SHF and UHF communications for NATO maritime and land forces. The spacecraft are hardened against nuclear effects and resistant to signal jamming. Built by British Aerospace and Matra Marconi Space, the spacecraft have a design operational life of seven years. The NATO SATCOM system currently consists of the NATO IVA and NATO IVB satellites and a previous-generation NATO IIID satellite. The satellites are operated in inclined geosynchronous orbits, with the NATO IVA satellite carrying operational traffic at 17.8° West, the NATO IVB as the primary spare at 20.2° West and the NATO IIID as the final spare at 18° West. The NATO Communications and Information Systems (CIS) Operating and Support Agency (NACOSA) at the Supreme Headquarters Allied Powers Europe in Mons, Belgium, coordinates and authorizes access to NATO satellites. The NATO SATCOM post-2000 study is evaluating options for a replacement for the NATO IV satellites after 2004. The proposed NATO Satellite Broadcast Service (SBS) is intended to be a counterpart to the US Global Broadcast Service.

[84] Gertz, B., 'China's military links forces to boost power', *Washington Times*, 16 Mar. 2000, p. A1.
[85] Long, W., 'China launches first navigation satellite', *Space Daily*, 31 Oct. 2000, URL <http://www.spacedaily.com/news/gps-00k.html>.
[86] *SPACEWARN Bulletin*, no. 566 (1 Jan. 2001), URL <http://nssdc.gsfc.nasa.gov/spacewarn/spx566.html>.
[87] The WAAS is described at URL <http://gps.faa.gov/Programs/WAAS/waas.htm>.

The Skynet 4 constellation consists of three spacecraft. The initial trio was launched in 1988 and 1990, with an expected operational life of seven years. Replacement spacecraft were launched in 1998, 1999 and most recently in February 2001. Skynet 4B, the oldest of the initial trio, was reportedly retired in June 1998, a few months after the launch of Skynet 4D.[88] The three new Skynet 4 spacecraft provide worldwide UHF and SHF communications, and two of the older Skynet spacecraft may remain on back-up status. The proposed Skynet 5 system, eventually to replace Skynet 4, includes two geosynchronous satellites to provide coverage over Europe, the Middle East, Africa, parts of Asia, the Atlantic Ocean and the eastern United States.[89]

Sicral (Satellite Italiano per Comunicazione Riservate), Italy's first military satellite, provides communications for the Italian Ministry of Defence. Launched with Skynet 4F on 7 February 2001, Sicral carries nine UHF, SHF and EHF transponders.

In November 2000 the French Ministry of Defence chose Alcatel Space as the prime contractor for the Syracuse III system, with the first launch to be carried out in late 2003. This new dedicated military spacecraft will augment the existing Telecom 2 hybrid civil/military communications satellites which host Syracuse II transponders. Syracuse III will provide significantly greater data throughput, operational flexibility and resistance to jamming.

The Spanish Hispasat is a dual-use system supporting civil, military and government communications requirements. It has provided X-band services to the Spanish Ministry of Defence since 1992. The Hispasat 1D spacecraft is scheduled for launch in late 2002, joining the Hispasat 1A, 1B and 1C satellites. On 13 July 2001 Loral Space & Communications was selected to build two new satellites. XTAR-EUR is scheduled to be launched in 2003, and SpainSat in 2004. The XTAR-EUR satellite will offer leased transponder services to government customers and provide a back-up to the Spanish Ministry of Defence. SpainSat, providing dedicated communications for the Spanish Ministry of Defence, will carry nine X-band transponders and a Ka-band payload.[90]

In December 1997 France, Germany and the UK signed an agreement for the joint project-definition phase of a future military satellite communications system known as TRIMILSATCOM. This project envisioned the joint development, manufacture and launch of a constellation of at least four SHF/EHF satellites, with the ownership and use of the constellation to be shared among the partner nations.[91] TRIMILSATCOM was initially the preferred approach for meeting the UK's SKYNET 5 requirement, although the UK withdrew

---

[88] URL <http://www.tbs-satellite.com/tse/online/sat_skynet_4b.html>.
[89] URL <http://www.mod.uk/dpa/projects/skynet5.htm>.
[90] 'Space Systems/Loral to build two new satellites to provide X-band satellite services to governments', Loral Space & Communications, 13 July 2001, URL <http://www.loral.com/inthenews/010713.html>.
[91] British Ministry of Defence, 'UK, France and Germany join forces for new generation of satellites', document 217/97, 16 Dec. 1997, available at URL <http://www.fas.org/news/uk/971216-uk.htm>.

from the project in August 1998. The BIMILSATCOM followed, with French and German participation; the first launch is planned for 2005.[92]

Helios is a military observation satellite programme developed by France, Italy and Spain. The Helios 1 programme was developed for a total investment of 10 billion francs (€1.52 million) shared among France (78.9 per cent), Italy (14.1 per cent) and Spain (7 per cent). Helios 1A was launched on 7 July 1995, followed by Helios 1B on 3 December 1999. The 1-metre resolution images from these satellites are also made available to the Western European Union.

The first of the improved Helios 2 series is scheduled for an initial launch in early 2004. Helios 2 is intended to provide significantly enhanced resolution, reduced access delay and a day/night capability. The total cost of the programme is estimated at $1.8 billion for the development and operation of two satellites over a 10-year period.[93] In contrast to Helios 1, initially only France committed substantial resources to Helios 2, although in early 2001 Belgium reportedly decided to contribute a 2.5 per cent share of the cost, the minimum contribution needed to gain access to imagery from the satellite.[94] Spain is expected to join the Helios 2 programme at a similar level.[95]

By 2006–2008, additional European imagery intelligence capabilities might include four German SAR-Lupe radar satellites, four Italian Cosmo/Skymed radar satellites and two French Pléiades high-resolution optical platforms.[96] In early 2001 the French and Italian governments reportedly agreed to jointly develop the Cosmo Pléiades, with launches planned in 2003–2006.[97] However, as of the end of 2001 the status of all these programmes remained uncertain, with no flight hardware under construction.

India

India embarked on high-resolution satellite imaging with the launch of the Indian Remote Sensing Satellite (IRS) IRS-1C in 1995 and IRS-1D in 1997. These satellites provided imagery with a resolution of 5.8 metres, which was the highest-resolution imagery publicly available prior to the launch of the Ikonos satellite in late 1999. While useful for mapping, this imagery had only modest national security applications.

According to published plans, the IRS-P5 (Cartosat-1) was intended to be India's first high-resolution imagery intelligence satellite system, with a ground sample distance (GSD) of 2.5 metres, a sigificant improvement over the 5.8-metre resolution of the IRS-1C earth resources satellite. The Cartosat

[92] McLean, A., 'European exploitation of space: when rather than if', *RUSI Journal*, Oct. 1999, pp. 47–50.
[93] Boucheron, J.-M., 'Rapport d'information sur le renseignement par l'image', Député, no. 3219, Assemblée Nationale, 4 July 2001.
[94] 'Belgium to join Helios-2 program', *Intelligence Newsletter*, no. 406 (23 May 2001).
[95] 'Belgium and Spain to join Helios 2', *Orbireport*, 12 July 2001.
[96] Western European Union (WEU), 'Space systems for Europe: observation, communications and navigation satellites', WEU Assembly document 1643, 18 May 1999.
[97] 'France and Italy to jointly develop Cosmo Pleiades remote sensing system', *SPACEandTECH Digest*, 5 Feb. 2001, available at URL < http://www.spaceandtech.com>.

programme was approved in 1997, with Cartosat-1 initially scheduled for launch in late 1999 and the follow-on Cartosat-2 planned for launch in 2002. Cartosat-2 was planned to offer imagery with a resolution of less than 1 metre.[98] By mid-2000 the Cartosat-1 remained under development, with a launch anticipated in 2001–2002.[99] By 2001 the Cartosat-1 was scheduled for 2002–2003, and the annual report of the Indian Space Research Organization simply indicated that 'work on a more advanced cartographic satellite, Cartosat-2, has also been initiated', with a launch target date in 2003–2004.[100]

The Technology Experiment Satellite (TES), a remote sensing and photo-reconnaissance satellite, was launched by a Polar Satellite Launch Vehicle PSLV-C3 rocket from the Sriharikota High Altitude Range on 22 October 2001. The 1100-kg satellite carried a high-resolution panchromatic (black and white image) camera, with a GSD variously reported as either 2.5 metres or 1 metre.[101]

Israel

Israel began working on satellite reconnaissance technology in the late 1980s and launched the Ofeq 1 (Horizon) satellite in 1988.[102] The Ofeq 2, launched in 1990,[103] also used the Shavit (Comet) booster based on the Jericho 2 ballistic missile. These small experimental spacecraft did not include intelligence collection capabilities, despite press reports to the contrary.[104] Israel launched the Ofeq 3 imagery intelligence satellite in 1995, although because it had an orbital inclination of 143.4° it could only cover the area of the earth between latitudes 36.6° North and South. Use of the Shavit booster required it to be launched across the Mediterranean Sea into a retrograde orbit in order to avoid overflight of Arab countries to the east. On 22 January 1998 Israel attempted to launch a new imagery intelligence satellite, but the Shavit booster failed after launch and was destroyed. Although sometimes described as 'Ofeq 4', it was the first commercial EROS launch. Ofeq 3 re-entered on 24 October 2000, having substantially exceeded its two-year life expectancy. The ImageSat International commercial satellite was launched soon afterwards. Future plans for the Ofeq programme are unclear, although some reports suggest that another launch may be planned for early 2002.

---

[98] 'Cabinet nod for two satellite proposals', *India World*, 25 June 1997.
[99] 'Plan to launch INSAT series', *The Tribune*, 7 June 2000.
[100] Indian Space Research Organization (ISRO), *Annual Report 2000–2001*, available at URL <http://www.isro.org/rep2002/Annual%20.htm>.
[101] Gopal, S., 'INDIA IN SPACE: launching of technology experiment satellite', South Asia Analysis Group Paper no. 349 (28 Oct. 2001), claims 1-metre resolution, while Roy, P. S. and Agarwal, V., 'Technology trends in remote sensing and data analysis', URL <http://www.gisdevelopment.net/technology/rs/techrs0018.htm>, claim 2.5-metre resolution.
[102] The authoritative transliteration appears to be 'Ofeq' although 'Offeq' and 'Ofek' are frequently encountered in the literature.
[103] Brinkley, J., 'Israel puts a satellite into orbit a day after threat by Iraqis', *New York Times*, 4 Apr. 1990.
[104] 'A new spy in the sky', *Time*, 2 Apr. 1990, p. 33.

Japan

While Japan has no dedicated military communications satellites, the Japan Self Defense Forces (JSDF) rely on the commercial Superbird spacecraft.[105] Space Communications Corporation (SCC), a Japanese satellite communications service company, was established in 1985 by Mitsubishi Corporation, Mitsubishi Electric Corporation and other Mitsubishi Group companies. SCC operates four Superbird communications satellites.[106]

Japan began to consider redefining its long-standing policy precluding the use of space for military purposes as early as 1994. North Korea's August 1998 Taepo Dong missile test prompted a national consensus in Japan for development of the Information Gathering Satellite (IGS) intelligence programme. On 6 November 1998 the Japanese Cabinet decided to develop and launch four satellites by 2002, including a pair with optical sensors of 1-metre resolution and a pair with imaging radar capabilities of somewhat lower resolution. On 29 September 1999 the Japanese and US governments signed an agreement facilitating the acquisition by Japan of remote sensing parts and components and related information for indigenous development of the IGS system. The spacecraft are projected to weigh about 1.5 metric tons and orbit at an altitude of about 500 km.[107]

By June 2001 the revised schedule called for the launch of two satellites in the winter of 2002 and a second pair in the summer of 2003, with a pair of second-generation spacecraft by the end of FY 2005 (the Japanese fiscal year begins on 1 April).[108] In August 2001 the Japanese Government requested funding of 70.7 billion yen in the FY 2002 state budget for the IGS project, a reduction of 8.5 per cent from the FY 2001 budget. At that time the launch of the four first-generation optical and radar satellites was projected for February and July 2002. The revised plan called for launching two identical replacement satellites in 2005–2006 and development of a pair of more advanced spacecraft that could be launched as early as 2008.[109] As of 20 November 2001 the National Space Development Agency (NASDA) of Japan launch schedule projected the launch of the first IGS satellellite on an H-IIA (Standard) booster in FY 2002 and the launch of the second IGS satellite on the same type of booster in FY 2003, but gave no indication of the timing of subsequent launches. However, other reports suggested that the first launch had in fact slipped to 2003.[110]

---

[105] Hirayama, [no initial], (Capt.), Communications Division Operations & Plans Department Maritime Staff Office, 'The view from Japan—current C4I in JMSDF & its future', 2 Feb. 2000.
[106] URL <http://www.superbird.co.jp/english/super_33.htm>.
[107] 'Japan aims to launch 4 intelligence satellites by 2002', *Kyodo News/Japan Weekly*, 13 Nov. 1998.
[108] Jiji Press Ticker Service, 'Japan delays info-gathering satellite plan', 13 June 2001.
[109] 'Gov't seeks 70 bil. yen for spy satellites', *Kyodo News*, 3 Sep. 2001.
[110] 'Australia to help Japan with spy satellite launch', *Kyodo News*, 22 Oct. 2001.

South Korea

The Korea Aerospace Research Institute (KARI) launched the Korea Multi-Purpose Satellite-1 (KOMPSAT-1) in December 1999 using the US commercial Orbital Sciences Corporation Taurus launch vehicle. This civilian remote sensing spacecraft, cooperatively developed by KARI and TRW, had several sensors, including one providing panchromatic imagery with a GSD of 6.6 metres—sufficient for mapping although not for military intelligence collection.[111]

Work started in 1999 on the Korea Multi-Purpose Satellite-2 (KOMPSAT-2), intended to provide 1-metre resolution imagery. By early 2000 ELOP Israel had begun assembly of the payload, similar to the Israeli Ofeq intelligence satellite, with a launch planned for 2003.[112] In March 2001 it was announced that KOMPSAT-2 would be launched on China's LM-2C rocket from Xichang in April 2004.[113]

Taiwan

Taiwan's ROCSAT-2 remote sensing mission, offering 2-metre resolution, was approved in October 1997. Taiwan's National Space Program Office (NSPO) signed a contract with France's Matra Marconi Space on 29 November 1999, to build the ROCSAT-2 satellite. A contract awarded to DASA Dornier of Germany in early 1999 had been cancelled because of opposition from China.[114] Initially scheduled to fly in 2002, in mid-2001 NSPO selected the Orbital Sciences Corporation Taurus rocket to launch ROCSAT-2 in mid-2003.[115]

Turkey

The Turkish Intelligence Satellite Supply Project called for a pair of high-resolution earth observation satellites.[116] In July 2000 the Turkish Government awarded Israel Aircraft Industries (IAI) a $270 million contract for an imagery intelligence satellite based on the Israeli Ofeq.[117] The competing French company Alcatel Space protested the award and in August 2000 won the contract. In January 2001 the Turkish Defence Ministry cancelled the contract, in retaliation for the French Parliament's condemnation of the Ottoman Empire's killing of its Armenian minority in the early 20th century. Following this can-

---

[111] 'Korea multi-purpose satellite', URL <http://kompsat.kari.re.kr>.
[112] Kelly, M., 'S Korea satellite assembly begins', *Flight International*, 2 May 2000, p. 30.
[113] Xinhua General News Service, 'China to launch S.Korean satellite', 21 Mar. 2001.
[114] 'Chinese opposition may halt German contract to build spacecraft', *China News*, 18 Aug. 1999.
[115] 'Taiwan drops India to launch ROCSAT 2 on Taurus', *SPACEandTECH Digest*, 15 June 2001.
[116] Petrie, G. and Buyuksalih, G., 'Aiming high—Turkey's efforts to develop a capability in space remote sensing', *Geoinfomatics*, Nov. 2001, pp. 34–39.
[117] 'Intelligence satellite', *Military Procurement International*, vol. 10, no. 16 (5 Aug. 2000).

cellation, IAI was the presumptive candidate to build the spacecraft,[118] although by the end of 2001 no contract award had been publicly announced.

V. Commercial operators

The precise division between imagery that is useful for national security planning and imagery that is useful for civil purposes is ambiguous and dependent on the specific applications. The French SPOT (Satellite pour l'observation de la terre) satellite, first launched in 1985, was the first to provide 10-metre resolution imagery, sufficient to depict airfields and other large installations. Within a decade, the Indian IRS-1C offered 5-metre imagery and a corresponding fourfold improvement in interpretability. Such imagery, however, was largely inadequate for most national security applications.

In 1992 the US Congress passed the Landsat Act, which authorized US companies to build and launch commercial imaging satellites. In 1994 President Clinton signed a Presidential Decision Directive (PDD-23) that further defined the government's remote sensing policies.[119]

By the late 1990s the impending introduction of commercial systems offering 1-metre resolution imagery marked the advent of commercial products with clear national security applications. These first-generation systems will soon be supplemented by commercial systems offering imagery of roughly 0.5-metre resolution.

Although not explicitly aimed at the national security market, other commercial systems are planning improved resolution capabilities. The French SPOT-5 spacecraft, to be launched in early 2002, will offer 2.5-metre panchromatic imagery.[120] The Canadian Radarsat-2 imaging radar satellite, with a launch planned in 2003, will offer all-weather imagery with a best resolution of 3 metres.[121]

Space Imaging

In April 1994 Space Imaging was granted a licence to offer 1-metre resolution satellite imagery. Lockheed Martin Commercial Space Systems built and launched the Ikonos satellite under contract to Space Imaging. On 27 April 1999 the first attempt to launch an Ikonos spacecraft failed because of a launch vehicle malfunction. On 24 September 1999 Space Imaging launched

---

[118] Demir, M., 'Turkey scraps spy satellite contract with France, opens door to Israel', *Jerusalem Post*, 25 Jan. 2001; and 'Turkey cancels spy satellite order', *SPACEandTECH Digest*, 29 Jan. 2001.

[119] Stout, M. and Quiggin, T., 'Exploiting the new high resolution satellite imagery: Darwinian imperatives?', *Commentary* (Canadian Security Intelligence Service), no. 75 (summer 1998); Dehqanzada, Y. A. and Florini, A. M., *Secrets for Sale: How Commercial Satellite Imagery Will Change the World* (Carnegie Endowment for International Peace: Washington, DC, 2000); and Baker, J. C., O'Connell, K. M. and Williamson, R. A. (eds), *Commercial Observation Satellites: At the Leading Edge of Global Transparency* (RAND and the American Society of Photogrammetry and Remote Sensing: Santa Monica, Calif., 2001).

[120] URL <http://www.spot.com/home/system/future/spot5/spot5.htm>.

[121] URL <http://radarsat.mda.ca/>; and URL <http://www.rsi.ca/rs2/home.htm>.

Ikonos-1, the world's first successful 1-metre resolution, commercial earth imaging satellite. The system provides 1-metre resolution black-and-white and 4-metre resolution colour imagery.

On 19 January 2001 Space Imaging announced that it had been awarded a licence by the National Oceanic and Atmospheric Administration (NOAA) to operate a spacecraft capable of providing 0.5-metre resolution imagery, with a launch anticipated in 2004.

Space Imaging has several regional affiliates able to task the Ikonos satellite and downlink high-resolution imagery directly to ground receiving stations.[122] Space Imaging Middle East, with its headquarters in Dubai, United Arab Emirates (UAE), was established as Dubai Space Imaging in November 1997 by a group of UAE investors. The company supplies imagery of Eastern Africa, Central Asia, the Middle East and the Persian Gulf. Japan Space Imaging Corporation was established in May 1988 by Mitsubishi Corporation for the East Asian region, including Japan. Space Imaging Europe S.A. (SIE) in Athens, Greece, has the exclusive right to a 12 million square kilometre territory that covers most of Europe, the Middle East and North Africa. On 21 November 2000 Space Imaging signed a contract with Turkey's Inta Space Systems Inc., a subsidiary of Cukurova Holdings, for the formation of a new commercial regional affiliate, Space Imaging Eurasia. In December 2000 Space Imaging Asia (e-HD.com) opened in Seoul, South Korea, with exclusive rights for coverage of Korea, and tasking rights over North-East Asia. On 31 August 2001 the Center for Remote Imaging, Sensing and Processing (CRISP) at the National University of Singapore began direct tasking and data collection of imagery from the Ikonos satellite.

DigitalGlobe

In 1993 the US Department of Commerce granted DigitalGlobe's predecessor, WorldView Imaging Corporation (WorldView), the first licence to operate an imagery satellite system. In January 1995 EarthWatch Incorporated (EarthWatch) was formed in a merger of WorldView and Ball Aerospace. The EarlyBird-1 satellite, designed to provide 3-metre resolution imagery, was launched on 24 December 1997 but failed in orbit four days later. On 21 November 2000 the QuickBird-1 satellite, designed to provide 1-metre imagery, was lost owing to a launch vehicle failure. In September 2001 EarthWatch became DigitalGlobe.[123]

The QuickBird-2 spacecraft was successfully launched on 18 October 2001. The spacecraft provides the highest-resolution satellite imagery available to the commercial market—61-cm panchromatic resolution at nadir. Originally,

---

[122] Space Imaging has affiliate companies in Greece—Space Imaging Europe S.A. (SIE); Japan—Japan Space Imaging Corporation; South Korea—Space Imaging Asia; Singapore—Center for Remote Imaging, Sensing and Processing (CRISP); Turkey—Space Imaging Eurasia; the UAE—Space Imaging Middle East (LLC) (SIME); and the USA, in Thornton, Colorado.
[123] URL <http://www.digitalglobe.com>.

the satellite was to collect and provide 1-metre imagery from a 600-km orbit, but the improved resolution was obtained by lowering the orbit to 450 km.

QuickBird satellite data are downlinked through two ground stations, in Norway and Alaska, which are linked to the control centre in Colorado. Eurimage is the exclusive distributor of DigitalGlobe products in Europe and North Africa, while Hitachi Software Engineering Co., Ltd is the exclusive distributor for customers in Asia. These two companies are also major investors in DigitalGlobe. Other international resellers include Sinclair Knight Merz in Australia, INTERSAT in Brazil, INCOM in Chile and BMP Geomatica in Peru.

Orbimage

Orbimage, an affiliate of Orbital Sciences Corporation, operates the OrbView-1 and OrbView-2 low-resolution satellites and plans to add a high-resolution satellite, OrbView-3. The attempted launch of the OrbView-4 imaging satellite on 21 September 2001 failed because of a malfunction in the Orbital Sciences Corporation's Taurus rocket. OrbView-4 carried sensors providing 1-metre resolution panchromatic and 4-metre resolution multi-spectral imagery. Following this loss, the company filed for bankruptcy, although plans apparently continued for future launches.[124]

OrbView-3, planned for launch in 2002, will have the same resolution as OrbView-4. OrbView-3 imagery will be downlinked in real time to ground stations located around the world or stored on-board the spacecraft and downlinked to Orbimage's master US ground stations. Orbimage ground stations are operational in Australia, Canada, Chile, Italy, South Korea and South Africa.

ImageSat International[125]

West Indian Space was incorporated in the Netherlands Antilles in 1997 as a joint venture of the Israeli companies IAI and Electro-Optics Industries (El-Op), and the US company Core Software Technologies. In August 2000 the company changed its name to ImageSat International, with ownership distributed between IAI (44 per cent), El-Op (12 per cent) and Core (44 per cent).[126] In late 2001 it was reported that the addition of other private investors

---

[124] Johnston, N., 'Va. satellite firm files plan to reorganize; Orbital Imaging's move follows satellite loss', *Washington Post,* 26 Sep. 2001.

[125] ImageSat has affiliate companies in Argentina—CONAE (National Commission on Space Activities); Israel—ImageSat Israel; Italy—IPT (Informatica per il Territorio); Japan—Hiroshima Earth Environment Information Center; South Korea—Satellite Technology Research Center; Russia—Sovinformsputnik; Singapore—Centre for Remote Imaging, Sensing and Processing (CRISP); South Africa—Satellite Applications Centre (SAC); Sweden—Metria Satellus; Taiwan—Center for Space and Remote Sensing Research; and the USA—Core Software Technology (CST).

[126] 'West Indian Space changes name to ImageSat, announces product offerings', *SPACEandTECH Digest,* 14 Aug. 2000.

from Europe and the USA had brought the non-Israeli holdings to about 60 per cent of the company.[127]

The EROS A1 satellite was launched by a Russian Start-1 booster on 5 December 2000. The EROS A1 provides either 1.8-metre resolution 'Standard' imagery or 1-metre 'Over-Sampled' imagery. In August 2001 ImageSat International initiated production of the follow-on EROS B1, designed to provide 0.5-metre resolution imagery, with a launch scheduled for late 2003.[128]

The lightweight EROS A1 spacecraft does not have on-board data storage, and imagery acquisition is restricted to footprints within a 2000-km radius of a ground station. The primary archive facility is located in Limassol, Cyprus, and as of early 2000 the company had reportedly signed agreements with 15 ground stations worldwide.[129] Satellite Operating Partners (SOP) provide dedicated regional coverage for the exclusive use of regional customers. SOP customers have complete control of the satellite over their area of coverage and retain a local archive of collected imagery. Alternatively, Priority Acquisition Service (PAS) provides confidential tasking of the ImageSat spacecraft while it is in a customer's coverage area (which appears to restrict the availability of the imagery to other customers).[130]

VI. Prevention of an arms race in outer space

The issue of the military uses of space has resurfaced on the arms control agenda. The USA's plans for an expansive BMD system architecture featuring space-based components, and the growing importance of 'space control' in US military strategy, have fuelled international concern about the militarization of outer space.[131] China, France and Russia have called for the negotiation of a new multilateral treaty prohibiting the deployment of weapons in space and restricting its use for peaceful purposes. These calls have been supported by other states, including Canada and Sri Lanka.

China has taken a leading role in advocating the creation of a non-militarized 'space sanctuary'. In the Conference on Disarmament, China proposed in 1999 the re-establishment of an ad hoc negotiating committee under item three of the CD agenda, 'Prevention of an arms race in outer space' (PAROS).[132] This proposal has been backed in principle by Russia.[133] In 2001

---

[127] Wagner, M. J., 'Evaluating the EROS-A1 satellite—an Israeli commercial high-resolution spy in the sky', *Geoinfomatics*, Nov. 2001, pp. 6–9.

[128] Opall, B., 'ImageSat initiates production of new craft', *Space News*, 13 Aug. 2001.

[129] Morring, F., Jr, 'Internet sales drive marketing for Israeli-built satellite', *Aviation Week's Space Business*, 3 Apr. 2000.

[130] Wagner (note 127).

[131] Referring to the USA's interest in the military uses of space, a senior Chinese representative warned at the United Nations that 'outer space is now faced with the danger of being weaponized, which manifests itself in two aspects, namely, the development of the missile defence programme and the "space control" plan'. Statement by Hu Xiaodi, Head of the Chinese delegation at the 2001 session of the United Nations Disarmament Commission, 10 Apr. 2001.

[132] China proposed the establishment of an ad hoc committee 'to negotiate the conclusion of an international legal instrument banning the testing, deployment and use of any weapons, weapon systems or

China intensified its diplomatic efforts to open substantive negotiations on space weapons in the light of the US administration's proposal to replace the 1972 Treaty on the Limitation of Anti-Ballistic Missile Systems (ABM Treaty) with a new US–Russian strategic framework.[134] Although it is not a party to the ABM Treaty, China has derived considerable security benefits from the treaty's prohibition of national missile defences. This created a strategic environment that allowed the Chinese deterrent to remain at a relatively modest number of warheads. Thus it was not surprising that China became one of the most vocal critics of US missile defence plans, suggesting that 'the crux of [the] matter lies in the attempt by one certain country aiming at absolute security to press ahead with a national missile defence system (NMD) covering the whole territory and to introduce weapons in outer space on the basis of its outstanding economic, scientific and technological capabilities'.[135]

In June 2001 the Chinese delegation introduced a proposed draft agreement. Under this proposal, states would agree 'not to test, deploy or use in outer space any weapons, weapon systems or their components. Not to test, deploy or use on land, in sea or atmosphere any weapons, weapon systems or their components that can be used for war-fighting in outer space. Not to use any objects launched into orbit to directly participate in combatant activities'. The draft defines 'weapon systems' as 'the collective of weapons and their indispensably linked parts that jointly accomplish battle missions' and components of weapon systems as 'subsystems that [are] directly and indispensably involved in accomplishing battle missions'.[136] Ambassador Hu candidly acknowledged that, owing 'to the complexity and sensitivity of the verification issue, the paper offers no specific ideas in this connection'. He also suggested that 'missile defense systems will undoubtedly incorporate space weapon systems. Some of these space weapon systems may be based in outer space, providing target information and guidance for weapon systems located on earth'.[137]

Prior to the introduction of the Chinese draft, progress on the PAROS issue in the CD had been almost non-existent, and the events of 2001 provide little encouragement for the future. Part of the problem is the Chinese draft itself, which is extraordinarily ambiguous in defining prohibited activities, possibly reflecting China's ambivalence as to which US activities are of most concern.

components thereof in outer space'. Re-establishment of an ad hoc committee on the prevention of an arms race in outer space and its mandate, Conference on Disarmament document CD/1576, 18 Mar. 1999. An ad hoc committee established in 1994 failed to reach agreement on a set of proposed confidence-building measures for outer space.

[133] Russia has suggested that the CD establish an ad hoc committee to negotiate a PAROS regime, which could potentially take the form of an international legal instrument. Statement by Vasily Sidorov, Ambassador of Russia to the CD, Conference on Disarmament document CD/PV.871, 22 Mar. 2001.

[134] See chapter 10 in this volume.

[135] Statement by Hu Xiaodi, Ambassador for Disarmament Affairs of China, at the Plenary of the Conference on Disarmament, Geneva, Conference on Disarmament document CD/PV.876, 7 June 2001.

[136] Delegation of China working paper, 'Possible elements of the future international legal instrument on the prevention of the weaponization of outer space', Plenary of the Conference on Disarmament, Geneva, Conference on Disarmament document CD/1645, 6 June 2001.

[137] Statement by Hu Xiaodi (note 135).

In contrast to some proposals of the 1980s, the Chinese draft is not simply a ban on anti-satellite weapons, nor is it simply a prohibition on space-based anti-missile interceptor weapons. The Chinese language would appear to correctly appreciate the extraordinarily critical role that space-based sensors play in proposed US ground-based anti-missile systems.

The Chinese proposal, with terms such as 'directly participate in combatant activities' and 'jointly accomplish battle missions', might be taken as directed against the increasingly intimate and indispensable connection between US military space systems and terrestrial war-fighting capabilities. The expressed Chinese anxieties concerning US hegemony enforced from space would be equally applicable to space-based lasers and space-directed 'smart bombs'.

Some difficulty would attend any effort to ban space-based interceptor systems or ground-based interceptors capable of destroying satellites. Such a ban would preclude all but the shortest-range interceptors currently projected in US anti-missile plans. Even greater difficulties would confront an effort to ban the space-based sensors that are the key component of wide-area anti-missile systems. The task of banning those space systems that provide targeting support for terrestrial weapons would appear even more daunting.

Conventionally, arms control analysts have spoken of the militarization of space as being an irreversibly accomplished fact and the weaponization of space as a future condition subject to further debate. As initially formulated, the Chinese draft does not appear to acknowledge this conventional distinction.

Problems of definition and drafting aside, a more fundamental problem confronts the PAROS project. Historically, arms control regimes have reflected existing power relationships. Cold war arms control agreements between the superpowers were agreements between equals, as were many multilateral agreements. Some other arms control regimes, notably those in the non-proliferation area, have encompassed restrictions on the weak by the powerful. The annals of arms control are devoid of precedent for a regime imposed on the strong by the weak, as is proposed under PAROS.

VII. Ballistic missile defence[138]

In May 2001 President Bush stated that 'today's most urgent threat stems . . . from a small number of missiles in the hands of . . . states for whom terror and blackmail are a way of life. They seek weapons of mass destruction to intimidate their neighbors, and to keep the USA and other responsible nations from helping allies and friends in strategic parts of the world'.[139]

The Bush Administration's new programme, unlike that of the Clinton Administration, was not focused on a single architecture, but included parallel architectures with air-, land-, sea- and space-based components, as well as

---

[138] For more on ballistic missile defence, see chapter 10 in this volume.
[139] 'Remarks by the president to students and faculty at National Defense University', Fort Lesley J. McNair, Washington, DC, 1 May 2001.

multiple technologies in each configuration. Unlike the Clinton approach, under the new Bush plan there was no commitment to specific dates for production and deployment other than for the lower-tier terminal defence elements. As Defense Secretary Rumsfeld noted, 'We don't have a proposed architecture. All we have is a series of, a couple of handfuls of very interesting research and development and testing programs that we believe need to be tested'.[140]

The Terminal Defense Segment

The Terminal Defense Segment (TDS) supports the development of capabilities to intercept ballistic missiles in the terminal phase of their trajectory, as the missile or warhead approaches and re-enters the atmosphere. This provides only a brief opportunity for interception—from a few minutes before re-entry to a minute or less after re-entry.

The Patriot is a mobile US Army air and missile defence system, which was first used against ballistic missiles during the Gulf War, with limited success. The Patriot Advanced Capability 3 (PAC-3) was designed from the outset as an anti-missile system. It integrates an entirely new interceptor missile with improved radars and tracking computers. Unlike the earlier versions of the Patriot, which used an explosive warhead, PAC-3 features a high-velocity hit-to-kill interceptor.[141] As of mid-2001 the PAC-3 programme had demonstrated seven successes in eight attempts in hit-to-kill intercepts against ballistic missile targets and four successes in four intercepts against cruise missiles and air-breathing targets. PAC-3 interceptor missiles were first delivered to training battalions in 2001.[142]

The Medium Extended Air Defense System (MEADS) is intended to defend against short- and medium-range ballistic missiles, cruise missiles and aircraft. The MEADS acquisition strategy included competition between two transatlantic industrial teams. As presently structured, the programme will integrate the Patriot PAC-3 missile with a lightweight launcher, a 360° coverage radar and a new tactical operation centre, including technology from Germany, Italy and the USA.[143] With funding approved by the German Parliament in June 2001, the trilateral MEADS programme initiated a three-year Risk Reduction Effort in July 2001. A $216 million contract was awarded to MEADS International, a joint venture including Lockheed Martin and EuroMEADS (consisting of the Anglo-Italian Alenia Marconi Systems and the European Aeronautic Defence and Space Co.).[144]

[140] Press Conference with Defense Secretary Donald Rumsfeld at the Frontiers of Freedom Institute Conference, Washington, DC, 12 July 2001.
[141] 0604865C PAC3—EMD, BMDO, RDT&E, Budget Item Justification (R-2 Exhibit), June 2001.
[142] Kadish, R. T. (Lt-Gen.), USAF, Director, Ballistic Missile Defense Organization, 'The Ballistic Missile Defense Program', Statement before the House Armed Services Committee, Amended Fiscal Year 2002 Budget, 19 July 2000.
[143] 0603869C MEADS—DEM/VAL (PD-V), BMDO, RDT&E Budget Item Justification (R-2 Exhibit), June 2000.
[144] Gildea, K., 'NATO awards $216 million MEADS risk reduction contract', *Defense Daily International*, 13 July 2001.

The Arrow system (developed jointly by the USA and Israel) consists of the jointly developed Arrow II interceptor and launcher, integrated with the Israeli-developed fire control radar (Green Pine), battle management centre (Citron Tree) and launch control centre (Hazelnut Tree). In January 1998 Israel requested $169 million to fund the procurement of a third Arrow battery, and the US Congress provided the additional funding between 1998 and 2001. Since 1988, when the Arrow programme was initiated, Israel has improved the performance of its pre-prototype Arrow I interceptor to the point where it achieved a successful intercept in June 1994. The first integrated intercept flight test was successfully conducted in Israel on 1 November 1999. The Green Pine radar detected a Scud class ballistic target, and the Citron Tree battle management centre commanded the launch of the Arrow II interceptor. The Israeli Air Force declared the Arrow operational on 16 October 2000.[145]

The Theater High-Altitude Area Defense (THAAD) system is intended to intercept medium-range theatre ballistic missiles at long ranges and high altitudes. THAAD uses hit-to-kill technologies, can operate in both the endo- and exo-atmosphere, and has a much longer range than the PAC-3 system. In the mid-1990s pressure regarding the schedule led to a string of six flight test failures. The Engineering and Manufacturing Development (EMD) contract was awarded to Lockheed Martin Space Systems Company, with Raytheon as the major subcontractor.[146] After programme restructuring, THAAD achieved successful intercepts in the summer of 1999.[147] THAAD will be deployed in three incremental blocks, and a limited contingency capability could be available in 2005. By 2007 an improved version is intended to defeat all expected threats from short- and medium-range missiles. By 2012 an upgraded version of THAAD is intended to counter more advanced threats.[148]

The Navy Area Theater Ballistic Missile Defense (TBMD) is built on the existing capabilities of the Aegis Weapon Systems (AWS) on the CG-47 Ticonderoga Class cruisers, the DDG-51 Arleigh Burke Class destroyers and the Navy Standard Missile II (SM-2) Block IV missile. This medium-range system was intended to provide lower-tier protection to ports, coastal airfields, amphibious operations and other coastal sites.[149] As with the PAC-3 and THAAD, the Navy Area interceptor featured a direct hit guidance to provide hit-to-kill intercepts a large percentage of the time, along with a blast fragmentation warhead to ensure lethality if a direct hit is not achievable. Although the Navy Area Program had experienced various technical, cost and schedule problems, as of mid-2001 flight tests were expected to lead to an operational

---

[145] 0603875C International Cooperative Programs, BMDO, RDT&E Budget Item Justification (R-2A Exhibit), June 2001.
[146] 0604861C THAAD System—EMD, BMDO, RDT&E Budget Item Justification (R-2 Exhibit), June 2001.
[147] Statement of Lieutenant General Ronald T. Kadish, USAF, Director, Ballistic Missile Defense Organization, Before the House Armed Services Committee, Subcommittee on Military Research & Development, 14 June 2001.
[148] 0603881C Terminal Defense Segment, BMDO, RDT&E Budget Item Justification (R-2A Exhibit), June 2001.
[149] 0604867C Navy Area—EMD, BMDO, RDT&E Budget Item Justification (R-2 Exhibit), June 2001.

capability by about 2005. Somewhat surprisingly, in December 2001 the Pentagon cancelled the programme when it emerged that the project was more than 50 per cent over budget and had fallen over two years behind the planned schedule.[150]

The Midcourse Defense Segment

In mid-2001 the Bush Administration redesignated the NMD programme as the Missile Defense System, or the Midcourse Defense Segment (the MDS acronym is defined differently by different officials). The MDS includes capabilities for countering ballistic missiles in the mid-course stage of flight, including the Ground-Based Midcourse Project and the Sea-Based Midcourse Project, successors to the NMD and Navy Theater Wide (NTW) programmes.

A major focus of the MDS is the construction of a missile defence test bed that provides a short-term option to employ the test facility's radars, command and control, and interceptor missiles as an operational capability. The system could be put on alert status to provide a contingency capability in FY 2004.[151] The MDS amended budget request of $3941 million represents an increase of $1455 million over FY 2001 enacted funds, and an increase of $1237 million over the FY 2002 initial budget submission.[152]

The NTW programme, also known as Upper Tier, is intended to intercept medium- to long-range theatre ballistic missiles. NTW, also known as the Aegis Light-weight ExoAtmospheric Projectile (LEAP) Intercept (ALI) programme, is intended to conduct boost-phase intercepts when a ship is positioned near the missile launch site, mid-course intercepts, as well as terminal-phase intercepts near the defended area. As with the shorter-range Navy Area system that was cancelled in December 2001, MTW builds on the AWS and the Standard missile. However, the SM-2 Block IV is modified with a new third stage and an exo-atmospheric kinetic warhead (KW). Testing has included the Flight Test Round (FTR)-1 in July 2000, which experienced a failure in the new third-stage system. Three additional flight tests were planned for 2001.[153] The programme successfully executed FTR-1A in January 2001 and was scheduled to conduct an additional flight test, Flight Mission (FM)-2, in the fourth quarter of FY 2001. An additional five flight tests, FM 3–7, were scheduled for FY 2002.[154]

Under the Bush Administration's restructured Sea-Based MDS project, the NTW programme is focused on developing a contingency sea-based ascent and mid-course intercept capability that could be deployed by FY 2005 to

---

[150] Dao, J., 'Navy missile defense plan is canceled by the Pentagon', *New York Times,* 16 Dec. 2001.

[151] 0603882C Midcourse Defense Segment, BMDO, RDT&E Budget Item Justification (R-2 Exhibit), June 2001.

[152] 'The Ballistic Missile Defense Program—Amended Fiscal Year 2002 Budget', Prepared Statement of Lieutenant General Ronald T. Kadish, USAF, Director, Ballistic Missile Defense Organization, United States Senate, Committee on Armed Services, 12 July 2001.

[153] 0603868C Navy Theater Wide—DEM/VAL, BMDO, RDT&E Budget Item Justification (R-2 Exhibit), June 2001.

[154] 0603882C Midcourse Defense Segment, BMDO, RDT&E Budget Item Justification (R-2A Exhibit), June 2001.

provide a limited capability against medium-range ballistic missile threats. The project is also intended to provide a sea-based ascent mid-course phase hit-to-kill capability against intermediate- and intercontinental-range missiles in FYs 2008–10.[155] The sea-based boost programme is also considering an entirely new high-speed, high-acceleration booster using a boost-phase kill vehicle. This booster will also be evaluated (with a different kill vehicle) for the sea-based mid-course requirement.[156]

The Clinton Administration's NMD system elements comprised a Ground-Based Interceptor (GBI), ground- and space-based sensors, and a Battle Management, Command, Control and Communication (BM/C3) system. The ground-based sensors included a new X-Band Radar (XBR) and upgrades to existing early-warning radars (EWRs). The NMD system would also use satellites for warning and tracking. Initially, these would be limited to the Defense Support Program (DSP) and subsequently the Space-Based Infrared System (SBIRS). The programme was structured to field an initial capability (IC) by the end of FY 2006 and an expanded capability by the end of FY 2008. The IC included up to 20 ground-based interceptors at a single site in Alaska, a single ground-based XBR, upgraded early-warning radars (UEWR) and DSP satellite support. The expanded capability extended to 100 ground-based interceptors, an upgraded XBR, the upgrading of five EWRs and the SBIRS warning and tracking satellites. However, on 1 September 2000 President Clinton decided to delay a deployment decision and continue testing.[157]

The Bush Administration has announced plans for an MDS Test Bed Facility that includes these components at several locations. Five GBIs with supporting infrastructure at Fort Greely, Alaska, would demonstrate the design of a GBI deployment site, although the plans did not call for the launching of interceptors from Fort Greely during the test programme since the missiles would have to fly over populated areas. Two GBI test launch silos would be located at the Kodiak Launch Complex (KLC) in Alaska, a commercial space launch centre owned by the Alaska Aerospace Development Corporation. The upgrades to the existing COBRA DANE phased array radar on Shemya Island would include upgraded software, refurbishment of power plant there, and test-support infrastructure. The UEWR at Beale AFB in California would consist of software upgrades to the existing PAVE PAWS radar.[158]

The Boost Defense Segment

The objective of the Boost Defense Segment (BDS) of the Bush Administration's restructured programme is to develop the capability to intercept ballistic missiles shortly after launch. The administration's amended request of

---
[155] 0603882C Midcourse Defense Segment, BMDO (note 154).
[156] 'The Ballistic Missile Defense Program—Amended Fiscal Year 2002 Budget' (note 152).
[157] 0603871C NMD, BMDO, RDT&E Budget Item Justification (R-2 Exhibit), June 2001.
[158] Statement of Dr Patricia Sanders, Deputy for Test, Simulation and Evaluation, Ballistic Missile Defense Organization, Amended Fiscal Year 2002 Military Construction Budget, 31 July 2001.

$685 million for the BDS programme was an increase of $313 million over the FY 2001 enacted funding. The programme includes both directed energy (DE) and kinetic energy (KE) systems. Early activities include an intercept demonstration in 2003 using the Airborne Laser (ABL) and a KE intercept in 2006 under the new Space-Based Interceptor Experiment (SBX) programme.

The ABL programme is intended to develop an air-based laser weapon to intercept ballistic missiles in their boost phase. This weapon system integrates a laser and associated equipment into a modified commercial Boeing 747-400F aircraft. The ABL programme definition and risk reduction contract was awarded to the Boeing/TRW/Lockheed Martin team in November 1996. The initial prototype ABL aircraft, with a laser providing about half of the projected power of the production version, is intended to culminate in boost-phase missile intercepts in 2003. This half-power ABL will be available for deployment as an emergency capability, and two full-power aircraft are to be delivered by FY 2009. Procurement of additional full-power aircraft will be completed by 2011.[159]

The Space-Based Laser (SBL) project is intended to accomplish boost phase intercept prior to the deployment of mid-course countermeasures.[160] The Defense Advanced Research Projects Agency (DARPA) began work on the SBL programme in 1977, and in 1984 the programme was transferred to the Strategic Defense Initiative Organization (SDIO). In 1997 the programme was again transferred, from the Ballistic Missile Defense Organization (BMDO) to the Air Force. Despite a major injection of funding during the late 1980s, as of 1997 the programme remained no closer to an orbital flight demonstration than when it was initiated in 1977. On 8 February 1999 a contract was awarded to a team comprised of Lockheed Martin, TRW and Boeing.[161] The project is initially focused on a ground demonstration at a new test facility at the Stennis Space Center, Mississippi. In 1999 the programme was structured to support an in-space Integrated Flight Experiment (IFX) in 2012–13. Under the revised Bush Administration plan, the programme will proceed from a component development phase in 2002–2006, to an integrated ground test phase in 2007–10, to an on-orbit test phase in 2011–13.[162] Proponents claim that each SBL platform would be capable of destroying dozens of missiles during their boost phase. A 12-satellite constellation could intercept about 95 per cent of theatre missile threats, and a 24-satellite constellation could provide nearly complete national missile defence.[163]

The Space-Based Kinetic Energy Experiment is the only major initiative of the Bush Administration's missile defence programme that was not included in the Clinton Administration plans. The programme represents a revival of

---

[159] 0603883C Boost Defense Segment, BMDO, RDT&E Budget Item Justification (R-2 Exhibit), June 2001.
[160] More information is available on the Space-Based Laser-Integrated Flight Experiment Internet site at URL <http://www.sbl.losangeles.af.mil>.
[161] 0603174C Space Based Laser, BMDO, RDT&E Budget Item Justification (R-2 Exhibit), June 2001.
[162] 0603883C Boost Defense Segment BMDO (note 159).
[163] BMDO, 'Space-Based Laser', Fact Sheet 301-00-11, Nov. 2000.

the Space-Based Kinetic Kill Vehicle (KKV) and 'Brilliant Pebbles' projects of the Reagan and first Bush administrations. The initial objective of this project is to conduct a single test (in 2005–2006) in which a KKV engages a thrusting target against a below-the-horizon background. Such a test is intended to demonstrate the feasibility of intercepting missiles in their boost phase. The programme plan is to develop a space-based kill vehicle that would be launched on an existing booster and fired against a representative missile target (the interceptor itself would be launched on a ballistic trajectory, rather than being placed into orbit). The development of other space-based interceptor components would be initiated following initial ground-launched tests. [164]

The SBIRS is a network of several types of satellites to provide detection and tracking of long-range and tactical ballistic missiles. The SBIRS network includes the SBIRS–Low satellites, in low earth orbits, and SBIRS–High satellites (developed by the US Air Force), in geosynchronous and highly elliptical orbits. An integrated centralized ground station supports all SBIRS space elements and Defense Support Program (DSP) satellites. [165]

The SBIRS–GEO programme is intended to replace the current Defense Support Program early-warning satellites in geostationary orbit, while the SBIRS–HEO will replace the HERITAGE intelligence collection sensors currently hosted on other spacecraft in highly elliptical orbits. Unlike the scanning sensors of DSP, the SBIRS will use large focal plane array staring sensors, which provide continuous coverage of the earth. SBIRS will use a common spacecraft design for both HEO and GEO, with four spacecraft in GEO and a pair of spacecraft in HEO. By late 2001 it appeared that the $2 billion SBIRS–High programme was facing a $2.2 billion cost overrun and a three-year schedule delay, with the first spacecraft slipping from 2002 to 2004 or later. [166]

While SBIRS–High supports both traditional early-warning and missile defence applications, SBIRS–Low is primarily for missile defence. It provides initial warning of a ballistic missile attack against the USA, its deployed forces or its allies. SBIRS–Low satellites provide continuous tracking of targets, from launch to impact or intercept. Their capabilities include booster detection, mid-course tracking, target discrimination and intercept hit/kill assessment. This satellite network will pass data to boost, mid-course and terminal defence interceptors. The planned tracking and discrimination capabilities are essential components for the projected multi-layer defence system. [167] SBIRS-Low began programme definition activities in August 1999 with the award of two contracts. In 2000 the US Air Force delayed the launch

---

[164] 0603883C Boost Defense Segment BMDO (note 159).
[165] Kadish (note 142).
[166] Wall, R. and Fulghum, D. A., 'New sensor targets high-profile missions', *Aviation Week & Space Technology,* 21 Dec. 2001, p. 41; and *Report to Accompany H.R. 3338, Department of Defense Appropriation Bill, 2002 and Supplemental Appropriations, 2002,* 107th Congress, 1st Session, Senate Report 107-109, 5 Dec. 2001.
[167] Robbins, C. A., 'Troubled system shows hurdles missile-defense plans will face', *Wall Street Journal,* 15 June 2001, p. 1.

of the first SBIRS–Low spacecraft from 2004 to 2006.[168] In response, at the direction of Congress, in early 2001 responsibility for SBIRS–Low was transferred from the Air Force to the BMDO. The SBIRS–Low contract is intended to support full-constellation deployment by FY 2011.[169] The estimated life-cycle cost of the programme, to consist of 27–30 satellites, is as much as $23 billion (as against the official $10 billion estimate in early 2001).[170] While the FY 2003 request for SBIRS–Low was $385 million, Congress appropriated only $250 million.

VIII. Conclusions

Space-based systems are becoming an increasingly important component of military power, above all for the United States. The USA is currently investing billions of dollars annually in development and deployment of a wide range of new precision-guided weapons which are revolutionizing the conduct of warfare. These weapons rely heavily on an integrated 'system of systems' that combines intelligence, communications, navigation and other military space systems.

At present no country can rival or contest US space dominance or the advantages that this provides to its terrestrial military operations. At the end of 2001, the USA had nearly 110 operational military-related satellites, accounting for well over two-thirds of all military satellites orbiting the earth; Russia had about 40 and the rest of the world about 20.

While it is difficult to overstate the singular advantages of US military space systems relative to those of the rest of the world, it would be a mistake to underestimate the rapidity with which other states are beginning to use space-based systems to enhance their security. Although commercial satellite imagery provides capabilities that are almost trivial compared to those of advanced US systems, these capabilities are revolutionary compared to what was available a decade ago.

The issue of the 'weaponization' of outer space has reappeared on the arms control agenda. There is growing international concern that the USA's quest for 'full-spectrum dominance'—a key dimension of which is the USA's ability to dominate space and to deny its use to other countries—will give rise to a destabilizing arms race in space. This concern has become more urgent in the light of the Bush Administration's plans for an expansive ballistic missile defence system architecture featuring space-based components. China and Russia have taken the lead in calling for the negotiation of a new multilateral treaty prohibiting the deployment of weapons in space and restricting its use for peaceful purposes. For its part, the USA has shown little interest in agreements that would constrain its military activities in space, where it enjoys unrivalled superiority.

[168] Newman, R. J., 'Space watch, high and low', *Air Force Magazine,* July 2001, p. 35.
[169] 0603884C Sensors Segment, BMDO, RDT&E Budget Item Justification (R-2A Exhibit), June 2001.
[170] *Report to Accompany H.R. 3338* (note 166).

IX. Tables of operational military satellites

Conventions

( )	Uncertain data or information
. .	Data not available or not applicable
+	Payload in addition to the payload for the primary mission

Abbreviations and acronyms

CCAFS	Cape Canaveral Air Force Station
Comm.	Communications satellite
deg.	Degrees
design.	Designation
ELINT	Electronic intelligence
GeoSat	Satellite in geostationary orbit
Incl.	Orbital inclination
Intl	International
km	Kilometres
KSC	Kennedy Space Center
min.	Minutes
Nav.	Navigation satellite
NORAD	North American Aerospace Defense Command
poss.	Possibly
prob.	Probably
SHAR	Sriharikota High Altitude Range
VAFB	Vandenberg Air Force Base

Table 11.1. US operational military satellites, as of 31 December 2001[a]

Common name	Official name	Intl name	NORAD design.	Launched (date)	Launcher	Launch site	Perigee (km)	Apogee (km)	Incl. (deg.)	Period (min.)	Comments
Navigation satellites in medium earth orbit											
GPS 2-02	SVN 13/USA 38	1989-044A	20061	10 June 89	Delta 6925	CCAFS	19 576	20 752	53.4	717.9	
GPS 2-04[b]	SVN 19/USA 47	1989-085A	20302	21 Oct 89	Delta 6925	CCAFS	21 195	21 232	55.0	760.1	(Retired, not in SEM Almanac)
GPS 2-05	SVN 17/USA 49	1989-097A	20361	11 Dec 89	Delta 6925	CCAFS	19 821	20 543	56.2	717.9	
GPS 2-08	SVN 21/USA 63	1990-068A	20724	2 Aug 90	Delta 6925	CCAFS	19 699	20 666	55.1	717.9	
GPS 2-09	SVN 15/USA 64	1990-088A	20830	1 Oct 90	Delta 6925	CCAFS	19 975	20 389	56.0	717.9	
GPS 2A-01	SVN 23/USA 66	1990-103A	20959	26 Nov 90	Delta 6925	CCAFS	19 749	20 607	55.0	717.9	USA 066, moved to slot E5
GPS 2A-02	SVN 24/USA 71	1991-047A	21552	4 July 91	Delta 6925	CCAFS	19 925	20 439	56.3	717.9	
GPS 2A-03[b]	SVN 28/USA 79	1992-009A	21890	23 Feb 92	Delta 7925	CCAFS	19 925	20 435	53.8	717.9	(Retired, not in SEM Almanac)
GPS 2A-04[b]	SVN 25/USA 80	1992-019A	21930	10 Apr 92	Delta 7925	CCAFS	20 096	20 260	55.0	717.9	(Decommissioned May 1997)
GPS 2A-05	SVN 26/USA 83	1992-039A	22014	7 July 92	Delta 7925	CCAFS	19 835	20 529	55.6	718.0	
GPS 2A-06	SVN 27/USA 84	1992-058A	22108	9 Sep 92	Delta 7925	CCAFS	19 760	20 604	54.0	717.9	
GPS 2A-07	SVN 32/USA 85	1992-079A	22231	22 Nov 92	Delta 7925	CCAFS	20 037	20 328	55.4	717.1	
GPS 2A-08	SVN 29/USA 87	1992-089A	22275	18 Dec 92	Delta 7925	CCAFS	19 939	20 385	55.4	717.9	
GPS 2A-09	SVN 22/USA 88	1993-007A	22446	3 Feb 93	Delta 7925	CCAFS	19 787	20 577	53.4	717.9	
GPS 2A-10	SVN 31/USA 90	1993-017A	22581	30 Mar 93	Delta 7925	CCAFS	19 901	20 464	54.0	717.9	
GPS 2A-11	SVN 37/USA 91	1993-032A	22657	13 May 93	Delta 7925	CCAFS	19 859	20 502	54.1	717.9	
GPS 2A-12	SVN 39/USA 92	1993-042A	22700	26 June 93	Delta 7925	CCAFS	19 850	20 514	54.2	717.9	
GPS 2A-13[b]	SVN 35/USA 94	1993-054A	22779	30 Aug 93	Delta 7925	CCAFS	20 091	20 273	53.6	718.0	(Retired)
GPS 2A-15	SVN 34/USA 96	1993-068A	22877	26 Oct 93	Delta 7925	CCAFS	20 023	20 338	55.7	717.9	
GPS 2A-15	SVN 36/USA 100	1994-016A	23027	10 Mar 94	Delta 7925	CCAFS	19 999	20 361	54.0	717.9	
GPS 2A-16	SVN 33/USA 117	1996-019A	23833	28 Mar 96	Delta 7925	CCAFS	20 109	20 256	53.5	717.9	
GPS 2A-17	SVN 40/USA 126	1996-041A	23953	16 July 96	Delta 7925	CCAFS	20 059	20 305	56.1	717.9	
GPS 2A-18	SVN 30/USA 128	1996-056A	24320	12 Sep 96	Delta 7925	CCAFS	20 021	20 339	54.0	717.9	
GPS 2A-19	SVN 38/USA 134	1997-067A	25030	6 Nov 97	Delta 7925	CCAFS	19 967	20 398	55.0	717.9	
GPS IIR-02	SVN 43/USA 132	1997-035A	24876	23 July 97	Delta 7925	CCAFS	20 134	20 230	55.7	717.9	
GPS IIR-03	SVN 46/USA 145	1999-055A	25933	7 Oct 99	Delta 7925	CCAFS	20 156	20 209	52.7	717.9	
GPS IIR-04	SVN 51/USA 150	2000-025A	26360	11 May 00	Delta 7925	CCAFS	20 136	20 228	55.2	717.9	
GPS IIR-05	SVN 44/USA 151	2000-040A	26407	16 July 00	Delta 7925	CCAFS	20 033	20 331	55.0	717.9	

THE MILITARY USES OF OUTER SPACE 657

GPS IIR-06	SVN 41/USA 154	2000-071A	26605	10 Nov 00	Delta 7925	CCAFS	20 121	20 243	55.3	717.9	Replacing USA 066
GPS IIR-07	SVN 54/USA 156	2001-004A	26690	30 Jan 01	Delta 7925	CCAFS	20 121	20 244	55.1	717.9	

Communications satellites in elliptical orbit

SDS II-1[c]	USA 40	1989-061B	20167	8 Aug 89	STS	KSC	500	39 850	63.4	717.7	Prob. replaced by USA 137
SDS II-2[c]	USA 89	1992-086B	22518	2 Dec 92	STS 53	KSC	500	39 850	63.4	717.7	Prob. replaced by USA 162
SDS II-3	USA 125	1996-038A	23945	3 July 96	Titan IVA	CCAFS	500	39 850	63.4	717.7	
SDS III-1	USA 137	1998-005A	25148	29 Jan 98	Atlas IIA	CCAFS	500	39 850	63.4	717.7	+ COBRA BRASS (Capricorn)
SDS III-2	USA 162	2001-046A	26948	10 Oct 01	Atlas 2AS	CCAFS	500	39 850	63.4	717.7	+ COBRA BRASS (Aquila)
SLDCOM 1	USA 59	1990-050A	20641	8 June 90	Titan IV	VAFB	1 200	1 200	61.0	109.4	
SLDCOM 2	USA 72	1991-076A	21775	8 Nov 91	Titan IV	VAFB	1 200	1 200	61.0	109.4	
SLDCOM 3	USA 119	1996-029A	23893	12 May 96	Titan IV	VAFB	1 200	1 200	61.0	109.4	

Communications satellites in geostationary orbit

LES 9[c]	..	1976-023B	8747	15 Mar 76	Titan 3C	CCAFS	35 745	35 825	17.0	..	Lincoln Experimental Satellite
FLTSAT 4[c]	..	1980-087A	12046	31 Oct 80	Atlas	CCAFS	35 681	35 894	12.3	1436.0	172°E, Pacific, spare
FLTSAT 7[c]	USA 20	1986-096A	17181	5 Dec 86	Atlas	CCAFS	35 766	35 809	4.9	1436.1	100°W, CONUS, spare
FLTSAT 8[c]	USA 46	1989-077A	20253	25 Sep 89	Atlas-G	CCAFS	35 770	35 792	2.6	1435.9	23°W, Atlantic, spare
UFO 02[c]	USA 95	1993-056A	22787	3 Sep 93	Atlas I	CCAFS	35 166	36 409	2.8	1436.0	74°E, Indian Ocean, spare
UFO 03	USA 104	1994-035A	23132	24 June 94	Atlas I	CCAFS	35 775	35 799	3.3	1435.2	80°W
UFO 04 EHF	USA 108	1995-003A	23467	29 Jan 95	Atlas II	CCAFS	35 762	35 774	3.4	1436.0	2°E
UFO 05 EHF	USA 111	1995-027A	23589	31 May 95	Atlas II	CCAFS	35 767	35 808	3.5	1436.0	76°E
UFO 06 EHF	USA 114	1995-057A	23696	22 Oct 95	Atlas II	CCAFS	35 775	35 799	3.7	1436.1	104°W
UFO 07 EHF	USA 127	1996-042A	23967	25 July 96	Atlas II	CCAFS	35 770	35 805	5.2	1435.9	108°W
UFO 08 GBS	USA 138	1998-016A	25258	16 Mar 98	Atlas II	CCAFS	35 770	35 792	5.6	1436.0	44°W
UFO 09 GBS	USA 140	1998-058A	25501	20 Oct 98	Atlas II	CCAFS	35 770	35 805	4.6	1436.1	132°W
UFO 10 GBS	USA 146	1999-063A	25967	24 Nov 99	Atlas II	CCAFS	35 772	35 802	0.0	1436.1	79°E
DSCS III B-5[c]	USA 12/F3	1985-092A	16115	3 Oct 85	STS 51	KSC	35 786	35 786	0.0	1436.1	150°W, W. Pacific Residual #1
DSCS III A-2[c]	USA 44/F4	1989-069B	20203	4 Sep 89	Titan 34D	CCAFS	35 786	35 786	0.0	1436.1	130°W, E. Pacific Residual
DSCS III B-14[c]	USA 78/F5	1992-006A	21873	11 Feb 92	Atlas II	CCAFS	35 786	35 786	0.0	1436.1	45°W, W. Atlantic Residual
DSCS III B-12	USA 82/F6	1992-037A	22009	2 July 92	Atlas II	CCAFS	35 786	35 786	0.0	1436.1	60°E, Indian Ocean
DSCS III B-9[c]	USA 93/F7	1993-046A	22719	19 July 93	Atlas II	CCAFS	35 786	35 786	0.0	1436.1	180°E, W. Pacific Residual #2
DSCS III B-10[c]	USA 97/F8	1993-074A	22915	28 Nov 93	Atlas II	CCAFS	35 786	35 786	0.0	1436.1	57°E, Indian Ocean Residual
DSCS III B-7	USA 113/F9	1995-038A	23628	31 July 95	Atlas IIA	CCAFS	35 786	35 786	0.0	1436.1	52°W, W. Atlantic
DSCS III B-13	USA 135/F10	1997-065A	25019	24 Oct 97	Atlas IIA	CCAFS	35 786	35 786	0.0	1436.1	135°W, E. Pacific

Common name	Official name	Intl name	NORAD design.	Launched (date)	Launcher	Launch site	Perigee (km)	Apogee (km)	Incl. (deg.)	Period (min.)	Comments
DSCS III B-8	USA 148/F11	2000-001A	26052	21 Jan 00	Atlas IIA	CCAFS	35 786	35 786	0.0	1436.1	175°E, W. Pacific
DSCS III B-11	USA 153/F12	2000-065A	26575	20 Oct 00	Atlas IIA	CCAFS	35 786	35 786	0.0	1436.1	12°W, E. Atlantic
Milstar-1 1	USA 99/DFS-1	1994-009A	22988	7 Feb 94	Titan IVA	CCAFS	35 786	35 786	0.0	1436.1	177.5°, E. Pacific
Milstar-1 2	USA 115/DFS-2	1995-060A	23712	6 Nov 96	Titan IVB	CCAFS	35 786	35 786	0.0	1436.1	4°, E. Europe
Milstar-II 1	USA 157	2001-009A	26715	27 Feb 01	Titan IVB	CCAFS	35 786	35 786	0.0	1436.1	90°, W. Americas
GeoLITE	USA 158	2001-020A	26770	18 May 01	Atlas IIA	CCAFS	35 786	35 786	0.0	1436.1	
Weather satellites											
GFO		1998-007A	25157	10 Feb 98	Taurus	VAFB	785	787	108.1	100.5	GeoSat follow-on
DMSP 5D-2 F12	USA 106	1994-057A	23233	29 Aug 94	Atlas E	VAFB	836	854	98.6	101.8	
DMSP 5D-2 F13	USA 109	1995-015A	23533	24 Mar 95	Atlas E	VAFB	839	851	98.9	101.8	
DMSP 5D-2 F14[c]	USA 131	1997-012A	24753	4 Apr 97	Titan 2	VAFB	836	852	98.7	101.8	Back-up
DMSP 5D-3 F15	USA 147	1999-067A	25991	12 Dec 99	Titan 2	VAFB	835	847	98.8	101.7	
Oscar 29[c]	NNSS 30290	1987-080B	18361	16 Sep 87	Scout G	VAFB	1 013	1 179	90.3	107.1	Stored
Oscar 27	NNSS 30270	1987-080A	18362	16 Sep 87	Scout G	VAFB	1 013	1 177	90.3	107.1	Operational
Oscar 23	NNSS 30230	1988-033A	19070	26 Apr 88	Scout G	VAFB	1 015	1 301	90.4	108.5	Operational
Oscar 32	NNSS 30320	1988-033B	19071	26 Apr 88	Scout G	VAFB	1 015	1 300	90.4	108.5	Operational
Oscar 25[c]	NNSS 30250	1988-074A	19419	25 Aug 88	Scout G	VAFB	1 032	1 175	89.8	107.3	Stored
Oscar 31[c]	NNSS 30310	1988-074B	19420	25 Aug 88	Scout G	VAFB	1 033	1 073	89.8	107.3	Stored
Early-warning satellites in geostationary orbit											
DSP F15 DSP 15[b]	USA 65	1990-095A	20929	13 Nov 90	Titan IVA	CCAFS	35 786	35 786	0.0	1436.1	Prob. replaced by USA 149
DSP F16	USA 72	1991-080B	21805	24 Nov 91	STS 44	KSC	35 786	35 786	0.0	1436.1	
DSP F17 DSP 17	USA 107	1994-084A	23435	22 Dec 94	Titan IVA	CCAFS	35 786	35 786	0.0	1436.1	
DSP F18 DSP 20	USA 130	1997-008A	24737	23 Feb 97	Titan IVB	CCAFS	35 786	35 786	0.0	1436.1	
DSP F19 DSP 18	USA 142	1999-017A	25669	9 Apr 99	Titan IVB	CCAFS	35 786	35 786	0.0	1436.1	
DSP F20	USA 149	2000-024A	26356	18 May 00	Titan IVB	CCAFS	35 786	35 786	0.0	1436.1	
Electronic intelligence satellites											
Orion/Magnum 3	USA 67	1990-097B	20963	15 Nov 90	STS 38	KSC	35 786	35 786	3.0	1436.1	
Orion/Mentor 4	USA 110	1995-022A	23567	14 May 95	Titan IVA	CCAFS	35 786	35 786	3.0	1436.1	Poss. replacement for USA 8
Orion/Mentor 5	USA 139	1998-029A	25336	9 May 98	Titan IVA	CCAFS	35 786	35 786	3.0	1436.1	Poss. replacement for USA 48

THE MILITARY USES OF OUTER SPACE

Name	Designation	COSPAR	Cat#	Launch date	Launcher	Site	Perigee	Apogee	Incl.	Period	Notes
Mercury 1	USA 105	1994-054A	23223	27 Aug 94	Titan IVA	CCAFS	35 786	35 786	3.0	1436.1	
Mercury 2	USA 118	1996-026A	23855	24 Apr 96	Titan IVA	CCAFS	35 786	35 786	3.0	1436.1	
Trumpet 1	USA 103	1994-026A	23097	3 May 94	Titan IVA	CCAFS	500	39 850	63.4	717.7	+ HERITAGE
Trumpet 2	USA 112	1995-034A	23609	10 July 95	Titan IVA	CCAFS	500	39 850	63.4	717.7	+ HERITAGE
Trumpet 3	USA 136	1997-068A	25034	8 Nov 97	Titan IVA	CCAFS	500	39 850	63.4	717.7	+ EHF Polar Adjunct
Electronic ocean surveillance satellites											
NOSS 2-1 SSU 1	USA 60	1990-050E	20642	8 June 90	Titan IV	VAFB	883	1 332	63.4	107.4	
NOSS 2-1 SSU 2	USA 61	1990-050C	20691	8 June 90	Titan IV	VAFB	883	1 332	63.4	107.4	
NOSS 2-1 SSU 3	USA 62	1990-050D	20692	8 June 90	Titan IV	VAFB	885	1 330	63.4	107.4	
NOSS 2-2 SSU 1	USA 74	1991-076C	21799	8 Nov 91	Titan IV	VAFB	919	1 296	63.4	107.4	
NOSS 2-2 SSU 2	USA 76	1991-076D	21808	8 Nov 91	Titan IV	VAFB	918	1 297	63.4	107.4	
NOSS 2-2 SSU 3	USA 77	1991-076E	21809	8 Nov 91	Titan IV	VAFB	919	1 296	63.4	107.4	
NOSS 2-3 SSU 1	USA 119	1996-029D	23862	12 May 96	Titan IV	VAFB	1 026	1 189	63.4	107.4	
NOSS 2-3 SSU 2	USA 120	1996-029C	23908	12 May 96	Titan IV	VAFB	1 027	1 188	63.4	107.4	
NOSS 2-3 SSU 3	USA 121	1996-029E	23936	12 May 96	Titan IV	VAFB	1 026	1 189	63.4	107.4	
NOSS 3-1 SSU 1	USA 160	2001-040A	26905	8 Sep 01	Atlas 2AS	VAFB	1 015	1 200	63.0	107.4	
NOSS 3-1 SSU 2	..	2001-040C	26907	8 Sep 01	Atlas 2AS	VAFB	1 014	1 202	63.0	107.4	
Imagery intelligence satellites											
Improved Crystal 2	USA 116	1995-066A	23728	5 Dec 95	Titan IV	VAFB	317	956	97.9	97.4	
Improved Crystal 3	USA 129	1996-072A	24680	20 Dec 96	Titan IV	VAFB	284	989	97.9	97.5	
Improved Crystal 4	USA 161	2001-044A	26934	5 Oct 01	Titan IV	VAFB	300	975	97.9	97.5	
8X 2	USA 144	1999-028A	25744	22 May 99	Titan IV	VAFB	2 687	3 133	63.4	148.5	
Lacrosse/Onyx 2	USA 69	1991-017A	21147	8 Mar 91	Titan IV	VAFB	649	655	68.0	97.8	
Lacrosse/Onyx 3	USA 133	1997-064A	25017	24 Oct 97	Titan IV	VAFB	658	668	57.0	98.0	
Lacrosse/Onyx 4	USA 152	2000-047A	26473	17 Aug 00	Titan IV	VAFB	681	687	68.0	98.4	Replaced USA 86
Minor military/technology development satellites											
Stacksat–TEX	P87-2/USA 57	1990-031B	20561	4 Nov 90	Atlas E	VAFB	616	747	89.8	98.3	Transceiver experiment
RADCAL	P92-1	1993-041A	22698	25 June 93	Scout	VAFB	753	881	89.6	101.2	Radar calibration
REX II	P89-1A	1996-014A	23814	9 Mar 96	Pegasus XL	VAFB	799	829	90.0	101.1	Radiation Experiment II
FORTE	P92-A	1997-047A	24920	28 Aug 97	Pegasus XL	VAFB	793	828	70.0	101.1	d
ARGOS	P91-1	1999-008A	25634	24 Feb 99	Delta 2	VAFB	822	838	98.9	101.5	e
MTI	P97-3	2000-014A	26102	12 Mar 00	Taurus	VAFB	551	584	97.4	96.0	f

Common name	Official name	Intl name	NORAD design.	Launched (date)	Launcher	Launch site	Perigee (km)	Apogee (km)	Incl. (deg.)	Period (min.)	Comments
TSX-5	P95-2	2000-030A	26374	7 June 00	Pegasus	VAFB	404	1 660	68.9	105.8	g
MightySat II.1	P99-1	2000-042A	26414	19 July 00	Minotaur	CCAFS	483	507	97.8	94.5	

Note: In the column *Launched (date)*, '00' = 2000, '01' = 2001.

[a] Figures in italics = orbital elements are estimated.
[b] Inactive/retired spacecraft.
[c] Back-up/spare spacecraft.
[d] Fast on-orbit recording transient events.
[e] Advanced research and global observation satellite.
[f] Multi-spectral thermal imager.
[g] Tri-service experiments mission.

Table 11.2. Russian operational military satellites, as of 31 December 2001

Common name	Official name	Intl name	NORAD design.	Launched (date)	Launcher	Launch site	Perigee (km)	Apogee (km)	Incl. (deg.)	Period (min.)	Comments
Navigation satellites in low earth orbit											
Parus 76[a]	Cosmos 2195	1992-36A	22006	1 July 92	Cosmos-3M	Plesetsk	955	1 010	82.9	104.7	Plane 1, off status unclear
Parus 77[a]	Cosmos 2218	1992-73A	22207	29 Oct 92	Cosmos-3M	Plesetsk	967	1 013	82.9	104.9	Replaced by Cosmos 2366
Parus 78[a]	Cosmos 2233	1993-08A	22487	9 Feb 93	Cosmos-3M	Plesetsk	952	1 008	82.9	104.6	Plane 5, active 1999
Parus 79[a]	Cosmos 2239	1993-20A	22590	1 Apr 93	Cosmos-3M	Plesetsk	964	998	82.9	104.7	Plane 4, off active 1999
Parus 80[a]	Cosmos 2266	1993-70A	22888	2 Nov 93	Cosmos-3M	Plesetsk	964	998	82.9	104.7	(Plane 1)
Parus 81[a]	Cosmos 2279	1994-24A	23092	26 Apr 94	Cosmos-3M	Plesetsk	948	1 017	82.9	104.7	Replaced by Cosmos 378
Parus 84[a]	Cosmos 2327	1996-04A	23773	16 Jan 96	Cosmos-3M	Plesetsk	950	1 020	83.0	104.8	Replaced by Cosmos 2334
Parus 85[a]	Cosmos 2334	1996-52A	24304	5 Sep 96	Cosmos-3M	Plesetsk	967	1 009	82.9	104.8	Plane 2, standby active 1999
Parus 86	Cosmos 2336	1996-71A	24677	20 Dec 96	Cosmos-3M	Plesetsk	978	1 011	82.9	105.0	Plane 4
Parus 87	Cosmos 2341	1997-17A	24772	17 Apr 97	Cosmos-3M	Plesetsk	975	1 014	82.9	105.0	Plane 2
Parus 88	Cosmos 2346	1997-52A	24953	24 Sep 97	Cosmos-3M	Plesetsk	938	994	82.9	104.4	Plane 1

THE MILITARY USES OF OUTER SPACE   661

Name	Cosmos	Int'l desig.	Cat #	Launch date	Vehicle	Site	Perigee	Apogee	Incl.	Period	Notes
Parus 89	Cosmos 2361	1998-76A	25590	24 Dec 98	Cosmos-3M	Plesetsk	966	1 015	82.9	104.9	Plane 5
Parus 90	Cosmos 2366	1999-45A	25892	26 Aug 99	Cosmos-3M	Plesetsk	960	1 009	82.9	104.7	Plane 3, replaced Cosmos 2218
Parus 91	Cosmos 2378	2001-23A	26818	29 May 01	Cosmos-3M	Plesetsk	961	1 012	82.9	104.8	Plane 6, replaced Cosmos 2279

Navigation satellites in medium earth orbit

Name	Cosmos	Int'l desig.	Cat #	Launch date	Vehicle	Site	Perigee	Apogee	Incl.	Period	Notes
GLONASS 72 766[a]	Cosmos 2308	1995-09B	23512	7 Mar 95	Proton-K	Baikonur	19 042	19 218	63.8	675.7	Withdrawn 31 Oct. 2001
GLONASS 75 781[a]	Cosmos 2317	1995-37B	23621	24 July 95	Proton-K	Baikonur	19 086	19 174	64.9	675.7	Withdrawn 31 Oct. 2001
GLONASS 76 785[a]	Cosmos 2318	1995-37B	23622	24 July 95	Proton-K	Baikonur	19 052	19 208	64.9	675.7	Withdrawn 6 Apr. 2001
GLONASS 78 778[a]	Cosmos 2324	1995-68B	23735	13 Dec 95	Proton-K	Baikonur	19 117	19 143	64.9	675.7	Withdrawn 30 Dec. 2001
GLONASS 80 786	Cosmos 2362	1998-77A	25593	30 Dec 98	Proton-K	Baikonur	19 105	19 155	65.2	675.7	Replaced Cosmos 2235
GLONASS 81 784	Cosmos 2363	1998-77B	25594	30 Dec 98	Proton-K	Baikonur	19 104	19 156	65.1	675.7	
GLONASS 82 779	Cosmos 2364	1998-77C	25595	30 Dec 98	Proton-K	Baikonur	19 126	19 134	65.2	675.7	
GLONASS 83 783	Cosmos 2374	2000-63A	26564	13 Oct 00	Proton-K	Baikonur	19 120	19 140	64.7	675.7	
GLONASS 84 787	Cosmos 2375	2000-63B	26565	13 Oct 00	Proton-K	Baikonur	19 126	19 134	64.7	675.7	
GLONASS 85 788	Cosmos 2376	2000-63C	26566	13 Oct 00	Proton-K	Baikonur	19 030	19 230	64.7	675.7	
GLONASS 86 790	Cosmos 2380	2001-53C	26987	1 Dec 01	Proton-K	Baikonur	19 114	19 146	64.8	675.7	
GLONASS 87 789	Cosmos 2381	2001-53B	26988	1 Dec 01	Proton-K	Baikonur	19 069	19 191	64.8	675.7	
GLONASS 88 711	Cosmos 2382	2001-53A	26989	1 Dec 01	Proton-K	Baikonur	19 115	19 145	64.8	675.7	

Communications satellites in low earth orbit

Name	Cosmos	Int'l desig.	Cat #	Launch date	Vehicle	Site	Perigee	Apogee	Incl.	Period
Strela-3	Cosmos 2337	1997-06D	24728	14 Feb 97	Tsyklon 3	Plesetsk	1 402	1 412	82.6	113.9
Strela-3	Cosmos 2338	1997-06E	24729	14 Feb 97	Tsyklon 3	Plesetsk	1 408	1 413	82.6	114.0
Strela-3	Cosmos 2339	1997-06F	24730	14 Feb 97	Tsyklon 3	Plesetsk	1 411	1 415	82.6	114.0
Strela-3	Cosmos 2352	1998-36A	25363	16 June 98	Tsyklon 3	Plesetsk	1 312	1 872	82.6	118.0
Strela-3	Cosmos 2353	1998-36B	25364	16 June 98	Tsyklon 3	Plesetsk	1 302	1 867	82.6	117.8
Strela-3	Cosmos 2354	1998-36C	25365	16 June 98	Tsyklon 3	Plesetsk	1 308	1 870	82.6	117.9
Strela-3	Cosmos 2355	1998-36D	25366	16 June 98	Tsyklon 3	Plesetsk	1 304	1 866	82.6	117.8
Strela-3	Cosmos 2356	1998-36E	25367	16 June 98	Tsyklon 3	Plesetsk	1 299	1 865	82.6	117.8
Strela-3	Cosmos 2357	1998-36F	25368	16 June 98	Tsyklon 3	Plesetsk	1 295	1 861	82.6	117.7
Strela-3	Cosmos 2384	2001-58D	27058	27 Dec 01	Tsyklon 3	Plesetsk	1 411	1 418	82.5	114.1
Strela-3	Cosmos 2385	2001-58E	27059	27 Dec 01	Tsyklon 3	Plesetsk	1 416	1 417	82.5	114.1
Strela-3	Cosmos 2386	2001-58F	27060	27 Dec 01	Tsyklon 3	Plesetsk	1 404	1 417	82.5	114.0

Communications satellites in elliptical orbit

Name		Int'l desig.	Cat #	Launch date	Vehicle	Site	Perigee	Apogee	Incl.	Period
Molniya-M 3-44	..	1993-25A	22633	21 Apr 93	Molniya-M	Plesetsk	904	39 457	64.6	717.8

Common name	Official name	Intl name	NORAD design.	Launched (date)	Launcher	Launch site	Perigee (km)	Apogee (km)	Incl. (deg.)	Period (min.)	Comments
Molniya-M 3-45	..	1993-49A	22729	4 Aug 93	Molniya-M	Plesetsk	2 299	38 082	64.5	718.3	
Molniya-M 3-46	..	1994-51A	23211	23 Aug 94	Molniya-M	Plesetsk	1 770	38 586	64.9	717.8	
Molniya-M 3-47	..	1995-42A	23642	9 Aug 95	Molniya-M	Plesetsk	1 112	39 236	64.1	717.6	
Molniya-M 3-48	..	1996-60A	24640	24 Oct 96	Molniya-M	Plesetsk	1 634	38 726	64.8	717.9	
Molniya-M 3-49	..	1998-40A	25379	1 July 98	Molniya-M	Plesetsk	496	39 848	63.7	717.6	
Molniya-M 3-50	..	1999-36A	25847	8 July 99	Molniya-M	Plesetsk	1 547	38 809	63.4	717.8	
Molniya-M 3-51	..	2001-30A	26867	20 July 01	Molniya-M	Plesetsk	670	39 678	62.9	717.7	K-1 improved
Molniya-M 3-52	..	2001-50A	26970	25 Oct 01	Molniya-M	Plesetsk	627	39 717	62.9	717.5	
Communications satellites in geostationary orbit											
Luch-5	Luch	1994-82A	23426	16 Dec 94	Proton	Baikonur	35 738	35 772	3.1	1434.4	110°W
Luch-6	Luch 1	1995-54A	23680	11 Oct 95	Proton	Baikonur	35 757	35 818	2.1	1436.1	74°E
Potok/Geizer 6	Cosmos 2085	1990-61A	20693	18 July 90	Proton	Baikonur	35 773	35 802	7.0	1436.0	70°E
Potok/Geizer 7	Cosmos 2172	1991-79A	21789	22 Nov 91	Proton	Baikonur	35 760	35802	6.1	1435.7	4°W
Potok/Geizer 8	Cosmos 2291	1994-60A	23267	21 Sep 94	Proton	Baikonur	35 790	35 848	4.1	1437.7	95°W
Potok/Geizer 9	Cosmos 2319	1995-45A	23653	30 Aug 95	Proton	Baikonur	35 780	35 807	3.4	1436.4	20°W
Potok/Geizer 11	Cosmos 2371	2000-36A	26394	4 July 00	Proton	Baikonur	35 770	35 802	0.2	1436.0	79°E
Raduga 29	..	1993-13A	22557	25 Mar 93	Proton	Baikonur	35 772	35 803	5.1	1436.0	12°E
Raduga 30	..	1993-62A	22836	30 Sep 93	Proton	Baikonur	35 782	35 792	4.7	1436.0	69°E
Raduga 1-3	..	1994-08A	22981	5 Feb 94	Proton	Baikonur	35 766	35 783	4.5	1435.5	52°E
Raduga 1-4	..	1999-10A	25642	28 Feb 99	Proton K	Baikonur	35 783	35 792	2.1	1436.0	35°E
Raduga 1-5	Cosmos 2372	2000-49A	26477	28 Aug 00	Proton K	Baikonur	35 775	35 787	0.3	1435.9	52°E
Raduga 1-6	..	2001-45A	26936	6 Oct 01	Proton K	Baikonur	35 773	35 802	1.3	1436.2	83°E
Early-warning satellites in elliptical orbit											
Oko-77	Cosmos 2340	1997-15A	24761	9 Apr 97	Molniya-M	Plesetsk	3 022	37 391	64.6	718.9	Replaced Cosmos 2217
Oko-78	Cosmos 2342	1997-22A	24800	14 May 97	Molniya-M	Plesetsk	2 303	38 029	66.4	717.3	
Oko-79	Cosmos 2351	1998-27A	25327	8 May 98	Molniya-M	Plesetsk	2 558	37 818	63.3	718.2	
Oko-80	Cosmos 2368	1999-73A	26042	28 Dec 99	Molniya-M	Plesetsk	1 319	39 017	61.2	717.4	

			Launched (date)							
Early-warning satellites in geostationary orbit										
Prognoz 12	Cosmos 2350	1998-25A	25315	28 Apr 98	Proton	Baikonur	35 760	35 802	1	1 436	82°E, Asian coverage
Prognoz 13	Cosmos 2379	2001-37A	26892	25 Aug 01	Proton	Baikonur	35 782	35 793	2.0	1436.2	24°W, USA coverage
Electronic intelligence satellites										
Tselina-D[a]	Cosmos 2221	1992-80A	22236	24 Nov 92	Tsyklon 3	Plesetsk	619	643	82.5	97.3	
Tselina-D[a]	Cosmos 2228	1992-94A	22286	25 Dec 92	Tsyklon 3	Plesetsk	614	648	82.5	97.3	
Tselina-D[a]	Cosmos 2242	1993-24A	22626	16 Apr 93	Tsyklon 3	Plesetsk	615	647	82.5	97.3	
Tselina-2	Cosmos 2219	1992-76A	22219	17 Nov 92	Zenit-2	Baikonur	834	863	71.1	101.9	Status?
Tselina-2	Cosmos 2227	1992-93A	22284	25 Dec 92	Zenit-2	Baikonur	847	853	71.0	101.9	Status?
Tselina-2	Cosmos 2237	1993-16A	22565	26 Mar 93	Zenit-2	Baikonur	844	858	70.9	101.9	Status?
Tselina-2	Cosmos 2263	1993-59A	22802	16 Sep 93	Zenit-2	Baikonur	845	856	70.9	101.9	
Tselina-2[a]	Cosmos 2278	1994-23A	23087	23 Apr 94	Zenit-2	Baikonur	842	856	71.0	101.9	Status?
Tselina-2	Cosmos 2297	1994-77A	23404	24 Nov 94	Zenit-2	Baikonur	841	857	71.0	101.9	
Tselina-2	Cosmos 2322	1995-58A	23704	31 Oct 95	Zenit-2	Baikonur	846	852	71.0	101.9	
Tselina-2	Cosmos 2333	1996-51A	24297	4 Sep 96	Zenit-2	Baikonur	845	856	70.9	101.9	
Tselina-2	Cosmos 2360	1998-45A	25406	28 July 98	Zenit-2	Baikonur	844	856	70.9	101.9	
Tselina-2	Cosmos 2369	2000-06A	26069	3 Feb 00	Zenit-2	Baikonur	846	853	71.0	101.9	New heavy ELINT class
Electronic ocean surveillance satellites										
US-PM 10	Cosmos 2367	1999-72A	26040	26 Dec 99	Tsyklon 2	Baikonur	399	470	65.0	93.2	
US-PM 11	Cosmos 2383	2001-57A	27053	20 Dec 01	Tsyklon 2	Baikonur	405	417	65.0	92.7	
Imagery intelligence satellites										
Yantar-4K1[a]	Cosmos 2377	2001-22A	26775	29 May 01	Soyuz	Baikonur	170	260	67.1	88.8	Re-entered 10 Oct. 2001
Arkon 1	Cosmos 2344	1997-28A	24827	6 June 97	Proton	Baikonur	1 469	2 772	63.4	129.9	

Note: In the column *Launched (date)*, '00' = 2000, '01' = 2001.

[a] Inactive or back-up.

Table 11.3. Rest of the world, operational military satellites, as of 31 December 2001

Mission	Common name	Official name	Intl name	NORAD design.	Launched (date)	Launcher	Launch site	Perigee (km)	Apogee (km)	Period (min.)	Incl. (deg.)	Comments
Australia												
Comm.	Leasat 5	SYNCOM IV-05	1990-002B	20410	9 Jan 90	STS	KSC	35 779	35 795	4.3	1436.1	156°E
China												
Comm.	Zhongxing-22	ChinaSat-22	2000-003A	26058	25 Jan 00	LM-3A	Xichang	35 778	35 796	0.1	1436.1	52°E
Nav	Beidou-01A	BNTS-1A	2000-069A	26599	30 Oct 00	LM-3A	Xichang	35 781	35 794	0.1	1436.1	46°E
Nav	Beidou-01B	BNTS-1B	2000-082A	26643	20 Dec 00	LM-3A	Xichang	35 773	35 802	0.1	1436.1	70.5°E
Europe												
Comm.[a]	NATO 3D	..	1984-115A	15391	14 Nov 84	Thor Delta	CCAFS	35 776	35 798	5.3	1436.1	18°W, spare
Comm.	NATO 4A	..	1991-001A	21047	8 Jan 91	Delta 7925	CCAFS	35 779	35 796	3.7	1436.1	17.8°W
Comm.	NATO 4B	..	1993-076A	22921	7 Dec 93	Delta 7925	CCAFS	35 772	35 802	1.9	1436.1	20.2°W, spare
Comm.[a]	Skynet 4A	..	1990-001A	20401	1 Jan 90	Titan 3	CCAFS	35 784	35 790	5.2	1436.1	113°W, spare
Comm.[a]	Skynet 4C	..	1990-079A	20776	30 Aug 90	Ariane-4	Kourou	35 776	35 790	4.0	1436.0	67°E, spare
Comm.	Skynet 4D	..	1998-002A	25134	9 Jan 98	Delta 7925	CCAFS	35 784	35 790	1.8	1436.1	113°W
Comm.	Skynet 4E	..	19990-09B	25639	26 Feb 99	Ariane-4	Kourou	35 101	36 461	2.2	1435.9	97°E
Comm.	Skynet 4F	..	2001-005B	26695	7 Feb 01	Ariane-4	Kourou	35 776	35 798	3.3	1436.0	86°E
Comm.	Sircal 1	..	2001-005A	26694	7 Feb 01	Ariane-4	Kourou	35 775	35 800	0.1	1436.0	16.2°E
Imagery	Helios 1A	..	1995-033A	23605	7 July 95	Ariane-4	Kourou	680	681	98.1	98.3	
Imagery	Helios 1B	..	1999-064A	25977	4 Dec 99	Ariane-4	Kourou	679	682	98.1	98.3	
India												
Imagery	TES	..	2001-049A	26957	22 Oct 01	PSLV	SHAR	551	579	97.8	96.0	
Israel: commercial												
Imagery	EROS-A1	..	2000-079A	26631	5 Dec 00	START-1	Plesetsk	465	487	97.3	94.1	
USA: commercial												
Imagery	Ikonos 2	..	1999-051A	25919	24 Sep 99	Athena 2	VAFB	678	679	98.2	98.3	
Imagery	Quickbird 2	..	2001-047A	26953	18 Oct 01	Delta 2	VAFB	444	448	97.2	93.5	

Note: In the column *Launched (date),* '00' = 2000 and '01' = 2001.

[a] Back-up/spare spacecraft.

12. Chemical and biological weapon developments and arms control

JEAN PASCAL ZANDERS, JOHN HART and FRIDA KUHLAU

I. Introduction

The 1972 Biological and Toxin Weapons Convention (BTWC) and the 1993 Chemical Weapons Convention (CWC) regimes were seriously challenged in 2001.[1]

In the first half of 2001 it became increasingly clear that the negotiation of a protocol to strengthen the BTWC would not be achieved in time for the Fifth Review Conference of the States Parties, which was held at the end of the year. While it was known that the United States was critical of the draft protocol, few states anticipated that it would reject both the document and the negotiation format. Because of the US rejection of the multilateral negotiation process both the Ad Hoc Group (AHG) of states parties negotiating the protocol and the Fifth Review Conference were unable to agree on final declarations.

One of the reasons for the rejection of the protocol by the United States was its claim that the protocol would be inadequate to deal with the growing threat of chemical and biological weapon (CBW) proliferation to so-called states of concern and terrorists. For several years the USA had expected that violent and destructive attacks might be carried out against it. The terrorist attacks of 11 September confirmed US concerns.[2] Less than one month after the attacks, letters filled with concentrated anthrax spores were mailed to members of the news media and politicians. As of January 2002, the sender of the letters remained unknown. The difficulties encountered in the criminal inquiry, the treatment of people exposed to anthrax spores and the disinfection of contaminated offices added to the sense of vulnerability.

As a consequence of the growing threat perception, the USA has been conducting several secret projects in an attempt to improve its defence against biological weapons (BW). However, the legality of these projects under the

---

[1] The Convention on the Prohibition of the Development, Production and Stockpiling of Bacteriological (Biological) and Toxin Weapons and on Their Destruction is reproduced on the SIPRI Chemical and Biological Warfare Project Internet site at URL <http://projects.sipri.se/cbw/docs/bw-btwc-text.html>. The Convention on the Prohibition of the Development, Production, Stockpiling and Use of Chemical Weapons and on their Destruction (corrected version), 8 Aug. 1994, is reproduced at URL <http://projects.sipri.se/cbw/docs/cw-cwc-texts.html>. The 31 Oct. 1999 amendment to Part VI of the Verification Annex of the CWC is reproduced at URL <http://projects.sipri.se/cbw/docs/cw-cwc-verannex5bis.html>.

[2] On 11 Sep. 2001 terrorists flew hijacked commercial airliners into the twin towers of the World Trade Center in New York City and into the Pentagon across the Potomac River from Washington, DC. A fourth plane crashed in Pennsylvania while flying to an unknown destination. Approximately 3000 people were killed.

BTWC is questionable, and international concern has been expressed because the USA is unwilling to open them to international scrutiny, which was an additional reason for the US rejection of the draft protocol.

The Organisation for the Prohibition of Chemical Weapons (OPCW) is the body which oversees the implementation of the CWC. In 2001 it faced a serious budgetary shortfall. As a result the number of inspections was reduced and some technical assistance and cooperative activities were postponed or cancelled. The budget problems will continue in 2002. Destruction operations continued in three of the four declared chemical weapon (CW) possessor states (India, South Korea and the USA), and there is increased hope that Russia will start large-scale destruction of its chemical weapon stockpile in 2002.

Section II of this chapter describes the negotiation of the draft protocol to the BTWC prior to the Fifth Review Conference, the suspension of the AHG as a negotiation forum and the actions that led to the suspension of the Fifth Review Conference until November 2002. The impact of biological defence research in the USA on the BTWC is analysed in section III. The implementation of the CWC and the operation of the OPCW are discussed in section IV. The terrorist attacks with mail-delivered anthrax spores are examined in section V. Section VI summarizes the concerns regarding offensive CBW state programmes and the terrorist acquisition of chemical and biological (CB) agents. Section VII presents the conclusions.

II. Biological weapon disarmament

The BTWC entered into force in 1975. In 2001 Algeria became the 144th party to it. As of 1 January 2002, 18 states are signatories to the convention, and 31 countries have not yet signed the BTWC.[3]

On 22 October 2001 the United States concluded an agreement for aid worth up to $6 million with Uzbekistan as part of its Cooperative Threat Reduction (CTR) programme. The project will support the dismantlement of former Soviet BW facilities in Uzbekistan and the 'safe and secure' destruction of any 'residual pathogens' on Vozrozhdenie Island in the Aral Sea.[4]

---

[3] Review Conference of the States Parties to the Convention on the Prohibition of the Development, Production and Stockpiling of Bacteriological (Biological) and Toxin Weapons and on Their Destruction, List of States Parties to the Convention on the Prohibition of the Development, Production and Stockpiling of Bacteriological (Biological) and Toxin Weapons and on Their Destruction as at October 2001, Fifth Review Conference document BWC/CONF.V/INF.1, 26 Oct. 2001. Complete lists of parties, signatory and non-signatory states are available at the SIPRI CBW Project Internet site at URL <http://projects.sipri.se/cbw/docs/bw-btwc-mainpage.html>. See also annex A in this volume. Taiwan also ratified the BTWC in 1973 but is no longer a member state of the United Nations.

[4] US Department of State, Daily Press Briefing, 23 Oct. 2001, Washington, DC, URL <http://www.state.gov/r/pa/prs/dpb/2001/index.cfm?docid=5509>. The lead agency on the US side is the Defense Threat Reduction Agency (DTRA). The programme is scheduled to begin in Mar. or Apr. 2002. US official, Private communication with J. Hart, 26 Dec. 2001. Vozrozhdenie Island was the site of a BW field testing facility that was established in 1936. Some of its employees worked with the former Soviet Union's Anti-plague System, which was set up to combat naturally occurring plague outbreaks. It established stations which carried out basic research, developed vaccines and treatments, treated infected individuals and destroyed diseased animals.

Developments in the Ad Hoc Group

Between January 1995 and August 2001 the AHG met in regular session 24 times and elaborated a draft protocol text.[5] Initially, at a Special Conference in 1994, the AHG had been mandated only to further develop the potential verification measures listed by the Ad Hoc Group of Governmental Experts to Identify and Examine Potential Verification Measures from a Scientific and Technical Standpoint (VEREX) and to explore the possibility of creating a legally binding instrument to strengthen the BTWC.[6] Following the Fourth Review Conference, in 1996, the AHG received a mandate to negotiate a legally binding protocol and was requested to complete its work as soon as possible. The Fifth Review Conference (19 November–7 December 2001) became accepted as a possible target date.

In early 2001 some states or groups of states held strong positions on sensitive issues, and there were few indications that they were willing to compromise in order to conclude the negotiation. Iran and some other members of the Non-Aligned Movement (NAM) rejected the continued existence of multilateral export control arrangements (such as the Australia Group, AG) outside the BTWC regime, arguing that such arrangements caused restrictions on the transfer of biological materials and equipment to developing countries and are therefore discriminatory. Russia continued to have strong views on definitions and thresholds. China had serious concerns regarding the type and scope of information to be declared. Some members of the Western European and Other States Group (WEOG) resisted far-reaching measures to implement Article X of the BTWC (related to technology transfers) and many countries, including those from the West, had reservations regarding the criteria for declarations.[7]

Consequently, many delegations felt that the usefulness of the negotiation process—with the rolling text and the use of the Friends of the Chair (FoC) to discuss particular issues—had been exhausted.[8] This prompted the chairman of the AHG, Ambassador Tibor Tóth of Hungary, to draft a 'composite text' to break the impasse. The document, which he distributed on 30 March, included the consensus items from the rolling text and suggested compromise language for outstanding issues on the basis of Tóth's informal bilateral con-

---

[5] In 2001 the AHG met 3 times: 22nd session (12–23 Feb.), 23rd session (23 Apr.–11 May) and 24th session (23 July–17 Aug.).

[6] VEREX was created by the Third Review Conference of parties to the BTWC in 1991. The group met 4 times between Mar. 1992 and Sep. 1993 and listed 21 potential measures in its final report. Ad Hoc Group of Governmental Experts to Identify and Examine Potential Verification Measures from a Scientific and Technical Standpoint, BTWC Third Review Conference report, BWC/CONF.III/VEREX/9, Geneva, 24 Sep. 1993, pp. 132–33. For an overview of the proposed measures and the review conferences see Zanders, J. P., Hart, J. and Kuhlau, F., 'Biotechnology and the future of the Biological and Toxin Weapons Convention', SIPRI Fact Sheet, Nov. 2001, URL <http://projects.sipri.se/cbw/research/biotechnologyfactsheet.pdf>.

[7] 'Washington and the BWC protocol negotiation', *CBW Conventions Bulletin*, no. 51 (Mar. 2001), p. 1.

[8] Pearson, G. S., 'Strengthening the Biological and Toxin Weapons Convention', *CBW Conventions Bulletin* (note 7), p. 19. The rolling text is the draft text of the protocol.

sultations with individual delegations and the work of the FoC.⁹ The initial reactions at the 23rd session were encouraging. Nevertheless, China and some NAM countries expressed their preference for continuing the negotiation using the rolling text (which was viewed by many as a tactic to delay the negotiations).¹⁰ The US delegation welcomed the chairman's initiative but noted that the document did not address some fundamental US concerns regarding the substance of the draft protocol.¹¹

At the 24th (and final) session, on 23 July–17 August, Tóth's opening remarks indicated that it was time for the participating states to make the necessary compromises in order to secure common benefits in the long run. Most participants (including some NAM states which had previously expressed reservations to the chairman's initiative) now endorsed the composite text as a basis for final negotiation.¹²

On 25 July, however, the United States announced its rejection of the rolling text, the composite text and further efforts to negotiate such a document. The complete rejection of both the product and the multilateral disarmament process effectively ended efforts to complete the draft protocol prior to the Fifth Review Conference. The harshest criticism came from China, Cuba and Iran. Most states, however, reacted with restraint. This may have reflected the fact that the USA had not dismissed the AHG mandate and continued to support the objective of strengthening the BTWC within the multilateral framework, which allowed for a less pessimistic assessment of the US rejection as a temporary setback. In the WEOG, however, the divisions deepened.¹³ The European Union (EU) and Australia, in particular, disagreed with the US assessment that the cost of the protocol outweighed its benefits and that nothing could make the composite text acceptable.¹⁴

The delegates next turned to the sole remaining task of writing the report for consideration at the Fifth Review Conference, which, among other things,

---

⁹ Protocol to the Convention on the Prohibition of the Development, Production and Stockpiling of Bacteriological (Biological) and Toxin Weapons and on Their Destruction, BTWC Ad Hoc Group document BWC/AD HOC GROUP/CRP.8, 30 Mar. 2001. The (corrected) composite text is also reproduced as Annex B to Procedural Report of the Ad Hoc Group of the States Parties to the Convention on the Prohibition of the Development, Production and Stockpiling of Bacteriological (Biological) and Toxin Weapons and on Their Destruction, AHG document BWC/AD HOC GROUP/56-1, 18 May 2001.

¹⁰ The NAM members with this view included Cuba, Indonesia, Iran, Libya, Pakistan and Sri Lanka. Pearson, G. S., 'Strengthening the Biological and Toxin Weapons Convention', *CBW Conventions Bulletin*, no. 52 (June 2001), p. 17.

¹¹ The US position and its rejection of the draft protocol are discussed below.

¹² Rissanen, J., 'A turning point to nowhere? BWC in trouble as US turns its back on verification protocol', *Disarmament Diplomacy*, no. 59 (July/Aug. 2001), p. 12.

¹³ The WEOG as a group did not submit a joint AHG working paper as a consequence of internal differences of opinion. Federation of American Scientists, 'US public positions on the BTWC protocol', Washington, DC, Aug. 2001, URL <http://www.fas.org/bwc/news/USPublicPositionsOnProtocol.htm>.

¹⁴ Statement by the presidency on behalf of the European Union at the 24th meeting of the Special Group of States Parties to the Convention on the Prohibition of Biological and Toxin Weapons, in Response to the speech given by the US representative on 25 July, Geneva, 26 July 2001, URL <http://www.eu2001.be>; and 'Powell, Rumsfeld press conference in Canberra July 30', *Washington File*, International Information Programs, US Department of State, Washington, DC, 30 July 2001, URL <http://usinfo.state.gov/>.

would have laid out the future agenda of the AHG.[15] They were unable to reach final agreement on the document because some countries (notably Cuba and Iran) wanted to explicitly blame the USA for the failure of the protocol negotiations, and subsequent compromise formulations were unacceptable to one group of countries or another. As a consequence of these developments, the AHG did not produce a procedural report of the 24th session. While some delegates acted as if the rolling and composite texts no longer existed, the AHG mandate was not suspended by these events and the participating states may yet readopt the documents as a starting point for future negotiation.[16]

The suspended protocol

The BTWC protocol, as envisaged by the AHG, would have been implemented by an Organization for the Prohibition of Biological Weapons (OPBW).[17] Confidence in compliance would have been generated by means of declarations, visits and investigations.

The parties to the BTWC protocol would have been required to submit initial and, subsequently, annual declarations. In the initial declaration, they would have had to provide information on past offensive and defensive BW programmes.[18] Annual declarations would have been based on certain 'triggers', including: national biological defence programme(s) and/or activities against bacteriological (biological) and toxin weapons conducted in the previous year; certain maximum biological-containment facilities, high biological-containment facilities and plant pathogen-containment facilities that work with pathogens or toxins listed in the protocol; and certain production facilities.[19] Since biological agents can usually be grown quickly over a short period of time using small initial quantities, basing declarations on quantitative thresholds is of limited value. Partly for this reason the protocol was structured with a view towards ascertaining the capability of each state party.

The composite text envisaged three types of visits by OPBW teams to 'protocol-relevant' facilities: randomly selected transparency and voluntary assistance visits as well as declaration clarification procedures.[20] The main purpose of these on-site visits was to provide assurance of the completeness and correctness of the declarations submitted and thereby generate confidence in the compliance of the other parties. The OPBW would have conducted a

---

[15] Some countries argued that the report should be addressed by a special conference of states parties to the BTWC to be held in the week before the Fifth Review Conference. Other countries argued strongly against the idea because such a special conference could call into question the continuation of the AHG's work. (In any case, only the states party to the BTWC, not an ad hoc group of parties, can call a special conference.)

[16] Zanders, J. P., Discussions with delegates from several countries, Glion, Switzerland, 11–12 Oct. and Geneva, 19–25 Nov. 2001; Pearson, G. S., 'Strengthening the Biological and Toxin Weapons Convention', *CBW Conventions Bulletin*, no. 53 (Sep. 2001), pp. 18–23; and Rissanen (note 12), pp. 13–19.

[17] BWC/AD HOC GROUP/CRP.8 (note 9).

[18] BWC/AD HOC GROUP/CRP.8 (note 9), Article 4, part B, paras 3–5.

[19] BWC/AD HOC GROUP/CRP.8 (note 9), Article 4, part C, paras 6–5.

[20] BWC/AD HOC GROUP/CRP.8 (note 9), Article 6, part B, paras 15–48; Article 6, part C, paras 49–54; and Article 6, part D, paras 55–104.

maximum of 120 randomly selected transparency visits each year with a maximum of 7 such visits per state party.

The draft protocol also provided for two types of investigations to address cases of suspected non-compliance with the BTWC: field investigations and facility investigations. Investigation-related provisions dealing with the timing, degree of access and procedures for the Executive Council to allow or disallow an investigation were complicated and never fully resolved.

Each party would have been required to provide the Technical Secretariat of the OPBW with information on its domestic implementation legislation and other regulations governing the transfer of agents, toxins, equipment and technologies relevant to Article III of the BTWC.[21] The transfer guidelines included a requirement for end-user certificates, written commitments by receiving parties not to re-transfer the item, and the provision of relevant information regarding the receiving party's laws and regulations. The guidelines would also have been restricted to certain types of equipment, such as 'fermentors or bioreactors designed to prevent the release of aerosols with a total internal volume of 100 litres or more' or 'aerosol analytical equipment designed to determine the size of aerosol particles up to 20 microns in diameter that contain micro-organisms or toxins'.[22] In addition to transfer guidelines, Article VII contained provisions for voluntary notification among parties of aggregate data on certain exports or authorization for export of select equipment for prophylactic, protective or other peaceful purposes in order to promote transparency and act as a confidence-building measure (CBM).[23] The parties would have been permitted to consult among themselves on transfer-related questions and to exclude other parties, the Executive Council and the Director-General of the OPBW.[24]

With respect to scientific and technological exchange for peaceful purposes and technical cooperation, the parties would have been required to promote and support a list of activities, including: (a) the publication, exchange and dissemination of information on conferences, training programmes, research and development (R&D) relating to biotechnology, Good Laboratory Practice (GLP) and Good Manufacturing Practice (GMP); (b) the work of certain laboratories, such as those working on disease prevention and surveillance; and (c) assistance to parties to improve laboratory capabilities in certain areas.[25] In order to avoid hampering economic and technological development the parties to the protocol would have had to ensure that, individually or collectively, they did not take discriminatory measures that were incompatible with the obligations of the BTWC.[26] The draft protocol envisaged the establishment of a cooperation committee within the OPBW to oversee the implementation of Article X of the BTWC on technical and scientific cooperation

---

[21] BWC/AD HOC GROUP/CRP.8 (note 9), Article 7, part A, paras 1–2.
[22] BWC/AD HOC GROUP/CRP.8 (note 9), Article 7, part B, para. 5.
[23] BWC/AD HOC GROUP/CRP.8 (note 9), Article 7, paras 7–9.
[24] BWC/AD HOC GROUP/CRP.8 (note 9), Article 7, para. 16.
[25] BWC/AD HOC GROUP/CRP.8 (note 9), Article 14, para. 4.
[26] BWC/AD HOC GROUP/CRP.8 (note 9), Article 14, para. 6(a).

US objections to the projected protocol

Since the 1991 Third Review Conference the USA has, in contrast to many other parties to the BTWC, consistently expressed its conviction that the BTWC is not verifiable.[28] In 1995, when the AHG began its activities, the USA stated its requirements for the protocol in terms of transparency and measures to enhance confidence and compliance with the BTWC. The Clinton Administration viewed the protocol as a tool to deter proliferators that would increase the cost and risk of violating the BTWC, rather than as an instrument of verification.[29] US contributions to the negotiation process were not always constructive, mostly as a consequence of the deadlock in the inter-agency consultations within the US administration, and on several occasions the USA introduced working documents that blunted proposed measures.[30] Following a comprehensive policy review of the BW threat involving all relevant government agencies in the spring of 2001, the George W. Bush Administration concluded that the protocol as formulated in the composite text would not achieve the AHG's mandate and strengthen confidence in compliance with the BTWC.[31] It also opposed the negotiation process, which led to the development of a policy strategy to stop 'the momentum of [the] seven-year long process'.[32]

At the 23rd session of the AHG, on 23 April–11 May, at which participants discussed the composite text for the first time, the US delegation indicated that many of its national positions on the substance of the draft protocol were not reflected in the text.[33] While there were many early indications that the Bush Administration did not prefer negotiated multilateral disarmament treaties,[34] some independent institutes in Washington also criticized the draft protocol for its weaknesses. On the basis of exercises involving various experts, one

---

[27] BWC/AD HOC GROUP/CRP.8 (note 9), Article 14, paras 7–32.

[28] For an overview of US compliance and verification concerns and preferred policy options see US General Accounting Office (GAO), *US and International Efforts to Ban Biological Weapons*, GAO/NSIAD-93-113 (GAO: Washington, DC, Dec. 1992).

[29] Harris, E. D., 'Bioweapons treaty still a good idea', *Christian Science Monitor*, 24 Aug. 2001, URL <http://www.csmonitor.com/2001/0824/p11s3-coop.htm>.

[30] Rosenberg, B., 'US policy and the BWC protocol', *CBW Conventions Bulletin*, no. 52 (June 2001), p. 1. According to a review prepared by the Federation of American Scientists (FAS), the USA submitted or co-submitted only 18 working papers to the AHG between 1995 and 2001, in contrast to South Africa (86), the United Kingdom (66) and Russia (29). Federation of American Scientists, 'US public positions on the BTWC protocol', Washington, DC, Aug. 2001, URL <http://www.fas.org/bwc/news/USPublicPositionsOnProtocol.htm>.

[31] Statement by the United States to the Ad Hoc Group of Biological Weapons Convention States Parties, Geneva, 25 July 2001, p. 2.

[32] Remarks by John Bolton, Under Secretary of State for Arms Control, at the Monterey Institute of International Studies, Washington, DC, 11 Jan. 2002, as distributed by Federal News Service, 11 Jan. 2002, URL <http://www.fnsg.com>.

[33] Pearson (note 10), p. 19.

[34] MacKenzie, D., 'Bugs of war', *New Scientist* (Internet edn), 9 May 2001, URL <http://www.newscientist.com/news/news.jsp?id=ns9999714>.

report noted that the envisaged industry monitoring routines, in particular, would create compliance ambiguities that would be detrimental to the reputation of the companies involved. It attributed this outcome to the fact that the draft protocol was designed to minimize the inconvenience and intrusiveness of inspections to host facilities and concluded that the companies might find an ambiguous compliance report more damaging than full-blown inspections.[35] Another report judged the text inadequate as a 'cost-effective, sharply focused, useful instrument in the fight against BW proliferation'.[36] The widespread negative attitude towards the protocol in Washington complicated efforts by other countries before the 24th session of the AHG to convince the USA to endorse the document.[37]

On the basis of an assessment that biological weapons pose a unique threat and that therefore the arms control approaches to other weapon categories (in particular, the CWC model) do not offer a workable structure to deal with the BW threat, the USA maintains that the protocol would be unlikely to deter states seeking a BW capability. Its adoption would consequently put US national security at risk. The core US objections to the protocol can be summarized as follows:[38]

1. The protocol (or any other verification regime) could not improve the ability to verify compliance with the BTWC or make the BTWC enforceable. The information that parties would receive under the protocol would not be of a type that would enable a country to judge compliance. As BW facilities can be small, temporary and without distinguishing features, it is unlikely that clandestine weapon projects would be detected.

2. The protocol would harm legitimate activities in the field of biotechnology by increasing expenses, risking the loss of confidential business or proprietary information, and limiting certain types of research. The measures in the protocol to reduce such risks would be insufficient (especially with respect to frivolous allegations of non-compliance that could force companies to release confidential information in order to refute the allegations).

3. The nature of the biotechnology industry would make it almost impossible to take inventories of activity in a party to the BTWC as the basis of a national declaration. In the USA there are possibly tens of thousands of relevant

---

[35] Smithson, A. E., *House of Cards: The Pivotal Importance of a Technically Sound BWC Monitoring Protocol*, Report no. 37 (Henry L. Stimson Center: Washington, DC, May 2001), p. 98.

[36] Moodie, M., *The BWC Protocol: A Critique*, Special Report no. 1 (Chemical and Biological Arms Control Institute: Washington, DC, June 2001), p. 37.

[37] Comment by a diplomat at a discussion meeting between representatives of EU member states and non-governmental organizations at the Fifth Review Conference, Geneva, 23 Nov. 2001.

[38] Statement by the United States to the Ad Hoc Group of Biological Weapons Convention States Parties (note 31); Press conference by US Ambassador Donald A. Mahley, Geneva, 25 July 2001 as transcribed in 'Mahley says absence of protocol will not undercut BWC', International Information Programs, US Department of State, Washington, DC, 25 July 2001, URL <http://usinfo.state.gov/topical/pol/arms/stories/01072502.htm>; 'Wolfowitz cites importance of biological weapons treaty', *Washington File*, International Information Programs, US Department of State, Washington, DC, 30 July 2001, URL <http://usinfo.state.gov/>; and Bailey, K. C., 'The biological and toxin threat to the United States', Paper published by the National Institute for Public Policy, Fairfax, Va., Oct. 2001, p. 10, URL <http://www.nipp.org>.

facilities whose number and location change on an irregular but frequent basis. The on-site inspection of a subset of such sites (arbitrarily selected for declaration by each party) would require an international organization far larger than the OPCW, and no AHG participant was willing to contemplate so extensive an organization. In addition, most such facilities are located in the countries that are the least likely to conduct activities which are prohibited under the BTWC.

4. If the declarations under the protocol with respect to the BW defence programmes were sufficiently comprehensive, they would risk compromising legitimate and sensitive national security information to an extent that would be unacceptable to the USA. If, in contrast, exclusions in such declarations were allowed, a potential proliferator might be able to conceal significant aspects of its BW programme in legitimately undeclared facilities. The USA viewed the dilemma as irreconcilable.

5. The USA was not prepared to have the other tools it uses to deal with BW proliferation (e.g., export controls and non- and counter-proliferation policies) degraded by the protocol. In particular, it objected to the proposed measures to implement Article X of the BTWC (technology transfers and international cooperation for peaceful purposes) and the concomitant demands by certain countries for so-called parallel export control arrangements such as the Australia Group to be abolished.[39]

6. Some parts of the protocol might limit the scope of the prohibitions or fix the meaning of terms in the BTWC.

7. The US Senate would be unlikely to ratify the protocol.

Several other countries shared some of the US concerns. However, in contrast to the USA they were still prepared to negotiate further on the basis of the draft documents in order to achieve a satisfactory result.

The Fifth Review Conference of the Parties to the BTWC

The Fifth Review Conference met in Geneva on 19 November–7 December 2001. Formally, the draft protocol was not on the agenda: the review conferences evaluate the functioning of the BTWC and confirm or expand the obligations of states in the light of international developments and technological advances made in the fields of the biological sciences and biotechnology. However, the failure of the AHG to achieve an agreed document prior to this conference placed the burden of conceptualizing the future regime for the prohibition of BW and devising the mechanisms to strengthen it on the review conference. Furthermore, it was inevitable that the issues that had complicated the AHG negotiation would also affect the review of the convention (particularly regarding Article III on non-proliferation and Article X on international

---

[39] The AG is discussed in Anthony, I., 'Multilateral weapon and technology export controls', *SIPRI Yearbook 2001: Armaments, Disarmament and International Security* (Oxford University Press: Oxford, 2001), p. 619. Australia Group documents are accessible on the Internet at URL <http://www.australia group.net>.

cooperation and technology transfers).⁴⁰ In addition, there was a general expectation that the USA would submit its alternative proposals at the forum and that most parties to the BTWC would confirm the continued validity of the AHG mandate.

The US alternative proposals

In his plenary address on the opening day of the review conference, Under Secretary of State for Arms Control and International Security Affairs John Bolton reiterated the US opposition to the composite text and formally presented alternative proposals for mechanisms to implement specific articles of the BTWC. With respect to Article IV (national implementation) the USA suggested that parties enact strict national criminal legislation against prohibited BW activities with strong extradition requirements; establish sound national oversight mechanisms for the security and genetic engineering of pathogenic organisms; promote responsible conduct in the study, use, modification and shipment of pathogenic organisms; and devise a solid framework for bioscientists in the form of a code of ethical conduct that would have universal recognition. Regarding Article V (consultation and cooperation) the USA proposed that an effective UN procedure be established to investigate suspicious outbreaks or allegations of BW use. States parties would be required to accept international inspectors following the determination by the UN Secretary-General that an inspection should take place. The United States also supported the setting up of a voluntary cooperative mechanism for clarifying and resolving compliance concerns by mutual consent. The procedures would include exchanges of information, voluntary visits and other measures to clarify and resolve doubts about compliance. With regard to Article VII (assistance to victims) and Article X (technical and scientific cooperation) the USA proposed the adoption of strict biosafety procedures, improved international disease surveillance and enhanced mechanisms for sending expert response teams to cope with outbreaks.⁴¹ Bolton requested that these proposals be endorsed in the Final Declaration of the review conference.

US non-compliance allegations

To the surprise of many countries, the USA also formally accused Iran, Iraq, Libya, North Korea, Sudan and Syria of maintaining offensive BW programmes and expressed its grave concern about possible BW use by the

---

⁴⁰ One of the alternatives mentioned by the USA in July was expansion of the Australia Group in terms of both participation and the scope of equipment and material covered in the export control lists. Press conference by US Ambassador Donald A. Mahley (note 38).

⁴¹ 'Strengthening the international regime against biological weapons', Statement by the President, Office of the Press Secretary, Washington, DC, 1 Nov. 2001, URL <http://www.whitehouse.gov/news/releases/2001/11/20011101.html>; and Statement of John R. Bolton, Under Secretary of State for Arms Control and International Security, US Department of State, to the Fifth Review Conference of the Biological Weapons Convention, United States Mission, Office of Public Affairs, Geneva, 19 Nov. 2001.

Afghanistan-based al-Qaeda terrorist organization.[42] Iran, Iraq and Libya (all of which are parties to the BTWC) rejected the accusations.[43] Iran condemned the United States for duplicity with respect to arms control and disarmament and referred to US support for Israel (which is not a party to the BTWC, CWC or the 1968 Non-Proliferation Treaty, NPT), the secret US BW defence programmes and the reservations to the CWC formulated by the US Congress with respect to challenge inspections on US territory.[44] Iraq and Libya also referred to the US support for Israel; Iraq added that it had terminated its BW programme in 1991 and that the items which the international inspectors considered to be related to the programme had been destroyed.[45] The US statement was also noteworthy for the countries it did not name.[46]

The review process

Following two days of general debate, the parties reviewed the operation of the convention article by article in the Committee of the Whole. Proposals for clarification and expansion of the articles were suggested in response to the terrorist attacks of 11 September and the subsequent delivery of anthrax spores by post. Among the topics discussed were: assistance following incidents involving the use of BW or in the case of a natural outbreak of disease, the establishment of an oversight mechanism on genetic manipulation and a code of conduct. Other suggestions were made for regime-building activities between the quinquennial review conferences. These included proposals to hold annual meetings and create a scientific advisory panel to advise on new biotechnology developments. Chairman of the Fifth Review Conference

---

[42] Statement of John R. Bolton (note 41). In the weeks preceding the review conference Ambassador Donald Mahley and Assistant Secretary for Arms Control Avis Bohlen visited the capitals of the other members of the WEOG to seek support for the US proposals. Rissanen, J., 'Preparations for the review conference, US lobbies its proposals', *BWC Protocol Bulletin*, 1 Nov. 2001, distributed by the Acronym Institute, URL <http://www.acronym.org.uk/bwc/index.htm>. In the week immediately preceding the review conference the USA informed the WEOG of its intention to name the proliferators. The other WEOG members strongly opposed the move but were unable to dissuade the USA and only succeeded in having the relevant sections shortened. EU delegate, Geneva, Private communication with J. P. Zanders, 22 Nov. 2001.

[43] North Korea is also a party to the BTWC (but did not participate in the review conference) and Syria is a signatory state. Sudan has not signed the convention. Complete lists of parties, signatory and non-signatory states are available at the SIPRI CBW Project Internet site at URL <http://projects.sipri.se/cbw/docs/bw-btwc-1>. See also annex A in this volume.

[44] Reply of the Islamic Republic of Iran to the US statement, Document distributed at the Fifth Review Conference, Geneva, 19 Nov. 2001. The US BW defence programme is discussed below.

[45] Fifth Review Conference for SIPRI, Geneva, 19 Nov. 2001. Many unresolved issues remain with respect to Iraq's BW programme. See Wahlberg, M., Leitenberg, M. and Zanders, J. P., 'The future of chemical and biological weapon disarmament in Iraq: from UNSCOM to UNMOVIC', *SIPRI Yearbook 2000: Armaments, Disarmament and International Security* (Oxford University Press: Oxford, 2000), pp. 560–75.

[46] See section VI of this chapter. The identification of suspected violators is not unprecedented at the BTWC review conferences. E.g., Iraq and Russia were named by several countries at both the Third and the Fourth review conferences. However, as one observer noted, in 1996 there were mechanisms to address compliance concerns (the trilateral process and the United Nations Special Commission on Iraq, UNSCOM), whereas it does not appear that the USA has used or intends to use the procedures in the BTWC and those agreed at previous review conferences to resolve the current concerns. Pearson, G. S., 'The Biological and Toxin Weapons Convention', *Chemical and Biological Weapons Conventions Bulletin*, no. 54 (Dec. 2001), p. 25.

Ambassador Tóth remarked that these ideas came from many participating states, not just the United States.[47] However, the USA preferred voluntary measures to be adopted by individual states, whereas other states (most notably the NAM) wanted such measures to be adopted as part of a legally binding instrument to be negotiated in the AHG.[48] The EU, in particular, was reported to be trying to bridge the gap between the USA and other parties.[49]

In the final week, following the review by the Committee of the Whole, the delegates discussed compromise language for the final declaration. However, several issues (notably, export control arrangements, such as the Australia Group, and technology transfers) that had proved impossible to resolve in the AHG sessions were raised again. No country rejected the US proposals out of hand, but views on how to implement them varied greatly, even between the USA and the EU.[50] The United States insisted on having the UN Security Council (where it has veto power) rather than another international body determine the need for an investigation in the event of a suspected outbreak of disease. It also objected to including facilities in such investigations, although this may be critical, as was proved by the 1979 anthrax outbreak in Sverdlovsk (now Yekaterinburg) following its accidental release from an illegal Soviet military research facility. The issues of past compliance (and whether or not to include the countries named by the USA as violators of the BTWC) and the status of the AHG mandate also remained to be resolved.[51]

On 7 December, the final day of the review conference, the delegates reached agreement on most of the text in the final declaration that addressed articles I–XI and XIII–XV of the BTWC. However, two hours before the scheduled end of the review conference the USA unexpectedly tabled new language for Article XII which proposed terminating the AHG mandate in exchange for US acceptance of annual meetings to review the progress in implementing the new measures adopted at the Fifth Review Conference and to consider measures or mechanisms to effectively strengthen the BTWC. An expert group could meet to examine matters identified by the parties at the annual meetings, but it would not have a negotiation mandate.[52] The US submission was received with shock and anger. EU representatives, who had not been informed in advance, were particularly disturbed by the US action. During a brief recess, in which the regional groups considered their responses,

---

[47] Highlights of press conference held on developments in the Fifth Review Conference of States Parties to the Biological Weapons Convention, United Nations, Geneva, 27 Nov. 2001, URL <http://www.unog.ch/news/documents/newsen/pc011127.html>.

[48] Rissanen, J., 'Differences and difficulties as delegates consider wide range of proposals', *BWC Protocol Bulletin*, 30 Nov. 2001, URL <http://www.acronym.org.uk/bwc/index.htm>.

[49] Highlights of press conference held on developments in the Fifth Review Conference of States Parties to the Biological Weapons Convention, United Nations, Geneva, 30 Nov. 2001, URL <http://www.unog.ch/news/documents/newsen/pc011130.html>.

[50] The USA does not appear to have elaborated its proposals in any document or statement prior to these discussions. This led to different interpretations by the other delegations. The proposals by the various parties for the language of the final declaration are contained in Fifth Review Conference, Committee of the Whole document BWC/CONF.V/COW/1, 13 Dec. 2001, annex.

[51] Rissanen, J., 'Endgame in earnest: first draft of final declaration issued on penultimate day', *BWC Protocol Bulletin*, 6 Dec. 2001, URL <http://www.acronym.org.uk/bwc/index.htm>.

[52] The full text of the US proposal is reproduced in Pearson (note 46), p. 20.

the EU representatives refused to participate in a WEOG meeting with the USA and instead met as an EU group. As a consequence of the actions by the USA and some other participating state parties a final declaration was not adopted, and the review conference was adjourned until 11–22 November 2002.[53]

III. Biotechnology, biological defence research and the BTWC

Biotechnology and genetic engineering have the potential to improve the quality of life. However, such knowledge can also be used for hostile purposes, to increase the stability and virulence of existing warfare agents or to create new agents based on the components of an organism.[54] This inherent duality is reinforced by the growing possibility of the chance discovery in non-military biotechnological research of a new pathogen, or a new expression of a pathogen, with characteristics that could make it attractive for military use. Such discoveries confront scientists with the dilemma of whether or not to publish results or pursue a line of research. Publication makes the results available to governments with hostile intentions or terrorists, but it also allows the scientific community to devise countermeasures (e.g., pharmaceuticals or detectors) and policy makers to reinforce the norms against the misapplication of biology (e.g., the BTWC).[55]

The number of institutes, government agencies and countries engaging in BW defence activities is increasing rapidly. The focus is on traditional and potential biological warfare agents, but research is also being conducted on the evolution of microbes and genetic modification in order to increase pathogenicity, stability or resistance to various types of medical treatment.[56] Such modifications were carried out in the offensive Soviet BW programme. Genome studies into traditional warfare agents offer opportunities for new prophylaxis and pretreatments and might contribute to the development of generic biological warfare agent detectors, but the results could also lead to the development of enhanced warfare agents.[57] Other research involves the behaviour of

---

[53] Rissanen, J., 'Anger after the ambush: review conference suspended after US asks for AHG's termination', *BWC Protocol Bulletin*, 9 Dec. 2001, URL <http://www.acronym.org.uk/bwc/index.htm>.

[54] These issues are discussed in greater detail in Zanders, Hart and Kuhlau (note 6).

[55] E.g., in 1994 a report was published on research with a strain of *E. coli* that had been rendered 32 000 times less sensitive to a certain antibiotic using an engineering technique of DNA shuffling. In the technique, copies of a particular bacterial gene are first broken up in fragments and then reassembled so that the fragments are ordered in subtly different sequences. After reintroduction of these genes into the bacteria, the specimens exhibiting the desired traits are selected for further development. Following a request by the American Society of Microbiology, which expressed its concern about potential misuse, the researcher destroyed the strain. Dennis, C., 'The bugs of war', *Nature*, vol. 411 (17 May 2001), p. 234.

[56] The term 'microbe' encompasses all microbial agents, such as viruses, bacteria, fungi, protozoa and microalgae. US National Science and Technology Council, *The Microbe Project* (Executive Office of the President, Office of Science and Technology Policy: Washington, DC, Jan. 2001), p. 2.

[57] By Oct. 2001 all the genes of the plague bacterium (*Yersinia pestis*) had been mapped. In Nov. 2001 scientists reported that they had decoded the genome of the anthrax bacterium (*Bacillus anthracis*). Reuters Medical News, '*Yersinia pestis* genome sequenced', 3 Oct. 2001, URL <http://id.medscape.com/reuters/prof/2001/10/10.04/20011003scie004.html>; and Broad, W. J., 'Genome offers "fingerprint" for

agents in the environment in order to test experimental detectors, but these activities could also contribute to the development of enhanced dissemination devices. As was demonstrated by revelations in 2001, such work may reach the limits of what is interpreted as being permissible under the BTWC. In the future the calls for secrecy in order to prevent proliferation will increasingly conflict with the need for scientific freedom and increased transparency with regard to BW defence programmes to maintain confidence in compliance with the BTWC.

The US BW defence research programme

The USA has the world's largest BW defence programme. In the light of the growing threat perception regarding BW proliferation and terrorism involving biological agents, the USA drastically expanded its efforts to prevent and counter the effects of BW use in the second half of the 1990s.[58] It was argued that the USA should adopt a more proactive policy and that research should be conducted on biological and toxin agents in order to understand what is possible even if this entailed criticism that the USA was violating the BTWC.[59] There was also criticism that the existing programmes focused on agents, not on the technology and functioning of delivery systems.[60]

In September 2001 *New York Times* journalists Judith Miller, Stephen Engelberg and William J. Broad revealed three secret BW threat assessment activities: Project Clear Vision, Project Bacchus and Project Jefferson. All arguably test the limits of the BTWC as they are closely related to activities that might be undertaken as part of an offensive BW development programme. Other projects exist, but little is known about them.[61]

In 1997 the Central Intelligence Agency (CIA) started Project Clear Vision to analyse delivery systems for biological warfare agents in countries of proliferation concern. A copy of a Soviet-designed biological bomblet that disperses the agent in aerosol form was built and its performance was assessed under various atmospheric conditions in two sets of tests at the Battelle Memorial

anthrax', *New York Times* (Internet edn), 28 Nov. 2001, URL <http://www.nytimes.com/2001/11/28/health/28GENE.html>.

[58] Zanders, J. P., French, E. M. and Pauwels, N., 'Chemical and biological weapon developments and arms control', *SIPRI Yearbook 1999: Armaments, Disarmament and International Security* (Oxford University Press: Oxford, 1999), pp. 594–95. In 2001 US government spending on preparedness activities against terrorist attacks with chemical, biological, radiological or nuclear materials was $1.7 billion per year, an increase of 310% since fiscal year (FY) 1998. For FY 2000 funding specifically related to terrorism with biological agents was estimated to be in the range of $35–40 million. As a result of the $40 billion emergency supplemental appropriation after the 11 Sep. terrorist attacks these figures are expected to rise further in FY 2002. Committee on Science, US House of Representatives, *The Science of Bioterrorism: Is the Federal Government Prepared?*, Hearing held on 5 Dec. 2001, URL <http://www.house.gov/science/full/dec05/full_charter_120501.htm>.

[59] Bailey (note 38).

[60] Miller, J., Engelberg, S. and Broad, W., *Germs: Biological Weapons and America's Secret War* (Simon & Schuster: New York, 2001), pp. 287–89.

[61] Project Bite Size, Back Star and Druid-Tempest are mentioned in the context of secret US Department of Defense programmes to develop defences against BW and other unconventional weapons. Miller, Engelberg and Broad (note 60), pp. 296–97.

Institute, a military contractor in Columbus, Ohio. The tests were completed in mid-2000 and were deemed successful.

In addition to investigating the delivery systems of the former Soviet Union and other BW proliferators, it was argued that the Clear Vision Project ought to investigate the military implications of gene splicing, a genetic engineering technique that could be used to increase the lethality of microbes. The work would have involved the creation of new strains of pathogens, but it was reportedly halted before genes were inserted into a pathogen.[62]

Project Clear Vision ended in early 2001. Following briefings of White House officials (President Bill Clinton was reportedly never informed) and congressional intelligence committees, the programme became increasingly controversial and the CIA did not seek new appropriations to fund it. In addition, the project did not satisfy one of the CIA's conditions for a particular research item: namely, the availability of credible intelligence that an adversary country was developing or deploying a particular BW.[63]

In Project Bacchus experts from the Defense Threat Reduction Agency (DTRA), a Department of Defense (DOD) agency, assembled a production plant that would be capable of producing biological warfare agents at a former nuclear test site in the Nevada desert using only commercially available materials that were procured in the USA and Europe.[64] Anthrax simulants were used in the production runs. It was hoped that distinctive patterns (signatures) of purchase of equipment would emerge, but none was detected. Project Bacchus received $1.6 million in funding. According to the participants, the project demonstrated the ease with which a state or a terrorist organization could acquire significant amounts of a biological warfare agent. The DTRA followed Project Bacchus with Operation Divine Junker, which simulated an attack by military commandos on a plant in order to neutralize it. The simulation was intended to assess whether the fermentor and milling machine could be disabled without disseminating any of the agent. It was also deemed a success.[65]

The Bush Administration reportedly plans to expand the BW defence projects because of the growing BW threat.[66] The DOD Defense Intelligence Agency (DIA) has taken over the CIA project to genetically engineer a more potent strain of the anthrax bacterium, similar to one first created by Soviet

---

[62] Miller, Engelberg and Broad (note 60), p. 296.
[63] Miller, Engelberg and Broad (note 60), p. 296.
[64] Miller, J., Engelberg, S. and Broad, W. J., 'US germ warfare research pushes treaty limits', *New York Times* (Internet edn), 4 Sep. 2001, URL <http://www.nytimes.com>. 'Bacchus' is reportedly an acronym for Biotechnology Activity Characterization by Unconventional Signatures.
[65] Miller, Engelberg and Broad (note 60), pp. 297–98.
[66] Miller, Engelberg and Broad (note 64). According to the Director of the National Institute of Allergy and Infectious Diseases at the National Institutes of Health, the USA is primarily concerned with the 6 so-called 'Category A' biological agents: anthrax, botulinum toxin, haemorrhagic fevers (e.g., Ebola), plague, smallpox and tularaemia. The criteria for focusing on these agents are ease of dispersal, impact versus mortality rate and the availability of an adequate therapy. Anthony Fauci, Director of the National Institute of Allergy and Infectious Diseases at the National Institutes of Health, as quoted in Nartker, M., 'US response: bioterrorism differs from biowarfare, official says', *Global Security Newswire*, 17 Jan. 2002, URL <http://www.nti.org/d_newswire/issues/newswires/2002_1_17.html#3>.

scientists.[67] Its purpose is to assess the effectiveness of the anthrax vaccine currently being administered to US military personnel against the modified strain. The DIA included the anthrax programme in its Project Jefferson, a government effort to assess the BW threat. In October 2001 the DOD approved the project, and the Battelle Memorial Institute will probably be chosen to engineer the anthrax strain and develop the new vaccine.[68]

In December 2001 another classified aspect of the US biological defence programme came to light during the search for the domestic source of the anthrax spores used in the letters. In the early 1990s US Army scientists at the Life Sciences Division of Dugway Proving Ground in Utah had made small quantities of dried anthrax. It was milled into respirable particles and aerosolized in order to test decontamination techniques and biological agent detection systems. The anthrax spores were milled to a concentration of the range of 1 trillion spores per gram, which reportedly exceeds that of the anthrax produced in the US and Soviet BW programmes. Production batches reportedly rarely accumulate more than 10 grams at any given time.[69] The project was launched in the early 1990s after Iraq's BW threat in the 1991 Persian Gulf War. The dried anthrax batches may have been the first produced since President Richard M. Nixon renounced BW in 1969.

The BTWC and BW defence

The BTWC is governed by the general purpose criterion in Article I: all activities that may contribute to the acquisition or retention of any type of biological warfare agent, however created or manufactured, are prohibited. Exception is made for those activities that benefit prophylactic, protective or other peaceful purposes.[70] Even experimentation involving the open-air release of pathogens or toxins can be justified if it supports one or more of these non-prohibited purposes.[71] The BTWC does not specify any quantitative or qualitative limitations for the biological agents that are used in the non-prohibited activities. Consequently, there is a potential ambiguity, and judgement of compliance with the BTWC with respect to biological defence

[67] 'DoD news briefing Victoria Clarke, ASD PA', DefenseLINK, 4 Sep. 2001, URL <http://www.defenselink.mil/news/Sep2001/t09052001_t0904asd.html>.

[68] Miller, J., 'US agrees to clean up anthrax site in Uzbekistan', *New York Times* (Internet edn), 23 Oct. 2001, URL <http://www.nytimes.com>; Borger, J., 'Pentagon approves super strain', *The Guardian*, 24 Oct. 2001, URL <http://www.guardian.co.uk/Archive/Article/0,4273,4283710,00.html>; and 'News briefs', *Arms Control Today*, Nov. 2001, URL <http://www.armscontrol.org/act/2001_11/briefsnov01.asp>.

[69] Shane, S., 'Bioterror: organisms made at a military laboratory in Utah are genetically identical to those mailed to members of Congress', *Baltimore Sun* (Internet edn), 12 Dec. 2001, URL <http://www.baltimoresun.com>; Shane, S., 'Army confirms making anthrax in recent years', *Baltimore Sun* (Internet edn), 13 Dec. 2001, URL <http://www.baltimoresun.com>; and Harvey, M. and Zither, A., 'Army defends its anthrax-making program', *Los Angeles Times* (Internet edn), 13 Dec. 2001, URL <http://www.cmonitor.com/stories/front1101/1213army_2001.shtml>.

[70] Article I(1) of the BTWC.

[71] Final Document of the Fourth Review Conference of the Parties to the Convention on the Prohibition of the Development, Production and Stockpiling of Bacteriological (Biological) and Toxin Weapons and on their Destruction, document BWC/IV/9, Part II, Final Declaration, Article I, para. 7, p. 16.

activities depends largely on the judgement of intent (in which perceptions of the enemy inevitably play a significant role).

With the exception of rudimentary consultation mechanisms the BTWC contains no provisions to deal with compliance concerns. The parties have therefore adopted a set of CBMs through the process of the review conferences. States are requested, among other things, to submit information on relevant national defence R&D programmes and on research centres and laboratories that specialize in permitted biological activities of direct relevance to the BTWC. The CBMs are only politically binding, and the extent and quality of the annual responses have generally been poor.[72]

Despite the misgivings of some individuals, US officials have consistently argued that the BW threat evaluation projects fall within the scope of BW defence activities permitted under the BTWC. In May 2000 a microbiologist at the US Army Medical Research Institute of Infectious Diseases (USAMRIID) was infected with glanders. Research on its causative agent, *Burkholderia mallei*, was justified on the grounds that there is suspicion that attempts are being made to develop an aerosolized and antibiotic-resistant form of the pathogen. No case of human glanders had been recorded in the USA since 1945.[73] Similarly, Project Clear Vision was defended by CIA officials who insisted that the research activities were within the scope of activities for protective purposes that are permitted by the BTWC.[74] The bomblet developed by the project lacked a fuse and other parts required to make it operational, and US officials argued that there was no intent to develop it into a complete weapon. The project was further justified by the claim that such items were being sold on the international market.[75] CIA lawyers were reportedly also convinced that the BTWC permitted the creation of new microbe strains in order to assess their military implications.[76] In the case of Project Bacchus, CIA legal reviews concluded that the construction of a plant was permitted under the BTWC because of its defensive nature and the fact that no actual biological warfare agent was intended to be produced. The Bush Administration considers projects such as the one designed to create a genetically modified strain of anthrax in Project Jefferson as fully consistent with the BTWC. Administration representatives have argued that the convention allows such research on both pathogens and delivery systems for protective or defence pur-

---

[72] Sims, N. A., 'The regime of compliance: the addition of confidence-building measures', *The Evolution of Biological Disarmament*, SIPRI Chemical & Biological Warfare Studies no. 19 (Oxford University Press: Oxford, 2001), pp. 61–81.

[73] Srinivasan, A. et al., 'Glanders in a military research microbiologist', *New England Journal of Medicine*, vol. 345, no. 4 (26 July 2001), p. 256; Khan, A. S. and Ashford, D. A., 'Ready or not—preparedness for bioterrorism', *New England Journal of Medicine*, vol. 345, no. 4 (26 July 2001), pp. 287–89; and DeShazer, D. et al., 'Laboratory-acquired human glanders', *Case Studies*, Center for the Study of Bioterrorism & Emerging Infections, Saint Louis University School of Public Health, May 2000, URL <http://bioterrorism.slu.edu/case_studies/laboratory>. Glanders was used by German agents to infect livestock and draught animals in the USA during World War I.

[74] This does not imply that the legal reviews produced consensus advice. Miller, J., 'When is a bomb not a bomb? Germ experts confront the US', *New York Times* (Internet edn), 5 Sep. 2001, URL <http://www.nytimes.com>.

[75] Miller, Engelberg and Broad (note 64).

[76] Miller, Engelberg and Broad (note 60), p. 296.

poses.⁷⁷ Similar arguments were used to justify the generation of dried anthrax aerosols at the Dugway Proving Ground.

In contrast to many other parties to the BTWC, the USA has consistently made detailed declarations of its BW defence activities (including activities conducted by the Battelle Memorial Institute facility in West Jefferson, Ohio).⁷⁸ However, the various programmes that were revealed in the second half of 2001 had never been declared, nor was the site for tests in the Nevada desert.⁷⁹ The surprise and concern that have been expressed about these programmes are motivated less by the fact that the USA conducts a wide variety of BW defence projects in order to deal with its perceived security threats than by the intent that motivates and justifies the secrecy and non-disclosure. As two BW disarmament experts wrote, CBMs are 'an assortment of activities that states engage in with the primary aim to become more sure that each understands the actions and/or intentions of the others'.⁸⁰ These doubts about intent have led to serious questions about the permissibility of these activities under the BTWC among members of the arms control community and governments.⁸¹

The BTWC is unclear about when a particular activity should be considered defensive or offensive, and 'purpose' is determined by the judgement of intent. Nevertheless, on the basis of analyses of past programmes and proliferation allegations, certain activities have become widely accepted as potential indicators of an offensive programme: certain kinds of vaccine research (especially if the disease is not indigenous), large-scale vaccinations of troops against certain agents, the creation of non-naturally occurring disease strains (especially those with heightened pathogenicity), the development of agent delivery systems, agent production installations, open-air release of pathogens, the presence of an explosive chamber inside a research establishment, and so on.

US officials have justified the projects on the grounds that the activities, installations and equipment are part of a defensive programme and have argued that secrecy is necessary in order not to provide potential adversaries

---

⁷⁷ 'DoD news briefing Victoria Clarke, ASD PA' (note 67)

⁷⁸ There is no official systematic presentation of the data submitted under the CBMs. According to 1 analysis by independent researchers based on 1998 data, the USA declared 18 facilities and 1063 (full- and part-time) personnel participating in its BW defence programmes. Annual financing amounted to c. $88.3 million. Chevrier, M. I. and Hunger, I., 'Confidence-building measures for the BTWC: performance and potential', *Nonproliferation Review*, vol. 7, no. 3 (fall/winter 2000), p. 33.

⁷⁹ 'Verification watch', *Trust & Verify*, Sep./Oct. 2001, URL <http://www.vertic.org/tnv/septoct01/watch.html>; and Private communication with J. P. Zanders, 6 Sep. 2001. Because the decision to go forward with the anthrax research under Project Jefferson was not taken until Oct. 2001, the programme would not yet have been part of the CBM submissions.

⁸⁰ Chevrier and Hunger (note 78), p. 25.

⁸¹ Miller, Engelberg and Broad (note 64); 'USA abetting proliferation of arms of mass destruction', *El Pais* (Madrid), 6 Sep. 2001, in Foreign Broadcast Information Service, *Daily Report–West Europe (FBIS-WEU)*, FBIS-WEU-2001-0906, 7 Sep. 2001; 'DPRK radio urges US to drop research on biological weapons', Central Broadcasting Station (Pyongyang), 14 Sep. 2001, Foreign Broadcast Information Service, *Daily Report–East Asia (FBIS-EAS)*, FBIS-EAS-2001-0914, 17 Sep. 2001; and Kempf, H., 'Les Américains ont réactivé leur recherche sur les armes biologiques' [The Americans have reactivated their research into biological weapons], *Le Monde* (Internet edn), 11 Dec. 2001, URL <http://www.lemonde.fr>.

with information about weaknesses in US BW defence. In so doing, the USA has argued that a wide range of activities which could contribute directly to an offensive programme falls outside the core prohibitions of the BTWC. The US biological defence activities also risk undermining the non-proliferation norms which the USA seeks to establish by enabling countries of proliferation concern to plausibly deny that certain suspicious activities they may be undertaking are connected to an offensive BW programme. Several such countries denied their interest in BW in the wake of the disclosures about the US projects.[82]

IV. Chemical weapon disarmament

The Chemical Weapons Convention entered into force on 29 April 1997. As of 1 January 2002, 145 states had ratified or acceded to the convention and a further 29 states had signed it.[83] Eighteen members of the United Nations have neither signed nor ratified the CWC.[84]

Implementing the CWC

As of 31 December 2001 the OPCW had conducted 1117 inspections from the date of entry into force of the CWC and overseen the destruction of 6374 tonnes (of a declared total of 69 869 tonnes) of chemical agent and 2 098 013 munitions (of a declared total of 8 624 493 munitions).[85]

In 2001 the OPCW faced a deficit of approximately 10 per cent of the 2001 budget. The Sixth Conference of the States Parties (CSP) to the CWC, which met on 14–19 May 2001, was therefore mainly preoccupied with seeking to clarify the nature of the budgetary deficit, finding ways to address its consequences and preventing future shortfalls. As a consequence of the budgetary deficit, the OPCW was only able to conduct 197 of 293 planned inspections for 2001.[86] In addition, some provisions of the OPCW's financial regulations were suspended, and technical assistance and cooperation activities were post-

---

[82] Reply of the Islamic Republic of Iran to the US statement (note 44); and 'Minister: Russia not developing biological weapons', Interfax (Moscow), 6 Sep. 2001, in Foreign Broadcast Information Service, *Daily Report–Central Eurasia (FBIS-SOV)*, FBIS-SOV-2001-0906, 7 Sep. 2001.

[83] Dominica, Nauru, Uganda and Zambia became parties in 2001.

[84] They are Andorra, Angola, Antigua and Barbuda, Barbados, Belize, Egypt, Iraq, Korea (North), Lebanon, Libya, Palau, Sao Tome and Principe, Solomon Islands, Somalia, Syria, Tonga, Tuvalu and Vanuatu.

[85] In 2001 there were 2 inspections of abandoned CW sites (40%), 62 of CW destruction facilities (98%), 26 of CW production facilities (57%), 28 of CW storage facilities (70%), 37 of old CW sites (43%), 19 of Schedule 1 facilities (100%, plus 1 additional inspection), 28 of Schedule 2 plant sites (70%), 12 of Schedule 3 plant sites (29%) and 17 of DOC/PSF plant sites (53%). Percentages indicate the number of planned inspections actually completed. OPCW official, Private communication with J. Hart, Jan. 2002.

[86] By 31 Dec. 2001, 98% of CWDFs, 57% of CWPFs, 70% of CWSFs, 43% of OCW sites, 40% of ACW sites, 100% of Schedule 1 facilities, 70% of Schedule 2 plant sites, 29% of Schedule 3 plant sites and 53% of DOC/PSF plant sites scheduled to be inspected during 2001 had actually received inspections. OPCW official, Private communication with J. Hart, Jan. 2002.

poned, curtailed or cancelled.[87] Director-General José Bustani warned that the OPCW was not in a position to adequately assist victims of CW use.[88] He also warned that, without additional funding, the Technical Secretariat (TS) of the OPCW would be forced to further reduce the number of inspections and that it would only be able to carry out some 60 per cent of its planned verification and international cooperation activities in 2002.[89]

The problem was largely caused by considerable delays between the inspections of CW (Article IV of the CWC) and CW production facilities (Article V) and the reimbursement by parties for direct costs related to these inspections.[90] Other factors were the late annual contributions to the OPCW by some parties, internal budgetary procedures, and the delays in CW destruction facilities becoming operational, so that there were fewer 'reimbursable inspections' than projected in the budget calculations.[91]

In order to remedy the situation, the Sixth CSP decided to apply the OPCW's 1999 cash surplus, totalling €2 709 614 (c. $2.4 million), towards the deficit incurred in 2000.[92] It also authorized the application of accrued interest in special accounts to offset the 2001 cash deficit.[93] In addition, a voluntary fund was established. The parties have provided some cost-free services, such as preparation of samples for the annual analytical laboratory proficiency tests. In December 2001 the TS nevertheless projected a €6 million deficit (c. $5.3 million) if no steps are taken to address the underlying reasons for the financial shortfall.[94] The budgetary discussions are complicated by a number of broader, somewhat more philosophical questions with political elements such as whether the size of the TS needs to be expanded to meet its objectives

---

[87] Opening statement by the Director-General to the Executive Council at its Twenty-Seventh Session, OPCW document EC-XXVII/DG.10, 4 Dec. 2001, para. 66; and 'The OPCW: twenty-seventh session of the Executive Council, OPCW establishes anti-terrorism working group', OPCW Press Release no. 31/2001, 13 Dec. 2001.

[88] Simons, M., 'Money short for battle on chemicals used in war', *New York Times* (Internet edn), 5 Oct. 2001, URL <http://www.nytimes.com>.

[89] 'The OPCW: twenty-seventh session of the Executive Council, OPCW establishes anti-terrorism working group' (note 87); Opening statement by the Director-General to the Executive Council at its Twenty-Seventh Session (note 87), para. 81; and 'Executive Council concludes its twenty-seventh session', *Secretariat Brief*, no. 30 (18 Dec. 2001), p. 2.

[90] At least €1.3 million ($1.25 million) of the estimated cost of reimbursement under Articles IV and V were not actually incurred in 2001 because the inspection costs were not generated. The costs were, however, incorporated into the 2001 budget when it was prepared in 2000. 'Organisation for the Prohibition of Chemical Weapons', *Secretariat Brief*, no. 28 (20 July 2001). Over the next 2–3 years the problem may worsen as the number of inspectable facilities, mainly CW destruction facilities, will increase.

[91] Bustani, J., Opening statement by the Director-General to the Conference of the States Parties at its sixth session, The Hague, Netherlands, 14 May 2001, para. 44; and Statement by the Director-General at its twenty-sixth session, OPCW document EC-XXVI/DG.11, 25 Sep. 2001, para. 76.

[92] Decision: withholding the distribution of the cash surplus for 1999, OPCW document C-VI/DEC.18, 19 May 2001. The OPCW financial regulations do not allow the organization to spend more money than is allocated in its annual programme and budget. Any surpluses are either to be refunded to parties or applied to parties' contributions for the following year. This decision was made on an exceptional, one-off basis.

[93] Decision: authorisation to use accrued interest in special accounts to offset the 2001 cash deficit, OPCW document C-VI/DEC.19, 19 May 2001.

[94] 'Executive Council concludes its twenty-seventh session', *Secretariat Brief*, no. 30, 18 Dec. 2001, p. 2.

or whether it is large enough already and the need to reallocate or better manage existing funds.

The Sixth CSP also approved an authentication and certification procedure for the central OPCW analytical database and on-site databases,[95] a relationship agreement between the UN and the OPCW,[96] privileges and immunities agreements with five parties,[97] a request by Russia to use a former chemical weapon production facility (CWPF) for non-prohibited purposes,[98] an obligation for parties to require end-user certificates from non-CWC parties receiving transfers of Schedule 3 chemicals,[99] a decision to allow the Executive Council (EC) to consider and conclude negotiated texts of cooperation agreements between the OPCW and other international organizations as may be required for the effective implementation of the CWC,[100] and a scale of assessment for 2002 in which the US contribution was reduced from 25 to 22 per cent.[101]

---

[95] Decision: authentication and certification procedure for the central OPCW analytical database and on-site databases, OPCW document C-VI/DEC.4, 17 May 2001.

[96] Decision: relationship agreement between the United Nations and the OPCW, OPCW document C-VI/DEC.5, 17 May 2001.

[97] Decision: agreement between the Republic of Austria and the Organisation for the Prohibition of Chemical Weapons on the privileges and immunities of the OPCW, OPCW document C-VI/DEC.12, 17 May 2001; Decision: agreement between the Government of the Republic of Belarus and the Organisation for the Prohibition of Chemical Weapons on the privileges and immunities of the OPCW, OPCW document C-VI/DEC.13, 17 May 2001; Decision: agreement between the Organisation for the Prohibition of Chemical Weapons and the Government of the Republic of Panama on the privileges and immunities of the OPCW, OPCW document C-VI/DEC.7, 17 May 2001; Decision: agreement between the Republic of the Philippines and the Organisation for the Prohibition of Chemical Weapons on the privileges and immunities of the OPCW, OPCW document C-VI/DEC.14, 17 May 2001; and Decision: agreement between the Organisation for the Prohibition of Chemical Weapons and the Portuguese Republic on the privileges and immunities of the OPCW, OPCW document C-VI/DEC.6, 17 May 2001.

[98] Decision: request by the Russian Federation to use the chemical weapons production facility, (filling of hydrocyanic acid into munitions) at OJSC 'Orgsteklo', Dzerzhinsk, for purposes not prohibited under the Convention, OPCW document C-VI/DEC.8, 17 May 2001.

[99] Decision: provisions on transfers of Schedule 3 chemicals to states not party to the Convention, OPCW document C-VI/DEC.10, 17 May 2001. The CWC categorizes chemical compounds of particular concern in schedules depending on their importance for the production of chemical warfare agents and for legitimate civilian manufacturing processes. Each list has different reporting requirements. Schedule 1 contains toxic chemicals that can be used as CW and that have few uses for permitted purposes. They are subject to the most stringent controls. Schedule 2 includes toxic chemicals and precursors to CW but which generally have greater commercial application. Schedule 3 chemicals can be used to produce CW but are also used in large quantities for non-prohibited purposes. The CWC also places reporting requirements on firms which produce certain discrete organic chemicals (DOC) that are not on any of the schedules and may contain phosphorus, sulphur or fluorine (DOC/PSFs). The CWC requires parties to adopt the necessary measures to ensure that Schedule 3 chemicals transferred to non-parties are not used for purposes prohibited by the CWC (para. 26, Part VIII of the Verification Annex). Products containing 30% or less of a Schedule 3 chemical or products identified as 'consumer goods packaged for retail sale for personal use, or packaged for individual use' are exempted from the end-user requirement. Decision: provisions on transfers of Schedule 3 chemicals to states not party to the Convention, OPCW document C-VI/DEC.10, 17 May 2001, para. 2.

[100] Decision: cooperation with international organisations, OPCW document C-VI/DEC.15, 17 May 2001.

[101] Decision: scale of assessments for 2002, OPCW document C-VI/DEC.20, 19 May 2001. The USA has also reduced its scale of assessment to the UN by the same percentage. The OPCW took this decision because the CWC requires that the OPCW's activities be paid for by members in accordance with the UN scale of assessment adjusted to take into account certain factors such as differences in membership between the two organizations. CWC, Article VIII, para. 7.

In response to the terrorist attacks of 11 September 2001, the EC of the OPCW established an anti-terrorism working group during its 27th session, held on 4–7 December.[102] The group will cooperate with the UN Security Council's Counter-Terrorism Committee, which was established on 28 September in accordance with UN Security Council Resolution 1373.[103]

A meeting of the OPCW working group that was mandated to prepare for the First Review Conference of the States Parties to the CWC was held on 29 November.[104] Other bodies, including the Scientific Advisory Board (SAB), are also preparing contributions for the conference.[105]

Destruction of chemical weapons and related facilities

The states that are declared possessors of CW are India, South Korea, the USA and Russia.[106] At the end of 2001, *India* had destroyed approximately 29 per cent of its Category 1 CW and over 39 per cent of its Category 2 CW.[107] It uses a neutralization-based destruction technology at a converted CW production facility. *South Korea* resumed its destruction operations after upgrading its CW destruction facility. It is expected to meet its Phase 2 (and possibly Phase 3) intermediate destruction deadlines for Category 1 CW.[108] The exact size and composition of the Indian and South Korean stockpiles are not publicly known.[109]

US CW destruction

At the end of 2001 approximately 25 per cent of the USA's 31 279.74-tonne CW stockpile as declared to the OPCW had been destroyed.[110] The US Army reportedly expects the cost of destroying the stockpile to reach $24 billion, a

---

[102] 'The OPCW: twenty-seventh session of the Executive Council, OPCW establishes anti-terrorism working group' (note 87).

[103] UN Security Council Resolution 1373, 28 Sep. 2001.

[104] Opening statement by the Director-General to the Executive Council at its Twenty-Seventh Session (note 87), paras 73–74.

[105] Mills, P., 'Progress in the Hague', *CBW Conventions Bulletin*, no. 51 (Mar. 2001), p. 17.

[106] Zanders, J. P., Hersh, M., Simon, J. and Wahlberg, M., 'Chemical and biological weapon developments and arms control', *SIPRI Yearbook 2001* (note 39), pp. 517–26.

[107] Opening statement by the Director-General to the Executive Council at its Twenty-Seventh Session (note 87), para. 14. Category 1 CW consists of Schedule 1 chemicals and their parts and components. Category 2 CW consists of all other (non-Schedule 1 chemicals) and their parts and components. CWC, Part IV(A), para. 16.

[108] Opening statement by the Director-General to the Executive Council at its Twenty-Seventh Session (note 87), para. 13; and CWC, Part IV(A), Verification Annex, para. 17 (a). The destruction of Category 1 CW is divided into 4 'phases' spread over 10 years starting from the CWC's entry into force on 29 Apr. 1997. CWC, Part IV(A), Verification Annex, para. 17(a).

[109] However, the size of India's and South Korea's combined CW stockpiles can be established by subtracting the US and Russian CW stockpiles from previous editions of the *OPCW Annual Report*, which contains aggregate stockpile amounts and compositions.

[110] US delegation to the CWC Preparatory Commission, Private communication with the authors, 6 Feb. 1996. Types and quantities of the CW stockpile are given in Zanders, J. P., Eckstein, S. and Hart, J., 'Chemical and biological weapon developments', *SIPRI Yearbook 1997: Armaments, Disarmament and International Security* (Oxford University Press: Oxford, 1997), p. 450.

rise of more than $9 billion over previous budget projections.[111] The reasons for the cost increase include delays in obtaining environmental permits, the rising cost of contractors and costs associated with containing and stabilizing leaking munitions.[112] In the wake of the 11 September terrorist attacks, the US Congress approved $40 million in supplementary funding to enhance the security of US CW stockpiles in addition to the approximately $120 million per year which the US Army was already spending on safeguarding CW stockpiles.[113] It was also announced that, partly as a result of the 11 September attack, the US Army would accelerate the schedule of its destruction operations at its Aberdeen facility and complete destruction operations there by the end of 2002, some three years ahead of schedule.[114]

The US Army Program Manager for Chemical Demilitarization (PMCD) is responsible for the destruction of CW. The PMCD consists of the Chemical Stockpile Disposal Program, the Alternative Technologies and Approaches Program (ATAP) and the Non-Stockpile Chemical Materiel Program (NSCMP).[115] A separate programme for Assembled Chemical Weapons Assessment (ACWA) was established in 1997 by Public Law 104-208. In 2001 it continued to review alternative, non-incineration-based technologies.[116] The NSCMP is responsible for disposing of former CW production facilities as well as recovered CW and CW *matériel*.[117] Non-stockpiled CW are reportedly located at over 100 locations, and the CW stockpile is stored at eight locations.[118] Incineration is the 'baseline' CW destruction technology, while alternative, neutralization-based destruction technologies will be used to

---

[111] Miller, A. C. and Levin, M., 'Disposal of chemical arms in US lags as costs mount', *Los Angeles Times*, 29 Sep. 2001, p. 1, URL <http://www.latimes.com>.

[112] 'Rocket containing chemical weapon found leaking', *MSNBC*, 3 May 2001, URL <http://www.msnbc.com/local/wvtm./m41407.asp?cp1=1>.

[113] Firestone, D., 'Army tightens security at nation's 8 chemical arms depots', *New York Times* (Internet edn), 2 Oct. 2001, URL <http://www.nytimes.com>; Westphal, J., US congressional testimony before the Senate Committee on Appropriations Subcommittee on Defense, 25 Apr. 2001, URL <http://www.senate.gov/~appropriations/defense/testimony/westfall42501.htm>; and *Baltimore Sun*, 11 Dec. 2001, quoted in *Chemical & Biological Arms Control Dispatch*, Dispatch no. 163, 1–15 Dec. 2001 (Chemical and Biological Arms Control Institute: Washington, DC, Dec. 2001).

[114] Brown, L. H., 'Army speeds APG plan: Aberdeen stockpile will be destroyed by end of the year', *Baltimore Sun*, 10 Jan. 2002, URL <http://www.sunspot.net/news/local/bal-md.mustard10jan10.story>.

[115] Westphal (note 113).

[116] The DOD must certify to Congress that the alternative technologies are as 'safe and cost-effective' as incineration, that implementing them will not take longer than implementing incineration and that the technologies satisfy relevant state and federal environmental and safety laws. Parker, M., US congressional testimony before the Senate Committee on Appropriations' Subcommittee on Defense, 25 Apr. 2001, URL <http://www.senate.gov/~appropriations/defense/testimony/ACWA.htm>. Additional information is presented in Zanders, French and Pauwels (note 58), pp. 515–16.

[117] A number of mobile, neutralization-based treatment systems continue to be developed to handle and destroy non-stockpiled CW. In addition, a fixed destruction facility for the disposal of non-stockpiled CW will operate at Pine Bluff Arsenal. 'Washington Group International to destroy chemical weapons', *Defence Systems Daily*, 13 Dec. 2001.

[118] The CW stockpiles are located at Aberdeen Proving Ground, Md.; Anniston Army Depot, Alabama; Blue Grass Army Depot, Ky.; Deseret Chemical Depot, Utah; Newport Chemical Depot, Ind.; Pine Bluff Arsenal, Ark.; Pueblo Depot, Colo.; and Umatilla Chemical Depot, Oreg.. 'Chemical demilitarization', US Army Corps of Engineers (Huntsville Center), Fact sheet, Sep. 2001, URL <http://www.hnd.usace.army.mil>. Non-stockpiled CW consist of (*a*) binary CW; (*b*) miscellaneous chemical warfare items, including unfilled munitions, support equipment and devices to be employed in conjunction with the use of CW; (*c*) recovered chemical weapons; (*d*) former production facilities; and (*e*) buried chemical warfare *matériel*.

destroy CW in at least two locations: Aberdeen and Newport.[119] Destruction operations are currently taking place only at Tooele, Utah.[120] In 2001 closure operations, which are expected to be completed in September 2003,[121] were begun at the Johnston Atoll Chemical Agent Disposal System (JACADS), located south-west of Hawaii.[122] Construction of chemical weapon destruction facilities (CWDFs) at the Pueblo Chemical Depot and Blue Grass Chemical Activity remained suspended pending the results of congressionally mandated alternative destruction technology studies.[123] Construction of the Anniston and Umatilla CWDFs was completed in June 2001 and August 2001, respectively.[124] Both are scheduled to begin operating in 2002.[125] Construction of the Pine Bluff, Aberdeen, and Newport CWDFs continued in 2001.[126] At a December 2001 OPCW Executive Council meeting the USA indicated that it may have to request an extension of the 2007 deadline for destroying its stockpiled CW.[127]

Russian CW destruction

The declared Russian CW stockpile consists of about 40 000 agent tonnes and is stored at seven locations.[128] Russia conducted limited destruction operations

[119] 'Chemical demilitarization' (note 118). About 1269 tonnes of VX are stored in bulk at Newport, while *c*. 1623 tonnes of sulphur mustard are stored in bulk at Aberdeen. A pilot CW destruction technology consisting of neutralization followed by supercritical water oxidation is to be tested on the VX. A second pilot CW destruction technology to be tested at Aberdeen consists of neutralization using hot water followed by biodegradation of the neutralization products. If successful, the pilot destruction technologies will be scaled up. Westphal (note 113).
[120] The CW is stockpiled at Deseret Chemical Depot.
[121] Westphal (note 113).
[122] CW destruction operations at JACADs were completed on 29 Nov. 2000. Secondary waste products and chemical agent identification sets (CAIS) will be disposed of as part of the closure activities. 'Chemical weapons destruction complete on Johnston Atoll', Press release distributed by the Office of the Assistant Secretary of Defense (Public Affairs), Washington, DC, 30 Nov. 2000, URL <http://www.defenselink.mil/news/Nov2000/b11302000_bt715-00.html>; 'Mission accomplished: JACADS safely destroys over 400,000 chemical weapons on Johnston Island', US Environmental Protection Agency information sheet, URL <http://www.epa.gov/region09/features/jacads>; and 'Chemical demilitarization' (note 118).
[123] 'Chemical demilitarization' (note 118); Committee on Review and Evaluation of Alternative Technologies for Demilitarization of Assembled Chemical Weapons: Phase II, and US Board on Army Science and Technology Division on Engineering and Physical Sciences (US National Research Council), *Analysis of Engineering Design Studies for Demilitarization of Assembled Chemical Weapons at Pueblo Chemical Depot* (National Academy Press: Washington, DC, 2001); and Miller and Levin (note 111). The destruction technologies to be used at both sites should be selected in 2002. Miller, A. and Levin, M., 'US to step up arms disposal', *Los Angeles Times*, 3 Oct. 2001, URL <http://www.latimes.com>.
[124] 'Chemical demilitarization' (note 118); Gillespie, K., 'Completion ceremony held for Anniston chemical agent disposal facility', *Bulletin* (US Army Corps of Engineers, Huntsville Center), vol. 22, no. 5 (June 2001), pp. 1, 4; and Gillespie, K., 'Umatilla chemical agent disposal facility completed', *Bulletin* (US Army Corps of Engineers, Huntsville Center), vol. 22, no. 7 (Aug. 2001), p. 3.
[125] Westphal (note 113).
[126] 'Chemical demilitarization' (note 118).
[127] Executive Council concludes its twenty-seventh session, *Secretariat Brief*, OPCW document, no. 30 (18 Dec. 2001), p. 3.
[128] Kambarka, Udmurtia Republic; Gorny, Saratov oblast; Kizner, Udmurtia Republic; Maradikovsky, Kirov oblast; Pochep, Bryansk oblast; Leonidovka, Penza oblast; and Shchuchye, Kurgan oblast. For background on Russian CW destruction see Hart, J. and Miller, C. D. (eds), *Chemical Weapon Destruction in Russia: Political, Legal and Technical Aspects*, SIPRI Chemical & Biological Warfare Studies, no. 17 (Oxford University Press: Oxford, 1998).

in 2001. In April it began destroying Category 2 CW at Shchuchye and Category 3 CW at Leonidovka and Maradikovskiy.[129] Russia completed the destruction of its Category 3 CW, consisting of over 4300 unfilled munitions and devices, by mid-November 2001.[130] On 27 September Russia completed the transfer of some 10 tonnes of phosgene (Category 2 CW) from approximately 4000 projectiles stored at Shchuchye into bulk containers.[131] All Category 2 CW are to be destroyed by 29 April 2002.[132]

However, systematic destruction of Category 1 CW was delayed again, principally as a consequence of Russia's generally weakened economy since the CWC's entry into force and a lack of high-level political commitment to resolve political, legal and technical difficulties associated with the destruction programme. In 2001 several developments indicated that Russia will begin large-scale destruction operations in 2002. Funding for CW destruction was increased from 500 million roubles (*c.* $16 million) in FY 2000 to 3 billion roubles (*c.* $96.6 million) in FY 2001.[133] Russian government officials estimated that the destruction of the CW stockpile will cost about $3 billion (other sources estimate the total cost to be in the range of $6–7 billion) and hope to meet 20–30 per cent of the cost with foreign assistance.[134]

Russia's revised destruction plan, which supersedes the 1996 comprehensive destruction plan, specifies that 1 per cent of the stockpile is to be destroyed by 2003, 20 per cent by 2007, 45 per cent by 2008 and 100 per cent by 2012.[135] Russia therefore requested the EC to approve a five-year extension of the final destruction deadline.[136] The plan also calls for the construction of three full-scale CWDFs, to be located at Gorny, Kambarka and Shchuchye.[137] Small-scale detoxification facilities are to constructed at Leonidovka, Maradikovskiy and Pochep.[138] Work on the construction of CWDFs at Gorny and

[129] Mills, P., 'Progress in The Hague, developments in the Organization for the Prohibition of Chemical Weapons', Quarterly review no. 34, *CBW Conventions Bulletin*, no. 52 (June 2001), p. 4.

[130] Reuters, 'Russia destroys chemical arms', 22 Nov. 2001, URL <http://www.iht.com/articles/39631.htm>; and Opening statement by the Director-General to the Executive Council at its Twenty-Seventh Session (note 87), para. 9.

[131] Opening statement by the Director-General to the Executive Council at its Twenty-Seventh Session (note 87), para. 10. Russia intends to destroy the phosgene at the Prikladnaya Khimiya Research Centre, located in Perm. Opening statement by the Director-General to the Executive Council at its Twenty-Seventh Session (note 87), para. 11.

[132] 'Twenty-sixth session of the Executive Council, 25–28 September 2001', *OPCW Synthesis* (winter/Dec. 2001), p. 31.

[133] Pikayev, A., 'Russian implementation of the CWC', ed. J. B. Tucker, *The Chemical Weapons Convention, Implementation Challenges and Solutions* (Monterey Institute of International Studies: Monterey, Calif., Apr. 2001), p. 35.

[134] 'Russia to destroy all chemical weapons stockpiles by 29 April 2012', Interfax (Moscow), 11 Dec. 2001, in FBIS-SOV-2001-1211, 12 Dec. 2001; 'Twenty-sixth session of the Executive Council, 25–28 September 2001' (note 132), p. 31; ITAR-TASS (Moscow), 30 Nov. 2001, in 'Russia revises chemical weapon destruction program', FBIS-SOV-2001-1130, 3 Dec. 2001; and Military News Agency, 'Russia resumes construction of chemical weapons scrapping facilities', 6 Apr. 2001, URL <http://www.military news.ru/fcl_l/eanews.asp?is=62925>.

[135] On making amendments and additions to the Resolution by the Government of the Russian Federation of March 21, 1996 (no. 305) on approving the Federal Special Program Chemical Weapons Stockpiles Destruction in the Russian Federation, Resolution no. 510, 5 July 2001.

[136] 'Twenty-sixth session of the Executive Council, 25–28 September 2001' (note 132), p. 31.

[137] Federal Special Program (note 135).

[138] Federal Special Program (note 135).

Shchuchye was intensified in 2001.[139] CW destruction is expected to begin at Gorny in 2002 and continue until 2005,[140] and at Kambarka and Shchuchye by 2005 and continue until 2011.[141] The building which will house the Gorny pilot CW destruction facility has been constructed, and equipment and infrastructure are currently being installed.[142] The plant will destroy 220 tonnes of lewisite using alkaline hydrolysis followed by electrolysis.[143] Legal uncertainties, including the effect of local and regional laws banning the transport of CW, have apparently not been fully resolved.[144]

The principal bodies involved in Russian CW destruction are the Munitions Agency, the Interdepartmental Scientific Council on Chemical and Biological Weapons Convention Problems, and the State Commission on Chemical Disarmament. The stockpile is under the jurisdiction and control of the Munitions Agency, headed by Zinovy Pak. The agency acts as Russia's National Authority to the OPCW and is responsible for the implementation of Russia's CW destruction programme, including the allocation of funds and the conversion or destruction of former CWPFs.[145] On 8 February 2001 the Russian Prime Minister's Cabinet issued Directive no. 87 outlining the mandate of the Directorate for the Safe and Secure Storage and Destruction of Chemical Weapons, located within the Munitions Agency.[146] The directorate is responsible for the safe storage, transport and destruction of CW, ensuring that CW is not diverted, letting contracts and R&D for CW destruction technologies.[147] The Interdepartmental Scientific Council, which is responsible to the Munitions Agency and headed by Academician Anatoliy Kuntsevich, is mandated to provide relevant CW-related scientific expertise, especially with regard to the selection of CW destruction technologies.[148] On 4 May 2001 a 22-member State Commission on Chemical Disarmament, whose chairman is Sergey Kiriyenko and deputy chairman is Zinovy Pak, was established to improve cooperation between the various bodies involved in CW destruction.[149]

---

[139] 'Russia resumes construction of chemical weapons scrapping facilities' (note 134).

[140] Federal Special Program (note 135); and 'European Union non-proliferation and disarmament actions in the Russian Federation', Statement provided by the Embassy of the Federal Republic of Germany, Stockholm, 3 Dec. 2001.

[141] Federal Special Program (note 135); and 'Twenty-sixth session of the Executive Council, 25–28 September 2001' (note 132), p. 31.

[142] 'European Union non-proliferation and disarmament actions in the Russian Federation' (note 140).

[143] 'European Union non-proliferation and disarmament actions in the Russian Federation' (note 140). The arsenic will be removed by electrolysis and later purified for industrial use.

[144] The governor of Bryansk oblast, e.g., is apparently opposed to the construction of any destruction facility anywhere in the oblast. Pikayev (note 133).

[145] Russia's payment for direct costs of inspection owed to the OPCW is made by the Munitions Agency. 'News chronology', *CBW Conventions Bulletin*, no. 53 (Sep. 2001), p. 28.

[146] Averre, D. and Khripunov, I., 'Chemical weapons disposal: Russia tries again', *Bulletin of the Atomic Scientists*, vol. 57, no. 5 (Sep./Oct. 2001), p. 60 ; and Pikayev (note 133).

[147] Pikayev (note 133).

[148] 'News chronology' (note 145), p. 45.

[149] 'On the formation of a State Commission on Chemical Disarmament', *Ministry of Foreign Affairs of the Russian Federation Daily News Bulletin*, 7 May 2001; and 'News chronology' (note 145), p. 26.

CW destruction assistance

CW destruction assistance is provided to Russia by Canada, the EU, Finland, France, Germany, Italy, the Netherlands, Norway, Sweden, Switzerland, the USA and the UK (table 12.1).[150] Approximately $374.4 million worth of assistance has been offered, while $237.4 million has been allocated to specific projects.[151]

Conversion of Russian CWPFs

One of the key concerns of the Russian Government regarding CWC implementation has been the fact that the convention requires each party to request Executive Council approval for conversion.[152] For Russia this issue is more important than for other parties to the CWC because its CWPFs were generally part of larger industrial complexes, while CWPFs in other countries were usually dedicated facilities. Russia has declared a total of 24 CWPFs of which at least 6 have been destroyed and 6 converted.[153] In 2001, Russia had seven conversion requests awaiting action by the OPCW Executive Council.[154]

Old and abandoned chemical weapons

According to the OPCW, the states which have officially declared that they possess old chemical weapons (OCW) are Belgium, France, Germany, Italy, Japan and the UK.[155]

On 13 April 2001 *French* authorities evacuated over 12 000 residents of the town of Vimy for more than a week while 173 tonnes of World War I-vintage explosive materials, including munitions containing sulphur mustard and phosgene, were transported to the Suippes military camp, located in the Marne

---

[150] Russian Embassy statement, Stockholm, 6 Mar. 2002; 'Russia to destroy all chemical weapons stockpiles by 29 April 2012', Interfax (note 134); 'Norway assists Russia in destroying chemical weapons', Interfax (Moscow), 30 Nov. 2001, in FBIS-SOV-2001-1130, 3 Dec. 2001; 'As if by magic: London and Rome earmark almost $20 million for destruction of Russian chemical weapons', *Rossiyskaya Gazeta* (Moscow), 22 Sep. 2001, in 'Putin's envoy secures funds for chemical weapon destruction on trip to Europe', FBIS-WEU-2001-0925, 22 Sep. 2001; and Interfax (Moscow), 7 Jan. 2002, 'Russia: new chemical weapons processing plant to be built this year', FBIS-SOV-2002-0107, 8 Jan. 2002.

[151] Russian Embassy statement, Stockholm, 6 Mar. 2002.

[152] CWC, Verification Annex, Verification Annex, paras 64–72, CWC; Gilbert, J. A. *et al.*, 'Destruction or conversion of Russian chemical weapon production facilities', eds Hart and Miller (note 128), pp. 55–74; and Kalinina, N., 'The problems of Russian chemical weapon destruction', eds Hart and Miller (note 128), pp. 8–9.

[153] 'List of chemical weapons production facilities subject to conversion or destruction', appendix no. 3, Federal Special Program (note 135); and 'Verification of destruction of chemical weapons and chemical weapons production facilities', *OPCW Annual Report 2000*, p. 17.

[154] Russian Embassy statement, Stockholm, 6 Mar. 2002.

[155] 'Old chemical weapons' are defined by the CWC as either (*a*) CW produced before 1925, or (*b*) CW produced between 1925 and 1 Jan. 1946 which have deteriorated to such an extent that they can no longer be used as CW. CWC, Article II, para. 5.

Table 12.1. Type, location and amount of Russian CW destruction assistance
Figures are in US $m.

Country/entity	Type[a]	Location(s)	Allocated or offered
Canada	..	..	..
European Union[b]	..	Gorny, Shchuchye	c. 1.76
Finland[c]	Detection equipment	Karmbarka	c. 0.89
France	..	..	..
Germany[d]	Pilot destruction plant	Gorny	c. 30.8
Italy[e]	Gas pipeline construction	Shchuchye	c. 7
Netherlands[f]	Electricity transformer	Gorny	c. 10
Norway[g]	Electricity transformer	Shchuchye	c. 1
Sweden[h]	Analytical equipment	Kambarka	..
Switzerland[i]	..	..	18–30
United Kingdom[j]	Water and power supply	Shchuchye	c. 18
United States[k]	Pilot destruction facility	Shchuchye	c. 866

[a] The listed types of assistance are not comprehensive.

[b] Some aid is coordinated and/or channelled through member states. A total of €700 000 (c. $617 000) is intended for Munitions Agency administrative tasks. The EU is also involved in assisting with the destruction of a former mustard production facility at Dzerzhinsk.

[c] The Finnish assistance programme was apparently begun in 1997. Detection equipment was scheduled to be delivered by the end of 2001.

[d] Germany has provided assistance since 1993 and currently provides CW destruction assistance within the framework of a 17 Dec. 1999 agreement between the EU and Russia on cooperation in the area of non-proliferation of nuclear, chemical and biological weapons. Germany's assistance consists of the construction of a facility for draining CW agent stored in bulk, equipment for a mobile and a stationary analytical laboratory, destruction equipment for initial hydrolysis of lewisite followed by arsenic extraction by electrolysis, an incinerator for liquid and solid CW residues, equipment for purification of off gases and liquid waste, and engineering and technical support for these activities. A mobile and a fixed laboratory are operational; various CW destruction equipment is on-site; and CW destruction equipment provided by Germany is being installed.

[e] On 17 Feb. 2001 an agreement between Italy and Russia on CW destruction assistance for the period 2000–2002 entered into effect.

[f] The Netherland's assistance is scheduled to operate from autumn 2001 to spring 2002. A total of €4.55 million (c. $4 million) has been allocated for the first phase of the assistance— the financing of an electric transformer at Gorny. The Netherlands is currently prepared to offer assistance totalling €11.34 million (c. $10 million).

[g] Norway agreed to provide c. $1 million to pay for an electrical transformer for the CWDF at Shchuchye. Norway's assistance will be coordinated with assistance from the UK.

[h] Sweden conducted a risk assessment of the CW stockpile at Kambarka. Sweden's assistance is currently focused on the delivery of analytical medical equipment to a local Kambarka hospital. The equipment is to be used to measure the levels of arsenic in the population before

and during CW destruction operations, detect any health effects of destruction operations on workers and promote an effective hospital response in case of CW-related accidents.

i The Swiss Parliament is reportedly considering an expenditure of CHF 30–50 million ($18–30 million) in 2007–2009.

j On 20 Dec.2001 the UK signed a 3-year agreement worth £12 million (*c.* $18 million) to assist with the establishment and maintenance of water and power supplies for Shchuchye's CWDF.

k In 2001 the USA completed a high-level review of non-proliferation and threat reduction assistance programmes to Russia, including CW destruction assistance provided within the framework of the CTR programme. Work on renovating and equipping a US-funded Central Analytical Laboratory was completed in Jan. 2001. The laboratory, located at the State Scientific Research Institute for Organic Chemistry and Technology (GosNIIOKhT), is important as it will be used to provide quality assurance for Russia's CW destruction programme. On 10 July 2001 Russia approved a second protocol to the US–Russian Agreement on the Safe and Secure Transportation, Storage and Destruction of Weapons and the Prevention of Weapons Proliferation. A total of $35 million in assistance has been allocated for FY 2002. The US plans to contribute assistance eventually totalling *c.* $866 million towards construction of the CWDF at Shchuchye, the main focus of US assistance. Preparatory site work was begun at Shchuchye in January 2001, and it is estimated that construction of the CWDF should be completed in 2006.

Sources: Russian Embassy statement, Stockholm, 6 Mar. 2002; *CBW Conventions Bulletin*, no. 53 (Sep. 2001), pp. 40–41, 44; Könberg, M., 'Rysslands destruktion av kemiska stridsmedel' [Russia's destruction of chemical warfare agents], FOI [Swedish Defence Research Agency], *BC-bulletin*, no. 6 (May 2001), pp. 1–10; 'Finland support for Russian CW storage facility', *Disarmament Diplomacy*, no. 52 (Nov. 2001), p. 63; Russian Ministry of Foreign Affairs, 'Note sent to German side on consent to distribution of gratuitous allocations made by the European Union for the destruction of chemical weapons in the Russian Federation', Press Release no. 0264, 6 Mar. 2001; 'European Union non-proliferation and disarmament actions in the Russian Federation', German Embassy statement, Stockholm, 3 Dec. 2001; OPCW Press Release 001/00, 20 Apr. 2000; 'Areas and volumes of international assistance extended to the Russian Federation for implementing the Program', appendix no. 6, The Federal Special Program Chemical weapons stockpiles destruction in the Russian Federation, revised plan approved by Resolution no. 510, 5 July 2001; ITAR-TASS (Moscow), 'Italy helping to scrap Russia's chemical weapons', 9 June 2001, in Foreign Broadcast Information Service, *Daily Report–Central Eurasia (FBIS-SOV)*, FBIS-SOV-2001-0609, 11 June 2001; and RIA (Moscow), 17 Sep. 2001, in 'Russia, Italy to cooperate in chemical weapons destruction', FBIS-SOV-2001-0918, 19 Sep. 2001; 'European Union non-proliferation and disarmament actions in the Russian Federation', Statement from the Embassy of the Netherlands, Stockholm, 28 Nov. 2001; Interfax (Moscow), 30 Nov. 2001, in 'Norway assists Russia in destroying chemical weapons', FBIS-SOV-2001-1130, 3 Dec. 2001; 'UK, Norway join forces with Russia to destroy chemical weapons', *Defence Systems Daily*, 7 Dec. 2001, URL <http://defence-data.com>; Private communication with J. Hart, 19 June 2001; ITAR-TASS (Moscow), 20 Dec. 2001, in 'UK to help fund construction of Russian chemical weapons elimination facility', FBIS-SOV-2001-1220, 21 Dec. 2001; 'UK, Russia sign treaty on CW weapon destruction', *Defence Systems Daily*, 21 Dec. 2001, URL <http://defence-data.com>; Koch, A., 'US review to aid Russia's WMD legacy programmes', *Jane's Defence Weekly*, vol. 37, no. 2 (9 Jan. 2002), p. 8; Gillespie, K., 'Huntsville Center team completes Russian chemical laboratory', *Bulletin* (US Army Corps of Engineers, Huntsville Center), vol. 22, no. 1 (Feb. 2001), pp. 1, 8; Zanders, J. P., Eckstein, S. and Hart, J., 'Chemical and biological weapon developments and arms control', *SIPRI Yearbook 1997: Armaments, Disarmament and International Security* (Oxford University Press: Oxford, 1997), p. 448; and US Army Corps of Engineers (Huntsville Center), 'Chemical demilitarization', Fact Sheet, Sep. 2001, URL <http://www.hnd.usace.army.mil>.

region.[156] The munitions were refrigerated during transport and finally deposited in underground nuclear missile silos to await destruction.[157] A CWDF is planned to be ready for operation at Suippes in 2005.[158] Approximately 250 tonnes of World War I munitions are reportedly uncovered in France annually, 10–15 per cent of which are CW.[159]

The United States' Army Corps of Engineers and the Environmental Protection Agency (EPA) continued systematically surveying and testing approximately 1600 properties to locate CW-contaminated soil and any remaining World War I-era CW munitions in Spring Valley in the north-west section of Washington, DC.[160] During World War I the US Army had rented the area from American University in order to develop and test CW.[161] Containers filled with sulphur mustard and lewisite as well as mortar shells were uncovered in 2001.[162] The current activities are expected to continue for at least two years.[163]

The countries which have officially declared to the OPCW the presence of abandoned chemical weapons (ACW) on their territory are China, Italy and Panama.[164]

On 27 June 2001 the *Japanese* Government reportedly approved a plan to remove ACW in Jilin province, China, for which it will provide the necessary

---

[156] 'Un risque d'explosion de munitions provoque une évacuation massive près d'Arras' [A risk of explosion of munitions provokes a massive evacuation near Arras], *Le Monde*, 15–16 Apr. 2001; and 'Les "évacués" de Vimy pourraient rentrer chez eux en fin de semaine' [The Vichy 'evacuees' will be able to return home at the end of the week], *Le Monde*, 19 Apr. 2001, Ministry of the Interior communiqué, n.d. [2001].

[157] BBC News Online, 'WWI shells reach army base', 16 Apr. 2001, URL <http://news.bbc.co.uk/low/english/world/europe/newsid_1279000/1279852.stm>.

[158] 'News chronology' (note 145), p. 28; and 'Itinéraire, commandement, coordination et composition du convoi itinéraire' [Itinerary, command, coordination and composition of the convoy itinerary], Mise en sécurité du centre de munitions de Vimy, 15 Apr. 2001, URL <http://www.pas-de-calais.pref.gouv.fr/details.asp?table=news&ID=79>. The CWDF was scheduled to be ready for operation in 2002. 'News chronology', *CBW Conventions Bulletin*, no. 40 (June 1998), p. 21.

[159] 'News chronology' (note 158), p. 21.

[160] Tucker, J., 'Chemical weapons: buried in the backyard', *Bulletin of the Atomic Scientists*, vol. 57, no. 5 (Sep./Oct. 2001), pp. 51–56; Vogel, S., 'Evidence of DC toxins unheeded: in 1986, US failed to act on warnings of buried munitions', *Washington Post*, 9 July 2001, p. A01; Holly, D., 'Old army chemicals plague DC', *Philadelphia Inquirer* (Internet edn), 15 Apr. 2001, URL <http://inq.philly.com>; and Argetsinger, A. and Vogel, S., 'Excavation by military forces some AU [American University] closings', *Washington Post*, 8 Jan. 2001, p. B01. OCW were previously discovered in the area in Jan. 1993. In Feb. 1998 more munitions were discovered in the backyard of the South Korean Ambassador to the USA in what was apparently a former disposal pit. Elevated levels of arsenic in soil samples have also been reported.

[161] The role of American University in the USA's World War I CW programme is discussed in Jones, D. P., 'The role of chemists in research on war gases in the United States during World War I', PhD thesis, University of Wisconsin, 23 May 1969, pp. 115–65; and Brophy, L. P., Miles, W. D. and Cochrane, R. C., *The Chemical Warfare Service: from Laboratory to Field* (Office of the Chief of Military History, US Army: Washington, DC, 1959), pp. 5–8, 24–25. In July 2001 American University filed a damage claim in the amount of $87 million against the US Army. Vogel, S., 'AU seeks $87 million in burial of weapons: claim alleges Army mishandled cleanup', *Washington Post* (Internet edn), 14 July 2001, p. B01, URL <http://www.washingtonpost.com>.

[162] Vogel, S., 'WW I chemicals removed from Spring Valley yard, army unearths mustard gas variant', *Washington Post* (Internet edn), 6 July 2001, p. B01, URL <http://www.washingtonpost.com>.

[163] Tucker (note 160), p. 55.

[164] The CWC defines 'abandoned chemical weapons' as CW abandoned by a state after 1 Jan. 1925 on the territory of another state without the consent of the latter. CWC, Article II, para. 6.

funding.¹⁶⁵ Further information regarding Japanese BW and CW activities was made public.¹⁶⁶ There were also reports that German and Russian companies may cooperate with China and Japan in destruction of the ACW.¹⁶⁷

At the invitation of the *Panamanian* Government, the OPCW carried out a fact-finding mission on 12–19 July to three locations in Panama, including San José Island, where the USA had operated a field-test facility during and after World War II.¹⁶⁸ The team was shown conventional munitions, fragments and the remnants of what appeared to be CW cylinders, rockets and air bombs.¹⁶⁹ A total of four intact CW air bombs were reportedly found on the island.¹⁷⁰ However, the origin and nature of many of the items shown to the team could not be positively identified. On 4 September, Panama's Foreign Minister gave the findings to the US Department of State and requested that the USA formally declare whether it is aware of the existence of CW abandoned on Panamanian territory and provide technical and financial assistance for the disposal of any such weapons.¹⁷¹ An inter-agency US government group is reportedly studying the OPCW report.¹⁷² Conflicting information on the types and quantities of CW shipped to Panama, their dates of production, the total number of sites where CW may be located and questions regarding the possible effects of contamination have been reported.¹⁷³

---

¹⁶⁵ 'News chronology' (note 145), p. 42.

¹⁶⁶ Chu, H., 'China haunted by WW II chemical weapons', *Los Angeles Times* (Internet edn), 27 Dec. 2001, URL <http://www.latimes.com>.

¹⁶⁷ 'German company plans using Russia's experience in weapons destruction', ITAR-TASS (Moscow), 24 May 2001, in 'German company plans using Russia's experience in weapons destruction', FBIS-SOV-2001-0524, 29 May 2001.

¹⁶⁸ Statement from the Panamanian Embassy, Stockholm, 7 Mar. 2002. The facility, which is located c. 97 km from Panama City, was established to test the characteristics of sulphur mustard in a tropical environment. This was believed to be necessary in part because of Japanese use of CW agents on mainland China and the climate and terrain where combat between Japanese and Allied forces was occurring. Brophy, L. P. and Fisher, G. J. B., *The Chemical Warfare Service: Organizing for War* (Office of the Chief of Military History, US Army: Washington, DC, 1959), pp. 106, 135–38; and Brophy, Miles and Cochrane (note 161), pp. 41, 411. The facility was used by personnel from the Canadian Army and Air Force, US Army and Navy and the British Army. Brophy and Fisher (note 168), p. 137.

¹⁶⁹ Lindsay-Poland, J., 'Panama calls for US chemical cleanup', Nov. 2001, URL <http://www.forusa.org/panama/1101_panamacleanup.html>.

¹⁷⁰ Four intact CW were identified: 3 1000-lb (c. 455 kg) AN-M79 bombs and 1 500-lb (c. 228 kg) AN-M78 bomb. All were originally filled with non-persistent agent. Statement from the Panamanian Embassy (note 168). According to an official US Army military historical source, 1000-lb AN-M79 bombs filled with phosgene and cyanogen chloride were tested at San José Island. Brophy, Miles and Cochrane (note 161), p. 41. In general, either the chemical fill of a munition left in the field has leaked out and hydrolysed or it is in nearly the same condition as it was the day the munition was filled. Prior to transferring US military bases to the Panamanian Government, the USA carried out a programme to locate munitions (mostly conventional) and render them harmless. However, it is difficult to ascertain whether every munition fired has been located, especially in view of the fact that much of the land consists of jungle. In addition, under the CWC a chemical weapon may consist of a munition body only. CWC, Article II, para. 1.

¹⁷¹ Statement from the Panamanian Embassy (note 168); and Lindsay-Poland (note 169).

¹⁷² US Embassy statement, 20 Nov. 2001, Stockholm.

¹⁷³ Hernandez, S., 'Panama–US: chemical weapons in Canal Zone sour relations', 9 Aug. 2001, URL <http://www.oneworld.org>; Pugliese, D., 'DND fears toxic legacy in Panama', *Ottawa Citizen Online*, 5 Aug. 2001, URL <http://www.ottawacitizen.com/national/010805/5043507.htm>; Pugliese, D., 'Deadly American legacy lingers in Panama', *Ottawa Citizen*, 23 Apr. 2001; Pugliese, D., 'Canada's toxic wartime secret', *Ottawa Citizen*, 22 Apr. 2001; and Pugliese, D., 'Canada may be part of Panama bomb menace', *Ottawa Citizen*, 24 Apr. 2001. Hernandez reports that the USA may have tested CW at 15 other sites in addition to San José Island.

A total of 1420 munitions, including CW, dating from the Italian–Ethiopian war in the 1930s were reportedly discovered at a construction site in the town of Amba Alage, located in the Tigray province of northern Ethiopia.[174] The Ministry of Trade and Industry, Ethiopia's CWC national authority, reportedly estimated that Italy transported some 80 000 tonnes of chemical agent during the war.[175] The Italian Government indicated that it was prepared to assist with the destruction of any weapons, including CW, that it may have left on Ethiopian territory.[176] However, at the end of 2001 Ethiopia had not declared to the OPCW that it possessed ACW and a joint Italian–Ethiopian investigative team found no chemical munitions among the recovered items examined.[177]

V. Terrorism with mail-delivered anthrax spores

Anthrax bacteria as a biological warfare agent

For many decades the spores of the anthrax bacterium have been considered a prime agent of biological warfare. Several countries—including Iraq, Japan, the former Soviet Union, the United Kingdom and the USA—have prepared them as a weapon. Their hardiness, wide availability and potential lethality also make them a potential candidate for biological terrorism.[178] However, the underlying mechanisms for anthrax virulence are still incompletely understood. Scientists have only begun to understand the biochemical causes of the virulence of the anthrax bacterium. In November 2001 scientists reported that they have decoded the anthrax bacterium genome.[179]

Bacillus anthracis, the causative agent of anthrax, occurs naturally worldwide. It can persist in the environment for decades in sporulated form. Anthrax primarily affects grazing animals and is encountered chiefly among livestock

---

[174] United Nations Office for the Coordination of Humanitarian Affairs (OCHA), Integrated Regional Information Network for the Horn of Africa, IRIN, 'Ethiopia: buried armaments discovered in Tigray', 3 May 2001, URL <http://www.irinnews.org/report.asp?ReportID=6206&SelectRegion=Horn_of_Africa&SelectCountry=ETHIOPIA>; 'Workers stumble on Italian chemical weapons cache in Ethiopia', *African Environmental Newsletter*, 6 May 2001; and 'Buried armaments discovered in Tigray', UN Integrated Regional Information Network/All Africa Global Media via COMTEX, 4 May 2001, URL <http://library.nothernlight.com/FD20010503310000062.html?cb=0&dx=1006&sc=0#doc>. See also *The Problem of Chemical and Biological Warfare: The Rise of CB Weapons*, vol. 1 (Almqvist & Wiksell: Stockholm, 1971), pp. 142–46, 257–58.

[175] BBC News Online, 'Ethiopia accuses Italy over weapons', 3 May 2001, URL <http://news.bbc.co.uk/low/english/world/africa/newsid_1310000/1310932.stm>; and IRIN, 'Ethiopia: buried armaments discovered in Tigray', 3 May (note 174). It is unclear whether the figure includes munition weight. The main CW agent used by Italy was sulphur mustard, which was generally dispersed using spray tanks attached to aircraft. It is unclear how much agent remained prior to Italy's withdrawal from Ethiopia in 1941. *New Encyclopædia Britannica*, vol. 4, Micropædia, 15th edn (Encyclopædia Britannica Inc.: London, 1985), p. 580.

[176] Italian Ministry of Foreign Affairs statement, 3 May 2001, unofficial translation of statement provided by the Italian Embassy, Stockholm.

[177] Private communication with J. Hart, 29 Jan. 2002.

[178] Atlas, R. M., 'The medical threat of biological weapons', ed. R. M. Atlas, 'Special issue: biological weapons', *Critical Reviews in Microbiology*, vol. 24, no. 3 (1998), p. 160.

[179] Broad, W. J., 'Genome offers "fingerprint" for anthrax', *New York Times* (Internet edn), 28 Nov. 2001, URL <http://www.nytimes.com>.

like cattle, sheep, goats and horses. It is most common in agricultural regions with inadequate control programmes for anthrax in livestock. *Bacillus anthracis* has a low incidence of infection in humans and the disease is mostly associated with agricultural, horticultural or industrial exposure. There are almost no known cases of human-to-human transmission.[180]

Anthrax bacteria in the vegetative (multiplying) state are rod-shaped and measure 1–1.2 to 3–5 microns; in sporulated form their size is approximately 1 micron. Anthrax spores cluster together to form particles. However, particle sizes are usually too large for the spores to reach the terminal alveoli of the lungs. In order for particles containing anthrax spores to be able to reach the terminal alveoli, where infection leading to the inhalational form of the disease can be initiated, they should ideally be no bigger than 5 microns.[181] Military programmes therefore try to deliver anthrax spores as an aerosol of small particles.

On the basis of experiments involving non-human primates, the LD_{50} dose (the amount required to kill 50 per cent of the people exposed) for (untreated) inhalational anthrax has been determined to vary enormously—from 2500 to 760 000 spores. The individual susceptibility among humans may vary greatly, as was suggested by the analysis of the anthrax outbreak in Sverdlovsk in the former Soviet Union in 1979.[182] For instance, none of the 66 documented fatal cases involved a person younger than 24 years of age. The human ID_{50} (the amount required to infect 50 per cent of the people exposed) of anthrax is usually set at 8 000–50 000 spores; however, infection may occur at far lower doses. It cannot be excluded that a single spore can cause the disease.[183]

Depending on the point of entry into the body, the disease can manifest itself as inhalational anthrax if the spores settle in the lungs; gastrointestinal (with two distinct syndromes—abdominal and the (rare) oropharyngeal) anthrax after ingestion; or cutaneous anthrax if they penetrate the skin. In the body the spores germinate into vegetative cells. Initially, cell damage is local. However, if the bacteria succeed in entering the bloodstream the disease may become

---

[180] Franz, D. R. *et al.*, 'Clinical recognition and management of patients exposed to biological warfare agents', *Clinics in Laboratory Medicine*, vol. 21, no. 3 (Sep. 2001), p. 437; Friedlander, A. M., 'Anthrax', eds F. R. Sidell, E. T. Takajufi and D. R. Franz, *Medical Aspects of Chemical and Biological Warfare* (Office of the Surgeon General, Department of the Army: Washington, DC, 1997), p. 469; 'Use of anthrax vaccine in the United States; recommendations of the Advisory Committee on Immunization Practices (ACIP)', *Morbidity and Mortality Weekly Report*, vol. 49, no. RR-15 (15 Dec. 2000), p. 1; and Cymet, T. C. *et al.*, 'Symptoms associated with anthrax exposure: suspected "aborted" anthrax', *Journal of the American Osteopathic Association*, vol. 102, no. 1 (Jan. 2002), pp. 41, 42.

[181] A micron is 1 millionth of a metre. Large particles, like pollen, are on average 20 micron in size and are stopped by the hairs in the nose. Particles of the range of 5–15 micron (e.g., fly ash and some pollution) can enter the respiratory tract but are caught by the mucous and ciliary (hairlike) cells of the bronchial walls. Particles smaller than 1 micron are also usually trapped in the upper respiratory tract because the air molecules push them against the bronchial walls. World Health Organization, *Hazard Prevention and Control in the Work Environment: Airborne Dust*, document WHO/SDE/OEH/99.14 (1999), pp. 4–7, URL <http://www.who.int/peh/Occupational_health/dust/dusttoc.htm>.

[182] For a detailed analysis of the Sverdlovsk incident see Guillemin, J., *Anthrax: The Investigation of an Outbreak* (University of California Press: Berkeley, Calif., 1999).

[183] Matthew Meselson, Harvard biologist, quoted in Broad, W. J. *et al.* 'Excruciating lessons in the ways of a disease', *New York Times* (Internet edn), 31 Oct. 2001, URL <http://www.nytimes.com>.

systemic and rapidly lethal. The mortality rate of untreated inhalational anthrax approaches 100 per cent. Death occurs within a few days of the onset of symptoms.[184] The mortality rate of gastrointestinal anthrax varies depending on the outbreak, but it may also approach 100 per cent. Cutaneous anthrax, which in most cases remains localized, is usually curable.[185]

Two factors characterize the virulence of the anthrax bacterium. First, the vegetative cells are encased in a polypeptide capsule that prevents the so-called scavenger cells (phagocytes, such as macrophages and neutrophils) from ingesting the invading bacteria.[186] Second, the vegetative anthrax bacterium releases a potent toxin that attacks the macrophages, thereby wiping out the first line of defence of the immune system. The anthrax toxin consists of three proteins: protective antigen, oedema factor and lethal factor. The protective antigen binds to the surface of the cell, where an enzyme trims off molecules, seven of which then combine to form a ring-shaped structure, or heptamer. The heptamer captures the two factors, which are then transported through the membrane of the attacked cell. Through biochemical action inside the cytosol the oedema factor and lethal factor catalyse different molecular reactions that lead to the destruction of the phagocyte and the release of cytokines.[187]

There are no reliable estimates of the number of human anthrax cases worldwide, but it is believed that over 95 per cent of the cases are cutaneous.[188] According to statistics published before the anthrax attacks in the autumn of 2001, the annual incidence of human anthrax cases in the USA dropped from approximately 130–200 in the early 20th century to none in the period 1992–99; a single case of cutaneous anthrax involving a 67-year-old farmer in

---

[184] Early symptoms of anthrax infection, which can last for several days, are nondescript (often described as flu-like). Dixon, T. C. et al., 'Anthrax', *New England Journal of Medicine*, vol. 341, no. 11 (9 Sep. 1999), p. 815; and Bush, L. M. et al., 'Index case of fatal inhalational anthrax due to bioterrorism in the United States', *New England Journal of Medicine*, vol. 345, no. 22 (29 Nov. 2001), p. 1608. The anthrax attacks in the United States in the autumn of 2001 led to a mortality rate for inhalational anthrax of around 44% (see below). This is primarily due to the high state of alert after the confirmation of the first anthrax cases, the accelerated diagnostic procedures and the application of multiple antibiotics as soon as an anthrax infection was suspected. It cannot be excluded, however, that in the event of an attack on a larger scale than the ones in the USA in which the emergency services could be overwhelmed that the mortality rate for inhalational anthrax would be much higher.

[185] Inglesby, T. V. et al., 'Anthrax as a biological weapon', *Journal of the American Medical Association*, vol. 281, no. 18 (12 May 1999), p. 1737; Dixon et al. (note 184), p. 815; and Pannier, A. D. et al., 'Crystal structure of the anthrax lethal factor', *Nature*, vol. 414 (8 Nov. 2001), URL <www.nature.com>.

[186] Upon entering the body the spores are actually enveloped by the macrophages, where they germinate and become vegetative. After being released by the macrophages, which have meanwhile taken the invading bacteria to the regional lymph nodes, the anthrax bacteria multiply in the lymphatic system and subsequently enter the bloodstream. Once released from the macrophages, there is no indication of immune response to the vegetative cells. Dixon, T. C. et al. (note 184), p. 815.

[187] Young, J. A. T. and Collier, R. J., 'Attacking anthrax', *Scientific American*, vol. 286, no. 3 (Mar. 2002), pp. 38–40.

[188] 'Use of anthrax vaccine in the United States: recommendations of the Advisory Committee on Immunization Practices (ACIP)' (note 180), p. 1. In industrialized states inhalational anthrax may make up the remaining 5%; developing countries, where animals are not always vaccinated, may suffer from a higher incidence of gastrointestinal anthrax as a consequence of the higher risk of consumption of contaminated meat. Pile, J. C. et al., 'Anthrax as a potential biological warfare agent', *Archives of Internal Medicine*, vol. 158 (9 Mar. 1998), p. 430.

North Dakota was diagnosed in August 2000. Only 18 of all US cases in the 20th century involved inhalational anthrax (most of the cases were among wool or goat hair workers); the most recent report dates back to 1976. The remainder were cutaneous anthrax, and there were no reports of gastrointestinal anthrax.[189]

The manufacture of a lethal anthrax aerosol is believed by some to be beyond the capability of individuals or groups without access to advanced biotechnology.[190] This is in part related to difficulties in producing sufficient quantities of particles of the right size.

Letters as a means of delivering anthrax bacteria

On 5 October 2001 a 63-year-old man died from inhalational anthrax in Florida. The same day, a nasal swab taken from a co-worker, who had been admitted to a local hospital for pneumonia, tested positive for anthrax spores.[191] Given the extremely low incidence of anthrax the cases were an early indicator of an unnatural outbreak.[192] The Federal Bureau of Investigation (FBI) and public health officials began the search for the source of the spores.[193]

As of 7 December 2001, a total of 22 cases of confirmed or suspected anthrax exposure had been reported to the Centers for Disease Control and Prevention (CDC). Half of the victims contracted inhalational anthrax and five of them died. There were 7 confirmed and 4 suspected cases of cutaneous anthrax. The casualties occurred in Florida (2 inhalational and 0 cutaneous), New York (1 and 7), Washington, DC (5 and 0), New Jersey (2 and 4) and Connecticut (1 and 0).[194] The age of the people who developed inhalational anthrax ranged from 43 to 94 years; the incubation period from the time of exposure to the onset of the symptoms, when known, was 5–11 days. The incubation period for cutaneous anthrax was estimated to be 1–10 days.[195] No

---

[189] 'Use of anthrax vaccine in the United States: recommendations of the Advisory Committee on Immunization Practices (ACIP)' (note 180), pp. 1–2; Shirley, L. et al., 'Human anthrax associated with an epizootic among livestock North Dakota, 2000', *Morbidity and Mortality Weekly Report*, vol. 50, no. 32 (17 Aug. 2001), pp. 677–78; and Editorial note to the article, p. 678.

[190] Inglesby et al. (note 185), p. 1736.

[191] 'Update: investigation of anthrax associated with intentional exposure and interim public health guidelines, October 2001', *Morbidity and Mortality Weekly Report*, vol. 50, no. 41 (19 Oct. 2001), p. 890.

[192] Inglesby et al. (note 185), p. 1737.

[193] 'Update: investigation of anthrax associated with intentional exposure and interim public health guidelines, October 2001' (note 191).

[194] 'Update: investigation of bioterrorism-related anthrax—Connecticut, 2001', *Morbidity and Mortality Weekly Report*, vol. 50, no. 48 (7 Dec. 2001), p. 1077. Earlier, an additional case of suspected cutaneous anthrax was listed, but the person was removed from the statistic as he no longer met the CDC surveillance case definition for anthrax following the negative results of biopsies of the skin lesion. No new cases were reported between Dec. 2001 and Feb. 2002.

[195] 'Update: investigation of bioterrorism-related anthrax and interim guidelines for clinical evaluation of persons with possible anthrax', *Morbidity and Mortality Weekly Report*, vol. 50, no. 43 (2 Nov. 2001), p. 944; and Goldman, J. J. and Garvey, M. J., '5th person dies of anthrax: case baffles investigators', *Los Angeles Times*, 22 Nov. 2001 (Internet edn), URL <http://www.latimes.com>.

one claimed responsibility for these covert attacks or articulated particular demands.[196]

One or more mailed letters or packages were suspected as the source of exposure in Florida, and several environmental samples taken from regional and local postal centres tested positive for anthrax bacteria. Postal workers in Florida tested negative.[197] With regard to the cases outside Florida, the investigators were able to identify four letters which had been contaminated with anthrax spores. They had all been mailed from Trenton, New Jersey. Two letters, sent on 18 September, were addressed to the National Broadcasting Company (NBC) and *The New York Post*. The two other letters, postmarked 9 October, were sent to Democratic Senate Majority Leader Tom Daschle and Senator Patrick Leahy (the letter to Leahy was found among quarantined mail on 16 November). The victims included no addressee but several people responsible for opening the mail for the addressees were infected.[198] Other letters that contained powdery substances proved to be hoaxes.[199]

It has been established that significant amounts of the anthrax spores leaked from the four envelopes as a consequence of mechanical agitation by the high-speed sorting machines. The letters contained approximately 1 gram of spores, each in a concentration of approximately 1 trillion spores per gram. The pores in the paper of the envelopes are about 10 microns, whereas the anthrax spores were about 1 micron. The very fine particles were alleged to have been treated with a chemical that enables the particles to float in the air by eliminating the static charge that would make them clump.[200] This claim cannot be confirmed on the basis of public sources. All the letters contained anthrax spores of the same Ames strain, but it is unclear whether they were prepared in the same way.[201] The leaking spores not only infected the postal workers closest to the sorting machines, but also contaminated large sections of the postal facilities, and (re)aerosolization may have infected personnel further away.[202] They also

---

[196] Caruso, J. T., Testimony before the Subcommittee on Technology, Terrorism and Government Information of the Committee on Judiciary, US Senate, Washington, DC, 6 Nov. 2001, URL <http://judiciary.senate.gov/110601f-caruso.htm>.

[197] 'Update: investigation of bioterrorism-related anthrax and interim guidelines for exposure management and antimicrobial therapy, October 2001', *Morbidity and Mortality Weekly Report*, vol. 50, no. 42 (26 Oct. 2001), p. 909.

[198] 'Update: investigation of bioterrorism-related anthrax and interim guidelines for exposure management and antimicrobial therapy, October 2001' (note 197), p. 910; and Meyer, J., 'Tainted letter opened with care', *Los Angeles Times*, 6 Dec. 2001 (Internet edn), URL <http://www.latimes.com>.

[199] Garvey, M., 'Fugitive on FBI wanted list suspected in hoax letters', *Los Angeles Times*, 30 Nov. 2001 (Internet edn), URL <http://www.latimes.com>; and Eggen, D., 'Marshals arrest suspect in hoax anthrax mailings', *Washington Post*, 6 Dec. 2001, p. A26.

[200] In early 2002 the issue of silica (or another chemical preventing clustering) has been clouded with disinformation as part of the efforts by some US policy shapers to deny any direct or indirect involvement by US laboratories or to implicate a foreign government in the attacks.

[201] According to one assessment, the clumping noticed in Florida and in the letters sent to the New York media may have resulted from exposure to humidity during the mail processing or delivery. Hatch Rosenberg, B., 'Analysis of the source of the anthrax attacks', Federation of American Scientists, Washington, DC, 17 Jan. 2002, URL <http://www.fas.org/bwc/news/anthraxreport.htm>. The Ames strain is one of 89 known genetic varieties of the anthrax bacterium. Warrick, J., 'One anthrax answer: Ames strain not from Iowa', *Washington Post*, 29 Jan. 2002, p. A02.

[202] 'Evaluation of *Bacillus anthracis* contamination inside the Brentwood Mail Processing and Distribution Center—District of Columbia, October 2001', *Morbidity and Mortality Weekly Report*, vol. 50, no. 50 (21 Dec. 2001), pp. 1129–33.

contaminated other letters passing through the sorting machines, which may explain the infection of the individuals who became ill but who did not work in the vicinity of one of the addressees or in or near a postal facility.[203] The cross-contamination in the US postal system also explains the presence of anthrax spores in other countries, such as Lithuania and Peru.[204] Whether the intention of the sender was cross-contamination and infection of people is uncertain because he or she ensured that the envelopes were tightly sealed and included warnings of the presence of the anthrax bacteria and the need to take antibiotics.[205]

The ease with which the anthrax spores became aerosolized meant that, in addition to the mail facilities, large sections of the US Senate buildings and newspaper offices were contaminated, requiring extensive and costly clean-up operations. (The Brentwood Mail Processing and Distribution Center in the District of Columbia was still closed in the second quarter of 2002.) These were also hampered by the lack of consensus about what constitutes a safe environment following decontamination.[206]

Proliferation implications of the anthrax attacks

In the wake of the 11 September attacks there was intense speculation as to whether members of the al-Qaeda network had also been preparing for a chemical or biological attack. These fears were heightened by the discovery of a crop duster manual among the belongings of a man being held in FBI custody in connection with the 11 September attacks and by the subsequent letters filled with anthrax spores.

Initially, the Bush Administration assumed that the anthrax letters had been sent by al-Qaeda members in connection with the 11 September attacks.[207] The

---

[203] The case of a 94-year-old Connecticut woman remains a mystery as environmental samples taken from her home, local businesses and other areas, as well as the nasal swabs of 460 employees working in postal centres that process the mail for the victim's town all proved negative for anthrax spores. According to calculations based on detailed analysis of weather conditions, wind patterns and air turbulence, Martin Furmanski concluded that it was possible that anthrax spores from Trenton, N.J., had travelled all the way to Connecticut. She died on 21 Nov. 2001. 'Update: investigation of bioterrorism-related inhalational anthrax—Connecticut, 2001', *Morbidity and Mortality Weekly Report*, vol. 50, no. 47 (30 Nov. 2001), pp. 1050–51; and MacKenzie, D., 'Wind may explain mystery anthrax cases', *New Scientist* (Internet edn), 14 Dec. 2001, URL <http://www.newscientist.com>. In Jan. 2002 the source of her infection was still unknown.

[204] 'Update: investigation of bioterrorism-related anthrax and interim guidelines for exposure management and antimicrobial therapy, October 2001' (note 197), pp. 914–15; and Knight, W., 'Resilient anthrax spreads far and wide', *New Scientist* (Internet edn), 2 Nov. 2001, URL <http://www.newscientist.com/news/news.jsp?id=ns99991516>.

[205] 'Update: investigation of bioterrorism-related anthrax and interim guidelines for exposure management and antimicrobial therapy, October 2001' (note 197), p. 912; and Hatch Rosenberg (note 201).

[206] 'The science of bioterrorism: is the federal government prepared?', Charter to the hearing organized by Committee on Science, US House of Representatives, Washington, DC, 5 Dec. 2001, URL <http://www.house.gov/science/full/dec05/full_charter_120501.htm>.

[207] Johnston, D., 'In shift, officials look into possibility anthrax cases have bin Laden ties', *New York Times* (Internet edn), 16 Oct. 2001, URL <http://www.nytimes.com>; and Weiss, R. and Eggen, D., 'Additive made spores deadlier', *Washington Post*, 25 Oct. 2001, p. A01.

al-Qaeda network had been mentioned as possibly possessing CBW, but no such weapons were found following the capture of al-Qaeda sites in Afghanistan. Some literature on chemical and biological warfare was retrieved, but it was similar to that which can be downloaded from the Internet. In addition, for several years jihad war manuals have reportedly contained sections devoted to chemical and biological warfare and instructions on how to prepare toxins, toxic agents and drugs.[208]

The fear of an attack using a crop duster filled with a chemical or biological agent led the US Federal Aviation Administration (FAA) to impose a nationwide flying ban on such aircraft. Furthermore, Mohammed Atta, a central figure in the 11 September attacks, reportedly rented an aeroplane four times from an airfield less than 2 km from the residence of the first anthrax victim in Florida and questioned workers at a second airfield about crop dusters.[209]

However, crop dusters may not be suitable for the dissemination of CB agents. They have spray tanks in the typical range of 1514–1892 litres, although some aircraft can carry up to 3028 litres.[210] Considering that Aum Shinrikyo manufactured 6–7 litres of sarin for its attacks in the Tokyo underground and its production capability was in the range of tens of litres (although the sect's plans called for 70 tonnes of agent),[211] from the perspective of a potential terrorist the volume of spray tanks is huge. As agent production inside the country against which the terrorist attacks are planned appears unfeasible, the alternative is importation from abroad. This scenario, however, places high demands on the maintenance of the stability and viability of the agent during transport and storage. Considering the difficulties Iraq experienced in these respects, this may prove to be a significant challenge. Furthermore, the nozzles of the spray installation would typically produce droplets too large for optimal results or for inhalation of the agent.[212] Major modifications would have to be made to the nozzles in order to produce finer mists.[213] Crop dusters are designed for spraying at very low levels in order to achieve the right concentration, and much of the agent would evaporate or be

[208] Harris, E. D., Testimony before the Committee on International Relations, US House of Representatives, Washington, DC, 5 Dec. 2001, URL <http://www.house.gov/international_relations/harr1205.htm>.
[209] 'FBI imposes new restrictions on crop-dusters', Cable News Network, URL <http://www.cnn.com/2001/US/0923/inv.crop.dusters/index.html>; and Associated Press, 'Florida anthrax victim dies', ABC News, 5 Oct. 2001, URL <http://abcnews.go.com/sections/us/DailyNews/anthraxvictim 011005.html>.
[210] For technical specifications of the Air Tractor models of crop dusters see URL <http://www.airtractor.com/models/ATmodels.html>.
[211] Tu, A. T., 'Anatomy of Aum Shinrikyo's organization and terrorist attacks with chemical and biological weapons', *Archives of Toxicology, Kinetics and Xenobiotic Metabolism*, vol. 7, no. 3 (autumn 1999), pp. 51, 55.
[212] Agricultural crop dusters typically dispense materials with a particle size of 100 microns or more, whereas the particle size of a biological agent must be in the 1–10 micron range in order to penetrate the human lung. Smithson, A. E., Prepared statement before the Subcommittee on Oversight and Investigations of the Committee on Energy and Commerce, US House of Representatives, Washington, DC, 10 Oct. 2001, available at URL <http://www.stimson.org/cbw/?sn=CB20011221144>.
[213] UNSCOM found that Iraq experienced great difficulties in modifying the nozzles for spray tanks and was forced to buy the required components abroad. Richard Spertzel, former head of the UNSCOM BW inspections, cited in Broad, W. J., 'Experts call for better assessment of threats', *New York Times* (Internet edn), 2 Oct. 2001, URL <http://www.nytimes.com/>.

destroyed if the aircraft were to fly at higher altitudes.[214] These elements, taken together, do not make a crop duster ideal for disseminating CB agents for terrorist purposes.[215] By the end of 2001 the issue of the crop dusters had virtually disappeared from the discussion on CB attacks in the USA.

The purity and high concentration of the mail-delivered anthrax spores and the fact that they aerosolized easily became a cause of major concern, because the characteristics appeared to point to an origin in a military BW programme. This gave rise to speculation about state involvement in the attacks. Hardliners in the USA who are seeking a pretext to remove Iraqi President Saddam Hussein tried to link Iraq to the attacks.[216] The initial reports suggested that bentonite had been used as a chemical additive. According to former United Nations Special Commission on Iraq (UNSCOM) inspectors, the Iraqi BW programme used bentonite as part of a unique system to create anthrax spore powders that are light and easily airborne.[217] The claim that the spores had been chemically treated has not been publicly confirmed. The purity and concentration of the anthrax spores in the letter to Senator Daschle were later described as better than that produced in the Soviet, US or Iraqi BW programmes.[218]

By the end of October 2001 there were a growing number of indications that the source of the anthrax spores might be within the USA. The US administration objected to a French-sponsored UN Security Council resolution condemning the letter attacks on the grounds that the UN Security Council only deals with matters of international security.[219] In early November the FBI released a profile of the sender of the letter: a lone, Western individual who has scientific expertise and access to anthrax samples and a well-equipped laboratory and is used to working with highly hazardous substances.[220] In December, Professor Barbara Hatch Rosenberg of the Federation of American Scientists charged that the perpetrator is a US citizen working in the US biological defence programme.[221]

---

[214] Ron Manley, head of the Verification Division of the OPCW, cited in MacKenzie, D., 'Invisible enemies', *New Scientist*, vol. 172, no. 2311 (6 Oct. 2001), p. 6.

[215] However, crop dusters filled with a chemical toxicant could be used to great effect in a terrorist attack against agricultural produce.

[216] Rose, D. and Vulliamy, E., 'Iraq "behind US anthrax outbreaks"', *The Observer* (Internet edn), 14 Oct. 2001, URL <http://www.observer.co.uk/international/story/0,6903,573893,00.html>. There were also numerous allegations that Iraq had assisted al-Qaeda with the 11 Sep. attacks.

[217] Spertzel, R. O., Russia, Iraq, and other potential sources of anthrax, Testimony before the Committee on International Relations, US House of Representatives, Washington, DC, 5 Dec. 2001, URL <http://www.house.gov/international_relations/sper1205.htm>. Spertzel added that UNSCOM found evidence that Iraq was seeking a supply of pharmaceutical silica in 1988 and 1989, but it did not find definitive proof that the acquisition had actually occurred. Bentonite is also commercially produced in the USA and an Internet search produces several company names.

[218] Spertzel (note 217); and Hatch Rosenberg (note 201).

[219] US Department of State, Daily Press Briefing, 1 Nov. 2001, URL <http://www.state.gov/r/pa/prs/dpb/2001/5880.htm>.

[220] Federal Bureau of Investigation, 'Linguistic/behavioral analysis of anthrax letters', 9 Nov. 2001, URL <http://www.fbi.gov/majcases/anthrax/amerithrax.htm>.

[221] Hatch Rosenberg (note 201). Hatch Rosenberg made her claim public during her address as an NGO representative to the Fifth Review Conference of the States Parties to the BTWC in Nov. 2001 and circulated the first version of her paper on the SIPRI Internet CBW Discussion Forum. Its contents were

VI. CBW proliferation

The debate about the threat posed by the proliferation of CBW intensified in the latter part of 2001 following the terrorist attacks of 11 September, the mailing of anthrax-contaminated letters in the USA and reports of such letters (most of which were hoaxes) elsewhere in the world.

US proliferation allegations

Even before the September attacks the growing concern in the USA that an adversary, whether a state or a non-state actor, might use chemical or biological agents against it had contributed to massive resource allocation to defence and protection programmes at the national, state and local levels. The threat perception undoubtedly contributed to the US preparedness to name states that it perceives to be in contravention of the prohibitions of the BTWC and the CWC in unclassified reports and at diplomatic meetings, such as the Fifth Review Conference of the BTWC.[222] In his first State of the Union Address, on 29 January 2002, President George W. Bush described Iran, Iraq and North Korea as constituting an 'axis of evil'.[223]

In 2001 the US Secretary of Defense, the DOD Chemical and Biological Defense Program and the CIA each released reports on CBW proliferation and the implementation of measures to counter the threat. Together they named 10 countries as seeking CBW or as having the necessary infrastructure to start such programmes: China, India, Iran, Iraq, Libya, North Korea, Pakistan, Russia, Sudan and Syria. Some country assessments focus on the different stages of progress of the CBW programmes, and others appear to address regional instabilities and the possibility that governments might renege on their commitments to the BTWC and CWC should the regional security environment deteriorate.[224]

---

subsequently made public by Broad, W. J. and Miller, J., 'Anthrax inquiry looks at US labs', *New York Times* (Internet edn), 2 Dec. 2001, URL <http://www.nyt.com>.

[222] In 2000 the USA formally accused Iran of violating its CWC commitments, although the USA did not demand that a challenge inspection be conducted as it could have done under the provisions of the convention. Zanders, Hersh, Simon and Wahlberg (note 106), pp. 533–34.

[223] 'The President's State of the Union Address', The White House, Washington, DC, 29 Jan. 2002, URL <http://www.whitehouse.gov/news/releases/2002/01/print/20020129-11.html>.

[224] Office of the Secretary of Defense, *Proliferation: Threat and Response* (Department of Defense: Washington, DC, Jan. 2001); Department of Defense Chemical and Biological Defense Program, *Annual Report to Congress and Performance Plan* (Department of Defense: Washington, DC, July 2001), pp. 5–9; and Central Intelligence Agency, 'Unclassified report to Congress on the acquisition of technology relating to weapons of mass destruction and advanced conventional munitions, 1 January through 30 June 2001', Jan. 2002, URL <http://www.cia.gov/publications/bian/bian_jan_2002.htm>.

Iraq's CBW programmes and their elimination

There is serious concern about the status of Iraq's CBW programmes. In 2001 the US DOD reported that Iraq may be reconstituting its CBW capability.[225] A January 2001 press report quoted allegations by senior US government officials that Iraq had rebuilt several factories for the production of CBW in an industrial complex in al-Fallujah, west of Baghdad.[226]

When the inspections by UNSCOM ended in December 1998 there were many unresolved questions regarding Iraq's CBW programmes.[227] The UN Monitoring, Verification and Inspection Commission (UNMOVIC) succeeded UNSCOM in 1999, but as of January 2002 it had not yet conducted any inspections inside Iraq. UNMOVIC nonetheless continues to prepare for such inspections should Iraq allow the return of international inspectors.[228] The UN Security Council also placed Iraq under an international sanctions regime in order to compel it to comply with the conditions of Resolution 687, which includes the destruction of its CBW and the termination of the CBW-related programmes under international supervision.[229] In November 2001 the Security Council extended the sanctions regime, which had previously been modified in an attempt to gain Iraqi cooperation.[230] Earlier, UNMOVIC revised and refined the list of items and materials whose transfer to Iraq is controlled.[231]

The past South African CBW programme

Project Coast, under which the various components of South Africa's CBW programme were coordinated, was officially launched in 1981 and funded from 1982 until 1993. The trial of Brigadier Wouter Basson, the key figure in Project Coast, began in November 1999.[232] The criminal indictments are not directly connected to Basson's CBW activities, but information about Project Coast has emerged throughout the court hearings.[233]

Agent production in Project Coast appears to have focused on chemicals intended for crowd and riot control such as CR (Dibenz(b,f)-1:4-oxazepine)

---

[225] Office of the Secretary of Defense (note 224), pp. 40, 41; and Department of Defense Chemical and Biological Defense Program (note 224), p. 7.

[226] Lee Myers, S. and Schmitt, E., 'Iraq rebuilt weapons factories, officials say', *New York Times* (Internet edn), 22 Jan. 2001, URL <http://www.nytimes.com>. The details are not included in the DOD reports released in 2001, but the CIA biannual report contains a general statement regarding Iraq's reconstruction of its CBW programme. Central Intelligence Agency (note 224).

[227] Wahlberg, Leitenberg and Zanders (note 45), pp. 560–75.

[228] Note by the Secretary-General, UN Security Council document S/2001/1126, 29 Nov. 2001.

[229] UN Security Council Resolution 687, 3 Apr. 1991.

[230] Wahlberg, Leitenberg and Zanders (note 45), p. 565.

[231] Letter dated 1 June 2001 from the Executive Chairman of the United Nations Monitoring, Verification and Inspection Commission addressed to the President of the Security Council, UN document S/2001/560 [reissued for technical reasons], 15 Oct. 2001. The reissue corrected some clerical errors.

[232] Weekly reports of the Basson Trial are available from the Centre for Conflict Resolution, University of Capetown at URL <http://ccrweb.ccr.uct.ac.za/cbw/cbw_index.html>. Basson trial, week 45 report, 2–3 May 2001, prepared and distributed by Chandré Gould and Marlene Burger, Centre for Conflict Resolution, University of Capetown.

[233] Zanders, Hersh, Simon and Wahlberg (note 106), pp. 536–37. Some charges relate to Basson's illegal possession of documents pertaining to the CBW activities.

and BZ (3-Quinuclidinyl benzilate). At the trial Basson confirmed that from the end of 1991 until the beginning of 1993 weaponization of incapacitants was accelerated and that the programme was set to be completed in 1994. According to his testimony, BZ was acquired through Abdul Razak, a Libyan, who had acquired the agent from Hong Kong. Five tonnes of BZ were delivered and, except for 980 kg, all of it was weaponized by the South African Defence Force (SADF), the forerunner of the South African National Defence Forces (SANDF), between June and December 1992.[234]

Basson testified that the CR was produced at the Delta G Scientific facility in Midrand (located between Johannesburg and Pretoria) and claimed that it had been used once by the SADF during the final attack at Tumpo in Angola.[235] He also testified that a few hundred (or perhaps as many as 1000) 81-mm mortar shells had been imported from Israel and then filled with CR at Swartklip Products in 1987–88.[236] Roelf Louw, an employee of the arms manufacturer Armscor, provided similar information.[237]

The possession and use of incapacitants for purposes other than riot control are prohibited under the CWC, which opened for signature in January 1993. According to Basson, Minister of Defence Eugene Louw therefore ordered all incapacitants apart from tear gas to be destroyed. The Co-ordinating Management Committee (CMC) then decided to remove the CR canisters from the shells and to store them separately. A subsequent decision was made to dump the chemicals into the sea.[238]

The Basson trial has provided insight into foreign involvement in South Africa's CBW programme despite the UN sanctions regime. Basson's claims that 500 kg of methaqualone had been purchased with the assistance of Swiss intelligence chief Peter Regli prompted the Swiss Ministry of Defence to investigate the allegation in August 2001. Another 500 kg was allegedly obtained from Croatia through Swiss intelligence agent Jurg Jacomet and the Swiss intelligence services.[239] Basson confirmed that he signed a deal in 1992 with Franjo Kajfe, then Croatian Minister of Energy, concerning the manufacture of methaqualone, which was later used to produce mandrax. There are discrepancies in the testimony about the Croatian deal: the state prosecution claims that despite the money transfer the goods were never delivered. Basson maintains that they were delivered.[240]

The true extent of the CBW-relevant exchanges between Basson and Libya is unclear. Basson claimed that the only direct Project Coast transaction with

---

[234] Basson trial, week 49 report, 27 July 2001; and Gould, C., Centre for Conflict Resolution, University of Capetown, Private communication with F. Kuhlau, 22 Jan. 2002.

[235] Basson trial, week 48 report, 23 July 2001. Although Basson did not give a date in court, it is believed that the use of CW occurred during the 2nd attack on 23 Mar. 1988. Gould, C., Centre for Conflict Resolution, University of Capetown, Private communication with F. Kuhlau, 13 Feb. 2002.

[236] Basson trial, week 55 report, 10 Sep. 2001.

[237] 'The continuing trial of Wouter Basson', *CBW Conventions Bulletin*, no. 52 (June 2001), p. 32.

[238] Basson trial, week 49 report, 27 July 2001. The chemicals that were allegedly dumped in the sea included cocaine, ecstasy, methaqualone and BZ.

[239] 'The continuing trial of Wouter Basson', *CBW Conventions Bulletin*, no. 53 (Sep. 2001), p. 25.

[240] Babic, J., 'HDZ sold chemical weapons to the Republic of South Africa', *Nacional*, no. 310, 25 Oct. 2001, URL <http://www.nacional.hr/htm/310052.en.htm>.

Libya was the purchase of BZ and certain 'cultures' supplied by the University of Tripoli.[241] Dr David Chu, managing director of Medchem Forschungs (a company specifically set up by Basson to promote the Roodeplaat Research Laboratory, RRL, in Europe) testified that Libya was a potential buyer for RRL.[242] Basson also claimed that he had made a series of deals on behalf of East German, Libyan and Russian financial actors with the consent of General Kat Liebenberg, then chief of the SADF.[243] Basson informed the court about a group of CBW experts, led by Libyan Abdul Razak, which met regularly to exchange information and discuss developments in the field. It included 'Russians, Libyans, East Germans, Chinese, Americans and Swiss', and Basson admitted to having supplied the group with the results of research conducted at Protechnik.[244] The trial concluded in April 2002; Basson was acquitted.

VII. Conclusions

The process of strengthening the BTWC suffered a serious setback in 2001 with the suspension of the AHG as an appropriate forum to negotiate measures to reinforce the BTWC regime. The Fifth Review Conference has been suspended until November 2002. These developments leave the BTWC a weak disarmament treaty that lacks compliance, enforcement and verification provisions at a time when rapid technological advances in the fields of biology and biotechnology are straining the convention. In addition, the anthrax-contaminated letters in the USA underscore the reality of the use of biological agents for terrorist purposes. Despite the evident urgency of these developments the international community is not united in its approach to them. The US preference for addressing the proliferation threat by means of national policy initiatives and technology development programmes may lead other states to adopt a similar policy, which in turn might lead to international competition in BW defence. Many such activities are similar to those for the development of an offensive BW programme. In the absence of international instruments to generate transparency with respect to these activities, suspicion about the intent of other states could easily lead to an international biological armament competition. This would be more likely if the reconvened Fifth Review Conference fails to reach a final declaration in November 2002.

In contrast to the BTWC, the CWC has a functioning verification regime and all parties appear committed to its fundamental principles. The four states that are declared possessors of CW are moving forward with the destruction of their stockpiles and CW-related installations, although it appears increasingly likely that the final destruction deadlines will need to be extended for Russia

---

[241] Basson trial, week 49 report, 1 Aug. 2001.
[242] 'The continuing trial of Wouter Basson' (note 237), pp. 30–31.
[243] Dispatch Online, 27 Sep. 2001, URL <http://www.dispatch.co.za/2001/09/27/southafrica/MBASSON.htm>.
[244] 'The continuing trial of Wouter Basson' (note 239), pp. 23–24. Protechnik's activities were in the field of CW defence. It is currently South Africa's single small-scale facility, as defined by the CWC.

and the USA. The OPCW faces financial difficulties because some of its operational procedures require modification and because not all parties prioritize their obligations, such as the reimbursement of inspection costs. In accordance with the obligations of the CWC, the OPCW must consider how and where to devote its inspection resources, particularly as regards the chemical industry. Consequently, further financial shortfalls could seriously hamper some of its core activities. The OPCW also has a responsibility to assist parties to implement the CWC and in emergencies, such as the use or threat of use of CW. Failure to meet these obligations could lead states to question the relevance of the convention.

The First Review Conference of the States Parties to the CWC will be held in 2003. The quality of the preparatory work and the selection of implementation issues will be critical to its success. A high-level political commitment will be necessary to prevent inconclusive discussions on outstanding implementation issues. Failure to address fundamental issues could jeopardize the long-term viability of a relatively strong disarmament treaty.

The attacks of 11 September 2001 increased the sense of vulnerability to indiscriminate mass-casualty terrorism throughout the world. This sense of vulnerability was further augmented by a series of letters containing very high-grade anthrax particles that were sent to representatives of the US media and politicians. Five people died and another 17 contracted the disease. However, despite their coincidence in time the two events appear unrelated (except perhaps that the sender of the letters wanted to exploit the anxiety already present). The al-Qaeda attacks were driven by fanaticism opposed to the values of the dominant power in world politics. The anthrax-contaminated letters were possibly sent by a highly qualified scientist whose motives remain obscure. Despite the difference in scale, both events demonstrated the potential for social and economic disruption.

The BTWC and the CWC are not a panacea for dealing with CB terrorism, but they establish a core set of norms that govern the behaviour of states, companies and individuals. They offer a first line of defence against the terrorist use of CB agents by complicating the efforts of terrorists to acquire such weapons. In this context, too, the failure to achieve a protocol to the BTWC signifies that in the foreseeable future it will not be possible to establish an emergency assistance set-up in the event of the use of biological agents similar to the one being developed under the CWC.

13. Conventional arms control

ZDZISLAW LACHOWSKI*

I. Introduction

In 2001 there were a number of positive changes in the multilateral and regional conventional arms control regimes. The general trend was a focus by the international community on regional and domestic sources of conflict and relevant arms control measures, particularly operational measures. In Europe the focus was on the implementation of agreed measures and, after the 11 September terrorist attacks on the United States, the search for new approaches to the politico-military dialogue.

The second conference to review the operation of the 1990 Treaty on Conventional Armed Forces in Europe (CFE Treaty) was held in 2001. Russia's continued non-compliance with its agreed flank levels has hindered the entry into force of the 1999 Agreement on Adaptation of the CFE Treaty (Agreement on Adaptation). However, Russia has met its commitments regarding troop withdrawals from Moldova. In Georgia the future of one Russian military base and the continued presence of Russian forces remain to be resolved. The Balkan arms control regimes worked well, and the agreement on regional stabilization 'in and around Yugoslavia' was successfully concluded. Regional and bilateral confidence- and security-building measures (CSBMs) continued to work smoothly, and new bilateral CSBMs were introduced in Europe. The Organization for Security and Co-operation in Europe (OSCE) military doctrine seminar evaluated the new threats and challenges and identified possible additional directions for the work of the OSCE in the CFE zone of application. After years of deadlock the 1992 Treaty on Open Skies entered into force on 1 January 2002, after Belarus and Russia ratified it in 2001.

The number of parties to the Convention on the Prohibition of the Use, Stockpiling, Production and Transfer of Anti-Personnel Mines and on their Destruction (APM Convention) continued to increase. The Second Review Conference of the 1981 Convention on Prohibitions or Restrictions on the Use of Certain Conventional Weapons which may be deemed to be Excessively Injurious or to have Indiscriminate Effects (CCW Convention or 'Inhumane Weapons Convention') extended the application of the convention to domestic armed conflicts.

---

* Maaike Reijlink contributed to the subsections on the fourth Vienna military doctrine seminar and the OSCE–Korea CSBM seminar.

SIPRI Yearbook 2002: Armaments, Disarmament and International Security

Table 13.1. CFE and CFE-1A ceilings and holdings in the ATTU zone, as of 1 January 2002[a]

State	Tanks Ceilings	Tanks Holdings	ACVs Ceilings	ACVs Holdings	Artillery Ceilings	Artillery Holdings	Aircraft Ceilings	Aircraft Holdings	Helicopters Ceilings	Helicopters Holdings	CFE 1A Manpower Ceilings	CFE 1A Manpower Holdings
Armenia	220	110	220	146	285	229	100	6	50	7	60 000	44 618
Azerbaijan	220	220	220	210	285	282	100	48	50	15	70 000	69 966
Belarus	1 800	1 608	2 600	2 507	1 615	1 471	294	212	80	58	100 000	79 870
Belgium	334	146	1 005	743	320	270	232	135	46	46	70 000	39 123
Bulgaria	1 475	1 475	2 000	1 885	1 750	1 738	235	232	67	43	104 000	54 495
Canada	77	0	263	0	32	0	90	0	13	0	10 660	
Czech Rep.	957	622	1 367	1 241	767	585	230	112	50	34	93 333	49 491
Denmark	353	238	336	311	503	479	106	68	18	12	39 000	25 293
France	1 306	1 084	3 820	3 339	1 292	764	800	588	374	284	325 000	184 988
Georgia	220	90	220	114	285	109	100	7	50	3	40 000	40 000
Germany	4 609	2 460	3 281	2 382	2 445	1 725	900	386	280	202	345 000	271 806
Greece	1 735	1 735	2 498	2 176	1 920	1 901	650	523	65	20	158 621	158 621
Hungary	835	743	1 700	1 478	840	834	180	92	108	49	100 000	33 408
Italy	1 348	1 253	3 339	2 934	1 955	1 404	650	497	142	133	315 000	173 522
Kazakhstan	50	0	200	0	100	0	15	0	20	0	0	0
Moldova	210	0	210	209	250	148	50	0	50	0	20 000	7 227
Netherlands	743	328	1 040	689	607	392	230	143	50	14	80 000	37 981
Norway	170	141	275	245	491	184	100	72	24	0	32 000	14 733
Poland	1 730	1 144	2 150	1 392	1 610	1 482	460	207	130	111	234 000	162 693
Portugal	300	187	430	353	450	363	160	101	26	0	75 000	36 751
Romania	1 375	1 258	2 100	2 051	1 475	1 384	430	204	120	22	230 000	109 143
Russia	6 350	5 066	11 280	9 647	6 315	5 874	3 416	2 406	855	523	1 450 000	650 802
Slovakia	478	272	683	534	383	374	100	79	40	19	46 667	32 366
Spain	891	698	2 047	1 002	1 370	1 054	310	191	80	28	300 000	160 372
Turkey	2 795	2 445	3 120	2 831	3 523	2 990	750	343	130	28	530 000	515 749
Ukraine	4 080	3 895	5 050	4 725	4 040	3 705	1 090	855	330	205	450 000	305 000
UK	1 015	608	3 176	2 344	636	459	900	511	356	267	260 000	206 762
USA	4 006	657	5 152	1 639	742	327	784	228	396	132	250 000	98 232

[a] Iceland and Luxembourg have no TLE in the application zone.

Source: Conventional Armed Forces in Europe (CFE): A Review and Update of Key Treaty Elements (US Department of State: Washington, DC, Jan. 2002).

II. European arms control

European arms control remains the most advanced regime of its type worldwide. In the past 12 years it has evolved remarkably, embracing pan-European, regional, structural and operational measures and mechanisms to address the emerging threats and challenges in Europe. European arms control has reduced the threat of large-scale military attack and has enhanced confidence, cooperation and mutual reassurance in Europe.

The Treaty on Conventional Armed Forces in Europe

The 1990 CFE Treaty set equal ceilings within its Atlantic-to-the-Urals (ATTU) zone of application on the major categories of heavy conventional armaments and equipment of the groups of states parties—originally the members of the North Atlantic Treaty Organization (NATO) and the Warsaw Treaty Organization (WTO). There are 30 parties to the CFE Treaty.[1] The main reduction of excess treaty-limited equipment (TLE) was carried out in three phases from 1992 to 1995. In January 2002 Russia appeared not to object to the Baltic states joining NATO, provided that they first become parties to the CFE Treaty.[2]

The 1999 Agreement on Adaptation introduced a new regime of arms control that discards the bipolar concept of a balance of forces. It is based on national and territorial ceilings, codified in the agreement's protocols as binding limits, and opens the CFE Treaty to European states which are not yet parties.[3] The agreement has not entered into force, mainly because of the refusal of the NATO and other states to ratify it in the face of Russia's continuing violation, in the North Caucasus, of the provisions of the CFE Treaty. Only Belarus has ratified the Agreement on Adaptation and deposited its instrument of ratification with the depositary, the Netherlands. The CFE Treaty and the associated documents and decisions therefore continue to be binding on all parties, and the Joint Consultative Group (JCG)—established to monitor implementation, resolve issues arising from implementation and consider measures to enhance the viability and effectiveness of the CFE Treaty—continues to prepare for the entry into force of the Agreement on Adaptation.

By 1 January 2001 more than 63 500 pieces of conventional armaments and equipment within and outside the ATTU zone had been scrapped or converted to civilian use by the parties, with many parties reducing their holdings to

---

[1] The parties to the CFE Treaty are listed in annex A in this volume. For discussion of conventional arms control in Europe before 1999 see the relevant chapters in previous editions of the SIPRI Yearbook. For the text of the CFE Treaty and Protocols see Koulik, S. and Kokoski, R., SIPRI, *Conventional Arms Control: Perspectives on Verification* (Oxford University Press: Oxford, 1994), pp. 211–76; and the OSCE Internet site at URL <http://www.osce.org/docs/english/1990-1999/cfe/cfetreate.htm>.

[2] Interfax (Moscow), 11 Jan. 2002, in 'Moscow sources: Russia may quit arms treaty if Baltic states join NATO', Foreign Broadcast Information Service, *Daily Report–Central Eurasia* (*FBIS-SOV*), FBIS-SOV-2002-0111, 11 Jan. 2002. As of Apr. 2002 there was no official Russian position on this issue. The view of the NATO states is that no formal linkage can exist between NATO enlargement and the CFE.

[3] For the text of the Agreement on Adaptation see *SIPRI Yearbook 2000: Armaments, Disarmament and International Security* (Oxford University Press: Oxford, 2000), pp. 627–42.

lower levels than required. Data on CFE ceilings and holdings in the treaty application zone as of 1 January 2002 are presented in table 13.1.

The Second CFE Review Conference

In accordance with Article XXI of the CFE Treaty, the Second Conference to Review the Operation of the Treaty on Conventional Armed Forces in Europe and the Concluding Act of the Negotiation on Personnel Strength (CFE-1A) was held in Vienna on 28 May–1 June 2001. The aim of the conference was to assess the implementation of the CFE Treaty and its associated documents since the first review conference, which was held in 1996.[4] The general assessment by the parties was that the regime has operated in a satisfactory manner. The participants reaffirmed all of the obligations and commitments undertaken at the 1999 OSCE Istanbul Summit Meeting.[5] They also noted that some issues required further consideration and resolution in the JCG, including treaty operation and implementation, unaccounted-for and uncontrolled treaty-limited equipment (UTLE), other non-compliance matters and arrangements for the entry into force of the Agreement on Adaptation of the CFE Treaty.[6]

Treaty operation and implementation issues

The second review conference addressed issues related to entry into force and focused on updating the Protocol on Existing Types of Conventional Armaments and Equipment (POET) before entry into force of the Agreement on Adaptation. The parties have modernized their arsenals, removing various types, models and versions of equipment and introducing new ones. Since some treaty definitions are unclear or ambiguous, different national interpretations of the definitions have developed, resulting in different national implementation practices. If the list of weapons to be covered by the adapted treaty is not clarified this will cause political and legal problems, hamper the work of inspectors and cause ambiguity and friction. It will also complicate the accession of new parties after entry into force of the adapted treaty, since they would probably possess new types of equipment. The second review confer-

---

[4] The Final Document of the First Conference to Review the Operation of the Treaty on Conventional Armed Forces in Europe and the Concluding Act of the Negotiation on Personnel Strength, Vienna, 31 May 1996; and Annex A, Document agreed among the States Parties to the Treaty on Conventional Armed Forces in Europe of 19 November 1990 (the Flank Document) are reproduced in *SIPRI Yearbook 1997: Armaments, Disarmament and International Security* (Oxford University Press: Oxford, 1997), pp. 511–17.

[5] OSCE, Final Act of the Conference of the States Parties to the Treaty on Conventional Armed Forces in Europe, Istanbul, 17 Nov. 1999. The text is reproduced as appendix 10B in *SIPRI Yearbook 2000* (note 3), pp. 642–46.

[6] Formal Conclusions of the Second Conference to Review the Operation of the Treaty on Conventional Armed Forces in Europe and the Concluding Act of the Negotiation on Personnel Strength, CFE document CFE.DOC/1/01, 1 June 2001. The conference was chaired by Italy. For a detailed analysis of the Second CFE Review Conference see Dunay, P., 'Der KSE-Prozess nach der Zweiten Überprüfungskonferenz des Vertrags' [The CFE process after the Second Review Conference of the Treaty], *OSZE-Jahrbuch 2001: Jahrbuch zur Organisation für Sicherheit und Zusammenarbeit in Europa (OSZE)* (Nomos Verlagsgesellschaft: Baden-Baden, 2001), pp. 321–40.

ence therefore urged the JCG to update POET in line with the agreement reached at the first review conference: to correct any inaccuracies, including removal of types, models and versions of equipment that do not meet the treaty criteria;[7] to discuss an annual update by the JCG, if appropriate; and to consider the creation of an electronic version of the lists in all official languages.

Other implementation issues were discussed, including limitations and related treaty obligations, interpretation of treaty counting rules, notifications and exchange of information, verification, and preparation for entry into force of the Agreement on Adaptation and its implementation.[8]

Special emphasis was put on the issue of UTLE. This type of equipment is present in several places in the area of application: in Nagorno-Karabakh (Armenia/Azerbaijan), the Abkhazia and Tshinkvali region and South Ossetia (Georgia), and the Trans-Dniester region (Moldova). Resolution of the UTLE issue lies in achieving a political settlement in these regions rather than in military–technical arrangements. The parties noted that this situation adversely affects the CFE regime and promised to continue to address the issue in the JCG, as tasked by the first review conference. In 2001, as in previous years, there remained an unresolved discrepancy of 1970 TLE items between actual levels and the aggregate amount of TLE that the eight former Soviet republics were committed to destroy or convert based on Soviet data submitted at the signing of the CFE Treaty in 1990. Most of the UTLE is believed to be derelict or not under government control in the Caucasian states.

Although abiding by its overall treaty limitations,[9] since 31 May 1999 Russia had been in breach of the 1996 Flank Document.[10] Its holdings of TLE in the flank zone exceeded the agreed limits, although the excess of equipment has been progressively diminishing. The Second CFE Review Conference acknowledged that Russia had met its obligations regarding equipment east-of-the-Urals under Annex E of the 1996 Flank Document.[11] Russia had completed the destruction of the total quantity of TLE necessary to meet its commitments and continued to scrap 2300 tanks, as required.

---

[7] According to the chairman of the POET Working Group, the goal should be a comprehensive update of the protocol including: adding new types of equipment, deleting possible types that do not meet treaty definitions, renaming types that are listed under the wrong names, reclassifying types that are listed under the wrong categories/subcategories, and identifying types that are no longer in operational service. Joint Consultative Group document JCG.TOI/10/01, 30 Oct. 2001.

[8] Denmark cited the occasional lack of common understanding of treaty definitions, instances of refusal to allow inspection of areas associated with declared sites, the taking of photographs during inspections, etc. The problem of the status of some TLE at Russian repair facilities as 'not combat capable' also remains unsettled. Denmark declared that its aim was not to criticize but to 'make practice correspond better' to the spirit and letter of the CFE Treaty. Delegation of Denmark to the OSCE, Opening Statement, Second Review Conference of the CFE Treaty, Second Review Conference document RC.DEL/22/01, 28 May 2001.

[9] In 2000 and 2001 Russia was requested to clarify its movements of artillery systems in order to alleviate concern that it might be exceeding its treaty limits.

[10] See note 4.

[11] CFE, Final Document (note 4), Annex E: Statement of the Representative of the Russian Federation to the Treaty on Conventional Armed Forces in Europe. Annex E is reproduced in *SIPRI Yearbook 1997* (note 4), pp. 515–17. Under its 14 June 1991 political commitment, the Soviet Union was to reduce a total of 14 500 items outside the ATTU zone.

In October Ukraine announced the completion of its reduction obligations under the extended deadline for conversion of some of its TLE. In 2001 it converted the remaining 131 ACVs.[12] There was also a dispute between Azerbaijan and Armenia about an alleged inconsistency in the Armenian data furnished to the JCG concerning the number of 'recovered' tanks.[13]

The JCG Group on Treaty Operation and Implementation (TOI) continued to work on such issues as the distribution of inspection costs between the inspected and inspecting states parties, establishing the notification formats that will be required to implement the Agreement on Adaptation,[14] declared site diagrams and access within declared sites. The TOI Group agreed on formats for scheduled and ad hoc notification and exchange of information, and formats for certain verification activities were actively considered at the end of 2001. However, although some progress was made, the problem of 'paid' (i.e., conducted at the expense of the inspecting or observing party) inspections was not resolved.[15]

Russian non-compliance with flank limitations

Since the autumn of 1999 the Russian equipment in Chechnya have exceeded the numbers allowed by the CFE Treaty's flank limitations. Russia has also sought to ensure its CFE partners that the increase in equipment was of a temporary nature and has gradually reduced its TLE quantities. On 22 January 2001, President Vladimir Putin announced a plan to hand over responsibility for operations in Chechnya to the Federal Security Service (Federal'naya Sluzhba Bezopasnosti, FSB) and to reduce Russia's armed forces in Chechnya to a 15 000-man army division and 7000 internal security troops. Neither a timetable nor details of equipment reduction were provided.[16] However, in the light of the hostilities in Chechnya, Putin stopped the troop withdrawals in early May.

At the second review conference Russia provided new data indicating that the quantity of its equipment in the flank zone had decreased considerably to 100 armoured combat vehicles (ACVs) in excess of the agreed level (see table 13.2). Russia referred to the 'obviously tangible tendency towards full implementation of the flank obligations' and called on other states to follow

---

[12] Ukraine converted 121 BMP-1 AIFVs and 10 BTR-60 APCs to civilian use. Joint Consultative Group document JCG.Jour/438, 23 Oct. 2001.

[13] In Oct. 2001 Armenia informed the JCG about 8 additional T-54/55 tanks 'recovered from various parts retrieved from the scene of border clashes' in 1992–94. Azerbaijan claimed that the notified loss of tanks by Armenia concerned T-72s exclusively. Armenia replied that it had recovered the tanks from the Azerbaijani losses. Joint Consultative Group documents JCG.DEL/29/01, 30 Oct. 2001; JCG.DEL/30/01, 6 Nov. 2001; and JCG.DEL/32/01, 13 Nov. 2001.

[14] The CFE adaptation process resulted in additional inspections equal to 25% of the passive declared site inspection quota of the states, which are to be conducted at the expense of the inspecting state.

[15] Letter from the Chairman of the Joint Consultative Group to the Minister of Foreign Affairs of Romania, Chairman of the Ninth Meeting of the Ministerial Council of the OSCE, Joint Consultative Group document JCG.DEL/37/01/Rev.1, 27 Nov. 2001.

[16] 'Putin scaling down war despite new fighting', *New York Times* (Internet edn), 23 Jan. 2001, URL <http://www.nytimes.com/2001/01/23/world/23RUSS.html>.

the example of Belarus and ratify the Agreement on Adaptation; Russia also reiterated its intention to do so.[17] In addition, Russia warned against harming the CFE Treaty by withdrawal from key non-proliferation and nuclear arms control agreements, upsetting the complex balance of military capabilities in Europe as a whole or in specific regions, using or threatening the use of force without the sanction of a United Nations Security Council resolution or conducting an ill-advised bloc policy. Russia cautioned against admitting the Baltic states to NATO because of the potentially adverse effect on the key provisions of the CFE Treaty, especially those concerning the flank and the Central European stability zone.[18]

The parties welcomed Russia's provision of new information. However, most delegations demanded that Russia supply more data and be more transparent in a manner consistent with CFE counting rules and procedures, including additional inspections to monitor the TLE withdrawals.[19] The NATO countries insisted that the often repeated Russian commitments concerning the flank zone (prompt reduction of Russian holdings to agreed levels) must be met before they initiate national ratification processes.

The change in the Russian–US relationship in the wake of the 11 September terrorist attacks and the US decision, on 13 December, to withdraw from the 1972 Treaty on the Limitation of Anti-Ballistic Missile Systems (ABM Treaty) affected the evolution of Russia's position on European security.[20] In November, the Russian delegate to the JCG announced that the Russian Defence Ministry had approved plans to finalize reductions of the Russian forces in North Caucasus in line with the agreed flank ceilings.[21] In January 2002 Russia announced that it had complied with the agreed limitations and renewed its call for the NATO states to ratify the Agreement on Adaptation.[22]

Withdrawal of Russian TLE from Georgia

At the Istanbul OSCE Summit Meeting Russia pledged that it would reduce the level of its heavy ground weapons on Georgian territory to the equivalent of a brigade.[23] The Russian TLE located at Vaziani and Gudauta (Abkhazia)

---

[17] Statement by the Director of the Department for Security Affairs and Disarmament Issues, Russian Ministry of Foreign Affairs Yuri S. Kapralov, delivered at the Second CFE Review Conference, Second Review Conference document RC.DEL/23/01, 28 May 2001. Ukraine also ratified the Agreement on Adaptation but did not deposit its instrument of ratification.

[18] Kapralov (note 17).

[19] E.g., the US representative demanded that 'the excess Russian holdings and "temporary presence" must be eliminated in a way that Treaty partners can readily understand and verify'. Opening remarks. Assistant Secretary of State Avis. T. Bohlen, Second Review Conference document RC.DEL/1/01/Rev.1, 28 May 2001.

[20] See chapter 10 in this volume.

[21] Joint Consultative Group document JCG.DEL/38/01, 28 Nov. 2001.

[22] Russia also reportedly suggested further changes in the adapted CFE Treaty regime, including tougher restrictions on combat aircraft and more stringent regulations regarding temporary deployments of NATO forces in areas adjacent to Russia. Interfax (Moscow), 11 Jan. 2002, in 'Russia expects NATO to ratify adapted treaty on conventional forces', FBIS-SOV-2002-0111, 11 Jan. 2002.

[23] OSCE (note 5); and *SIPRI Yearbook 2000* (note 3), p. 646. The basic temporary deployment is 153 tanks, 241 ACVs and 140 artillery pieces.

Table 13.2. Russian entitlements and holdings in the flank zone under the 1999 Agreement on Adaptation, 1999–2002

	Tanks	ACVs	Artillery
Territorial sub-limits for revised flank zone[a]	1 300	2 140	1 680
Holdings in the revised flank			
Oct. 1999	1 493	3 534	1 985
July 2000	1 442	3 017	1 857
Nov. 2000	1 327	2 790	1 746
May 2001	1 304	2 246	1 609
Jan. 2002	1 294	2 044	1 557

ACVs = armoured combat vehicle

[a] In the Leningrad military district (MD), excluding the Pskov *oblast* (region); and in the North Caucasus MD, excluding: the Volgograd *oblast*; the Astrakhan *oblast*; that part of the Rostov *oblast* east of the line extending from Kushchevskaya to the Volgodonsk *oblast* border, including Volgodonsk; and Kushchevskaya and a narrow corridor in Krasnodar *kray* (territory) leading to Kushchevskaya.

were scheduled to be removed, and those two bases as well as the repair facilities at Tbilisi were to be closed by 1 July 2001. Georgia agreed that Russia could temporarily deploy TLE at the Batumi and Akhalkalaki bases. The OSCE established a voluntary fund to help Russia finance the withdrawal of forces, and several OSCE states have contributed to it.[24] By the end of 2000 Russia had completed the scheduled reductions and destroyed additional quantities of heavy ground weapons.[25]

The withdrawal is complicated by the volatile situation in Georgia and near its borders. Progress was slowed in 2001 by accusations by Georgia and Russia against each other in the JCG and by a lack of dialogue between the two states. In the first half of 2001 the future use of the Vaziani and Gudauta bases remained unresolved, as did the issue of the long-term presence of Russian forces in Batumi and Akhalkalaki. Russia handed over control of its Vaziani base to Georgia on 29 June 2001, but it failed to pull out of the Gudauta base by 1 July. The failure was alleged to be 'beyond the control of the Russian side' and because of the 'opposition of [the] local Abkhaz and Russian population and [the] lack of conditions for a safe withdrawal of the Russian personnel and military equipment, which the Abkhaz armed forces could take possession of and use in new hostilities against Georgia'.[26] Russia

---

[24] OSCE, Final Act (note 5), para. 19.

[25] Lachowski, Z., 'Conventional arms control', *SIPRI Yearbook 2001: Armaments, Disarmament and International Security* (Oxford University Press: Oxford, 2001), pp. 556–57.

[26] Interfax (Moscow), 2 July 2001, in 'Russia defends itself against Georgian criticism of slow troop withdrawal', FBIS-SOV-2001-0702, 2 July 2001; and Radio Free Europe/Radio Liberty, 'Russia hands over one military base to Georgia on schedule' and '. . . But not a second', Radio Free Europe/Radio Liberty, *RFE/RL Newsline*, 2 July 2001, URL <http://www.rferl.org/newsline/2001/07/020701.asp>. Russian Foreign Minister Igor Ivanov suggested that 'with the support of Chechen and international terrorists, the Georgian side provoked hostilities in Abkhazia, which clearly made it even more difficult

proposed that 300 Russian military personnel be permitted to remain at the base to conduct peace operations and guard equipment. Georgia rejected the demand as unacceptable, and the two governments held talks to find a compromise, such as allowing Russian troops to remain temporarily at Gudauta. In early November Russia declared that the military base had been dismantled and the troops withdrawn. However, Georgia alleged that Russia had not complied with transparency measures regarding the remaining Russian military personnel at Gudauta and the schedule for the pullout. Georgia refused to confirm the Russian withdrawal pending the resolution of these and related issues. In December the OSCE called for the resumption of Georgian–Russian negotiations concerning transparency measures with regard to the closure of the base at Gudauta.[27] The terms of the Russian withdrawal from Batumi and Akhalkalaki have not been agreed. Georgia has proposed a three-year withdrawal period; Russia has suggested a 14-year withdrawal schedule.[28] The situation was complicated in October by allegations that Chechen military forces were fighting alongside Georgian partisans in the breakaway province of Abkhazia.[29] Talks between Georgian and Russian experts were resumed in the JCG in February 2002.

The issue of Russian TLE in Moldova

Under its 1994 constitution, Moldova is permanently neutral and refuses to host foreign forces on its territory. However, the 1994 agreement with Russia on the withdrawal of Russian troops has not entered into force. At the 1999 Istanbul OSCE Summit Meeting, Russia pledged to withdraw and/or destroy Russian treaty-limited conventional armaments and equipment by the end of 2001 and to pull out its troops by the end of 2002.[30] A decision was taken to facilitate the withdrawal and destruction of Russian armaments and to establish an OSCE-administered fund for that purpose.[31]

The OSCE and other Western states have repeatedly criticized Russia for lack of progress in the withdrawal of its troops and armaments from the Trans-Dniester region, noting that the last shipment of Russian arms and military equipment from the region was in November 1999. Concerns have been expressed regarding Russia's ability to meet the schedule for withdrawals. However, Russia has consistently stated that it will carry out its CFE/OSCE commitments (weapon destruction or withdrawal plus disposal of some 42 000 tonnes of Soviet-vintage munitions). Russia blamed the delay on inadequate funds, the unsettled relations between Moldova and its separatist Trans-

to reach agreement with Sukhumi'. Letter of the Minister of Foreign Affairs of the Russian Federation, 12 Nov. 2001, OSCE document SEC.DEL/29/01, 15 Nov. 2001 (in Russian).

[27] Decision no. 2, Statements by the Ministerial Council, (2), OSCE Ministerial Council, Bucharest, 2001, OSCE document MC(9).DEC/2, 4 Dec. 2001.

[28] Interfax (Moscow), 9 June 2001, 'Georgia insists on withdrawal of Russian bases within three years', FBIS-SOV-2001-0609, 9 June 2001.

[29] 'Tensions between Russia and Georgia reach new heights', *Financial Times*, 12 Oct. 2001, p. 7.

[30] OSCE, Final Act (note 5), para. 19. The *c.* 42 000 tonnes of ammunition stored in the Trans-Dniester region pose a grave threat to this unstable region.

[31] OSCE, Final Act (note 5), para. 19.

Dniester region, the obstruction of the Trans-Dniester administration and its economic demands. Following confirmation that the OSCE would fund the disposal of the Russian armaments remaining in eastern Moldova, the destruction of heavy weapons at the Operative Group of Russian Forces (the former 14th Army) base in Tiraspol began in June, under the supervision of the OSCE Mission to Moldova.[32] In July and August significant progress was made in the dismantling and withdrawal of 108 T-64 tanks, 131 ACVs and 125 heavy artillery pieces. After a brief pause in early September the reduction process was renewed, and the opposition of the Tiraspol authorities was overcome by a deal with Russia for the reduction of the Trans-Dniester region's debt. In mid-November Russia announced the completion of the withdrawal of its TLE from Moldova.[33] The announcement was welcomed by the OSCE, which expressed its expectation that the withdrawal would be officially confirmed at the Bucharest OSCE ministerial meeting in December 2001.[34] The Ministerial Council in Bucharest commended Russia on accomplishing the withdrawal ahead of schedule, which it hailed as a model for constructive and fruitful cooperation in dealing with other issues.[35]

Regional arms control in Europe

Arms control in the Balkans is designed to play an important stabilizing role in post-conflict security building.[36] The 1996 Agreement on Sub-Regional Arms Control (Florence Agreement, also known as the Article IV Agreement)—signed by Bosnia and Herzegovina and its two entities (the Muslim–Croat Federation of Bosnia and Herzegovina and the Republika Srpska) Croatia and the Federal Republic of Yugoslavia (FRY, Serbia and Montenegro)—remains the only structural (i.e., dealing with arms reductions and limitations) regional arms control arrangement still operating below the European level.[37] The characteristic feature of this arms control agreement is that compliance with its terms is both monitored and assisted from outside by the international com-

---

[32] 'OSCE submits plan to dispose of 40 000 tons of Russian munitions in Moldova', OSCE Press Release, 4 Oct. 2001, URL <http://www.osce.org/news/generate.php3?news_id=2026>

[33] According to the Russian military, all the equipment was to be transferred beyond the Urals. ITAR-TASS (Moscow), 14 Nov. 20001, in 'Russia fulfils obligation to withdraw weapons from Dniester region', FBIS-SOV-2001-1114, 14 Nov. 2001.

[34] 'US statement on removal of Russian equipment from Moldova', *Washington File*, 23 Nov. 2001, URL <http://usinfo.state.gov/cgi-bin/washfile/display.pl?p=/products/washfile/topic/intrel&f=01112301.wpo&t=/products/washfile/newsitem.shtml>.

[35] These issues include the withdrawal from the Trans-Dniester region of the Russian equipment not limited by the CFE Treaty and the withdrawal or destruction of ammunition belonging to Russia. Decision no. 2 (note 27).

[36] Under the terms of the General Framework Agreement for Peace in Bosnia and Herzegovina (Dayton Agreement), 21 Nov. 1995, Annex 1-B, Agreement on Regional Stabilization, negotiations were launched with the aim of agreeing on CSBMs in Bosnia and Herzegovina (Article II), reaching an arms control agreement for the former Yugoslavia (Article IV) and establishing 'a regional balance in and around the former Yugoslavia' (Article V). The Agreement on Regional Stabilization is reproduced in *SIPRI Yearbook 1996: Armaments, Disarmament and International Security* (Oxford University Press: Oxford, 1996), pp. 241–43.

[37] In this section 'regional' in the OSCE context refers to areas below the CFE/OSCE level. The text of the Florence Agreement is reproduced in *SIPRI Yearbook 1997* (note 4), pp. 517–24.

munity. In contrast to the general situation in Europe, the military security of the subregion is built on a balance of forces among the local powers, which have not developed a satisfactory degree of security cooperation.

The Florence Agreement

In 2001 stability prevailed following two years of problems related to the implementation of the Florence (Article IV) Agreement. The quality of the annual exchange of information improved. The parties focused on the difficult issues of inspections by Bosnia and Herzegovina and exempted equipment (under Article III) and on voluntary reductions of holdings in the five categories of weapon covered by the agreement. Inspections, with one group of exceptions, were carried out with relative ease, and OSCE assistants took part as observers in most inspections.

Two major issues are unresolved. The first issue concerns inspections by Bosnia and Herzegovina (the agreement gives all parties the right to carry out inspections). It has not been able to conduct such inspections because the Republika Srpska is blocking the inspections. The second issue is how to encourage the parties to reduce their levels of agreement-limited armaments (ALA) that are exempt from the counting rules and to lower their ceilings for ALA.[38]

III. European CSBMs

In 2001 the implementation of the Vienna Document 1999 of the Negotiations on Confidence- and Security-Building Measures in Europe proceeded smoothly.[39] The eleventh Annual Implementation Assessment Meeting (AIAM) took place in Vienna on 26–28 February 2001. The delegations emphasized the need to adapt the document to changed circumstances, if necessary, rather than to renegotiate it. Many delegations stressed the importance of the 2000 OSCE Document on Small Arms and Light Weapons (SALW) and called for its early and full implementation.[40] The Forum for Security Cooperation (FSC) held extensive discussions on the SALW document in order to assist states in their preparation for the first information exchange, held on 30 June 2001. The FSC later decided to hold a SALW workshop in Vienna on 4–5 February 2002.[41]

---

[38] Annual Report on the Implementation of the Agreement on Confidence- and Security-Building Measures in Bosnia and Herzegovina (Article II, Annex 1-B), and the Agreement on Sub-Regional Arms Control (Article IV, Annex 1-B, Dayton Peace Accords), 1 Jan.–30 Nov. 2001. Major General Claudio Zappulla (Italian Air Force), Personal Representative of the OSCE CIO, OSCE document CIO.GAL/71/01, 26 Nov. 2001.

[39] Vienna Document 1999 of the Negotiations on Confidence- and Security-Building Measures, OSCE document FSC.DOC/1/99, 16 Nov. 199

[40] The text of the document is reproduced in *SIPRI Yearbook 2001* (note 25), pp. 590–98. The 2001 UN conference on the illicit trade of SALW is discussed in appendix 13A in this volume.

[41] The decisions are presented in OSCE document FSC.DEC/5/01, 17 Oct. 2001; and OSCE document FSC.DEC/8/01, 28 Nov. 2001.

Workshops were held on the 1994 OSCE Code of Conduct on Politico-Military Aspects of Security[42] (in Switzerland and Ukraine) and the Global and Annual Exchanges of Military Information (GEMI and AEMI, the latter within the purview of the Vienna Document 1999, held in Vienna).

As of 1 July 2001, the Netherlands no longer hosts the OSCE Communications Network, which the Dutch Ministry of Foreign Affairs has maintained for 10 years. The network, which has been relocated to Vienna, is vital to the CSBM/CFE notification and information exchange system. The FSC established a back-up procedure for the operation of the Communications Network until work on phase II of its modernization has been completed.[43] With Yugoslavia having joined the network, the number of connected participating states rose to 39.

The FSC decided that by December 2002 an information exchange of updated data relating to major weapons and equipment systems should be conducted in electronic form (on CD-ROM). The Conflict Prevention Centre (CPC) will be provided with a copy of the data exchange and will report on the implementation of the decision as well as coordinate the provision of technical assistance to participating states.[44]

The fourth Vienna military doctrine seminar

The Vienna Document 1999 encouraged the participating states to 'hold periodic high-level military doctrine seminars similar to those already held'.[45] On 11–13 June 2001 the FSC held the fourth Vienna seminar on military doctrines and defence policies. It was attended by experts and high-level representatives from defence ministries, the military, other state institutions, policy-oriented organizations and academics. The seminar focused on: (*a*) changes and challenges in the security environment relevant to defence policy and military doctrine; (*b*) military doctrine and reforms of the armed forces; and (*c*) multilateral security approaches in the OSCE area.

The seminar addressed such general issues as the difference between threats and risks, the methodology of risk assessment and the causes of change in the security environment. Despite different perceptions and interpretations the participating states acknowledged the low threat of global military conflict and the greater potential for local wars and regional military conflicts as the result of inter-ethnic, territorial and religious differences. The seminar participants emphasized the threat of low-level regional conflicts, the proliferation of weapons of mass destruction, international and national terrorism, organized

---

[42] Budapest Document 1994, Budapest Decisions IV, Code of Conduct on Politico-Military Aspects of Security, URL <http://www.osce.org/docs/english/summite.htm>.

[43] The central electronic mail server was relocated in early Sep., and operations resumed on 1 Oct. 2001. For details of the back-up procedure see OSCE document FSC.DEC/3/01, 20 June 2001.

[44] OSCE document FSC.DEC/6/01, 14 Nov. 2001.

[45] Vienna Document 1999 (note 39), chapter II, para. 15.7.

crime, and ecological and humanitarian disasters.[46] Most countries see the need to adapt their military doctrines and forces. Smaller, more flexible, more rapidly deployable, mobile, interoperable and self-sustained forces are considered better able to carry out missions with multiple functions and cooperate with civilian agencies (e.g., police) and organizations.

Four major directions of the future work of the FSC were outlined as a result of the seminar: (*a*) discussion of new risks and challenges with a view to agreeing a comprehensive OSCE approach to counter them; (*b*) the growing role of multinational structures in the OSCE area; (*c*) continued discussion of the evolution of military doctrines and security policies including presentation of the military doctrines by states at the FSC plenaries, further clarification of the nature and objectives of military doctrines and defence policies, the FSC contribution to the process of bringing closer military doctrines and the existing arms control regimes, the impact of the military and technological revolution on the possible use of force, strategic stability and arms control prospects, elaboration of a unified technology for the OSCE states used in their military doctrines and so on; and (*d*) adaptation of the FSC activities to the evolving OSCE security environment through better coordination of FSC and Permanent Council activities, adjustment to the new political and military activities (CSBMs for crisis situations, entrusting the Code of Conduct with some operational functions, such as with regard to peacekeeping operations and the like) and the elaboration of new CSBMs.[47] However, the chairman's conclusions are not binding, and in the latter part of 2001 they were overtaken by dramatic events. In the light of the terrorist attacks on the USA, the OSCE is certain to reassess the relationship between military doctrine and multilateral security cooperation with regard to prioritizing the tasks and missions of armed forces.

Enhancing security cooperation after 11 September

After the 11 September attacks discussion was initiated on how to combat terrorism using existing FSC instruments and documents, including CSBMs, and on the kind of new measures that ought to be developed.[48] There was discussion of the relevance of such documents as the OSCE Document on SALW, the Code of Conduct, the 1994 Principles Governing Non-Proliferation and the Conventional Arms Transfers Questionnaire.[49] The inclusion of the Vienna

[46] OSCE Forum for Security Co-operation, 4th Seminar on Military Doctrines and Defence Policies in the OSCE Area, Vienna, 11–13 June 2001, Report of the Working Session Rapporteur, OSCE document FSC.GAL/66/01, 13 June 2001.

[47] Summary Report by the Chairman of the 4th Seminar on Military Doctrines and Defence Policies in the OSCE Area to the OSCE Forum for Security Co-operation, OSCE document FSC.DEL/338/01, 4 July 2001.

[48] These issues were addressed in various proposals made at the FSC in the autumn of 2001 concerning the future role of the FSC and a new agenda for it. E.g., the Russian proposal, OSCE document FSC.JOUR/340, 19 Sep. 2001; the Slovenian chairmanship proposal, OSCE document FSC.DEL/441/01, 3 Oct. 2001; the EU proposal, OSCE document FSC.DEL/450/01/Rev. 1, 17 Oct. 2001; and the Swedish chairmanship draft proposal, OSCE document FSC.DEL/483/01/Rev 1,. 20 Nov. 2001.

[49] OSCE Document on SALW (note 40); Code of Conduct (note 42), Principles Governing Non-Proliferation, CSCE Forum for Security Co-Operation, Vienna, 3 Dec. 1994, reproduced at URL <http://

Document 1999 was also suggested. The list, from which the FSC will need to select key documents, covers most agreed documents.

The OSCE Bucharest Ministerial Meeting decided to strengthen the role of the OSCE by making the Permanent Council the permanent forum for political dialogue among participating states. As the body responsible for reviewing the implementation of OSCE commitments and negotiating arms control and CSBMs, the FSC is to address the aspects of the new security environment which fall within its mandate and act to strengthen the politico-military dimension. It is intended to retain its autonomy and decision-making capacity but to work more closely with other OSCE bodies on security issues, provide expert advice to the Permanent Council and OSCE field operations and advise the Permanent Council or the Chairman-in-Office (CIO). In order to facilitate interaction between the Permanent Council and the FSC, the CIO will be represented at its Troika meetings, and the chairman of the FSC will be represented at Troika meetings on matters of FSC concern.[50] The Troika has proposed a 'road map' for the FSC in its efforts to combat terrorism.[51]

In the autumn of 2001 a Russian proposal that an expert meeting on terrorism be held under the auspices of the FSC was supported by many delegations. At the end of 2001 the FSC took a step towards reform by extending the duration of the chairmanship (to a four-month term).[52]

The CSBM Agreement in Bosnia and Herzegovina

The 1996 Agreement on Confidence- and Security-Building Measures in Bosnia and Herzegovina—negotiated under Article II of Annex 1-B of the 1995 General Framework Agreement for Peace in Bosnia and Herzegovina (Dayton Agreement)—outlines a set of measures to enhance mutual confidence and reduce the risk of conflict in the country. The parties to the agreement are Bosnia and Herzegovina and its two entities. Stability and peace remain dependent on a strong international engagement and presence. Several domestic factors also determine the level of military security. Formally, two separate armed forces exist, but in reality there are three because two components (the Croats and Bosnian Muslims) of the Federation of Bosnia and

---

www.fas.org/nuke/control/osce/text/NONPROLE.htm>; 'Questionnaire on participating states' policy and procedures for the export of conventional arms and related technology', OSCE document FSC.DEC/10/95, Annex 2, 26 Apr. 1995, reproduced at URL <http://projects.sipri.se/expcon/oscefsc116.htm>. The participating states pledged, among other things, to use the FSC to strengthen their efforts to combat terrorism, to enhance the implementation of existing politico-military commitments and agreements (including the SALW and the Code of Conduct), and to examine the relevance of other documents. The Bucharest Plan of Action for Combating Terrorism, Annex to MC(9).DEC/1, Ninth Meeting of the Ministerial Council, Bucharest, 3–4 Dec. 2001.

[50] Decision no. 3, Fostering the Role of the OSCE as a Forum for Political Dialogue, MC(9).DEC/1, Ninth Meeting of the Ministerial Council, Bucharest, 3–4 Dec. 2001. The OSCE Ministerial Troika is composed of the current CIO, the CIO of the preceding year and the incoming CIO.

[51] Decision no. 1, Combating Terrorism, MC(9).DEC/1, Ninth Meeting of the Ministerial Council, Bucharest, 3–4 Dec. 2001.

[52] OSCE document FSC.DEC/9/01, 12 Dec. 2001.

Herzegovina have not been integrated. There is insufficient transparency in military budgets and the joint institutions are very weak.[53]

In 2001 the greatest success was achieved under Measure XI, the programme of military contacts and cooperation. Three seminars were held by Finland, Germany and Romania for mid-level officers and senior non-commissioned officers, with the aim of introducing the OSCE Code of Conduct and the ways in which OSCE states have integrated it into their military doctrines and practice. Aerial observation exercises also proved successful. The parties have developed a protocol and measures that allow them to use military aircraft to support humanitarian missions. The original mandate was broadened from a mechanism to support a risk-reduction measure to encompass humanitarian assistance requirements within Bosnia and Herzegovina.

The issue of transparency in military budgets has also made considerable headway. The development of realistic defence budgets and defence postures is essential in this regard. Consequently, in December 2000 the Joint Consultative Commission (JCC) decided to proceed with an audit of the military budget of the Croat–Muslim Federation. The Federation completed its report and submitted it to both the CIO Personal Representative and the Head of the OSCE Mission to Bosnia and Herzegovina. Republika Srpska chose to abstain until the Federation's budget was completed, prepared its own internal audit and provided a report on it to the CIO Personal Representative. The process of improving budget transparency will continue in the future.

In contrast to the preceding two years, inspections were carried out almost without incident. One inspection was interrupted because of political turbulence, and one was not fully completed because of a coordination problem with the Stabilization Force (SFOR). As a result, the OSCE and SFOR changed their guidelines to allow greater flexibility. The parties therefore regarded these interruptions as minor, and the spirit of the agreement was maintained.

For several years, the parties have improved their exchanges of information. In mid-2001, they demilitarized equipment held in historical collections. The Protocol on Existing Types of Conventional Armaments and Equipment was reviewed and adopted at the third review conference, held on 19–21 February 2001. Recommendations were made to improve the Protocol on Notification and Exchange of Information. The parties also approved a new Protocol on Aerial Observation and updated the Protocol on Visits to Weapon Manufacturing Facilities. In addition, they agreed to update the Agreement on CSBMs by the summer of 2002 by incorporating the decisions taken since January 1996 by the JCC and the three review conferences. The parties agreed to maintain the CIO Personal Representative as the chairman of the JCC until the next review conference.[54]

---

[53] The defence budget of Bosnia and Herzegovina amounts to about 6% of its gross domestic product. The country 'can in fact barely afford an army half of its current size'. 'Awareness raising campaign on military expenditure starts in Bosnia and Herzegovina', OSCE Press Release, 11 Oct. 2001.

[54] Annual Report (note 38); Annual Report 2001 on OSCE Activities (1 Nov. 2000–31 Oct. 2001), pp. 64–65; and Final Document of the Third Conference to Review the Implementation of the Agree-

Negotiations under Article V of the Agreement on Regional Stabilization

On 18 July 2001, the 20 states participating in the negotiations under Article V of the Agreement on Regional Stabilization, which aims to find lasting solutions for the regional stabilization of South-Eastern Europe ('in and around the former Yugoslavia'), reached consensus on a politically binding joint document. The Concluding Document of the negotiations ended a long and, at times, difficult negotiating process, which began in 1996 with the Article II and IV agreements. The original mandate of the negotiation—to bridge the arms control obligations of the parties to the Florence Agreement with the obligations of the neighbouring parties to the CFE Treaty—was not fully attained. Structural arms control measures were not addressed since certain participating states did not wish to address future arms control limitations until the time of their accession to the CFE Treaty. The participants also could not agree on a binding information exchange which would go beyond the existing obligations. Moreover, since many countries in the region are already parties to various bilateral and multilateral agreements and consider that their participation in the numerous organizations and initiatives effectively ensures their security, there was no major incentive to expand such measures. Finally, the admission of the FRY to the OSCE in November 2001 helped change the political situation and relax tensions and fears in the region.

As a result, the Concluding Document provides a list of voluntary CSBMs, for the most part inspired by Chapter X (regional measures) of the Vienna Document 1999. They cover defence-related information, expanded military contacts and cooperation, military activities, inspections and evaluations visits, demining and destruction of anti-personnel mines, and small arms and light weapons.

A commission of participating states was established to review the implementation of the measures, and the states undertook to cooperate closely with the 1999 Stability Pact for South Eastern Europe.[55] The measures took effect on 1 January 2002.[56]

The Sub-Table on Defence and Security of the Stability Pact's Working Table III on Security Issues welcomed the conclusion of the Article V negotiation and stressed the need to establish close links between Working Table III and the Article V Commission, particularly through coordinated, regular joint meetings. The Regional Arms Control, Verification and Implementation

---

ment on Confidence- and Security-Building Measures in Bosnia and Herzegovina, Vienna, 19–21 Feb. 2001.

[55] The Stability Pact for South Eastern Europe is reproduced in *SIPRI Yearbook 2000* (note 3), pp. 214–20, and is available at URL <http://www.stabilitypact.org/official%20Texts/PACT.HTM>. A brief summary of the pact and the partners are listed in the glossary in this volume.

[56] The Concluding Document of the Negotiations under Article V of Annex 1-B of the General Framework Agreement for Peace in Bosnia and Herzegovina, ArtV.DOC/1/01, Vienna, 18 July 2001 is reproduced as appendix 13B in this volume. Statement by Ambassador Henry Jacolin, Special Representative of the OSCE for Article V (Regional Stability), at the Joint PC/FSC Meeting, Vienna, 19 July 2001.

Assistance Centre (RAVIAC) also declared its desire to contribute to the implementation of the accord.[57]

New bilateral CSBM accords

The Vienna Document 1999 committed the participating states to pursue regional CSBM arrangements. In 2001 three new CSBM agreements were created: between Lithuania and Russia, Belarus and Lithuania, and Belarus and Ukraine.

The first agreement, established by the exchange of diplomatic notes on 19 January between the Lithuanian mission to the OSCE and the Russian delegation on Military Security and Arms Control in Vienna, provides for one additional evaluation visit to formations or units in Lithuania and Russia's Kaliningrad *oblast* and annual exchange of additional information about military forces on Lithuanian territory and the Kaliningrad *oblast*.[58] This accord symbolizes Russia's changing attitude towards its Baltic Sea neighbours and the will to allow more insight into the military activities of this formerly closed and heavily armed area.

The second agreement, in the form of a 19 July 2001 joint statement to the FSC,[59] envisages: one additional evaluation visit to assess military information and one additional inspection of the specified area in Belarus and Lithuania, above the quotas under the Vienna Document; exchange of additional information on the armed forces; provision of information concerning the 'most extensive military activity' being carried out on the territories of both countries which do not reach the Vienna Document notification thresholds; holding meetings of experts to assess implementation of these CSBMs; and possible automatic extension of the CSBM implementation after the end of the calendar year.[60]

The third agreement was signed by Belarus and Ukraine on 16 October. It envisages: notification and observation of military activities in the border areas of both states;[61] bilateral exchange of military information (deployments and activities at and above the level of regiment/brigade) in an agreed format;

---

[57] Working Table III on Security Issues. Fifth Meeting of the Working Table on Security Issues, Chairman's Conclusions, State Secretary Kim Travik, Budapest, 27–28 Nov. 2001. The Sub-Table on Defence and Security Issues has a supporting, facilitating and coordinating role in providing funding and resources for Article V measures. Arms control-related cooperation between Working Table III and the Article V forum is discussed in Lachowski (note 25), pp. 567–68.

[58] OSCE document FSC.DEL/20/01, 24 Jan. 2001. Russia was initially reluctant to respond to the 1998 Finnish–Swedish proposal to adopt the bilateral CSBM arrangements agreed by the other 8 states around the Baltic Sea. In 2000 Russia implemented bilateral CSBM accords (extra evaluation visit plus exchange of information) with Estonia and Finland.

[59] It also took the form of an exchange of notes between the 2 respective missions to the OSCE in Vienna.

[60] Joint Statement by the Permanent Delegation of the Republic of Belarus to the OSCE and the Permanent Representation of the Republic of Lithuania to the OSCE concerning additional confidence- and security-building measures at the Forum for Security Cooperation, OSCE document FSC.DEL/384/01, 19 July 2001.

[61] The relevant parameters are: 42 days in advance; at least 5000 men; 100 tanks or 150 ACVs or 75 artillery pieces of 100 mm and above; or 50 sorties of combat aircraft and/or attack helicopters.

and a greater number of inspections and visits (one above the Vienna Document quotas in each instance). In addition, each side will inform the other about the use of military units exceeding the agreed levels in emergencies caused by a natural or technological disaster. They will meet at least once a year to assess the implementation of the agreement.[62]

Poland and Ukraine have announced the negotiation of complementary CSBMs on the expansion of the scope of military information on planned activities, prior notification of certain military activities, observation, a greater number of evaluations and a joint assessment of the implementation of the measures.[63]

The Treaty on Open Skies

The 1992 Treaty on Open Skies, based on a 1955 initiative by US President Dwight D. Eisenhower, was signed on 24 March 1992 by the members of NATO and the former Warsaw Treaty Organization.[64] The entry into force of the treaty was long blocked by the failure of Belarus and Russia to ratify this confidence-building instrument. It was 'held hostage' to other outstanding political and military issues (e.g., missile defence; the 1993 Treaty on Further Reductions and Limitations of Strategic Offensive Arms, START II Treaty; and the Comprehensive Nuclear Test-Ban Treaty, CTBT). Owing to Russia's change of policy in 2001, both states ratified the treaty and deposited their instruments of ratification on 2 November 2001. The treaty entered into force on 1 January 2002.[65]

In the run-up to the entry into force of the treaty, the Open Skies Consultative Commission (OSCC) decided to establish three informal working groups on certification, sensors, and rules and procedures. The working group on certification started its work in November 2001.[66]

Now that the treaty has entered into force additional OSCE states may apply for participation in the Open Skies regime until 1 July 2002, and after that any

---

[62] Agreement between the Cabinet of Ministers of Ukraine and the Government of the Republic of Belarus on additional confidence- and security-building measures, Kiev, 16 Oct. 2001, submitted to the FSC, OSCE document FSC.DEL/476/01, 7 Nov. 2001.

[63] Statement by Poland and Ukraine at the AIAM, Vienna, 27 Feb. 2001. In early 2002 the agreement, which is an intergovernmental accord, was reported to be in the final stage of negotiation.

[64] For the terms of the treaty and the list of parties and signatories see annex A in this volume.

[65] Kyrgyzstan has signed but not yet ratified the treaty, but it does not belong to either of the categories of states whose ratification is necessary for its entry into force.

[66] The Aug. 2001 joint certification exercise in Fürstenfeldbruck, Germany, demonstrated that the certification of aircraft and sensors is the main priority of the states parties. Ukraine has proposed to arrange a joint certification of observation aircraft as a first practical approach to future activities under the treaty. The 2001 session of the Working Group on Certification agreed several issues, subsequently adopted by the OSCC plenary on 17 Dec. 2001: (*a*) the Decision on provisions for the initial certification period; and (*b*) the Chairperson's Statements on: issues related to certification of observation aircraft and sensors, the use of a standard CD-ROM format for distribution of certification documentation, the OSCC determination of the number of individuals participating in a certification, principles for the conduct of the C-130 H/POD-system certification, principles for joint certifications and the use of one calibration target for certification. It was not possible to agree on provisions for the use of a standard signature page for the Certification Report. The OSCC also adopted rules of procedure and working methods for the OSCC. *OSCC Journal*, no. OSCC.XXVI.JOUR/74 (17 Dec. 2001).

country may request to accede to the treaty.[67] On 5 November 2001 Sweden and Finland announced their intention to accede to the treaty.[68] Reciprocal voluntary unarmed reconnaissance overflights continued as in previous years. Since 1996 more than 350 such trial flights have taken place. These flights have resulted in increased interest in sustaining the Open Skies regime.[69]

IV. Non-European CSBM arrangements

Outside Europe there is interest in the European CSBM experience and the possibility of applying it to the problems of lack of transparency, unpredictability and other military concerns. The Association of South-East Asian Nations (ASEAN) countries are continuing their confidence-building measure (CBM) dialogue, and China, Kazakhstan, Kyrgyzstan, Russia and Tajikistan are working to streamline the operation of their arms control agreements.[70] The OSCE states have recently activated a programme to share their CSBM experience with interested countries. In March 2001 the OSCE and South Korea discussed the applicability of CSBMs to the Korean peninsula.[71]

The ASEAN Regional Forum

The political and security dialogue within the ASEAN Regional Forum (ARF) covers both military and defence-related measures and non-military issues which have a significant impact on regional security.[72] The ARF's flexible, step-by-step military security-related process is characterized by various national and international voluntary CBM undertakings (such as seminars and workshop meetings of defence and military officials, visits to defence facilities, documents and briefings on regional security concerns, etc.) carried out within the Track I (official) and Track II (unofficial) dialogues. It attempts to develop incremental confidence building through preventive diplomacy by the elaboration of approaches to conflicts that are designed to create the premises for agreement. Two meetings of the Intersessional Support Group on CBMs

---

[67] For 6 months after entry into force of the Treaty on Open Skies (until 1 July 2002), any other OSCE participating state may apply for accession by submitting a written request to 1 of the depositaries for consideration by the OSCC. Applications are subject to consensus agreement by the OSCC. After 1 July 2002 the OSCC will consider the application for accession of any other state. See annex A in *SIPRI Yearbook 2001* (note 25) and Annex A in this volume.

[68] Cyprus and Lithuania have also expressed interest in joining the regime.

[69] 'All conditions fulfilled for Open Skies Treaty to enter into force', OSCE Press Release, 5 Nov. 2001. The information seminar on the treaty, held in Oct. 2001, underlined the potential for application in such areas as environmental protection, humanitarian crises and natural disasters. Report on the Information Seminar on the Treaty on Open Skies, Vienna, 1 and 2 Oct. 2001, Vienna, 4 Oct. 2001, OSCE document OSCE.DEL/35/01, 5 Oct. 2001.

[70] China, Kazakhstan, Kyrgyzstan, Russia and Tajikistan, the 'Shanghai Five', established the Shanghai Forum in 2000. It was replaced in 2001 by the Shanghai Cooperation Organization (SCO). The Shanghai Forum is discussed in Lachowski (note 25), p. 569. For a brief description of the SCO see the glossary in this volume.

[71] OSCE–Korea Conference 2001: Applicability of OSCE CSBMs in Northeast Asia, Seoul, 19–21 Mar. 2001.

[72] The members of ASEAN and ARF are listed in the glossary in this volume.

(ISG on CBMs) are usually held between the annual meetings of the ARF. The ASEAN foreign ministers review annually the recommendations made by ISG and ARF senior officials. The ARF notes that the process continues to develop 'at a pace comfortable to all ARF participants' and on the basis of consensus and non-interference in internal affairs. However, this approach is often criticized as seeking the lowest common denominator by harmonizing common positions rather than striving for a more ambitious agenda.

In the intersessional period 2000–2001 the regional security situation was assessed by the ASEAN as relatively stable, and 'remarkable' progress was noted in the adoption by the ARF participants of papers concerning the three basic confidence-building areas: on the enhanced role of the ARF Chairman; the terms of reference for ARF experts/eminent persons; and on the concept and principles of preventive diplomacy. The second volume of the *Annual Security Outlook* (ASO), a regional document outlining security issues and the concerns of ARF members, was also published. The meeting of ARF foreign ministers meeting in Hanoi also agreed that the ASO will no longer be confidential.[73]

Arms control in Central Asia

In 1996 and 1997 the Shanghai Five, the four CIS states and China agreed on CBMs and arms reductions in the 100 km-wide areas adjacent to the borders.[74] The agreements differ from the European solutions. The information exchanged under both agreements is confidential. For example, one of the CBMs requires notification of the temporary entry of river-going combat vessels into the border areas. The agreements have injected a measure of stability, cooperation and confidence in this part of Central Asia.

Following the completion of the ratification processes, both agreements began to be implemented in 1999 when trial inspections began. Verification differs from the CFE Treaty regime and has been adapted to meet regional needs (e.g., separate inspections of armed forces and border units, not more than two inspections annually in the Eastern and Western sectors, and so on). The parties have encountered some difficulties in their inspection activities because of the need first to coordinate inspections among the four CIS states and then with China. A Joint Control Group, headed by two co-chairmen from the four CIS states and China, addresses the implementation issues of the 1997 agreement on arms reductions. It has also recently been instructed to supervise

---

[73] See Co-Chairmen Summary Report of the Meeting of the ARF Intersessional Support Group (ISG) on Confidence-Building Measures (CBMs), Seoul, 1–3 Nov, 2000, and in Kuala Lumpur, 18–20 Apr. 2001, URL <http://www.aseansec.org/print.asp?file=/amm/arf8doc.3.htm>; and Chairman's Statement of the 8th ASEAN Regional Forum, Hanoi, 25 July 2001, URL <http://www.aseansec.org/print.asp?file=/amm/hanoi05.htm>. For ASEAN CBM developments in previous years see the *SIPRI Yearbooks 1998–2000*.

[74] The 1996 Agreement between Russia, Kazakhstan, Kyrgyzstan and Tajikistan (as a joint party) and China on Confidence Building in the Military Field in the Border Area; and the 1997 Agreement between Russia, Kazakhstan, Kyrgyzstan and Tajikistan (as a joint party) and China on the Mutual Reduction of Armed Forces in the Border Area.

the implementation of the 1996 CBM agreement. The modalities for this multilateral supervision are still being worked out because until recently monitoring has been conducted on a bilateral basis.[75]

The OSCE–Korea CSBM seminar

In the wake of the June 2000 summit meeting between President Kim Dae-Jung of South Korea and President Kim Jong Il of North Korea, South Korea took the initiative to hold an OSCE–Korea Conference on the 'Applicability of OSCE CSBMs in Northeast Asia'. Its focus was on information and experience sharing between the OSCE and North-East Asia, notably CBMs and CSBMs.[76]

Three lessons were drawn from the OSCE experience. The first lesson was the need for a 'gradual approach', starting with a Helsinki-type first generation of CBMs that are acceptable to both sides.[77] Second, the success of CBMs is dependent on the political will of the parties. There are currently incentives for both sides to enter into and implement CBMs. Advances in military technology and strategy are bound to favour South Korea over time, whereas in North Korea the defence industry is a heavy economic drain. South Korea is interested in reducing the danger of a surprise attack.[78] However, there is discernible opposition to CBM arrangements among the South Korean military.

The third lesson to be learned from both Korean and European history is the importance of consolidating the process through a comprehensive and institutional approach. Europe's former division into two blocs differs from the situation in North-East Asia, where bilateral alliance networks dominate. Moreover, the relative homogeneity in Europe is non-existent in North-East Asia, which until recently has lacked the tradition of multilateral dialogue or cooperation. Both North and South Korea are parties to the ARF. Positive results in the ARF have led to a call for the creation of a special forum between North Korea and South Korea as a way to address various security issues.

A fundamental issue is the form and role of the CBMs. North Korea perceives CBMs as a 'top–down' approach, that is, political agreements among

---

[75] Russian Federation delegation, 'Some specific features and implementation experience of the agreements on strengthening confidence in [the] military sphere and on mutual armed forces reduction in the framework of [the] "Shanghai Five group"', OSCE–Korea Conference 2001 (note 71), Conference document no. 022.

[76] 'Co-chairmen's summary', OSCE–Korea Conference 2001 (note 71), Conference document no. 026.

[77] It was remarked that in the context of North-East Asia it would be easier and more appropriate to start from other CBMs (not military ones) so that confidence-building would not become a hostage to the lack of or slow progress in the military field. Tasanen, A., 'OSCE CSBMs and Asian (ARF) CBMs—an attempt to synthesis between European and Asian views based on impressions from [the] ARF (Track One) seminar "Approaches to Confidence-Building", held on 2–4 Oct. 2000 in Helsinki, Finland', OSCE–Korea Conference 2001 (note 71), Conference document no. 015. A somewhat similar conclusion was drawn from the CSBM seminar for the Mediterranean Partners for Co-operation held in Portoroz, Slovenia, in 2000.

[78] Kyongsloo Lho, a South Korean scholar, noted: 'North Korea's extreme arrogance and South Korea's hubris no longer constitute the impediments they once did to a potential confidence-building process on the peninsula'. Kyongsloo Lho, 'Confidence- and security-building measures (CSBMs) for the Korean peninsula', OSCE–Korea Conference 2001 (note 71), Conference document no. 019.

leaders to be followed by an implementation process in which the parties change their behaviour in accordance with a new political perception.[79] South Korea and the West prefer a 'bottom–up', incremental, step-by-step approach, in which progress in implementation builds confidence among the parties concerned. Part of the confidence-building process consists therefore of agreeing on terminology and definitions.

V. Landmines and certain conventional weapons

Landmines

According to recent estimates, 230–245 million anti-personnel mines (APMs) are stored in the arsenals of approximately 100 countries.[80] Some 215–225 million landmines are possessed by countries which have not signed the 1997 Convention on the Prohibition of the Use, Stockpiling, Production and Transfer of Anti-Personnel Mines and on their Destruction (APM Convention). These estimates are 5–20 million lower than those for 2000. The largest stockpiles of APMs are alleged to be in China (110 million), Russia (60–70 million), the USA (11.2 million), Pakistan (6 million), India (4–5 million) and Belarus (4.5 million). Ukraine has revised its stockpile disclosure from 10 million to more than 6 million. Belarus disclosed the size of its stockpiles of APMs as between one-half and one-third of the estimated figure of 10–15 million APMs. Eight of the 12 largest producers and exporters of landmines are now parties to the APM Convention.[81] Fourteen producers of landmines—including major producers such as China, India, Pakistan and Russia as well as several dozen mine-using countries involved in conflicts—have not signed the APM Convention.

Two multilateral agreements deal with landmines. The 1997 APM Convention aims at the elimination of *all* anti-personnel mines, but it is hampered by the absence of strong monitoring and enforcement provisions. The amended (landmine) Protocol II of the CCW Convention is a hybrid, combining humanitarian and arms control measures. The parties to the amended Protocol II of the CCW Convention include most major producer and user countries (e.g., China, India and Pakistan, but not Russia) and the convention

---

[79] Snyder, S., 'Which CBMs for the Korean peninsula?', OSCE–Korea Conference 2001 (note 71), Conference document no. 018.

[80] International Campaign to Ban Landmines, Landmine Monitor, *Landmine Monitor Report 2001: Towards a Mine-Free World*, 2001. The recent US estimates of the number of fatalities caused by landmines and unexploded ordnance were lowered from the previously reported 26 000 casualties annually to less than 10 000 a year. The number of buried APMs has fallen from 80–110 million to 45–50 million APMs in *c.* 60 countries. *To Walk the Earth in Safety: the United States Commitment to Humanitarian Demining* (US Department of State: Washington, DC, 2001), as quoted in 'Fatalities from landmines drop dramatically worldwide', US Department of State, International Information Program, 4 Dec. 2001, URL <http://usinfo.state.gov/topical/pol/arms/stories/01120502.htm>.

[81] The 8 states are: Belgium, Bosnia and Herzegovina, Bulgaria, the Czech Republic, France, Hungary, Italy and the UK. *Landmine Monitor Report 2001* removed 2 countries, Turkey and Yugoslavia, from its list of 16 producers.

includes other 'inhumane' weapons not covered by the APM Convention, such as delayed-action weapons, anti-vehicle mines and booby traps.

The APM Convention

As of 1 January 2002, there were 122 parties to the APM Convention, and another 20 states had signed.[82] However, 51 states have not acceded to the convention. These include three of the five permanent members of the UN Security Council (China, Russia and the USA), other major landmine producers, such as India and Pakistan, all but four former Soviet republics and many states in the Middle East and Asia. The signatories include all the states of the western hemisphere except Cuba and the USA, all the NATO nations except Turkey and the USA, all the EU member states except Finland, most of the African countries and numerous states in the Asia–Pacific region (the regional distribution is shown in table 13.3).

In March 2001 the US National Research Council issued a report commissioned by the Department of Defence which concluded that while some of alternative technologies could be ready by the 2006 deadline, 'in certain situations, some alternatives will not be ready until later, and anti-personnel landmines will need to be retained'.[83] Assistant Secretary of State for Legislative Affairs Paul Kelly stated in July 2001 that the US Administration had to 'examine the need for landmines on the modern battlefields of the future' and 'cannot undercut the effectiveness of [the US] military on the way to that future'.[84] On 19 December 2001, 124 members of Congress sent a bipartisan letter to President George W. Bush urging him to stick to the May 1998 pledge by President Bill Clinton and direct the ongoing US landmine policy review towards the goal of eliminating APMs from the US arsenal.[85]

In 2001 it was claimed that Ugandan forces used APMs in the Democratic Republic of the Congo (DRC). Uganda, a party to the convention, denied the charge. There were also allegations that the following signatories used APMs in 2001: Angola, Burundi, Ethiopia, Rwanda (in the DRC) and Sudan. Landmines were used in armed conflicts by both rebel and government forces. Except for Angola, the governments of these countries denied the accusations. Most instances of the use of APMs were in ongoing conflicts (e.g., Russia in Chechnya, along the border between Tajikistan and Kyrgyzstan, and Nepal, the Former Yugoslav Republic of Macedonia (FYROM) and the FRY).[86]

[82] A summary of the convention, the parties and signatories are given in annex A in this volume.
[83] In May 1998 the USA indicated its willingness to join the APM Convention on certain conditions. Lachowski, Z., 'The ban on anti-personnel mines', *SIPRI Yearbook 1999: Armaments, Disarmament and International Security* (Oxford University Press: Oxford, 1999), pp. 656–57. Committee on Alternative Technologies to Replace Antipersonnel Landmines, Commission on Engineering and Technical Systems, Office of International Affairs, *Alternative Technologies to Replace Antipersonnel Landmines* (National Research Council, National Academy Press: Washington, DC, 2001), pp. 77–78, URL <http://bob.nap.edu/books/0309073499/html>.
[84] 'Bush team shies from Clinton landmine policy', *Arms Control Today*, News Briefs, Sep. 2001, pp. 38.
[85] 'Bush urged to redirect landmine policy review', *Arms Control Today*, News Briefs, Jan./Feb. 2002, p. 42.
[86] *Landmine Monitor Report 2001* (note 80).

Table 13.3. The status of the APM Convention, as of 1 January 2002

Region	Signed but not ratified	Ratified or acceded	Not signed	Total
Africa	7	37	4	48
Americas	3	30	2	35
Asia–Pacific	5	15	19	39
Europe/Central Asia	5	35	13	53
Middle East/North Africa	–	5	13	18
Total	**20**	**122**	**51**	**193**

Source: Based on Mines Action Canada, 1 Jan. 2002, URL <http://www.minesaction canada.com>.

The CCW Convention

In the run-up to the Second CCW Convention Review Conference, held on 11–21 December 2001, its Preparatory Committee held three sessions: in December 2000 and in April and September 2001. In addition, informal open-ended consultations were convened in August 2001. Sixty-three states parties, 4 signatory states and 13 states not parties to the CCW Convention as well as representatives of the International Committee of the Red Cross (ICRC) and the United Nations Children's Fund (UNICEF) took part in the conference. Five issues were identified for consideration at the review conference.[87]

1. *Expanding the scope of the convention to non-international armed conflicts.* By expanding the scope of the original Protocol II to cover civil wars and domestic conflicts, the amended Protocol II had broken new ground for changes in international practice (e.g., in the Hague International Criminal Tribunal). The aim of the new initiative was that all of the other protocols (I, III and IV) should apply to all types of conflicts—international and internal.[88] The conference extended the application of the convention to non-international armed conflicts (amended Article I). This is another important broadening of the trend towards making international humanitarian law applicable to all parties to an internal conflict, including non-state actors.[89]

2. *Explosive remnants of war* (ERW—also called unexploded ordnance, UXO). The ICRC proposed that 'explosive remnants of war', especially cluster munitions, be included on the agenda of the review conference. However, because of the technical complexities, the importance of such weapons

---

[87] See also Matheson, M. J., 'Filling the gaps in the Conventional Weapons Convention', *Arms Control Today*, Nov. 2001, pp. 12–16.

[88] The US proposal of 14 Dec. 2000 is available at URL < http://www.ccwtreaty.com/article1.html>.

[89] This, however, does not constitute a rule for future protocols which may 'apply, exclude or modify the scope of their application' in relation to Article I. Final Declaration, Second Review Conference of the States Parties to the Convention on Prohibitions or Restrictions on the Use of Certain Conventional Weapons which may be deemed to be Excessively Injurious or to have Indiscriminate Effects, Geneva, doc. CCW/CONF.II/MC.I/1, 11–21 Dec. 2001, para 7, p. 6.

and the fact that the discussion of the issue is at an early stage, the 2001 conference was unable to adopt restrictions on ERW.[90] The review conference established an open-ended group of governmental experts with a coordinator to 'discuss ways and means' of addressing the issue of ERW, including: factors and types of ammunition; technical improvements which could reduce the risks of munitions becoming ERW; strengthening existing international humanitarian law to minimize post-conflict risks of ERW; ERW-related warning, assistance and clearance steps; and so on. The coordinator is to submit recommendations, adopted by consensus, 'at an early stage' (to the meeting in December 2002, if possible) for consideration by the states parties.[91]

3. *Mines other than anti-personnel mines.* These types of mines endanger the civilian population as well as civilian traffic and humanitarian relief operations and other peace missions. In April 2001 Denmark and the USA proposed a new, fifth CCW protocol to address anti-vehicle mines.[92] The proposal would not prohibit anti-vehicle mines, but mines would be furnished with self-destructing and back-up self-deactivating mechanisms, making them easier to detect. In addition, the deployment of mines would be regulated. However, several major mine-using countries, such as China, India, Pakistan and Russia, oppose the proposal because anti-vehicle mines constitute an essential element of their national defence. Nevertheless, the review conference demonstrated considerable support for balanced restrictions on the use of anti-vehicle mines. In effect, the conference gave a mandate to the group of governmental experts to further explore the issue, and its coordinator is to submit a report, adopted by consensus, to the states parties.[93]

4. *Stronger compliance mechanism.* Although the amended Protocol II provides for some monitoring and enforcement measures (such as sanctions against violators), it lacks regular procedures to address non-compliance. The USA proposed adding a provision for investigating allegations of non-compliance and conducting on-site inspections to be added to the amended Protocol II and the proposed anti-vehicle mines protocol.[94] China and other non-aligned countries felt that the US proposal would infringe on their national sovereignty and would also cause revision of the amended Protocol II. The US proposal therefore failed.

The EU states and South Africa presented proposals that were less far-reaching than the US proposal. The conference was unable to synchronize the proposals and settle the sensitive issues of sovereignty and security, cost and other matters as well as address the concerns of the developing countries. As a result, it was agreed that the 'Chairman-designate' should undertake consulta-

---

[90] While commending the ICRC proposal the USA claimed that, compared to traditional unitary bombs, improved, more reliable cluster bombs will cause less destruction, can shorten conflicts and benefit friendly forces, while reducing the harm to civilian populations during armed conflicts. Statement of Edward Cummings, Head of the US delegation to the Second Preparatory Conference of the 2001 CCW Review Conference, 5 Apr. 2001, URL<http://www.ccwtreaty.com/ccw0405.html>

[91] Final Declaration (note 89), p. 7

[92] The Danish–US proposal was revised in July 2001, RL <http://www.ccwtreaty.com/usdan1.htm>. Since then an additional 10 co-sponsors have been added.

[93] Final Declaration (note 89), p. 7.

[94] The US proposal can be accessed at URL <http://www.ccwtreaty.com/comply1.htm>.

tions on 'possible options' to promote compliance with the convention and its annexed protocols and report on this to the parties.[95]

5. The Swiss Government proposed prohibiting the use of small-calibre weapons and ammunition that cause excessive damage inside the human body. This proposal was also not adopted by the parties.[96] It was given lukewarm support but no UN financing, and experts from interested states parties were given the task of dealing with the various aspects of the proposal.

Viewed in the light of the failures or stalemates in other areas of arms control in 2001 (e.g., those associated with the CTBT, missile defence and the proposed protocol to the 1972 Biological and Toxin Weapons Convention[97]), the outcome of the Second CCW Convention Review Conference should be regarded as a success. The work of the conference is being continued by a group of governmental experts with two coordinators: one for explosive remnants of war and one for mines. The EWR issue seems to be gaining in importance, and the group may be able to produce a report for the meeting of the parties to be held in December 2002. However, there remains a risk that states parties which are not interested in further strengthening the CCW will obstruct this work.

VI. Conclusions

Measures aimed at promoting confidence, transparency, openness and security regained prominence in Europe after the conclusion of the 1999 Agreement on Adaptation of the CFE Treaty. The new qualitative changes that took place in Europe's security environment in 2001—the NATO–Russia rapprochement in the wake of the 11 September terrorist attacks and the forthcoming enlargements of NATO and the EU—will affect military cooperation in Europe, including arms control endeavours such as the extension of the CFE Treaty area of application, Russia's possible enhanced influence on the development of the European arms control regime, new regional challenges and CSBMs. Although the entry into force of the Agreement on Adaptation is being implemented, it is still stalled by Russia's insufficient progress towards compliance with the commitments which it made at the OSCE Istanbul Summit Meeting with regard to the CFE Treaty.

Arms control in the new security environment differs from arms control in the cold war period. The changed situation allows states to take a more balanced approach that lessens the risk of the irrational and disproportionate responses of the past. Efforts to reduce, limit and monitor armaments are less important, although they retain their unique role as an international 'insurance policy'. Against this backdrop, the building of confidence and security is changing in character, context, scope and function.

---

[95] Final Declaration (note 89), p. 7
[96] Final Declaration (note 89), p. 7.
[97] See chapter 12 in this volume.

There are four characteristic features of the process of controlling weapons and consolidating military security in Europe today. First, the 'hard' (structural) steps to regulate armaments are being replaced by 'soft' (operational) arrangements, such as CSBMs, risk reduction, transparency and other cooperative mechanisms. Second, the new measures are increasingly becoming region-oriented—moving from the pan-European to the regional, subregional, bilateral and even domestic level. Third, there is debate as to whether CSBMs are applicable in times of crisis or conflict. There is no consensus on this issue, and one view is that new arrangements, mechanisms and institutions are needed. Others believe that the necessary instruments exist but that the political will is lacking. Fourth, the autonomous role of CSBMs in regulating relations between states is increasingly constrained by their inclusion in synergistic packages of military and non-military measures for crisis management, conflict prevention and post-conflict rehabilitation (e.g., the Stability Pact for South Eastern Europe) or in counter-terrorism arrangements. 'Soft measures' may be effective in resolving security problems in volatile regions and combating terrorism in Europe.

The Forum for Security Co-operation plays an important role in the proposed reform of the political and military dimension of the OSCE. In the wake of 11 September the agenda of the FSC is being adapted to meet new security challenges, involve the FSC more closely in operational security issues (e.g., by giving it an additional advisory role within the OSCE), make better use of its political and military expertise, and improve its organizational efficiency.

Although the European example of conventional arms control measures is seen as a positive model, conventional arms control remains a low security priority elsewhere in the world. The hopes for progress and the adoption of arms control measures on the Korean peninsula have not been fulfilled. The 11 September attack on the USA focused attention on the problem of international terrorism.

The regulation of excessively injurious conventional weapons or those that have an indiscriminate effect has gained prominence as concern has grown in the international community about the suffering of both civilians and combatants. The 2001 Second Review Conference of the CCW Convention extended the application of the convention to domestic armed conflicts and expressed support for additional work on other issues of humanitarian concern.

Appendix 13A. The UN conference on the illicit trade in small arms and light weapons

PIETER D. WEZEMAN

I. Introduction

Since the early 1990s there has been a growing realization in the arms control and disarmament policy-making community that, while major conventional weapons and weapons of mass destruction have previously received great attention, small arms and light weapons (SALW) play a significant role in most armed conflicts.[1] Control of the proliferation and availability of small arms is therefore considered an important instrument for conflict prevention and resolution and is the subject of a number of recent multilateral initiatives.[2] Activities in 2001 included the start of the implementation of the 2000 Organization for Security and Co-operation in Europe (OSCE) Document on Small Arms and Light Weapons and the extension of the 1998 Economic Community of West African States (ECOWAS) Moratorium on the Importation, Exportation and Manufacture of Small Arms and Light Weapons.[3]

The global high point of the discussion on small arms was the UN Conference on the Illicit Trade in Small Arms and Light Weapons in All Its Aspects, held in July 2001. The conference and the document it produced—the UN Programme of Action to Prevent, Combat and Eradicate the Illicit Trade in Small Arms and Light Weapons in All Its Aspects—reflect the principal elements of other multilateral initiatives and form the global framework for the further development of these initiatives.

II. The UN conference

Since the mid-1990s, small arms have received growing attention within the United Nations.[4] Based on a 1997 recommendation by the UN Panel of Governmental

---

[1] In this appendix the term 'small arms' refers to 'small arms and light weapons', for which there is as yet no agreed definition. The most common definition used is given in the UN Report of Governmental Experts on Small Arms, UN document A/52/298, 27 Aug. 1997, p. 11. The definition essentially categorizes small arms as weapons designed for personal use, and light weapons as designed for use by several persons serving as a crew. See also Wezeman, P. D., Wezeman, S. T. and Chipperfield, N., 'Transfers of small arms and other weapons to armed conflicts', *SIPRI Yearbook 2001: Armaments, Disarmament and International Security* (Oxford University Press: Oxford, 2001), pp. 410–20.

[2] For further discussion of developments regarding multilateral initiatives in the field of small arms up until 2001 see Small Arms Survey, *Tackling the Small Arms Problem: Multilateral Measures and Initiatives, Small Arms Survey 2001* (Oxford University Press: Oxford, 2001), pp. 251–91; and Small Arms Survey, *Small Arms Survey 2002* (Oxford University Press: Oxford, 2002), The small arms debate also includes the use of weapons by criminals and individual citizens. It is not always possible to make a clear distinction between large-scale crime and conflict. Some multilateral efforts are aimed specifically at controlling the use of weapons to combat crime, such as the 2001 UN Protocol against the Illicit Manufacturing of and Trafficking in Firearms, Their Parts and Components and Ammunition, UN document A/RES/255, 8 June 2001.

[3] See OSCE Forum for Security Co-operation document FSC.DOC/1/00, 24 Nov. 2000; and, for the ECOWAS Moratorium, UN document A/53/763, 18 Dec. 1998.

[4] The first concrete outcome of this debate took shape in 1995 when, in Resolution 50/70B, the UN General Assembly requested the Secretary-General to prepare a report on small arms and light weapons with the assistance of a group of governmental experts.

Experts on Small Arms, in December 1999 the UN General Assembly decided to organize the Conference on the Illicit Trade in Small Arms and Light Weapons in All Its Aspects, which was held in New York on 9–20 July 2001.[5] It was preceded by two meetings in 2000 and one in 2001 in which a draft Programme of Action was prepared to form the basis for the conference deliberations. Several groups of states, including the European Union (EU), the Organization of African Unity (OAU) and the Organization of American States (OAS), held regional conferences and meetings to prepare common positions at the UN conference.[6]

The aim of the conference was to agree a common Programme of Action to Prevent, Combat and Eradicate the Illicit Trade in Small Arms and Light Weapons in All Its Aspects.[7] In the Programme of Action, which is a political statement and not legally binding, the states participating in the conference announced they would undertake a series of measures to prevent, combat and eradicate the illicit trade in small arms and light weapons.

At the national level these measures include: creating and enforcing controls over the production and international transfer of small arms; marking small arms during production; maintaining records of all holdings and transfers of small arms; using end-user certificates and notifying the original exporting state when small arms are re-transferred; developing regulations for arms brokering; establishing standards and procedures related to the management and security of weapons held by authorized bodies; developing disarmament, demobilization and reintegration of combatants; and destroying confiscated small arms.

At the regional level the measures include: working towards regional legally binding instruments aimed at combating the illicit trade in small arms, encouraging moratoria or similar initiatives on the transfer and manufacture of small arms, establishing trans-border customs cooperation and networks for information sharing, developing transparency to combat illicit trade in small arms, and addressing the special needs of children affected by armed conflict.

At the global level the measures include: cooperating in the UN system in order to ensure the implementation of UN arms embargoes, encouraging disarmament and demobilization of ex-combatants and their reintegration into civilian life, encouraging states to enhance cooperation with Interpol, and promoting dialogue and a culture of peace by encouraging education and public awareness programmes.

During the debate on the Programme of Action, the USA was the most vocal opponent to a number of provisions[8] and stated that it would not support the Programme of Action unless certain provisions were removed. The US Government viewed these

---

[5] For information and documents on the conference see the UN Department for Disarmament Affairs Internet site, URL <http://www.un.org/Depts/dda/CAB/smallarms/>. For a categorized database see the Internet site of the Small Arms Survey, URL <http://www.smallarmssurvey.org/Database.html>.

[6] The EU adopted the Programme for the Preventing and Combating of Illicit Trafficking in Conventional Arms in June 1997; the OAU adopted the Bamako Declaration on an African Common Position on the Illicit Proliferation, Circulation and Trafficking of Small Arms and Light Weapons in Dec. 2001, UN document A/CONF.192/PC/23, 10 Jan. 2001; and the Latin American and Caribbean states adopted the Brasilia Declaration in Nov. 2000, UN A/CONF.192/PC/19, 19 Dec. 2000. For a list of international activities in preparation of the UN Conference see Report of the United Nations Conference on the Illicit Trade in Small Arms and Light Weapons in All Its Aspects, UN document A/CONF.192/15, 20 July 2001, pp. 17–19.

[7] Report of the United Nations Conference on the Illicit Trade in Small Arms and Light Weapons in All Its Aspects (note 5).

[8] For a detailed overview of the main points of discussion at the conference see Batchelor, P., 'The 2001 UN conference on small arms: a first step?', *Disarmament Diplomacy*, Sep. 2001, URL <http://www.acronym.org.uk/dd/dd60/60op1.htm>.

provisions as diverting attention from practical measures to cope with the illicit small arms trade or as falling under the sovereignty of state governments and therefore not legitimate areas for international cooperation and action.[9] After the text had been amended, the USA accepted the provisions that called for: (*a*) constraining the legal trade and manufacture of small arms; (*b*) international advocacy activities by non-governmental organizations (NGOs); and (*c*) a mandatory review conference. Discussions on two other issues resulted in last-minute negotiations and the extension of the conference by one day.[10]

The draft Programme of Action contained a paragraph which stated that countries should supply small arms and light weapons only to governments, thereby excluding supplies to non-state actors without authorization from the government in the recipient country.[11] The USA insisted that the paragraph be removed, arguing that it would preclude assistance to oppressed non-state groups and that '[d]istinctions between governments and non-governments are irrelevant in determining responsible and irresponsible end-users of arms'.[12] The African states opposed the US proposal, arguing that the supply of small arms to non-state actors was the most important way in which the proliferation of small arms in Africa is exacerbated. The EU also supported the proposed ban on small arms supplies to non-state actors.

Another paragraph called for consideration of a prohibition on the unrestricted trade in and private ownership of small arms specifically designed for military purposes. The USA objected to this provision as well, which it perceived as possibly restricting the right of private persons to own arms, a prohibition that would be opposed by the strong pro-gun lobby in the USA. Because the USA stated that it would not support the Programme for Action if these paragraphs remained in the document, even in altered form, both paragraphs were deleted in the final version, following agreement by the African delegations in order for it to be possible for the conference to achieve consensus.

The US position was strongly criticized.[13] However, the assertiveness of the USA made it possible for other states, which might otherwise have spoken out against certain provisions, to 'hide behind' the USA and thus avoid strong public and government criticism.

The conference was attended by a large number of NGOs and other representatives of civil society.[14] The NGOs included both organizations that were strongly in favour of strict controls on small arms trade and possession, organized mainly in the Inter-

---

[9] Under Secretary of State for Arms Control and International Security Affairs John R. Bolton, US Statement at Plenary Session of the UN Conference on the Illicit Trade in Small Arms and Light Weapons in All Its Aspects, 9 July 2001, URL <http://www.un.org/Depts/dda/CAB/smallarms/statements/usE.html>.

[10] DDA 2001 Update, United Nations Department for Disarmament Affairs (DDA), June/July 2001, p. 4.

[11] Draft Programme of Action to Prevent, Combat and Eradicate the Illicit Trade in Small Arms and Light Weapons in All Its Aspects, Working paper by the Chairman of the Preparatory Committee, A/CONF.192/PC/L.4/Rev.1, 12 Feb. 2001.

[12] Under Secretary of State for Arms Control and International Security Affairs John R. Bolton (note 9).

[13] The president of the conference criticized the USA in expressing his disappointment at the Conference's inability to agree 'due to the concerns of one state—on language recognizing the need to establish and maintain controls over private ownership of these deadly weapons, and the need for preventing sales of such weapons to non-state groups'. Batchelor (note 8). For similar criticism see 'Small arms win', *International Herald Tribune*, 1 July 2001, p. 6; and Lynch, C., 'US fights small arms controls, stance on UN pact not shared by allies', *Washington Post*, 10 July 2001, p. A01.

[14] A total of 119 registered NGOs and 380 representatives attended the conference. DDA 2001 Update (note 10), p. 3.

national Action Network on Small Arms (IANSA), and organizations that were in favour of liberal controls on the civilian possession of small arms, organized mainly in the World Forum on the Future of Sport Shooting Activities (WFSA).[15] The conference stressed the role of civil society in combating the illicit trade in small arms and the Programme of Action encouraged regional organizations and states to facilitate the appropriate cooperation of civil society, including NGOs.

III. Conclusions

The Programme of Action adopted by the conference has no legal status and does not create a regime. Most of the provisions are of a general nature. They do not present new norms regarding the possession of and trade in arms but only call on governments to take proper action to prevent those forms of arms possession and trade that already have been declared illicit under national laws. Moreover, there are no specific guidelines for how to operationalize the agenda. The Programme of Action can therefore be perceived as a weak outcome of the conference.

On the other hand, it can be argued that the Programme of Action is a clear declaration of the political will of the international community to act against the proliferation of small arms and an important first step towards building norms and implementing collective measures against the illicit trade in small arms.[16] It also represents the first global framework to guide the work of national governments, international organizations, especially regional organizations, and civil society in combating the illicit trade in small arms.

There is a risk that the international momentum created by the conference may whither away. In order to maintain the momentum the UN General Assembly decided to convene a conference in 2006 to review the progress of the implementation of the Programme of Action, preceded by biennial meetings of states.[17]

---

[15] For an assessment of the influence of NGOs on the conference see Batchelor, P., 'NGO perspectives: NGOs and the small arms issue', *Disarmament Forum*, no. 1 (2002), URL <http://www.unog.ch/unidir/2-01-e7%20batchelor.pdf>.

[16] A view taken by, e.g., the UN Secretary-General. DDA 2001 Update (note 10), p. 1.

[17] UN General Assembly Resolution 56/24, 10 Jan. 2002.

Appendix 13B. Documents on conventional arms control

CONCLUDING DOCUMENT OF THE NEGOTIATIONS UNDER ARTICLE V OF ANNEX 1-B OF THE GENERAL FRAMEWORK AGREEMENT FOR PEACE IN BOSNIA AND HERZEGOVINA

Vienna, 18 July 2001

Representatives of the twenty States referred to in the mandate of 23 November 1998 for negotiations under Article V of Annex 1-B of the General Framework Agreement for Peace in Bosnia and Herzegovina (hereafter 'the participating States'), have engaged in negotiations under the auspices of the OSCE Forum for Security Co-Operation (FSC) in Vienna from 8 March 1999. They participated in the process as sovereign and independent States, on the basis of full equality.

The recent democratic changes in South-East Europe and the admission of the Federal Republic of Yugoslavia to the OSCE, as well as its commitment to the principles and standards of the OSCE and the Vienna Document 1999, have been of special relevance to these negotiations.

General

1. The participating States acted in accordance with the mandate for negotiations under Article V of Annex 1-B of the General Framework Agreement for Peace in Bosnia and Herzegovina and were guided by the relevant OSCE Summit and Ministerial Council decisions.

2. The participating States underscore the importance of strict compliance with the provisions of the United Nations Charter and, in particular, of the full implementation of Security Council resolutions relevant to these negotiations.

3. The participating States reaffirm their adherence to the Helsinki Final Act, the Charter of Paris for a New Europe, and the Charter for European Security, and in particular recognize the indivisibility of security in Europe, and the inherent right of each and every participating State to be free to choose or change its security arrangements, including treaties of alliance, as they evolve. They reaffirm their commitment to full implementation of the provisions of the Vienna Document 1999, the Code of Conduct on Politico-Military Aspects of Security, and other FSC-agreed instruments.

4. The participating States underline their support for the aims and objectives and for the full implementation of the General Framework Agreement for Peace in Bosnia and Herzegovina. They recall the achievements reached through the Florence and Vienna Agreements.

5. The participating States are resolved to enrich their broad security dialogue and to further co-operation and good neighbourly relations, based on the principles of the Helsinki Final Act: sovereign equality and respect for the rights inherent in sovereignty; refraining from the threat or use of force; inviolability of frontiers; territorial integrity of States; peaceful settlement of disputes; non-intervention in internal affairs; respect for human rights and fundamental freedoms, including the freedom of thought, conscience, religion or belief; equal rights and self-determination of peoples; co-operation among States; and fulfilment in good faith of obligations under international law.

6. The participating States recall that they are committed to take appropriate measures in preventing their respective territories from being used for the preparation, organization or commission of acts of extremist violence, including terrorist activities, against other participating States and their citizens.

7. The participating States, noting the existing initiatives for co-operation, emphasize the significant contribution of regional bilateral and multilateral agreements and arrangements to OSCE-wide confidence- and security- building. They underline their commitment to the aims and objectives of the Stability Pact for South Eastern Europe.

8. The participating States reaffirm the significance of the Open Skies Treaty.

9. The participating States recall that the adapted CFE Treaty, upon its entry into force, will be open to voluntary accession by other OSCE participating States in the area between the Atlantic Ocean and the Ural Mountains and thereby will provide an important additional contribution to European stability and security.

Defence-related information

10. The participating States note the particular importance of defence budget transparency. They encourage the exchange of information on the actual yearly expenditures (in terms of the relevant currency). They further encourage the provision of information about financial or other forms of contribution received from any other State and applied to its defence budget, including financial donations to any defence or defence-related budget; donations of armaments and equipment; and defence-related loans, leases or sales on preferential terms.

11. Those participating States who so wish may consider, on a voluntary basis, the exchange of information with regard to their national holdings of conventional armaments, in a bilateral and reciprocal framework.

Expanded military contacts and co-operation

12. Recognizing the need for further developing friendly relations between States throughout Europe, the participating States will intensify their efforts to promote and facilitate military contacts and co-operation in accordance with Chapter IV of the Vienna Document 1999. Pursuing the goals described in Chapter X of the Vienna Document 1999, they will, on a voluntary basis and as appropriate, promote and facilitate:

Military contacts

12.1. The establishment of a regular security dialogue at the appropriate political and military levels; the establishment of points of contact at different levels between Ministries of Defence, General Staffs, military schools and academies and between regional commands and units, particularly in border areas; the establishment of contacts between military formations, units and institutions, particularly in border areas ('partnerships'), including sporting and cultural events; the organization of seminars/workshops on military–political matters; the reservation of places in mid-level (i.e., Command and General Staff College) and top-level (i.e., War College) military schools and academies for members of the armed forces from the participating States; the establishment of contacts between national verification units.

Military co-operation and risk reduction

12.2. The establishment of joint training for peacekeeping, search and rescue or disaster relief; the establishment of joint forces/headquarters for peacekeeping or disaster relief; the arrangement of visits for other participating States to military facilities or military formations in addition to the provisions set forth in the Vienna Document 1999; the arrangement of visits for other participating States to air bases in addition to the provisions set forth in the Vienna Document 1999; the establishment of hotlines between regional military commanders, particularly in border areas; the development of consultative mechanisms in case of unusual military activities, disaster relief, etc.; the use of the Regional Arms Control Verification and Implementation Assistance Centre (RACVIAC) for arms control-related instruction and training.

Military activities

13. Recognizing the significance of certain military activities, in particular in border areas, and taking into account existing bilateral and multilateral agreements, participating States may consider, on a voluntary basis and as appropriate, reducing the thresholds for military activities of their own forces subject to prior notification and observation to lower levels than those set out in Chapters V and VI of the Vienna Document 1999. Subject to the security needs of participating States they may develop additional criteria for notification and observation.

Inspections and evaluation visits

14. In accordance with paragraph 144.9 of the Vienna Document 1999, participating States may, on a voluntary basis and as appropriate, consider offering supplementary inspections and evaluation visits of their own forces, particularly in border areas. Other States, whose forces are present in the area, may voluntarily agree to participate in such supplementary inspections. The relevant provisions for inspections or evaluation visits set forth in Chapter IX of the Vienna Document 1999 will apply, unless otherwise agreed.

Antipersonnel mines

15. The participating States may provide, on a voluntary basis and as appropriate, financial and technical support in response to requests by other participating States for the de-mining of areas on their territory where antipersonnel mines are emplaced and for the destruction of antipersonnel mines.

Small arms and light weapons

16. The participating States recognize that the excessive and destabilizing accumulation and uncontrolled spread of small arms are problems that have contributed to the intensity and duration of the majority of recent armed conflicts. They are of concern because they pose a threat and a challenge to peace, and undermine efforts to ensure an indivisible and comprehensive security. The participating States reaffirm their commitment to the OSCE Document on Small Arms and Light Weapons. They remain determined to implement fully the measures agreed therein, noting their relevance to the objectives of Article V. The participating States will co-operate as appropriate in combating illicit trafficking in all its aspects; in safe and effective management of stockpiles; in reduction and destruction of surpluses; and in early warning, conflict prevention; crisis management, and post-conflict rehabilitation issues related to small arms and light weapons. They will also seek to provide, as appropriate, financial and technical support for activities in this field.

Commission

17. A Commission of the participating States is established to review the implementation of the measures contained in this Concluding Document. It will be chaired by a participating State. The Chairmanship will rotate alphabetically in the French language, beginning with Albania. Unless otherwise agreed, the Chairmanship will change every year. Unless otherwise agreed, the Commission will meet once a year. Extraordinary meetings may be convened at the request of any participating State following appropriate consultations with all participating States by the Chairman.

18. Decisions will be taken by consensus. The Commission will define its own procedures and working methods.

19. The Commission will meet under the auspices of the OSCE. It will inform the FSC and the Permanent Council (PC) of its activities and will liaise with the sub-table on Defence and Security Issues of Table III of the Stability Pact for South Eastern Europe.

Final

20. The measures contained in this Concluding Document are of a voluntary nature. This Concluding Document is politically binding and will become effective on 1 January 2002.

Source: Organization for Security and Co-operation in Europe, 'Regional arms control agreements, Article V', URL <http://www.osce.org/representatives/arms/article5/article5.pdf>.

14. Multilateral export controls

IAN ANTHONY

I. Introduction

This chapter describes identified changes in the guidelines and procedures of five multilateral export control regimes: the Australia Group (AG), the Zangger Committee, the Missile Technology Control Regime (MTCR), the Nuclear Suppliers Group (NSG) and the Wassenaar Arrangement on Export Controls for Conventional Arms and Dual-Use Technologies (WA).

In 2001 Bulgaria joined the Australia Group, and South Korea joined the MTCR, in each case bringing the number of participating states to 33. There are now 41 states that participate in one or more of the regimes while 27 states participate in all of them. The European Commission also participates in the Australia Group and the Zangger Committee and is represented in the NSG as an observer.[1] Table 14.1 lists the members of each regime.

In 2001 the MTCR completed work on a draft International Code of Conduct against Ballistic Missile Proliferation. The draft will be discussed among states with a view to adopting the code in 2002.

Multilateral export control will play a role in counter-terrorism measures. The annual plenary meeting of the MTCR was one of the first opportunities at which officials could discuss the implications of the attacks on the United States that occurred on 11 September 2001.[2] In early October the AG participating states discussed the role of export controls in reducing the threat of terrorist attacks using chemical and biological weapons (CBW). The group underlined that their objectives include preventing the acquisition of CBW by non-state actors.[3]

In December 2001 the participating states agreed to modify the initial elements of the Wassenaar Arrangement to make clear their commitment to prevent the acquisition of conventional arms and dual-use goods and technologies by terrorist groups and organizations as well as by individual terrorists.

---

[1] The Zangger Committee is an informal group of states that meet to discuss how to interpret their obligations under Article 3.2 of the 1968 Non-Proliferation Treaty (NPT). The committee is not part of the NPT. For additional information see URL <http://projects.sipri.se/expcon/NSG_documents.html>.

[2] The meeting took place in Ottawa on 25–28 Sep. 2001.

[3] *The Australia Group: Tackling the Threat of Chemical and Biological Weapons*, Media Release 1 Oct. 2001, Document AG/Oct01/Press/Chair/24. Australia Group documents are available on the Internet at URL <http://www.australiagroup.net>. The Australia Group is an informal network of countries that consult on and harmonize national export licensing measures that apply to lists of items agreed among the group. The participating states have agreed 6 lists of items and have made a political commitment to ensure that all items on these lists are subject to national export controls. The objective is to prevent trade and international cooperation from contributing to CBW programmes.

Table 14.1. Membership of multilateral weapon and technology export control regimes, as of 1 January 2002

State	Zangger Committee[a] 1974	NSG[b] 1978	Australia Group[a] 1985	MTCR[c] 1987	Wassenaar Arrangement 1996
Argentina	x	x	x	x	x
Australia	x	x	x	x	x
Austria	x	x	x	x	x
Belarus		x			
Belgium	x	x	x	x	x
Brazil		x		x	
Bulgaria	x	x	x[d]		x
Canada	x	x	x	x	x
China	x				
Cyprus		x	x		
Czech Republic	x	x	x	x	x
Denmark	x	x	x	x	x
Finland	x	x	x	x	x
France	x	x	x	x	x
Germany	x	x	x	x	x
Greece	x	x	x	x	x
Hungary	x	x	x	x	x
Iceland			x	x	
Ireland	x	x	x	x	x
Italy	x	x	x	x	x
Japan	x	x	x	x	x
Korea, South	x	x	x	x[d]	x
Latvia		x			
Luxembourg	x	x	x	x	x
Netherlands	x	x	x	x	x
New Zealand		x	x	x	x
Norway	x	x	x	x	x
Poland	x	x	x	x	x
Portugal	x	x	x	x	x
Romania	x	x	x		x
Russia	x	x		x	x
Slovakia	x	x	x		x
Slovenia	x	x			
South Africa	x	x		x	
Spain	x	x	x	x	x
Sweden	x	x	x	x	x
Switzerland	x	x	x	x	x
Turkey	x	x	x	x	x
UK	x	x	x	x	x
Ukraine	x	x		x	x
USA	x	x	x	x	x
Total	**35**	**39**	**33**	**33**	**33**

Note: The years in the column headings indicate when the export control regime was formally established, although the groups may have met on an informal basis before then.

[a] The European Commission participates in this regime.

[b] The Nuclear Suppliers Group. The European Commission is represented in this regime as an observer.

[c] The Missile Technology Control Regime.

[d] Became a member of the regime in 2001.

II. The Missile Technology Control Regime

The MTCR is an informal, voluntary association of countries that share the goal of non-proliferation of unmanned delivery systems for weapons of mass destruction and seek to coordinate national export licensing efforts aimed at preventing their proliferation. It was established by seven states in 1987. In 2001 South Korea participated fully in the MTCR, bringing the number of participating states to 33.[4]

The full participation of South Korea had been under discussion for several years. However, the South Korean Government did not submit a formal request to participate until January 2001—after determining the future of its own ballistic missile programme.[5] Until January 2001 South Korea was bound by a 1979 bilateral understanding with the USA according to which it would not develop missiles with ranges in excess of 180 km. In response to the development of ballistic missiles by North Korea, South Korea has expressed an interest in developing missiles with ranges up to 500 km.[6] Under an agreement with the USA reached in January 2001, the South Korean Government adopted new guidelines that enabled it to develop and produce guided missiles able to deliver a 500-kg payload to a range of up to 300 km.[7] This cleared the way for South Korea to participate in the MTCR, which requires a consensus among current participants.

The plenary meeting of the MTCR took place after the 11 September terrorist attacks on the USA. During the general information exchange, the possession of Scud missiles by the Taliban forces in Afghanistan and the possible implications was one of the issues taken up by participating states.

MTCR compliance issues

In 2001 the MTCR continued to discuss the issue of compliance with agreed measures. The national approaches of Russia and the United States to implementing their MTCR obligations have attracted particular attention.

Russia continued to modify its national export control system, partly in response to allegations that it did not comply with its MTCR commitments.

Allegations related to Russia focus on two different issues. First, the allegation has been made that Russian entities continue to supply missile-related items to missile programmes of concern—including Iran and North Korea, whose nuclear programmes cause proliferation concerns to the USA, in par-

---

[4] Plenary Meeting of the Missile Technology Control Regime, Ottawa, Canada, 25–28 Sep. 2001, Press Release, 28 Sep. 2001.

[5] S. Korea's New Missile Guidelines: Guidelines Balance Security, Non-Proliferation, Statement by Richard Boucher, Spokesman, US Department of State, 17 Jan. 2001, URL <http://usinfo.state.gov/topical/pol/arms/stories/01011702.htm>.

[6] It is discussed in Anthony, I., 'Responses to proliferation: the North Korean ballistic missile programme', *SIPRI Yearbook 2000: Armament, Disarmament and International Security* (Oxford University Press: Oxford, 2000), pp. 647–66.

[7] 'ROKG official notes ROK to be full member of MTCR in March', Seoul Yonhap, 21 Mar. 2001, in Foreign Broadcast Information Service, *Daily Report–East Asia* (*FBIS-EAS*), FBIS-EAS-2001-0321, 21 Mar. 2001.

ticular. Second, Russia has continued to have the more aggressive export orientation in its aerospace and arms industries that was observed in 2000. President Vladimir Putin has, in effect, annulled political agreements about military–technical cooperation with several countries of concern reached bilaterally with the USA by President Boris Yeltsin.[8] Of particular concern has been Russian marketing of missiles such as the Yakhont cruise missile and the Iskander-E land-based missile in the Middle East and South Asia.[9]

At the same time, developments in Russia's export control system in 2001 were expected to reduce the probability that missile technologies could be exported without the consent of the responsible Russian authorities. Through a Presidential Decree issued in April 2001 transfers of items on the Russian national control list developed for missiles and missile-related technologies using intangible means required a licence.

As a result, any operation or transaction resulting in the transfer of controlled items either to a foreign country or to a foreign person (including so-called 'deemed exports' of cases where a foreign person in Russia gained access to such items) became an activity subject to licence. This includes transfers via electronic means.[10] In order to assist with enforcement of these controls, in particular enforcement of intangible technology transfers,[11] Russia revised its export control reporting system to include the Ministry of Education and the Russian Academy of Science in the system of reporting to the Russian Federation Export Control Commission. These bodies would be required to create systems to ensure that the activities of Russian scientists with access to items and technologies subject to control were consistent with the export control laws and, similarly, to ensure compliance with the regulations by foreign students studying in Russia.[12]

In the case of the United States, governments in several other countries continued to complain about the use of US national legislation to control not only US exports but also activities taking place in other states. The National Defense Authorization Act for Fiscal Year 1991 requires mandatory US sanctions against foreign persons who export an item in the MTCR Annex to a country that is not a member of the MTCR.[13]

---

[8] Anthony, I., 'Multilateral weapon and technology export controls', *SIPRI Yearbook 2001: Armament, Disarmament and International Security* (Oxford University Press: Oxford, 2001), pp. 615–39.

[9] The Russian Agency for Conventional Armaments (RAV) has described the Iskander-E missile as a weapon of deterrence in local conflicts and a strategic weapon for small countries. The Iskander-E, a conventionally armed version of a missile being developed in Russia for a range of missions, including delivery of tactical nuclear weapons, is marketed as MTCR-compliant in that it has a range lower than 300 km. The RAV is a government agency to which Russian enterprises report and which, in effect, represents their interests within government.

[10] 'Russia endorses export rules to prevent development of chemical weapons', Interfax (Moscow), 28 Sep. 2001, in Foreign Broadcast Information Service, *Daily Report–Central Eurasia* (*FBIS-SOV*), FBIS-SOV-2001-0928, 1 Oct. 2001.

[11] Intangible technology transfers are discussed in Anthony (note 8), pp. 631–35.

[12] In recent years high-profile cases have been reported of foreign students gaining access to controlled technologies through participation in international scientific projects in Russia or with Russian partners.

[13] National Defense Authorization Act for Fiscal Year 1991, Public Law 101-510, Nov. 5, 1990, sections 1702 and 1703.

Missile-related sanctions have been applied against Chinese entities on several occasions in the past (in 1991 and 1993). China is not a member of the MTCR and has complained that the USA is using sanctions to apply US laws in cases where China has not broken any commitment or undertaking that it has given. The question of how to avoid a situation arising in which the USA would be compelled to introduce missile-related sanctions against China became an important issue in Chinese–US bilateral relations. In 1994 President Bill Clinton lifted the sanctions after China issued a statement agreeing not to export ground-to-ground missiles inherently capable of delivering at least a 500-kg payload with a range of at least 300 km. In 2000 China issued a more specific statement about how it would translate this commitment into its national export control system.[14]

The George W. Bush Administration has paid close attention to missile proliferation in bilateral talks with China, in particular Chinese implementation of commitments given in November 2000.[15] In August 2001 officials held talks intended 'to clarify China's willingness to implement fully the terms of the November 2000 missile agreement'. The talks were described as 'inconclusive'.[16] Prior to the meeting public reports suggested that the United States was still concerned about Chinese exports to Pakistan of items considered to be for use in Pakistan's missile programme.[17]

At the beginning of September the USA determined that Chinese and Pakistani entities had engaged in missile technology proliferation activities.[18] Accordingly, US law required the denial for two years of export licence applications authorizing the export of controlled missile technology items to entities found to have been engaged in missile-related transfers.

In China this decision led to sanctions being applied to the China Metallurgical Equipment Company, and in Pakistan sanctions were applied to the National Development Complex. In each case no new US Government contracts may be concluded with either entity for MTCR Annex-controlled equipment or technology for two years, while any licence applications to export MTCR Annex-controlled equipment or technology to either entity will be denied for two years.[19] Under the US Arms Export Control Act the Chinese Government was also subject to sanctions. Accordingly, export licences will

---

[14] China agreed to elaborate a list of goods and technologies that could contribute to missile development and production and to ensure that exports of these items would be subject to control. See Anthony (note 8); and chapter 5 in this volume.

[15] In Nov. 2000 the Chinese Foreign Ministry issued a statement that China would shortly introduce into its export control legislation a comprehensive list of missile-related items and dual-use items that could not be exported without authorization.

[16] State Department spokesman Philip Reeker, 'US and China wrap up missile talks in Beijing', 24 Aug. 2001, URL <http://www.nautilus.org/napsnet/dr/0108/AUG24.html#item8>.

[17] 'Sino-US missile talks officially halted', Hong Kong Agence France Press, 24 Aug. 2001, reproduced in Foreign Broadcast Information Service, *Daily Report–China* (*FBIS-CHI*), FBIS-CHI-2001-0824, 24 Aug. 2001.

[18] 'Bureau of Nonproliferation: imposition of missile proliferation sanctions against a Chinese entity and a Pakistani entity', *Federal Register*, vol. 66, no. 176 (11 Sep. 2001), p. 47256.

[19] 'Pakistan, China's principled stand' in Islamabad Khabrain (in Urdu) 4 Sep. 2001, reproduced as 'Daily hopes world to oppose "unjustified" US sanctions on China, Pakistan', FBIS-CHI-2001-0905, 4 Sep. 2001.

be denied for MTCR Annex-controlled equipment or technology to 'all activities of the Chinese government relating to the development or production of missile equipment or technology and all activities of the Chinese government affecting the development or production of electronics, space systems or equipment, and military aircraft'.[20] In addition, no US Government contracts may be placed involving the activities described above for a two-year period.[21] These sanctions may impact on the ability of US companies to use Chinese satellite launch facilities.

The imposition of the sanctions placed a question mark over the resumption of Chinese–US talks on how to implement the November 2000 agreement.[22]

The International Code of Conduct and efforts to control ballistic missile proliferation

While the ongoing proliferation of ballistic missiles capable of delivering nuclear, biological and chemical (NBC) weapons creates a security challenge, states have not put in place a system of international legal control. During 2000 and 2001 discussions within the MTCR aimed to develop an International Code of Conduct against Ballistic Missile Proliferation (ICOC) and to bring about the adoption of such a code.[23] At the plenary meeting in Ottawa in September 2001 a final draft code was agreed among the MTCR participating states.[24]

The draft ICOC contains a set of broad principles against ballistic missile proliferation, in favour of peaceful uses of space and supporting existing non-proliferation regimes. The draft also contains some confidence-building measures (CBMs) in the form of annual disclosures of information on ballistic missile and space launch vehicle (SLV) programmes and advance notification of ballistic missile and SLV launches.

The ICOC is only one of several initiatives currently taking place that is intended to put in place a system of international control for missiles. The United Nations has on its agenda the question of 'missiles' in all their aspects, while Russia has stimulated discussion of missile proliferation by proposing the creation of a Global Control System for Non-Proliferation of Missiles and Missile Technologies (GCS).

These processes suggest that many states see a need for an international instrument addressing the security impact of missiles, but there is no agree-

---

[20] 'Bureau of Nonproliferation' (note 18).
[21] 'Bureau of Nonproliferation' (note 18).
[22] He Yafei, an official from the Chinese embassy in Washington, is quoted as saying 'we want to engage in dialogue with the United States to find a way out, but sanctions have to be lifted first. The US side cannot expect, as with other countries, to continue with China on nonproliferation consultations while sanctions are in place'. Agence France Press (Hong Kong), 18 Sep. 2001, reproduced in 'China warns US to lift sanctions before resuming proliferation talks', FBIS-CHI-2001-0918, 18 Sep. 2001.
[23] The code is based on a Canadian proposal put forward in the 1999 plenary meeting of the MTCR in Noordwijk, the Netherlands.
[24] The final draft is available at URL <http://projects.sipri.se/expcon/draficoc.htm>.

ment on the purpose of such an instrument, its scope and legal form or the details of how it might operate.

Recent experience in other arms control processes raises doubts that such disagreements could be sufficiently narrowed in open-ended discussions in global forums to permit the adoption of any text. However, the objective of the MTCR participating states is to have a code adopted by as many states as possible and within a reasonable time.

Achieving a multilateral agreement on missile proliferation could have been pursued through the United Nations. In November 2000 the UN General Assembly requested the Secretary-General to prepare a report on missiles with the assistance of a panel of governmental experts for consideration in 2002.[25] The General Assembly also sought the views of member states on this question and, as of August 2001, had received nine replies.[26]

A comparison of the contents of the replies points to some of the difficulties in agreeing on a single approach in the UN context. The Russian response makes clear that the UN focus should be on missile proliferation, with a particular emphasis on 'the political instability in individual regions of the world' and the efforts of states 'to stimulate industrial and economic development through access to missile and space technologies'.[27] The Russian view is that other missile-related issues are better addressed through bilateral arrangements between states and, in particular, between Russia and the USA.

Similarly, the European Union (EU) member states were critical of the UN process, which 'lacks sufficient focus, in particular regarding what we see as the overriding problem in the field of missiles, that is, the proliferation of ballistic missiles, and in particular those capable of carrying weapons of mass destruction'.[28]

The reply by Pakistan explicitly rejects this focus, arguing that 'considering the issue of missiles in the limited context of "horizontal proliferation" will inevitably lead to partial, iniquitous and controversial solutions'.[29]

China introduced another point of potential disagreement by underlining the need for any agreement to promote international cooperation on the peaceful use of outer space.[30]

A Russian suggestion to create a GCS was announced at the opening of the 2000 Review Conference of the 1968 Non-Proliferation Treaty (NPT). The GCS would have three main elements. First, there would be a multilateral transparency regime applied to missile launches described by Russian officials

---

[25] Resolutions adopted by the General Assembly, 55/33. General and complete disarmament: missiles, UN document A/Res/55/33, 12 Jan. 2001.
[26] Replies were received from Belarus, Bolivia, China, El Salvador, Mexico, Pakistan, Russia, Saudi Arabia and Sweden (on behalf of the European Union).
[27] Russian Federation response contained in Missiles: Report of the Secretary General, UN document A/56/136, 5 July 2001.
[28] Sweden (on behalf of the states members of the United Nations that are members of the European Union) response contained in Missiles: Report of the Secretary General (note 27).
[29] Pakistan response contained in Missiles: Report of the Secretary General, UN document A/56/136/Add.2, 6 Sep. 2001.
[30] Chinese response contained in Missiles: Report of the Secretary General, UN document A/56/136/Add.1, 15 Aug. 2001.

as a CBM.³¹ This mechanism would be based on existing bilateral Russian–US arrangements in the area of missile launch notification. Under these arrangements a Joint Data Exchange Center (JDEC) would be established in Moscow. According to the Russian proposal, the establishment of the JDEC would create the technical capacity to establish a repository for data on launches by other states in the framework of a multilateral arrangement. Second, under the GCS positive security assurances would be provided to states that renounce national missile programmes. Third, multilateral consultations would be arranged on the problem of missile proliferation.³²

Russian officials have been invited to explain in more detail how the GCS would function and have briefed their counterparts about it in, for example, the NATO–Russian Permanent Joint Council (PJC) and in two seminars for officials organized in Moscow in March 2000 and February 2001. However, at these meetings no draft text was proposed, although texts were produced that explained the GCS in general terms. The USA decided not to participate in additional meetings in Moscow and also stayed away from an international conference on the peaceful uses of space organized in Moscow in May 2001.³³

Representatives of the MTCR participating states took part in the UN deliberations and attended the GCS-related meetings in Moscow. These processes are not considered to be incompatible with or to exclude the need for the ICOC. The MTCR participating states consider the ICOC to be the most advanced and the most promising of the current initiatives in that a text has been prepared and a process for its adoption has been decided upon.

The MTCR participating states have agreed to use their own diplomatic channels to develop the greatest possible support for the draft text of the ICOC as agreed in the Ottawa MTCR plenary. In July 2001 the European Union adopted a common position on the fight against ballistic missile proliferation, pledging to support the universalization of the ICOC and to 'actively support an ad hoc international negotiating process, leading to an International Conference for its adoption no later than 2002'.³⁴

At the beginning of February 2002, 78 states endorsed the draft ICOC at a meeting in Paris.³⁵ European Union states will coordinate and facilitate preparations for an international conference that is expected to take place at the end of 2002.

The draft ICOC as released from the MTCR is seen as a politically binding measure that can be modified by consensus (i.e., unless consensus is obtained the text will not be changed). Aware of the criticism of unfairness levelled

---

³¹ Described in Kile, S. N., 'Nuclear arms control and ballistic missile defence', *SIPRI Yearbook 2001* (note 8), pp. 439–40.

³² Missiles: Report of the Secretary-General (note 27).

³³ 'Russian Foreign Ministry official bewildered by no US official presence at space forum', ITAR-TASS, 10 Apr. 2001, in Foreign Broadcast Information Service, *Daily Report–Central Eurasia (FBIS-SOV)*, FBIS-SOV-2001-0410, 10 Apr. 2001.

³⁴ Council Common Position of 23 July 2001 on the fight against ballistic missile proliferation (2001/567/CFSP), *Official Journal of the European Communities*, L 202, 27 July 2001, p. 1.

³⁵ US Department of State, *Draft International Code of Conduct Against Ballistic Missile Proliferation*, Feb. 11 2002; and Nartker, M., 'International response: code of conduct ineffective, experts say', *Global Security Newswire*, Feb. 15 2001.

against the MTCR by non-participating states, it was proposed that the 'universalization of the draft code should take place through a transparent and inclusive negotiating process open to all states on the basis of equality'.[36]

Some consultation on the draft code took place during its elaboration. The MTCR participating states reported the existence of the ICOC to the UN General Assembly and to the Conference on Disarmament (CD), and circulated the preliminary draft text to all CD participating states.

Distributing the document gave states an opportunity to consider its contents and to pass their views to the MTCR participating states should they wish to do so, although it was not proposed that the CD should take up the ICOC as an element of its agenda.[37] In addition, a roundtable discussion of the draft code in Warsaw in May 2001 was attended by a number of critical states that do not participate in the MTCR, including China and India.

The draft ICOC is seen as a step towards the development of globally accepted norms in support of ballistic missile non-proliferation. The first section elaborates principles that states will abide by when subscribing to the code. The second section contains general measures to be implemented by states, including a commitment to reduce, where possible, national holdings of ballistic missiles capable of delivering weapons of mass destruction. The third section addresses the issue of cooperation with states which eliminate existing ballistic missile programmes or space launch vehicle programmes.

The fourth section of the draft ICOC describes transparency measures that subscribing states agree to implement with respect to ballistic missile programmes and SLV programmes. The measures consist of annual declarations and information exchange on national policies, on the number and generic class of ballistic missiles and SLVs launched during the preceding year, and pre-launch notification for ballistic missile and SLV launches and test flights.

A fifth section describes organizational aspects of the code consisting of a schedule of meetings, a mechanism for information exchange and a mechanism for voluntary resolution of questions arising from declarations.

The development of the ICOC has been managed by MTCR states in a way that is both flexible and innovative, although whether this is sufficient to lead to the successful adoption of a text remains to be seen.

III. The Nuclear Suppliers Group

The Nuclear Suppliers Group was established in 1978 following three years of discussion among seven nuclear supplier countries (Canada, France, the Federal Republic of Germany, Japan, the UK, the USA and the USSR). It is an informal arrangement of nuclear supplier states that seek to prevent the acqui-

---

[36] Plenary Meeting of the Missile Technology Control Regime, Ottawa, Canada, 25–28 Sep. 2001, Department of Foreign Affairs and International Trade Press Release, 28 Sep. 2001, available at URL <http://projects.sipri.se/expcon>.

[37] Russia has made reference in national statements to the possibility that the United Nations would have a primary role in the practical elaboration of an agreement or agreements, while Pakistan has insisted that discussing the question in the United Nations is essential.

sition of nuclear weapons by states other than those recognized as nuclear weapon states in the framework of the NPT.

The NSG has developed Guidelines for Nuclear Transfers and Guidelines for Nuclear-Related Dual-Use Equipment, Materials, Software and Related Technology that participating states apply in making national decisions about what kinds of exports to authorize. It has also drawn up lists of items to which these guidelines apply. These guidelines and lists are published by the International Atomic Energy Agency (IAEA) as INFCIRC/254.[38]

Apart from questions of membership and list development, in 2001 the NSG established a Consultative Group. This is a standing body that facilitates consultations among participating states on, for example, the interpretation of agreed guidelines for nuclear supply.[39]

The NSG participating states decided to establish an Internet site to facilitate access to public documents.[40] This decision was one more measure within a transparency initiative launched several years ago to explain the objectives and procedures of the NSG. The Internet site will be managed by the German Government on behalf of the NSG.

In 2001 the European Union withdrew an offer to finance the establishment of a secure fax network connecting NSG participating states. The offer, made in 1999 and to be supported using common funds, was revoked because no countries had taken it up. These funds were released for other purposes.

Another set of issues concerned how to interpret Russian nuclear cooperation with India in the context of NSG guidelines. This is not a new issue for the NSG to consider. In 1998 the NSG tried, unsuccessfully, to persuade Russia not to supply two nuclear reactors to India.[41]

This issue was raised again when Russia agreed to sell 58 tonnes of low-enriched uranium fuel pellets to India's nuclear power station at Tarapur. This agreement was reached in October 2000, at which time Indian reports suggest that India and Russia also discussed the question of additional supplies of reactors.[42] Indian reports quote the chief of the Indian Atomic Energy Commission, Anil Kadodkar, as saying that Russia offered to supply four new reactors for the Kudamkulam power plant in Tamil Nadu during a meeting of the Indo-Russian joint commission in Moscow in January 2001.[43]

Under the NSG guidelines nuclear suppliers have committed themselves not to supply controlled items to any end-user unless the recipient country has

---

[38] Communications Received from Certain Member States Regarding Guidelines for the Export of Nuclear Material, Equipment and Technology, INFCIRC/254/Rev.4/Part 1, 15 Mar. 2000; and Communications Received from Certain Member States Regarding Guidelines for Transfers of Nuclear-Related Dual-Use Equipment, Materials, Software and Related Technology, INFCIRC/254/Rev.4/Part 2*, 9 Mar. 2000. IAEA documents are available at URL <http://www.iaea.org/worldatom/Documents/>.

[39] Press Statement, Nuclear Suppliers Group Plenary Meeting, Aspen, Colo., 10–11 May 2001.

[40] Nuclear Suppliers Group, URL <http://www.nuclearsuppliersgroup.org>.

[41] Hibbs, M., 'Russia–India: West may pressure IMF on Russian reactor sales', *Global Beat: Nuclear Watch*, June 26 1998, URL <http://www.nyu.edu/globalbeat/nucwatch/nucwatch062698.html>.

[42] Chellaney, B., 'Russia steps in to save Tarapur N-plant', *Hindustan Times* (Internet edn), 12 Oct. 2000, URL <http://www.hindustantimes.com/nonfram/121000/detEXC01.asp/>.

[43] Malhotra, J., 'Russia offers 4 more N-reactors to India', *India Express* (Internet edn), 20 Feb. 2001, URL <http://www.indian-express.com/ie/daily/20010220/iin20044.html>.

placed all of its nuclear activities under full-scope IAEA safeguards. This commitment was adopted in Warsaw in 1992.[44] India has many nuclear facilities that are not under full-scope safeguards.

The United States has argued that both supply of the reactors and the supplies of nuclear fuel are inconsistent with Russia's NSG commitments. Decisions reached by Russia in 1998 and any offers made in 2001 to supply reactors to India would be in conflict with the 1992 Warsaw Statement on Full Scope Safeguards.[45] Russia has argued that specific contracts to supply reactors to India, agreed in 1997 and 1998, were implementing a bilateral Memorandum of Understanding (MOU) signed with India in 1988. In the Russian view, commitments made prior to 1992 are not governed by the Warsaw Statement.

Russia does not claim that agreements on nuclear fuel are 'grandfathered' since they were reached in 1998. Before the Indian nuclear tests of 1998 the Tarapur reactor purchased nuclear fuel from China (which is not a member of the NSG). However, after the Indian tests China stopped supplying this fuel. Russia has argued that nuclear supply arrangements do not prohibit transfers made on the grounds of safety and that Tarapur should be seen as a special case. According to the Russian argument, the reactors will become unsafe if they continue to burn existing fuel. Moreover, it is argued that the non-proliferation arguments against nuclear supply are weak because India has already demonstrated its capability to manufacture nuclear weapons using resources that are not related to the Tarapur facility.[46]

The safety exemption (contained in paragraph 4 of the NSG guidelines) states that transfers may be made to a non-nuclear weapon state without a safeguards agreement 'only in exceptional cases when they are deemed essential for the safe operation of existing facilities'. In these cases the nuclear supplier should 'inform and, if appropriate, consult in the event that they intend to authorize or deny such transfers'.[47]

Individual participating states take national licensing decisions according to their own interpretation of their commitments under the NSG. However, in 1994 the NSG suppliers agreed on how this safety exemption should be interpreted for licensing purposes. Transfers should be authorized 'only when deemed to be essential in order to prevent or correct a radiological hazard pos-

---

[44] Statement on full-scope safeguards, agreed at the Meeting of Adherents to the Nuclear Suppliers Guidelines, Warsaw, 31 Mar.–3 Apr. 1992. The statement was subsequently published by the IAEA as INFCIRC/405, May 1992, URL <http://www.iaea.org/worldatom/Documents/Infcircs/Others/inf405.shtml>.

[45] 'Russian shipment of low enriched uranium fuel to India', Statement by Philip T. Reeker, US Department of State, 16 Feb. 2001 reproduced in *Washington File*, 16 Feb. 2001, URL <http://usinfo.state.gov/topical/pol/arms/stories/01021601.htm>.

[46] The arguments are laid out in Stratford, R. K., 'Starting over: building a non-proliferation regime from scratch', Paper delivered to the Non-Proliferation Symposium How to Harmonize Peaceful Uses of Nuclear Energy and Nonproliferation Policy organized by the Japan Atomic Industrial Forum, Tokyo, 7–8 Mar. 2001. Stratford is the Director of the Office of Nuclear Energy Affairs, US Department of State. The papers are archived at URL <http://www.jaif.or.jp/english/npsympo/sympo_2nd.html/>.

[47] Guidelines for Nuclear Transfers, Article 4(b). The most recent version of the guidelines have been published by the IAEA as INFCIRC/254/Rev. 4/Part , 15 Mar. 2000. The document is archived at URL <http://www.iaea.org/worldatom/Documents/Infcircs/2000/infcirc254r4p1a1.pdf>.

ing a significant danger to public health and safety and which cannot be realistically met with any other means'.[48]

At successive NSG meetings the Russian arguments were rejected by all the participating states except Belarus. Most representatives agreed that the Russian transfer of nuclear fuel could not reasonably be said to fall within the 1994 interpretation. One unnamed official said that the Russian action was a 'flagrant violation' of NSG agreements and that 'if the reactors are unsafe, then they shouldn't operate'.[49] In its response the US State Department noted that the transfers were part of a pattern of Russian nuclear export activity that 'raises serious questions about Russia's support for the goal of preventing nuclear proliferation'.[50] Other states apparently argued that the NSG depended on solidarity among its members for success. If one state was able to carry out commercial activities of this kind then other participating states might have to review their national positions.

In 2001 there were developments in Russia's nuclear establishment and national export control system that were of relevance to analysis of Russia's implementation of its NSG commitments.

In March 2001 President Putin removed the Minister of Atomic Energy, Yevgeniy Adamov, from his position.[51] In general, Adamov had lobbied hard within the Russian Government for steps to increase nuclear exports and international industrial cooperation in the field of nuclear energy. Adamov was a strong supporter of agreements with India and Iran that were contentious both in the context of Russia's international obligations in regard to nuclear non-proliferation and in bilateral relations with the United States.

It is not clear whether the reasons for Adamov's removal were related to non-proliferation concerns. Adamov was mentioned in a report of the Anti-Corruption Commission of the Russian Parliament released just before President Putin took the decision.[52] Moreover, it is not clear that the dismissal signals a change in policy. In April 2001 the new minister, Alexander Rumyantsev, announced that the agreement to ship nuclear fuel to India would be fulfilled.[53]

In June 2001 the Russian Government issued a Federal Decree containing Regulations on Control over Foreign Economic Activity in Respect of Nuclear-Related Dual Use Equipment and Materials and Related Tech-

---

[48] Thorne, C. E., *A Guide to Nuclear Export Controls* (Proliferation Data Services: Burke, Va., 1997), pp. 1–3.
[49] Quoted in Hibbs, M., 'NSG objects again after Russia says LEU exports to India are proceeding', *Nuclear Fuel*, vol. 26 no. 3 (5 Feb. 2001).
[50] 'Russian shipment of low enriched uranium fuel to India' (note 45).
[51] 'Adamov's dismissal is a good sign for nuclear nonproliferation', PIR Center, 28 Mar. 2001, reproduced at URL <http://www.pircenter.lrg/board/article.php3?artid=639>.
[52] According to the report, Adamov was linked to at least 10 companies inside and outside Russia, mostly consulting and import/export companies managing aspects of nuclear trade. Employees of Minatom are forbidden to have private business interests.
[53] 'New nuclear minister pushes for spent nuclear fuel imports', *Russian Journal Online*, Daily News Report: Energy, 6 Apr. 2001, URL <http://www.russiajournal.com/news/rj_news.shtml?nd=578>. Rumyantsev has also made it clear that Russian nuclear cooperation with Iran will continue.

nology.⁵⁴ The regulations prohibited transfers of controlled items under four conditions: (*a*) for use in carrying out activities for the creation of nuclear explosive devices; (*b*) for use in states not possessing nuclear weapons in carrying out activities in the field of the nuclear fuel cycle not placed under IAEA safeguards; (*c*) in the case of the existence of an unacceptable risk of their being used for purposes indicated in a and b; and (*d*) when the transfer is contrary to the purpose of the non-proliferation of nuclear weapons.

IV. The Wassenaar Arrangement

The Wassenaar Arrangement is an informal arrangement in which the participating states intend to contribute to regional and international security by promoting transparency and greater responsibility with regard to transfers of conventional arms and dual-use goods and technologies, thus preventing destabilizing accumulations.

Through national policies the participating states seek to prevent transfers of agreed items from contributing to the development or enhancement of military capabilities that undermine regional and international security, and to ensure that transferred items are not diverted to support such capabilities. The arrangement mainly provides a mechanism for information exchange and does not attempt to develop common controls. However, under its initial elements the arrangement is intended 'to enhance cooperation to prevent the acquisition of armaments and sensitive dual-use items for military end-uses, if the situation in a region or the behaviour of a state is, or becomes, a cause for serious concern to the participating states'.⁵⁵

In 2001 the main issues of contention within the WA concerned disagreements among the participating states about how to enhance transparency in reporting on conventional arms transfers. This issue had two elements. First, the contents and use of information reported informally by states in papers describing national perspectives on the armament dynamic in particular regions and subregions form part of the general information exchange between participating states within the Wassenaar Arrangement. Second, the question is addressed of how to advance the more specific information exchange, which currently consists of exchanges of information every six months on deliveries of conventional arms to states that do not participate in the WA. Conventional arms have the same definition for reporting purposes as the original categories used in the UN Register of Conventional Arms.⁵⁶

During 2001 an increasing number of states submitted papers for consideration during the general information exchange. Some countries submitted mul-

---

⁵⁴ These regulations are available in English on the Internet site of the SIPRI Export Control Project at URL <http://projects.sipri.se/expcon/dualuse/russiadu.htm>.

⁵⁵ The Wassenaar Arrangement on Export Controls for Conventional Arms and Dual-Use Goods and Technologies: Initial Elements, Vienna, 12 July 1996, URL <http://www.wassenaar.org/docs/docindex.html>.

⁵⁶ The background to these issues is provided in Anthony, I., 'Multilateral weapon and technology export controls', *SIPRI Yearbook 2000* (note 6), pp. 667–84; and Anthony, I., 'Multilateral weapon and technology export controls', *SIPRI Yearbook 2001* (note 8), pp. 615–39.

tiple papers—for example, the Russian Federation submitted papers on six different regions. These national papers generated significant discussion among participating states. For example, the papers submitted by Japan and South Korea presented different conclusions about the implications for regional security of deliveries of arms to North Korea. Russia disagreed with the evaluation submitted by the United States of the implications of arms deliveries to India for regional and international security.

Disagreement about the particular content of papers notwithstanding, these exchanges indicate a positive evolution of the Wassenaar Arrangement. However, they also highlight some shortcomings in the current procedures.

The papers submitted by participating states and the discussion generated by them go some way to addressing the criticism that the WA pays too little attention to the way in which the norms established in the initial elements are implemented. The development of the general information exchange is made more difficult by two features of current reporting procedures.

First, the reporting is confined to the armament dynamic in non-participating states. This can lead to important matters being excluded from discussions within the WA. For example, at least one state submitted a paper addressing the impact of arms deliveries on regional security in the Caucasus. However, the paper could not take into account the impact of developments in the North Caucasus for the Caucasus as a whole because the North Caucasus forms part of the territory of Russia, a WA participating state.

Second, states still tend to confine their reporting to systems contained in the seven categories listed in appendix 3 of the document Initial Elements of the Wassenaar Arrangement on Export Controls for Conventional Arms and Dual-Use Goods and Technologies. In 2001 a small group of military experts from the main exporting countries met to discuss what kinds of equipment could be included in an expanded information exchange without any compromise to the security or commercial interests of exporters. In addition, two additional subcategories of military items were added to the mandatory reporting of transfers/licences granted. These were armoured bridge-launching vehicles and gun-carriers specifically designed for towing artillery.[57]

V. The impact of the 11 September terrorist attacks on multilateral export control

The multilateral export control regimes were not designed to address the issue of terrorist access to weapons of different kinds. However, after the terrorist attacks against the United States it has become more obvious that counteracting transnational terrorist networks requires international cooperation.[58] After 11 September 2001 all of the multilateral regimes have taken notice in

[57] Public Statement, Seventh Plenary of the Wassenaar Arrangement, Vienna, Dec. 7 2001. This statement and the revised categories for reporting purposes are both archived at URL <http://www.wassenaar.org/>.

[58] For more information see 'Counterterrorism and the nonproliferation regime', a special issue of *The Monitor: International Perspectives on Nonproliferation*, vol. 8, no. 1 (winter 2002).

their meetings of the need to examine how their activities could contribute to eliminating terrorism.

The risk that terrorist groups would acquire non-conventional weapons became a focus of particular attention.[59] In testimony before the US Senate the Director of Central Intelligence noted that

as early as 1998, Bin Ladin publicly declared that acquiring unconventional weapons was 'a religious duty' . . . we know that al-Qa'ida was working to acquire some of the most dangerous chemical agents and toxins. Documents recovered from al-Qa'ida facilities in Afghanistan show that Bin Ladin was pursuing a sophisticated biological weapons research program. We also believe that Bin Ladin was seeking to acquire or develop a nuclear device. Al-Qa'ida may be pursuing a radioactive dispersal device—what some call a 'dirty bomb'.[60]

Issues that have been under discussion in export control regimes in recent years are relevant to combating terrorist groups.[61] The implementation of end-use or 'catch-all' controls against groups and individuals identified as terrorists by the United Nations may be one feasible approach.

The more widespread use of end-use controls has increased the need for information sharing among regime members. In response to the attacks of 11 September the Australia Group is currently discussing enhanced information sharing within the group and expanding its scope to cover dual-use equipment and technology. The Wassenaar Arrangement decided to amend its initial elements for the first time to include the commitment that participating states 'will continue to prevent the acquisition of conventional arms and dual-use goods and technologies by terrorist groups and organisations, as well as by individual terrorists'.[62]

Export control regimes lack a common risk assessment that can be the basis for national decisions about whether to authorize a given export and the conditions to attach to an authorization. Measures to help identify the actual end-user of controlled items and to reduce the risk of unauthorized re-export of controlled items are likely to remain a focal point of discussion. Since the dissolution of the Coordinating Committee on Multilateral Export Controls (COCOM) arrangements states that participate in regimes have emphasized national decision making and insist that regimes do not target any particular state or group of states. If risk assessment procedures suggested that programmes of concern are in fact concentrated in a small number of states this approach might be called into question.

---

[59] The impact of the distribution of anthrax using the postal service in Sep. 2001 is discussed in chapter 12 in this volume.

[60] *Worldwide Threat: Converging Dangers in a Post 9/11 World*, Testimony of Director of Central Intelligence George J. Tenet before the Senate Select Committee on Intelligence Feb. 6 2002.

[61] For further discussion see Anthony, I., 'Combating terrorism: the role of regimes', *The Monitor: International Perspectives on Nonproliferation*, vol. 8, no. 1 (winter 2002), pp. 7–11.

[62] The Wassenaar Arrangement on Export Controls for Conventional Arms and Dual-Use Goods and Technologies, Initial Elements as adopted by the Plenary of 11–12 July 1996 and amended by the Plenary of 6–7 Dec. 2001, URL <http://www.wassenaar.org/docs/docindex.html>.

Regimes are also examining the conditions under which simplified procedures might be applied between states that have a high degree of confidence regarding compliance with arms control agreements and the effectiveness of their export control systems. The realization that groups planning terrorist acts may already be located within the territories of regime participants is causing a reassessment of the wisdom of simplified procedures.

VI. Conclusions

A significant number of states have developed common rules and habits of cooperation in the framework of the multilateral export control regimes. Nevertheless, there has been a growing sense that the momentum established within the regimes in the first part of the 1990s was not maintained.

Prior to the attacks of 11 September 2001, however, the experience of the regimes was that there remain significant disagreements between participating states over important issues. Disagreements often stem from the fact that licensing decisions are based on national interpretations of regime rules. These are in turn steered by the interests of participating states rather than a common norm or a common perception of the risks posed by particular transfers.

After 11 September certain decisions that were difficult to take in the framework of the regimes may have become possible.

Particular attention is being paid in this regard to the following issues: the development of procedures for sharing information related to licensing and enforcement; the development of a more harmonized approach to risk assessment and the identification of programmes of concern; the development of common approaches to end-user controls in countries where programmes of concern are located; and the question of how to apply controls to new types of commercial practices in a changing market.

Annexes

Annex A. Arms control and disarmament agreements

Annex B. Chronology 2001

Annex A. Arms control and disarmament agreements

CHRISTER BERGGREN

Notes

1. The agreements are listed in the order of the date on which they were opened for signature (multilateral agreements) or signed (bilateral agreements); the date on which they entered into force and the depositary for multilateral treaties are also given. Information is as of 1 January 2002 unless otherwise indicated. Where confirmed information on entry into force or new parties became available in early 2002, this information is given in notes.

2. The main source of information is the lists of signatories and parties provided by the depositaries of the treaties.

3. For a few major treaties, the substantive parts of the most important reservations, declarations and/or interpretive statements made in connection with a state's signature, ratification, accession or succession are given in footnotes below the list of parties.

4. States listed as parties have ratified, acceded or succeeded to the agreements. Former non-self-governing territories, upon attaining independence, sometimes make general statements of continuity to all agreements concluded by the former colonial power. This annex lists as parties only those new states which have made an uncontested declaration on continuity or have notified the depositary about their succession.

5. The Russian Federation, constituted in 1991 as an independent state, confirmed the continuity of international obligations assumed by the Soviet Union. In order to become signatories/parties, the other former Soviet republics which were constituted in 1991 as independent sovereign states subsequently signed, ratified or acceded to agreements.

6. Czechoslovakia split into two states, the Czech Republic and Slovakia, in 1993. Both states have succeeded to all the agreements listed in this annex to which Czechoslovakia was a party.

7. The Socialist Federal Republic of Yugoslavia (SFRY) split into several states in 1991–92. In accordance with the information received from the depositaries, 'Yugoslavia' is listed for the agreements which the SFRY (former Yugoslavia) signed or ratified. 'Yugoslavia (FRY)' is listed for the agreements to which the Federal Republic of Yugoslavia (FRY) has acceded. On 6 March 2001 the FRY notified the United Nations that it intends to succeed to a number of treaties deposited with the Secretary-General. (In March 2002 it was proposed that the name of the country be changed to 'Serbia and Montenegro', subject to parliamentary approval.) The former Yugoslav republics of Bosnia and Herzegovina, Croatia, Macedonia and Slovenia have succeeded, as independent states, to several agreements.

8. Taiwan, while not recognized as a sovereign state by some nations, is given as a party to those agreements which it has ratified.

9. Unless stated otherwise, the multilateral agreements listed in this annex are open to all states for signature, ratification, accession or succession.

10. A complete list of UN member states, with the year in which they became members, appears in the glossary at the front of this volume. Not all the states listed in this annex are UN members.

Protocol for the Prohibition of the Use in War of Asphyxiating, Poisonous or Other Gases, and of Bacteriological Methods of Warfare (Geneva Protocol)

Signed at Geneva on 17 June 1925; entered into force on 8 February 1928; depositary French Government

The protocol declares that the parties agree to be bound by the prohibition on the use in war of these weapons.

Parties (133): Afghanistan, Albania, Algeria[1], Angola[1], Antigua and Barbuda, Argentina, Australia, Austria, Bahrain[1], Bangladesh[1], Barbados, Belarus, Belgium, Benin, Bhutan, Bolivia, Brazil, Bulgaria, Burkina Faso, Cambodia[1], Cameroon, Canada, Cape Verde, Central African Republic, Chile, China[1], Côte d'Ivoire, Cuba, Cyprus, Czech Republic, Denmark, Dominican Republic, Ecuador, Egypt, Equatorial Guinea, Estonia, Ethiopia, Fiji[1], Finland, France, Gambia, Germany, Ghana, Greece, Grenada, Guatemala, Guinea-Bissau, Holy See, Hungary, Iceland, India[1], Indonesia, Iran, Iraq[1], Ireland, Israel[2], Italy, Jamaica, Japan, Jordan[3], Kenya, Korea (North)[1], Korea (South)[1], Kuwait[1], Laos, Latvia, Lebanon, Lesotho, Liberia, Libya[1], Liechtenstein, Lithuania, Luxembourg, Madagascar, Malawi, Malaysia, Maldives, Malta, Mauritius, Mexico, Monaco, Mongolia, Morocco, Nepal, Netherlands, New Zealand, Nicaragua, Niger, Nigeria[1], Norway, Pakistan[1], Panama, Papua New Guinea[1], Paraguay, Peru, Philippines, Poland, Portugal[1], Qatar, Romania, Russia, Rwanda, Saint Kitts and Nevis, Saint Lucia, Saint Vincent and the Grenadines, Saudi Arabia, Senegal, Sierra Leone, Slovakia, Solomon Islands, South Africa, Spain, Sri Lanka, Sudan, Swaziland, Sweden, Switzerland, Syria, Tanzania, Thailand, Togo, Tonga, Trinidad and Tobago, Tunisia, Turkey, Uganda, UK[4], Uruguay, USA[4], Venezuela, Viet Nam[1], Yemen, Yugoslavia[1]

Signed but not ratified: El Salvador

[1] The protocol is binding on this state only as regards states which have signed and ratified or acceded to it. The protocol will cease to be binding on this state in regard to any enemy state whose armed forces or whose allies fail to respect the prohibitions laid down in it.

[2] The protocol is binding on Israel only as regards states which have signed and ratified or acceded to it. The protocol shall cease to be binding on Israel in regard to any enemy state whose armed forces, or the armed forces of whose allies, or the regular or irregular forces, or groups or individuals operating from its territory, fail to respect the prohibitions which are the object of the protocol.

[3] Jordan undertakes to respect the obligations contained in the protocol with regard to states which have undertaken similar commitments. It is not bound by the protocol as regards states whose armed forces, regular or irregular, do not respect the provisions of the protocol.

[4] The protocol shall cease to be binding on this state with respect to use in war of asphyxiating, poisonous or other gases, and of all analogous liquids, materials or devices, in regard to any enemy state if such state or any of its allies fails to respect the prohibitions laid down in the protocol.

Treaty for Collaboration in Economic, Social and Cultural Matters and for Collective Self-defence among Western European states (Brussels Treaty)

Signed at Brussels on 17 March 1948; entered into force on 25 August 1948; depositary Belgian Government

The treaty provides for close cooperation of the parties in the military, economic and political fields.

Parties (7): *Original parties:* Belgium, France, Luxembourg, Netherlands, UK

Germany and Italy acceded through the 1954 Protocols.

See also Modified Brussels Treaty and Protocols of 1954.

Convention on the Prevention and Punishment of the Crime of Genocide (Genocide Convention)

Adopted at Paris by the UN General Assembly on 9 December 1948; entered into force on 12 January 1951; depositary UN Secretary-General

Under the convention any commission of acts intended to destroy, in whole or in part, a national, ethnic, racial or religious group as such is declared to be a crime punishable under international law.

Parties (133): Afghanistan, Albania*, Algeria*, Antigua and Barbuda, Argentina*, Armenia, Australia, Austria, Azerbaijan, Bahamas, Bahrain*, Bangladesh*, Barbados, Belarus*, Belgium, Belize, Bosnia and Herzegovina, Brazil, Bulgaria*, Burkina Faso, Burundi, Cambodia, Canada, Chile, China*, Colombia, Congo (Democratic Republic of the), Costa Rica, Côte d'Ivoire, Croatia, Cuba, Cyprus, Czech Republic, Denmark, Ecuador, Egypt, El Salvador, Estonia, Ethiopia, Fiji, Finland, France, Gabon, Gambia, Georgia, Germany, Ghana, Greece, Guatemala, Guinea, Haiti, Honduras, Hungary*, Iceland, India*, Iran, Iraq, Ireland, Israel, Italy, Jamaica, Jordan, Kazakhstan, Korea (North), Korea (South), Kuwait, Kyrgyzstan, Laos, Latvia, Lebanon, Lesotho, Liberia, Libya, Liechtenstein, Lithuania, Luxembourg, Macedonia (Former Yugoslav Republic of), Malaysia*, Maldives, Mali, Mexico, Moldova, Monaco, Mongolia*, Morocco*, Mozambique, Myanmar (Burma)*, Namibia, Nepal, Netherlands, New Zealand, Nicaragua, Norway, Pakistan, Panama, Papua New Guinea, Paraguay, Peru, Philippines*, Poland*, Portugal*, Romania*, Russia*, Rwanda*, Saint Vincent and the Grenadines, Saudi Arabia, Senegal, Seychelles, Singapore*, Slovakia, Slovenia, South Africa, Spain*, Sri Lanka, Sweden, Switzerland, Syria, Tanzania, Togo, Tonga, Tunisia, Turkey, Uganda, UK, Ukraine*, Uruguay, USA*, Uzbekistan, Venezuela*, Viet Nam*, Yemen*, Yugoslavia*, Zimbabwe

Signed but not ratified: Bolivia, Dominican Republic

* With reservation and/or declaration.

Geneva Convention IV Relative to the Protection of Civilian Persons in Time of War

Signed at Geneva on 12 August 1949; entered into force on 21 October 1950; depositary Swiss Federal Council

The convention establishes rules for the protection of civilians in areas covered by war and on occupied territories.

Parties (189): Afghanistan, Albania*, Algeria, Andorra, Angola*, Antigua and Barbuda, Argentina, Armenia, Australia*, Austria, Azerbaijan, Bahamas, Bahrain, Bangladesh, Barbados*, Belarus*, Belgium, Belize, Benin, Bhutan, Bolivia, Bosnia and Herzegovina, Botswana, Brazil, Brunei Darussalam, Bulgaria*, Burkina Faso, Burundi, Cambodia, Cameroon, Canada, Cape Verde, Central African Republic, Chad, Chile, China*, Colombia, Comoros, Congo (Democratic Republic of the), Congo (Republic of), Costa Rica, Côte d'Ivoire, Croatia, Cuba, Cyprus, Czech Republic*, Denmark, Djibouti, Dominica, Dominican Republic, Ecuador, Egypt, El Salvador, Equatorial Guinea, Estonia, Eritrea, Ethiopia, Fiji, Finland, France, Gabon, Gambia, Georgia, Germany*, Ghana, Greece, Grenada, Guatemala, Guinea, Guinea-Bissau*, Guyana, Haiti, Holy See, Honduras, Hungary*, Iceland, India, Indonesia, Iran*, Iraq, Ireland, Israel*, Italy, Jamaica, Japan, Jordan, Kazakhstan, Kenya, Kiribati, Korea (North)*, Korea (South)*, Kuwait*, Kyrgyzstan, Laos, Latvia, Lebanon, Lesotho, Liberia, Libya, Liechtenstein, Lithuania, Luxembourg, Macedonia (Former Yugoslav Republic of)*, Madagascar, Malawi, Malaysia, Maldives, Mali, Malta, Mauritania, Mauritius, Mexico, Micronesia,

Moldova, Monaco, Mongolia, Morocco, Mozambique, Myanmar (Burma), Namibia, Nepal, Netherlands, New Zealand, Nicaragua, Niger, Nigeria, Norway, Oman, Pakistan*, Palau, Panama, Papua New Guinea, Paraguay, Peru, Philippines, Poland*, Portugal*, Qatar, Romania*, Russia*, Rwanda, Saint Kitts and Nevis, Saint Lucia, Saint Vincent and the Grenadines, Samoa (Western), San Marino, Sao Tome and Principe, Saudi Arabia, Senegal, Seychelles, Sierra Leone, Singapore*, Slovakia*, Slovenia, Solomon Islands, Somalia, South Africa, Spain, Sri Lanka, Sudan, Suriname*, Swaziland, Sweden, Switzerland, Syria, Tajikistan, Tanzania, Thailand, Togo, Tonga, Trinidad and Tobago, Tunisia, Turkey, Turkmenistan, Tuvalu, Uganda, UK, Ukraine*, United Arab Emirates, Uruguay*, USA*, Uzbekistan, Vanuatu, Venezuela, Viet Nam*, Yemen*, Yugoslavia*, Zambia, Zimbabwe

* With reservation and/or declaration.

In 1989 the Palestine Liberation Organization (PLO) informed the depositary that it had decided to adhere to the four Geneva Conventions and the two Protocols of 1977.

See also Protocols I and II of 1977.

Treaty of Economic, Social and Cultural Collaboration and Collective Self-Defence (Modified Brussels Treaty); Protocols (Paris Agreements)

Signed at Paris on 23 October 1954; entered into force on 6 May 1955; depositary Belgian Government

The 1948 Brussels Treaty was modified by four protocols which amended the original text to take account of political and military developments in Europe, allowing the Federal Republic of Germany and Italy to become parties in return for controls over German armaments and force levels (annulled, except for weapons of mass destruction, in 1984). The Western European Union (WEU) was created through the Modified Brussels Treaty. The treaty contains an obligation for collective defence of its members.

Members of the WEU: Belgium, France, Germany, Greece, Italy, Luxembourg, Netherlands, Portugal, Spain, UK

Antarctic Treaty

Signed at Washington, DC, on 1 December 1959; entered into force on 23 June 1961; depositary US Government

Declares the Antarctic an area to be used exclusively for peaceful purposes. Prohibits any measure of a military nature in the Antarctic, such as the establishment of military bases and fortifications, and the carrying out of military manoeuvres or the testing of any type of weapon. The treaty bans any nuclear explosion as well as the disposal of radioactive waste material in Antarctica.

In accordance with Article IX, consultative meetings are convened at regular intervals to exchange information and hold consultations on matters pertaining to Antarctica, as well as to recommend to the governments measures in furtherance of the principles and objectives of the treaty.

The treaty is subject to ratification by the signatories and is open for accession by UN members or by other states invited to accede with the consent of all the parties entitled to participate in the consultative meetings provided for in Article IX.

Parties (45): Argentina†, Australia†, Austria, Belgium†, Brazil†, Bulgaria, Canada, Chile†, China†, Colombia, Cuba, Czech Republic, Denmark, Ecuador†, Estonia, Finland†, France†, Germany†, Greece, Guatemala, Hungary, India†, Italy†, Japan†, Korea (North), Korea (South)†,

Netherlands†, New Zealand†, Norway†, Papua New Guinea, Peru†, Poland†, Romania, Russia†, Slovakia, South Africa†, Spain†, Sweden†, Switzerland, Turkey, UK†, Ukraine, Uruguay†, USA†, Venezuela†

† Party entitled to participate in the consultative meetings.

The Protocol on Environmental Protection to the Antarctic Treaty (**1991 Madrid Protocol**) entered into force on 14 January 1998.

Treaty Banning Nuclear Weapon Tests in the Atmosphere, in Outer Space and Under Water (Partial Test Ban Treaty, PTBT)

Signed at Moscow by three original parties on 5 August 1963, opened for signature by other states at London, Moscow and Washington, DC, on 8 August 1963; entered into force on 10 October 1963; depositaries British, US and Russian governments

The treaty prohibits the carrying out of any nuclear weapon test explosion or any other nuclear explosion: (*a*) in the atmosphere, beyond its limits, including outer space, or under water, including territorial waters or high seas; and (*b*) in any other environment if such explosion causes radioactive debris to be present outside the territorial limits of the state under whose jurisdiction or control the explosion is conducted.

Parties (125): Afghanistan, Antigua and Barbuda, Argentina, Armenia, Australia, Austria, Bahamas, Bangladesh, Belarus, Belgium, Benin, Bhutan, Bolivia, Bosnia and Herzegovina, Botswana, Brazil, Bulgaria, Canada, Cape Verde, Central African Republic, Chad, Chile, Colombia, Congo (Democratic Republic of the), Costa Rica, Côte d'Ivoire, Croatia, Cyprus, Czech Republic, Denmark, Dominican Republic, Ecuador, Egypt, El Salvador, Equatorial Guinea, Fiji, Finland, Gabon, Gambia, Germany, Ghana, Greece, Guatemala, Guinea-Bissau, Honduras, Hungary, Iceland, India, Indonesia, Iran, Iraq, Ireland, Israel, Italy, Jamaica, Japan, Jordan, Kenya, Korea (South), Kuwait, Laos, Lebanon, Liberia, Libya, Luxembourg, Madagascar, Malawi, Malaysia, Malta, Mauritania, Mauritius, Mexico, Mongolia, Morocco, Myanmar (Burma), Nepal, Netherlands, New Zealand, Nicaragua, Niger, Nigeria, Norway, Pakistan, Panama, Papua New Guinea, Peru, Philippines, Poland, Romania, Russia, Rwanda, Samoa (Western), San Marino, Senegal, Seychelles, Sierra Leone, Singapore, Slovakia, Slovenia, South Africa, Spain, Sri Lanka, Sudan, Suriname, Swaziland, Sweden, Switzerland, Syria, Taiwan, Tanzania, Thailand, Togo, Tonga, Trinidad and Tobago, Tunisia, Turkey, Uganda, UK, Ukraine, Uruguay, USA, Venezuela, Yemen, Yugoslavia, Zambia

Signed but not ratified: Algeria, Burkina Faso, Burundi, Cameroon, Ethiopia, Haiti, Mali, Paraguay, Portugal, Somalia

Treaty on Principles Governing the Activities of States in the Exploration and Use of Outer Space, Including the Moon and Other Celestial Bodies (Outer Space Treaty)

Opened for signature at London, Moscow and Washington, DC, on 27 January 1967; entered into force on 10 October 1967; depositaries British, Russian and US governments

The treaty prohibits the placing into orbit around the earth of any objects carrying nuclear weapons or any other kinds of weapons of mass destruction, the installation of such weapons on celestial bodies, or the stationing of them in outer space in any other manner. The establishment of military bases, installations and fortifications, the testing of any type of weapons and the conduct of military manoeuvres on celestial bodies are also forbidden.

Parties (102): Afghanistan, Algeria, Antigua and Barbuda, Argentina, Australia, Austria, Bahamas, Bangladesh, Barbados, Belarus, Belgium, Benin, Brazil, Brunei Darussalam, Bulgaria, Burkina Faso, Canada, Chile, China, Cuba, Cyprus, Czech Republic, Denmark, Dominican Republic, Ecuador, Egypt, El Salvador, Equatorial Guinea, Fiji, Finland, France, Germany, Greece, Guinea-Bissau, Hungary, Iceland, India, Iraq, Ireland, Israel, Italy, Jamaica, Japan, Kazakhstan, Kenya, Korea (South), Kuwait, Laos, Lebanon, Libya, Madagascar, Mali, Mauritius, Mexico, Mongolia, Morocco, Myanmar (Burma), Nepal, Netherlands, New Zealand, Niger, Nigeria, Norway, Pakistan, Papua New Guinea, Peru, Poland, Portugal, Romania, Russia, Saint Kitts and Nevis, Saint Lucia, Saint Vincent and the Grenadines, San Marino, Saudi Arabia, Seychelles, Sierra Leone, Singapore, Slovakia, Solomon Islands, South Africa, Spain, Sri Lanka, Sweden, Swaziland, Switzerland, Syria, Taiwan, Thailand, Togo, Tonga, Tunisia, Turkey, Uganda, UK, Ukraine, Uruguay, USA, Venezuela, Viet Nam, Yemen, Zambia

Signed but not ratified: Bolivia, Botswana, Burundi, Cameroon, Central African Republic, Colombia, Congo (Democratic Republic of the), Ethiopia, Gambia, Ghana, Guyana, Haiti, Holy See, Honduras, Indonesia, Iran, Jordan, Lesotho, Luxembourg, Malaysia, Nicaragua, Panama, Philippines, Rwanda, Somalia, Trinidad and Tobago, Yugoslavia

Treaty for the Prohibition of Nuclear Weapons in Latin America and the Caribbean (Treaty of Tlatelolco)

Original treaty opened for signature at Mexico, Distrito Federal, on 14 February 1967; entered into force on 22 April 1968. The treaty was amended in 1990, 1991 and 1992; depositary Mexican Government

The treaty prohibits the testing, use, manufacture, production or acquisition by any means, as well as the receipt, storage, installation, deployment and any form of possession of any nuclear weapons by Latin American and Caribbean countries.

The parties should conclude agreements with the IAEA for the application of safeguards to their nuclear activities. The IAEA has the exclusive power to carry out special inspections.

The treaty is open for signature by all the independent states of the region.

Under *Additional Protocol I* states with territories within the zone (France, the Netherlands, the UK and the USA) undertake to apply the statute of military denuclearization to these territories.

Under *Additional Protocol II* the recognized nuclear weapon states (China, France, Russia (at the time of signing, the USSR), the UK and the USA) undertake to respect the statute of military denuclearization of Latin America and not to contribute to acts involving a violation of the treaty, nor to use or threaten to use nuclear weapons against the parties to the treaty.

Parties to the original treaty (32): Antigua and Barbuda, Argentina, Bahamas, Barbados, Belize, Bolivia, Brazil, Chile, Colombia, Costa Rica, Dominica, Dominican Republic, Ecuador, El Salvador, Grenada, Guatemala, Guyana, Haiti, Honduras, Jamaica, Mexico, Nicaragua, Panama, Paraguay, Peru, Saint Kitts and Nevis, Saint Lucia, Saint Vincent and the Grenadines, Suriname, Trinidad and Tobago, Uruguay, Venezuela

Signed but not ratified: Cuba

Amendments ratified by: Argentina, Barbados, Belize, Brazil, Chile, Colombia, Costa Rica, Dominican Republic, Ecuador, El Salvador, Grenada, Guatemala, Guyana, Jamaica, Mexico, Panama, Paraguay, Peru, Suriname, Uruguay, Venezuela

Note: Not all the countries listed had ratified all three amendments by 1 Jan. 2002.

Parties to Additional Protocol I: France[1], Netherlands, UK[2], USA[3]

Parties to Additional Protocol II: China[4], France[5], Russia[6], UK[2], USA[7]

[1] France declared that Protocol I shall not apply to transit across French territories situated within the zone of the treaty, and destined for other French territories. The protocol shall not limit the participation of the populations of the French territories in the activities mentioned in Article 1 of the treaty, and in efforts connected with the national defence of France. France does not consider the zone described in the treaty as established in accordance with international law; it cannot, therefore, agree that the treaty should apply to that zone.

[2] When signing and ratifying Protocols I and II, the UK made the following declarations of understanding: The signing and ratification by the UK could not be regarded as affecting in any way the legal status of any territory for the international relations of which the UK is responsible, lying within the limits of the geographical zone established by the treaty. Should any party to the treaty carry out any act of aggression with the support of a nuclear weapon state, the UK would be free to reconsider the extent to which it could be regarded as bound by the provisions of Protocol II.

[3] The USA ratified Protocol I with the following understandings: The provisions of the treaty do not affect the exclusive power and legal competence under international law of a state adhering to this Protocol to grant or deny transit and transport privileges to its own or any other vessels or aircraft irrespective of cargo or armaments; the provisions do not affect rights under international law of a state adhering to this protocol regarding the exercise of the freedom of the seas, or regarding passage through or over waters subject to the sovereignty of a state. The declarations attached by the USA to its ratification of Protocol II apply also to Protocol I.

[4] China declared that it will never send its means of transportation and delivery carrying nuclear weapons to cross the territory, territorial sea or airspace of Latin American countries.

[5] France stated that it interprets the undertaking contained in Article 3 of Protocol II to mean that it presents no obstacle to the full exercise of the right of self-defence enshrined in Article 51 of the UN Charter; it takes note of the interpretation by the Preparatory Commission for the Denuclearization of Latin America according to which the treaty does not apply to transit, the granting or denying of which lies within the exclusive competence of each state party in accordance with international law. In 1974, France made a supplementary statement to the effect that it was prepared to consider its obligations under Protocol II as applying not only to the signatories of the treaty, but also to the territories for which the statute of denuclearization was in force in conformity with Protocol I.

[6] On signing an ratifying Protocol II, the USSR stated that it assumed that the effect of Article 1 of the treaty extends to any nuclear explosive device and that, accordingly, the carrying out by any party of nuclear explosions for peaceful purposes would be a violation of its obligations under Article 1 and would be incompatible with its non-nuclear weapon status. For states parties to the treaty, a solution to the problem of peaceful nuclear explosions can be found in accordance with the provisions of Article V of the NPT and within the framework of the international procedures of the IAEA. It declared that authorizing the transit of nuclear weapons in any form would be contrary to the objectives of the treaty.

Any actions undertaken by a state or states parties to the treaty which are not compatible with their non-nuclear weapon status, and also the commission by one or more states parties to the treaty of an act of aggression with the support of a state which is in possession of nuclear weapons or together with such a state, will be regarded by the USSR as incompatible with the obligations of those countries under the treaty. In such cases it would reserve the right to reconsider its obligations under Protocol II. It further reserves the right to reconsider its attitude to this protocol in the event of any actions on the part of other states possessing nuclear weapons which are incompatible with their obligations under the said protocol.

[7] The USA signed and ratified Protocol II with the following declarations and understandings: Each of the parties retains exclusive power and legal competence, to grant or deny non-parties transit and transport privileges. As regards the undertaking not to use or threaten to use nuclear weapons against the parties, the USA would consider that an armed attack by a party, in which it was assisted by a nuclear weapon state, would be incompatible with the treaty.

Treaty on the Non-proliferation of Nuclear Weapons (Non-Proliferation Treaty, NPT)

Opened for signature at London, Moscow and Washington, DC, on 1 July 1968; entered into force on 5 March 1970; depositaries British, Russian and US governments

The treaty prohibits the transfer by nuclear weapon states (defined in the treaty as those which have manufactured and exploded a nuclear weapon or other nuclear explosive device prior to 1 January 1967) to any recipient whatsoever, of nuclear weapons or other nuclear explosive devices or of control over them, as well as the assistance, encouragement or inducement of any non-nuclear weapon state to manufacture or otherwise acquire such weapons or devices. It also prohibits the receipt by non-nuclear weapon states from any transferor whatsoever, as well as the manufacture or other acquisition by those states, of nuclear weapons or other nuclear explosive devices.

The parties undertake to facilitate the exchange of equipment, materials and scientific and technological information for the peaceful uses of nuclear energy and to ensure that potential benefits from peaceful applications of nuclear explosions will be made available to non-nuclear weapon parties to the treaty. They also undertake to pursue negotiations in good faith on effective measures relating to cessation of the nuclear arms race at an early date and to nuclear disarmament, and on a treaty on general and complete disarmament.

Non-nuclear weapon states undertake to conclude safeguard agreements with the International Atomic Energy Agency (IAEA) with a view to preventing diversion of nuclear energy from peaceful uses to nuclear weapons or other nuclear explosive devices. A Model Protocol additional to the safeguards agreements, strengthening the measures, was approved in 1997; Additional Safeguards Protocols are signed by states individually with the IAEA.

A Review and Extension Conference, convened in 1995 in accordance with the treaty, decided that the treaty should remain in force indefinitely.

Parties (188): Afghanistan†, Albania, Algeria†, Andorra, Angola, Antigua and Barbuda†, Argentina†, Armenia†, Australia†, Austria†, Azerbaijan†, Bahamas†, Bahrain, Bangladesh†, Barbados†, Belarus†, Belgium†, Belize†, Benin, Bhutan†, Bolivia†, Bosnia and Herzegovina†, Botswana, Brazil†, Brunei Darussalam†, Bulgaria†, Burkina Faso, Burundi, Cambodia†, Cameroon, Canada†, Cape Verde, Central African Republic, Chad, Chile†, China†, Colombia, Comoros, Congo (Democratic Republic of the)†, Congo (Republic of), Costa Rica†, Côte d'Ivoire†, Croatia†, Cyprus†, Czech Republic†, Denmark†, Djibouti, Dominica†, Dominican Republic†, Ecuador†, Egypt†, El Salvador†, Equatorial Guinea, Eritrea, Estonia†, Ethiopia†, Fiji†, Finland†, France†, Gabon, Gambia†, Georgia, Germany†, Ghana†, Greece†, Grenada†, Guatemala†, Guinea, Guinea-Bissau, Guyana†, Haiti, Holy See†, Honduras†, Hungary†, Iceland†, Indonesia†, Iran†, Iraq†, Ireland†, Italy†, Jamaica†, Japan†, Jordan†, Kazakhstan†, Kenya, Kiribati†, Korea (North)†, Korea (South)†, Kuwait, Kyrgyzstan, Laos, Latvia†, Lebanon†, Lesotho†, Liberia, Libya†, Liechtenstein†, Lithuania†, Luxembourg†, Macedonia (Former Yugoslav Republic of), Madagascar†, Malawi†, Malaysia†, Maldives†, Mali, Malta†, Marshall Islands, Mauritania, Mauritius†, Mexico†, Micronesia, Moldova, Monaco†, Mongolia†, Morocco†, Mozambique, Myanmar (Burma)†, Namibia†, Nauru†, Nepal†, Netherlands†, New Zealand†, Nicaragua†, Niger, Nigeria†, Norway†, Oman, Palau, Panama, Papua New Guinea†, Paraguay†, Peru†, Philippines†, Poland†, Portugal†, Qatar, Romania†, Russia†, Rwanda, Saint Kitts and Nevis†, Saint Lucia†, Saint Vincent and the Grenadines†, Samoa (Western)†, San Marino†, Sao Tome and Principe, Saudi Arabia, Senegal†, Seychelles, Sierra

Leone, Singapore†, Slovakia†, Slovenia†, Solomon Islands†, Somalia, South Africa†, Spain†, Sri Lanka†, Sudan†, Suriname†, Swaziland†, Sweden†, Switzerland†, Syria†, Taiwan, Tajikistan, Tanzania, Thailand†, Togo, Tonga†, Trinidad and Tobago†, Tunisia†, Turkey†, Turkmenistan, Tuvalu†, Uganda, UK†, Ukraine†, United Arab Emirates, Uruguay†, USA†, Uzbekistan†, Vanuatu, Venezuela†, Viet Nam†, Yemen, Yugoslavia†, Zambia†, Zimbabwe†

† Party with safeguards agreements in force with the International Atomic Energy Agency (IAEA), as required by the treaty, or concluded by a nuclear weapon state on a voluntary basis.

Additional Safeguards Protocols are in force for 23 states: (Australia, Azerbaijan, Bangladesh, Bulgaria, Canada, Croatia, Ecuador, Holy See, Hungary, Indonesia, Japan, Jordan, Lithuania, Monaco, New Zealand, Norway, Panama, Peru, Poland, Romania, Slovenia, Turkey and Uzbekistan.) Taiwan, although it has not concluded a safeguards agreement, has agreed to apply the measures contained in the Model Additional Safeguards Protocol.

Treaty on the Prohibition of the Emplacement of Nuclear Weapons and other Weapons of Mass Destruction on the Seabed and the Ocean Floor and in the Subsoil thereof (Seabed Treaty)

Opened for signature at London, Moscow and Washington, DC, on 11 February 1971; entered into force on 18 May 1972; depositaries British, Russian and US governments

The treaty prohibits implanting or emplacing on the seabed and the ocean floor and in the subsoil thereof beyond the outer limit of a 12-mile seabed zone any nuclear weapons or any other types of weapons of mass destruction as well as structures, launching installations or any other facilities specifically designed for storing, testing or using such weapons.

Parties (94): Afghanistan, Algeria, Antigua and Barbuda, Argentina[1], Australia, Austria, Bahamas, Belarus, Belgium, Benin, Bosnia and Herzegovina, Botswana, Brazil[2], Bulgaria, Canada[3], Cape Verde, Central African Republic, China, Congo (Republic of), Côte d'Ivoire, Croatia, Cuba, Cyprus, Czech Republic, Denmark, Dominican Republic, Ethiopia, Finland, Germany, Ghana, Greece, Guatemala, Guinea-Bissau, Hungary, Iceland, India[4], Iran, Iraq, Ireland, Italy[5], Jamaica, Japan, Jordan, Korea (South), Laos, Latvia, Lesotho, Libya, Liechtenstein, Luxembourg, Malaysia, Malta, Mauritius, Mexico[6], Mongolia, Morocco, Nepal, Netherlands, New Zealand, Nicaragua, Niger, Norway, Panama, Philippines, Poland, Portugal, Qatar, Romania, Russia, Rwanda, Saint Vincent and the Grenadines, Sao Tome and Principe, Saudi Arabia, Seychelles, Singapore, Slovakia, Slovenia, Solomon Islands, South Africa, Spain, Swaziland, Sweden, Switzerland, Taiwan, Togo, Tunisia, Turkey[7], UK, Ukraine, USA, Viet Nam[8], Yemen, Yugoslavia[9], Zambia

Signed but not ratified: Bolivia, Burundi, Cambodia, Cameroon, Colombia, Costa Rica, Equatorial Guinea, Gambia, Guinea, Honduras, Lebanon, Liberia, Madagascar, Mali, Myanmar (Burma), Paraguay, Senegal, Sierra Leone, Sudan, Tanzania, Uruguay

[1] Argentina precludes any possibility of strengthening, through this treaty, certain positions concerning continental shelves to the detriment of others based on different criteria.
[2] Brazil stated that nothing in the treaty shall be interpreted as prejudicing in any way the sovereign rights of Brazil in the area of the sea, the seabed and the subsoil thereof adjacent to its coasts. It is the understanding of Brazil that the word 'observation', as it appears in para. 1 of Article III of the treaty, refers only to observation that is incidental to the normal course of navigation in accordance with international law.
[3] Canada declared that Article I, para. 1, cannot be interpreted as indicating that any state has a right to implant or emplace any weapons not prohibited under Article I, para. 1, on the seabed and ocean floor, and in the subsoil thereof, beyond the limits of national jurisdiction, or as constituting any limitation on the principle that this area of the seabed and ocean floor and the subsoil thereof shall be reserved for

exclusively peaceful purposes. Articles I, II and III cannot be interpreted as indicating that any state but the coastal state has any right to implant or emplace any weapon not prohibited under Article I, para. 1 on the continental shelf, or the subsoil thereof, appertaining to that coastal state, beyond the outer limit of the seabed zone referred to in Article I and defined in Article II. Article III cannot be interpreted as indicating any restrictions or limitation upon the rights of the coastal state, consistent with its exclusive sovereign rights with respect to the continental shelf, to verify, inspect or effect the removal of any weapon, structure, installation, facility or device implanted or emplaced on the continental shelf, or the subsoil thereof, appertaining to that coastal state, beyond the outer limit of the seabed zone referred to in Article I and defined in Article II.

[4] The accession by India is based on its position that it has full and exclusive rights over the continental shelf adjoining its territory and beyond its territorial waters and the subsoil thereof. There cannot, therefore, be any restriction on, or limitation of, the sovereign right of India as a coastal state to verify, inspect, remove or destroy any weapon, device, structure, installation or facility, which might be implanted or emplaced on or beneath its continental shelf by any other country, or to take such other steps as may be considered necessary to safeguard its security.

[5] Italy stated, *inter alia*, that in the case of agreements on further measures in the field of disarmament to prevent an arms race on the seabed and ocean floor and in their subsoil, the question of the delimitation of the area within which these measures would find application shall have to be examined and solved in each instance in accordance with the nature of the measures to be adopted.

[6] Mexico declared that the treaty cannot be interpreted to mean that a state has the right to emplace weapons of mass destruction, or arms or military equipment of any type, on the continental shelf of Mexico. It reserves the right to verify, inspect, remove or destroy any weapon, structure, installation, device or equipment placed on its continental shelf, including nuclear weapons or other weapons of mass destruction.

[7] Turkey declared that the provisions of Article II cannot be used by a state party in support of claims other than those related to disarmament. Hence, Article II cannot be interpreted as establishing a link with the UN Convention on the Law of the Sea. Furthermore, no provision of the Seabed Treaty confers on parties the right to militarize zones which have been demilitarized by other international instruments. Nor can it be interpreted as conferring on either the coastal states or other states the right to emplace nuclear weapons or other weapons of mass destruction on the continental shelf of a demilitarized territory.

[8] Viet Nam stated that no provision of the treaty should be interpreted in a way that would contradict the rights of the coastal states with regard to their continental shelf, including the right to take measures to ensure their security.

[9] In 1974, the Ambassador of Yugoslavia transmitted to the US Secretary of State a note stating that in the view of the Yugoslav Government, Article III, para. 1, of the treaty should be interpreted in such a way that a state exercising its right under this article shall be obliged to notify in advance the coastal state, in so far as its observations are to be carried out 'within the stretch of the sea extending above the continental shelf of the said state'. The USA objected to the Yugoslav reservation, which it considers incompatible with the object and purpose of the treaty.

Convention on the Prohibition of the Development, Production and Stockpiling of Bacteriological (Biological) and Toxin Weapons and on their Destruction (Biological and Toxin Weapons Convention, BTWC)

Opened for signature at London, Moscow and Washington, DC, on 10 April 1972; entered into force on 26 March 1975; depositaries British, Russian and US governments

The convention prohibits the development, production, stockpiling or acquisition by other means or retention of microbial or other biological agents, or toxins whatever their origin or method of production, of types and in quantities that have no justification of prophylactic, protective or other peaceful purposes, as well as weapons, equipment or means of delivery designed to use such agents or toxins for hostile purposes or in armed conflict. The destruction of the agents, toxins, weapons, equipment and means of delivery in the possession of the parties, or their diversion to peaceful purposes, should be effected not later than nine months after the entry into force of the convention. According to a mandate from the 1996 BTWC Review Conference, verification and other measures to strengthen the convention are being discussed and considered in an Ad Hoc Group.

Parties (145): Afghanistan, Albania, Algeria, Argentina, Armenia, Australia, Austria, Bahamas, Bahrain, Bangladesh, Barbados, Belarus, Belgium, Belize, Benin, Bhutan, Bolivia, Bosnia and Herzegovina, Botswana, Brazil, Brunei Darussalam, Bulgaria, Burkina Faso, Cambodia, Canada, Cape Verde, Chile, China, Colombia, Congo (Democratic Republic of the), Congo (Republic of), Costa Rica, Croatia, Cuba, Cyprus, Czech Republic, Denmark, Dominica, Dominican Republic, Ecuador, El Salvador, Equatorial Guinea, Estonia, Ethiopia, Fiji, Finland, France, Gambia, Georgia, Germany, Ghana, Greece, Grenada, Guatemala, Guinea-Bissau, Honduras, Hungary, Iceland, India, Indonesia, Iran, Iraq, Ireland, Italy, Jamaica, Japan, Jordan, Kenya, Korea (North), Korea (South), Kuwait, Laos, Latvia, Lebanon, Lesotho, Libya, Liechtenstein, Lithuania, Luxembourg, Macedonia (Former Yugoslav Republic of), Malaysia, Maldives, Malta, Mauritius, Mexico, Monaco, Mongolia, Netherlands, New Zealand, Nicaragua, Niger, Nigeria, Norway, Oman, Pakistan, Panama, Papua New Guinea, Paraguay, Peru, Philippines, Poland, Portugal, Qatar, Romania, Russia, Rwanda, Saint Kitts and Nevis, Saint Lucia, Saint Vincent and the Grenadines, San Marino, Sao Tome and Principe, Saudi Arabia, Senegal, Seychelles, Sierra Leone, Singapore, Slovakia, Slovenia, Solomon Islands, South Africa, Spain, Sri Lanka, Suriname, Swaziland, Sweden, Switzerland*, Taiwan, Thailand, Togo, Tonga, Tunisia, Turkey, Turkmenistan, Uganda, UK, Ukraine, Uruguay, USA, Uzbekistan, Vanuatu, Venezuela, Viet Nam, Yemen, Yugoslavia, Zimbabwe

Signed but not ratified: Burundi, Central African Republic, Côte d'Ivoire, Egypt, Gabon, Guyana, Haiti, Liberia, Madagascar, Malawi, Mali, Morocco, Myanmar (Burma), Nepal, Somalia, Syria, Tanzania, United Arab Emirates

* With reservation.

Treaty on the Limitation of Anti-Ballistic Missile Systems (ABM Treaty)

Signed by the USA and the USSR at Moscow on 26 May 1972; entered into force on 3 October 1972

The parties undertake not to build nationwide defences against ballistic missile attack and limits the development and deployment of permitted strategic missile defences. The treaty prohibits the parties from giving air defence missiles, radars or launchers the technical ability to counter strategic ballistic missiles and from testing them in a strategic ABM mode.

The **1974 Protocol** to the ABM Treaty introduces further numerical restrictions on permitted ballistic missile defences.

Belarus, Kazakhstan, Russia and Ukraine signed the **1997 Memorandum of Understanding on Succession (MOUS)**, in which they assumed the obligations of the former USSR regarding the treaty. Russia and the USA signed a set of Agreed Statements, including the **1997 Demarcation Agreement,** specifying the demarcation line between strategic missile defences, which are not permitted under the treaty, and non-strategic or theatre missile defences (TMD), which are permitted under the treaty. The set of 1997 agreements on anti-missile defence were ratified by Russia in April 2000, but because the USA did not ratify them they did not formally enter into force.

Note: On 13 December 2001 the USA, with six months' notice, announced its withdrawal from the ABM Treaty.

Treaty on the Limitation of Underground Nuclear Weapon Tests (Threshold Test Ban Treaty, TTBT)

Signed by the USA and the USSR at Moscow on 3 July 1974; entered into force on 11 December 1990

The parties undertake not to carry out any individual underground nuclear weapon test having a yield exceeding 150 kilotons.

Treaty on Underground Nuclear Explosions for Peaceful Purposes (Peaceful Nuclear Explosions Treaty, PNET)

Signed by the USA and the USSR at Moscow and Washington, DC, on 28 May 1976; entered into force on 11 December 1990

The parties undertake not to carry out any underground nuclear explosion for peaceful purposes having a yield exceeding 150 kilotons or any group explosion having an aggregate yield exceeding 150 kilotons.

Convention on the Prohibition of Military or Any Other Hostile Use of Environmental Modification Techniques (Enmod Convention)

Opened for signature at Geneva on 18 May 1977; entered into force on 5 October 1978; depositary UN Secretary-General

The convention prohibits military or any other hostile use of environmental modification techniques having widespread, long-lasting or severe effects as the means of destruction, damage or injury to states party to the convention. The term 'environmental modification techniques' refers to any technique for changing—through the deliberate manipulation of natural processes—the dynamics, composition or structure of the earth, including its biota, lithosphere, hydrosphere and atmosphere, or of outer space. The understandings reached during the negotiations, but not written into the convention, define the terms 'widespread', 'long-lasting' and 'severe'.

Parties (66): Afghanistan, Algeria, Antigua and Barbuda, Argentina, Australia, Austria, Bangladesh, Belarus, Belgium, Benin, Brazil, Bulgaria, Canada, Cape Verde, Chile, Costa Rica, Cuba, Cyprus, Czech Republic, Denmark, Dominica, Egypt, Finland, Germany, Ghana, Greece, Guatemala, Hungary, India, Ireland, Italy, Japan, Korea (North), Korea (South)*, Kuwait, Laos, Malawi, Mauritius, Mongolia, Netherlands*, New Zealand, Niger, Norway, Pakistan, Papua New Guinea, Poland, Romania, Russia, Saint Lucia, Saint Vincent and the Grenadines, Sao Tome and Principe, Slovakia, Solomon Islands, Spain, Sri Lanka, Sweden, Switzerland, Tajikistan, Tunisia, UK, Ukraine, Uruguay, USA, Uzbekistan, Viet Nam, Yemen

Signed but not ratified: Bolivia, Congo (Democratic Republic of the), Ethiopia, Holy See, Iceland, Iran, Iraq, Lebanon, Liberia, Luxembourg, Morocco, Nicaragua, Portugal, Sierra Leone, Syria, Turkey, Uganda

* With declaration.

Protocol I Additional to the 1949 Geneva Conventions, and Relating to the Protection of Victims of International Armed Conflicts
Protocol II Additional to the 1949 Geneva Conventions, and Relating to the Protection of Victims of Non-International Armed Conflicts

Opened for signature at Bern on 12 December 1977; entered into force on 7 December 1978; depositary Swiss Federal Council

The protocols confirm that the right of the parties to international or non-international armed conflicts to choose methods or means of warfare is not unlimited and that it is prohibited to use weapons or means of warfare which cause superfluous injury or unnecessary suffering.

Parties to Protocol I (158) and Protocol II (151): Albania, Algeria*, Angola[1]*, Antigua and Barbuda, Argentina*, Armenia, Australia*, Austria*, Bahamas, Bahrain, Bangladesh, Barbados, Belarus, Belgium*, Belize, Benin, Bolivia, Bosnia and Herzegovina, Botswana, Brazil, Brunei Darussalam, Bulgaria, Burkina Faso, Burundi, Cambodia, Cameroon, Canada*, Cape Verde, Central African Republic, Chad, Chile, China*, Colombia, Comoros, Congo (Democratic Republic of the)[1], Congo (Republic of), Costa Rica, Côte d'Ivoire, Croatia, Cuba, Cyprus, Czech Republic, Denmark*, Djibouti, Dominica, Dominican Republic, Ecuador, Egypt*, El Salvador, Equatorial Guinea, Estonia, Ethiopia, Finland*, France*, Gabon, Gambia, Georgia, Germany*, Ghana, Greece, Grenada, Guatemala, Guinea, Guinea-Bissau, Guyana, Holy See, Honduras, Hungary, Iceland*, Ireland, Italy*, Jamaica, Jordan, Kazakhstan, Kenya, Korea (North)[1], Korea (South)*, Kuwait, Kyrgyzstan, Laos, Latvia, Lebanon, Lesotho, Liberia, Libya, Liechtenstein*, Lithuania, Luxembourg, Macedonia (Former Yugoslav Republic of), Madagascar, Malawi, Maldives, Mali, Malta*, Mauritania, Mauritius, Mexico[1], Micronesia, Moldova, Monaco, Mongolia, Mozambique[1], Namibia, Netherlands*, New Zealand*, Nicaragua, Niger, Nigeria, Norway, Oman, Palau, Panama, Paraguay, Peru, Philippines[2], Poland, Portugal, Qatar*[1], Romania, Russia*, Rwanda, Saint Kitts and Nevis, Saint Lucia, Saint Vincent and the Grenadines, Samoa (Western), San Marino, Sao Tome and Principe, Saudi Arabia*, Senegal, Seychelles, Sierra Leone, Slovakia, Slovenia, Solomon Islands, South Africa, Spain*, Suriname, Swaziland, Sweden*, Switzerland*, Syria*[1], Tajikistan, Tanzania, Togo, Trinidad and Tobago*, Tunisia, Turkmenistan, Uganda, UK, Ukraine, United Arab Emirates*, Uruguay, Uzbekistan, Vanuatu, Venezuela, Viet Nam[1], Yemen, Yugoslavia*, Zambia, Zimbabwe

In 1989 the Palestine Liberation Organization (PLO) informed the depositary that it had decided to adhere to the four Geneva Conventions and the two Protocols.

* With reservation and/or declaration.

[1] Party only to Protocol I.
[2] Party only to Protocol II.

Convention on the Physical Protection of Nuclear Material

Opened for signature at Vienna and New York on 3 March 1980; entered into force on 8 February 1987; depositary IAEA Director General

The convention obligates the parties to protect nuclear material for peaceful purposes while in international transport.

Parties (67): Albania, Antigua and Barbuda, Argentina*, Armenia, Australia, Austria, Belarus, Belgium, Bolivia, Bosnia and Herzegovina, Botswana, Brazil, Bulgaria, Canada, Chile, China*, Croatia, Cuba, Cyprus, Czech Republic, Denmark, Ecuador, Estonia, Euratom*, Finland, France*, Germany, Greece, Grenada, Guatemala, Hungary, Indonesia*, Ireland, Israel*, Italy, Japan, Kenya, Korea (South)*, Lebanon, Liechtenstein, Libya, Lithuania, Luxembourg, Macedonia (Former Yugoslav Republic of), Mexico, Moldova, Monaco, Mongolia*, Netherlands*, Norway, Panama, Pakistan*, Paraguay, Peru*, Philippines, Poland*, Portugal[†], Romania, Russia*, Slovakia, Slovenia, Spain*, Sudan, Sweden, Switzerland, Tajikistan, Trinidad and Tobago, Tunisia, Turkey*, UK, Ukraine, USA, Uzbekistan, Yugoslavia

Signed but not ratified: Dominican Republic, Haiti, Morocco, Niger, South Africa

Note: Grenada acceded to the convention on 9 January 2002. Israel deposited its ratification, with reservation, to the convention on 22 January 2002. Bolivia acceded on 24 Jan. 2002. Yugoslavia succeeded on 5 Feb. 2002. Kenya acceded on 11 Feb. 2002. Albania acceded on 5 Mar. 2002.

* With reservation and/or declaration.

Convention on Prohibitions or Restrictions on the Use of Certain Conventional Weapons which may be Deemed to be Excessively Injurious or to have Indiscriminate Effects (CCW Convention, or 'Inhumane Weapons' Convention)

The convention, with protocols I, II and III, was opened for signature at New York on 10 April 1981; entered into force on 2 December 1983; depositary UN Secretary-General

The convention is an 'umbrella treaty', under which specific agreements can be concluded in the form of protocols. To become a party a state must ratify a minimum of two of the original protocols.

Protocol I prohibits the use of weapons intended to injure by fragments which are not detectable in the human body by X-rays.

Protocol II prohibits or restricts the use of mines, booby-traps and other devices.

Amended Protocol II, which entered into force on 3 December 1998, reinforces the constraints regarding landmines.

Protocol III restricts the use of incendiary weapons.

Protocol IV, which entered into force on 30 July 1998, prohibits the employment of laser weapons specifically designed to cause permanent blindness to unenhanced vision.

Parties to the convention and original protocols (88): Argentina*, Australia, Austria, Bangladesh, Belarus, Belgium, Benin[1], Bolivia, Bosnia and Herzegovina, Brazil, Bulgaria, Cambodia, Canada, Cape Verde, China, Colombia, Costa Rica, Croatia, Cuba, Cyprus*, Czech Republic, Denmark, Djibouti, Ecuador, El Salvador, Estonia[1], Finland, France*[2], Georgia, Germany, Greece, Guatemala, Holy See, Hungary, India, Ireland, Israel[2], Italy, Japan, Jordan[1], Korea (South)[3], Laos, Latvia, Lesotho, Liechtenstein, Lithuania[1], Luxembourg, Macedonia (Former Yugoslav Republic of), Maldives[1], Mali, Malta, Mauritius, Mexico, Moldova, Monaco[3], Mongolia, Nauru, Netherlands*, New Zealand, Nicaragua[1], Niger, Norway, Pakistan, Panama, Peru[1], Philippines, Poland, Portugal, Romania, Russia, Senegal[4], Seychelles, Slovakia, Slovenia, South Africa, Spain, Sweden, Switzerland, Tajikistan, Togo, Tunisia, Uganda, UK, Ukraine, Uruguay, USA[2], Uzbekistan, Yugoslavia

Signed but not ratified the convention and original protocols: Afghanistan, Egypt, Iceland, Morocco, Nigeria, Sierra Leone, Sudan, Turkey, Viet Nam

* With reservation and/or declaration.
   [1] Party only to Protocols I and III.
   [2] Party only to Protocols I and II.
   [3] Party only to Protocol I.
   [4] Party only to Protocol III.

Parties to Amended Protocol II (63): Argentina, Australia, Austria, Bangladesh, Belgium, Bolivia, Bosnia and Herzegovina, Brazil, Bulgaria, Cambodia, Canada, Cape Verde, China, Colombia, Costa Rica, Czech Republic, Denmark, Ecuador, El Salvador, Estonia, Finland, France, Germany, Greece, Guatemala, Holy See, Hungary, India, Ireland, Israel, Italy, Japan, Jordan, Korea (South), Liechtenstein, Lithuania, Luxembourg, Maldives, Mali, Moldova, Monaco, Nauru, Netherlands, New Zealand, Nicaragua, Norway, Pakistan, Panama, Peru, Philippines, Portugal, Senegal, Seychelles, Slovakia, South Africa, Spain, Sweden, Switzerland, Tajikistan, UK, Ukraine, Uruguay, USA

Parties to Protocol IV (61): Argentina, Australia, Austria, Bangladesh, Belarus, Belgium, Bolivia, Bosnia and Herzegovina, Brazil, Bulgaria, Cambodia, Canada, Cape Verde, China, Colombia, Costa Rica, Czech Republic, Denmark, El Salvador, Estonia, Finland, France, Germany, Greece, Holy See, Hungary, India, Ireland, Israel, Italy, Japan, Latvia, Liechtenstein, Lithuania, Luxembourg, Maldives, Mali, Mexico, Moldova, Mongolia, Nauru, Netherlands, New Zealand, Nicaragua, Norway, Pakistan, Panama, Peru, Philippines, Portugal, Russia, Seychelles, Slovakia, South Africa, Spain, Sweden, Switzerland, Tajikistan, UK, Uruguay, Uzbekistan

South Pacific Nuclear Free Zone Treaty (Treaty of Rarotonga)

Opened for signature at Rarotonga, Cook Islands, on 6 August 1985; entered into force on 11 December 1986; depositary Director of the South Pacific Bureau for Economic Co-operation (from 1988, South Pacific Forum Secretariat)

The treaty prohibits the manufacture or acquisition by other means of any nuclear explosive device, as well as possession or control over such device by the parties anywhere inside or outside the zone area described in an annex. The parties also undertake not to supply nuclear material or equipment, unless subject to IAEA safeguards, and to prevent in their territories the stationing as well as the testing of any nuclear explosive device and undertake not to dump, and to prevent the dumping of, radioactive wastes and other radioactive matter at sea anywhere within the zone. Each party remains free to allow visits, as well as transit, by foreign ships and aircraft.

The treaty is open for signature by the members of the South Pacific Forum.

Under *Protocol 1* France, the UK and the USA undertake to apply the treaty prohibitions relating to the manufacture, stationing and testing of nuclear explosive devices in the territories situated within the zone, for which they are internationally responsible.

Under *Protocol 2* China, France, Russia, the UK and the USA undertake not to use or threaten to use a nuclear explosive device against the parties to the treaty or against any territory within the zone for which a party to Protocol 1 is internationally responsible.

Under *Protocol 3* China, France, the UK, the USA and Russia undertake not to test any nuclear explosive device anywhere within the zone.

Parties (13): Australia, Cook Islands, Fiji, Kiribati, Nauru, New Zealand, Niue, Papua New Guinea, Samoa (Western), Solomon Islands, Tonga, Tuvalu, Vanuatu

Parties to Protocol 1: France, UK; **signed but not ratified:** USA
Parties to Protocol 2: China, France[1], Russia, UK[2]; **signed but not ratified:** USA
Parties to Protocol 3: China, France, Russia, UK; **signed but not ratified:** USA

[1] France declared that the negative security guarantees set out in Protocol 2 are the same as the CD declaration of 6 Apr. 1995 referred to in UN Security Council Resolution 984 of 11 Apr. 1995.

[2] The UK declared that nothing in the treaty affects the rights under international law with regard to transit of the zone or visits to ports and airfields within the zone by ships and aircraft. The UK will not be bound by the undertakings in Protocol 2 in case of an invasion or any other attack on the UK, its territories, its armed forces or its allies, carried out or sustained by a party to the treaty in association or alliance with a nuclear weapon state or if a party violates its non-proliferation obligations under the treaty.

Treaty on the Elimination of Intermediate-Range and Shorter-Range Missiles (INF Treaty)

Signed by the USA and the USSR at Washington, DC, on 8 December 1987; entered into force on 1 June 1988

The treaty obligates the parties to destroy all land-based missiles with a range of 500–5500 km (intermediate-range, 1000–5500 km; and shorter-range, 500–1000 km) and their launchers by 1 June 1991. The treaty was implemented by the two parties before this date.

Treaty on Conventional Armed Forces in Europe (CFE Treaty)

Original treaty signed at Paris on 19 November 1990; entered into force on 9 November 1992; depositary Netherlands Government

The treaty sets ceilings on five categories of treaty-limited equipment (TLE)—battle tanks, armoured combat vehicles, artillery of at least 100-mm calibre, combat aircraft and attack helicopters—in an area stretching from the Atlantic Ocean to the Ural Mountains (the Atlantic-to-the-Urals, ATTU, zone).

The treaty was negotiated and signed by the member states of the Warsaw Treaty Organization (WTO) and NATO within the framework of the Conference on Security and Co-operation in Europe (from 1995 the Organization for Security and Co-operation in Europe, OSCE).

The **1992 Tashkent Agreement**, adopted by the former Soviet republics (with the exception of the three Baltic states) with territories within the ATTU zone, and the **1992 Oslo Document** (Final Document of the Extraordinary Conference of the States Parties to the CFE Treaty) introduced modifications to the treaty required because of the emergence of new states after the break-up of the USSR.

Parties (30): Armenia, Azerbaijan, Belarus, Belgium, Bulgaria, Canada, Czech Republic, Denmark, France, Georgia, Germany, Greece, Hungary, Iceland, Italy, Kazakhstan, Luxembourg, Moldova, Netherlands, Norway, Poland, Portugal, Romania, Russia, Slovakia, Spain, Turkey, UK, Ukraine, USA

The first Review Conference of the CFE Treaty adopted the **1996 Flank Document**, which reorganized the flank areas geographically and numerically, allowing Russia and Ukraine to deploy more TLE along their borders.

The **1999 Agreement on Adaptation of the CFE Treaty** replaces the CFE Treaty bloc-to-bloc military balance with individual state limits on TLE holdings and provides for a new structure of limitations and new military flexibility mechanisms, flank sublimits and enhanced transparency; it opens the CFE regime to all the other European states. It will enter into force when it has been ratified by all the signatories. The **1999 Final Act**, with annexes, contains politically binding arrangements with regard to the North Caucasus and Central and Eastern Europe, and withdrawals of armed forces from foreign territories.

1 ratification of the Agreement on Adaptation deposited: Belarus

Concluding Act of the Negotiation on Personnel Strength of Conventional Armed Forces in Europe (CFE-1A Agreement)

Signed by the parties to the CFE Treaty at Helsinki on 10 July 1992; entered into force simultaneously with the CFE Treaty; depositary Netherlands Government

The agreement limits the personnel of the conventional land-based armed forces of the parties within the ATTU zone.

Treaty on the Reduction and Limitation of Strategic Offensive Arms (START I Treaty)

Signed by the USA and the USSR at Moscow on 31 July 1991; entered into force on 5 December 1994

The treaty requires the USA and Russia to make phased reductions in their offensive strategic nuclear forces over a seven-year period. It sets numerical limits on deployed strategic nuclear delivery vehicles (SNDVs)—ICBMs, SLBMs and heavy bombers— and the nuclear warheads they carry. In the Protocol to Facilitate the Implementation of the START Treaty (**1992 Lisbon Protocol**), which entered into force on 5 December 1994, Belarus, Kazakhstan and Ukraine also assumed the obligations of the former USSR under the treaty. They pledged to eliminate all the former Soviet strategic weapons on their territories within the seven-year reduction period and to join the NPT as non-nuclear weapon states in the shortest possible time.

Treaty on Open Skies

Opened for signature at Helsinki on 24 March 1992; entered into force 1 January 2002; depositaries Canadian and Hungarian governments

The treaty obligates the parties to submit their territories to short-notice unarmed surveillance flights. The area of application stretches from Vancouver, Canada, eastward to Vladivostok, Russia.

The treaty was negotiated between the member states of the Warsaw Treaty Organization (WTO) and NATO. It was opened for signature by the NATO states, the new states of the former WTO, and the new states of the former Soviet Union (except the three Baltic states). For six months after entry into force of the treaty (until 1 July 2002), any other OSCE member state may apply for accession, and after that any country may request to accede to the treaty.

26 ratifications deposited: Belarus, Belgium, Bulgaria, Canada, Czech Republic, Denmark, France, Georgia, Germany, Greece, Hungary, Iceland, Italy, Luxembourg, Netherlands, Norway, Poland, Portugal, Romania, Slovakia, Spain, Russia, Turkey, UK, Ukraine, USA

Signed but not ratified: Kyrgyzstan

Treaty on Further Reduction and Limitation of Strategic Offensive Arms (START II Treaty)

Signed by the USA and Russia at Moscow on 3 January 1993; not in force as of 1 January 2002

The treaty requires the USA and Russia to eliminate their MIRVed ICBMs and sharply reduce the number of their deployed strategic nuclear warheads to no more than 3000–3500 each (of which no more than 1750 may be deployed on SLBMs) by 1 January 2003 or no later than 31 December 2000 if the USA and Russia reach a formal agreement committing the USA to help finance the elimination of strategic nuclear weapons in Russia.

On 26 September 1997 the two parties signed a *Protocol* to the treaty providing for the extension until the end of 2007 of the period of implementation of the treaty.

Note: The US Senate ratified the treaty on 26 January 1996; the Russian Duma and Federation Council approved ratification on 14 and 19 April 2000, respectively.

Convention on the Prohibition of the Development, Production, Stockpiling and Use of Chemical Weapons and on their Destruction (Chemical Weapons Convention, CWC)

Opened for signature at Paris on 13 January 1993; entered into force on 29 April 1997; depositary UN Secretary-General

The convention prohibits both the use of chemical weapons (also prohibited by the 1925 Geneva Protocol) and the development, production, acquisition, transfer and stockpiling of chemical weapons. Each party undertakes to destroy its chemical weapons and production facilities within 10 years after the treaty enters into force.

Parties (145): Albania, Algeria, Argentina, Armenia, Australia, Austria, Azerbaijan, Bahrain, Bangladesh, Belarus, Belgium, Benin, Bolivia, Bosnia and Herzegovina, Botswana, Brazil, Brunei Darussalam, Bulgaria, Burkina Faso, Burundi, Cameroon, Canada, Chile, China, Colombia, Cook Islands, Costa Rica, Côte d'Ivoire, Croatia, Cuba, Cyprus, Czech Republic, Denmark, Dominica, Ecuador, El Salvador, Equatorial Guinea, Eritrea, Estonia, Ethiopia, Fiji, Finland, France, Gabon, Gambia, Georgia, Germany, Ghana, Greece, Guinea, Guyana, Holy See, Hungary, Iceland, India, Indonesia, Iran, Ireland, Italy, Jamaica, Japan, Jordan, Kazakhstan, Kenya, Kiribati, Korea (South), Kuwait, Laos, Latvia, Lesotho, Liechtenstein, Lithuania, Luxembourg, Macedonia (Former Yugoslav Republic of), Malawi, Malaysia, Maldives, Mali, Malta, Mauritania, Mauritius, Mexico, Micronesia, Moldova, Monaco, Mongolia, Morocco, Mozambique, Namibia, Nauru, Nepal, Netherlands, New Zealand, Nicaragua, Niger, Nigeria, Norway, Oman, Pakistan, Panama, Papua New Guinea, Paraguay, Peru, Philippines, Poland, Portugal, Qatar, Romania, Russia, Saint Lucia, San Marino, Saudi Arabia, Senegal, Seychelles, Singapore, Slovakia, Slovenia, South Africa, Spain, Sri Lanka, Sudan, Suriname, Swaziland, Sweden, Switzerland, Tajikistan, Tanzania, Togo, Trinidad and Tobago, Tunisia, Turkey, Turkmenistan, Uganda, UK, Ukraine, United Arab Emirates, Uruguay, USA, Uzbekistan, Venezuela, Viet Nam, Yemen, Yugoslavia (FRY), Zambia, Zimbabwe

Signed but not ratified: Afghanistan, Bahamas, Bhutan, Cambodia, Cape Verde, Central African Republic, Chad, Comoros, Congo (Democratic Republic of the), Congo (Republic of), Djibouti, Dominican Republic, Grenada, Guatemala, Guinea-Bissau, Haiti, Honduras, Israel, Kyrgyzstan, Liberia, Madagascar, Marshall Islands, Myanmar (Burma), Rwanda, Saint Kitts and Nevis, Saint Vincent and the Grenadines, Samoa (Western), Sierra Leone, Thailand

Treaty on the Southeast Asia Nuclear Weapon-Free Zone (Treaty of Bangkok)

Signed at Bangkok on 15 December 1995; entered into force on 27 March 1997; depositary Government of Thailand

The treaty prohibits the development, manufacture, acquisition or testing of nuclear weapons inside or outside the zone area as well as the stationing and transport of nuclear weapons in or through the zone. Each state party may decide for itself whether to allow visits and transit by foreign ships and aircraft. The parties undertake not to dump at sea or discharge into the atmosphere anywhere within the zone any radioactive material or wastes or dispose of radioactive material on land. The parties should conclude an agreement with the IAEA for the application of full-scope safeguards to their peaceful nuclear activities.

The zone includes not only the territories but also the continental shelves and exclusive economic zones of the states parties.

The treaty is open for signature by all the states in South-East Asia.

Under a *Protocol* to the treaty China, France, Russia, the UK and the USA are to undertake not to use or threaten to use nuclear weapons against any state party to the treaty. They should further undertake not to use nuclear weapons within the Southeast Asia nuclear weapon-free zone. The protocol will enter into force for each state party on the date of its deposit of the instrument of ratification.

Parties (10): Brunei Darussalam, Cambodia, Indonesia, Laos, Malaysia, Myanmar (Burma), Philippines, Singapore, Thailand, Viet Nam

Protocol: no signatures, no parties

Agreement on Confidence- and Security-Building Measures in Bosnia and Herzegovina between Bosnia and Herzegovina, the Federation of Bosnia and Herzegovina and the Republika Srpska

Signed at Vienna on 26 January 1996, entered into force on 26 January 1996

The agreement is largely based on the Vienna Document 1994 but includes additional restrictions and restraints measures on military movements, deployments and exercises and provides for exchange of information and data relating to major weapon systems.

African Nuclear-Weapon-Free Zone Treaty (Treaty of Pelindaba)

Signed at Cairo on 11 April 1996; not in force as of 1 January 2002; depositary Secretary-General of the Organization of African Unity

The treaty prohibits the research, development, manufacture and acquisition of nuclear explosive devices and the testing or stationing of any nuclear explosive device. Each party remains free to allow visits, as well as transit by foreign ships and aircraft. The treaty also prohibits any attack against nuclear installations. The parties undertake not to dump or permit the dumping of radioactive wastes and other radioactive matter anywhere within the zone. The parties should conclude an agreement with the IAEA for the application of comprehensive safeguards to their peaceful nuclear activities.

The zone includes the territory of the continent of Africa, island states members of the OAU and all islands considered by the OAU to be part of Africa.

The treaty is open for signature by all the states of Africa. It will enter into force upon the 28th ratification.

Under *Protocol I* China, France, Russia, the UK and the USA are to undertake not to use or threaten to use a nuclear explosive device against the parties to the Treaty.

Under *Protocol II* China, France, Russia, the UK and the USA are to undertake not to test nuclear explosive devices anywhere within the zone.

Under *Protocol III* states with territories within the zone for which they are internationally responsible are to undertake to observe certain provisions of the treaty with respect to these territories. This protocol is open for signature by France and Spain.

The protocols will enter into force simultaneously with the treaty for those protocol signatories that have deposited their instruments of ratification.

16 ratifications deposited: Algeria, Botswana, Burkina Faso, Côte d'Ivoire, Gambia, Guinea, Kenya, Mali, Mauritania, Mauritius, Nigeria, South Africa, Swaziland, Tanzania, Togo, Zimbabwe

Signed but not ratified: Angola, Benin, Burundi, Cameroon, Cape Verde, Central African Republic, Chad, Comoros, Congo (Democratic Republic of the), Congo (Republic of), Djibouti, Egypt, Eritrea, Ethiopia, Gabon, Ghana, Guinea-Bissau, Lesotho, Liberia, Libya, Malawi, Morocco, Mozambique, Namibia, Niger, Rwanda, Sao Tome and Principe, Senegal, Seychelles, Sierra Leone, Sudan, Tunisia, Uganda, Zambia

Protocol I: ratifications deposited: China, France[1], UK[3]; **signed but not ratified:** Russia[2], USA[4]

Protocol II: ratifications deposited: China, France, UK[3]; **signed but not ratified:** Russia[2], USA[4]

Protocol III: ratifications deposited: France

[1] France stated that the Protocols did not affect its right to self-defence, as stipulated in Article 51 of the UN Charter. It clarified that its commitment under Article 1 of Protocol I was equivalent to the negative security assurances given by France to non-nuclear weapon states parties to the NPT, as confirmed in its declaration made on 6 Apr. 1995 at the Conference on Disarmament, and as referred to in UN Security Council Resolution 984.

[2] Russia stated that as long as a military base of a nuclear state was located on the islands of the Chagos archipelago these islands could not be regarded as fulfilling the requirements put forward by the Treaty for nuclear-weapon-free territories. Moreover, since certain states declared that they would consider themselves free from the obligations under the Protocols with regard to the mentioned territories, Russia could not consider itself to be bound by the obligations under Protocol I in respect to the same territories. Russia interpreted its obligations under Article 1 of Protocol I as follows: It would not use nuclear weapons against a state party to the Treaty, except in the case of invasion or any other armed attack on Russia, its territory, its armed forces or other troops, its allies or a state towards which it had a security commitment, carried out or sustained by a non-nuclear-weapon state party to the treaty, in association or alliance with a nuclear-weapon state.

[3] The UK stated that it did not accept the inclusion of the British Indian Ocean Territory within the African nuclear weapon-free zone without its consent, and did not accept, by its adherence to Protocols I and II, any legal obligations in respect of that territory. Moreover, it would not be bound by its undertaking under Article 1 of Protocol I in case of an invasion or any other attack on the United Kingdom, its dependent territories, its armed forces or other troops, its allies or a state towards which it had security commitment, carried out or sustained by a party to the treaty in association or alliance with a nuclear-weapon state, or if any party to the treaty was in material breach of its own non-proliferation obligations under the treaty.

[4] The USA stated, with respect to Protocol I, that it would consider an invasion or any other attack on the USA, its territories, its armed forces or other troops, its allies or on a state toward which it had a

security commitment, carried out or sustained by a party to the treaty in association or alliance with a nuclear-weapon state, to be incompatible with the treaty party's corresponding obligations. The USA also stated that neither the treaty nor Protocol II would apply to the activities of the UK, the USA or any other state not party to the treaty on the island of Diego Garcia or elsewhere in the British Indian Ocean Territories. No change was, therefore, required in US armed forces operations in Diego Garcia and elsewhere in these territories.

Agreement on Sub-Regional Arms Control concerning Yugoslavia (Serbia and Montenegro), Bosnia and Herzegovina, and Croatia (Florence Agreement)

Adopted at Florence on 14 June 1996; entered into force upon signature

The agreement was negotiated under the auspices of the OSCE in accordance with the mandate in the 1995 General Framework Agreement for Peace in Bosnia and Herzegovina (Dayton Agreement). It sets numerical ceilings on armaments of the former warring parties: Bosnia and Herzegovina and its two entities, Croatia and the Federal Republic of Yugoslavia. Five categories of heavy conventional weapons are included: battle tanks, armoured combat vehicles, heavy artillery (75 mm and above), combat aircraft and attack helicopters. The reductions were completed by 31 October 1997. It is confirmed that 6580 weapon items were destroyed by that date.

Comprehensive Nuclear Test-Ban Treaty (CTBT)

Opened for signature at New York on 24 September 1996; not in force as of 1 January 2002; depositary UN Secretary-General

The treaty prohibits the carrying out of any nuclear weapon test explosion or any other nuclear explosion, and urges each party to prevent any such nuclear explosion at any place under its jurisdiction or control and refrain from causing, encouraging, or in any way participating in the carrying out of any nuclear weapon test explosion or any other nuclear explosion.

The treaty will enter into force 180 days after the date of the deposit of the instrument of ratification of the 44 states listed in an annex to the treaty but in no case earlier than two years after its opening for signature. All the 44 states possess nuclear power reactors and/or nuclear research reactors.

The 44 states whose ratification is required for entry into force are Algeria, Argentina, Australia, Austria, Bangladesh, Belgium, Brazil, Bulgaria, Canada, Chile, China, Colombia, Congo (Democratic Republic of the), Egypt, Finland, France, Germany, Hungary, India, Indonesia, Iran, Israel, Italy, Japan, Korea (North), Korea (South), Mexico, Netherlands, Norway, Pakistan, Peru, Poland, Romania, Russia, Slovakia, South Africa, Spain, Sweden, Switzerland, Turkey, UK, Ukraine, USA and Viet Nam.

90 ratifications deposited: Argentina, Australia, Austria, Azerbaijan, Bangladesh, Belarus, Belgium, Benin, Bolivia, Brazil, Bulgaria, Cambodia, Canada, Chile, Costa Rica, Croatia, Czech Republic, Denmark, Ecuador, El Salvador, Estonia, Fiji, Finland, France, Gabon, Germany, Greece, Grenada, Guyana, Holy See, Hungary, Iceland, Ireland, Italy, Jamaica, Japan, Jordan, Kenya, Kiribati, Korea (South), Laos, Latvia, Lesotho, Lithuania, Luxembourg, Macedonia (Former Yugoslav Republic of), Maldives, Mali, Malta, Mexico, Micronesia, Monaco, Mongolia, Morocco, Namibia, Nauru, Netherlands, New Zealand, Nicaragua, Nigeria, Norway, Panama, Paraguay, Peru, Philippines, Poland, Portugal, Qatar, Romania, Russia, Saint Lucia, San Marino, Senegal, Sierra Leone, Singapore, Slovakia, Slovenia, South

Africa, Spain, Sweden, Switzerland, Tajikistan, Turkey, Turkmenistan, Uganda, UK, Ukraine, United Arab Emirates, Uruguay, Uzbekistan

Signed but not ratified: Albania, Algeria, Andorra, Angola, Antigua and Barbuda, Armenia, Bahrain, Belize, Bosnia and Herzegovina, Brunei Darussalam, Burkina Faso, Burundi, Cameroon, Cape Verde, Central African Republic, Chad, China, Colombia, Comoros, Congo (Democratic Republic of the), Congo (Republic of), Cook Islands, Côte d'Ivoire, Cyprus, Djibouti, Dominican Republic, Egypt, Equatorial Guinea, Ethiopia, Georgia, Ghana, Guatemala, Guinea, Guinea-Bissau, Haiti, Honduras, Indonesia, Iran, Israel, Kazakhstan, Kuwait, Kyrgyzstan, Liberia, Libya, Liechtenstein, Madagascar, Malawi, Malaysia, Marshall Islands, Mauritania, Moldova, Mozambique, Myanmar (Burma), Nepal, Niger, Oman, Papua New Guinea, Samoa (Western), San Marino, Sao Tome and Principe, Seychelles, Solomon Islands, Sri Lanka, Suriname, Swaziland, Thailand, Togo, Tunisia, USA, Vanuatu, Venezuela, Viet Nam, Yemen, Yugoslavia, Zambia, Zimbabwe

Note: San Marino ratified the treaty on 12 March 2002.

Joint Statement on Parameters on Future Reductions in Nuclear Forces

Signed by the USA and Russia at Helsinki on 21 March 1997

In the Joint Statement the two sides agree that once the 1993 START II Treaty enters into force negotiations on a START III treaty will begin. START III will include lower aggregate levels of 2000–2500 nuclear warheads for each side.

Convention on the Prohibition of the Use, Stockpiling, Production and Transfer of Anti-Personnel Mines and on their Destruction (APM Convention)

Opened for signature at Ottawa on 3–4 December 1997 and at the UN Headquarters, New York, on 5 December 1997; entered into force on 1 March 1999; depositary UN Secretary-General

The convention prohibits anti-personnel mines, which are defined as mines designed to be exploded by the presence, proximity or contact of a person and which will incapacitate, injure or kill one or more persons.

Each party undertakes to destroy all its stockpiled anti-personnel mines as soon as possible but not later that four years after the entry into force of the convention for that state party. Each party also undertakes to destroy all anti-personnel mines in mined areas under its jurisdiction or control not later than 10 years after the entry into force of the convention for that state party.

Parties (122): Albania, Algeria, Andorra, Antigua and Barbuda, Argentina, Australia, Austria, Bahamas, Bangladesh, Barbados, Belgium, Belize, Benin, Bolivia, Bosnia and Herzegovina, Botswana, Brazil, Bulgaria, Burkina Faso, Cambodia, Canada, Cape Verde, Chad, Chile, Colombia, Congo (Republic of), Costa Rica, Côte d'Ivoire, Croatia, Czech Republic, Denmark, Djibouti, Dominica, Dominican Republic, Ecuador, El Salvador, Equatorial Guinea, Eritrea, Fiji, France, Gabon, Germany, Ghana, Grenada, Guatemala, Guinea, Guinea-Bissau, Holy See, Honduras, Hungary, Iceland, Ireland, Italy, Jamaica, Japan, Jordan, Kenya, Kiribati, Lesotho, Liberia, Liechtenstein, Luxembourg, Macedonia (Former Yugoslav Republic of), Madagascar, Malawi, Malaysia, Maldives, Mali, Malta, Mauritania, Mauritius, Mexico, Moldova, Monaco, Mozambique, Namibia, Nauru, Netherlands, New Zealand, Nicaragua, Niger, Nigeria, Niue, Norway, Panama, Paraguay, Peru, Philippines, Portugal, Qatar, Romania, Rwanda, Saint Kitts and Nevis, Saint Lucia, Saint

Vincent and the Grenadines, Samoa (Western), San Marino, Senegal, Seychelles, Sierra Leone, Slovakia, Slovenia, Solomon Islands, South Africa, Spain, Swaziland, Sweden, Switzerland, Tajikistan, Tanzania, Thailand, Togo, Trinidad and Tobago, Tunisia, Turkmenistan, Uganda, UK, Uruguay, Venezuela, Yemen, Zambia, Zimbabwe

Signed but not ratified: Angola, Brunei Darussalam, Burundi, Cameroon, Cook Island, Cyprus, Ethiopia, Gambia, Greece, Guyana, Haiti, Indonesia, Lithuania, Marshall Islands, Poland, Sao Tome and Principe, Sudan, Suriname, Ukraine, Vanuatu

Vienna Document 1999 on Confidence- and Security-Building Measures

Adopted by the participating states of the Organization for Security and Co-operation in Europe at Istanbul on 16 November 1999; entered into force on 1 January 2000

The Vienna Document 1999 builds on the 1986 Stockholm Document on Confidence- and Security-building Measures (CSBMs) and Disarmament in Europe and previous Vienna Documents (1990, 1992 and 1994). The Vienna Document 1990 provided for military budget exchange, risk reduction procedures, a communication network and an annual CSBM implementation assessment. The Vienna Documents 1992 and 1994 introduced new mechanisms and parameters for military activities, defence planning and military contacts.

The Vienna Document 1999 introduces regional measures aimed at increasing transparency and confidence in a bilateral, multilateral and regional context and some improvements, in particular regarding the constraining measures.

Annex B. Chronology 2001

CHRISTER BERGGREN

For the convenience of the reader, key words are indicated in the right-hand column, opposite each entry. They refer to the subject areas covered in the entry. Definitions of the acronyms can be found on page xix. The dates are according to local time.

16 Jan.	The President of the Democratic Republic of the Congo (DRC), Laurent-Desiré Kabila, is assassinated. His son, Joseph Kabila, is appointed president of the DRC on 26 Jan.	DRC
17 Jan.	Under an agreement reached with the USA, the South Korean Government adopts new guidelines that will enable it to develop ballistic missiles capable of delivering a 500-kg payload to a range of up to 300 km. This allows South Korea to join the Missile Technology Control Regime (MTCR).	South Korea; Missiles; MTCR
17 Jan.	India successfully carries out the second flight test of the Agni-II medium-range ballistic missile.	India; Missiles
19 Jan.	Lithuania and Russia exchange notes confirming an agreement on the application of CSBMs, providing for one additional evaluation visit to military formations or units in Lithuania and the Russian Kaliningrad oblast and annual exchanges of additional information about military forces on Lithuanian territory and in the Kaliningrad oblast.	Kaliningrad; Russia; Lithuania; CSBMs
20 Jan.	George W. Bush is inaugurated as the 43rd President of the United States.	USA
22 Jan.	The Council of the European Union decides to establish the Political and Security Committee (PSC), the EU Military Committee (EUMC) and the EU Military Staff (EUMS) as permanent political and military structures. The EUMC becomes permanent on 9 Apr. and the EUMS on 11 June.	EU; ESDP
25 Jan.	The Council of Europe restores Russia's voting rights, initially suspended in Apr. 2000 because of criticism of Russia's human rights record in the Chechen Republic.	Russia/ Chechnya; Human rights; Council of Europe
25 Jan.	At a meeting in Yamoussoukro, Côte d'Ivoire, delegations from Benin, Burkina Faso, Côte d'Ivoire, Mali, Niger and Togo pledge not to host rebel forces on their territories.	West Africa
6 Feb.	Likud Party candidate Sharon is elected Israeli Prime Minister.	Israel
8 Feb.	Colombian President Pastrana and Revolutionary Armed Forces (FARC) leader Marulanda Vélez agree to resume peace negotiations and to extend for a further eight months the FARC-controlled demilitarized zone. In further talks on 14 and 23 Feb. the two sides agree on confidence-building measures.	Colombia

SIPRI Yearbook 2002: Armaments, Disarmament and International Security

16 Feb.	Indonesian government negotiators and separatist rebel leaders from the Aceh province agree to extend their ceasefire indefinitely and to begin peace negotiations. Violence nonetheless breaks out in Mar.	Indonesia
16 Feb.	The US and British air forces attack Iraqi military command, control and communication sites outside Baghdad to destroy Iraqi air defence infrastructure. They carry out several attacks on Iraqi defence sites during 2001.	Iraq/USA; UK
18 Feb.	Ethnic violence erupts between indigenous Dayaks and immigrant groups on the island of Borneo, Indonesia, resulting in hundreds of deaths and thousands of refugees.	Indonesia; Ethnic conflicts
20 Feb.	Russian Defence Minister Sergeyev presents to NATO Secretary General Lord Robertson the outline of a proposal for a European theatre missile defence system, to be developed in cooperation between Russia, NATO and the USA.	NATO; Russia; Missile defence
22 Feb.	The UN Security Council unanimously adopts Resolution 1341, demanding that the parties to the conflict in the Democratic Republic of the Congo (DRC) begin disengaging on 15 Mar. and that the signatories to the 1999 Lusaka Agreement adopt plans for the complete withdrawal of all foreign troops by 15 May. Forces of the UN Mission in the DRC (MONUC) will verify the troop withdrawal.	DRC; UN
26 Feb.	Armed fighting between ethnic Albanian rebels and the Former Yugoslav Republic of Macedonia (FYROM) security forces breaks out close to the border between the Yugoslav province of Kosovo and FYROM. The ethnic Albanians demand greater political and cultural rights in FYROM. After further attacks by ethnic Albanian rebels, FYROM closes its border and on 4 Mar. holds consultations with UN and NATO officials.	FYROM
7 Mar.	The UN Security Council unanimously adopts Resolution 1343, reimposing an arms embargo on Liberia in order to stop its weapon supplies to Sierra Leonean rebels. The resolution also imposes an embargo on the export of Liberian diamonds. The embargoes will take effect on 7 May and will last one year.	Liberia; Sierra Leone; UN; Arms embargo
12 Mar.	At a meeting between Russian President Putin and Iranian President Khatami in Moscow, Russia agrees to resume the export of conventional arms and nuclear technology to Iran.	Russia; Iran; Arms trade; Nuclear technology
13 Mar.	North Korea cancels cabinet-level talks with South Korea on reconciliation between the two countries.	North Korea/South Korea
14 Mar.	The Federal Republic of Yugoslavia (FRY) troops enter, with UN approval, the Ground Safety Zone established between the Yugoslav province of Kosovo and the Republic of Serbia, for the first time since June 1999, as a consequence of actions by armed ethnic Albanian rebels. The zone remains under the authority of the Kosovo Force (KFOR) Commander.	Yugoslavia; Kosovo; UN

16 Mar.	The Senegalese Government and Casamance Movement of Democratic Forces (MDFC) rebel leaders sign a peace accord at Ziguinchor, Senegal. A follow-up agreement on military issues is signed on 23 Mar.	Senegal
21 Mar.	The UN Security Council unanimously adopts Resolution 1345, strongly condemning ethnic Albanian extremist violence in the Former Yugoslav Republic of Macedonia (FYROM).	FYROM; UN
21 Mar.	The UN confirms the withdrawal of most of the forces of the warring parties from the Democratic Republic of the Congo (DRC). On 29 March UN military observers move into the areas cleared of foreign and rebel forces in the eastern DRC (see *22 Feb.*).	DRC; UN
27 Mar.	The USA vetoes a UN Security Council resolution urging the creation of an observer force to protect civilians in the Palestinian West Bank and Gaza territories.	UN; USA; Palestinians
30 Mar.	The Chairman of the 1972 Biological and Toxin Weapons Convention Ad Hoc Group, Tóth, issues a compromise 'composite text' to serve as the basis for a final agreement on a protocol to strengthen the BTWC.	BTWC
1 Apr.	In Belgrade, former President of the Federal Republic of Yugoslavia (FRY) Milosevic is arrested by Yugoslav authorities. He is accused of theft of state funds and abuse of power. He is later handed over to the International Criminal Tribunal for the Former Yugoslavia (ICTY) in The Hague (see *28 June*).	Yugoslavia
12 Apr.	The presidents of 10 West African countries hold an emergency summit meeting in Abuja, Nigeria, on a series of rebel offensives and national rivalries that have drawn in Guinea, Liberia and Sierra Leone and caused a refugee crisis. The leaders urge the UN Security Council to approve the deployment of peacekeeping forces to the war zone.	West Africa
17 Apr.	The interior ministers of Iran and Saudi Arabia sign, in Teheran, a security agreement, constituting a framework for cooperation on regional stability and other issues of common interest.	Iran; Saudi Arabia
18 Apr.	The Russian State Duma ratifies the 1992 Open Skies Treaty. On 27 May President Putin signs the ratification bill into law. On 17 May the Parliament of Belarus ratifies the treaty, thereby allowing the treaty to enter into force.	Russia; Belarus; Open Skies Treaty
20 Apr.	MONUC peacekeeping forces arrive in the Democratic Republic of the Congo (DRC) to protect UN bases, equipment and 500 unarmed observers monitoring the ceasefire (see *22 Feb.* and *21 Mar.*).	DRC; UN; Peacekeeping
27 Apr.	In Moscow, Russian Defence Minister Ivanov and North Korean People's Armed Forces Minister Kim Il-chol sign an agreement on military cooperation.	Russia; North Korea

28 Apr.	At a Shanghai Forum meeting in Moscow, the foreign ministers of China, Kazakhstan, Kyrgyzstan, Russia and Tajikistan call on the world community to lift sanctions against Iraq as soon as possible in accordance with UN Security Council resolutions.	Shanghai Forum; Iraq; Sanctions
1 May	US President Bush calls for the building of an expansive missile defence system and pledges unilateral reductions in the US nuclear arsenal. He urges Russia to cooperate on replacing the 1972 ABM Treaty with a new strategic framework.	USA; Missile defence
5 May	A fact-finding committee, led by former US Senator Mitchell, issues a call for an Israeli freeze on the building of new settlements in the West Bank and Gaza Strip and for strict Palestinian measures against terrorism as a requirement for the resumption of peace talks.	Israel/ Palestinians
15 May	Revolutionary United Front (RUF) rebels and the pro-government militia Civil Defence Force agree to a ceasefire during peace talks with the Sierra Leone Government aimed at ending the civil war in Sierra Leone. A disarmament, demobilization and reintegration programme is implemented.	Sierra Leone
15 May	Defence ministers of the Western European Armaments Group (WEAG) member states sign, in Brussels, a Memorandum of Understanding on European Undertakings for Research Organization Programmes and Activities (EUROPA), a framework agreement for bilateral and multilateral arms research and technology cooperation in Europe.	WEAG; Arms cooperation
21 May	At a ceremony in Moscow, the inspections under the 1987 Treaty on the Elimination of Intermediate-Range and Shorter-Range Missiles (INF Treaty) are completed. The treaty has resulted in the elimination of 2692 medium- and shorter-range missiles equipped with 4000 nuclear warheads.	INF Treaty
21 May	NATO Secretary General Lord Robertson and Minister of Foreign Affairs of the Former Yugoslav Republic of Macedonia (FYROM) Casule sign an agreement providing the legal basis for the presence of Kosovo Force (KFOR) troops on the territory of FYROM (see *31 Mar.*).	FYROM; KFOR
24 May	Security forces from the Federal Republic of Yugoslavia (FRY) and the Republic of Serbia return to the remaining sector of the Ground Safety Zone between Kosovo and Serbia. Ethnic Albanian guerrillas in the area disarm and disband.	Yugoslavia; Kosovo
25 May	The presidents of the member states of the CIS Collective Security Council (Armenia, Belarus, Kazakhstan, Kyrgyzstan, Russia and Tajikistan) agree, at their meeting in Yerevan, Armenia, to form a joint rapid reaction force of 3000 troops in order to enhance regional security in Central Asia.	Central Asia; CIS

CHRONOLOGY 2001   789

26 May	One month after the 36th country deposits its instrument of ratification of the Constitutive Act of the African Union, the act enters into force. The African Union (AU) will replace the OAU and will become operational in 2002. The Organization of African Unity (OAU) holds its last summit meeting on 9–11 July in Lusaka, Zambia.	Africa; African Union; OAU
30 May	During a meeting in Budapest between NATO's North Atlantic Council and the European Union's General Affairs Council, foreign ministers of NATO and the EU meet officially for the first time.	EU; NATO
31 May	Iran successfully tests a new class of Iranian-made surface-to-surface guided medium-range missiles, Fateh-110, the latest step in its missile programme.	Iran; Missiles
31 May	The UN General Assembly adopts, without a vote, the Protocol Against the Illicit Manufacturing of and Trafficking in Firearms, their Parts and Components and Ammunition, supplementing the 2000 UN Convention Against Transnational Organized Crime.	UN; Small arms; Arms control
31 May	The NATO Parliamentary Assembly, meeting in Vilnius, Lithuania, approves a declaration on NATO enlargement, calling on the North Atlantic Council to invite new members by 2002.	NATO; Enlargement
5 June	US Secretary of State Powell and Uzbek Foreign Minister Kamilov sign, in Washington, DC, an agreement providing the 'legal framework' for cooperation between the two countries in threat reduction and defence.	USA; Uzbekistan
6 June	The Russian Duma passes a bill allowing the import of spent nuclear fuel from other nations. The bill is signed by President Putin on 11 July.	Russia; Nuclear fuel
6 June	US President Bush announces the resumption of negotiations with North Korea on the implementation of the 1994 Agreed Framework, relating to missile programme verification and missile export. The first preliminary discussions are held on 13 June, in New York.	USA; North Korea
7 June	At the Conference on Disarmament (CD) in Geneva, China presents a draft treaty to ban weapons in outer space.	China; Space; CD
12 June	Russia and India successfully test the jointly developed PJ-10, a supersonic, 280 km-range, air-, land- and sea-launched cruise missile.	Russia; India; Missiles
13 June	At a meeting in Brussels with NATO heads of state and government, US President Bush presents a new US concept of deterrence, based on ballistic missile defences. European NATO leaders insist that a new security strategy must respect existing arms control pacts.	USA; Europe; NATO; Security policy

14 June   The Former Yugoslav Republic of Macedonia (FYROM) formally asks NATO to send troops to help it disarm ethnic Albanian rebels.   FYROM; NATO

14–15 June   Uzbekistan is admitted as a new member of the Shanghai Forum, which at the same meeting, in Shanghai, formally becomes the Shanghai Cooperation Organization (SCO), open to all countries. The SCO will continue to strengthen military cooperation to guarantee that member countries can coordinate their fight against terrorism, separatism and extremism.   Central Asia; Shanghai Cooperation Organization

16 June   At the meeting of EU heads of state and government in Gothenburg, Sweden, the EU leaders agree to complete negotiations with the first applicant nations by the end of 2002. The European Council adopts the Declaration on Prevention of Proliferation of Ballistic Missiles.   EU; Enlargement; Missiles; Non-proliferation

28 June   Former President of the Federal Republic of Yugoslavia (FRY) Milosevic is extradited for trial at the International Criminal Tribunal for the Former Yugoslavia (ICTY) in The Hague. In May 1999 he was charged with crimes against humanity and violations of the laws of war in Kosovo. On 9 Oct. he is indicted for war crimes committed in Croatia during 1991 and 1992. On 23 Nov. he is charged with responsibility for Serb atrocities in the 1992–95 war in Bosnia and Herzegovina.   Yugoslavia; War crimes; ICTY

29 June   The Russian Army completes the withdrawal from Vaziani, a Russian military base in Georgia, which is transferred to Georgian military authorities. The Russian Army does not withdraw from the military base at Gudauta, Georgia.   Russia; Georgia; Military bases

29 June   The Russian Defence Ministry states that the reduction of 365 000 servicemen is to be completed by 2003–2004 in order to reach the optimal size of the army and navy of 1 million servicemen.   Russia; Force reductions

1 July   First Minister Trimble of Northern Ireland resigns from his post because of lack of action in the Irish Republican Army (IRA) disarmament process.   Northern Ireland

5 July   Ethnic Albanian rebels and the Government of the Former Yugoslav Republic of Macedonia (FYROM) sign separate ceasefire agreements with NATO.   FYROM

5 July   Russian Prime Minister Kasyanov approves Resolution 510, amending Russia's 1996 Comprehensive Chemical Weapon Destruction Programme. According to the new plan, Russia will complete the destruction of its chemical weapon stockpile by 2012.   Russia; CW

5 July   The heads of state and government of the Economic Community of West African States (ECOWAS), meeting in Lusaka, Zambia, decide to renew the 1998 Moratorium on Importation, Exportation and Manufacture of Light Weapons in West Africa for a second three-year period.   ECOWAS; Small arms

14 July   The USA successfully tests a prototype missile interceptor, hitting a mock warhead launched 7700 km away.   USA; Missile defence

16 July	Russian President Putin and Chinese President Jiang Zemin sign, in Moscow, a 25-year Treaty on Good-neighbourliness, Friendship and Cooperation, including a pledge not to use nuclear or other forces against each other.	Russia/China
18 July	The negotiations under Article V of Annex 1-B of the 1995 General Framework Agreement for Peace in Bosnia and Herzegovina (Dayton Agreement) on the regional stabilization of South-Eastern Europe ends with the Concluding Document, a joint politically binding document comprising a list of voluntary CSBMs for its participating states, to take effect on 1 Jan. 2002.	South-East Europe; CSBMs; Dayton Agreement
21 July	The states participating in the UN Conference on the Illicit Trade in Small Arms and Light Weapons in All Its Aspects adopt the Programme of Action to Prevent, Combat and Eradicate the Illicit Trade in Small Arms and Light Weapons in All Its Aspects.	UN; Small arms
22 July	At the Group of Eight (G8) summit meeting in Genoa, US President Bush and Russian President Putin agree to hold simultaneous talks on nuclear arsenal reductions and the US plans to deploy a missile defence system.	USA/Russia; Nuclear weapons; Missile defence; G8
23 July	Burundian President Buyoya signs, in Arusha, Tanzania, a power-sharing agreement with 19 Hutu and Tutsi political parties that is designed to end the civil war in Burundi between the two ethnic groups. A three-year transitional government, including members from the Hutu and Tutsi ethnic groups, enters office on 1 Nov.	Burundi
23 July	The Council of the European Union adopts a Common Position on the fight against ballistic missile proliferation, strongly supporting the 1999 Draft International Code of Conduct (ICOC), elaborated by the members of the Missile Technology Control Regime (MTCR).	EU; MTCR
25 July	At the 24th session of the Ad Hoc Group of the States Parties to the 1972 Biological and Toxin Weapons Convention (BTWC), held in Geneva, the parties fail to agree on a protocol or a final declaration (see *30 Mar.* and *7 Dec.*).	BTWC
2 Aug.	In the first conviction by the International Criminal Tribunal for the Former Yugoslavia (ICTY) on a charge of genocide (in the wars that broke up Yugoslavia), former Bosnian Serb general Krstic is found guilty of genocide committed in Srebrenica in July 1995.	ICTY
4 Aug.	At a meeting with Russian President Putin in Moscow, North Korean leader Kim Jong-il declares that North Korea is suspending its ballistic missile launches until 2003.	North Korea; Missiles
7 Aug.	Philippine President Arroyo and Murad Eibrahim, chief of staff of the Moro Islamic Liberation Front (MILF), sign a ceasefire agreement in Kuala Lumpur.	Philippines

8 Aug.	Russian President Putin signs a decree confirming the list of toxic agents, equipment and technology that must be put through customs control, in order to ensure that Russia fulfils its international obligations under the 1972 Biological and Toxin Weapons Convention (BTWC). He also signs a decree confirming the list of equipment, materials and technologies that may be used in the construction of missiles and must be under export control for the purpose of ensuring the implementation of Russia's international obligations in the non-proliferation of missiles.	Russia; BTWC; Missiles; Non-proliferation
10 Aug.	Following a Palestinian suicide bomber attack killing 15 people, Israeli forces occupy Palestinian institutions, most significantly the Orient House, the national headquarters for the Palestinians in Jerusalem.	Palestinians/ Israel
13 Aug.	Former Yugoslav Republic of Macedonia (FYROM) Prime Minister Georgievski and Albanian Democratic Party leader Xhaferi sign a ceasefire accord, brokered by the USA and the EU. On 14 Aug. the ethnic Albanian National Liberation Army (NLA) agrees to hand over its weapons to NATO forces in exchange for the constitutional and legislative reforms that will expand cultural and political rights for the ethnic Albanian minority in the FYROM.	FYROM
15 Aug.	NATO authorizes the deployment of troops in 'Task Force Harvest' to disarm the ethnic Albanian rebels in Macedonia. A British NATO team arrives in Skopje on 17 Aug., and on 22 Aug. the full contingent of 3500 troops begins to arrive.	Macedonia; NATO
20 Aug.	Political leaders of different factions of the Democratic Republic of Congo (DRC), including President Kabila, open the Inter-Congolese National Dialogue in Gaborone, Botswana. The talks are aimed at agreeing on the establishment of a transitional government and free elections.	DRC
21 Aug.	The China Arms Control and Disarmament Association (CACDA), the first Chinese national non-governmental organization in the field of arms control, disarmament and international security, is founded.	China; NGO
7–11 Sep.	Religious and ethnic violence between Christians and Muslims spreads in central Nigeria, claiming hundreds of lives. The fighting continues through October.	Nigeria
10 Sep.	The UN Security Council unanimously votes to lift the arms embargo on the Federal Republic of Yugoslavia (FRY).	Yugoslavia; UN
11 Sep.	Two hijacked passenger planes deliberately hit and destroy the World Trade Center towers in New York. Another hijacked passenger plane hits the Pentagon, in Arlington, Virginia, near Washington, DC. A fourth passenger plane, heading for Washington, DC, crashes near Pittsburgh, Pennsylvania. The attacks kill nearly 3000 people. US President Bush denounces the attacks as an 'act of war'.	USA

12 Sep.	The UN Security Council adopts Resolution 1368, determined to combat by all means threats to international peace and security caused by terrorist acts. The North Atlantic Council agrees that if it is determined that the 11 Sep. attack was directed from abroad (thus defined as an 'act of war') it shall be regarded as an act covered by Article 5 of the 1949 North Atlantic Treaty (Washington Treaty), which states that an armed attack against one or more of its members shall be considered an attack against them all. US federal authorities claim that the hijackers who carried out the 11 Sep. attacks are followers of the Islamic militant leader Usama bin Laden and his al-Qaeda terrorist network, operating out of Afghanistan. Taliban officials in Afghanistan deny the accusations. US military forces in the USA and around the world are put on the highest state of alert.	USA; Terrorism; UN; NATO
14 Sep.	Pakistan agrees to open its air space, to give logistical support and to share its intelligence with US forces for an attack on Afghanistan, where the Taliban regime is harbouring bin Laden. Domestic unrest erupts in Pakistan. Russia rejects participation in a retaliatory strike and the use by the United States of countries in Central Asia as the staging ground for attacks on Afghanistan. On 22 Sep. US President Bush decides to lift the sanctions, barring economic military and economic assistance, on India and Pakistan. (The authorization bill is signed on 27 Oct.)	USA/ Afghanistan; Russia; Pakistan
15 Sep.	US and British special operations forces enter Afghanistan from Pakistan, Tajikistan and Uzbekistan for surveillance and the search for bin Laden. On 28 Sep. US and British military officials confirm that special military forces are conducting reconnaissance missions inside Afghanistan.	USA/UK/ Afghanistan
19 Sep.	US President Bush orders the deployment of additional US air and naval forces to the Persian Gulf area, within striking distance of Afghanistan.	USA/ Afghanistan
20 Sep.	Iran declares that it will not offer its air territory for attacks on Afghanistan. US President Bush demands that the Taliban regime in Afghanistan promptly deliver bin Laden and members of his al-Qaeda network to US authorities and immediately and permanently close down terrorist training camps. On 25 Sep. Kyrgyzstan opens up its air space for the US Air Force.	Russia/Central Asia/Iran/ Afghanistan/ USA
21 Sep.	Pakistan withdraws its diplomats from Afghanistan. The United Arab Emirates, and on 25 Sep. Saudi Arabia, sever diplomatic ties with Afghanistan because of the Taliban regime's harbouring of terrorists.	Pakistan; United Arab Emirates; Saudi Arabia/ Afghanistan
21 Sep.	At a European Council meeting in Brussels, the chairman, Belgian Prime Minister Verhofstadt, declares that the EU member states are prepared to join military actions against states harbouring or supporting terrorists.	EU
22 Sep.	British Northern Ireland Secretary Reid orders the suspension of Northern Ireland's home-rule government (see *1 July*).	Northern Ireland

25 Sep.	The ethnic Albanian National Liberation Army (NLA) in the Former Yugoslav Republic of Macedonia (FYROM) meets the NATO demand that it hand over 3300 weapons within a month. On 27 Sep. NATO's Permanent Council activates the 'Amber Fox' operation with a 700-troop follow-up Task Force, aimed at ensuring the safety of international civilian monitors from the EU and the OSCE during the internal peace-consolidation phase in FYROM. The operation is extended until Mar. 2002.	FYROM
26 Sep.	Pakistan seals off its border with Afghanistan. Iranian religious leader Ayatollah Khamenei declares that Iran will maintain neutrality in the event of a military campaign against Afghanistan. It will only take part in a campaign led by the UN.	Pakistan; Iran; Afghanistan
26 Sep.	Meeting in Brussels, defence ministers of NATO and Russia agree on closer intelligence cooperation on terrorism and to hold enhanced senior-level consultations.	Russia; NATO; Terrorism
26 Sep.	Meeting at Gaza International Airport, Palestinian Chairman Arafat and Israeli Foreign Minister Peres agree on a series of confidence-building measures and to form a joint committee to deal with disputes that arise over enforcing measures recommended in the Mitchell Report (see *5 May*).	Israel/ Palestinians
27 Sep.	Ali Ahmeti, political leader of the ethnic Albanian National Liberation Army (NLA) in the Former Yugoslav Republic of Macedonia (FYROM), declares that his forces have formally disbanded and pledges to cooperate in the peace effort.	FYROM
28 Sep.	The UN Security Council unanimously adopts Resolution 1373, obliging member states to stop all financial and other support to suspected terrorists. The UN will also expand information sharing among member states to combat terrorism, and there will be a Security Council mechanism to monitor implementation on a continuous basis.	UN Security Council; Terrorism; Afghanistan
28 Sep.	At the 16th Plenary Meeting of the Missile Technology Control Regime (MTCR), held in Ottawa, the participating states agree on the 1999 Draft International Code of Conduct (ICOC) against ballistic missile proliferation.	MTCR
2 Oct.	The USA presents evidence regarding the responsibility of the 11 Sep. attacks to the NATO Permanent Council, which activates Article 5 of the 1949 North Atlantic Treaty (see *12 Sep.*). Two days later the evidence is presented to Pakistan.	USA; NATO; Afghanistan
2 Oct.	US Under Secretary of Defense for Acquisition, Technology, and Logistics Aldridge announces that the USA will not meet the 2007 deadline for the destruction of its chemical weapon stockpiles.	USA; Chemical weapons
3 Oct.	The EU–Russia Summit in Brussels adopts a decision to strengthen dialogue and cooperation on political and security issues, with regular meetings and consultations between the EU Political and Security Committee and Russian representatives.	EU; Russia

CHRONOLOGY 2001 795

4 Oct.	The opposition forces in Afghanistan (the anti-Taliban United Islamic Front for the Salvation of Afghanistan, UIFSA, or Northern Alliance) receive offers of military aid from Russia and Iran and hold a meeting with US representatives. The USA contributes $320 million in additional humanitarian assistance to the people of Afghanistan. On 5 Oct. Uzbekistan offers the use of airbases for US military operations in Afghanistan.	Afghanistan
6 Oct.	The first victim of pulmonary anthrax from contaminated letters dies in Florida, USA. Spores of the anthrax bacteria are distributed by mail to several other media producers. On 15 Oct. parts of the US Senate Office Building in Washington, DC, are shut down because of an anthrax-contaminated letter addressed to a senator. The House of Representatives is adjourned on 17–23 Oct.	USA; Terrorism; BW
7 Oct.	US and British air forces attack military targets of the Taliban regime and al-Qaeda network training camps in Afghanistan.	Afghanistan/ USA/UK
10 Oct.	At a meeting called by Iran in Doha, Qatar, the Organization of the Islamic Conference (OIC) rejects the targeting of any Islamic or Arab state under the pretext of fighting terrorism.	OIC; Afghanistan
12 Oct.	In a joint statement, the USA and Uzbekistan announce their decision to establish a long-term mutual commitment to advance security and regional stability, including consultations about appropriate steps to address the situation in the event of a direct threat to the security or territorial integrity of Uzbekistan.	USA; Uzbekistan
14 Oct.	Afghan Taliban officials declare their willingness to negotiate a hand-over of bin Laden to a third country if the USA stops the air attacks. The offer is rejected by US President Bush.	Afghanistan/ USA
16 Oct.	The Belarussian–Ukrainian Intergovernmental Agreement on Additional CSBMs is signed in Kiev, Ukraine.	Belarus/ Ukraine; CSBMs
17 Oct.	Israeli Tourism Minister Zeevi, leader of the National Union Party, is assassinated. The Popular Front for the Liberation of Palestine (PFLP) claims responsibility. After the Palestinian Authority refuses to hand over the assassins, the Israeli Army enters several West Bank towns and fighting erupts. On 22 Oct. US President Bush calls for the withdrawal of Israel from the West Bank.	Israel/ Palestinians; USA
20 Oct.	US ground forces begin to conduct military operations in southern Afghanistan. On 31 Oct. US Secretary of Defense Rumsfeld acknowledges that ground forces are also cooperating with Afghan opposition forces.	USA; Afghanistan
22 Oct.	The US–Uzbekistan Cooperative Agreement on Elimination of Biological Weapons is signed in Tashkent. It contains a project on the decontamination of a biological weapons development and test site on Vozrozhdeniye Island in the Aral Sea. Up to $6 million will be spent under the terms of the agreement.	BW; Uzbekistan; USA
23 Oct.	The Irish Republican Army (IRA) declares that it has started to dismantle its arsenal. The dismantlement is confirmed by the Independent International Commission on Decommissioning.	Northern Ireland; IRA

24 Oct.	Belarus, China and Russia submit to the First Committee of the UN General Assembly a resolution on the preservation of and compliance with the 1972 ABM Treaty. It is adopted on 29 Nov. as UN General Assembly Resolution 56/24.	ABM Treaty
27 Oct.	Nur Misuari, Governor of the Philippine Autonomous Region for Muslim Mindanao (ARMM) and head of the Moro National Liberation Front (MNLF), reaches an agreement with the Muslim separatist Moro Islamic Liberation Front (MILF) rebel group to promote the establishment of an independent Islamic state in the southern Philippines.	Philippines
3 Nov.	Tajikistan offers military cooperation with the USA and its airfields as bases for carrying out US air strikes against Taliban and al-Qaeda forces in Afghanistan.	Tajikistan; USA; Afghanistan
5 Nov.	The governments of the depositary nations, Canada and Hungary, announce that Russia and Belarus have deposited their instruments of ratification of the 1992 Open Skies Treaty on 2 Nov. The treaty will enter into force on 1 Jan. 2002 (see *18 Apr.*).	Open Skies Treaty
13 Nov.	Meeting with Russian President Putin in Washington, DC, US President Bush announces that the USA will reduce its operationally deployed strategic nuclear warheads to a level between 1700 and 2200 by the year 2012. In a joint statement on cooperation against bioterrorism, both countries confirm their strong support for the 1972 Biological and Toxin Weapons Convention (BTWC).	USA; Russia; Nuclear weapons; BTWC
12–13 Nov.	Backed by a coordinated US bombing campaign and military advisers, the Northern Alliance forces enter Kabul, the Afghan capital. Taliban troops retreat to the south.	Afghanistan
14 Nov.	Weapons and equipment are withdrawn from the Operative Group of Russian Armed Forces (the former 14th Army) base in Tiraspol, capital of the Trans-Dniester region of Moldova. Russia thereby fulfils its obligations under the 1999 OSCE Istanbul Summit Declaration and is in compliance with the 1990 CFE Treaty.	Russia; Moldova; Trans-Dniester
14 Nov.	The UN Security Council unanimously adopts Resolution 1378, affirming that the UN should play a central role in supporting the efforts of the Afghan people to urgently establish a new, broad-based, multi-ethnic transitional administration, leading to the formation of a new government.	UN; Afghanistan
16 Nov.	The Parliament of the Former Yugoslav Republic of Macedonia (FYROM) adopts amendments to its constitution granting the ethnic Albanian minority more rights (see *13 Aug.*). FYROM President Trajkovski declares an amnesty for former ethnic Albanian guerrilla fighters, with the exception of those indictable by the UN war crimes tribunal.	FYROM
21 Nov.	US President Bush calls for Iraq to let UN weapon inspectors return to the country. Iraq rejects the call on 27 Nov.	USA/Iraq

22 Nov.	At a NATO–Russia summit meeting in Moscow, NATO Secretary General Robertson proposes to give Russia a decision-making role within the Alliance on some security issues. These include anti-terrorism measures, crisis management, non-proliferation and arms control, and civil emergencies (see *7 Dec.*).	Russia; NATO
29 Nov.	Meeting in Gaborone, Botswana, government and industry negotiators from countries within the so-called Kimberley Process agree on a framework for a system to certify legitimately traded diamonds and stop the market for diamonds mined by rebels and others in war zones.	Arms trade; Diamonds
2 Dec.	Because of the terrorist attacks carried out in Israel by the Harakat al-Muqawama al-Islamiyya (Islamic Resistance Movement, Hamas), the Palestinian Authority declares a state of emergency. On 4 Dec. Israeli forces attack security offices of the Palestinian Authority and Chairman Arafat. Other areas in the West Bank and Gaza Strip are also attacked and blocked off.	Palestinians/ Israel
5 Dec.	The USA and Russia announce the completion of reductions in their strategic nuclear arsenals to 6000 accountable warheads each, as required under the 1991 START I Treaty. All nuclear warheads and strategic offensive weapons have been removed from Belarus, Kazakhstan and Ukraine.	Nuclear disarmament; START
5 Dec.	Meeting in Petersberg near Bonn, Germany, representatives of various political and ethnic factions in Afghanistan sign the Agreement on Provisional Arrangements in Afghanistan Pending the Re-establishment of Permanent Government Institutions. An Interim Authority headed by a Chairman, Pashtun tribal leader Karzai, becomes the internationally recognized government of Afghanistan for six months as of 22 Dec. An independent commission will convene an Emergency Loya Jirga, a constituent assembly of provincial leaders, within six months. According to the agreement, the UN Security Council is to authorize the deployment of a multinational peacekeeping force. On 6 Dec. the UN Security Council unanimously adopts Resolution 1383, endorsing the agreement and declaring its willingness to take further action.	Afghanistan; UN
6 Dec.	Taliban forces abandon Kandahar, their last stronghold, completing the Taliban's fall from power in Afghanistan. On 16 Dec. US Secretary of State Powell states that the al-Qaeda network is effectively destroyed. However, the US military search for al-Qaeda forces continues.	Afghanistan
7 Dec.	At a meeting in the framework of the NATO–Russia Permanent Joint Council (PJC) in Brussels, NATO and Russian foreign ministers decide to engage in the formation of a new cooperation mechanism within a council of 20 (instead of 19 plus Russia), which should elaborate and implement joint decisions and actions, replacing the PJC. The procedures of this new council will be worked out during the spring of 2002.	NATO; Russia; PJC

7 Dec.	At the Fifth Review Conference of the States Parties to the 1972 Biological and Toxin Weapons Convention (BTWC), the US delegation proposes the termination of the mandate for the Ad Hoc Group negotiating a protocol to the convention (see 30 Mar.). The conference is suspended until 11 Nov. 2002.	BTWC
9 Dec.	Kazakhstan agrees to provide the USA access to its airbases during the military operations in Afghanistan. On 11 Dec. a Tajik airbase begins to be used for military and humanitarian operations.	Afghanistan; Kazakhstan; Tajikistan; USA
11–21 Dec.	At the Second Review Conference, held in Geneva, the parties to the 1981 Convention on Prohibitions or Restrictions on the Use of Certain Conventional Weapons which may be Deemed to be Excessively Injurious or to have Indiscriminate Effects (CCW Convention, or 'Inhumane Weapons' Convention) adopt an amendment to Article 1 of the convention, making the prohibitions and restrictions of the convention apply to internal (e.g., insurrections, rebel movements and terrorism) as well as international conflicts.	CCW Convention;
12 Dec.	China becomes a member of the World Trade Organization (WTO). Taiwan becomes a member of the WTO on 1 Jan. 2002.	China; Taiwan; WTO
13 Dec.	US President Bush announces the US withdrawal from the 1972 ABM Treaty on six months' notice, in order to carry out a testing programme to develop a missile defence.	USA; ABM Treaty; Missile defence
13 Dec.	After further terrorist attacks by the Harakat al-Muqawama al-Islamiyya (Hamas) organization, Israeli forces attack targets in the West Bank and the Gaza Strip. The Israeli Government breaks all diplomatic and security contacts with the Palestinian Authority and Chairman Arafat. On 21 Dec. Hamas suspends its attacks on Israelis.	Israel/ Palestinians
13 Dec.	The Indian Parliament House in New Delhi is attacked by five gunmen; 12 people die in the attack. Indian authorities blame Pakistan-based Muslim militant groups for the attack. The two countries move large numbers of troops to the Line of Control in Kashmir.	India/Pakistan
14–15 Dec.	Meeting in Laeken, Belgium, the European Council adopts the Declaration on the Operational Capability of the Common European Security and Defence Policy (ESDP), declaring that the EU is now able to conduct some crisis management operations, even though it lacks assured EU access to NATO's operational planning capabilities.	EU; Crisis management; ESDP
20 Dec.	The UN Security Council unanimously adopts Resolution 1386, authorizing the establishment for six months of an International Security Assistance Force (ISAF) in Afghanistan to assist the Afghan Interim Authority in the maintenance of security in Kabul and its surrounding area.	UN; Peacekeeping; Afghanistan

About the contributors

Dr Ian Anthony (United Kingdom) is the Leader of the SIPRI Internet Database on European Export Controls Project. In 1992–98 he was Leader of the SIPRI Arms Transfers Project. His most recent publication for SIPRI is *A Future Arms Control Agenda: Proceedings of Nobel Symposium 118, 1999* (2001), for which he is co-editor (with Adam Daniel Rotfeld). He is also editor of the SIPRI volumes *Russia and the Arms Trade* (1998), *Arms Export Regulations* (1991) and SIPRI Research Report no. 7, *The Future of Defence Industries in Central and Eastern Europe* (1994), and author of *The Naval Arms Trade* (SIPRI, 1990) and *The Arms Trade and Medium Powers: Case Studies of India and Pakistan 1947–90* (1991). He has written or co-authored chapters for the SIPRI Yearbook since 1988.

Christer Berggren (Sweden) was a SIPRI librarian with special responsibilities for the Arms Control and Disarmament Documentary Survey Project. He is now a librarian at the Swedish National Defence College.

Oleg Bukharin (United States) is a Research Scientist with the Program on Science and Global Security at Princeton University's Woodrow Wilson School of Public and International Affairs. He conducts research and writes on the Russian and US nuclear weapon programmes, nuclear arms control, and safeguards and security of nuclear materials and facilities.

Professor George Bunn (United States) is a Consulting Professor at the Stanford University Center for International Security and Cooperation. He has worked for the US Atomic Energy Commission, the US Nuclear Regulatory Commission and the US Arms Control and Disarmament Agency (ACDA). At ACDA he was General Counsel and a member of the delegation that negotiated the 1968 Non-Proliferation Treaty and Ambassador to the Geneva Disarmament Conference. He is the author of *Arms Control by Committee: Managing Negotiations with the Russians* (Stanford University Press, 1992), and reports and articles on nuclear arms control topics including the physical protection of nuclear material from theft and sabotage. He has also served as professor and dean at the University of Wisconsin Law School.

Nicholas Chipperfield (United Kingdom) is a Research Assistant on the SIPRI Arms Transfers Project and has contributed to the SIPRI Yearbook since 2000. Before joining SIPRI he worked for the British American Security Information Council (BASIC) in Washington, DC and London.

Dr Renata Dwan (Ireland) is the Leader of the SIPRI Project on Conflict Prevention, Management and Resolution. Previously, she was Deputy Director of the EastWest Institute (EWI) European Security Programme at the EWI Budapest Centre. She was Hedley Bull Junior Research Fellow in International Relations at the University of Oxford. Her most recent publications include the edited volume *Building Security in the New States of Eurasia: Subregional Cooperation in the Former Soviet Space* (2000) and, for SIPRI, *Preventing Violent Conflict: The Search for Political Will, Strategies and Effective Tools* (2000) and SIPRI Research Report no. 16, *Executive*

Policing: Enforcing the Law in Peace Operations (2002). She is currently on leave from SIPRI to serve as Special Adviser to the European Union Police Mission in Bosnia and Herzegovina planning in the Secretariat of the Council of the European Union.

Mikael Eriksson (Sweden) is a Research Assistant on the Uppsala Conflict Data Project at the Department of Peace and Conflict Research, Uppsala University. He is currently working on the Stockholm Process on the Implementation of Targeted Sanctions.

Dr David Gold (USA) is Visiting Fellow at the Center for Global Change and Governance, Rutgers University, Newark, New Jersey. He was previously Senior Economic Affairs Officer in the United Nations Secretariat in New York. His research is focused on trends in US defence spending, economics of the peace dividend and economics of the arms trade. He has published in professional journals and books, including, most recently, 'Defence spending and the US economy', *Survival* (2001) and 'Could we have done better? a retrospective on the 1990s peace dividend in the United States', in ed. A. Markusen, *America's Peace Dividend* (Council of Foreign Relations and Columbia International Affairs Online, 2000).

Dr Björn Hagelin (Sweden) is the Leader of the SIPRI Arms Transfers Project. Before joining SIPRI in 1998 he was a Researcher and Associate Professor at the Department of Peace and Conflict Research, Uppsala University. His recent publications include a chapter on Sweden's defence industry in eds A. Eriksson and J. Hallenberg, *The Changing European Defence Industry Sector: Consequences for Sweden?* (2000). He also contributed to Gummett, P. and James, A. (eds), *The European Defence Industry and the New Arms Economy* (Palgrave, forthcoming), the final report of the international project Managing European Technology: Defence and Competitiveness Issues.

Joshua Handler (United States) is a Ph.D. Candidate at the Woodrow Wilson School of Public and International Affairs, Princeton University. In 1988–96 he served as research coordinator for Greenpeace's Nuclear Free Seas Campaign and then as coordinator of the Disarmament Campaign. He is co-author of the Neptune Papers, a series of monographs on the naval nuclear arms race, and a number of books, articles and papers on the US and Russian military and nuclear weapon programmes.

John Hart (United States) has been a Researcher on the SIPRI Chemical and Biological Warfare (CBW) Project since 2001. Previously, he worked as an On-Site Inspection Researcher at the London-based Verification Research, Training and Information Centre (VERTIC) and as a Research Associate at the Monterey Institute of International Studies (MIIS) Center for Nonproliferation Studies. In 1996–97 he worked as a Research Assistant on the SIPRI CBW Project. He is co-author of the SIPRI Fact Sheets 'The Chemical Weapons Convention' (1997) and 'Biotechnology and the future of the Biological and Toxin Weapons Convention' (2001). He co-edited *Chemical Weapon Destruction in Russia: Political, Legal and Technical Aspects*, SIPRI Chemical & Biological Warfare Studies, no. 17 (1998) and contributed to the SIPRI Yearbooks in 1997 and 1998.

ABOUT THE CONTRIBUTORS 801

Dylan Hendrickson (United States) is a Senior Research Fellow at the International Policy Institute, King's College London. His work focuses on international responses to armed conflicts and security problems in Africa and South-East Asia. For the past three years he has worked closely with the British Government, helping to develop and operationalize its security sector reform policy.

Dr Andrzej Karkoszka (Poland) is Head of the Think Tank at the Geneva Centre for the Democratic Control of Armed Forces (DCAF). He was a research fellow at SIPRI in 1973–75 and 1977–81. He has previously served as Adviser in the Chancellery of the Office of the President of Poland, Director of the Ministry of Defence Department of International Security, First Deputy Minister of National Defence, Deputy Chairman of the Polish Team for the NATO accession negotiations and Professor of Central European Security at the George C. Marshall European Center for Security Studies. His most recent publications include 'Verification of implementation of arms control agreements', *Polish Quarterly of International Affairs* (2001) and 'NATO enlargement: 2002 and beyond', eds J. Bozo and F. Beltram, *NATO a New Alliance* (IFRI, 2001).

Shannon N. Kile (United States) is a Researcher on the SIPRI Project on Military Technology and International Security. He is the author of chapters in the SIPRI volume *A Future Arms Control Agenda: Proceedings of Nobel Symposium 118, 1999* (2001) and SIPRI Research Report no. 7, *The Future of the Defence Industries in Central and Eastern Europe* (1994) and a co-author (with Adam Daniel Rotfeld) of a chapter in the Organization for Security and Co-operation in Europe (OSCE) *OSCE Yearbook* (1997). He has contributed to two SIPRI books on Russian security policy: *Russia and Europe: The Emerging Security Agenda* (1997) and *Russia and Asia: The Emerging Security Agenda* (1999) and is the author of the SIPRI Fact Sheet 'Missile defence and the ABM Treaty: a status report' (2001). He has contributed to the SIPRI Yearbook since 1995 on nuclear arms control.

Hans M. Kristensen (Denmark) is a Senior Researcher with the Nautilus Institute in Berkeley, California, where he directs nuclear weapons strategy research. He has been a member of the Danish Defence Commission and a senior researcher with the Military Information Unit of Greeenpeace International, Washington, DC. He is co-author of the Nuclear Notebook series in *The Bulletin of the Atomic Scientists*. Other recent publications include *The Matrix of Deterrence: US Strategic Command Force Structure Studies* (Nautilus Institute, 2001), *The Post Cold War SIOP and Nuclear Warfare Planning: A Glossary, Abbreviations, and Acronyms* (with William M. Arkin, National Resources Defense Council, 1999), and 'The nuclear war files', an on-line repository of US nuclear planning documents hosted by the Nautilus Institute Internet site.

Frida Kuhlau (Sweden) joined the SIPRI Chemical and Biological Warfare Project in 2001. She is co-author of the SIPRI Fact Sheet 'Biotechnology and the future of the Biological and Toxin Weapons Convention' (2001). She assists in the development of the Internet-based Educational Module on Chemical & Biological Weapons Non-proliferation (http://cbw.sipri.se), a joint project with the Centre for Peace and Security Studies at the Free University of Brussels and the International Relations and Security Network (ISN) at the Swiss Federal Institute of Technology in Zurich.

Dr Zdzislaw Lachowski (Poland) is Leader of the SIPRI Project on Conventional Arms Control. He formerly worked at the Polish Institute of International Affairs in Warsaw. He has published extensively on the problems of European military security and arms control as well as issues concerning European political integration. He is the author of chapters in the SIPRI volume *A Future Arms Control Agenda: Proceedings of Nobel Symposium 118, 1999* (2001) and in *Between the Balance of Forces and Cooperative Security in Europe: Adapting the CFE Regime to the New Security Environment* (1999, in Polish) and has contributed to the SIPRI Yearbook since 1992.

Evamaria Loose-Weintraub (Germany) is a Research Assistant on the SIPRI Military Expenditure and Arms Production Project. She is responsible for data on military expenditure in Europe and Central and South America. She is the author of chapters in the SIPRI volume *Arms Export Regulations* (1991) and co-author of a chapter in SIPRI Research Report no. 7, *The Future of the Defence Industries in Central and Eastern Europe* (1994) and of 'Overview of world military expenditure' in the *UNESCO Encyclopedia of Life Support Systems* (forthcoming 2002). She has contributed to most editions of the SIPRI Yearbook since 1984.

Wuyi Omitoogun (Nigeria) is a Researcher on the SIPRI Military Expenditure and Arms Production Project. He is the coordinator of a new project on the Defence Budgeting Process in Africa. He is the author of 'Arms control and conflict in Africa' in *Arms Control and Disarmament: A New Conceptual Approach* (UN Department for Disarmament Affairs, 2000) and the forthcoming SIPRI Research Report no. 17, *Military Expenditure in Africa*.

Thomas Papworth (United Kingdom) was Research Assistant on the SIPRI Projects Conflicts and Peace Enforcement, and Conflict Prevention, Management and Resolution. Before joining SIPRI he was an intern at the International Institute for Strategic Studies (IISS) in London. He contributed to the SIPRI Yearbooks in 2000 and 2001.

John Pike is Director of GlobalSecurity.org, a non-profit public policy group focused on defence, space and intelligence issues, which he founded in 2000.

Professor Adam Daniel Rotfeld (Poland) is Director of SIPRI and Leader of the SIPRI Project on Building a Cooperative Security System in and for Europe. He has participated in many multilateral negotiations and served as the Personal Representative of the Conference on Security and Co-operation in Europe (CSCE) Chairman-in-Office to examine the settlement of the conflict in the Trans-Dniester region (1992–93). In 2001 he was appointed by the President of Poland to Poland's National Security Council and as the Deputy Minister for Foreign Affairs with responsibility for international security (2002–). He is the author or editor of over 20 books and more than 300 articles on the legal and political aspects of relations between Germany and the Central and East European states after World War II (recognition of borders, the Munich Agreement and the right of self-determination), human rights, confidence- and security-building measures, European security and the CSCE/OSCE process. He has written chapters on global and regional security systems and European and transatlantic security structures for the SIPRI Yearbook since 1991.

ABOUT THE CONTRIBUTORS 803

Dr Taylor B. Seybolt (United States) was the Leader of the SIPRI Conflicts and Peace Enforcement Project in 2000–2002. He is now a program officer at the United States Institute of Peace. Prior to joining SIPRI he was a research fellow at the Harvard University Kennedy School of Government. He has written articles and chapters on conflicts worldwide, humanitarian military intervention and the regional spread of intra-state conflicts. He contributed to the SIPRI Yearbooks in 2000 and 2001.

Elisabeth Sköns (Sweden) is the Leader of the SIPRI Military Expenditure and Arms Production Project. She is the author of chapters on the economics of arms production and the internationalization of arms production for the SIPRI volume *Arms Industry Limited* (1993) and other publications. She is also the author of chapters on military expenditure and their determinants and economic impact, including in *New Millennium, New Perspectives* (UN University, 2000). She has contributed to most editions of the SIPRI Yearbook since 1983.

Margareta Sollenberg (Sweden) is a Research Assistant on the Uppsala Conflict Data Project at the Department of Peace and Conflict Research, Uppsala University. She has been editor of *States in Armed Conflict* since 1994 and has contributed to the SIPRI Yearbook since 1995.

Petter Stålenheim (Sweden) is a Researcher on the SIPRI Military Expenditure and Arms Production Project. He is responsible for data on military expenditure in Asia and Oceania and for the maintenance of the SIPRI Military Expenditure Database. He has contributed to the SIPRI Yearbook since 1998.

Professor Peter Wallensteen (Sweden) has held the Dag Hammarskjöld Chair in Peace and Conflict Research since 1985 and was Head of the Department of Peace and Conflict Research, Uppsala University in 1972–99. He has most recently published *Understanding Conflict Resolution: War, Peace and the Global System* (SAGE, 2002) and *Conflict Prevention through Co-Operation Development* (Uppsala, 2001). He has co-authored chapters in the SIPRI Yearbook since 1988.

Reinhilde Weidacher (Italy) is a Researcher on the SIPRI Military Expenditure and Arms Production Project. She is the author of a report for the Swedish Defence Research Establishment on the Italian arms industry (1998) and co-author (with Elisabeth Sköns) of a chapter on the economics of arms production for the *Encyclopedia of Violence, Peace and Conflict* (1999).

Pieter D. Wezeman (Netherlands) is a Researcher on the SIPRI Arms Transfers Project. He has authored or co-authored several articles and papers on arms export issues. From 2000 he has focused on the issue of small arms transfers to areas of conflict. He has contributed to the SIPRI Yearbook since 1995.

Siemon T. Wezeman (Netherlands) is a Researcher on the SIPRI Arms Transfers Project. He is co-author (with Edward J. Laurance and Herbert Wulf) of SIPRI Research Report no. 6, *Arms Watch: SIPRI Report on the First Year of the UN Register of Conventional Arms* (1993), (with Bates Gill and J. N. Mak) of *ASEAN Arms Acquisitions: Developing Transparency* (1995) and (with Pieter D. Wezeman) of a paper for the Bonn International Center for Conversion (BICC) on Dutch surplus weapon exports (1996). He has contributed to SIPRI Research Report no. 13, *Arms, Transparency and Security in South-East Asia* (1997) and to the SIPRI Yearbook since 1993.

Sharon Wiharta (Indonesia) is a Research Assistant on the SIPRI Projects Conflicts and Peace Enforcement, and Conflict Prevention, Management and Resolution. Prior to joining SIPRI she worked at the Center for International Affairs at the University of Washington in Seattle.

Lyudmila Zaitseva (Kyrgyzstan) is a visiting researcher at the Stanford University Center for International Security and Cooperation. She retains a position on the staff of the National Nuclear Center of Kazakhstan. Her research at Stanford focuses on maintaining physical security for nuclear and other radioactive materials as well as for the facilities in which they are kept. She established the Stanford University Database on Nuclear Smuggling, Theft and Orphan Radiation Sources.

Dr Jean Pascal Zanders (Belgium) is the Leader of the SIPRI Chemical and Biological Warfare Project and editor of the SIPRI Chemical & Biological Warfare Studies series. He was previously Research Associate at the Centre for Peace and Security Studies at the Free University of Brussels. He has published extensively on chemical and biological weapon issues in English, Dutch and French since 1986 and has edited *Chemical Weapons Proliferation* (1991, with Eric Remacle) and *The 2nd Gulf War and the CBW Threat* (1995). He has contributed to the SIPRI Yearbook since 1997 and to the SIPRI volume *The Challenge of Old Chemical Munitions and Toxic Armament Wastes* (1997). He is co-author of the SIPRI Fact Sheets 'The Chemical Weapons Convention' (1997), 'Iraq: the UNSCOM experience' (1998) and 'Biotechnology and the future of the Biological and Toxin Weapons Convention' (2001). He is also the principal author of the Internet-based educational module on Chemical and Biological Weapon Non-proliferation (http://cbw.sipri.se). His most recent paper is 'Challenges to disarmament regimes: the case of the Biological and Toxin Weapons Convention', *Global Society* (2001). He is currently preparing a book on managing dual-use technology transfers in a proliferation environment.

Dr Nicholas Zarimpas (Greece) was the Leader of the SIPRI Project on Military Technology and International Security. Previously, he worked on civilian plutonium management and nuclear fuel cycle issues as an administrator at the Nuclear Energy Agency of the Organisation for Economic Co-operation and Development (OECD). He has also held appointments with the Joint Research Centre Ispra and the Environment Directorate-General of the European Commission. During the past 10 years he has acted as scientific secretary to several international multidisciplinary research, technological and policy committees. He is the editor of *Transparency in Nuclear Warheads and Materials: The Political and Technical Dimensions* (SIPRI, forthcoming 2002).

SIPRI Yearbook 2002: Armaments, Disarmament and International Security
Oxford University Press, Oxford, 2002, 845 pp.
(Stockholm International Peace Research Institute)
ISBN 0-19-925176-2

ABSTRACTS

ROTFELD, A. D., 'Introduction: Global security after 11 September 2001', in *SIPRI Yearbook 2002*, pp. 1–18.

The terrorist attacks on the USA marked a watershed in the international security process. The policies and relations of the USA, Russia and other states changed—combating terrorism became a high priority. The transatlantic community disagrees whether to focus on defeating the al-Qaeda network or eliminating the roots of terrorism with broader policies. Expectations of a global response to terrorism fell flat. The interventions in Afghanistan and Kosovo reflect the aspiration to establish international principles and norms of international order. However, there is a lack of legal instruments to tackle situations in which states have traditionally exercised discretionary power or justified their actions as self-defence. The transformation and enlargement of NATO and the EU accelerated. The states of Central Asia have gained in significance. The USA has an unprecedented position in terms of military, economic and technological capabilities. Official statements reflect the view that security is based on interdependence, but the US tendency towards unilateralism prevails.

ERIKSSON, M., SOLLENBERG, M. and WALLENSTEEN, P., 'Patterns of major armed conflicts, 1990–2001', in *SIPRI Yearbook 2002*, pp. 63–76.

In 2001 there were 24 major armed conflicts in 22 locations throughout the world. The number of major armed conflicts and the number of conflict locations in 2001 were slightly lower than in 2000, when there were 25 major armed conflicts in 23 locations. The only interstate conflict that was active in 2001 occurred between India and Pakistan. The vast majority of the conflicts in 2001 occurred in Africa and in Asia. During the period 1990–2001 there were 57 different major armed conflicts in 45 different locations. The number of major armed conflicts in 2001 was lower than in most years since 1990.

SEYBOLT, T. B., 'Major armed conflicts', in *SIPRI Yearbook 2002*, pp. 21–62.

All of the 15 most deadly conflicts in 2001 were intra-state conflicts. Despite their intra-state nature, all of them were directly influenced by external actors—some who prolonged and intensified conflicts and others who promoted conflict resolution. Of the 15 conflicts, 11 spilled over international borders, in some cases causing conflicts in neighbouring states to intensify. In other cases, states in the region were not significantly affected. Although the general pattern of conflict worldwide in 2001 was consistent with previous years, the campaign against terrorism by the USA and its allies after 11 September directly influenced a small number of conflicts and had a much wider indirect impact.

SEYBOLT, T. B., 'Measuring violence: an introduction to conflict data sets', in *SIPRI Yearbook 2002*, pp. 81–96.

The systematic study of violent conflict was pioneered by Quincy Wright in *A Study of War*. Since the 1980s a multitude of conflict data-collection projects have emerged. They often disagree on some of the most basic questions. Is the world more or less violent today than in the past? Are wars more or less destructive than they used to be? What are the causes of conflict initiation, continuation and termination? The world's primary data-collection projects are presented, as are methodological, theoretical and policy-related reasons for the differences among them. They offer researchers a vast array of good data with which to develop academic theories and policy-related arguments.

DWAN, R., 'Conflict prevention', in *SIPRI Yearbook 2002*, pp. 97–150.

Conflict prevention was high on the agendas of both the EU and the UN in 2001, resulting in the release of major reports and ensuing high-level debates. The reports document the efforts of the EU and the UN to practically implement their new commitment to conflict prevention. Both the EU and the UN have emphasized long-term preventive efforts focused on the root causes of armed conflicts. This calls for an integrated approach and improved coordination among actors. In practice, however, this is difficult as preventive efforts in West Africa and Zimbabwe illustrate. The annex lists figures for current multilateral peace operations in 2001. For the first time since 1996 no new UN peace operation was launched. The limited size and scope of the five new multilateral missions meanwhile illustrate the increasing caution of the international states and organizations towards new operations as the costs of such commitments are realized.

LACHOWSKI, Z., 'The military dimension of the European Union', in *SIPRI Yearbook 2002*, pp. 151–73.

Events in 2001 served as a mid-course test for the pursuit of the European Security and Defence Policy (ESDP). The post-11 September developments brought home to the EU the reality of its role and potential in the transatlantic relationship. In 2001 the ESDP was declared operational, but the issue of EU access to NATO's assets remained unresolved. The question of duplication of efforts by the EU and NATO has also not been sufficiently addressed. Defining the ESDP and building public support for increased spending will pose a challenge in the coming years.

HENDRICKSON, D. and KARKOSZKA, A., 'The challenges of security sector reform', in *SIPRI Yearbook 2002*, pp. 175–201.

The international community is seeking to respond in a more integrated manner to the violent conflicts and security problems facing partner states. Restoration of a viable national capacity in the security domain, based on mechanisms that ensure transparency and accountability, is a vital element of the overall effort to strengthen governance. Security sector reform is nonetheless a new area of activity for international actors, and there is still not a shared understanding of what the priorities and objectives should be. The response of states to the terrorist attacks in the United States on 11 September 2001 may impede security sector reform by increasing the political influence and institutional autonomy of security services.

ANTHONY, I., 'Sanctions applied by the European Union and the United Nations', in *SIPRI Yearbook 2002*, pp. 203–28.

EU and UN initiatives have been made to improve the effectiveness of sanctions in managing international security problems. The UN sanctions regime against Iraq was modified in 2001. UN mandatory sanctions have been part of the international response to acts of terrorism. In September 2001 UN Security Council Resolution 1373 established counter-terrorism measures aimed at all entities or persons engaged in supporting or carrying out terrorist acts. The EU has been developing sanctions to advance its objectives on democratization and human rights. EU sanctions achieved some success in South-Eastern Europe when used as part of a broader set of security-building measures.

SKÖNS, E., LOOSE-WEINTRAUB, E., OMITOOGUN, W. and STÅLENHEIM, P., 'Military expenditure', in *SIPRI Yearbook 2002*, pp. 231–65.

World military expenditure increased by about 2 per cent in real terms in 2001 to $839 billion (in current prices) according to adopted budgets. Since 1998, when it reached its lowest point since the end of the cold war, it has increased by 7 per cent. When the actual expenditure figures for 2001 become available, the increase is likely to be greater because of additional expenditure generated by the 11 September terrorist attacks on the USA and the ensuing 'war on terrorism'. Five countries accounted for more than half of world military spending in 2001: the USA (36 per cent), Russia (6 per cent) and France, Japan and the UK (5 per cent each). The high-income countries have the highest per capita spending, while the countries where military spending imposes the heaviest economic burden are located in the Middle East and Africa.

SKÖNS, E. and WEIDACHER, R, 'Arms production', in *SIPRI Yearbook 2002*, pp. 323–72.

The arms industry has changed profoundly in size and structure since the early 1990s. Until the mid-1990s these changes were driven by the need to adjust arms-producing capabilities to the significant reduction in the demand for military equipment associated with the end of the cold war. Since the late 1990s arms procurement has stabilized and in a number of countries there had even been slight growth. The current restructuring of the arms industry is to a significant extent driven by large companies trying to secure their already strong positions on the world market through continued acquisitions and international cooperation arrangements. Increased concentration and internationalization in the production of armaments may require international measures to ensure transparency and accountability in arms production in the future.

GOLD, D., 'US military expenditure and the 2001 Quadrennial Defense Review, in *SIPRI Yearbook 2002*, pp. 309–22.

The USA is beginning a major military buildup following the terrorist attacks of 11 September 2001, the fourth since World War II and the continuation of a pattern where military spending goes through cycles but exhibits no long-term real growth. The long-run decline in the military burden reflects, among other causes, the political priority given to civilian objectives. The buildup continues almost all Clinton-era programmes and fails to resolve conflicts over military transformation. The Quadrennial Defense Review articulated broad objectives for transformation but contained no budgetary guidance. Thus, disputes will continue to dominate the budgetary process.

HAGELIN, B., WEZEMAN, P. D., WEZEMAN, S. T. and CHIPPERFIELD, N., 'International arms transfers', in *SIPRI Yearbook 2002*, pp. 373–465.

Since 1997 arms transfers have declined by about 33 per cent. For the period 1997–2001 the USA was the dominant supplier, accounting for nearly half of all deliveries. Russia was the second largest supplier for the period, but for 2001 Russia was the largest supplier. Taiwan was the largest recipient for 1997–2001. For 2001 China was by far the largest recipient. For 1997–2001 the most important arms transfers in terms of volume were between Taiwan and the USA, and China and Russia. The list of main suppliers and main recipients has changed little over the past 10 years. India and Pakistan are both major recipients of arms. They have received weapons or have weapon acquisition plans that could be destabilizing. On the other side of the scale, small imports of weapons have strongly influenced the course of the war in Sierra Leone and relations between West African countries. Competition has increased the importance of offsets, including technology transfers. The Joint Strike Fighter project may point to an unbalanced future transatlantic market.

ANTHONY, I., 'Arms control after the attacks of 11 September 2001', in *SIPRI Yearbook 2002*, pp. 469–88.

During its first year the George W. Bush Administration reassessed the role of arms control within foreign, security and defence policy. Arms control discussions were used to organize a dialogue about trends in international security. As part of the process of the adaptation of arms control there may be a loss of coherence in the position of some states, but this may be a transitory phenomenon as new norms and principles develop in a changing security environment. The proliferation of nuclear, biological and chemical munitions and the missile delivery systems could place the continental USA at risk in future. The emphasis on non-proliferation was strengthened after the terrorist attacks on the USA in September 2001. The focus on state-sponsored military programmes was supplemented by an examination of whether arms control could help to manage threats from non-state groups and individuals.

ZARIMPAS, N., 'Tactical nuclear weapons', in *SIPRI Yearbook 2002*, pp. 568–84.

More than a decade after the end of the cold war and successive nuclear weapon reductions, thousands of tactical nuclear weapons continue to be stockpiled in the Russian and US arsenals. Tactical nuclear weapons remain outside formal arms control agreements and, because of their small size, mobility and decentralized command and control arrangements, pose unique dangers. Recent developments have raised concerns that increased reliance on, and new missions for, these weapons can be expected. Such concerns are exacerbated by the lack of transparency surrounding their numbers and operational status. Tactical nuclear weapons deserve urgent attention. Russia and the USA should ensure that their tactical nuclear weapons are safely and securely stored; jointly reaffirm their commitment to the 1991–92 Presidential Nuclear Initiatives; provide updates on progress towards the elimination of these weapons; pursue increased transparency; reassess the perceived utility of such weapons in military and deterrence doctrines; halt further weapon modernization; unilaterally proceed with additional reductions; and consider ways of constructing a cooperative framework that drastically limits the numbers and locations of such weapons.

KILE, S. N., 'Ballistic missile defence and nuclear arms control', in *SIPRI Yearbook 2002*, pp. 489–524.

In 2001 the controversy over missile defence and the future of the 1972 ABM Treaty came to a head. President George W. Bush gave notice that the USA would withdraw from the treaty in order to build a large-scale national missile defence system. Bush's announcement had been expected and elicited a restrained response from China and Russia. It reflected his administration's rejection of treaty-based approaches to arms control. While 'agreeing to disagree' on missile defence, Bush and Russian President Vladimir Putin pledged to make deep cuts in strategic nuclear forces. However, there was a disagreement over whether these cuts should be made as informal initiatives or as part of a treaty. There was also a disagreement over whether all warheads removed from missiles would have to be dismantled, as advocated by Russia, or could be placed in storage.

BUKHARIN, O., 'The changing Russian and US nuclear warhead production complexes', in *SIPRI Yearbook 2002*, pp. 585–97.

The end of the cold war called for radical downsizing and restructuring of the Russian and US nuclear warhead production complexes. The USA has already concluded the first phase of its infrastructure reductions. However, the US policies of strengthening its warhead design capability and maintaining a large reserve of stored warheads could have a negative impact on future arms reductions. The Russian complex, although smaller compared to its cold-war size, remains oversized. International cooperation could significantly speed up the contraction of the Russian complex. Deep nuclear arms cuts in Russia and the USA would necessitate further infrastructure reductions.

BUNN, G. and ZAITSEVA, L., 'Efforts to improve nuclear material and facility security', in *SIPRI Yearbook 2002*, pp. 598–612.

There is a need for higher standards and stronger practices for the protection of nuclear facilities from terrorists and thieves in the aftermath of the attacks of 11 September 2001. Information from the International Atomic Energy Agency's Illicit Trafficking Database and from Stanford University's Database on Nuclear Smuggling, Theft and Orphan Radiation Sources shows that illicit trafficking in nuclear and other radioactive materials has taken place over the past decade not just from Russia but also from many other places around the world. The existing national and international protection standards and practices for protection of such materials need to be improved.

PIKE, J., 'The military uses of outer space', in *SIPRI Yearbook 2002*, pp. 613–64.

The 'weaponization' of outer space has resurfaced on the arms control agenda, with growing international concern that the USA's ability to dominate space and to deny its use to other countries will give rise to a destabilizing arms race in space. At the end of 2001, the one remaining superpower—the United States—had nearly 110 operational military spacecraft, well over two-thirds of all the military spacecraft orbiting the earth. Russia was a distant second, with about 40. The rest of the world had only about 20 military satellites in orbit.

ZANDERS, J. P., HART, J. and KUHLAU, F., 'Chemical and biological weapon developments and arms control', in *SIPRI Yearbook 2002*, pp. 665–708.

The major events in 2001 in the field of biological weapon control were responses to the anthrax attacks in the USA, the US rejection of a draft protocol to strengthen the 1972 Biological and Toxin Weapons Convention (BTWC), and its proposal to terminate the negotiating mandate of the Ad Hoc Group which had drafted the protocol. Numerous political and technical questions were raised during the anthrax investigation, including whether non-state actors are technically capable of producing the high-grade anthrax spores used and the nature of the anthrax strain and its provenance. The major issue facing the Organisation for the Prohibition of Chemical Weapons (OPCW), the body that oversees the implementation of the 1993 Chemical Weapons Convention (CWC), was an unprecedented budgetary shortfall mainly caused by how certain inspection costs are estimated and subsequently reimbursed to the OPCW.

LACHOWSKI, Z., 'Conventional arms control', in *SIPRI Yearbook 2002*, pp. 709–35.

The general trend in 2001 was a focus by the international community on regional and domestic sources of conflict and relevant arms control measures, particularly operational measures. In Europe the focus was on the implementation of agreed measures and, after the 11 September terrorist attacks on the USA, the search for new approaches to the politico-military dialogue. Russia's continued non-compliance with its agreed flank levels hindered the entry into force of the 1999 Agreement on Adaptation of the 1990 Treaty on Conventional Armed Forces in Europe (CFE Treaty). However, Russia has met its commitments regarding troop withdrawals from Moldova. In Georgia the future of one Russian military base and the continued presence of Russian forces remain to be resolved. Regional and bilateral confidence- and security-building measures (CSBMs) continued to work smoothly, and new bilateral CSBMs were introduced in Europe. On 1 January 2002 the 1992 Treaty on Open Skies entered into force. The Second Review Conference of the CCW ('Inhumane Weapons') Convention extended the application of the convention to domestic armed conflicts.

ANTHONY, I., 'Multilateral export controls', in *SIPRI Yearbook 2002*, pp. 743–58.

Changes in the membership, guidelines and procedures of the Australia Group (AG), Missile Technology Control Regime (MTCR), Nuclear Suppliers Group (NSG) and Wassenaar Arrangement are surveyed. In response to the September 2001 terrorist attacks on the United States new ideas on the role of export controls in counter-terrorism were put forward in the AG, the NSG and the Wassenaar Arrangement. MTCR participating states agreed the text of an International Code of Conduct (ICOC) aimed at discouraging ballistic missile proliferation. A decision on implementing the ICOC is expected in 2002. The NSG continued to discuss how to respond to decisions by Russia related to nuclear supply that are considered to violate the NSG guidelines.

Errata

SIPRI Yearbook 2001: Armaments, Disarmament and International Security

Page 356, table 5A.1, note a, second line	Should read: '(pp. 368–71)'.
Page 553, first paragraph of text, line 3	Should read: '14 500 items (table 8.3) which were inherited from the former Soviet Union'.
Page 555, table 8.4, column 'ACVs', the number in the row for Oct. 1999	Should read: '3 534'.
Pages 556, last line	'Guduata' should read 'Gudauta'.

INDEX

A400M transport aircraft 257
A-4 aircraft 565
A-5 aircraft 563
A-6 aircraft 397
A-10 aircraft 397
A-109 helicopter 393
Abkhazia 55, 137, 713, 715, 716, 717
ABM (Anti-Ballistic Missile) Treaty (1972) 487, 513, 579, 771, 796:
  US change of attitude towards 500–502, 646
  US withdrawal from 14, 472, 482, 484–85, 489, 490, 491, 506–11, 523, 715, 798
Abu Dhabi: military expenditure 248
ACSS 334
Adamov, Yevgeniy 754
ADI 361
ADM (atomic demolition mine) 557, 570, 571, 575
Advanced Technology Vessel 561
Advanced Wideband System (AWS) 538
Aegis Weapon System 649
Aero Vodochody 339
Aerospace Corp. 360
Afghanistan:
  Afghanistan Interim Authority (AIA) 43, 44, 125, 126, 241, 796, 797, 798
  arms embargo 39, 40, 221, 387–89
  arms trade and 40
  Bonn Agreement (2001) 43–44, 125, 241
  casualties 44
  caves attacked 42–43
  Chechens in 40, 52, 55
  civilians 44
  conflict in 2, 4, 21–22, 39–44, 66, 72, 121, 125–26, 163, 224, 239–42, 245, 253–54, 264, 508, 795
  drug trafficking 39–40
  EU and 223–24
  Europe's weakness and 166
  France and 549, 552
  GPS used in 620
  humanitarian aid 44, 126, 240
  International Security Assistance Force (ISAF) 12, 44, 124, 125–26, 133, 241, 798
  intervention in 2, 3, 4, 10
  Japan and 240
  Kabul 39, 42, 43, 44, 125, 126
  military expenditure 273, 279, 285
  NATO and 2
  Northern Alliance 39, 40–41, 43, 55, 67, 125, 389, 795, 796
  Pakistan and 41, 45, 241, 793
  peace mission in 124, 125
  reconstruction 44, 239, 240–41
  refugees 44
  Russia and 42, 508, 795
  sanctions against 39–40, 206, 221–24
  security sector reform 190
  Tajikistan and 796
  Taliban regime 4, 10, 21, 39, 40, 41, 45, 55, 67, 125, 221–22, 793, 795:
    collapse of 42, 796, 797
    foreign fighters 40
    sanctions against 39–40, 206, 222–24
    terrorism and 39, 221
  UK in war against 4, 41, 43, 125–26, 224, 240, 793, 795
  UN and 39, 41, 43–44, 125, 222–23, 797, 798
  USA in war against 2, 4, 21, 23, 39–44, 121, 125, 126, 163, 239–42, 253–54, 264, 313, 316, 625, 508, 793, 795
  USSR and 39, 43
Africa:
  aid to 185
  arms imports 376
  conflict prevention mechanisms 186–87
  conflicts in 24–39, 65, 69–71, 188
  Great Lakes region 22, 24, 32, 61, 65, 106, 119
  military expenditure 13, 188–89, 231, 233, 245–46, 264, 266, 276, 282–83, 306
  militia groups 189
  peace missions in 128–29
  Regional Peace Initiative 27
  security sector 179–80:
    investment in 189
    privatization 189
    reform 188–90, 194, 201
    state weaknesses 189
  UN and 114 *see also under names of individual countries*
  UN conference on illicit trade in small arms and 737–38
  see also under names of individual countries
African Union 30 fn. 47, 99 fn. 11, 789
Agni missile 559–61, 563, 785

INDEX 813

Agreement on Adaptation of the CFE Treaty (1999) 709, 711, 712, 713, 714
Agreement on Confidence- and Security-Building Measures in Bosnia and Herzegovina (1996) 722, 779
Agreement on Regional Stabilization (2001) 724–25
Agusta (Finmeccanica) 359
Agusta Westland 333, 334
Ahmeti, Ali 794
air security 10
Airborne Laser 652
Airbus 257
Airbus A400M transport aircraft 161
Airbus Military Co. 258
aircraft carriers: France 552
airliners: security for 607
Akash SAM system 383
AKUF (Arbeitsgemeinschaft Kriegsursachenforschung) 84
Alaska Aerospace Development Corporation 651
Albania:
  EU in 137
  MAPE 124, 137
  military expenditure 273, 279, 285, 307
  OSCE in 135
  see also Macedonia (FYROM): Albanian rebels, Albanian rights
Alcatel Space 637, 641
Alcoa 360
Aldridge, Edward, Jr 794
Alenia 335
Alenia Aerospazio 335, 340
Alenia Marconi Systems 333, 358, 648
Algeria:
  casualties 25
  conflict in 24–25, 69, 246
  military expenditure 246, 270, 276, 282
Alliant Tech Systems 329, 358
Alvis 361
AM General Corporation 361
Amber Fox, Operation 166 *see also* Macedonia (FYROM): TFF
America:
  arms imports 376
  conflicts in 65, 71–72
  see also under names of countries
Amnesty International 118
Amsterdam Treaty (1997) 152
Angola:
  casualties 27
  chemical weapons used in 706
  civilians 27
  conflict in 22, 25–27, 65, 69
  diamonds 25–26
  DRC and 26, 30, 31, 33
  military expenditure 270, 276, 282
  Namibia and 26–27
  oil 25
  refugees 26
  UN and 25–26
  UNITA 25, 26, 27, 205, 208, 390
  Zambia and 26
Annan, Kofi:
  conflict prevention and 99, 100
  DRC and 30
  Israel and 56
  Zimbabwe and 117
Antarctic Treaty (1959) 473, 764–65
Anteon 360
anthrax 14, 679–80, 681, 682:
  as BW agent 696–99
  proliferation implications 701–3
  terrorist attack, 2001 665, 696–701, 707, 708, 795
Anti-Personnel Mines (APM) Convention (1997) 473, 709, 730–31, 782–83
Antonov-24 aircraft 161
APA 334
Apache helicopter 332
Arafat, Yasser 55, 58, 794, 797, 798
Arbatov, Alexei G. 8 fn. 24, 508
Argentina:
  military expenditure 272, 278, 284, 304, 307
  sanctions against 210
Ariana Afghana Airlines 222
armed conflicts *see* conflicts
Armenia: military expenditure 274, 280, 285
arms control:
  adaptation 487, 488
  aims 470, 471
  characteristics 470–76
  compliance with 470, 471–72, 474–76, 485
  confidence in, loss of 487
  constraints 470
  definition 470–71
  effectiveness of 470, 472, 474–76
  failures of 469
  forms of 471, 472–74
  international security and 470
  linkage between processes 484, 485
  NBC weapons and 479
  non-compliance with 471, 472
  non-state actors and 487

sanctions and 480
states' differences on 483
terrorist attacks and 469, 470, 485–87, 488
treaties, scepticism about 491, 524
usefulness of 469, 471, 476
violation 472, 476, 478
arms production:
accountability 366, 367
changes 324, 325, 326
cold war 326
commercial confidentiality 372
companies' size 327–28, 330
company specialization 330–31
competition 330, 343
concentration 13–14, 325, 326–31, 352, 353
corruption 343
cost 343
data on 323, 354–72
decline in 323, 325
definitions 354–55
diversification 323, 330
downsizing 323
foreign investment 332, 337–38, 393
foreign ownership 325, 353
foreign subsidiaries 332
governments and 332, 371–72
history 325–26
internationalization 325, 331–40, 353, 393
joint ventures 13, 328, 333, 334, 336
mergers and acquisitions 326, 328, 329, 330, 333, 334, 336, 338
minor arms-producing countries 337–40
privatization 325, 341, 342, 343, 352, 353
rationalization 323, 330
reporting practices 367–68
restructuring 13–14
SIPRI data on 367
SIPRI's top 100 companies 326, 356–65
terrorist attacks and 323
transatlantic 13
transatlantic cooperation 396, 397, 402
transatlantic links 335–36
trends 323–25
arms trade:
arms embargoes 35, 39, 40, 109–10, 205–206, 208, 211–12, 215, 218–19, 221, 226, 375, 382, 384, 385–86, 387–90, 402
arms production investment and 337–38
changes in 326
co-production arrangements 393
cold war and 326

commercial confidentiality 366, 372
competition 391–95, 402
conflict areas and 380–90
data on 390, 460–65
decline in 325, 326
illicit 11, 17
internationalization and 331–32
military interoperability and 400
military technology transfers and 392–93, 400, 402
offsets 337, 339, 340, 341, 353, 392–95, 402
recipients 14, 379–80, 403–406
SIPRI Arms Transfers Project 373
SIPRI register of 413–65
SIPRI sources and methods 456–59
small transfers, significance of 384–87
suppliers 14, 374–79, 407–408
transparency 400–401
value of 390–91
volume 13, 403–10
world 390, 402
arms trade, control of 16
Arrow interceptor system 649
Arroyo, Gloria Macapagal 48, 49
ASEAN (Association of South-East Asian Nations):
military expenditure 267
Regional Forum (ARF) 727–28
security sector reform and 186
Asia:
arms imports 376
conflicts in 39–52, 65–66, 72–74, 190
military expenditure 264, 266, 272–73, 278–79, 284–85
security sector reform 188–90, 190–91, 194, 201
USA and 191
ASML 329
Aspin, Les 315
Astrium 334
Atlas booster 619, 624
Atta, Mohammed 702
Australia:
arms imports 377
arms production 332, 341, 380
biological weapons and 668
military expenditure 240, 273, 279, 285, 307
Australia Group 473–74, 475, 667, 743, 744, 757
Austria 171:
military expenditure 274, 280, 286

INDEX 815

AV-8B aircraft 397
AVIA Systems Group 339
Ayres, Bill 94
Azerbaijan:
  conflict and 66
  military expenditure 274, 280, 286, 307
  OSCE in 135

Babcock Borsig 359
Babcock International Group 361
BAE Systems 328, 330, 332, 333, 335, 336, 339, 340, 344, 353, 357, 393, 394, 396, 397, 398, 559, 636
BAE Systems Capital Limited 259
Bahadhur aircraft 558–59
Bahrain: military expenditure 238, 275, 281, 287
Ball Aerospace 643
ballistic missile defence *see under* United States of America
ballistic missiles, long-range: proliferation 491, 503
Baltiyskiy Zavod 348, 349, 350, 364
Bangkok Treaty (1995) 473, 779
Bangladesh 249:
  military expenditure 249, 273, 279, 285
Barak, Ehud 55
Basson, Wouter 705, 706
Behavioral Correlates of War 91–92
Belarus 210:
  arms exports 376
  Lithuania and 725
  military expenditure 274, 280, 286, 307
  nuclear reactor 604
  Ukraine and 725–26, 795
  USSR's nuclear weapons in 512, 797
Belgium:
  chemical weapons 691
  conflict prevention and 105, 112
  Europe's military capability 155, 156, 162, 164, 165
  military expenditure 274, 280, 286, 292
  Zimbabwe and 119
Belize: military expenditure 271, 277, 283
Benin: military expenditure 270, 276, 282
Berlusconi, Silvio 154 fn. 13
BF Goodrich 329, 360
Bharat Electronics 361
BIMILSATCOM 638
Biological and Toxin Weapons Convention (BTWC, 1972) 14–15, 473, 475–76, 480, 485, 486, 487, 770–71:

Ad Hoc Group (AHG) 665, 666, 667–71, 673, 707
  biotechnology and 672
  BW defence and 680–83
  non-compliance allegations 674–75
  protocol to 665, 666, 667–73, 787, 791
  review conferences 665, 667, 668, 673–77, 675–77, 707, 798
  Russia and 792, 796
  status of 666
  USA and 665, 668, 671–73, 674–75, 676, 677, 681–83, 798
  verification 671, 672, 681
biological weapons 14–15, 707:
  defence against 665, 677–78, 678–83
  proliferation 704–707
  terrorism and 743
biotechnology 672, 677
Black Sea 726
Blair, Tony 154, 167
Boeing 328, 330, 331, 332, 336, 344, 353, 357, 393, 396, 619, 620, 626
Boeing/TRW/Lockheed Martin team 652
Bofors 336
Bolivia:
  drugs 252
  military expenditure 272, 278, 284, 307
Bolton, John 477, 478, 479–80, 487 fn. 79, 505, 674
Bombardier 344, 360
bombers *see* nuclear bombers
Bond, Doug 94
Bond, Joe 94
Bosnia and Herzegovina:
  Concluding Document (2001) 740–42
  conflict in 66
  EU and 127
  Florence Agreement 718, 719
  military expenditure 236, 238, 274, 280, 286
  OSCE and 127, 135, 304, 723
  sanctions against 212
  security sector reform 196
  SFOR 127, 136, 723
  UN and 99
Botswana: military expenditure 246, 270, 276, 282
Brahimi, Lakhdar 43, 101
Brahimi Report (2000) 101, 102 fn. 20
Brazil:
  arms production 394
  military expenditure 233, 234, 235, 252, 272, 278, 284, 305–306, 307

816 SIPRI YEARBOOK 2002

Brecher, Michael 90
Bremer, Stuart 88, 91
'Brilliant Pebbles' 494, 653
British Nuclear Fuels 548
Brunei: military expenditure 273, 279, 285, 307
Brussels Treaty (1948/1954) 762, 764
Bulgaria:
  military expenditure 274, 280, 286, 307
  NATO and 9
Burghardt, Günter 484
Burkina Faso: military expenditure 270, 276, 282, 306
Burundi:
  Arusha Peace and Reconciliation Agreement (2000) 27
  casualties 27
  conflict in 22, 24, 27–29, 69, 791
  DRC and 29, 31, 32, 61
  EU and 27
  Implementation Monitoring Commission 28
  military expenditure 237, 238, 245, 270, 276, 282
  peace mission in 138
  protection and support force in 124, 139
  transitional government 27, 28, 29, 791
  refugees 27
  South Africa and 29, 124, 129
  UN and 27, 28–29
Bush, President George: Presidential Nuclear Initiatives 571–72, 574
Bush, President George W.:
  ABM Treaty and 14, 482, 489, 501, 504, 508, 511, 523
  anthrax attack and 701
  armed conflicts and 469–70
  arms control and 469, 470, 476–85, 488, 516, 517, 523–24, 580, 671
  arms exports 379
  BMD and 490, 491–92, 501–2, 503
  BRAC process 320
  BTWC and 671
  BW defence and 679, 681
  China's missile exports and 747
  EU and 166, 167
  inaugurated president 785
  Iraq and 217, 218, 796
  Israel and 57, 795
  military doctrine and 315
  military expenditure and 240, 309, 316, 317, 321
  military pay 319

  NATO and 7
  North Korea and 476–77
  nuclear security programmes 519–20
  nuclear weapons, reductions in 489, 515–17
  Putin, meetings with 8, 503–504, 506, 515, 791, 796
  satellites and 614
  terrorism and 226, 240
  weapons, skipping a generation 315
Bustani, José 684
Buyoya, President Pierre 28, 791
BZ 706

C^3I (command, control, communications and intelligence):
  EU 153, 160, 161
  Kazakhstan 249
  transatlantic 253
  USA 538
C-17 aircraft 331
C-160H Astart, aircraft 552
CACI International 361
CAE 329, 361
Cambodia 190:
  military expenditure 273, 279, 285
  security services 190, 191
Cameroon: military expenditure 270, 276, 282
Canada:
  EU and 165
  Flying Training 344
  military expenditure 272, 278, 284, 292
  terrorism and 243
Cape Verde: military expenditure 270, 276, 282
Caribbean countries: aid to 185
Carlyle Group 329
Carnegie Commission on Preventing Deadly Conflict 98
CASA 339
Castaño, Carlos 59
CCW (Certain Conventional Weapons) Convention (1981) 473, 709, 732–34, 735, 774–75, 798
CEA 358
CEE (Central and East European) states:
  arms exports 401
  arms imports 338
  arms production 341
  assistance to 256
  civil–military relations 178
  internal security forces 179

INDEX 817

military expenditure 232, 234, 267, 307, 308
NATO and 176
offsets and 395
security sector reform 176, 185, 196, 198, 201
social services 177
Centaur boosters 616
Center for Remote Imaging, Sensing and Processing 643
Central African Republic: military expenditure 270, 276, 282
Central America: military expenditure 266, 271–72, 277–78, 283–84, 307
Central Asia:
  military expenditure 248–49, 266, 272, 278, 284, 307
  security sector reform 19
  significance increased 2, 10
  terrorism and 10
  USA and 9, 191
  West and 10–11
CFE Treaty (Treaty on Conventional Armed Forces in Europe, 1990) 472, 709, 710, 711–18, 776–77:
  POET 712, 713
  review conferences 709, 712, 713–15
  Russia's non-compliance 709, 711, 713, 714–15
  Ukraine and 714
CFE-1A Agreement (1992) 710, 712, 777
Chad: military expenditure 270, 276, 282
Chechnya, conflict and 52–55, 66, 74:
  casualties 54
  CFE Treaty and 711, 714
  civilians and 54
  human rights abuses in 54
  OSCE and 54, 135
  radioactive landmine in 611
  Russian troop reductions 714
chemical weapons 14:
  declared possessors 686
  destruction of 666, 686–91
  old and abandoned 691–96
  proliferation 704–707
  terrorism and 743
Chemical Weapons Convention (CWC, 1993) 473, 476, 485, 486, 683–86, 707–708, 778:
  review conference 686, 708
Chevaline warhead 547
Chidambaram, Rajgopal 558
Chile:
  arms imports 252

military expenditure 252, 272, 278, 284, 304, 307
'reserved copper law' 252
China:
  ABM Treaty and 501, 509–10
  arms control and 481, 484, 485, 792
  arms exports 376, 382
  arms imports 14, 251, 349, 375, 379, 402
  biological weapons and 667, 668, 704
  chemical weapons 694–95, 704
  IDC and 523
  military expenditure 233, 234, 235, 249, 250–51, 252, 264, 273, 279, 285, 307
  military expenditure on R&D 251
  missiles and 381, 480, 485, 747
  nuclear exports 753
  nuclear forces 552–57
  nuclear modernization 526
  nuclear weapons:
    non-strategic 557
    numbers of 516, 525, 526, 552–53, 557
    tests 562
  outer space arms control 645–47, 789
  Russia and 791
  sanctions against 210, 747
  strategic nuclear forces 510
  Tibet 553
  West and 10
  WTO and 798
China Metallurgical Co. 747
Chinook helicopter 332
Chirac, President Jacques 548–49
Chowdhury, Anwarul Karim 207
Chowdhury Report 208
Chu, David 707
Chubais, Anatoliy 348
CIS (Commonwealth of Independent States):
  military expenditure 267, 307, 308
  peace missions 137
  rapid reaction force 788
Citron Tree battle management centre 649
Clinton, President Bill 309, 316, 319, 320, 343, 477, 478, 479, 481:
  ABM Treaty and 500–501
  BMD and 492, 494, 498, 647, 651, 652
  BTWC and 671
  BW defence and 679
  nuclear reductions, surplus warheads in reserve 517, 529
Cobham 360
CODELCO 252
cold war:
  arms production 326

arms trade 326
cold war's end:
  arms production and 326
  conflict and 98
  effects of 98, 175, 176, 323, 596
Colombia:
  AUC (United Self-Defence Forces of Colombia) 59, 60, 61
  casualties 60
  conflicts in 22, 58–61, 62, 71
  drugs 59, 252
  ENL (National Liberation Army) 58, 61
  FARC (Revolutionary Armed Forces of Colombia) 58, 59, 60, 61
  human rights abuses 59
  military aid to 59
  military expenditure 272, 278, 284, 307
  peace talks 785
  USA and 59
Commonwealth of Nations 119, 120
Comoros: OAU and 138
Computer Science Corporation 345, 358
computers 81, 263, 585, 648
Conference on Disarmament (CD) 645, 789
Conference on Facilitating the Entry into Force of the CTBT 522
Conference on Stability, Security, Development and Co-operation in Africa (CSSDCA) 186–87, 306
Conflict Data Project 84, 85, 89
conflict prevention 12:
  coordination and 101–102
  definition 98
  development aid and 101
  development and 101
  regional organizations and 106–107
  rise of 97
  terrorism and 121–23
conflict prevention, management and resolution 12
conflict resolution 13
Conflict Simulation Model (KOSIMO) 84, 89–90
conflicts:
  arms purchases 21, 22
  casualties 68 *see also under names of countries involved*
  causes of 98
  civilians 21
  conclusions on 61–62
  costs of 98–99
  definitions 23–24, 77–78
  development and 99
  economic gain and 61
  external interventions 175
  intensity, changes in 68
  intra-state 11, 21–22, 23, 65
  intra-state and interstate 22
  methods 79–80
  natural resources and 21, 22, 61–62
  neighbouring states and 11
  patterns of 11–12, 63–68
  rebels 21
  refugees 22
  regional patterns 64, 65–66
  sources 78–79
  'spillover' 21, 22, 61, 183
  study of 81
  trends 63–68
  see also following entry
conflicts, data sets:
  casualties 85
  causes and processes 90–94
  coding rules 86–88
  costs of wars 94–96
  definitions 84–86
  differences in 81, 82–88
  duration of 87
  leading sets 88–96
  methods 86–88
  occurrence patterns 88–90
  purposes 82–83
Congo, Democratic Republic of the (DRC):
  Angola and 26, 30, 31, 33
  Burundi rebels and 29, 31, 32, 61
  casualties 33
  conflict in 22, 24, 29–34, 61, 65, 70, 786, 787
  human rights abuses 34
  Inter-Congolese Dialogue 29, 31, 32–33, 792
  landmines used in 731
  Lusaka Ceasefire Accord (1999) 28, 29, 30–31, 32, 786
  military expenditure 245, 270, 276, 282
  Namibia and 30, 31
  natural resources 33
  OAU and 138
  president assassinated 29, 785
  Rwanda and 30, 31, 32, 33
  Uganda and 30, 31, 32, 33
  UN and 29, 786, 787
  uranium stolen 603
  warring parties withdraw 787
  Zimbabwe and 30, 119

INDEX 819

Congo, Republic of 65:
  military expenditure 270, 276, 282
Conté, President Lansana 35
Convention on the Physical Protection of
  Nuclear Material (CPPNM, 1980) 598 fn. 2,
  609–10, 773–74
conventional arms control 709–39:
  documents on 740–42
Cooperative Threat Reduction (CTR)
  programme 474, 477 fn. 44, 591
Coordinating Mechanism for Sierra Leone
  111
Cordant Technologies 360
Core Software Technologies 644
Correlates of War 81, 84, 88–89
Correlates of War–Militarized Interstate
  Disputes 91
Costa Rica: military expenditure 271, 277,
  283, 307
Côte d'Ivoire 108, 385:
  military expenditure 270, 276, 282
Cotonou Agreement (2000) 106, 113, 117,
  185
Council of Europe 106, 785
counter-terrorism measures 232
Cox Report 555
CR 705, 706
crime, organized 16, 17
Croatia 718:
  arms exports 385
  conflict in 66
  military expenditure 236, 238, 274, 280,
    286, 307
  OSCE and 127, 135
  sanctions against 212
  WEU Demining Assistance Mission 124,
    137
crop dusters 702–703
Crouch, Jack Dyer II 518
cruise missiles:
  China 557
  France 548, 549, 550–51, 552
  Israel and 567
  USA 528, 530, 531, 535, 537, 567, 574
Crusader mobile artillery system 314
CTBT (Comprehensive Nuclear Test-Ban
  Treaty, 1996) 473, 522–23, 578, 592, 596,
  781–82:
  CTBTO (Comprehensive Nuclear Test-Ban
    Treaty Organization) 522–23
Cuba:
  as 'backlash state' 479
  chemical weapons and 668, 669
  missile crisis 569 fn. 7
  Russian bases 9
Cubic Corporation 361
Cukurova Holdings 643
Cyprus: military expenditure 274, 280, 286
Czech Republic:
  arms exports 401
  arms imports 393, 394
  arms production 338, 339
  military expenditure 274, 280, 286, 292,
    307
  nuclear smuggling 603
  security sector reform 198
Czechoslovakia: military expenditure 274,
  280, 286

Dahlgren, Hans 112–13
Daimler-Chrysler 333
Danforth, John 37
DASA Dornier 641
Daschle, Tom 482–83, 703
Dassault 393, 394, 549, 566
Dassault Aviation 340, 358
Davenport, Christian 93
Davies, John 95
Dayton Agreement (1995) 722, 791
DCN 341, 358
DD-21 destroyer (DD-X) 318
DD-X destroyer (DD-21) 318
De-Targeting Agreement, US–Russian (1994)
  532
Declaration of San Salvador on Confidence-
  and Security-Building Measures (1998) 304
Defense Satellites Communications System
  (DSCS) 538
Defense Support Program 622–23
Delta rockets 619
democracy 7, 17, 18
Denel 360
Denmark 118, 153, 154, 200:
  arms exports 400, 401
  military expenditure 274, 280, 286, 292
developing countries:
  aid to 177
  military expenditure 231, 235, 237–38
  security sector reform 176, 178, 187
  social services 177
Devonport Management 359
DF missiles 553–54, 555, 557
Dhanush (Bow) SLBM 561
diamonds 21, 797 *see also under* Angola;
  Liberia; Sierra Leone
Diego Garcia 41

Diehl 359
Diehl Avionik Systeme 334
Diehl, Paul 91, 92
DigitalGlobe 643–44
Divine Junker, Operation 679
Djibouti: military expenditure 270, 276, 282
DRS Technologies 329, 360
drug trafficking 183, 191, 193
DSCS (Defense Satellite Communications System) 617–18
Dyncorp 346, 359

E-2C Hawkeye aircraft 552
EADS 258, 329, 332, 333, 334, 335, 336, 339, 340, 353, 357, 358, 550
earth-penetrating bombs 529, 530, 577–78
EarthWatch Inc. 643
East Asia: military expenditure 232, 234, 249–51, 264, 266, 273, 279, 284–85, 307
East Timor:
    Australia and 125
    INTERFET 125
    security sector reform 190
Eastern Europe: security sector reform 192
EC-135 Airborne National Command Post 538–39
ECOMOG (ECOWAS Monitoring Group) 111–12
ECOWAS (Economic Community of West African States):
    conflict prevention and 99, 109, 110, 111
    EU and 113, 115–16
    security sector reform and 186
    small arms and light weapons, moratorium 386–87, 736, 790
    UN and 115
Ecuador 60, 61:
    drugs 252
    military expenditure 272, 278, 284
EDA 329
EDS 346, 358, 359
EG&G 361
Egypt:
    arms imports 377
    military expenditure 247, 275, 281, 287
    peace mission in 138
Einhorn, Robert 218
El Salvador: military expenditure 272, 278, 283, 307
Elbit Systems 359
Electro-Optics Industries 644
Elisra 361

Elop Israel 641
Embraer 394
Emergency Action Messages 538, 539
Enduring Freedom, Operation 4, 126, 253
Engels AB 539
Engineered Support Systems 361
Enmod (Environmental Modification) Convention (1977) 772
Equatorial Guinea: military expenditure 270, 276, 282
Eritrea 63, 65, 68, 387:
    military expenditure 237, 238, 270, 276, 282
Estonia:
    military expenditure 274, 280, 286, 307
    NATO and 9
    OSCE in 134
Estrada, Joseph 48
ET Marinesysteme 334
Ethiopia 37, 63, 65, 68, 387:
    chemical weapons in 696
    military expenditure 237, 238, 270, 276, 282
EU (European Union):
    arms control and 483–84, 580
    arms embargoes 211–12, 387
    arms exports 372, 379, 400
    arms production 379
    biological weapons and 668, 676–77
    Charter of Fundamental Rights of the European Union 210
    Code of Conduct for Arms Exports 372, 400
    coercive instruments 115
    Commission 106, 107, 113, 211, 333
    Commission Communication on Conflict Prevention 100, 103–104, 105–7, 112
    Common Foreign and Security Policy (CFSP) 9–10, 13, 106 see also under EU (European Union), military dimension
    Common Position concerning conflict prevention, management and resolution in Africa 112
    conflict prevention and 97, 99–100, 103–107, 112–16
    Council of Europe and 106
    Council Policy Unit (CPU) 105, 106
    Country Strategy Papers (CSPs) 105, 106
    development aid 105, 106, 112, 115, 117
    ECOWAS and 113
    enlargement 2, 790

European Council 100, 107, 122, 157, 790
European Political Cooperation (EPC) 152, 210, 226
external assistance programmes 185
Inter-service Quality Support Group 106
military expenditure 267
missile proliferation and 749, 750, 790, 791
NATO and 8, 789
OSCE and 104, 106
Parliament 107
peace missions 137
Programme for the Prevention of Violent Conflicts 100, 103–5
reform 2, 9–10
Russia and 165, 794
sanctions 13, 203, 210–14, 226–27, 228
terrorism and 9–10, 122, 226–27, 227, 793
UN and 104, 106, 112, 113:
UN conference on illicit trade in small arms and 737–38
See also following entry; for relations with individual countries and organizations see under names of
EU (European Union), military dimension 12:
arms procurement 173
C^3I 153, 160, 161
Capabilities Commitment Conference 153, 160, 165
Capabilities Improvement Conference (CIC) 156, 158, 165, 256–57
Capability Action Plan (ECAP) 160, 162–63, 257
Combined Joint Task Force 152
Common Foreign and Security Policy (CFSP) 151, 152, 153, 154, 157, 162, 163, 173, 185, 203, 227, 228
crisis management 151, 155, 156, 158, 159, 164, 166, 170, 203, 798
Defence Ministers Council 158
ERRF 154, 155, 157, 158, 159, 166, 168, 169, 170, 172, 255, 256–57
ESDP 10, 12, 151, 153, 154, 155, 156–60, 163, 166, 168, 170, 171, 172, 173, 244, 256, 332, 484, 798
EUMC 157, 158, 159, 162, 164, 785
EUMS 157, 158–59, 165, 168, 171, 785
exercises 159–60, 164
financial hurdle 170–71
force capabilities 161–62
Force Catalogue 154, 162
forces interoperability 170
Headline Goal 151, 152, 156, 160, 162, 164, 172, 256

Headline Task Force 163
Helsinki Council meeting 151, 153, 156, 256
High Representative for the CFSP 153, 157, 159, 163, 164
history of 152–55
international organizations, cooperation with 166
ISTAR 160, 161
Laeken Council meeting 164, 165, 170, 257, 798
military capabilities 160–63
NATO and 156, 157, 159–60, 162, 163–64, 165, 167, 167–70, 172
Nice Council meeting 154, 156, 165, 168
non-EU countries and 156, 162, 164, 164–66, 167
operational capability 155–66
Petersberg tasks 151, 152, 153, 155, 161, 163, 172
Political and Security Committee (PSC) 157, 158, 159, 163, 164, 165, 168, 785
rapid reaction capability 153, 154
situation centre 157, 163
strategic capabilities 161–62, 170
terrorism and 163, 793
UN and 166
USA:
 relationship with 166–67
 reliance on 152
 wants Europe to do more 152
WEU and 124, 153
see also European Community
Eurimage 644
Euro-Atlantic Partnership Council (EAPC) 10
Eurocopter 340
Eurocopter Group *see* EADS
Eurofighter 399
Eurofighter International 393
Eurofighter Simulation Systems 334
EuroMEADS 648
Europe:
 ABM Treaty and 510–11
 arms control and 484, 711–19
 arms production 13, 332, 336, 353, 788:
  integration 332, 333–35, 353
 conflict in 52–55, 65, 66, 74
 CSBMs 719–27:
  new bilateral 725
 military expenditure 171, 273–75, 279–81, 285–86, 292–93
 see also following entries

Europe, Western:
  arms imports 376–77, 395
  arms production 325
  arms exports 372
  interoperability 253, 255
  military expenditure 232, 253–59, 264, 267, 307, 324
  military reform 253
  military weaknesses 253
  standardization 253
  terrorism and 253
  see also EU (European Union)
European Aeronautic Defence and Space Co. 648
European Community 152, 210, 211, 212
European Defence Industries Group 395
European Military Aircraft Co. 333–35
European Technology Acquisition Programme (ETAP) 335, 399
European Union *see under* EU (European Union)
Exocet missile 382
export controls 15, 743–58:
  terrorism and 743, 756–58
ExxonMobil 47

F-4 aircraft 565
F-5 aircraft 248
F-15 aircraft 248, 322, 393, 537, 566
F-16 aircraft 248, 252, 330, 331–32, 340 fn. 50, 374, 379, 397, 537, 563, 566
F-18 aircraft 314
F-22 aircraft 311, 314, 318, 322, 331
F-35 aircraft *see* JSF aircraft
F-117 aircraft 577
F/A-18 aircraft 397
F/A-18E/F aircraft 331
Falkland Islands 210
Fangataufa 549
Fateh-110 missile 789
Ferrostaal 340
FIAT 341, 359, 361
Fiji: military expenditure 273, 279, 285, 307
Financial Action Task Force on Money Laundering 227
Fincantiera 341
Fincantieri Gruppo 361
Finland 154, 340:
  arms exports and 401
  arms production 393
  military expenditure 274, 280, 286
  Open Skies Treaty and 727
Finmeccanica 333, 335, 341, 357

fissile material:
  production of 586:
    ending 476, 589
  weapon dismantlement and 584
FLIR Systems 329
Florence Agreement (1996) 472, 718, 719, 781
FLTSATCOM (Fleet Satellite Communication) system 618
Fowler Report 208 fn. 14
Framework Agreement Concerning Measures to Facilitate the Restructuring and Operation of the European Defence Industry (1998) 332, 379
France:
  ABM Treaty and 510
  arms exports 376, 378, 379, 381, 382, 391, 392, 393, 394, 401, 559
  arms industry 13–14
  arms production 324, 333, 335, 341, 371
  chemical weapons 691–94
  ERRF and 154, 155
  EU military capacity and 167, 171
  highly enriched uranium production stopped 549
  Marcoule reprocessing plant 549
  military expenditure 13, 233, 235, 253, 274, 280, 286, 293, 295, 548
  NATO and 154, 167
  nuclear forces 548–52
  nuclear modernization 526
  nuclear smuggling 605
  nuclear test facilities dismantled 549
  nuclear weapons 525:
    numbers of 526, 548, 549, 551–52
  Pierrelatte enrichment plant 549
  plutonium production stopped 549
  terrorism and 244
  Zimbabwe and 118, 119
Future Imagery Architecture (FIA) Program 625–26
Future Strategic Tanker Aircraft 258–59

G8 *see* Group of Eight (G8)
Gabon 29:
  military expenditure 270, 276, 282
Gaidar, Yegor 348
Gambia: military expenditure 270, 276, 282
Ganesh, R. N. 561
GEC 336
Geller, Dan 91
Gencorp 359
General Dynamics 328, 329, 336, 353, 357

General Electric 358
General Motors 359
genetic engineering 677, 679
Geneva Conventions (1949) 763–64:
   Protocols to 773
Geneva Protocol (1925) 762
Genocide Convention (1948) 763
Geo-Synchronous Satellite Launch Vehicle 561
Georgia:
   Afghanistan war and 241
   CFE and 713
   Chechens in 52, 55
   CIS in 137
   military expenditure 274, 280, 286, 307
   OSCE in 134
   peace missions in 137
   Russian bases in 137:
      withdrawal from 715–17, 790
   uranium missing 604
   see also Abkhazia
Georgievski, Ljubco 792
Geosynchronous Lightweight Technology Experiment (GeoLITE) 619
Germany:
   A400M aircraft 257–58
   ABM Treaty and 510
   arms exports 376, 378, 379, 381, 391, 392, 401
   arms production 324, 325, 333, 335, 340
   European security capability and 155, 171
   military expenditure 233, 235, 243, 253, 257, 274, 280, 286, 293
   military outsourcing 344
   plutonium smuggling 605
   sanctions reform and 210
Ghana: military expenditure 270, 276, 282
Ghauri missile 563
GIAT Industries 341, 359
Gibler, Doug 91
GKN 329, 357
glanders 681
Global Broadcast Service (GBS) 616, 636
Global Event-Data System 95
'Global Guardian' nuclear exercises 538
globalization:
   Americanization and 3
   decision making and 6–7
   democracy and 18
   governance of 18
   influence on all spheres 3, 7
   IT and 6
   poor states and 6
   terrorism and 3
   war and 1
GLONASS satellite system 628
GM Canada 359
GMP 644
Gochman, Chuck 91
Goertz, Gary 92
Golan Heights 567
Goldstone, Jack A. 93
Gorbachov, President Mikhail 572
Gore, Al 477
Gottemoeller, Rose 476
GPS (Global Positioning System) 620–21, 623, 636
Greece 154, 171, 340:
   Araxos AB 575
   arms production 340
   EU and NATO 165, 170, 172
   military expenditure 238, 274, 280, 286, 293
   Turkey and 168, 170
Green Pine radar 383, 649
Group of Eight (G8) 8, 98, 504, 791
Guatemala: military expenditure 272, 278, 284
guerrillas 21
Guinea:
   arms imports 384–85
   conflicts and 24, 34, 35, 36, 61, 108, 385
   ECOMOG and 111
   EU and 113
   Liberia and 35, 108, 114
   military expenditure 245, 271, 277, 282
   refugees 113
   Sierra Leone and 108
Guinea-Bissau 65, 108, 111:
   military expenditure 271, 277, 283
Gulf War *see* Persian Gulf War
Gurr, Ted Robert 93
Guyana: military expenditure 272, 278, 284

H-IIA booster 640
Harff, Barbara 93
Harpoon missile 382
Harrier aircraft 397
Harris 359
Hazelnut Tree launch control centre 649
HDW (Babcock Borsig) 360
HDW/Ferrostaal 329
Helgeson, Vidor 52
Helios satellites 638
Hellenic Aerospace Industries 340
Hellenic Shipyards 340

Helsinki summit meeting 579
Henry Dunant Centre for Humanitarian
 Dialogue 47
Hensel, Paul 91, 92
HERITAGE sensors 653
HEU Agreement, US–Russian (1993) 591
Hindustan Aeronautics 359, 558–59
Hitachi Software Engineering Co. 644
Honduras: military expenditure 272, 278, 284
Honest John missile 570
Honeywell International 358
Hoon, Geoff 125, 163, 547
Howaldtswerke Deutsche Werft 340
Hu Xiaodi 646
Hughes Global Services 635
Hughes Space and Communications Co. 635
human rights: respect for 7, 13, 17
Human Rights Watch 118
humanity, crimes against 17, 790
Hungary:
 military expenditure 274, 280, 286, 293
 security sector reform 198
Hunting 359

IAEA (International Atomic Energy Agency) 218:
 funding problems 608
 Illicit Trafficking Database 601–605
 International Basic Safety Standards for Protection against Ionizing Radiation and for the Safety of Radioactive Sources 611
 nuclear security and 598, 599, 600, 606, 608–10, 611–12
IAIA 349
Iceland: military expenditure 274, 280, 286
Il-78 aircraft 382
ImageSat International 644–45
INCOM 644
Incombank 348
India:
 Afghanistan and 241
 air defence 383
 Arab states and 382
 arms control and 481
 arms embargo on 375
 arms imports 14, 349, 375–78, 379, 380–83, 402
 arms production 372
 biological weapons and 704
 chemical weapons 666, 704
 China, nuclear imports from 753
 conflict in 21, 23, 44–46, 63, 65–66, 73, 85
 fissile material production 557
 Indira Gandhi Centre for Atomic Research 561
 Magazon Dockyard 561
 military expenditure 233, 234, 235, 249, 273, 279, 285, 307
 missiles 382, 383, 785, 789
 NPT and 475
 nuclear forces 557–61
 nuclear no-first-use policy 558
 nuclear policy 558
 nuclear weapons 380, 382, 481:
  numbers of 525, 526, 557
  tests 557–58
 Pakistan and 21, 23, 44–46, 63, 65–66, 249, 381, 558, 798
 Parliament attacked 381, 798
 Russia, nuclear cooperation with 752–55
 sanctions on 481, 793
 space programme 561
 terrorism and 23, 558
 West and 10
Indonesia:
 Aceh 46, 47, 48, 786
 casualties 47
 ceasefire 47
 conflict in 22, 46–48, 73, 786
 gas 47, 48
 Irian-Jaya 46, 48
 Java 47, 48
 military expenditure 273, 279, 285
 Moluccas 46
 sanctions against 210
 security forces 190–91
 terrorism and 242
INF Treaty (1987) 528, 571 fn. 18, 586, 776, 788
Inhumane Weapons Convention (1981) 473, 709, 732–34, 735, 774–75, 798
Inter-Governmental Authority on Development (IGAD) 37
Internal Wars and Failures of Governance: State Failure Data Set 93
International Action Network on Small Arms (IANSA) 738–39
International Campaign to Ban Landmines (ICBL) 23
International Code of Conduct against Ballistic Missile Proliferation (ICOC) 477, 748–51, 791, 794
International Committee of the Red Cross (ICRC) 732
International Criminal Tribunal for the Former Yugoslavia (ICTY) 787, 790, 791

INDEX   825

International Crisis Behavior Project 90–91
International Data Centre 523
international financial institutions (IFIs) 177, 197
International Monetary Fund (IMF) 106:
  conditionality 191, 197
  security sector reform and 177, 178
International Monitoring System 523
International Narcotics Control (INC) 252
International Security Assistance Force (ISAF) *see under* Afghanistan
Internet 86
INTERSAT 644
IR 361
IRA (Irish Republican Army) 790, 795
Iran:
  Afghanistan and 241, 793, 794
  arms imports 377
  arms production 247
  as 'backlash state' 479
  biological weapons and 667, 668, 669, 674, 675
  conflict in 74
  IDC and 523
  military expenditure 233, 247, 275, 281, 287
  missiles and 482, 745, 789
  nuclear weapons alleged 565
  Russian nuclear technology and 521, 786
  Saudi Arabia 787
  USA and 41, 704
Iraq:
  arms control and 480
  arms embargo 215, 218–19
  as 'backlash state' 479
  biological weapons and 674, 675, 680, 696, 702, 704–5
  chemical weapons and 704–705
  conflict in 74
  distrust of 219
  humanitarian imports 215
  Kuwait and 63
  military expenditure 275, 281, 287
  NBC programmes 480
  oil for food programme 215
  sanctions on 203, 209, 215–21, 705, 788
  UK and 217, 219
  UK/US attack on 786
  UN and 215–21
  USA and 23, 209, 217–18, 219, 480, 704, 786, 796
Ireland 171:
  military expenditure 274, 280, 286

Iridium satellites 620
Ishakawajima-Harima 359
Iskander-E missile 746
Islamic fundamentalism 5
Islamic world:
  autocracy in, challenges to 11
  civil war in 5, 11
  conflict in 11
  terrorism and 2, 5–6
  West and 10–11
Israel:
  arms embargo 567
  arms exports 377, 381, 383
  arms imports 377
  casualties 57
  ceasefire agreements 56
  criticisms of 56
  Dimona facility 565
  Eilabun facility 567
  Gaza Strip 56, 57, 787, 788, 795, 797, 798
  Hamas 797, 798
  Interim Agreement on the West Bank and the Gaza Strip (1995) 55–56
  Jewish settlements 56
  military expenditure 235, 238, 247, 275, 281, 287
  Mitchell Report 56, 794
  NPT and 475
  nuclear forces 565–67
  nuclear weapons 485, 525, 565–67
  numbers of 526, 565
  Palestinian areas occupied 792
  Palestinian Authority 55, 56, 57, 58, 797, 798
  Palestinians, conflict with 22, 55–58, 62, 74, 247, 788, 794, 795, 797, 798
  Palmikhim AFB 565
  peace mission in 139
  plutonium separation 565
  Sdot Micha AFB 565, 566
  Sharm el-Sheikh Fact-Finding Committee 56
  Sharon elected prime minister 785
  Soreq Center 565
  suicide bombings in 792
  Tel Nof AB 565, 566
  terrorism 56, 57
  USA and 57–58, 247, 787
  West Bank 56, 57, 787, 788, 795, 797, 798
  Yodefat Rafael 565
Israel Aircraft Industries (IAI) 358, 641, 644
Israel Military Industries 359
Issue Correlates of War 92

IST Group 349
IT (information technology) 4, 6, 316, 321, 328, 345, 353
Italy:
 arms exports 376, 377, 379
 arms production 324, 325, 333, 335, 341
 chemical weapons 696
 EU military capacity and 171
 military expenditure 233, 235, 253, 274, 280, 286, 293
 nuclear smuggling 603
ITT Industries 358
Ivanov, Igor 503, 504, 572
Ivanov, Sergey 503, 504, 507, 787
IVOCO (FIAT) 361
IZAR 360

Jacomet, Jurg 706
Jaguar aircraft 558, 559
Jammu and Kashmir *see* Kashmir
Japan:
 Afghanistan and 240
 arms production 372
 chemical weapons 694–95
 military expenditure 13, 233, 234, 235, 249–50, 273, 279, 285, 307
 sarin attack in 702
 terrorism and 240
Japan Electronic Computer 361
Japan Space Imaging Corporation 643
JAS-39 Gripen aircraft 325, 393, 394
JAST (Joint Advanced Strike Technology) 395–96
Jenkins, J. Craig 94
Jericho missile 565, 566–67
Joint Convention on the Safety of Spent Fuel Management and on the Safety of Radioactive Waste Management (1997) 611
Joint Statement on nuclear reductions, US–Russian 504
Joint Statement on Parameters on Future Reductions in Nuclear Forces (1997) 782
Joint Strike Fighter *see* JSF
Jordan:
 arms imports 377
 military expenditure 238, 275, 281, 287, 307
JSF aircraft 13, 311, 314, 330, 331, 335, 336, 395–400, 402, 537
Julang SLBM 555
Jumpseat satellites 624

Kabbah, President Ahmed Tejan 34, 128
Kabila, President Joseph 29, 32, 33, 128, 785, 792
Kabila, President Laurent 29, 785
Kadish, Ronald 497, 499
Kadodkar, Anil 752
Kajfe, Franjo 706
Kamilov, Abdulaziz 789
Kansas Events Data System 96
Karzai, Hamid 39, 43, 797
Kashmir 23, 44–46, 249, 381:
 casualties in 46
Kashol Group 349
Kasyanov, Mikhail 790
Kawasaki Heavy Industries 358
Kazakhstan:
 Afghanistan and 798
 conflicts 249
 military doctrine 249
 military expenditure 248, 249, 272, 278, 284
 military reform 248
 sanctions against 210
 significance increased 2
 USSR's nuclear weapons in 512, 797
Kazan Gorbunov plant 540
Kelly, Paul 731
Kenya 37, 61, 67:
 military expenditure 271, 277, 283
KH-11 KENNAN satellites 625
Khamenei, Ayatollah 794
Khatami, President Mohammad 786
Kim Dae-Jung 729
Kim Il-chol 787
Kimberley Process 797
Kim Jong Il 729, 791
Kiriyenko, Sergey 690
KnAAPO 352
Komatsu 360
Koor Industries 361
Korea, North:
 arms control and 480
 as 'backlash state' 479
 biological weapons and 674
 Korea, South and 786
 military expenditure 264, 273, 279, 285, 307
 missiles and 381, 482, 745, 785, 791
 Neutral Nations Supervisory Commission 138
 OSCE CSBM seminar and 729–30
 Russia and 787, 791
 USA and 476–77, 704, 789

Korea, South:
  arms exports 393
  arms imports 379, 393
  arms production 372
  chemical weapons 666
  Korea, North and 786
  military build-up 250
  military expenditure 235, 249, 264, 273, 279, 285, 307
  MTCR and 745
  Neutral Nations Supervisory Commission 138
  OSCE CSBM seminar and 729–30
Korean War 311
Kosovo 3, 66, 117, 135, 136, 152, 196, 213, 214, 604, 786, 788
Kostunica, President Vojislav 214
Krasnoe Sormovo shipbuilding plant 348
Krauss-Maffei Wegmann 359
Krstic, Radislav 791
Krupp 325
Kumaratunga, President Chandrika Bandaranaike 51, 52
Kuntsevich, Anatoly 690
Kursk: sinking of 543
Kuwait 41, 63, 247:
  arms imports 247, 377
  military expenditure 238, 275, 281, 287
Kwajalein Atoll 500
Kyrgyzstan:
  Afghanistan war and 793
  military expenditure 249, 272, 278, 284
  significance increased 2

L-3 Communications 329, 346, 358
L-159 aircraft 339, 393
Laden, Usama bin 41, 42, 43, 55, 222, 387, 600, 757, 793, 795
Lagos, President Ricardo 252
Lake, Anthony 479
landmines 730–31, 733 *see also* Anti-Personnel Mines Convention
Laos: military expenditure 273, 279, 285
laser weapons 626, 652
Latin America:
  conflicts in 192–93
  drugs 193
  military expenditure 233, 306–307
  security sector reform 192–94
  USA and 193–94
Latvia:
  military expenditure 274, 280, 286, 307
  NATO and 9
  OSCE in 134
law: international order and 6, 7
Lebanon: military expenditure 275, 281, 287, 307
Leng, Russell 91
Lesotho: military expenditure 271, 277, 283
Lexis-Nexis Academic Universe 86
Liberia:
  arms embargo on 35, 109–10, 226, 389, 390, 786
  arms imports 385, 387
  conflict and 34, 35, 61, 65, 107, 108, 114, 385
  development aid to 113
  diamonds 35, 109, 786
  ECOMOG and 111
  EU and 113
  Guinea and 35, 108, 114
  military expenditure 271, 276, 277, 282, 283
  refugees 113, 114
  sanctions against 109–10, 114, 115, 208, 226
  Sierra Leone and 21, 24, 34, 35, 107–108, 109, 114, 786
  UN and 208
Libya:
  arms control and 480
  as 'backlash state' 479
  biological weapons and 674, 675, 704
  chemical weapons and 704, 706
  military expenditure 270
  sanctions against 210, 225
Liebenberg, Kat 707
Lincoln Experimental Satellites 619–20
lithium-6 562
Lithuania:
  Belarus and 725
  military expenditure 274, 280, 286, 307
  NATO and 9
  Russia and 725, 785
Litton 357
Ljubljana summit meeting 8
LM-2C rocket 641
Lockheed Martin 311, 328, 329, 330–31, 336, 344, 353, 357, 396, 535, 548, 617, 642, 648, 649, 652
Long March booster 636
Loral Space and Communications 637
Louw, Eugene 706
Luxembourg 154, 171:
  military expenditure 274, 280, 293

M4 SLBM 551–52
M5 SLBM 551
M45 SLBM 551
M51 SLBM 548, 551
M-9 missile 564
M-11 missile 564
Maastricht Treaty (Treaty on European Union, 1992) 152, 211
Macedonia (FYROM):
  Albanian rebels 127, 786, 787, 790, 792, 794
  Albanian rights 796
  ceasefire 792
  EU and 164, 166
  KFOR and 788
  military expenditure 274, 280, 286, 307
  NATO and 124, 127, 164, 794
  OSCE in 127, 134
  peace missions in 124, 127, 136
  TFF 127, 136, 166
  TFH 127, 136
Madagascar: military expenditure 271, 277, 283
Major Episodes of Political Violence 85, 90
Malawi: military expenditure 271, 277, 283
Malaysia: military expenditure 273, 279, 285
Mali 111:
  military expenditure 271, 277, 283
Malta: military expenditure 274, 280, 286
Mandela, President Nelson 27, 28
Mano River Union (MRU) 108, 110, 111, 112, 114, 115
Maoz, Zeev 91
Marconi 359
Marconi Mobile 360
Marshall, Monty G. 90, 93
Marulanda Vélez, Manuel 60, 785
Masire, President Ketumile 30, 32
Massoud, Ahmad Shah 39
Matra BAe Dynamics 333
Matra Marconi Space 636, 641
Mauritania: military expenditure 271, 277, 283
Mauritius: military expenditure 271, 277, 283, 306
MBDA 333, 334
Medchem Forschungs 706
Médecins sans Frontières 23
Medium Extended Air Defense System (MEADS) 495, 648
Mercury spacecraft 624
Mexico:
  drugs 252
  military expenditure 272, 278, 284, 307
MICAH (International Civilian Support Mission in Haiti) 133
Middle East:
  arms control 484
  arms imports 377
  conflicts in 55–58, 66, 74
  military expenditure 13, 231, 232, 234, 238–39, 246–48, 264, 267, 275, 281, 287, 294, 307
  USA and 67
MiG-27 aircraft 558–59
Mikhailov, Vladimir 540
Mil helicopter plant 348
military expenditure:
  accountability 304
  changes, 1992–2001 234
  concentration of 231, 233
  CSBMs 302, 303, 304
  data on 302–308
  definition 298, 302
  determinants of 264
  economic burden 234–39, 302
  economic constraints 264
  GDP and 231, 234
  global pattern 233–34, 237
  GNI and 236, 237
  official data on 302–308
  per capita 231, 234
  private financing 258
  regional 233
  regional trends 245–63
  reporting of 302–306
  security sector reform and 197
  SIPRI and 232, 302, 305, 306–7
  SIPRI's sources and methods 297–301
  tables of 266–87
  terrorist attacks and 231, 232, 239–44
  transparency and 186, 232, 302, 304, 306, 308
  trends 231, 232–39, 245–63, 264
  world 13, 231, 232
  see also under names of countries
military functions: outsourcing 343, 346
military services: outsourcing 342–46
military technology 323, 392–93
Milosevic, President Slobodan 203 fn. 1, 213, 214, 787, 790
MILSATCOM (Military Satellite Communications) 618
MILSATCOM satellites 538
Milstar satellites 538, 616, 617
mines 733 *see also* landmines

INDEX 829

Minorities at Risk 93
MINUGUA (UN Verification Mission in Guatemala) 133
MINURSO (UN Mission for the Referendum in Western Sahara) 131
Minuteman III missile 518, 532–33, 536
Mirage IVP aircraft 549
Mirage V aircraft 563
Mirage 2000 aircraft 340 fn. 50, 382, 549, 550, 559
MIRVs:
  Chinese 555
  Russian 539, 540, 541
Missile Technology Control Regime (MTCR) 473–74, 475, 743, 744, 745–51, 785, 791, 794
missiles, ballistic 470, 477, 481–82
Misuari, Nur 50
Mitchell Report 56, 794
Mitchell, Sara McLaughlin 92
Mitchell, George 788
Mitre 360
Mitsubishi Corporation 640, 643
Mitsubishi Electric 358
Mitsubishi Heavy Industries 357
Mitsui Shipbuilding 360
Mitterrand, President François 549–50
MKEK 361
Moldova:
  arms exports 385
  CFE and 713
  conflict in 717
  military expenditure 274, 280, 286, 307
  OSCE and 134, 718
  peace missions in 137
  Russian withdrawals from 708, 717–18, 796
Molniya satellites 629
money laundering 10, 191, 244
Mongolia: military expenditure 273, 279, 285, 307
Montenegro 214
MONUC (UN Observation Mission in the Democratic Republic of the Congo) 30, 32, 128, 132, 786, 787
Mordechai, Yitzhak 565
Morocco:
  military expenditure 238, 246, 270, 276, 282
  Western Sahara and 246
Moskovskiy, Aleksey 263
Motorola 361

Mozambique 121, 196:
  military expenditure 271, 277, 283
MPRI 346
Mubarak, President Hosni 225
Mugabe, President Robert 116, 117, 119
Munich Conference on Security Policy 4
MUOS (Mobile User Objective System) 618
Mururoa 549, 551
Museveni, President Yoweri 245–46
Musharraf, President General Pervez 41, 45, 46, 562
Mustafa, Ghulam 561
MX/Peacemaker missile 528, 532, 533, 536
Myanmar (Burma) 190:
  Bangladesh and 249
  conflict 73, 249
  military expenditure 273, 285
  sanctions against 210

Nagorno-Karabakh 713
Namibia:
  Angola and 26–27
  DRC and 31
  military expenditure 246, 271, 277, 283
NAPA 352
nationalism, revival of 16
NATO (North Atlantic Treaty Organization):
  Baltic states and 711, 715
  burden sharing 255–56, 303
  command structure 7
  Defence Capabilities Initiative (DCI) 160, 255, 257, 484
  Economics Directorate 303
  enlargement 2, 7, 9, 576, 789
  ESDI 166
  EU and 8, 156, 157, 159–60, 162, 163–64, 165, 167–70, 172, 789
  Flying Training in Canada 344
  inclusive security and 8
  Macedonia and 790, 792
  Military Committee 164
  military expenditure 244, 255–56, 264, 267, 292–95, 324, 325
  military expenditure reporting 303
  North Atlantic Council (NAC) 157, 164, 168
  nuclear weapons mix 570
  nuclear weapons stored 574
  Partnership for Peace (PFP) 160, 185
  peace missions 136
  peacekeeping operations 255
  rapid reaction forces and 8

reform 2, 7–8
Russia and 7, 8–9, 172, 797
satellites 636–37
Study on Enlargement 185
terrorism and 7, 8, 244–45, 255, 793, 794
transatlantic capability gap 253, 255–56
US costs 255–56
NATO–Russian Permanent Joint Council (PJC) 750
Naval Area Theater Ballistic Missile Defense 649–50
Navy Air Defense (NAD) 495, 496
Navy Ionospheric Monitoring System 620
NBC (nuclear, biological and chemical) weapons:
 responses to proliferation of 491
 see also under arms control; Iraq; terrorism
Ndayizeye, Domitien 28
NEC 359
Nepal: military expenditure 249, 273, 279, 285, 307
Netherlands 154, 162, 171:
 arms exports 376, 377
 arms production 324, 372
 military expenditure 274, 280, 286, 293
Network Centric Warfare (NCW) 345
New Partnership for Africa's Development (NEPAD) 187, 190, 201
New Programs and Concepts 349
New Zealand:
 arms imports 377
 military expenditure 273, 279, 285, 307
Newport News Shipbuilding 328, 357
NGOs (non-governmental organizations) 98, 99, 738–39
NH-90 helicopter 340, 393
Nicaragua: military expenditure 272, 278, 284
Nice Treaty (2000) 153, 154, 173
Niger 111:
 military expenditure 271, 277, 283
Nigeria 111, 115, 233, 792:
 military expenditure 246, 271, 277, 283
Nissan Motor 361
Nixon, President Richard M. 680
No Dong missile 563
Non-Aligned Movement (NAM): biological weapons and 667, 668
Nordic Standard Helicopter Programme (NSHP) 393
North Africa: military expenditure 246, 266, 276, 282
North America: military expenditure 232, 264, 266, 272, 278, 284, 292

Northern Ireland 790, 793, 795
Northrop Grumman 328, 329, 336, 353, 357, 396
Norway 200, 340:
 arms exports 377
 arms production 393
 military expenditure 275, 281, 286, 294
NPO Saturn 352
NPT (Non-Proliferation Treaty, 1968) 473, 475, 578, 768–69:
 Review Conference, 2000 503, 580, 749
 Review and Extension Conference, 1995 580
 tactical nuclear weapons and 580
NSCC 334
nuclear arms control 14, 489, 506, 508, 511, 511–18:
 disagreements about 14
 weapons dismantling 517, 543
 see also nuclear weapons: reductions in *under* Russia; United States of America
nuclear artillery shells 543, 572
nuclear bombers:
 China 553
 France 549–51
 India 558–59
 Israel 565–66
 Pakistan 563
 Russia 539–40
 USA:
  B-1 528, 531
  B-2 530, 530–31, 536
  B-52 530, 531, 536
  numbers of 531
nuclear bombs 528, 530, 531, 535, 537, 543, 546, 547, 570, 574, 580, 581
Nuclear Detonation System 621, 623
nuclear facilities:
 numbers of 598
 physical protection of 608–12
 threat to 598–600, 609
nuclear material:
 illegal trafficking in 601–606, 612:
  destinations 606
 physical protection of 608–12
 theft of 598, 601, 602, 603, 609
Nuclear Material, Convention on the Physical Protection of (1987) 598 fn. 2, 609–10, 773–74
nuclear mines 543, 572
nuclear research reactors 598, 599, 604
nuclear security initiatives, cooperative 518–22

Nuclear Suppliers Group (NSG) 473–74, 475, 743, 744, 751–55
nuclear warhead production complexes:
  downsizing of 586, 587, 591–92, 593, 596, 597:
  cooperation over 591
  purpose of 585
  restructuring of 586, 587, 591–92, 596
  Russia 586, 587–93
  USA 585, 586, 593–96
nuclear warheads 528–29, 531, 533, 535, 545, 546, 551:
  design 585
  design theft allegations 555
  dismantlement 517, 543, 583, 586, 590
  reliability 586
  safety 586
  storage of 592, 596
  transparency and 582–84
  W62 528
nuclear weapons:
  earth-penetrating 529, 530, 577–78
  non-strategic 543
  numbers of 525
  reductions in 483, 489, 504
  reliability testing 529
  see also under names of possessor countries and tactical nuclear weapons
Nuclear Weapons Council 529
nuclear weapons tests: ending of 585
Nye, Joseph 19

OAS (Organization of American States):
  conflict prevention and 99
  military expenditure and 304–305
  security sector reform and 186
OAU (Organization of African Unity):
  conflict prevention and 99
  DRC and 30
  EU and 113
  peace missions 138
  replaced 789
  security sector reform and 186
  sovereignty principle 119
  Zimbabwe and 119
  see also African Union
Oceania:
  arms imports 377
  military expenditure 232, 266, 273, 279, 285, 307
OECD (Organisation for Economic Co-operation and Development) 1, 185:
  military expenditure 267

Oh, Churl 94
oil prices 247
Oman: military expenditure 233, 238, 247, 275, 281, 287
Omar, Mullah Mohammad 39, 42
ONCAP 329
ONEXIMbank 348, 349
ONYX spacecraft 625
OPEC (Organization of the Oil Exporting Countries): military expenditure 267
Open Skies Treaty (1992) 709, 726–27, 777, 787, 796
Orbimage 644
Orbital Sciences Corporation 641, 644
Ordnance Factories 358
Organisation for the Prohibition of Chemical Weapons (OPCW) 666, 683–84, 685, 707
Organisme Conjoint de Coopération en Matière d'Armement (OCCAR) 257
Organization for Security and Co-operation in Europe (OSCE):
  Code of Conduct (1994) 720, 721, 723
  Conflict Prevention Centre 720
  Document on Small Arms and Light Weapons (2000) 719, 721, 736
  Forum for Security Co-operation (FSC) 719, 720–21, 735
  Istanbul Summit (1999) 712, 715, 717, 734
  Korea CSBM seminar 729–30
  military doctrine seminar 709
  military expenditure 267
  military expenditure reporting 303–304
  peace missions 134–36
  Permanent Council 721–22
  security sector reform and 186
  for relations with individual states see under names of states
Organization of the Islamic Conference 795
Organization for the Prohibition of Biological Weapons (OPBW) 669, 670
Orion spacecraft 624
Oscar navigation satellites 620, 622
Oshkosh Truck 361
outer space:
  arms race in 14–15, 510, 645–47, 654
  terrestrial war-fighting and 647, 654
  US dominance in 654
  see also satellites
Outer Space Treaty (1967) 473, 474–75, 765–66

Pacific countries: aid to 185
Pakistan:
    Abdul Qadeer Khan Research Laboratories 562
    Afghanistan and 41, 45, 241, 793, 794
    air defence 383
    arms control and 481
    arms embargo 382, 563
    arms imports 378, 380–83, 402
    CBW and 704
    conflict in 21, 44–46, 63, 65–66, 73
    debts 241
    economy 241
    fissile material produced 562
    India and 21, 23, 44–46, 63, 65–66, 249, 381, 798
    Islam and 5
    Islamic extremists in 562
    Khushab reactor 562
    military expenditure 238, 273, 279, 285
    missile technology and 480
    missiles 381, 382, 383, 747, 749
    NPT and 475
    nuclear forces 562–64
    nuclear weapons 380, 382, 481:
        numbers of 526, 562
    nuclear weapon tests 562
    PINSTECH 562
    plutonium reprocessing 562
    sanctions on 481, 747, 793
    Sargodha AB 563, 564
    significance increased 2, 10
    Taliban and 382
    terrorism and 123, 794
    uranium enrichment 562
    USA and 41, 241:
        aid to 793
Palestine Human Rights Monitoring Group 57
Palestinian Authority (PA) 55, 56, 57, 58
Palmer, Glenn 91
Panama 193:
    chemical weapons 695
    military expenditure 272, 278, 284
Papandreou, George A. 169
Papua New Guinea 139:
    military expenditure 273, 279, 285
Paraguay: military expenditure 272, 278, 284, 304–305
Paris Agreements (1954) 764
Partial Test Ban Treaty (PTBT, 1963) 473, 765
Pastrana, President Andrés 60, 61, 785

Pathfinder satellite 538
Patria Industry 340
Patriot PAC-3 495, 496, 648
peace missions, multilateral 124–39:
    table of 129–39
    reluctance to commit resources 124
Peaceful Nuclear Explosions Treaty (PNET, 1976) 473, 772
Peacemaker missile *see* MX/Peacemaker missile
Pearl Harbor 5
Pelindaba Treaty (1996) 473, 779–81
Peres, Shimon 58, 794
Performance Diesels Co. 334
Perle, Richard 478–79
Pershing missiles 567, 570
Persian Gulf War (1991) 549, 614, 623, 680
Peru:
    conflict in 71
    drugs 252
    military expenditure 272, 278, 284, 304–305, 307
Pfetsch, Frank R. 89
Philippines 796:
    ceasefire agreement 791
    conflict in 22, 48–50, 62, 73
    military expenditure 273, 279, 285, 307
    terrorism and 242
PJ-10 missile 789
Plowshare Program 577
plutonium:
    ageing of 596
    production of 585, 596:
        ending of 587, 589, 593
    theft, diversion of 519
    widespread existence of 608
Poland 210:
    arms exports 401
    arms production 338–39
    military expenditure 275, 281, 287, 294, 307
    security sector reform 198
    Ukraine and 726
Polar Adjunct programme 616
Polar Space Launch Vehicle 561
Pollins, Brian 91
Portugal 154, 171:
    military expenditure 275, 281, 287, 294
post-cold war period: term's inadequacy 1
Powell, Colin 46, 122, 217–18, 315, 483, 517, 789, 797
Prabhakaran, Vellupillai 50, 52
Pratt & Whitney 339, 358

Presidential Nuclear Initiatives (PNIs), US–Russian (1991–92) 543, 571–73, 578, 580, 584
Primex Technologies 360
Priority Acquisition Service 645
Prithvi missile 558, 559
Project Bacchus 678, 681
Project Clear Vision 678–79, 680, 681
Project Jefferson 678, 680, 681
Protocol for the Assessment of Nonviolent Direct Action (PANDA) 94–95
Proton booster 631
Pugwash Conferences of Science and World Affairs 23
Putin, President Vladimir:
   ABM Treaty and 484–85, 490, 504, 507, 508, 509
   arms industry and 262, 349
   Bush, meetings with 8, 503–4, 506, 515, 791, 796
   Chechnya and 714
   foreign policy, domestic motives 509
   military expenditure and 259
   nuclear weapons, reductions in 489, 515–17, 577
   tactical nuclear weapons and 579
PZL Rzeszow 339
PZL Swidnik 339
PZL Warszawa-Okecie 339
PZL-130 Orlik aircraft 339

al-Qaeda 67–68, 72, 187, 222, 600, 675, 701, 702, 708, 797
Qatar: arms imports 377
Quintao, Geraldo 305–306

Rabbani, Burhanuddin 39
Racal 332
radioactive material 601, 610–12 *see also* nuclear material
Rafael 359
Rafale aircraft 393, 548, 549, 551, 552
RAND Corporation 398
Rarotonga Treaty (1985) 472, 775–76
Ray, Jim 91
Raytheon 328, 336, 353, 357, 626, 649
Razak, Abdul 706, 707
Reagan, President Ronald 311, 313, 319, 321, 653
Regan, Patrick 91, 94
Regional Arms Control, Verification and Implementation Assistance Centre 724–25
Regli, Peter 706

Reid, John 793
Republika Srpska 304, 718, 723
Rheinmetall 358
Rheinmetall DeTec 358
Rice, Condoleezza 478
Richard, Alain 244
Rivalry Data Set 92
Robertson, Lord 9, 16 fn. 48, 164, 244
Robust Nuclear Earth Penetrator 529
Rockwell International 359
'rogue states' 479 fn. 49, 490, 511
Rolls Royce 357
Rolls-Royce Snecma 334
Romania:
   arms exports 385
   military expenditure 275, 281, 287, 307
   NATO and 9
Rome, Treaty of (1957) 211
Rosenberg, Barbara Hatch 703
Rotorism 334
RUAG SUISSE 359
Ruggiero, Renato 154 fn. 12
Rumyantsev, Alexander 754
Rumsfeld, Donald 2, 309, 311, 316, 321, 493, 501, 614, 648–50, 795
Russia:
   ABM Treaty and 485, 502–503, 507–508, 513
   Aleysk 540
   Anadyr AB 540
   armaments programme, 2002 263
   armed forces reductions 790
   arms control 484
   arms exports 13, 14, 323, 352, 375, 376, 377, 378, 379, 381, 383, 385, 391, 392, 402
   arms industry 262, 263
   arms procurement 262, 263, 323
   arms production 14, 346–52, 364–65, 375:
     companies list 350, 364–65
     concentration 346, 352
     conversion 346, 587, 589, 591, 592
     industry structure 351–52
     overcapacity 351, 352
     privatization 348–51, 352
   Association of Soldiers' Mothers 54
   Baikonur, launches from 541, 632
   Bershet 541
   biological weapons and 667, 704
   BMD and 502–503
   chemical weapon production facility, use of 685
   chemical weapons and 704

chemical weapons destruction 666, 688–91, 790:
    aid for 691, 692
closed cities 587, 589, 590 fn. 9, 592
conventional forces weakness 576, 579
Dombarovskiy 540
economic problems 592
EU and 165, 794
European theatre defence and 786
financial crisis, 1998 349
Iran, nuclear technology and 521, 786
Kaliningrad 579–80, 725, 785
Kamchatka Peninsula 542
Kartaly 540
Kola Peninsula 542
Kostroma 541
Kozelsk 541
Krasnoyarsk 541
Lesnoy 593
Lyulka-Saturn design bureau 349
Mayak storage site 584
military expenditure 13, 233, 234, 235, 259–63, 264, 275, 281, 287, 307, 308
military expenditure on R&D 263, 375
military industrial reorganization 375
military technology cooperation 746
Minatom 587, 589–92
missile development with India 789
missile proliferation and 748, 749–50
Molnia plant 589, 590
Moscow 591
MTCR and 745–46
National Security Council 262
NATO and 8–9
Novouralsk 589
nuclear exports 752–55
nuclear forces 514, 539–45:
    affording 516
    modernizing 525
nuclear imports 789
nuclear material in 519
nuclear materials, security of 599, 602, 607
nuclear scientists, occupying 519, 521, 591
nuclear security assistance to 519–22, 608, 612
nuclear security in 607–8, 612
nuclear smuggling 603, 604–606
nuclear weapons:
    numbers of 525, 526, 539, 543, 544, 586
    reductions in 489, 504, 508, 511–18, 791, 797
    safety and security 575–76
peace missions 137

Plesetsk 541
Rosoboronexport 375
Sarov 590, 591, 593
security policy 8
security sector reform 192
Sever 590
Seversk 591
Sheremetyevo airport 604
Snezhinsk 591, 593
Southern Tier and 605–606, 612
State Armaments Programme 262
strategic forces, decline in 503
strategic stability talks 595
Sukhoi design bureau 348, 352
Tatishchevo 541
terrorism and 2, 8, 794
Trekhgorny 590
Ukraine AB 539
Urals Electromechanical Plant 590
USA and 8–9, 477, 715
Uzhur 540
Western security structures and 8–9
Zarechny 590, 591
Zelenogorsk 589
Zheleznogorsk 589
see also Chechnya, conflict in
Russian Avionics 349
Rwanda:
    arms embargo 208
    conflict in 24, 70
    DRC and 30, 31, 32, 33
    military expenditure 197, 237, 238, 245, 271, 277, 283
    UN and 99, 208
Ryazan AB 539
Rybinsk Motors 349

S-300/SA-10 SAM system 383
S-300V/SA-12 383
S-400 missile system 383
Saab 339, 358, 394
Sagarika SLBM 561
SAGEM Groupe 359
Saint Malo Declaration (1998) 152–53, 154, 166
SALT treaties (1972, 1979) 528
sanctions:
    arms embargoes 205–6
    declaratory element 205
    definition 204
    effectiveness 227–38
    human rights and 210–11
    individuals and 205

non-state entities and 203, 205
objective 205
terrorism and 225–27, 228
types of 205
wider strategy and 228
see also under EU (European Union); United Nations; *and under names of countries concerned*
Sankoh, Foday 35, 36
Santa Barbara 336
sarin 702
Satellite Operating Partners (SOP) 645
satellites:
general references
commercial 616, 642–45
communication through 538
integration with military forces 614
interceptor system 494
lifetime 614
numbers 613
Persian Gulf War and 614, 623
individual countries and organizations
Australia 635, 664
Belgium 638
Canada 642
China 635–36, 664
Europe 636–38, 664
France 637, 638, 642
Germany 637, 638
India 638–39, 642, 664
Israel 639, 664
Italy 637, 638
Japan 640:
  navigation 627–28, 660–61
  ocean-surveillance 632–33, 663
  weather 630
Korea, South 641, 643
NATO 636–37
Russia:
  anti-satellite systems 634
  communications 628–30, 661–62
  early-warning 630–31, 662–63
  economic problems and 628
  electronic intelligence 631–32, 663
  imagery intelligence 633
Spain 637, 638
Taiwan 641, 643
Turkey 641–42
UK 637
USA:
  anti-satellite systems 626
  commercial 642, 664
  communications 615–20, 657–58

dominance and 614–15, 654
early-warning 622–23, 658
electronic intelligence 658–59
imagery intelligence 625–26, 659
military operations and 615, 625
navigation 620–21, 622, 656–57
number of 654
ocean-surveillance 623–24, 659
signals intelligence 624–25
technology development 626–27
weather 621–22, 658
USSR 627, 628–29, 630, 631, 632, 633, 634
Saudi Arabia 67:
  Afghanistan and 41, 793
  arms imports 14, 374, 377, 379, 402
  Iran and 787
  military expenditure 233, 234, 235, 238, 247–48, 275, 281, 287
  oil revenue 248
Savimbi, Jonas 26
SBIRS programme 653, 654
Scharping, Rudolph 155, 258, 484
Schröder, Joseph W. 167
Schrodt, Philip A. 96
Science Applications International 345, 357
Scientific Production Center Tekhnokompleks 351–52
Scranton, USS 538
Scud missile 563, 623
SDS (Satellite Data System) 619
Seabed Treaty (1971) 473, 769–70
security:
  civil–military relations 178
  democratic control of 12, 177, 185, 186
  domestic and external 6
  freedom and 7
  militarization of 177
  non-military 6
  rethinking 178
security assistance:
  civil–military relations 177
  democracy and 177
  development aid and 176
  rethinking of 177
  security sector reform and 180, 182, 199–200
  shortcomings 177
security policies:
  reform of 1, 6–11, 12
  terrorism and 6
security sector:
  accountability 180

civil society and 180
defining 178–80
elements of 179
private actors 179–80, 180
security sector reform:
  approaches to 182
  assistance for 184
  challenges for 194–200
  civil oversight of 180, 181, 182, 185
  confidence- and security-building and 186
  conflict and 196
  conflict prevention and 180, 183
  counter-terrorism and 187
  defining 180–81
  definition 176
  democratization 180, 182
  development aid and 175, 177, 184–85
  development and 180
  donor countries and 184
  expense of 182
  governance and 183
  human rights and 180, 181
  human rights training and 195
  IFIs and 177
  instruments of 183–87
  integrated programmes 198–200
  international aid 195, 196
  international security and 175, 183–88
  military expenditure and 197
  objectives, conflicting 197
  operationalizing concepts 195
  policy agenda 177–82
  receptivity to varies 194
  regional approaches 181
  regional contexts 188–94
  regional organizations 185–87
  state weakness and 200
  support for 182
  support for weak 194, 195
  terrorism and 183, 187–88, 201
  terrorist attacks (11 September 2001) and 187–88, 201
  transparency 180, 181, 186
  unfavourable environments 196–97
security system, global 3, 9, 17
Senegal 108, 111, 787:
  military expenditure 246, 271, 277, 283
SEPI 360
Serbia:
  EU and 164
  NATO and 164
Serbia and Montenegro *see* Yugoslavia, Federal Republic of (FRY)

Serco 344, 346, 548
Sergeyev, Igor 786
Severnaya shipyard 348
Seychelles: military expenditure 271, 277, 283
Shaheen missile 564
Shalikashvili, John 315
Shanghai Cooperation Organization 790 *see also following entry*
Shanghai Forum 728–29, 788, 790
Sharon, Ariel 55–56, 57, 58, 785
Shavit (Comet) booster 639
Shemya Island 651
Shin Maywa 361
Shrharikota High Altitude Range Centre 561
Sicral 637
Sierra Leone:
  Abuja Agreement (2000) 68
  casualties in 36
  ceasefire agreements, 2001 35–36, 68, 788
  conflicts in 21, 22, 24, 34–36, 65, 107–108, 245, 385
  diamonds 21, 35, 36, 109, 114
  ECOMOG and 111
  elections 34, 36
  EU and 112–13
  Guinea and 35, 36
  Liberia and 21, 34, 107–108, 109, 114, 786
  Lomé Peace Agreement (1999) 34, 35
  military expenditure 245, 271, 277, 283
  peace accord 24
  peace missions in 34, 35, 36, 110, 111, 114, 128
  rebels 24
  refugees 108, 113
  RUF 34, 35, 36, 108, 109, 128, 384, 385, 389, 390, 788:
    arms embargo 384, 385–86
    sanctions 206, 208
  security sector reform 189–90, 196, 198–99
  UK and 34, 35, 189, 198–99, 385
  UN and 34, 35, 36, 68, 109, 110, 111, 114, 128, 132, 199, 208, 245
Silicon Graphics 360
Sinclair Knight Merz 644
Singapore: military expenditure 238, 273, 279, 285
Singapore Aerospace 361
Singapore Technologies 359
Singer, J. David 81, 88
Single Integrated Operational Plan (SIOP) 530, 531, 539, 553
'Six-plus-Two' states 223

Skynet Systems 637
Slovak Republic:
  military expenditure 275, 281, 287
  NATO and 9
Slovenia:
  military expenditure 275, 281, 287, 307
  NATO and 9
SMA 341
Small, Melvin 81
Smart Acquisition 258
Smiths Industries 359
Snecma 341, 358
Sneh, Ephraim 567
Sokol 349
Solana, Javier 9, 105, 153, 157, 163, 170, 171, 203 fn. 2
Solheim, Erik 51
Sollenberg, Margareta 89
Solomon Islands 139
Somalia:
  conflicts in 36–37, 61, 65, 70
  military expenditure 271, 277, 283
SOSTAR 334
South Africa:
  arms imports 338
  arms production 332, 338, 393
  Burundi and 24, 124, 129
  CBW programme 705–707
  land problems 121
  military expenditure 246, 271, 277, 283, 306
  sanctions on 210
  Zimbabwe and 120, 121
South America:
  conflicts in 59–61
  drugs 252
  military expenditure 234, 252, 266, 272, 278, 284, 307
  USA and 252
South Asia: military expenditure 232, 233, 249, 264, 266, 273, 279, 285, 307
South East European Cooperation Process (SEECP) 186
Southern African Development Community (SADC) 117, 119, 120–21, 186
Southern Rhodesia: sanctions on 210
Space Communications Corporation 640
Space Imaging 642–43
Space Imaging Asia 643
Space Imaging Eurasia 643
Space Imaging Europe S.A. 643
space launch vehicle (SLV) 748
Space Shuttle 619

Space-Based Interceptor Experiment programme 652
Space-Based Kinetic Energy Experiment 652–53
Space-Based Kinetic Kill Vehicle 653
Space-Based Laser project 652
Spain 154, 171:
  arms exports 377, 379
  arms production 324, 333, 335, 341, 372
  military expenditure 235, 275, 281, 287, 294
Srebrenica 99
Sri Lanka:
  conflict in 22, 50–52, 62, 74, 85
  Liberation Tigers (LTTE) 50, 51, 52
  military expenditure 273, 279, 285
Srpska, Republika 304
SS-18 ICBM 539, 540–41
SS-19 ICBM 541, 541–42
SS-24 ICBM 541
SS-25 ICBM (Topol) 541
SS-27 ICBM 507, 541
SS-N-18 SLBM 542–43
SS-N-20 SLBM 542, 543
SS-N-23 SLBM 542
SSBNs:
  China 555–56
  France 548, 551–52
  India 561
  Russia 539, 542–43
  UK 545–47
  USA 528, 533–35, 538
SSNs:
  France 552
  USA 537–38
ST Engineering 359
Stability Pact for South Eastern Europe (1999) 214, 304, 724
Standard Missile 650
Stanford University Database on Nuclear Smuggling, Theft and Orphan Radiation Sources (DSTO) 599, 601, 603, 604
START I Treaty (1991) 353, 511–12, 517, 525, 526, 532, 534, 539, 540, 541, 777, 797:
  Lisbon Protocol (1992) 512
START II Treaty (1993) 512–13, 516, 517, 528, 529–30, 540–41, 541, 542, 579, 595, 778
START III treaty 513, 528, 578–79
Start-1 booster 645
states:
  failed 6, 183
  threats from poor 6

stealth technology 399, 531
Stewart & Stevenson 361
Stoll, Ric 91
STRATCOM (US Strategic Command) 528, 529, 530
Straw, Jack 220–21
Su-27 aircraft 351
Su-30MKI aircraft 382, 559
submarines *see* SSNs; SSBNs
Sudan:
   biological weapons and 480, 674, 704
   casualties in 37
   chemical weapons and 704
   conflicts in 37–39, 61, 71
   human rights abuses 38
   military expenditure 38, 245, 271, 277, 283
   oil 38
   refugees 38
   sanctions against 225
   terrorism and 123
Sukarnoputri, Megawati 48
Sukhoi Aviation 3522
Super Étendard aircraft 549, 550, 552
Suriname 85
Surya programme 561
Swaziland: military expenditure 271, 277, 283
Sweden:
   arms exports 377, 379, 400
   arms imports 393
   arms production 325, 335, 336, 339, 340, 393
   conflict prevention and 99, 104, 112, 155
   crisis management and 154
   Europe's military capability 155, 165
   military expenditure 275, 281, 287
   Open Skies Treaty and 727
   Zimbabwe and 118
Switzerland 210:
   arms exports 377
   military expenditure 275, 281, 287
Syria:
   arms control and 480
   biological weapons and 674, 704
   chemical weapons and 704
   military expenditure 238, 247, 275, 281, 287

TACAMO aircraft 538–39
tactical nuclear weapons:
   arms control agreements and 568, 576, 578–79, 582–84
   current status 573–75
   dangers of 575
   definitions 568–70
   history of 570–73
   numbers of 573
   Presidential Nuclear Initiatives and 543, 571–73, 579, 580, 584
   proliferation and 575, 576
   recent developments 576–78
   reductions 580–84
   Russian weakness and 576–77, 581
   safety and security 575–76
   security threats 568
   terrorists and 581–82
   transparency lacking 568
   use of 576
Taepo Dong missile 640
Taiwan:
   arms imports 14, 374, 377, 379–80, 402
   arms production 372
   military expenditure 249, 273, 279, 285, 307
   WTO and 798
Tajikistan:
   Afghanistan and 796, 798
   conflict in 66
   military expenditure and 272, 278, 284
   OSCE in 134
   significance increased 2
   terrorism and 123
Tanzania:
   Burundi rebels and 28, 29
   military expenditure 271, 277, 283
   radar system bought 199 fn. 68
   USA attacked in 67
Task Force Fox (TFF) 127, 136, 166
Task Force Harvest (TFH) 127, 136
Taurus launch vehicle 641, 644
Taylor, Charles Louis 94
Taylor, President Charles 35, 109
Teledyne Technologies 361
Tenix 360
terrorism:
   aim of 5–6
   arms control and 486, 488
   attitudes to 22–23
   authoritarian regimes and 187, 188
   broad policies and 2
   causes 122
   combating 2, 3, 4–5, 12, 16–17, 22–23, 44, 122, 123, 231, 232, 239, 253, 264, 311, 316, 321, 486:
      militarization of 23, 123
   definition 2 fn. 5

drug trafficking 191
financing 10, 12
global nature of 16
global responsibility and 1–4, 16
intelligence and 187, 188
international relationships changed by 2, 12
military capacities, denying 486
military expenditure and 231, 232, 239–44
money laundering 191
NBC weapons and 470, 486–87
nuclear facilities and 598–600, 610, 612
nuclear materials and 521, 600, 609, 612
radioactive material and 611
sanctions against 225–27
scale of 5
sponsoring states 486
weapons, denying 486
see also following entry
terrorist attacks (11 September 2001) 1, 5–6, 8, 14, 15, 16, 22, 39, 167:
 aeronautics market and 341
 arms control and 469, 470, 485–87, 488
 BMD and 491
 conflict prevention and 121–23
 effectiveness of 23
 effects of 22, 67, 172, 239, 311, 317, 322, 540, 686
 export controls and 743
 foreign policy changes and 67
 FSC and 721–22
 intra-state conflict and 22
 Israel and 55
 nuclear facilities and 598–600, 606–12
 Russian nuclear weapons and 521
 sanctions and 203
 security sector reform and 183, 187–88
 Turkey and 169
 UN and 122
Textron 358
Thailand: military expenditure 273, 279, 285, 307
Thales 328, 329, 332, 336, 353, 357, 358
Thales Avionics 361
Thales Nederland 360
Thales Optronics 361
Thales Raytheon Systems 334, 336
Theater High-Altitude Area Defense (THAAD) 496, 649
Third Party Interventions in Intrastate Conflict 94
Threshold Test Ban Treaty (TTBT, 1974) 473, 772

ThyssenKrupp 360
TI Group 360
Tibet 553
Tipnis, A. Y. 559
Titan 329, 359, 624
Titan booster 616, 622
TK Werften 360
Tlatelolco Treaty (1967) 473, 766–67
Togo 111, 113:
 military expenditure 271, 277, 283
Topol-M missile 350
Tornado aircraft 397, 537, 546
Toshiba 360
Tóth, Tibor 667–68, 676, 787
Tracking and Data Relay Satellite (TDRS) spacecraft 619
Tracor 336
Trajkovski, President Boris 796
transatlantic relationships 2–3, 7–8, 13–14, 17
Trans-Dniester 713, 717–18
TRANSIT satellite system 620, 627, 628
Trident I missile 534, 535, 536
Trident II missile 518, 525, 533, 534–35, 536, 546, 547
Trilateral Initiative (1996) 584
Trimble, David 790
TRIMILSATCOM 637–38
Trishul SAM system 383
tritium 562, 595:
 production of 585, 586, 593
Trumpet spacecraft 624
TRW 357, 617, 652
Tunisia: military expenditure 270, 276, 282, 306
Turboprop International 334
Turin International Training Camp 102
Turkey:
 Armenians killed by 641
 arms imports 14, 377, 378, 379, 402
 conflict in 75
 EU and NATO 165, 167–70, 172
 Greece and 168, 170
 JSF 398, 399
 military expenditure 235, 238, 275, 281, 287, 294, 307
 nuclear smuggling and 606
 nuclear weapons in 574
 significance increased 2
Turkmenistan: military expenditure 248, 272, 278, 284
Type 214 submarines 340
Typhoon aircraft 399

UFO satellites 616, 618
Uganda:
　DRC and 30, 31, 32, 33, 61
　military expenditure 197, 245, 271, 277, 283
　security sector reform and 196
Ukraine:
　arms exports 376, 378–79, 385
　Belarus and 725–26, 795
　EU and 166
　military expenditure 275, 281, 287
　nuclear reactor 604
　Poland and 726
　Ukrspetsexport 375
　USSR's nuclear weapons in 512, 797
Ultra Electronics 361
UNAMSIL (UN Assistance Mission to Sierra Leone) 34, 35, 36, 110, 111, 114, 128, 132
UNBIH (UN Mission in Bosnia and Herzegovina) 127
UNDOF (UN Disengagement Observer Force) 130
UNFICYP (UN Peacekeeping Force in Cyprus) 130
UNICEF (UN Children's Fund) 732
UNICOI (UN International Commission of Inquiry) 208
UNIFIL (UN Interim Force in the Lebanon) 130
UNIKOM (UN Iraq/Kuwait Observer Mission) 130
Union of Soviet Socialist Republics (USSR):
　break-up of 66
　BW and 696
　KGB 191
　nuclear forces 514
　nuclear security assistance to 518–22
　nuclear weapons mix 571
　sanctions against 210
　scientists, occupying 519, 521
　Sverdlovsk (Yekaterinburg) anthrax outbreak 676
　see also Russia
United Arab Emirates (UAE):
　Afghanistan and 41, 793
　arms imports 377
　military expenditure 247, 248, 275, 281, 287
United Defense 336, 358
United Kingdom:
　arms exports 376, 377, 379, 381, 391, 392, 401
　arms industry, PFI and 259
　arms production 324, 335, 336, 339, 371, 397
　Atomic Weapons Establishment 548
　BW and 696
　Defence Ministry 184, 346, 390–91
　Department for International Development 184
　EU military capacity and 167
　EU and NATO and 167, 168
　Europe and 154
　European security capability and 155
　JSF and 398–99, 402
　Lockerbie bombing 225
　military expenditure 13, 233, 235, 244, 253, 275, 281, 287, 294:
　　private financing 258
　military outsourcing 343
　NATO and 155, 157
　nuclear smuggling and 605
　nuclear weapons, numbers of 525, 526, 547
　nuclear forces 545–48
　Private Finance Initiative 258
　RAF Lossiemouth 546
　RAF Marnham 546
　security sector reform and 184, 195, 200
　terrorism and 244
　for relations with other states see under names of those states
United Nations:
　arms embargoes 35, 384, 385, 387–89, 390, 792
　Brahimi Report (2000) 101, 102 fn. 20
　Charter 125, 206:
　　conflict prevention and 101
　coercive instruments 115
　Conference on the Illicit Trade in Small Arms and Light Weapons 736–39, 791
　conflict prevention and 97, 99, 100–103, 108–12, 113, 114–16:
　　report on 97, 100–3, 107
　Counter-Terrorism Committee 226, 227, 686
　Department of Peacekeeping Operations (DPKO) 184–85
　Department of Political Affairs (DPA) 102, 109, 184
　Development Programme (UNDP) 117, 120, 184
　ECOWAS and 111
　General Assembly, conflict prevention and 103
　General Assembly Resolution 56/24 796
　Human Development Index 108

INDEX 841

institutional changes 102
Inter-Agency Task Force on West Africa 109, 110, 114, 115
International Training Camp, Turin 102
military expenditure reporting 303, 305, 306–307
Millennium Summit 98
missiles and 748, 749
Office for the Coordination of Humanitarian Affairs (OCHA) 27, 110–11
Office of the Iraq Programme 219, 220
Office for West Africa 110, 114
peace missions 12, 130–33
peace operations, wariness over new 111, 114, 124, 129
Petersberg tasks 104
reform 102
Register of Conventional Arms 372, 400
sanctions 12–13, 109, 115, 203, 204–10, 212, 215–24:
  effectiveness 204
  legitimacy questioned 206
  lifting 209
  terrorism and 225–27
  Working Group on 206–10, 225, 228
  see also arms embargoes
Security Council, conflict prevention and 103
Security Council Resolution 661 218
Security Council Resolution 687 216, 218, 220, 705
Security Council Resolution 747 225
Security Council Resolution 986 215
Security Council Resolution 1013 208
Security Council Resolution 1051 218, 219
Security Council Resolution 1054 225
Security Council Resolution 1076 221
Security Council Resolution 1193 221
Security Council Resolution 1267 222, 225
Security Council Resolution 1284 215, 216, 221
Security Council Resolution 1295 208
Security Council Resolution 1333 222–23
Security Council Resolution 1341 786
Security Council Resolution 1343 109, 389, 786
Security Council Resolution 1345 787
Security Council Resolution 1352 217
Security Council Resolution 1363 223
Security Council Resolution 1368 793
Security Council Resolution 1373 225, 226, 227, 794
Security Council Resolution 1378 796
Security Council Resolution 1382 203, 217, 219–21
Security Council Resolution 1386 798
security sector reform and 184
small arms control and 789
sovereignty principle 117
terrorism and 122, 793, 794 *see also* Counter-Terrorism Committee
Trust Fund for Preventive Action 102
for relations with individual countries and organizations see under names of
United States of America:
  Aberdeen facility 687, 688
  Alaska 494, 499, 500, 501
  Anniston chemical facility 688
  Area Missile Defense programme 318
  Arms Control and Disarmament Agency (ACDA) 567
  arms control policy 469–70, 477–82, 485–87, 488, 516, 517, 523–24
  Arms Export Control Act 747
  arms exports 13–14, 247, 252, 331–32, 374–75, 376, 377, 379, 382, 390, 391, 392, 394, 402
  arms industry, excessive profits 313
  arms production 13, 326, 328–31, 332, 336, 343, 371–72
  ballistic missile defence:
    architectures of 647–54
    arms control and 481–82, 484, 645
    Boost Defense Segment 651–54
    Bush calls for 788, 789
    China and 251
    contracts for 331
    costs 493–94
    debate on 490–93
    Europe and 166
    funding for 490, 495, 497
    Ground-Based Interceptor (GBI) missile 498, 648–50
    Ground-based Midcourse Defense 499
    Midcourse Defense Segment 650–51
    military expenditure and 314, 318
    name changed 650
    plans altered 492–93, 495
    R&D 490
    'rogue states' and 647
    space-based sensors, importance of 647
    test bed facility 499–500, 651
    testing 498–99, 506, 790

US dominance and 615
Ballistic Missile Defense Organization (BMDO) 482, 495
ballistic missile threats to 482
Bangor 534, 537
Barksdale AFB 531
Base Force Structure 315
Base Realignment and Closure (BRAC) process 319, 320
Battelle Memorial Institute 678–79, 680, 682
biological weapon defence 665–66, 675, 678–80, 681–83
biological weapons and 668, 696, 699–701, 703, 707
Blue Grass Chemical Activity 688
Bottom–Up Review 315
BTWC and 665, 668, 671–73, 674–75, 676, 677, 681–83
California 500
CBW proliferation allegations 704
chemical weapons 694, 695:
  destruction 666, 686–88, 794
  security 687
CIA 543, 552, 553, 554, 555, 556, 561, 564, 567, 678, 679
Coast Guard 317, 320
Commission on Offsets in International Trade 395
Congress 239, 317, 319, 379:
  ABM Treaty and 502
  BMD and 490–91, 494, 498
Congressional Budget Office 240, 255, 309, 493–94
Congressional Research Service 329, 390
Cooperative Threat Reduction project 521, 522, 591, 666
CTBT and 522
Davis-Monthan AFB 563
defence planning changed 2
Defense Department (DOD) 239, 240, 242, 311, 313, 314, 317, 318, 320, 321, 328, 343, 477, 492
Defense Emergency Reserve Fund 242
Defense Intelligence Agency 679
Defense Threat Reduction Agency (DTRA) 679
Device Assembly Plant 596
DIA 557, 635
dominance of 3–4, 469
Dugway Proving Ground 680, 682
economic dominance 3–4
Ellsworth AFB 532

embassies bombed 67, 222
Energy Department (DOE) 519, 521, 522, 591, 593, 595, 596
EU and NATO 166–67, 168
Fissile Material Disposition programme 522
fissile material physical control and accounting regime (MPC&A) 519, 520, 521, 522
Foreign Military Sales Program 392
Fort Greely 500, 651
General Accounting Office 255
gun lobby 738
Hanford Reservation 593, 595
health care 319
homeland security 320
International Science and Technology Centers 521
international security and 18
Johnston Atoll Chemical Agent Disposal System 688
Justice Department 320
Kansas City Plant 593
Kings Bay Weapon Facility 537
Kirtland AFB 537
Kodiak Island launch complex 500, 651
landmine policy 731
Los Alamos National Laboratory 535, 596
Malstrom AFB 532, 533
military doctrine and 315–16
military expenditure 13, 188, 232, 233, 235, 239–40, 242–43, 264–65, 272, 278, 284, 292, 309–22, 324, 491, 494
military expenditure on R&D 242–43
military outsourcing 343, 344
Military Sealift Command 346
Minot AFB 531, 532, 533
Missile Defense Agency 318
Missile Defense Program 493
missile proliferation and 746–47, 750
Modern Pit Facility 535
Mound Laboratory 593
MTCR and 745, 746–47
multilateralism 4
NASA 619
National Command Authority (NCA) 538, 539
National Defense Authorization Act (1991) 746
National Missile Defense Act (1999) 490
National Reconnaissance Office (NRO) 618, 619, 625–26
naval command and control 538–39

INDEX 843

Navy–Marine Corps Intranet 346
Nellis AFB 537
Nevada Test Site 529, 585, 596
Newport facility 688
Norfolk Naval Station 573
Nuclear Cities Initiative 591
nuclear forces 514, 526–39
Nuclear Posture Review (NPR, 2001) 513, 518, 525, 527–28, 528, 529, 530, 531, 538, 586 fn. 4
Nuclear Regulatory Commission 599, 605
nuclear reserves 526, 529
nuclear security in 607–608
nuclear stockpile, enduring 528–29
Nuclear Threat Initiative 609
nuclear weapons:
  numbers of 525, 526, 527–28, 574, 586
  reductions in 489, 506, 508, 511–18, 791, 796, 797
  variety of 570
nuclear weapons tests 529
Oak Ridge Plant 526, 596
Office of Management and Budget 309
Pantex Plant 526, 595
Pentagon building, attack on 1, 792 *see also* terrorist attacks (11 September 2001)
Pine Bluff facility 688
Pinellas Plant 593
plutonium pits 526, 535
Pressler Amendment (1984) 563
Pueblo Chemical Depot 688
Quadrennial Defense Review (QDR, 1997) 314, 315, 322
Quadrennial Defense Review (QDR, 2001) 309, 314–16, 321, 322
Redirection of Biotechnical Scientists programme 521
Rocky Flats Plant 535, 593
Russia no longer an enemy 482, 513, 516, 715
Russian Transition Initiative 521
Russia's nuclear exports and 753, 754
San Diego Naval Air Station 537
Sandia National Laboratories 599
Savannah River 593
Science-Based Stockpile Stewardship (SBSS) 585, 596
security, perception changes 4–5
Security Assistance Act 401
security sector reform and 195
security system and 16, 18
Sequoyah Plant 595

social security 312
spacecraft, numbers of 14
Stennis Space Center 652
strategic stability talks 595
tax cuts 311, 312
terrorism and 1, 2, 4, 10, 18, 22, 187, 188, 495:
  threat perception grows 665
terrorist attacks on, September 2001 1, 5–6, 8, 14, 15, 16, 22, 55, 62, 67–68, 242, 264, 311, 507, 792, 794 *see also* Afghanistan: USA in war against
Theater High-Altitude Area Defense (THAAD) 496, 649
theatre missile defence 314, 318, 495, 649
Tooele facility 688
Transportation Department 320
Umatilla chemical facility 688
UN conference on illicit trade in small arms 737–38
unilateral action 3, 4
Uzbekistan and 789
Vandenberg AFB 622
Warhead and Fissile Material Transparency programme 521
Warren AFB 532, 533
Washington 1 fn. 1, 792
Watts Bar plant 595
White Sands 626
Whiteman AFB 530, 532
World Trade Centre, attack on 1, 85, 792 *see also* terrorist attacks on
Yorktown Naval Weapon Station 537
for relations with other states, see under names of states concerned
United States Marine Repair 361
United Technologies 357
UNMEE (UN Mission in Ethiopia and Eritrea) 133
UNMIBH (UN Mission in Bosnia and Herzegovina) 131
UNMIK (UN Interim Administration in Kosovo) 131
UNMOGIP (UN Military Observer Group in India and Pakistan) 130
UNMOP (UN Mission of Observers in Preklava) 131
UNMOVIC (UN Monitoring, Verification and Inspection Commission) 216, 218, 221, 705
UNOMIG (UN Observer Mission in Georgia) 131

UNSCOM (UN Special Commission on Iraq) 216, 218, 703, 705
UNSMA (UN Special Mission in Afghanistan) 133
UNTAET (UN Transitional Administration in East Timor) 132
UNTSO (UN Truce Supervision Organization) 130
uranium, highly enriched:
　HEU Agreement 591
　production of 585:
　　ending of 587, 589, 593
　research reactors and 599, 604
　theft, diversion of 519, 601, 603
　widespread existence of 608
uranium, low-enriched 602
Uruguay: military expenditure 272, 278, 284, 307
Uzbekistan:
　Afghanistan war and 241
　biological weapons 795
　military expenditure 248, 272, 278, 284
　military reform 248
　Russia and 795
　significance increased 2
　USA and 666, 789, 795
　USSR's BW facilities in 666

V-22 Osprey aircraft 314, 318
Vajpayee, Atal Bihari 45
Venezuela: military expenditure 272, 278, 284
Verhofstadt, Guy 793
Veridian 359
Veritas Capital 329
Vienna Document 1994 303
Vienna Document 1999 719, 720, 721–22, 725, 783
Viet Nam:
　military expenditure 273, 279, 285
　Russian bases 9
Viet Nam War 311, 317
Violent Intrastate Nationalist Conflicts 94
Vito test 529
Vosper Thornycroft 360

Wade, President Abdoulaye 246
Wahid, President Abdurrahman 48
Wallensteen, Peter 89
Warsaw Pact *see following entry*
Warsaw Treaty Organization 571: military expenditure 264

Wassenaar Arrangement on Export Controls for Conventional Arms and Dual-Use Goods and Technologies 474, 743, 744, 755–56
wealth: disproportion in 1
weapons of mass destruction: non-proliferation of 8
West Africa:
　Abuja summit 787
　arms imports 402
　conflict in 107–116, 787
　EU and 108, 112–16
　mineral wealth in 61
　small arms and light weapons 386–87
　UN and 108, 109–12, 114–16
West Indian Space 644
Western European Armaments Group 788
Western European and Other States Group (WEOG) 667, 668, 677
WEU (Western European Union) 124, 137, 152, 168
WGS (Wideband Gapfiller Satellite) System 617, 618
Wheeler's Island 560
Wickremesinghe, Ramil 52
Wilkenfeld, Jonathan 90
Wolfowitz, Paul 4
World Bank:
　conditionality 191, 197
　conflict prevention and 102, 106
　economic conditionality 177–78
　security sector reform and 177–78
　Tanzanian radar and 199 fn. 68
World Event/Interaction Survey Project 96
World Food Programme (WFP) 116, 117
World Forum on the Future of Sport Shooting Activities 739
World War II 326
WorldView Imaging Corporation 643
Wright, Quincy 81

Xhaferi, Arben 792

Yakhont missile 746
Yeltsin, President Boris 543, 572, 577, 747
Yemen 67:
　military expenditure 238, 275, 281, 287
Yugoslavia: break-up of 66
Yugoslavia, Federal Republic of (FRY):
　Florence Agreement and 718
　Ground Safety Zone 786, 788
　KFOR 786

military expenditure 237, 238, 275, 281, 287, 307
nuclear research reactor 604
OSCE in 127, 136
peace missions in 124, 127
sanctions against 212–14, 792
Yugoslavia, former: sanctions 203

Zambia:
  Angola and 26
  military expenditure 271, 277, 283
Zangger Committee 743, 744
Zeevi, Rehavam 795
Zenit booster 631, 632
Zimbabwe:
  Abuja Agreement (2001) 120, 121
  Africa, status in 119
  aid to 117, 118
  Burundi rebels and 29
  conflict in 116–21
  development 116
  DRC and 30, 31, 119
  economic collapse 116
  elections 118, 119
  EU and 117–21
  farm invasions 116, 118
  food 117
  France and 118, 119
  government oppression 117, 118, 121
  land reform 116, 117, 118, 119, 120, 121
  military expenditure 246, 271, 277, 283
  Movement for Democratic Change 116, 119
  neighbouring countries' dependence on 121
  refugees 116
  sanctions against 118, 119, 210
  South Africa and 120, 121
  UK and 116, 117, 118, 119, 120
  UN and 117
  ZANU–PF Party 116